PHYSICAL METALLURGY

PART II

LIST OF CONTRIBUTORS

E. Arzt

H. Biloni

J.L. Bocquet

G. Brébec

R.W. Cahn

G.Y. Chin

R.D. Doherty

H.E. Exner

G. Frommeyer

D.R. Gaskell

K. Girgis

H. Gleiter

P. Haasen

J.P. Hirth

E.D. Hondros

E. Hornbogen

W.L. Johnson

H.W. King

G. Kostorz

Y. Limoge

J.D. Livingston

F.E. Luborsky

T.B. Massalski

R.F. Mehl [†]

P. Neumann

A.D. Pelton

D.G. Pettifor

M. Rühle

M.P. Seah

J.-L. Strudel

R.M. Thomson

C.M. Wayman

J. Weertman

J.R. Weertman

M. Wilkens

H.J. Wollenberger

PHYSICAL METALLURGY

Third, revised and enlarged edition

Edited by

R.W. CAHN

Université de Paris-Sud, France

P. HAASEN

Universität Göttingen, Germany

PART II

1983

NORTH-HOLLAND PHYSICS PUBLISHING

AMSTERDAM – OXFORD – NEW YORK – TOKYO

ISBN: 0444 86 628 0 Set
        0444 86 786 4 Part I
        0444 86 787 2 Part II

PUBLISHED BY:

North-Holland Physics Publishing

a division of

Elsevier Science Publishers B.V.

P.O. Box 103
1000 AC Amsterdam
The Netherlands

SOLE DISTRIBUTORS FOR THE USA AND CANADA:

Elsevier Science Publishing Company Inc.
52, Vanderbilt Avenue
New York, N.Y. 10017
U.S.A.

Library of Congress Cataloging in Publication Data
Main entry under title:

Physical metallurgy.

  Includes bibliographies and indexes.
  1. Physical metallurgy.   I. Cahn, R.W. (Robert W.), 1924–
II. Haasen, P. (Peter)
TN690.P44   1983   669′.94   83-17292
ISBN 0–444–86628–0 (Elsevier : set)
ISBN 0–444–86786–4 (Elsevier : pt. 1)
ISBN 0–444–86787–2 (Elsevier : pt. 2)

Printed in The Netherlands

PREFACE TO THE THIRD EDITION

The first edition of this book was published in 1965 and the second in 1970. The book continued to sell well during the 1970s and, once it was out of print, pressure developed for a new edition to be prepared. The subject had grown greatly during the 1970s and R.W.C. hesitated to undertake the task alone. He is immensely grateful to P.H. for converting into a pleasure what would otherwise have been an intolerable burden!

The second edition contained twenty-two chapters. In the present edition, eight of these twenty-two have been thoroughly revised by the same authors as before, while the others have been entrusted to new contributors, some being divided into pairs of chapters. In addition, seven chapters have been commissioned on new themes. The difficult decision was taken to leave out the chapter on superpure metals and to replace it by one focused on solute segregation to interfaces and surfaces – a topic which has made major strides during the past decade and which is of great practical significance. A name index has also been added.

Research in physical metallurgy has become worldwide and this is reflected in the fact that the contributors to this edition live in no fewer than seven countries. We are proud to have been able to edit a truly international text, both of us having worked in several countries ourselves. We would like here to express our thanks to all our contributors for their hard and effective work, their promptness and their angelic patience with editorial pressures!

The length of the book has inevitably increased, by 50% over the second edition, which was itself 20% longer than the first edition. Even to contain the increase within these numbers has entailed draconian limitations and difficult choices; these were unavoidable if the book was not to be priced out of its market. Everything possible has been done by the editors and the publisher to keep the price to a minimum (to enable readers to take the advice of G.CHR. LICHTENBERG [1775]: "He who has two pairs of trousers should pawn one and buy this book".).

Two kinds of chapters have been allowed priority in allocating space: those covering very active fields and those concerned with the most basic topics such as phase transformations, including solidification (a central theme of physical metallurgy), defects and diffusion. Also, this time we have devoted more space to

v

experimental methods and their underlying principles, microscopy in particular. Since there is a plethora of texts available on the standard aspects of X-ray diffraction, the chapter on X-ray and neutron scattering has been designed to emphasize less familiar aspects. Because of space limitations, we regretfully decided that we could not include a chapter on corrosion.

This revised and enlarged edition can properly be regarded as to all intents and purposes a new book.

Sometimes it was difficult to draw a sharp dividing line between physical metallurgy and process metallurgy, but we have done our best to observe the distinction and to restrict the book to its intended theme. Again, reference is inevitably made occasionally to nonmetallics, especially when they serve as model materials for metallic systems.

As before, the book is designed primarily for graduate students beginning research or undertaking advanced courses, and as a basis for more experienced research workers who require an overview of fields comparatively new to them, or with which they wish to renew contact after a gap of some years.

We should like to thank Ir. J. Soutberg and Drs. A.P. de Ruiter of the North-Holland Publishing Company for their major editorial and administrative contributions to the production of this edition, and in particular we acknowledge the good-humoured resolve of Drs. W.H. Wimmers, former managing director of the Company, to bring this third edition to fruition. We are grateful to Dr. Bormann for preparing the subject index. We thank the hundreds of research workers who kindly gave permission for reproduction of their published illustrations: all are acknowledged in the figure captions.

Of the authors who contributed to the first edition, one is no longer alive: Robert Franklin Mehl, who wrote the introductory historical chapter. What he wrote has been left untouched in the present edition, but one of us has written a short supplement to bring the treatment up to date, and has updated the bibliography. Robert Mehl was one of the founders of the modern science of physical metallurgy, both through his direct scientific contributions and through his leadership and encouragement of many eminent metallurgists who at one time worked with him. We dedicate this third edition to his memory.

April 1983

Robert W. CAHN, Paris
Peter HAASEN, Göttingen

CONTENTS

Chapter 3. Electron theory of metals, by D.G. Pettifor 73

Chapter 4. Structure of solid solutions, by T.B. Massalski 153

Chapter 5. Structure of intermetallic compounds, by K. Girgis   219

Chapter 13. *Interfacial and surface microscopy, by E.D. Hondros and M.P. Seah*   855

PART II

Chapter 14. Diffusive phase transformations in the solid state, by Roger D. Doherty 933

Chapter 15. Phase transformations, nondiffusive, by C.M. Wayman 1031

Chapter 19. Mechanical properties, mildly temperature-dependent, by J. Weertman and J.R. Weertman ... 1259

Chapter 22. Mechanical properties of multiphase alloys, by Jean-Loup Strudel . . 1411

Chapter 26. Magnetic properties of metals and alloys, by F.E. Luborsky, J.D. Livingston and G.Y. Chin

CHAPTER 14

DIFFUSIVE PHASE TRANSFORMATIONS IN THE SOLID STATE

Roger D. DOHERTY

*Department of Materials Engineering,
Drexel University,
Philadelphia, PA 19104, USA*

R.W. Cahn and P. Haasen, eds.
Physical Metallurgy; third, revised and enlarged edition
© *Elsevier Science Publishers BV, 1983*

1. General considerations

1.1. Introduction

Solid-state phase transformations are a central topic in physical metallurgy, since almost all industrial alloys are heat-treated after casting to improve their properties. The heat-treatment changes the microstructure of the alloy, either by a recovery and recrystallization process (ch. 25), or by some type of phase transformation. Two main types of phase transformation are found: polymorphous changes and precipitation reactions. In a *polymorphous change,* in for example iron, cobalt or titanium, there is a change of crystal structure. This affects *all* the atoms in the alloy and has a tremendous scope for changing the microstructure of the alloy (see ch. 16). In *precipitation reactions,* which are crucial in alloys based on aluminium, copper and nickel, the main method of modifying the microstructure is to alloy with elements that are soluble in the base metal at high temperatures but precipitate out of solution at lower temperatures.

In both polymorphous and precipitation reactions, there is a migration of an interface between two crystalline phases, and there are two possible modes of interfacial migration. In the first of these modes, atoms make thermally activated random jumps across the interface, a "diffusive" mechanism. In the second mode, the daughter crystal grows into the parent, by a coordinated shear-type motion of all the atoms at the interface. The first type of transformation is the subject of this chapter while the second type is the subject of ch. 15. Some transformations, such as those in iron-based alloys that are described as bainitic, appear to have both a diffusive and a martensitic character, and will be briefly considered here.

1.2. Driving forces – free energy changes

All structural transformations are driven, at constant temperature and pressure, by the possible reduction in Gibbs free energy, G, from the original to the final structure. The definition of G is (see also ch. 6):

$$G = H - TS, \tag{1}$$

where T is the absolute temperature, S is the entropy, and H, the enthalpy, is given by:

$$H = U + PV + W. \tag{2}$$

U is the internal energy, P is the pressure, V the volume and W all the other work done on the material. In transformations between condensed phases the second of these terms can be neglected with respect to the first and for most transformations the work term in eq. (2) can also be neglected. Under these circumstances the difference between G, the Gibbs, and F, the Helmholtz free-energy is insignificant, and the symbol F will be used throughout this chapter for the function that must be minimized during a structural transformation:

$$F = U - TS. \tag{1a}$$

For a transformation from a parent α-phase to a daughter β-phase, the driving "force", in Joules per unit volume, or more accurately pressure, in Newtons per unit area, is given in terms of the changes in internal energy, $\Delta U_{\alpha\beta}^{v}$, and entropy, $\Delta S_{\alpha\beta}^{v}$, by:

$$\Delta F_{\alpha\beta}^{v} = \Delta U_{\alpha\beta}^{v} - T\Delta S_{\alpha\beta}^{v}. \tag{3}$$

At the equilibrium transformation temperature, T_e, $\Delta F_{\alpha\beta}^{v}$ is of course zero so if $\Delta U_{\alpha\beta}^{v}$ and $\Delta S_{\alpha\beta}^{v}$ are constant, at a finite undercooling, $\Delta T = (T_e - T)$, the driving force is given by:

$$\Delta F_{\alpha\beta}^{v} = \Delta U_{\alpha\beta}^{v} \Delta T / T_e \approx \Delta H_{\alpha\beta}^{v} \Delta T / T_e. \tag{4}$$

For small undercoolings the constancy of $\Delta U_{\alpha\beta}^{v}$ and $\Delta S_{\alpha\beta}^{v}$ are reasonable assumptions but for larger undercoolings a correction should be made to eq. (4) by the standard thermodynamic methods, see for example SWALIN [1972]. Table 1 gives some typical values of the latent-heat changes, from which it can be seen that the driving force for solid-state phase changes will be much smaller than for solidification at similar undercoolings.

For a precipitation reaction in a binary alloy AB, such as the one described by the free-energy–composition curves in fig. 1, the free-energy changes can be readily determined. The overall molar free-energy change of the alloy is IJ, where J lies on the common tangent to the two free-energy curves. The free energy per mole of precipitate is KL where K lies on the tangent to the α-phase at I. The intercepts M and N at $C_B = 1$ are the partial molar free energies (chemical potentials) of component B in the α-phase at compositions C_0 and C_α, respectively. From standard solution thermodynamics, we have:

$$\bar{F}_B(C_0) = F_B^0 + RT \ln a_B(C_0), \tag{5a}$$

$$\bar{F}_B(C_\alpha) = F_B^0 + RT \ln a_B(C_\alpha), \tag{5b}$$

where F_B^0 is the free energy of pure B in its standard state, R is the gas constant and a_B is the activity of B in the α-phase at the two compositions C_0 and C_α. MN is then given by:

$$MN = \bar{F}_B(C_0) - \bar{F}_B(C_\alpha) + RT \ln \frac{a_B(C_0)}{a_B(C_\alpha)}. \tag{6}$$

Table 1
Latent heat changes.

Element	Transformation	Latent heat change (kJ/g-mol)	T_e (K)
Fe	liquid to solid	-15.5	1809
Ti	liquid to solid	-18.9	2133
Fe	austenite to ferrite	-0.9	1183
Ti	bcc to hcp	-3.5	1155

References: p. 1024.

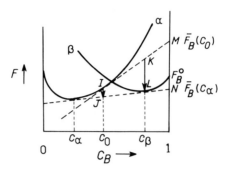

Fig. 1. Free-energy–composition curve of two phases α and β showing conditions for nucleation of β from superaturated α of composition C_0.

If α is a dilute solution with C_0 and $C_\alpha \ll 1$, such as in the phase diagram shown in fig. 2, then the activity coefficient f_α of B in the α-phase is constant (Henry's Law) so eq. (6) becomes:

$$MN = RT \ln(C_0/C_\alpha).\tag{6b}$$

If, again as in fig. 2, β is a dilute terminal solid solution of A in B, then KL in fig. 1 is effectively MN and the driving force per mole of precipitate is given by:

$$\Delta F^m_{\alpha\beta} = RT \ln(C_0/C_\alpha).\tag{6c}$$

In general C_α is not close to zero and C_β not almost unity so that a correction must be made, giving:

$$\Delta F^m_{\alpha\beta} = \frac{C_\beta - C_\alpha}{1 - C_\alpha} RT \ln(C_0/C_\alpha).\tag{6d}$$

For the precipitation of an intermetallic phase such as $CuAl_2$ from an aluminium-rich solid solution, the effect of the modification to eq. (6d) is to decrease the driving force very significantly, by 66% for $CuAl_2$. An extreme version of this effect is found for the growth of ferrite, α, from low-carbon austenite, γ, in iron–carbon alloys. Since the precipitate is solute-depleted, the form of the equation

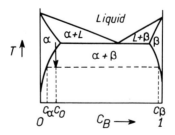

Fig. 2. Simple phase diagram derived from fig. 1.

for the driving force is inevitably changed; so here:

$$\Delta F_{\gamma\alpha}^{m} = \frac{C_{\gamma} - C_{\alpha}}{C_{\gamma}} RT \ln\left(\frac{1 - C_0}{1 - C_{\gamma}}\right)$$ (7)

where the compositions are given in atomic fractions of carbon.

The low driving force for this reaction even when the alloy is cooled well into the two-phase region, $\alpha + \gamma$, for example when the volume fraction of α will be about 0.5, makes this reaction very different from normal precipitation reactions. The difference arises since $\Delta F_{\gamma\alpha}^{m}$ is the driving force for nucleation, while the volume fraction of new phase, $(C_{\gamma} - C_0)/(C_{\gamma} - C_{\alpha})$ is the supersaturation, Ω, that drives the diffusional growth processes. That is, for ferrite formation nucleation is much more difficult but growth is easier than in normal precipitation reactions. This is considered in more detail in § 2.2.6.2 *.

Finally it should be remembered that if the matrix phase is not a dilute terminal solid solution the activity coefficient, f_{α}, is not likely to be constant, so the correction for this effect, derived by PURDY [1971], needs to be included, giving:

$$\Delta F_{\alpha\beta}^{m} = \left(\frac{C_{\beta} - C_{\alpha}}{1 - C_{\alpha}}\right) \frac{RT}{\epsilon_{\alpha}} \ln(C_0/C_{\alpha}),$$ (8)

where ϵ_{α} is the "Darken" non-ideality factor, $\epsilon_{\alpha} = 1 + \mathrm{d} \ln f_{\alpha}/\mathrm{d} \ln C_{\alpha}$.

1.3. Stable and unstable free-energy curves

The free-energy–composition curves shown in fig. 1 are both stable, since any alloy on a curve where the curvature, $\mathrm{d}^2F/\mathrm{d}C^2$, is positive, cannot spontaneously reduce its free energy, except by nucleating a new phase. In a phase diagram such as that sketched in fig. 3, where an α-phase below a critical temperature T_c decomposes into solute-rich and solute-poor phases α' and α'', the free-energy–composition curve is unstable below T_c, fig. 4. This instability arises in the part of the curve where the curvature, $\mathrm{d}^2F/\mathrm{d}C^2$, is negative. Within this *spinodal* region an alloy of composition i can split into j and k and reduce its free energy accordingly (CAHN [1968]). This can occur not only if α' and α'' are the stable phases but also when, as in fig. 4, there is an even more stable β-phase present, whose formation, however, requires the system to overcome a nucleation barrier. In this case α' and α'' would be metastable.

1.4. Gibbs's two types of transformation

From figs. 3 and 4 it is possible to see the distinction between the two types of transformation first described by the father of this subject, Willard Gibbs. In the first of these reactions, now usually described as a *nucleation and growth* reaction, a

---

* Just above the eutectoid temperature in Fe–C, C_{γ} is 0.8 wt% C ($C_{\gamma} = 0.0017$), C_{α} is almost zero, and so for an alloy with C_0 of 0.4 wt% C ($C_0 = 0.0008$), $\Omega = 0.5$, the driving force for nucleation is only 6.5 J/g-mole, while for precipitation of a B-rich phase from an aluminium solid solution at say 150°C with $C_{\alpha} = 0.0001$ and with $C_0 = 0.01$, $\Omega = 0.01$ but the driving force for formation of the β is 16000 J/g-mole!

References: p. 1024.

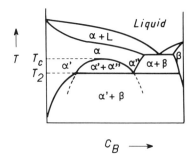

Fig. 3. More complex phase diagram than in fig. 2, showing the development of solid-state immiscibility in the α-phase below a critical temperature T_c.

small region of a new phase forms from within the matrix phase. The new region, the *nucleus*, has, in general, a different composition and structure from the parent phase. The nucleus is separated from the parent matrix by an interface which will have an interfacial energy, σ, in J/m^2. This change is described as being large in magnitude but localized in extent. Other names for this mode of transformation are "heterogeneous" or "discontinuous". However these latter terms have other meanings in different parts of this subject and are best not used as alternative names for this type of Gibbsian transformation. ("Heterogeneous" nucleation and "discontinuous" precipitation have specific meanings within the area of nucleation-and-growth reactions.) The topic of nucleation-and-growth reactions is discussed in § 2.

The second mode of transformation is typified by a composition fluctuation within an unstable region of a free-energy curve. The result of such a fluctuation is . that a region of initially uniform composition evolves a composition wave whose amplitude grows with time (fig. 5). This type of reaction, where the disturbance is limited in magnitude but delocalized in space, is described as *spinodal decomposition*

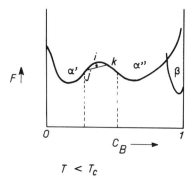

Fig. 4. Free-energy–composition curve for the phase diagram of fig. 3 at a temperature below the eutectoid temperature T_2. The curvature, d^2F/dC^2, is negative in the part between the dashed lines, positive elsewhere.

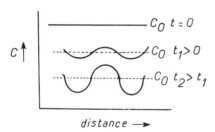

Fig. 5. Variation in solute distribution with time during a continuous reaction, Gibbs type-2: *spinodal decomposition.*

when the wavelength is significantly larger than the atomic dimensions. This type of change is known to occur in many precipitation reactions, for example in Al–Zn alloys (RUNDMAN and HILLIARD [1967]). It may also occur in ordering reactions (COOK *et al.* [1969] and SOFFA and LAUGHLIN [1983]). The term *continuous transformation* is now the usual general name for this mode of structural change *.

For a continuous ordering reaction the wavelength of the composition fluctuation is that of the interatomic spacing, see § 3.2.

CHRISTIAN [1970] has pointed out, in an earlier version of this chapter, that it would seem impossible for there to be a continuous transformation unless the parent and product phases share a common crystal lattice or, if both are fluid, a lack of crystal lattice. Reactions can, and frequently do, however, start in a continuous manner with, subsequently, a *different* structure nucleating in a solute-rich or solute-poor region of the common parent lattice.

1.5. First-order and higher-order transformations

In most structural transformations in physical metallurgy there is, at equilibrium, a discontinuous change in the slope of the free-energy–temperature curve at the equilibrium temperature (fig. 6). This is described as a first-order transformation as there is a discontinuity in the first derivative, dF/dT (EHRENFEST [1933]). In second- and higher-order transformations, there is a discontinuity in the second or higher derivative, d^nF/dT^n. Examples of second-order transformations are some ordering transformations, for example in β-brass where, as the temperature rises, the order steadily falls to zero. In a first-order transformation the derivative of free energy with temperature, dF/dT, at constant pressure is the entropy. A discontinuous change in entropy then requires a finite latent heat, ΔH: $\Delta S = \Delta H/T$. In second order phase transformations there is a discontinuity in the second derivative,

* There is a possible confusion here since continuous precipitation is sometimes used for nucleation-and-growth reactions which do not involve the motion of a high-angle grain boundary. The context of the discussion will usually avoid any difficulty since the term is used only as a distinction from discontinuous precipitation which does involve the movement of a high-angle boundary (§2.5).

References: p. 1024.

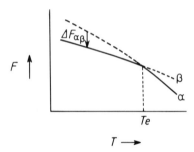

Fig. 6. Variation of free energy with temperature, showing a first-order reaction at an equilibrium transition temperature T_e.

d^2F/dT^2, and therefore in the derivative of enthalpy, the specific heat.

CHRISTIAN [1979] has recently discussed the thermodynamic classification of structural changes and indicated the problems of linking phase transformations as studied by physical metallurgists with the study of critical-point phenomena of interest to theoretical physicists.

1.6.   Short-range and long-range diffusion

In polymorphous changes, in solidification and in recrystallization, all in pure materials, the only atomic process is the motion of atoms across the interface: the *interfacial* process. For nucleation-and-growth reactions where the parent and product phases have different compositions, there are then two *successive* processes; firstly, long-range diffusion over distances of many atomic spacings, commonly described as the *diffusional step* and secondly, atomic transport across the interface, normally a thermally activated short-range diffusional process which is described, by contrast, as the *interfacial step* or *reaction*. The long-range diffusion only involves a fraction of the total number of atoms in the new phase, those required to change the composition of matrix to new phase. The fraction may be nearly 100%, for example in the precipitation of a B-rich phase from a supersaturated A-rich solid solution (fig. 2). In cases like the precipitation of bcc ferrite from dilute fcc austenite in low-carbon steels, the fraction of atoms taking part in the diffusional step may be very much less than one. By contrast *all* the atoms must take part in the interfacial reaction.

Since the two reactions, the diffusional step and the short-range interfacial step, are *successive* reactions, the *slowest* of the two will be rate-controlling (§ 2.3). This may be contrasted with the common situation where there are two *alternative* reactions, such as the precipitation of a stable or a metastable phase. Where reactions are alternatives then the *fastest* reaction will control, not only the kinetics of the process but also the final microstructure. This situation occurs, for example, in precipitation-hardening reactions at low temperatures and high driving forces, where metastable precipitates with a low barrier to nucleation can form in preference to the, more difficult-to-nucleate, stable phase. The initial metastable reaction then removes most of the driving force for the formation of the stable phase, so the metastable product may then remain for the lifetime of the sample.

It is possible that in some reactions there can be a diffusive long-range reaction to allow a required change of composition followed by a nondiffusive, that is, a martensitic interface reaction. The bainitic reaction (§2.6) appears to fall into this category but this suggestion is still controversial (AARONSON [1979]).

1.7. Techniques for studying phase transformations

Since phase transformations produce a change of microstructure, any structural investigation technique can be used to study phase transformations. EDINGTON [1979] has recently reviewed this topic and detailed references to the tremendous range of published work can be found there.

The techniques used include direct observation using optical, electron, and more recently field-ion microscopical methods. The increased resolution of the last two techniques involves a penalty in that the regions studied are highly localized and potentially unrepresentative. Diffraction techniques are also of vital importance, both those involving general diffraction and also selected-area diffraction in both transmission microscopy and, at even higher spatial resolution, scanning transmission electron microscopy (STEM). At lower spatial resolution, electron channelling patterns and X-ray Kossel techniques have been of value in investigations involving bulk samples. Of increasing importance, particularly for many quantitative investigations, are analytical studies of highly localized regions. Conventional electron-probe microanalysis has allowed chemical analysis down to $1–2$ μm, analytical electron microscopy by either X-ray emission or by electron energy loss has allowed routine analysis down to a resolution of 10 nm. Ultimate or near-ultimate resolution has been achieved by the atom-probe field-ion microscopy technique, the determination of individual atoms by time-of-flight measurements of field evaporated atomic layers. Other successful analytical techniques include high-resolution Auger electron spectroscopy and ion probe mass spectrometry using focussed ion beam erosion. For a detailed comparison of the techniques, the recent conference on microanalysis at high spatial resolution provides a detailed source of information (LORIMER [1981]).

Many of these techniques have only recently become available and are all expensive in equipment costs, so much less use has been made of them than of the very widely used transmission electron microscopy which revolutionized this subject after its introduction in the late 1950s. However, as EDINGTON [1979] has shown, these latter techniques, introduced in the 1970s, have made extensive contributions to the subject and will continue to do so. Use of these techniques may modify the present situation, which is that in many areas, the *theories* of phase transformations are well ahead of rigorous *tests* of those theories, at least, as far as quantitative experimental studies are concerned.

In conclusion, it may be noted that there are two distinct modes of investigation of microstructural change which can be used. In most cases the microstructure is

References: p. 1024.

investigated in samples where the reaction has been halted by rapid quenching, normally to room temperature. This allows detailed study, at the highest resolution, of the material. However, the development with time of the reaction, in the particular region studied, cannot then be followed by this technique. The alternative technique, of *in-situ* observation of change in a hot-stage microscope, allows continuous investigation of the transformations in a selected region but usually at poorer resolution and with severe time limitations. An additional problem for the in-situ studies is the worry about modification of transformation by the presence of near-by surfaces and by the illumination, usually a beam of electrons. The use of high voltage electron microscopy has reduced the first of these difficulties but the introduction of point defects by the high-voltage beam can be a difficulty in some cases (e.g., WEAVER et al. [1978] and DOHERTY and PURDY [1983].

2. Nucleation-and-growth transformations

2.1. Theory of nucleation

The basis of the theory of nucleation is that when a new phase nucleates within a parent phase, an interface is formed between the two phases. The interface creates a local increase of free energy when the first few atoms assemble in the new structure. The theory of this process, which describes the interfacial energy barrier to nucleation, was originally developed for vapour to liquid condensation by VOLMER and FLOOD [1934] and BECKER and DÖRING [1935]. The model was applied by Turnbull to solidification of metals and subsequently solid-to-solid metallic phase transformations in the late 1940s (TURNBULL and FISHER [1949]) and since that date has been fundamental to the qualitative and quantitative understanding of structural changes in physical metallurgy. The theory appears to be, qualitatively at least, highly successful both for solidification (ch. 9, §3) and in describing the nucleation step in all but one types of solid-state nucleation-and-growth reactions. The subject of solid-state nucleation has been reviewed on many occasions, for example by KELLY and NICHOLSON [1963], RUSSELL [1970], NICHOLSON [1970], CHRISTIAN [1975] and RUSSELL [1980], with similar conclusions to those advanced here – that the theory is in good agreement, qualitatively and semi-quantitatively, with a large amount of experimental data. There are, however, very few fully quantitative experimental tests of the theory; but these do support the theory to a surprisingly high degree. The one exception where the theory has been found not to apply is to the nucleation of new grains in the recrystallization of deformed materials, see DOHERTY [1978], where the nuclei pre-exist in the deformed state, or can develop by processes that do not require a local increase in free energy. This is discussed in ch. 25, §3.3.

The model assumes that a new phase, differing from the parent phase in structure and often in composition, is built up, atom by atom, by thermally activated atom transfer across the interface *. There is, during nucleation, a change in the free

---

* This picture may not be satisfactory for the nucleation of martensitic transformations (ch. 15, §2.5.4).

energy, which is given in the usual continuum description as a decrease in the volume free energy, due to transfer from a less stable to a more stable phase, and an increase in the interfacial free energy due to the increase of the area of the interface between the two phases. In the conventional model of nucleation the macroscopic values of these parameters are used even though it is unlikely that a cluster of only a few atoms would have the macroscopic properties that the new phase will show when it has grown to a respectable size. The change of free energy, ΔF, when a new phase of volume V_β and interface area A_β forms, causing an increase in elastic strain energy per unit volume of precipitate, ΔF_E^v, is given as:

$$\Delta F = V_\beta \Delta F_{\alpha\beta}^v + A_\beta \sigma + V_\beta \Delta F_E^v, \tag{9}$$

where σ is the specific energy of the $\alpha-\beta$ interface and $\Delta F_{\alpha\beta}^v$ is the free-energy change per unit volume, given by the macroscopic free energy change per gram-molecule divided by the volume of a g-mole of the new phase V_m. The first and last terms in eq. (9) are both dependent on the volume and may be treated together. Only if the first term is larger (more negative) than the third term, which is almost inevitably positive, will the reaction proceed. The volume of the nucleus increases as the third power of the nucleus size ($\frac{4}{3}\pi r^3$ if the nucleus is a sphere of radius r), while the area term, which is positive and acts as the barrier to the process, increases only as the second power of the nucleus size ($4\pi r^2$). Figure 7 plots eq. (9) for a spherical nucleus, and shows that at small radii the nucleus is unstable and tends to redissolve in the matrix; only when the precipitate has reached the *critical radius, r^*,* and the critical free energy increase, ΔF^*, does the nucleus become capable of metastable existence, and after addition of one atom ($r > r^*$) it can grow with a continuous decrease of free energy. For a spherical nucleus the values of the critical parameters are given by:

$$r^* = -2\sigma/\left(\Delta F_{\alpha\beta}^v + \Delta F_E^v\right), \tag{10a}$$

$$\Delta F^* = \tfrac{16}{3}\pi\sigma^3/\left(\Delta F_{\alpha\beta}^v + \Delta F_E^v\right)^2 \tag{10b}$$

Value of r^* and ΔF^* can be obtained by substitution of appropriate expressions

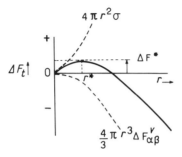

Fig. 7. The nucleation problem; free energy of a sphere of new phase in a supersaturated matrix, as a function of the radius of the sphere.

References: p. 1024.

for $\Delta F_{\alpha\beta}^{v}$, for example eqs. (4), (6c), (6d) or (8), with the value of $\Delta F_{\alpha\beta}^{v}$ obtained by dividing the value of $\Delta F_{\alpha\beta}^{m}$ by the volume of a g-mole of the new phase, V_{m}. For most metallic phases the volume of a g-mole of pure metal or substitutional solid solution is close to 10^{-5} m^3. The problem is to find the appropriate value for ΔF_{E}^{v}; this is discussed in §2.1.2, but for the rest of §2.1 it will be assumed that the strain term is negligible, as it will be for example in solidification from most metallic melts, at least above the glass transition temperatures.

The rate of nucleation per unit volume, $I_{\alpha\beta}^{v}$, can be obtained from the simple application of statistical mechanics to show that if the critical formation energy of a nucleus is ΔF^* then at equilibrium the concentration of critical-sized nuclei, N^*, in a unit volume is given by:

$$N^* = N^{v} \exp(-\Delta F^*/kT), \tag{11}$$

where N^{v} is the number of atomic sites per unit volume on which the nucleation could have started. The rate of nucleation is then the product of this concentration of critical nuclei and the rate of atomic addition to the nuclei to make them just supercritical:

$$I_{\alpha\beta}^{v} = N^* A^* v \exp(-\Delta F_{a}/kT) \exp(-\Delta F^*/kT), \tag{12}$$

where A^* is the number of surface atomic sites on the critical nucleus to which an atom can join after overcoming a growth barrier ΔF_{a} (§2.2.1), and v is the atomic vibration frequency of an atom in the matrix at the interface, usually assumed to be of the order of 10^{13}/s. In most cases A^* is assumed equal to the total number of atomic sites on the interface but again as discussed in §2.2.3 this may not always be valid and atoms may be able to add on only at *ledges* on the interface.

The theory needs to be modified to take into account the expected loss of critical-sized nuclei by their growth, during nucleation, into supercritical regions. This loss reduces the value of N^* to less than the value expected at equilibrium, see eq. (11), by a factor which is usually about 0.05 (RUSSELL [1970]). The near constancy of this correction arises since the rate of formation of critical-sized nuclei and their growth to become supercritical both occur at a similar rate determined by atomic diffusion onto the growing cluster of atoms.

There are many problems with the use of the theory, given by eq. (12), to predict experimentally measured quantities such as the density of the new grains of the product phase. These difficulties include the usual lack of knowledge of the interfacial energy σ which plays such a vital part in determining $I_{\alpha\beta}^{v}$ ($I_{\alpha\beta}^{v}$ being proportional to the exponential of σ cubed), the difficulties in calculating the strain energy term in eq. (10b) and in calculating how the driving force for the phase change, $\Delta F_{\alpha\beta}^{v}$, falls as the reaction proceeds, so that the rate of nucleation can be integrated to give a total number of growing nuclei. (See, for example, SERVI and TURNBULL [1966], and LANGER and SCHWARTZ [1980]). Finally, as discussed in §2.4.2, a significant number of nuclei will dissolve up by a process of Ostwald ripening during the precipitation reaction itself. However, despite these quantitative difficulties the theory is extremely useful in describing a large volume of experimental observations. A major success of this theory is its ability to account for the observation that nucleation, by this mechanism which is loosely but universally described as *homogeneous nucleation*, increases from a rate that is almost undetectable to rates that are too fast to measure over a narrow range of *undercooling*. This is readily shown by use of eq. (11) for a matrix phase change driven by a free-energy function of eq. (4). Under these circumstances the concentration of critical nuclei is given by

$$\log N^* = 28 - K/\Delta T^2 \tag{13}$$

since the number of atomic sites in a cubic metre is about 10^{28} and K is a constant given by

$$K = 16\pi\sigma^3 T_c^2 V_m^2 / 3\Delta H_{\alpha\beta}^v kT. \tag{14}$$

If the undercooling which gives a density of nuclei of $1/m^3$ is $\Delta T(0)$ and that which gives a density 10^6 times as high is $\Delta T(6)$, then substitution in eq. (13) shows: $\Delta T(6)/\Delta T(0) = (28/22)^{1/2} = 1.1$. This shows that a 10% increase in supercooling will increase the density of critical-sized nuclei, and therefore the rate of nucleation, by a million times. The constant K in eq. (14) can be easily evaluated if the undercooling that gives a measurable rate of nucleation is known, by substitution of a value of N^* of about $10^3/m^3$ in eq. (13). This allows an estimate of the effective interfacial energy to be easily made. In many cases, for example in solidification of pure metals, the determination of the interfacial energy by use of eq. (13) has given reasonable values of this energy even when the undercooling appears to have been underestimated (CANTOR and DOHERTY [1979]).

Another major success of the theory of homogeneous nucleation is its ability to understand a large range of experimental experience which shows that during many precipitation reactions the phase that forms is not the equilibrium structure. It is frequently found that a metastable phase, characterized by good atomic fit with the matrix and therefore a low value of interfacial energy, σ, is the one that forms. The theory is compatible with this phenomenon since the rate of nucleation is so sensitive to σ; the successful reaction will often be the one that nucleates fastest, and this is usually the reaction that shows the lowest interfacial energy, even though it does not yield the most stable phase. Similarly, nucleation theory can account for the fact that when a new crystalline phase forms in a matrix of a crystalline parent phase there is always a particular orientation-relationship between the phases. This appears to be the relationship that minimizes the value of the crucial parameter, the interfacial energy σ. This idea is discussed further in §2.1.2.

Finally it should be recognized that additional qualitative support for this picture of nucleation is provided by an apparent failure of the model. The failure arises since in many cases nucleation occurs at much smaller undercooling than expected. In these cases it is found that the new phase forms on some defect in the parent phase, a grain boundary for example. The second term in eq. (9) will then be significantly reduced by the fact that the new phase has consumed some of the existing energy of the defect, thereby reducing the barrier to nucleation. This topic of *heterogeneous nucleation* is discussed in §2.1.4.

2.1.1. Interfacial structure and energy

A major difference between the interfaces produced by nucleation during solidification from a liquid or glassy phase, and by nucleation in solid-state reactions, is in the possibility of atomic matching across an interface between two crystals. For crystal–crystal interfaces the atomic matching ranges from perfect, as in fully *coherent* interfaces, through less perfect matching in *semicoherent* interfaces, to random matching in fully *incoherent* interfaces. The atomic structure of such interfaces is discussed in ch. 10B, §2.4.

A simple example of a fully coherent interface is one between two phases that have the same crystal structure, lattice parameter and orientation, often loosely described as "cube–cube". For fully coherent interfaces the only contribution to the interfacial energy, σ, comes from the higher energy of the unlike A–B bonds across the interface. Since the system is showing immiscibility it is expected that the unlike bond energy $h(AB)$ will be greater than the average energies of the bonds, $[h(AA) + h(BB)]/2$, as described in any discussion of the so-called *quasi-chemical* bond model (e.g., SWALIN [1972]).

Such completely perfect matching is unlikely to be found and deviations from coherency can develop as the extreme conditions described above are relaxed. In most reactions there will be a small difference in the lattice parameters of the two phases, a_α and a_β, which gives rise to a misfit, δ:

$$\delta = 2(a_\alpha - a_\beta)/(a_\alpha + a_\beta). \tag{15}$$

For small misfits and small precipitates, the difference in atomic spacing can be taken up by elastic strain in the two phases; but at larger precipitate sizes the system can lower its elastic strain energy by concentrating the strain in a cross-grid of edge dislocations spaced a distance d apart (fig. 8). Over the dislocation spacing there will be n atomic planes in the phase with the larger parameter and $n + 1$ planes in the other phase. When the misfit is much less than unity, as it usually is, then:

$$n = 1/\delta, \tag{16}$$

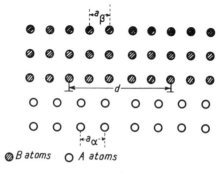

$\oslash$ *B atoms* $\bigcirc$ *A atoms*

Fig. 8. Semicoherent interface between two phases with the same structure but different atomic spacing and containing a set of edge dislocations a distance d apart.

and

$$d = (a_\alpha + a_\beta)/2\delta. \tag{17}$$

For an semicoherent interface the interfacial energy has two components; a "chemical" term as discussed for the coherent interface and an additional "structural" term, due to the cross-grid of edge dislocations. For small precipitates, where the diameter of the nucleus may be less than the dislocation spacing, d, the nucleus will be expected to remain coherent, though elastically strained, and consequently will have a higher solubility because the elastic energy raises the energy of the precipitate phase (CAHN [1968]).

A fully incoherent phase boundary will have an atomic structure rather like that of a high-angle grain boundary, whose misorientation is well away from any possible coincidence-site boundary relationship (ch. 10B, §2.2.1). Fully incoherent interfaces are likely to arise in two main ways. The first way is if after nucleation, the orientation of the surrounding matrix phase is changed by the passage of a high-angle grain boundary in a recrystallization or grain-growth reaction. This phenomenon has been discussed by DOHERTY [1982] with examples of the resulting change in interfacial properties and consequent microstructural changes. The other main origin of incoherent interfaces arises when the two crystals, α and β, do not have a common crystal structure – which is the usual situation in most precipitation reactions and is inevitable in all polymorphous changes, though not in all ordering reactions. As is discussed in the next section, under these conditions part at least of the interface is almost certain to be incoherent.

2.1.2. Equilibrium shape

When the nucleus and the matrix have different crystal structures, it in invariably found that there is a definitive orientation relationship between the phases. BARRETT [1952] has given a very complete listing of the orientation relationships for many reactions. The orientation relationship appears to allow a good fit to develop between the two crystals either along a particular plane or occasionally along a particular direction. Typical examples of these orientation relationships are provided by fcc–hcp matrix–precipitate pairs in systems such as Al–Ag_2Al and in the Cu–Si system, which show a good fit between (111) fcc and (0001) hcp. These planes are parallel and provide the *habit plane* of the flat plate-like precipitate crystals that form in these reactions (e.g. fig. 24 below). The atomic fit between the two structures is shown in fig. 9, where it can be seen that the two structures have identical atomic arrangements on the habit plane, so allowing full coherence in two dimensions. The two equilibrium phases have slightly different lattice parameters but the intermediate metastable precipitate, γ', has the same atomic spacing as the aluminium matrix, in the close-packed plane, and so shows full coherence, and consequently very low energy, for the habit plane (the structural component being zero). Precipitates are however, three-dimensional and the interface must lie at all possible orientations in space around the crystal; consequently the other parts of the interface, the rims of the plate-like crystals, must have much poorer atomic matching and a higher

References: p. 1024.

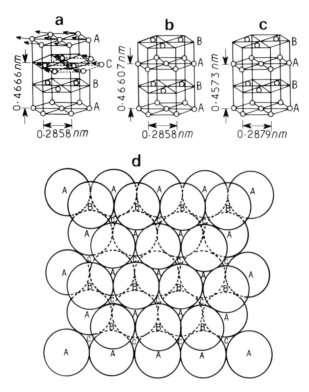

Fig. 9. Atomic structure of fcc–cph matrix–precipitate structures in the Al–Ag_2Al system: (a) the fcc matrix; (b) γ' intermediate precipitate, fully coherent; (c) equilibrium Ag_2Al, partially coherent; (d) the fcc ⟨111⟩ stacking sequence.

interfacial energy. In other words, the magnitude of σ will vary rapidly around the precipitate. This can be represented in a section through the σ *plot* shown in fig. 10. The σ plot is a radial plot of the interfacial energy as a vector as a function of the orientation of the interface of the precipitate. The deep cusps in the plot occur at the orientations of good fit, corresponding to the habit plane. AARONSON *et al.* [1968] have calculated that the ratio of energy between the good fit and the rest of the interface is $1:2.7$ for the equilibrium Ag_2Al phase; similar calculations for the intermediate phase give a higher ratio of $1:10$ (FERRANTE [1978]). For anisotropic σ plots such as the one seen in fig. 11, the Gibbs–Wulff theorem (MULLINS [1963] and MARTIN and DOHERTY [1976]) predicts the equilibrium shape of the precipitate, for the orientation relationship that gives the particular anisotropic σ plot. This shape is the inner envelope of so-called *Wulff planes* that are planes drawn perpendicular to the vector from the origin at the intersection with the σ plot. Figure 11 shows as an example the Wulff plane LBM for the vector OB and also the equilibrium shape of the precipitate, which provides for facet planes at the cusps (which represent good-fit interfaces).

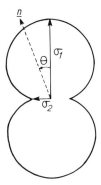

Fig. 10. The σ plot, the variation of interface energy with the orientation of the boundary plane for the particular orientation that allows good fit at the cusp plane, σ_2.

The *aspect ratio, A,* of the precipitates, the length-to-thickness ratio, is equal to the energy ratio $\sigma_{incoherent}/\sigma_{coherent}$. The cusp gives a good-fit *plane* where there is good match of crystal planes as in the fcc–hcp system discussed and in many other precipitation reactions, for example Al–θ' in the aluminium–copper system which has habit planes on the matrix (100) planes (fig. 15, below). For systems like the fcc–bcc system in, for example, the iron–copper system, there is only a good-fit *direction,* which is the close-packed direction ⟨110⟩ fcc ⟨111⟩ bcc. This form of matching of directions provides the well-developed *needle morphology,* with an aspect ratio of 5:1, reported by SPEICH and ORIANI [1965]. For good-fit directions, the σ plot, schematically given in two dimensions by fig. 10, must be rotated about the vertical axis to produce the appropriate three-dimensional form: for plate-like crystals the three-dimensional plot is obtained by rotation about the horizontal axis.

Since the ends of needle crystals, and the rims of plate-like crystals, have poorer atomic fit than the coherent needle axis or habit planes of the plates, it is commonly assumed that these ends and rims are fully incoherent. This may be usually the case but it need not necessarily be true, since in the fcc–hcp pair Al–Ag_2Al it has been seen that the rims of the plates are also faceted. This implies that the rim is not without some coherency.

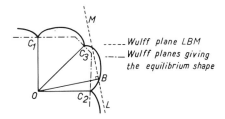

Fig. 11. The Wulff construction. The Wulff plane at B is LBM, that is normal to OB at B. The equilibium shape is the inner envelope of Wulff phases, in this case, the cusps C_1, C_2 and C_3.

It is usually assumed that during nucleation the nucleus forms with the minimum interfacial energy, since any other assumption increases the free-energy barrier to nucleation and strongly reduces the rate of nucleation. The assumption of minimum interfacial energy requires the nucleus to have the equilibrium shape: this will be plate-like or needle-like if the σ plot shows the normal anisotropy for two crystals with different crystal lattices. If the plot is, however, fully isotropic the expected shape of the nucleus will be a sphere. In two examples of "cube–cube" relative orientations between crystals with the same structure, nearly spherical equilibrium shapes have been reported. In the bcc–bcc beta–gamma brass system studied by STEPHENS and PURDY [1975] and in the all-fcc system of Ni–Ni$_3$X (where X denotes a mixture of several atoms) (see, e.g., RICKS et al. [1983]), equilibrium shapes that were nearly spherical were reported. In the latter case it seems that the precipitate develops the usually reported cuboid morphology only with growth in systems when there is some small misfit, δ. Both the brass and the nickel systems can show an evolution into a dendritic morphology during growth. This type of shape instability is discussed in §2.5.

For plate-like precipitates with an equilibrium value of the aspect ratio A, that is, with $A = \sigma_1/\sigma_2$, the critical values of the plate radius, $a(y)^*$, and nucleation barrier, ΔF^*, are given by:

$$a(y)^* = -8\sigma_1/3\left(\Delta F_{\alpha\beta}^{\mathrm{v}} + \Delta F_{\mathrm{E}}^{\mathrm{v}}\right), \tag{18}$$

$$\Delta F^* = 256\pi\sigma^3/27A\left(\Delta F_{\alpha\beta}^{\mathrm{v}} + \Delta F_{\mathrm{E}}^{\mathrm{v}}\right)^2. \tag{19}$$

σ, is the interfacial energy of the incoherent rim and σ_2 that of the coherent habit plane.

2.1.3. Strain energy

Strain energy plays a vital role in the nucleation of solid-state phase changes, as shown by eqs. (9)–(19). There appear to be two different origins of strain that can be developed by the formation of a new phase in a matrix phase, when both are crystalline. The first type of strain is that caused by a misfit between two coherent phases before the precipitate has grown to a sufficient size to induce a dislocation to take up the distortion. The second type of strain is that caused when the new phase occupies a different volume from the region of the matrix that it has replaced. This second type of strain can arise in various ways. One of these is when the volume per atom is different in the two structures and the precipitate grows without a change in the number of atom sites. An example would be the nucleation of the less dense bcc ferrite from austenite in pure iron. A further method for the development of this volume strain is if there is a significant difference in the rates of diffusion of two components of the alloy. An example is when zinc-rich γ-brass precipitates from β-brass when there is a more rapid inward flux of the faster diffusing zinc atoms than the compensating outward movement of copper atoms. This increases the *number of atomic sites* in the region where γ-brass is growing, giving a dilational strain. This effect has been directly observed on the surface of a brass alloy, by

CLARK and WAYMAN [1970] who showed that the γ-phase stood above the surrounding surface. In the interior of the sample this displacement would produce an elastic strain unless relieved by plastic deformation.

A local strain around a precipitate will produce a large energy increase if the strain is taken up elastically since the elastic moduli of metals, both tensile and bulk, are large. For aluminium where Young's modulus is 70 GPa, a 1% strain would give an elastic energy of 3.5 MJ/m^3 (35 J/g-mole), a 5% strain would give an energy 25 times larger. This energy can be reduced plastically by the introduction of dislocations into the interface in the first case, raising the interfacial free energy. For the second, dilational, type of strain the elastic energy can be relaxed either plastically, by dislocation motion, or by diffusional motion of point defects. Both of these release mechanisms are easier at high temperatures, especially the motion of point defects.

Two limiting conditions are likely to apply. The first condition would be for precipitation at low temperatures where there will usually be a high free-energy driving force (eq. 6d), C_α being extremely small in a precipitation reaction at low temperatures. The second condition would occur for high-temperature reactions, usually with lower driving forces. With such small driving forces then the strain energy opposing nucleation can easily be larger than the free-energy decrease driving the reaction. For example, precipitation of AlCu$_2$ from an aluminium–copper alloy at 800 K with a typical high-temperature supersaturation ratio, C_0/C_α, of 1.1 gives a $\Delta F_{\alpha\beta}^m$ of only about 200 J/g-mole which would be offset by an elastic strain of only 2.5%. For low-temperature precipitation-hardening reactions, the supersaturation ratios are much larger and the elastic strain energy more easily provided by the large free-energy change driving the precipitation. Electron microscopy has shown that such elastic distortion does occur under these low-temperature, high-supersaturation conditions (see, e.g., KELLY and NICHOLSON [1963]).

NEMOTO [1974] has shown by direct observation, using in-situ high-voltage electron microscopy, that during dissolution of cementite by decarburization at 700°C with high rates of dissolution, part of the change of volume of the reaction was provided by intense localized dislocation motion. For slower rates of dissolution such dislocation movement was not observed and the strain was assumed to be accommodated by movement of iron vacancies. The analysis of the coupled diffusion of substitutional iron atoms and rapid interstitial carbon atoms to relax localized dilational strain has been provided by ORIANI [1966] for the case of precipitate coarsening in the same, iron–cementite, system.

Recently, MAKENAS and BIRNBAUM [1980] have shown that there is very significant plastic deformation during hydride precipitation in the niobium–hydrogen system. They charged hydrogen electrolytically into previously prepared thin foils of niobium and observed precipitation by TEM as the foils were continuously studied during cooling. Precipitation of hydride occurs by rapid interstitial diffusion of hydrogen causing a large expansion ($\Delta V/V$ of 18%) at temperatures too low for diffusional relaxation. Around each precipitate there was an intense tangle of dislocations with additional prismatic dislocation loops punched out in the $\langle 111 \rangle$

References: p. 1024.

slip directions. During resolution the plastic strain was not reversible and this led to a large temperature hysteresis between precipitation and resolution. Similar examples of plastic deformation have been obtained during other low-temperature hydride precipitation reactions reviewed by MAKENAS and BIRNBAUM [1980].

LEE *et al.* [1980,1983] have reviewed the current theoretical and experimental results on the influence of elastic and plastic strains on precipitation in general, and nucleation in particular. For the *elastic* situation the detailed theoretical analysis shows that the elastic energy is minimized when the precipitates are formed with a thin disc- or plate-like shape. This is the morphology revealed by electron-microscopic studies in all cases where Guinier–Preston (GP) zones form with the solute-rich zone having atomic size difference from the lattice atoms that the solute replaces. Disc-like zones are found for example in aluminium–copper and copper–beryllium alloys where there are significant size differences between the atoms; however in systems like aluminium–silver, where the atoms are almost the same size, spherical zones are found (KELLY and NICHOLSON [1963]). The precipitate shapes found in the same system when the intermediate and equilibrium structures are produced are plate-like but this is due to the fact that the later precipitates, being of hexagonal symmetry, have a different structure from the fcc matrix. The observations on GP zones are therefore much more relevant to understanding the influence of elastic strain on precipitate shape than are observations on shapes, where there is a complication due to differences in precipitate structure. Finally it should be noted that, as expected, the habit plane of misfitting disc-like zones appears to be in the matrix direction with the lowest elastic modulus, $\langle 100 \rangle$ in most cubic metals (WERT [1976]).

LEE *et al.* [1980] have discussed the expected *plastic* yielding that can occur around a spherical nucleus under the assumption of isotropic elasticity. They assumed that for precipitates larger than about 1 μm the macroscopic yield stress will operate. But for smaller precipitates they assumed that the yield stress will be raised by the lack of dislocations in the sub-micron regions. By use of a model proposed by ASHBY and JOHNSON [1969], in which dislocations can be nucleated with the help of a supersaturation of point defects, LEE *et al.* [1980] were able to derive an expected yield stress for flow, as a function of the precipitate radius. Their analysis gave results that agreed well with a range of experimental results, as regards the critical size of precipitate which went from coherent, elastically strained, to semi-coherent.

2.1.4. Heterogeneous nucleation

In many examples of nucleation in solids it is found that the nucleation sites are not distributed randomly, but are concentrated at particular sites in the matrix, usually at some crystal defect. The defects that act as the *heterogeneous* sites for nucleation are grain boundaries, grain edges, grain corners, dislocations, stacking faults etc., as described in some detail by NICHOLSON [1970]. CAHN [1956,1957] provided the currently accepted analysis for nucleation of incoherent precipitates on grain boundaries and on dislocations. The analysis shows that the nucleus forms on the pre-existing defect so that the energy of formation of the nucleus is reduced by that proportion of the defect energy that is consumed by the nucleus as it is formed. For nucleation on the boundary between two matrix grains (fig. 12) the grain boundary that has been destroyed is shown as the dashed line within the new crystal. The critical parameter is the semi-angle θ determined by the ratio $\sigma_{\alpha\alpha}/2\sigma_{\alpha\beta}$, where $\sigma_{\alpha\alpha}$ is the grain-boundary energy and $\sigma_{\alpha\beta}$ is the incoherent energy of the interface:

$$\cos \theta = \sigma_{\alpha\alpha}/2\sigma_{\alpha\beta}. \tag{20}$$

When the new phase "wets" the grain boundary, which occurs when $2\,\sigma_{\alpha\beta} < \sigma_{\alpha\alpha}$, then there is no barrier to nucleation. This can often arise when the second phase is a liquid, as occurs for example in the *hot shortness* of steels in the presence of molten iron sulphide, and in other cases of liquid-metal embrittlement. In general, grain boundaries in metals are not "wetted" by crystalline second phases and so some fraction of the barrier to homogeneous nucleation remains in the case of heterogeneous nucleation, as given by:

$$\Delta F^*(\text{het}) = \Delta F^*(1 - \cos \theta)^2 (2 + \cos \theta)/2 \tag{21}$$

For nucleation at grain *edges* (the lines where three grain boundaries meet) and at grain *corners* (where four grains, and four grain edges, meet), there is an even larger reduction in the barrier to nucleation. Such reductions in the barrier to nucleation do not automatically lead to a large increase in the nucleation rate since there is a dramatic reduction in the number of suitable sites for nucleation to start [the term N^v in eq. (11)]. If the mean grain diameter is d, the thickness of the grain boundary is δ_b and the volume of an atomic site is V_s then the number of nucleating sites per grain declines as $d^3/V_s : d^2\delta_b/V_s : d\delta_b^2/V_s : \delta_b^3/V_s$ as we consider homogeneous nucleation, grain-boundary nucleation, grain-edge and grain-corner nucleation. For a grain size of 50 μm and a boundary thickness of 0.5 nm the ratios are $10^{15} : 10^{10} : 10^5 : 1$. These are very large reductions. The result of the two effects, the reduction in ΔF^* and in N^v, together, is that for nucleation under large driving

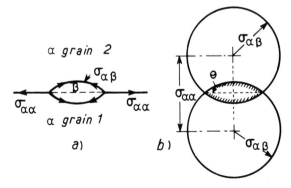

Fig. 12. (a) The double hemispherical cap model for the expected shape of an incoherent precipitate at an α–α grain boundary showing the destroyed α–α grain boundary as the dashed line. (b) The Wulff construction for this situation. (After LEE and AARONSON [1975].)

References: p. 1024.

forces and with small interfacial energies, homogeneous nucleation will be expected to dominate, giving a very high density of new precipitates. With smaller driving forces or with higher values of the interfacial energy (if no good atomic matching between the two phases is possible) then the consequent large value of ΔF^* in eq. (10) requires heterogeneous nucleation, at a suitable site, for any significant rate of nucleation to be possible. This conclusion is in good qualitative agreement with a huge amount of experimental evidence for the sites of nucleation in many systems (NICHOLSON [1970]). It is always found, for example, that with a small supercooling into a two-phase field giving a low driving force, nucleation occurs solely at grain corners and edges and only at somewhat higher driving forces is nucleation found at grain-boundary sites and eventually grain-interior sites. Moreover, it is very commonly found that grain-interior nucleation is not homogeneous, since dislocation sites are also strong heterogeneous nucleation sites (CAHN [1957]). It appears that only for very large driving forces indeed, such as by diffusion of a third element into a binary alloy to give, for example, internal oxidation or nitriding (with a solute whose oxide or nitride has extremely low solubility) can homogeneous nucleation be expected to occur as the fastest process for largely incoherent interfaces. In other examples there is never enough driving force for homogeneous nucleation except with largely coherent interfaces, as in GP zone formation. Most intermediate precipitates appear to nucleate either on dislocations or by some two-stage heat treatments, on pre-existing GP zones (LORIMER and NICHOLSON [1969] and JACOBS and PASHLEY [1969].

The classical theory of heterogeneous nucleation, outlined above, assumed that the energy of the interface between the two phases has a constant value. This is a reasonable assumption for systems where there is no significant atomic matching but this is only likely if one of the two phases is non-crystalline, for example the internal oxidation of copper–silicon to precipitate amorphous silica. In most metallic precipitation reactions, two crystalline phases are involved with a very high likelihood of

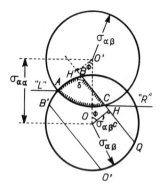

Fig. 13. The Wulff construction for a precipitate with a low-energy coherent interface BH, $\sigma_{\alpha\beta}{}^c$, between the grain-boundary precipitate and the upper grain at a particular angle ϕ between the habit plane and the boundary plane (after LEE and AARONSON [1975]).

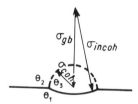

Fig. 14. The equilibrium shape for a precipitate "cube–cube" oriented with the upper grain, with a small value of the coherent energy, σ_{coh}, and a larger value of the incoherent energy, σ_{incoh} (after DOHERTY [1982]).

some atomic matching. This may not make a very great difference to the theory of nucleation on dislocations but for nucleation on grain boundaries it is known that a significant difference from the classical theory is produced. This arises since, if the precipitate has an orientation relationship giving good atomic fit with one crystal, then it is very unlikely to have any atomic matching with the other grain at a grain-boundary nucleation site. The assumption of a constant interfacial energy is then invalid.

The situation of grain-boundary nucleation with lattice matching has been discussed by LEE and AARONSON [1975], who have developed a modification to the Gibbs–Wulff theorem for grain-boundary nucleation (figs. 13 and 14). They show that the expected shapes are very different from that of fig. 12 and consequently the effectiveness of grain-boundary nucleation sites is very different from that given by eq. (21). Lee and Aaronson then found that the barrier to nucleation is reduced still further, by factors of up to 100, depending on the ratio $\sigma_{coh}/\sigma_{incoh}$, and the orientation of the facet plane with respect to the grain-boundary plane. The largest reduction occurs when the facet plane is nearly parallel to the grain boundary, ϕ tending to zero (fig. 13).

2.1.5. Experimental studies of nucleation

Despite the very extensive range of experimental investigations on precipitation processes in metals, as described for example in the recent conference on Precipitation Processes (RUSSELL and AARONSON [1978]), there are very few investigations that have attempted to test the various theories of nucleation quantitatively. As previously mentioned, however, there are numerous experimental observations that are in agreement, at least in outline, with the main predictions of the theories. Amongst the qualitative observations are the following:

(i) There is invariably an orientation relationship between the nucleated phase and the parent crystal lattice, and this orientation relationship is apparently always one that allows for good atomic matching between the two crystals, so that the interfacial energy will be minimized.

(ii) At high supersaturations it is commonly found that a metastable precipitate structure is formed rather than the stable phase. In every case where this has been studied, the metastable phase has better atomic matching with the matrix than has the stable phase. This has already been discussed in the aluminium–silver system but is perhaps better known with reference to the aluminium–copper system (fig. 15). The phases that are precipitated before the equilibrium θ-phase, $CuAl_2$, are the disc-like GP zones and the tetragonal θ' which is fully coherent with the aluminium matrix on the cube planes. This pattern of precipitation is found in a wide range of metallic and nonmetallic solid-state precipitation reactions, as reviewed by RUSSELL and AARONSON [1978].

(iii) Owing to the strong barrier to homogeneous nucleation provided by the interfacial energy and the potency of many heterogeneous nucleation sites in solids, homogeneous nucleation is only found with either extremely high free-energy driving forces or with precipitates with very low values of the interfacial energy. Typical examples of homogeneous nucleation are found for the precipitation of GP zones or intermediate phases which have the same crystal structure as the matrix. In addition to the many different zones formed in precipitation-hardened aluminium alloys (LORIMER [1978]) other examples are fcc cobalt precipitated in copper–cobalt alloys (SERVI and TURNBULL [1966]) and ordered fcc precipitates of Ni_3Al in nickel–aluminium alloys (KIRKWOOD [1970], HIRATA and KIRKWOOD [1977] and WENDT [1981]). These three experiments will be discussed in more detail here and in § 2.4.2, since they also have provided some of the few examples of quantitative tests

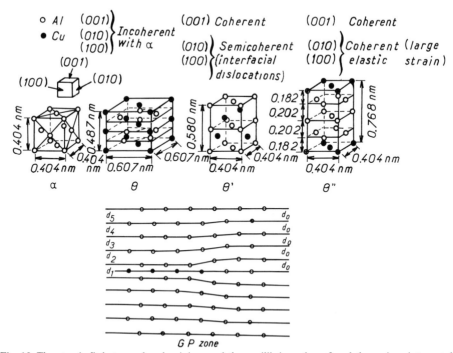

Fig. 15. The atomic fit between fcc aluminium and the equilibrium phase θ and the various intermetallic phases θ' and θ'' and the GP zones in Al–Cu system (after HORNBOGEN [1967]).

of the kinetic theory of homogeneous nucleation.

(iv) As described in §2.1.4., the variation of heterogeneous nucleation sites with supersaturation seems also to be in good agreement with the predictions of the classical theory of heterogeneous nucleation, but with the exception of Aaronson's investigation of grain-boundary nucleation [AARONSON [1979]) there again seems to have been no published quantitative tests of the theory for grain-boundary nucleation and none at all for nucleation on dislocations.

The classic test of *homogeneous nucleation* theory in metallic solids was provided by SERVI and TURNBULL [1966]. They studied the precipitation of fcc cobalt precipitates by an in-situ technique: electrical resistivity measurements on supersaturated copper–cobalt alloys that had been directly quenched to the precipitation temperature after solution treatment. The overall kinetics of the reaction were close to those expected for a nucleation-and-growth reaction with nuclei forming only in the earliest stages of the reaction. The bulk of the transformation appeared to show diffusion-controlled growth of a near-constant number of nuclei. Servi and Turnbull reported however that the agreement with the predictions of this model was not exact. As described in various reviews (for example, TURNBULL [1956]), the fraction $f(x)$ of the supersaturated solute that remains in the solution during diffusion-controlled growth with a constant number of growth centres is given by

$$f(x) = 1 - \exp(-t/\tau)^n, \tag{22}$$

with the exponent $n = 3/2$.

Over a range of cobalt contents, from 1.0 wt% to 2.69 wt% Co, the observed value of the exponent n fell from 1.95 to 0.97. The investigators ignored this discrepancy and calculated the density of nuclei from the time constant τ. From these precipitate densities, which were soon afterwards confirmed by electron microscopy (see TANNER and SERVI [1966]) Servi and Turnbull were able to determine initial rates of nucleation. They were then able to show that these, experimental, nucleation rates agreed closely with the predictions of steady-state nucleation theory [eq. (12)]. The observed apparent interfacial energy was then found to be 0.20 J/m^2 which compared very closely to a calculated value, based on the quasi-chemical bonding model, of 0.23 J/m^2. They reported a difference in the calculated and observed pre-exponential terms in eq. (12) of a factor of 100, the observed value being smaller than the calculated value. Two corrections to this have been found – a mis-placed π in Servi and Turnbull's equation 2 and the incorrect insertion of the atomic fraction of cobalt in their equation 6 (RUSSELL [1970]). The effect of these corrections is that the steady-state theory apparently underestimates the rate of nucleation by a factor of 10^5.

Despite this failure, the ability of the theory to come close to giving the right variation of nucleation rate with temperature and supersaturation and to give a good value for the interfacial energy seems to indicate that the theory is rather successful. A similar success has been reported by STRUTT [1974] for nucleation of a coherent metastable phase in the LiF–MgF$_2$ ionic system studied by ionic conductivity.

References: p. 1024.

RUSSELL (1970) has described a similar successful test of nucleation theory, in a precipitation reaction in an oxide glass system.

More recent tests of nucleation theory have been carried out in the nickel–aluminium system. These studies take into account the potential loss of precipitates that have nucleated, by the process of Ostwald ripening (§ 2.4.2).

LANGER and SCHWARTZ [1980] have reconsidered nucleation theory in its application to the intrinsically simpler case of unmixing of liquids at small supercoolings. In this type of reaction there is no complications due either to the anisotropy of interfacial energy or to strain energy effects. Despite earlier reports, for example by HEADY and CAHN [1973], of a marked failure of the theory in a case of liquid unmixing, Langer and Schwartz were able to show that the basic theory of nucleation remained valid, or rather that there was no unambiguous evidence for a significant failure of the theory. In the unmixing experiments in liquids close to the unmixing temperature the experimental problem appears to be of observing an initial rate of nucleation rather than the onset of a significant amount of unmixing. Observable unmixing, the onset of "cloudyness", involves diffusion close to the unmixing temperature and this is an inherently slow process.

It should be recognized of course that the current theory of nucleation cannot be expected to be really quantitatively reliable since, as has already been mentioned, the use of bulk parameters like the interfacial energy and free energy for very small nuclei must be a doubtful procedure – although essential in the absence of an accurate way of describing the energies of small clusters of atoms. Such atomic-energy calculations might be better made for ionic materials, or ones bonded by Van der Waals forces, for which more reliable interatomic calculations can be made.

The only reports, known to the author of this chapter, on quantitative testing of the theory of *heterogeneous nucleation* are that of LANGE [1979] as presented by AARONSON [1979] and that of PLICHTA et al. [1980]. The first investigation was of the grain-boundary nucleation rate of ferrite from austenite in high-purity iron–carbon alloys. Considerable care was taken to distinguish grain-edge from grain-boundary nucleation: only data for nucleation on the boundaries between two austenite grains were considered. In order to make a comparison with the experimental results, an equilibrium shape for the ferrite nucleus on the grain boundary was needed. The only model that was at all successful was a pill-box nucleus shape for β on the boundary between two α grains, where all the interfaces are partially or fully coherent. It was reported that all other nucleus shapes predicted nucleation rates orders of magnitude smaller than the rates observed. As Aaronson has pointed out, a pillbox model requires the habit plane of the precipitate to be accurately parallel to the appropriate plane in both the austenite grains, a rather unlikely situation. However, since austenite is unstable in the iron–carbon system, room-temperature crystallographic studies to test this hypothesis could not be carried out.

The other study on heterogeneous nucleation kinetics was the observations by PLICHTA et al. [1980] on nucleation of the massive transformation in two Ti alloys. In the *massive transformation*, discussed further in § 2.2, the driving force for nucleation is determined from the under-cooling, ΔT, below the temperature where two phases with the same composition have the same free energy. As in the earlier study in iron, the spherical cap model underestimated the observed nucleation rates by, in this case, thousands of orders of magnitude! Agreement could again be achieved only by use of the pillbox model with a low-energy coherent interface, again with only a small fraction of potential grain-boundary nucleation sites appearing to be active.

Two ideas seem to follow from these two sets of very interesting results. Firstly there is the need for experiments on a system where the orientations of both phases can be studied after precipitation in a system with reliable thermodynamic data, for example precipitation in a dilute alloy. Secondly, there is the possibility that heterogeneous nucleation at grain boundaries may not depend on the general grain-boundary structure but may occur mainly at grain-boundary defects such as the intrinsic and extrinsic dislocations known to exist in metallic grain boundaries (ch. 10B, §2.2). It has been reported, for example by DESALOS *et al.* [1981], that after controlled rolling of low-alloy steels there is an enhanced density of nucleation of ferrite from austenite (fig. 16). This could indicate that nucleation is enhanced on dislocations absorbed on austenite grain boundaries.

PLICHTA and AARONSON [1980] attempted to provide an answer to the first problem by observing the orientation relationships for boundary-nucleated massive precipitates in Ag–26 at% Al which has the same bcc–hcp pair of phases as in the Ti alloys studied. Plichta and Aaronson showed that 46 out of 47 boundary precipitates did indeed have the "Burgers" relationship, $(0001)\|(110\}$, $[11\bar{2}0]\|\langle1\bar{1}1\rangle$, with one of the two matrix grains, but rarely with both. They were able to provide evidence that for about half of the non-Burgers-related orientations some indication of good atomic fit was possible. Unfortunately no nucleation-rate experiments were carried out in the crystallographically investigated system.

Additional evidence for the failure of the classical model of boundary nucleation has been described by BUTLER and SWANN [1976]. They studied by TEM the density of heterogeneously nucleated precipitates in Al–Mg–Zn alloys. They studied precipitates forming under higher driving forces, giving much higher precipitate densities than studied in the observations of LANGE [1979] and PLICHTA *et al.* [1980]. Butler and Swann showed direct evidence for preferential nucleation on defects at boundaries, notably at grain-boundary steps (fig. 17) and at dislocations in low-angle boundaries. It is therefore possible, in experiments such as those described by

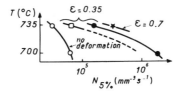

Fig. 16. Observed rate of nucleation, after 5% transformation of γ to α in a low-carbon steel, with no deformation and with two levels of prior hot deformation in the austenite (after DESALOS *et al.* [1981]).

References: p. 1024.

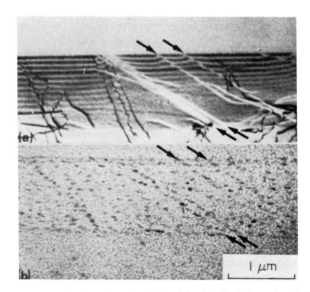

Fig. 17. Precipitation on a grain boundary in Al–Zn–Mg: (a) Grain-boundary ledges, arrowed, and dislocations in the as-quenched alloy; (b) the same area after 75 s ageing at 240°C, with preferential nucleation on the ledge and on some dislocations. (From BUTLER and SWANN [1976].)

LANGE [1979] and PLICHTA et al. [1980]), that the precipitates seen may indeed have come from *defects on the grain boundaries* and that this accounts for the high rate of nucleation. Other aspects of Butler and Swann's observations are in agreement with the analysis of LEE and AARONSON [1975]. Butler and Swann showed that the density of precipitates varied from boundary to boundary and along an individual boundary when the boundary plane varied; this was considered by Butler and Swann as being due to the change of angle ϕ between the boundary plane and the precipitate habit plane. In some cases as the boundary plane varied a different precipitate orientation was found. In most TEM studies of grain-boundary precipitation, as for example that of WILLIAMS and EDINGTON [1976], the precipitates at a boundary are always found to be one of the variants of the precipitates nucleated inside one of the two grains that join at the boundary.

   In the absence of additional evidence, particularly on the role of grain-boundary dislocations, it is probably safest at present to regard the existing models of heterogeneous nucleation as being not yet fully established. However, the analysis of Lee and Aaronson seems likely to be valid for general boundary nucleation but to need revision to take into account the higher level of heterogeneities – those on the grain boundary itself. These heterogeneities probably will dominate the reaction at low driving forces where nucleation is studied by optical microscopy.

   In addition, there is a tremendous lack of reliable interfacial-energy information needed to make use of the theory (HONDROS [1978]). BUTLER and SWANN [1976], for example, report evidence that higher solution-treatment temperatures that will reduce segregation of impurities and so raise the boundary energy, cause the expected *increase* of nucleation rate. Such qualitative ideas, though valuable, do not of course substitute for quantitative evaluation of the interfacial energy under various conditions.

2.2. Growth processes

2.2.1. Growth without change of composition

When the growing precipitate phase and the matrix have the same composition then the growth process merely requires atom transfer across the interface. This situation is equivalent to the movement of a high-angle grain boundary in pure metals (ch. 25, §3.4). One complication in polymorphous phase changes, as with grain boundary migration, it the possibility of *solute drag* (CAHN [1963] and LÜCKE and STÜWE [1963]). The solute drag phenomenon arises if a trace of an alloy addition is adsorbed onto the interface between the growing and shrinking phases. The rate of motion of the interface can then be greatly retarded unless the reaction is sufficiently fast for the interface to "escape" from the solute atmosphere. For grain-boundary movement it is difficult to acquire a sufficient driving force for such escape, but in solid-state phase changes the driving force is likely to be sufficiently large for solute drag not to be a significant problem in many situations, for example with supercoolings of more than about 50 K [eq. (4)].

For a phase transformation where the growing α crystal is a different phase from the β matrix, a new situation arises which is not present when the two crystals are the same phase, as in grain-boundary migration. The distinction is that any impurity will almost always prefer to concentrate in one or other of the two phases, that is, the partition coefficient,

$$K_{\beta\alpha} = C_\alpha/C_\beta, \tag{23}$$

will not be unity.

This is illustrated by fig. 18 which shows a phase transformation in a dilute binary alloy where the solute segregates to the high-temperature β-phase and $k_{\beta\alpha} < 1$.

The free-energy–composition curve at a temperature, T_0, below the equilibrium transformation temperature, T_e, is also seen in fig. 18, and at this temperature a critical composition $C(T_0)$ is indicated. For an alloy with less solute than this critical concentration, the β-phase is allowed to transform directly to α without change of composition. This is often considered in terms of the temperature to which an alloy must be cooled to allow this *diffusionless transformation* to be thermodynamically possible, region II of fig. 18a. It is clear that a diffusionless transformation to α of the same composition is not the most stable product in region II, but is a metastable product whose formation will take place only if it can occur significantly faster than the formation of solute-depleted α of composition C_α. In region I of the two-phase region, formation of solute-depleted α is the only possible product. At a temperature

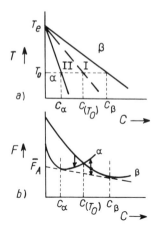

Fig. 18. (a) Phase diagram showing equilibrium below an equilibrium transition temperature, T_e; (b) the free-energy–composition curve to show the possibility of a diffusionless transformation, β to α, in region II. In region I only a reaction to give solute-depleted α can lower the free energy.

below the two-phase field the diffusionless product is then the most stable product. Since the diffusionless transformation does not require long-range solute transport, it will generally occur faster than any reaction requiring long-range diffusion (§ 2.2.2). Diffusionless transformation is known to occur in the class of transformations described as *massive* after the shapes of the product grains often found (MASSALSKI [1970]). Since no metal can ever be absolutely pure, any polymorphous reaction, if it is to occur without long-range diffusion of the impurity, must also be treated as an example of a massive transformation. In the following simple analysis it is assumed that the product and parent phases have the same composition and moreover that there is no solute build-up or depletion at the interface, in other words there is no solute drag effect considered.

  Figure 19 shows the expected variation of free energy per atom across a β–α interface below a transition temperature, when the low-temperature α-phase is more stable. There is a difference $\Delta F_{\beta\alpha}^{a}$ in free energy and an activation barrier ΔF_a to atomic migration across the interface. In this treatment the interface is assumed to be incoherent, which has two important consequences; firstly that the barrier to atomic transport is likely to be of the order of the activation energy for grain-boundary migration rather than the higher energy for bulk diffusion; the second effect is that there should be no difficulty for atoms to find a site for them to "sit" on after crossing the interface (§ 2.2.3).

  The flux of atoms from β to α, $J_{\beta\alpha}$, and the reverse flux, $J_{\alpha\beta}$, are given by:

$$J_{\beta\alpha} = \nu C_{ac} \exp(-\Delta F_a/kT), \tag{24}$$

$$J_{\alpha\beta} = \nu C_{ac} \exp\left[-\left(\Delta F_a + \Delta F_{\beta\alpha}^{a}\right)/kT\right], \tag{25}$$

where ν is the atomic vibration frequency of atoms at the interface, and C_{ac} is the

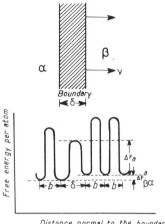

Fig. 19. Thermally activated boundary motion; there is an activation barrier ΔF_a for atoms to cross the boundary and a driving force, $\Delta F_{\beta\alpha}^a$ per atom, for growth of α at the expense of β.

accommodation coefficient, the fraction of atoms that, having left one of the crystals, can immediately find a site in the other crystal. C_{ac} for an incoherent interface is usually assumed to be constant and nearly unity or, more logically, about 0.5. The net flux of atoms is given by the difference of the fluxes $J_{\beta\alpha}$ and $J_{\alpha\beta}$:

$$J_{\beta\alpha}(\text{net}) = \nu C_{ac}\left[\exp(-\Delta F_a/kT)\right]\left[1 - \exp(-\Delta F_{\beta\alpha}^a/kT)\right]. \tag{26}$$

For small undercoolings $\Delta F_{\beta\alpha}^a \ll kT$, so the difference term in eq. (26) becomes $\Delta F_{\beta\alpha}^a/kT$. The velocity of the interface, v, is the product of the net flux per atom site and the atomic spacing, b, in the product α-phase:

$$v = bJ_{\beta\alpha}(\text{net}) = M\Delta F_{\beta\alpha}^a. \tag{27}$$

The mobility, M, is obtained from eq. (26) and (27) as:

$$M = \nu b C_{ac}\exp(-\Delta F_a/kT)/kT. \tag{28}$$

There are alternative definitions of interfacial mobility based on the replacement of the free-energy difference, $\Delta F_{\beta\alpha}^a$, in eq. (27) by either a temperature difference, ΔT_i, an interface supercooling (ch. 9, §4.1), or a composition difference in a phase change involving composition changes (§2.2.2).

The simple linear relationship between the interface velocity and the driving force [eq. (27)] fails if the accommodation coefficient, C_{ac}, is not a constant. This may occur for certain faceted interfaces where there is a difficulty in starting a new atomic layer. This is often found in solidification, particularly of high-entropy-of-freezing materials (ch. 9, §4.2). For solid-state transformations the same situation may arise for coherent interfaces between phases of different structure or orientation (§2.2.3).

2.2.2.   Transformations involving long-range diffusion

In region I of the phase diagram in fig. 18, the formation of the low-temperatu...
phase requires the diffusion of solute away from the low-solute α-phase. On the
other hand, for the typical precipitation reaction, such as the growth of solute-rich β
from supersaturated α (fig. 2), the reaction proceeds by the diffusion of solute to the
growing β-precipitates; the two situations are shown in fig. 20. The composition of
the matrix phase at the interface is C_i, which is shown as lying somewhere between
the equilibrium value (C_β for the polymorphous reaction, C_α for the precipitation
reaction) and the bulk matrix composition well away from the new phase, C_0. For a
spherical, solute-rich β-precipitate of radius r the composition outside the precipitate
is a function of radial distance R and time t since the nucleation of the precipitate.
The flux J_R across a sphere of radius R is given by:

$$J_R = -4\pi R^2 D \, (dC'/dR)_R \tag{29}$$

where C' is given as moles of solute per unit volume and the relationship between C'
and the atomic fraction of solute C is:

$$C' = C/V_m. \tag{30}$$

V_m is the volume of one g-mole of alloy, which for simplicity is assumed independent
of composition. With this change eq. (29) becomes:

$$J_R = -4\pi R^2 D \, (dC/dR)_R/V_m. \tag{31}$$

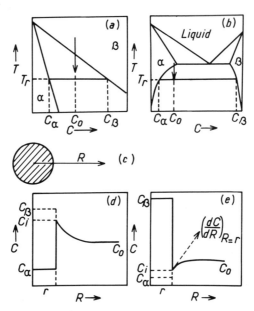

Fig. 20. Diffusion around a growing precipitate. In the matrix phase-change (a and d), solute is diffusion
away from the low-solute α, while in (b) and (e) solute is diffusing towards the solute-rich β.

The volume of the precipitate is $\frac{4}{3}\pi r^3$, and the *increase* in the number of solute moles, n, in the precipitate above that of the matrix from which it forms, is:

$$n = \tfrac{4}{3}\pi r^3 (C_\beta - C_i)/V_m, \tag{32}$$

$$dn/dr = 4\pi r^2 (C_\beta - C_i)/V_m. \tag{33}$$

The flux to the precipitate, dn/dt is then given by:

$$dn/dt = 4\pi r^2 (C_\beta - C_i)(dr/dt)/V_m. \tag{34}$$

This flux will be that provided by diffusion [eq. (31)], with $R = r$. Provided that there is a low supersaturation $\Omega = (C_0 - C_i)/(C_\beta - C_i)$, then during the time that it takes a solute atom to diffuse across the solute-depleted region around the precipitate, the solute distribution and the instantaneous growth rate, dr/dt, can both be regarded as constant. This means that flux-balancing eqs. (34) and (31) will give:

$$\frac{4\pi r^2}{V_m}(C_\beta - C_i)\frac{dr}{dt} = \frac{4\pi R^2 D}{V_m}\frac{dC}{dR}. \tag{35}$$

The change of sign in the flux equation (31) arises since the negative flux is one in the $-R$ direction, which is a *positive* flux for precipitate growth. Rearranging eq. (35) yields:

$$\frac{dR}{R^2} = \frac{D\,dC}{r^2(dr/dt)(C_\beta \text{-} C_i)}. \tag{36}$$

Equation (36) can then be integrated from $R = r$, $C = C_i$, to $R = \infty$, $C = C_0$, and this gives:

$$\frac{1}{r} = \frac{D(C_0 - C_i)}{r^2(C_\beta - C_i)(dr/dt)}. \tag{37}$$

Equation (37) can then be rearranged to give the growth rate, that is, the interface velocity, v:

$$v = dr/dt = \frac{D}{r}\frac{(C_0 - C_i)}{(C_\beta - C_i)}. \tag{38}$$

The interface velocity is also given by a relationship derived from the local departure from solute equilibrium $(C_i - C_\alpha)$, by use of a slightly modified definition of mobility [eq. (28)]. The velocity is then given by:

$$v = dr/dt = M'\frac{(C_i - C_\alpha)}{(C_\beta - C_i)}. \tag{39}$$

There is a straightforward relationship between the two mobilities, M and M', which can be obtained by relating the local departure from solute equilibrium at the interface, $(C_i - C_\alpha)$, to the free energy per atom ($\overrightarrow{IJ}$ in fig. 1) with $C_i = C_0$.

The two equations for the growth rate can be combined to determine the

References: p. 1024.

previously undefined parameter, the interface composition in the matrix phase, C_i:

$$rM'/D = (C_0 - C_i)/(C_i - C_\alpha). \tag{40}$$

The dimensionless parameter, rM'/D, which is the composition equivalent of the Nusselt number in heat flow, determines the value of C_i. With $rM'/D \gg 1$, we have $C_i = C_\alpha$, the condition of *local equilibrium* (HILLERT [1975]). This is the situation normally described as that of *diffusion-controlled growth*, since here the interface reaction is so fast that any further increase of mobility does not further accelerate the reaction. The rate of growth is then fully described by eq. (38) with, of course, $C_i = C_\alpha$. The opposite extreme, with $rM'/D \ll 1$, gives $C_i = C_0$. This arises when the diffusion is so rapid, and atom transport across the interface so slow, that the solute-depleted region around the precipitate is eliminated. We then have the condition usually described as *interface-controlled growth*. The intermediate case, where $rM'/D \approx 1$ is usually described as *mixed control*.

Diffusion control. With $C_i = C_\alpha$, eq. (38) becomes:

$$r \, dr = D \frac{(C_0 - C_\alpha)}{(C_\beta - C_\alpha)} \, dt = D\Omega \, dt, \tag{41}$$

where Ω is the dimensionless supersaturation. Equation (41) can be integrated to give the square-root relation expected for diffusion control:

$$r = (2 D\Omega t)^{1/2}. \tag{42}$$

Interface control. From eq. (39) with $C_i = C_0$, we obtain, by integration, the linear relationship

$$r = M'\Omega t. \tag{43}$$

It should be clear from the role of the parameter rM'/D in controlling the reaction, that as $r \to 0$ the situation should tend towards interface control. This may only be detectable, however, for very small values of the critical radius of nucleation, r^* [eq. (10a)], since growth cannot occur until a critical nucleus has formed by thermal fluctuation. Even after nucleation, precipitates close to the critical radius will have their growth conditions modified by interfacial effects, with the solubility C_α increased by the Gibbs–Thomson equation (§2.2.6). The only conditions for which this type of interface control [eq. (43)] or even mixed control can be realistically expected would be for very low values of mobility under conditions of faceted growth. This is discussed in §2.2.3.

More general solutions for diffusion-controlled growth of both spheres and flat plates have been provided by AARON et al. [1970], including the exact solution for the general case without the approximations involved in eq. (35). For a sphere the radius is given by:

$$r = (\alpha D t)^{1/2}. \tag{44}$$

The dimensionless coefficient α is defined by:

$$\Omega = \frac{\alpha}{2} \exp \alpha \left[\exp(-\alpha) - (\alpha \pi)^{1/2} \mathrm{erf}(\alpha^{1/2}) \right]. \tag{45}$$

The approximate relationship of eq. (42) is valid for many precipitation reactions with a low supersaturation $\Omega < 0.1$, for which $\alpha = 2 \, \Omega$.

For growth of plates, the half-thickness a_x is given by:

$$a_x = 2\alpha'(Dt)^{1/2}, \tag{46}$$

$$\Omega = \alpha' \pi^{1/2} \exp\left(\frac{\alpha'^2}{2} \right) \mathrm{erfc}\left(\frac{\alpha'}{2} \right). \tag{47}$$

For oblate and prolate spheroids, that are reasonable approximations to the frequently found plate-like and needle-like precipitate shapes, HAM [1958] and HORVAY and CAHN [1961] have shown that the half-lengths, a_y, and half-thicknesses, a_x, are given by:

$$a_x = 2(\beta Dt)^{1/2}, \tag{48}$$

$$a_y = 2A(\beta Dt)^{1/2}. \tag{49}$$

The diffusion parameters β are obtained from the supersaturation and the aspect ratio $A(A = a_y/a_x)$ by complex equations of a form similar to eqs. (45) and (47) (see HORVAY and CAHN [1961], who provide figures showing how β increases with the supersaturation at different values of the aspect ratio). The assumptions that lead to eqs. (48) and (49) were merely that the precipitate was formed with a given shape, a plate or a needle, and from then on, growth occurred at local equilibrium all round the precipitate, that is, with $C_i = C_\alpha$. The analysis then predicts that the shape is preserved, that is, the precipitate grows with a constant aspect ratio. This result has been challenged following the application of stability theory to the growth process (MULLINS and SEKERKA [1963]; §2.2.6). However, as will be discussed in §2.2.6, the criticisms of the Ham, Horvay and Cahn analysis do not rule out its usefulness for the highly elongated "Widmanstätten" shapes. In fact, the Ham, Horvay and Cahn analysis seems to have very wide validity, except at high Ω ($\Omega \geqslant 0.5$).

2.2.3. Role of interface structure in growth processes

AARONSON [1962] pointed out that for the plate-like *Widmanstätten precipitates* in systems like Al–Ag$_2$Al (fig. 9, above), the broad faces of the precipitates have a very good atomic fit (§2.1.2). This is because the atomic positions on (0001) hexagonal planes are the same as on the {111} fcc planes in the matrix. For growth of either phase by single-atom transfer to the other structure, a very considerable rise in energy results (fig. 21a, b). This continues with a second atom (fig. 21c) and only becomes a reasonable structure with either three, five or six atoms added (fig. 21d, e). The obvious conclusion of this idea is that it will be very difficult for individual atoms to join the new precipitate on the broad face: in other words the accommodation coefficient C_{ac} in eqs. (24) and (25) will be close to zero, giving a

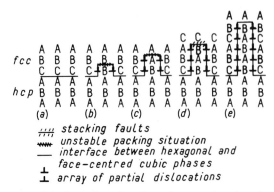

stacking faults
unstable packing situation
interface between hexagonal and
face-centred cubic phases
array of partial dislocations

Fig. 21. Problem of growth of the lower hcp phase from the upper fcc phase by random atomic jumps: intermediate structures (b) and (c) are very unstable compared to (d), a three-layer ledge, or (e), a five- (or six)layer ledge. (After MARTIN and DOHERTY [1976].)

very low mobility M or M'. The common solution to this problem is the formation of a *ledge* on the broad face of the precipitate. Atoms can then add to the *riser* of the ledge which will be incoherent, so the interface propagates forward by lateral motion of the ledges. When this model was first put forward by AARONSON [1962] there was little direct evidence to support the suggestion though a strong need, from the observed shapes of the Widmanstätten precipitate, for an explanation of the very high aspect ratios found; these are often of the order of 100 or more (FERRANTE and DOHERTY [1979]). AARONSON [1962] and AARONSON et al. [1970] argued that the high aspect ratio of the Widmanstätten plates and of needles was due to the need for ledge growth at the coherent parts of the precipitate interface. Following the work of WEATHERLY and SARGENT [1970] and WEATHERLY [1971] it has become clear that the growth ledges postulated by Aaronson do almost invariably actually occur on the broad faces of Widmanstätten precipitates, and that the coherent areas of these interfaces do often show reduced mobility compared to incoherent parts of the interface (e.g. PURDY [1978]). Figure 22 shows the dislocation-like contrast shown by the ledges on faceted Mg_2Si precipitates in an aluminium alloy.

It was implicitly assumed by Aaronson that ledged interfaces would always show interface-controlled motion, that is, ledged interfaces would be *inhibited* and not grow at the rates expected for diffusion control. This does not follow automatically however, as will be seen in the next section.

The structure of growth ledges, on good-fit interfaces, need not be based on the ledge moving only by individual random atomic addition to the growing phase, usually by atomic addition at the risers. As was shown by NICHOLSON and NUTTING [1961] and by HREN and THOMAS [1963], growth of a hexagonal precipitate such as Ag_2Al from a fcc matrix can occur by glide of an $a/6 \langle 112 \rangle$ dislocation on a $\{111\}$ habit plane (fig. 23). The glide of a single such dislocation will cause two atomic layers of the matrix fcc structure to transform to the hexagonal structure, and to cause shear of the matrix structure. This idea raises a fascinating problem, since this

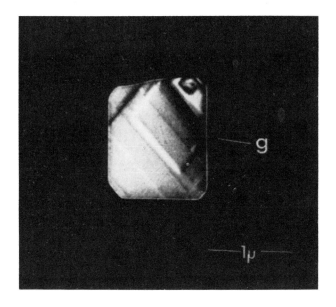

(a)

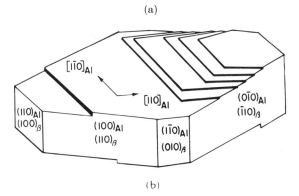

(b)

Fig. 22. Growth ledges on a Mg₂Si plate-like precipitate seen by dark-field electron microscopy (from WEATHERLY [1971]).

```
B  B │A  A  A  A│B  B
C  C │B  B  B  B│C  C
A  A │C  C  C  C│A  A
B  B │A  A  A  A│B  B
C  C │B  B  B  B│C  C
A  A  A  A  A  A  A  A
B  B  B  B  B  B  B  B
A  A  A  A  A  A  A  A
```
*Perfect crystal after $\frac{a}{6}[\bar{1}1\bar{2}]$
displacement of partial
dislocation.*

Fig. 23. Growth of hcp phase from fcc by glide of partial dislocation: two-layer ledge. (After DOHERTY [1982].)

References: p. 1024.

type of transformation when occurring in cobalt and in cobalt–nickel alloys (GAUNT and CHRISTIAN [1959]) is apparently fully martensitic (see ch. 15). However, in dilute aluminium–silver alloys it is not possible thermodynamically for diffusionless precipitation of the hexagonal phase to occur, since the fcc solid solution will not contain enough silver for a diffusionless transformation. There is in addition overwhelming evidence that the precipitates are indeed silver-rich compared to the solid solution (e.g., LAIRD and AARONSON [1969] and FERRANTE and DOHERTY [1976]). Further evidence for the dual martensitic and diffusional character of the fcc to hcp transformation in aluminium–silver alloys is provided by the observations of AARONSON et al. [1977] on the heterogeneous nucleation of the intermediate γ'-phase on dislocations. The precipitates were observed to form by the glide of a partial Shockley dislocation, away from the other Shockley partial which together with the first partial had made up a perfect lattice dislocation in the supersaturated solid solution before precipitation (fig. 24). By assuming that the transformation was a diffusional one, that is, that the precipitates were indeed the expected silver-rich metastable γ'-phase, the authors were able to predict successfully the required undercooling for this type of nucleation to occur. It is clear that this reaction at least

Fig. 24. Al–15 wt% Ag, solution-treated at 500°C, directly quenched to 360°C and held there for 60 s, showing the γ' plates forming behind moving partial dislocations (from AARONSON et al. [1977]).

has the shear characteristics of a martensitic reaction, while also involving a change of composition. This combination seems perfectly acceptable for a case where the reaction requires diffusion but the interface process can occur by a shear process. Shear by dislocation glide parallel to the habit plane gives the invariant-plane strain characteristic of the martensitic transformation (GAUNT and CHRISTIAN [1959]).

Irrespective of the mechanism of ledge movement across the coherent interface, the occurrence of ledged interfaces (see for example the review by AARONSON [1974]) indicates that that this is an important phenomenon and that the diffusional mobility of these interfaces will be controlled by the nucleation and growth of the ledges (§ 2.2.4).

It should be recognized, however, that growth ledges of any form should only be expected for coherent interfaces between phases which have *different* structures or orientations. The difficulty identified in fig. 21 only arises since the atomic fit is restricted to isolated atomic planes. If the two phases have identical structure and orientation as in, for example, the so-called "cube–cube" relationship, then there is no difficulty in adding atoms onto the growing phase – in fact there is no interface step at all since the arrival of the required solute at the interface changes the structure automatically from that of the matrix to that of the precipitate. Examples of this type of cube–cube relationship, already cited in § 2.1.2, are β–γ brass and Ni–Ni$_3$Al precipitates, a further example of growing interest is the similar structure observed in aluminium–lithium alloys containing Al$_3$Li (e.g., WILLIAMS and EDINGTON [1975]). As expected, despite a large quantity of experimental observations, particularly in the nickel alloys, there are no unambiguous reports of ledge structures at the coherent interfaces between cube–cube matrix interface pairs.

2.2.4. Growth of ledged interfaces

JONES and TRIVEDI [1971,1975] have analyzed the growth kinetics expected for a ledged interface. They considered ledges of height h and spacing λ. If the ledges have an instantaneous velocity v_s then the overall interface velocity, v_i is:

$$v_i = v_s h / \lambda. \tag{50}$$

Jones and Trivedi started by assuming a constant ledge velocity and then determined a solution to the diffusion equation that would describe the solute field feeding solute to the ledge. They produced a relationship between the dimensionless Peclet number, $p = v_s h / 2D$, and the dimensionless supersaturation, Ω, for an isolated ledge. Their solution, for the case when there is diffusion-controlled motion of the riser of the ledge, is:

$$\Omega = 2 p \alpha(p). \tag{51}$$

The function $\alpha(p)$ can be evaluated numerically from data provided by JONES and TRIVEDI [1971]. Typical values of p are about 0.03 at $\Omega = 0.1$, 0.08 at $\Omega = 0.2$ and 0.24 at $\Omega = 0.4$. It was reported by JONES and TRIVEDI [1971] that the solute field about a moving ledge has a finite extent (fig. 25), and that this field becomes more localized around the ledge as the supersaturation increases and the ledge velocity

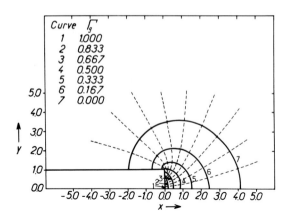

Fig. 25. Analytical model of ledge growth: Predicted contours of equal solute content about a ledge moving with a constant velocity. Contours are for different values of the normalized supersaturation, $\Gamma_g = 1 - \Omega/\Omega_0$. With $\Omega_0 = (C_0 - C_\alpha)/(C_\beta - C_\alpha)$ the initial supersaturation and $\Omega = (C - C_\alpha)/(C_\beta - C_\alpha)$ the remaining supersaturation, the value of Γ_g is 0.0 for zero supersaturation (equilibrium solubility), and 1.0 for full supersaturation. (After JONES and TRIVEDI [1971].)

rises. An important consequence of this analysis is that, provided ledges are more widely spaced than the extent of the diffusion field, each ledge can move independently and with a constant ledge velocity, and so the interface velocity v_i itself is a constant:

$$v_i = 2 Dp/\lambda. \tag{52}$$

DOHERTY and CANTOR [1983] have drawn attention to a striking problem raised by this result when considered in relation to the expected velocity for the *diffusion-controlled* growth of a plate, which is obtained from eq. (46) as:

$$v_i(\alpha') = \alpha'(D/t)^{1/2}. \tag{53}$$

The difficulty is that since eq. (53) contains time in the denominator of the velocity function, the velocity of the *mobile*, diffusion-controlled interface falls steadily with time, eventually becoming smaller than the velocity predicted for the ledge model of the inhibited interface. This happens for any constant value of the ledge spacing, λ, however large it might be, provided only that it is finite. Moreover, the velocity advantage of the ledged interface continues to grow with longer and longer times, becoming, at infinite time, infinitely larger. Although fixed ledge spacings and infinite growth times are not experimentally realistic, the "thought experiment' described above indicates a major difficulty with the Jones and Trivedi analysis. DOHERTY and CANTOR [1983] were unable to find an error in the analysis, but tested the results by computer simulation of the diffusion field around a pair of ledges. The simulation caused ledges to be formed in the fully supersaturated matrix and to grow by solute flux to the ledge riser, which was assumed to show local equilibrium, $C_i = C_\alpha$. The rest of the interface was not allowed to accept solute. There was inevitably a higher flux to the top of the ledge than to the base, but the ledge shape was held stable by distributing this arriving flux uniformly along the riser; this makes the implicit assumption of rapid interface diffusion along the incoherent riser interface. This seems a reasonable assumption for a model which involves the diffusion-controlled growth of the individual riser.

The results of the finite-difference computer simulation were very different from the analytical results provided by Jones and Trivedi. As would be expected, the high local flux from a fully supersaturated matrix gave a high initial ledge velocity but this velocity fell steadily with time. For very widely separated ledges the total ledge displacement, X_s, showed an initial time exponent of about 0.76, which increased to 0.88 at longer times for supersaturations up to 0.1 (fig. 26). The origin of this ever-decreasing ledge velocity was the steady movement of isoconcentrates away from the ledge as time increased (fig. 27). The only condition for which a constant velocity of the ledge was found was for high supersaturations, $\Omega = 0.5$, and long times, and here the solute distribution was highly asymmetric (fig. 27). The asymmetry appears to be physically reasonable since the ledge is growing rapidly and therefore "catching up" with the isoconcentrates in front of the ledge but leaving a huge "trail" of depleted matrix behind it.

An additional difference between the computer model and the earlier analytical result was found when the interledge spacing λ was reduced (fig. 28). When the growth times reached values (λ^2/D) where the diffusion fields were found to overlap, the time exponent of the ledge displacement fell to 0.5, which, by eq. (52), gives the same time exponent for the interface displacement, a_x. This is the time exponent for full diffusion-controlled growth of the interface [eq. (46)]. The computer results also showed, after the overlap of the diffusion fields from adjacent ledges, that the interface displacement not only had the correct time exponent but

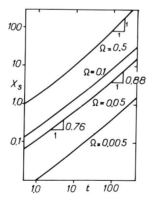

Fig. 26. Computer simulation of ledge growth. The total ledge displacements X_s in units of ledge height h, for dimensionless time of growth, $t = h^2/D$, for various supersaturations. Only for $\Omega = 0.5$ was a constant velocity found at long times. Isolated ledge. (After DOHERTY and CANTOR [1983].)

References: p. 1024.

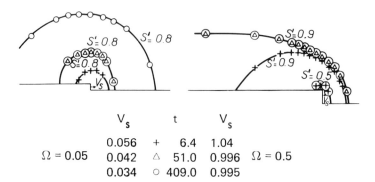

$$V_s \qquad t \qquad V_s$$

	0.056	+	6.4	1.04	
$\Omega = 0.05$	0.042	△	51.0	0.996	$\Omega = 0.5$
	0.034	○	409.0	0.995	

Fig. 27. Computer simulation of ledge growth. Predicted contours of equal solute supersaturation at growing ledge for three values of t in units of h^2/D, with the corresponding values of the ledge velocity, V_s, in units of D/h, for two different supersaturations, 0.05 and 0.5. S' a normalized supersaturation, defined by $S' = (C - C_\alpha)/(C_0 - C_\alpha)$; for the full initial supersaturation $S' = 1$. Isolated ledges. (After DOHERTY and CANTOR [1983].)

also, to within a few percent, the displacement had the correct value [from eq. (46)] for the *slope* of a plot of a_x versus $t^{1/2}$. The plots had time delays corresponding to the initial periods of slow growth, where solute close to the interface could not join the precipitate except at the ledge. At times where the diffusion fields had over-lapped, the behaviour was as though the interface could accept solute everywhere. This occurs, apparently, since the diffusion fields around interacting ledges are very similar to the diffusion field about a flat interface which can accept solute at all points. It appears therefore that the computer model is giving physically reasonable results while the analytical model at present does not. This is an unsatisfactory situation that needs to be resolved.

Unsatisfactory as the theory is for ledge growth, there is no effective theory at all for ledge *formation*. WEATHERLY [1971] has applied conventional nucleation theory to the formation of a "pillbox" disc on the surface of a precipitate, but the predicted

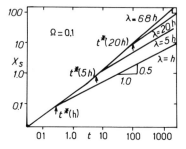

Fig. 28. Computer simulation of ledge growth: closely spaced ledges, $\lambda = h$, $\lambda = 5h$, $\lambda = 20h$ and $\lambda = 68h$. Total ledge displacement as a function of time of diffusion. For $t > t^*$ the diffusion fields have interacted and a $a_x \sim (t)^{1/2}$ relationship is found. (After DOHERTY and CANTOR [1983].)

rates of nucleation are too low for any reasonable rate of ledge formation. This happens even during initial growth when the interface has full solute supersaturation and therefore a high driving force for pillbox nucleation. At later stages of growth, and especially during Ostwald ripening (§ 2.4), there is so little supersaturation that "homogeneous" ledge nucleation is impossible (FERRANTE and DOHERTY [1979]). Although various heterogeneous nucleation sites have been described (AARONSON [1974]), for example on screw dislocations and on intersecting precipitates, there is apparently no theory for these processes, and few reliable measurements of ledge nucleation rates.

2.2.5. Quantitative experimental observations of growth rates

As with nucleation studies there is a considerable lack of quantitative data on growth rates, despite the large number of qualitative observations on growth of product phases by diffusional processes. There are a few such quantiative measurements however and these will be briefly described.

2.2.5.1. Interface-controlled growth rates, without change of composition. The best data for this type of reaction are probably those for grain-boundary migration in high purity alloys (ch. 25, § 3.4.1), but for the purposes of the present chapter it can be deduced from the current state of knowledge, from the work of AUST and RUTTER [1963] to the recent review by HAESSNER and HOFMANN [1978], that the single-atom model of eq. (27) does appear to describe the process very well, with activation energies of the order of those expected for boundary diffusion. However, in grain-boundary migration the experiments are often complicated by solute-drag effects and by the problem of knowing what is the magnitude of the free energy driving force. However, it seems safe to conclude from the present state of knowledge that the basic model for mobility of an incoherent boundary is in quite good shape. Low-angle grain boundaries and *coherent* twin boundaries show low mobility (ch. 25, §§ 2.4.2 and 3.4.2).

For polymorphous phase transformations there are considerable difficulties in determining growth rates, since in many cases the kinetics of the process are apparently controlled by nucleation. A clear example of this is provided by the numerous studies of the fcc–bcc phase transition in iron with a range of trace alloy additions. WILSON *et al.* [1979] briefly reviewed the field and presented additional data. They report a series of thermal arrests at different cooling rates for high-purity iron and mild steel; similar results have been previously found. Two transformations occurring at the highest temperatures, approximately 840 and 735°C, are both described by the authors as occurring by the massive transformation mechanism, the product in each case being apparently ferrite of the same composition as the austenite. The distinction is due apparently to different nucleation mechanisms. The high-temperature transformation product is described as equiaxed, having a polygonal shape, and was treated by Wilson *et al.* as occurring with the volume stress relaxed, while the lower-temperature product is described as massive, and consists of small grains with an irregular outline.

A detailed review of the general topic of *massive transformations* has been provided by MASSALSKI [1970]; details of the various structural features in this type of transformation can be obtained from this source. One early investigation that did provide quantitative data on the diffusionless type of transformation is the study of the massive transformation in copper–zinc alloys by KARLYN et al. [1969]. The authors quenched bcc brass containing 38 at% Zn to room temperature and observed the transformation products on briefly up-quenching samples by pulse-heating. They only obtained the diffusionless fcc product when the sample was held in the single-phase α region and it was not found when samples were held in region II of the phase diagram (fig. 18, above) where it is expected that the diffusionless transformation *should* occur. The explanation offered for this failure to see the transformation in the two-phase region was that of a nucleation difficulty. It was hypothesized that only pre-existing nuclei would have had time to develop, and since these pre-existing nuclei had been held in the two-phase region, they were all surrounded by rejected solute. The effect of this rejected solute would be to cause the local concentration at the interface to be outside region II of the phase diagram. When the samples were heated into the single-phase field, however, the nuclei would, after a brief time delay to reabsorb the excess zinc solute, be able to grow diffusionlessly; this was what actually was seen to happen and the authors were able to describe the growth rate of the α grains by eq. (27) with the *low* activation energy (for copper alloys) of about 60 kJ/mole.

More recently, PLICHTA et al. [1978] compared the growth rates found in a series of massive reactions in titanium binary alloys with previous studies of massive transformation in other systems. They showed, for all the systems, that the activation energy was significantly less than expected for bulk diffusion. This result was obtained by use of eq. (27) and again provides evidence that this equation does indeed adequately describe interface-controlled diffusionless transformations.

2.2.5.2. Reactions involving long-range solute diffusion. The range of quantitative experimental data available prior to 1968 was reviewed by AARONSON et al. [1970]. They reported that in most cases the experimental studies showed good agreement with diffusion-controlled growth models, such as were described in §2.2.2. These authors also considered additional mechanisms such as the growth of grain-boundary precipitates, the theories for which were not included in §2.2.2. For several important examples of thickening of Widmanstätten plates, for example in Fe–C, Cu–Si and Al–Cu, they reported evidence for ledged growth; moreover, the growth data showed signs of growth slower than expected for diffusion control. Later investigations of precipitation of the hexagonal phase from the fcc matrix phase in Cu–Si by KINSMAN et al. [1971,1973] and by TIEN et al. [1971,1973] again showed in this case apparent control not by silicon diffusion to the growing hexagonal phase but control by interface processes that allowed the ledge mechanism to operate. Tien et al. used an electrical-resistivity technique and at small driving forces suggested a dislocation pole mechanism similar to that used to interpret twinning in fcc metals (e.g., KELLY and GROVES [1970]). The twinning dislocation was an $a/6 \langle 112 \rangle$ partial, and $c[00.1]$ and $a/2 \langle 112 \rangle$ type were the pole dislocations in the hexagonal and cubic phases respectively. The Cu–Si phase diagram is an interesting example of the type of system which has a very low driving force for nucleation since the two phases, cubic α and hexagonal κ, are very similar in composition [eq. (6d)]. The microstructure of the precipitated two-phase structure shows the hexagonal plates running completely across the matrix grains, so this is a rather unusual precipitation reaction, though somewhat similar to Widmanstätten ferrite in the iron-carbon system in having a low driving force for nucleation at a large supersaturation for growth. The precipitates in Cu–Si nucleate apparently at stacking faults while the iron system requires grain-boundary nucleation.

In the more usual precipitation reaction where a high density of metastable plate-like precipitates form, for example the θ'-phase in Al–Cu or Ag_2Al in Al–Ag, there has been a considerable amount of work and even more controversy. SANKARAN and LAIRD [1974] extended the observations of AARONSON and LAIRD [1968] and claimed that the results showed inhibited thickening compared to the rates predicted by the diffusion-control model. They also reported that the lengthening reaction was faster than that expected for linear unstable lengthening (§ 2.2.6). CHEN and DOHERTY [1977] reanalyzed the earlier data on the growth kinetics of θ' in Al–Cu and suggested that the results appeared to fit quite well to the HAM [1958,1959] and HORVAY and CAHN [1961] diffusion-controlled model for growth with a constant aspect ratio. This lead, not unexpectedly, to a considerable controversy between the various authors (SANKARAN and LAIRD [1977,1978], AARONSON [1977] and DOHERTY, FERRANTE and CHEN [1977,1978b]). It is probably impossible for someone deeply involved in a scientific argument to give a really objective account of the situation but it appears that it is reasonable to say that the data presented agree surprisingly well with the diffusion-controlled model, that is, with no significant inhibition of the growth rate, but with very large aspect ratios. These high length-to-width shapes appear to indicate some significant *initial* inhibition of thickening.

The situation is perhaps easier to understand in the studies of precipitate growth in the Al–Ag system (LAIRD and AARONSON [1969], FERRANTE and DOHERTY

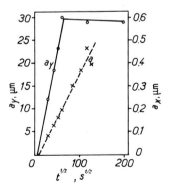

Fig. 29. Half-lengths, a_y, and half-thicknesses, a_x, of plate-like precipitates of Ag_2Al growing in aluminium, as a function of time$^{1/2}$ (after FERRANTE and DOHERTY [1979]).

References: p. 1024.

[1976,1979], and RAJAB and DOHERTY [1983]). Figure 29 shows the observed growth kinetics in a particular experiment and indicates that the results show the root-time relationship of diffusion control. Moreover, the slopes of the plots are close to those predicted by HORVAY and CAHN [1961] for diffusion control [eqs. (48) and (49)]. However, in this and other experiments the value of the aspect ratio, A, varies widely with the precipitate growth conditions and in every case is very much greater than the value expected for precipitates with the equilibrium shape determined by interfacial energies. Ferrante and Doherty postulated that the aspect ratio was increased by immobility in the thickening reaction at the *earliest stages of the reaction*, before the measurements shown by fig. 29 could be made. Under these conditions the aspect ratio would increase rapidly as had been previously argued by Aaronson.

The details of this hypothesis are easily given; the diffusion-controlled thickening rates are given by:

$$da_x/dt = (D\beta/t)^{1/2}, \tag{54}$$

$$da_y/dt = A(D\beta/t)^{1/2}; \tag{55}$$

when integrated these give the growth equations (48) and (49) quoted above.

The rate of change in aspect ratio, dA/dt is derived from the definition of A, $A = a_y/a_x$:

$$dA/dt = \frac{1}{a_x}\left(\frac{da_y}{dt} - A\frac{da_x}{dt}\right). \tag{56}$$

Substitution of eqs. (54) and (55) into eq. (56) shows, as had been pointed out by HAM [1958], that with diffusion-controlled lengthening and thickening from an initial aspect ratio A, that shape will be preserved ($dA/dt = 0$). However, if the thickening reaction is inhibited by some factor f, with $f < 1$, eq. (54) becomes:

$$da_x/dt = f(D\beta/t)^{1/2}. \tag{54a}$$

Therefore, under the conditions of inhibited thickening there will be an *increase* of the aspect ratio, given by:

$$dA/dt = \frac{A}{a_x}\left(\frac{D\beta}{t}\right)^{\frac{1}{2}}(1 - f). \tag{57}$$

FERRANTE and DOHERTY [1979] showed direct evidence for this effect by observations using TEM at the earliest stages of precipitation, at shorter times than appear on fig. 29. The observations were of thickening slower than predicted, $f < 1$, and a steadily rising value of the aspect ratio, A [eq. (57)]. RAJAB and DOHERTY [1983] extended the observations in the Al–Ag system to a wider range of temperatures and supersaturations. They also found diffusion-controlled growth rates. More significantly, they found that the value of the aspect ratio, though constant for a given reaction condition, varied systematically with the *density* of precipitates. The aspect ratio fell as the density increased. The authors argued that this variation in A was due to initial inhibition that was caused by a shortage of growth ledges. However, when precipitates on different habit planes intersected each other, which occurs earlier for a higher density of precipitates, then inhibition ceased and the precipitates retained the value of A at the time of intersection. The microstructural origin of this effect was assumed to be easier ledge nucleation at the intersection of two precipitates. Structural studies by RAJAB [1982] gave experimental support for this hypothesis.

The current position on observation of precipitate growth kinetics suggests therefore that the theoretical ideas described in §2.2 appear to be well founded; most reactions are diffusion-controlled except those with a coherent interface, between different structures, when a ledge mechanism is required. Even with a ledged growth, however, diffusion-controlled growth is possible provided that there is a sufficiently active source of ledges. The major problem that remains is the lack of understanding about the mechanisms and kinetics of ledge formation. A similar conclusion is arrived at for Ostwald ripening (§2.4).

Finally, mention should be made again of the studies on nucleation in the copper–cobalt system by SERVI and TURNBULL [1966]. Their observations showed clear evidence for diffusion-controlled growth of metastable fcc cobalt precipitates in the fcc matrix. This is a "cube–cube" orientation relationship which should not need growth ledges. The agreement with diffusion control in this system is very satisfactory confirmation of the mobility of interfaces between coherently connected structures when they have the same orientation and structure. The precipitate growth was studied in this system from the *earliest* stages of growth, at the smallest values of radius r, by virtue of the electrical resistivity technique used.

2.2.6. Growth instabilities

This section is divided into two parts; firstly, consideration of the onset of instabilities, and secondly, the linear growth expected for well developed instabilities, the lengthening of dendrites and of Widmanstätten plates and needles with a constant tip or rim radius.

2.2.6.1. Initial instability. For a spherical precipitate growing in a solute field (fig. 20, above) with either full diffusion control or even with mixed control, $C_i \neq C_0$, the solute gradient sets up a potentially unstable situation, first analyzed by MULLINS and SEKERKA [1963]. If the sphere is perturbed by any oscillation, a spherical harmonic, then those parts of the sphere with a locally larger value of the radius will penetrate into more supersaturated matrix and grow faster, while parts of the sphere with a smaller value of r will grow more slowly, and as a result the instability will grow. This situation had been long recognized in solidification and had been treated formally by TILLER *et al.* [1963] as *constitutional supercooling* (ch. 9, §6). In solidification, a temperature gradient provides some stability against growth of perturbations into the common growth form, a highly branched *dendrite*. For an isothermal precipitation reaction no such effect exists, but Mullins and Sekerka showed how the interfacial energy could partially stabilize the growing sphere.

In the following discussion, only the case of the growth of a solute-rich precipitate

References: p. 1024.

by flux towards the growing precipitate will be considered. The opposite case, of growth of a solute-poor phase, for example, the growth of alpha-iron in austenite in Fe–C alloys, merely requires an appropriate change of notation.

To describe this stabilization, the Gibbs–Thomson equation for the increase of solubility of a solute-rich spherical precipitate of radius r is needed. This is given in the general case by PURDY [1971] (see also MARTIN and DOHERTY [1976]) as:

$$C_\alpha(r) = C_\alpha \left(1 + \frac{(1 - C_\alpha)}{(C_\beta - C_\alpha)} \frac{2\sigma V_m}{RTr\epsilon_\alpha} \right), \tag{58}$$

where ϵ_α is the Darken non-ideality factor (§ 1.2). For dilute terminal solid solutions this reduces to the usual form:

$$C_\alpha(r) = C_\alpha \left(1 + \frac{2\sigma V_m}{RTr} \right). \tag{58a}$$

The effect of surface energy on solubility is such that when the precipitate nucleates at the critical radius r^*, $C_\alpha(r) = C_0$ and so there is zero supersaturation, and the growth rate dr/dt is zero. The general equation for growth is

$$dr/dt = \frac{D}{r} \frac{C_0 - C_\alpha(r)}{C_\beta - C_\alpha(r)}. \tag{59}$$

Substitution of eq. (58a) into eq. (59) gives, with $C_\beta - C_\alpha(r) \approx C_\beta - C_\alpha$:

$$dr/dt = \frac{2D\sigma V_m C_\alpha}{(C_\beta - C_\alpha)RT} \frac{1}{r} \left(\frac{1}{r^*} - \frac{1}{r} \right). \tag{60}$$

This growth rate is plotted out in fig. 30, which shows that the growth rate rises to a maximum value at $r = 2r^*$, and then declines, becoming that given by eq. (41) only for $r \gg r^*$.

Mullins and Sekerka showed that for this type of diffusion-controlled growth the spherical shape was stable against all perturbations until it reached a limiting radius

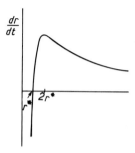

Fig. 30. Variation of growth rate of a spherical precipitate with precipitate radius at a fixed supersaturation. At $r = r^*$, the critical radius for nucleation, the growth rate is zero.

for absolute stability, $r = 7r^*$. The physical origin of the stability is that while an unperturbed sphere has a constant solubility, at the maximum of the perturbation the radius of curvature, which is the radius used in eq. (58), *decreases*, while at the minimum in the perturbation the radius of curvature increases. So the solubility rises at the maxima and decreases at the minima. This creates a concentration difference that opposes the development of the perturbation. The radius of curvature of a surface is the mean of any two radii that are perpendicular to each other.

The condition of *absolute instability* is that a perturbation of amplitude δ_s is growing, that is, $d\delta_s/dt$ is positive, and a spherical harmonic with two maxima, $l = 2$, is the first to show this type of instability. CORIELL and PARKER [1967] extended Mullins and Serkerka's analysis and showed that absolute stability was less important than relative stability. The criterion for a sphere to show *relative instability* is that

$$\dot{\delta}_s/\delta_s \geqslant \dot{r}/r. \tag{61}$$

This criterion arises since if the relative increase $\dot{\delta}_s/\delta_s < \dot{r}/r$, then although the sphere is unstable, the observed shape will become apparently *more spherical*, rather than less, with time. The smallest radius that shows relative instability is the $l = 3$ spherical harmonic with three maxima; this occurs when $r > 21r^*$.

The analysis above is a limiting case since, as discussed by SHEWMON [1965], there are other stabilizing factors that must be considered, particularly for solid-state reactions. SHEWMON [1965] developed this analysis since he could not find any examples of solid-state dendritic growth, and wished to understand why. However, unstable growth shapes are known, and have been studied, in a few metallic solids (MALCOLM and PURDY [1967], BAINBRIDGE and DOHERTY [1969] and RICKS *et al.* [1983]. Despite these examples the problem remained why there were so *few* examples of solid-state dendritic precipitates.

The additional stabilizing factors that inhibit growth perturbations include the following:

   (i) interface immobility;
   (ii) interfacial diffusion;
   (iii) diffusion within the precipitate;
   (iv) strong anisotropy in the interfacial energy;
   (v) closely spaced precipitates;
   (vi) high initial aspect ratios.

The first four of these effects have been discussed by SHEWMON [1965] and each was shown to be strongly stabilizing, though no detailed analysis was made for (iv), the stability of an interface at a cusp orientation (figs. 10 and 11, above). The physical basis of the enhanced stability is easily seen, however, for a flat interface, at a cusp orientation – such as the habit plane of a Widmanstätten precipitate. Any perturbation would have to increase not only the interfacial *area*, the usual problem for a perturbation, but a perturbation at a cusp also requires the interface to rotate *away* from the cusp plane and thereby greatly increase the specific interfacial *energy*. Partial immobility (i), mixed control, will stabilize the precipitate since it reduces the

References: p. 1024.

concentration gradients that cause instability. High diffusivity at the interface (ii), or within the precipitate (iii), relative to the diffusivity in the matrix phase is stabilizing since flux through these pathways can reduce the interfacial area without contributing to precipitate growth. The analysis of surface smoothing by the various diffusion pathways was given by MULLINS [1959]. For metallic *solidification*, where dendrites are ubiquitous, there are none of the effective stabilizing influences, while many if not most of the solid-state diffusional reactions do show these stabilizing influences. The additional stabilizing influence of closely spaced precipitates (v) is easily understood, though not yet apparently properly analyzed. The effect arises since closely spaced precipitates show an overlap of the diffusion fields of adjacent particles (so-called *soft impingement*); this effect reduces the concentration gradients about each precipitate, $C_0 \rightarrow C_\alpha$. The result of this is seen for solidification either when there is a high rate of primary nucleation (KATTAMIS *et al.* [1967]) or secondary nucleation in stir-casting (see VOGEL [1977]); under these conditions of closely spaced growth centres, the solidification growth form becomes spherical, not dendritic.

The effect of an anisotropic precipitate shape (vi) on shape stability has not been formally analyzed with a perturbation technique, but DOHERTY [1982] indicated why a shape with a high aspect ratio should be stable by use of the equations developed in §2.2.2 for stable growth by comparison with the expected linear growth rate of the end of a Widmanstätten plate (§2.2.6.2).

The combination of all these effects appears to explain the lack of shape instabilities, at least those that might give dendritic shapes, in all but a few solid-state reactions. The few dendritic shapes that are found occur with cube–cube oriented precipitates that have low and isotropic interfacial energies and diffusivities; in addition, the precipitates have ordered structures, which have low rates of diffusion within the precipitates and also show low densities of precipitates. The clearest examples are γ-brass in β-brass (MALCOLM and PURDY [1967] and BAINBRIDGE and DOHERTY [1969]) and Ni_3Al in Ni at small undercoolings that give low precipitate densities (e.g., RICKS *et al.* [1983]). Deliberate attempts to produce shape instabilities in other cube–cube systems, for example on silver precipitates in copper (DOHERTY [1976]), failed, presumably owing to the high rates of interfacial diffusion expected in the semicoherent interface between silver and copper, with the large misfit between the two phases, and also owing to the significant rate of diffusion within the precipitate.

2.2.6.2. Linear growth models. For both dendrites and plate-like and needle-like Widmanstätten growth forms, there are well developed models for linear growth of the tips which are assumed to have a constant tip radius. For the tip of the needle the radius is the same in both directions; for the rim of the plate the constant radius is the small one between the two parallel habit planes and not the large radius of the plate itself. The early model for lengthening of a plate was originated by ZENER [1946] and modified by HILLERT [1957]. The essential features of the model are the assumption of a constant tip radius which will give, for a needle crystal, a growth rate given by eq. (41). However, since there has to be an effect of the surface energy, the growth rate changes into that given by eq. (59) and shown by fig. 30. For a plate-like crystal, the equations are slightly different, but the physical reasoning and the resultant form of the relationship, between velocity and plate-tip radius are very similar to that seen for a simple needle crystal.

Having produced such a relationship it is necessary for the system, and the theoretician, to select some particular value of the radius for the plate-like or needle-like crystal. For convenience and for lack of any other simple choice the initial theory chose the maximum-velocity criterion (HILLERT [1957]). As explained in recent reviews for plate-lengthening (e.g., HILLERT [1975]) and for needle- and dendritic growth by DOHERTY [1980], the simple theory has proved inadequate and has had to be modified. Two major modifications have been required: the first is for a more accurate diffusion model and the second is the adoption of a more successful determining principle for tip radius. Both of these will be briefly described.

(1) *A better diffusional model.* For the plate-like crystal the required diffusional solution, neglecting interfacial energy considerations, is that due to IVANTSOV [1947]. The Peclet number p', here defined as $vr/2D$ with r the radius of the rim of the plate then the peclet number is given by:

$$\Omega = (\pi p')^{1/2} \exp(p') \operatorname{erf}(p')^{1/2}. \tag{62}$$

This gives a unique value of p' for a given supersaturation Ω but still leaves the choice of the radius r and therefore the growth velocity v undetermined. TRIVEDI [1970a] introduced the influence of surface energy; his rather complex equation can be well approximated at low-to-medium values of Ω by the equation due to BOSZE and TRIVEDI [1974]:

$$vr^*/D = \frac{27}{256\pi} \left(\frac{\Omega}{1 - 2\Omega/\pi - \Omega^2/2\pi} \right) \tag{63}$$

The value of the velocity v in eq. (63) is the maximum velocity, r^* is the critical radius of nucleation.

HILLERT [1975] has shown that this equation gives good agreement with the observed rates of linear lengthening of Widmanstätten α-iron in Fe–C alloys, and PURDY [1971] showed equivalent agreement for the needle-like α-brass precipitates forming from β-brass in Cu–Zn alloys. In both these cases the precipitates grow from grain-boundary nucleation sites since, the composition diferences between the two phases being small, there is insufficient driving force for nucleation within grains [eqs. (6d) and (7)].

(2) *Determining principle for tip radius.* Despite the obvious attractions to the theoretician of choosing the radius that gives the maximum growth rate, there is no reason why nature should share his outlook. Various other determining principles have been suggested in the past, mainly for the linear growth of two-phase products, eutectics, eutectoids and cellular precipitation reactions, see §2.5 (PULS and KIRKALDY [1972] and LIVINGSTON and CAHN [1974]). In her studies, on dendritic growth of γ dendrites in β-brass, FAIRS [1977] measured both the linear growth rate and tip radius for a range of precipitation conditions. She observed that the tip

References: p. 1024.

radius was larger than that expected to give the maximum growth rate, and that the growth rate was, in turn, less than that expected for maximum velocity. These observations were duplicated by the detailed studies by GLICKSMAN et al. [1976] on the growth of thermal dendrites in undercooled succinonitrile, a transparent organic analogue for metals with a low entropy of melting. In this study, the authors reported that the growth velocities were significantly *less* than expected for the maximum velocity; they reported the tip radii but did not comment on the values. The data on succinonitrile were reanalyzed by DOHERTY, CANTOR and FAIRS [1978a] who showed that the effect observed by Fairs was occurring in the organic solidification dendrites, the tip radius was significantly larger than that which would give the maximum velocity but the data lay on the plot predicted by the Trivedi equation (TRIVEDI [1970b]) for dendritic lengthening.

Fairs showed that in both her solid-state dendrites and the organic dendrites the operating tip radius seems to be determined by a *stability* criterion. This was originally thought to be relative stability but, as argued by HUANG and GLICKSMAN [1981], absolute stability is more likely to be appropriate for a perturbation on a tip with a constant tip radius. It is, however, still not obvious, at least to the present author, why stability rather than the maximum-velocity criterion controls the tip radius and therefore the growth velocity. LANGER and MÜLLER-KRUMBHAAR [1978] have developed a detailed theory for dendritic growth with the tip radius determined by a stability criterion and this general theory appears to be very successful. According to this description, the dendritic or needle growth rate is determined from the Ivantsov equation for a needle crystal:

$$\Omega = p' \exp(p') E_1(p'), \tag{64}$$

where E_1 is the integral exponential function, and the Peclet number, p', is determined by:

$$p' = vr/2D = (vd_0/2D\sigma')^{1/2}, \tag{65}$$

where d_0 is a "capillary length", the last term, without r, in the Gibbs–Thomson equation, and σ' a stability constant that is given as 0.025.

A similar analysis may be feasible for determining the operating rim radius for the growth of plate-like precipitates.

In this analysis there is still no adequate treatment for describing how an initial instability in a spherical precipitate develops into a branched dendrite, whose linear growth can be described by stationary diffusion fields around the constant tip shape [eqs. (64) and (65)]. An even more striking problem is the lack of a stability theory to describe how a plate-like Widmanstätten precipitate initially growing, like a sphere with a size–root-time relationship, can evolve a perturbation at the rim, and then subsequently adopt the constant growth rate and tip radius experimentally found (HILLERT [1975]). The present author (DOHERTY [1982]) has suggested that it is likely that the change from a stable growth relationship, $a_y = 2A(\beta Dt)^{1/2}$, to an unstable *linear* growth, given for example by eq. (63), might be expected to occur when the linear growth rate becomes *faster* then the lengthening of the stable shape. If this is

Table 2.
Plate half-length for which eqs. (49) and (63) predict the same lengthening rate.

Supersaturation Ω	Supersaturation Ω^*	Precipitate tip radius r, nm	Growth parameter β	Plate half-length a_y^*, μm
0.01	0.01	$1060r_c = 6.0 \times 10^4$	7×10^{-5}	2.3×10^6
0.03	0.0306	$306r_c = 7 \times 10^3$	3×10^{-4}	0.12×10^6
0.1	0.107	$100r_c = 6 \times 10^2$	4×10^{-3}	1.2×10^4
0.3	0.38	$28r_c = 60$	5×10^{-2}	1.2×10^3
0.6	1.07	$10r_c = 10$	1×10^{-1}	50

For $a_r > a_y^*$ linear lengthening would be expected.

correct, then the value of the aspect ratio, A, is crucial since large values of A will greatly delay the transition to linear lengthening, eq. (4a). The other factor of crucial importance is the supersaturation, Ω. Table 2 shows the expected half-length a_y^* for which the Bosze and Trivedi relationship [eq. (63)] gives the same lengthening rate as the stable Horvay and Cahn result. The table was obtained for a typical value of the aspect ratio seen for precipitates nucleated within a grain, $A = 100$, and a capillary length of 0.6 nm.

The huge values of a_y^* at the low supersaturations typical of age-hardening precipitations of solute-rich precipitates from solution-treated alloys, $\Omega < 0.1$–0.3, suggest why linear lengthening is not seen in such precipitation reactions as Ag_2Al from Al and θ' in Al–Cu (§2.2.5). However, matrix phase transformations such as the formation of ferrite from austenite can occur at large supersaturations but low nucleation densities (§1.3), so the observation of linear unstable growth of Widmanstätten plates in this system is understandable. This can only be a tentative description, since the analysis is approximate and the operating value of A in the stable growth regimes of Widmanstätten ferrite is not known. However, experimental observation on the early stages of the formation of Widmanstätten plates at grain boundaries should be relatively straightforward.

2.3. Precipitate dissolution

There has been much less interest in the dissolution of precipitates than in their growth for the obvious reason that, provided an alloy is heated into a single-phase region, all the precipitates will dissolve, given sufficient time. However, for precipitates that are not completely soluble after an increase of temperature, there will be a period of dissolution that will affect the microstructure because of the different shrinkage rates of different precipitates. In addition, observation of precipitate dissolution can provide insight into the amount of diffusion control. This has been exploited by HALL and HAWORTH [1970] and ABBOTT and HAWORTH [1973] in studies of the dissolution of plate-like precipitates in the Al–Cu and Al–Ag systems by microprobe analysis of the solute fields in the matrix normal to the habit plane of the precipitates. In the first example apparent diffusion-controlled dissolution was found, while in the Al–Ag alloy some evidence for mixed control was reported with the solute content, C_i, being less than the equilibrium value. For such systems, where a ledge mechanism would be expected, it should be remembered that for *dissolution* of isolated plates the incoherent rims of the plates are good sources of ledges (MARTIN and DOHERTY [1976]).

A detailed analysis of precipitate dissolution for spherical precipitates and for diffusion away from the broad faces of plate-like precipitates has been published by AARON and KOTLER [1971]. Their review provides a very detailed summary of the theory and need not be reproduced here. Only some major points will be described. Aaron and Kotler compare the processes of growth and dissolution and show that one is not the simple opposite of the other, particularly for spherical precipitates. The distinction arises since for dissolution the rate, for a diffusion-controlled reaction, is initially infinite and then falls as the concentration gradient in front of the interface becomes shallower. However as the process continues, the radius of the sphere shrinks and the volume providing the solute falls so the dissolution rate rises again. The effect is due to the spherical symmetry of the matrix absorbing the solute. Plate-like dissolution by *thinning* however does not show this acceleration, and the rate of thinning continues to decline (fig. 31). Aaron and Kotler also point out that the details of the diffusional analysis are more difficult for dissolution than for growth since in the dissolution case the source of solute, the interface, is retreating from the surrounding matrix so that the solute distribution close to the interface undergoes a more complex behaviour during dissolution than during growth (fig.32). This problem becomes very severe for the case of dissolution of a plate-like or needle-like precipitate growing back from the tip. In this case the tip will move into a region of matrix already enriched with solute due to the thinning process, and no analytical solution of the problem appears to be available.

In the analysis provided by AARON and KOTLER [1971] it was shown that, although capillarity effects are expected to increase the rate of dissolution of spherical particles owing to the rise of solubility at small radii, these modifications appear to be small for the 1 μm-sized and larger particles of interest in dissolution reactions above the equilibrium solvus. Although this was not considered by Aaron and Kotler, the acceleration due to capillarity is likely to be of greater significance for the dissolution of the very much smaller precipitates of concern in precipitation

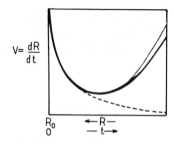

Fig. 31. Dissolution of an isolated sphere (heavy solid line) and a plate (dashed line) by diffusion-controlled dissolution. The influence of the Gibbs–Thomson effect on the dissolution rate of small spheres is shown by the thin solid line. (After AARON and KOTLER [1971].)

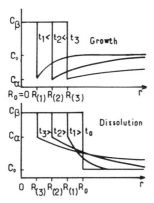

Fig. 32. Distinction between growth and dissolution of a solute-rich β precipitate: the composition is C_α at the interface and C_0 in the matrix a long way from the precipitate. (After AARON and KOTLER [1971].)

reversion. This is the dissolution of very fine metastable precipitates when the ageing temperature is raised during a precipitation-hardening reaction. SARGENT and PURDY [1974] pointed out the possibility that a fine dispersion of precipitates may redissolve completely even when the raised temperature remains within the two-phase field, so long as the radii of the precipitates are less than the critical radius for nucleation of the alloy at the higher temperature.

2.4. Competitive growth – Ostwald ripening

2.4.1. After precipitation

There are two subsequent stages to the free growth of isolated precipitates described in §2.2. The first of these occurs when the diffusion fields around adjacent precipitates start to overlap, a process usually described as *soft impingement*. As described in various reviews of overall transformation kinetics (for example TURN-BULL [1956], FINE [1964] and CHRISTIAN [1970,1975]) the rate of the precipitation reaction will be given by eq. (22) during the period after soft impingement. The references cited here include a general discussion of what is often called either a Johnson–Mehl or an Avrami equation of the type given by eq. (22). In the general case the equation gives the amount of new phase formed divided by the amount of the new phase present at equilibrium, which is the change of solute quoted for a precipitation reaction in eq. (22). The values of the exponent n are related in table 3 to the particular nucleation conditions. Table 3 was originally published by CHRISTIAN [1970]. The time constant, τ, is determined form the nucleation and growth rates and the form of the equation, $1 - \exp(-t/\tau)^n$, comes from competition between the growth centers either by "soft" impingement of diffusion fields or by "hard" impingement between growing grains. A full discussion of this topic has not been provided here, firstly since the earlier reviews are rather complete and,

Table 3
Value of n in kinetic law $f = 1 - \exp(- kt^n)$.

(a) Polymorphous changes, discontinuous precipitation, eutectoidal reactions, interface controlled growth, etc.

Conditions	n
Increasing nucleation rate	> 4
Constant nucleation rate	4
Decreasing nucleation reate	3–4
Zero nucleation rate (saturation of point sites)	3
Grain-edge nucleation after saturation	2
Grain-boundary nucleation after saturation	1

(b) Diffusion-controlled growth (early stages of reaction only)

Conditions	n
All shapes growing from small dimensions, increasing nucleation rate	$> 2^{1/2}$
All shapes growing from small dimensions, constant nucleation rate	$2^{1/2}$
All shapes growing from small dimensions, decreasing nucleation rate	$1^{1/2}-2^{1/2}$
All shapes growing from small dimensions, zero nucleation rate	$1^{1/2}$
Growth of particles of appreciable initial volume	$1-1^{1/2}$
Needles and plates of finite long dimensions, small in comparison with their separations	1
Thickening of long cylinders (needles), e.g. after complete edge impingement	1
Thickening of very large plates, e,g. after complete edge impingement	1/2
Segregation to dislocations (very early stage only)	2/3

secondly, because (with a few exceptions) use of the overall reaction kinetics has played little role in understanding the process since similar values of n can be obtained for different mechanisms. In the opinion of the present author, use of eq. (22) is of much less value in understanding reaction mechanisms than is the determination of the separate nucleation and growth rates. However, the equation is of value in providing a physical picture of the overall process and in confirming that individual nucleation and growth rates have been correctly measured. A nice illustration of the use of the equation is its application in biology to the two-dimensional spread of colonies of lichen on rocks or of oysters in freshly established oyster beds.

Following the soft impingement of diffusion fields, the precipitates drain the remaining solute from the matrix and have near to the equilibrium volume fraction of the new phase.

This does not mark the end of the reaction, however, since the microstructure of a distribution of precipitates has a high area of interface, and this raises the free energy of the system. The process of reduction of this energy involves the growth of larger precipitates at the expense of smaller ones. In a similar way, after primary recrystallization of deformed metals the resulting grain structure usually has a significant amount of grain boundary area that drives the process of grain growth (ch. 25).

The process of competitive precipitate growth in a near-equilibrium matrix composition is commonly described as *Ostwald ripening* after the discoverer of the same process in salt precipitating from aqueous solutions (OSTWALD [1900]). The mechanism of this process is provided by the Gibbs–Thomson equation (58). This shows that with a range of precipitates the solubility will be least with the largest precipitates and largest with the smallest precipitates. This produces a solute gradient from small precipitates to large ones and diffusion will then cause the smallest precipitates to shrink and disappear. The solute from the shrinking precipitates feeds the larger ones so the mean precipitate size rises. The rate of growth is described by fig. 30 but with the critical radius of nucleation replaced by the mean radius, $\bar{r}$; $\bar{r} = r^*$ in fact.

The topic has been reviewed by MARTIN and DOHERTY [1976] and is discussed also in chapter 10B of this volume, so full details of the process will not be given here. For diffusion-controlled growth the classic (LSW) theory of the process due to WAGNER [1961] and LIFSHITZ and SLYOZOV [1961], predicts that, even starting from a narrow size distribution, the ripening process will yield a steady-state distribution of particle sizes that will cause the mean precipitate radius to coarsen with time as:

$$\bar{r}_t^3 = \bar{r}_0^3 + \tfrac{8}{9}\frac{\sigma D C_\alpha V_{\mathrm{m}} t}{RT}. \tag{66}$$

This result has been extensively tested both qualitatively and quantitatively many times and in general it appears to be very successful. However, there are a number of problems. The main problem is that although almost all tests on spherical and near-spherical precipitates give the expected $\bar{r}^3$–t relationship, the experimental slopes are often larger than expected from eq. (66) and correspondingly there is a somewhat wider size distribution than predicted. A particular point frequently noted is that the simple diffusion theory [eq. (60) and fig. 30] predicts that the fastest growth rate will be for particles *twice* the size of the average, yet the LSW-predicted spread of particle sizes for diffusion-limited growth cuts off at a maximum particle size of only $1.5\bar{r}$. It is not at all clear *why* particles are theoretically limited to this size, since fig. 30 suggests that after extended growth the maximum precipitate size is likely to be $2\bar{r}$. Experimental size distributions, even after very long periods of coarsening, do show precipitates up to $2\bar{r}$ in size (see for example BOWER and WHITEMAN [1969] and fig. 33). That is, the experimental results fit better with the simple analysis than with the more rigorous LSW analysis. It is possible that in most precipitation reactions the particles start from a wider size distribution than the steady-state one – a situation that has not apparently been theoretically analyzed.

Another problem with the analysis is that it was developed for *point* particles, that is, for a volume fraction of second phase that is close to zero, so that individual particles are only aware of the *average* solute content, $\bar{C}_\alpha = C_\alpha(\bar{r})$, and they are unaware of the local concentrations introduced by neighbouring particles. ARDELL [1972] introduced an approximate, but physically reasonable, correction for the

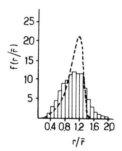

Fig. 33. Experimental results for the distribution of precipitate reduced radii ($r/\bar{r}$) against reduced radius after extensive diffusion-controlled coarsening of Fe_3Si in Fe at 700°C, aged 312 h. The dotted line shows the expected distribution in the limit of a low volume fraction ($f \to 0$). (After BOWER and WHITEMAN [1969].)

effect of finite volume fractions and showed that when the volume fraction of second phase was 0.25, the rate of coarsening implied by eq. (66) was expected to increase *tenfold*. The experimental results however, particularly in the intensively studied $Ni-Ni_3X$ systems, show no significant rise in coarsening rate with volume fraction (CHELLMAN and ARDELL [1974]). BRAILSFORD and WYNBLATT [1979] have produced a modified theory that suggests that the acceleration of coarsening with volume fraction is much smaller than predicted by Ardell. For a full discussion of the various theoretical modifications to the initial LSW theory, to take account of interface immobility, and alternative diffusion paths along grain boundaries and dislocations, the discussion by MARTIN and DOHERTY [1976] can be consulted. More recent developments in both theory and experiment are discussed in chapter 10B.

2.4.2. During initial nucleation and growth

It was pointed out by KAMPMANN and KAHLWEIT [1967,1970] that competitive dissolution of small precipitates would not be restricted to periods *after* precipitation was complete. This is because during all stages of precipitation the average solute content of the matrix will be falling steadily owing to nucleation, and more significantly, to growth of precipitates – for example as measured by the electrical resistivity studies of cobalt precipitation in Cu–Co alloys (SERVI and TURNBULL [1966]). The fall in solute content causes the critical radius of nucleation, r^*, to rise, and any precipitate whose radius falls below the size will redissolve. The process is shown schematically in fig. 34. The alloy is quenched below the solubility limit, becoming supersaturated at $t = t_s$. The supersaturation, S, here defined as $\bar{C}/C_\alpha$, first rises with time (because C_α falls) so r^* first falls with time. The first nucleus forms at t', when S is still rising, r^* falling; When, at t'', the last nucleus forms, S is falling already, and, consequently, $r^*(t)$ is rising: the smallest precipitates are dissolved up. Numerical solutions of this problem were published by KAMPMANN and KAHLWEIT [1970], assuming diffusion-controlled growth and dissolution, using

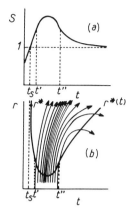

Fig. 34. Competitive coarsening during precipitation. An alloy is quenched becoming supersaturated at $t = t_s$. The first nucleation occurs at t' and the last at t''. As the supersaturation, $S = \bar{C}/C_\alpha$, falls, $r^*(t)$ rises and small precipitates redissolve. (After KAMPMANN and KAHLWEIT [1970].)

homogeneous nucleation theory and assuming that the supersaturation builds up exponentially with time. Figure 35 shows the result of one set of calculations with fig. 35a showing a change of supersaturation, S; curve 1 is the change in the absence of precipitation and curve 2 the actual supersaturation that is determined by the precipitation, and in turn determines the nucleation and dissolution rate. The total number of precipitates larger than the critical size is seen in fig. 35b as the number of precipitates, normalized by dividing by the final solute atomic concentration Z/N_t. The number of precipitates rises during the precipitation reaction to a maximum value and then starts to fall while the supersaturation is still large, approximately 4. Figure 36 shows the computed change of i^*, the critical precipitate size in terms of the number of atoms of solute, during coarsening; this is compared to the "steady-state" loss of precipitates predicted by an analysis equivalent to the

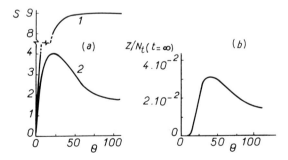

Fig. 35. Numerical solution to competitive coarsening: (a) the variation in supersaturation S with time θ, curve 1 with no precipitation, curve 2 with precipitation; (b) the variation in the density of the precipitates Z, scaled with the equilibrium solute content $N_t(t = \infty)$. (After KAMPMANN and KAHLWEIT [1970].)

References: p. 1024.

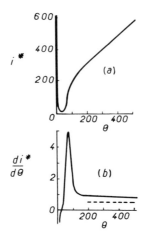

Fig. 36. Competitive coarsening during precipitation: (a) the variation in the critical size i^* with coarsening time, where i^* is the number of atoms in the critical cluster; (b) the rate of change of the critical size with time. (After KAMPMANN and KAHLWEIT [1970].)

LSW analysis described in §2.4.1, shown as the dashed line. KAMPMANN and KAHLWEIT [1970] suggest, however, that the discrepancy is probably within the "experimental error" of the numerical analysis used. The dimensionless time parameter, θ, scales with the diffusion coefficient, so that with the diffusion coefficient expected for liquids, $D = 10^{-9}$ m^2/s, $\theta = 10$ correspond to only 10^{-6} s, but with a substitutional diffusion coefficient of $D = 10^{-16}$ m^2/s this would be 100 s. It is clear from these computations, therefore, that the number of growing precipitates seen at the later stages of the reaction may seriously *under-estimate* the original density of nuclei, for example in the determination of the nucleation rate in the experiments of Servi and Turnbull discussed in §2.1.5.

Recently, LANGER and SCHWARTZ [1980] evaluated analytically the effect of surface-energy-driven coarsening on precipitate density. The results reported show very similar trends to those described here but a detailed comparison of the two evaluations has not been attempted by the present author, or by Langer and Schwartz.

There are several experimental studies that have attempted to test the idea of competitive precipitate dissolution during precipitation in the Ni–Ni$_3$Al system. These experiments used both transmission electron microscopy (KIRKWOOD [1970], and HIRATA and KIRKWOOD [1977]) and, more recently, field ion microscopy/atom probe techniques (WENDT [1981]). In all these studies a fall of at least an order of magnitude in the precipitate density was found while the reaction was proceeding. The results of WENDT [1981] on precipitation in a Ni–14 at% Al alloy at 550°C are shown in figs. 37 and 38. The supersaturation Δc falls with time and reaches about zero between 600 and 1400 min (fig. 37a). The relationship between number density and time is expected to be a simple reciprocal relationship during classical LSW

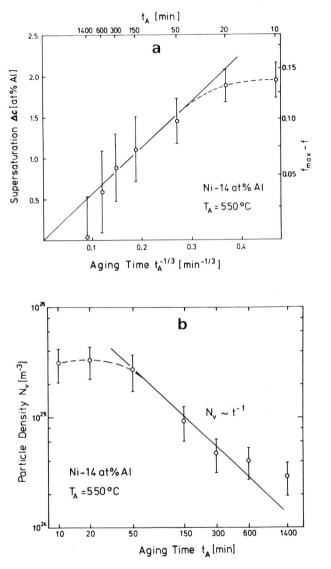

Fig. 37. Precipitation in a Ni–14 at% Al alloy, aged at 550°C: (a) super-saturation, Δc, versus ageing time, t_A (from right to left); f is the precipitated volume fraction; (b) particle density, N_v, versus ageing time, compared with the theoretical slope (solid line). (From WENDT [1981].)

coarsening ($N_v \propto 1/r^3 \sim 1/t$) at $\Delta c \approx 0$. The results shown in fig. 37b indicate that the functional relationship still applies even when $\Delta c > 0$. The results by TEM show similar effects, though the study by Hirata and Kirkwood showed a two-stage fall in precipitate density with time; the initial, more rapid, fall was ascribed to an

References: p. 1024.

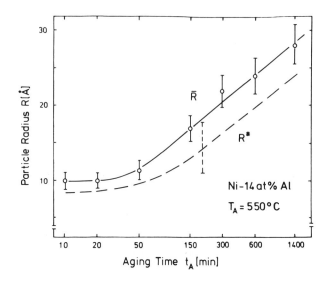

Fig. 38. Measured mean particle radius, $\bar{R}$, and calculated radius of critical nuclei, R^*, versus ageing time for the alloy of fig. 37 (from WENDT [1981]).

anomalously high diffusion coefficient immediately after the quench due to trapped vacancies.

HIRATA and KIRKWOOD [1977] used their maximum precipitate densities and were able to get close agreement with the classical results described in §2.1. Wendt used his experimental results to test the theory proposed by LANGER and SCHWARTZ [1980]. Reasonable agreement was obtained.

2.4.3. Coarsening of Widmanstätten precipitates

One of the early studies of precititate coarsening was of the growth of the needle-like precipitates of fcc copper in bcc iron (SPEICH and ORIANI [1965]). Despite the anisotropic shape indicating, together with the orientation relationship, a good-fit direction between the close-packed $\langle 110 \rangle$ fcc and $\langle 111 \rangle$ bcc, the precipitates showed apparent diffusion-controlled growth with a constant aspect ratio of about 4. The authors used a modified version of the HAM [1958,1959] analysis to describe the diffusion-controlled coarsening process and obtained reasonable agreement with their experimental results. FERRANTE and DOHERTY [1979], as part of their investigation of the role of ledged interfaces in the growth of plate-like precipitates in Al–Ag, studied the coarsening of these precipitates after the completion of precipitation. The plate-like precipitates in this system had much larger aspect ratios than expected for equilibrium shapes: $A > 100$ while $A_{eq} = 2.7$ (AARON-SON et al. [1968]). Ferrante and Doherty showed that with $A \gg A_{eq}$ the modified Gibbs–Thomson equations for the solubility at the incoherent rim, C_r, and that at the semicoherent facet plane, C_f are given by:

$$C_r = C_\alpha \left[1 + \left(\frac{1 - C_\alpha}{C_\gamma - C_\alpha} \right) \left(1 + \frac{A}{A_{eq}} \right) \frac{\sigma_r V_m}{RTa_\gamma} \right],$$ (67a)

$$C_f = C_\alpha \left[1 + \left(\frac{1 - C_\alpha}{C_\gamma - C_\alpha} \right) \frac{2\sigma_r V_m}{RTa_\gamma} \right].$$ (67b)

Despite the difference in solubility, the precipitates showed a reluctance to achieve the expected equilibrium shape, even with very extended coarsening times. The conclusion reached by Ferrante and Doherty was that this reluctance in the Al–Ag$_2$Al system to change shape was due to thickening inhibited by the lack of growth ledges. Subsequent work by RAJAB [1982] has confirmed that hypothesis for Al–Ag$_2$Al. An unexpected result from the coarsening study with these out-of-shape Ag$_2$Al precipitates was that despite inhibited thickening, the coarsening reaction, measured by the rate of *lengthening*, was very fast. However, further analysis of the results by use of the Ham, Horvay and Cahn analysis, and by use of eq. (67a), showed that the intuitive expectation, that inhibited thickening would slow coarsening, had been wrong. With inhibited thickening, $A \gg A_{eq}$, there is a much higher driving force for lengthening, as the only means of ridding the system of the excess interfacial energy. Under the highly approximate condition of a constant aspect ratio, the analysis yielded the following prediction for the rate of growth of the mean half-length of the precipitate:

$$\bar{a}_y^3(t) = \bar{a}_y^3(0) + \frac{3AD\sigma}{2\pi RT} \left(1 + \frac{A}{A_{eq}} \right) \frac{C_\alpha(1 - C_\alpha)}{(C_\gamma - C_\alpha)^2} \cdot t.$$ (68)

This gave surprisingly close agreement with the experimental results, despite the fact that the aspect ratio was not constant but showed a strong correlation with precipitate size.

More detailed investigation of the coarsening of plate-like precipitates with $A \gg A_{eq}$ due to the effects of thickening inhibition during the prior precipitation has been provided by MERLE and FOUQUET [1981] and MERLE and MERLIN [1981]. They studied θ' in Al–Cu and found in that system that the precipitates formed with high and variable aspect ratios; with the *highest* aspect ratios for those conditions with the *lowest* precipitate densities. During coarsening the aspect ratio fell towards what appears to be the equilibrium value of about 20, the rate of fall being *fastest* in those microstructures which had the *highest* precipitate density (fig. 39). The implication of these results is clearly that the highest densities lead to the greatest chance of precipitate intersection and thus a source of growth ledges. MERLE and MERLIN [1981] and subsequently MERLE and DOHERTY [1982] showed that the kinetics of the change of shape can be described by a ledge model with solute transport at 225°C by *interface diffusion* around the incoherent edge of the precipitate. However, the agreement required the *experimental* measurement of ledge spacing and there

References: p. 1024.

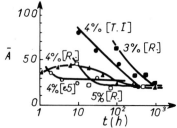

Fig. 39. The variation of the mean aspect ratio of the plate-like θ' precipitates in Al–Cu at 225°C. T.I. is an interrupted quench from the solution-treatment temperature, R is a room-temperature quench followed by the 225°C age, and $\epsilon 5$ is a 5% room-temperature strain prior to ageing at 225°C. 3%, 4% and 5% are the Cu contents. Note that $\bar{A}$ falls *fastest* with the *highest density* of precipitates. (After MERLE and FOUQUET [1981].)

remains no way, at present, of *predicting* this vital parameter, owing to the lack of a successful model for ledge nucleation.

2.5. Discontinuous reactions – reactions at a moving interface

The reactions discussed in this section are all those occurring at a *moving* interface, giving a *two-phase* product. The moving boundary may be a phase boundary, as in eutectic and eutectoidal decomposition, or it may be a matrix grain boundary as in the precipitation reaction conventionally described as *discontinuous precipitation* (GUST [1979]. This name is an unfortunate one since the general name for the second class of transformations considered by Gibbs, those that do *not* involve nucleation and growth, is *continuous transformations* (SOFFA and LAUGHLIN [1983]). The term "discontinuous" should then apply to *all* Gibbs type-I, nucleation-and-growth, transformations and not just the subset of them that occur at a moving boundary. The term "discontinuous transformations" will be used here in the narrow meaning even though an alternative term, such as MB("Moving Boundary" reactions) would be preferable at least for the purpose of precision if not for elegance. This review is not the place to make this unilateral change of notation, but care should be taken to avoid the expression "continuous precipitation" for reactions that either are nucleated entirely within a matrix grain or for those that are nucleated at grain boundaries and then grow *away* from the boundary, for example Widmanstätten ferrite. The term continuous transformations should be restricted to reactions which do *not* involve nucleation and growth.

2.5.1. Eutectoidal decomposition

In *eutectoidal decomposition,* the general reaction involves the change $\gamma \rightarrow \alpha + \beta$, and the most important example is the "pearlite" reaction, the decomposition of austenite, γ, in an iron–carbon alloy containing 0.8 wt% C, to give ferrite, α, and cementite, Fe_3C: $\gamma \rightarrow \alpha + Fe_3C$. The description *pearlite* comes from the similarity between the optical effect produced by light diffraction from the lamellar spacing

(~ optical wavelengths) of the two-phase eutectoid and the optical effect produced by the "mother of pearl" biological composite.

The process can be understood from the schematic microstructure (fig. 40), for the transformation occurring isothermally at an undercooling ΔT below the eutectoid temperature (fig. 41). The extrapolations in the phase diagram give the compositions of γ in equilibrium with α and β as $C_\gamma(\alpha)$ and $C_\gamma(\beta)$; the equivalent values in α are $C_\alpha(\gamma)$ and $C_\alpha(\beta)$. If the main diffusion path is along the moving boundary, as it often is, then the compositions in the boundary are given as $K_{\gamma\alpha}^b C_\gamma(\alpha)$ and $K_{\gamma\beta}^b C_\gamma(\beta)$, where $K_{\gamma\alpha}^b$ and $K_{\gamma\beta}^b$ are the partition coefficients for solute between the boundary and the decomposing austenite. Since the separate values of the different coefficients are rarely known an average value, K_γ^b, is often used.

From figs. 40 and 41 it can be seen that there is a composition difference ΔC_γ within the γ-phase and equivalent composition differences ΔC_α in the α-phase and ΔC_b in the boundary. In the initial treatment of the diffusional problem given by ZENER [1946], it was assumed that diffusion took place down a solute gradient dC/dy, given by $2\Delta C_\gamma/\lambda_s$, in the austenite, where λ_s is the lamellar spacing. The solute diffusion was assumed to occur within a thickness of approximately $\lambda_s/2$ away from the interface. By a mass balance of the diffusing solute needed for the reaction, the following relationship between the velocity and the spacing was obtained:

$$v = \frac{2\,D\Delta C_\gamma}{f_\alpha f_\beta} \cdot \frac{1}{C_\beta - C_\alpha} \cdot \frac{1}{\lambda_s}, \tag{69}$$

where f_α and f_β are the relative widths of the two phases, assuming that the molar volumes are the same in both phases (HILLERT [1972]). Equation (69) fails to take into account, however, the interfacial energy, $\sigma_{\alpha\beta}$, of the interface between the product phases. The origin of the lamellar equivalent to the Gibbs–Thomson effect is shown in fig. 42. The interfacial energy per unit volume is $2\sigma_{\alpha\beta}/\lambda_s$, and so the two-phase product is raised in molar free energy by $2\sigma_{\alpha\beta}V_m/\lambda_s$, with a consequent reduction of ΔC_γ to $\Delta C_\gamma(\lambda_s)$. If the spacing that reduces ΔC_γ to zero, $\lambda_s(\text{rev})$, is included, eq. (69) can be changed to give the velocity corrected for the interfacial energy effects. The description of $\lambda_s(\text{rev})$ is used since at that spacing, both the

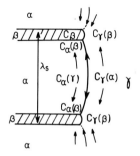

Fig. 40. Growth of an $\alpha + \beta$ lamellar structure, showing the variation in composition, $C_\gamma(\alpha)$, $C_\gamma(\beta)$ in the γ-phase, and $C_\alpha(\gamma)$, $C_\alpha(\beta)$ in the α-phase, that drive the reaction.

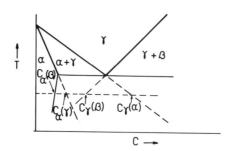

Fig. 41. The equilibrium diagram showing the origin of the composition differences seen in fig. 40.

two-phase product and the high-temperature austenite phase have the same free energy and so the reaction would be reversible at $\lambda_s = \lambda_s(\text{rev})$. With this effect included the velocity becomes:

$$V = \frac{2D\Delta C_\gamma(\lambda_s)}{f_\alpha f_\beta (C_\beta - C_\alpha)} \frac{1}{\lambda_s} = \frac{2D\Delta C_\gamma}{f_\alpha f_\beta (C_\beta - C_\alpha)} \frac{1}{\lambda_s} \left(1 - \frac{\lambda_s(\text{rev})}{\lambda_s}\right). \tag{70}$$

In these diffusional equations the compositions must, as always, be given in units of atoms per unit volume.

In the iron–carbon system, the diffusion coefficient in α is significantly faster than in the close-packed austenite, γ, so α diffusion is an alternative pathway which can be described by appropriate modification to eq. (70). For diffusion in the interface the analysis by TURNBULL [1955] showed that not only must the compositions and the diffusion coefficient be changed but the diffusion thickness is reduced

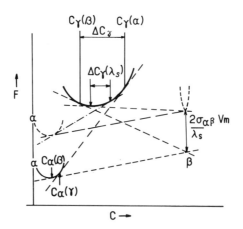

Fig. 42. The free-energy–composition diagram that gives the diagram seen in fig. 41. The composition difference ΔC_γ is reduced to $\Delta C_\gamma(\lambda_s)$ by the rise in free energy per mole of $\alpha + \beta$, $2\sigma_{\alpha\beta}V_m/\lambda_s$, caused by the lamellar spacing.

from the spacing $\lambda_s/2$ to the boundary thickness δ_b, so the resultant equation is:

$$v(b) = \frac{8D_b k_\gamma^b \Delta C_\gamma \delta_b}{f_\alpha f_\beta (C_\beta - C_\alpha)} \frac{1}{\lambda_s^2} \left(1 - \frac{\lambda_s(\text{rev})}{\lambda_s} \right). \tag{71}$$

HILLERT [1969,1972,1982] has discussed these equations and the more rigorous solutions to the diffusion equations in these cases and also for the possibilities that there may be some significant supersaturation left in the product phases and that some free energy or equivalently composition difference will be required to drive the interface, that is, there may some element of interface control. For full analytical details and analysis of experimental results, the reviews by Hillert should be consulted.

It should be apparent from eqs. (70) and (71) that the problem is not fully specified, even in the simple treatments, until some means of determining the operating lamellar spacing can be found. Experiment shows that there is a constant spacing at any undercooling and that this spacing decreases as the undercooling increases and consequently $\lambda_s(\text{rev})$ decreases. The discussion of the choice of spacing is given in §2.5.4.

The fall in spacing, λ_s, and critical spacing, $\lambda_s(\text{rev})$, together with the usual exponential fall of the diffusion coefficient with temperature, means that the eutectoid growth velocity will initially *increase* with undercooling, reach a maximum and then decrease rapidly in a temperature range where diffusion becomes very slow. This qualitative pattern is always found and is of great importance in for example the heat-treatment of steels. In outline, the carbon-containing austenite needs to be quenched to martensite and to do this, the diffusional decomposition to pearlite, and to the other diffusional product, bainite (§2.6), must be avoided. To achieve this, the transformation to pearlite needs to be slowed down by inhibiting either the nucleation of pearlite or its growth. The diffusional changes are usually reported on time–temperature–transformation (TTT) diagrams showing the time for isothermal tranformation; these diagrams normally show "C-curve" behaviour (fig. 43), the "nose" of the C-curve being that for the maximum overall rate of transformation derived from a Johnson–Mehl equation (§2.4.1).

Although the transformation can be slowed down by an increase in austenite grain size, which reduces the density of boundary nucleation sites, this is not the best solution, for reasons related to optimizing mechanical properties. The best method of improving the "hardenability" of steels is by *slowing* down the pearlite *growth*. This is achieved by addition of small amounts of substitutional metallic contents (nickel, manganese, chromium, etc.) which are soluble in the austenite but which at equilibrium will *partition* between the two product phases, but do so by slow substitutional diffusion. This the basis of *low-alloy* steels. At high temperature the solute must partition since pearlite with uniformly distributed solute will have a higher free energy than the parent austenite. At lower temperatures, and higher undercoolings, there will be sufficient free-energy difference to allow transformation with "no partitioning", but with an inevitable decrease in driving force and therefore

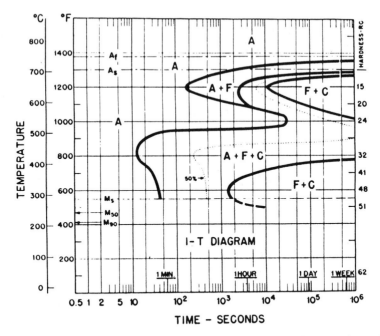

Fig. 43. Isothermal TTT diagram for a low-alloy steel, containing 0.4% C, 0.3% Mo, 0.8% Cr, 1% Mn and 1.8% Ni. The high-temperature transformation is firstly to precipitate ferrite (F) followed by pearlite (F + C), while the lower-temperature diffusional reaction is the formation of bainite (§2.6). (From the Atlas of Isothermal Transformation and Cooling Curves, ASM, Metals Park, OH, 1977.)

an increase in the critical spacing, λ_s(rev). The "no partitioning" of ternary solute in eutectoids is comparable to the diffusionless precipitation of single-phase products discussed in §2.2.1. The reduction in growth velocity caused by these ternary solute effects increases with the amount of the low-alloy solute additions.

There has been considerable experimental interest in studies of solute partitioning in steels (see, e.g., WILLIAMS et al. [1979], RIDLEY and LORIMER [1981], SMITH et al. [1981] and RICKS [1981]). Although it is clearly established that at high temperatures there is full solute partitioning in low-alloy steel pearlites, and that with a fall in temperature the partioning becomes incomplete, experimental difficulties have so far prevented demonstration of the full no-partitioning of solute during transformation. The difficulty is spatial resolution of the analysis, since at higher undercoolings the experimental spacings become very fine and exceed the current analytical resolution.

2.5.2. Discontinuous precipitation

An equation similar to eq.(71) describes the growth of a two-phase product behind a moving matrix grain boundary that relieves a supersaturated α matrix, the reaction being $\alpha \rightarrow \alpha' + \beta$. This process occurs in many low-temperature precipitation reactions and arises by grain-boundary nucleation of the precipitate phase β, which can then grow by grain-boundary diffusion of solute at temperatures where the competitive process involves *very slow bulk* diffusion. In order to provide the solute the boundary must migrate into the supersaturated grain interior. GUST [1979] has published a very detailed review of the literature on this topic, starting from the original observation of the precipitation of copper at moving grain boundaries in Ag–Cu alloys (AGEEW and SACHS [1930]). There are several points of interest in this type of transformation; they include the problem of how the reaction initiates and in particular how it is possible for different parts of the same boundary to migrate in different directions, as reported for example by WILLIAMS and EDINGTON [1976]; the question of what determines the residual supersaturation in the product, since in many cases this is reported to be appreciable (e.g., PORTER *et al.* [1974]); there is also the general problem of determining which reaction, the discontinuous one or growth by volume diffusion, will occur under any given circumstances. At present there are no definite answers to any of these questions though some general observations can be made.

As regards the amount of remaining supersaturation there is extensive evidence for significant residual supersaturation. This comes from lattice-parameter measurements and more recently from very high resolution microanalysis by electron energy loss techniques applied to Mg–Al alloys (PORTER *et al.* [1974]) and to Al–Li alloys (WILLIAMS and EDINGTON [1976]). Figure 44 shows the analysis results across a matrix grain boundary in Mg–Al, showing the discontinuous change at the boundary, characteristic of a process proceeding by boundary diffusion, and that the resulting solid solution is still supersaturated in the aluminium solute.

The initiation of the discontinuous reaction has been studied by TU [1972] in the Pb–Sn, solder, system at room temperature. His micrographs show regular grain-boundary nucleation of a colony of Sn-rich precipitates which grow apparently with an orientation relationship with the grain that is *not* consumed, when the boundary

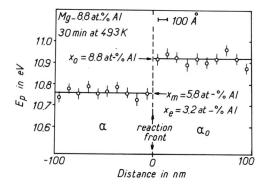

Fig. 44. Microanalysis by electron energy loss, across the migrating grain boundary during discontinuous precipitation in a Mg–Al alloy (after PORTER *et al.* [1974]).

References: p. 1024.

starts to move. As the boundary starts to move, further precipitates form to give the operating lamellar spacing.

The final question concerns the conditions required for the discontinuous reaction to dominate. This will presumably require that the discontinuous process occurs faster than either frequent grain-*interior* precipitation (presumably of a metastable, easy to nucleate, phase) or the alternative reaction of the growth of a boundary-nucleated phase *into* the grain away from the boundary, by a process determined by bulk diffusion. It might be expected that the discontinuous process will take over from growth of boundary precipitates as the temperature falls, since the relative advantage of boundary diffusion over bulk diffusion increases as the temperature falls. Boundary diffusion has a lower activation energy. Results reviewed by GUST [1979] confirm this description with the discontinuous process in, for example, copper-rich Cu–In (PREDEL and GUST [1975]) dominating the precipitation process as the temperature of the reaction fell below 300°C. However, if the temperature falls very far, general grain-interior nucleation dominates most precipitation reactions, at least in alloys selected for precipitation hardening (KELLY and NICHOLSON [1963]). For such alloys, high temperatures of precipitation frequently cause discontinuous precipitation of the stable phase in place of the general precipitation of the strengthening metastable phases. On this basis there should be only a narrow "window" of conditions that allows the discontinuous reaction to proceed. This is just as well since discontinuously precipitated alloys are unlikely to have good mechanical properties, having neither the high strength of precipitation-hardened alloys nor the good toughness of single-phase alloys.

2.5.3. Discontinuous coarsening

The ordinary coarsening process, described in §2.4, involves the reduction of interfacial energy by growth of large precipitates, at the expense of smaller ones, by diffusion either through the lattice or – for a set of precipitates all linked by grain boundaries – by boundary diffusion in *stationary* boundaries. LIVINGSTON and CAHN [1974] discovered a *discontinuous* coarsening reaction in polycrystalline eutectoids in Co–Si, Cu–In and Ni–In, when annealed close to the eutectoid temperature. Figure 45 shows a typical microstructure in which the upper grain is growing into the lower grain at the right while the opposite is happening on the left. The grain boundary is only visible in the grey Co_2Si phase. The origin of this effect is, of course, the reduction in interfacial area and was successfully analyzed by LIVINGSTON and CAHN [1974] on the basis of boundary diffusion to give a predicted boundary velocity:

$$v = \frac{8C_b D_b \delta_b}{f_\alpha^2 f_\beta^2 (C_\beta - C_\alpha)} \frac{\sigma_{\alpha\beta}}{\lambda_2^2 \lambda_1 RT} \left(1 - \frac{\lambda_1}{\lambda_2}\right), \tag{72}$$

where λ_1 and λ_2 are the spacings before and after coarsening, C_b is the grain-boundary composition, $\sigma_{\alpha\beta}$ is the energy of the interface between the two phases, which is assumed to have the same properties on either side of the migrating

Fig. 45. Discontinuous coarsening of eutectoid in Co–Si alloys; a polycrystalline sample annealed at 1000°C for 96 h. ×1875. (From LIVINGSTON and CAHN [1974].)

boundary, and the other terms have their previous meanings. This analysis was consistent with all the observations made. Of particular interest was the clear demonstration of the means of determining which way the boundary must move. As seen in fig. 45, the grain whose lamellae lie nearly parallel to the initial boundary plane is able to grow, since it automatically has a larger effective spacing along the boundary. The only difficulty that arises here is the inevitable change of interfacial plane that occurs with this process, even though the two phases have the same relative orientations; the properties of the interface may vary with boundary orientation unless the interface is fully incoherent.

A further very important point discussed by Livingston and Cahn was the magnitude of the spacing ratio, λ_2/λ_1, which was shown to be between 5 and 7 for Cu–In and Co–Si and between 10 and 20 for Ni–In. This is discussed in more detail in §2.5.4.

Other examples of dicontinuous coarsening have been given by GUST [1979], though without discussion. One further example has already been mentioned, the work of WILLIAMS and EDINGTON [1976] on Al–Li alloys. At low ageing temperatures close to room temperature, the discontinuous reaction apparently involved formation by *precipitation* of the two-phase lamellar structure from a fully super-saturated solid solution, but at higher ageing temperatures the discontinuous reaction was one of *coarsening* when a lamellar Al + Al_3Li, cube–cube oriented, product grew out from grain boundaries to consume the fine coherent precipitate nucleated in the adjacent grain. The reaction halted after a while when the spherical Al_3Li

References: p. 1024.

distribution had coarsened by the usual LSW process, and also presumably when the aluminium supersaturation had been removed by completion of the grain-interior reaction. An interesting result reported by Williams and Edington was that grain boundaries close to coincidence-site orientations (ch. 10B, §2.2) were less able to show the discontinuous reaction than general high-angle grain boundaries.

DOHERTY [1982] has discussed the formation of similar coarsened lamellar structures by deformation-induced boundary migration in nickel alloys containing coherent precipitates.

2.5.4.   Determination of lamellar spacing in discontinuous reactions

In the initial ZENER [1946] analysis it was assumed that the system adopted the spacing that allowed the interface to migrate at the *maximum velocity*, which for volume-diffusion mechanisms gives $\lambda_s = 2\lambda_s(\text{rev})$ and for interface diffusion mechanisms $\lambda_s = 1.5\lambda(\text{rev})$. PULS and KIRKALDY [1972] discussed alternative optimization criteria, including the spacing that gives the *maximum rate of entropy production*. This yields somewhat larger optimum spacings, which are $3\lambda_s(\text{rev})$ and $2\lambda_s(\text{rev})$ for volume and interface diffusion mechanisms, respectively. There is, however, considerable difficulty in estimating the appropriate values for $\lambda_s(\text{rev})$ since its evaluation requires knowledge of the interfacial energy as well as the free energy driving the eutectoid reactions. Free-energy calculations for discontinuous precipitation will be easier, particularly in dilute alloys. However, as LIVINGSTON and CAHN [1974] pointed out, the discontinuous coarsening reaction does not have either of these difficulties, since the relevant parameter is the spacing ratio which was *directly* measured. The spacing ratios in coarsening were found by Livingston and Cahn to be considerably bigger than predicted by either of the proposed optimization criteria, maximum velocity or maximum rate of entropy production. The coarsening results therefore suggest growth well away from the maximum-velocity condition, at larger spacings. The same conclusion is also suggested by various experimental studies on discontinuous precipitation (RUSSEW and GUST [1979] and SPEICH [1968]), recently reviewed by HILLERT [1982]. It was shown by Hillert that the experimental results lay close to the line, predicted by his diffusional model, of a plot of the residual supersaturation against lamellar spacing, but at *larger values of the spacing* than was predicted for the maximum-velocity criterion.

By comparison with the equivalent problem in dendritic growth, it would appear that there are *experimental* reasons, firstly, for expecting growth on the *large* side of the spacing predicted by the maximum-velocity criterion, and secondly, for the potential successful application of some type of stability criterion for lamellar spacing. Unfortunately, there seem to be currently no successful theoretical stability analyses for discontinuous reactions. This is a major theoretical problem that needs a solution.

2.5.5.   Diffusion-induced grain-boundary migration

In discontinuous reactions the available free energy of the precipitation, for example IJ in fig. 1, can be applied either to the growing β-phase alone which then pulls the depleted α matrix along, or the energy can be applied to *both phases*. HILLERT [1972, 1982] in his analysis of discontinuous precipitation has used this idea of the partition of the free energy between the two phases. A simpler experimental situation was devised by HILLERT and PURDY [1978], in a process that is now called by the title of this section, *diffusion-induced grain-boundary migration*, DIBM. This can be achieved by alloying, or dealloying, a polycrystalline sample from the surface at a temperature at which bulk diffusion has been "frozen out", but boundary diffusion is still rapid. HILLERT and PURDY [1978] introduced zinc into thin polycrystalline iron samples from a vapour source, an Fe–11.3 wt% Zn alloy, at temperatures between 545 and 600°C. They observed boundary migration which left a layer of zinc-enriched solid solution behind the migrating boundary. The driving force for this reaction is the free energy of mixing per atom, ΔF_m^a. The effective mobilities of the boundaries could be estimated from

$$v = M\Delta F_m^a. \tag{73}$$

Mobilities of grain boundaries in iron, at 580°C, of about 10^{-17} m^4/Js were found which compare with much higher values found from ordinary boundary-mobility experiments and from SPEICH's [1968] work on discontinuous precipitation in iron-rich Fe–Zn alloys ($M = 10^{-15}$ and 10^{-12} m^4/Js respectively). It is not clear whether the discrepancy arises from extraneous sources such as solute drag or if the use of eq. (73) is not justified.

CAHN *et al.* [1978] and BALLUFFI and CAHN [1981] have reviewed several other examples of DIBM and have shown that it occurs under a whole range of conditions, including both alloying and dealloying in systems with either a positive or a negative deviation from ideal solution behaviour, that is, with positive or negative heats of mixing. To the highest available resolution there is a discontinuity in the solute content ahead of the migrating boundary, as was previously shown for discontinuous precipitation by fig. 44. Cahn *et al.* draw attention to a significant theoretical problem in DIBM and equivalently in discontinuous precipitation. This is how the overall free-energy decrease can *couple* with the individual atomic motions required for boundary migration. The problem is shown in the schematic migrating-boundary structure in fig. 46. The boundary contains solute in equilibrium with the solute content of the growing grain on the left but, apart from the boundary plane, there is no solute in the grain on the right. If an atomic layer of solute-rich alloy is formed on the grain on the *left* by atom transfer from the right, together with solute addition *down* the boundary, a decrease of free energy of mixing will have occurred. But this difference in free energy is apparently *not available* to the atoms in the right grain, that must jump across the boundary. BALLUFFI and CAHN [1981] make the interesting prediction that DIBM *could not occur* if the solute were an *isotope* of the matrix, that is, if the solution were *ideal* and the rate of boundary diffusion of the two types of atom were the same. This seems unlikely since DIBM can occur with both positive and negative deviations from ideality, so it is not obvious that it cannot occur for a truly ideal solution. CAHN *et al.* [1978] propose a model for the process based on *differences in the rate of boundary diffusion*, the grain-boundary equivalent of the

References: p. 1024.

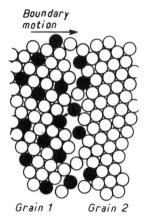

Fig. 46. Diffusion-induced grain-boundary motion; solute atoms, solid circles, have diffused down the boundary and are causing growth of grain 1, which is being solute-enriched and therefore causing a reduction in free energy by the free energy of mixing. What is the driving force causing a net transfer of atoms across the boundary?

Kirkendall effect. This has the result that extra lattice sites must be created, or lost, at the boundary which provides, via the climb of grain-boundary edge dislocations, a means of moving the grain boundary. There is clear evidence, for example in the Fe–Zn alloy experiments of Hillert and Purdy, that there is surface tilting at the migrating grain-boundary regions, characteristic of volume changes predicted by the boundary Kirkendall effect. It is not clear, to this author at least, whether there might not be a means of providing some boundary driving force from the entropy effects of an atomically sharp concentration gradient at the edge of the right-hand grain in fig. 46. However, no analysis that predicts this effect is available, so further development awaits the result of the crucial experiment to see if DIBM can or cannot be produced by isotope diffusion.

2.5.6. Experimental results on discontinuous reactions

PULS and KIRKALDY [1972] have published a detailed review on the experimental results of eutectoidal decompositions. For the eutectoidal reaction in binary iron–carbon alloys, the reactions appear to go at the rates expected for volume diffusion of carbon, though with some indication of rate enhancement by additional boundary diffusion. In the observation on eutectoid growth in binary substitutional alloys several studies, such as the fcc Al–78% Zn eutectoid (CHEETHAM and RIDLEY [1971]) show *boundary* diffusion to be dominant while in the Cu–Al eutectoid where the decomposing "austenite" phase is bcc with a high bulk diffusivity, *volume* diffusion control was reported (ASUNDI and WEST [1966]) *. In all these eutectoid

---

* This result is interesting for the theory of precipitate shape stability (§2.2.6). γ-brass and also γ in Cu–Al show solid-state dendrites and it was argued in §2.2.6.1 that a high rate of bulk diffusion relative to interface diffusion is required for instabilities to form. That the eutectoidal decomposition of bcc β in Cu–Al is by bulk rather than interface diffusion is in agreement with this idea, since the bcc β-phase shows γ dendrites.

systems, as in eutectic solidification where the reviewed results were considered along with solid-state reactions by KIRKALDY and SHARMA [1980], the results were tested against models which assumed a maximum-velocity, or the equivalent, a minimum-undercooling criterion, and showed physically sensible behaviour. The data on discontinuous precipitation have already been mentioned but it could be argued that the lack of full details renders that analysis perhaps not definitive.

From their study on Fe–C–X ternary alloys, PULS and KIRDALDY [1972] suggested that the data are apparently explicable on the reasonable basis that there is no partitioning of the substitutional solute and hence carbon diffusion control operates at low temperatures; at high temperatures, however, there is evidence for a shift to control by boundary diffusion of the substitutional elements, which are then partitioning.

Unfortunately there is lack of full information on the crucial parameters, such as the surface energies and the relevant diffusion coefficients, to allow a really satisfactory test of the theory to be made. However, at the moment the overall situation appears to be one of major doubt about the operating criterion that determines spacing, but apart from this problem the theories seem in good shape. The data on growth velocities appear to be satisfactory, but the discontinuous-coarsening results cast real doubt on the spacing-determining criteria, at least in that reaction. There is an obvious need for more data on the spacing ratios in discontinuous coarsening and for observations on the spacings in systems where the interfacial energies and the driving forces are well established.

2.6. Bainitic transformations

On isothermal TTT diagrams for iron–carbon alloys there is an important transformation at temperatures below the "nose" of the pearlite reaction (fig. 43, above). The product of this lower-temperature transformation is called *bainite* and appears to be a mixture of α-Fe and finely dispersed iron carbide, Fe_3C. Two varieties of bainite, known as *upper bainite* and *lower bainite*, have been identified, corresponding to their respective temperatures of formation. The subject has been reviewed many times (AARONSON [1969], HEHEMAN [1970], HEHEMAN *et al.* [1972] and BHADESHIA and EDMONDS [1980]), and full details of the experimental results and differing views on the transformation can be obtained from these reviews. There is a considerable dispute as to the interpretation of the results, particularly concerning the significance of the surface relief that is known to accompany the reaction, and hence as to the martensitic character, or otherwise, of the reaction. Following the initial ideas of KO and COTTRELL [1952] it has been commonly accepted that a martensitic shear step is involved in the bainitic reaction (KELLY and NUTTING [1961] and CHRISTIAN [1965]). There is also no doubt from the kinetics (HILLERT [1975]) and from the identification of carbides within the structure (e.g. BHADESHIA [1980]) that long-range carbon diffusion is part of the process. However, Bhadeshia and Edmonds propose that the diffusion of carbon occurs after the reaction is complete. The more conventional view is that the reaction involves the removal of

References: p. 1024.

carbon by rapid interstitial diffusion to allow carbide precipitation and the carbon-depleted matrix then undergoes a martensitic transformation. This would make thermodynamic sense, since carbon stabilizes the fcc austenite structure so that the martensitic transformation temperature, M_s, is strongly depressed by carbon. The removal of carbon could then allow the austenite, now carbon-depleted, to transform to martensite. AARONSON [1969] and AARONSON et al. [1978] do not accept this model as fully demonstrated but there appears to be no reason, at least in principle, why the Ko and Cottrell picture could not be right (see §2.2) but the subject is certainly not settled. It is difficult to accept that bainite could form martensitically and then decompose by precipitating carbides since in this case the initial reaction should be the normal martensite reaction, which kinetically it clearly is not. Martensite in medium-carbon iron alloys forms "athermally" at lower temperatures than the partially isothermal bainite (fig. 43). In an athermal reaction there is a given fraction transformed which changes with temperature but not isothermally with time at a given temperature (ch. 15, §2.5.2).

A confusing problem is raised by the observation that Widmanstätten ferrite formation at high temperature also shows evidence for surface relief (CLARK and WAYMAN [1970] and BHADESHIA [1981]) but following the review by HILLERT [1975] it is difficult to understand how this can have any martensitic character, since almost all the driving force appears to be taken up in the carbon diffusion reaction. Martensite formation in iron alloys seems always to require a very significant driving force. Considering their industrial importance, these various reactions in iron alloys seem surprisingly little understood.

There are other reactions which are also called bainitic by virtue of showing both diffusional changes in composition as well as surface relief effects characteristic of martensitic shear, for example Cu–Al (SMITH [1962]), Cu–Sn (DE BONDT and DERUYTERRE [1967]) and Cu–Zn, where both rods and plates of fcc (α) precipitation from bcc (β) shows surface relief (CLARK and WAYMAN [1970]).

This is clearly an area in which there is no clearly accepted qualitative model, let alone a full quantitative analysis. Unusually for a topic in metallic solid-state phase transformations, the problem is a lack of agreed theory rather than a shortage of quantitative data.

3. Continuous transformations

Although the vast majority of phase transformations in physical metallurgy are of Gibbs' first type – nucleation and growth, Gibbs' second type of transformation, now usually called *continuous transformations*, are amongst the most interesting and important reactions. A continuous transformation can only occur if the initial and final structures share a common lattice (or lack of a lattice, as in liquids and glasses). A particularly important version of this is the *spinodal decomposition.*

The reaction is solely a *diffusion* reaction involving atoms exchanging lattice sites. Although binary equilibrium diagrams indicate only a few potential examples of *equilibrium spinodal decomposition* in metals, there are numerous metastable reactions that could occur by the continuous mechanism (see CAHN [1968] and SOFFA and LAUGHLIN [1983]) and in addition the process of *continuous ordering* is also to be included as a continuous reaction, following COOK *et al.* [1969] and DE FONTAINE [1975].

3.1. Spinodal decomposition

Two excellent early reviews of this subject have been provided by CAHN [1968] and HILLIARD [1970]; these give a clear account of the basic theory and a review of relevant experimental observations up to the late sixties. The following will therefore give an outline of the initial theory, following CAHN [1968], and will then try to bring out important advances made in the last fifteen years. For a very complete account of the experimental results the detailed review by SOFFA and LAUGHLIN [1983] should be consulted.

In a region of a single phase where the curvature of the free-energy composition plot, d^2F/dC^2, is negative (fig. 4, above) *uphill* diffusion can occur, as this lowers the free energy. In the literature on spinodal decomposition the discussion is usually in terms of this second differential, while in the literature on diffusion, the discussion is based on the "Darken term", $1 + d(\ln f_\alpha)/d(\ln C_\alpha)$ (SHEWMON [1963], also ch. 8, §5.3). If D_{A*} and D_{B*} are the tracer diffusion coefficients of the two components of atomic fraction C_A and C_B, then the interdiffusion coefficient, $\tilde{D}$, is given by:

$$\tilde{D} = (C_A D_{B*} + C_B D_{A*}) \left(1 + \frac{d \ln f_\alpha^B}{d \ln C_B}\right). \tag{74}$$

It is readily shown (e.g., MARTIN and DOHERTY [1976]) that:

$$\frac{d^2F}{dC^2} = RT \left(\frac{1}{1-C_B} + \frac{1}{C_B}\right) \left(1 + \frac{d \ln f_\alpha^B}{d \ln C_\alpha^B}\right). \tag{75}$$

From here onwards, the atomic fraction C will stand, as elsewhere in this chapter, for the atomic fraction of the second component, B. So $C_A = 1 - C_B = 1 - C$.

From eqs. (74) and (75) we obtain:

$$\tilde{D} = \left[(1-C)D_{B*} + CD_{A*}\right] \frac{C(1-C)}{RT} \frac{d^2F}{dC^2}, \tag{76}$$

$$\tilde{D} = M_D \frac{d^2F}{dC^2}. \tag{76a}$$

Equations (76) and (76a) define the diffusional mobility, M_D.

As a consequence of this effect, an initially homogeneous solid solution is unstable to an initial infinitesimal fluctuation (fig. 5), so that some sinusoidal perturbations can grow. The crucial problem concerns the *wavelength* of the perturbation that can grow, and to discuss this, the thermodynamics of inhomogeneous solid solutions are needed (CAHN and HILLIARD [1958]).

References: p. 1024.

For an homogeneous AB alloy in which we exchange a unit amount of A ($\Delta F = -\mu_A$, where μ_A is the chemical potential of A, $\mu_A = \bar{F}_A$) for a unit amount of B ($\Delta F = \mu_B$), the change of free energy, ΔF, is:

$$\Delta F = (\mu_B - \mu_A)/V_m = \frac{dF^v}{dC}, \tag{77}$$

where F^v is the free energy of the number of atomic sites that occupies unit volume of the homogeneous solid solution. In this simple treatment it is assumed that the atoms are of equal size; a correction to account for the change of lattice parameter with composition in a real system will be introduced later.

The diffusive flux, $\tilde{J}$, is then given by the usual equation, the first law of diffusion:

$$\tilde{J} = -\tilde{D}\nabla C', \tag{78}$$

with $\tilde{D}$ given by eq. (76), and $C' = C/V_m$, the concentration expressed in atoms of B per unit volume.

From the thermodynamics of inhomogeneous solid solutions, a correction for inhomogeneity is introduced as:

$$\left(\frac{dF^v}{dC}\right)_{inh} = \frac{dF^v}{dC} - \frac{2K\nabla^2 C}{V_m}, \tag{79}$$

where K is the *gradient-energy coefficient* which is determined by the difference in number of like atomic neighbours between an atom in a homogeneous alloy and an atom in an alloy which has a variation in composition. CAHN [1968] pointed out that the first differential, ∇C, will have no effect on the energy since under a fixed gradient, ∇C, any excess of one component in one direction will be balanced by an equal depletion of that component in the opposite direction. The value of K is given by:

$$K = N^v k T_c \psi^2, \tag{80}$$

where T_c is the critical temperature, below which homogeneous alloys will wish to *unmix* into A-rich and B-rich regions, and ψ is the "interaction distance" of atoms (CAHN [1968]).

Substitution of eqs. (76a) and (79) into the diffusion equation (78) gives:

$$\tilde{J} = -M_D \frac{d^2 F}{dC^2} \nabla C + 2M_D K \nabla^3 C. \tag{81}$$

The change of composition with time, dC/dt, is obtained in the normal way for the derivation of the second law of diffusion (e.g., SHEWMON [1963]), as:

$$\frac{dC}{dt} = V_m \frac{dC'}{dt} = \frac{M_D d^2 F}{dC^2} \nabla^2 C - 2M_D K \nabla^4 C. \tag{82}$$

CAHN [1961] showed that this differential equation has the following solution:

$$C = C_0 + \exp(R(\beta)t)\cos(\beta r) \cdot \text{const.}, \tag{83}$$

with $C_0 =$ initial composition; $\beta =$ wave number $2\pi/\lambda_w$; $\lambda_w =$ wavelength of the particular fluctuation. The "amplification" factor $R(\beta)$ is given by:

$$R(\beta) = -M_D\beta^2\left(\frac{d^2F}{dC^2} + 2K\beta^2\right). \tag{84}$$

Since M_D is inherently positive, it can be seen, from the term in brackets in eq. (84), that in a system showing unmixing, when K will be positive, short-wavelength fluctuations will decay since $K\beta^2 > -d^2F/dC^2$ for large values of β, but fluctuations can grow below a critical wave number, β^*:

$$\beta^* = \left(-\frac{d^2F}{dC^2}\frac{1}{2K}\right)^{1/2},$$

with the corresponding critical wavelength, $\lambda^* \doteq 2\pi/\beta^*$. The fastest growing wavelength, λ_{max}, is determined from eqs. (83) and (84) as the maximum amplification at $\beta^*/\sqrt{2}$, owing to the double effect of increasing wave number in reducing the *diffusion distance*, as well as its effect in reducing the *driving force*, a similar result to that previously seen in other areas of phase transformation (§§ 2.2.6 and 2.5). In the present case, all wavelengths will be found, to some extent, in any perturbation; in addition, at very small amplitudes, all wavelengths can grow *independently* so the fastest-growing wavelength will be seen and should dominate the spinodal decomposition.

An important modification to the theory comes from considerations of elastic strains that arise when there is a change of lattice parameter with change of alloy content. With η defined as the unit strain per unit composition difference, E as Young's Modulus and ν as Poisson's ratio, eq. (84) is changed by the elastic strain to:

$$R(\beta) = -M_D\beta^2\left(\frac{d^2F}{dC^2} + \frac{2\eta E}{1-\nu} + K\beta^2\right). \tag{84a}$$

The strain term, in eq. (84a), acts in addition to the gradient-energy term to *inhibit* the reaction. It is more convenient, however, to include the second term with the first and consider a *coherent spinodal* region as one in which the *sum* of the first two terms in the parentheses of eq. (84a) is negative. Only within the coherent spinodal region can fluctuations develop while the crystal remains fully coherent across the composition variation.

RUNDMAN and HILLIARD [1967] tested the model for spinodal decomposition by small-angle X-ray scattering experiments with an Al–Zn alloy. Al–Zn has a phase diagram very similar to that shown in fig. 3, above. Their results (fig. 47) show very well the behaviour expected for spinodal decomposition. The alloy, Al–22 at% Zn, had been quenched from the single-phase, α, region and annealed at 65°C for the times indicated. The critical value of wave number, β^*, where there is no change of scattered intensity is seen, together with the maximum rate of increase of intensity at a wave number about 0.7 of the "cross-over" value of β^*. The interpretation of small-angle diffraction is discussed in ch. 12, §5.2. This was not the only clear experimental demonstration of the importance of the gradient term in an unstable

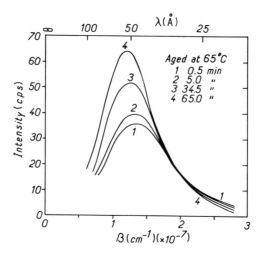

Fig. 47. Small-angle X-ray spectra for an Al–22 at% Zn alloy quenched from 425°C and annealed at 65°C for the times indicated (after RUNDMAN and HILLIARD [1967]).

solid solution; diffusion experiments in stable solid solutions, made very inhomogeneous by repeated depositions of alloys of different compositions, also demonstrated the importance of the gradient energy term (COOK and HILLIARD [1969] and PHILOFSKY and HILLIARD [1969].

Other early experimental studies that supported the linear spinodal model described by CAHN [1968] are discussed very clearly by HILLIARD [1970] whose review nicely complements that of Cahn.

Following the initial development of Cahn's theory of spinodal decomposition there have been various modifications to the theory to try to deal with later stages in the reaction when the linear model of independent growth of all fluctuations is no longer valid. A further important modification was consideration of the influence of *random thermal fluctuations*, "Brownian Motion", on the process.

Random thermal fluctuations will give local increases of energy paid for by the increase of entropy that comes from the "disorder" introduced by the small groups of atoms having the increased energy. In nucleation theory this process gives the vital equation (13) for the probability of creating a nucleus with local increase of energy ΔF^*.

COOK [1970] introduced this idea of random solute movements into a treatment of spinodal decomposition and showed that the effects seen in experiments on small-angle scattering could be fitted to the changes proposed by his model. The effect of the thermal fluctuations is essentially that for alloys close to the spinodal boundary the distinction between spinodal decomposition and the nucleation and growth of coherent "zones" becomes much less clear-cut than in the original model of spinodal decomposition.

LANGER [1975] has discussed this in more detail and shown that the initial model of spinodal decomposition, as described above, predicts that at the spinodal point, when $d^2F/dC^2 = 0$, the critical wavelength becomes infinite and the only mechanism for transformation will then be by the nucleation of a solute-rich "zone". However, nucleation of such a zone will occur frequently with a small critical radius and with the very low-energy, "diffuse" interface expected in these circumstances *. In other words, nucleation will result in a structure very much like that of spinodal decomposition. The same picture is given in recent models of spinodal decomposition modified by thermal fluctuations. Figure 48a shows the structure factor S computed by LANGER *et al.* [1975] as a function of q, a temperature-modified wave number and τ, a modified annealing time, for an alloy held at the spinodal boundary. It can be seen that a spinodal-like fluctuation builds up even with $d^2F/dC^2 = 0$, but the maximum intensity shifts to larger wavelengths with time.

The other modification to the theory deals with the later time development of the fluctuations. The results, also computed by LANGER *et al.* [1975] for the expected development of the structure factor for an alloy at the centre of the spinodal region are shown in fig. 48b; $q = 1$ corresponds to the critical wave number β^*. The value q_p is that for $\beta^*/\sqrt{2}$, the fastest growing fluctuation in the linear theory. It can be seen that for short times this is where the maximum actually appears, but at longer times, the peak in the structure factor moves to smaller wave numbers, larger wavelengths. In addition, the peak intensity does not continue to grow *exponentially* with time. This appears to indicate the movement towards the LSW-type *coarsening*, with larger wavelengths growing at the expense of shorter wavelengths. Even within spinodal decomposition a process that is equivalent to interfacial-energy-induced coarsening appears to be occurring *during* the decomposition and is not restricted to the last stages of the reaction when the solute-rich and solute-poor regions have discrete interfaces and are close to the equilibrium form of α' and α''.

Experimental data on the later stages of spinodal decomposition are shown in fig. 49 for Al–22 at% Zn (HILLIARD [1970]). The data here are for a higher temperature, 150°C, than in fig. 47 which was obtained at 65°C, so the diffusional unmixing is much more advanced than in the earlier study. It is seen in fig. 49 that the peak of intensity is shifted to much smaller wave numbers as the reaction proceeds; this is exactly the effect produced in the computer model of LANGER *et al.* [1975]. Assuming an activation energy for diffusion of Zn in Al of 120 kJ/g-mole gives the ratio of diffusion coefficients at the two temperatures, $\tilde{D}(150°C)/\tilde{D}(65°C)$, as 5000. The much higher intensities in fig. 49 than in fig. 47 can be seen, confirming that the reaction is, indeed, further advanced.

TZAKALAKOS [1977] and TZAKALAKOS and HILLIARD [1980] have been able to

---

* Simple application of the "quasi-chemical bond" model to a coherent interface between solute-rich and solute-poor zones shows that the interfacial energy σ, which is a *free energy*, is reduced at relatively high temperatures by making the interface extend over several atomic layers at some increase in energy but with a more than compensating increase in entropy. Atom-probe FIM has shown, however, that at temperatures much lower than T_c the coherent interface (in Cu–1.9 at% Ti) is atomically sharp (V. ALVENSLEBEN [1982]).

References: p. 1024.

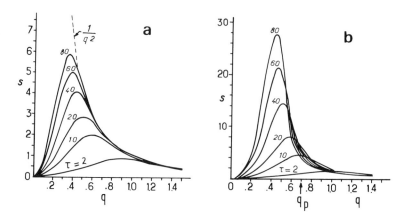

Fig. 48. Computed structure factor as a function of modified wavenumber for increasing times, τ, for (a) an alloy at the edge of the spinodal region and (b) an alloy at the centre of the spinodal region. (After LANGER et al. [1975].)

provide some analytical insight into the later stages of spinodal decomposition, when the compositional amplitude of the fluctuation is no longer small, but begins to approach the difference in composition $\Delta C_{\alpha'\alpha''}$ between the solute-rich and solute-poor regions of the two phase systems (fig. 3). The difficulty is readily seen from the free-energy–composition curve (fig. 4), since, as the composition fluctuation reaches the spinodal points, $d^2F/dC^2 = 0$, the driving force for further unmixing then vanishes. DITCHEK and SCHWARTZ [1980] discuss the theory and extend it beyond the single wavelength considered by Tzakalakos, to a range of wavelengths. For a single wavelength the amplitude grows until it reaches a *critical* wave form which has

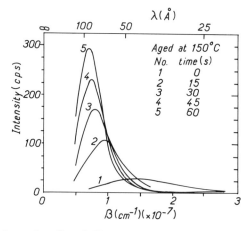

Fig. 49. Experimentally observed small-angle X-ray spectra for an Al–22 at% Zn alloy annealed at 150°C for the times indicated (after RUNDMAN et al. [1970]).

a tanh(βr) function (CAHN and HILLIARD [1958,1959]).

The system can then continue to lower its free energy only by perturbing the wavenumber to smaller values; this will *reduce* the gradient energy which is opposing further decomposition. Ditchek and Schwartz's extension, to consideration of a range of wavelengths, allows the reaction to continue by a growth in amplitude of waves with smaller wavenumbers. In the application of the analysis to experimental results in spinodally decomposing alloys, several fitting parameters are required. These take into account the initial composition waves produced during the quench after the solution treatment in the single-phase region. This is a weakness in their test of the theory, but nevertheless the comparison of the experimental results and the theory, shown as the points and the solid lines, respectively, in fig. 50, is very satisfactory indeed. The results were obtained for a Cu, 10.8 at% Ni, 3.2 at% Sn alloy, solution-treated at 800°C, quenched to room temperature and then aged at 350°C for various times, before examination of the composition fluctuation; $\epsilon = 0.015$ corresponds to the metastable equilibrium value of the composition difference of about 2.4 at% Sn between the two phases; the modulation is almost entirely in the tin content. Figure 50 shows very satisfactory agreement with the theory and in addition various features of the theory and the results can be seen. These features include the initial growth of amplitude at a fixed wavelength, $\lambda_w = 5$ nm; the departure from exponential growth of the amplitude, which occurs before the onset of the increase in the dominant wavelength. Finally it can be seen that the faster quench gives an initially smaller modulation in composition than does the slower

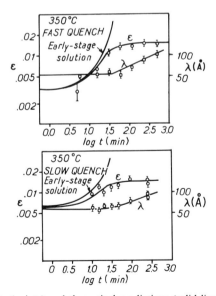

Fig. 50. Experimental results (points) and theoretical predictions (solid lines) for the amplitude, ϵ, and dominant wavelength, λ, for the spinodal decomposition of Cu–Ni–Sn after slow and fast quenches (after DITCHEK and SCHWARTZ [1980]).

References: p. 1024.

quench. In the Cu–Ni–Sn alloys the tin-rich regions show an ordering reaction to give a DO_{22}, Al_3Ti, superlattice at later stages of the decomposition.

In a recent, high-resolution, study by a field-ion microscopy/atom-probe technique, BIEHL [1980] was able to study the details of spinodal decomposition in an alloy of Cu–2.7 at% Ti. This showed the expected steady rise in the titanium content of the Ti-rich regions as the reaction progressed at 350°C. The rise in solute content saturated at 20 at% Ti, the composition of the ordered phase, Cu_4Ti. The dominant wavelength grew with time both during, and after, the time that the modulation was increasing in magnitude, with the dominant wavelength of the modulation increasing with reaction time to the 1/4 power. The diameter of individual titanium-rich clusters also grew, in this case with time to the 1/3 power, the time exponent expected for LSW coarsening both *during* and after the change in solute *content*.

It is striking that in the study of decomposition of Ni–14 at% Al by WENDT [1981], previously discussed in §2.2.6, it was clear that the reaction in the nickel alloy was much better described as a *nucleation-and-growth* reaction, rather than as a spinodal reaction. This was seen since the solute-rich regions, when first detected, had the required composition of Ni_3Al. Wendt had used the same techniques, FIM/AP, as BIEHL [1980]. It seems surprising that an alloy whose composition, 14 at% Al, is quite close to the composition of the final precipitate, should show a clear nucleation behaviour, while the Cu–2.7 at% Ti alloy that was much further from the composition of the ordered phase, Cu_4Ti, should show all the indications of spinodal decomposition. The difference is probably due to there being a very different form of the free-energy–composition curve in Cu–Ti at 350°C than in Ni–Al at 550°C (their homologous temperatures, T/T_m, are 0.46 and 0.48, respectively); this is discussed further in §3.2. In the meantime, V. ALVENSLEBEN [1982] has shown clearly by FIM/AP that a Cu–1.9 at% Ti alloy decomposes by nucleation and growth. So the spinodal point must be between 1.9 and 2.7 at% Ti at 350°C.

It is interesting that despite the apparent loss of distinction between spinodal decomposition and nucleation of zones with diffuse and coherent interface as argued by LANGER [1975] and earlier in this section, the atom-probe technique is able to demonstrate a distinct difference in the reactions on the basis of the time behaviour of the *composition of the centre of the zone*.

These atom-probe results suggest that further clarification of the theory of spinodal decomposition is going to be needed, if it is ever to be possible to predict which type of reaction will be expected to occur *faster* in a particular alloy, spinodal decomposition or nucleation and growth.

The classic electron microscopy studies of spinodal decomposition were provided by BUTLER and THOMAS [1970] and by LIVAK and THOMAS [1971] using ternary Cu–Ni–Fe alloys. In the first investigation, a symmetrical composition (51.5 at% Cu, 33.5 at% Ni, 15 at% Fe) near the centre of the spinodal region, was investigated. In the second study, an asymmetrical composition (32 at% Cu, 45.5 at% Ni, 15 at% Fe) nearer to the edge of the spinodal region was used. In both studies, Curie-temperature measurements, which are very sensitive to composition of the Fe–Ni rich phase, supplemented the microscopy. In the "as-quenched" alloys no sign of decomposition could be seen but, since the scattering factors of the components are similar in the Cu–Ni–Fe alloys, any initial perturbations will be very difficult to detect, unlike the case in the Cu–Ni–Sn alloy used by DITCHEK and SCHWARTZ [1980]. Butler and Thomas showed that the waves developed along $\langle 100 \rangle$ directions, which are the elastically softest directions. They also found that the two-phase structure was initially composed of rod-like particles with diffuse interfaces, that developed planar interfaces with extended coarsening. At long times of ageing, the interfaces lost their coherency by the usual formation of dislocation cross grids. There are faults in the periodicity and from the theories of coarsening of lamellar structures (see, e.g., MARTIN and DOHERTY [1976]) it would be expected that coarsening will occur by the *migration* of such faults to remove the excess interfacial area.

BUTLER and THOMAS [1970] report that the wavelength of the modulations coarsened, with a time to the power $1/3$ dependency, both during the later stages of the reaction and also where the Curie temperature was varying. This change indicated a variation in composition of the copper-depleted regions, that is, spinodal decomposition was occurring. For alloys aged at 625°C these simultaneous changes can be clearly seen (fig. 51a), and these results confirm that the coarsening reaction is occurring during the initial decomposition as well as after the decomposition.

In the study on the asymmetrical Cu–Ni–Fe alloys, LIVAK and THOMAS [1971] gave similar results, but there were some important distinctions. One of these distinctions was the slower development of the composition fluctuations; the change of Curie temperature continued for 100 h at 625°C for the asymmetrical alloy (fig. 51b) as distinct from only 1–5 h (fig. 51a) for the symmetrical alloy. This distinction is expected since the value of d^2F/dC^2 is smaller in alloys away from the symmetry point. In the asymmetrical alloy there was, in addition, no sign of growth in the dominant wavelength for nearly 10 h at 625°C (fig. 51b) so that coarsening appears to develop at a *later* stage in the reaction in the asymmetrical alloy than in the symmetrical alloy.

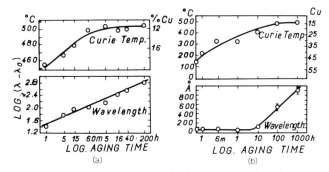

Fig. 51. Variation of the Curie temperature and the dominant wavelength with time on ageing at 625°C, for two Cu–Ni–Fe alloys: (a) the symmetrical alloy (after BUTLER and THOMAS [1970]); (b) the symmetrical alloy with 32 at% Cu (after LIVAK and THOMAS [1971]).

References: p. 1024.

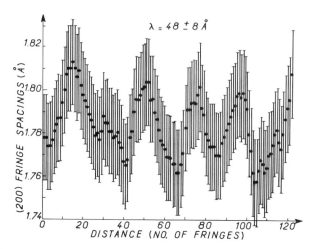

Fig. 52. The variation in fringe spacing from lattice images of a spinodally decomposed Cu–Ni–Cr alloy aged at 700°C for 10 min (after Wu *et al.* [1978]).

In an important extension of the application of TEM to spinodal decomposition, Sinclair *et al.* [1976] and Wu *et al.* [1978] used *lattice imaging techniques* to demonstrate the change in lattice parameter on a very fine scale, produced in spinodally decomposing alloys. Figure 52 shows the measured variation in lattice spacing in a 68 at% Cu, 29 at% Ni, 3 at% Cr alloy aged at 700°C for 10 min. The apparent wavelength of the fluctuation, 4.8 ± 0.8 nm, is very close to that obtained from electron diffraction, 5 ± 0.5 nm, in the same sample. In this lattice-imaging study, it was clearly shown that, as would be expected, the interface was *diffuse* at the early stages of the reaction and became much sharper at the later stages; the lattice was then no longer continuous.

3.2. Continuous ordering

In the discussion of nucleation-and-growth reactions in § 2, no specific discussion was offered for the precipitation of *ordered phase* from a disordered matrix, since there seems no real distinction between growth of an ordered phase from a disordered matrix and the growth of any other phase, in a Gibbs type-I transformation. For continuous reactions, however, continuous ordering is dictinctly different from spinodal decomposition and will therefore be briefly considered here as separate reaction. Much of what follows is derived from the original paper by Cook *et al.* [1969] and the detailed review by Soffa and Laughlin [1983]; reference should be made to these publications and to reviews by De Fontaine [1975,1981] and by Cook [1976] for more detailed accounts of the topic.

It has already been mentioned that in Cu–Ti and in Cu–Ni–Sn the end-result of spinodal decomposition reactions can be the formation of an ordered phase in solute-enriched regions of the structure, that is, the spinodal decomposition is the first continuous reaction and then order can develop, presumably continuously, as the composition increases towards the ideal value for a particular long-range-ordered structure. However, continuous ordering can occur without change of composition if the alloy is close to the correct composition. This has been analyzed by COOK *et al.* [1969].

The spinodal decomposition model discussed in §3.1 was based on a *continuum* analysis (CAHN [1961,1962]). However, HILLERT [1956,1961] had already considered a *discrete lattice* model, albeit in one dimension only. Hillert's approach was extended by COOK, DE FONTAINE and HILLIARD [1969] to a three-dimensional, discrete lattice model. Their model considered not only spinodal decomposition, the unmixing reaction with a negative d^2F/dC^2 and a positive gradient energy parameter K, as discussed previously, but also the opposite case, a strongly *positive* d^2F/dC^2, and a *negative K*, a situation expected for a system that can go, without a change in composition, from a disordered high-temperature structure to an *ordered* low-temperature structure, as the temperature falls below a critical temperature T_c.

The results of the discrete lattice calculation are very similar to the continuum model in the spinodal case. Figure 53 shows the amplification factor here called α, as a function of a/λ where a is the simple cubic lattice parameter, for three situations. Figure 53a is for a spinodal decomposition, that is, with negative d^2F/dC^2 and positive K. The critical wavelength, λ_c, and the fastest-growing wavelength, λ_m, are shown. Figure 53b shows the amplification factor for a *continuous ordering reaction*, with positive d^2F/dC^2 and negative K, where the maximum amplification occurs at $a/\lambda = 1$, that is, the maximum and minimum in composition occur on *adjacent lattice sites*, giving an ordered structure. Figure 53c shows the situation with a negative amplification at all values of a/λ, which will occur above the critical instability temperature, T_{instab}. This temperature is predicted by this treatment to occur when:

$$d^2F/dC^2 = -32K/a^2. \tag{85}$$

In a real situation, in any one region of the lattice, various perturbations will conflict with each other, but eventually one perturbation will dominate the rest and establish a particular *domain* of order in that region. The order in the domain will then steadily increase. Different regions will be dominated by waves with differing *phases* of modulation, giving the multiple domain structure of the ordered alloy. (For a discussion on domain structure see ch. 10B, §2.6) COOK *et al.* [1969] extend the analysis to bcc and fcc lattices. As in the original continuum model of spinodal decomposition, the analysis of COOK *et al.* [1969] is restricted to small perturbations where d^2F/dC^2 and K remain constant, so again only the early stages of the reaction are described by this theory. Extensions to the initial discussion on continuous ordering have been provided by DE FONTAINE [1975,1981] and by COOK [1976], who also briefly review some of the early experimental investigations on continuous ordering.

SOFFA and LAUGHLIN [1983] have reviewed the considerable amount of published

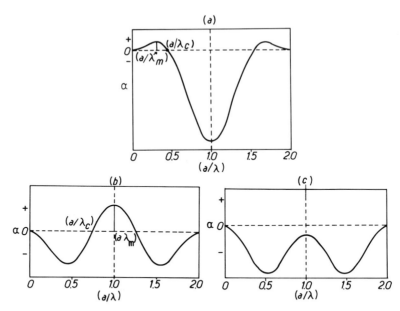

Fig. 53. The amplification factor, α, for the discrete lattice model of continuous reactions: (a) for spinodal decomposition; (b) for ordering below the instability temperature; (c) above the instability temperature. a is the lattice parameter of the simple cubic structure. (After COOK et al. [1969].)

work on continuous ordering reactions but they concentrate mainly on the more numerous cases of decomposition of metastable solid solutions in which composition fluctuations occur on a large scale, spinodal decomposition, with ordering developing in the solute-rich regions. They discuss only a few studies of continuous ordering occurring without a change of compositions. However, the two-stage reactions discussed by Soffa and Laughlin are probably more important technologically.

CHEN and COHEN [1979] have investigated the process of continuous ordering, without changes of composition, by studies on the intensities of scattered X-rays at superlattice positions in various alloys at temperatures *above* the ordering temperatures. For Cu_3Au, apart from some additional, heterophase, fluctuations just above the critical temperature, the data behave in the expected way for this type of theory (COOK [1976]). Extrapolation indicates an instability temperature, below which continuous ordering is expected. Between T_c and T_{instab}, ordering will be expected to occur but by a *nucleation* mechanism. Similar results were reported by Chen and Cohen for a higher copper alloy, for Co–Pt and for Fe–Al. Continuous ordering has also been demonstrated in β-brass (KUBO et al. [1980]) and in Ni–Mo alloys as reviewed by DE FONTAINE [1975,1981]. It appears from the present data that continuous ordering can occur in the manner proposed and that to date the proposed model appears to be reasonably successful.

Figure 54, from SOFFA and LAUGHLIN [1983], illustrates various possible free-en-

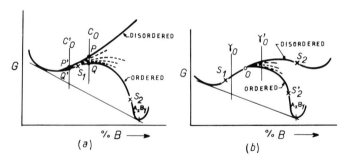

Fig. 54. Free-energy–composition curves of alloys that show various types of spinodal decomposition and continuous ordering, see the text (after SOFFA and LAUGHLIN [1983]).

ergy–composition curves that can give ordered second phases by continuous reactions, involving a change of composition. In fig. 54a the disordered curve is stable everywhere. An alloy of composition C_0 can lower its free energy by ordering, whereupon the ordered phase can then spinodally decompose; this type of process is described as a "conditional spinodal" (ALLEN and CAHN [1976]). An alloy of lower solute content, C'_0, even after ordering will not decompose spinodally but will precipitate the ordered phase only via a nucleation process. In fig. 54b, however, the disordered curve also shows an instability so that a composition γ'_0 can initially decompose in the normal spinodal manner, with the solute-rich cluster becoming ordered in the normal course of the reaction when the composition of the solute-rich regions has reached a value where the ordered phase has a lower free energy than the disordered phase. The distinction between a C'_0 and a γ'_0 alloy could be the difference between the systems, Ni–Al and Cu–Ti, discussed in § 3.1. It was reported there that experiments had shown that in the nickel alloy the ordered phase was precipitated with the correct composition right from the start of the decomposition, while in the copper alloy there was steady increase in the titanium content of the titanium-rich regions for $C_0 > 1.9$ at% Ti. The schematic diagrams of fig. 54 appear to describe, qualitatively, the complex processes that can take place in various important alloys that precipitate ordered phases.

4. Application of phase transformation theory to specific alloy systems

The application of phase transformation theory to specific systems is a huge topic and will only be treated in the briefest outline here by giving references to the various accounts of the field. The main alloy system is of course the one based on iron, since 50 times as much steel is produced than that of the next most common alloy family, based on aluminium, though the rate of growth of aluminium is considerably faster than that of iron (FINNISTON [1978]). Chapter 16 of this volume deals with the physical metallurgy of steels and gives the most appropriate discussion of the topic, and further references. For completeness however, two additional references to this topic can be given: PICKERING [1978] and HONEYCOMBE [1981].

For aluminium an excellent review of the various alloy reactions is provided by POLMEAR [1981] whose book also includes a discussion of the phase transformations in magnesium and titanium alloys. More detailed accounts of aluminium are provided by the three-volume book edited by VAN HORN [1967]; the extremely detailed review by MONDOLFO [1976] provides exceptional detail on the equilibrium diagrams and microstructures of all known aluminium alloys. LORIMER [1978] reviewed the physical metallurgy of aluminium precipitation reactions. The same volume on precipitation processes, edited by RUSSELL and AARONSON [1978], includes accounts of precipitation in nickel alloys by MERRICK [1978] and in titanium alloys by WILLIAMS [1978] as well as EDMONDS and HONEYCOMBE [1978] on iron. For the nickel alloys, SIMS and HAGEL [1972] and BETTERIDGE and HESLOP [1974] provide detailed additional information and for titanium alloys there are also the conference reports on titanium published by PERGAMON PRESS [1971] and PLENUM PRESS [1973].

As discussed in §1 for metals such as iron and titanium, the change of matrix crystal structure offers tremendous scope for many heat-treatments while in the alloys with a single matrix crystal structure, aluminium, copper, nickel, magnesium etc., only precipitation is readily available as a means of modifying the microstructures. However, although this is not discussed in this chapter, various additional microstructural modifying processes are available and are of increasing importance; these include rapid quenching from the liquid (chs. 9 and 28) which gives a much wider range of starting microstructure for subsequent treatment to work on than is provided by the conventional ingot casting route; the production of composite structures by powder metallurgy (ch. 29) and also the wide control of microstructure made possible by *thermomechanical* treatments in which phase transformations are combined with deformation. This has been widely exploited for grain-size control in low-carbon alloys of iron and for ausformed iron, see ch. 16, but is now of increasing interest for modification of microstructure by precipitation during deformation. HORNBOGEN and KÖSTER [1978] have shown something of the interesting results obtained by cold deformation, followed by precipitation, in supersaturated solid solutions. The next step is the combination of warm deformation with precipitation, an area only just being explored in alloys other than iron. An important review of the field was provided in the book edited by MORRIS [1979], on thermomechanical processing of aluminium alloys. There are also excellent reviews by FERTON [1981] on thermomechanical processing of aluminium alloys and by LUETJERING and PETERS [1981] on thermomechanical processing of titanium alloys.

5.   *Problems in phase transformations*

A very interesting and unusual approach to science in general and materials science in particular was the challenge offered by MADDIN [1976] in the 25th volume of Materials science and Engineering to 45 scientists to write a brief account of what was *not known* in the 45 topics of major interest to the authors. Many of the topics were from the area of phase transformations. 40 of those invited accepted the challenge and produced an important series of articles. From the previous pages some of the important questions raised may be listed and there are no doubt many more not revealed by the review in this chapter because of insufficient knowledge.

The questions that seem most important include the following:

(1) The need for more data to test the theories of heterogeneous nucleation on both grain boundaries and dislocations, to answer the questions raised by *coherency* of the interface and the higher-level heterogeneities in grain boundaries.

(2) What is the origin of the growth ledges on coherent interfaces between phases of different structures, both during precipitation and during coarsening?

(3) The rationalization of the role and significance of surface relief in diffusional transformations, particularly the formation of Widmanstätten ferrite and the bainite reaction in steels.

(4) Understanding the determining principle of the lamellar microstructures in discontinuous reactions.

(5) What are the conditions in which discontinuous precipitation at grain boundaries occurs in preference to nucleation and growth without the interference of a moving grain boundary?

(6) What is the mechanism of diffusion-induced grain-boundary migration?

(7) What are the conditions in which spinodal decomposition occurs in preference to nucleation and growth of coherent phases?

(8) The problem of shape stability for non-spherical precipitates.

(9) The solution to the problem of what effect, if any, the volume fraction of second-phase particles has on the rate of Ostwald ripening.

(10) The need for a great deal more data on free energies, interfacial energies and the role of impurities in modifying interfacial energies, which is needed for accurate prediction of phase-transformation kinetics and thus of resultant microstructures.

(11) The problem of the complications to all aspects of phase transformations produced by external variables such as deformation, radiation (HUDSON [1978]) and temperature gradients (McLEAN [1979]) and the effects both on microstructures and on the phase transformations that produce these microstructures.

References: p. 1024.

It is clear from this list that despite the great successes achieved in the past 30 years in understanding phase transformations in metals, much remains unknown and therefore fascinating in this field.

Acknowledgements

The author is very grateful to many scientists whose discussions have provided a great deal of such insight as this chapter contains. In particular I would like to acknowledge Dr. H.I. Aaronson, Dr. J.W. Cahn, Dr. B. Cantor, Prof. J.W. Christian, Prof. M. Hillert, Dr. G.R. Purdy as well as my former doctoral students Dr. B.G. Bainbridge, Dr. S.J.M. Fairs, Dr. M. Ferrante and Dr. K. Rajab. The mistakes are however all my responsibility. I would like to thank Prof. P. Haasen and Prof. R.W. Cahn for their critical reading of the manuscript and for their suggested modifications.

References

AARON, H.B., and G.R. KOTLER, 1971, Metallurg. Trans. **2**, 393.

AARON, H.B., D. FAINSTEIN and G.R. KOTLER, 1970, J. Appl. Phys. **41**, 4404.

AARONSON, H.I., 1962, in: The Decomposition of Austenite by Diffusional Processes (Interscience, New York) p. 387.

AARONSON, H.I., 1969, in: The Mechanisms of Phase Transformations in Crystalline Solids (The Metals Society, London) p. 270.

AARONSON, H.I., 1974, J. Microsc. **102**, 275.

AARONSON, H.I., 1977, Scripta Metall. **11**, 731 and 741.

AARONSON, H.I., 1979, in: Phase Transformations, York Conference (The institution of Metallurgists, London) p. II-1.

AARONSON, H.I., and C. LAIRD, 1968, TMS AIME **242**, 1437.

AARONSON, H.I., J.B. CLARK and C. LAIRD, 1968, Met. Sci. **2**, 155.

AARONSON, H.I., C. LAIRD and K.R. KINSMAN, 1970, in: Phase Transformation (ASM, Metals Park, OH) p. 313.

AARONSON, H.I., K.C. RUSSELL and G.W. LORIMER, 1977, Metallurg. Trans. **8A**, 1644.

AARONSON, H.I., J.K. LEE and K.C. RUSSELL, 1978, in: Precipitation Processes in Solids, eds. K.C. Russell and H.I. Aaronson (Met. Soc. AIME, Warrendale, PA) p. 31.

ABBOTT, K., and C. HAWORTH, 1973, Acta Metall. **21**, 951.

AGEEW, N., and G. SACHS, 1930, Z. Phys. **66**, 293.

ALLEN,S., and J.W. CAHN, 1974, Acta Metall. **24**, 425.

ARDELL, A.J., 1972, Acta Metall. **20**, 61.

ASHBY, M.F., and L. JOHNSON, 1969, Phil. Mag. **20**, 1009.

ASUNDI, M.P., and D.R.F. WEST, 1966, J. Inst. Metals **94**, 327.

AUST, K.T., and J.W. RUTTER, 1963, in: Recovery and Recrystallization of Metals, ed. L. Himmel (Interscience, New York) p. 131.

BAINBRIDGE, B.G., and R.D. DOHERTY, 1969, in: Quantitative Relationship between Properties and Microstructure, eds. D.G. Brandon and A. Rosen (Israel Univ. Press, Jerusalem) p. 427.

BALLUFFI, R.W., 1982, Metallurg. Trans. **13A**, 2069.

BALLUFFI, R.W., and J.W. CAHN, 1981, Acta Metall. **29**. 493.

BARRETT, C.S., 1952, Structure of Metals, 2nd Ed. (McGraw–Hill, New York) p. 548.

BECKER, R., and W. DÖRING, 1934, Ann. Phys. **24**, 719.

BETTERIDGE, W., and J. HESLOP, eds., 1974, The Nimonic Alloys (Crane and Russak, New York).

BHADESHIA, H.K.D.H., 1980, Acta Metall. **28**, 1103.

BHADESHIA, H.K.D.H., 1981, Acta Metall. **29**, 1117.

BHADESHIA, H.K.D.H., and D.V. EDMONDS, 1980, Acta Metall. **28**, 1265.

BIEHL, K.E., 1980, Ph.D. thesis, Univ. of Göttingen; E. BIEHL and R. WAGNER, in preparation.

BOSZE, W.R., and R. TRIVEDI, 1974, Metallurg. Trans. **5**, 511.

BOWER, E.N., and B.M. WHITEMAN, 1969, in: The Mechanisms of Phase Transformations in Crystalline Solids (The Metals Society, London) p. 119.

BRAILSFORD, I.N., and P. WYNBLATT, 1979, Acta Metall. **27**, 489.

BUTLER, E.P., and P. SWANN, 1976, Acta Metall. **24**, 343.

BUTLER, E.P., and G. THOMAS, 1970, Acta Metall. **18**, 347.

CAHN, J.W., 1956, Acta Metall. **4**, 441.

CAHN, J.W., 1957, Acta Metall. **5**, 168.

CAHN, J.W., 1961, Acta Metall. **9**, 795.

CAHN, J.W., 1962, Acta Metall. **10**, 179.

CAHN, J.W., 1963, Acta Metall. **10**, 789.

CAHN, J.W., 1968, TMS AIME **242**, 166.

CAHN, J.W., and J.E. HILLIARD, 1958, J. Chem. Phys. **28**, 259.

CAHN, J.W., and J.E. HILLIARD, 1959, J. Chem. Phys. **31**, 3 and 788.

CAHN, J.W., J.D. PAN and R.W. BALLUFFI, 1978, Scripta Metall. **13**, 2069.

CANTOR, B., and R.D. DOHERTY, 1979, Acta Metall. **27**, 33.

CHEETHAM, D., and N. RIDLEY, 1971, J. Inst. Metals **99**, 371.

CHELLMAN, D.J., and A.J. ARDELL, 1974, Acta Metall. **22**, 577.

CHEN, H., and J.B. COHEN, 1979, Acta Metall. **27**, 603.

CHEN, H., and R.D. DOHERTY, 1977, Scripta Metall. **11**, 725.

CHRISTIAN, J.W., 1965, in: The Physical Properties of Martensite and Bainite (The Metals Society, London) p. 1.

CHRISTIAN, J.W., 1970, in: Physical Metallurgy, 2nd Ed., ed. R.W. Cahn (North-Holland, Amsterdam) p. 471.

CHRISTIAN, J.W., 1975, The Theory of Transformations in Metals and Alloys, 2nd Ed. (Pergamon Press, Oxford).

CHRISTIAN, J.W., 1979, in: Phase Transformations, York Conference (The Institution of Metallurgists, London) p. 1.

CLARK, H.M., and C.M. WAYMAN, 1970, in: Phase Transformations ASM, Metals Park, OH) p. 59.

COOK, H.E., 1970, Acta Metall. **18**, 297.

COOK, H.E., 1976, Mater. Sci. Eng. **25**, 127.

COOK, H.E., and J.E. HILLIARD, 1969, J. Appl. Phys. **40**, 2192.

COOK, H.E., D. DE FONTAINE and J.E. HILLIARD, 1969, Acta Metall. **17**, 765.

CORIELL, S.R., and R.L. PARKER 1967, in: Crystal Growth, ed. H.S. Peiser (Pergamon Press, Oxford) p. 703.

DE BONDT, M., and A. DERUYTERRE, 1977, Acta Metall. **15**, 993.

DE FONTAINE, D., 1975, Acta Metall. **23**, 553.

DE FONTAINE, D., 1981, Metallurg. Trans. **12A**, 559.

DESALOS, Y., A. LEBON and R. LOMBRY, 1981, in: Les Traitements Thermomechaniques, Comptes Rendus, 24ᵉ Colloque de Metallurgie Saclay, France (CEN-Saclay) p. 137.

DITCHEK, B., and L.H. SCHWARTZ, 1980, Acta Metall. **28**, 807.

DOHERTY, R.D., 1976, unpublished research.

DOHERTY, R.D., 1978, in: Recrystallization in Metallic Materials, ed. F. Haessner (Riederer-Verlag, Stuttgart) p. 23.

DOHERTY, R.D., 1980, in: Crystal Growth, ed. B. Pampkin (Pergamon Pres, Oxford) p. 485.

DOHERTY, R.D., 1982, Met. Sci. **16**, 1.

DOHERTY, R.D., and B. CANTOR, 1983, in: Solid-State Phase Transformations, the Pittsburg Conference, ed. H.I. Aaronson (Met. Soc. AIME, Warrendale, PA) p. 547.

DOHERTY, R.D., and G.R. PURDY, 1983, in: Phase Transformations, the Pittsburg Conference, ed. H.I. Aaronson (Met. Soc. AIME, Warrendale, PA) p. 561.

DOHERTY, R.D., M. FERRANTE and Y.H. CHEN, 1977, Scripta Metall. 11, 733.

DOHERTY, R.D., B. CANTOR and S.J.M. FAIRS, 1978a, Metallurg. Trans. 9A, 621.

DOHERTY, R.D., M. FERRANTE and Y.H. CHEN, 1978b, Scripta Metall. 12, 885.

EDINGTON, J.W., 1979, in: Phase Transformations, York Conference (The Institution of Metallurgists, London) p. I-1.

EDMONDS, D.V., and R.W.K. HONEYCOMBE, 1978, in: Precipitation Processes in Solids, eds. K.C. Russell and H.I. Aaronson (Met. Soc. AIME, Warrendale, PA) p. 121.

EDMONDS, D.V. and R.W.K. HONEYCOMBE, 1979, in: Phase Transformations, York Conference (The Institution of Metallurgists, London) p. 121.

EHRENFEST, P., 1933, Proc. Acad. Sci. Amsterdam 36, 153.

FAIRS, S.J.M., 1977, D. Phil. thesis, Susses Univ.

FERRANTE, M., 1978, D. Phil. thesis, Sussex Univ.

FERRANTE, M., and R.D. DOHERTY, 1976, Scripta Metall. 10, 1059.

FERRANTE, M., and R.D. DOHERTY, 1979, Acta Metall. 27, 1603.

FERTON, D., 1981, in: Les Traitements Thermomechaniques, Comptes Rendus, 24e Colloque de Metallurgie, Saclay, France (CEN-Saclay) p. 243.

FINE, M.E., 1964, Introduction to Phase Transformations in Condensed Systems (Macmillan, New York).

FINNISTON, M., 1978, Metallurg. Trans. 9A, 327.

GAUNT, P., and J.W. CHRISTIAN, 1959, Acta Metall. 7, 529.

GLICKSMAN, M.E., R.J. SCHAEFER and A.J. AYERS, 1976, Metallurg. Trans. 7A, 1747.

GUST, W., 1979, in: Phase Transformations, York Conference (The Institution of Metallurgists, London) p. II-27.

HAESSNER, F., and S. HOFMANN, 1978, in: Recrystallization in Metallic Materials, ed. F. Haessner (Riederer-Verlag, Stuttgart) p. 63.

HALL, M.G., and C. HAWORTH, 1970, Acta Metall. 18, 331.

HAM, F.S., 1958, J. Phys. Chem. Solids 6, 335.

HAM, F.S., 1959, Quart. Appl. Maths. 17, 135.

HEADY, R.B., and J.W. CAHN, 1973, J. Chem. Phys. 58, 896.

HEHEMAN, R.F., 1970, in: Phase Transformations (ASM, Metals Park, OH) p. 379.

HEHEMAN, R.F., K.R. KINSMAN and H.I. AARONSON, 1972, Metallurg. Trans. 3, 1077.

HILLERT, M., 1956, D.Sc. Thesis MIT.

HILLERT, M., 1957, Jernkontorets Ann. 141, 757.

HILLERT, M., 1961, Acta Metall. 9, 525.

HILLERT, M., 1969, in: The Mechanisms of Phase Transformation in Crystalline Solids (The Metals Society, London) p. 231.

HILLERT, M., 1972, Metallurg. Trans. 3, 2729.

HILLERT, M., 1975, Metallurg. Trans. 6A, 5.

HILLERT, M., 1982, Acta Metall. 30, 1689.

HILLERT, M., and G.R. PURDY, 1978, Acta Metall. 26, 333.

HILLIARD, J.E., 1970, in: Phase Transformations (ASM, Metals Park, OH) p. 497.

HIRATA, T., and D.H. KIRKWOOD, 1977, Acta Metall. 25, 1425.

HONDROS, E., 1978, in: Precipitation Processes in Solids, eds. K.C. Russell and H.I. Aaronson (Met. Soc. AIME, Warrendale, PA) p. 1.

HONEYCOMBE, R.W.K., 1981, Steels: Microstructure and properties (Edward Arnold, London).

HORNBOGEN, E., 1967, Aluminium 43, 115.

HORNBOGEN, E., and U. KÖSTER, 1978, in: Recrystallization in Metallic Materials, ed. F. Haessner (Riederer-Verlag, Stuttgart) p. 159.

HORVAY, G., and J.W. CAHN, 1961, Acta Metall. 9, 695.

HREN, J.A., and G. THOMAS, 1963, TMS AIME 221, 304.

HUANG, S.-C., and M.E. GLICKSMAN, 1981, Acta Metall. 29, 701.

HUDSON, B., 1978, in: Precipitation Processes in Solids, eds. K.C. Russell and H.I. Aaronson (Met. Soc. AIME, Warrendale, PA) p. 284.

IVANTSOV, P.G., 1947, Dokl. Akad. Nauk. SSSR **58**, 567.

JACOBS, M.H., and D.W. PASHLEY, 1969, in: The Mechanisms of Phase Transformation in Crystalline Solids (The Metals Society, London) p. 43.

JONES, G.J., and R. TRIVEDI, 1971, J. Appl. Phys. **42**, 4299.

JONES, G.J., and R. TRIVEDI, 1975, J. Cryst. Growth **29**, 155.

KAMPMANN, L., and M. KAHLWEIT, 1967, Ber. Bunsen Gesell. Phys. Chem. **71**, 78.

KAMPMANN, L., and M. KAHLWEIT, 1970, Ber. Bunsen Gesell. Phys. Chem. **74**, 456.

KARLYN, D., J.W. CAHN and M. COHEN, 1969, TMS AIME **245**, 1971.

KATTAMIS, T.Z., U.T. HOLMBERG, and M.C. FLEMINGS, 1967, J. Inst. Metals **9**, 343.

KELLY, A., and G.W. GROVES, 1970, Crystallography and Crystal Defects (Addison–Wesley, Reading, MA) p. 310.

KELLY, A., and R.B. NICHOLSON, 1963, Prog. Mater. Sci. **10**, 151.

KELLY, P.M., and J. NUTTING, 1961, J. Iron Steel Inst. **197**, 199.

KINSMAN, H.A., H.I. AARONSON and E. EICHEN, 1971, Metallurg. Trans. **2**, 1041.

KINSMAN, H.A., H.I. AARONSON and E. EICHEN, 1973, Metallurg. Trans. **4**, 369.

KIRKALDY, J.S., and R.C. SHARMA, 1980, Acta Metall. **28**, 1009.

KIRKWOOD, D.H., 1970, Acta Metall. **18**, 563.

KO, T., and S.A. COTTRELL, 1952, J. Iron Steel Inst. **172**, 307.

KUBO, H., I. CORNELIS and C.M. WAYMAN, 1980, Acta Metall. **28**, 395.

LAIRD, C., and H.I. AARONSON, 1969, Acta Metall. **17**, 505.

LANGE, W.F., 1979, Ph.D. Thesis, Michigan Technol. Univ., Houghton.

LANGER, J.S., 1971, Ann. Phys. **65**, 53.

LANGER, J.S., 1975, in: Fluctuations, Instabilities and Phase Transitions, ed. T. Riste (Plenum, New York) p. 19.

LANGER, J.S., and H. MÜLLER-KRUMBHAAR, 1978, Acta Metall. **26**, 61.

LANGER, J.S., and A.J. SCHWARTZ, 1980, Phys. Rev. **A21**, 948.

LANGER, J.S., M. BAR-ON and H.D. MILLER, 1975, Phys. Rev. **A11**, 1417.

LEE, J.K., and H.I. AARONSON, 1975, Acta Metall. **23**, 799 and 809.

LEE, J.K., and W.C. JOHNSON, 1983, in: Solid-State Phase Transformations, the Pittsburg Conference, ed. H.I. Aaronson (Met. Soc. AIME, Warrendale, PA) p. 127.

LEE, J.K., Y.Y. EARMME, H.I. AARONSON and K.C. RUSSELL, 1980, Metallurg. Trans. **11A**, 1837.

LIFSHITZ, I.M., and V.V. Slyozov, 1961, J. Phys. Chem. Solids **19**, 35.

LIVAK, R.J., and G. THOMAS, 1971, Acta Metall. **19**, 497.

LIVINGSTON, J.D., and J.W. CAHN, 1974, Acta Metall. **22**, 495.

LORIMER, G.W., 1978, in: Precipitation Processes in Solids, eds. K.C. Russell and H.I. Aaronson (Met. Soc. AIME, Warrendale, PA) p. 87.

LORIMER, G.W., ed., 1981, Quantitative Analysis with High Spatial Resolution (The Metals Society, London).

LORIMER, G.W., and R.B. NICHOLSON, 1969, in: The Mechanisms of Phase Transformations in Crystalline Solids, (The Metals Society, London) p. 36.

LÜCKE, K., and H.P. STÜWE, 1963, in: Recovery and Recrystallization of Metals, ed. L. Himmel (Wiley, New York) p. 171.

LUETJERING, G., and M. PETERS, 1981, in: Les Traitements Thermomechaniques, Comptes Rendus, 24^e Colloque de Metallurgie, Saclay, France (CEN-Saclay) p. 225.

MADDIN, R., 1976, Mater. Sci. Eng. **25**, 1.

MALCOLM, J.A., and G.R. PURDY, 1967, TMS AIME **239**, 1391.

MAKENAS, B.J., and H.K. BIRNBAUM, 1980, Acta Metall. **28**, 979.

MARTIN, J.W., and R.D. DOHERTY, 1976, Stability of Microstructure in Metallic Systems (Cambridge Univ. Press).

MASSALSKI, T.B., 1970, in: Phase Transformations (ASM, Metals Park, OH) p. 433.

MCLEAN, M., 1979, in: Phase Transformations, York Conference (The Institution of Metallurgists, London) p. II-54.

MERLE, P., and R.D. DOHERTY, 1982, Scripta Metall. **16**, 357.

MERLE, P., and F. FOUQUET, 1981, Acta Metall. **29**, 1919.

MERLE, P., and J. MERLIN, 1981, Acta Metall. **29**, 1929.

MERRICK, H.F., 1978, in: Precipitation Processes in Solids, eds. K.C. Russell and H.I. Aaronson (Met. Soc. AIME, Warrendale, PA) p. 161.

MONDOLFO, L.F., 1976, Aluminium Alloys: Structures and Properties (Butterworths, London).

MORRIS, J.G., ed., 1979, Thermomechanical Processing of Aluminum Alloys (AIME, New York).

MULLINS, W.W., 1959, J. Appl. Phys. **30**, 77.

MULLINS, W.W., 1963, in: Metal Surfaces (ASM, Metals Park, OH) p. 17.

MULLINS, W.W., and R.F. SEKERKA, 1963, J. Appl. Phys. **34**, 323.

NEMOTO, M., 1974, Acta Metall. **22**, 847.

NICHOLSON, R.B., 1970, in: Phase Transformations (ASM, Metals Park, OH) p. 269.

NICHOLSON, R.B., and J. NUTTING, 1961, Acta Metall. **9**, 331.

ORIANI, R.A., 1966, Acta Metall. **14**, 84.

OSTWALD, W., 1900, Z. Phys. Chem. **34**, 495.

PERGAMON PRESS (Oxford), 1971, Science, Technology and Application of Titanium.

PHILOFSKY, E.M., and J.E. HILLIARD, 1969, J. Appl. Phys. **40**, 2198.

PICKERING, F.B., 1978, Physical Metallurgy and the Design of Steels (Appl. Science Publ., London).

PLENUM PRESS (New York), 1973, Titanium Science and Technology.

PLICHTA, M.R., and H.I. AARONSON, 1980, Acta Metall. **28**, 1041.

PLICHTA, M.R., H.I. AARONSON and J.H. PEREPEZKO,, 1978, Acta Metall. **26**, 1293.

PLICHTA, M.R., J.H. PEREPEZKO, H.I. AARONSON and W.F. LANGE, 1980, Acta Metall. **28**, 1031.

POLMEAR, I.J., 1981, Light Alloys (Edward Arnold, London).

PORTER, D.A., D.B. WILLIAMS and J.W. EDINGTON, 1974, in: Electron Microscopy 1974, vol. 1, eds. J.V. Sanders and D.J. Goodchild (Australian Academy of Science, Canberra) p. 656.

PREDEL, B., and W. GUST, 1975, Mater. Sci. Eng. **17**, 41.

PULS, M.P., and J.S. KIRKALDY, 1972, Metallurg. Trans. **3**, 2779.

PURDY, G.R., 1971, Met. Sci. **5**, 81.

PURDY, G.R., 1978, Acta Metall. **26**, 477.

RAJAB, K., 1982, D. Phil. thesis, Univ. of Sussex.

RAJAB, K., and R.D. DOHERTY, 1983, in: Solid-State Phase Transformations, the Pittsburg Conference, ed. H.I. Aaronson (Met. Soc. AIME, Warrendale, PA) p. 555.

RICKS, R.A., 1981, in: Quantitative Analysis with High Spatial Resolution, ed. G.W. Lorimer (The Metals Society, London) p. 85.

RICKS, R.A., A.J. PORTER and R.C. ECOB, 1983, Acta Metall. **31**, 43.

RIDLEY, N., and G.W. LORIMER, 1981, in: Quantitative Analysis with High Spatial Resolution, ed. G.W. Lorimer (The Metals Society, London) p. 80.

RUNDMAN, K.B., and J.E. HILLIARD, 1967, Acta Metall. **15**, 1025.

RUNDMAN K.B., D. DE FONTAINE and J.E. HILLIARD, 1970, quoted by HILLIARD [1970].

RUSSELL, K.C., 1970, in: Phase Transformations (ASM, Metals Park, OH) p. 219.

RUSSELL, K.C., 1980, Adv. Colloid and Interface Sci. **13**, 205.

RUSSELL, K.C., and H.I. AARONSON, eds., 1978, Precipitation Processes in Solids (Met. Soc. AIME, Warrendale, PA).

RUSSEW, K., and W. GUST, 1979, Z. Metallk. **70**, 523.

SANKARAN, R., and C. LAIRD, 1974, Acta Metall. **22**, 957.

SANKARAN, R., and C. LAIRD, 1977, Scripta Metall. **11**, 385.

SANKARAN, R., and C. LAIRD, 1978, Scripta Metall. **12**, 877.

SARGENT, C.M., and G.R. PURDY, 1974, Scripta Metall. **8**, 569.

SERVI, I.S., and D. TURNBULL, 1966, Acta Metall. **14**, 161.

SHEWMON, P.G., 1963, Diffusion in Solids (McGraw–Hill, New York).

SHEWMON, P.G., 1965, TMS AIME **233**, 736.

SIMS, C.T., and W.C. HAGEL, 1972, The Superalloys (Wiley, New York).

SINCLAIR, R., R. GRONSKY and G. THOMAS, 1976, Acta Metall. **24**, 789.

SMITH, C.S., 1962, in: Decomposition of Austenite by Diffusional Processes, eds. V.F. Zackay and H.I. Aaronson (Interscience, New York) p. 237.

SMITH, G.D.W., A.J. GARRATT-REED and J.B. VANDER SANDE 1981, in: Quantitative Analysis with High Spatial Resolution, ed. G.W. Lorimer (The Metals Society, London) p. 238.

SOFFA, W.A., and D.E. LAUGHLIN, 1983, in: Solid-State Phase Transformations, the Pittsburg Conference, ed. H.I. Aaronson (Met. Soc. AIME, Warrendale, PA) p. 159.

SPEICH, G.R., 1968, TMS AIME **242**, 1359.

SPEICH, G.R., and R.A. ORIANI, 1965, TMS AIME **233**, 623.

STEPHENS, D.E., and G.R. PURDY, 1975, Acta Metall. **23**, 1343.

STRUTT, J., 1974, D. Phil thesis, Univ. of Sussex.

SUNDQUIST, B.E., 1973, Metallurg. Trans. **4**, 1919.

SWALIN, R.A., 1972, Thermodynamics of Solids, 2nd Ed. (Wiley, New York).

TANNER, L., and I.S. SERVI, 1966, Acta Metall. **14**, 231.

TIEN, J.K., P.G. SHEWMON and J.S. FOSTER, 1971, Metallurg. Trans. **2**, 1193.

TIEN, J.K., P.G. SHEWMON and J.S. FOSTER, 1973, Metallurg. Trans. **4**, 370.

TILLER, W.A., K.A. JACKSON, J.W. RUTTER and B. CHALMERS, 1953, Acta Metall. **1**, 428.

TRIVEDI, R., 1970a, Metallurg. Trans. **1**, 921.

TRIVEDI, R., 1970b, Acta Metall. **18**, 287.

TU, K.N., 1972, Metallurg. Trans. **3**, 2769.

TURNBULL, D., 1955, Acta Metall. **3**, 55.

TURNBULL, D., 1956, in: Solid State Physics, eds F. Seitz and D. Turnbull (Academic, New York) vol. 3, p. 226.

TURNBULL, D., and J.C. FISHER, 1949, J. Chem. Phys. **17**, 71.

TZAKALAKOS, T., 1977, Ph.D. thesis, Northwestern Univ..

TZAKALAKOS, T., and J.E. HILLIARD, 1980, quoted by DITCHEK and SCHWARTZ [1980].

V. ALVENSLEBEN, L., 1982, Dipl. thesis, Göttingen; L. V. ALVENSLEBEN and R. WAGNER, to be published.

VAN HORN, K., ed., 1967, Aluminum, vols. 1–3 (ASM, Metals Park, OH).

VOGEL, A., 1977, D. Phil thesis, Univ. of Sussex.

VOLMER, M., and H. FLOOD, 1934, Z. Phys. Chem. **170**, 273.

WAGNER, C., 1961, Z. Elektrochem. **65**, 581.

WEATHERLY, G.C., 1971, Acta Metall. **19**, 181.

WEATHERLY, G.C., and C.M. SARGENT, 1970, Phil. Mag. **22**, 1049.

WEAVER, L., B. HUDSON and J. NUTTING, 1978, Met. Sci. **12**, 257.

WENDT, H., 1981, Ph.D. thesis, Univ. of Göttingen; H. WENDT and P. HAASEN, Acta Metall. 31 (1983) in press.

WERT, J.A., 1976, Acta Metall. **24**, 65.

WILLIAMS, D.B., and J.W. EDINGTON, 1975, Met. Sci. **9**, 529.

WILLIAMS, D.B., and J.W. EDINGTON, 1976, Acta Metall. **24**, 323.

WILLIAMS. J.C., 1978, in: Precipitation Processes in Solids, eds. K.C. Russell and H.I. Aaronson (Met. Soc. AIME, Warrendale, PA) p. 191.

WILLIAMS, P.R., M.K. MILLER, P.A. BEAVAN and G.D.W. SMITH, 1979, in: Phase Transformations, York Conference (The Institution of Metallurgists, London) p. II-98.

WILSON, E.A., S.M.C. VICKERS, C. QUIXALL and A. BRADSHAW, 1979, in: Phase Transformations, York Conference (The Institution of Metallurgists, London) p. II-67.

WU, C.K., R. SINCLAIR and G. THOMAS, 1978, Metallurg. Trans. **9A**, 381.

ZENER, C., 1946, TMS AIME **167**, 550.

ZETTLEMOYER, A.C., 1969, Nucleation (M. Dekker, New York).

Further reading

The Mechanism of Phase Transformations in Crystalline Solids (The Metals Society, London, 1969).

A.C. Zettlemoyer, Nucleation (M. Dekker, New York, 1969).

Phase Transformations (ASM, Metals Park, OH, 1970).

G.A. Chadwick, The Metallography of Phase Transformation (Butterworths, London, 1972).

H. Warlimont, ed., Order–Disorder Transformations in Solids (Springer, Berlin, 1974).

J.W. Christian, The Theory of Transformation in Metals and Alloys, 2nd Ed., 1975; also 1st Ed., 1965 (Pergamon Press, Oxford).

M. Hillert, Diffusion and Interface Control of Reactions in Alloys, Metallurg. Trans. 6A (1975) 5.

J.W. Martin and R.D. Doherty, The Stability of Microstructure in Metallic Systems (Cambridge Univ. Press. 1976).

K.C. Russell and H.I. Aaronson, eds., Precipitation Processes in Solids (Met. Soc. AIME, Warrendale, PA, 1978).

Phase Transformations, York Conference (The Institution of Metallurgists, London, 1979).

I.N. Easterling and O.A. Porter, Phase Transformations in Metals and Alloys (Van Nostrand, London, 1981).

H.I. Aaronson, ed., Solid-State Phase Transformations, the Pittsburg Conference (Met. Soc. AIME, Warrendale, PA, 1983).

CHAPTER 15

PHASE TRANSFORMATIONS, NONDIFFUSIVE

C.M. Wayman

Department of Metallurgy and Mining Engineering
and Materials Research Laboratory
University of Illinois at Urbana-Champaign
Urbana, IL 61801, USA

R.W. Cahn and P. Haasen, eds.
Physical Metallurgy; third, revised and enlarged edition
© *Elsevier Science Publishers BV, 1983*

1. Overview

The title of this chapter indicates that we will be concerned with phase changes where there is no long range movement of atoms in the sense of, say, their diffusional flow down a concentration gradient according to Fick's Law. Indeed, in most cases to be considered the atoms move less than an interatomic distance and retain their relative relationship with their neighbors during the phase change. The prototype of this behavior is a martensitic transformation where the transformation process *per se* is equivalent to the deformation of the parent crystal lattice into the product (martensite) crystal lattice. Because of their nature, martensitic transformations are frequently referred to as "displacive" or "shear-like" and "diffusionless." Over the years a vast amount of geometrical or crystallographic information on martensitic transformations has accrued, and indeed, as we shall see, all martensitic transformations have certain common crystallographic characteristics, which in many ways form the basis for identifying a martensitic transformation as such. From a consideration of these characteristics a largely successful phenomenological crystallographic theory has evolved, to which some attention will be devoted in the present chapter, along with a consideration of other features of martensitic transformations including thermodynamics and kinetics. As will be seen, there are also other phase changes, not commonly recognized as martensitic (e.g., bainite formation), which are also well described by the martensite crystallography theory, and these may be rationalized. And finally, some other types of diffusionless, displacive solid-state transformations such as omega phase and charge density wave formation are considered, both of which involve lattice distortions with correlated atomic displacements.

2. Martensitic transformations

2.1. Introduction and general characteristics

The name *martensite* (to honor the German scientist Martens) was originally used to describe the hard microconstituent found in quenched steels. Since, materials other than steel have been found to exhibit the same type of solid-state phase transformation, known as a *martensitic transformation* – frequently also called a *shear* or *displacive transformation*. Martensite, the description given to the product phase of a martensitic transformation, now transcends its original meaning considering that martensitic transformations have also been found in nonferrous alloys, pure metals, ceramics, minerals, inorganic compounds, solidified gases and polymers.

It has long been recognized that martensitic transformations are diffusionless, a characteristic easily seen by noting that many martensitic transformations can readily occur below 100 K where diffusional atomic movements are insignificant. It is not implied that all martensitic transformations occur at low temperatures; indeed, many of them occur at comparatively high temperatures, but these too are diffusionless. When a new phase is formed from its parent martensitically, discrete regions of the solid typically transform at a high velocity which is independent of temperature. In most cases the amount of transformation resulting is characteristic of the temperature and does not increase with time. As pointed out by CHRISTIAN [1970], the overall kinetics of a martensitic reaction depend on both the nucleation and growth stages and will largely be dominated by the slower of the two. For example, slow thermal nucleation may give rise to isothermal transformation characteristics. As will be seen later, the nucleation of martensitic transformations does not obey classical nucleation theory.

The interface boundary between the martensite and its parent is intimately related to the transformation growth process. Such an interface is highly glissile, and from low-temperature experiments is known not to require thermal activation for its movement. The martensite–parent interface can be totally coherent or semi-coherent, depending upon the particular material undergoing transformation. In most cases, e.g. the formation of martensite in ferrous alloys, the interface is semi-coherent and the parent and product lattices are coherently accommodated only over local regions of the boundary, leading to an accumulating misfit which must be periodically alleviated by an auxiliary deformation process. On the other hand, in, e.g., the fcc → hcp transformation in Co and its alloys, the martensite–parent interface is fully coherent, and when the interface moves (normal to itself) the complete structural deformation specified by the relation between the two lattices is accomplished. In contrast, in the case of a semi-coherent martensite interface, the region (martensite) generated by the advancing interface does not undergo the simple lattice change implied by the respective unit cells of the two phases. More discussion of interfaces will follow later, but for now it is emphasized that both semi-coherent and coherent martensite–parent interfaces must be glissile.

Martensitic transformations are most readily distinguished from other solid-state phase changes on the basis of their crystallographic characteristics, which imply a "military" (as opposed to "civilian") mode of atomic transfer from the parent phase to the product. They feature a coordinated structural change involving a lattice correspondence and an ideally planar parent–product interface which during movement (transformation) produces an invariant plane strain shape deformation. A fine-scale inhomogeneity in the martensite, such as slip, twinning or faulting, is usually observed at the electron microscope scale. This secondary deformation, an intrinsic part of the transformation process, produces the invariant plane condition at the macroscopic scale and provides a semi-coherent glissile interface between the martensite and the parent phase. Crystallographic features between the martensite and the parent phase such as the habit (invariant) plane and orientation relationship are usually not expressible in terms of exact relations involving simple Miller indices. The various crystallographic features of martensitic transformations will be discussed and illustrated with representative experimental examples. After this, a brief account of the development of the phenomenological crystallographic theory will be presented, followed by an algebraic analysis.

References: p. 1073.

2.2. Experimental observations of crystallographic features

Figure 1 is an optical micrograph of a polycrystalline Fe–24.5% Pt alloy taken after transformation of the parent phase into martensite by cooling. Prior to transformation (and when in the parent phase condition) the specimen was polished flat. The contrast observed results from a change in shape of the transformed regions. The light and dark shades correspond to regions of martensite which have undergone distortion in different senses with respect to the initial surface, and this surface tilting or macroscopic distortion is known as the *shape deformation* or *shape strain*. If successive serial sections of the specimen were removed and the specimen observed at each stage, it would be seen that the shape of the martensite units is that of lenticular plates. An X-ray analysis would reveal that the plates showing different optical contrast also feature different lattice orientations with respect to the initial parent grain. X-ray analysis would also show that a structural change had occurred, which for this Fe–Pt alloy is an fcc → bcc transformation. The differently oriented plates are different crystallographic variants of the habit plane and orientation relationship.

Referring again to fig. 1, if straight scratches were purposely abraded on a flat specimen still in the parent phase condition, it would be noted that after the martensitic transformation occurred the scratches would be displaced in a characteristic manner. After analyzing several nonparallel scratches crossing a given plate and noting the initial position of the same scratches in the untransformed parent phase a distortion matrix which describes the shape deformation can be derived, analysis of

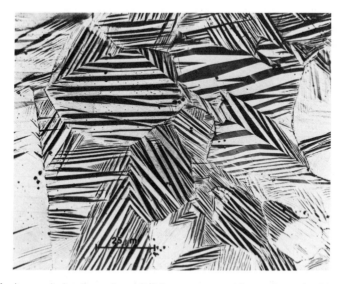

Fig. 1. Optical micrograph showing surface relief due to a martensitic transformation in an Fe–24.5% Pt alloy. The specimen was polished to a flat condition at room temperature and then cooled to produce the martensite.

which shows that a homogeneous deformation (at the optical microscope scale) has taken place. Further consideration shows that this distortion (shape deformation) resembles a simple shear, but in general is not exactly a shear; rather, an *invariant-plane strain* (fig. 2) has taken place with the plane of reference being the undistorted and unrotated habit plane. Mathematically, an invariant plane strain is a homogeneous distortion such that the displacement of any point is in a common direction and proportional to the distance from a fixed (i.e. not influenced by the strain) plane of reference, which is the invariant plane. In most martensitic tranformations a volume change accompanies the structural change which produces a normal component to the invariant-plane strain.

The above discussion indicates that the martensitic transformation shape deformation can be represented as an invariant-plane strain, a homogeneous deformation. Although this is well established, a past difficulty in the understanding of the geometry of martensitic transformations was the recognition that the homogeneous shape deformation matrix, when "applied" to the parent structure, does not in general generate the known martensite crystal structure. In iron alloys the martensitic transformation converts an fcc parent into a bcc (or bct) product, but the measured shape deformation matrix when applied to the parent will not produce a bcc structure. This apparent inconsistency will be explained later.

Martensite plates in a given alloy usually possess a unique habit plane, as shown for example in fig. 3. The stereographic projection shows the experimentally determined habit plane poles for different martensite plates, which, as can be seen, cluster near the plane in the parent phase with Miller indices $(3\ 10\ 15)_P$ (subscripts P and M, respectively, are used to designate the parent and martensite).

Early work suggested that, apart from a small relative rotation of corresponding unit cells in the parent and product phases, a homogeneous, lattice distortion would account for the known structural change in martensitic transformations. As early as 1924, BAIN suggested that the austenite (parent phase) $\rightarrow$ martensite transformation

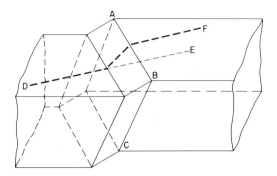

Fig. 2. Schematic representation of the invariant-plane strain shape deformation characteristic of martensitic transformations. The initially straight scratch DE is displaced to the position DF when the martensite plate with habit plane ABC is formed. The plane ABC is invariant (undistorted and unrotated) as a result of the martensitic transformation.

References: p. 1073.

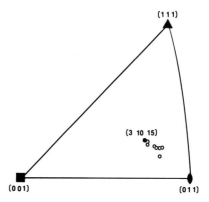

Fig. 3. Unit triangle of stereographic projection showing experimental habit planes (open circles) for martensite formed in an Fe–7% Al–1.5% C alloy. The average habit plane cannot be expressed in terms of simple Miller indices, and is near the plane $\langle 3\ 10\ 15\rangle_P$ of the parent phase.

in steels could be explained by a homogeneous "upsetting" of the parent fcc lattice into the required bcc (or bct) lattice, as shown schematically in fig. 4. This simple, intuitively derived deformation was later justified mathematically in that the particular correspondence between lattices envisioned by Bain when compared to others was found to involve the smallest principal strains. There are many ways (*correspondences*) to generate a bcc product from an fcc parent by means of a homogeneous distortion, and another possible correspondence is shown in fig. 5, but analysis

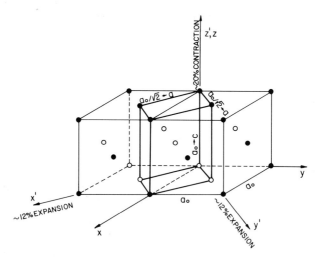

Fig. 4. Lattice distortion and correspondence proposed by BAIN [1924] for the fcc–bcc (bct) martensitic transformation in iron alloys. The correspondence related cell in the parent phase (indicated by bold lines) becomes a unit cell in the martensite as a consequence of a homogeneous "upsetting" with respect to the z' axis.

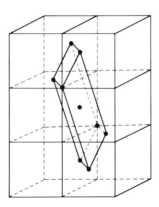

Fig. 5. Alternative lattice correspondence for the fcc–bcc (bct) martensitic transformation which involves larger principal distortions than the Bain correspondence.

shows that the Bain deformation involves the smallest principal strains. The *lattice correspondence*, a unique relationship between any lattice point in the initial lattice and the point it becomes in the final lattice, implied by the Bain distortion has also been verified experimentally using an ordered Fe_3Pt alloy which undergoes nominally a fcc to bcc transformation. By observing corresponding superlattice reflections in the parent and martensitic phases by means of transmission electron diffraction it was deduced that the Bain correspondence actually applies (TADAKI and SHIMIZU [1970]).

From the lattice correspondence shown in fig. 4 one would expect for example, $[001]_M \parallel [001]_P$, $[010]_M \parallel [110]_P$, $(112)_M \parallel (101)_P$, $(011)_M \parallel (111)_P$, etc. However, such exact parallelisms are not observed. For the Fe–Pt alloy shown in fig. 1 the following orientation relationship was observed:

$[001]_P - [001]_M$:   9.10° apart,

$[\bar{1}01]_P - [\bar{1}\bar{1}1]_M$:   4.42° apart,

$(111)_P - (011)_M$:   0.86° apart,

which is typical of the orientation relationships found in iron alloys and steels. Further consideration of the above orientation relationship shows that the correspondence cell is not only distorted ("upset") but also is rotated (about 10° from $[001]_P$ towards $[110]_P$). This is termed a *rigid body rotation*. The combined distortion–rotation is known as the *lattice deformation*. Additional examination of the above orientation relationship shows that the close-packed planes $(111)_P$ and $(011)_M$ are nearly parallel to each other and the close-packed directions $[\bar{1}01]_P$ and $[\bar{1}\bar{1}1]_M$ are a few degrees apart. This is an example of the *Greninger–Troiano* orientation relationship. When the close-packed planes and directions in two coexisting structures are parallel, this situation is frequently termed the *Kurdjumov–Sachs* type of orientation relationship.

References: p. 1073.

2.3.   The phenomenological crystallographic theory of martensitic transformations

The feasibility of the Bain lattice correspondence and distortion has been presented as one involving minimal atomic displacements. However, the Bain distortion itself is inconsistent with the experimentally established invariant plane strain shape deformation. The reason for this is as follows. Referring to fig. 4 again, it is noted that the correspondence cell is contracted ($\sim 20\%$) along the z' axis and expanded the same amount ($\sim 12\%$) along the x' and y' axes. Such a homogeneous distortion will leave no plane invariant (i.e. undistorted and unrotated). But suppose that the distortion along y' vanished. In other words, one of the principal distortions is less than unity (along z'), one is greater than unity (along x') and the remaining one (along y') is unity. These conditions imply the distortion of an initial sphere (parent phase) into a triaxial ellipsoid (martensite) following which the sphere and ellipsoid can fit together along an undistorted plane of contact (habit plane) as shown in fig. 6. But this special set of conditions is not generally found in practice; the principal distortions are determined by the lattice correspondence and observed lattice parameters of the two phases.

The apparent inconsistency described above is resolved by envisioning a *lattice-invariant deformation* (involving no structural change) such as slip or twinning to occur in conjunction with the Bain distortion. The additional deformation must be lattice invariant because the necessary structural change is effected by the Bain distortion alone. The role of the additional deformation is essentially to shear (distort) the ellipsoid resulting from the Bain distortion into tangency with the initial sphere; then one of the principal distortions (OX in fig. 6) becomes unity and an undistorted contact plane exists. This additional deformation (slip, twinning or faulting) being a shear is known as the *inhomogeneous shear* or *complementary shear* of the crystallography theory.

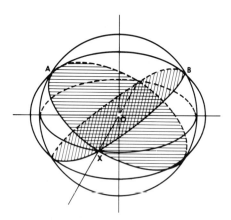

Fig. 6. The Bain distortion has a mathematical analog wherein an initial unit sphere is distorted into an ellipsoid. If one of the principal distortions is unity and the other two are of opposite sign, then the sphere and ellipsoid can fit together along the undistorted planes AOX or BOX. See text for discussion.

One problem still remains. Even though the required structural change has been effected (by the Bain distortion) and an undistorted contact plane has been provided for (by the inhomogeneous shear), the habit plane is still not unrotated, as required from observation. Further consideration of the sphere–ellipsoid analog shows that although an undistorted plane now exists, it has been rotated from its initial unrotated position. Consequently, a rigid body rotation is additionally incorporated along with the Bain distortion and inhomogeneous shear. These are the three phenomenological steps describing the total transformation crystallography. There is no time sequence implied as to which step occurs when. Of course the combined effect of these three operations must be equivalent to the shape deformation.

Working with the theory just described, different crystallographic features such as the habit plane and orientation relationship can be predicted by supposing the inhomogeneous shear to occur on different crystallographic planes and directions. For example, in most iron alloys (where similar lattice parameters for the parent and martensite phase are found) an inhomogeneous shear on the (112)–$[\bar{1}\bar{1}1]_M$ twinning system will predict a habit plane near $(3\ 15\ 10)_P$, but assuming the inhomogeneous shear system to be (011)–$[\bar{1}\bar{1}1]_M$ predicts a habit plane near $(111)_P$. It should be noted that since the lattice parameters and correspondence are usually known, the Bain distortion for a given transformation is specified, and the flexibility in the theory comes through different suppositions concerning the plane and direction of the inhomogeneous shear. Once the inhomogeneous shear system is assumed, a shear of a certain magnitude will produce an undistorted plane, which when rigidly rotated to its original position becomes the invariant habit plane.

The previous discussion and examples cited above have centered around iron alloys. Because of the importance of steels there has been substantial work on them and thus an abundance of experimental data exists. However, the principles presented are quite general and apply to all martensitic transformations.

2.3.1. The inhomogeneous shear and martensite substructure
The inhomogeneous shear was introduced in the crystallographic description of martensite transformations to ensure that the habit plane is macroscopically undistorted. Figure 7 is a schematic representation of the appearance of internally twinned and internally slipped martensite plates. Although there are localized distortions at the interface, the "saw tooth" effect because of alternating twins (called transformation twins) or slip lamellae prevents accumulation of any strain at the interface over large distances. For internally twinned martensite the Bain distortion is envisioned to occur along different contraction axes in the two regions, and the twinning plane in the martensite is derived from a mirror plane in the parent. In the case of internally slipped martensite the Bain distortion is the same in all regions of a plate. Figure 8 shows that the same effective shear angle γ can be accomplished by slip or twinning. Note that the relative thickness of the two twin components determines the angle γ.

Because of the inhomogeneous shear mentioned above, one would expect to observe some kind of substructure in the martensite. Such observations have been

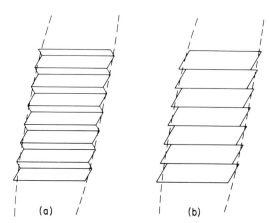

Fig. 7. Schematic representation of the inhomogeneous shear in a martensitic transformation involving (a) internally twinned, and (b) internally slipped plates of martensite. The serrated effect at the interface (habit plane) in each case prevents the long-range accumulation of strain and consequently the interface remains macroscopically undistorted.

made since the introduction of the theory. Figure 9 is a transmission electron micrograph of a martensite plate in an Fe–Ni–C alloy. The fine striations crossing the plate are transformation twins ($\langle 112 \rangle_M$ twinning plane). Regions adjacent to the martensite are retained austenite. The electron diffraction pattern from the martensite shows twin related reflections; if the twin plane is indexed specifically as $(112)_M$, the habit plane trace becomes specifically $(3\ 15\ 10)_P$. If (112)–$[\bar{1}\bar{1}1]_M$ twinning is used in the habit plane calculations, the predicted habit plane is in fact $(3\ 15\ 10)_P$. Thus the particular variant of the twin plane is found to be consistent with the particular variant of the habit plane, and for the Fe–Ni–C alloy the experimental observations are in excellent agreement with those features which are predicted using the phenomenological crystallographic theory. This is also the case for many other

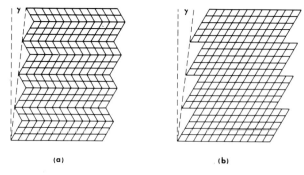

Fig. 8. As shown for twinning (a) and slip (b) the same magnitude of the inhomogeneous shear, as given by the angle γ, can be accomplished by either.

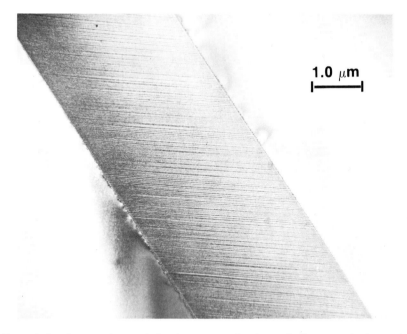

Fig. 9. Transmission electron micrograph showing a martensite plate and adjacent retained austenite in an Fe–30% Ni–0.4% C alloy. The striations within the plate running from left to right are transformation twins.

martensitic transformations which have been studied in some detail using transmission electron microscopy and diffraction. Notable exceptions, however, are certain steels which transform to martensite with a $\{225\}_P$ habit plane. The martensite in these materials has a complex substructure, as seen in the electron microscope, and is not very well explained using the theory described above. In addition, there is no adequate crystallographic description of the lath martensites found in ferrous alloys.

2.3.2. Mathematical description of the phenomenological crystallographic theory

The basic equation of the crystallography theory just described is:

$$P_1 = R\bar{P}B, \tag{1}$$

where B represents the Bain distortion, $\bar{P}$ is a simple shear (following in a mathematical sense the Bain distortion), R is the rigid body rotation previously mentioned, and P_1 is the invariant-plane strain shape deformation. B, $\bar{P}$, R and P_1 are all (3×3) matrices. The matrix product $R\bar{P}B$ is equivalent to the shape deformation P_1 and the rotation matrix R rotates the plane left undistorted by $\bar{P}B$ to its original position. That is, $P_1 = R\bar{P}B$ is an invariant-plane strain.

Although eq. (1) indicates that the inhomogeneous shear $\bar{P}$ follows the Bain distortion, it should be noted that the same end result can be obtained by "allowing"

References: p. 1073.

the shear to occur in the parent phase prior to the Bain distortion. By following this alternative procedure, there are certain computational simplifications to be gained. In this case the basic equation becomes:

$$P_1 = RBP, \tag{2}$$

where P as before represents a simple shear.

With reference to the previous discussion, the invariant-plane strain shape deformation can be expressed as:

$$P_1 = I + m d p' = \begin{pmatrix} 1 & 0 & 0 \\ 0 & 1 & 0 \\ 0 & 0 & 1 \end{pmatrix} + m[d_1 d_2 d_3](p_1 p_2 p_3)$$

$$= \begin{pmatrix} 1 + m d_1 p_1 & m d_1 p_2 & m d_1 p_3 \\ m d_2 p_1 & 1 + m d_2 p_2 & m d_2 p_3 \\ m d_3 p_1 & m d_3 p_2 & 1 + m d_3 p_3 \end{pmatrix}, \tag{3}$$

where p' (prime meaning transpose) being a plane normal is written as a (1×3) row matrix, in contrast to d, a lattice vector, which is a (3×1) column matrix. Considering fig. 4 again, the Bain distortion for the fcc $\rightarrow$ bcc (bct) case can be written as:

$$B = \begin{pmatrix} \sqrt{2}\, a/a_0 & 0 & 0 \\ 0 & \sqrt{2}\, a/a_0 & 0 \\ 0 & 0 & c/a_0 \end{pmatrix}. \tag{4}$$

When typical values for a, a_0 and c are substituted for the fcc $\rightarrow$ bct transformation in iron alloys, a representative B matrix is

$$B = \begin{pmatrix} 1.12 & 0 & 0 \\ 0 & 1.12 & 0 \\ 0 & 0 & 0.8 \end{pmatrix}. \tag{5}$$

It is seen from eq. (5) that two of the principal distortions are greater than unity (1.12) and the third is less than unity (0.8). Thus the necessary conditions mentioned earlier (§2.3) for an invariant-plane strain do not exist insofar as the Bain distortion itself is concerned.

Referring back to eq. (2), it is noted that P is a simple shear and therefore of the invariant-plane strain form $(I + m d p')$. Also, the inverse of P, $P^{-1} = I - m d p'$, is a simple shear of the same magnitude on the same plane, but in the opposite direction. That is, both P and P^{-1} are invariant-plane strains. It is thus convenient to rewrite eq. (2) as:

$$P_1 P_2 = RB, \tag{6}$$

where $P_2 = P^{-1}$. Since both P_1 and P_2 are invariant-plane strains, their product RB is an invariant-*line* strain, S. The invariant-line strain is defined by the *planes* (i.e., their intersection) which are invariant to P_1 and P_2. If the invariant-line strain S is known, all crystallographic features of a given martensitic transformation can be predicted. It is not within the scope of the present account to go into the details of the invariant-line strain analysis (BOWLES and MACKENZIE [1954]), but certain highlights should be mentioned. The Bain correspondence and distortion B are known from the lattice parameters of the parent and martensite phases. R is determined once the plane p_2' and direction d_2 of P_2 are assumed. It should be noted that $RB = S$ constitutes the lattice deformation of the transformation.

The important results of the invariant-line strain analysis, noting that the shape strain is $P_1 = I + m_1 d_1 p_1'$ and that the simple shear ("mathematically" preceding the Bain distortion) is $P_2 = I + m_2 d_2 p_2'$, are as follows (where the magnitudes, directions and planes of the component invariant-plane strains are given respectively by m, d and p'):

$$d_1 = [Sy_2 - y_2]/p_1' y_2, \tag{7}$$

$$p_1' = (q_2' - q_2'S^{-1})/q_2'S^{-1}d_1, \tag{8}$$

where y_2 is any vector lying in p_2' (except the invariant line x) and q_2' is any normal (other than n', the row eigenvector of S^{-1}, i.e., $n'S^{-1} = n'$) to a plane containing d_2. The normalization factor for d_1 in eq. (7) is $1/m_1$ and therefore P_1, m_1, d_1 and p_1' are all determinable. The matrix R is determined from the requirement that x and n' which are displaced by the Bain distortion must be totally invariant. R defines the orientation relationship within any small region of the martensite plate not involving P_2. Thus, the assumed correspondence and lattice parameters determine B, the assumption of p_2' and d_2 allows R to be determined, and $RB = S$ defines the elements of P_1.

The previous description of the crystallographic theory parallels the analysis given by BOWLES and MACKENZIE [1954] but the treatments of WECHSLER *et al.* [1953] and BULLOUGH and BILBY [1956] are equivalent.

Some variations in the basic theory just presented include the introduction of a dilatation parameter, which relaxes slightly the requirement that the habit plane be undistorted, and the incorporation of two inhomogeneous shear systems such that

$$P_1 = RBS_2S_1 \tag{9}$$

where S_2 and S_1 are the two inhomogeneous shear systems involved. Neither of these modified approaches is without criticism, and it has been argued that the double shear approach loses generality.

2.3.3. Some other crystallographic observations

As pointed out earlier in this discussion, the martensite habit plane is near {3 10 15}$_P$ for an Fe–24.5% Pt alloy. A similar habit plane is also found for martensite formed in Fe–Ni alloys containing approximately 30% Ni. However, in some steels, e.g. Fe–8% Cr–1.1% C, the habit plane is near {225}$_P$, and the habit plane of lath martensites typical of low-carbon steels and dilute iron alloys is near {111}$_P$ However, the martensite–parent orientation relationship is almost the same in the {3 10 15}, {225} and {111} transformations.

References: p. 1073.

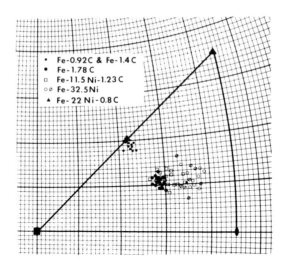

Fig. 10. Unit triangle of a stereographic projection showing the results of habit plane measurements on martensite formed in five different Fe-alloys.

It is unclear why different steels transform to martensite with a different habit plane, but it is generally thought that the different habit planes result from different operative modes of the inhomogeneous shear, i.e., different planes and directions, or even the operation of two or more modes within a given martensite plate or lath as suggested by the multiple shear theories. Figure 10 shows some experimental habit plane determinations for a variety of iron alloys.

Techniques are improving for measuring the shape strain associated with martensitic transformations and the information so obtained is leading to a clearer understanding of the transformation crystallography. In addition, the use of transmission electron microscopy has greatly enhanced the understanding of the crystallography of martensite transformations.

2.3.4. Further observations on the martensite morphology and substructure

In general, martensite in ferrous alloys takes the form of *lenticular plates* as seen in fig. 1 for an Fe–Pt alloy. This morphology is readily understood because a lenticular plate, as a mechanical twin, is a low strain energy shape in a "shear" transformation. However, other morphologies are not uncommonly found in ferrous alloys, and these are briefly mentioned here.

In low-carbon steels (up to 0.4% C) and in some other ferrous alloys the martensite takes the form of *laths* rather than plates. A typical lath has dimensions $0.3 \times 4 \times 200 \ \mu m$. As mentioned above, the habit plane of such laths is near $\{111\}_P$. In Fe–Ni–C alloys, three distinct morphologies are found and are classified as to *lenticular, thin plate,* and *butterfly* morphologies. In one well studied case it has been shown that butterfly martensite (fig. 15, below) forms at the highest temperatures, lenticular martensite forms at intermediate temperatures, and thin-plate martensite forms at the lowest temperatures. In some materials, particularly ferrous alloys, a needle-like morphology known as surface martensite is formed under some conditions. Apparently this morphology is easily nucleated at surfaces and can develop into a "needle" form because matrix constraints are relaxed near a free surface.

Whereas the plate types of ferrous martensite frequently form in a self-accommodating manner involving several habit plane variants, lath martensites do not appear to do so. Instead, the laths tend to cluster together into packets, each adjacent lath having the same habit plane, orientation relationship, and shape deformation. The intrinsic lath morphology (as opposed to plate morphology) is thought to be related to the larger shape strain associated with lath martensite. Unlike the plate morphology which typically contains a high density of transformation twins as shown in fig. 9, martensite laths are typically untwinned and contain a high density of internal dislocations as shown in fig. 11. Since the inhomogeneous shear in ferrous lath martensite is not twinning, it is expected that slip is the mode involved and that the lattice-invariant deformation would be effected by the movement of interface dislocations. An example of such interface dislocations in an Fe–Ni–Mn martensite lath is shown in fig. 12. The high density of dislocations usually observed within lath martensites may not be directly related to the transformation crystallography. It may result from the constraints a lath experiences during growth, i.e., the internal dislocations probably arise from accommodation distortion.

Another martensite morphology found in certain steels, notably austenitic stain-

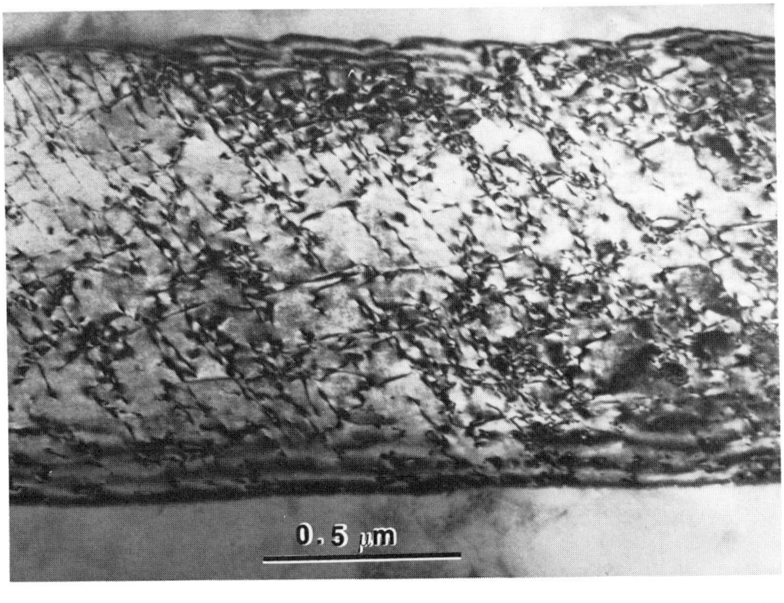

Fig. 11. Transmission electron micrograph showing dislocations (rather than twins) in a martensite lath formed in an Fe–20% Ni–5% Mn alloy.

References: p. 1073.

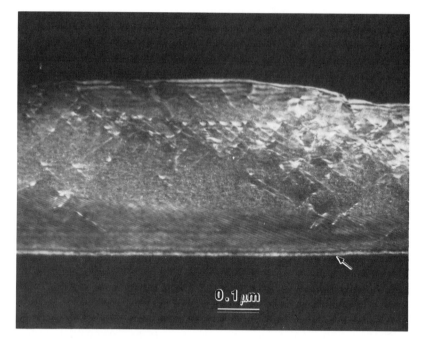

Fig. 12. Transmission electron micrograph showing interface dislocations (parallel to the arrow) in a lath of martensite in an Fe–20% Ni–5% Mn alloy.

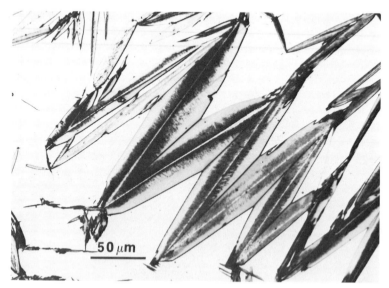

Fig. 13. Optical micrograph showing lenticular martensite formed at −79°C in an Fe–31% Ni–0.23% C alloy. The more heavily etched linear central region of the plates is known as a midrib.

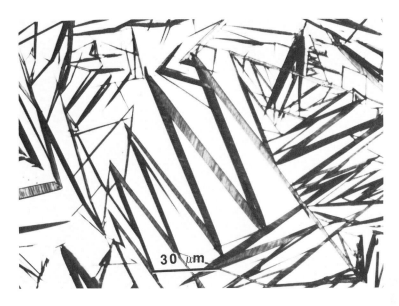

Fig. 14. Optical micrograph showing the "thin-plate" morphology of martensite, observed in an Fe–31% Ni–0.29% C alloy transformed at −160°C.

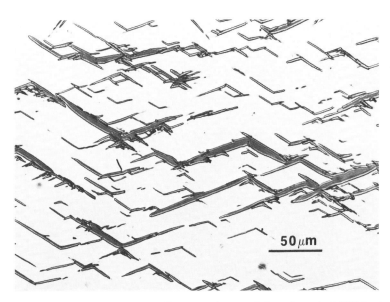

Fig. 15. Optical micrograph showing butterfly martensite morphology formed at 0°C in an Fe–20% Ni–0.7% C alloy.

References: p. 1073.

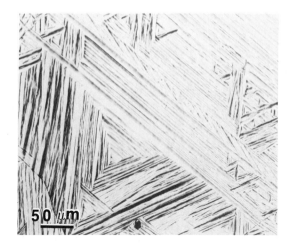

Fig. 16. Optical micrograph showing surface relief following the formation of lath martensite in an Fe–0.2% C alloy.

less steels and Fe–Mn–C alloys, is the *banded morphology* (fig. 17, below). However in these alloys, the transformation is not fcc → bcc, but fcc → hcp instead. This transformation is accomplished by distorting the fcc parent phase on {111} planes by the passage of partial dislocations which generate the hcp martensite. As would be expected, the bands delineate {111}$_P$ planes. This type of fcc → hcp martensite transformation is similar to that which occurs in cobalt and cobalt alloys, and, as mentioned earlier, involves a fully coherent interface and no lattice-invariant deformation in the usual sense.

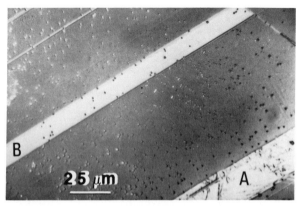

Fig. 17. Optical micrograph showing bands of hcp martensite (A and B) formed in the fcc austenite matrix in an Fe–12% Mn–0.48% C alloy.

Figures 13–17 are optical micrographs showing typical martensite morphologies in ferrous alloys.

2.4. Martensite–parent interfaces

The phenomenological crystallographic theory of martensitic transformations was developed to provide a description of the overall atomic displacements involved in these transformations, consistent with the observed geometrical features. The theory describes the displacements of lattice points required to transform the parent lattice into the martensite lattice in its observed orientation and to generate the observed shape change. In martensitic transformations between structures whose primitive unit cells contain a single atom, the displacements of atoms and lattice points are the same, but in other cases shuffling movements of some of the atoms are required. The phenomenological theory does not specify the order in which the individual atoms are displaced during transformation and thus does not prescribe a particular transformation mechanism. Rather, it simply describes a set of displacements with which the growth mechanisms must be compatible.

Growth of martensite plates involves both the edgewise propagation of the plate across the parent crystal and the thickening of the plate. In the usual type of martensitic transformation where the shape and lattice deformations differ, a glissile interface parallel to the habit plane which can produce thickening by moving in the direction of its normal, contains areas of coherence separated either by matching dislocations or twin boundaries. Such a glissile boundary can move by homogeneous transformation of the coherent regions combined with gliding of the matching dislocations in their slip planes, or by the extension of the twin boundaries. The movement of such boundaries is conservative (CHRISTIAN [1982]) because all the atoms in the parent swept by the boundary are incorporated in the martensite. An interface which moves conservatively is glissile, as with dislocation motion. There is no net flux of vacancies or other point defects. The individual atom displacements relative to their neighbors are less than an interatomic distance, and any activation energy is appreciably smaller than that for self-diffusion.

If the interface is not parallel to the habit plane, as in a lenticular or tapered plate, it must contain steps, the migration of which parallel to the habit plane provides an alternative mechanism for thickening as well as a mechanism for edgewise growth. Since the gliding of the steps produces the displacements of the matrix which generate the shape deformation, they are transformation dislocations with Burgers vector mdh, where md is the displacement vector of the shape strain and h is the step height. Alternatively, the Burgers vector b of the transformation dislocations is given by:

$$b = (I - S^{-1})h, \tag{10}$$

where S is the homogeneous lattice deformation. For irrational habit planes it is uncertain what value h will take.

When the transformation dislocations generate the final lattice, as in the fcc → hcp

References: p. 1073.

cobalt transformation, no other defects are required in the interface. More usually, however, the lattice that would be generated by the transformation dislocations, i.e., the lattice obtained by applying the shape strain homogeneously to the parent lattice, differs from the martensite lattice. It is then necessary to propose that additional displacements occur, effectively within the transformation dislocations. The additional displacements are those given by the inhomogeneous shear of the phenomenological theory. They are suitably small and contribute no additional change of shape if the interface contains matching dislocations or twin boundaries as well as transformation dislocations.

In a martensitic transformation occurring as described above, edgewise growth in the direction of the matching dislocations or along the intersection of the twinning plane and the habit plane involves only extension of existing defects. However, growth at right angles requires the periodic generation of new dislocations or twin boundaries.

2.5.   Energetics of martensitic transformations

2.5.1.   Transformation hysteresis and the reverse martensite-to-parent transformation

Two representative but widely different cases are considered here: the martensitic transformation (non-thermoelastic) in Fe–Ni (~ 30%) alloys *, and that (thermoelastic) in the typical β-phase (3/2 electron/atom Hume-Rothery phases) alloys such as β-brass. In both cases the shear component of the macroscopic shape strain is ~ 0.2 (shear strain component ~ 12°) but the accommodation by the parent phase is very different, comparing the two cases. The transformation hysteresis is also very different. In the Fe–Ni alloys the martensite–parent interface appears to become immobilized after a plate of martensite has thickened to a certain extent, and when the martensite is heated, the interface will not move backwards, apparently being pinned by its damaged environment. Instead, the reverse martensite → parent (austenite) transformation takes place by the nucleation of small platelets of the parent phase within the martensite plates (KESSLER and PITSCH [1967]). And in ferrous alloys containing carbon, i.e., steels, the usual stages of martensite tempering occur (KRAUSS [1980]) within which the martensite decomposes through diffusional processes. On the other hand, in alloys of the β-phase type, the matrix or parent phase accommodation of the martensite plates appears to be essentially elastic, with no dislocation or "debris" generation, and the interface remains glissile, capable of "backwards" movement and the shrinkage of martensite plates during heating to effect the reverse transformation. In addition, in such *thermoelastic transformations* (KURDJUMOV and KHANDROS [1949]), stored elastic energy apparently predisposes the reverse martensite–parent transformation, allowing it to occur "prematurely"

---

* Some aspects of martensitic transformations in steels, and especially the characteristics of the M_s temperature, are discussed in ch. 16, § 3.2.

(KAUFMAN and COHEN [1958]). In such cases, the stored elastic energy contributes to the driving force for the reverse transformation.

2.5.2. Thermoelastic and non-thermoelastic martensitic transformations

Thermoelastic and non-thermoelastic martensitic transformations may be contrasted by those observed in Au–Cd and Fe–Ni alloys, respectively (OTSUKA and WAYMAN [1977]), as indicated in fig. 18 (KAUFMAN and COHEN [1958]) where it can be seen that a substantial difference in transformation hysteresis, as given by the difference between the A_f (completion temperature for the reverse transformation) and M_s temperatures, exists. The magnitude of the hysteresis is related to the transformation driving force. The thermoelastic transformation (Au–Cd) is characterized by a small hysteresis. Another difference, comparing these two transformation modes, is in the manner of the martensite–parent reverse transformation. In the thermoelastic case, the martensitic transformation on cooling proceeds by the continuous growth of martensite plates and the nucleation of new plates. When cooling is stopped, growth ceases, but resumes by further growth upon subsequent cooling, until the martensite plates impinge with each other or with grain boundaries. The reverse transformation upon heating occurs by the backwards movement of the martensite–parent interface, and the plates of martensite "shrink" and revert completely to the initial parent phase orientation.

In contrast, for a non-thermoelastic transformation, once a plate of martensite grows to a given size during cooling, it will not grow further upon subsequent cooling because the interface apparently becomes immobilized. The immobilized interface does not undergo inverse movement during heating, but instead the parent phase is nucleated within the immobilized martensite plates, and a given plate as a whole does not revert to the original parent phase orientation (KESSLER and PITSCH [1967]).

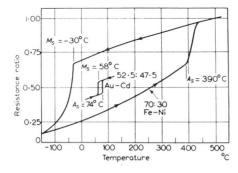

Fig. 18. Graph showing electrical resistance change during heating and cooling for Fe–Ni and Au–Cd alloys, indicating the hysteresis for the martensitic transformation on cooling and the reverse transformation on heating for non-thermoelastic and thermoelastic transformation, respectively (after KAUFMAN and COHEN [1957]).

References: p. 1073.

2.5.3. Free energy change of a martensitic transformation

Figure 19 shows schematically the change in chemical free energies of martensite and austenite (parent phase) with temperature. T_0 is the temperature at which the austenite and martensite are in thermodynamic equilibrium and M_s is the temperature at which the transformation starts upon cooling. The difference in free energies between austenite (γ) and martensite (α'), $\Delta G_{M_s}^{\gamma \to \alpha'}$, at the M_s temperature is the critical chemical driving force for the onset of the martensitic transformation (other features of fig. 19 will be described later).

In general, the free energy change associated with a martensitic transformation is given by:

$$\Delta G^{P \to M} = \Delta G_C^{P \to M} + \Delta G_{NC}^{P \to M}, \tag{11}$$

where $\Delta G_C^{P \to M}$ is the chemical free energy change associated with the transformation from parent to martensite (proportional to the amount of martensite formed), and $\Delta G_{NC}^{P \to M}$ is the non-chemical energy opposing the transformation [consisting of elastic (strain) and surface energy]. Since the martensite–parent interface is essentially coherent in a thermoelastic transformation, the surface energy term should be small, and the elastic energy term will dominate, increasing rapidly with the martensite plate size.

For a thermoelastic transformation, under a given set of conditions, $\Delta G^{P \to M}$ for the system takes a minimum, so if the temperature is decreased, $\Delta G_C^{P \to M}$ decreases, and thus a martensite plate grows until another minimum is reached. But if the temperature is raised, $\Delta G_C^{P \to M}$ increases and the martensite shrinks until another minimum is reached, if the interface remains mobile. Accordingly, when the size of a martensite plate corresponds to a free energy minimum at a given temperature, the system can be described to be in a state of thermoelastic equilibrium. This is the origin of the terminology *thermoelastic transformation*.

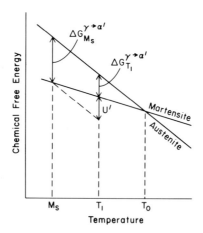

Fig. 19. Schematic diagram showing the free energy change for a martensitic transformation. See text for details.

In view of the above discussion, a transformation which involves a mobile interface (during formation and reversal) is a thermoelastic martensitic transformation. Because of the mobile, reversible nature of such an interface, the transformation is associated with crystallographic reversibility, another important characteristic of thermoelastic martensitic transformations. In most thermoelastic transformations the parent phase is ordered and thus has a rather high elastic limit. This would favor transformation reversibility because less transformation "debris" would form in a stronger matrix.

2.5.4. Nucleation of martensite

If a martensite transformation were able to nucleate in an ideally perfect crystal of the parent phase, the attendant free energy change could be written (COHEN and WAYMAN [1981]) as:

$$\Delta G = (4\pi/3)r^2 c \Delta g_{\text{ch}} + (4\pi/3)r^2 c(Ac/r) + 2\pi r^2 \sigma. \tag{12}$$

Here, the "particle" of martensite is assumed to be a thin oblate spheroid of volume $(4\pi/3)r^2 c$ and the surface area $\approx 2\pi r^2$; r is the radius, c is the semithickness, Δg_{ch} is the theoretical chemical free energy change per unit volume of the martensite particle, Ac/r ($= \Delta g_{\text{el}}$) is the elastic strain energy per unit volume of the martensite particle, and σ is the interfacial energy per unit area.

The strain-energy factor A is given by:

$$A = [\pi(2 - \nu)/8(1 - \nu)]\mu\gamma_0^2 + (\pi/4)\mu\epsilon_n^2, \tag{13}$$

where the Poisson ratio (ν) and the shear modulus (μ) are each taken to be the same for the parent and product phases. It is assumed that the martensitic shape deformation is an invariant-plane strain, having a shear component γ_0 and a normal component ϵ_n. With typical values for the martensitic reaction in ferrous alloys, A is about 2.4×10^3 MJ/m^3.

Most martensitic transformations, and certainly the fcc → bcc (or bct) structural change in iron-base alloys, involve semi-coherent rather than fully coherent interfaces, and so the interfacial energy is estimated to be about 100–200 mJ/m^2 (100–200 erg/cm^2) rather than to lie in the coherent range of 10–20 mJ/m^2 (10–20 erg/cm^2). Δg_{ch} is approximately -170 MJ/m^3 (-300 cal/mole) for a typical ferrous alloy at its martensite-start temperature (M$_s$), and when this quantity is substituted into eq. (12) along with the above values of A and σ, the resulting function of ΔG versus r and c is a saddle-shaped surface with a barrier on the order of 8×10^{-16} J per nucleation event ($\sim 5 \times 10^3$ eV). This is $\sim 10^5$ kT at temperatures where the transformation is known to nucleate spontaneously. Clearly, the available thermal energy is much too small for homogeneous nucleation by heterophase fluctuations, and so special nucleation sites must be considered. It is frequently thought that dislocation arrays serve as martensite nuclei.

Martensitic reactions in ferrous alloys can occur either isothermally or athermally (i.e., on continuous cooling). Most transformations in steels have been observed athermally as a function of dropping temperature, but the best quantitative insights

References: p. 1073.

into the reaction kinetics are obtained from isothermal transformations in which case nucleation and transformation rates can be determined experimentally. In any event (i.e., athermal or isothermal reaction kinetics) the transformation kinetics are controlled by the nucleation rate and not the growth rate. It is the relatively rapid growth rate which leads to this situation.

The important but not well understood problem of martensite nucleation was previously addressed by CHRISTIAN [1970] who concluded that the critical stage in martensite nucleation involves the attainment of a condition from which non-thermally activated growth can begin, which implies the formation of a semi-coherent interface. He further suggested a framework within which to consider the martensite nucleation process involving (a) classical nucleation utilizing thermal fluctuations (either homogeneously or at randomly distributed preferred sites), (b) the quenching-in of a pattern of embryos or compositional fluctuations which are subcritical at high temperatures but become supercritical at lower temperatures, and (c) the non-thermally activated rearrangement of the structure of a defect, or the interaction of different defects to produce a configuration resembling the martensitic structure. Recent thinking (CLAPP [1973], OLSON and COHEN [1976]) would tend to favor the last of these categories.

2.6. Mechanical effects in non-thermoelastic martensitic transformations

2.6.1. Introductory comments

A martensitic transformation proceeds by a displacive or "shear" process, which on a macroscopic (optical microscope) scale may be described as an invariant-plane strain. Thus, in general, when a martensitic transformation is coupled with an externally applied stress, other phenomena such as stress- and strain-induced martensitic transformations, the generation of new nucleation sites, and transformation-induced plasticity (TRIP) may occur.

It is generally characteristic that non-thermoelastic martensite in most iron and nonferrous alloys exhibits a large transformation volume change and a large thermal hysteresis during cooling and heating (i.e., a large difference between M_s and A_s) in comparison with thermoelastic martensite. The latter was shown in fig. 18 where the A_s temperature of the Fe–Ni alloy is much higher than M_s. Thus, in the case of non-thermoelastic martensite, the strain- or stress-induced martensite formed above M_s is stable at the stressing temperature and is not retransformed to austenite when the specimen is unloaded. Furthermore, since the austenite and martensite are somewhat damaged by plastic deformation because of the large shape deformation and volume change during transformation, a crystallographically reversible transformation of the type parent phase $\rightleftarrows$ martensite becomes difficult. In iron alloys, martensite can be induced during plastic deformation of the austenite because the parent phase usually exhibits good ductility.

Since the martensitic transformation is achieved by a cooperative shear movement of atoms, it is easily envisioned that an applied elastic stress aids the transformation.

However, the role of plastic strain in the martensitic transformation is quite complex. Here, the view is taken that a deformation-induced martensitic transformation can be more clearly understood considering the effect of an applied stress rather than the effect of strain.

A remarkable increase in elongation is obtained when ferrous martensite is formed during deformation. This phenomenon is called *transformation-induced plasticity* (TRIP). An explanation of the TRIP phenomenon and various controlling factors will be discussed.

2.6.2. Chemical and mechanical driving forces

The chemical free energy change associated with a martensitic transformation was depicted in fig. 19. Referring to fig. 19 again, when a stress is applied to the austenite at T_1 (between M_s and T_0), the mechanical driving force, U, due to the stress is added to the chemical driving force, $\Delta G_{T_1}^{\gamma \to \alpha'}$, and the martensitic transformation starts at the critical stress where the total driving force is equal to $\Delta G_{M_s}^{\gamma \to \alpha'}$. $U'(= \Delta G_{M_s}^{\gamma \to \alpha'} - \Delta G_{T_1}^{\gamma \to \alpha'})$ in fig. 19 is the critical mechanical driving force necessary for the stress-induced martensitic transformation at T_1. The mechanical driving force, U, is a function of stress and the orientation of a transforming martensite plate, and can be expressed (PATEL and COHEN [1953]) as:

$$U = \tau \gamma_0 + \sigma \epsilon_n,\tag{14}$$

where τ is the shear stress resolved along the transformation shear direction in the martensite habit plane, γ_0 is the transformation shear strain along the shape-strain shear direction in the habit plane, σ is the normal stress resolved perpendicular to the habit plane, and ϵ_n is the dilational component of the transformation shape

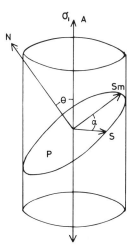

Fig. 20. Schmid-factor diagram for the influence of an applied stress σ_1 along stress-axis A in inducing a martensitic transformation on habit plane P with normal N. S is the direction of the shape strain for the martensite and S_m is the maximum shape strain elongation parallel to the habit plane.

References: p. 1073.

strain. When a specimen is stressed by amount σ_1 as shown in fig. 20, τ and σ may be expressed as follows for any given orientation of a martensite plate (ONODERA *et al.* [1976]);

$$\tau = (1/2)\sigma_1(\sin 2\theta)\cos\alpha, \tag{15}$$

$$\sigma = \pm(1/2)\sigma_1(1 + \cos 2\theta), \tag{16}$$

where σ_1 is the absolute value of the applied stress (tension or compression), θ is the angle between the applied stress axis and the normal to the habit plane, and α is the angle between the transformation shear direction and the maximum shear direction of the applied stress in the habit plane. The plus and minus signs in eq. (16) correspond to tension and compression, respectively. From eqs. (14)–(16) the mechanical driving force from the applied stress σ_1 is

$$U = (1/2)\sigma_1[\gamma_0(\sin 2\theta)\cos\alpha \pm \epsilon_0(1 + \cos 2\theta)]. \tag{17}$$

2.6.3. Critical stress to induce martensitic transformation

When a martensitic transformation starts by the application of stress to a polycrystalline austenite in which the orientation of grains is distributed at random, a martensite plate whose orientation yields a maximum value of U, eq. (17), will be formed first. The maximum value of U is obtained under the conditions $\alpha = 0$, and $dU/d\theta = 0$, from which the critical mechanical driving force (U') is obtained as:

$$U' = (1/2)\sigma_1'[\gamma_0\sin 2\theta' \pm \epsilon_0(1 + \cos 2\theta')], \tag{18}$$

where σ_1' is the critical applied stress for the start of the martensite transformation. In the case of Fe–Ni alloys, for example, θ' is calculated from the known component of the transformation shape strain ($\gamma_0 = 0.20$, $\epsilon_n = 0.04$) as 39.5° for tension and 50.5° for compression.

If the chemical driving force ($\Delta G^{\gamma \to \alpha'}$) decreases linearly with increase in temperature above M_s, it is expected that the critical applied stress for the start of martensite formation increases linearly with an increase in stressing temperature. Actually, it has been observed (OLSON and COHEN [1972], ONODERA *et al.* [1976]) that the critical stress for martensite formation increases linearly with an increase in temperature in the range between M_s and M_s^σ as shown in fig. 21. However, above M_s^σ (e.g., at T_2), martensite is induced at a stress σ_b after plastic deformation of the austenite occurs.

In general, it has been observed (OTSUKA and WAYMAN [1977]) that only a single (of usually 24) variant of martensite is induced when an elastic stress is applied. This is the variant which gives rise to maximum elongation in the direction of the tensile axis.

When the austenite is deformed at temperatures above M_s^σ (e.g., at T_2 in fig. 21), it begins to deform plastically at a stress σ_a, and is strain-hardened up to σ_b. Then the martensitic transformation starts to take place. σ_b is considerably lower than σ_c, which is obtained by extrapolating the critical stress–temperature line between M_s and M_s^σ. This decrease (i.e., $\sigma_c - \sigma_b$) in the critical applied stress for martensite

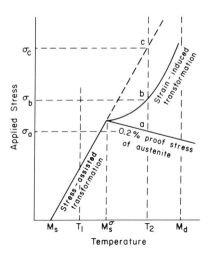

Fig. 21. Schematic stress–temperature diagram showing the critical stress to initiate the formation of martensite as a function of temperature. The various regimes of behavior are described in the text.

formation is due to plastic deformation of austenite. As to the role of plastic deformation of austenite on the deformation-induced martensitic transformation, two different views have been expressed. One is the strain-induced martensite nucleation hypothesis proposed by OLSON and COHEN [1972, 1976, 1979], and the other holds that the stress is locally concentrated at obstacles (e.g. grain boundaries, twin boundaries, etc.) by plastic deformation of austenite, and thus the concentrated stress becomes equivalent to σ_c in fig. 21, the latter proposed by ONODERA and TAMURA [1979] and others.

In contrast to the M_s temperature where a martensitic transformation will begin as a consequence of the chemical driving force, reference is frequently made to the M_d temperature. M_d is the temperature above which the chemical driving force becomes so small that nucleation of martensite cannot be mechanically induced, even in the plastic strain regime.

2.6.4. Transformation-induced plasticity

As an example of TRIP behavior (ZACKAY *et al.* [1967]), we consider an austenitic Fe–29% Ni–0.26% C alloy ($M_s = -60°C$, $M_d = 25°C$) deformed in tension at various temperatures. The relationship between tensile properties and test temperature for this alloy is shown in the upper part of fig. 22 (TAMURA *et al.* [1970]). Note that the elongation is nearly 100%. An inverse temperature dependence of the yield stress (0.2%) is observed between M_s^σ and M_s. This results from the shape strain of martensite formed before yielding of the austenite occurs. The tensile strength is increased with decreasing test temperature just above M_s^σ. Such an elongation enhancement is attributed mainly to the suppression of necking, due to the increase in work-hardening rate by the formation of martensite. Furthermore, it is thought

References: p. 1073.

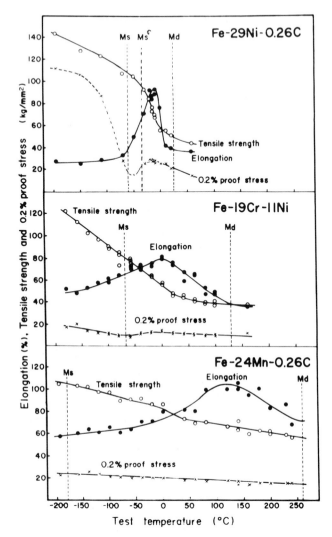

Fig. 22. Experimental data showing the effect of test temperature on the tensile properties of three metastable austenitic alloys strained at the rate $\dot{\varepsilon} = 5.5 \times 10^{-4}/s$ (after TAMURA et al. [1970]).

that the initiation and propagation of cracks are suppressed by the formation of martensite during deformation, because the stress concentrations may be released by the formation of preferential variants of martensite at regions of stress concentration. Similar results are seen in fig. 22 for two other alloys.

With an increase in carbon content, the martensite hardness is increased substantially, but the austenite hardness only slightly increases. Accordingly, in the case of high-carbon steels, since a larger work-hardening rate can be obtained even with a small amount of martensite formation, the uniform elongation by TRIP becomes much larger (TAMURA *et al.* [1972, 1973]).

2.7. Mechanical effects in thermoelastic martensitic transformations

2.7.1. General

Thermoelastic martensitic transformations are accompanied by very unusual mechanical effects in both the parent and martensite phases. In addition to the now popular *shape memory effect* (SME) and the mechanical behavior of alloys exhibiting this type of behavior, closer inspection shows that SME alloys also show a variety of other kinds of interesting mechanical behavior, such as superelasticity associated with the formation of a reversible stress-induced martensite, the rubberlike effect, "training" and the two-way shape memory effect, extensive deformation resulting from stress-induced martensite-to-martensite transformations, unusual damping behavior, and finally, high stresses generated during the reverse martensite-to-parent transformation.

The number of materials exhibiting the shape memory effect is now extensive, including many Cu-based alloys, those of the noble metals based on Ag and Au, the classic Ni–Ti alloys (NITINOL), ternary variations of the same such as Ni–Ti–Cu and Ni–Ti–Fe, Ni–Al alloys, and Fe–Pt alloys, to mention a few. In these various alloy systems, in general, the martensite forms thermoelastically, is crystallographically reversible, and the parent phase (and hence the martensite) is ordered. These generalities were first suggested by SHIMIZU and WAYMAN [1972]. The SME martensites are either internally twinned or internally faulted as a consequence of the inhomogeneous shear of the phenomenological crystallographic theory. In addition, the parent phase is usually of the B2 or DO3 type and the martensite crystal structure is given by the stacking sequence of prior (parent phase) {110} planes, e.g., 2H, 3R, 9R and 18R.

2.7.2. The shape memory effect

A brief description of the SME is as follows. An object in the low temperature martensitic condition when deformed and then unstressed (usually below M_f, the temperature at which the martensitic transformation is completed) will regain its original shape when heated through the A_s–A_f temperature interval to cause the reverse martensite-to-parent transformation. Strains (i.e., deformed martensite) on the order of 6–8% are completely recovered. The process of regaining the original shape is clearly associated with the reverse transformation of the deformed martensitic phase.

Substantial progress has recently been made in understanding the nature of the SME (SCHROEDER and WAYMAN [1977], SABURI *et al.* [1979]), and a brief summary of the state of the art is as follows. A single crystal of the parent phase will usually transform into 24 orientations of martensite (variants of $\langle hkl \rangle$). But when this multi-orientation martensite configuration is deformed, a single orientation of

References: p. 1073.

martensite eventually results as a consequence of twinning and the movement of certain martensite interfaces. The twins which form in the (deformed) martensite are simply other orientations (variants) of martensite, and it has been shown (SCHROEDER and WAYMAN [1977a]) that twinning can convert one orientation (variant) of martensite to an other. A similar conversion results when martensite–martensite boundaries move under stress whereby one plate orientation grows at the expense of another. The end result is that there is only one remaining orientation (of the original 24) of martensite, and this single remaining orientation is the variant whose shear component of the shape deformation will permit maximum elongation of the specimen as a whole in the direction of the tensile axis (SABURI *et al.* [1979]). The entire deformation is mediated by martensite variant reorientation; if dislocation slip intervenes, the shape memory becomes imperfect.

Continuing, although the original parent phase single crystal transforms into many (up to 24) orientations of martensite, the reverse situation does not occur. Rather, the single crystal of martensite resulting from deformation below the M_f temperature, transforms during heating to a single orientation of parent phase. This is a consequence of the martensite and parent lattice symmetries involved and the

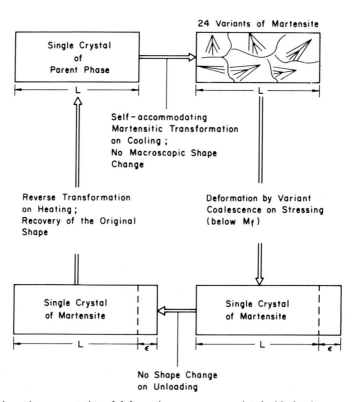

Fig. 23. Schematic representation of deformation processes associated with the shape memory effect.

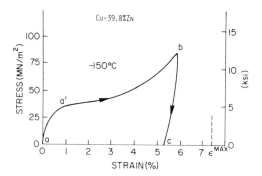

Fig. 24. Stress–strain curve for a Cu–Zn shape-memory alloy deformed below its M_f temperature. The low flow stress a' associated with the martensite deformation is to be noted.

necessity to maintain atomic ordering during the reverse transformation. In this connection we note that the highly symmetric parent phase (usually cubic) has many equivalent ways for the parent–martensite lattice change (Bain distortion) to occur, thus leading to the many variants of martensite actually observed. But on the other hand, the relatively unsymmetric martensite (e.g., monoclinic in a currently commercial Cu–Zn–Al alloy) does not have such a multiplicity of choices, the consequence of which is that only a single variant of the parent phase (in fact, the original variant) is usually nucleated in the course of the reverse martensite (deformed)-to-parent transformation. Thus, in a way of speaking, the single crystal of martensite unshears to form a single crystal of the parent phase, such "unshearing" during the reverse transformation restoring the specimen to its original shape. A schematic representation of the shape memory process is shown in fig. 23 (SABURI *et al.* [1979]). The processes just described appear to be universal in nature, irrespective of the particular alloy system or crystal structure of the martensite.

Figure 24 (SCHROEDER and WAYMAN [1977a]) shows a stress–strain curve for a Cu–Zn SME alloy single crystal specimen deformed below the M_f temperature. It is seen that the martensite begins to deform (a') at a relatively low stress, 35 MN/m² (5 ksi). The residual strain at point c was completely recovered during heating.

2.7.3. The two-way shape memory effect

With the shape memory effect, as previously described, a specimen "deformed" in the martensitic condition will "undeform" and regain its original shape after heating from A_s to A_f. In contrast, the *two-way shape memory* (TWSM) involves a reversible deformation: a specimen will spontaneously deform during cooling from M_s to M_f and then undeform during heating from A_s to A_f. For some time it has been realized (WASILEWSKI [1975]) that such behavior occurs as a result of deformation in either the parent phase or the martensite. Research using Cu–Zn single crystals (SCHROEDER and WAYMAN [1977b]) has shown conclusively that two types of two-way shape

References: p. 1073.

memory can occur, depending on the particular manner of stress-cycling or "training," as described below.

SME cycling involves cooling a specimen below the M_f temperature, deforming it to produce a preferential martensite variant, as described earlier, and heating it to above the A_f temperature. This procedure is repeated several times, the manner of martensite deformation (e.g., tension, compression, bending) remaining unchanged.

SIM cycling involves deformation of a specimen above the M_s temperature in order to produce *stress-induced martensite* (SIM), followed by reversal of the SIM when the applied load is released. This process is also repeated several times using the same means of stressing each time.

The two-way shape memory is observed after both SME and SIM cycling, and the terms "SME training" and "SIM training" have been suggested (SCHROEDER and WAYMAN [1977b]). In either case, the two-way behavior comes from the preferential formation and reversal of a "trained" variant of martensite formed after either SME or SIM cycling and cooling the specimen from M_s to M_f. That is, the "training" of a specimen to form a preferential variant of martensite upon cooling to M_f can result from prior deformation of the martensite formed thermally (upon cooling) followed by heating, or SIM cycling above the M_s temperature. In both cases the first part of the TWSM occurs upon cooling. In either case, the result is that a major portion of the martensite in a given specimen corresponds to a preferred variant, which produces a spontaneous strain during cooling, the amount of the strain corresponding to the shape strain of the preferred variant. It has been observed that the effectiveness of SME training is inferior to that of SIM training, but nevertheless, in both cases the preferential formation of a single variant (orientation) of martensite may be explained by the programming effect of the generation of an inbuilt pattern of stresses (or defects) as a consequence of the cycling processes. Apparently, these stresses (or defects) become more inbuilt and effective with increased cycling (SCHROEDER and WAYMAN [1976]).

2.7.4. The engine effect in shape memory alloys

Although SME materials generally deform in the martensitic condition at comparatively low stresses, as seen in fig. 24, surprisingly large stresses are generated when the deformed martensite is heated from A_s to A_f. As a case in point (JACKSON et al. [1972]), a nearly equiatomic Ni–Ti alloy when martensitic will undergo deformation (as at point a' of fig. 24) at a stress of about 70 MN/m^2 (10 ksi). Yet when constrained and heated to A_f it is found that a "thermomechanical recovery stress' of about 700 MN/m^2 (100 ksi) is generated. In other words, heat can be used to create a mechanical force which can do work. This is the basis of numerous heat engines based on shape memory alloys which are presently being experimented with (GOLDSTEIN and MCNAMARA [1979]).

2.7.5. Pseudoelastic effects

Pseudoelasticity refers to a situation where comparatively large strains as a result of loading to well above the elastic limit are completely recovered upon unloading at a constant temperature. It is convenient to subdivide pseudoelastic behavior into two categories (OTSUKA and WAYMAN [1977]), "superelasticity" and the "rubberlike" effect, depending on the nature of the driving forces and mechanisms involved. If the behavior is caused by SIM formation and its subsequent reversion, the terminology *superelasticity* is used. Here, there is a mechanical contribution to the transformation driving force in a thermoelastic martensitic transformation. In contrast, *rubberlike* behavior involves deformation behavior of the martensite phase itself and involves no phase transformation.

2.7.5.1. Superelasticity. When a stress is applied above the M_s temperature, a mechanically elastic martensite is stress-induced in most thermoelastic alloys, and this SIM will disappear when the stress is released. This behavior could be described as a mechanical type of shape memory. A stress–strain curve for a Cu–Zn alloy showing typical superelastic behavior is shown in fig. 25. Two plateau regions are seen. The upper plateau corresponds to the formation of SIM plates. Using a single crystal specimen, as for fig. 25, it was observed that only one (of 24) variant of SIM was formed. The preferred variant was found to be the one whose shape strain will permit maximum specimen elongation in the direction of the tensile axis (SCHROEDER and WAYMAN [1979]). The parallel plates will nucleate, lengthen and coalesce, eventually leading to a single crystal of martensite at the end of the upper plateau region. Upon unloading the specimen, the stress–strain curve follows the lower plateau region, which involves the nucleation etc., of parallel plates of only one variant of the parent phase. The plateau stresses as shown in fig. 25 depend on the test temperature, and in fact the upper plateau stress is zero at the M_s temperature, as it should be. The temperature dependence of the stress to produce SIM is shown in fig. 26. The slope of the M_s variation with temperature may be used with a modified version of the Clausius–Clapeyron equation,

$$d\sigma/dM_s = -\Delta H/\epsilon T, \tag{19}$$

to determine the transformation latent heat, ΔH (OTSUKA and WAYMAN [1977]).

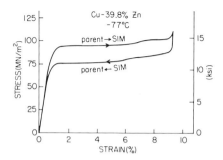

Fig. 25. Stress–strain curve for a thermoelastic Cu–Zn shape-memory alloy deformed above the M_s temperature, showing superelastic behavior as a consequence of the formation and reversion of a reversible, stress-induced martensite.

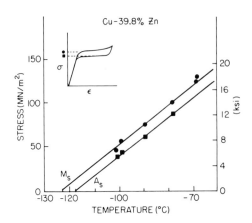

Fig. 26. Stress–temperature plot of experimental data showing the temperature dependence of the applied stress required to produce stress-induced martensite in a Cu–39.8% Zn alloy.

Stress–strain curves of the type shown in fig. 25 are frequently referred to as superelastic loops.

2.7.5.2. Rubberlike behavior. The type of behavior described in the previous section is indeed "rubberlike" in nature, but to avoid confusion it is considered best to reserve the term *rubberlike* to describe a type of behavior characteristic of a fully martensitic (formed by cooling) specimen. Rubberlike behavior, also a pseudoelastic effect, is found in the martensite phase of some alloys which undergo thermoelastic transformation. Rubberlike behavior does not involve a phase transformation, and appears to be related to the reversible movement of transformation twin boundaries or martensite boundaries, but the nature of the restoring force still remains unclear.

2.7.6. Martensite-to-martensite transformations

Figure 27 shows the deformation behavior of a Cu–39.8 wt% Zn single crystal strained at $-88°C$, some 35° above the M_s temperature (SCHROEDER and WAYMAN [1978]). The first upper plateau corresponds to the formation of SIM as described previously, but the second plateau, starting at about 9% strain, corresponds to a second martensitic transformation which is stress-induced from the first martensite "mother." The two lower plateaux are a result of the reverse transformations, occurring in an inverse sequence. By means of these successive martensite-to-martensite transformations, completely recoverable strains as high at 17% can be realized. Figure 27 shows a double superelastic loop. It should also be noted that the second stress-induced martensite in the Cu–Zn alloy can be formed from the first SIM, or from the normal thermally formed (upon cooling) martensite.

In the case of the Cu–Zn alloy discussed above, the initial martensite has a 9R structure, and additional deformation deforms the 9R structure into a 3R structure. This occurs by shearing on the basal (close-packed) plane of the former such that the

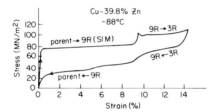

Fig. 27. Stress–strain diagram showing the deformation behavior of a Cu–39.8% Zn alloy single crystal specimen deformed above the M$_s$ temperature. A double superelastic loop is observed which is the result of two successive stress-induced martensitic transformations. The first stress-induced transformation is that depicted by fig. 25, while the second stress-induced transformation originates from the first-formed (under stress) martensite.

structural stacking sequence is changed from ... ABCBCACAB... to ... ABCABCABC... .

3. Crystallographically similar transformations

3.1. The bainite transformation in steels

It is widely agreed that the bainite reaction in steels has certain martensitic characteristics, yet it is universally known that the austenite → bainite reaction is controlled by the diffusion rate of carbon atoms. Nevertheless, several workers have shown that when bainite plates or laths are formed, the surface relief produced is that characteristic of an invariant-plane strain. There are also other martensitic-like features such as an irrational habit plane and transformation substructure, and various studies of the austenite–bainite orientation relationship imply that the Bain correspondence between the austenite and bainitic ferrite exists.

Clearly, the bainitic transformation cannot be classified as a diffusionless transformation because of the attendant diffusion of interstitial carbon atoms. On the other hand, it appears very likely that the iron atoms undergo "military," martensite-like movements, and that the rapid diffusion of carbon atoms from one interstice to another does not destroy the lattice correspondence held by the iron atoms. In view of this, one would expect the phenomenological crystallographic theory to apply to the bainite transformation. Several instances of such an application have been reported, with varying degrees of success.

3.2. Oxides and hydrides

Various metallic oxides and hydrides form with a plate morphology and exhibit an invariant plane strain shape deformation. Such plates are also internally twinned or internally slipped. An example of an internally twinned plate in a tantalum oxide is shown in fig. 28. In this case (VAN LANDUYT and WAYMAN [1968]) the observed

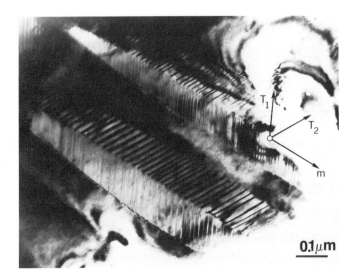

Fig. 28. Transmission electron micrograph showing plates of the sub-oxide TaO$_y$ following the oxidation of tantalum. The striations within the plates are twins, and the general crystallography of the process is well described by the phenomenological crystallographic theory developed for martensitic transformations.

crystallographic features of the oxide plates are excellently predicted by the phenomenological crystallographic theory developed for martensitic transformations. As with the formation of bainite, discussed in the previous section, it appears that a lattice correspondence between the metallic atoms in the two phases exists and remains intact despite the occurrence of interstitial diffusion (of oxygen) during the phase change.

3.3. The CuAu II ordering reaction

The formation of the ordered othorhombic CuAu II phase from the disordered cubic phase parent at ~ 390°C generates an invariant-plane strain relief and conforms well with the phenomenological crystallographic theory of martensitic transformations. These findings led SMITH and BOWLES [1960] to conclude that the mechanism of this transformation, which leads to internally twinned plates of the CuAu II, is similar to that of a martensitic transformation, even though place changes between neighboring atoms are necessarily involved during the ordering reaction. But AARONSON and KINSMAN [1977] dispute that the ordering reaction is similar to a martensitic transformation, claiming that the atomic jumps required for ordering are inconsistent with a shear transformation and that conformity with the crystallographic theory is not a sufficient criterion to identify a shear transformation.

The CuAu II ordering reaction was later considered by BOWLES and WAYMAN

[1979] who suggested that the place exchanges between neighboring atoms, required to achieve the ordering, are redundant in the sense that they do not contribute to the transformation shape change so that the theory still prescribes the displacements involved in producing the lattice change. They proposed that the growth of CuAu II plates occurs by the gliding of transformation dislocations at a rate determined by the rate of ordering at the dislocations. They also emphasized that there are no known examples of transformations involving substitutional diffusion to or from an interface which have all the crystallographic features of a martensitic transformation.

4.   Omega phase formation

The $\beta \to \omega$ phase change (cubic–hexagonal) involves correlated atomic displacements and occurs in Ti, Zr and Hf alloys and in β-phase alloys of the noble metals. It has been extensively investigated in recent years. The reader is referred to two reviews: SASS [1972] and WILLIAMS *et al.* [1973]. In general, some controversial points concerning the $\beta \to \omega$ transformation remain, but certain of its features have come into focus. The $\beta \to \omega$ phase change occurs both athermally and isothermally, but there appears to be no fundamental basis for differentiating between athermally formed and isothermally formed omega (WILLIAMS *et al.* [1973]).

Omega phase formation can occur reversibly and without diffusion at quite low temperatures, and the ω-phase cannot be suppressed by rapid quenching. These characteristics parallel those found in martensitic transformations, but the ω-phase "particles" exhibit a cuboidal or ellipsoidal morphology, depending upon the relative parent and product lattice parameters. At least during the early stages of transformation, the ω particles are quite small (~ 15 Å in diameter), and they are aligned along $\langle 111 \rangle$ directions. Their observed density is very high (10^{18}–10^{19} particles/cm^3). The smallness of the particles and their high density suggest that their nucleation is not as serious a barrier as their growth, in contrast to martensitic transformations where the opposite is the case. Figure 29 shows an example of omega particles.

A number of alloys exhibiting the $\beta \to \omega$ transformation feature the following orientation relationship:

$$(111)_\beta \parallel (0001)_\omega,$$

$$[1\bar{1}0]_\beta \parallel [2\bar{1}10]_\omega.$$

Interestingly, these same alloys (although at different compositions) also undergo a $\beta \to \alpha'$ martensitic transformation in which case the orientation relationship is of the Burger's type:

$$(110)_\beta \parallel (0001)_{\alpha'},$$

$$[1\bar{1}1]_\beta \parallel [11\bar{2}0]_{\alpha'}.$$

It should be noted that the β–ω orientation relationship shows a multiplicity of only

References: p. 1073.

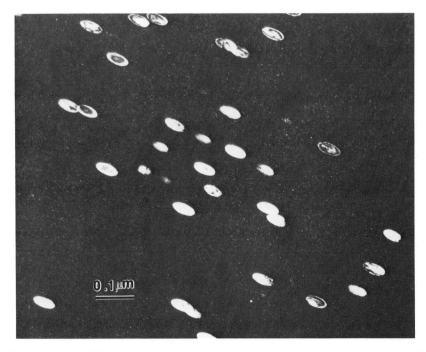

Fig. 29. Transmission electron micrograph (dark field) showing ω-phase particles formed at 480°C in a Ti–Mo–Sn–Zr alloy. Only one of the four variants of the ω-phase is in contrast.

four, which, when compared to that of martenstic transformations (where the multiplicity is much higher) suggests that the respective lattice correspondences are different.

Another comparison can be made between the diffusionless $\beta \rightarrow \omega$ and $\beta \rightarrow \alpha'$ (martensitic) transformations. When martensite forms, the atomic displacements are mostly homogeneous, as given by a Bain-type distortion. Each atom (apart from additional shuffles in some cases) is homogeneously transported to its final position according to the Bain deformation and correspondence. But in the $\beta \rightarrow \omega$ transformation, some of the corresponding atoms undergo movements whereas others do not. Figure 30 (SASS [1972]) shows a $(\bar{1}01)$ section through a β unit cell. Atoms A, B, E and F remain in place during the $\beta \rightarrow \omega$ transformation, whereas atoms C and D undergo shuffle movements in opposite sense along $[\bar{1}\bar{1}\bar{1}]_\beta$, in the end reaching the position 1/2[111], initially having been situated at the 2/3 and 1/3 positions. Accordingly, the 1st, 4th, 7th, 10th, etc., planes remain unchanged while 2 and 3, 5 and 6, 8 and 9, etc., shuffle past each other in opposite senses. The $\beta \rightarrow \omega$ atomic movements are formally consistent with atom movements given by a $2/3\langle 111 \rangle$ displacement wave having nodes at $n\langle 111 \rangle_\beta$ positions (n = any integer).

Advances in understanding the $\beta \rightarrow \omega$ transformation have come from the application of transmission electron microscopy and diffraction. Diffraction patterns

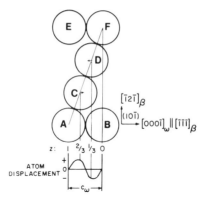

Fig. 30. Schematic diagram showing atomic movements involved during ω-phase formation. This is a $(10\bar{1})$ section through the parent bcc unit cell. The $2/3\langle111\rangle$ longitudinal displacement waves cause atomic movements (in directions indicated by the two arrows) which are those needed to form the ω-structure (after Sass, [1972]).

frequently exhibit networks of diffuse intensity (sheets of intensity on $\langle111\rangle_\beta$ planes) which obscure the identification of diffraction maxima from the ω-phase. These diffuse diffraction effects are observed at temperatures above which those from the ω-phase are clearly identified, and the former have been ascribed to pretransformation "linear defects".

Impurities have a marked influence on the $\beta \rightarrow \omega$ transformation. For example, the presence of 1200 ppm of oxygen can lower the ω-start temperature by about 600 K, presumably because oxygen stiffens the matrix and depresses the transformation start temperature. Apparently the oxygen atoms somehow interact with the pretransformation linear defects along $\langle111\rangle_\beta$ and impede their ordering.

Several problems related to ω-phase formation are still outstanding, including the possible role of lattice vacancies in promoting linear defects and hence nucleation, the high nucleation frequency and low particle growth rate, and the extreme embrittlement of the matrix β-phase after the ω-phase forms.

5. *Phase changes and charge density waves*

A *charge density wave* (CDW) is a static modulation of conduction electrons and is a Fermi-surface driven phenomenon usually accompanied by a periodic lattice distortion. In essence, the electronic energy of the solid is lowered as a consequence of the lattice distortion, the attendant strain energy of which is more than compensated by the reduction in electronic energy. The present understanding of CDWs in conductors follows from the pioneering work of Peierls [1955], Kohn [1959] and Overhauser [1968, 1971]. A detailed description of CDW phenomena has been published by Wilson et al. [1975].

References: p. 1073.

Considering a simple one-dimensional metal, PEIERLS [1955] suggested that such a solid is susceptible to a periodic lattice distortion which can lower the total energy of the system by means of a distortion which changes the lattice periodicity. In elementary form, the basic ideas are shown in fig. 31. Figure 31a shows the familiar $E-k$ relationship for a one-dimensional monovalent metal. The first Brillouin zone with boundary at $\pm\pi/a$ is half occupied. The upper part of the figure shows that the electronic charge density is of periodic form, being maximum in the vicinity of the ion cores. In fig. 31b the one-dimensional lattice has been periodically distorted as shown in the upper part of the figure and the resulting period $2a$ causes the formation of a charge density wave of period $2a$ associated with the new super-lattice. Since the lattice has been doubled in real space, a new band gap will appear at $\pm\pi/2a$ in k-space which permits some electrons to "spill down," thus lowering the Fermi energy. Thus the (periodic) lattice distortion has a rather simple driving force which is purely electronic in origin. Numerous examples of a CDW phase change have recently been found in quasi-one-dimensional organic conductors (e.g. TTF–TCNQ) and quasi-two-dimensional layered compounds (e.g. transition-metal dichalcogenides).

A favorable Fermi surface geometry is necessary for the formation of a CDW, which will most likely occur when the shape of the Fermi surface permits a connection by the same wavevector Q, i.e., $Q = 2k_f$. This modulation with wavevector Q will modify the Fermi surface by creating gaps at these *nested* positions. If the nested (i.e. dimpled) portion of the Fermi surface is significant, the energy "gain" by creating energy gaps may overcome the energy "cost" arising from the periodic lattice distortion, thus allowing the formation of a CDW. In other words, a structural change will occur when the CDW formation is accompanied by ion

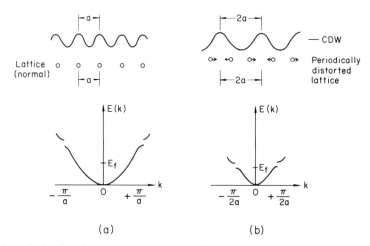

Fig. 31. Schematic drawing showing a periodic lattice distortion, charge density waves, and energy band gaps before and after a periodic lattice distortion in a hypothetical one-dimensional metal. Note the overall lowering of the Fermi level following the periodic lattice distortion. See text.

displacements which stabilize the charge perturbation. Another requirement for forming a CDW is a strong electron–phonon coupling, required to permit ionic displacements to reduce the otherwise prohibitive Coulomb energy, and precursor phenomena such as a soft phonon mode (CHAN and HEINE [1973]) might occur above the transition temperature to assist the CDW instability. The normal crystalline periodicity is altered in the presence of a CDW. However, it should be noted that the wavevector of a CDW is determined by the Fermi surface and is therefore not necessarily an integral fraction of a reciprocal lattice vector of the undistorted parent phase. Consequently an "incommensurate" phase may result, which is considered to have lost its translational symmetry (AXE [1977]) and is thus a quasi-crystalline phase.

The incommensurate state described above may not actually correspond to the lowest possible energy state and, accordingly, the CDW or the lattice may undergo a further distortion which makes the two "commensurate" in which case the CDW wavevector is an integral fraction of the underlying lattice. The commensurate state is usually referred to as a "locked-in" state. Thus, there can be two phase changes associated with a CDW formation: the normal (parent)-to-incommensurate transition (usually second order) and the incommensurate-to-incommensurate transformation (first order). Since a CDW is accompanied by a lattice distortion, diffraction techniques (electron, neutron, X-ray) can be used to reveal satellite reflections appearing near the Bragg reflections of the parent phase as a consequence of the new superlattice associated with the formation of the CDW. These satellites are separated from the associated Bragg reflections by a reciprocal lattice vector determined by the CDW wavevector.

In lower-dimensional systems (1-d and 2-d materials) the possibility of CDW formation is enhanced because the simple structures involved lead to a high probability for favorable Fermi surface nesting. But in three-dimensional materials CDW phenomena should, in principle, be rather rare because of the unlikelihood of favorable Fermi surface nesting.

Besides the incommensurate state mentioned above, MCMILLAN [1976] introduced the concept of *discommensurations*, which are narrow out-of-phase regions (defects) between large in-phase (commensurate) regions. Accordingly, the incommensurate-to-commensurate change is viewed as a defect (discommensuration) "melting" transition.

A recent study (HWANG *et al.* [1983]) of the so-called "premartensitic effects" in Ti–Ni alloys has been carried out using Ti–Ni–Fe alloys containing a few percent of Fe in order to suppress the martensitic transformation (lower M_s) while at the same time leaving the temperature regime of the "premartensite" phenomena essentially unchanged. Space does not permit a detailed presentation of this work, but some of the highlights are significantly new and these are briefly considered. The "premartensitic" resistivity anomaly (increase with decreasing temperature) in Ti–Ni type alloys has been ascribed (HWANG *et al.* [1983]) to the formation of a three-dimensional charge density wave which appears to evolve in two stages. In the first, the B2(CsCl) parent is gradually (second order) distorted into an incommensurate

References: p. 1073.

phase with decreasing temperature. The incommensurate phase is most simply described as one with "distorted cubic" symmetry. Its gradual transition involves the appearance of new $\frac{1}{3}\langle 110 \rangle_{B2}$ superlattice reflections which intensify with decreasing temperature. At this stage, the $\frac{1}{3}\langle 110 \rangle$ superlattice reflections are in slightly irrational (incommensurate) positions. As the temperature is further decreased these superlattice reflections "lock in" to commensurate positions which are precise multiples of $\frac{1}{3}\langle 110 \rangle_{B2}$, and the resulting structure is rhombohedral (designated the *R-phase*) as a consequence of a homogeneous distortion which involves an expansion along $\langle 111 \rangle$ cube diagonals (and a small contraction in all directions in the plane normal to $\langle 111 \rangle$). This incommensurate-to-commensurate transformation is first-order in nature. Thus, the events observed during the so-called *premartensitic state* of Ti–Ni–X alloys identify with the sequence: B2(CsCl) → distorted cubic (incommensurate) → R, rhombohedral (commensurate). In addition, microdomains typical of anti-phase domains in ordered alloys have been identified using dark field electron microscope images formed by using the new superlattice reflections.

Finally, it should be mentioned that the incommensurate-to-commensurate transformation in Ti–Ni–Fe alloys is also associated with the appearance of twin-like domains, as shown in fig. 32. The twin plane has been identified as a {110} plane. It would appear that the only difference between the "twins" and the "matrix" within which they exist is that the sense of the $\langle 111 \rangle$ expansion (Bain distortion) occurs along different c axes in each. In this respect, the formation of twins during the incommensurate-to-commensurate change appears to be a result of overall strain compensation, not unlike that occurring in a martensitic transformation. But in the case of the Ti–Ni–Fe alloys, when the actual martensite is formed at a lower

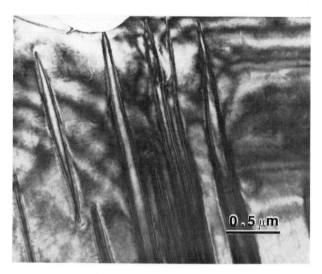

Fig. 32. Twin-related domains as a consequence of an incommensurate-to-commensurate phase change in a $Ti_{50}Ni_{47}Fe_3$(at%) alloy. See text for explanation.

temperature, the {110} twins and the commensurate R-phase are completely destroyed by the advancing martensite interface, and, seemingly, the "premartensitic" commensurate twinned structure presents no obstacle to the growth of the martensite plates.

In the above description of charge density waves and related phenomena, a measure of consistency has been attempted by referring to the *second order* parent → incommensurate (gradual) change as a *transition*, as opposed to the *first-order* incommensurate → commensurate (abrupt) change which is viewed as a *transformation*.

Acknowledgement

I would like to acknowledge many helpful comments from numerous colleagues and members of my own research group. In particular, I wish to thank Professor G. Krauss for providing fig. 16, Professor K. Shimizu for fig. 17, Dr. M. Umemoto for figs. 13–15, and Professor J.C. Williams for fig. 29.

References

AARONSON, H.I., and K.R. KINSMAN, 1977, Acta Metall. **25**, 367.

AXE, J.D., 1977, in: Electron-Phonon Interactions and Phase Transitions, ed. T. Riste (Plenum, New York) p. 50.

BAIN, E.C., 1924, Trans. AIME **70**, 25.

BOWLES, J.S., and J.K. MACKENZIE, 1954, Acta Metall. **2**, 129, 138 and 224.

BOWLES, J.S., and C.M. WAYMAN, 1979, Acta Metall. **27**, 833.

BULLOUGH, R., and B.A. BILBY, 1956, Proc. Phys. Soc. **B 69**, 1276.

CHAN S.K., and V. HEINE, 1973, J. Phys. **F3**, 795.

CHRISTIAN, J.W., 1970, Phase Transformations, in: Physical Metallurgy, 2nd Ed., ed. R.W. Cahn (North-Holland, Amsterdam) ch. 10.

CHRISTIAN, J.W. 1982, Metallurg. Trans. **13A**, 509.

CLAPP, P.C., 1973, Phys. Stat. Sol. (b) **57**, 561.

COHEN, M., and C.M. WAYMAN, 1981, Fundamentals of Martensitic Reactions, in: Metallurgical Treatises, eds. J.K. Tien and J.F. Elliott (Metallurgical Society AIME, Warrendale, PA) p. 445.

GOLDSTEIN, D.M., and L.J. McNAMARA, 1979, Report NSWC MP 79-441, Proc. of Heat Engine Conference (U.S. Naval Surface Weapons Center, Silver Springs, MD).

HWANG, C.M., M. MEICHLE, M.B. SALAMON and C.M. WAYMAN, 1983, Phil. Mag. **A47**, 9 and 31.

JACKSON, C.M., H.J. WAGNER and R.J. WASILEWSKI, 1972, National Aeronautics and Space Agency Report SP-5110.

KAUFMAN, L., and M. COHEN, 1958, Prog. Met. Phys. **7**, 165.

KESSLER, H., and W. PITSCH, 1967, Acta Metall. **15**, 401.

KOHN, W., 1959, Phys. Rev. Lett. **2**, 393.

KRAUSS, G., 1980, Principles of the Heat Treatment of Steels (ASM, Metals Park, OH) ch. 8.

KURDJUMOV, G., and L.G. KHANDROS, 1949, Dokl. Akad. Nauk SSSR **66**, 211.

McMILLAN, W., 1976, Phys. Rev. **B14**, 1496.

OLSON, G.B., and M. COHEN, 1972, J. Less-Common Met. **28**, 107.

OLSON, G.B., and M. COHEN, 1976, Metallurg. Trans. **7A**, 1897, 1905 and 1915.

OLSON, G.B., and M. COHEN, 1979, Proc. US–Japan Symp. on Mechanical Behavior of Metals and Alloys Associated with Displacive Transformations (Rensselaer Polytechnic Inst., Troy, NY) p. 7.

ONODERA, H., and I. TAMURA, 1979, Proc. US-Japan Symp. on Mechanical Behavior of Metals and Alloys Associated with Displacive Transformations (Rensselaer Polytechnic Inst., Troy, NY) p. 24.

ONODERA, H., H. GOTO and I. TAMURA, 1976, Suppl. Trans. Japan Inst. Met. **17**, 327.

OTSUKA, K., and C.M. WAYMAN, 1977, Reviews on the Deformation Behavior of Materials * **2**, 81.

* A quarterly, published by Freund, Tel Aviv.

OVERHAUSER, A.W., 1968, Phys. Rev. **167**, 691.

OVERHAUSER, A.W., 1971, Phys. Rev. **B3**, 3173.

PATEL, J.R., and M. COHEN, 1953, Acta Metall. **1**, 531.

PEIERLS, R.E., 1955, Quantum Theory of Solids (Oxford Univ. Press, Oxford).

SABURI, T., S. NENNO and C.M. WAYMAN, 1979, Proc. ICOMAT-79, Int. Conf. on Martensitic Transformations, Cambridge, Massachusetts (Alpine Press, Boston, MA) p. 619.

SASS, S.L., 1972, J. Less-Common Met. **28**, 157.

SCHROEDER, T.A., and C.M. WAYMAN, 1977a, Acta Metall. **25**, 1375.

SCHROEDER, T.A., and C.M. WAYMAN, 1977b, Scripta Metall. **11**, 225.

SCHROEDER, T.A., and C.M. WAYMAN, 1978, Acta Metall. **26**, 1745.

SCHROEDER, T.A., and C.M. WAYMAN, 1979, Acta Metall. **27**, 405.

SHIMIZU, K., and C.M. WAYMAN, 1972, Met. Sci. J. **6**, 175.

SMITH, R., and J.S. BOWLES, 1960, Acta Metall. **8**, 405.

TADAKI, T., and K. SHIMIZU, 1970, Trans. Japan Inst. Met., **11**, 44.

TAMURA, I., T. MAKI, H. HATO, Y. TOMOTA and M. OKADA, 1970, Proc. 2nd Int. Conf. on Strength of Metals and Alloys, Asilomar, CA, vol. 3, p. 894.

TAMURA, I., T. MAKI, S. SHIMOOKA, M. OKADA and Y. TOMOTA, 1972, Proc. 3rd Conf. on High Strength Martensitic Steels, Havirov, Czechoslovakia, p. 118.

TAMURA, I., Y. TOMOTA and M. OZAWA, 1973, Proc. 3rd Int. Conf. on Strength of Metals and Alloys, Cambridge, UK, vol. 1, p. 611 and vol. 2, p. 540.

VAN LANDUYT, J., and C.M. WAYMAN, 1968, Acta Metall. **16**, 803 and 815.

WASILEWSKI, R., 1975, in: Shape Memory Effects in Alloys, ed. J. Perkins (Plenum, New York) p. 245.

WECHSLER, M.S., D.S. LIEBERMAN and T.A. READ, 1953, Trans. AIME **197**, 1503.

WILLIAMS, J.C., D. DE FONTAINE and N.E. PATON, 1973, Metallurg. Trans. **4**, 2701.

WILSON, J.A., F.J. DISALVO and S. MAHAJAN, 1975, Adv. Phys. **24**, 117.

ZACKAY, V.F., E.R. PARKER, D. FARH and R. BUSCH, 1967, Trans. ASM **60**, 252.

General bibliography

Williams, J.C., D. de Fontaine and N.E Paton, 1973, Metallurg. Trans. **4**, 2701. The ω-phase as an Example of an Unusual Shear Transformation.

Christian, J.W., 1964, The Theory of Transformation in Metals and Alloys (Pergamon Press, Oxford).

Wayman, C.M., 1964, Introduction to the Crystallography of Martensitic Transformations (Macmillan, New York).

Wilson, J.A., F.J. DiSalvo and S. Mahajan, 1975, Charge-density Waves and Superlattices in the Metallic Layered Transition Metal Dichalcogenides, Adv. Phys. **24**, 117.

J. Perkins, ed., 1975, The Shape Memory Effect in Alloys (Plenum, New York).

Proceedings JIMIS-1, 1976, Japan Inst. Met. Int. Symp. on Martensitic Transformations, Suppl. Trans. Japan Inst. Met., vol. 17.

Proceedings ICOMAT-77, 1977, International Conference on Martensitic Transformations, Kiev USSR.

Phase Transformations, 1979, Series 3, No. 11, vols. 1 and 2 (The Institution of Metallurgists, London).

Proceedings ICOMAT-79, 1979, International Conference on Martensitic Transformations, Cambridge, MA (Alpine Press, Boston, MA).

Nishiyama, Z., 1979, Martensitic Transformation, eds. M.E. Fine, M. Meshii and C.M. Wayman (Academic, New York).

Barrett, C.S., and T.B. Massalski, 1980, Structure of Metals, 3rd Ed. (Pergamon Press, Oxford) ch. 18.

Proceedings of International Conference on Solid–Solid Phase Transformations, 1981, Pittsburgh, PA (Metallurgical Society AIME, Warrendale, PA, 1983).

Cohen, M., and C.M. Wayman, 1981, Fundamentals of Martensitic Reactions, in: Metallurgical Treatises (Metallurgical Society AIME, Warrendale, PA) p. 445.

CHAPTER 16

PHYSICAL METALLURGY OF STEELS

E. HORNBOGEN

Institut für Werkstoffe
Ruhr-Universität Bochum
4630 Bochum 1, FRG

R.W. Cahn and P. Haasen, eds.
Physical Metallurgy; third, revised and enlarged edition
© *Elsevier Science Publishers BV, 1983*

1. Iron and steel

1.1. Introduction

Steels are alloys of the element iron with carbon and usually several other elements. There are three major groups of alloying elements, those that form interstitial and substitutional solutions, and those that are immiscible with the crystal lattices of iron (table 1). We shall not deal with the last category in the following. Commercial steels always contain a large number of both types of alloying elements, some added intentionally, others unintentionally during production and processing of the steel. Unwanted elements are for example: N (from the air), S (from the coke), P, As (from some ores), and H (introduced during pickling). Examples for the composition of some commercial steels are given in table 2.

Steels are the most important group of alloys used by mankind since about 2000 BC to the present date. There are several reasons for this long-lasting popularity of the alloys of iron:

a) The crust of the earth contains about 4.2 wt% Fe, which can be rather easily reduced to the metallic state.

b) Iron has a melting temperature (1540°C) that allows thermally activated processes at not too high temperatures ($T > 400°C$) while these processes (except those involving interstitial diffusion exclusively) are very slow at room temperature.

c) There are two phase transformations in solid iron that allow the formation of a

Table 1
Solubility of the elements in the crystal lattices of iron.

H 1																	He 3
Li 3	Be 2											B[a] 3	C 1	N 1	O 1	F 3	Ne 3
Na 3	Mg 3											Al 2	Si 2	P[b] 2	S 3	Cl 3	Ar 3
K 3	Ca 3	Sc 3?	Ti 2	V 2[c]	Cr 2[c]	Mn 2[c]	Fe	Co 2[c]	Ni 2[c]	Cu 2	Zn 2	Ga 2	Ge 2	As 2	Se 3	Br 3	Kr 3
Rb 3	Sr 3	Y 3?	Zr 2	Nb 2	Mo 2	Tc 2	Ru 2	Rh 2[c]	Pd 2[c]	Ag 3	Cd 3	In 3	Sn 2	Sb 2	Te 3	I 3	Xe 3
Cs 3	Ba 3	La 3?	Hf 2?	Ta 2	W 2	Re 2	Os 2	Ir 2[c]	Pt 2[c]	Au 2	Hg 3	Tl 3	Pb 3	Bi 3	Po 3?	At 3	Rn 3
Fr 3	Ra 3	Ac 3?	Th 3?	Pa 3?	U 3	Np 3?	Pu 3?	Am 3?	Cm 3?	Bk 3?	Cf 3?						

1. interstitial elements
2. substitutional elements
3. practically no solubility

[a] soluble in grain boundaries
[b] P may be interstitial in the fcc structure
[c] complete solubility at certain temperatures.

Table 2
Important alloying elements in some commercial steels.

no.	Composition (wt%)								Structure	Application
	C	Mn	Cr	Ni	Mo	W	Si	V		
1	0.12	0.45	–	–	–	–	–	–	ferrite plus pearlite	construction steel
2	0.65	0.70	–	–	–	–	–	–	tempered martensite	tool steel
3	0.50	<1.0	13.0	–	–	–	–	–	tempered martensite	knife blade steel (stainless)
4	0.10	0.40	5.0	–	0.5	–	–	–	bainite	high temperature steel
5	<0.15	0.40	–	6.0	–	–	–	–	ferrite	low temperature steel
6	0.80	0.30	4.0	–	–	18.0	–	2.0	tempered martensite and tungsten-carbide	high speed steel
7	0.10	<1.0	18.0	8.0	–	–	–	–	austenite	stainless steel
8	<0.07	–	–	–	–	–	4.0	–	ferrite	transformer steel

wide variety of microstructures. As a consequence a large range of different physical properties can be obtained.

d) Iron is ferromagnetic at low temperatures ($T_c = 768°C$) which adds a new useful physical property and leads to many anomalies of other properties as a consequence of ferromagnetism (§§ 1.2, 2.2, 6.3).

There is a strong scientific interest in iron and its alloys. Nevertheless steel is still the (inorganic) material with the largest number of unsolved problems. For example there is not yet a completely satisfactory explanation for the ($\gamma \rightarrow \alpha$) phase transformation of iron. The advent of dual-phase steel after 1975 has shown that new developments even of mass-produced structural steel are not impossible. The large amount of empirical knowledge that has been accumulated during four thousand years is only of limited value for a basic understanding. Progress has been made during recent years in the quantitative understanding for example of phase diagrams, crystallography of transformations and of some mechanical properties of steels. Frequently we must however be satisfied with a qualitative understanding of structure and properties of steels. One major experimental difficulty is that the properties of pure iron are very difficult to determine. Even the purest iron obtained by zone-refining contains enough (especially interstitial) impurities to strongly affect many structure-sensitive properties, such as the onset of plastic yielding.

1.2. Some properties of pure iron

The iron atom is situated towards the end of the first transition series with an electron configuration $Ar3d^64s^2$. The neighbouring elements with larger atomic

References: p. 1136.

number than iron (Co, Ni, Cu) crystallize in close-packed structures until the 3d states are filled in Cu. The transition elements with less-filled 3d states (β-Ti, V, Cr, δ-Mn) have a bcc structure. The maximum of the binding energy for Cr with its half-filled 3d band indicates that a contribution of covalent bonding by 3d-states may stabilize this structure (fig. 1).

Iron is found between these two groups of elements. It crystallizes in both the fcc ($906° < T_\gamma < 1401°C$) and the bcc ($1401 > T_\alpha < 609°C$) lattices. The re-occurrence of the bcc structure at high temperatures can be rationalized by dividing the difference in free energy of bcc and fcc iron, $\Delta F_{\alpha\gamma}$, into a magnetic and a nonmagnetic term (ZENER [1955]). The magnetic term is negligible up to temperatures at which magnetization of α-Fe does not change much. At higher temperatures (depending on T_c) the bcc structure is increasingly stabilized due to the entropy of demagnetization. Without this effect, α-Fe should transform to γ-Fe at about 700°C without another transformation at 1401°C. In fig. 2, $\Delta F_{\alpha\gamma} = F_\alpha - F_\gamma$ is plotted as a function of temperature. The contribution that is due to magnetic uncoupling in the α-phase is shown as a dashed line. This term produces the curvature of $\Delta F_{\alpha\gamma}(T)$ as determined by JOHANNSON [1937] and the re-occurrence of the bcc structure above 1401°C.

The increasing value of $\Delta F_{\alpha\gamma}$ with decreasing temperature indicates that an increasing pressure is needed to produce the closest packed iron at temperatures below 906°C. At 20°C this pressure is 130 kBar as has been determined by both static and dynamic measurements. It is uncertain whether the high-pressure phase is fcc or hcp. In the latter case there should be a triple point between the bcc, fcc, and hcp phases in the p-T-diagram (figs. 3 and 4).

There is no TEM experimental evidence for the existence of thermal vacancies in quenched α-Fe. The reason is probably the high ratio of energy of formation u_F to energy of motion u_M: $u_F/u_M \approx 4$ as compared to about 1 for the fcc lattice (ch. 17 § 2.2.1). Vacancies probably anneal out during quenching of α-Fe (JOHNSON [1960]). It is very likely that these results are strongly influenced by interstitial elements, especially by carbon.

The Burgers vector of the minimum energy dislocation in α-Fe is $(a/2)\langle 111 \rangle$.

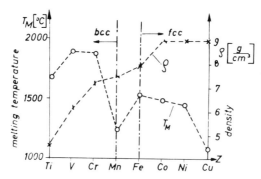

Fig. 1. Melting temperature, T_M, and density, ϱ, versus atomic number Z of the elements of the first transition period.

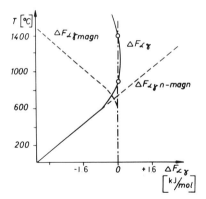

Fig. 2. Free energy difference $\Delta F_{\alpha\gamma}$ resolved into magnetic and nonmagnetic components (after ZENER [1955]).

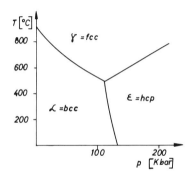

Fig. 3. Pressure–temperature diagram of iron (after BUNDY [1965]).

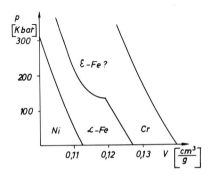

Fig. 4. Dynamic pressure–volume curves of Cr, Fe and Ni (after MINSHALL [1961]).

References: p. 1136.

Segments of $a\langle 100\rangle$ can form by the reaction $(a/2)[1\bar{1}1]+(a/2)[11\bar{1}]\to a[100]$. The stacking-fault energy of α-Fe (as of the other bcc transition metals) is very high. Stacking faults have therefore not been observed in α-Fe even by subtle weak beam electron microscopy. Details of the core structure of dislocations in the bcc lattice and their consequences on crystal plasticity are discussed in ch. 18. The probability for the occurrence of annealing twins is low. In γ-Fe the Burgers vector of the complete dislocations is $(a/2)\langle 110\rangle$. The stacking-fault energy (at 1000°C) determined from the frequency of annealing twins is 75 mJ/m² (NUTTING and CHARNOCK [1967]) (table 7). Stacking faults and annealing twins are therefore probable to be found in γ-Fe and its alloys.

Many physical properties, like specific volume or electrical conductivity, change discontinuously during the phase transformations of iron. They are often used to determine the transformation temperatures in steels. There are discontinuities of the self-diffusion coefficient of iron at the transformation temperatures. The diffusivity of α-Fe is about 10^2 times higher than that of γ-Fe at the same temperatures (fig. 5). This can qualitatively be explained by the closer packing of γ-Fe (BIRCHENALL [1951]). Below T_c there is an anomalous decrease of the diffusion coefficient of about 60% during the transition to complete magnetic order. A change in equilibrium vacancy concentration, not magnetostriction, is used as an explanation for this behaviour (BIRCHENALL and BORG [1960]).

An amorphous structure can be obtained for pure iron neither by vapour deposition nor by splat cooling down to temperatures close to 0 K. Such a structure is exclusively found in certain alloys of iron.

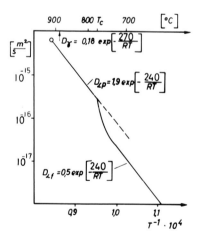

Fig. 5. Self-diffusion coefficient of paramagnetic ($D_{\alpha p}$) and ferromagnetic ($D_{\alpha f}$) α-iron; Q in kJ. (After BIRCHENALL [1960].)

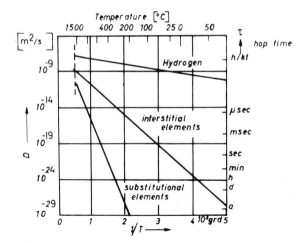

Fig. 6. Average diffusion coefficients D of substitutional (s) and interstitial elements in iron.
$Q_s = 250 \pm 40$ kJ/mol; in α-Fe: $Q_C = 83$ kJ/mol; $Q_N = 74$ kJ/mol; $Q_O = 95$ kJ/mol.

2.   Alloys of iron

2.1.   Interstitial alloys

The elements H, C, N and O can occupy interstitial sites in the lattices of iron (ch. 4). All these elements fullfil the condition $r_i \ll r_{Fe}$ (table 3, fig. 6). The amount of solubility can however not be explained on the basis of atomic-size ratios. Increased filling of the 3d states reduces the solubility of all interstitials as shown for the first transition period and C in fig. 7. Iron has the lowest carbon solubility of the bcc metals. The solubility in the fcc iron is larger, owing to the more favourable geometry. With increasing atomic number the solubility decreases to very low values for copper.

The solubilities depend on the phase with which the α-iron interstitial solid

Table 3
Maximum solubility of interstitials in α-Fe; $r_{\alpha\text{-Fe}} = 1.24$ Å.

element	atomic radius (for coord. no. 12) (nm)	max. solubility (at%)	at temperature (°C)	in equil. with
H	< 0.04	$1-2 \times 10^{-3}$	905	H_2
B	0.098	< 0.02	915	$Fe_2 B$
C	0.091	0.095	723	$Fe_3 C$
N	0.092	0.40	590	$Fe_4 N$
O	~ 0.092	$0.7-13 \times 10^{-4}$	906	FeO

References: p. 1136.

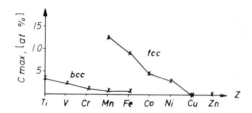

Fig. 7. Maximum solubility of carbon in the elements of the first transition series.

solution is in equilibrium: stable equilibria would be for example those with nitrogen gas or with graphite. There is a high activation energy of nucleation of these pure elements because of large surface energy terms. Metastable phases form instead, with lower activation energy and a higher solubility of the iron-rich solid solution. Phases that form in Fe–C and Fe–N alloys are listed in table 4. In fig. 8 the solubilities of some of these phases are shown. Which phase is formed depends on temperature and time of the heat treatment and on the imperfection structure of the solid solution. The probability of formation of nitrogen gas or graphite is so low that these phases rarely precipitate in steels.

In Fe–C alloys, graphite can form from the fcc solid solution at grain boundaries only after very long periods of ageing. Normally Fe_3C (cementite, ch. 5) or the less stable carbides form instead of graphite (§ 5.10). There are as many Fe–C diagrams as there are stable and metastable phases. In fig. 9 the metastable Fe–Fe_3C diagram is shown in addition to the stable Fe-graphite diagram. The temperature dependence of metastable solubility of C [wt%] in α-Fe is

$$C_{\alpha\text{-Fe}-\text{Fe}_3\text{C}}(T) = 2.55 \exp\left(\frac{40.7 \text{ kJ/mol}}{RT}\right).$$

The corresponding value for the heat of solution in γ-Fe is smaller. DARKEN and GURRY [1953] give 31.5 kJ/mol for γ-Fe-Fe_3C and 42 kJ/mol for γ-Fe-graphite.

There are several important composition ranges in the Fe–Fe_3C diagram (fig. 9):

Table 4
Metastable compounds that precipitate in Fe–C and Fe–N alloys.

	Composition	Crystal structure	Nucleation sites	Formation temperatures
cementite	Fe_3C	fig. 9	dislocations, grainb.[2]	> 200°C
ϵ-carbide [3]	$Fe_{2-3}C$	hcp [1]	matrix, splat cooling	< 250°C
α'-carbide [3]	Fe_8C	fig. 16	matrix	< 100°C
γ'-nitride [4]	Fe_4N	fcc [1]	dislocations, grainb. [2]	> 250°C
α'-nitride	Fe_8N	fig. 16	matrix	< 300°C

[1] Arrangement of iron atoms, bc N-atom.
[2] Only at high temperatures.
[3] There is not complete certainty about these structures.
[4] In stable equilibrium with γ-Fe.

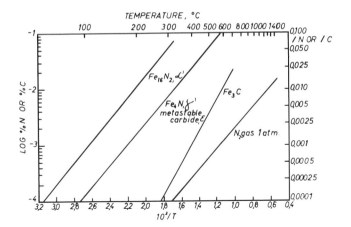

Fig. 8. Solubilities of C and N in α-Fe in equilibrium with different phases (average values from different investigations after LESLIE and KEH [1965]).

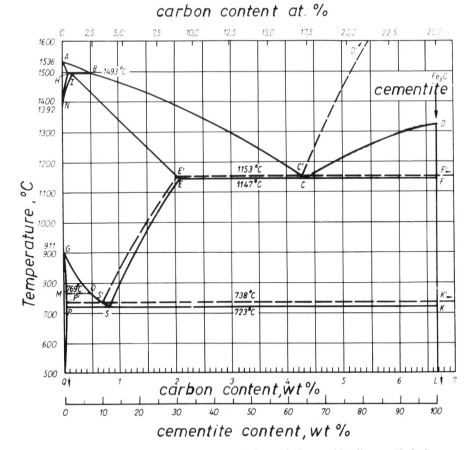

Fig. 9. The metastable Fe–Fe₃C diagram and the stable Fe–graphite diagram (dashed).

References: p. 1136.

a. The maximum solubility in γ-iron (austenite) $C_{\gamma-Fe_3C} = 2.0$ wt% $= 8.9$ at% is the limiting composition separating steel and cast iron (§ 7.2).

b. Cast iron has usually hypo-eutectic or eutectic composition: $C_E = 4.3$ wt% $= 17.3$ at%,

$$1_E \rightarrow \gamma\text{-Fe}(C) + Fe_3C.$$

c. The steels can be subdivided in hypo-eutectoid and hypereutectoid depending on whether their composition is smaller or larger than $C_p = 0.80$ wt% $= 3.61$ at% C. The microstructure that originates from the eutectoid reaction

$$\gamma\text{-Fe}(C) \rightarrow \alpha\text{-Fe}(C) + Fe_3C$$

is known as pearlite, if it is formed by lamellar growth.

d. The maximum solubility of carbon in α-Fe is $C_{\alpha-Fe_3C} = 0.02$ wt% $= 0.095$ at%. In alloys with smaller carbon contents, carbides can only form by precipitation from α-Fe(C) (ferrite).

The lower the carbon content, the lower is the temperature below at which precipitation can start. Below 250°C the less stable ϵ-carbides (LESLIE [1961]) or probably even the coherent α'-carbide form (PHILLIPS [1963]) (fig. 16a). The smallest carbon contents that can be obtained at the present date, $(1-2) \times 10^{-4}$ wt%, correspond to the solubility at about $+50°C$.

Nitrogen has a relatively large solubility in α-Fe. The solubility of oxygen is very small (ENGELL et al. [1967]). The difference between the solubilities of N, C and O in α-iron cannot be explained on the basis of atomic-size ratios (table 3). These solubilities are only valid for perfect crystals of iron in equilibrium with a certain phase. If lattice defects are present, the solubility will be increased due to segregation. Grain boundaries, dislocations, radiation defects in α-Fe, and in addition stacking faults in γ-Fe have to be considered because they offer interstitial sites energetically favourable to those in the perfect lattice. These defects are not in thermal equilibrium, their density can vary over many orders of magnitude due to mechanical-, heat-, or radiation-treatments of steel. The concentration of segregated interstitial atoms depends on the binding energy u_B (i.e. the energy difference between a site in the perfect lattice and at the defect), the temperature and the concentration of interstitials in the lattice, c_L. MCLEAN [1957] has proposed the following equation for the concentration at a grain boundary, c_B, with an average binding energy of the grain boundary sites $\bar{u}_B$:

$$c_B = \frac{a\, c_L \exp \bar{u}_B/RT}{1 + a\, c_L \exp \bar{u}_B/RT}.$$

Figure 10 gives the results of the calculation for carbon in α-Fe. Taking $\bar{u}_B = 84$ kJ, $\alpha = 0.005$. $c_B = 1$ if all sites with $\bar{u}_B$ in the grain boundary structure are occupied i.e. at low temperatures or high values of c_L. Segregation to grain boundaries is important for the understanding of grain-size dependence of mechanical properties (figs. 30 and 32, and ch. 19) and grain-boundary embrittlement (§ 5.11). Segregation to dislocations is the reason of strain ageing (§ 5.1), segregation to stacking faults the

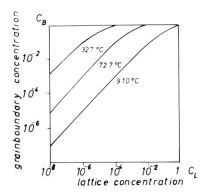

Fig. 10. Portion C_B of grain-boundary sites ($u_B = 84$ kJ/mol) occupied at different temperatures and concentration c_L of carbon (after MᶜLᴇᴀɴ [1957]).

first step of nucleation of carbides in some austenitic steels, segregation to radiation defects one reason for embrittlement of radiated steels. Precipitation of interstitials from α-Fe(C) is also highly affected by lattice defects. In table 4 the preferred nucleation sites have been added. The energy contributed from the defect leads to a minimum of the activation energy of nucleation at the site of the defect (chs. 14 and 21), so that the distribution of precipitates is determined by the distribution of defects in such cases.

2.2. Substitutional alloys

The elements which have been marked as 2 (table 1) are completely or partially soluble as substitutional elements in the iron lattices. The Hume-Rothery rules for solubility are well fullfilled for alloys with the transition elements: Large solubility of the elements in the neighbourhood of Fe (Ni, Co, Mn, Cr and V), limited solubility of Ti, and practically no solubility of Ca and K. The alloying elements have different energies of solution in α- and γ-iron. The $\Delta F_{\alpha\gamma}(T)$ curve (fig. 2) is shifted in different directions depending on whether $\Delta F_{\alpha\gamma}$ is increased or decreased by the alloying element (fig. 11). This leads to an increase or decrease of the temperature range in which austenite is stable. For simplicity the curves are drawn for α- and γ-phase of the same composition (i.e. the martensitic T_0-temperatures are shown for $\Delta F_{\alpha\gamma} = 0$, §3.2). We can divide the alloying elements into two groups depending whether they close or open the γ-field. Typical phase diagrams are shown in figs. 9, 12 and 13. The following rules tell how the alloying elements behave in this respect:
a) the γ-field is opened by all interstitial elements, because the fcc lattice provides sites of smaller strain energy than does the bcc lattice;
b) the γ-field is opened by elements with a fcc or hcp structure (Cu, Au, Ni, Pt, γ-Mn);
c) the γ-field is closed by elements that form Hume-Rothery phases with Cu, Ag

References: p. 1136.

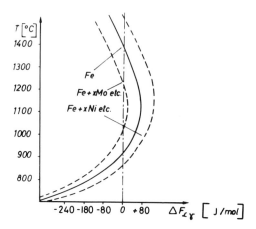

Fig. 11. The effect of solute elements with larger heat of solution in γ-Fe than in α-Fe (e.g. Mo) and vice versa (e.g. Ni) on the transformation temperatures ($\Delta F_{\alpha\gamma} = 0$), assuming constant concentrations of α- and γ-solid solution and no change of T_c in α-Fe (schematic).

and Au, even if they form a fcc lattice (Al, Si, P, Zn, Ga, Ge, As etc.);

d) the γ-field is closed by the bcc transition metals (V, Ti, Mo, W, Cr etc.);

The phase diagram of Fe–Cr alloys shows that small additions of chromium lower the $\gamma \rightarrow \alpha$ transformation temperature while the loop is closed by larger Cr-additions. ZENER [1955] has pointed out that this behaviour can be explained by the magnetic term of the free energy of α-Fe (fig. 2). The addition of Cr to Fe has two effects: It shifts the $\Delta F_{\alpha\gamma}(T)$ curve as shown in fig. 11 and it lowers the Curie temperature T_c. Consequently the term $\Delta F_{\alpha\gamma\ \text{magn}}$ (fig. 2) appears at a lower temperature and changes the shape of the curve to that shown in fig. 13b. This T_c-effect should play a role for transformation temperatures of all alloys of iron. In Fe–Cr alloys it is evident because the difference in heat of solution of Cr in α-Fe and γ-Fe is relatively small.

In many alloys the α- and γ-field is bounded by a *miscibility gap* (fig. 13c). The α- or γ-solid solutions are then in equilibrium with an intermetallic compound or a solid solution. Special attention is drawn to the ordered bcc structures Fe_3Al (ch. 4, fig. 33d), and Fe_3Si that can form as stable coherent precipitates in α-iron. In addition metastable equilibria occur frequently in substitutional alloys of Fe which forms the matrix of a microstructure. Such coherent phases have been found in binary bcc solid solutions with Cu, Au, Al, Mo, W, Be (HORNBOGEN et al. [1966]). They play an important role in maraging steels (§5.6).

The crystal structure of the solid solution of Fe at room temperatures, as it follows from the phase diagrams (figs. 12, 13), is used to classify steels. If they consist predominantly of α-Fe solid solutions they are known as *ferritic* steels, solid solutions of γ-Fe are the base of *austenitic* steels. A metastable fcc structure can be obtained for example from the ternary alloy indicated in fig. 14. This metastable

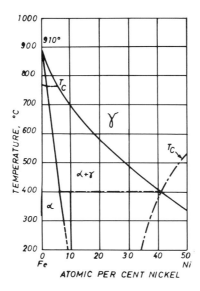

Fig. 12. Fe–Ni phase diagram; T_c is also shown for the metastable γ-alloys with 34–41 at% Ni (see fig. 47).

phase is the well-known austenitic stainless steel. Precipitation of stable or metastable coherent, ordered phases (γ') is very common in the fcc iron alloys. Such (γ + γ') microstructures provide the base for creep resistance of precipitation-hardened stainless steels and Ni-base superalloys.

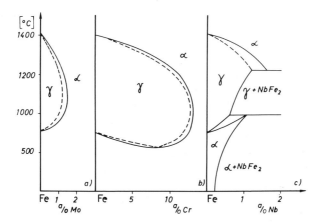

Fig. 13. Closed γ-fields in Fe–Mo, Fe–Cr, and Fe–Nb phase diagrams.

References: p. 1136.

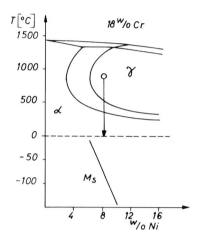

Fig. 14. Section of the Fe–Cr–Ni diagram showing the composition and heat treatment of austenitic stainless steel. Below M_s the metastable fcc alloys become instable and transform to bcc martensite, while at $M_s < T < M_d$ strain-induced transformation can take place.

2.3. Interstitial plus substitutional alloys

Even plain carbon steel contains a certain amount of substitutional elements (especially Mn). Therefore interstitials are always in interaction with atoms substituted in the iron lattice. The interaction energy as compared to that with iron, Δu, can be positive or negative. $\Delta u = (u_{\mathrm{Fe}-i}) - (u_{s-i})$. The lattice of iron with interstitial (i) and substitutional (s) atom in solutions can be expected to show a distribution of the i-atoms that depends on that of the s-atoms (fig. 15). A qualitative idea of the sign of Δu can be obtained from the stability of the carbides of the s-atoms and from their effect on the solubility of carbon (fig. 7, table 5). For the first long period we find a decreasing stability of carbides and of carbon solubility with increasing atomic number:

$$\gamma\text{-Fe} \rightarrow \text{Co} \rightarrow \text{Ni} \rightarrow \text{Cu and } \beta\text{-Ti} \rightarrow \text{V} \rightarrow \text{Cr} \rightarrow (\text{Mn}) \rightarrow \alpha\text{-Fe.}$$

We can divide the substitutional transitional elements in two groups depending on the sign of Δu, which will lead to the following effects on Fe–C alloys:
a) the solubility of carbon is increased by elements with negative Δu and vice versa.
b) there is a tendency to form alloy carbides if Δu is negative and to form alloy free carbides if Δu is positive.
A very large number of carbides can form in Fe–s–C alloys in stable or metastable equilibria. In table 5 the crystal structures of carbides of the elements of the first transition series are listed. There are basically the following possibilities for the composition of carbides that form in Fe–s–C alloys:
a) $\mathrm{Fe}_x\mathrm{C}_y$ solute free carbide;
b) $\mathrm{Fe(s)}_x\mathrm{C}_y$ carbide with substitutional composition of the alloy;

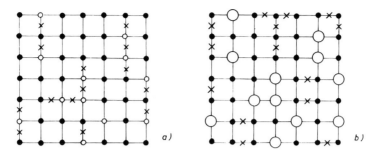

Fig. 15. Schematic drawing of the structure of interstitial (×) plus substitutional (○) solid solutions, in iron (●): (a) attraction, (b) repulsion between interstitial and substitutional solute.

c) $[Fe_x s_z]C_y$ carbide with Fe replaced by a definite amount of s;
d) $s_x C_y$ iron-free carbide.

Many $[Fe_x s_z]C_y$-type carbides occur in alloy steels, still more in rapidly solidified alloys of iron, in addition to those listed in tables 4 and 5.

η-carbide is cubic with 96 metal and 16 carbon atoms in the elementary cell. The metal atoms are iron and group V, VI transition metals. The composition is M_6C, a typical representative $[Fe_4W_2]C$ forms in high speed steels (§5.8, table 6). W, Mo and Cr can be substituted in small, Mn in large amounts in Fe_3C (cementite, ch. 4, fig. 9). If a higher concentration of group V, VI elements is present a mixed carbide of the type $Cr_{23}C_6$ forms (table 5). Fe, Mn, V, Nb, Mo and W can replace Cr in this carbide in a wide range of concentrations. In tungsten steels it has a composition $[Fe_{21}W_2]C_6$. Another important group of mixed carbides are the χ-phases. Their structure can be derived from substitutional compounds with β-Mn structure. It forms in Cr–Mo and Cr–W steels. C dissolves in the β-Mn structure and stabilizes

Table 5
Composition and melting temperatures [°C] of carbides that form in equilibrium with transition metals
(after GOLDSCHMIDT [1967]).

TiC	V_2C	$Cr_{23}C_6$	$Mn_{23}C_6$	(Fe_3C)	(Co_3C)	(Ni_3C)
3140	2200	1580	1010	*g	*g	*g
ZrC	Nb_2C	Mo_2C	Tc	Ru	Rh	Pd
3550	3100	2410	g?	g	g	g
HfC	Ta_2C	W_2C	Re	Os	Ir	Pt
3890	3400	2800	g	g	g	g

TC   cubic (B1)
T_2C hcp (L′3)
$T_{23}C$ cubic (D8$_4$)
T_3C orthorhombic (fig. 4.8, DO$_{11}$)
*   carbide not stable in liquid state
g   stable equilibrium with graphite.

References: p. 1136.

it. A pure χ-phase has, for example, a composition $Fe_{28}Cr_{23}W_7$ which is shifted to higher W-contents by C additions.

Metastable carbides that form in ferrite are characterized by greater similarity to α-iron as compared to the more stable carbides. This allows them to form semicoherent or coherent interfaces with the α-Fe. Figure 16a shows the structure which fullfils the requirements for full coherency if the particles are small. An ordered arrangement of interstitials that only distorts the bcc lattice (α') has been found by JACK [1951] in Fe–N alloys. This structure forms during ageing of Fe–N alloys at a low temperature. It is likely that C can be substituted for N (PHILLIPS [1963]) which is not possible in Fe_3C or Fe_4N (table 4). Because substitutional elements are almost immobile at temperatures at which α'-phase forms (fig. 6) in alloy steels a composition $[Fe(s)]_8(C, N)$ can be expected for this metastable phase. The special carbides (Fe-free: TiC, VC, NbC) with a B1-structure (fig. 16b) are coherent with the γ-Fe solid solutions. As a consequence they can form as ultra-fine dispersions, which raise the yield stress of micro-alloyed steel (HSLA-steel, i.e. high-strength low-alloy steel). An excellent survey of interstitial phases in steels has been made by GOLDSCHMIDT [1967]. Besides activation energy of nucleation, the diffusion conditions determine which carbide forms in a steel. In fig. 6, above, the temperature-dependence of diffusion coefficients of substitutional and interstitial elements and H are plotted. We can distinguish three important temperature ranges for steels:

a) $T < -100°C$,   interstitials except H become immobile;
b) $-100°C < T < +350°C$,   interstitials are mobile, substitutionals immobile;
c) $T > 350°C$,   interstitials and substitutionals are mobile.

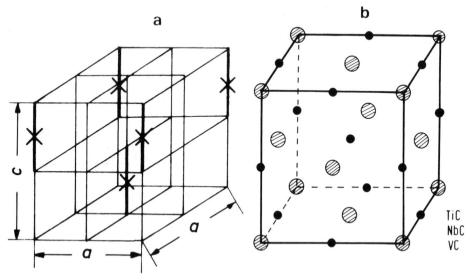

Fig. 16. (a) Structure of the metastable compound $Fe_{16}N_2$ (α'), crosses indicate N- or C-atoms, $a = 5.72$ Å, $c = 6.29$ Å (after JACK [1951]). (b) Structure of the carbides TiC, VC, NbC and carbonitride as Ti(C, N).

3.   Transformation reactions

3.1.   Pearlite

The iron-rich side of the Fe–Fe$_3$C phase diagram (fig. 9) provides the background for a large number of solid state reactions. A summary of some heat treatments and reactions is provided in fig. 17. Several metastable equilibrium lines are added to this schematic diagram as dashed lines. There are diffusion-controlled and diffusion-less reactions. In alloy steels, reactions can be controlled by substitutional or interstitial diffusion. The microstructure of a steel can only be understood if all the individual reactions and their mutual effects are known.

If an alloy of eutectoid composition is cooled below the eutectoid temperature T_p the pearlite reaction can occur. Nucleation of two new types of crystals inside the austenite crystal is not probable because of the high surface- and strain energy terms of nucleation. It is more probable that an autocatalytic reaction starts at grain boundaries and moves into the austenite crystal with a reaction front that leaves an aggregate of lamellar- or rod-shaped particles behind (ch. 14, and fig. 36d). Four parameters are important for the microstructure of pearlitic steel: the velocity of the reaction front G, the lamellar spacing, the rate of nucleation v (fig. 18) and the austenite grain size d (ch. 14). The first two factors are interrelated. Many experiments indicate that the lamellar spacing decreases in proportion to undercooling (ch. 14). If the velocity of the reaction front G is constant at a temperature $T < T_p$, the pearlite reaction starting after an incubation period t_s is completed at a time t_f that depends on the austenite grain diameter d (§ 3.4):

$$t_f - t_s = d/2G.$$

If in the theories (ch. 14) the diffusion coefficient of C in α-Fe and the supersaturation of carbon is used, the rate of formation of pearlite in steels can only be

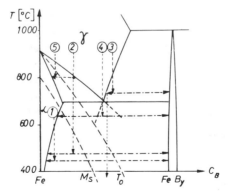

Fig. 17. Some heat treatments in Fe alloy with extended γ-field: (1) quench ageing; (2) marageing or tempering of martensite; (3) ausageing; (4) eutectoid formation; (5) intercritical annealing to obtain dual-phase or duplex structures.

References: p. 1136.

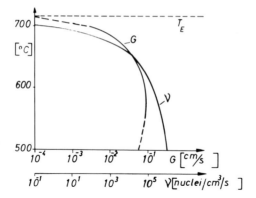

Fig. 18. Rate of nucleation v and velocity of the reaction front G for the pearlite reaction (Fe, 0.78 wt% C, 0.03 Mn; ASTM grain size 5) after MEHL and HAGEL [1956]. These factors and the grain size determine the beginning and end of the pearlite reaction as shown in the TTT-diagrams figs. 19, 23.

calculated if the carbide has the same substitutional concentration as the austenite and if the substitutional elements do not have a considerable effect on the activity of dissolved carbon. There are two reasons for a decrease of the rate of formation of

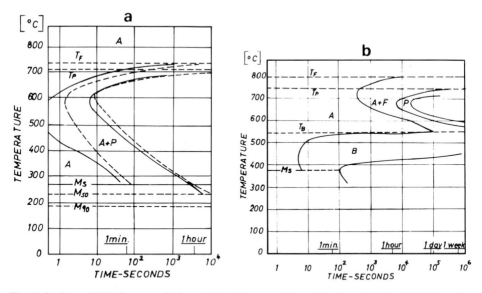

Fig. 19. Isothermal TTT-diagrams which are obtained by quenching and isothermal ageing: (a) High-carbon steel; small additions of B (dashed lines) move the beginning of the transformation to longer times (the lines separating proeutectoid ferrite and pearlite are not shown); Fe, 0.63 wt% C, 0.87 wt% Mn, with 0.0000 wt% B and 0.0018 wt% B, respectively. (b) Alloy steel; formation of ferrite and pearlite is shifted to longer times. There are two maxima of the reaction rates of pearlite (~ 700°C) and bainite (~ 400°C). Fe, 0.3 wt% C, 0.84 wt% Mn, 0.60 wt% Ni, 0.73 wt% Cr, 0.90 wt% Mo, 0.11 wt% V.

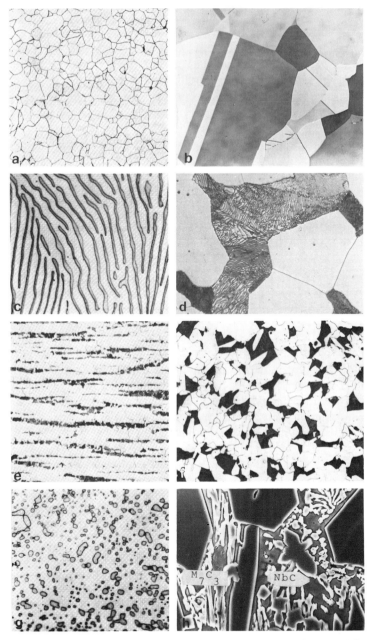

Fig. 20. a) Single-phase microstructures; ferrite. 100×. b) Austenite, 1.2 wt% C, 12 wt% Mn, 1100°C water quenched. 200×. c) Lamellar structure: pearlite, 0,8 wt% C. 2500×. d) Ferrite–pearlite structure, 0.4 wt% C. 500×. e) Anisotropic ferrite–pearlite structure (banding in rolled steel, dark areas: pearlite). 100×. f) ferrite–martensite structure, dual-phase, 0.11 C, 0.15 Mo, 0.24 P wt%, intercritical annealing: 60 min, 725°C. 150×. g) Dispersed Fe$_3$C-particles, 1 wt% C. 1000×. h) Hard facing Fe-base alloy, high volume fraction of NbC and M$_7$C$_3$, M = Cr, Fe, 5.5 C, 30 Cr, 7 Nb wt%. 1000×.

Figs. 20a, c, e, g, courtesy J.R. VILELLA †, U.S. Steel Corp.

Fig. 20b, h, courtesy H. BERNS, Ruhr-University.

References: p. 1136.

pearlite by the addition of third elements (to Fe and C):
a) Elements that segregate to grain boundaries influence nucleation and can slow down the beginning of the lamellar reaction. Small additions of B are most effective in this respect.
b) Substitutional elements that, in equilibrium, are present in different concentrations in the carbide and in the ferrite must diffuse in addition to carbon (fig. 6). Diffusion of the s-atoms then becomes rate-controlling. The rate of formation of pearlite is therefore decreased by all alloying elements, but least by those that can be substituted into Fe_3C (Ni, Co, Mn). Examples for the effect of alloying elements on the rate of the pearlite reaction are shown in fig. 19.

In steels with a composition different from the eutectoid, transformation of the austenite starts with formation of ferrite or cementite. The phases start to form at grain boundaries at small undercooling, and grow with a large variety of morphologies depending on composition and cooling rate. The structure of the interface determines the morphology. Plates with $\{111\}_{\gamma\text{-Fe}}$ habit plane form frequently (AARONSON [1962]). Growth of such proeutectoid ferrite plates seems to be not exclusively diffusion-controlled. It takes place partially by shear as the jerky growth curves indicate. Figure 20 shows some of the typical microstructures which are observed in Fe–C steels of different composition.

A heat treatment of a hypo-eutectoid steel in the $(\alpha + \gamma)$-field (intercritical anneal) will produce a duplex or dual-phase (DP) microstructure. DP-steels have been recently developed for applications for which a high uniform elongation and a high tensile strength is required. The microstructure of such steels consists of two types of solid solutions (ferrite and austenite). The austenite may transform during cooling from the annealing temperatures (see fig. 20).

3.2.   Martensite *

If austenite is cooled fast enough so that formation of pearlite cannot start (critical cooling rate), a degree of undercooling can be reached at which the fcc structure loses its mechanical stability and forms a (distorted) bcc lattice by shear (ch. 14, §8). The slower the pearlite reaction takes place, the lower is the rate of cooling that is needed to obtain undercooled austenite. The temperature at which diffusionless transformation starts is known as M_s. The M_s-temperature is determined by thermodynamic and mechanical factors. It is always lower than the temperature at which α- and γ-solid solution of the same composition are in equilibrium (T_0, $\Delta F_{\alpha\gamma} = 0$, fig. 11). The M_s-temperature can be above or below room temperature, and steels can be classified accordingly. For austenitic steels $M_s <$ room temperature, martensitic steels $M_s >$ room temperature; both must of course be cooled faster than the critical rate to avoid formation of pearlite. Between T_0 and M_s metastable austenite transforms partially into martensite under external stress.

Substitutional solutions of γ-iron transform to a bcc structure. For interstitial

---

* See also ch. 15 for a more general discussion of this type of transformation.

solid solutions an ordering of the interstitial atoms takes place during the transformation. In one shear system only one of the three types of interstitial positions in the body-centred lattice are occupied, so that a tetragonal distortion results (fig. 21a; compare also fig. 16).

In addition to the shear that leads to the change in crystal structure, a lattice-invariant shear of the martensite takes place to reduce the elastic strain in the neighbourhood of a martensite crystal (ch. 14). The amount of shear for the $\gamma \to \alpha$ transformation is about 0.12 (WAYMAN [1964]). The internal plastic deformation takes place by twinning or slip (KELLY and NUTTING [1961]. It leads to martensite crystals that contain dislocations and thin twins. The crystallography and the mechanisms of plastic deformation correspond to those that take place in ferrite of the same composition, temperature and strain rate (§4.1).

The M_s-temperature of steel depends predominantly on three factors:
a) the equilibrium temperature T_0 (fig. 11);
b) the strain- and the surface-energy of nucleation, F_N, and the existence of sites for heterogeneous nucleation, i.e., grain boundaries;
c) the critical resolved stress which has to be overcome to shear the austenite into martensite requiring additional energy (if heterogeneous nuclei are available), F_τ. The energy needed for the start of martensite formation is

$$\Delta F_{\alpha\gamma} \geq F_N + F_\tau;$$

this corresponds to an undercooling

$$\Delta T = T_0 - M_s = \frac{F_N + F_\tau}{\Delta S_{\alpha\gamma}}.$$

(With $\Delta S_{\alpha\gamma}$ the transformation entropy and the differences in specific heat neglected.)

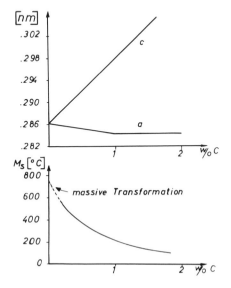

Fig. 21a. c/a ratio and M_s for bct Fe–C martensite.

References: p. 1136.

The value of F_N can be assumed to be about constant because grain boundaries as nucleation site are always present in steels, the volume change of the transformation does not vary much with composition and the dimensions of the specimen are macroscopic. The energy term F_τ is determined by the critical shear stress τ which has to be overcome by the partial dislocations that move during shear while the martensite crystal is formed:

$$F_\tau = \tfrac{1}{2}(V\tau\varphi_{\gamma\alpha}),$$

where V is the molar volume, and $\varphi_{\gamma\alpha}$ the angle of shear for the $\gamma \to \alpha$ transformation). τ can be increased by solid-solution-, precipitation-, work- and radiation-hardening. All obstacles to dislocation motion increase F_τ and therefore decrease M_s. Alloying elements also lower or raise T_0. The effect which these elements have on M_s can only be understood if their influence on F_τ and T_0 is considered. There are many measurements that indicate that not only the alloying elements that lower T_0: C, Ni, Mn, but also those which raise T_0: Al, Si, Mo, W, V, do lower M_s. This is shown with an empirical equation which is used to calculate martensite temperatures of alloy steels (the chemical symbols indicating the values for wt%):

$$M_s = (561 - 474C - 33Mn - 17Ni - 17Cr - 21Mo, W, Si)°C.$$

These are elements which produce strong solid-solution-hardening (§5.1), so that it is likely that the rise of T_0 is overcompensated by the increase of F_τ. The largest effect on M_s in alloys of constant macroscopic composition can be brought about by precipitation of small coherent particles in the austenite. F_τ passes through a maximum as precipitation-hardening occurs while M_s can be lowered several hundred °C, depending on density and structure of the particles (MEYER and HORNBOGEN [1968]).

Figure 21b shows the variation of the M_s temperature of an Fe–Ni–Al alloy which had been preaged to produce ordered coherent particles in the austenite prior to transformation. Superimposed on the effect of varying F_τ is an increase in T_0 due to precipitation of alloying elements which leads to the increase of M_s after long periods of time if particles become large and F_τ becomes small. Figure 21c shows the microstructures corresponding to two different states of heat-treatment of the alloy.

In order to obtain complete transformation to martensite the steel has to be cooled considerably below M_s (fig. 21d). Retained austenite is found in the spaces between the martensite. This austenite becomes work-hardened by the martensite crystals as they form, and thus stabilized against transformation. If cooling is interrupted at a temperature below M_s, then resumed after a period of time, the transformation does not continue until an additional undercooling is reached (*stabilization* of austenite). This will be due to strain-ageing of deformed austenite and the thermally activated reduction of transformation stresses that would have aided further transformation (e.g., GLOVER [1955]).

The size of the martensite crystals is limited by the austenite grain size. In steels, small martensite crystals are wanted because isotropic properties and small internal stresses are wanted. Heat-treatments that lead to grain growth of austenite before cooling below M_s are therefore avoided.

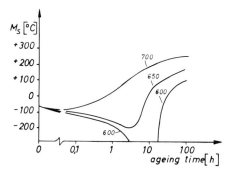

Fig. 21b. Variation of the M_s-temperature of an Fe, 27.4 at% Ni, 12.5 at% Al, 0.06 at% C alloy by ageing the austenite at different temperatures, as indicated (fig. 17, no. 3) to form coherent precipitates.

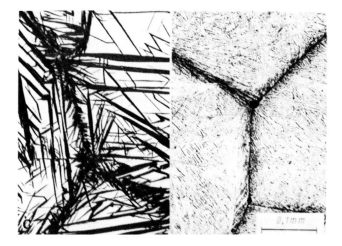

Fig. 21c. Incomplete (40%) and complete transformation of austenite to martensite of an 0.06 at% C, 27.4 at% Ni, 12.5 at% Al steel ($M_s = -69°C$). The transformation starts at austenite grain boundaries. Austenite grain size determines the maximum dimension of martensite crystals. Former austenite grains can be revealed even in the completely transformed alloy. Carbide precipitates are visible at the grain-boundary nodes.

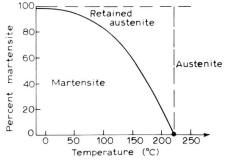

Fig. 21d. Martensite that has formed below M_s in a steel with 1.1 at%C and 2.8 wt% Cr after austenitizing at 1040°C. $M_s = 230°C$, the transformation is almost complete at 20°C. (After COHEN [1949].)

References: p. 1136.

3.3. Bainite

During a bainitic reaction, shear transformation is cooperating with short-range diffusion. The reaction takes place above M_s. Iron and the substituted elements rearrange by shear, while carbon precipitates preferentially at the lattice defects (dislocations, twin boundaries) that have been produced by lattice invariant shear. The microstructure of bainite consists of lense-shaped crystals, like martensite, that contain metastable carbides (usually ϵ-carbide) in a fine dispersion. The process can proceed as shown schematically in fig. 22a, and (figs. 22b and c) so create the microstructure of lower bainite. Because of the necessity of diffusion of carbon the reaction is time dependent. The model implies that diffusion takes place in ferrite. Precipitation in the austenite before shear has occurred is not likely because austenite is stabilized against shear by small particles (fig. 21b), is free of imperfections, and has a lower diffusivity than ferrite. The bainite reaction takes place if a steel has been cooled fast enough to a temperature $500°C > T > M_s$ that the pearlite reaction could not start.

A distinction is needed between upper and lower bainite. Lower bainite forms above M_s with the same habit plane as martensite and ϵ- or a less stable carbide

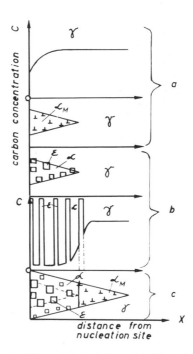

Fig. 22a. Formation and growth of lower bainite (schematic). (a) nucleation at a site for heterogeneous nucleation (grain boundary) and a zone of low carbon concentration (i.e. high M_s-temperature), formation of dislocations by lattice invariant shear; (b) precipitation of ϵ-carbide preferentially at the dislocations, carbon depletion raises the M_s-temperature at the interface with the austenite; (c) new nucleation at the tip as in a), sidewise growth by diffusion through the interface.

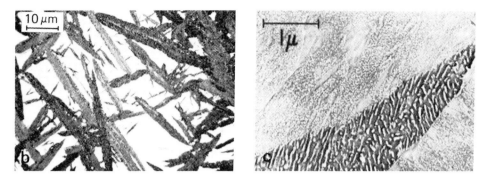

Fig. 22b and c. Microstructure of lower bainite. 3.23 wt% Cr; 30 min at 350°C. Partially transformed 0.66 wt% C (SPEICH [1962]). (b) Optical micrograph. (c) Electron micrograph (replica).

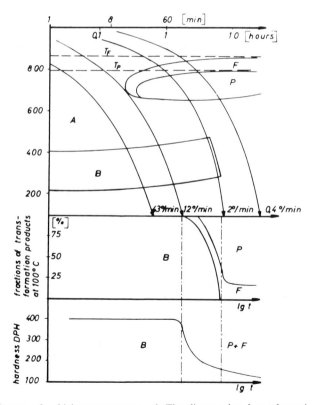

Fig. 23. TTT diagram of a high-temperature steel. The diagram has been determined by continuous cooling. It has to be read along the cooling curves. The additional diagrams indicate microstructure and hardness of this steel cooled from 1010°C at different rates. A – austenite, F – ferrite, P – pearlite, B – bainite, Fe, 0.10 wt% C, 0.40 wt% Mn, 5.39 wt% Cr, 0.51 wt% Mo.

References: p. 1136.

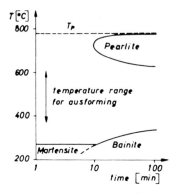

Fig. 24. Isothermal TTT-diagram of an ausform steel (only starting temperatures are shown). Fe, 0.40 wt% C, 5.0 wt% Cr, 1.3 wt% Mo, 1.0 wt% Si, 0.5 wt%V (see fig. 42b).

precipitates with the substitutional solute distribution of the austenite. In the upper bainite range (300–500°C) the mobility of insterstitials becomes large and substitutional elements can move to a limited extend. Consequently Fe₃C or other more stable alloy carbides can form and thermally activated relaxation of the interfaces leads to crystals with no defined habit planes (PICKERING [1968]).

In carbon steels or steels with alloying elements that form no alloy carbides (Mn, Co, Ni) there is no gap between bainite and pearlite. In steels containing alloying elements that form alloy carbides (V, Mo, W) there is a large difference in the nature of the carbide phases that form by the bainite and pearlite reaction (figs. 19, 23, 24). In these alloys the carbide of the bainite is relatively less stable than the carbide of the pearlite containing the alloying element in high concentration. Therefore the temperature up to which bainite can form is lowered and no direct transition exists to the temperature range in which pearlite forms (fig. 19b). The basic characteristics of the three reactions that start from metastable austenite are summarized in table 6.

Table 6
Characteristics of different transformation reactions of Fe–C austenite.

	Pearlite	Bainite	Martensite
Free energy F of the transf. product	F_P	$< F_B$	$< F_M$
av. diffusion paths of carbon S [nm]	$10 < S_P$	$10 > S_B > 0$	$S_M = 0$
velocity of the reaction front G_{max}	G_P	$\approx G_B$	$\ll G_M$
lattice invariant plastic shear	–	+	+
temperature of formation, T	$720°C > T_P > \sim 500°C$	$\sim 500°C > T_B > M_s$	$T_M \leq M_s$
strength of transformation product, σ	σ_P	$< \sigma_B$	$< \sigma_M$

There is a fourth type of reaction in iron alloys with M_s temperatures above 400°C, i.e., with relatively low alloy contents. Even at extremely high cooling rates (500°C/s) transformation occurs above M_s by a mechanism in which the rapid motion of the γ–α interface produces ferrite with the same composition as the austenite. The morphology of the transformation product is different from martensite because thermally activated motion of the γ–α interface occurs by individual atomic hops across the reaction front in a direction opposite to that of its motion. The transformed structure is almost free of the defects which are due to lattice-invariant shear in martensite (GILBERT and OWEN [1962]). This reaction is known as *massive transformation* (see also ch. 14).

3.4. Time–Temperature–Transformation diagrams

The microstructure of steel can be understood if the conditions are known under which *all* the above-mentioned reactions take place. A survey can be given in a diagram which indicates beginning and end of these reactions at different temperatures. In addition information can be provided on the fraction of austenite that has been transformed. This is for example important for $T < M_s$. The martensite reaction is frequently not complete even at large undercooling (residual austenite, fig. 21b). The time–temperature–transformation- (TTT-) diagram is determined for an alloy which has been heat-treated in the γ-field under defined conditions and with a known austenite grain size. TTT-diagrams can be determined in two ways:

a) the specimen is quenched from the austenite to temperatures T at which austenite is metastable. At these temperatures, beginning and end of the reactions is measured, for example in a dilatometer or metallographically. The diagram is read along the time axes (figs. 19, 24).

b) The specimen is cooled continuously at different cooling rates. The temperatures at which reaction takes place are measured and similar diagrams are obtained which have to be read along the cooling curves. The curves of the continuous-cooling diagrams are shifted a little to lower temperatures and longer time as compared to the isothermal diagrams. They are especially useful to guide practical heat treatments because steels are usually cooled continuously during processing (fig. 23). Continuous-cooling diagrams can be supplemented by diagrams which give information on microstructure and mechanical properties of the steel after cooling at a certain rate to room temperature (ROSE and PETER [1953]), fig. 23.

Figure 19b is a isothermal TTT-diagram of a steel of hypo-eutectoid composition. Proeutectoid ferrite forms at temperatures below T_F, pearlite below T_P. Both pearlite and bainite show a maximum reaction rate at about 700°C and 450°C respectively. The M_s-temperature is a straight horizontal line in all diagrams, i.e. independent of the cooling rate. The volume portion of martensite increases with undercooling, that of bainite and pearlite with time. The shapes of the curves in the TTT-diagrams can be explained in the following way. Increasing driving force below the equilibrium with decreasing temperature leads to the maximum reaction rate, which depends on nucleation rate and velocity of the reaction front (fig. 18). Both are diffusion controlled and therefore retarded with decreasing temperature. There are two maxima if a less stable carbide (than Fe_3C) will form in bainite. For this carbide a metastable phase diagram with a lower equilibrium temperature T_B is valid. The reaction can only start below this temperature. A high reaction rate of bainite at relatively low temperatures follows from the smaller diffusion path in bainite as compared to pearlite. A large gap between the pearlite and bainite formation ranges (fig. 24) indicates a large difference in stability between the two carbides. This gap is used in ausform steels (§ 5.7).

The TTT-diagram indicates the effects alloying elements have on the pearlite reaction and on martensite temperatures (figs. 19, 23, 24). The critical cooling rate for martensite formation can be read directly from continuous-cooling diagrams. These diagrams are also essential for a proper understanding of hardenability (§ 5.3).

3.5. Tempering of martensite

Martensite is a supersaturated solid solution with a high density of defects. The equilibrium structure of an Fe–C alloy would consist of graphite in defect-free ferrite. If martensite is heated to temperatures at which atoms and defects become mobile a large number of intermediate microstructure form by the following processes:
a) reduction of elastic internal stresses that have not been reduced by lattice invariant plastic shear;
b) rearrangement of the dislocations to networks and annihilation of dislocations;
c) grain growth;
d) precipitation of metastable or stable phases in the martensite lattice or at the defects;
e) decomposition or transformation of residual austenite.

If Fe–C martensite is heated from room temperature the following processes can be observed (TURKALO [1961]; TEKIN and KELLY [1965]):
α) 20–250°C, segregation of C at the dislocation, precipitation of ϵ-carbide at dislocations and probably of α'-carbide coherently in the lattice. The tetragonality of the martensite decreases and disappears. The structure becomes similar to that of lower bainite. Dislocations are prevented from annealing out by the particles.
β) 200–300°C, decomposition of retained austenite to bainite.
γ) 250–400°C, Fe_3C starts to form at former austenite grain boundaries and by transformation of some ϵ-carbide particles, dislocations anneal out as ϵ-particles dissolve.
δ) 400–720°C, coarsening of Fe_3C particles which approach spherical shape. Grain growth is controlled by dissolution of particles.
ϵ) $T > 720$°C, the $\alpha \to \gamma$ transformation starts. The rate of grain growth increases as soon as all carbide particles are dissolved.

Heating a hypo-eutectic steel to a temperature at which it retransforms completely to austenite and cooling it for ferrite–pearlite formation is known as *normalizing*. It leads to formation of a completely new microstructure and to a loss of textures that have existed before the heat treatment. Corresponding reactions take place in martensites containing substitutional elements. If long-range diffusion of these elements is needed for the formation of precipitates these will form only above 350°C. This is found for alloy carbides and for carbon-free martensite in which intermetallic compounds form (i.e. maraeging steels). Approach to equilibrium is slower, i.e., takes place at higher temperatures if the martensite is heated. This is used if a steel has to be stabilized against loss of strength at elevated temperatures. The elements V, Cr, Mo are added for example to increase temper resistance of steel. Mo has the strongest effect because of its low diffusivity and its tendency to form alloy carbides. Elements that form stable alloy carbides (V, Nb, W) are often added to steels to make them less sensitive to rapid grain growth during heat treatment in the austenite range. Such carbides dissolve only above 1300°C. Alloys containing these particles (high speed steels, §5) can be well homogenized while a small austenite grain size is still preserved.

4. Deformation and recrystallization

4.1. Microstructure of deformed steel

The distribution of dislocations is basically different in ferritic and austenitic steel because of their large difference in stacking fault energy. Plastic deformation of ferrite below 20°C leads to wavy slip lines and to evenly distributed dislocations parallel to $\langle 111 \rangle_{bcc}$ (screw dislocations), at 20°C dislocations rearrange to tangles and cell walls, formation of regular networks starts at 300–400°C (depending on dislocation density, KEH [1961]). The sources for dislocations are grain boundaries or non-coherent phase boundaries. Fe_3C deforms plastically only at high stresses or high temperatures (KEH [1963]). High stresses can be provided by zones of high dislocation density in ferrite in which Fe_3C is embedded. A limited amount of plastic deformation of Fe_3C can take place in ferrite–cementite composites like pearlite. At large amounts of plastic deformation fracture of Fe_3C crystallites and flow of ferrite around the Fe_3C particles takes place. A dispersion of non-coherent carbide particles leads to a large number of dislocation sources (ferrite–Fe_3C-interface) and keeps dislocations from rearranging; the dislocation density increases in such steels much more rapidly with amount of deformation than shown in fig. 25d. Extremely high dislocation densities ($N = 5 \times 10^{13}/cm^2$) can be obtained for example in an alloy of iron with 1 wt% carbon precipitated as Fe_3C and deformed by wire drawing.

Important for the mechanical properties of steels is the distribution of slip at a given amount of plastic deformation. The distribution is determined by the number of active sources and the possibility for cross-slip. In austenitic steel the dislocations are confined to their slip planes because of the low stacking-fault energy (table 7, and fig. 25c). Many dislocations move in the same slip system and straight steps in the surface occur. In α-Fe with low alloying content slip takes place in many slip

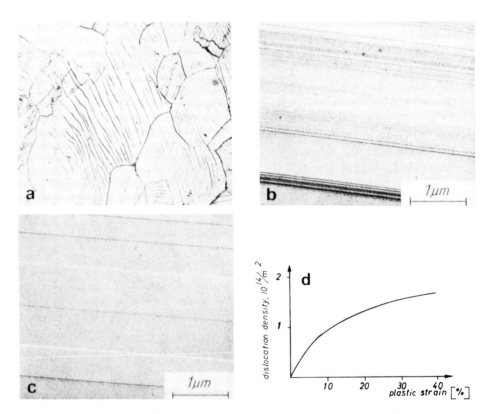

Fig. 25. Different types of slip distribution at small amounts of plastic deformation: (a) dislocations can leave slip planes easily, a large number of slip systems available; 'wavy' slip bands in low-alloy ferrite at 20°C, 3% deformed, light micrograph 200×; Fe, 0.91 wt% Cu, 0.003 wt% C; (b) only {111} slip planes, difficulty to cross-slip because of low SFE (table 7): slip bands of predominantly one slip system; in austenitic steels (Ni, 18.8 at% Cr, 62 at% Al, 2% deformed, replica); (c) coherent particles are present, that are sheared by dislocation; high single-slip steps that are widely spaced, for example austenitic steels hardened by γ′-precipitates at 20°C, or ferrite containing small α′-carbides at −200°C (Ni 18.8 at% Cr, 62 at% Al, particle size 400 Å, 2% deformed, replica); (d) dislocation density N versus plastic strain (at −78 to +200°C) in polycrystalline α-Fe; The dislocation density is relatively independent of dislocation arrangement and deformation temperature (KEH [1961]).

planes, because the dislocations can easily leave a certain slip plane (fig. 25a). Highly localized slip can be produced in ferrite (at low temperatures) and austenite by precipitation of small coherent particles that are sheared by dislocations (fig. 25c). Coarse slip makes steels sensitive to crack-initiation in fatigue, stress corrosion, and to low-temperature brittleness.

The macroscopic distribution of slip depends on the density of sources that can produce dislocations at a stress smaller than that needed to propagate them through a crystal. If there are a large number of suitable grain boundaries and dislocations, slip is distributed evenly over the specimen volume. This is the case in austenitic

Table 7
Stacking-fault energy [a] of some austenitic steels [b].

No.	Composition (wt%)				γ_N[mJ/m^2]	γ_T[mJ/m^2]
	Ni	Cr	Si	Co		
1 [c]	–	–	–	–		75
2	30	–	–	–		40
3	60	–	–	–		70
4	16	15	–	–	23	14
5	23	16	–	–	28	22
6	16	16	1.4	–	11	3.4
7	10	16	–	9.5	8	2.2
8	12	18	–	–	11	
9	15	–	–	–		57

γ_N dislocation node measurements; γ_T twin frequency measurements.

[a] The accuracy of such measurements is still not very high as indicated by the difference between γ_N and γ_T.

[b] After SILCOCK *et al.* [1966] and NUTTING and CHARNOCK [1967].

[c] γ-Fe.

steel because of the relatively small tendency of interstitials to segregate. In carbon steels these sources are usually blocked at room temperature because C-atoms (and N) are mobile and have a high tendency to segregate to the sources. In the extreme case dislocations are then produced only by one source, and a zone of high dislocation density spreads discontinuously into the material producing new dislocations by an autocatalytic process. The deformed zone, known as a *Lüders band*, sweeps through the whole specimen. This inhomogeneous deformation is followed by a homogeneous multiplication process of the Lüders-dislocations (see fig. 30b).

Mechanical twinning can take place in ferritic steels and, under special circumstances in austenite steels (HORNBOGEN [1964b]). Twinning in steel is favoured by high crystal perfection, low temperature, high strain rate, solid-solution hardening (fig. 45, below).

4.2. Recovery and recrystallization

Thermally activated processes like dislocation-climb and grain-boundary motion become rapid in pure iron above 350°C. This temperature is shifted to higher values by dissolved second elements. The rate of formation and growth of recrystallized grains can be changed by several orders of magnitude even by small additions of second elements (VENTURELLO *et al.* [1963]; LESLIE *et al.* [1963]; figs. 26 and 27a). If the atoms are precipitated the effect can be still stronger. LESLIE [1965] has shown that precipitation of Cu at dislocation cells in α-Fe prevents their rearrangement and formation of recrystallization nuclei. If the formation and motion of a recrystallization front is inhibited, "in situ" recrystallization (i.e. continuous growth of subgrains) occurs at a very slow rate. Figure 27 shows that the recrystallization

References: p. 1136.

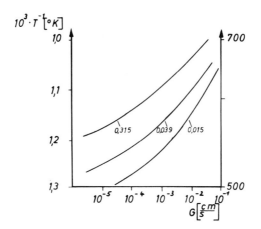

Fig. 26. The effect of small Mo-additions on the velocity of the recrystallization front of iron at 5% recryst., 60% deformation, at% Mo as indicated near curves (after LESLIE et al. [1963]).

behaviour of steels can only be understood if distribution of alloying elements is well known. Two effects of second-phase particles have to be distinguished in steels. Particles that are precipitating or atoms that are segregating while recrystallization takes place slow down this reaction. Particles that had been present before recrystallization, which are in a coarse distribution, and which have depleted the matrix, can accelerate recrystallization. They provide their interfaces as nucleation sites (ch. 25 and figs. 27b and 28). Such particles have a similar effect on nucleation of pearlite (fig. 20b) as pearlite aggregates can initiate recrystallization. Owing to the phase transformation, effects of segregation, and particles there are many ways of influencing annealing textures in steels. If a steel is heated above the temperature of

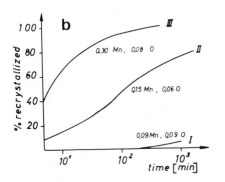
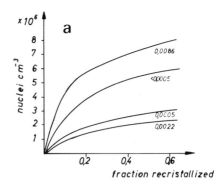

Fig. 27. (a) The effect of C-additions (in wt%), smaller and larger (0.0086 wt%) than the solubility, on formation of recrystallization nuclei between 400 and 500°C after 80% deformation (after VENTURELLO et al. [1963]). (b) Acceleration of recrystallization if alloying elements are precipitated as MnO particles (curves II and III), and not dissolved and segregated (curve I and fig. 26) (after LESLIE et al. [1965]).

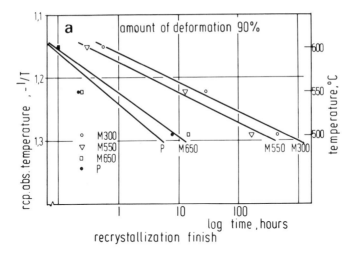

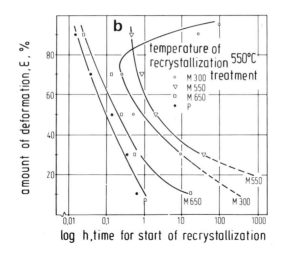

Fig. 28. Recrystallization of a steel with 0.22 wt% C and different distributions of Fe$_3$C-particles: P: Ferrite–Pearlite; M: Martensite tempered at 300, 550, 650°C. (a) $T-t$-diagram for complete recrystallization. (b) T-ϵ-diagram for start of recrystallization.

$\alpha \rightarrow \gamma$ transformation and then cooled, any texture will disappear, because a large number of orientations of the ferrite originate from any one austenite grain. Below the transformation temperature there are the following possibilities for the textures that can form if a cold-worked steel is recrystallized:
a) the rolling texture is preserved if "in situ" recrystallization occurs due to precipitation of particles (LESLIE [1961b];
b) the normal recrystallization texture of α-Fe develops if recrystallization takes

References: p. 1136.

place by motion of a recrystallization front that is not much impeded by particles or segregation (HAESSNER and WEIK [1956]);

c) special textures can develop either during secondary recrystallization or else during primary recrystallization in the presence of a coarse distribution of particles which can act as nucleation sites for a large variety of orientations, including some that grow with particular ease.

Of practical importance are the annealing textures of soft magnetic Fe–Si alloys (ch. 26). The two types of textures (ch. 26) contain one or two $\langle 100 \rangle$ directions in the plane of a sheet. They form during secondary recrystallization. Inclusions of MnS, FeS, Fe_4N etc. play an important, but not quite clarified role in the formation of these textures. These phenomena have been summarized by WASSERMANN and GREWEN [1962]. Because of the high purity which is needed to obtain low watt losses (§6.2) the alloys have to be free of second phases. Therefore impurities may have to be added to control texture formation and then to be removed by a final treatment to obtain the wanted magnetic properties (WASSERMANN and GREWEN [1962]).

The effects of particles are also responsible for the differences of the textures that develop in rimmed and killed steel, which are important for the plastic anisotropy which determines their deep drawing properties (DILLAMORE and FLETCHER [1963]).

Further details of recrystallization of two-phase alloys, including steels, can be found in ch. 25, §3.7.

5. Mechanical properties

5.1. Strength of ferrite

α-iron has a shear modulus in $\langle 111 \rangle$ directions of 61.000 MPa. We can estimate a theoretical shear stress $\tau_{th} \approx 2000$ MPa for single crystals and $\sigma_{th} \approx 6000$ MPa for polycrystals of random orientation. Single crystals of α-iron of very high purity yield at room temperature at a CRSS of $\tau_0 \approx 10$ MPa, polycrystals yield at $\sigma_0 \approx 30$–40 MPa subtracting the grain size effect. The yield stress of steels can assume values in between σ_0 and σ_{th}, depending on composition and microstructure (fig. 29).

The most important components of the microstructure of many steels are the solid solutions of α-iron (ferrite). The mechanical properties of ferrite are determined by the stress necessary to create and move dislocations. The latter depends on the following factors (MCLEAN [1967]):

a) lattice-friction stress of the bcc lattice;

b) friction stress due to dissolved interstitial atoms;

c) friction stress due to substitutional atoms;

d) stress due to work hardening;

e) stress due to interaction of dislocations with grain boundaries.

There is uncertainty about the lattice-friction stress σ_p. A theoretical treatment by FRIEDEL [1964] results in $\sigma_p \approx 400$ MPa at 0 K and of about zero at room temperature. The effect of interstitials is problematic because their distribution is

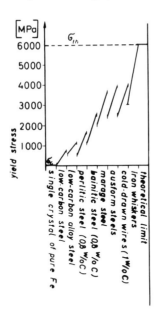

Fig. 29. Strength ranges of different iron base materials.

frequently not known. The solubility of carbon and nitrogen in α-iron at room temperature (fig. 8) is smaller than the C- and N-content of the highest purity iron available. In addition, there is some mobility of these atoms (1 jump per sec at room temperature) down to $-200°C$. Therefore we can expect precipitation of α'-phase in ferritic steels at these temperatures, as well as thermally activated interaction of interstitials with the stress field of moving dislocations (GLEITER [1968]). These processes will cause a strong temperature dependence of the yield stress (fig. 30a). Substitutional elements not only lead to solution hardening themselves (FLEISCHER [1962]) but also change solubility and precipitation kinetics of interstitials (fig. 15). The role of dislocations and grain boundaries (ch. 19) in strengthening of ferrite depends strongly on the state of segregation of interstitials. Attempts have been made to separate the different components of the CRSS of iron–interstitial alloys (MCLEAN and DINGLEY [1967]) and of alloys that contain substitutional elements in addition (LÜTGERING and HORNBOGEN [1968]). The strengthening effect due to dislocations σ_D has been investigated by KEH [1961]. It follows the relationship $\sigma_D = \alpha Gb\sqrt{N}$ (fig. 31) which is rather independent of the local distribution of the dislocations in the ferrite grains. The influence of the segregation condition on the constant k_y of the Hall–Petch relationship (ch. 19) indicates that grain boundaries become stronger obstacles the more atoms are segregated (FISHER [1962]; COTTRELL [1963], and fig. 32). In fig. 30a typical stress–strain curves of Fe–C alloys are shown, measured at different temperatures. The stress–strain curves are changed drastically if the ferrite is heat-treated differently (fig. 33). The heat treatments lead

References: p. 1136.

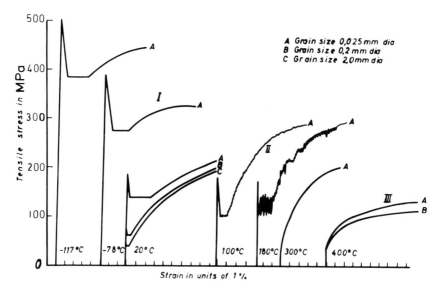

Fig. 30a. Stress–strain curves of a slowly cooled α-Fe alloy at different temperatures (after McLean and Dingley [1967]), Fe, 0.004 at% C, 0.003 at% N, 0.03 at% substit. impurities.

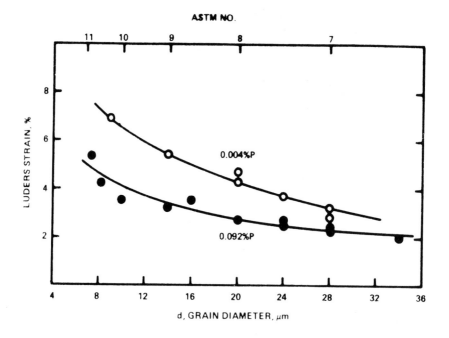

Fig. 30b. Grain-size dependence of the Lüders strain at 20°C, for two P-contents (after Hu [1983]).

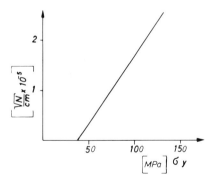

Fig. 31. Yield stress σ_y versus dislocation density N in α-Fe (after KEH [1961]).

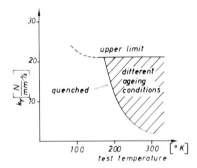

Fig. 32. Variation of k_y (from the Hall–Petch relationship for grain-size dependence of lower yield stress) with the heat treatment of the steel because of segregation at grain-boundaries, α-Fe, 0.001 wt% C+N (after FISHER [1962]).

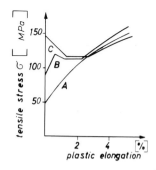

Fig. 33. Variation of the shape of the stress–strain curve measured at room temperature, cooled from 600°C with rates that lead to different segregation at grain boundaries: (A) quenched; (B) 1°C/min; (C) 6°C/h. Fe, 0.006 at% C, 0.012 at% N. (After LÜTGERING and HORNBOGEN [1968].)

References: p. 1136.

to different segregation conditions at grain boundaries and dislocations and to different dispersions of precipitated metastable carbides. The typical stress–strain curves of fig. 30a can be interpreted as follows: Curve I: at the upper yield point the formation of dislocations starts at one grain boundary. These dislocations are able to propagate through other grain boundaries as a Lüders front at the lower yield stress. At the end of the plateau the front has moved through the sample and grains are filled with dislocations ($N \approx 10^{10}/\mathrm{cm}^2$), which cause the Lüders strain (1–8%) (fig. 30b). Plastic deformation continues by multiplication of these dislocations, leading to a positive work-hardening coefficient. Curve III: there is no discontinuous yielding. Dislocations form at many grain boundaries and move into the grains where they interact with solute atoms or particles and, as deformation continues, with other slip dislocations. Curve II: yielding occurs as described for curve I, however the velocity of the dislocations is of the same order of magnitude as the diffusing interstitial atoms. Dislocations are caught up by interstitials, pinned, and become mobile again after considerable increase of stress (dynamic strain ageing).

Type-I-behaviour is typical for segregated (hardened) grain boundaries, type III-behaviour occurs if grain boundaries are soft (in rapidly quenched ferrite or at high temperatures or in austenite), so that nucleation of dislocations in a large number of slip systems and grains occurs simultaneously. Jerky flow (type II) in steels can be expected for certain strain rates in a certain temperature range (fig. 30a). Continuous jerk-free flow reoccurs at very high temperatures when the tendency to segregation and the pinning force become small.

It follows from fig. 33 that only the lower yield stress of steels with the same segregation conditions in the grain boundaries can be compared. This was confirmed by FISHER [1962], who showed in addition that there is an upper limit for the k_y value (fig. 32). This can be explained by the grain-boundary dislocation model of the passage of a crystal-lattice dislocation through a grain boundary, fig. 34 (GLEITER et al. [1968]). It can be expected from this model, that grain-boundary dislocations become immobile at the upper limit of k_y. Lower k_y-values in less segregated grain

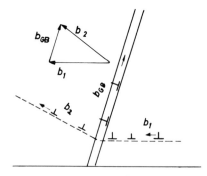

Fig. 34. Passage of a grain boundary by dislocation as it has to take place at the lower yield stress in the plateau of the stress–strain curve. Relaxation of stress by slip in the grain boundary depends on the segregation condition, figs. 10, 32.

boundaries are correlated with higher mobility of these dislocations. In specimens that yield continuously, grain-boundary dislocations can be assumed to be more mobile than crystal-lattice dislocations, and therefore move before grain-boundary sources are producing lattice dislocations.

If the effect of grain boundaries and dislocations on the yield stress is known, the influence of interstitial and substitutional atoms dissolved in ferrite can be investigated. Because of the tetragonality of its stress field there is a very strong solution-hardening effect of carbon in ferrite. The interstitials in octahedral sites of austenite have a relatively small effect. Measurements with small solute concentrations have indicated the following values for C in Fe:

$$\frac{d\sigma}{dc_{bcc}} \approx 3.5 \times 10^5 \text{ MPa}; \qquad \frac{d\sigma}{dc_{fcc}} \approx 7 \times 10^3 \text{ MPa}.$$

For large concentrations $\sigma_i \sim c^{1/2}$ is found as predicted by FLEISCHER [1962] (fig. 35).

If the effect of solid-solution hardening by substitutional elements is investigated, the measured yield stresses have to be corrected for segregation, solution, and precipitation hardening of the interstitials which are present even in the purest iron alloys. The two groups of alloying elements (§2.3) either accelerate or decelerate segregation and precipitation. Steels containing different substitutional elements (for example Mo or Ni) having the same interstitial contents and having been heat treated in the same way can therefore show completely different stages of segregation and precipitation and therefore completely different stress strain curves.

Fig. 35 shows that solution-hardening in iron can be explained by a general theory of solution hardening (FLEISCHER [1962]). The results of fig. 35 have been obtained by correcting for the grain-boundary, dislocation and interstitial interactions. The anomalies known for the solution-hardening behaviour of important alloying elements in steel can be explained if the above-mentioned factors are all considered. The strengthening effect of Ni in ferrite is due to a high work-hardening

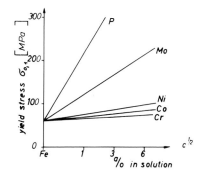

Fig. 35. Solid-solution hardening of α-Fe by different elements. The values have been corrected for the strengthening effects caused by the interstitials.

References: p. 1136.

rate caused by an enhancement of precipitation and segregation of interstitials: thus strengthening is only produced indirectly by the dissolved Ni atoms. A minimum in the yield stress versus Cr concentration curve (HORNE et al. [1963]) is a consequence of the opposite effect of Cr on C. Increasing concentrations of Cr lead to a decreasing rate of segregation at grain boundaries and dislocations, and precipitation of interstitial carbon. Therefore alloys heat-treated in the same way show a decreasing yield stress for small additions of Cr, which only at higher Cr concentrations is overcompensated by substitutional solution hardening.

5.2. Strength of ferrite–pearlite steels and dual-phase steel

The metallographic structure of most structural steels consists of proeutectoid ferrite and pearlite in different volume portions and distributions (figs. 20a, 20e). In coarse aggregates the yield strength of the alloy is linearly dependent on strength and volume portions of ferrite and pearlite (fig. 36a). The strength of pearlite depends primarily on the spacing of its lamellae (fig. 20c) or on the spacing between spherical carbides (MANNING et al. [1968], fig. 20d). Experimental results indicate that yield strength of pearlite increase proportional by the undercooling (fig. 36b). Empirically a proportionality between yield stress of pearlite and the logarithm of the lamellar spacing was determined (GENSAMER et al. [1942]). A quantitative interpretation of the strength of pearlite based on dislocation mechanisms is difficult and has not as yet been achieved.

It is evident from fig. 36b that the strengthening mechanism changes as the bainitic transformation starts. The strength originates from a fine dispersion of carbides and from defects introduced by the shear transformation.

An equiaxed microstructure of ferrite and pearlite aggregates is usually aimed for in structural steels because quasi-isotropic properties are achieved with such a microstructure provided that no texture exists for the ferritic matrix. Anisotropic

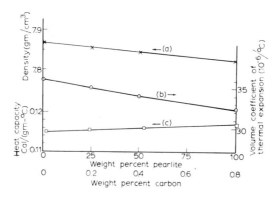

Fig. 36a. Properties versus microstructure; example: pearlite content of annealed steels. (a) Density: read left. (b) Volume coefficient of thermal expansion: read right. (c) Heat capacity: read left. The variables of both coordinates depend on composition.

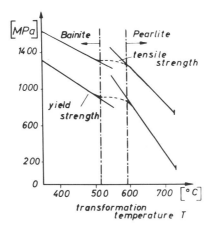

Fig. 36b. Strength of a steel with almost eutectoid composition (Fe, 0.78 wt% C, 0.63 wt% Mn) in which decomposition of austenite took place at different temperatures (after GENSAMER *et al.* [1942]).

microstructures can originate if the pearlite nucleates at the non-coherent interfaces of inclusions that have been spread out during hot rolling (fig. 20e). Mechanical properties parallel and perpendicular to such pearlite bands are different.

Hardenability is an important technological property of steels. It can be defined for example as the depth to which a steel specimen is hardened at maximum possible rate of cooling (brine quench). The measurements are made with a standardized cylindrical specimen, which is heated to form austenite. Then a water jet is directed against one cross-sectional surface until the whole specimen has been cooled to room temperature by thermal conduction. Hardness and microstructure is then investigated along the axis of the cylinder. Hardenability of different steels can be compared or conclusions can be drawn from microstructure and mechanical properties of work-pieces of known cross-sections that have been cooled at certain rates. It may be emphasized that a small austenite grain size is of great importance for good technological properties of a hardened steel (§3.2). Excessive growth of austenite grains is avoided by using the lowest possible temperatures and shortest times for complete austenitization or by adding particles (alloy carbides) that do not dissolve at temperature of austenitization, or boron atoms (§5.5, 5.8) to decrease grain-boundary mobility. The production of ultrafine austenite grain size to obtain high-strength low-carbon steel is discussed in §5.7.

In addition to the established microstructures of structural steels (ferrite–pearlite, tempered martensite, and finally dispersed special carbides in HSLA-steels), steels with Dual-Phase structure have recently provided a new group with challenging mechanical properties. The matrix of DP-steels is formed by ferrite with not too high interstitial content (< 0.1 wt% C), but considerable amounts of substitutional elements (Si, P, Mo, V). By discontinuous or continuous annealing in the $(\alpha + \gamma)$ field (fig. 17) or controlled rolling from the austenite a coarse dispersion of 20–30

References: p. 1136.

vol% martensite in ferrite is formed (fig. 20). For the continuous mass production of steel sheet with such a microstructure an advanced technology is required – the continuous annealing line. Exact values of temperature and duration of intercritical annealing treatments are necessary to obtain reproducible microstructure and mechanical properties. Still more delicate is the process of controlled rolling, for which the parameters are temperature, duration, and the amount of deformation.

The stress–strain curve shows a small yield stress, continuous yielding, a high rate of work-hardening and therefore a combination of a large tensile strength and uniform elongation (fig. 41b). Plastic deformation takes place almost exclusively in the ferrite component of the microstructure. Its composition and the indirect effect of the martensite component of the microstructure determine the properties of the bulk alloy (fig. 20f). There are good perspectives for the application of these steels as automobile sheets.

5.3. Strength of martensite

The carbon atoms which are highly soluble but have little strengthening effect in austenite (§ 5.1) are forced into the bcc lattice by the martensitic transformation. Solution-hardening by carbon alone can explain a large portion of the strength of martensite (fig. 37). The following additional factors contribute to strengthening (COHEN and WINCHELL [1962]; LESLIE [1966]; SPEICH and WARLIMONT [1968]):

a) lattice defects introduced by lattice-invariant shear, i.e., dislocations, twin boundaries and interfaces of martensite crystals;
b) transformation stresses (which have not been reduced by plastic shear);
c) particles precipitated in martensite, segregation zones at dislocations and boundaries;
d) lattice defects inherited from the austenite;

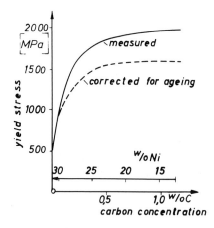

Fig. 37. Strength of martensite as a function of carbon content. The Ni concentration has been varied to obtain comparable alloys with the same M_s-temperature (after COHEN and WINCHELL [1962]).

e) particles that have formed in the austenite and that have either sheared into a
   metastable structure coherent with the martensite or that have lost coherency
   during the transformation.

In most steels the major contribution to the yield strength comes from solid-solution
hardening by carbon (fig. 37). There can, however, be considerable hardening by
lattice defects alone in pure iron transformed by high-pressure shock waves (fig. 4,
LESLIE [1961]) or by small particles precipitated from substitutional solution (Cu,
Ni$_3$Al) that have formed in austenite and are sheared into martensite (MEYER and
HORNBOGEN [1967]).

The grain size of the austenite limits the size of the martensite crystals and
therefore the amount of boundary-hardening. Special hotrolling conditions of the
austenite or in the $\alpha \rightarrow \gamma$ transformation range can lead to extremely small austenite
grain size (grain diameter $< 1 \mu$m) and consequently additional strength of the
martensite, especially in low-carbon steels (GRANGE [1966]). The heat-treatment for
hardening of a steel consists in heating to the austenite temperature until a
homogeneous solid solution is obtained and then cooling at a rate larger than the
critical rate below M$_s$. The cooling rate varies inside a sample. Therefore sometimes
martensitic transformation takes place in the outer zones and formation of pearlite
in the inner portions of a specimen. Alloying elements that slow down the pearlitic
reaction increase *hardenability* of pieces with larger cross-sections. In steel with M$_s$
not much above room temperature, residual austenite will be present which can be
transformed by further cooling to obtain full hardness. High-carbon steels are
frequently not heated into the homogeneous γ-field before hardening but only into
the austenite-plus-carbide field. After cooling, an aggregate of carbide and martensite is obtained. The residual carbide has the function of inhibiting grain-growth of
austenite and increases wear resistance, for example in cutting tools (§ 5.8).

5.4. Strength and ductility of tempered martensite

Above room temperature, dissolved carbon precipitates as carbides and therefore
solution-strengthening is replaced by particle-strengthening. The different precipitate
phases (α', ϵ, Fe$_3$C, alloy carbides) lead to maxima of yield strength. Simultaneously,
healing out of dislocations, twin boundaries and grain boundaries takes place and
causes a continuous decrease of strength. The sum of these effects determines the
mechanical properties of tempered martensite.

For Fe–C alloys precipitation hardening is only effective below $\sim 100°$C. At
higher temperatures softening is dominant, even though at 200–250°C precipitation
of ϵ-carbide leads to some hardening. The structure and the mechanical properties of
bainite and tempered martensite in this stage are almost identical. Quenched
martensite is tempered to increase ductility of the steel. Uniform elongation of a
carbon steel (0.8 wt% C) is zero up to 150°C and it reaches full ductility (30–50%)
by tempering above 350°C. The heat treatment is chosen as a compromise between
strength which is lost and ductility which is gained (LEMENT *et al.* [1954]). The
tempering behaviour of alloy steels which contain elements that form alloy carbides

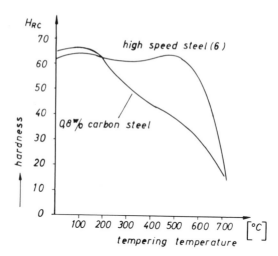

Fig. 38. Strength of a carbon steel and a high-alloy steel (table 2, no. 6) during tempering of martensite. Secondary hardening takes place at 100°C and 500°C respectively.

(V, Mo, W) is quite different. Strength is preserved to much higher temperatures because carbides can only precipitate if the substitutional elements become mobile (> 350°C). Below this temperature martensite defects are pinned by segregation, so that strength decreases only moderately. Maximum precipitation-hardening is found in such steels at about 500°C (fig. 38). Because such steels have the property of preserving strength at elevated temperatures (retention of hardness) they are used as tool steels or high-temperature steels (HOBSON and TYAS [1968]).

5.5 Precipitation-hardened ferritic steel

Precipitation-hardening of a steel with a ferritic matrix occurs if coherent or partially coherent particles form during slow cooling or ageing of a supersaturated solid solution. There are three basic types of such precipitates that form in a fine dispersion in ferrite and act as obstacles to the motion of dislocations:

a) The purely interstitial α'- or ϵ-phase (table 4, and fig. 16) forms coherently between $-100°C$ and $+250°C$ in the bcc lattice; the interstitial atoms become ordered. Precipitation-hardening is due to the interaction of slip dislocations with the stress field outside and with order inside the particle. This type of precipitation (quench-ageing) is not utilized as a strenthening mechanism; it has however consequences on low-temperature ductility of steel (§5.11; DAVENPORT and BAIN [1930]; LESLIE and KEH [1965]).

b) Considerable strengthening can be achieved by precipitation of substitutional elements from iron, i.e. elements that do not form carbides and have a solubility decreasing with temperature—for example Cu or Fe_3Al. Cu precipitates from ferrite as a metastable bcc phase which transforms to the stable fcc structure above a particle diameter of ~ 10 nm. Precipitation takes place above 400°C. The range of the ageing temperature for precipitation-hardening by substitutional phases is 450–600°C (HORNBOGEN *et al.* [1966]). Maximum precipitation-hardening occurs in copper steels if the bcc copper-rich particles transform and lose coherency (fig. 39).

c) Most effective strengthening can be expected if both processes of precipitation mentioned above are combined, i.e. if particles that contain substitutional elements, with interstitials bonded strongly to those elements, are precipitated. Because of the strong directionality of the bonds with the strong carbide formers (§2.3) only small particles or clusters can stay coherent with the bcc matrix (< 3 nm). This combination of elements is achieved by adding small amounts of Nb, Ti, or V to carbon steel. The increase in strength of steels of the same composition and heat-treatment is indicated in table 8. The interstitial compounds that may be responsible for this effect are α' in the case of coherency, and $(Nb, Fe)_6(C,N)_2$ or $Nb(C, N)$, with NaCl-structure, for non-coherency (fig. 16); $Nb(C, N)$ has been identified at grain boundaries. The same phase appears finely dispersed in the ferrite matrix and is responsible for the major portion of the hardening. The grain-boundary precipitate has an indirect hardening effect by

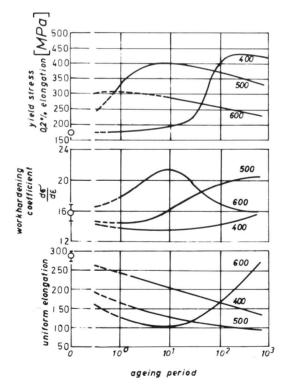

Fig. 39. Precipitation-hardening of an Fe–Cu alloy, quenched from 840°C and aged at different temperatures [°C]. Fe, 0.91 wt% Cu, 0.003 wt% C.

References: p. 1136.

Table 8

The effect of small Nb-additions on strength of a hot-rolled and air-cooled low-carbon steel [a].

	Composition (wt%)				Yield point (MPa)	Tensile strength (MPa)	Elongation (%)
	Nb	C	N	Mn			
1	< 0.001	0.16	0.004	0.94	270	450	32
2	0.02	0.17	0.004	0.96	380	500	27
3	0.04	0.17	0.004	0.94	410	540	24

[a] After LESLIE [1963].

inhibiting grain growth and therefore leading to a higher grain-boundary strengthening as compared to the Nb-free steel (HSLA-steels).

5.6. Marageing steel

The heat-treatment termed *marageing* is identical with tempering of martensite (fig. 17). Marageing steels are given a composition which provides for precipitation-hardening from a relatively soft martensite, i.e., one with a low interstitial concentration. The ageing treatments are done at temperatures at which substitutional elements can move to a limited extent ($300 < T_A < 700°C$, fig. 17). These materials belong to the ductile ultrahigh-strength steels (i.e., $\sigma_y > 1500$ MPa). A typical marageing steel contains iron (with low carbon content), 10–25% Ni to lower the martensite temperature, and the substitutional elements which produce the precipitates (DECKER et al. [1962]; DECKER [1963]). Cooling below M_s leads to a martensite strengthened mainly by dislocations produced by lattice-invariant shear (fig. 40a). Precipitate particles that form during ageing must be finely dispersed and effective obstacles for dislocations. Particles with ordered bcc structure such as $(Fe, Ni)_3Al$ or $(Fe, Ni)Al$ (ch. 4) fullfil this condition because they can form coherently in the matrix as a stable phase. Al alone does not harden sufficiently; it can be replaced by transition elements, for example Ti, V, Nb, Mo. These elements increase the lattice parameter of the metastable structure of the particles and therefore lead to strain-field hardening (fig. 40b) (SPEICH and FLOREEN [1963]). Intermetallic compounds which permit no coherency with the matrix form directly at dislocations or grain boundaries. Most important are the fcc (Cu, Ni_3Al) and the hcp (Ni_3Ti) structure (MEYER and HORNBOGEN [1967]).

In addition to their contribution to precipitation-hardening, the particles pin the dislocations of the martensite and prevent them from healing out. Recovery and recrystallization in marageing steels is synchronized with the coarsening of the particles. The following processes have to be considered for overageing:

a) transformation of metastable coherent precipitates to more stable non-coherent ones;

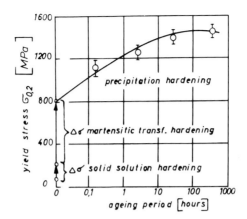

Fig. 40a. Components of yield stress of an experimental marage alloy. Fe, 27 at% Ni, 12 at% Al, 0.10 at% C.

b) dissolution of small particles in favour of bigger ones that have formed at dislocations;

c) coarsening of particles at dislocations, which are released so that recovery can take place.

Overageing will be accelerated if transformation to the austenite takes place during ageing. This reaction determines the upper limit of the ageing temperature and therewith the amount of elements that are added to lower the M_s temperature. The most remarkable property of marageing steels is the combination of high yield stress with high fracture toughness in a rather wide temperature range (BRUCH [1978], and

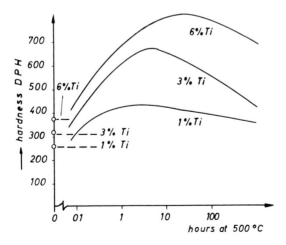

Fig. 40b. The effect of Ti on hardness of aged Fe, 20 wt% Ni martensite, 1.1, 3.0, 5.8 wt% Ti, 0.01–0.2 wt% C (after SPEICH and FLOREEN [1963]).

References: p. 1136.

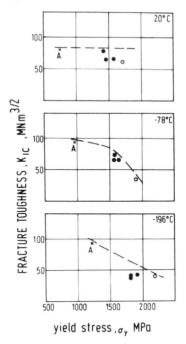

Fig. 40c. Fracture toughness as a function of yield stress of a 0.1 C, 18 Ni, 8 Co, 5 Mo (wt%) marageing steel (after BRUCH and HORNBOGEN [1978]).

fig. 40c). Fracture toughness is not very sensitive to the ageing treatment, so that the state of maximum yield stress is associated with considerable toughness (fig. 40c).

Not all marageing steels are hardened by the formation of substitutional phases. If alloy carbides are used for strengthening, ageing must take place at the same temperature as for substitutional phases. The sequence martensitic transformation → precipitation in ferrite can be reversed to: precipitation in austenitic → martensitic transformation (fig. 17). The effect of "ausageing" on M_s was represented in fig. 21b. The strength of martensite produced from aged austenite increases about in proportion to the decrease of M_s. This process is very effective in strengthening martensite crystals. The decrease of M_s will, however, lead to a decreasing volume fraction of martensite, so that cooling below room temperature is necessary to obtain high strength (MEYER and HORNBOGEN [1968]).

5.7. Thermo-mechanical treatments

Marageing is one approach by which ultrahigh strength is obtained by different hardening mechanisms (fig. 40a), i.e., the introduction of different types of obstacles to dislocation motion into the iron lattice. Mechanical treatments that lead to increased defect density can be applied in addition   and in different sequences together with the thermal treatments (KULA and RADCLIFFE [1963]). The oldest thermo-mechanical treatment is the production of high-strength piano wire from carbon steel. This material is tempered to produce Fe_3C particles and then work-hardened by wire drawing. The dispersed hard particles reduce the rearrangement of dislocations. Therefore a higher work-hardening rate than in a homogeneous iron alloy is reached.

Hot-rolling of low carbon steel to produce very small austenite grain size before

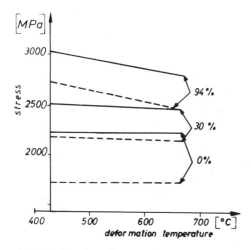

Fig. 41a. The yield strength (dashed) and tensile strength of the ausform steel of fig. 24, after different amounts of deformation at different temperatures in the pearlite–bainite gap (after SCHMATZ *et al.* [1963]).

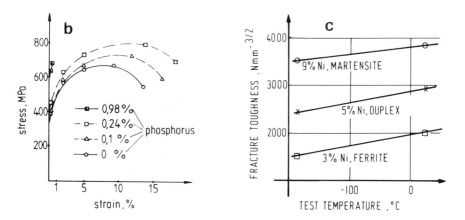

Fig. 41b, c. Mechanical properties resulting from microstructures obtained by thermomechanical treatments such as controlled rolling. (b) Stress–strain curves of dual-phase steels, ferrite +25 vol% martensite (0,11 C, 0.15 Mo wt% 4) with different P-contents. (c) Fracture toughness of a 5 wt% Ni-steel with duplex structure, 50 vol% ferrite and martensite, compared with that of the individual phases.

References: p. 1136.

transformation is another thermo-mechanical treatment. The additional grain-boundaries lead to an increase in strength because of their effect on the microstructure after transformation (GRANGE [1966]).

A process which is used to produce very high strength is known as *ausforming*. Ausform steels have a TTT-diagram with a wide gap between pearlite and bainite formation temperatures (fig. 24). The steel is cooled – avoiding pearlite formation – to a temperature in the gap, rolled, and cooled below M_s. A typical composition and the corresponding mechanical properties of such a material are shown in fig. 24 and 41a (SCHMATZ *et al.* [1963]). There are the following reasons for the additional strength obtained by ausforming:

a) dislocations produced in austenite are inherited by the martensite;

b) particles or segregation zones are sheared from the austenite into the martensite;

c) the size of the martensite crystals is decreased.

The strength that can be obtained cannot be explained by an increased dislocation density alone, so that contributions of precipitation-hardening (ausageing) must play a role. *Ausform steels are the strongest ductile iron-base materials that can be obtained at present besides some metallic glasses* (figs. 29, 41). If temperature, time, and amount of deformation are well defined this process is known as *controlled rolling*. A lowering of the temperature of the steel during plastic deformation makes recovery and recrystallization interfere with carbide precipitation and $\gamma \rightarrow \alpha$ phase transformation. Combinations of these reactions take place, by which favourable microstructures will form (HORNBOGEN [1979]). Controlled rolling is applied, for example, to a HSLA-steel for which a small grain size and an ultra-fine dispersion of NbC-, TiC-, or VC-particles is desired. A new application is the production of dual-phase steels

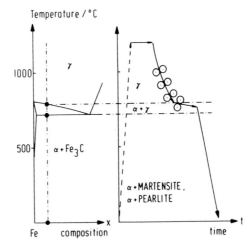

Fig. 42. Semi-schematic representation of controlled rolling as an example for thermomechanical processing of steel.

(fig. 42). The steel is rolled while it is cooled into the $(\alpha + \gamma)$-field, and then cooled fast enough for martensitic transformation of the γ-component of this microstructure.

5.8. High-speed steels

These steels are required to operate as cutting tools at high speed. They have to maintain high strength (and wear resistance) at high temperatures (600°C, red hardness). A typical composition of a high-speed steel is given in table 2. The most important alloying elements are 1 wt% C, 18 wt% W or the equivalent amount of Mo. The composition indicates that stable alloy carbides (types M_6C, MC) are required to form. The heat treatment given to these steels assures that these carbides stay partly undissolved while the matrix can undergo martensitic transformation:

The steel is heated into the austenite field (1250–1300°C) until all of the $M_{23}C_6$-carbide has dissolved. This carbide contains the Cr and Mn which is needed in solution to reduce the critical cooling rate for martensitic transformation. The structure obtained after cooling consists of martensite, retained austenite (10–20%), and undissolved alloy carbides (5–10%). The tempering treatment has the purpose to relieve internal stresses produced by the transformation, transform residual austenite and precipitation-harden the martensite.

The strength of a tempered high-speed steel remains almost unchanged up to 450°C, then increases and falls off above 600°C (fig. 38). The hardness-maximum at ~ 500°C is known as secondary hardening. The decrease in hardness above 600°C is due to coarsening of the carbide precipitated in the martensite. Some strength is preserved at higher temperatures from the undissolved alloy carbides. These carbides are responsible for the wear resistance (HOBSON and TYAS [1968]).

Other ferritic steels for elevated-temperature use have a similar microstructure. Hardened carbon steels can be used only up to 200°C. For moderate requirements Cr is added to form the $(Fe, Cr)_{23}C_6$-carbide at higher temperatures and strength is then preserved up to 350°C. In steels used up to 550°C the elements V, Mo, W are added. Ferritic steels are not suitable for higher temperatures: either precipitation-hardening austenitic steels or Ni-base alloys are used for such requirements.

5.9. Austenitic steel

The following properties of fcc iron can be correlated with mechanical properties that are basically different from that of α-iron: low diffusion coefficient and stacking-fault energy, high solubility of interstitials, metastability.

The low stacking-fault energy of γ-iron and many of its alloys (table 7) leads to planar dislocation arrangement and therefore to the higher work-hardening rate of stainless steel as compared to ferritic steels at room temperature and consequently a large uniform elongation in the tensile test. Slip lines produced by small amounts of deformation are shown in fig. 25b. This makes them sensitive for crack-initiation under fatigue and stress corrosion conditions. If the austenite is deformed at temperatures $T_0 > T > M_s$, deformation-induced martensite increases the work-hardening rate still more. This effect is used if high wear resistance is needed. An austenitic steel with such a deformation behaviour is known as *Hadfield steel*, which contains about 18 wt% Mn and 1 wt% C (the Fe–Mn phase diagram is similar to Fe–Ni, fig. 12).

Because of the lower interaction energy $\bar{u}$ (§2.1), the tendency for interstitials to segregate, and therefore for discontinuous yielding and strain-ageing, is much less pronounced in austenitic steels.

The diffusion coefficient of interstitial and substitutional elements is 10–1000 times smaller than in ferrite (at the same temperature). Therefore, particle growth leading to overageing takes place at higher temperatures. Austenitic steels can be used above the limit for ferritic high-temperature steels. Such steels are usually precipitation-hardened by a coherently precipitated ordered phase, γ', i.e. (Fe, Ni)$_3$(Al, Ti) and can be used up to 750°C.

Precipitation of interstitial compounds at grain boundaries is unwanted in some austenitic stainless steels. In such cases alloy carbide formers (§2.3) Nb, Ta are added in amounts equivalent to the carbon contents, which then is removed from the solution and concentrated in large carbide particles. For still higher service temperatures iron is partially or completely replaced by Ni or Co. These superalloys can be used up to 1050°C.

5.10. Low-temperature steels

Structural steels are expected to break in a ductile manner, i.e., associated with a large expenditure of strain energy (fig. 43a). In brittle fracture plastic strain is negligable and the crack propagates either along grain boundaries (figs. 43b, 44, 45b), twin boundaries (fig. 45a), or {100}-planes of the bcc lattice (fig. 45a). Plastic separation along slip planes can, for example, play a role in fatigue-crack propagation. This is shown for an austenitic steel and {111} planes in fig. 43c. Low-temperature brittleness is a general property of the bcc lattice which can be modified by chemical composition and microstructure. Therefore all ferritic steels become brittle at low temperatures (§5.11). Low-temperature steels must fulfill the condition that the transition to brittleness (ch. 23) does not occur above the temperature of use, for example not above −200°C for steels in contact with liquid nitrogen, while normal carbon steel becomes brittle at −50°C. An addition of Ni to low-carbon steels has the effect of preserving ductility at such low temperatures (fig. 43a). The Ni concentration is limited by the extend of the α-solid solutions. The higher the Ni concentration, the lower is the temperature to which the alloy can be heated without $\alpha \rightarrow \gamma$ transformation (fig. 12). Steels with 5–9 wt% Ni are used for low-temperature applications (−200°C). For use at higher temperatures the Ni content can be reduced.

Interstitials are mobile down to −50°C and in supersaturation in all ferritic steels (figs. 6, 8). At a constant heat-treatment (i.e. rate of cooling, or ageing treatment) additions of Ni increase the activity of carbon and therefore the tendency for

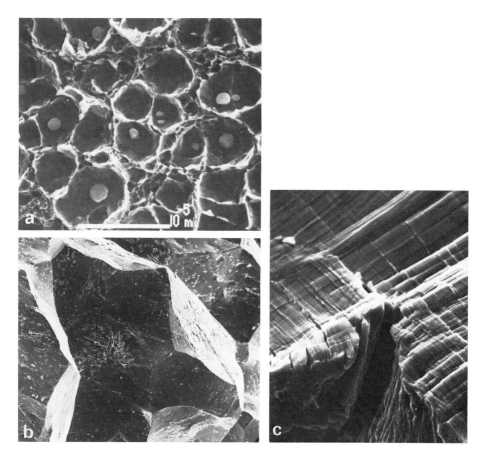

Fig. 43. Scanning electron microscopy of fracture mechanism in steel: (a) tough dimple fracture initiated by pore formation at inclusions; Ni, Cr, Mo-steel, 0.25 wt% C, 2500×; (b) brittle intercrystalline fracture at former austenite grain boundaries; Fe 0.6 wt% C, quenched from 1280°C, 150×; (c) fatigue-crack tip in austenitic steel propagating along {111} slip bands; 40 at% Ni, 6 at% Al, 5000×.

Fig. 43a: courtesy J. ALBRECHT, BBC, Baden; fig. 43b: H. BERNS, Ruhr-University; fig. 43c: K.H. ZUM GAHR, Siegen University.

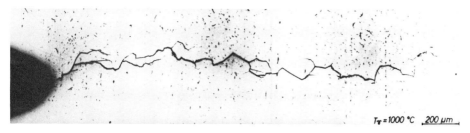

Fig. 44. Brittle inter- and transcrystalline fracture in notched specimen of hardened tool steel, 0.96 C, 2.0 Mn, 0.3 Cr wt%, quenched from 1000°C.

References: p. 1136.

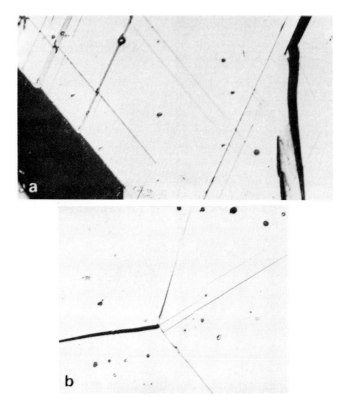

Fig. 45. Low temperature brittleness of ferrite: (a) crack nuclei at twin intersections, cleavage along the {112} twin interfaces (left) and transcrystalline cracks along {100}; (b) cracking along a grain boundary. Fe, 1.78 wt% P, 0.003 wt% C, 0.003 wt% N, deformed at −178°C, light micrographs 200×.

segregation and precipitation of interstitials. At −200°C the interstitials (C, N) are practically immobile in their ordered positions in the precipitates. If such clusters or particles are small they are sheared by dislocations and strong pile-ups are formed because the shear resistance decreases in the slip plane in which the dislocations have moved by shearing-off particles. If, at higher temperatures, sheared particles can reform before the next dislocation has passed, this effect disappears. Above a critical size the particles are by-passed by cross-slip. Under such conditions an increased resistance to shear after passage of a dislocation through a slip plane can be expected at all temperatures. Localized or homogeneous slip distribution follows from these different behaviours (fig. 25b, c, §4.1). An increasing Ni content in steel has the effect of producing particles larger than the critical size at all cooling rates. In Ni-free steel this condition can only be obtained by ageing treatments at low temperatures. Elements with negative Δu (§2.3) have the effect of slowing down precipitation.

The steel is brittle at low temperature in the presence of a planar dislocation distribution (localized slip), which produces high stress concentrations by pile-ups (ch. 19) and large surface steps. Transition to ductility occurs if dislocations can easily cross-slip, particles are by-passed.

Low-temperature brittleness is a property of the bcc structure of ferrite. In addition to modifications of this phase by suitable solute atoms, microstructure is a factor which affects the transition to brittle behaviour. As an example the fracture toughness of a Ni-steel with a duplex structure is shown in fig. 41c. The microstructure consists of an equal number of grains with high and low nickel content. Even if the low-nickel phase alone would behave in a brittle manner, a propagating crack is repeatedly stopped by the high-nickel component so that the macroscopic properties indicate a ductile behaviour even at low temperatures. A similar effect is obtained by duplex or dual-phase microstructures with ferrite and untransformed austenite as microstructural components.

Austenitic steels, like all fcc alloys, do not show low-temperature brittleness of this type because of their much smaller tendency for interstitials to segregate and precipitate. Their applicability at low temperatures is limited by the M_s temperature (fig. 14). Suitable compositions or heat-treatments (fig. 21b) can be found to allow their use at the lowest temperatures.

In summary, brittleness of steel can be due to different causes:

a) low-temperature brittleness by transcrystalline cleavage, caused by a change in glide behaviour;

b) *temper-embrittlement*, intercrystalline, caused by segregation of certain elements at grain boundaries; temper-embrittlement can be aided by coherent precipitates which cause localized slip (ch. 13, §6.3);

c) pseudo-intercrystalline brittleness due to highly localized but ductile separation along particle-free zones at former austenite boundaries (for example relaxation-embrittlement in weldments);

d) embrittlement due to transformation-induced microcracks in high-carbon tool steels;

e) intercrystalline brittleness caused by the formation of liquid films or of graphite during high-temperature heat treatment.

f) embrittlement from hydrogen which entered into the steel from the environment (for example H_2O from the atmosphere, during pickling).

6. *Other physical properties*

Steels are normally used because of their mechanical properties. There are, however, iron base alloys for which other physical properties are of primary interest. Steels with high neutron-absorption cross-section, low thermal-expansion coefficient and soft magnetic steel may serve as examples.

References: p. 1136.

6.1. Absorber steels

· Steels which are used in the construction of atomic reactors, for example as canning materials, must fulfill the special requirement not to contain elements with isotopes that have a high absorption cross-section for thermal and epi-thermal neutrons. B, Co, Ta, Nb, Ti, Al, N are such elements that are normally contained in steels.

The opposite is true for steels that are used as absorber materials or for shielding. Among the alloying elements in question, B has the best ability to absorb neutrons over a wide energy range (fig. 46). ($B = 0.8^{11}B + 0.2^{10}B$; $B : \sigma_a = 750$ barn; $^{10}B : \sigma_\alpha = 3470$ barn). The macroscopic neutron-absorption coefficient depends on the B-concentration in the steel. Because B is hardly soluble in α-Fe it is either present segregated or as particles ($Fe_2 B$). Above 2 wt% B ferritic steels become so brittle that they cannot be worked (grain-boundary brittleness). Therefore austenitic steels are often used that contain up to 5 wt% B as a dispersion of $(Fe, Cr)_2 B$. The absorption coefficient of a steel is structure-sensitive, because self-shielding leads to a decrease of the macroscopic absorption coefficient above a certain particle size. Dispersions of B_4C are sometimes used because the lattice of this compound is able to absorb a portion of the He atoms that form during the (n, α) reaction of ^{10}B, so that the tendency to form gas bubbles is reduced.

6.2. Transformer steel

An alloy for use as transformer steel should fulfill the following conditions:
a) it should consist of atoms and a crystal structure with the maximum possible magnetization;
b) it should have a microstructure that allows easy motion of magnetic domain walls;
c) the crystals should be oriented with their direction of easy magnetization (ch. 26) in the direction of the desired magnetization.

The condition a) is fulfilled best by iron *, b) implies that the material should be free of particles, dislocations and grain boundaries. This is best fulfilled in a well-annealed pure single crystal. For iron such a microstructure is difficult to obtain. The $\gamma \rightarrow \alpha$ transformation produces grain boundaries and dislocations which are difficult to remove by annealing below 900°C. If an alloying element is added which closes the γ-loop (fig. 13) this difficulty can be avoided. Alloys with more than 2.2 wt% Si can be heat-treated up to the melting temperature. Commercial transformer steel contains 3–5 wt% Si. A loss in maximum magnetization has, however, to be tolerated if Si is added. A beneficial effect follows from the increased electrical resistivity due to solute atoms, which leads to a decrease of eddy-currents that are induced by the changes in magnetization. The condition c) is fulfilled for an iron crystal with a $\langle 100 \rangle$ direction in the direction of magnetization or for a sheet with a

---

* Except Fe–Co alloys.

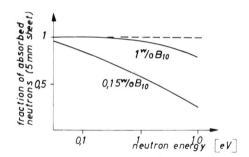

Fig. 46. Neutron absorption of steels with different ^{10}B contents. B is present as BFe_2 particles (table 3). (After STEVENS [1958].)

Goss (i.e. one $\langle 100 \rangle$ in the plane of the sheet) or a cube texture (i.e. two $\langle 100 \rangle$ in plane of sheet, §4.2).

New materials which soon may compete with transformer steel are iron-base metallic glasses. The homogeneity of the glass structure allows easy motion of Bloch walls and provides very low energy losses. (See also ch. 26, §§4.2 and 5.)

6.3. Invar alloy

A consequence of the particular concentration dependence of the Curie temperature T_c of Fe–Ni alloys (fig. 12) is an anomalous behaviour of the coefficient of thermal expansion α_T. There is a minimum of $\alpha(0°C)$ in the range of the fcc solid solutions (fig. 47) at 36 wt% Ni. The minimum shifts to higher concentrations at higher temperatures and is always just below T_c (36 wt% Ni, $T_c = 260°C$). If the alloy is cooled, the decrease in volume is partially compensated by magnetostriction of the austenite. The small value of $\alpha(0°C)$ shown in fig. 47 can be reduced to about zero by rapid cooling and plastic deformation of the alloy, which produces complete

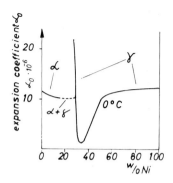

Fig. 47. Thermal expansion coefficient α_0 (at 0°C) for the Fe–Ni alloys. The low-Ni austenitic alloys are metastable. (Compare with fig. 12; after CHEVENARD [1921].)

References: p. 1136.

disorder. The extremely small expansion coefficient of these alloys is utilized in steels for measuring instruments and in watch springs; for this purpose, the fact that the expansion coefficient is zero only over a small temperature range is no obstacle.

7. Solidification

7.1. Rimming steel, killed steel

The microstructure of a steel is not only determined by its composition and the solid state reactions (§2–4) but also by the mode of its solidification. The two extremes of complete mixing and no mixing (ch. 9) are approximately verified for the conditions under which rimming steel and killed steel solidify.

The different behaviour is a consequence of the discontinuous decrease in solubility of O during the transition from liquid to solid iron (fig. 48). The oxygen can either react with carbon to form a *gaseous* reaction product, $O + C \rightarrow CO$, with O produced by liquid → solid reaction, C dissolved in liquid iron and CO not soluble in liquid iron, or with elements (with larger reaction enthalpy with O than with carbon) such as Si, Al, Mg, Zr, Mn, to form a *solid* reaction product, $3O + 2Al \rightarrow Al_2O_3$, with Al dissolved in liquid steel and Al_2O_3 dispersed as particles in liquid steel.

Above a sufficient supersaturation of oxygen, CO-bubbles will form and grow at an increasing rate in the liquid steel. They will cause convection of the melt ahead of the freezing metal (ch. 9). By deoxidation, the O-concentration is lowered and solid particles are produced that stay in the melt as a dispersion, or move to the surface. Transport of atoms during solidification takes place predominantly by liquid diffu-

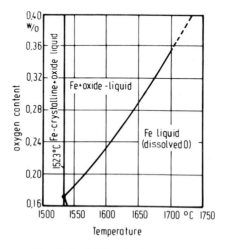

Fig. 48. Partial Fe−O phase diagram.

sion (ch. 9). If there is convection and the k_0 values of the alloying elements are much smaller than unity (ch. 9) there will be strong segregation of these elements to the interior of the ingot. The ingot forms a rim of rather pure iron (hence the name, *rimming steel*), and alloying elements, especially C, N, S and P, are concentrated in the centre in a melt with lowered melting temperature. There is an optimum amount of Al added that is just sufficient to avoid bubble formation (*killed steel*). If more is added, surplus Al will dissolve in the ferrite and change its properties. If less Al is added, moderate bubble formation and consequently convection will occur (semi-killed steel).

Most low carbon steels (< 0.25 wt% C) are rimming steels. The high-purity zone in the surface is preserved during rolling. All steels which have to have an even distribution of alloying elements, i.e. most alloyed steels, are killed steels. They contain some of the deoxidation products dispersed as small particles which sometimes have an favourable effect on low-temperature brittleness (§5.11, and fig. 48).

A different method of removing gases dissolved in liquid steel is vacuum-casting. The equilibrium concentrations of H and N are proportional to $\sqrt{p}$, where p is the partial pressure of the diatomic gas. O is removed as CO. The technological limit of low pressures in steel production, given by the vacuum systems, is 10^{-1} Torr. This is, however, sufficient to reduce the hydrogen content so far that vacuum-cast steels never show flaws and hydrogen embrittlement (§5.1).

7.2. Cast iron

Cast irons are alloys of iron with more than 2 wt%, usually 2.6–3.6 wt%, C. The eutectic and the eutectoid reaction (fig. 9) can follow the stable phase diagram

$$\text{liquid} \rightarrow \gamma + \text{graphite}; \quad \gamma \rightarrow \alpha + \text{graphite (fig. 49a)}$$

or the metastable phase diagram

$$\text{liquid} \rightarrow \gamma + \text{Fe}_3\text{C}: \quad \gamma \rightarrow \alpha + \text{Fe}_2\text{C (fig. 49c)}.$$

Third alloying elements can be subdivided in those who favour carbide formation (as compared to the Fe–C alloy), the elements with less occupied 3d shells than iron (Mn, V, Cr, Ti or corresponding transition elements of the other periods, for example Mo, W), and those who favour graphite formation (i.e., elements with more complete 3d shells of the long period, Co, Ni, Cu, and the elements with complete shells, of which Al, Si, P are of practical importance). The different behaviour can again be related to an increase or decrease of activity of carbon by these alloying elements, i.e., to the interaction energy between carbon, iron and the alloying elements. Additions of Si to commercial cast iron leads to formation of graphite during the eutectic reaction. Such cast iron is known as 'grey' because of the colour of its fracture surface. It usually contains 1–3 wt% Si. The eutectoid reaction takes place according to the metastable conditions so that the microstructure consists of pearlite and graphite (fig. 49b). The microstructure of cast iron depends, however, much on composition and cooling rate so that many intermediate stages between

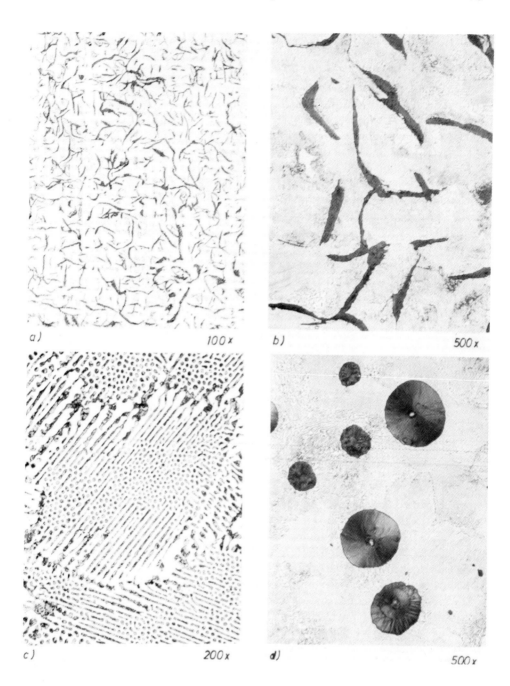

Fig. 49. Microstructures of grey, white, and nodular cast iron: (a) graphite and ferrite; (b) graphite and pearlite; (c) cementite and pearlite; (d) nodular graphite and pearlite. (Courtesy J. MOTZ.)

stable and metastable equilibrium can be obtained. Addition of Mn or an increased cooling rate shifts the microstructure towards the metastable conditions: 'white' cast iron (fig. 49c).

There is a wide variety of morphologies of graphite. Most frequently it is lamellar. A special case is nodular graphite. Nodules form for example if cast iron is desoxidized with Ce or Mg. It is probable that the desoxidation products CeO, MgO act as heterogeneous nuclei (MOTZ [1959]). The nodules are built from segments, so that always the low-surface-energy (0001) planes of graphite form the interface with iron. Nodules can form in liquid and solid (γ)-iron. This particular morphology of graphite makes cast iron malleable (fig. 49d). The mechanism by which additives modify the morphology of cast iron is treated in detail in ch. 9, § 11.1.5.2.

Cast iron with lamellar graphite has a very low tensile strength, because of the low tensile strength of graphite. It is widely used, however, because of its strength under compression and its ability to damp mechanical vibrations.

Graphite-forming iron-base materials approach closely full thermodynamic equilibrium. The opposite is true for alloys in which crystallization can be totally avoided during rapid cooling. Binary iron–carbon alloys do not belong to the compositions with a good glass-forming ability. Figure 50 shows a semi-schematic TTT-diagram for a Fe–C alloy of about eutectic composition and for cooling from the liquid phase. Only at very high cooling rates (band thickness $< 0.2 \, \mu$m) an Fe–C-glass is formed. At lower cooling rates the metastable ϵ-solid solution is formed, which at still slower cooling is replaced by the metastable (γ + Fe$_3$C) eutectic. There are two ways to improve glass-forming ability of iron alloys (HORNBOGEN and SCHMIDT [1982], and fig. 50):

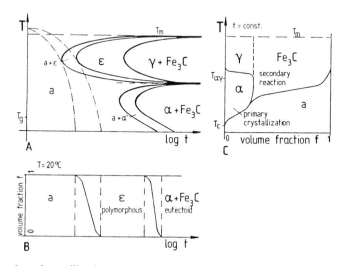

Fig. 50. Formation of metallic glasses (a) and metastable crystal structures from Fe–C-alloys: (A) temperature–time crystallization diagram; (B) structures formed after cooling; (C) structures which form during reheating of a glass.

References: p. 1136.

a) to increase the stability (i.e., lower the free energy) of the amorphous alloy;
b) to slow down the rates of the various reactions by which crystallization proceeds. All alloying elements with positive interaction energy with carbon stabilize the melt. Still more effective is, however, an addition of boron. There exists no crystal structure of iron with any appreciable solubility for this element. Together with crystallization, decomposition either by primary crystallization or by an eutectic reaction has to take place. This leads to low crystallization rates. The critical cooling rates for glass formation are therefore lowered by additions of large amounts of boron (~ 20 at%) to steels or cast irons. Alloys in the compositional range of steels cannot be obtained as glasses. Metastable crystalline phases and microstructures as obtained by rapid cooling, increasingly play a role as starting materials (powders) for the production of high-duty alloys such as high-speed steels or superalloys.

References

AARONSON, H.I., 1962, in: Decomposition of Austenite by Diffusional Processes (Interscience, New York) p. 387.

BIRCHENALL, C.E., 1951, in: Atom Movements (ASM, Metals Park, OH) p. 112.

BIRCHENALL, C.E. and R.J. BORG, 1960, Trans. AIME **218**, 980.

BRUCH, U., and E. HORNBOGEN, 1978, Arch. Eisenhüttenw. **49**, 357 and 409.

BUNDY, F.P., 1965, J. Appl. Phys. **36**, 619.

CHEVENARD, P. 1921, Acad. Sci. (Paris) **172**, 594.

COHEN, M., 1949, Trans ASM **41**, 35.

COHEN, M., and P.G. WINCHELL, 1962, Trans. ASM **55**, 347.

COTTRELL, A.H., 1963, in: The Relation between Structure and Mechanical Properties of Metals (HMSO, London) p. 456.

DARKEN, L.S., and R.W. GURRY, 1953, Physical Chemistry of Metals (McGraw–Hill, New york) p. 397.

DAVENPORT, E.S., and E.C. BAIN, 1930, Trans. AIME **90**, 117.

DECKER, R.F., 1963, in: The Relation between Structure and Mechanical Properties of Metals (HMSO, London) p. 647.

DECKER, R.F., J.T. EASH and A.J. GOLDMAN, 1962, Trans. Quart. ASM **55**, 58.

DILLAMORE, I.L., and S.F.H. FLETCHER, 1966, in: Recrystallisation, Grain Growth and Textures (ASM, Metals Park, OH) p. 448.

ENGELL, H.J., W. FRANK and A. SEEGER, 1967, Z. Metallk. **58**, 452.

FISHER, R.M., 1962, Ph. D. thesis, Cambridge.

FISHER, R.M., E.J. DULIS and K.G. CAROLL, 1953, Trans. AIME **197**, 690.

FLEISCHER, R.L., 1962, Acta Metall. **10**, 835.

FRIEDEL, J., 1964, Dislocations (Pergamon Press, London) p. 67 and 121.

GENSAMER, M., E.B. PEARSALL, W.S. PELLINI and J.R. LOW, 1942, Trans. ASM **30**, 983.

GILBERT, A., and W.S. OWEN, 1962, Acta Metall. **10**, 45.

GLEITER, H., E. HORNBOGEN and G. BÄRO, 1968, Acta Metall., **16**, 1053.

GLOVER, S.G., 1956, in: The Mechanism of Phase Transformation in Metals (The Institute of Metals, London).

GOLDSCHMIDT, H.J., 1967, Interstitial Alloys (Butterworths, London).

GRANGE, R.A., 1966, Trans. ASM **59**, 26.

HAESSNER, and H. WEIK, 1956. Arch. Eisenhüttenw., **27**, 153.

HOBSON, G., and P.S. TYAS, 1968, Metals and Materials **2**, 144.

HORNBOGEN, E., 1960, Trans. AIME **218**, 634.

HORNBOGEN, E., 1964a, Trans. ASM **57**, 120.

HORNBOGEN, E., 1964b, in: High Energy Rate Working of Metals (Central Institute for Industrial Research, Oslo) p. 345.

HORNBOGEN, E., 1979, Metallurg. Trans. **10A**, 947.

HORNBOGEN, E., and I. SCHMIDT, 1982, in: Rapidly Solidified Amorphous and Crystalline Alloys, eds. B.H. Kear, B.C. Giessen and M. Cohen (North-Holland, Amsterdam) p. 199.

HORNBOGEN, E., G. LÜTGERING and M. ROTH, 1966, Arch. Eisenhüttengew. **37**, 523.

HORNE, G.T., R.B. ROY and H. PAXTON, 1963. J. Iron Steel Inst. **201**, 161.

HU, HSUN, 1983, Metallurg. Trans. **14A**, 85.

JACK, K.H., 1951, Proc. Roy. Soc. **A208**, 200.

JOHANNSON, H., 1937, Arch. Eisenhüttenw. **11**, 241.

JOHNSON, J.W., 1960, J. Less-Common Met. **2**, 241.

KEH, A.S., 1961, in: Imperfections in Crystals (Interscience, New York) p. 213.

KEH, A.S., 1962, Acta Metall. **11**, 1101.

KEH, A.S., and W.C. PORR, 1960, Trans. ASM **53**, 81.

KELLY, P.M., and J.M. CHILTON, 1968, Acta Metall. **16**, 637.

KELLY, P.M., and J. NUTTING, 1961. J. Iron Steel Inst. **197**, 199.

KULA, E.B., and S.V. RADCLIFFE, 1963, J. Metals **12**, 755.

LEMENT, B.C., B.C. AVERBACH and M. COHEN, 1954, Trans. ASM **46**, 851.

LESLIE, W.C., 1961a, Acta Metall. **9**, 1004.

LESLIE, W.C.., 1961b, Trans. AIME **221**, 752.

LESLIE, W.C., 1963, in: The Relation Between Structure and Mechanical Properties in Metals (HMSO, London) p. 333.

LESLIE, W.C., 1966, in: Strengthening Mechanisms in Metals and Ceramics (Syracuse Univ. Press) p. 43.

LESLIE, W.C. and A.S. KEH, 1965, AIME Met. Soc. Conf., 337.

LESLIE, W.C., E. HORNBOGEN and G.E. DIETER, 1962, J. Iron Steel Inst. **200**, 622.

LESLIE, W.C., J.T. MICHALAK and F.W. AUL, 1963, in: Iron and its Dilute Solid Solutions (Interscience, New York) p. 119.

LESLIE, W.C., J.T. MICHALAK, A.S. KEH and R.J. SOBER, 1965, Trans. ASM **58**, 672.

LÜTGERING, G., and E. HORNBOGEN, 1968, Z. Metallk. **59**, 29.

MCLEAN, D., 1957, Grain Boundaries (Oxford Univ. Press. London) p. 118.

MCLEAN, D., and D.J. DINGLEY, 1967, Acta Metall. **15**, 885.

MANNING, R.D., H.M. REICHOLD and J.M. HODGE, 1968, in: Transformation and Hardenability in Steels (Climax Molybdenum Co.) p. 169.

MEHL, R.F., and W.C. HAGEL, 1956, Prog. Met. Phys. **6**, 74.

MEYER, W., and E. HORNBOGEN, 1967, Z. Metallk. 58, 297, 372 and 445.

MEYER, W., and E. HORNBOGEN, 1968, Arch. Eisenhüttenw. **39**, 73.

MINSHALL, F.S., 1961, AIME Met. Soc. Conf. **9**, 249.

MOTZ, J., 1959, Giesserei **46**, 953.

NUTTING, J.W., and W. CHARNOCK, 1967, Met. Sci. J. **1**, 77 and 123.

PHILLIPS, V.A., 1963, Trans. ASM **56**, 600.

PICKERING, F.B., 1968, in: Transformation and Hardenability in Steels (Climax Molybdenum Co.) p. 109.

ROSE, A., and W. PETER, 1952, Stahl und Eisen **72**, 1063.

SCHMATZ, D.J., F.W. SCHALLER and V.F. ZACKEY, 1963, The Relation Between Structure and Mechanical Properties of Metals (HMSO, London) p. 647.

SILCOCK, J.M., R.W. ROOKES and J. BARFORD, 1966, J. Iron Steel Inst. **204**, 623.

SPEICH, G.R., 1962, in: Decomposition of Austenite by Diffusional Processes (Interscience, New York) p. 353.

SPEICH, G.R., and S. FLOREEN, 1963, Trans. ASM **57**, 714.

SPEICH, G.R., and H. WARLIMONT, 1968, J. Iron Steel Inst. **204**, 385.

STEVENS, H.E., 1958, Nucl. Sci. Eng. **4**, 373.

TEKIN, E., and P.M. KELLY, 1965, Precipitation from Iron Base Alloys (Gordon and Breach, New York) p. 173.

TURKALO, A.M., 1961, Trans. ASM **54**, 344.

VENTURELLO, G., C. ANTONIONE and F. BONACCORSO, 1963, TRANS. AIME **227**, 433.

WASSERMANN, G., and J. GREWEN, 1962, Texturen metallischer Werkstoffe (Springer, Berlin) p. 712.

WAYMAN, C.M., 1964, Introduction to the Crystallography of Martensitic Transformations (McMillan, London) p. 138.

WERT, C., 1950, Trans. AIME **180**, 1242.

ZENER, C., 1955, Trans. AIME **203**, 619.

Further reading

Bain, E.C. and H.W. Paxton, Alloying Elements in Steel (Amer. Soc. Metals, 1961).

Cohen, M., The Martensite Transformation, in: Phase Transformations in Solids (Wiley, New York, 1951) p. 588.

Duckworth, W.E., The Heat-treatment of Low-Alloys Steels, Metallurgical Reviews **13** (1968) 145.

Houdremont, E., Handbuch der Sonderstahlkunde (Springer, Berlin, 1956).

Hume-Rothery, W., The Structures of Alloys of Iron (Pergamon Press, London, 1966).

Jaffee, R.J., and B.A. Wilcox, eds., Fundamental aspects of structural alloy design (Plenum, New York, 1977).

Leslie, W.C., The Physical Metallurgy of Steel (McGraw–Hill, New York, 1981).

Kot, R.A., and B.L. Bamfitt, eds., Fundamentals of Dual-Phase Steel (Met. Soc. AIME, New York, 1981).

Kot, R.A., and J.W. Morris, eds., Structure and Properties of Dual-Phase Steels (Met. Soc. AIME, New York, 1979).

Mackenzie, I.M., Recent Developments in Structural Steels, Metallurgical Reviews **13** (1968) 189.

Microalloying 75 (Union Carbide Corp., 1977).

Speich, G.R., and J.B. Clark, eds., Precipitation from Iron-Base Alloys (Gordon and Breach, New York, 1965).

Zackay, V.F., and H.I. Aaronson, eds., Decomposition of Austenite by Diffusional Processes (Interscience, New York, 1962).

Transformation and Hardenability of Steels (Climax Molybdenum Co., 1968).

Strengthening Mechanisms in Steels (Climax Molybdenum Co., 1970).

Physical Properties of Martensite and Bainite, British Iron and Steel Inst. Special Report **93** (1965).

CHAPTER 17

POINT DEFECTS

H.J. WOLLENBERGER

Hahn–Meitner-Institut für
Kernforschung Berlin Gmbh
1000 Berlin 39, FRG

R.W. Cahn and P. Haasen, eds.
Physical Metallurgy; third, revised and enlarged edition
© *Elsevier Science Publishers BV, 1983*

1. Introduction

Point defects are lattice defects of zero dimensionality, i.e., they do not possess lattice structure in any dimension. Typical point defects are *impurity atoms* in a pure metal, *vacancies* and *self-interstitials*. This chapter covers the properties of vacancies and self-interstitials, their interaction with other lattice defects, the production mechanisms and their importance for radiation damage of materials. Properties of atomic solutes per se are of interest mainly with respect to thermodynamics of alloys which are treated in ch. 4.

Vacancies are produced simply by heating, in concentrations sufficiently high for quantitative investigations. Interstitials must be produced by doing external work on the crystal. Such work is done on an atomic scale by energetic particle irradiation. Collisions between the projectiles and lattice atoms cause displacements of atoms from substitutional sites to interstitial sites. Thus vacancies and interstitials are produced in equal numbers. As one vacancy and one interstitial form a Frenkel defect, irradiation is essentially a Frenkel defect production process. This is disadvantageous with respect to experimental research on interstitial properties. Property changes of crystals caused by irradiation also involve vacancies. Vacancy formation is treated in §2.2.1 and Frenkel defect production in §3.1.

Plastic deformation also produces vacancies and interstitials. Although its application is much less expensive than particle irradiation, it has not yet become a common procedure for point defect production. There are many reasons. Point defect production during deformation is not well understood and does not allow a controlled defect production. The significance of point defects in plastic deformation is treated in ch. 20, and their recovery behaviour after plastic deformation in ch. 23.

Anomalously high point defect concentrations occur in some non-stoichiometric intermetallic compounds. There, vacancies and interstitials clearly play the role of additional alloying elements and are of thermodynamic significance in this sense. This topic is treated in ch. 4.

Other methods such as rapid quenching, evaporation on cold substrates or laser annealing depend on the thermally activated production, which is treated in the present chapter for controlled conditions which allow successful vacancy investigation. The above three kinds of treatment are even less developed than plastic deformation and do not give new information about point defects.

In pure metals and in the majority of alloys vacancies provide thermally activated atom transport and, hence, vacancy properties directly influence atomic transport (ch. 8). Vacancy properties also provide information on interatomic forces by means of specific perturbations due to the vacant lattice site.

The interstitial is a very interesting defect because of the large lattice perturbation it causes. A number of important properties result from this effect. It also plays a fundamental role as one constituent of the Frenkel defect generated by radiation. The strong lattice perturbation seems to play a key role in the evolution of typical damage structures. It might also be of significant importance for the stability of alloy phases under particle irradiation.

The properties of vacancies and interstitials are described in §§2 and 3 on a quantitative basis wherever possible. Sometimes the reader will notice outstanding deviations of data from different authors. This problem goes back to the late sixties, when conferences of point-defect researchers became exciting events also for non-specialists in this field. A controversy had arisen on the interpretation of the generally observed radiation-damage recovery. The question was whether a satisfactory explanation requires the simultaneous existence of two interstitial species annihilating at quite different temperatures or needs only one interstitial species. The reader is referred to the report on the panel discussion held at the Jülich Conference on Vacancies and Interstitials in Metals (see further reading: Vacancies and Interstitials in Metals, 1970) in order to get the flavour of the atmosphere of that period. Many papers in the literature give a lively impression. In the meantime a large number of experimental results was reported and even completely new methods were applied. They certainly narrowed the scope of the controversy considerably. It would be impossible to outline within this article the controversy in terms of *all* experimental results which were claimed at the time to be of relevance. Instead of this, the scientific content and today's state of the controversy is briefly outlined in a dedicated section (§2.3).

The application of alloy materials in nuclear reactors, particularly in the core region with its high neutron flux density, causes radiation damage to be a technological problem. This is, in particular, true for the fast breeder reactor and the future fusion reactor technology. Important property changes such as swelling, radiation-induced creep and radiation-induced atom redistribution are briefly outlined in §4 of the present chapter.

2. Vacancy properties

2.1. Theoretical background

The entropy of a defect-containing crystal is larger than that of a perfect crystal. Therefore the Gibbs free energy change resulting from changing the atomic-defect concentration c by δc is given by:

$$\delta G = \left(\Delta H^f - T\Delta S^f + k_B T \ln c \right) \delta c, \tag{1}$$

where ΔH^f is the activation enthalpy of formation, $k_B \ln c$ is the ideal entropy of mixing, and ΔS^f is that entropy change which arises in addition to $k_B \ln c$ (excess entropy of mixing). (k_B is the Boltzmann constant.) In the close-packed metals, formation of vacancies requires the smallest amount of energy when compared with other lattice defects. Hence, they are dominant in thermal equilibrium.

The entropy change ΔS^f is mainly due to the change of the phonon spectrum of the crystal by the introduced defects. For the high-temperature harmonic approximation one obtains:

$$\Delta S_v^f = k_B \sum_i \ln(\omega_{0i}/\omega_i), \tag{2}$$

where ω_{0i} and ω_i are the eigenfrequencies of the crystal without and with vacancies, respectively. The phonon spectrum is changed by the change of the atomic coupling for the nearest neighbours of the vacancy and by the overall lattice volume change. The entropy change ΔS^f is positive if on average the ω_i are lowered and negative if they are enhanced.

The simplest model assumes nearest-neighbour interaction by a "spiral spring" with force constant f and calculates the change of the Einstein frequencies of the nearest neighbours of the vacancy by removing one of its twelve coupling springs (in the fcc lattice). For the vibration towards the vacancy, $\omega_{0E}^2 = 4f/M$ (M = atomic mass) is changed to $\omega_E^2 = 3f/M$. Vibrations perpendicular to the removed spring are unchanged. For all other atoms all eigenfrequencies are unchanged. Hence we obtain $\Delta S_v^f = \frac{1}{2}(12k_B \ln 4/3) = 1.73 \, k_B$.

Refinement of the calculation must take into account the static atomic relaxations around the vacancy. These change the force constants for many more atoms than the nearest neighbours. Correct consideration of this effect seems to be difficult. A negative sign of the relaxation-contribution to the formation-entropy change was obtained by SCHOTTKY et al. [1964] and a positive one by BURTON [1971]. In total, ΔS_v^f was obtained by SCHOTTKY et al. [1964] to be $0.49k_B$ for Cu, $0.92k_B$ for Ag and $1.22k_B$ for Au, whereas BURTON [1971] obtained $1.8k_B$–$2.0k_B$ for fcc and $2.2k_B$–$2.4k_B$ for bcc metals.

The sensitivity of the result to the number of atomic shells considered as contributing to the frequency redistribution has been investigated systematically for different pair-interaction potentials by HATCHER et al. [1979]. In the model crystal, about 5×10^3 atoms around the vacancy were allowed to relax statically and for 10^2 to 5×10^2 atoms in the central region dynamical displacements (vibrations) were allowed. The pair interaction was simulated by the *Born–Mayer potential* introduced by GIBSON et al. [1960] for computer simulation of Cu (see also §3.3.1) and by the *Morse potential* introduced by COTTERILL and DOYAMA [1966]. This last potential was smoothly cut off at $r = 1.2a$ by HATCHER et al. [1979]. It then fits the observed lattice constant, bulk modulus and enthalpy of vacancy formation for Cu. The entropy of formation was found to depend approximately linearly on the inverse of the number of atoms allowed to vibrate. The relationship was followed up to the 19th atomic shell around the vacancy. By extrapolation, $\Delta S_{1v}^f = 2.3k_B$ was obtained for the Morse potential and $1.6k_B$ for the Born–Mayer potential. The influence of the static lattice relaxation on the magnitude of ΔS_{1v}^f is demonstrated by comparing the above with $2.28k_B$ for the Morse potential and $2.52k_B$ for the Born–Mayer potential; both values are for the unrelaxed atoms (perfect lattice around the vacancy). HATCHER et al. [1979] have also found $\Delta S_{1v}^f = 2.1k_B$ for a potential simulating α-Fe. The cut-off Morse potential describes the formation and migration properties of mono- and di-vacancies in Cu reasonably well. The results are unsatisfactory for the Born–Mayer and the cited long-range Morse potential (COT-TERILL and DOYAMA [1966]). For the hcp Mg, $\Delta S_{1v}^f = 1.5k_B$–$2k_B$ was found by using an empirical potential (MONTI and SAVINO [1981]).

Discrepancies between the results from different potentials are particularly large for the *relaxation volume* ΔV_{1v}^{rel}, i.e. the volume change of a crystal by removing one atom from the interior and withdrawing it from the crystal. The quantity is measured by the lattice-parameter change. The Born–Mayer potential for Cu yields -0.47 atomic volume, whereas the Morse potential yields -0.02 atomic volume for both the extended and the cut-off version (DEDERICHS *et al.* [1978]). With $\Delta H_{1v}^f = -0.41$ eV, the Born–Mayer potential fails to give a reasonable enthalpy of formation. With the corresponding values $+1.17$ eV and $+1.29$ eV for the extended and the cut-off Morse potential, respectively, satisfactory agreement with experimental values is obtained.

Ab-initio calculations of vacancy formation enthalpies are obviously problematic. Reviews are by FRIEDEL [1970], EVANS [1977], HEALD [1977], and STOTT [1978]. The problem consists of calculating the energetics of the electron system for a vacancy-containing crystal. Only so-called simple metals could be treated. It had to be assumed that the core electrons are rigidly confined to the nucleus and any charge redistribution caused by the missing atom core only affects the conduction electrons. For transition metals the vacancy effect on the outermost d electrons has not been described successfully as yet. For the simple metals two entirely different approaches have been applied.

The first consists in replacing the vacancy by a repulsive impurity potential acting in a free electron gas (jellium). The second approach is based on pseudo-potential theory and assumes the vacancy effect to be weak enough to follow a linear response formalism. The results of both approaches are unsatisfactory in terms of agreement with experimental data. Reasons are the improper treatment of the effect of the local charge redistribution as well as of that of the lattice relaxation around a vacancy on the electron system. More promising seems to be the combination of self-consistent cluster calculations (quantum-chemical approach) with lattice defect calculations, such as have been applied successfully to calculations of the energetics of rare-gas interstitials in metals (MELIUS *et al.* [1978]).

2.2. Experimental methods and results

2.2.1. Enthalpy and entropy of formation

2.2.1.1. The role of di-vacancies. For thermal equilibrium one obtains from eq. (1):

$$c_v^0 = \exp\left(\Delta S_v^f / k_B\right) \cdot \exp\left(-\Delta H_v^f / k_B T\right). \tag{3}$$

The determination of enthalpy and entropy of formation according to this equation requires the determination of equilibrium vacancy concentrations c_v^0. Two different procedures are applied: either measurement of c_v at high temperatures such that c_v is indeed in thermal equilibrium during the measurement, or quenching of the sample from high temperatures to low temperatures such that the defects are immobile. The latter method avoids difficult measurements at high temperatures but requires considerable efforts to avoid vacancy losses during quenching. Review articles are by HOCH [1970], BALLUFFI et al. [1970] and SIEGEL [1978].

Even the most careful measurement of c_v^0 at high temperatures cannot yield the wanted formation enthalpy ΔH_{1v}^f and entropy ΔS_{1v}^f of the monovacancy because c_v^0 does not necessarily involve only monovacancies. Actually it is given by $c_v^0 = c_{1v}^0 + 2c_{2v}^0 + \cdots = \sum_{n=1}^{N} nc_{nv}^0$, where c_{nv}^0 is the equilibrium concentration of clusters consisting of n vacancies. For equilibrium concentration measurements, higher aggregates than di-vacancies need not be considered owing to their negligibly small concentration in the normal cubic metals. We then have in the dilute solution approximation:

$$c_v^0 = c_{1v}^0 + 2c_{2v}^0 = \exp\left(\Delta S_{1v}^f/k_B\right)\exp\left(-\Delta H_{1v}^f/k_B T\right)$$
$$+ 2g_{2v}\exp\left[\left(2\Delta S_{1v}^f - \Delta S_{2v}^b\right)/k_B\right]\exp\left[\left(\Delta H_{2v}^b - 2\Delta H_{1v}^f\right)/k_B T\right], \qquad (4)$$

where ΔS_{2v}^b and ΔH_{2v}^b are entropy and enthalpy of binding of the di-vacancy, respectively. For the di-vacancy formed by two vacancies at nearest-neighbour sites in the fcc structure, the geometry factor becomes $g_{2v} = 6$. In bcc and hcp structures different di-vacancy configurations may exist simultaneously with non-degenerate values of g_{2v}, ΔS_{2v}^b and ΔH_{2v}^b. Up to now, there is no reliable method of measurement which separates c_{2v}^0 and c_{1v}^0. Hence, the formation enthalpy as derived from the temperature dependence of c_v^0 is an effective one:

$$\Delta H_v^f = -\frac{\partial \ln c_v^0(T)}{\partial(1/k_B T)}$$
$$= \Delta H_{1v}^f + 2g_{2v}\frac{\left(\Delta H_{1v}^f - \Delta H_{2v}^b\right)\exp\left[\left(G_{2v}^b - G_{1v}^f\right)/k_B T\right]}{1 + g_{2v}\exp\left[\left(G_{2v}^b - G_{1v}^f\right)/k_B T\right]}, \qquad (5)$$

where G denotes the Gibbs free energy. For bcc and hcp metals, eq. (5) becomes much more complicated due to the possible occurrence of different di-vacancy configurations (SEEGER [1973]). Equation (5) introduces a temperature dependence of ΔH_v^f which must be considered whenever ΔH_v^f is determined over a large temperature range. (See also ch. 8, §3.1.).

2.2.1.2. Differential dilatometry. The classic method of direct c_v^0 determination is *differential dilatometry* (SIMMONS and BALLUFFI [1960a, b, 1962, 1963], BIANCHI et al. [1966], FEDER and CHARBNAU [1966], FEDER and NOWICK [1967, 1972], FEDER [1970], JANOT et al. [1970], V. GUÉRARD et al. [1974], JANOT and GEORGE [1975]). It is based on a theory by ESHELBY [1955, 1956], which relates the microscopic volume change ΔV^{rel} of randomly distributed dilatation centers (point defects) to the total volume change of the defect-containing crystal as compared to the perfect crystal. The total crystal volume change consists of the Eshelby contribution plus that caused by adding to the total crystal volume the atomic volumes of those atoms which are taken out of the interior of the crystal in order to produce vacancies. The crystal volume change can be measured by the length change $\Delta l/l$ of the sample for cubic crystals and the average microscopic (lattice cell) change $\Delta a/a$ by the lattice parameter change of the same sample. One obtains:

$$c_v = 3(\Delta l/l - \Delta a/a), \tag{6}$$

which is correct for cubic crystals and the small vacancy concentrations as usual in such metals.

Precise measurements of this type are extremely difficult. For illustration of the orders of magnitude to be measured, the Al data by SIMMONS and BALLUFFI [1960] are shown in fig. 1. Data referring to vacancy formation are collected in table 1.

2.2.1.3. Positron-annihilation spectroscopy. The second common method of measuring vacancy concentrations in thermal equilibrium is *positron-annihilation spectroscopy* (PAS). Review articles on PAS application for studies of vacancy properties are by SEEGER [1973], DOYAMA and HASIGUTI [1973], TRIFTSHÄUSER [1975a], SIEGEL [1978], MIJNARENDS [1979] and WEST [1979]. High-energy positrons injected in metal crystals are rapidly thermalized by electron–hole excitations and interactions with phonons. The thermalized positron diffuses through the lattice and ends its life by annihilation with an electron. The lifetime depends on the total electron density occurring along the diffusional path of the positron. Vacancies obviously trap positrons in a bound state, and because of the missing core electrons at the vacant lattice site, the local electron density is significantly reduced. This condition causes the lifetime of trapped positrons to be enhanced by 20–80% as compared to that of free positrons in the perfect lattice. Consequently, positrons in a vacancy-containing crystal end their lifes by annihilation either as free positrons or as trapped positrons. The lifetimes for both fates are different and the probability of trapping is proportional to the vacancy concentration. Lifetime measurements are possible as γ-quanta are emitted at the birth of a positron as well as at its decay. Fortunately, thermalization happens within about one picosecond whereas the average lifetime in the metal crystal is in the order of 200 ps.

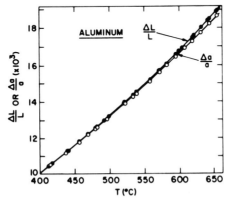

Fig. 1. Differential dilatometry in vacancy equilibrium conditions for aluminium. $\Delta L/L$ is the relative length change of a sample, $\Delta a/a$ is the relative lattice-constant change. (After SIMMONS and BALLUFFI [1960].)

References: p. 1208.

Table 1
Properties of vacancy formation.

Glossary:

Property symbol	Definition and units
$c_v(T_M)$	Vacancy concentration at the melting temperature ($\times 10^4$)
ΔS_v^f	Activation entropy of formation (units of k_B), referring either to single-vacancy values or to effective values, depending on data evaluation
ΔH_v^f	Activation enthalpy of formation (in eV) referring to single-vacancy- or effective values, depending on data evaluation
ΔV_v^{rel}	Relaxation volume of the vacancy (in atomic-volume units)
ρ_v	Electric resistivity contribution per unit concentration of vacancies (in units of 10^{-4} Ωcm)

Measurement symbol	Method of measurement for data evaluation
DD	Differential dilatometry
$\Delta \rho_q, \Delta \rho_{irr}$	Electric resistivity of quenched and irradiated samples, respectively
TEM	Transmission electron microscopy
PAS	Positron-annihilation spectroscopy
DXS	Diffuse X-ray scattering
ΔV^{SD}	activation volume of self-diffusion
ΔV_{1v}^m	activation volume of vacancy migration
ΔV_{1v}^f	activation volume of vacancy formation

Vacancy formation properties

Property	Value	Measurement	Reference
Aluminium (fcc)			
$c_v(T_M)$	9.4	DD, $\Delta \rho_q$	SIEGEL [1978]
ΔS_v^f	1.76	DD	BIANCHI et al. [1966]
	0.6	DD	V. GUÉRARD et al. [1974]
	1.69	$\Delta \rho_q$	FURUKAWA et al. [1976]
	0.7	DD & $\Delta \rho_q$	SIEGEL [1978]
ΔH_v^f	0.77	DD	FEDER and NOWICK [1958]
	0.76	DD	SIMMONS and BALLUFFI [1960a]
	0.71	DD	BIANCHI et al. [1966]
	0.73	$\Delta \rho_q$	BASS [1976]
	0.66	PAS	McKEE et al. [1972a]
	0.71	PAS	McKEE et al. [1972b]
	0.62	PAS	SNEAD et al. [1972]
	0.62, 0.69	PAS	HALL et al. [1974]
	0.67	PAS	KIM et al. [1974]
	0.66	DD	V. GUÉRARD et al. [1974]
	0.66	PAS	TRIFTSHÄUSER [1975b]
	0.69	$\Delta \rho_q$	TZANETAKIS et al. [1976]
	0.70	$\Delta \rho_q$	FURUKAWA et al. [1976]
	0.60, 0.65, 0.68	PAS	DLUBEK et al. [1977a]
	0.66	$\Delta \rho_q$	BERGER et al. [1978]
	0.66	PAS	FLUSS et al. [1978]
	0.66	DD	SIEGEL [1978]
	0.73	DD	HOOD and SCHULTZ [1978]

Table 1 (continued)

Property	Value	Measurement	Reference
Aluminium (fcc)			
$\Delta V_{\mathrm{v}}^{\mathrm{rel}}$	0.35	$\Delta V_{1\mathrm{v}}^{\mathrm{f}}$	EMRICK and MCARDLE [1969]
	−0.05	DXS	HAUBOLD [1972]
	0.05	$\Delta V^{\mathrm{SD}} - \Delta V_{1\mathrm{v}}^{\mathrm{m}}$	SCHILLING [1978]
ρ_{v}	3.0	DD	SIMMONS and BALLUFFI [1960a]
	0.8	$\Delta \rho_{\mathrm{q}}$ & TEM	LEE and SIEGEL [1975]
	2.0	$\Delta \rho_{\mathrm{q}}$ & $\Delta \rho_{\mathrm{irr}}$	RIZK et al. [1976]
	1.1	theory	FUKAI [1969]
			BENEDEK and BARATOFF [1971]
Cadmium (hcp)			
$c_{\mathrm{v}}(T_{\mathrm{M}})$	5	DD	FEDER and NOWICK [1972]
			JANOT and GEORGE [1975]
$\Delta H_{\mathrm{v}}^{\mathrm{f}}$	0.52	PAS	CONNORS et al. [1971]
	0.41	DD	FEDER and NOWICK [1972]
	0.39	PAS	MCKEE et al. [1972a]
	0.42	PAS	WEST [1973]
	0.41	DD	JANOT and GEORGE [1975]
	0.42	$\Delta \rho_{\mathrm{q}}$	SIMON et al. [1975]
	0.47	PAS	SINGH and WEST [1976]
	0.39, 0.42, 0.45	PAS	MASCHER et al. [1981]
Cobalt (hcp)			
$\Delta H_{\mathrm{v}}^{\mathrm{f}}$	1.34	PAS	MATTER et al. [1979]
Copper (fcc)			
$c_{\mathrm{v}}(T_{\mathrm{M}})$	2	DD	SIMMONS and BALLUFFI [1963]
	1.5–2	$\Delta \rho_{\mathrm{q}}$	BOURASSA and LENGELER [1976]
$\Delta S_{\mathrm{v}}^{\mathrm{f}}$	1.5	DD	SIMMONS and BALLUFFI [1963]
	2.2	$\Delta \rho_{\mathrm{q}}$	BOURASSA and LENGELER [1976]
	2.6	PAS, $\Delta \rho_{\mathrm{q}}$	MANTL and TRIFTSHÄUSER [1978]
	2.8	$\Delta \rho_{\mathrm{q}}$	BERGER et al. [1979]
$\Delta H_{\mathrm{v}}^{\mathrm{f}}$	1.17	DD	SIMMONS and BALLUFFI [1963]
	1.14	$\Delta \rho_{\mathrm{q}}$	WRIGHT and EVANS [1966]
	1.20	PAS	SUEOKA [1974]
	1.29	PAS	TRIFTSHÄUSER and MCGERVEY [1975]
	1.27	$\Delta \rho_{\mathrm{q}}$	BOURASSA and LENGELER [1976]
	1.21	PAS	FUKUSHIMA and DOYAMA [1976]
	1.26	PAS	RICE-EVANS et al. [1976]
	1.28	PAS	NANAO et al. [1977]
	1.04, 1.19, 1,22	PAS	DLUBEK et al. [1977]
	1.22	PAS	CAMPBELL et al. [1978]
	1.30	PAS	BERGER et al. [1979]
	1.31	PAS	FLUSS et al. [1979]
$\Delta V_{\mathrm{v}}^{\mathrm{rel}}$	−0.2	DXS	HAUBOLD and MARTINSEN [1978]
ρ_{v}	0.62	DD & $\Delta \rho_{\mathrm{q}}$	BOURASSA and LENGELER [1976]
			BERGER et al. [1979]

References: p. 1208.

Table 1 (continued)

Property	Value	Measurement	Reference
Gold (fcc)			
$c_v(T_M)$	7.2	DD	SIMMONS and BALLUFFI [1962]
	6.4	$\Delta\rho_q$ & $\Delta l/l$ & V_{1v}^f	MORI *et al.* [1962]
ΔS_v^f	1.1	$\Delta\rho_q$ & $\Delta l/l$ & V_{1v}^f	MORI *et al.* [1962]
ΔH_v^f	0.94	DD	SIMMONS and BALLUFFI [1962]
	0.97	$\Delta\rho_q$ & $\Delta l/l$ & V_{1v}^f	MORI *et al.* [1962]
	0.98	$\Delta\rho_q$	FLYNN *et al.* [1965]
	0.95	$\Delta\rho_q$	KINO and KOEHLER [1967]
	0.95	$\Delta\rho_q$	CAMANZI *et al.* [1968]
	0.94	$\Delta\rho_q$	WANG *et al.* [1968]
	0.98, 1.00	PAS	HALL *et al.* [1974]
	0.97	PAS	TRIFTSHÄUSER and McGERVEY [1975]
	0.97	$\Delta\rho_q$	LENGELER [1976]
	0.92	PAS	HERLACH *et al.* [1977]
	0.96	PAS	DLUBEK [1977a]
	0.89–0.96	DD	SAHU *et al.* [1978]
ΔV_v^{rel}	−0.5	DD	SIMMONS and BALLUFFI [1962]
	−0.44	$\Delta a/a$ & $\Delta\rho_q$	HERTZ *et al.* [1973]
	−0.15	DXS & $\Delta\rho_q$	EHRHART *et al.* [1979]
α-iron (bcc)			
ΔH_v^f	1.60	PAS	SCHAEFER *et al.* [1977]
	1.4	PAS	KIM and BUYERS [1978]
	1.60	PAS	MATTER *et al.* [1979]
Indium (hcp)			
ΔH_v^f	0.55	PAS	McKEE *et al.* [1972a]
	0.48	PAS	TRIFTSHÄUSER [1975b]
	0.39	PAS	SINGH *et al.* [1975]
	0.59	PAS	RICE-EVANS *et al.* [1977]
	0.43	PAS	PUFF *et al.* [1982]
Magnesium (hcp)			
ΔH_v^f	0.58	DD	JANOT *et al.* [1970]
Molybdenum (bcc)			
ΔH_v^f	3.24	$\Delta\rho_q$	SUEZAWA and KIMURA [1973]
	3.0	PAS	MAIER *et al.* [1979b]
	3.2	$\Delta\rho_q$	SCHWIRTLICH and SCHULTZ [1980a]
Niobium (bcc)			
ΔH_v^f	2.6	PAS	MAIER *et al.* [1979b]
	$\geqslant 2.7$	$\Delta\rho_q$	SCHWIRTLICH and SCHULTZ [1980b]
Nickel (fcc)			
ΔH_v^f	1.60	$\Delta\rho_q$	MAMALUI *et al.* [1969]
	1.45, 1.73	PAS	CAMPBELL *et al.* [1977]
	1.65, 1.74	PAS	NANAO *et al.* [1977]
Nickel (fcc)	1.6, 1.72, 1.8	PAS	DLUBEK *et al.* [1977]

Table 1 (continued)

Property	Value	Measurement	Reference
Nickel (fcc)			
ΔH_v^f	1.58–1.63	$\Delta\rho_q$ & ΔS_{1v}^f	Wycisk and Feller-Kniepmeier [1978]
	1.7	PAS	Maier et al. [1979a]
	1.55	PAS	Matter et al. [1979]
	1.54	PAS	Snead et al. [1979]
ΔV_v^{rel}	−0.22	DXS	Bender and Ehrhart [1982]
ρ_v	2.3–3.6	$\Delta\rho_q$ & ΔS_{1v}^f	Wycisk and Feller-Kniepmeier [1978]
Lead (fcc)			
$c_v(T_M)$	1.7	DD	Feder and Nowick [1967]
ΔS_v^f	2.6	$\Delta\rho_q$ & DD	Leadbetter et al. [1966]
	0.7 ± 2.0	DD	Feder and Nowick [1967]
	1.6	$\Delta\rho_q$	Knodle and Koehler [1978]
ΔH_v^f	0.58	$\Delta\rho_q$ & DD	Leadbetter et al. [1966]
	0.49	$\Delta a/a$	Feder and Nowick [1967]
	0.54	PAS	Triftshäuser [1975b]
	0.58–0.65	PAS	Sharma et al. [1976]
	0.62	PAS	Hu et al. [1979]
ρ_v	5.0	$\Delta\rho_q$ & DD	Leadbetter et al. [1966]
	2.8	$\Delta\rho_q$	Knodle and Koehler [1978]
Platinum (fcc)			
$c_v(T_m)$	100	c_p & $\Delta a/a$	Kraftmakher and Strelkov [1970]
	8.4	$\Delta\rho_q$ & Q^{SD}	Schumacher et al. [1968]
ΔS_v^f	1.3	$\Delta\rho_q$ & Q^{SD}	Schumacher et al. [1968]
	4.5	c_p & $\Delta a/a$	Kraftmakher and Strelkov [1970]
ΔH_v^f	1.51	$\Delta\rho_q$	Jackson [1965]
	1.49	$\Delta\rho_q$	Schumacher et al. [1968]
	1.6	c_p & $\Delta a/a$	Kraftmakher and Strelkov [1970]
	1.3	$\Delta\rho_q$	Zetts and Bass [1975]
	1.31	$\Delta\rho_q$	Mišek [1979]
	1.32	PAS	Maier et al. [1979b]
	1.15	$\Delta\rho_q$	Emrick [1982]
ΔV_v^{rel}	−0.42	$\Delta a/a$ & $\Delta\rho_q$	Hertz et al. [1973]
	−0.33	$\Delta\rho_q$ & press. dep.	Charles et al. [1975]
	−0.28	$\Delta\rho_q$ & press. dep.	Emrick [1982]
ρ_v	5.75	DD & $\Delta\rho_q$	Berger et al. [1973]
Silver (fcc)			
$c_v(T_M)$	1.7	DD	Simmons and Balluffi [1960b]
ΔH_v^f	1.09	DD & ΔS_v^f	Simmons and Balluffi [1960b]
	1.10	$\Delta\rho_q$	Cuddy and Machlin [1962]
	1.10	$\Delta\rho_q$	Doyama and Koehler [1962]
	1.16	PAS	Triftshäuser and McGervey [1975]
	1.19	PAS	Campbell et al. [1978]
ρ_v	1.3	DD	Doyama and Koehler [1962]

References: p. 1208.

Table 1 (continued)

Property	Value	Measurement	Reference
Tantalum (bcc)			
$\Delta H_{\rm v}^{\rm f}$	2.8	PAS	Maier *et al.* [1979b]
	3.1	$\Delta \rho_{\rm q}$	Tietze *et al.* [1982]
Tungsten			
$c_{\rm v}(T_{\rm M})$	1.0	$\Delta \rho_{\rm q}$ & TEM	Rasch *et al.* [1980]
$\Delta S_{\rm v}^{\rm f}$	2.3	$\Delta \rho_{\rm q}$ & TEM	Rasch *et al.* [1980]
$\Delta H_{\rm v}^{\rm f}$	3.6	$\Delta \rho_{\rm q}$	Gripshover *et al.* [1970]
	3.1	$\Delta \rho_{\rm q}$	Kunz [1971]
	4.0	PAS	Maier *et al.* [1979]
	3.67	$\Delta \rho_{\rm q}$	Rasch *et al.* [1980]
$\rho_{\rm v}$	6.3	$\Delta \rho_{\rm q}$ & TEM	Rasch *et al.* [1980]
Zinc (hcp)			
$\Delta H_{\rm v}^{\rm f}$	0.54	PAS	McKee *et al.* [1972a]
	0.51	PAS	West [1973]
$\Delta V_{\rm v}^{\rm rel}$	−0.60	$Q^{\rm SD}$ & press. dep.	Chhabildas and Gilder [1972]

    A critical aspect of the applicability of PAS for vacancy concentration measurements is the existence of positron bound states in vacancies. For some metals such as Al, Fe, Cu, Zn, Nb, Ag, Cd, In, W, Au, and Pd their existence must be concluded from the experimental observations and was established theoretically (Seeger [1973], Triftshäuser [1975], West [1973], Maier *et al.* [1978]). For other metals, like Li, Na, Ga, Sb, Hg, and Bi, bound states could not be found (Kusmiss and Stewart [1967], Brandt and Waung [1968], Mackenzie *et al.* [1971], Kim and Stewart [1975]) although theoretical calculations predicted them (Hodges [1970], Brandt [1974], Manninen *et al.* [1975], Tam [1977]). Reasons for this discrepancy are still under discussion. From PAS, information is extracted not only from the positron lifetimes but also from the angular correlation between the directions of the emitted annihilation γ-rays and from the Doppler broadening of these γ-rays. These quantities yield information on the net momenta of the annihilating electron–positron pairs. They allow a distinction to be made between annihilations with the higher-momentum core electrons and the lower-momentum valence or conduction electrons.

    Vacancy concentrations are determined by means of lifetime spectrum measurements as well as momentum techniques (angular correlation and Doppler broadening). The former avoids additional assumptions on positron-annihilation parameters but requires high-resolution measuring techniques and expanded data deconvolution. Although the deconvolution of momentum-distribution data requires critical assumptions on the temperature dependence of positron-annihilation parameters, these methods have become more popular for $\Delta H_{\rm v}^{\rm f}$ determinations. In fig. 2, normalized trapping rates derived from angular correlation measurements are shown for Cu and Au. Indicated are the range of measurement of differential dilatometry and the range of statistical significance for the PAS data. One realizes that the reliable PAS data extend to about two orders of magnitude lower vacancy concentration than the differential dilatometry data. As monovacancies certainly predominate in this concentration range, PAS studies are of great importance as a complement to differential dilatometry on the one side, and to resistivity measurements in quenched samples on the other side. More results are given in table 1.

2.2.1.4. Resistivity measurements after quenching. All problems of measurements at temperatures near the melting point can be avoided if equilibrium vacancy concentrations can quantitatively be quenched from such temperatures T_q to temperatures which immobilize the vacancies. Properties which are proportional to the vacancy concentration could then be measured on quenched and unquenched samples for direct comparison. This idea appeared so challenging that incredibly large efforts were put into the development of quenching methods since the pioneering work by KAUFFMAN and KOEHLER [1952]. Methods and problems are reviewed by BALLUFFI et al. [1970] and SIEGEL [1978]. There are indeed problems

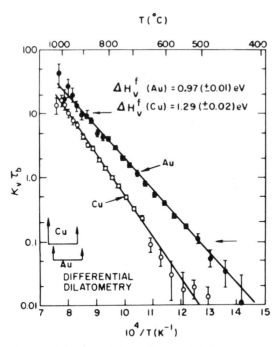

Fig. 2. Arrhenius plot of a quantity measured by positron-annihilation spectroscopy in gold and copper according to TRIFTSHÄUSER and McGERVEY [1975]. The product $k_v \tau_b$ is proportional to the vacancy concentration according to common models. The horizontal arrows indicate the boundaries of the range of highest statistical data significance. The range measured by means of differential dilatometry (SIMMONS and BALLUFFI [1962, 1963]) is indicated also.

References: p. 1208.

inherent to the quenching process per se. During quenching the vacancies are still highly mobile in a substantial fraction of the total temperature interval being passed. The migrating vacancies are able to react with other defects or with one another, with the following consequences: (i) Vacancy losses to sinks as dislocations, grain boundaries and surfaces. The quenched concentration c_v will therefore be smaller than $c_v^0(T_q)$. The inhomogeneous distributions of sinks will cause locally varying concentrations. (ii) Vacancy clustering, which causes repartitioning of the cluster size distribution existing at T_q. High-order clusters are favoured in comparison to the equilibrium distribution at T_q.

By modelling the vacancy reaction scheme for the conditions of quenching, vacancy loss and repartitioning of cluster sizes were studied in great detail (BAL-LUFFI et al. [1970]). With the aid of such calculations, quenching results obtained for systematically varied quenching rates could be corrected and extrapolated to infinite quenching rates in the case of Au (FLYNN et al. [1965]). Systematic studies of this kind were performed also on Al (BASS [1967] and BERGER et al. [1978]).

An entirely different approach was successfully followed by LENGELER [1976] and LENGELER and BOURASSA [1976]. They were able to rapidly quench single crystals of Au and Cu with dislocation densities so low that vacancy losses are negligible even for high T_q values. The effect is demonstrated in fig. 3. More resistivity quenching data are incorporated in table 1.

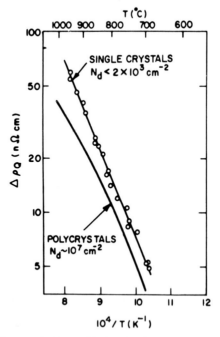

Fig. 3. Arrhenius plot of the quenched-in residual resistivity of gold. The influence of the dislocation density N_d is shown by the curve for quenched polycrystals. (After LENGELER [1976].)

2.2.2. Activation enthalpy of migration

2.2.2.1. Problems of methodology. The common way of investigating the migration properties of vacancies consists of quenching a sample from high temperatures and subsequent annealing at increasing temperatures in order to induce the diffusion-controlled annihilation of the excess vacancies at the annealing temperature. Review articles are due to KOEHLER [1970], CHIK [1970], SCHILLING *et al.* [1970] and BALLUFFI [1978]. The residual resistivity has mostly been taken as a measure of the vacancy concentration in the lattice. This property is certainly preferable to all others in view of its convenient application and high sensitivity. The isochronal recovery behaviour of a pure metal containing point defects is sketched in fig. 4. The resistivity recovers in distinct steps which are labelled according to VAN BUEREN [1955]. Stage I is observed in irradiated samples only. Stage II occurs in irradiated and plastically deformed samples. Quenched samples show resistivity recovery at temperatures varying from stage III to stage IV, depending on the individual metal. As the first recovery stage of quenched samples does not generally coincide with stage III of irradiated samples, a straightforward classification of this stage is not possible. This situation created the *stage III controversy* (§2.3).

The resistivity method is not appropriate to give information on details of the underlying defect reactions. Its inability to indicate secondary vacancy reactions beside annihilation as clustering and trapping at impurities is one of the origins of

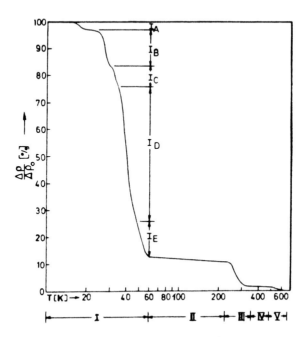

Fig. 4. Isochronal recovery curve of the electric resistivity of Cu electron-irradiated at 4 K.

References: p. 1208.

the stage III controversy. But at that time no other method existed which allowed a quantitative separation of different vacancy reactions. Stage V is caused by recrystallization (see ch. 25) and occurs in all samples which contain larger defect agglomerates.

A real advance was achieved by applying PAS to the question at which temperature vacancies become mobile. Another significant step was made when the ability of nuclear probe atom methods as Mössbauer spectroscopy and perturbed γ–γ angular correlation was detected to distinguish between interstitials and vacancies. Because of the problems described in §2.2.2.3, a satisfactory interpretation of stage III in irradiated samples restricts the boundary conditions for data evaluation of the recovery of quenched samples.

As many as 21 experimental results are listed by BALLUFFI [1978] which favour stage III interpretation by vacancy migration. Since then new methods such as the perturbed γ–γ angular correlation measurements have fruitfully been applied. Their results also favour the vacancy interpretation (see, for example, PLEITER and HOHENEMSER [1982]).

2.2.2.2. Two selected pieces of evidence for vacancy migration in stage III. Among the older results, the positron-annihilation behaviour in stage III seems especially conclusive with respect to vacancy migration in that stage. The important quantity is the so-called *lineshape parameter* $R = |(I_v^t - I_v^f)/(I_c^t - I_c^f)|$. The energy of the positron-annihilation γ quanta is Doppler-broadened because of the non-vanishing momenta of the annihilating electron–positron pairs. The centre of the curve (intensity versus energy) is caused by low-momentum pairs, i.e., involves conduction and valence electrons, whereas the sides stem from high-momentum pairs, i.e., involve core electrons. The quantity I is simply a suitably broad section of the Doppler line integral, one below the maximum (I_v) and the other below the branches (I_c). The upper subscripts t and f in the expression for R refer to trapped and free positrons, respectively, as obtained from the irradiated and unirradiated state.

The lineshape parameter R was found to be considerably larger for positrons trapped by voids than for those trapped by single vacancies (MANTL and TRIFTSHÄUSER [1978]). For trapping by dislocation loops its size is close to that for single vacancy trapping. The result from measurements in electron-irradiated Cu through stage III recovery is shown in fig. 5. The corresponding resistivity-decrease between 200 K and 300 K amounted to about 85% of its value at 200 K. The increase of R indicates a further fractional decrease of core electron density as compared to the presence of merely single vacancies below 200 K. This can only be visualized by vacancy agglomeration in three-dimensional arrangements. The effect of interstitial-type dislocation loops which could be formed in stage III according to the two-interstitial model was found to produce $R = 0.62 \pm 0.03$ in neutron-irradiated aluminium (GAUSTER et al. [1975]). Interstitial agglomeration can therefore be excluded as an origin of the observed R enhancement. MANTL and TRIFTSHÄUSER [1978] assume the three-dimensional agglomerates to be the nuclei of the vacancy loops observed at higher temperatures by electron microscopy (OHR [1975]). The

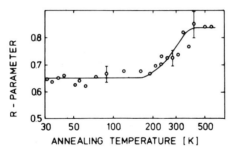

Fig. 5. Lineshape parameter R of low-temperature electron-irradiated Cu upon isochronal annealing (after MANTL and TRIFTSHÄUSER [1978]).

same behaviour of R as in Cu has been observed in the bcc Mo whereas in Al the observed increase in R did not significantly exceed the error limit (MANTL and TRIFSHÄUSER [1978]).

Fortunately, the *perturbed γ–γ angular correlation* (PAC) technique indicated the arrival of a vacancy-type defect in stage III at the probe atom Cd in Al (RINNEBERG *et al.* [1978], RINNEBERG and HAAS [1978]). The PAC technique essentially measures the hyperfine interaction of the probe-atom nucleus with the electric crystal field. Interstitials or vacancies situated at the nearest-neighbour sites of such probe atoms obviously cause sufficiently large field gradients to produce measurable signals. Review articles on this type of application of PAC are by PLEITER and HOHENEMSER [1982], WICHERT [1982] and NIESEN [1981]. The PAC signals are determined by the electric field gradient at the site of the probe-atom nucleus and are in that sense specific for the type of defect which is placed at a nearest-neighbour site. Although the electric field gradient cannot be calculated yet for a given defect with sufficient accuracy the method has already been applied with great success for defect identification in many metals. If it happens that the probe atom in a given host metal traps the vacancy, a quenched sample containing probe atoms will exhibit the PAC signal caused by the vacancy. Similarly one obtains the signal caused by the interstitial when one irradiates the sample at low temperatures and anneals through stage I or irradiates in stage II. In this way one obtains the "fingerprints" of vacancy and interstitial. Clear evidence for the arrival of vacancies at the probe atoms in stage III has been obtained for Al, Ag (BUTT *et al.* [1979], DEICHER *et al.* [1981a]), Au (DEICHER *et al.* [1981b]), Cu (WICHERT *et al.* [1978]), Pt (MÜLLER [1979]), Cd and Zn (SEEBOECK *et al.* [1982]).

The PAC signal also gives information on the deviation of the electric field-gradient tensor from axial symmetry. When measurements are performed in single crystals, the tensor orientation with respect to the crystal orientation can be obtained. Knowledge of orientation and deviation from axial symmetry provides valuable criteria for answering the question whether mono-, di-, or tri-vacancies decorate the probe atom and cause the observed signal. Indeed, the PAC technique has resolved a number of different configurations for the metals quoted above. The variety of defect arrangements trapped at In in Au after different treatments is illustrated in fig. 6. From the field-gradient tensor properties, it has been concluded that defect 3 is a monovacancy at a nearest-neighbour site, defect 4 is a planar $\langle 111 \rangle$ vacancy loop and defects 1 and 2 are multiple vacancies in different arrangements (DEICHER et al. [1981b]). It should be kept in mind that the spectrum of visible vacancy arrangements does depend on the nature of the probe atom because of the required trap property. Nevertheless, the variety of arrangements to be seen in fig. 6 gives a good impression of the complexity of the stage III processes which, in the earlier controversy, were often identified with either monovacancy or mono-interstitial reactions.

2.2.2.3. Experimental determination of ΔH_v^m. The primary problem of experiments involving quenching and subsequent annealing consists in the large variety of defect reactions which are likely to occur simultaneously. In addition, questions arise with regard to the actual boundary conditions produced during the quenching and annealing procedure. The following aspects must be considered: (i) inhomogeneous local distributions of vacancies, due to losses to sinks or cluster formation (§2.2.3), (ii) time-dependent sink concentrations and configurations during the annealing process, e.g. by nucleation and growth of vacancy clusters, (iii) sink efficiencies which depend on both instantaneous vacancy concentration and temperature, (iv) vacancy–impurity interactions, (v) drift diffusion effects caused by elastic interaction of vacancies and sinks. Computer simulation by means of the rate-equation approach shows that the main defect parameters can hardly be inferred from a standard analysis of isothermal and isochronal data if, for example, the actual clustering kinetics are unknown (JOHNSON [1968]).

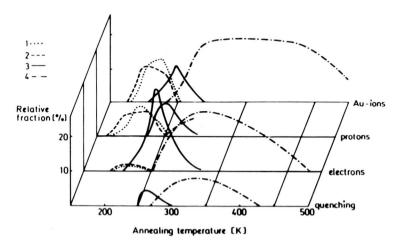

Fig. 6. Fraction of In atoms in gold which emit perturbed $\gamma - \gamma$ angular correlation signals upon isochronal annealing after quenching and low-temperature irradiation with the particles indicated. The different types of lines indicate different types of signals and hence, different defects trapped by the probe atoms (see text).

The experimentally observed influence of clustering is illustrated in fig. 7 which shows for Au and Al the observed temperatures T_a of annealing stages as function of the quenching temperature T_q, according to BALLUFFI [1978]. Within the temperature range shown for T_a, vacancies either annihilate at fixed sinks or form immobile clusters which dissociate at considerably higher temperatures than 450 K. The behaviour of T_a for Al is illustrated in more detail in fig. 8 which shows the temperature-differentiated isochronal recovery curve for the T_q's indicated, according to LEVY *et al.* [1973]. Peak B was quantitatively interpreted by a single diffusion-controlled annihilation process with an activation enthalpy of 0.65 eV, which was ascribed to monovacancy migration. Peak A could not be interpreted by a single process and is characterized by an effective activation enthalpy of 0.44–0.5 eV. Obviously, at low vacancy concentrations as quenched from low temperatures, monovacancies are the dominant defects. For higher vacancy concentrations as quenched from temperatures above 500°C, multiple vacancies are formed which migrate faster than monovacancies and accelerate the annihilation process. This feature describes, in general, the annihilation behaviour of quenched-in vacancies in most metals, although the separation between the processes involving monovacancies and those involving mobile multivacancies is not so obvious as in Al.

In order to obtain reliable data, the boundary conditions must carefully be controlled in the experiments (LENGELER and BOURASSA [1976], §2.2.1.3) and/or systematically varied solutions of clustering models must be fitted to the data for evaluation (SAHU *et al.* [1978]). The results of such fits are not single sets of enthalpy and entropy changes for migration of single vacancies but "fields of existence" for possible parameter sets which include also the vacancy-binding parameters up to higher clusters. The enthalpies of vacancy migration are listed in table 2 either as effective quantities ΔH_v^m or as specific quantities ΔH_{1v}^m valid for the monovacancies, depending on the type of data evaluation performed.

The stage III controversy arose as the recovery stage III of irradiated samples always occurred at lower temperatures than the main recovery stage in quenched

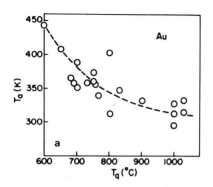
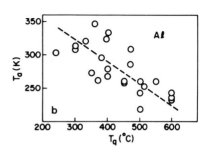

Fig. 7. Dependence of the temperature T_a, characterizing the temperatures of vacancy-annihilation stages in isochronal annealing experiments on quenching temperature T_q for Au and Al (after BALLUFFI [1978]).

References: p. 1208.

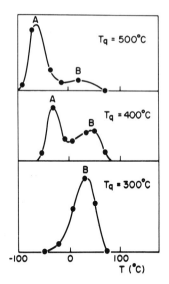

Fig. 8. Temperature-differentiated isochronal resistivity recovery curve of Al for different quenching temperatures T_q (after LEVY *et al.* [1973]).

samples. A systematic comparison is by BALLUFFI [1978]. As an example, the temperatures of recovery T_a measured by more than 10 groups of authors in Pt are compared in fig. 9. The widespread data-scatter for both types of treatment as well as the systematic difference between irradiation and quenching becomes obvious. It is quite suggestive to take the discrepancy of 200 K between the most often observed recovery temperatures in irradiated and quenched samples as overwhelming evidence for the migration of two entirely different defects, as is done by the advocates of the two-interstitial model. According to this model, stage III in irradiated metals is caused by the migration and subsequent recombination of the so-called stable interstitial which arises from the metastable one by conversion (SEEGER [1958, 1975]). Its enthalpy of migration is supposed to equal the effective enthalpy of migration found for stage III, whereas the vacancy-migration enthalpy follows from the respective quantity found in quenched samples. The behaviour of prequenched and irradiated samples has been studied for Al (BAUER [1969]), Au (LEE and KOEHLER [1968]) and Pt (DUESING and SCHILLING [1969], SONNENBERG *et al.* [1972]). Typical curves for Pt are shown in fig. 10. The splitting of the single stage III in the unquenched state into two stages in the prequenched state has been interpreted by SCHILLING *et al.* [1970] in terms of a mono–di-vacancy model which takes into account the different concentrations and local distributions of sinks in the quenched and unquenched samples. Very careful measurements and data evaluation revealed a monotonic variation of $\Delta H_{\mathrm{eff}}^m$ when proceeding through the stage III–IV region. The data could quantitatively be interpreted by a mono–di-vacancy model with $\Delta H_{1v}^m = 1.45$ eV, $\Delta H_{2v}^m = 1.00$ eV and $\Delta H_{2v}^b = 0.15$ eV (SONNENBERG *et al.* [1972]).

Table 2
Activation enthalpy of vacancy migration and self-diffusion.

Glossary

Symbol	Meaning
$\Delta\rho_q$, $\Delta\rho_{irr}$, $\Delta\rho_{pd}$	electric resistivity of quenched, irradiated and plastically deformed samples, respectively
PAC_{irr}	perturbed $\gamma - \gamma$ angular correlation of irradiated samples
PAS_{irr}, PAS_{pd}	positron-annihilation spectroscopy of irradiated and plastically deformed samples, respectively

Property (in eV)		Measurement	Reference
ΔH_v^m	Q^{SD}		
Aluminium (fcc)			
0.65		$\Delta\rho_q$	DeSorbo and Turnbull [1959]
0.59		$\Delta\rho_{irr}$	Ceresara et al. [1963]
0.62		$\Delta\rho_q$	Federighi [1965]
0.62		$\Delta\rho_q$	Budin and Lucasson [1965]
0.57		$\Delta\rho_{pd}$	Ceresara et al. [1965]
0.62		$\Delta\rho_{irr}$	Federighi et al. [1965]
0.58		$\Delta\rho_{irr}$	Isebeck et al. [1966]
0.58		$\Delta\rho_{irr}$	Brugière and Lucasson [1967]
0.61		$\Delta\rho_{irr}$	Garr and Sosin [1967]
0.62		$\Delta\rho_{irr}$	Lwin et al. [1968]
0.59		$\Delta\rho_{irr}$	Dimitrov-Frois and Dimitrov [1968]
0.58		$\Delta\rho_{irr}$	Bauer [1969]
0.60		$\Delta\rho_{irr}$	Kontoleon et al. (1973)
0.65		$\Delta\rho_q$	Levy et al. [1973]
	1.28		Peterson [1978]
Cadmium (hcp)			
0.38		$\Delta\rho_q$	Simon et al. [1975]
0.23 ‖ c, 0.37 ⊥ c		$\Delta\rho_{irr}$	Roggen et al. [1978]
	0.81 ‖ c, 0.85 ⊥ c		Peterson [1978]
Copper (fcc)			
0.67		$\Delta\rho_{irr}$	Meechan et al. [1960]
0.71		$\Delta\rho_{irr}$	Dworschak et al. [1964]
0.77		$\Delta\rho_q$	Lucasson et al. [1964]
0.71		$\Delta\rho_{irr}$	Dworschak and Koehler [1965]
1.06		$\Delta\rho_q$	Ramsteiner et al. [1965]
0.85		$\Delta\rho_q$	Wright and Evans [1966]
0.69		$\Delta\rho_{irr}$	Brugière and Lucasson [1968]
0.72		$\Delta\rho_{irr}$	Schilling et al. [1970]
0.74		$\Delta\rho_q$	Bourassa and Lengeler [1976]
0.70		$\Delta\rho_{irr}$	Wienhold et al. [1978]
0.69		PAC_{irr}	Wichert et al. [1978]
	2.07		Peterson [1978]
Gold (fcc)			
0.83		$\Delta\rho_q$	Schüle et al. [1962]
0.80		$\Delta\rho_{irr}$	Dworschak et al. [1964]
0.80		$\Delta\rho_{irr}$	Bauer and Sosin [1964]
0.84		$\Delta\rho_q$	Ytterhus and Balluffi [1965]

References: p. 1208.

Table 2 (continued)

Property (in eV)		Measurement	Reference
ΔH_v^m	Q^{SD}		
Gold (fcc)			
0.70		$\Delta \rho_q$	De Laplace et al. [1966]
0.70		$\Delta \rho_q$	Siegel [1966b]
0.85		$\Delta \rho_{irr}$	Sharma et al. [1967]
0.94		$\Delta \rho_q$	Kino and Koehler [1967]
0.78		$\Delta \rho_1$	Camanzi et al. [1968]
0.85		$\Delta \rho_{irr}$	Lee and Koehler [1968]
0.90		$\Delta \rho_q$	Wang et al. [1968]
0.77		$\Delta \rho_{irr}$	Wittmaack [1970]
0.69, 0.78, 0.81		$\Delta \rho_q$	Das and Dawson [1971]
0.70		$\Delta \rho_{irr}$	Antesberger et al. [1978]
0.83–0.89		TEM, Q^{SD}	Sahu et al. [1978]
0.62–0.72		PAC_{irr}	Deicher et al. [1981]
0.71		$\Delta \rho_{irr}$	Sonnenberg and Dedek [1982]
	1.76		Peterson [1978]
α-Iron (bcc)			
0.55		$\Delta \rho_{pd}$	Cuddy [1968]
1.24		TEM	Kiritani et al. [1979]
1.22		PAS_{irr}	Schaefer [1981]
0.55		$\Delta \rho_{irr}$	Kugler et al. [1982]
0.53		PAS_{pd}	Segers et al. [1982]
0.55		PAS_{irr}	Vehanen et al. [1982]
	2.88		Hettich et al. [1977]
Molybdenum (bcc)			
1.25		$\Delta \rho_{irr}$	Peacock and Johnson [1963]
1.29		$\Delta \rho_{irr}$	Afman et al. [1970], Afman [1972], Stals [1972]
1.62		$\Delta \rho_q$	Suezawa and Kimura [1973]
1.6		$\Delta \rho_q$	Yoshioka et al. [1974]
1.30		$\Delta \rho_{irr}$	Kugler et al. [1982]
1.3		TEM	Phillipp et al. [1982]
	4.53		Maier et al. [1979c]
Niobium (bcc)			
0.56		$\Delta \rho_{irr}$	Faber et al. [1974]
0.55		$\Delta \rho_{irr}$	Faber and Schultz [1977]
0.95		TEM	Phillipp et al. [1982]
	3.96		Lundy et al. [1965]
	3.64		Peterson [1978]
Nickel (fcc)			
1.1.		$\Delta \rho_{irr}, \Delta \rho_{pd}$	Sosin and Brinkman [1959]
0.92, 1.1		$\Delta \rho_{pd}$	Simson and Sizmann [1962]
1.03, 1.46		$\Delta \rho_{irr}$	Mehrer et al. [1965]
1.4		$\Delta \rho_q$	Mughrabi and Seeger [1967]
1.32–1.27		$\Delta \rho_q$	Wycisk and Feller-Kniepmeier [1978]
1.2		TEM	Kiritani and Takata [1978]
1.1		$\Delta \rho_{irr}$	Antesberger et al. [1978]
1.04		$\Delta \rho_{irr}$	Khanna and Sonnenberg [1981]
	2.88		Peterson [1978]

Table 2 (continued)

Property (in eV)		Measurement	Reference
ΔH_v^m	Q^{SD}		
Lead (fcc)			
0.54		$\Delta \rho_q$	KNODLE and KOEHLER [1978]
	1.13		PETERSON [1978]
Platinum (fcc)			
1.48		$\Delta \rho_q$	BACCHELLA et al. [1959]
1.46		$\Delta \rho_{irr}, \Delta \rho_q, \Delta \rho_{pd}$	PIERCY [1960]
1.38		$\Delta \rho_q$	JACKSON [1965]
1.36		$\Delta \rho_{irr}$	BAUER and SOSIN [1966]
1.45		$\Delta \rho_q$	POLAK [1967]
1.38		$\Delta \rho_q$	SCHUMACHER et al. [1968]
1.3–1.4		$\Delta \rho_q$	RATTKE et al. [1969]
1.45		$\Delta \rho_{irr}$	DOYAMA et al. [1971]
1.51		$\Delta \rho_q$	HEIGL and SIZMANN [1972]
1.4		$\Delta \rho_{irr}$	SONNENBERG et al. [1972]
1.31		PAC_{irr}	MÜLLER [1979]
	2.96		PETERSON [1978]
Silver (fcc)			
0.58, 0.86		$\Delta \rho_q$	QUÉRÉ [1961]
0.60, 0.88		$\Delta \rho_q, \Delta \rho_{pd}$	RAMSTEINER et al. [1962]
0.83		$\Delta \rho_q$	DOYAMA and KOEHLER [1962]
0.58, 0.67		$\Delta \rho_{irr}$	DWORSCHAK et al. [1964]
0.65		$\Delta \rho_{irr}$	ANTESBERGER et al. [1978]
0.64		$\Delta \rho_q$	HILLAIRET et al. [1982]
	1.76		PETERSON [1978]
Tantalum (bcc)			
0.7		$\Delta \rho_{irr}$	FABER et al. [1974]
0.70		$\Delta \rho_{pd}$	PERETTI et al. [1980]
1.1		TEM	PHILLIPP et al. [1982]
	4.1		PETERSON [1978]
Tungsten (bcc)			
1.7		$\Delta \rho_{irr}$	KINCHIN and THOMPSON [1958]
1.1		$\Delta \rho_{irr}$	NEELY et al. [1968]
	5.45		MUNDY et al. [1978]
Zinc (hcp)			
0.44		$\Delta \rho_q$	SIMON et al. [1974]
0.34 ∥ c, 0.42 ⊥ c		$\Delta \rho_{irr}$	ROGGEN [1978]
	0.95 ∥ c, 1.00 ⊥ c		PETERSON [1978]

BALLUFFI (1978) compared the reaction processes in stage III for the irradiated and quenched state by plotting ΔH_{eff}^m against $\overline{T}_a$, the T_a values averaged over all reported measurements, as shown in fig. 11. The data for the different metals lie well on straight lines through the origin, but quenched and irradiated states are characterized by different lines. The equality of $\Delta H_{eff}^m / k_B T_a$ for the different metals

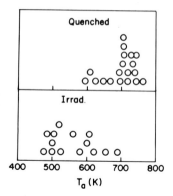

Fig. 9. Temperatures T_a (definition see fig. 7) as measured in Pt between 400 K and 800 K after quenching or irradiation, by different authors. Each circle represents one published isochronal recovery stage. The ordinate gives the observation frequency of a given T_a. (After BALLUFFI [1978].)

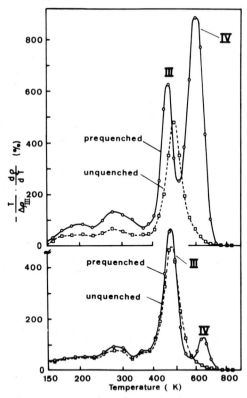

Fig. 10. Temperature-differentiated isochronal recovery curves of Pt after low-temperature electron-irradiation. Shown is the influence of pre-quenching and that of initial defect concentration (the ratio of quenched-in resistivity to irradiation-induced resistivity is large for the upper curves, small for the lower curves). The curves are normalized by different values of $\Delta\rho_{III}$ (see ordinate). Merely for that reason a deviation between dashed and solid curves occurs below 350 K. (After SCHILLING et al. [1970].)

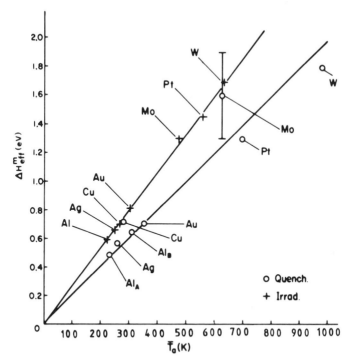

Fig. 11. Effective enthalpies of defect migration versus T_a (definition see fig. 7) averaged over the results by different authors for quenched and irradiated samples (after BALLUFFI [1978].

indicates the close similarity of the parameters determining migration and annihilation, such as frequency factor, sink geometry and density, and initial defect concentration. The different slopes for quenched and irradiated states indicate deviation
of at least one of these parameters. The difference can be explained by a number of
jumps-per-vacancy until annihilation which is larger by a factor of 10^3 in the
quenched state than in the irradiated state. This order of magnitude had already
been derived earlier (SONNENBERG et al. [1972], JOHNSON [1970], SCHILLING et al.
[1970], LEE and KOEHLER [1968], LWIN et al. [1968]).

2.2.3. Agglomeration

Vacancy agglomerates which are large enough to be observable in the electron
microscope have been studied in great detail (ch. 11, § 6). The topology of configurations involves dislocation loops, stacking-fault tetrahedra and voids. Review articles
are by EYRE et al. [1977], KIRITANI [1982], and SHIMOMURA et al. [1982]. Since the
resolution of common electron microscopy is limited to 1–2 nm, the observable
agglomerates certainly contain more than ten vacancies. About the same resolution
limit holds for diffuse X-ray scattering (EHRHART et al. [1982] and LARSON and
YOUNG [1982]). Field ion microscopy does allow imaging of agglomerates consisting

References: p. 1208.

of less than ten vacancies (WAGNER [1982]). It has, however, not been applied yet to questions like shape and size distribution of small vacancy agglomerates as formed by the encounter of migrating vacancies.

While experimental information on configuration and stability of di-, tri- and tetra-vacancies is lacking, computer simulations give some information by using a special nearest-neighbour interaction potential constructed to similate the bcc α-Fe (JOHNSON [1964]). This potential was also used for simulating a fcc model crystal which behaves similar to Ni with respect to the elastic properties. This model yielded for the single vacancy: $\Delta H_{1v}^f = 1.49$ eV, $\Delta H_{1v}^m = 1.32$ eV and $V_{1v}^{rel} = -0.15$ atomic volume. The stable di-vacancy consisting of the vacancies at nearest-neighbour sites is characterized by $\Delta H_{2v}^b = 0.25$ eV and $\Delta H_{2v}^m = 0.9$ eV. The most stable tri-vacancy, shown in fig. 12a, yielded $\Delta H_{3v}^b = 0.75$ eV and $\Delta H_{3v}^m = 1.02$ eV. The migration includes an intermediate dissociation step. Reorientation of the tri-vacancy with a lower activation enthalpy than ΔH_{3v}^m occurs by jump of the atom which forms a tetrahedron together with the tri-vacancy (fig. 12b). The stable tetravacancy is that shown in fig. 12d with $\Delta H_{4v}^b = 1.51$ eV. The stability of the three-dimensional tetrahedron as compared to the two-dimensional rhombic configuration (fig. 12c) might be a direct consequence of the nearest-neighbour interaction potential (DEDE-RICHS *et al.* [1978]).

For the α-Fe model crystal $\Delta H_{1v}^m = 0.68$ eV was found for the nearest-neighbour jump. The most stable di-vacancy configuration places the vacancies at second-neighbour distances $\Delta H_{2v}^b = 0.2$ eV. Migration happens either via the metastable nearest-neighbour configuration or via a fourth-neighbour configuration, both with an activation enthalpy being about the same as for the mono-vacancy migration. The most stable tri-vacancy has two nearest-neighbour and one second-neighbour spacings with $\Delta H_{3v}^b = 0.49$ eV (JOHNSON and BEELER [1977]). The first immobile aggregate was found to be the tetravacancy. The local vibration densities of states of the neigbouring atoms of vacancy clusters were calculated for Cu, α-Fe and α-Ti by

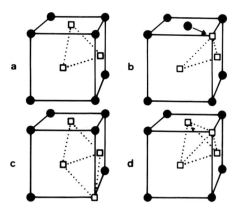

Fig. 12. Tri- and tetra-vacancy configurations in the fcc structure (see text).

YAMAMOTO [1982]. Formation and migration enthalpies of di-vacancies have frequently been derived from experiments as described in §§2.2.1 and 2.2.2. But the derived numbers result from fitting models to measured data by assuming that eq. (4) holds. In no case has the observed deviation of $c_v^0(1/T)$ from linearity been proven by direct evidence to be caused by di-vacancies. Another reason could be a temperature dependence of the activation quantities of vacancy formation (PETERSON [1978], STOTT [1978], FUKAI [1978], AUDIT and GILDER [1978], AUDIT [1982], GANNE and QUÉRÉ [1982], VAROTSOS and ALEXOPOULOS [1982]).

2.2.4. Interaction with other defects

Vacancies interact not only with one another, but also with lattice defects like interstitials, dislocations, surfaces, solutes, Bloch walls etc. The interaction with interstitials governs the Frenkel pair recombination (§3.2.2.4). The interaction with dislocations and surfaces influences the vacancy annihilation and is therefore of great importance for the interpretation of annealing experiments of irradiated and quenched samples (§§2.2.2 and 4). The interaction with solutes governs the solute diffusion, also known as impurity diffusion, and its relationship to the self-diffusion of the solvent atoms. This topic is treated in ch. 8.

"Constitutional vacancies" in ordered binary solid solutions are treated in ch. 4 §9.1 and from the point of view of diffusion in ch. 8, §5.2. Complex interactions of vacancies, vacancy clusters and solutes with dislocations cause such phenomena as *dislocation pinning* (NOWICK and BERRY [1972], and SOSIN [1970]) or *pipe diffusion* (chapter 8, §7). The interaction with Bloch walls has been used to determine the activation enthalpy for vacancy migration by measurements of the magnetic diffusion after-effect (KRONMÜLLER [1970]).

2.3. The controversy over the recovery-stage III interpretation

The controversy on the recovery-stage III interpretation resulted from the fact that the main recovery stage of quenched samples did not exactly coincide with stage III in irradiated samples (§2.2.2.3). Unfortunately, in the early days the difference was particularly large for Cu, the standard research material of the metal physicists. The lack of total recovery at the end of stage III also played a significant role in the dispute. The latter problem, however, is of less significance than the former and is omitted here.

The primary source of disagreement is the interpretation of the stage III recovery. The advocates of the "one-interstitial-" or (a better expression) *vacancy model* explain stage III by vacancy migration and annihilation. The temperature difference of the recovery-rate maxima of isochronal annealing curves were ascribed to the different number of jumps the vacancies had to perform in order to reach the sinks which are of different density and configuration after quenching and irradiation, respectively (§2.2.2.2).

The advocates of the *two-interstitial model* solved the problem by ascribing stage III after irradiation to migration and annihilation of the stable interstitial, by calling the first stage after quenching stage IV and ascribing it to migrating vacancies. The interstitial migrating in stage I (fig. 4) was thought to be a metastable configuration. Through 20 years, the open question of the stage III–stage IV origin energized research in the point-defect field. Considerable improvement of experimental methods was achieved in three directions: (i) determination of interstitial properties like configuration, dynamic behaviour, migration, recombination and clustering behaviour, (ii) improvement and understanding of the quenching procedure, especially with regard to impurity effects, (iii) detection of vacancy migration and clustering (positron annihilation, perturbed angular correlation, see §2.2.2.2). It is impossible to outline the many stages of the debate here. As an introduction to the controversy the reader is referred to SEEGER [1975], YOUNG [1978] and BALLUFFI [1978]. The two latter authors conclude that in many metals the experimental observations can satisfactorily be explained by the one-interstitial model, i.e., by the vacancy migration in stage III. The experimental basis of the papers by SCHINDLER et al. [1978] and FRANK et al. [1979], which was taken as a strong support of the two-interstitial model by the authors just at the time of YOUNG's [1978] and BALLUFFI's [1978] reports could not be confirmed later (LAUPHEIMER et al. [1981]).

In recent years, the intensity of the debate has diminished although the advocates of the two-interstitial model report on new deficiencies of the one-interstitial model and successful applications of the two-insterstitial model (KIENLE et al. [1983], SEEGER [1982], SEEGER and FRANK [1983]). The general argumentation regarding interstitial properties is, as before, based on observations which cannot be explained by the well-known stage I interstitial properties directly. The second set of parameters introduced by the second interstitial of the two-interstitial model generally allows easy explanation of such observations.

The central question of the controversy is the consistent explanation of stage III by vacancy migration. To the present author it seems to be very difficult to contradict the experimental results reported in the literature which favour vacancy migration in stage III (§2.2.2). Very characteristic is the situation for Au for which recent measurements in specially pre-treated electron-irradiated samples yielded $\Delta H_{\rm eff}^{\rm m} = 0.71$ eV in stage III (SONNENBERG and DEDEK [1982]). The treatment consisted of high-dose α-particle irradiations which resulted in a very high density of dislocation loops. This high loop density acting as a high fixed sink density for the migrating stage III defect suppresses all other possible reactions of this defect.

The model fitted to equilibrium, quenching and self-diffusion data by SAHU et al. [1978] yielded $\Delta H_{\rm 1v}^{\rm m} = 0.83$ eV (§2.2.2). According to SEEGER and FRANK [1982] the differing migration enthalpies are the evidence for the migration of an interstitial in stage III with $\Delta H_{\rm i}^{\rm m} = 0.71$ eV. To the present author this conclusion, although suggestive, seems to be premature. The primary quenching data yield $\Delta H_{\rm eff}^{\rm m}$ values between 0.55 eV and 0.63 eV, depending on the initial vacancy concentration. The result $\Delta H_{\rm 1v}^{\rm m} = 0.83$ eV is derived by fitting a model which assumes a substantial di-vacancy contribution. Di-vacancies are responsible (i) for the curvature of the $c_{\rm v}^0(1/T)$ curve which determines $\Delta H_{\rm 2v}^{\rm b}$ and (ii) for the experimental $\Delta H_{\rm v}^{\rm m} = 0.62$ eV. This model is certainly not unique with regard to its basic assumptions. The question whether the above curvature could be due to temperature dependence of ΔH_{1v}^{f} (VAROTSOS and ALEXOPOULOS [1982]), for example, must carefully be reinvestigated as well as the atomic processes during annealing of quenched resistivity samples. The electron-microscopic work on quenched gold clearly demonstrates the strong tendency for vacancy agglomeration. At the same time, there are reliable indications that the electric resistivity does not change during the formation of small clusters containing not more than five vacancies or so. Hence, resistivity recovery experiments can show only a later state of a complex vacancy-annihilation regime with certainly time-dependent sink concentrations.

3. Self-interstitials

3.1. Production of interstitial atoms

3.1.1. Questions of relevance in the present context

The interstitial formation energy of a few eV can easily be provided by irradiating a crystal with energetic particles. For example, an electron of 300 keV energy transfers 19 eV recoil energy to a Cu nucleus via head-on collision. The maximum energy transferred to Cu by fission neutrons of 2 MeV amounts to 125 keV. The fundamental problems of radiation damage in materials intensively stimulated research regarding particle–lattice-atom interaction and the Frenkel-defect production which results. In the present section we deal with this matter just to the extent required for an understanding of the problems in answering two fundamental questions: (i) Which atomic defect concentration c_d is produced by a given *fluence* Φ (time-integrated flux density $\emptyset$) of particles penetrating a crystal? (ii) What is the difference in the Frenkel-defect production by different irradiating particles, electrons and neutrons for example? Among the numerous difficult questions on the Frenkel-defect production these two are particularly important for the determination of specific (microscopic) defect properties from irradiation-induced changes of macroscopic material properties.

3.1.2. Atomic displacement cross-section for electron irradiation

The atomic fraction of particle–target-atom collisions for a fluence Φ is given by $\sigma\Phi$. The term *collision* ought to be specified precisely in order to render σ a uniquely defined quantity. Electrons penetrating a crystal interact with the electrons as well as with the nuclei of the target atoms. In metals, the electron–electron interaction produces heat and hence determines the cooling conditions for irradiation facilities. At the same time it causes an angular spread of the electron beam around the incident beam direction. This spread is of relevance for the angular resolution of threshold-energy determinations and for the flux-density determination in any kind of electron-irradiation experiment except for very thin samples. Details are given below.

References: p. 1208.

The electron–nucleus interaction causes the displacement of atoms from regular lattice sites to interstitial positions. The necessary specification of this collision is given by the minimum recoil energy required for permanent displacement of an atom. Only collisions with recoil energies T larger than the displacement threshold T_d are of relevance for the Frenkel defect production. The total cross-section can be written as the recoil-energy integral of the differential cross-section $d\sigma$,

$$\sigma(E) = \int_{T_d}^{T_{max}} \frac{d\sigma(E, T)}{dT} \, dT, \tag{7}$$

taken from the displacement threshold energy to the maximum transferred energy (head-on collision), E is the electron energy. The differential cross section for the scattering of a relativistic (Dirac) electron by a point nucleus was calculated by MOTT [1932]. Relativistic electron scattering prefers mean recoil energies and suppresses high ones as compared to Rutherford scattering, for which $d\sigma/dT \propto 1/T^2 = T_{max}^{-2} \cos^{-4}\theta$ holds, where θ is the starting angle of the recoiling atom *with respect to the incident electron-irradiation* (for a review see *Further reading*: Corbett [1966]).

The lattice structure causes T_d to depend on the recoil impact direction *with respect to the lattice orientation, ϑ*. As is shown in fig. 13 for Cu, the orientation-dependent threshold energy spans a *threshold-energy surface* in the (T, ϑ, φ)-space, where the angular coordinates ϑ and φ are taken with respect to the lattice orientations.

Irradiation of textureless polycrystals means randomness of the recoil impact directions with respect to the lattice orientations. If for T_d in eq. (7) the absolute minimum $T_{d,min}$ of the threshold-energy surface were taken, σ would be calculated much larger than the experimental total displacement cross-section. All those collisions with $T > T_{d,min}$ but with impact directions for which $T < T_d$ holds do not produce Frenkel defects. This effect of the threshold-energy surface is often considered phenomenologically by introducing the *displacement probability, p(T)* into eq.

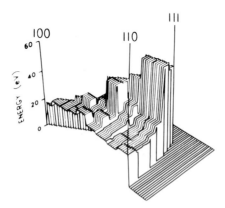

Fig. 13. Displacement threshold energy T_d as a function of the atomic recoil impact direction within the fundamental orientation triangle in Cu (after KING *et al.* [1981]).

(7), leading to the displacement cross-section

$$\sigma_d(E) = \int_{T_{d,\min}}^{T_{\max}(E)} p(T) \frac{d\sigma(E, T)}{dT} \, dT. \tag{8}$$

When the threshold-energy surface is known, $p(T)$ and hence σ_d can be calculated. This is true for all those particle irradiations for which $d\sigma/dT$ is known from nuclear physics. For the determination of the threshold-energy surface, such a particle irradiation of a single crystal would be desired which causes atomic displacements with just one recoil impact direction. By variation of the crystal orientation with respect to the recoil impact direction and by variation of the particle energy, the threshold energy surface could thus be scanned. Unfortunately, the statistical nature of particle interaction causes a distribution of recoil impact directions which fills the solid angle 2π (maximum angle between incident projectile and target-atom recoil impact is $\pi/2$) for any given incident particle direction. The distribution is governed by the angle-dependent differential cross-section of the particular type of particle–atom collision. Hence, any scanning of the threshold energy surface requires deconvolution of the measured data with respect to the angle-dependent differential cross-section. An additional problem arises for electron irradiation from the considerable beam spread upon passing a sample. It is due to the multiple scattering by target electrons. For electrons of 0.5 MeV, the angular distribution at the reverse side of a 25 μm thick Cu sample is such that about 40% of the electrons are scattered out of their incident direction by more than 40°. Review articles on earlier threshold-energy surface determinations are by SOSIN and BAUER [1968], VAJDA [1977] and JUNG [1981a].

The angular resolution of $T_d(\vartheta, \varphi)$ has been improved substantially during the last few years by applying high-voltage electron microscopy (HVEM) for in-situ defect production. Two measuring methods for defect production rates have been developed: (i) residual resistivity measurements in electron-microscope samples, i.e. samples of about 400 nm thickness and 0.1×0.1 mm^2 irradiated area (KING *et al.* [1980]) and (ii) rate measurements for nucleation and growth of interstitial-type dislocation loops (URBAN and YOSHIDA [1981]). The first method is applied at irradiation temperatures below 10 K, whereas the second one requires mobile interstitials, i.e., irradiation temperatures above 50 K (see §3.4). Both methods benefit by the small sample thickness, by the ease of sample tilting with respect to the beam direction and by the electron flux density which is some orders of magnitude larger than at common accelerator irradiations. The threshold energy surface shown in fig. 13 was obtained by resistivity-change rate measurements at six different electron energies and about 35 different crystal orientations, yielding about 200 data points altogether.

The displacement probability $p(T)$ derived from the threshold energy surface in fig. 13 is shown in fig. 14 with error bars. One realizes that calculations of the total displacement cross-section according to eq. (8) with such $p(T)$ data leads to uncertainties of 10–15%. A similarly exhaustive determination of $T_d(\vartheta, \varphi)$ as in Cu does not exist for any other metal, yet. As a consequence, defect production rates can generally not be calculated with satisfactory accuracy for given irradiation experiments by applying eq. (8). Another method often applied is the measurement of a standard quantity as the electric resistivity under the given irradiation condition and relating all measured property changes to that of the standard quantity. This procedure is described in § 3.2. The resistivity contribution per unit concentration of Frenkel defects, ρ_F, enters into the determination of $T_d(\vartheta, \varphi)$ as the displacement cross-section $\sigma_d(E, \theta)$ for the electron energy E and the incident direction θ is measured by the resistivity damage rate $d\rho/d\Phi = \rho_F\sigma_d$. If certain conditions regarding (i) the angular dependence of the displacement probability and (ii) the size of the available displacement energies T relative to $T_d(\vartheta, \varphi)$ are fulfilled, the value of ρ_F can be derived from such measurements as well as $T_d(\vartheta, \varphi)$ (ABROMEIT [1983]). The results in figs. 13 and 14 have been obtained with $\rho_F = 2 \times 10^{-4}$ Ωcm. But the optimal fit assuming the above mentioned conditions to be fulfilled yielded $\rho_F = 2.85 \times 10^{-4}$ Ωcm (KING and BENEDEK [1981], see also table 3). This result slightly deviates from that obtained from the second method of ρ_F determination, the Huang and diffuse X-ray scattering (see below) which yielded $\rho_F = 2.6 \times 10^{-4}$ Ωcm and 2.0×10^{-4} Ωcm, respectively (SCHILLING [1978]). This indicates that for Cu, ρ_F is still subject to about 30% uncertainty.

The accompanying measurement of a standard quantity, e.g., resistivity, as a measure of the defect concentration gains even more weight for irradiation experiments at high fluence levels or elevated temperatures. At low irradiation temperatures but high Frenkel-defect concentration in the sample not all defects produced according to the cross-section given by eq. (8) are stable. The spontaneous recombination reduces the effective production rate (review by WOLLENBERGER [1970]). It arises from the fact that vacancy and interstitial ought to be separated by a

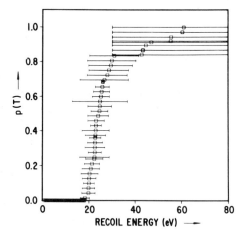

Fig. 14. Displacement probability versus recoil energy T_d as resulting from the threshold-energy surface in fig. 13. The error bars indicate the sensitivity of the threshold-energy surface fit to the measured damage rate data.

Table 3

Electric resistivity contribution per unit concentration of Frenkel defects $\rho_F[10^{-4}\,\Omega\,cm]$.

Metal	Constribution	Method [a]	Reference
Ag	2.1	$\Delta\rho_{irr}$	LUCASSON and WALKER [1962]
			ROBERTS et al. [1966]
Al	4.0	$\Delta\rho_{irr}$	JUNG [1981]
	3.9	HS & $\Delta a/a$	BENDER and EHRHART [1983]
Au	3.2	HS & $\Delta a/a$	EHRHART and SEGURA [1975]
Cd	5	$\Delta\rho_{irr}$	VAJDA [1977]
Co	15	$\Delta\rho_{irr}$	VAJDA [1977]
	16	HS & $\Delta a/a$	EHRHART and SCHÖNFELD [1982]
Cu	2.0	HS, DXS & $\Delta a/a$	SCHILLING [1978]
	2.85	$\Delta\rho_{irr}$	KING and BENEDEK [1981]
	2.8	HS & $\Delta a/a$	BENDER and EHRHART [1983]
α-Fe	30	$\Delta\rho_{irr}$	VAJDA [1977]
Mo	15	HS & $\Delta a/a$	EHRHART and SEGURA [1975]
	13	$\Delta\rho_{irr}$	VAJDA [1977]
Ni	6	$\Delta\rho_{irr}$	LUCASSON and WALKER [1962]
	7.1	HS & $\Delta a/a$	BENDER and EHRHART [1983]
Pt	9.5	$\Delta\rho_{irr}$	JUNG et al. [1973]
Ta	16	$\Delta\rho_{irr}$	JUNG et al. [1973]
			BIGET et al. [1979]
Zn	15	$\Delta\rho_{irr}$	VAJDA [1977]
	15	HS & $\Delta a/a$	EHRHART and SCHÖNFELD [1982]

[a] $\Delta\rho_{irr}$: electric resistivity of irradiated samples; HS: Huang scattering;
$\Delta a/a$: lattice parameter, DXS: diffuse X-ray scattering.

minimum distance in order to be stable even at $T = 0$. This minimum distance creates the volume of spontaneous recombination around the defects which must be subtracted from the total sample volume in order to get the volume available for defect production.

At elevated temperatures the thermally activated annihilation processes as treated in § 3.3.2.4 occur and reduce the primarily induced defect production rate (WOLLEN-BERGER [1970, 1978]). These complications cause large uncertainties in the determination of defect production rates by means of eq. (8).

3.1.3. Frenkel-defect production by neutrons and ions

Electron irradiation is the basic method of Frenkel-defect production when properties of single Frenkel defects are to be studied. Nuclear reactor technology introduced neutron-irradiated materials and raised the question concerning the underlying damage structure. Neutrons are "heavy" projectiles as compared with electrons. The energy transfer to target atoms for head-on collisions is given by the classical formula $T_{max} = 4\,MmE/(M + m)^2$, where E is the projectile energy, and m and M are the masses of projectile and target atom. (Reviews on radiation–solid interaction are by LEHMANN [1977], LEIBFRIED [1965] and SEITZ and KOEHLER

References: p. 1208.

[1956].) For Cu we have $T_{max} \approx E/16$ and the average recoil energy $\langle T \rangle = 5 \times 10^4$ eV for 1.6 MeV fission neutrons. This energy is three orders of magnitude larger than the displacement threshold energy, which leads to damage features completely different from those of electron irradiation. Before looking more closely at details of this type of damage, we realize that bombarding a sample with ions of the sample material (called self-ion irradiation) having 5×10^4 eV incident energy must have a similar effect as the fission neutron irradiation. Certainly self-ion irradiation is highly welcome as a form of simulation irradiation instead of true neutron irradiation because ion irradiation combined with any sample property measurement is much more easily achievable than the corresponding neutron irradiation (Ishino *et al.* [1977]; Gabriel *et al.* [1976]; *further reading*: Proc. workshop correlation of neutron- and charged particle damage 1976, Stiegler [1976]). In practice, a self-ion energy of 5×10^4 eV would not be appropriate for damage production in the bulk of samples as the penetration depth of such ions measures only a few atomic distances. With self-ions of 0.3–1 MeV energy, samples for electron microscopy (100–200 nm thickness) or electric resistivity measurements can be homogeneously damaged. If even thicker samples are required, as for mechanical property measurements for example, neutron irradiation is simulated by high-energy light-ion irradiation. Light ions are protons, deuterons and α-particles and energies are 20 MeV, for example. For such projectiles the penetration depth is of the order of 10–100 μm. The induced recoil energy spectra of the target atoms are still similar enough to those caused by reactor neutrons to assure satisfactory simulation conditions. The simulation aspect has made ion irradiation an important and frequently applied defect production method.

Lattice atoms which obtained recoil energies as large as 100 keV themselves act as projectiles in the lattice and by collisions with other lattice atoms they distribute their primary recoil energy. In this way a primary knocked-on-atom initiates a *collision cascade* which, according to the transferred energies, leads to a displacement cascade which, in turn, ends up in a defect cascade. The ranges in Cu of Cu^+ ions with energies 5×10^4 eV, 2×10^4 eV, and 5×10^3 eV are 11.5 nnd, 4.7 nnd, 1.1 nnd (nnd – nearest-neighbour distances), respectively. Hence, the density of collisions and defects strongly increases towards the end of the cascade.

As long as the collisions can be treated as binary collisions the process can analytically be described by a linear Boltzmann-type transport equation, as was originally done by Kinchin and Pease [1955]. These authors obtained the ratio $n = T/2T_d$, where T is the energy of the primary knock-on atom displacing n atoms in a cascade with recoil energies larger than the displacement threshold T_d. This treatment can only be a first-order approximation as the binary collision model breaks down as soon as the recoil energies become of the order of 10^4 eV. The multiple-collision events in this energy region form so-called *displacement spikes*, illustrated in figs. 15a, b. The corresponding result for n as obtained from computer simulation reads $n = 0.8\epsilon(T)/2T_{d,eff}$ for $T \geqslant 2.5T_{d,eff}$, $n = 1$ for $T_{d,eff} \leqslant T \leqslant 2.5T_{d,eff}$ and $n = 0$ for $T < T_{d,eff}$, where $\epsilon(T)$ is the damage energy, which is the total energy T diminished by the losses due to electron excitations (Norgett *et al.* [1974]). These excitations result from the Coulomb interaction of the moving ion with the electrons of the solid. By electron–phonon interaction the energy is lost as heat. In Ni, for example, the excitation losses become appreciable around 10^4 eV and measure about 60% of the total energy at 10^6 eV. By taking into account the number of Frenkel defects produced per primary recoil atom, we obtain for the total displacement cross-section:

$$\sigma_d(E) = \int_{T_{d,eff}}^{T_{max}(E)} \frac{d\sigma(E, T)}{dT} n(T)\, dT, \qquad (9)$$

where E is the energy of the incident particles (neutrons or ions) and $d\sigma/dT$ is the differential cross-section of a target atom to obtain energies T by collisions with the

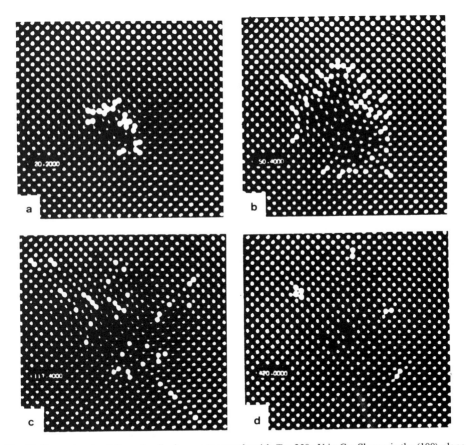

Fig. 15. Computer simulation of a displacement cascade with $T = 250$ eV in Cu. Shown is the $\langle 100 \rangle$ plane. The primary recoil impact direction was $\langle 110 \rangle$ given to the atom at that now vacant site in the cascade area which lies closest to the left lower corner of fig. 15a at $t = 0$. Atoms imaged by bright dots have more than 0.5 eV. In (b): $t = 6 \times 10^{-14}$ s (real time); in (c): $t = 3.5 \times 10^{-13}$ s; in (d), when $t = 10^{-12}$s a stable configuration is reached, consisting of five Frenkel defects. (From LEHMANN *et al.* [1981].)

References: p. 1208.

incident particles, and $T_{d,eff}$ an effective threshold energy suitably chosen to replace the anisotropic threshold-energy surface by a spherical surface. For neutrons and ions the differential cross-sections are entirely different functions of T.

The validity of eq. (9) for ion irradiation on Cu has extensively been studied by means of electric resistivity measurements for a wide range of recoil energies by AVERBACK et al. [1978]. The important finding is that the quantity $\sigma_{d,meas}/\sigma_{d,calc} = \xi$, called the *efficiency*, falls from unity around 10^3 eV to 0.5 at 10^4 eV and to 0.4 at 10^5 eV. A number of possible reasons for ξ being smaller than unity can be given. The importance of Frenkel defect recombination will certainly increase with increasing energy density of the cascade. Furthermore, the way recombination has been considered in deriving n in eq. (9) can of course not guarantee correct simulation of nature with respect to the many-body events among Frenkel defects of high density. The sequence of images in fig. 15 shows even in the computer model how drastically the number of Frenkel defects is reduced as the collision cascade decays.

Information on the structure and volume of defect cascades comes from computer simulation as well as from experiments. Figure 16 gives a three-dimensional impression of a computer simulation result. Two features are typical and important: (i) The vacancies are frequently clustered and are surrounded by single interstitials. This picture was postulated earlier by SEEGER [1958]. (ii) The defect density strongly fluctuates and the structure could be described as a complex of subcascades. This feature is typical for high-energy cascades.

The defect cascades can be imaged by electron microscopy (reviews are by JÄGER [1981], KIRITANI [1982] and SHIMOMURA et al. [1982]) and field-ion microscopy (reviews are by WAGNER [1982] and SEIDMAN [1978, 1976]). A particularly enlightening example of electron microscopy results is shown in fig. 17. By producing collision cascades in long-range ordered Cu_3Au, disordered volume regions are created which are the images of the collision cascades. By using the superlattice reflections and dark-field imaging, the disordered zones appear as dark contrast within the brightly imaged ordered matrix. Subcascade formation can clearly be seen in fig. 17b. The average cascade diameter was found to lie between 3 nm at $T = 10$ keV and 12 nm at $T = 100$ keV in satisfactory agreement with calculated values (SIGMUND [1972]) for $T \leqslant 50$ keV. This example shows the direct phase transformation by cascade-type damage. Another important aspect of the resulting defect cluster structure is their nature as point-defect sinks. Both aspects are treated in more detail in §4.

3.2. Determination of Frenkel-defect concentrations

Specific properties of Frenkel defects are determined by measuring a certain property change of a macroscopic sample upon irradiation and relating the measured change to the underlying atomic defect concentration, c_d. This procedure requires knowledge of c_d. Determination of c_d from the irradiation parameters as treated in the foregoing section is subject to substantial uncertainty. In order to avoid such uncertainties it has become more usual to measure the electric residual

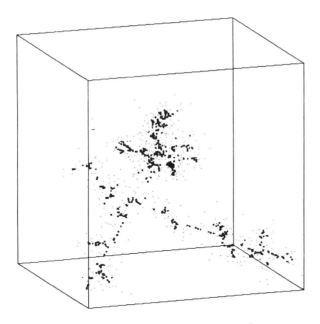

Fig. 16. Computer simulation of a displacement cascade with $T = 2 \times 10^5$ eV in a Cu model. The primary knock-on occurred in the lower left. The squares are vacancies and the (smaller) dots are interstitials. Cube edge measures 125 nearest-neighbour distances. (From HEINISCH [1981].)

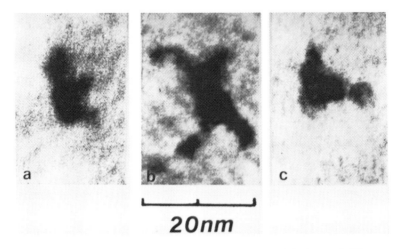

Fig. 17. Disordered zones in long-range ordered Cu_3Au irradiated with Cu^+ ions of different energy: (a) $T = 5 \times 10^4$ eV; (b) $T = 10^5$ eV; (c) separated subcascade at $T = 10^5$ eV. (From JENKINS and WILKENS [1976].)

References: p. 1208.

resistivity change $\Delta\rho$ of a sample made of a proper standard material for which the constant of proportionality ρ_F is known and to calculate $c_d = \Delta\rho/\rho_F$. Resistivity samples can generally be irradiated simultaneously with the sample under specific investigation. Thus ρ_F plays the role of a defect-concentration standard. As has been shown in the foregoing section, ρ_F can be derived from resistivity-damage rate measurements in single crystals under electron irradiation at low temperatures and under suitable variation of crystal orientation and electron energy. For the most extensively investigated Cu the error limits of ρ_F are certainly not less than $\pm 30\%$. For other metals the uncertainty of ρ_F is even larger. Values are given in table 3.

An entirely different method of determining c_d is the proper evaluation of the X-ray scattering intensity in the Huang range and in the range between the Bragg reflections (§3.2.3.2, also ch. 12, §3.3). The Huang scattering intensity is proportional to $c_d[(\Delta V_{1i}^{rel})^2 + (\Delta V_{1v}^{rel})^2]$ (HAUBOLD [1975], DEDERICHS [1973], TRINKAUS [1972]). By determining the lattice constant change, $\Delta a/a$, one obtains $c_d(\Delta V_{1i}^{rel} + \Delta V_{1v}^{rel})$. From both results, ΔV_{1i}^{rel} as well as c_d can be derived once ΔV_{1v}^{rel} is known from other measurements. Resistivity measurements on the same sample or under the same irradiation conditions then give ρ_F (table 3). The error limits amount to about $\pm 20\%$ (SCHILLING [1978]).

From the comparison of measured and calculated absolute intensities of the diffuse scattering between the Bragg peaks, c_d and hence ρ_F can be derived as well.

3.3. Interstitial properties

3.3.1. Results of model calculations

Methods for calculation of point-defect properties are provided by the theories of lattice statics and dynamics (reviews are by DEDERICHS and ZELLER [1980], LEIBFRIED and BREUER [1978] MARADUDIN *et al.* [1971], LUDWIG [1967]) and by computer simulation (reviews are by JOHNSON [1973], GEHLEN *et al.* [1972]). Computer simulation was introduced into the point-defect research field by GIBSON *et al.* [1960] and ERGINSOY *et al.* [1964]. They studied the fundamental displacement processes initiated by recoiling atoms and with that gave the first hints for an understanding of the shape of the threshold-energy surface (§3.1.2) on the basis of replacement-collision sequences (JUNG [1981b], URBAN and YOSHIDA [1981], TENEN-BAUM and DOAN [1977], VAJDA [1972], ROTH *et al.* [1975], BECKER *et al.* [1973], WOLLENBERGER and WURM [1965]).

Such computer simulations are based on the numerical solution of the equations of motion for a set of atoms arranged as in the crystal and coupled by spring forces which simulate the actual lattice–atom interaction. Critical aspects of the method are the choice of the the interaction potentials and the adjustment of the boundary conditions for the necessarily size-limited model crystal in a way which represents the correct embedding in an elastic continuum. The potentials have generally been chosen so as to generate two-body central forces and to match known physical properties, such as the elastic constants. Mostly, a certain functional form of the potential as the Morse, Lennard–Jones or Born–Mayer type is taken to be valid and the respective coefficients are determined by matching the model's properties to various physical properties of the modelled metal. In other cases, purely empirical potentials have been developed by the matching procedure. Potentials which result from pseudopotential theory have also been used. Unfortunately, the noble and transition metals for which most of the experimental defect research has been done are not favourable candidates for pseudopotential application. Computer simulations have been applied to study not only defect production mechanisms but also static and dynamic properties of defects (§2.1).

For analytical calculations of the interstitial properties the real space relaxation model is often replaced by the normal coordinate expansion model. The latter was introduced by KANZAKI [1957] and treats the defect–lattice interaction as a source function (*Kanzaki forces*) for the displacement in a harmonic lattice. The force equations for equilibrium are Fourier-transformed, yielding variables which are Fourier inverses of real space displacements. All energy changes are related to the first and second derivatives of the interactions, and these derivatives can directly be matched to the force constants as following from Born–v. Karman fits to the phonon dispersion curves. The properties of primary interest are the activation enthalpies of formation and of migration of the interstitial. The latter is determined as the difference between the interstitial formation enthalpy in the saddle point configuration and that in the equilibrium configuration, in accordance with common use in rate theory.

3.3.1.1. Activation enthalpies of formation in equilibrium and saddle-point configurations.
Specific feature of the self-interstitial is the strong lattice distortion with its large displacements in the neighbouring atom shells (0.14 nnd for the nearest neighbours of the two atoms forming a split interstitial configuration in Al (HAUBOLD [1976])). Consequently the lattice relaxation significantly influences the repulsive energy contribution. One easily realizes that the large capacity of computers is very helpful for handling just this contribution as accurately as possible.

The electronic contribution consists of the electron-energy change due to the volume change of the crystal upon insertion of the interstitial atom. Those results given in table 4 which were not obtained by computer simulation essentially deviate by different treatment of the electronic contribution.

In order to find the stable interstitial configuration, the formation enthalpy has been calculated by a number of authors for all configurations shown in fig. 18. In table 4 only minimum and maximum values are given. They indicate that with one exception the difference between minimum and maximum of the formation energy is less than 15% of the minimum value, and detailed comparison would show that there are always two or three different configurations for which the formation energy deviates by less than 5%. The origin of these small differences lies in the large relaxation effect. Without the lattice relaxation the formation enthalpy would be larger by nearly one order of magnitude (DEDERICHS et al. [1978]). Rearrangement of the interstitial atom or atom pair does not cause a significant change of the relaxation volume (see table 4) and, hence, changes the formation enthalpy only by a

References: p. 1208.

Table 4
Calculated properties of self-interstitials.

Metal	Configuration	ΔH_i^f (eV)	ΔV_i^{rel} (at. volume)	ΔH_i^m (eV)	Reference
Cu	⟨100⟩–split	5.07–5.82		< 0.24	HUNTINGTON [1942, 1953]
	Octahedral	5.14–6.09			
	Octahedral	2.5–2.6	1.67–2.01		TEWORDT [1958]
	⟨100⟩–split	3.2	1.10–1.25		
	Octahedral	2.73–3.42	1.21–1.78		SEEGER and MANN [1960]
	⟨100⟩–split	2.187	1.126	0.103	BENNEMANN and TEWORDT [1960]
	Octahedral	2.43–2.44	1.219–1.441		
	⟨100⟩–split	4.351		0.090	BENNEMANN [1961]
	⟨100⟩–split	4.139	2.20	0.05	JOHNSON and BROWN [1962] [a]
	Crowdion	4.840	2.57		
	⟨100⟩–split	2.47–2.84	1.37–1.44	0.46–0.57	SEEGER et al. [1962] [a]
	⟨111⟩–split	3.83–4.27	1.43–1.65		
	⟨100⟩–split	3.39	1.34	< 0.06	DOYAMA and COTTERILL [1967] [a]
	Tetrahedral	3.70	1.38		
	⟨100⟩–split	3.42	1.5	0.13	DEDERICHS et al. [1978] [a]
	Tetrahedral	3.89	1.48		
Ni	⟨100⟩–split	4.08	1.7	0.15	JOHNSON [1966]
	Crowdion	4.10		0.04	
	⟨111⟩–split	4.24		0.13	
α-Fe	⟨100⟩–split	4.6	2.34	0.21	JOHNSON [1965]
	Octahedral	5.73			DEDERICHS et al. [1978]
Al	⟨100⟩–split	2.89		0.15	LAM et al. [1980]
Mg	A–split	2.36			IMAFUKU et al. [1982]
	Hexahedral	2.66			

[a] Gives data on more configurations than listed in this table.

small amount. The data reported by DEDERICHS et al. [1978] are obtained by computer simulation with a Morse potential modified in order to soften the strong repulsive core of this potential and to fit (in addition to other properties – lattice constant, compression modulus, vacancy formation enthalpy) the relaxation volume of the interstitial to the experimental value (see table 2). For Al an inter-ionic potential as derived from pseudo-potential theory was used.

The activation enthalpy for migration of the Cu interstitial was accordingly obtained to be around 0.1 eV, with the exception of 0.5 eV by SEEGER et al. [1962]. This difference was one of the sources which energized the stage III controversy (§ 2.3). The small difference between the formation enthalpies of the saddle point configuration and that of the equilibrium position shows again that the activation enthalpy of formation depends rather weakly on the interstitial configuration.

Details of possible migration steps can conveniently be investigated by computer simulation. The most probable migrational step obtained for the fcc structure

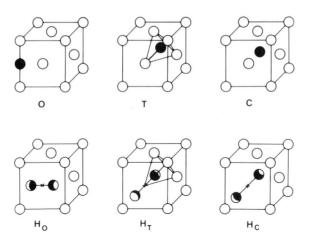

Fig. 18. Self-interstitial configurations in the free lattice. Symmetries: O – octahedral; T – tetrahedral, C – crowdion; H_O, H_T, H_C – dumbbell having axis along $\langle 100 \rangle$, $\langle 111 \rangle$, and $\langle 110 \rangle$, respectively.

(DEDERICHS *et al.* [1978], LAM *et al.* [1980]) is shown in fig. 19a. It consists of a translational motion of the centre of gravity of the dumbbell by one atomic distance and a rotational motion of the dumbbell axis by 90°. It should be noted that any other form of motion of the $\langle 100 \rangle$-split interstitial requires a considerably higher activation enthalpy. Especially the 90° axis rotation with fixed centre of gravity requires a four times larger activation enthalpy.

For the bcc α-Fe potential, the elementary jump is found again to be a translational step of the centre of gravity of one atomic distance and rotation of the axis by 60° as shown in fig. 19b. The activation enthalpy is 0.21 eV. The value of 0.33 eV according to JOHNSON [1964] could not be reproduced by DEDERICHS *et al.* [1978]. Contrary to the situation in the fcc structure, the pure 90° axis rotation requires only a little larger activation enthalpy, namely 0.25 eV.

3.3.1.2. Dynamic properties. Computer simulation furthermore revealed the occurrence of low-frequency resonant modes of the $\langle 100 \rangle$ dumbbell in the fcc structure besides high-frequency localized modes (SCHOLZ and LEHMANN [1972], IMAFUKU *et*

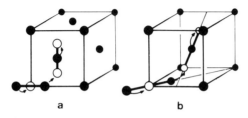

Fig. 19. Migrational steps (a) of the $\langle 100 \rangle$-split interstitial in the fcc lattice and (b) the $\langle 110 \rangle$-split in the bcc lattice, according to computer simulation results.

References: p. 1208.

al. [1982]). This unusual occurrence of both types of vibrational modes at one defect originates from the highly compressed lattice around the interstitial and its special configuration (DEDERICHS *et al.* [1973]). For the oppositely directed vibrations of the two dumbbell atoms along $\langle 100 \rangle$ as shown in fig. 20a, the small equilibrium separation between the two atoms (0.77 nnd) leads to a very strong force-constant which couples the two atoms. This constant leads to a localized mode and for modified Morse potential one obtains the mode A_{1g} shown in fig. 21, which lies well above the maximum lattice frequency, ω_{max}.

For the displacement directions of the two dumbbell atoms as shown in fig. 20c the strongly compressed spiral spring between the two atoms exhibits a negative bending spring component which acts perpendicular to the spiral spring axes. Quantitative evaluation shows that the force-constant of the negative bending spring becomes comparable to that of the restoring force of the perfect lattice. The resulting force-constant is small so that the librational mode in fig. 20c is a resonant mode with a very low frequency, as shown in fig. 21 for E_g. Another resonant mode (A_{2u}) is excited with the direction of atomic motion as shown in fig. 20b. As the motions in fig. 20b and c also strain the compressed springs with the nearest neighbours of the dumbbell, localized modes are excited as well (see A_{2u} and E_g beyond ω_{max} in fig. 21). The low-frequency resonances of the dumbbell lead to comparatively large thermal displacements of the interstitial atoms. In fig. 22 the mean squares of the atomic displacements are compared for dumbbell atoms and those in perfect lattice positions, as calculated for a Morse potential according to DEDERICHS *et al.* [1978]. It becomes obvious that the resonant modes are thermally populated already around 30 K, the beginning of stage I_D recovery (§ 3.3.2.4) in Cu. Indeed, large amplitudes of the resonant modes lead directly to the saddle point configuration for the migration jump. The flatness of the energy contour along the migration jump path is a direct consequence of the negative bending spring effect.

The presence of resonant modes should cause a low-temperature maximum in the temperature dependence of the specific heat (ZELLER and DEDERICHS [1976]), and some experimental indication for its occurrence has been found in electron-irradia-

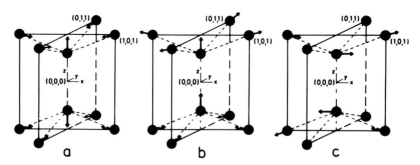

Fig. 20. Localized and resonance modes of the $\langle 100 \rangle$-split interstitial: (a) localized mode A_{1g} (see fig. 21); (b) resonance and localized mode A_{2u}; (c) resonance and localized mode E_g. (From DEDERICHS *et al.* [1978].)

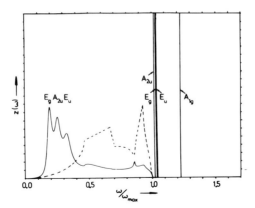

Fig. 21. Local frequency spectrum of the $\langle 100 \rangle$-split interstitial (averaged over all directions) for a modified Morse potential (from DEDERICHS *et al.* [1978]).

ted Cu (MONSAU and WOLLENBERGER [1980]). Experimental difficulties, however, prevented a reliable error analysis of the result.

The influence of the resonance modes on the temperature dependence of elastic moduli was observed by HOLDER *et al.* [1974]. They derived $\omega_R = \omega_{max}/8$ which is in good agreement with the calculated modes in fig. 20. The most effective consequence of the resonant mode is the considerable elastic polarizability. In fig. 23 the displacement directions of the dumbbell atoms and their nearest neighbours are shown for two different applied shear stresses and for uniform compression. Obviously the $\langle 100 \rangle$-shear stress excites the resonant mode E_g, and computer experiments show that the rotation angle of the dumbbell axes is by a factor of twenty larger than the shear angle, simply because of the bending spring action. The

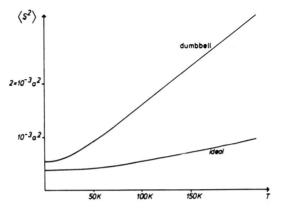

Fig. 22. Averaged square of the thermal displacements of one of the $\langle 100 \rangle$-split interstitial atoms compared with that of a regular lattice atom (from DEDERICHS *et al.* [1978]).

References: p. 1208.

presence of dumbbell interstitials in a crystal therefore must decrease the elastic constant c_{44} with respect to a perfect crystal. Large decreases of the shear moduli of irradiated polycrystals have indeed been observed many years ago (KÖNIG *et al.* [1964], WENZL *et al.* [1971]). This negative sign of the modulus change was a long-standing problem in theory and had not been solved earlier than the detection of the resonant mode in the computer experiments.

The existence of a para-elastic polarizability of interstitials was discussed by DEDERICHS *et al.* [1978]. For anisotropic defects it could be even one order of magnitude larger than the dia-elastic polarizability. It can directly be observed, however, only for such interstitials which are able to reorient in an external field without performing migration steps towards annihilation. On the other hand, migration in an external field will be influenced by the para-elastic polarizability. This effect might be of great importance for irradiation-induced creep (BULLOUGH and WILLIS [1975], see also §4).

3.3.1.3. Arrhenius behaviour of diffusion. The low activation enthalpy for interstitial migration and the low temperature at which it takes place (stage I_D recovery in Cu around 35 K, §3.3.2.4) raised the question whether Arrhenius behaviour, as it has always been assumed for stage I recovery interpretation, could actually be justified (FLYNN [1975]). Indeed, the majority of vibrational lattice modes occupy their ground states of zero-point motion at temperatures much lower than the Debye temperature ($T_{Deb} = 310$ K for Cu). Hence little justification can be given for a classical description of the migration process from this point of view. On the other hand, computer experiments clearly show that the migrational step is always a consequence of sufficient excitation of both resonant modes of the split interstitial (§§3.3.1.2 and 3.3.2.5). Excitation of these modes is possible only by phonons of this frequency. Since $\Delta H_i^m = 30\,\hbar\omega_R$, extremely high excitations must be provided by incoming phonons. The fluctuation possibility for this state is essentially classic, i.e., the migration velocity depends on $\exp(-\Delta H_i^m/kT)$. This is indeed the classic Arrhenius behaviour because the phonon states corresponding to ω_R are classically populated already at low temperatures. The motion of the atoms neighbouring the hopping dumbbell observed in computer simulations suggests a description in terms of the classic activated state.

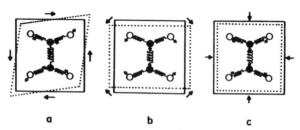

a   b   c

Fig. 23. Homogeneous deformation of a crystal containing a $\langle 100 \rangle$-split interstitial: (a) $\langle 100 \rangle$ shear (modulus c_{44}); (b) $\langle 110 \rangle$ shear (modulus $(c_{11} - c_{12})/2$); (c) compression (modulus $(c_{11} + 2c_{12})/3$).

3.3.1.4. Multiple interstitials. In the foregoing sections it has been shown that interstitials are dilatation centres with large relaxation volumes. Two interstitials approaching one another interact via these atomic displacement fields.

The interaction energy can be calculated by means of Kanzaki forces (KANZAKI [1957]) simulating the defect-induced displacement field. For distances large compared to the range of the Kanzaki forces (a few atomic distances), the energy can be calculated by means of the Green's function of the elastic continuum. From a multipole expansion of the forces one obtains a leading term of the direct interaction, the dipole–dipole interaction or the so-called first order size interaction. Here, the interaction energy decreases with increasing distance, r, as r^{-3}. It depends on the direction of the defect-connecting line with respect to the axes of the elastic anisotropy.

Therefore, the interaction of two-point defects generally is attractive and repulsive as well. For cubic crystals and isotropic defects one obtains ESHELBY's formula [1956] as first-order term of a perturbation expansion with respect to the anisotropy parameter $d = c_{11} - c_{12} - 2c_{44}$:

$$E_{int} = -\frac{15}{8\pi} d \left(\frac{5}{3c_{11} + 2c_{12} + 4c_{44}} \right) \frac{\Delta V_1^{rel} \Delta V_2^{rel}}{r^3} \left(\frac{3}{5} - \sum_i \frac{r_i^4}{r^4} \right). \qquad (10)$$

For $d > 0$ and interstitial–interstitial interaction, i.e., $\Delta V_1^{rel} = \Delta V_2^{rel} > 0$ we have an attractive interaction along $\langle 100 \rangle$ and repulsive ones along $\langle 110 \rangle$ and $\langle 111 \rangle$. For $d = 0$ the multipole interaction does not vanish. The influence of higher-order terms of the perturbation expansion was studied by DEDERICHS and POLLMANN [1972]. A numerical solution free of approximations was given by MASAMURA and SINES [1970].

Higher multipole interactions decrease with r^{-5} or even faster (SIEMS [1968], HARDY and BULLOUGH [1967]). Of special interest is the induced interaction based upon the polarizability of the defects. It depends on r as r^{-6} (SIEMS [1968]). For the interaction of two interstitials in Cu or Al with their high polarizabilities, the induced-interaction energy is comparable to the dipole interaction only for $r \leqslant 3$ atomic distances (TRINKAUS [1975]).

The above described relationships are certainly not applicable to interstitials approaching one another as close as necessary to form multiple interstitials. This range of interaction has been studied by computer simulation (SCHOBER [1977], SCHOBER and ZELLER [1978] and INGLE *et al.* [1981]). Quite detailed information on static and dynamic properties of di- and tri-interstitials for different Cu potentials were obtained. The stability of configurations was studied up to clusters consisting of no fewer than 37 interstitials. The activation enthalpies of binding were found to be larger than the activation enthalpies of migration for all cases. Dissociation of multiple interstitials must therefore not be expected according to this fcc model. The relaxation volume change per clustering interstitial amounts to -5% to -15% of ΔV_i^{rel}. The formation enthalpy per interstitial decreases by about 30% when going from two to ten interstitials.

References: p. 1208.

The stable di-interstitial consists of parallel dumbbells at nearest-neighbour sites, tilted by a small angle ($< 10°$) in the $\langle 110 \rangle$ plane. Stable tri-interstitials are formed by mutually orthogonal dumbbells at nearest neighbour sites. Adding one atom in the octahedral lattice position to the tri-interstitial configuration yields the stable four-interstitial. Larger clusters arise from the four-interstitial cluster by adding further dumbbells such that each is equidistant to the central octahedral interstitial and is aligned orthogonally to its nearest-neighbour dumbbell. Such three-dimensional structures are more stable than two-dimensional ones for fewer than nine interstitials and are less stable than these for more than 13 interstitials. The stable two-dimensional clusters are platelets of octahedral interstitials on $\langle 111 \rangle$ planes.

The activation enthalpy of di-interstitial migration was found to lie below or above that of mono-interstitial migration, depending on the details of the potential. For the higher-order clusters, the migration enthalpies increase rapidly with the number of interstitials in the cluster.

The dynamic behaviour of the di-interstitial is essentially similar to that for the mono-interstitial. A variety of resonant and localized modes occur. According to the resonant modes the thermal displacements of the dumbbell atom are nearly as large as for the single dumbbell. The change of the elastic constant c_{44} per interstitial is even larger than that of the single dumbbell. Although the quantitative results depend significantly on the details of the pair-interaction potential used for the simulation, the general dynamic behaviour makes at least di-interstitials a very interesting object for experimental studies.

3.3.2. Experimental methods and results

3.3.2.1. Relaxation volume.
Introduction of an additional atom into a perfect finite crystal enhances the volume of this crystal by the relaxation volume of the self-interstitial, ΔV_i^{rel}. Because of possible relaxation anisotropies, interstitials of concentration c_i must be introduced randomly with respect to the anisotropy axes in order to produce the correct total volume change $c_i \Delta V_i^{rel}$.

The method of determining ΔV_{1i}^{rel} is given by the appropriate evaluation of the Huang scattering (EHRHART [1978]) or the diffuse scattering between the Bragg peaks (HAUBOLD and MARTINSEN [1978], §§3.2 and 3.3.2.2 also ch. 12, §3.3). As the relaxation volumes of interstitial and vacancy enter quadratically into the Huang scattering intensity and ΔV_v^{rel} is considerably smaller than ΔV_i^{rel}, the former needs to be known only approximately. The experimental results are collected in table 5. The deviation of ΔV_i^{rel} from one atomic volume can be imaged as that additional volume which must be spent because of the perturbation of the regular packing of spheres by the inserted interstitial atom. Perturbation of the fcc structure obviously requires more additional volume than the less dense bcc structure.

3.3.2.2. Configuration.
Experimental information on interstitial configurations comes from diffuse X-ray scattering and mechanical (§3.3.2.5) or magnetic relaxation experiments. The open question of the potential existence of two interstitial configurations in one and the same metal as postulated by the two-interstitial model generated a strong impact on the development of both the theory of the point-

Table 5
Relaxation volume of self-interstitials [a].

Metal	ΔV_i^{rel} (at. volume)	Measurement	Reference
Al	1.9	$\Delta a/a$	WAGNER et al. [1970]
	1.9	DXS & $\Delta a/a$	HAUBOLD [1975]
	1.9	HS & $\Delta a/a$	BENDER and EHRHART [1983]
Cd	11–19	HS & $\Delta a/a$	EHRHART and SCHÖNFELD [1982]
Co	1.5	HS & $\Delta a/a$	EHRHART and SCHÖNFELD [1982]
Cu	1.3	$\Delta a/a$	DWORSCHAK et al. [1972]
	1.45	DXS & $\Delta a/a$	HAUBOLD and MARTINSEN [1978]
	1.7	HS & $\Delta a/a$	BENDER and EHRHART [1983]
Mo	1.1	HS & $\Delta a/a$	EHRHART [1978]
Ni	1.8	HS & $\Delta a/a$	BENDER and EHRHART [1983]
Pt	2.0	$\Delta a/a$	HERTZ and PEISL [1975]
Zn	3.5	HS & $\Delta a/a$	EHRHART and SCHÖNFELD [1982]
Zr	0.6	HS & $\Delta a/a$	EHRHART and SCHÖNFELD [1982]

[a] Glossary see tables 1–3.

defect-induced diffuse X-ray scattering (DEDERICHS [1973], TRINKAUS [1972]) and the measuring technique (EHRHART et al. [1974]).

As the result of these efforts, the configuration of the interstitial in Al, Cu, Ni, Mo, Fe and Zn has been determined. For the fcc metals the $\langle 100 \rangle$ split interstitial, for the bcc the $\langle 110 \rangle$ split and for the hexagonal Zn the $\langle 0001 \rangle$ split interstitial has been found.

The analysis of diffuse X-ray scattering makes use of the information which is included in the intensity profile scattered outside the Bragg reflections due to the imperfect crystal. Here the relevant imperfections are the atomic displacements around interstitial atoms and vacancies. An order-of-magnitude comparison of the different scattering contributions to the total diffuse scattering cross-section is shown in fig. 24, according to EHRHART et al. [1974]. The interesting intensity profile occurs on a background one to two orders of magnitude larger than this. Information can be obtained only by a careful subtraction of the intensity scattered by an interstitial-free crystal from that of an interstitial-containing one.

Owing to the minor lattice distortions around a vacancy (§2.1) the diffuse intensity caused by this defect is only a small correction to the interstitial contribution.

The diffuse intensity profile contains the symmetry of the atomic distortions as well as the distortion strength, i.e. the interstitial atom extension. The unknown configuration is determined by comparing calculated intensity profiles with the measured ones. Figure 25 shows calculated scattering profiles for three different interstitial configurations in Al and the comparison with measured data. The agreement between measured and calculated data for the $\langle 100 \rangle$ split interstitial is convincing. The profiles for this configuration have been calculated with a distance

References: p. 1208.

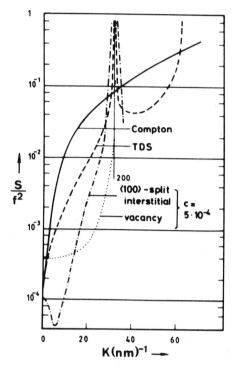

Fig. 24. Cross-sections for Compton scattering, thermal diffuse scattering (TDS) at 4 K and point-defect X-ray scattering by Al as a function of the scattering wave vector ($\lambda = 15.4$ nm) (after EHRHART *et al.* [1974]).

between the two dumbbell atoms of 0.85 nnd of Al. The four nearest neighbours of the dumbbell were found to be displaced outwards by about 0.07 nnd. The long-range atomic displacement field around lattice defects determines the scattering at small values of the scattering vector, q. Measurements close to the Bragg peaks (Huang scattering) can therefore resolve the symmetry of interstitials (EHRHART [1978]). Although high-angular-resolution techniques must be applied, the overall experimental expenditure and the requirements for sample quality are less exceptional than in the case of diffuse X-ray measurements between the Bragg peaks. This advantage originates from the much higher defect-induced scattering intensity. Huang scattering measurements are particularly powerful for point-defect clustering studies.

3.3.2.3. Formation enthalpy. The formation enthalpy of a Frenkel defect can be determined by measuring the heat release caused by annihilating Frenkel defects. In practice, Frenkel defects are generated in a sample by low-temperature irradiation. In order to obtain a reliable measure for the defect concentration generated, a resistivity sample is irradiated simultaneously. After sufficient irradiation, the calorimetric sample is then heated at a constant heating rate. The power input required for

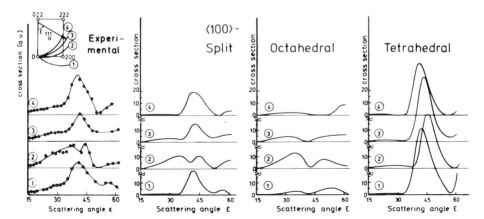

Fig. 25. Comparison of the diffuse X-ray-scattering intensity measured in Al electron-irradiated at 4 K with curves calculated for different interstitial configurations. The numbers 1 to 4 indicate the circles of the Ewald sphere shown in the insert (left upper corner) along which the intensity was measured and calculated. (After EHRHART *et al.* [1974].)

this heating condition varies as the stored energy releases. Different methods of measurement for this entity have been applied with varying success. One of the most serious problems in measuring stored energies in irradiated samples is the required decoupling of the sample from good thermal contact with a cooling stage during the irradiation to optimal thermal isolation during the calorimetric measurements. This procedure must be performed at temperatures below 10 K (stage I recovery). Another very inconvenient condition is the continuous nonlinear increase of the heat capacity of the sample with increasing temperature. The total enhancement in the stage I recovery interval exceeds one order of magnitude. The stored-energy release must be measured on this strongly varying background (review by WENZL [1970]).

In the case of stored-energy measurements, all authors have given the ratio $\Delta Q / \Delta \rho$, i.e. stored-energy release per resistivity recovery. The ΔH_F^f values in table 6 were calculated from the reported ratios by multiplying them by the ρ_F values given in table 3. The enthalpy of interstitial formation has been calculated according to $\Delta H_F^f - \Delta H_v^f = \Delta H_v^f = \Delta H_i^f$, with the average vacancy data from table 1.

3.3.2.4. Activation enthalpy of migration. Most experimental information on the migration enthalpy of interstitials comes from electric resistivity recovery measurements performed after low-temperature irradiation. The recovery curve as found in Cu after electron irradiation at 4.2 K is shown schematically in fig. 4. (More recovery measurements have been performed on Cu than on any other metal.) The curve can be taken as a prototype for many metals, although the detailed structure slightly varies from metal to metal. Substantially different curves are observed for Au, Cd and V.

The generally agreed interpretation of the stage I recovery consists of the so-called close pair recombination in stages I_A–I_C and the free migration of the

Table 6
Activation enthalpy of Frenkel defect and self-interstitial formation [a].

Metal	ΔH_F^f (eV)	ΔH_i^f (eV)	Measurement [b]	Reference
Al	4.3	3.6	n	SCHILLING and TISCHER [1967]
	3.7	3.0	e^-	WOLLENBERGER [1965]
Cu	2.9	1.7	n	BLEWITT [1962]
	4.1	2.9	n	LOSEHAND et al. [1969]
	4.2	3.0	e^-	MEECHAN and SOSIN [1959]
	5.5	4.3	d^{++}	GRANATO and NILAN [1965]
	5.4	4.2	e^-	WOLLENBERGER [1965]
α-Fe	6.3	4.7	e^-	MOSER [1966]
	6.6	5.0	e^-	BILGER et al. [1968]
	13.6	12.0	n	BILGER et al. [1968]
Pt	2.9	1.7	e^-	FEESE et al. [1970]
	2.4	1.2	d^{++}	JACKSON [1979]

[a] Denoted by ΔH_F^f (Frenkel defect) and ΔH_i^f (self interstitial). For evaluation procedure see text.
[b] Stored-energy release and resistivity recovery measured after irradiation with electrons (e^-), neutrons (n) and deuterons (d^{++}), respectively.

interstitial in stages I_D and I_E with interstitial–vacancy recombination upon random diffusion and eventual encounter with a vacancy. For an extensive review on earlier experimental work in fcc metals and its interpretation, see SCHILLING et al. [1970]. The local correlation of those vacancies and interstitials which were produced by one and the same displacement process (displacement distance r_p of a correlated Frenkel defect) must cause a recovery stage for which the significant average number of migrational jumps per interstitial is determined by r_p. For recombination with non-correlated vacancies the specific number of interstitial jumps depends on the concentration of the Frenkel defects. Hence, we expect two recovery stages which are well separated when the average distance of the Frenkel defect is large compared to r_p and overlap when both quantities are similar in size. These two stages are identified with I_D and I_E.

The curve in fig. 4 does not fall to zero in stage I, which means incomplete Frenkel-defect recombination in stage I_E. Among many possible reasons, two are mainly effective in those cases represented by fig. 4: (i) formation of immobile di-interstitials, tri-interstitials and so forth, (ii) trapping of interstitials by solutes. A complete description of the stage I_E recovery must consider these reactions. A review on the theory of diffusion-controlled defect reactions has been given by SCHROEDER [1980].

The first systematic quantitative study of stage I_D–I_E recovery was performed by CORBETT et al. [1959] for electron-irradiated Cu by applying the numerical description of a diffusion-controlled model reaction scheme (WAITE [1957]) to suitably measured electric resistivity recovery curves. Excellent agreement between measured data and theoretical prediction was obtained. The obtained ΔH_i^m values are collected in table 7.

Since then a large number of stage I resistivity recovery investigations have been performed. A critical review with respect to the experimental verification of interstitial configuration and migration models was published by YOUNG [1978]. The respective behaviour of bcc metals has been reviewed by SCHULTZ [1982].

As an example of data obtained by means of a rather elaborate measuring technique, fig. 26 shows differential isochronal resistivity-recovery curves for electron-irradiated Pt, according to SONNENBERG et al. [1972]. The influence of the initial defect concentration (given by the initial residual resistivity $\Delta\rho_0$ and corresponding to about 10 at-ppm to 1000 at-ppm) becomes quite obvious. The quality of the fit of appropriate rate-equation solutions to the stage $I_D–I_E$ recovery is shown in fig. 27, according to SONNENBERG [1972]. The series of figures clearly shows the increasing weight of both uncorrelated recovery and suppression of recovery by di-interstitial formation with increasing initial defect concentration. The influence of interstitial-trapping solutes is shown by i + f in the upper two figures.

Table 7 gives the migration enthalpies obtained by this type of investigations. The

Table 7
Activation enthalpy of interstitial migration [a].

Metal	ΔH_i^m (eV)	Measurement [b]	Reference
Ag	0.100	$\Delta\rho_{irr}$	HERSCHBACH [1963]
	0.088	$\Delta\rho_{irr}$	RIZK et al. [1977]
Al	0.115	$\Delta\rho_{irr}$	SIMPSON and CHAPLIN [1969]
	0.115	MR	SPIRIĆ et al. [1977]
	0.112	$\Delta\rho_{irr}$	RIZK et al. [1976]
Co	0.1	MA	COPE et al. [1968]
			SULPICE et al. [1968]
Cu	0.117	$\Delta\rho_{irr}$	CORBETT et al. [1959]
Ga	0.073	$\Delta\rho_{irr}$	MYHRA and GARDINER [1975]
α-Fe	0.3	MA	SCHAEFER et al. [1975]
Mo	0.083	$\Delta\rho_{irr}$	KUGLER et al. [1982]
Ni	0.15	$\Delta\rho_{irr}$	PERETTO et al. [1966]
			LAMPERT and SCHAEFER [1972]
			KNÖLL et al. [1974]
			FORSCH et al. [1974]
	0.14	MA	PERETTO et al. [1966]
			LAMPERT and SCHAEFER [1972]
			KNÖLL et al. [1974]
			FORSCH et al. [1974]
Pb	0.01	$\Delta\rho_{irr}$	BIRTCHER et al. [1974]
			SCHROEDER et al. [1975]
Pt	0.063	$\Delta\rho_{irr}$	SONNENBERG et al. [1972]
			DIBBERT et al. [1972]
Zr	0.26	$\Delta\rho_{irr}$	NEELY [1970]
	0.30	MR	PICHON et al. [1973]

[a] Denoted by ΔH_i^m.
[b] See glossary of foregoing tables; in addition MR and MA indicate mechanical relaxation and magnetic after-effect, respectively.

References: p. 1208.

table also includes data obtained not by resistivity recovery measurements but by elastic after-effect measurements in Al (SPIRIĆ *et al.* [1977]) and magnetic after-effect measurements in Ni (KNÖLL *et al.* [1974]) and in Fe (SCHAEFER *et al.* [1976], CHAMBRON *et al.* [1976]). These properties, measured in single crystal samples at a fixed temperature after isochronal annealing periods at increasing temperatures, give information not only on the recombination rate but also on the symmetry and a possible orientation relaxation of the migrating defect. In Al the results show that a defect with tetragonal symmetry reorients while migrating, until it recombines with a vacancy. This behaviour is exactly predicted by the computer experiments with the combined translational and rotational jump of the $\langle 100 \rangle$-dumbbell (§ 3.3.1). The annihilation kinetics of the after-effect follows closely that of the resistivity. At temperatures below stage I_D no reorientation relaxation has been observed which could be ascribed to free interstitials. Therefore a pure dumbbell rotation without migration can be excluded, which is again in agreement with the theoretical prediction.

For Ni reorientation relaxation of a defect with $\langle 100 \rangle$-symmetry in stage I_D has been derived from magnetic after-effect measurements.

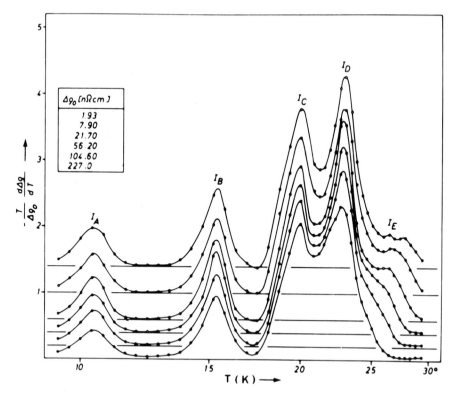

Fig. 26. Temperature-differentiated isochronal resistivity recovery curves of electron-irradiated Pt for different initial irradiation-induced resistivity increments $\Delta \rho$ (after SONNENBERG *et al.* [1972]).

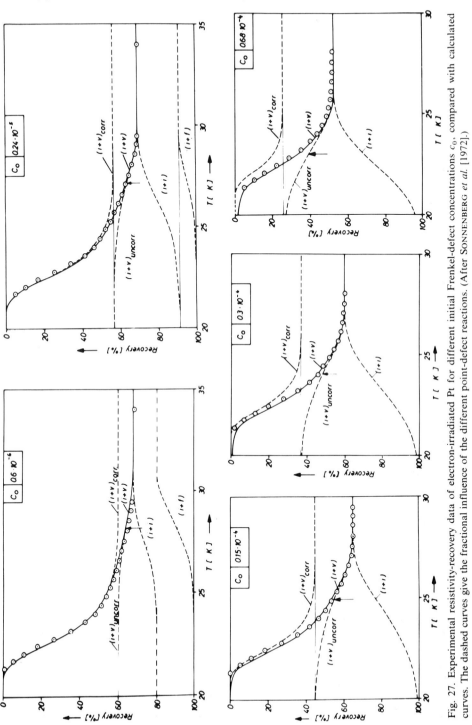

Fig. 27. Experimental resistivity-recovery data of electron-irradiated Pt for different initial Frenkel-defect concentrations c_0, compared with calculated curves. The dashed curves give the fractional influence of the different point-defect reactions. (After SONNENBERG *et al.* [1972].)

For α-Fe, magnetic alignment experiments have identified the ⟨110⟩-symmetry of the interstitial (CHAMBRON et al. [1976]) and magnetic after-effect measurements have shown the reorientation relaxation during annealing in stage I_D (SCHAEFER et al. [1976]).

For other metals, including Au (DWORSCHAK et al. [1981], SEGURA and EHRHART [1979], SCHROEDER and STRITZKER [1977], BIRTCHER et al. [1975], GWOZDZ and KOEHLER [1973]), Cd (COLTMAN et al. [1971], O'NEAL and CHAPLIN [1972]), Nb (SCHULTZ [1982]), Ta (SCHULTZ [1982]) and V (SCHULTZ [1982], COLTMAN et al. [1975]), recovery stages which could unambiguously be ascribed to correlated and uncorrelated recombination of freely migrating interstitials could not be detected. On the other hand, long-range mobility of the interstitials below 5 K has been concluded from Huang scattering experiments which do show interstitial cluster formation during electron irradiation at 5 K in Au, Cd, Mg and Nb (EHRHART [1978]). Many other observations give more indirect hints of long-range interstitial mobility. In particular, Au has been studied with respect to this problem in great detail (DWORSCHAK et al. [1981, 1975] and SEEGER [1970]). A stage I_E, if it exists, must occur below 0.3 K (BIRTCHER et al. [1975], SCHROEDER and STRITZKER [1977]. The unexpectedly large displacement distances (DWORSCHAK et al. [1975]) at low temperatures could be due to the special coupling of interstitials to the vibrational lattice modes (§ 3.3.1.3).

As to Mo and W, recovery stages observed around 35 K for Mo (MAURY et al. [1978]) and 27 K for W (DAUSINGER [1978]) do show a temperature shift as function of the initial defect concentration. But the kinetics are completely different from what is observed in Cu, for example. According to SCHULTZ [1982], the observed kinetics would be in accordance with a model that allows not only for vacancy–interstitial recombination but also for interstitial trapping at certain neighbouring sites of a vacancy. By this effect the amount of recovery in stage I_E will be reduced. At some higher temperature the close pair formed by the trapping event is allowed to recombine leading to the typical close-pair recovery stage, as indeed observed a few degrees above the quasi-I_E stage. Direct evidence for this interpretation is still lacking.

3.3.2.5. Dynamic properties. The dynamic behaviour of the self-interstitial is probably the property most sensitive to the configuration. As has been shown in § 3.3.1.2, the decrease of the elastic modulus with increasing interstitial concentration played a key role in the understanding of the nature of interstitial structure. After a number of modulus measurements in polycrystals which demonstrated the negative sign and the large absolute value of the modulus change (KÖNIG et al. [1964], WENZL et al. [1971]), the first measurements in single crystals containing randomly distributed single Frenkel defects have been reported by HOLDER et al. [1974] on thermal-neutron-irradiated Cu and by REHN and ROBROCK [1977] in electron-irradiated Cu. The data obtained by the former authors are shown in fig. 28, indicating the largest effect for c_{44}. An interstitial concentration of 300 at-ppm softens the crystal by a c_{44} change of 1%. If the concentration of single interstitials could be enhanced up to 3% (limiting factor: spontaneous recombination – see

§3.1.2) and the softening proceeded linearly, the crystal would be unstable against a $\langle 100 \rangle$ shear stress ($c_{44} = 0$). Quenched-in vacancies were found to produce much smaller changes of the elastic constants (FOLWEILER and BROTZEN [1959]). The sequence of decreasing softening for c' and κ can be understood in terms of the dumbbell interstitial. In the case of $\langle 110 \rangle$ shear stress which measures $c' = (c_{11} - c_{12})/2$, the bending spring of the dumbbell coupling is not activated, as can be seen in fig. 23, but the bending springs of the interaction with the nearest neighbours of the dumbbell are operating. As the equilibrium distance is larger than that between the dumbbell atoms the force-constants are smaller and hence the softening is less pronounced. This effect is even smaller for hydrostatic compression (bulk modulus κ). Data for Al are given by ROBROCK and SCHILLING [1976].

In the bcc Mo the largest softening has been observed for c' (OKUDA [1975]), which shows the $\langle 110 \rangle$ symmetry of the split interstitial in this metal.

The large change of the elastic constants should significantly influence the phonon dispersion curves (WOOD and MOSTOLLER [1975], SCHOBER et al. [1975]). Such an effect has indeed been observed by NICKLOW et al. [1979].

3.3.2.6.   Interstitial agglomeration. The formation of small multiple interstitials, even di- or tri-interstitials, has been concluded from measurements of the diffuse X-ray scattering, Huang scattering (§3.3.2.2) and mechanical relaxation measurements (§3.3.2.5). In fig. 29 diffuse X-ray scattering measurements on electron-irradiated Al subsequently annealed to 40 K are compared with calculations for two di-interstitial models according to EHRHART et al. [1974]. Obviously the parallel dumbbell di-interstitial is formed in Al at the end of recovery stage I. This configuration was predicted by computer simulation with a Cu potential (§3.3.1.4).

For the Huang scattering intensity the formation of clusters consisting of n

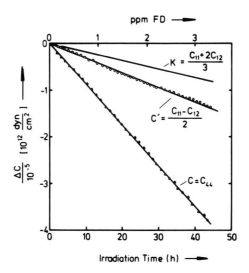

Fig. 28. Elastic moduli of Cu during electron irradiation at 4 K (after HOLDER et al. [1974].

References: p. 1208.

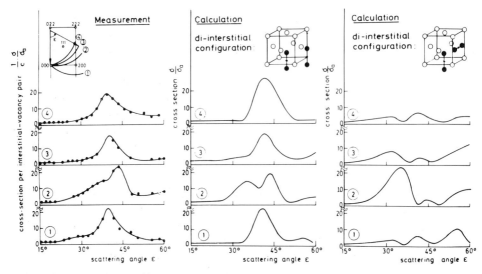

Fig. 29. Comparison of the diffuse X-ray-scattering intensity measured in Al electron-irradiated at 4 K and subsequently heated to 40 K, with calculated curves for two different di-interstitial configurations. Numbers 1 to 4 as in fig. 25. (After EHRHART et al. [1974].)

interstitials leads to $(c_d/n)(n\Delta V_i^{rel})^2 = nc_d(\Delta V_i^{rel})^2$ instead of $c_d(\Delta V_i^{rel})^2$ for single interstitials of concentration c_d. This enhancement is predicted for a linear super-position of the atomic displacements in the cluster. The effect of the actual relaxation upon clustering seems, however, to remain smaller than the gain by the factor n (EHRHART and SCHILLING [1974]). The cluster formation can easily be monitored, as the wavevector dependence of the intensity according to q^{-2} is substituted by a q^{-4} dependence in the upper q range. From the boundary of the q^{-2} validity the cluster radius can be estimated. For Al, with the same experimental conditions as in fig. 25, $n = 2$ was found in agreement with the above result. Between 40 K and 45 K, n was found to increase from 2 to 4. For Cu it has been found that already at the end of recovery stage I small loop-shaped clusters with $n = 5$–10 exist (EHRHART and SCHLAGHECK [1974]). The lack of clusters smaller than with $n = 5$ leads to the conclusion that in Cu, di- and tri-interstitials are mobile in the same temperature region as the mono-interstitial is. This behaviour was also found for di-interstitials by the computer simulations (§ 3.3.1.4). More information on small interstitials clusters in Au, Cd, Co, Nb, Mg, Zn and Zr from Huang scattering is given by EHRHART [1978], SÉGURA and EHRHART [1979] and EHRHART and SCHÖN-FELD [1982].

3.3.2.7. Interstitial–solute interaction. Comparison of the stage I recovery of pure metals with that of impure ones showed that solutes trap migrating self-intersti-tials within certain temperature ranges (SCHILLING et al. [1970]). If this trapping were due to a long-range interaction between solutes and interstitials this interaction should be of elastic nature, because of the effective screening of charges by the conduction electrons. The interaction potential would then be described to a first approximation by eq. (10). The overall effect of solutes on migrating interstitials is an attractive one, despite the existence of repulsive directions (PROFANT and WOLLENBERGER [1976]). The drift-diffusion of interstitials in the interaction potential has been treated in detail by SCHROEDER [1980].

The long-range interaction causing drift cannot reliably be applied at distances between solute and interstitial of three atomic distances or less. Answers to questions such as the configuration of an interstitial placed close to a solute or the binding energy of this pair must be derived from microscopic calculations. A first attempt was made by DEDERICHS et al. [1978], who simulated a solute in Cu in computer experiments by simply shifting the Cu–Cu interaction potential by a certain distance R_0. Enlarging the equilibrium distance of the neighbour simulates oversized solutes, diminishing it simulates undersized solutes. The model predicts a number of interesting results. For undersized solutes it shows the formation of a mixed dumbbell. Solute and one solvent atom share one lattice position in the $\langle 100 \rangle$ split configuration. By oversized solutes the self-interstitial dumbbell is trapped at different neighbouring lattice positions of the solute atom, with different binding energies, of the order of magnitude of 0.1 eV.

The interaction energies in saddle-point configuration of the self-interstitials have not been reported. The saddle-point energies for different types of jumps of the mixed dumbbell atoms have been investigated with the result shown in fig. 30. The three different kinds of jumps are illustrated in fig. 31. As $\Delta H_i^m = 0.1$ eV, activation enthalpies for cage motion are very small according to fig. 30. For volume size misfits larger than about 10%, the mixed dumbbell should be able to rotate before dissociation occurs. Since subsequent rotational and caging jumps allow long-range migration of the mixed dumbbell (see fig. 31) activation enthalpies of migration smaller than that of the vacancy are to be expected. Such conditions are indeed found experimentally with considerable practical implications for solute transport (§4.4).

Similar calculations have been performed for different solutes in Al by using inter-ionic potentials derived from first principles (LAM et al. [1980, 1981]). The pair potentials for the Al–Al and Al–solute interaction extended up to the nineth neighbour. For the solute Zn, mixed dumbbell formation was found with a binding energy of 0.38 eV. For Be, Ca, K and Li, maximum binding energies between 0.62 eV (K) and 1.75 eV (Li) were found for different configurations but in no case a split configuration with solvent atom and solute sharing a lattice site was found. For Mg no binding was found. The saddle-point energy for mixed-dumbbell migration was found to be much larger than this dissociation energy.

The experimental situation of the solute–interstitial interaction has been reviewed by WOLLENBERGER [1978]. Since then, a number of observations have been reported which indicate the existence of quite a number of different solute–interstitial complex configurations with different physical properties in one and the same irradiated dilute alloy. These results are reviewed by ROBROCK [1983].

Trapping of migrating interstitials by solutes prevents the interstitials from

References: p. 1208.

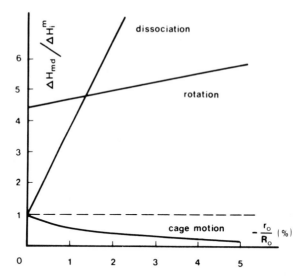

Fig. 30. Normalized enthalpies of mixed dumbbell dissociation, rotation and cage motion according to computer simulation of Cu (DEDERICHS et al. [1978]) versus size-misfit. The ratio r_0/R_0 of the potential shift r_0 divided by the equilibrium atom distance R_0 of Cu must be multiplied by about six in order to obtain the volume size-misfit comparable with experiments.

recombining with vacancies. This can clearly be seen in the isochronal resistivity recovery curve, as exemplified in fig. 32. The presence of 112 at-ppm Au in Cu suppressed recovery in stage I_E. The recovery is retarded and occurs stepwise around

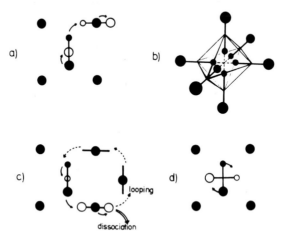

Fig. 31. Elementary jumps of a mixed dumbbell: (a) Jump of the solute; (b) by cage motion of the solute around the octahedral site the six mixed dumbbells shown are formed consecutively; (c) jump sequence of solvent atoms and solute which in case of looping leads to migration steps of the solute; (d) rotation. (From DEDERICHS et al. [1978].)

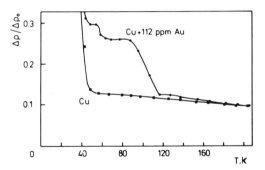

Fig. 32. Isochronal resistivity recovery of Cu and dilute Cu–Au electron-irradiated at 4 K (after CANNON and SOSIN [1975]).

60 K and between 90 K and 120 K. Beyond this the resistivities of dilute alloy and pure solvent are equal, indicating that the Au atoms no longer trap interstitials. By fitting diffusion models to the recovery curves in the stage $I_D–I_E$ range the reaction, capture or trapping radius of the solute for migrating interstitials is obtained. Appropriate evaluation of the final recovery stage yields the activation enthalpy for dissociation of the complex. Activation enthalpies obtained this way for different solutes in Cu are shown in table 8.

The final recovery stage could also be caused by the solute–interstitial pairs becoming mobile as a whole and migrating until the interstitial recombines with a vacancy. Then the activation enthalpy derived from this recovery stage would characterize the saddle point for the complex migration. The recovery experiments do not discriminate between these two types of reactions. This question is particularly interesting for the solutes Co, Fe, Ge, Ni and Zn in Cu for which only an upper limit for the binding energy can be deduced. In these cases trapping was not observed at all. This could, in principle, be due to migration of the complex already at the lowest temperature investigated (55 K).

Solutes in Al seem to cause even stronger binding since the recovery curves of dilute alloys do not coincide with that of the pure Al before stage III recovery, i.e., before vacancies start moving (WOLLENBERGER [1975], DWORSCHAK *et al.* [1976]). The quantitative evaluation of resistivity measurements with respect to reaction radii is somewhat handicapped by the fact that the resistivity does not give any hint on defect reactions occurring if they are not accompanied by resistivity change. But whenever the resistivity decreases, we simply interpret the change as a result of recombination. Reactions which must carefully be considered in this respect are formation of mobile or immobile di-interstitials which seem not to be accompanied by a measurable resistivity change but do change the defect reaction scheme significantly. Configurational changes of interstitial–solute pairs, on the other hand, could be accompanied by resistivity changes. Careful choice of the experimental conditions can to some extent help to exclude undesired reactions. The significance of di-interstitial formation in this respect has recently been discussed in some detail by MAURY *et al.* [1980].

References: p. 1208.

Table 8

Activation enthalpies of self-interstitial solute pair dissociation [a].

Metal	Ag	Au	Be	Co	Cr	Fe	Ge	In
ΔH^{diss} (eV)	0.40	0.33	>1	<0.15	0.36	<0.20	<0.20	0.44
	Mg	Mn	Ni	Pd	Sb	Si	Ti	Zn
	0.43	0.32	<0.15	0.30	0.32	0.44	0.40	<0.20

[a] Denoted by ΔH^{diss}; values according to WOLLENBERGER [1978].

The simplest reaction scheme in view of the solute–interstitial trapping is obtained by electron irradiation at temperatures corresponding to the recovery stages I_E and II. Complications can be confined to those inherently originating from the trapping reaction scheme (WOLLENBERGER [1975, 1978], LENNARTZ et al. [1977], and ABROMEIT and WOLLENBERGER [1983]).

A large number of capture radii have been determined by resistivity-change rate measurements in the stage II recovery range. For Cu and Al as solvents, the so determined capture radii of nearly all solutes are found to depend on temperature as T^{-2}. If the capture radii were determined by drift diffusion of the interstitial in the long-range interaction field, the potential energy should depend on distance R as $R^{-1/2}$ in order to explain the T^{-2} dependence (SCHROEDER [1980]). No physical reason can be given for such a long-range potential. As for the recombination radius, a $T^{-1/3}$ dependence was found (LENNARTZ et al. [1977]), as expected for the elastic dipole interaction. The curious T^{-2} dependence might arise from the fact that the trapped interstitials are detrapped from varying well depths during the period of measurement. By assuming the appropriate number of trapping sites and interstitial transition rates between them, the T^{-2} dependence was quantitatively explained for Au in Cu as an example (ABROMEIT and WOLLENBERGER [1983]).

The absolute size of the effective capture radii is derived in the order of a few atomic distances for Cu, and for Al as solvent about twice as large. The sizes found in Cu disagree with the elastic interaction theory by a factor of about two. For Al, however, which is nearly isotropic, there is a disagreement of approximately one order of magnitude. The disagreement in the case of Al indicates that this type of theoretical description is not correct. It should be noted that diffusion jumps of the dumbbell interstitial in the vicinity of a solute have not been investigated by computer experiments until now.

While resistivity measurements have yielded substantial information on the trapping behaviour of solutes in terms of capture rate constants, methods such as ion-channeling measurements, Mössbauer effect analysis, perturbed γ–γ angular correlation and mechanical relaxation measurements have extended our knowledge of configurations and motion characteristics of solute–interstitial pairs.

The *ion-channeling method* has been reviewed by SWANSON et al. [1978] and HOWE and SWANSON [1982]. It is the most direct method for localizing impurity atoms in a host lattice. Whenever solutes are displaced from regular lattice sites by trapped interstitials, ion channeling, in principle, can detect them and measure the displacement distance. Most of the solute–interstitial pair investigations have been performed by means of Rutherford backscattering of channeled ions. This method has the limitation that it has sufficient sensitivity only for solute masses large compared to the solvent mass. Consequently this method has been applied mainly to dilute Al alloys. The existence of ⟨100⟩ mixed dumbbells has been concluded for the investigated alloys with undersized solutes and no solute displacement was found for oversized ones. By using proton-induced X-ray emission as an indicator for the presence of impurities in interstitial positions (ECKER [1982]) the mass difference problem is avoided.

It should be kept in mind that channeling measurements can unambiguously identify the interstitial position of a solute only when the solute does not occupy more than one type of position. Data evaluation in case of simultaneous presence of more than one position becomes ambiguous and in practice very uncertain for even one of the assumed positions. Besides the different configurations possible for the solute–one-interstitial complex, the trapping of more than one interstitial per solute could lead to different interstitial positions of the solute. The latter problem can be solved, in principle, by careful investigations of the defect-concentration dependence of the solute position.

The *mechanical relaxation methods* measure the damping of applied elastic vibrations caused by the stress-induced reorientation of solute–interstitial complexes (see reviews by KRONMÜLLER [1970], NOWICK and BERRY [1972]). By investigating single crystals the symmetry of the complex can be derived. Complexes of different configurations are expected to show different activation enthalpies for reorientation and migration. For In in Cu, for example, not less than six different damping maxima are obtained (KOLLERS *et al.* [1981]), pointing to six different complex-configurations. Four of them annihilate close to the final resistivity recovery stage of this alloy (DWORSCHAK *et al.* [1978]) with small differences in annihilation temperature among each other. The remaining two annihilate at considerably lower temperatures.

A similar multiplicity of reorientation-damping peaks and annihilation temperatures has been found in dilute Al alloys (ROBROCK [1982], HULTMAN *et al.* [1981]). It has also been shown that the activation enthalpy for re-orientation significantly depends on the solute species (GRANATO [1981]).

For Fe in Al a re-orientation process occurs at 8 K with an activation enthalpy of about 13 meV (REHN *et al.* [1978]). It has been identified to arise from the cage motion detected earlier by Mössbauer studies (see below). Some disagreement regarding details of the cage geometry as concluded from internal friction (ROBROCK and SCHOBER [1981]) and Mössbauer spectroscopy (PETRY *et al.* [1982]) clearly shows progress and limits reached with the application of these methods.

By *Mössbauer effect studies*, the cage motion of a solute which has trapped a self-interstitial has been detected. In a dilute alloy of ^{57}Co in Al or Ag irradiated at low temperatures and subsequently annealed through recovery stage I or irradiated at temperatures within stage II, a second Mössbauer line of the ^{57}Co (^{57}Fe)

References: p. 1208.

transition was observed besides the well-known one arising from the solute dissolved substitutionally (VOGL and MANSEL [1976]). From the isomer shift of the new line it must be concluded that the Mössbauer isotope is placed closer to the next solvent atom than a nearest-neighbour distance. Hence, it must be situated in the close vicinity of an interstitial atom. The Debye–Waller factor f of this new line shows an unusual temperature dependence, as shown in fig. 33. It decreases between 10 K and 20 K (for Al) by about 80%. This decrease can be measured reversibly so long as the sample is not heated into recovery stage III (Al) or to 100 K (Ag), i.e., until the trapped interstitials are annihilated. This decrease has been quantitatively explained by a cage motion of the Mössbauer isotope in the cage shown in fig. 31b (VOGL et al. [1976]). The essential origin of the reversible Debye–Waller-factor drop is a locally restricted motion of the Mössbauer isotope within the lifetime of the Mössbauer quanta (10^{-7} s). The low-temperature range of its occurrence fits well to the theoretical result in fig. 30.

Mössbauer studies have also demonstrated the existence of more than one solute–interstitial-complex configuration. In irradiated Mo–^{57}Co not less than five different new Mössbauer lines were resolved (MANSEL et al. [1981, 1982]).

The perturbed γ–γ angular correlation technique introduced in §2.2.2.2 in connection with the identification of vacancies arriving at the probe atoms in the temperature range of stage III has also been applied to questions of interstitial–solute interaction (reviews quoted in §2.2.2.2). Again, a variety of different configurations of interstitial probe atom pairs must be concluded from the results for certain dilute alloys.

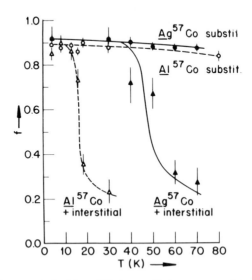

Fig. 33. Debye–Waller factor as derived from Mössbauer effect measurements in ^{57}Co in electron-irradiated Al and Ag. Irradiated and unirradiated states are indicated with " + interstitial" and "substit.", respectively. (After VOGL et al. [1976].)

The application of probe atom methods seems still to possess considerable potential even when the theoretical problem of field-gradient calculation remains unsolved. A systematic application to single crystal samples could give more information on the symmetry of the complexes and systematic studies of the kinetics could give valuable information on the interstitial population of the complexes (ABROMEIT and WOLLENBERGER [1983]).

4. Radiation damage at high particle fluences

4.1. General remarks

Reactor materials are generally used at temperatures well above recovery stage III (§ 2.2.2), i.e., vacancies are mobile during neutron irradiation. In a pure metal the interstitials are then mobile, too (§§ 2 and 3). For a given alloy the same cannot be said off-hand. What does an interstitial in a concentrated binary or ternary alloy look like? At which temperatures does it become mobile? Once it migrates, does it transport A and B atoms and even C atoms with similar probabilities? Or does it migrate via A atoms only? The answers to these questions are of great importance for the stability of alloys under irradiation (§ 4.3). These questions also reflect the basic problems of interstitial migration in alloys. Actually, interstitial mobility has been demonstrated directly by atomic ordering at relatively low temperatures or by defect recovery studies only for a few alloys (WOLLENBERGER [1981], WAGNER et al. [1982]). Nevertheless, interstitial mobility at technically relevant irradiation temperatures has generally been assumed for alloy materials in theories of void growth, for example.

The radiation-damage problem encountered in engineering materials at high fluences, which also occurs in pure metals, and in that sense originates from vacancy and interstitial properties, is the *dimensional instability*. It occurs as swelling, i.e. decrease of density, and as volume-conserving creep which happens at significantly lower temperatures than the thermally activated creep. The quantitative features of these phenomena do considerably vary from alloy to alloy. But this can be viewed as a manifestation of the different vacancy and interstitial properties in the different alloys.

4.2. Void formation and growth

Most materials irradiated at temperatures between $0.3T_M$ and $0.6T_M$ show a measurable and approximately linear volume increase with neutron fluence, for fluences larger than 10^{21}–10^{22} neutrons/cm^2. Such fluence levels are achieved in fuel-element claddings of fast breeder reactors, such as the French prototype PHENIX, in about one month. Electron microscopy has shown the existence of *voids*, the total volume of which matched the observed volume increase. As an example, voids in Ni and stainless steel are shown in fig. 34. The morphology of voids as three-dimensional vacancy arrangements is multiform (for reviews see *Further reading* list: Corbett and Ianiello [1972], Bleiberg and Bennett [1977] and Carpenter *et al.* [1980]), and void lattices have been observed, too (EVANS *et al.* [1972], WIFFEN [1972], BENOIST and MARTIN [1975], KRISHAN [1982]). Voids have also been studied by small-angle X-ray scattering, which can fruitfully be combined with other techniques (ch. 12, §5.3). Void formation was not expected before its detection (CAWTHORNE and FULTON [1967]), as the earlier investigations showed larger vacancy clusters to be stable only as planar aggregates (dislocation loops, stacking fault tetrahedra). Moreover, such aggregates were no longer stable at those temperatures at which voids are formed. Hence, their observation raised two questions: (i) What is the nucleation process for these three-dimensional vacancy arrangements? (ii) What are the stability criteria?

The first question is still open. There is no theoretical indication that vacancy clusters of more than ten to twenty vacancies could be energetically favourable in the form of voids compared to planar aggregates. Therefore, lattice heterogeneities as precipitations or homophase impurity clusters must be responsible for the nucleation. Extensive researches have been done to establish whether atomic He gas bubbles are the nuclei or not (see, for example, LANORE *et al.* [1975], GLASGOW *et al.* [1981]). It became quite obvious that implanted He significantly enhances the nucleation rate. But the limited sensitivity of the analytical methods does not yet allow proof of the complete absence of such impurities for all conditions for which void formation is observed. By reactor neutron irradiation, impurities are produced

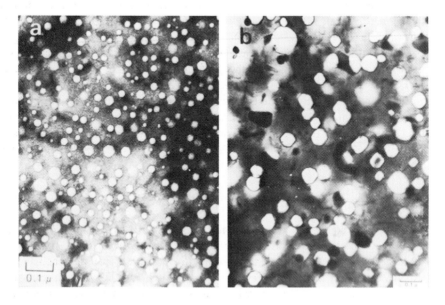

Fig. 34. Voids in (a) Ni irradiated by 200 keV Ni$^+$ ions at 500°C (ADDA [1972]) and (b) type 316 stainless steel irradiated by 7.1×10^{22} neutrons/cm^2 at 525°C (APPLEBY *et al.* [1972]).

due to nuclear transmutation. High cross-sections for (n, α) transmutations lead to He production which is sufficiently large to explain the observed void nucleation rates.

The relative volume increase and, thus, the void volume fraction can easily reach 10% or more. A corresponding amount of clustered interstitial atoms has never been detected. The vacancies contributing to void formation must therefore have escaped both recombination and annihilation at dislocations acting as sinks. As recombination annihilates vacancies and interstitials in equal numbers, the annihilation at sinks must be unmatched, in favour of vacancy escape. The common explanation of such *biased* vacancy and interstitial annihilation considers the elastic interaction between the migrating point-defects and dislocations as causing a drift diffusion towards the dislocations (reviews see, for example, BRAILSFORD and BULLOUGH [1978], BULLOUGH and WOOD [1980] and *Further reading* list: Wiffen *et al.* [1979], Nygren *et al.* [1981]). The interaction energy for a dilatation centre A with the volume change ΔV_A^m in the strain field of an edge dislocation is proportional to $\Delta V_A^m / r$, where r is the distance of A from the dislocation core. The proportionality to ΔV_A^m makes the interaction strength quite different for vacancy and interstitial (Cu, for example: $\Delta V_v^m = -0.2$, $\Delta V_i^m = 1.6$). In a given material, the topologies and concentrations of saturable sinks (e.g. sessile dislocations), unsaturable sinks and saturable traps composes a complex scheme of spatially varying defect reactions. By appropriate modelling of such diffusion and drift-controlled reaction systems, a bias of interstitial and vacancy annihilation of a few percent is obtained, in agreement with the experimental results.

The influence of pretreatment and irradiation temperature on the swelling is illustrated in fig. 35a. The pretreatment obviously controls the bias factor via dislocation arrangement and concentration. The temperature controls the defect mobility and, by this, the stationary defect concentration which in turn determines the relative influence of the recombination. At low temperatures the recombination rate is large because of the slowly migrating vacancies. The biased sink annihilation is suppressed. At higher temperatures the latter is favoured over recombination. At even higher temperatures, the stationary vacancy concentration approaches thermal equilibrium values and, hence, does not provide the supersaturation required for void formation. Slowing down of one of the migrating defect species can be achieved by adding solutes which temporarily trap interstitials (§ 3.3.2.7). In this way swelling is reduced (see, for example, POTTER *et al.* [1977]) as shown in fig. 35b.

Although the void growth phenomenon seems to be basically understood, the prediction of a growth rate for a material that has not yet been irradiated is difficult, as the atomic processes which determine the different reaction-rate constants are not understood quantitatively. It should be kept in mind also that the bias annihilation presumes interstitial mobility, the occurrence of which has not been demonstrated for many alloys.

References: p. 1208.

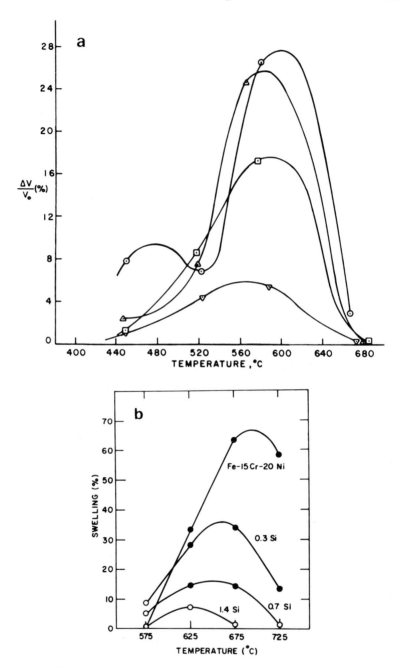

Fig. 35. Influence of pretreatment and solute additions on void-swelling: (a) 316 type stainless steel after 14.1×10^{22} neutrons/cm², ⊙ – annealed, ▲ – 10% cold worked (cw), ▣ – 20% cw, ▽ – 30% cw (from FLINN and KENFIELD [1976]); (b) Fe–15 at% Cr–20 at% Ni with quoted Si additions after a fluence corresponding to 140 displacements per atom (from JOHNSTON et al. [1976]).

4.3. Irradiation-induced creep

The thermally activated creep of materials is treated in ch. 20. Important feature of this phenomenon is the exponential temperature dependence of the creep rate, $\dot{\epsilon}$, for a fixed stress, σ. If the creep experiment is done under irradiation, $\dot{\epsilon}$ becomes nearly independent of temperature below about $T_M/2$. This behaviour originates from the high irradiation-induced vacancy- and interstitial-concentration which allows sufficient dislocation climb and deformation by atom transport, despite the small diffusivities of these defects (for reviews see *Further reading* list: Carpenter *et al.* [1980], Gittus [1978]). For not too large applied stresses σ, the creep rate $\dot{\epsilon}$ is found to be proportional to σ and to the defect production rate k_0. One of the successful models uses the *stress-induced preferential absorption* (SIPA) of defects at dislocations capable of climbing (HEALD and SPEIGHT [1974]), as illustrated in fig. 36. If again the elastic-energy-controlled drift diffusion is assumed to determine the defect current towards dislocation sinks, the orientation of the dislocations with respect to the external stress will govern the rate constants for defect annihilation. According to the model calculation, dislocations of type I (fig. 36) absorb more interstitials than vacancies, and vice versa for type II dislocations. The resulting climb along the directions indicated cause the volume-conserving elongation of the sample parallel to σ.

4.4 Radiation-induced phase transformation

Many multiphase alloys are not in thermodynamic equilibrium at application temperatures. When they are applied in radiation environments as in nuclear reactor cores, high concentrations of mobile vacancies and possibly interstitials provide an intense transport of matter, which drives the multiphase system towards equilibrium. The resulting phase transformations are radiation-accelerated. Other processes can superimpose on such phase transformations directed towards thermodynamic equilibrium. One of them results from the fact that the radiation-induced supersaturation

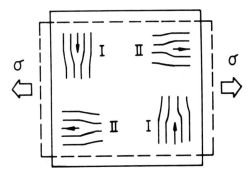

Fig. 36. Dislocation climb directions by preferential vacancy- and interstitial absorption under applied stress σ.

References: p. 1208.

of vacancies and interstitials causes defect fluxes towards internal sinks and surfaces. The situation in front of such a sink is sketched in fig. 37. If the partial diffusion coefficients D_A^v and D_B^v of the alloy-constituents A and B due to vacancy migration were different (see ch. 8) the vacancy flux J_v to the sink would cause different atom fluxes J_A and J_B, with the result of a depletion of the faster-transported component near the sink, as shown in fig. 37b. The process works as an inverse Kirkendall effect. If the interstitial flux J_i provided atom fluxes as shown in fig. 37c, component A would become enriched as shown in fig. 37d. Under such circumstances the steady-state concentration gradient ∇c_A was derived by WIEDERSICH et al. [1979] to be proportional to $[(D_A^v/D_B^v) - (D_A^i/D_B^i)]\nabla c_v$. Atomic redistribution also occurs if one of the defect species forms a mobile complex with one of the constituents and this complex is able to migrate large distances without dissociating. Then this complex diffuses down the concentration gradient of its own species. In such a case the complex-forming alloy constituent will be enriched at the defect sink.

There is no reason for these redistribution mechanisms to stop when the solubility

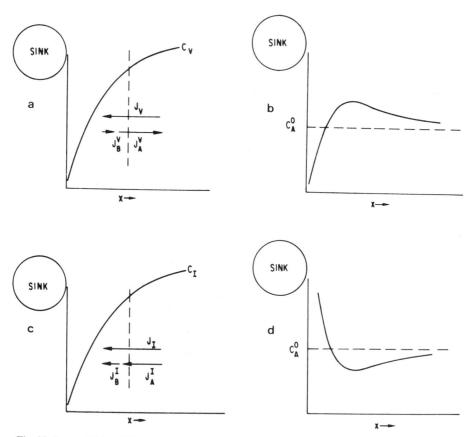

Fig. 37. Inverse Kirkendall effect produced by vacancy (a, b) and interstitial (c, d) fluxes to sinks.

limit of the enriched component is passed near the sink. If nucleation of the phase which is stable for the redistributed alloy composition is possible, precipitation will occur and will be driven as long as the irradiation-induced defect flux is maintained and no resolution mechanism counteracts the precipitation. In the temperature range in which the irradiation-induced steady-state vacancy concentration approaches the thermal-equilibrium concentration, thermally driven back-diffusion counteracts with increasing efficiency.

Atom redistribution as well as irradiation-driven precipitation has been studied in many alloy systems under a great variety of irradiation conditions (see *Further reading* list: Workshop on Solute Segregation and Phase Stability during Irradiation — Stiegler [1979]; Irradiation Phase Stability — Holland *et al.* [1981]). Redistribution was studied by using the sample surface as defect sink and measuring the depth profiles of the alloy constituents beneath that surface (see ch. 13, and, for example, WIEDERSICH *et al.* [1977], MARWICK *et al.* [1979] and WAGNER *et al.* [1983]). For studies of redistribution and precipitation near internal sinks, high-voltage electron microscopy has proved to be a very valuable tool. Electron energies from about 0.5 MeV upwards are sufficient to produce Frenkel defects during the imaging (§ 3.1.2). The beam intensity is extremely high, so that the number of atomic displacements produced by nuclear-reactor irradiation can be exceeded by orders of magnitude within minutes. Moreover, electron irradiation is advantageous as it avoids collision-cascade effects which give rise to additional complex phase instabilities (§ 4.4.2). The alloy system studied most extensively with respect to the fundamental problems is Ni–Si (see, for example, BARBU and MARTIN [1977] and OKAMOTO *et al.* [1982]). Precipitation of the ordered phase Ni_3Si was observed at the sample surface as well as at dislocation loops in undersaturated alloys containing a few atomic percent of Si.

Another interesting system is Cu–Be in which the ordered phase CuBe is precipitated at internal sinks by Cu^+ ion irradiation (KOCH *et al.* [1981]) but a Guinier–Preston-zone-type decomposition occurs upon electron irradiation (WAHI and WOLLENBERGER [1982]). For this alloy it could be shown that the interstitial forms a stable complex with the Be atom (mixed dumbbell) which migrates by an activation enthalpy which is slightly smaller than that of the vacancy (BARTELS *et al.* [1979]). This condition provides a rapid precipitation mechanism.

The essential origin of the radiation-induced atom redistribution is irreversible defect annihilation. The spatially inhomogeneous distribution of sinks causes defect fluxes and these in turn cause redistribution. The recombination reaction can also cause redistribution (MARTIN [1980], CAUVIN and MARTIN [1981]) and this aspect is briefly treated in ch. 8, § 8.3.2.

Atom redistribution, swelling and radiation-induced creep depend on each other and influence each other in a complex way. Extensive efforts have been devoted to the objective of undoing this knot for the technically interesting stainless steels (see, for example, GARNER [1981]), to achieve a general understanding of the important correlations. Quantitative treatments are still in the state of estimations.

A further atomic redistribution process is atomic mixing within the collision

References: p. 1208.

cascade volume, as caused by heavy particle irradiation. An example of disordering of a long-range ordered alloy is shown in fig. 17. As many γ-phases are long-range ordered, this collision-induced dissolution of ordered phases is an important elementary mechanism of radiation-induced phase transformations (NELSON et al. [1972]). Atomic mixing can influence the stability even of non-ordered precipitates (ABROMEIT [1979]). Displacement cascades matching a substantial fraction of the precipitate size can transport enough matter out of this precipitate into the surrounding matrix to make the precipitate unstable in this way.

Phase transformation could also be induced if the point defects created during irradiation were to change the free enthalpies of two equilibrating phases by different amounts such that the equilibrium is removed. According to BOCQUET and MARTIN [1979], this effect is too small to explain one of the experimentally observed radiation-induced phase transformations.

Radiation-induced segregation and phase transformations are further discussed in ch. 8, § 8.3.2.

Acknowledgement

Mrs. Barbara Standke perfectly phonotyped the manuscript despite the many duties she attended to simultaneously. Dr. Christian Abromeit carefully searched through the literature for the data given in the tables and critically read the manuscript.

References

ABROMEIT, C., 1979, A Model for Phase Stability under Irradiation, in: Irradiation Behaviour of Metallic Materials for Fast Reactor Core Components, Proc. of the Int. Conf. Ajaccio, Corse, France, 1978, eds. J. Poirier and J.M. Dupouy, p. 89.

ABROMEIT, C., 1983, to be published in J. Phys. F.

ABROMEIT, C., and H. WOLLENBERGER, 1983, Phil. Mag., to be published.

ADDA, Y., 1972, Radiation-Induced Cavities in Metals: Presentation of Some Results, in: Radiation-Induced Voids in Metals, eds. J.W. Corbett and L.C. Ianniello, (CONF-710601, Natl. Techn. Inf. Service, Springfield, VA 22161) p. 31.

AFMAN, H.B., 1972, Phys. Stat. Sol. (a) 13, 623.

AFMAN, H.B., J.H. MOOY and H. RADEMAKER, 1970, Scripta Metall. 4, 545.

ANTESBERGER, G., K. SONNENBERG and P. WIENHOLD, 1978, J. Nucl. Mater. 69/70, 660.

APPLEBY, W.K., D.W. SANDUSKY and U.E. WOLFF, 1972, Fluence and Temperature Dependence of Void Formation in Highly Irradiated Stainless Steels, in: Radiation-Induced Voids in Metals, eds. J.W. Corbett and L.C. Ianniello (Conf-710601, Natl. Techn. Inf. Service, Springfield, VA 22161) p. 156.

AUDIT, P., 1982, Vacancy-Thermal Expansion in Aluminium, in: Point Defects and Defect Interactions in Metals, eds. J. Takamura, M. Doyama and M. Kiritani (Univ. of Tokyo Press, 1982) p. 291.

AUDIT, P., and H.M. GILDER, 1978, J. Nucl. Mater. 69/70, 641.

AVERBACK, R.S., R. BENEDEK and K.L. MERKLE, 1978, Phys. Rev. B18, 4156.

BACCHELLA, G.L., E. GERMAGNOLI and S. GRANATA, 1959, J. Appl. Phys. 30, 748.

BALLUFFI, R.W., 1978, J. Nucl. Mater. 69/70, 240.

BALLUFFI, R.W., K.H. LIE, D.N. SEIDMAN and R.W. SIEGEL, 1970, Determination of Concentrations and

Formation Energies and Entropies of Vacancy Defects from Quenching Experiments, in: Vacancies and Interstitials in Metals, eds. A. Seeger, D. Schumacher, W. Schilling and J. Diehl (North-Holland, Amsterdam) p. 125.

BARBU, A. and G. MARTIN, 1977, Scripta Metall. **11**, 771.

BARTELS, A., F. DWORSCHAK, H.-P. MEURER, C. ABROMEIT and H. WOLLENBERGER, 1979, J. Nucl. Mater. **83**, 24.

BASS, J., 1967, Phil. Mag. **15**, 717.

BAUER, W., 1969, Rad. Eff. **1**, 23.

BAUER, W., and A. SOSIN, 1964, Phys. Rev. **136**, A474.

BAUER, W., and A. SOSIN, 1966, Phys. Rev. **147**, 482.

BECKER, D.E., F. DWORSCHAK and H. WOLLENBERGER, 1973, Rad. Eff. **17**, 25.

BENDER, O., and P. EHRHART, 1982, Frenkel Defects in Ni and Ni-base Alloys, in: Point Defects and Defect Interactions in Metals, eds. J. Takamura, M. Doyama and M. Kiritani (Univ. of Tokyo Press, 1982) p. 639.

BENDER, O., and P. EHRHART, 1983, to be published in J. Phys. **F**.

BENEDEK, R., and A. BARATOFF, 1971, J. Phys. Chem. Solids **32**, 1015.

BENNEMANN, K.H., 1961, Phys. Rev. **124**, 669.

BENNEMANN, K.H., and L. TEWORDT, 1960, Z. Naturf. **15a**, 772.

BENOIST, P., and G. MARTIN, 1975, Stability of Void Lattices under Irradiation, in: Proc. Int. Conf. on Fundamental Aspects of Radiation Damage in Metals, Gatlinburg, 1975, eds. M.T. Robinson and F.W. Young, Jr. (CONF-751006, Natl. Techn. Inf. Service, Springfield, VA 22161) p. 1236.

BERGER, A.S., D.N. SEIDMAN and R.W. BALLUFFI, 1973, Acta Metall. **21**, 123.

BERGER, A.S., S.T. OCKERS, M.K. CHASON and R.W. SIEGEL, 1978, J. Nucl. Mater. **69/70**, 734.

BERGER, A.S., S.T. OCKERS and R.W. SIEGEL, 1979, J. Phys. **F9**, 1023.

BIANCHI, G., D. MALLEJAC, C. JANOT, and G. CHAMPIER, 1966, Compt. Rend. Acad. Sci. (Paris) **B263**, 1404.

BIGET, M., F. MAURY, P. VAJDA, A. LUCASSON and P. LUCASSON, 1979, Phys. Rev. **B19**, 820.

BILGER, H., V. HIVERT, J. VERDONE, J.L. LEVEQUE and J.C. SOULIE, 1968, Point Defects in Iron, in: International Conference on Vacancies and Interstitials in Metals, Jülich (Jül-Conf-2, p. 751), unpublished.

BIRTCHER, R.C., J.W. LYNN and J.S. KOEHLER, 1974, Phys. Rev. Lett. **33**, 899.

BIRTCHER, R.C., W. HERTZ, G. FRITSCH and J.F. WATSON, 1975, Very Low Temperature Electron-Irradiation and Annealing of Gold and Lead, in: Proc. Int. Conf. on Fundamental Aspects of Radiation Damage in Metals, Gatlinburg, 1975, eds. M.T. Robinson and F.W. Young, Jr. (CONF-751006, Natl. Techn. Inf. Service, Springfield, VA 22161) p. 405.

BLEWITT, T.H., 1962, in: Radiation Damage in Solids, ed. D.S. Billington (Academic, New York) p. 630.

BOCQUET, J.L., and G. MARTIN, 1979, J. Nucl. Mater. **83**, 186.

BOURASSA, R.R., and B. LENGELER, 1976, J. Phys. **F6**, 1405.

BRAILSFORD, A.D., and R. BULLOUGH, 1978, J. Nucl. Mater. **69/70**, 434.

BRANDT, W., 1974, Appl. Phys. **4**, 1.

BRANDT, W., and H.F. WAUNG, 1968, Phys. Lett. **27A**, 100.

BRUGIÈRE, R., and P. LUCASSON, 1967, Phys. Stat. Sol. **24**, K77.

BRUGIÈRE, R., and P. LUCASSON, 1968, Phys. Stat. Sol. **30**, K139.

BUDIN, C. and P. LUCASSON, 1965, Phys. Lett. **16**, 229.

BULLOUGH, R., and J.R. WILLIS, 1975, Phil. Mag. **31**, 855.

BULLOUGH, R., and M.H. WOOD, 1980, J. Nucl. Mater. **90**, 1.

BURTON, J.J., 1971, Phys. Rev. **B5**, 2948.

BUTT, R., R. KEITEL and G.VOGL, 1979, HMI-Report, unpublished.

CAMANZI, A., N.A. MANCINI, E. RIMINI and G. SCHIANCHI, 1968, Vacancy–Impurity-atom Interaction in Au–Pt Dilute, in: International Conference on Vacancies and Interstitials in Metals, Jülich (Jül-Conf-2, p. 154), unpublished.

CAMPBELL, J.L, C.W. SCHULTE and J.A. JACKMAN, 1977, J. Phys. **F7**, 1985.

CAMPBELL, J.L., C.W. SCHULTE and R.R. GINGERICH, 1978, J. Nucl. Mater. **69/70**, 609.

CANNON, C.P., and A. SOSIN, 1975, Rad. Eff. **25**, 253.

CAUVIN, R., and G. MARTIN, 1981, Phys. Rev. **B23**, 3333.

CAWTHORNE, C., and E.J. FULTON, 1967, Nature **216**, 515.

CERESARA, S., T. FEDERIGHI, D. GELLI and F. PIERAGOSTINI, 1963, Nuovo Cim. **29**, 1244.

CERESARA, S., H. ELKHOLY and T. FEDERIGHI, 1965, Phil. Mag. **12**, 1105.

CHAMBRON, W., J. VERDONE and P. MOSER, 1975, Determination of Point Defects Symmetry in a Cubic Lattice by Magnetic Relaxation, in: Proc. Int. Conf. on Fundamental Aspects of Radiation Damage in Metals, Gatlinburg, 1975, eds. M.T. Robinson and F.W. Young, Jr. (CONF-751006, Natl. Techn. Inf. Service, Springfield, VA 22161) p. 261.

CHARLES, M., J. HILLAIRET, M. BEYELER and J. DE LAPLACE, 1975, Phys. Rev. **B11**, 4808.

CHHABILDAS, L.C., and H.M. GILDER, 1972, Phys. Rev. **B5**, 2135.

CHIK, K.P., 1970, Annealing and Clustering of Quenched-in Vacancies in Metals, in: International Conference on Vacancies and Interstitials in Metals, Jülich (Jül-Conf-2, p. 183), unpublished.

COLTMAN, R.R., C.E. KLABUNDE, J.K. REDMAN and A.L. SOUTHERN, 1971, Rad. Eff. **7**, 235.

COLTMAN, R.R., C.E. KLABUNDE, J.K. REDMAN and J.M. WILLIAMS, 1975, Rad. Eff. **24**, 69.

CONNORS, D.C., V.H.C. KRISP and R.N. WEST, 1971, J. Phys. **F1**, 355.

COPE, H., G. SULPICE, C. MINIER, H. BILGER and P. MOSER, 1968, Post-Irradiation Recovery of Two Hexagonal Ferromagnetic Metals: Cobalt and Gadolinium, in: International Conference on Vacancies and Interstitials in Metals, Jülich (Jül-Conf-2, p. 792), unpublished.

CORBETT, J.W., R.B. SMITH and R.M. WALKER, 1959, Phys. Rev. **114**, 1442 and 1460.

CORNELIS, J.L., P. STALS, P. DE MEESTER, J. ROGGEN and J. NIHOUL, 1978, J. Nucl. Mater. **69/70**, 704.

COTTERILL, R.M.J. and M. DOYAMA, 1966, Energies and Atomic Configurations of Line Defects and Plane Defects in fcc Metals, in: Lattice Defects and Their Interaction, ed. R.R. Hasiguti (Gordon and Breach, New York) p. 1.

CUDDY, L.J., 1968, Acta Metall. **16**, 23.

CUDDY, L.J., and E.S. MACHLIN, 1962, Phil. Mag. **7**, 745.

CURR, R.M., 1955, Proc. Roy. Soc. **A68**, 156.

DAS, K.B., and H.I. DAWSON, 1971, J. Phys. Soc. Japan **30**, 388.

DAUSINGER, F., 1978, Phil. Mag. **A37**, 819.

DEDERICHS, P.H., and J. POLLMANN, 1972, Z. Phys. **255**, 315.

DEDERICHS, P.H., and R. ZELLER, 1980, Dynamical Properties of Point Defects in Metals, in: Springer Tracts in Modern Physics, Vol. 87, eds. G. Höhler and E.A. Niekisch (Springer, Berlin) ch. 6.4.

DEDERICHS, P.H., C. LEHMANN and A. SCHOLZ, 1973, Phys. Rev. Lett. **31**, 1130.

DEDERICHS, P.H., C. LEHMANN, H.R. SCHOBER, A. SCHOLZ and R. ZELLER, 1978, J. Nucl. Mater. **69/70**, 176.

DEICHER, M., E. RECKNAGEL and TH. WICHERT, 1981a, Rad. Eff. **54**, 155.

DEICHER, M., O. ECHT, E. RECKNAGEL and TH. WICHERT, 1981b, Geometrical Structure of Lattice Defect–Impurity Configurations Determined by TDPAC, in: Nuclear and Electron Resonance Spectroscopies Applied to Materials Science, eds. E.N. Kaufmann and G.K. Shenoy (Elsevier–North-Holland, New York) p. 435.

DE LAPLACE, J., J. HILLAIRET, C. MAIRY and Y. ADDA, 1966, Mém. Sci. Rev. Mét. **63**, 282.

DESORBO, W., and D. TURNBULL, 1959, Phys. Rev. **115**, 560.

DIBBERT, H.J., K. SONNENBERG, W. SCHILLING and U. DEDEK, 1972, Rad. Eff. **15**, 115.

DIMITROV-FROIS, C., and O. DIMITROV, 1968, Interaction of Point Defects with Magnesium Impurity Atoms in Neutron-Irradiated Aluminium, in: International Conference on Vacancies and Interstitials in Metals, Jülich, (Jül-Conf-2, p. 290), unpublished.

DLUBEK, G., O. BRÜMMER and N. MEYENDORF, 1977a, Phys. Stat. Sol. (a) **39**, K95.

DLUBEK, G., O. BRÜMMER and N. MEYENDORF, 1977b, Appl. Phys. **13**, 67.

DOYAMA, M., and R.R. HASIGUTI, 1973, Cryst. Lattice Defects **4**, 139.

DOYAMA, M., and J.S. KOEHLER, 1962, Phys. Rev. **127**, 21.

DOYAMA, M., and R.M.J. COTTERILL, 1967, Energies and Atomic Configurations of Point Defects in fcc

Metals, in: Lattice Defects and Their Interactions, ed. R.R. Hasiguti (Gordon and Breach, New York) p. 79.

DOYAMA, M., J.S. KOEHLER, Y.N. LWIN, E.A. RYAN and D.G. SHAW, 1971, Phys. Rev. **B4**, 281.

DUESING, G., and W. SCHILLING, 1969, Rad. Eff. **1**, 65.

DWORSCHAK, F., and J. KOEHLER, 1965, Phys. Rev. **140**, A941.

DWORSCHAK, F., K. HERSCHBACH and J.S. KOEHLER, 1964, Phys. Rev. **133**, A293.

DWORSCHAK, F., H. WAGNER and P. WOMBACHER, 1972, Phys. Stat. Sol. (b) **52**, 103.

DWORSCHAK, F., H.E. SCHEPP and H. WOLLENBERGER, 1975, J. Appl. Phys. **46**, 1049.

DWORSCHAK, F., TH. MONSAU and H. WOLLENBERGER, 1976, J. Phys. **F6**, 2207.

DWORSCHAK, F., R. LENNARTZ, J. SELKE and H. WOLLENBERGER, 1978, J. Nucl. Mater. **69/70**, 748.

DWORSCHAK, F., G. HOLFELDER and H. WOLLENBERGER, 1981, Rad. Eff. **59**, 35.

ECKER, K.H., 1982, Verhandl. DPG (VI) **17**, 892.

EHRHART, P., 1978, J. Nucl. Mater. **69/70**, 200.

EHRHART, P., and W. SCHILLING, 1974, Phys. Rev. **B8**, 2604.

EHRHART, P., and U. SCHLAGHECK, 1974, J. Phys. **F4**, 1589.

EHRHART, P., and B. SCHÖNFELD, 1982, Self-Interstitial Atoms and their Agglomerates in hcp Metals, in: Point Defects and Defect Interactions in Metals, eds. J. Takamura, M. Doyama and M. Kiritani (Univ. of Tokyo Press, 1982) p. 47.

EHRHART, P., and E. SEGURA, 1975, X-ray Investigation of Interstitials and Interstitial Clusters after Low Temperature Electron-Irradiation and Thermal Annealing of Gold, in: Proc. Int. Conf. on Fundamental Aspects of Radiation Damage in Metals, Gatlinburg, 1975, eds. M.T. Robinson and F.W. Young, Jr. (CONF-751006, Natl. Techn. Inf. Service, Springfield, VA 22161) p. 295.

EHRHART, P., H.G. HAUBOLD and W. SCHILLING, 1974, Investigation of Point Defects and Their Agglomerates in Irradiated Metals by Diffuse X-ray Scattering, in: Festkörperprobleme XIV/ Advances in Solid State Physics, ed. H.J. Queisser (Vieweg, Braunschweig) p. 87.

EHRHART, P., H.D. CARSTANJEN, A.M. FATTAH and J.B. ROBERTO, 1979, Phil. Mag. **A40**, 843.

EHRHART, P., B. SCHÖNFELD and K. SONNENBERG, 1982, Agglomerates of Interstitial Atoms and Vacancies in Electron-Irradiated Copper, in: Point Defects and Defect Interactions in Metals, eds. J. Takamura, M. Doyama and M. Kiritani (Univ. of Tokyo Press, 1982) p. 687.

EMRICK, R.M., 1982, J. Phys. **F12**, 1327.

EMRICK, R.M., and P.B. MCARDLE, 1969, Phys. Rev. **188**, 1156.

ERGINSOY, C., G.H. VINEYARD and A. ENGLERT, 1964, Phys. Rev. **133**, A595.

ESHELBY, J.D., 1955, Acta Metall. **3**, 487.

ESHELBY, J.D., 1956, The Continuum Theory of Lattice Defects, in: Solid State Physics, 3, eds. F. Seitz and D. Turnbull (Academic, New York) p. 79.

EVANS, J.H., R. BULLOUGH and A.M. STONEHAM, 1972, The Observation and Theory of the Void Lattice in Molybdenum, in: Radiation-Induced Voids in Metals, eds. J.W. Corbett and L.C. Ianniello (CONF-710601, Natl. Techn. Inf. Service, Springfield, VA 22161) p. 522.

EVANS, R., 1977, Calculation of Point Defect Formation Energies in Metals, in: Vacancies '76, eds. R.E. Smallman and J.E. Harris (The Metals Society, London) p. 30.

EYRE, B.L., M.H. LORETTO and R.E. SMALLMAN, 1977, Electron Microscopy Studies of Point Defect Clusters in Metals, in: Vacancies '76, eds. R.E. Smallman and J.E. Harris (The Metals Society, London) p. 63.

FABER, K., and H. SCHULTZ, 1977, Rad. Eff. **31**, 157.

FABER, K., J. SCHWEIKHARDT and H. SCHULTZ, 1974, Scripta Metall. **8**, 713.

FEDER, R., 1970, Phys. Rev. **B2**, 828.

FEDER, R., and H.P. CHARBNAU, 1966, Phys. Rev. **149**, 464.

FEDER, R., and A.S. NOWICK, 1958, Phys. Rev. **109**, 1959.

FEDER, R., and A.S. NOWICK, 1967, Phil. Mag. **15**, 805.

FEDER, R., and A.S. NOWICK, 1972, Phys. Rev. **B5**, 1244.

FEDERIGHI, T., 1965, in: Lattice Defects in Quenched Metals, eds. R.M.J. Cotterill, M. Doyama, J.J. Jackson and M. Meshii (Academic, New York) p. 217.

FEDERIGHI, T., S. CERESARA and F. PIERAGOSTINI, 1965, Phil. Mag. **12**, 1093.

FEESE, K., D. HOFFMANN and H. WOLLENBERGER, 1970, Cryst. Lattice Defects **1**, 245.

FLINN, J.E., and T.A. KENFIELD, 1976, Neutron Swelling Observations on Austenitic Stainless Steels Irradiated in EBR-II, in: Correlation of Neutron and Charged Particle Damage, Proc. of the workshop in Oak Ridge, ed. J.O. Stiegler (CONF-760673, Natl. Techn. Inf. Service, Springfield, VA 22161) p. 253.

FLUSS, H.J., L.C. SMEDSKJAER, M.K. CHASON, D.G. LEGNINI and R.W. SIEGEL, 1978, J. Nucl. Mater. **69/70**, 586.

FLUSS, H.J., L.C. SMEDSKJAER, R.W. SIEGEL, D.G. LEGNINI and M.K. CHASON, 1979, Positron-Annihilation Measurement of the Vacancy Formation Enthalpy in Copper, in: Vth Conf. on Positron Annihilation, eds. R.R. Hasiguti and K. Fujiwara (Japan Inst. of Metals, Sendai, 1979) p. 97.

FLYNN, C.P., 1975, Thin Solid Films **25**, 37.

FLYNN, C.P., J. BASS and D. LAZARUS, 1965, Phil. Mag. **11**, 521.

FOLWEILER, R.C., and F.R. BROTZEN, 1959, Acta Metall. **7**, 716.

FORSCH, K., J. HEMMERICH, H. KNÖLL and G. LUCKI, 1974, Phys. Stat. Sol. (a) **23**, 223.

FRANK, W., A. SEEGER and R. SCHINDLER, 1979, Rad. Eff. **40**, 239.

FRIEDEL, J., 1970, Theory of Point Defects in Metals, in: Vacancies and Interstitials in Metals, eds. A. Seeger, D. Schumacher, W. Schilling and J. Diehl (North-Holland, Amsterdam) p. 787.

FUKAI, Y., 1969, Phil. Mag. **20**, 1277.

FUKAI, Y., 1978, J. Nucl. Mater. **69/70**, 573.

FUKUSHIMA, H., and M. DOYAMA, 1976, J. Phys. **F6**, 677.

FURUKAWA, K., J. TAKAMURA, N. KUWANA, R. TAHARA and M. ABE, 1976, J. Phys. Soc. Jap. **41**, 1584.

GABRIEL, T.A., J.D. AMBURGEY and N.M. GREEWE, 1976, Nucl. Sci. and Eng. **61**, 21.

GANNE, J.P., and Y. QUÉRÉ, 1982, Intrinsic Thermal Expansion of Point Defects in Metals, in: Point Defects and Defect Interactions in Metals, eds. J. Takamura, M. Doyama and M. Kiritani (Univ. of Tokyo Press, 1982) p. 232.

GARNER, F.A., 1981, Proc. Symp. on Phase Stability during Irradiation, eds. J.R. Holland, L.K. Mansur and D.I. Potter (Met. Soc. AIME, Warrendale, PA) p. 165.

GARR, K.R., and A. SOSIN, 1967, Phys. Rev. **162**, 681.

GAUSTER, W.B., S. MANTL, T. SCHOBER and W. TRIFTSHÄUSER, 1975, Annealing of Dislocation Loops in Neutron-Irradiated Copper Investigated by Positron Annihilation, in: Proc. Int. Conf. on Fundamental Aspects of Radiation Damage in Metals, Gatlinburg, 1975, eds. M.T. Robinson and F.W. Young, Jr. (CONF-751006, Natl. Techn. Inf. Service, Springfield, VA 22161) p. 1143.

GEHLEN, P.C., I.R. BEELER and R.I. JAFFEE, eds., 1972, Interatomic Potentials and Simulation of Lattice Defects (Plenum, New York).

GIBSON, J.B., A.N. GOLAND, M. MILGRAM and G.H. VINEYARD, 1960, Phys. Rev. **120**, 1229.

GLASGOW, B.B., A. SI-AHMED, W.G. WOLFER and F.A. GARNER, 1981, J. Nucl. Mater. **103/104**, 981.

GRANATO, A.V., 1981, private communication.

GRANATO, A.V., and T.G. NILAN, 1965, Phys. Rev. **137**, A1250.

GRIPSHOVER, R.J., M. KHOSHNEVISAN, J.S. ZETTS and J. BASS, 1970, Phil. Mag. **22**, 757.

GWOZDZ, P.S., and J.S. KOEHLER, 1973, Phys. Rev. **B8**, 3616.

HALL, T.M., A.N. GOLAND and C.L. SNEAD, 1974, Phys. Rev. **B10**, 3062.

HARDY, J.R., and R. BULLOUGH, 1967, Phil. Mag. **15**, 237.

HATCHER, R., D.R. ZELLER and P.H. DEDERICHS, 1979, Phys. Rev. **B19**, 5083.

HAUBOLD, H.-G., 1972, thesis, Aachen, unpublished.

HAUBOLD, H.-G., 1975, Study of Irradiation-Induced Point Defects by Diffuse Scattering, in: Proc. Int. Conf. on Fundamental Aspects of Radiation Damage in Metals, Gatlinburg, 1975, eds. M.T. Robinson and F.W. Young, Jr. (CONF-751006, Natl. Techn. Inf. Service, Springfield, VA 22161) p. 268.

HAUBOLD, H.-G., 1976, Rev. Physique Appl. **11**, 73.

HAUBOLD, H.-G., and D. MARTINSEN, 1978, J. Nucl. Mater. **69/70**, 644.

HEALD, P.T., 1977, Discrete Lattice Models of Point Defects, in: Vacancies '76, eds. R.E. Smallman and J.E. Harris (The Metals Society, London) p. 11.

HEALD, P.T., and M.V. SPEIGHT, 1974, Phil. Mag. **29**, 1075.

HEIGL, F. and R. SIZMANN, 1972, Cryst. Lattice Defects **3**, 13.

HEINISCH, K.L., 1981, J. Nucl. Mater. **103/104**, 1325.

HERLACH, D., H. STOLL, W. TROST, H. METZ, T.E. JACKMAN, K. MAIER, H.E. SCHAEFER and A. SEEGER, 1977, Appl. Phys. **12**, 59.

HERSCHBACH, K., 1963, PHYS. REV. **130**, 554.

HERTZ, W., and H. PEISL, 1975, J. Phys. **F5**, 2241.

HERTZ, W., W. WAIDELICH and H. PEISL, 1973, Phys. Lett. **43A**, 289.

HETTICH, G., H. MEHRER and K. MAIER, 1977, Scripta Metall. **11**, 795.

HILLAIRET, J., C. MAIRY, C. MINIER, P. HAUTOJÄRVI, A. VEHANEN and J. YLI-KAUPPILA, 1982, Vacancies in Silver: A Resistivity and Positron-Annihilation Study, in: Point Defects and Defect Interactions in Metals, eds. J. Takamura, M. Doyama and M. Kiritani (Univ. of Tokyo Press, 1982) p. 284.

HOCH, M., 1970, Equilibrium Measurements in High-Melting-Point Materials, in: Vacancies and Interstitials in Metals, eds. A. Seeger, D. Schumacher, W. Schilling and J. Diehl (North-Holland, Amsterdam) p. 81.

HODGES, C.H., 1970, Phys. Rev. Lett. **25**, 285.

HOLDER, J., A.V. GRANATO and L.E. REHN, 1974, Phys. Rev. **B10**, 363.

HOOD, G.M., and R.J. SCHULTZ, 1978, J. Nucl. Mater. **69/70**, 607.

HOWE, L.M., and M.L. SWANSON, 1982, Ion Channeling Investigations of the Interactions between Irradiation-produced defects and solute atoms in metals, in: Point Defects and Defect Interactions in Metals, eds. J. Takamura, M. Doyama and M. Kiritani (Univ. of Tokyo Press, 1982) p. 53.

HU, C.-K., S. BERKO, G.R. GRUZALKSI and W.K. WARBURTON, 1979, Positron-Annihilation Lifetime and Doppler Profile Studies in Pb, Pb (+Tl) and Pb (+Cd), in: Proc. 5th Conf. Positron Annihilation, eds. R.R. Hasiguti and K. Fujiwara (Japan. Inst. of Metals, Sendai).

HULTMAN, K.L., J. HOLDER and A.V. GRANATO, 1981, J. Physique, Colloque **C5-42**, Suppl. 10, 753.

HUNTINGTON, H.B., 1942, Phys. Rev. **61**, 325.

HUNTINGTON, H.B., 1953, Phys. Rev. **91**, 1092.

HUNTINGTON, H.B., and F. SEITZ, 1942, Phys. Rev. **61**, 315.

IMAFUKU, M., R. YAMAMOTO and M. DOYAMA, 1982, Computer Studies of Self-Interstitials in Magnesium, in: Point Defects and Defect Interactions in Metals, eds. J. Takamura, M. Doyama and M. Kiritani (Univ. of Tokyo Press, 1982) p. 145.

INGLE, K.W., R.C. PERRIN and H.R. SCHOBER, 1981, J. Phys. **F11**, 1161.

ISEBECK, K., R. MÜLLER, W. SCHILLING and H. WENZL, 1966, Phys. Stat. Sol. **18**, 467.

ISHINO, S., S. IWATA, Y. MATSUTANI and S. TANAKA, 1977, Computer Simulation and Neutron, Heavy-Ion and Electron Irradiation Correlation, in: Radiation Effects in Breeder Reactor Structural Materials, Scottsdale, eds. M.L. Bleiberg and J.W. Bennett (Met. Soc. AIME, Warrendale, PA) p. 879.

JACKSON, J.J., 1965, Point Defects in Quenched Platinum, in: Lattice Defects in Quenched Metals, eds. R.M.J. Cotterill, M. Doyama, J.J. Jackson and M. Meshii (Academic, New York) p. 467.

JACKSON, J.J., 1979, Phys. Rev. **B20**, 534.

JACQUES, H. and K.H. ROBROCK, 1981, J. Physique, Colloque **C5-42**, Suppl. 10, 723.

JÄGER, W., 1981, J. Microsc. Spectrosc. Electron. **6**, 437.

JANOT, C., and B. GEORGE, 1975, Phys. Rev. **B12**, 2212.

JANOT, C., D. MALLEJAC and B. GEORGE, 1970, Phys. Rev. **B2**, 3088.

JENKINS, M.L., and M. WILKENS, 1976, Phil. Mag. **34**, 1155.

JOHNSON, R.A., 1964, Phys. Rev. **134**, A1329.

JOHNSON, R.A., 1965, J. Phys. Chem. Solids **26**, 75.

JOHNSON, R.A., 1966, Phys. Rev. **145**, 423.

JOHNSON, R.A., 1968, Phys. Rev. **174**, 684.

JOHNSON, R.A., 1970, Phys. Rev. **B1**, 3956.

JOHNSON, R.A., 1973, J. Phys. **F3**, 295.

JOHNSON, R.A., and E. BROWN, 1962, Phys. Rev. **127**, 446.

JOHNSON, R.A., and J.R. BEELER, 1977, Phys. Rev. **156**, 677.

JOHNSTON, W.G., T. LAURITZEN, J.W. ROSOLOWSKI and A.M. TURKALO, 1976, J. Metals **28**, 19.

JUNG, P., 1981a, Phys. Rev. **B23**, 664.

JUNG, P., 1981b, Rad. Eff. **59**, 103.

JUNG, P., R.L. CHAPLIN, H.J. FENZL, K. REICHELT and P. WOMBACHER, 1973, Phys. Rev. **B8**, 553.

KANZAKI, H.J., 1957, J. Phys. Chem. Solids **2**, 24.

KAUFFMAN, J.W., and J.S. KOEHLER, 1952, Phys. Rev. **88**, 149.

KEYS, L.K., and J. MOTEFF, 1970, J. Nucl. Mater. **34**, 260.

KHANNA, S.K., and K. SONNENBERG, 1981, Rad. Eff. **59**, 91.

KIENLE, W., W. FRANK and A. SEEGER, 1983, Rad. Eff., to be published.

KIM, S.M., and W.J.L. BUYERS, 1978, J. Phys. **F8**, L103.

KIM, S.M., and A.T. STEWART, 1975, Phys. Rev. **B11**, 4012.

KIM, S.M., W.J.L. BUYERS, P. MARTEL and G.M. HOOD, 1974, J. Phys. **F4**, 343.

KIM, Y.-W., and J.M. GALLIGAN, 1978, J. Nucl. Mater. **69/70**, 680.

KINCHIN, G.H., and R.S. PEASE, 1955, Rep. Prog. Phys. **18**, 1.

KINCHIN, G.H., and M.W. THOMPSON, 1958, J. Nucl. Energy **6**, 275.

KING, W.E., and R. BENEDEK, 1981, Phys. Rev. **B23**, 6335.

KING, W.E., K.L. MERKLE and M. MESHII, 1981, Phys. Rev. **B23**, 6319.

KINO, T., and J.S. KOEHLER, 1967, Phys. Rev. **162**, 632.

KIRITANI, M., 1982, Nature of Point Defects and Their Interactions Revealed by Electron Microscope Observation of Their Clusters, in: Point Defects and Defect Interactions in Metals, eds. J. Takamura, M. Doyama and M. Kiritani (Univ. of Tokyo Press, 1982) p. 59.

KIRITANI, M., and H. TAKATA, 1978, J. Nucl. Mater. **69/70**, 277.

KIRITANI, M., H. TAKATA, K. MORIYAMA and F. EUCHI FUJITA, 1979, Phil. Mag. **A40**, 779.

KNODLE, W.C., and J.S. KOEHLER, 1978, J. Nucl. Mater. **69/70**, 620.

KNÖLL, H., U. DEDEK and W. SCHILLING, 1974, J. Phys. **F4**, 1095.

KOCH, R., R.P. WAHI and H. WOLLENBERGER, 1981, J. Nucl. Mater. **103/104**, 1211.

KOEHLER, J.S., 1970, Electrical Resistivity Measurements of Vacancies in Metals, in: Vacancies and Interstitials in Metals, eds. A. Seeger, D. Schumacher, W. Schilling and J. Diehl (North-Holland, Amsterdam) p. 169.

KÖNIG, D., J. VÖLKL and W. SCHILLING, 1964, Phys. Stat. Sol. **7**, 591.

KOLLERS, G., H. JACQUES, L.E. REHN and K.H. ROBROCK, 1981, J. Physique, Colloque **C5-42**, Suppl. 10, 729.

KONTOLEON, N., K. PAPATHANASSOPOULOS and K. CHOUNTAS, 1973, Rad. Eff. **20**, 273.

KRAFTMAKHER, YA.A., and P.G. STRELKOV, 1970, Equilibrium Concentration of Vacancies in Metals, in: Vacancies and Interstitials in Metals, eds. A. Seeger, D. Schumacher, W. Schilling and J. Diehl (North-Holland, Amsterdam) p. 59.

KRISHAN, K., 1982, Phil. Mag. **A45**, 401.

KRONMÜLLER, H., 1970, Studies of Point Defects in Metals by Means of Mechanical and Magnetic Relaxation, in: Vacancies and Interstitials in Metals, eds. A. Seeger, D. Schumacher, W. Schilling and J.Diehl (North-Holland, Amsterdam) p. 667.

KUGLER, H., I.A. SCHWIRTLICH, S. TAKAKI, K. YAMAKAWA, U. ZIEBART, J. PETZOLD and H. SCHULTZ, 1982, Stage III Recovery in Electron-Irradiated bcc Transition Metals, in: Point Defects and Defect Interactions in Metals, eds. J. Takamura, M. Doyama and M. Kiritani (Univ. of Tokyo Press, 1982) p. 520.

KUNZ, W., 1971, Phys. Stat. Sol. (b) **48**, 387.

KUSMISS, J.H., and A.T. STEWART, 1967, Adv. Phys. **16**, 471.

LAM, N.Q., N.V. DOAN and Y. ADDA, 1980, J. Phys. **F10**, 2359.

LAM, N.Q., N.V. DOAN, L. DAGENS and Y. ADDA, 1981, J. Phys. **F11**, 2231.

LAMPERT, G., and H.-E. SCHAEFER, 1972, Phys. Stat. Sol. (b) **52**, 475; Phys. Stat. Sol. (b) **53**, 113.

LANORE, J.M., L. GLOWINSKI, A. RISBET, P. REGNIER, J.L. FLAMENT, V. LEVY and Y. ADDA, 1975, Studies of Void Formation in Pure Metals, in: Proc. Int. Conf. on Fundamental Aspects of Radiation Damage in Metals, Gatlinburg, 1975, eds. M.T. Robinson and F.W. Young, Jr. (CONF-751006, Natl. Techn. Inf. Service, Springfield, VA 22161) p. 1169.

LARSON, B.C., and F.W. YOUNG, Jr., 1982, Vacancy and Interstitial Loops in Irradiated Metals, in: Point Defects and Defect Interactions in Metals, eds. J. Takamura, M. Doyama and M. Kiritani (Univ. of Tokyo Press, 1982) p. 679.

LAUPHEIMER, A., W. FRANK, M. RÜHLE, A. SEEGER and M. WILKENS, 1981, Rad. Eff. Lett. **67**, 95.

LEADBETTER, A.J., D.M.T. NEWSHAM and N.H. PICTON, 1966, Phil. Mag. **13**, 371.

LEE, C., and J.S. KOEHLER, 1968, Phys. Rev. **176**, 813.

LEE, C., and R.W. SIEGEL, 1975, Electrochem. Soc. Extend. Abstr. **75-1**, 252.

LEHMANN, C., 1977, Interaction of Radiation with Solids, in: Defects in Crystalline Solids, vol. 10, eds. S. Amelinckx, R. Gevers and J. Nihoul (North-Holland, Amsterdam).

LEHMANN, C., A. SCHOLZ and W. SCHILLING, 1981, Computer Simulation of Radiation-Induced Defects, film (16 mm, colour, 32 min) produced and distributed by EMW. Huschert Film-Studio (Weststr. 34-36, D-4000, Düsseldorf-Benrath, F.R. Germany).

LEIBFRIED, G., 1965, Bestrahlungseffekte in Festkörpern (Teubner, Stuttgart).

LEIBFRIED, G., and N. BREUER, 1978, Point Defects in Metals I, in: Springer Tracts in Modern Physics, vol. 81, eds. G. Höhler and E.A. Niekisch (Springer, Berlin).

LENGELER, B., 1976, Phil. Mag. **34**, 259.

LENGELER, B., and R.R. BOURASSA, 1976, J. Phys. **F6**, 1405.

LENNARTZ, R., F. DWORSCHAK and H. WOLLENBERGER, 1977, J. Phys. **F7**, 2011.

LEVY, V., J.M. LANORE and J. HILLAIRET, 1973, Phil. Mag. **28**, 373.

LOSEHAND, R., F. RAU and H. WENZL, 1969, Rad. Eff. **2**, 69.

LUCASSON, A., P. LUCASSON and C. BUDIN, 1964, J. Physique **25**, 2550M.

LUCASSON, P.G., and R.M. WALKER, 1962, Phys. Rev. **127**, 485.

LUDWIG, W., 1967, Recent Developments in Lattice Theory, in: Springer Tracts in Modern Physics, vol. 43, ed. G. Höhler (Springer, Berlin).

LUNDY, T.S., F.R. WINSLOW, R.E. PAWEL and C.J. MCHARGUE, 1965, Trans. Amer. Inst. Min. Engrs. **233**, 1533.

LWIN, Y.N., M. DOYAMA and J.S. KOEHLER, 1968, Phys. Rev. **165**, 787.

LYNN, K.G., C.L. SNEAD, J.J. HURST and K. FARRELL, 1979, Determination of the Vacancy Formation Enthalpy for High-Purity Ni, in: Vth Int. Conf. on Positron Annihilation, eds. R.R. Hasiguti and K. Fujiwara (Japan. Inst. of Metals, Sendai) p. 119.

MACKENZIE, J.K., T.W. CRAIG and B.T.A. MCKEE, 1971, Phys. Lett. **33A**, 227.

MAIER, K., H. METZ, D. HERLACH, H.E. SCHAEFER and A. SEEGER, 1978, J. Nucl. Mater. **69/70**, 589.

MAIER, K., G. REIN, B. SAILE, P. VALENTA and H.E. SCHAEFER, 1979a, Positron-Annihilation Measurements in Cu, Ni, Pd and Pt in Thermal Equilibrium, in: Vth Int. Conf. on Positron Annihilation, eds. R.R. Hasiguti and K. Fujiwara (Japan. Inst. of Metals, Sendai) p. 101.

MAIER, K., M. PEO, B. SAILE, H.E. SCHAEFER and A. SEEGER, 1979b, Phil. Mag. **A40**, 701.

MAIER, K., H. MEHRER and G. REIN, 1979c, Z. Metallk. **70**, 271.

MAMALUI, A.A., T.D. OSITINSKAJA, V.A. PERVAKOV and V.I. KHOMKEVICH, 1969, Sov. Phys. Solid State **10**, 2290.

MANNINEN, M., R. NIEMINEN, P. HAUTOJÄRVI and J. ARPONEN, 1975, Phys. Rev. **B12**, 4012.

MANSEL, W., J. MARANGOS and G. VOGL, 1981, Hyperfine Interactions **10**, 687.

MANSEL, W., J. MARANGOS and D. WAHL, 1982, J. Nucl. Mater. **108/109**, 137.

MANTL, S. and W. TRIFTSHÄUSER, 1978, Phys. Rev. **B17**, 1645.

MARADUDIN, A.A., E.W. MONTROLL, G.H. WEISS and I.P. IPATOVA, 1971, Theory of Lattice Dynamics in the Harmonic Approximation, Solid State Phys., Suppl. III, eds. H. Ehrenreich, F. Seitz and D. Turnbull (Academic, New York).

MARTIN, G., 1980, Phys. Rev. **B21**, 2122.

MARWICK, A.D., R.C. PILLER and P.M. SIVELL, 1979, J. Nucl. Mater. **83**, 35.

MASAMURA, R.A., and G. SINES, 1970, J. Appl. Phys. **41**, 3930.

MASCHER, P., L. BREITENHUBER and W. PUFF, 1981, Phys. Stat. Sol (b) **104**, 601.

MATTER, H., J. WINTER and W. TRIFTSHÄUSER, 1979, Appl. Phys. **20**, 135.

MAURY, F., M. BIGET, P. VAJDA, A. LUCASSON and P. LUCASSON, 1978, Rad. Eff. **38**, 53.

MAURY, F., A. LUCASSON, P. LUCASSON, J. LE HERICY, P. VAJDA, C. DIMITROV and O. DIMITROV, 1980, Rad. Eff. **51**, 57.

MCKEE, B.T.A., W. TRIFTSHÄUSER and A.T. STEWART, 1972a, Phys. Rev. Lett. **28**, 358.

MCKEE, B.T.A., A.G.D. JOST and I.K. MACKENZIE, 1972b, Can. J. Phys. **50**, 415.

MEECHAN, C.J., and A. SOSIN, 1959, Phys. Rev. **113**, 422.

MEECHAN, C.J., A. SOSIN and J.A. BRINKMAN, 1960, Phys. Rev. **120**, 411.

MEHRER, H., H. KRONMÜLLER and A. SEEGER, 1965, Phys. Stat. Sol. **10**, 725.

MELIUS, C.F., C.L. BISSON and W.P. WILSON, 1978, Phys. Rev. **B18**, 1647.

MIJNARENDS, P.E., 1979, Electron Momentum Densities in Metals and Alloys, in: Topics in Current Physics 12, ed. P. Hautojärvi (Springer, Berlin) ch. 2.

MIŠEK, K., 1979, Czech. J. Phys. **B29**, 1243.

MONSAU, TH., and H. WOLLENBERGER, 1980, unpublished results.

MONTI, A.M., and E.J. SAVINO, 1981, Phys. Rev. **B23**, 6494.

MORI, T., M. MESHII and J.W. KAUFFMAN, 1962, J. Appl. Phys. **33**, 2776.

MOSER, P., 1966, Mém. Sci. Rev. Métallurg. **63**, 431.

MOTT, N.F., 1932, Proc. Roy. Soc. **A135**, 429.

MÜLLER, H.-G., 1979, thesis, Univ. of Bonn, unpublished.

MUGHRABI, H., and A. SEEGER, 1967, Phys. Stat. Sol. **19**, 251.

MUNDY, J.N., S.J. ROTHMAN, N.Q. LAM, H.A. HOFF and L.J. NOWICKI, 1978, Phys. Rev. **B18**, 6566.

MYHRA, S. and R.B. GARDINER, 1975, Rad. Eff. **27**, 35.

NANAO, S., K. KURIBAYASHI, S. TANIGAWA and M. DOYAMA, 1977, J. Phys. **F7**, 1403.

NEELY, H.H., 1970, Rad. Eff. **3**, 189.

NEELY, H.H., D.W. KEEPER and A. SOSIN, 1968, Phys. Stat. Sol. **28**, 675.

NELSON, R.S., J.A. HUDSON and D.J. MAZEY, 1972, J. Nucl. Mater. **44**, 318.

NICKLOW, R.M., W.P. CRUMMETT and J.M. WILLIAMS, 1979, Phys. Rev. **B20**, 5034.

NIESEN, L., 1981, Hyperfine Interactions **10**, 619.

NORGETT, M.J., M.T. ROBINSON and I.M. TORRENS, 1974, Nucl. Eng. Design **33**, 50.

NOWICK, A.S., and B.S. BERRY, 1972, Anelastic Relaxation in Crystalline Solids (Academic, New York).

OHR, S.M., 1975, The Nature of Defect Clusters in Electron-Irradiated Copper, in: Proc. Int. Conf. on Fundamental Aspects of Radiation Damage in Metals, Gatlinburg, 1975, eds. M.T. Robinson and F.W. Young, Jr. (CONF-751006, Natl. Techn. Inf. Service, Springfield, VA 22161) p. 650.

OKAMOTO, P.R., L.E. REHN, R.S. AVERBACK, K.H. ROBROCK and H. WIEDERSICH, 1982, Mechanism and Kinetics of Radiation-Induced Segregation in Ni–Si alloys, in: Point Defects and Defect Interactions in Metals, eds. J. Takamura, M. Doyama and M. Kiritani (Univ. of Tokyo Press, 1982) p. 946.

OKUDA, S., 1975, Experimental Studies on Self-Interstitials in bcc Metals, in: Proc. Int. Conf. on Fundamental Aspects of Radiation Damage in Metals, Gatlinburg, 1975, eds. M.T. Robinson and F.W. Young, Jr. (CONF-751006, Natl. Techn. Inf. Service, Springfield, VA 22161) p. 361.

O'NEAL, T.N., and R.C. CHAPLIN, 1972, Phys. Rev. **B5**, 3810.

PEACOCK, G.E., and A.A. JOHNSON, 1963, Phil. Mag. **8**, 563.

PERETTI, H.A., M.A. MONDINO and A. SEEGER, 1980, Mater. Sci. Eng. **46**, 129.

PERETTO, P., J.L. OLDON, C. MINIER-CASSAYRE, D. DAUTREPPE and P. MOSER, 1966, Phys. Stat. Sol. (b) **53**, 113.

PETERSON, N.L., 1978, J. Nucl. Mater. **69/70**, 3.

PETRY, W., G. VOGL and W. MANSEL, 1982, Z. Phys. **B46**, 319.

PHILLIPP, F., B. SAILE and K. URBAN, 1982, Investigation of Vacancy Diffusion in Niobium, Molybdenum, Tantalum by Means of High Voltage Electron Microscopy, in: Point Defects and Defect Interactions in Metals, eds. J. Takamura, M. Doyama and M. Kiritani (Univ. of Tokyo Press, 1982) p. 261.

PICHON, R., E. BISOGNI and P. MOSER, 1973, Rad. Eff. **20**, 159.

PIERCY, G.R., 1960, Phil. Mag. **5**, 201.

PLEITER, F., and C. HOHENEMSER, 1982, Phys. Rev. **B25**, 106.

POLAK, J., 1967, Phys. Stat. Sol. **21**, 581.

POTTER, D.I., L.E. REHN, P.R. OKAMOTO and H. WIEDERSICH, 1977, Void Swelling and Segregation in Dilute Nickel Alloys, in: Radiation Effects In Breeder Structural Materials (AIME, New York) p. 377.

PROFANT, M., and H. WOLLENBERGER, 1976, Phys. Stat. Sol. (b) **71**, 515.

PUFF, W., P. MASCHER, P. KINDL and H. SORMANN, 1982, Positron Lifetime Measurements in Indium, in: VIth Int. Conf. on Positron Annihilation, Arlington, eds. R.R. Hasiguti and K. Fujiwara (Japan. Inst. of Metals, Sendai).

QUÉRÉ, Y., 1961, C.R. Acad. Sci. **252**, 2399.

RAMSTEINER, F., W. SCHÜLE and A. SEEGER, 1962, Phys. Stat. Sol. **2**, 1005.

RAMSTEINER, F., G. LAMPERT, A. SEEGER and W. SCHÜLE, 1965, Phys. Stat. Sol. **8**, 863.

RASCH, K.-D., R.W. SIEGEL and H. SCHULTZ, 1980, Phil. Mag. **A41**, 91.

RATTKE, R., O. HAUSER and J. WIETING, 1969, Phys. Stat. Sol. **31**, 167.

REHN, L.E., and K.H. ROBROCK, 1977, J. Phys. **F7**, 1107.

REHN, L.E., K.H. ROBROCK and H. JACQUES, 1978, J. Phys. **F8**, 1835.

RICE-EVANS, P., T. HLAING and D.B. REES, 1976, J. Phys. **F6**, 1079.

RICE-EVANS, P., T. HLAING and I. CHAGLAR, 1977, Phys. Lett. **A60**, 368.

RINNEBERG, H., and H. HAAS, 1978, Hyperfine Interactions **4**, 678.

RINNEBERG, H., W. SEMMLER and G. ANTESBERGER, 1978, Phys. Lett. **A66**, 57.

RIZK, R., P. VAJDA, F. MAURY, A. LUCASSON, P. LUCASSON, C. DIMITROV and O. DIMITROV, 1976, J. Appl. Phys. **47**, 4740.

RIZK, R., P. VAJDA, F. MAURY, A. LUCASSON and P. LUCASSON, 1977, J. Appl. Phys. **48**, 481.

ROBERTS, C.G., W.P. RICKEY and P.E. SHEARIN, 1966, J. Appl. Phys. **37**, 4517.

ROBROCK, K.H., 1982, Study of Self-Interstitial-Atom–Solute-Atom-Complexes by Mechanical Relaxation, in: Point Defects and Defect Interactions in Metals, eds. J. Takamura, M. Doyama and M. Kiritani (Univ. of Tokyo Press, 1982) p. 353.

ROBROCK, K.H., 1983, The Interaction of Self-Interstitial Atoms with Solute Atoms in Metals, in: Phase Stability and Solute Redistribution, eds. F. Nolfi and J. Gittus (Appl. Sci. Publ., London) p. 115.

ROBROCK, K.H., and W. SCHILLING, 1976, J. Phys. **F6**, 303.

ROBROCK, K.H., and H.R. SCHOBER, 1981, J. Physique, Colloque **C5-42**, Suppl. 10, 735.

ROGGEN, J., J. NIHOUL, J. CORNELIS and L. STALS, 1978, J. Nucl. Mater. **69/70**, 700.

ROTH, G., H. WOLLENBERGER, C. ZECKAU and K. LÜCKE, 1975, Rad. Eff. **26**, 141.

SAHU, R.P., K.C. JAIN and R.W. SIEGEL, 1978, J. Nucl. Mater. **69/70**, 264.

SCHAEFER, H.-E., 1981, Habilitation, Univ. of Stuttgart.

SCHAEFER, H.-E., D. BUTTEWEG and W. DANDER, 1975, Defects in High Purity Iron after 27 K Electron Irradiation, in: Proc. Int. Conf. on Fundamental Aspects of Radiation Damage in Metals, Gatlinburg, 1975, eds. M.T. Robinson and F.W. Young, Jr. (CONF-751006, Natl. Techn. Inf. Service, Springfield, VA, 22161) p. 463.

SCHAEFER, H.-E., K. MAIER, M. WELLER, D. HERLACH, A. SEEGER and J. DIEHL, 1977, Scripta Metall. **11**, 803.

SCHILLING, W., 1978, J. Nucl. Mater. **69/70**, 465.

SCHILLING, W., and P. TISCHER, 1967, Z. Angew. Phys. **22**, 56.

SCHILLING, W., G. BURGER, K. ISEBECK and H. WENZL, 1970, Annealing Stages in the Electrical Resistivity of Irradiated fcc Metals, in: Vacancies and Interstitials in Metals, eds. A. Seeger, D. Schumacher, W. Schilling and J. Diehl (North-Holland, Amsterdam) p. 255.

SCHINDLER, R., W. FRANK, M. RÜHLE, A. SEEGER and M. WILKENS, J. Nucl. Mater. **69/70**, 331.

SCHOBER, H.R., 1977, J. Phys. **F7**, 1127.

SCHOBER, H.R., and R. ZELLER, 1978, J. Nucl. Mater. **69/70**, 341.

SCHOBER, H.R., V.K. TEWARY and P.H. DEDERICHS, 1975, Z. Phys. **B21**, 255.

SCHOLZ, A., and C. LEHMANN, 1972, Phys. Rev. **B6**, 813.

SCHOTTKY, G., A. SEEGER and G. SCHMID, 1964, Phys. Stat. Sol. **4**, 439.

SCHROEDER, H., and B. STRITZKER, 1977, Rad. Eff. **33**, 125.

SCHROEDER, H., R. LENNARTZ and U. DEDEK, 1975, Recovery of Pure Lead after Electron Irradiation at 4.7 K and below 3 K, in: Proc. Int. Conf. on Fundamental Aspects of Radiation Damage in Metals, Gatlinburg, 1975, eds. M.T. Robinson and F.W. Young, Jr. (CONF-751006, Natl. Techn. Inf. Service, Springfield, VA 22161) p. 411.

SCHROEDER, K., 1980, Theory of Diffusion-Controlled Reactions of Point Defects in Metals, in: Springer Tracts in Modern Physics, vol. 87, eds. G. Höhler and E.A. Niekisch (Springer, Berlin) p. 171.

SCHÜLE, W., A. SEEGER, D. SCHUMACHER and K. KING, 1962, Phys. Stat. Sol. **2**, 1199.

SCHUMACHER, D., A. SEEGER and O. HÄRLIN, 1968, Phys. Stat. Sol. **25**, 359.

SCHULTZ, H., 1982, Atomic Defects in bcc Transition Metals, in: Point Defects and Defect Interactions in Metals, eds. J. Takamura, M. Doyama and M. Kiritani (Univ. of Tokyo Press, 1982) p. 183.

SCHWIRTLICH, I.A., and H. SCHULTZ, 1980a, Phil. Mag. **A42**, 601.

SCHWIRTLICH, I.A., and H. SCHULTZ, 1980b, Phil. Mag. **A42**, 613.

SEEBOECK, R., W. ENGEL, S. HOTH, R. KEITEL and W. WITTHUHN, 1982,Vacancies in Zinc and Cadmium Produced by Proton- and Electron-Irradiation, in: Point Defects and Defect Interactions in Metals, eds. J. Takamura, M. Doyama and M. Kiritani (Univ. of Tokyo Press, 1982) p. 271.

SEEGER, A., 1958, On the Theory of Radiation Damage and Radiation Hardening, in: Proc. 2nd U.N. Int. Conf. on the Peaceful Uses of Atomic Energy, vol. 6 (United Nations, Geneva) p. 250.

SEEGER, A., 1970, Rad. Eff. **2**, 165.

SEEGER, A., 1973a, Cryst. Lattice Defects **4**, 221.

SEEGER, A., 1973b, J. Phys. **F3**, 248.

SEEGER, A., 1975, The Interpretation of Radiation Damage in Metals, in: Proc. Int. Conf. on Fundamental Aspects of Radiation Damage in Metals, Gatlinburg, 1975, eds. M.T. Robinson and F.W. Young, Jr. (CONF-751006, Natl. Techn. Inf. Service, Springfield, VA 22161) p. 493.

SEEGER, A., 1982, Phys. Lett. **89A**, 241.

SEEGER, A., and W. FRANK, 1983, Rad. Eff., in press.

SEEGER, A. and E. MANN, 1960, J. Phys. Chem. Solids **12**, 326.

SEEGER, A., E. MANN and R. V. JAN, 1962, J. Phys. Chem. Solids **23**, 639.

SEGERS, D., F. VAN BRABANDER, L. DORIKENS-VAN PRAET and M. DORIKENS, 1982, Appl. Phys. **A27**, 129.

SEGURA, E., and P. EHRHART, 1979, Rad. Eff. **42**, 233.

SEIDMAN, D.N., 1976, Field-Ion Microscope Studies of the Defect Structure of the Primary State of Damage of Irradiated Metals, in: Radiation Damage in Metals, eds. N.L. Peterson and S.D. Harkness (ASM, Metals Park, OH).

SEIDMAN, D.N., 1978, Surf. Sci. **70**, 532.

SEITZ, F., and J.S. KOEHLER, 1956, Displacement of Atoms during Irradiation, in: Solid State Physics, vol. 2, eds. F. Seitz and D. Turnbull (Academic, New York) p. 305.

SHARMA, R.K., C. LEE and J.S. KOEHLER, 1967, Phys. Rev. Lett. **19**, 1379.

SHARMA, S.C., S. BERKO and W.K. WARBURTON, 1976, Phys. Lett. **58A**, 405.

SHIMOMURA, Y., K. YAMAHAWA, K. KITAGAWA and H. ODA, 1982, Studies of Point Defect Clusters with an Electron Microscope, in: Point Defects and Defect Interactions in Metals, eds. J. Takamura, M. Doyama and M. Kiritani (Univ. of Tokyo Press, 1982) p. 712.

SIEGEL, R.W., 1966a, Phil. Mag. **13**, 359.

SIEGEL, R.W., 1966b, Phil. Mag. **13**, 337.

SIEGEL, R.W., 1978, J. Nucl. Mater. **69/70**, 117.

SIEMS, R., 1968, Phys. Stat. Sol. **30**, 645.

SIGMUND, P., 1972, Rev. Roum. Phys. **17**, 969.

SIMMONS, R.O., and R.W. BALLUFFI, 1960a, Phys. Rev. **117**, 52.

SIMMONS, R.O., and R.W. BALLUFFI, 1960b, Phys. Rev. **119**, 600.

SIMMONS, R.O., and R.W. BALLUFFI, 1962, Phys. Rev. **125**, 862.

SIMMONS, R.O., and R.W. BALLUFFI, 1963, Phys. Rev. **129**, 1533.

SIMON, J.P., P. VOSTRY, J. HILLAIRET and P. VAJDA, 1974, Phys. Stat. Sol. (b) **64**, 277.

SIMON, J.P., P. VOSTRY, J. HILLAIRET and V. LEVY, 1975, Phil. Mag. **31**, 154.

SIMPSON, H.M., and R.L. CHAPLIN, 1969, Phys. Rev. **178**, 1166.

SIMSON, P., and R. SIZMANN, 1962, Z. Naturf. **17a**, 596.

SINGH, H.P., and R.N. WEST, 1976, J. Phys. **F6**, L267.

SINGH, H.P., G.S. GOODBODY and R.N. WEST, 1975, Phys. Lett. **A55**, 237.

SMOLUCHOWSKI, M., 1917, Z. Phys. Chem. (Leipzig) **92**, 129.

SNEAD, C.L., T.M. HALL and A.N. GOLAND, 1972, Phys. Rev. Lett. **29**, 62.

SONNENBERG, K., and U. DEDEK, 1982, Rad. Eff. **61**, 175.

SONNENBERG, K., W. SCHILLING, H.J. DIBBERT, K. MIKA and K. SCHROEDER, 1972a, Rad. Eff. **15**, 129.

SONNENBERG, K., W. SCHILLING, K. MIKA and K. DETTMANN, 1972b, Rad. Eff. **16**, 65.

SOSIN, A., 1970, Study of Point Defects by Means of Dislocation Pinning Effects, in: International Conference on Vacancies and Interstitials in Metals, Jülich (Jül-Conf-2, p. 729), unpublished.

SOSIN, A., and W. BAUER, 1968, Atomic Displacement Mechanism in Metals and Semi-Conductors, in: Studies in Radiation Effects in Solids, ed. G.J. Dienes, vol. 3 (Gordon and Breach, New York).

SOSIN, A., and J.A. BRINKMAN, 1959, Acta Metall. **7**, 478.

SPIRIĆ, V., L.E. REHN, K.H. ROBROCK and W. SCHILLING, 1977, Phys. Rev. **B15**, 672.

STALS, L., 1972, On the Kinetics of Point Defect Annealing in Solids, in: Defects in Refractory Metals, eds. R. de Batist, J. Nihoul and L. Stals (SCK–CEN, Mol) p. 191.

STOTT, M.J., 1978, J. Nucl. Mater. **69/70**, 157.

SUEOKA, O., 1974, J. Phys. Soc. Japan **36**, 464.

SUEZAWA, M., and H. KIMURA, 1973, Phil. Mag. **28**, 901.

SULPICE, G., C. MINIER, P. MOSER and H. BILGER, 1968, J. Physique **29**, 253.

SWANSON, M.L., L.M. HOWE and A.F. QUENNEVILLE, 1978, J. Nucl. Mater. **69/70**, 372.

TAKAMURA, J.I., 1970, Point Defects, in: Physical Metallurgy, ed. R.W. Cahn (North-Holland, Amsterdam) ch. 14.

TAM, S.W., 1975, J. Metals **27**, A23.

TENENBAUM, A., and N.V. DOAN, 1977, Phil. Mag. **35**, 379.

TEWORDT, L., 1958, Phys. Rev. **109**, 61.

TIETZE, M., S. TAKAKI, I.A. SCHWIRTLICH and H. SCHULTZ, 1982, Quenching Investigations on bcc Metals, in: Point Defects and Defect Interations in Metals, eds. J. Takamura, M. Doyama and M. Kiritani (Univ. of Tokyo Press, 1982) p. 265.

TRIFTSHÄUSER, W., 1975a, Festkörperprobleme XV/Adv. in Solid State Phys., ed. H.J. Queisser (Vieweg, Braunschweig) p. 381.

TRIFTSHÄUSER, W., 1975b, Phys. Rev. **B12**, 4634.

TRIFTSHÄUSER, W., and J.D. McGERVEY, 1975, Appl. Phys. **6**, 177.

TRINKAUS, H., 1972, Phys. Stat. Sol. **51**, 1972.

TRINKAUS, H., 1975, Theory of Polarization-Induced Elastic Interaction of Point Defects, in: Proc. Int. Conf. on Fundamental Aspects of Radiation Damage in Metals, Gatlinburg, 1975, eds. M.T. Robinson and F.W. Young, Jr. (CONF-751006, Natl. Techn. Inf. Service, Springfield, VA 22161) p. 254.

TZANETAKIS, P., J. HILLAIRET and G. REVEL, 1976, Phys. Stat. Sol. (b) **75**, 433.

URBAN K., and N. YOSHIDA, 1981, Phil. Mag. **A44**, 1193.

VAJDA, P., 1977, Rev. Mod. Phys. **49**, 481.

VAN BUEREN, H.G., 1955, Z. Metallk. **46**, 272.

VAROTSOS, P., and K. ALEXOPOULOS, 1982, Phys. Stat. Sol. (b) **110**, 9.

VEHANEN, A., P. HAUTOJÄRVI, J. JOHANSSON and J. YLI-KAUPPILA, 1982, Phys. Rev. **B25**, 762.

V. GUÉRARD, B., H. PEISL and R. ZITZMANN, 1974, Appl. Phys. **3**, 37.

VOGL, G., and W. MANSEL, 1975, Mössbauer Studies of Interstitials in fcc Metals, in: Proc. Int. Conf. on Fundamental Aspects of Radiation Damage in Metals, Gatlinburg, 1975, eds. M.T. Robinson and F.W. Young, Jr. (CONF-751006, Natl. Techn. Inf. Service, Springfield, VA 22161) p. 349.

VOGL, G., W. MANSEL and P.H. DEDERICHS, 1976, Phys. Rev. Lett. **36**, 1497.

VOGL, G., W. MANSEL, W. PETRY and V. GRÖGER, 1978, Hyperfine Interactions **4**, 681.

WAGNER, H., F. DWORSCHAK and W. SCHILLING, 1970, Phys. Rev. **B2**, 3856.

WAGNER, R., 1982, Analytical Field Ion Microscopy in Metallurgy, in: Crystal Growth, Properties and Applications, ed. H.C. Freyhardt (Springer, Berlin).

WAGNER, W., R. POERSCHKE and H. WOLLENBERGER, 1982, J. Phys. **F12**, 405.

WAGNER, W., L.E. REHN, H. WIEDERSICH and V. NAUNDORF, 1983, submitted to Phys. Rev.

WAHI, R.P. and H. WOLLENBERGER, 1982, J. Nucl. Mater. **113**, 207.

WAITE, T.R., 1957, Phys. Rev. **107**, 463 and 471.

WANG, G.G., D.N. SEIDMANN and R.W. BALLUFFI, 1968, Phys. Rev. **169**, 553.

WEIDINGER, A., M. DEICHER and J. BUSSE, 1982, Radiation Damage in Tantalum observed by PAC, in: Point Defects and Defect Interactions in Metals, eds. J. Takamura, M. Doyama and M. Kiritani (Univ. of Tokyo Press, 1982) p. 268.

WEINBERG, A.M., and E.P. WIGNER, 1958, The Physical Theory of Neutron Chain Reactors (Univ. of Chicago Press, Chicago).

WENZL, H., 1970, Physical Properties of Point Defects in Cubic Metals, in: Vacancies and Interstitials in Metals, eds. A. Seeger, D. Schumacher, W. Schilling and J. Diehl (North-Holland, Amsterdam) p. 363.

WENZL, H., F. KERSCHER, V. FISCHER, K. EHRENSPERGER and K. PAPATHANASSOPOULOS, 1971, Z. Naturf. **26**, 489.

WEST, R.N., 1973, Adv. Phys. **22**, 263.

WEST, R.N., 1979, Positron Studies of Lattice Defects in Metals, in: Topics in Current Physics 12, ed. P. Hautojärvi (Springer, Berlin) ch. 3.

WICHERT, TH., 1982, PAC Study of Point Defects in Metals, in: Point Defects and Defect Interactions in Metals, eds. J. Takamura, M. Doyama and M. Kiritani (Univ. of Tokyo Press, 1982) p. 19.

WICHERT, TH., M. DEICHER, O. ECHT and E. RECKNAGEL, 1978, Phys. Rev. Lett. **41**, 1659.

WIEDERSICH, H, P.R. OKAMOTO and N.Q. LAM, 1977, Solute Segregation During Irradiation, in: Radiation Effects in Breeder Reactor Structural Materials, Scottsdale, eds. M.L. Bleiberg and J.W. Bennett (AIME, New York) p. 801.

WIEDERSICH, H., P.R. OKAMOTO and N.Q. LAM, 1979, J. Nucl. Mater. **83**, 98.

WIENHOLD, P., K. SONNENBERG and A. ANTESBERGER, 1978, Rad. Eff. **36**, 235.

WIFFEN, F.W., 1972, The Effect of Alloying and Purity on the Formation and Ordering of Voids in bcc Metals, in: Radiation-Induced Voids in Metals, eds. J.W. Corbett and L.C. Ianniello (CONF-710601, Natl. Techn. Inf. Service Springfield, VA 22161) p. 386.

WITTMAACK, K., 1970, Phys. Stat. Sol. **37**, 633.

WOLLENBERGER, H., 1965, Frühjahrstagung DPG Freudenstadt, unpublished.

WOLLENBERGER, H., 1970, Production of Frenkel Defects during Low-Temperature Irradiations, in: Vacancies and Interstitials in Metals, eds. A. Seeger, D. Schumacher, W. Schilling and J. Diehl (North-Holland, Amsterdam) p. 215.

WOLLENBERGER, H., 1975, Defect interactions above stage I, in: Proc. Int. Conf. on Fundamental Aspects of Radiation Damage in Metals, Gatlinburg, 1975, eds. M.T. Robinson and F.W. Young, Jr. (CONF-751006, Natl. Techn. Inf. Service, Springfield, VA 22161) p. 582.

WOLLENBERGER, H., 1978, J. Nucl. Mater. **69/70**, 362.

WOLLENBERGER, H., 1981, Z. Metallk. **72**, 608.

WOLLENBERGER, H., and J. WURM, 1965, Phys. Stat. Sol. **9**, 601.

WOOD, R.F., and M. MOSTOLLER, 1975, Phys. Rev. Lett. **35**, 45.

WRIGHT, P., and J.H. EVANS, 1966, Phil. Mag. **13**, 521.

WYCISK, W., and M. FELLER-KNIEPMEIER, 1978, J. Nucl. Mater. **69/70**, 616.

YAMAMOTO, R., 1982, Lattice Vibrations around a Vacancy and Vacancy Clusters in Metals, in: Point Defects and Defect Interactions in Metals, eds. J. Takamura, M. Doyama and M. Kiritani (Univ. of Tokyo Press, 1982) p. 120.

YOSHIOKA, K., M. SUEZAWA and H. KIMURA, 1974, Scripta Metall. **8**, 111.

YOUNG, F.W., Jr., 1978, J. Nucl. Mater. **69/70**, 310.

YTTERHUS, J.A., and R.W. BALLUFFI, 1965, Phil. Mag. **11**, 707.

ZELLER, R., and P.H. DEDERICHS, 1976, Z. Phys. **B25**, 139.

ZETTS, J.S., and J. BASS, 1975, Phil. Mag. **31**, 419.

Further reading

Bleiberg, M.L., and J.W. Bennett, eds., 1977, Radiation Effects in Breeder Reactor Structural Materials, Proc. Int. Conf. Scottsdale 1977 (AIME, New York).

Cahn, R.W., ed., 1970, Physical Metallurgy (North-Holland, Amsterdam), ch. 14, by J. Takamura.

Carpenter, G.J.C., C.E. Coleman and S.R. MacEwen, eds., 1980, Fundamental Mechanisms of Radiation-Induced Creep and Growth, Proc. Int. Conf. Chalk River 1979, J. Nucl. Mater. 90.

Corbett, J.W., 1966, Electron Radiation Damage in Semiconductors and Metals, in: Solid State Physics Suppl. 7, eds. F. Seitz and D. Turnbull (Academic, New York).

Corbett, J.W., and L.C. Ianniello, eds., 1972, Radiation-Induced Voids in Metals, Proc. Int. Conf. Albany 1971 (CONF-710601, Natl. Techn. Inf. Service, Springfield, VA 22161).

Cundy, M.R., P. van der Hardt and R.H. Loelgen, eds., 1977, Measurements of Irradiation-Enhanced Creep in Nuclear Materials, Proc. Int. Conf. Petten 1976, J. Nucl. Mater. 65 (1977).

Flynn, C.P., 1972, Point Defects and Diffusion (Clarendon Press, Oxford).

Gehlen, P.C., J.R. Beeler, Jr., and R.I. Jaffee, eds., 1971, Interatomic Potentials and Simulation of Lattice Defects, Battelle Inst. Materials Science Colloquia, Seattle and Harrison, Hotsprings (Plenum, New York).

Gittus, J., 1978, Irradiation Effects in Crystalline Solids (Appl. Science Publ., London).

Hautojärvi, P., ed., 1979, Positrons in Solids, Topics in Current Physics, vol. 12 (Springer, Berlin).

Höhler, G., and E.A. Niekisch, eds., 1978, Point Defects in Metals I, Springer Tracts in Modern Physics, vol. 81 (Springer, Berlin).

Höhler, G., and E.A. Niekisch, eds., 1980, Point Defects in Metals II, Springer Tracts in Modern Physics, vol. 87 (Springer, Berlin).

Holland, J.R., L.K. Mansur and D.I. Potter, eds., 1981, Proc. Symp. on Phase Stability during Irradiation (Met. Soc. AIME, Warrendale, PA).

Krippner, M., ed., 1969, Radiation Damage in Reactor Materials, Proc. Symp. Vienna 1969 (IAEA, Vienna).

Lehmann, Chr., 1977, Interaction of Radiation with Solids and Elementary Defect Production, in: Defects in Crystalline Solids, vol. 10, eds. S. Amelinckx, R. Gevers and J. Nihoul (North-Holland, Amsterdam).

Nelson, R.S, 1968, The Observation of Atomic Collisions in Crystalline Solids, in: Defects in Crystalline Solids, vol. 1, eds. S. Amelinckx, R. Gevers and J. Nihoul (North-Holland, Amsterdam).

Nowick, A.S., and B.S. Berry, 1972, Anelastic Relaxation in Crystalline Solids, in: Mater. Science Series, eds. A.M. Alper, J.L. Margrave and A.S. Nowick (Academic, New York).

Nygren, R.E., R.E. Gold and R.H. Jones, eds., 1981, Fusion Reactor Materials, Proc. 2nd Topical Meeting on Fusion reactor Materials, Seattle 1981, J. Nucl. Mater. 103/104 (1981).

Peterson, N.L., and R.W. Siegel, eds., 1976, Properties of Atomic Defects in Metals, Proc. Int. Conf. Argonne 1976, J. Nucl. Mater. 69/70 (1978).

Robinson, M.T., and F.W. Young, Jr., eds., 1975, Proc. Int. Conf. on Fundamental Aspects of Radiation Damage in Metals, Gatlinburg 1975 (CONF-751006, Natl. Techn. Inf. Service, Springfield, VA 22161).

Seeger, A., D. Schumacher, W. Schilling and J. Diehl, eds., 1970, Vacancies and Interstitials in Metals (North-Holland, Amsterdam).

Stiegler, J.O., ed., 1976, Correlation of Neutron and Charged Particle Damage, Proc. Workshop Oak Ridge 1976 (CONF-760673, Natl. Techn. Inf. Service, Springfield, VA 22161).

Stiegler, J.O., ed., 1979, Workshop on Solute Segregation and Phase Stability During Irradiation, Gatlinburg 1978, J. Nucl. Mater. 83 (1979).

Wiffen, F.W., J.H. DeVan and J.O. Stiegler, eds., 1979, Fusion Reactor Materials, Proc. 1st Topical Meeting Miami 1979, J. Nucl. Mater. 85/86 (1979).

CHAPTER 18

DISLOCATIONS

J.P. HIRTH

Metallurgical Engineering Department
The Ohio State University
Columbus, OH 43210, USA

R.W. Cahn and P. Haasen, eds.
Physical Metallurgy; third, revised and enlarged edition
© *Elsevier Science Publishers BV, 1983*

1.   *Elementary geometrical properties*

The concept of crystal dislocations was introduced by POLANYI [1934], OROWAN [1934] and TAYLOR [1934], although the elastic properties of dislocations in isotropic continua had been known since 1905 (TIMPE, also VOLTERRA [1907]). Dislocations are defects whose motion produces plastic deformation of crystals at stresses well below the theoretical shear strength of a perfect crystal. In fig. 1a, b and c the glide motion of an edge dislocation is shown to cause plastic shear strain. One can imagine a virtual process of cutting the crystal on a glide plane, shearing the cut surface by a shear displacement vector *b*, the *Burgers vector,* and gluing the cut surfaces together, creating the edge dislocation in fig. 1b. The dislocation bounds a slipped area and is a line defect. It is characterized by the Burgers vector *b* and by a unit vector ξ tangent to the dislocation line at a point in question. The same dislocation could be formed by opening a cut under normal tractions, fig. 1d, and inserting a plane of matter.

In order that the deformation not produce a high energy fault on the cut surface, *b* is usually a perfect lattice vector as illustrated for an edge dislocation in fig. 2.

The choice of the ± sense of ξ is arbitrary, but once chosen, the ± sense of *b* is fixed by the following convention: imagine a perfect reference crystal, select a vector ξ in it, and construct a closed circuit in it, right-handed relative to ξ. Then construct the same Burgers circuit in the real crystal, as shown for example in fig. 2. The vector SF connecting the start of the circuit to the finish is the Burgers vector of a dislocation if it is contained within the circuit. In this operation, the circuit must not pass through the nonlinear core region within an atomic spacing or two of the dislocation line.

For the *edge dislocation*, *b* is seen to be perpendicular to ξ. In fig. 3 a *screw dislocation* can be imagined to have been created by a cut and displacement operation, or simply by shearing the slipped area by motion of the dislocation in

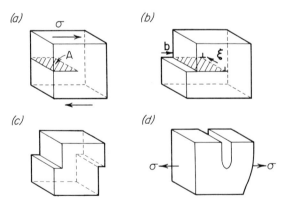

Fig. 1. (a–c) shear of a crystal under a shear stress σ by an amount *b* by passage of an edge dislocation; (d) creation of same dislocation under normal stress σ once added material is placed in the opened cut.

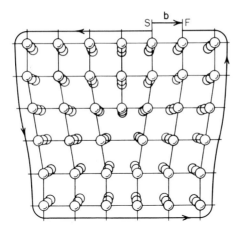

Fig. 2. An edge dislocation in a simple cubic crystal. A Burgers circuit is also shown, projected toward the viewer for clarity so that it passes through the center of atoms not shown. The sense vector ξ points out of the page.

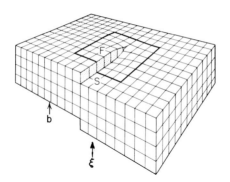

Fig. 3. A screw dislocation in a simple cubic crystal.

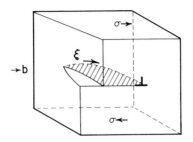

Fig. 4. A mixed dislocation.

References: p. 1257.

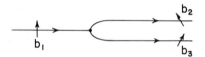

Fig. 5. Three dislocations meeting at a node. Arrows on lines indicate sense of ξ. Equivalent Burgers circuits also shown.

from the surface of a perfect crystal. The screw dislocation has b parallel to ξ. It is right-handed if b points in the same direction as ξ as in fig. 3; left-handed otherwise. In general, fig. 4, the slipped or displaced surface can be arbitrary, the dislocation line can be arbitrarily curved, and the dislocation is called *mixed* when the angle β between b and ξ is neither 0 nor an integer multiple of $\pi/2$.

Some other properties follow directly from the above definitions. If ξ is reversed, the sense of b is also reversed as seen from fig. 2 since the circuit is also reversed when ξ is reversed. Since the dislocation line bounds a displaced area, the line cannot end within otherwise perfect crystal but can only end at a free surface, a grain boundary, a second-phase interface, or a dislocation node. A *node* is a point where two or more dislocation lines join. Translation of a Burgers circuit along ξ without the circuit cutting through a dislocation core does not change the vector SF or thus the total b (imagine such an operation for fig. 2); such circuits are called *equivalent* Burgers circuits. Thus, if a dislocation denoted by its Burgers vector, b_1, splits into two dislocations b_2 and b_3, enclosed by equivalent circuits, fig. 5, an analog of Kirchhoff's law applies and $b_1 = b_2 + b_3$. If the ξ for dislocations 2 and 3 are reversed, then the signs of b_2 and b_3 change by the earlier axiom and: $\sum_i b_i = 0$ for dislocations meeting at a node if all sense vectors ξ are selected to point toward the node.

2. Elastic fields of dislocations

2.1. Displacements and stresses

Consider the right-handed screw dislocation with the geometry as indicated in fig. 6. Imagine that the cylindrical region is a portion of an infinite continuum. In cylindrical coordinates, the displacements can be deduced by inspection:

$$u_r = u_\theta = 0, u_z = b\theta/2\pi. \tag{1}$$

Differentiation and the use of Hooke's Law gives the strain and stress fields:

$$\varepsilon_{\theta z} = \frac{1}{2}\left(\frac{1}{r}\frac{\partial u_z}{\partial \theta} + \frac{\partial u_\theta}{\partial z}\right) = \frac{b}{4\pi r}, \tag{2}$$

$$\sigma_{\theta z} = 2\mu\varepsilon_{\theta z} = \mu b/2\pi r, \tag{3}$$

with μ the shear modulus.

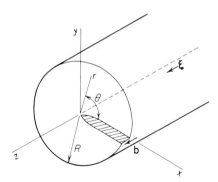

Fig. 6. A right-handed screw dislocation along the axis of a cylindrical region of radius R and length L.

The edge dislocation, fig. 7, has displacements u_x and $u_y \neq 0$ but $u_z = 0$. Thus, the solution for this plane-strain problem proceeds by deducing the Airy stress function Ψ. Guided by the screw dislocation result that $\sigma \sim 1/r$ and observing in fig. 2 that $(\sigma_{rr} + \sigma_{\theta\theta}) \sim -\sin\theta$ to give compression above, and tension below the glide plane $\theta = 0$, one finds:

$$\Psi = \frac{-\mu b}{\pi(1-\nu)} r \sin\theta \ln r, \tag{4}$$

with ν = Poisson's ratio. Differentiation of eq. (4) gives:

$$\sigma_{rr} = \frac{1}{r}\frac{\partial\Psi}{\partial r} + \frac{1}{r^2}\frac{\partial^2\Psi}{\partial\theta^2} = -\frac{\mu b \sin\theta}{2\pi(1-\nu)r},$$

$$\sigma_{\theta\theta} = \frac{\partial^2\Psi}{\partial r^2} = \sigma_{rr}, \qquad \sigma_{r\theta} = -\frac{\partial}{\partial r}\left(\frac{1}{r}\frac{\partial\Psi}{\partial\theta}\right) = \frac{\mu b \cos\theta}{2\pi(1-\nu)r},$$

$$\sigma_{zz} = \nu(\sigma_{rr} + \sigma_{\theta\theta}). \tag{5}$$

This field has the physical features required by fig. 2; compression above the glide plane, tension below it, and maximum shear stresses on the glide plane $\theta = 0$. The

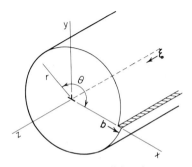

Fig. 7. An edge dislocation.

References: p. 1257.

strains are given by Hooke's Law and their integrals give the displacements, see HIRTH and LOTHE [1982].

For the screw dislocation, the strain energy density is $w = \sigma_{\theta z}^2/2\mu$. The total strain energy per unit length in the cylindrical region in fig. 6 is:

$$\frac{W}{L} = \int_{r_0}^{R} wr \, dr \, d\theta = \frac{\mu b^2}{4\pi} \ln \frac{R}{r_0}. \tag{6}$$

Here a core cutoff radius r_0 is introduced to avoid divergence. This is necessary because the use of linear elasticity gives an artifical divergence in $\varepsilon_{\theta z}$, eq. (2), as $r \to r_0 \approx b$.

For the edge dislocation, an alternate procedure is simpler. From thermodynamics, the strain energy is equal to the isothermal reversible work done on the system in creating the strain. For fig. 7, imagine creating the dislocation by making a cut on the surface $\theta = 0$, $r > 0$, and displacing it by b, applying stresses $\sigma_{r\theta}$ ($\theta = 0$) to the cut to accomplish the displacement. Since $\sigma_{r\theta} \propto b$ finally, the average stress in the process is $\frac{1}{2}\sigma_{r\theta}$ and the surface work on an element dA is $\frac{1}{2}\sigma_{r\theta}(\theta = 0)b \, dA$. The total strain energy per unit length is then given by the integral of this surface work over A:

$$\frac{W}{L} = \int_{r_0}^{R} \tfrac{1}{2}\sigma_{r\theta}(\theta = 0)b \, dr = \frac{\mu b^2}{4\pi(1 - \nu)} \ln \frac{R}{r_0}. \tag{7}$$

The displacement and stress field for a mixed dislocation is given by decomposition of b into edge and screw components b_e and b_s and use of the above expressions with $b_e = b \sin \beta$ and $b_s = b \cos \beta$ substituted for b. In general, the strain energy, being quadratic in σ, cannot be superposed, but the screw and edge components of the mixed dislocation have no common stress components, so W/L can also be added,

$$\frac{W}{L} = \frac{\mu b^2}{4\pi(1 - \nu)} \left[\sin^2\beta + (1 - \nu)\cos^2\beta\right] \ln \frac{R}{r_0} = E(\beta) \ln \frac{R}{r_0}, \tag{8}$$

where $E(\beta)$ is called the *energy factor*.

2.2. Peach–Koehler force

As just discussed, the strain energy is the isothermal work done on the system, which, for a system under constant external pressure, is the Gibbs free energy G of the system. For an isothermal, constant pressure system, ΔG for a hypothetical reversible displacement of the system in some reaction coordinate c_r is a criterion for whether the displacement in the actual case tends to occur spontaneously (forward if $\Delta G < 0$, backward if $\Delta G > 0$, not if $\Delta G = 0$). In a general sense, we define a virtual thermodynamic force as $F = -\partial G/\partial c_r$. Then if $F > 0$, the system tends to change spontaneously in the direction of the reaction coordinate.

We are interested in whether a dislocation will tend to move spontaneously in the

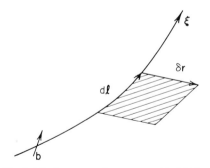

Fig. 8. Displacement of a segment d*l* of dislocation.

presence of a stress field, so the appropriate reaction coordinate is the position **r** of the dislocation. Imagine the process, fig. 8, of cutting an area $dA = n\,dA = dl \times \delta r$ $= dl(\xi \times \delta r)$, displacing the dislocation by δr, removing or adding matter if necessary, and gluing the surfaces together. If surface forces are needed to prevent the cut from relaxing, work can be extracted $(\Delta G < 0)$ as the displacement is accomplished in the hypothetical process. The self-stresses of the dislocation are symmetric and hence do no *net* work in the process. However, any other stresses σ acting in the core region *can* extract work. The stresses produce surface forces $\sigma \cdot dA$ and in the displacement **b** extract an amount of work $-\delta G = \delta W$, given by

$$\delta W = \boldsymbol{b} \cdot \boldsymbol{\sigma} \cdot dA = \boldsymbol{b} \cdot \boldsymbol{\sigma} \cdot (\boldsymbol{\xi} \times \delta \boldsymbol{r}) \delta l = \delta \boldsymbol{r} \cdot (\boldsymbol{b} \cdot \boldsymbol{\sigma} \times \boldsymbol{\xi}) \delta l.$$

But by the definition of the virtual thermodynamic force **F**, $\delta W = \boldsymbol{F} \cdot \delta \boldsymbol{r}$. Thus, the virtual force is (PEACH and KOEHLER [1950]):

$$\frac{\boldsymbol{F}}{L} = \boldsymbol{b} \cdot \boldsymbol{\sigma} \times \boldsymbol{\xi} \qquad \text{or} \qquad \frac{F_k}{L} = -\varepsilon_{ijk}\xi_i \sigma_{j\ell} b_\ell, \tag{9}$$

where ε_{ijk} is the permutation operator ($\varepsilon_{ijk} = 1$ or -1 for ε_{123} or ε_{321} or their cyclic permutations, 0 otherwise).

Consider the example of fig. 7, with only $\xi_3 = 1$ nonzero, and with only $b_1 = b$ nonzero. According to eq. (9), a shear stress σ_{21} produces a glide force $F_1/L = \sigma_{21}b_1$; a normal stress σ_{11} produces a climb force $F_2/L = -\sigma_{11}b_1$; other stresses produce no force. For the screw of fig. 6, with $\xi_3 = 1$, $b_3 = b$, a stress σ_{23} produces a glide force $F_1/L = \sigma_{23}b_3$ while a stress σ_{13} produces a (cross-slip) glide force $F_2/L = -\sigma_{13}b_3$.

Thus, the Peach–Koehler equation (9) reveals the direction of prospective spontaneous motion of a dislocation in the presence of stresses σ. *The force is a virtual thermodynamic one and must not be confused with a mechanical force.*

2.3. Dislocation interactions

Interaction forces and energies also can be developed from eq. (9). Consider first the interaction between parallel dislocations A and B in fig. 9, with $\xi_3 = 1$ and

References: p. 1257.

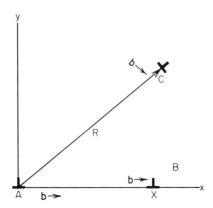

Fig. 9. Parallel dislocations; ξ points out of the page.

$b_1 = b$. Equation (9) then gives for the glide force on B:

$$\frac{F_1^B}{L} = \xi_3^B \sigma_{21}^A b_1^B = \frac{\mu b^2}{2\pi(1-\nu)X}, \tag{10}$$

since $\sigma_{21}^A = \sigma_{r\theta}(\theta = 0, \ x = X) = \mu b/2\pi(1-\nu)X$. The interaction force is repulsive between like-sign edge dislocations; attractive between opposite-sign dislocations. For more general parallel dislocations, such as A and C in fig. 9, the use of eq. (9) is more complicated, involving more terms. However, in cylindrical coordinates with $r = R$ at dislocation C, the radial interaction force has the simple form

$$\frac{F_r}{L} = \frac{\mu}{2\pi R}\left[(\boldsymbol{b}_A\cdot\boldsymbol{\xi}_A)(\boldsymbol{b}_C\cdot\boldsymbol{\xi}_C) + \frac{1}{(1-\nu)}(\boldsymbol{b}_A\times\boldsymbol{\xi}_A)\cdot(\boldsymbol{b}_C\times\boldsymbol{\xi}_C)\right]. \tag{11}$$

Equation (11) is seen to reproduce eq. (10) for the edge case and to give $F_r/L = \mu b^2/2\pi R$ for parallel like-sign screw dislocations. There is a weak oscillating F_θ/L component as well in the general case (NABARRO [1952]).

The interaction energy as a function of R can be determined by integration. For example, for the above screw case

$$W_{\text{int}} = \int_{R_a}^{R}\frac{F_r}{L}\,dr = \frac{\mu b^2}{2\pi}\ln\frac{R}{R_a}, \tag{12}$$

where R_a is a reference position and can be chosen to equal r_0, for example.

A procedure for more general interactions is exemplified by the perpendicular screw case of fig. 10. Coordinates are selected for dislocation A so that the stress field can be conveniently and simply expressed by eq. (3), $\sigma'_{\theta z} = \mu b/2\pi r'$. The stress tensor is then transformed to coordinates x, y, z fixed on dislocation B to make the use of eq. (9) easy, giving $\sigma_{yz} = -\mu b \cos^2\theta'/2\pi h$. Equation (9) then readily gives $(F_x^B/L) = -\mu b^2 \cos^2\theta'/2\pi h$ for the local value of the interaction force, the locus of

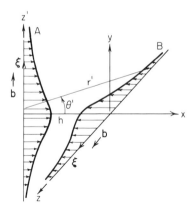

Fig. 10. Two perpendicular right-handed screw dislocations separated by a distance h; light arrows represent the interaction force.

which is plotted in fig. 10. This procedure works for defects other than dislocations. Defect A could be a solute atom, a precipitate, or the like, and, provided one can express its stress field, σ^A, the interaction force on B can be determined as above. The equal and opposite nature of the net forces in fig. 10 is fortuitous; in general, Newton's law does not apply for these virtual thermodynamic forces.

All of the interaction forces have the characteristic feature that they vary inversely with the spacing of dislocations. This result has important implications for work hardening, where the macroscopic stress scales with the dislocation "mesh-length," a quantity directly scaling with the dislocation spacing (KUHLMANN-WILSDORF [1966]).

The examples presented here are all for infinite straight dislocations. Stress, force, and energy expressions are available for generally curved dislocations in the form of line integrals, but, except for special cases such as circles, the integrals can only be solved numerically. Analytical results are available for straight-line segments, angular dislocations and infinitesimal loops. These results are directly applicable to some arrays, such as stacking-fault tetrahedra, and can be used to obtain fairly accurate approximate results for curved arrays such as extended dislocation nodes. The results are very lengthy, beyond the scope of the present work, and can be found in NABARRO [1967] and HIRTH and LOTHE [1982].

2.4. Surface effects

At a free surface, equilibrium requires that no forces act on a surface element. If the area element A is defined by a local normal unit vector n, the condition is $\sigma \cdot A = A\sigma \cdot n = 0$. For dislocations in finite media, this surface condition modifies the elastic field. As an example, suppose that the surfaces in fig. 6 are free surfaces bounding the cylinder. The stresses $\sigma_{\theta z}$ from eq. (3) would act on the end surface A_z and produce a net torque on the end. These extraneous stresses are removed by

References: p. 1257.

superposing equal and opposite tractions on the surface, solving the elastic problem for a cylinder under such tractions, and adding the resultant strain and displacement field to the infinite medium result. For the case of fig. 6, a net twist appears in the cylinder, the added stresses are $\sigma_{\theta z} = \mu b r / \pi R^2$, and the free energy of the system is decreased by $\Delta W / L = - \mu b^2 / 4 \pi$ (ESHELBY [1953]).

Similarly, for the edge dislocation in a finite cylinder, fig. 7, stresses σ_{rr} and $\sigma_{r\theta}$ must be imposed on the cylinder surface $r = R$. In this case the free energy decreases by $\Delta W / L = \mu b^2 / 8 \pi (1 - \nu)$. Surface stresses thus characteristically produce energy changes $\sim \mu b^2 / 10$. These changes are of the order of uncertainties in W/L because of uncertainties in core energy and in the use of isotropic elasticity, so they can consistently be neglected.

When a dislocation is parallel to a free surface, the superposed field often can be determined by an "image" construction. Consider the screw dislocation in fig. 11. Rotation of coordinates shows that eq. (3) gives stresses $\sigma_{xz} = - \mu b y / 2 \pi R^2$ acting on the free surface where $r = R$. A solution which gives opposite stresses which exactly cancel these extraneous stresses is that of an opposite-sign dislocation at the mirror-image position relative to the surface. Thus, the field in the body is that of the real dislocation $\sigma_{\theta z} = \mu b / 2 \pi r$ and that of the *image dislocation* $\sigma_{\theta' z'} = - \mu b / 2 \pi r'$. At the site of the real dislocation, the image field is $\sigma_{yz} = - \sigma_{\theta' z'} (\theta = \pi, r' = 2\lambda) = \mu b / 4 \pi \lambda$. Thus, by eq. (9), the free surface exerts an image force

$$F_1 / L = \mu b^2 / 4 \pi \lambda, \tag{13}$$

attracting the dislocation to the free surface.

For the edge dislocation, the simple image construction is insufficient and an added stress field is needed to satisfy the free surface boundary condition (HIRTH and LOTHE [1982]). However, the force associated with the free surface tractions is still that of the simple image dislocation. This result is general (LOTHE [1976]); for single dislocations inclined to the surface, the simple image construction gives the correct force produced by the surface tractions. For two or more dislocations near

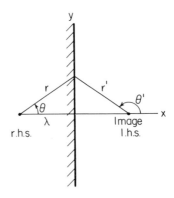

Fig. 11. A right-handed screw dislocation, parallel to and a distance λ from a free surface.

the surface, the simple image construction is insufficient. The added stress field of one dislocation, discussed above for the edge, would produce a force on the other dislocation.

MARCINKOWSKI [1979] has criticized the image method and instead recommends a method of a distribution of infinitesimal surface dislocations involving a solution of a Cauchy integral. Several methods, including the above two, are correct and self-consistent, so the selection of a method is a matter of taste and convenience.

2.5. Line tension

For arbitrary dislocations in a deformed crystal, the determination of W/L is a most formidable task. However, a rough approximation can be made. On the average, since like-sign dislocations repel while opposite-sign ones attract, a dislocation tends to be surrounded by opposite-sign dislocations at an average spacing λ. In accord with St. Venant's principle, a pair of dislocations of opposite sign are sources of equal and opposite stress fields which should cancel one another over distances greater than $\sim \lambda$ from the pair. Thus, the strain field of the dislocation is limited to a radius $R \approx \lambda$. Hence, since as already discussed, core and surface terms can be neglected, eq. (8) applies roughly for the arbitrary case with $R \approx \lambda$. Moreover, $\lambda \sim \rho^{-1/2}$ where ρ is the dislocation density (m/m^3). Typical values of ρ range from $\sim 10^{10}$ m/m^3 for annealed crystals to $\sim 10^{16}$ m/m^3 for cold worked crystals, corresponding to 10^{-8} m $< \lambda < 10^{-5}$ m. With $r_0 \cong b \cong 3 \times 10^{-10}$ m and $\nu \cong 0.3$, the range of W/L in eq. (8) is from 0.28 to 1.18 μb^2.

Since μ and b are constant for a given material, W/L is a constant with the above approximations. A stretched linear-elastic string has a constant energy per unit length W/L numerically equal to the *line-tension* force $\mathbb{S}$ restraining the string, since $\partial W/\partial L = W/L = \mathbb{S}$. In the above approximation the dislocation can be thought of similarly as a stretched line constrained by line tension forces acting parallel to the line:

$$\mathbb{S} = \frac{\partial W}{\partial L} = g\mu b^2. \tag{14}$$

Here g is a numerical factor varying from 0.3 to 1.2 with an average value of roughly 0.7, or in a cruder approximation, $1/2$.

The simple line tension analogy is qualitatively useful. For example, in this approximation a bowed-out dislocation or zig-zag dislocation is unstable with respect to a straight dislocation in the absence of stress. Also, the equilibrium angles at a node are determined by the line tensions.

In a more accurate representation, one must account for the variation of W/L with β in eq. (8). $E(\beta)$ is a minimum for the screw orientation $\beta = 0$ or π, so a torque exists tending to twist the dislocation into screw orientation. Including this factor, but retaining the concept that $\mathbb{S}$ is the change in W with change in the total line length of a dislocation, DE WIT and KOEHLER [1959] found:

$$\mathbb{S} = \frac{W}{L} + \frac{\partial^2(W/L)}{\partial\beta^2} = \frac{W}{L}\left[1 + \frac{1}{E(\beta)}\frac{\partial^2 E(\beta)}{\partial\beta^2}\right]. \tag{15}$$

References: p. 1257.

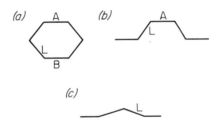

Fig. 12. Bowed-out configurations of an edge dislocation; b is normal to the page.

In the simple approximation that $\mathbb{S} = W/L$, the value of $\mathbb{S}$ for an edge exceeds that for a screw, eq. (8), by a factor $1/(1 - \nu) \approx 1.4$. With the inclusion of the factor $\partial^2(W/L)/\partial\beta^2$, which tends to stabilize screws and destabilize edges, the situation is reversed and $\mathbb{S}$ for a screw exceeds that for an edge by a factor $(1 + \nu)/(1 - \nu)(1 - 2\nu) \approx 4.6$.

Even more accurately, rather than use the rough St. Venant approximation, one should compute the self and interaction energy. This can be done by the straight line segment equations mentioned previously. For the configurations in fig. 12, HIRTH *et al.* [1965] find:

$$\mathbb{S} = \frac{\mu b^2}{4\pi(1 - \nu)} \ln \frac{gL}{r_0}, \tag{16}$$

with g a numerical factor equal to 6.41, 1.11, and 0.18, respectively, for cases a, b, and c. The result is understandable qualitatively. In case a, segments A and B are equal and opposite and so work must be done against their attractive interaction force in forming the loop. In case b, segment B is missing, so less work is needed and $\mathbb{S}$ is smaller. In case c, there is even less attractive segment length and $\mathbb{S}$ is lowest. Interestingly, as first noted by MOTT and NABARRO [1948], the line tension of a bowed segment depends on the bowout length L and *not* on the outer cutoff R. This result is important in the theory of bowout in solution-hardened or dispersion-hardened systems.

3.   Crystal lattice effects

3.1.   Peierls barrier

PEIERLS [1940], in a model elaborated by NABARRO [1947], modeled the periodic potential in the glide plane experienced by a straight dislocation lying in a low-index direction. In the model, the bonds across the glide plane were considered to interact by an atomic potential, while the remainder of the lattice was elastic. The model is useful in giving an analytical result that removes the artificial core divergencies of the Volterra dislocation, eqs. (2), (3) and (5). However, as discussed by HIRTH and

LOTHE [1982] and by BULLOUGH and TEWARY [1979], the model is physically unrealistic (compared to a cylindrical atomic model centered on the dislocation) and predicts that symmetry positions are positions of energy maxima instead of minima. Nevertheless, atomic calculations indicate that the potential is roughly equivalent to the sinusoidal form of the Peierls–Nabarro result. Thus, it is useful to retain their result as an empirical representation of the barrier.

For a dislocation displaced by a distance x from a symmetry position (such as in fig. 2), the potential is:

$$\frac{W}{L} = \frac{W_p}{L} \sin^2 \frac{\pi x}{a}, \tag{17}$$

where a is the period, usually equal to b, and W_p is the *Peierls energy*. The maximum slope of this oscillating potential gives the *Peierls stress*, σ_p, needed to overcome the barrier at 0 K, but the barrier is usually overcome instead by kink formation as discussed below. In the original model the explicit form for W_p was

$$\frac{W_p}{L} = \frac{\mu b^2}{2\pi(1-\nu)} \exp\left(-\frac{4\pi\zeta}{b}\right) = \frac{\mu b^2}{2\pi(1-\nu)} \exp\left(-\frac{2\pi d}{b(1-\nu)}\right), \tag{18}$$

where ζ is the *width* of the spreading of the core in the glide plane and d is the glide plane spacing. This form is not verified in atomic calculations, but the qualitative trends of a marked decrease in σ_p or W_p with an increase in core spreading or d are retained and are useful in predicting crystal slip systems. Because d and b appear in the exponent in eq. (18), the relation suggests that the smallest b and largest d should correspond to the observed slip system, in good agreement with experiment when dislocation extension is not an important phenomenon. As an indication of the change of Peierls energy with slip system, the ratio of W_p for the closest packed plane slip system, $\langle 110 \rangle$–$\langle 111 \rangle$, to that of the next closest packed plane system, $\langle 110 \rangle$–$\langle 100 \rangle$, is 0.36 in fcc crystals.

A dislocation line containing *kinks* is shown in fig. 13. The configuration of a kink such as B is determined by a balance between line tension forces, which tend to make w in fig. 13 large and hence straighten the line, and the forces associated with the Peierls barrier, which tend to make the kink abrupt. In the simple line-tension model of eq. (14), this force balance gives an equilibrium kink width w and kink formation energy W_k given in terms of the total elastic energy W/L by

$$w = a\left(W/2W_p\right)^{1/2}, \qquad W_k = (2a/\pi L)\left(2W\,W_p\right)^{1/2}. \tag{19}$$

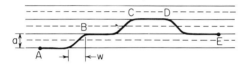

Fig. 13. Double kink CD and single kink B in a dislocation segment AE. Dashed and solid lines are maxima and minima in the varying lattice potential, respectively.

Typically, W increases successively for bcc metals, ionic crystals and covalent crystals with values $W_p \sim 10^{-4}$, 10^{-3} and $10^{-2}W$, $W_k \sim 0.1$, 0.05 and 0.01 μb^3 and $w \approx 70$, 20 and 7a, respectively. For fcc and hcp metals W_p is much smaller, so that Peierls barrier effects are only important well below room temperature.

Kinks are of three types: *geometric kinks* such as B in fig. 13 are present by geometric necessity since the segment AE spans one Peierls hill. At thermal equilibrium, a dislocation line contains a concentration of *thermal kinks* of $\sim (1/b)$ $\exp(- W_k/kT)$. Finally, since the motion of a dislocation produces an offset of b in the entire lattice, as illustrated in fig. 1, *intersection kinks* can be produced on each dislocation if two nonparallel dislocations intersect.

3.2. Core structure and energy

With the advent of fast computers, atomic calculations for the nonlinear core region have become prevalent. They still have the drawback of using atomic pair potentials which do not accurately represent real metals. However, they do approximate real metal behavior fairly well and they are most useful in delineating classes of core behavior for different crystals.

Such calculations have been performed using either lattice statics or lattice dynamics as reviewed by PULS [1981]. With the development of flexible boundary methods, in which a cylindrical atomic region concentric with the dislocation core is matched compatibly with a surrounding elastic continuum, the use of lattice dynamics in real space is very efficient according to Puls.

One important feature of these results is related to the volume of a dislocation. In the linear elastic approximation, the net volume change produced by a dislocation in a crystal is zero. The atomic calculations, both for metals and for ionic crystals, reveal that both edge and screw dislocations produce a volume expansion of about an atomic volume per close-packed atomic plane cut by a dislocation, i.e. the volume-equivalent of a row of interstitial atoms.

As also discussed by Puls, calculations of the core energy, while fewer, are of interest. In the reversible process of creating the edge dislocation, §2.1, in addition to the elastic energy of eq. (8) stored between r_0 and R, elastic energy flows into the core and is stored there as core energy. The core energy can only be estimated by atomic calculations. Rather than adding a core term to all energies, it is convenient to retain the results of eqs. (6) to (8) and to incorporate the core energy by adjusting the value of r_0. In terms of a parameter α defined by $r_0 = b/\alpha$, HIRTH and LOTHE [1982] find that $\alpha \approx 3$–5 for ionic crystals with screw values larger than those for edges, while $\alpha \approx 0.5$–2 for both partial and perfect dislocations in metals.

One of the successes of atomic calculations has been in the understanding of the properties of screw dislocation cores in bcc metals, reviewed by VITEK [1974] and by SEEGER and WÜTHRICH [1976]. Figure 14 illustrates the result which indicates splitting of a $\frac{1}{2}[111]$ screw dislocation into a dislocation $q[111]$ at the origin and three dislocations $(\frac{1}{6} - \frac{1}{3}q)[111]$ at B, C, and D. Because the splitting is only over atomic dimensions and because of the variable nature of q, these are called *fractional*

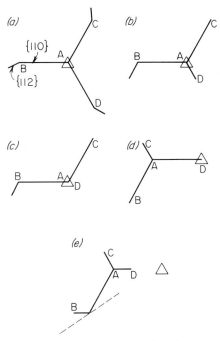

Fig. 14. (a) Core dissociation of screw dislocation; (b–e) Motion of screw under stress producing glide on the average on the dashed {112} plane shown in e.

dislocations. For the glide sequence in fig. 14, the value of σ_p is large since the fractional dislocations must contract before motion can occur. Also, because of nascent extensions on {112} planes, also indicated in fig. 14, σ_p is asymmetric, being smaller in the so-called easy twinning direction than in the opposite direction, in agreement with experiments.

3.3. Stacking faults and partial dislocations

In close-packed fcc and hcp crystals, stacking faults on {111} planes have relatively low energies and, consequently, perfect dislocations extend to form *partial dislocations* bounding a stacking fault. As an example, we consider the fcc case and the commonly observed intrinsic stacking fault, formed when a {111} plane is removed from the normal stacking sequence ABCABCABC and the crystal is collapsed together normally to form the fault ABCA:CABC (see discussion of fig. 17e). The notation of THOMPSON [1953] is useful. If one connects the atom centers of the four atoms near a corner of a unit cell, one forms the *Thompson tetrahedron*, fig. 15. The faces are the {111} glide planes and the edges are the $\frac{1}{2}\langle 110 \rangle$ Burgers vectors (glide directions). Then AB, etc., denote Burgers vectors and (d) denotes glide plane.

Now consider the view of the glide plane in fig. 16. A perfect dislocation AB

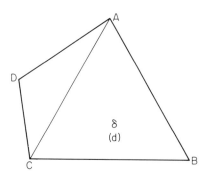

Fig. 15. Thompson tetrahedron.

would shear the crystal and move an atom from 1 to 2. Clearly, in a central-force atom model, the path along the potential "valleys" (1)–(3)–(2) would be more favourable energetically. Yet, at the midpoint (3) the configuration is that of an *intrinsic stacking fault*. Hence, motion occurs first by the partial δB, creating the fault, followed by the partial $A\delta$, annihilating the fault. The actual configuration is shown in fig. 17a. The equilibrium spacing is achieved when the repulsive forces between the partials, eq. (11), just balance the attractive force γ of the fault. Here γ is the stacking fault energy (J/m^2), numerically equal to the surface tension (N/m) (see ch. 4, § 9.2).

The partial dislocations $A\delta$, etc., of the glide type, are also given on the Thompson tetrahedron, fig. 15, and are called *Shockley partials*. The correct sequence to form an intrinsic fault is established if: viewing the fault along positive ξ and viewing the plane as if one were outside the tetrahedron [as for (d) in fig. 17], place the Greek letter outside the fault. An example is given in fig. 17b for the dislocation AB bending from plane (d) to (c). Use of the continuity axiom of § 1.1 reveals that a *stair-rod partial* $\gamma\delta$ must exist at the bend. Stair-rods are sessile, since their glide would produce a very high energy fault.

Some other important arrays are also shown in fig. 17. Figure 17c shows a

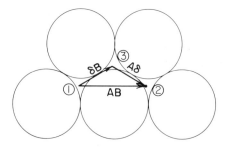

Fig. 16. Partial dislocation Burgers vectors in fcc crystals.

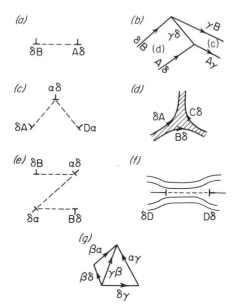

Fig. 17. Extended arrays in fcc. View is parallel to the $\{111\}$ planes in a, b, e, and f, and perpendicular to it in d.

Lomer–Cottrell barrier (TEUTONICO [1964]) formed by the interaction of BA(d) and DB(a), the barrier strength being associated with the attractive interaction of Bδ and αB to form $\alpha\delta$ *. Figure 17d is an extended node, and 17e is an extended dipole. Figure 17f shows the removal of a plane of atoms by vacancy condensation to form a fault bounded by *Frank partials* of the type Dδ, the vector magnitude equalling the glide plane spacing. Frank partials are also sessile, but, as indicated by their formation process, they can climb by vacancy absorption or emission. Fig. 17g is a stacking fault tetrahedron, formed from a Frank loop such as that in fig. 17f, with all faces corresponding to faults.

Extrinsic faults, equivalent to insertion of a plane of atoms into the normal sequence, ABCABACABC have energies about equal to those of intrinsic faults (GALLAGHER [1970]) but are rarely observed. Extension to form this fault involves pairs of partials bounding each side. The more complex movements to form these pairs probably hinder this type of extension, accounting for its rare appearance.

Analogous faults and partials form in hcp crystals, diamond cubic crystals, and in layer structure crystals such as mica and graphite (AMELINCKX [1979]). The fault energy is too large for meaningful extension to form a stacking fault in bcc crystals, so extension into fractional dislocations occurs instead, as already discussed.

---

* The array shown in fig. 17c is the original result. HAZZLEDINE [1982] has recently shown that this is actually a rather flat saddle point with the true equilibrium configuration being asymmetric with relative extensions of the two arms in the ratio 3.82 in the isotropic case.

References: p. 1257.

3.4. Slip systems

Together, the concepts of Peierls energy and stacking faults are sufficient to explain most observed slip systems, with core structure and image effects having an influence in a few special cases. Equation (18), or its equivalent in a more exact model, predicts that the Peierls stress σ_p is least for the slip system with the smallest b and largest interplanar spacing d, that is, b along a low-index direction and d normal to a closed-packed plane. The value of σ_p increases rapidly with an increase in b or decrease in d, so the Peierls energy criterion would predict only a few low-index slip systems for a given crystal. A low stacking-fault energy is another factor favoring a given glide system, both because the extension of dislocations tends to restrict glide to the fault plane and because σ_p is less for Shockley-type partials.

The above ideas are borne out for observed slip systems as reviewed by HIRTH and LOTHE [1982]. $\langle 110 \rangle$–$\{111\}$ is the predominant slip system in fcc crystals, fitting both criteria, while $\langle 110 \rangle$–$\{100\}$ slip, on the second closest packed plane, appears occasionally. In hcp crystals with large c/a ratios, giving a relatively large d for the basal plane, which is also a low-fault energy plane, the $\langle 11\bar{2}0 \rangle$–$(0002)$ basal slip system is favored and is observed. With low c/a ratios, pyramidal slip or prismatic slip become favored and are observed.

In bcc crystals, the low-index $\langle 111 \rangle$ direction is invariably the slip direction. This non-close-packed lattice has a smaller difference between the largest $\{110\}$ and next largest $\{112\}$ d spacing, and both slip planes are observed. As discussed in § 3.2, the pertinent glide plane for screw dislocations is determined by the behavior of fractional dislocations in the core. The situation for edge dislocations is unresolved, but there is some indication of extension on $\{112\}$ planes at low temperatures.

In thin crystals, where surface slip nucleation is important, unusual slip systems are observed in the sense that they do not have the maximum resolved shear stress (VESELY [1972]; LOHNE [1974]). These results are associated with image effects, either favoring glide by enhancing kink nucleation for screw dislocations oblique to the surface, or suppressing glide by pulling a dislocation more normal to the surface and creating jogs on it.

3.5. Jogs

An offset in a dislocation line which has a component normal to the glide plane is called a *jog*. A jog in a screw dislocation is simply a kink in the cross-slip plane and has properties as discussed in § 3.1. A jog in an edge dislocation is shown in fig. 18a. Rearrangement of the jog to a kink-like configuration would involve a very large misfit energy, so the jog remains sharp. Therefore, jogs have energies W_j of the order of the dislocation line energy multiplied by the jog height or $\sim \mu b^3$. These energies are 10–100 times larger than the formation energies for kinks, W_k. Consequently, the concentrations of thermal jogs are much less than those of thermal kinks at a given temperature. Intersection jogs and geometric jogs also exist, analogous to the equivalent kinks.

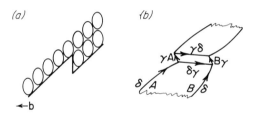

Fig. 18. Jogs in (a) a perfect edge dislocation and (b) an extended edge dislocation in fcc.

Climb occurs at jogs by vacancy formation or annihilation at jog sites. Self-interstitials could cause climb also, but their formation energy is so large in metals that they are only important in the case of radiation damage. For a perfect dislocation, the only constraint to double jog nucleation or jog motion, the processes which produce climb in an analogous way to glide by kink processes as shown in fig. 13, is the formation of the pertinent defect, the double jog or vacancy.

An *extended jog* in a fcc crystal is depicted in fig. 18b. For a *superjog*, where the jog height is a multiple of the interplanar spacing, the extended jog is composed of partial dislocations and stacking faults, as shown. For a unit jog with height equal to a single interplanar spacing, the *stair-rod dipole* γδ–δγ exists only over core dimensions. (For further details concerning dislocation dipoles see ch. 24, §§ 3.1 and 4.4). Rather than stair-rods and stacking faults, the jog is then better viewed as a jog-line, equivalent to a row of 1/3 vacancies for the present example (HIRSCH [1962]).

The extension of the jog evidently provides a constraint to both jog nucleation and motion. In early work, it was thought that the jog would have to constrict to a perfect dislocation to climb. However, recent concepts (see BALLUFFI and GRANATO [1979]) indicate that a vacancy itself can dissociate into the jog line and cause climb without constriction.

Observations by CHERNS *et al.* [1980] show that extended dislocations can also nucleate double jogs without constriction. Thus, the constraint to climb caused by extension, while present, is less than once expected.

4.   *Dislocation behavior at low homologous temperatures*

4.1.   Kink motion

In bcc metals at low temperatures, glide occurs by kink motion. For low stresses, kinks move and annihilate but nucleation easily replenishes them. The dislocation velocity v is the area swept per unit length per unit time or, fig. 13, the product of the equilibrium number of kinks of both geometric signs $2n_k$, the kink velocity v_k, and the distance between Peierls valleys $\sim b$. v_k is given by the kink mobility

D_k/kT, where D_k is the kink diffusivity, and the driving force from eq. (9), σb^2,

$$v_k = \frac{2\sigma b^2 D_k}{kT} \exp\left(-\frac{W_k}{kT}\right).$$
                                                                    (20)

For larger stresses, kinks are swept to the end of a segment of length L before they can be replenished, and double-kink nucleation is controlling. The formation energy $2W_k$ of a double-kink then enters the velocity expression, which becomes:

$$v_k = \frac{\sigma b L D_k}{kT} \exp\left(-\frac{2W_k}{kT}\right),$$
                                                                    (21)

see HIRTH and LOTHE [1982]. Further developments and detailed derivations are given by these authors and by SEEGER [1981], who also notes that the interesting phenomenon of soliton motion in one-dimensional conductors can be described by the same form of kinetics.

4.2. Point forces and bowout

For fcc and hcp metals or for bcc transition metals above about room temperature, $kT \geq W_k$ and the effects of the Peierls barrier are removed by thermal fluctuation. In principle, dislocations under a driving force could then accelerate until their motion was controlled by damping produced by phonon scattering or radiation. However, except in shock-loading, dislocation motion becomes hindered by obstacles and breakaway from the obstacles becomes rate-controlling. A review of this subject is given by KOCKS et al. [1975].

Bowout at obstacles is illustrated in fig. 19. For small bowout, the force on a dislocation segment $\sigma b \lambda$ is transmitted to pinning points via line tension forces acting on the pinning points. In the simple line tension case, eq. (14), the resultant force in the direction of bowout acting at the pinning point is $\mu b^2 \cos(\phi/2)$. The opposite stabilizing force F is the interaction force of the dislocation and a pinning point, often so localized that F can be considered a point force. At equilibrium, these forces are all equal and for a linear array of obstacles as in fig. 19:

$$\sigma = \frac{\mu b}{\lambda} \cos\left(\frac{\phi}{2}\right) = \frac{F}{b\lambda}.$$
                                                                    (22)

For stresses $\sigma < \sigma_c$, the critical stress for breakaway, corresponding to $\phi > \phi_c$ and

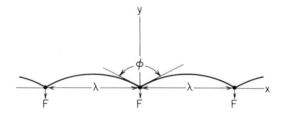

Fig. 19. Bowout from pinning points.

$F < F_c$, the bowout is stable. This case is relevant to the Granato–Lücke internal friction theory and to jog drag. For $\phi \geqslant \phi_c$, $F > F_c$; breakaway of dislocations from the pinning points occurs, relevant to solid-solution or precipitation-hardening and to dislocation–dislocation intersection. For obstacles so strong that $F < F_c$ when $\phi = 0$, complete bowout occurs at the particles, relevant to dispersion-hardening and the operation of dislocation sources. These several applications are now considered in succession.

4.3. Granato–Lücke internal friction theory

The vibration of a loop of length λ, such as that in fig. 19, produces energy dissipation and the GRANATO–LÜCKE [1956] internal friction peak. In discussing eq. (14), we mentioned the analogy of a dislocation of a stretched elastic string. The analogy is useful here and the vibrating loop under a driving shear stress σ is equivalent to a damped vibrating string segment. In this analogy, the dislocation has an effective mass per unit length $m = \mathcal{S}/C_t^2 = \mu b^2/2C_t^2$ with eq. (14) for the line tension and with C_t the transverse sound velocity. The damping force B is given by $\sigma b = Bv$, with v the dislocation velocity. As reviewed by ALSHITS and INDENBOM [1975], B is temperature dependent, but a rough value at room temperature for many materials is $B \cong 10^{-2} \mu b/C_t$.

With the above parameters, the equation of dynamic equilibrium for the vibrating loop in the coordinates of fig. 19 is

$$m\frac{\partial^2 y}{\partial t^2} + B\frac{\partial y}{\partial t} - \mathcal{S}\frac{\partial^2 y}{\partial x^2} = \sigma b. \tag{23}$$

The solution of eq. (23) is a standard problem in mechanics. The solution gives an internal friction (the relative energy loss per cycle $\Delta W/W$) for the fundamental mode of:

$$Q^{-1} = \frac{16\alpha^2\mu\rho b^2}{\pi m\Omega_1} \frac{\omega\tau_1}{\left[1 - (\omega/\Omega_1)^2\right]^2 + (\omega\tau_1)^2}, \tag{24}$$

where ρ is the *active dislocation length per unit volume*, α is the *Schmid stress resolution factor*, Ω_1 is the *fundamental frequency*, $\Omega_1 = \pi^2\mathcal{S}/m\lambda^2$, ω is the driving frequency and τ_1 is the relaxation time, $\tau_1 = B/m\Omega_1^2$. Inspection of eq. (24) shows that the internal friction peak, the maximum value of Q^{-1}, occurs near $\omega = \Omega_1$.

For large amplitudes, that is for large stresses, the pinning points can be dragged in either the x or y directions or breakaway from the pinning points can occur. In either situation, the internal friction becomes amplitude dependent, increasing at large stresses.

The double-kink formation process discussed in §4.1 also gives rise to internal friction phenomena, the so-called Bordoni and Niblett–Wilks peaks observed in bcc and fcc metals. These are somewhat specialized to low temperatures, so they are not discussed here. A review of this subject is presented by FANTOZZI *et al.* [1982].

References: p. 1257.

4.4. Dislocation sources

Suppose that the segment λ in fig. 19 is a small portion of a larger loop. Then $\phi/2$ can be chosen near $\pi/2$ and the radius of curvature of the loop is $r = \lambda \tan(\phi/2) \cong \lambda/\cos(\phi/2)$. Then the simple line-tension expression of eq. (22) becomes

$$\sigma = \mu b/r. \tag{25}$$

Consider now the segment λ in fig. 20, pinned in its glide plane in such a way that the dislocation leaves the glide plane at the pinning points. As the stress increases the dislocation bows out further and r decreases. At position B, $r = \lambda/2$, the radius of curvature is a minimum, and the loop can grow spontaneously with an increase in r since eq. (25) can no longer be satisfied. When the loop reaches position D, the parts which touch one another are equal and opposite dislocations and annihilate one another (the dashed line in fig. 20). A complete loop is formed and the portion constrained by the pinning points can again bowout and repeat the process. The segment λ so operating is called a *Frank–Read dislocation source*. The source is important in dislocation multiplication since many dislocation loops can be produced from one segment.

If only one end of a segment is pinned, with the other end reaching a free surface, a single-ended source forms, with the dislocation winding into a spiral. Both types of source can also operate in climb. They are then referred to as *Bardeen–Herring sources*.

There are other types of sources, but most of these can be thought of as variants of the above sources. Consider, for example, the double cross-slip of a screw dislocation in fig. 21. The screw dislocation can be imagined to cross-slip at A to avoid an obstacle and then to cross-slip back to the original glide plane at B. Provided that h is large enough that the Peach–Koehler force and the inertia of loop B can overcome its attractive interaction force with A (that is, if h is large enough, since the latter force $\propto 1/h$) the loop B can evidently act just like a Frank–Read source. The superjogs AB are sessile and pin the cross-slipped segment at its ends.

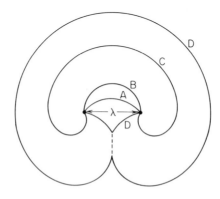

Fig. 20. Bowout of a segment λ to become a Frank–Read source.

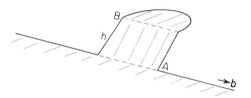

Fig. 21. Double cross-slip of a screw dislocation.

4.5. Dislocation pile-ups

Dislocations emanating from a source move until they encounter an obstacle. They then form an array called a *dislocation pile-up* as shown in fig. 22 for edges. At equilibrium the net force on each of the N edge dislocations is zero. Each dislocation is acted upon by the applied stress with a force σb. As shown for the second dislocation, this force is balanced by the summation of the repulsive interaction force for all but the lead dislocation. The lead dislocation only encounters the short-range stress field σ_0 of the obstacle which gives a repulsive force $\sigma_0 b$, balanced by the *sum* of σb and the interaction forces.

A simple way to deduce the stress field of the pile-up is to imagine an infinitesimal increase in σ, which, since each dislocation is at equilibrium, to first-order produces the same virtual displacement δx of each dislocation. Since the entire array translates rigidly, the interaction forces, which sum to zero for the entire array, do no work. Moreover, the change in total free energy W is zero to first order since $\partial W / \partial x = 0$ defines equilibrium. Thus, the change in free energy $N\sigma b \delta x$ produced by the applied stress acting on all of the dislocations is balanced by the free energy change $\sigma_0 b \delta x$ produced by the leading dislocation moving against the stress field of the obstacle. Thus,

$$N\sigma = -\sigma_0 = \sigma_e. \tag{26}$$

The *effective stress* (or local stress) σ_e at the lead dislocation of the pile-up, composed of the applied stress and the sum of the stress fields of the other dislocations, is $N\sigma$. This is an enormous stress concentration since N typically ranges from ~ 20 to ~ 500.

The dislocation next to the tip encounters an apparent obstacle consisting of the

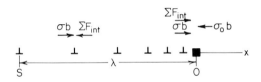

Fig. 22. View parallel to glide plane showing pile-up extending from source S to obstacle O.

first dislocation and its obstacle, so similar reasoning shows that the stress at this site is $(N-1)\sigma$. Thus, the stress decreases and the spacing increases with distance from the pile-up tip. An exact solution for the positions involves N equations in N unknowns, the positions of the dislocations, and can only be solved numerically as reviewed by CHOU and LI [1969] and HIRTH and LOTHE [1982]. As also discussed by those authors, an analytical integral solution can be obtained if the discrete dislocation array is replaced by a continuous distribution of infinitesimal dislocations with the same net Burgers vector Nb. The integral solution is precisely the continuum mechanics solution for a mode II shear crack. Thus, the infinitesimal dislocations make the connection between crystal plasticity and continuum plasticity.

Such solutions give the result for the number of dislocations in the pile-up $N = \pi(1-\nu)\lambda\sigma/\mu b$. Thus, the effective stress at the tip of the pile-up is:

$$\sigma_e = N\sigma = \pi(1-\nu)\lambda\sigma^2/\mu b. \tag{27}$$

This result is useful in explaining effects of microstructure on flow or fracture. A critical stress σ_e is required to propagate yield past an obstacle or to nucleate a crack there, with σ_e a material constant depending on surface energy, Frank–Read source size and the like. Then eq. (27) shows that the applied stress at yield or fracture varies as $\lambda^{-1/2}$. With the grain boundaries as barriers, the largest pile-ups have $\lambda = d/2$, with d the grain size, giving $\sigma \propto d^{-1/2}$, the form of the so-called *Hall–Petch relation*.

The stress field ahead of and near to the pile-up tip varies as $r^{-1/2}$, where r is the distance from the tip. The field of the infinitesimal dislocations gives, in fact, precisely the stress intensity $\sigma_e = K/(\pi r)^{1/2}$, identical to the continuum mechanics result. The far stress field ahead of a pile-up varies as $Nb/(r')^{1/2}$, where r' is the distance from a point $3/4$ of the way from the source to the obstacle. Thus, the field is equivalent to that of a superdislocation with Burgers vector Nb at $r' = 0$. Even if cross-slip or climb should rearrange the dislocations at the tip, *blunting* the pile-up, the effects on the field would be short-range according to St. Venant's principle. Thus, the stress concentration remains that of a superdislocation except near the tip. The superdislocation analogy is useful in qualitative reasoning.

4.6. Pinning in alloys

In alloys, a variety of structural entities can pin dislocations. For solute atoms with strong binding energy (interaction energy) with dislocations, such as interstitials in bcc metals, either the model of fig. 19 or variants of it are applicable. The concentration of solute in the two planes through which the dislocation glides is $c = 2/\lambda^2 d$, with d the interplanar spacing and λ now the mean spacing of solute in the plane. Together with eq. (22) this result gives the FLEISCHER [1964] solid solution hardening law $\sigma \propto c^{1/2}$, valid for dilute solutions. Some modification of eq. (22) might be appropriate, as discussed in connection with fig. 23, but this would not influence the $c^{1/2}$ dependence. For concentrated solutions, core interactions and short-range ordering or clustering effects are more important giving $\sigma \propto c$. For

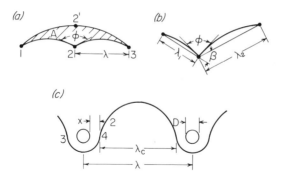

Fig. 23. (a, b) Bowout from pinning points; (c) bowout around particles.

weaker binding energy substitutional solutes or for intermediate concentrations with larger binding energies, a dislocation is more likely to bowout to break away from several solute pins simultaneously, giving a behavior $\sigma \propto c^{2/3}$ as discussed by HAASEN [1983] who elaborates on the theory of solid solution hardening.

For small coherent precipitates such as G–P zones in Al alloys, the obstacle strengths are such that the small-bowout eq. (22) should still be valid. The issue in this case is what produces the maximum resisting force F, although it is clear that dislocations cut the particles at breakaway. Possibilities are coherency strain interactions, the surface energy of the interface step formed when the particle is cut, incompatibility effects at the interface since the glide systems are not the same in the matrix and the precipitate, fault formation in the particle, or Peierls stress in the particle. Which of these is dominant is not yet known.

When the random distribution of pinning points in the glide plane, rather than the linear, regular array of fig. 19, is considered, modifications of eq. (22) appear. FRIEDEL [1956] supposed that at steady state the breaking of a pin at 2 in fig. 23a resulted in an area A being swept before the loop is repinned at 2′. In the small-bowout approximation and using eq. (25) he found for the area swept:

$$\frac{A}{\lambda^2} = \left(\frac{\mu b}{\sigma \lambda}\right)^2 \cos^3\left(\frac{\phi}{2}\right),\tag{28}$$

as can be verified readily for the geometry shown. If one associates $A \sim \lambda^2 \sim$ the mean area per pinning point in the plane, one can rearrange eq. (28) to give:

$$\sigma = \frac{\mu b}{\lambda} \cos^{3/2}\left(\frac{\phi}{2}\right) = h\frac{\mu b}{\lambda} \cos\left(\frac{\phi}{2}\right),\tag{29}$$

with $h = \cos^{1/2}(\phi/2)$ for the case shown. Except for the factor h, eq. (29) is identical to eq. (22).

More generally the critical bowout angle is expected to vary with length distribution λ_i, with pinning point strength distribution ϕ_i, and with the angle β_i as shown in fig. 23b. Equation (29) could then still be used but with $h = h(\phi_i, \lambda_i, \beta_i)$. For large

References: p. 1257.

bowouts a more complicated geometrical function would replace eq. (28). In computer simulations with random distributions of λ_i and β_i, and with the simple line-tension form, eq. (25), the simple result of eq. (29) was found to hold for $\phi \gtrsim \pi/3$, see KOCKS *et al.* [1975]. Since $h < 1$ for this case, the resisting force is less than that for a regular array. This can be rationalized in terms of easy breaking points with large λ_i or large β_i (fig. 23b) within the statistical distribution. Whether the form of eq. (29) holds for real arrays and other than in the approximation of eq. (25) remains an open question. For any version of breakaway from point obstacles, the dependence $\sigma \propto \lambda^{-1}$ of eqs. (22) and (29) is retained.

For discrete obstacles, other effects appear as shown in fig. 23c. If the particles are harder than the matrix, the usual case, image repulsion leads to a repulsion distance x, which, together with the finite size of the obstacles D, reduces the effective spacing from λ to λ_e. In a refined line-tension model, the dimension λ replaces L in eq. (16) and this expression replaces that of eq. (15) in eq. (29). Further, considering the interactions between adjacent loops, $\mathcal{S}$ would be lowered because of the attraction between opposite sign portions 1 and 2 in fig. 23. On the other hand, the harder particle would screen the interaction between segments 3 and 4 which would tend to offset the previous effect. Whether such refinements significantly influence eq. (29) in actual cases is not known and the use of this relation or the simpler form of eq. (22) is probably sufficient with the present state of understanding.

When the critical stress is exceeded for the case of fig. 23c, either the particles are cut or the loops bypass the particles, recombine, and leave a closed loop around each particle, in the Orowan process. Subsequent loops can form, giving rise to circular pile-ups (or their blunted versions) and attendant work-hardening in dispersion-hardened systems.

4.7. Work-hardening

In the absence of strong pinning points, work-hardening is dominated by dislocation–dislocation interactions. Early theories for work-hardening emphasized one single mechanism, but it now appears that a number of dislocation mechanisms can contribute. The loop length scaling of eq. (22) still applies for free segments, with dislocation interactions providing the pinning (see KUHLMANN-WILSDORF [1966]). Here, we briefly enumerate the hardening interactions.

For *forest* interactions of a dislocation with other dislocations threading its glide plane, the initial interaction is the elastic one of fig. 10. In the process of intersection, energy must be supplied to form intersection jogs and kinks. The latter process is more important for extended dislocations, but the relative importance of the two contributions is unresolved for perfect dislocations. Continued pinning is provided by sessile jogs on screw dislocations.

Forest dislocations can be provided by glide on secondary slip systems or by several special mechanisms. One of these is dipole formation, which can be analyzed using fig. 21. If after cross-slip, h is less than the critical value to operate the new

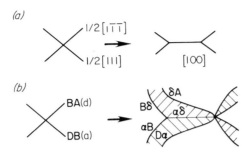

Fig. 24. Attractive junctions in (a) perfect bcc dislocations, (b) extended fcc dislocations.

segment as a source, the original dislocation and the new segment interact to form dislocation dipoles, pure edge in character. A dipole loop could also form by the condensation of vacancies into a disc which collapses. In materials of low stacking fault energy the latter process leads to the formation of stacking fault tetrahedra, fig. 17.

Another possibility in the process of intersection is the formation of *attractive junctions*, illustrated in fig. 24. The reacted part of fig. 24b is the Lomer–Cottrell barrier of fig. 17. Because W_{int} is negative between the reacting dislocations, energy must be supplied to remove the attractive junctions to complete the intersection process, and therefore their formation contributes to hardening.

Grain boundaries or Lomer–Cottrell barriers extensive in length are sufficiently strong barriers to sustain dislocation pile-ups (HIRTH [1972]; THOMPSON [1977]). These not only generate back stresses to stop the source producing the pile-up dislocations, but also generate long-range internal stresses, both effects leading to hardening.

The ideas discussed here provide the dislocation mechanisms to be considered in work-hardening. The integration of these models to provide a macroscopic work-hardening relation is discussed by WEERTMAN and WEERTMAN [1983].

5. Dislocation behavior at high homologous temperatures

5.1. Osmotic climb forces

At temperatures greater than about $0.4T_m$, vacancy contributions become sufficient for dislocation climb to be important in deformation and recovery. In the absence of stress, the *equilibrium vacancy atom fraction* is:

$$c_0^0 = \exp(-G_f/kT), \tag{30}$$

where G_f is the formation energy of a vacancy. In the presence of external pressure P, the formation of a vacancy as in fig. 25a requires extra work $P\delta V$. Here δV is the

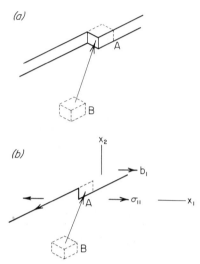

Fig. 25. Creation of a vacancy at B by depositing the atom from that site to (a) site A at a step on a free surface or (b) site A at a jog on an edge dislocation.

volume change on vacancy formation, equal to the atomic volume v_a minus the relaxation volume $\delta v + \delta v'$ arising from the partial collapse of neighboring atoms into the vacant site. The term δv is the local volume change and $\delta v'$ is an image contraction of the crystal associated with satisfying the free surface boundary condition. In the presence of an internal pressure p at site b there is extra work $p(\delta v + \delta v')$. Thus, the equilibrium atom fraction becomes

$$c = \exp\{-[G_f + Pv_a - (P+p)(\delta v + \delta v')]/kT\}$$
$$= c_0^0 \exp\{-[Pv_a - (P+p)(\delta v + \delta v')]/kT\}$$
$$= c^0 \exp(-Pv_a/kT). \tag{31}$$

It is convenient to include the image terms in c^0 and use this value as a reference concentration for dislocation problems so that first-order effects of stress can be separated from weak effects arising from internal pressure gradients (HIRTH and LOTHE [1982]).

For the creation of a vacancy by the process of fig. 25b, the net work is

$$\delta W = -\sigma_{11} b_1 ah + \overline{G}_v = -\sigma_{11} v_a + kT \ln(c/c^0). \tag{32}$$

Here $\sigma_{11} b_1$ is the Peach–Koehler force on the edge dislocation, eq. (9); a is the length and h the climb distance of an atomic unit of dislocation line, so that $b_1 ah = v_a$, $\overline{G}_v$ is the vacancy chemical potential referred to a reference state c^0, eq. (31), and c is the actual vacancy concentration. The net force in the x_2 climb direction per unit length is

$$\frac{F}{L} = \frac{1}{a}\frac{\partial W}{\partial x_2} = \frac{1}{ah}\delta W = -\sigma_{11} b_1 + \frac{kT b_1}{v_a} \ln\left(\frac{c}{c^0}\right) = \frac{F_{el}}{L} + \frac{F_{os}}{L}. \tag{33}$$

Here F_{el} and F_{os} are the elastic and osmotic climb force on the dislocation, respectively. More generally, F_{el} would be the total Peach–Koehler force in the x_2 direction and could include forces on screw components of mixed dislocations, see §2.2.

Let us consider some local equilibrium examples. At local equilibrium the total force $F/L = 0$ and eq. (33) gives:

$$c = c^0 \exp(\sigma_{11} v_a / kT). \tag{34}$$

For a dislocation under uniform hydrostatic pressure P, $\sigma_{11} = -P$, eqs. (31) and (34) give the same local equilibrium value of c at the dislocation and at a free surface, and no vacancy diffusion or climb occurs. A dislocation under stress σ_{11} with local concentration c near a stress free surface with concentration c^0 experiences a vacancy concentration gradient and hence a tendency to climb. Often $\sigma_{11} v_a < kT$ at high temperatures and the gradient can be linearized, $c - c^0 \cong c^0 \sigma_{11} v_a / kT$.

If local equilibrium does not obtain, either F_{el} or F_{os} can give a net climb force. For example, a quenched-in concentration c of vacancies producing a supersaturation relative to c^0 at the final temperature gives an osmotic climb force given by eq. (33). For quenches from near T_m to $T_m/2$ or less, the osmotic forces are large, corresponding to stresses of $\sim \mu/10$ or greater, of the order of theoretical strengths of perfect crystals. Thus, there is little constraint to climb or to nucleation of vacancy discs which collapse to dislocation loops after quenches from near T_m. A special case is the development of a helical dislocation from a straight screw dislocation in which local equilibrium is established by a balance of the line-tension forces, tending to return the helix to a straight line, by the osmotic force on the edge component of the helical dislocation.

5.2. Jog drag

A climb-type process that can contribute to an overall glide process is the drag of jogs by screw dislocations. A jog on a screw is sessile, as seen in fig. 26, since the jog itself has edge character. Yet, whether it is a unit jog or a superjog, it can move along with the screw dislocation by vacancy absorption or emission. A steady-state

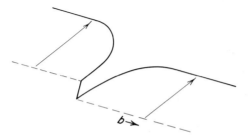

Fig. 26. Unstressed screw dislocation (dashed line) containing a jog, and glide of the dislocation under stress with pinning at the sessile jog.

References: p. 1257.

configuration could be that of fig. 19 with the osmotic forces at the jogs balancing line-tension forces from the loops bowed out by the applied glide force. With opposite-sign jogs, the local equilibrium concentrations at the jogs would be given by eq. (34) but with opposite signs of the exponent. Thus, concentration gradients would exist and the rate of climb would be controlled by vacancy diffusion from one sign jog to the other. Relevant to ch. 20, the concentration gradient in this case would involve a difference of exponentials, eq. (34), together giving a stress dependence $\sim 2 \sinh(\sigma_{11} v_a / kT)$.

For bowout forces giving point forces at the jog pinning points $F > F_{os}^* = G_f b_1 / v_a$, breakaway will occur in the sense that the dislocation will move rapidly in glide, generating a string of vacancies (or interstitials) in the wake of the jogs. This can happen because the energy released on glide by a unit atomic distance exceeds the formation energy G_f of the appropriate point defect.

5.3. Climb

In the absence of constraints, local equilibrium vacancy concentrations are maintained at jogs and the determination of the climb velocity of a dislocation becomes a diffusion boundary-value problem with the jogs as sources and sinks. Often the jog spacing is so small or the pipe diffusivity of vacancies along the core is so large that a dislocation acts as a perfect line source and sink and dislocation lines can be considered as the sources and sinks (BALLUFFI and GRANATO [1979]). The boundary-value solutions are beyond the scope of this work, but consideration of constraint is of interest.

In the presence of strong pinning, dislocations must climb to critical configurations as in fig. 20 to activate Bardeen–Herring sources. In the critical configuration, the sum of the elastic and osmotic forces, eq. (33), must balance the line tension forces $(\mu b / \lambda)$.

For straight perfect dislocations, there are no indications of constraints preventing the attainment of local equilibrium vacancy concentrations. For extended dislocations, jog nucleation and jog motion are complex and could provide constraint (§ 3.5). However, experiments on the relatively low stacking fault energy metal gold show that the climb efficiency is greater than or equal to 0.1 (SEIDMAN and BALLUFFI [1966]). Here the climb efficiency was defined as the climb rate of a dislocation, under moderate to small osmotic forces produced by quenching, divided by the maximum climb rate under pure diffusion control. In general, constraint would be negligible when the relaxation time for diffusion from source to sink over a distance L, $\tau \sim L^2 / D_{vac}$ with D_{vac} the diffusivity of vacancies, is large compared to the time to nucleate and move jogs over a typical segment length λ. In most cases τ is expected to be large compared to the nucleation time in the gold quenching experiment. Thus, generally, one expects constraint to be absent so that it is a good approximation to treat climb as a pure diffusional boundary-value problem.

5.4. Solute drag

In analogy with the vacancy case already discussed, there is an interaction energy $p\delta v$ between a dislocation with internal pressure field p and a solute atom with a characteristic local volume change δv, equivalent to a ball inserted into a hole of different size in an elastic medium. A straight screw dislocation has no p field and thus no interaction. For the edge dislocation of fig. 7, $p = -(\sigma_{rr} + \sigma_{\theta\theta} + \sigma_{zz})/3$. Hence, from eq. (5), the interaction energy is:

$$W_{int} = \frac{\mu b \delta v (1 + \nu)\sin\theta}{3\pi(1 - \nu)r}. \tag{35}$$

The solute concentration in local equilibrium in the presence of p, again analogous to § 5.1, is:

$$\frac{c}{1 - c} = \frac{c_0}{1 - c_0} \exp(-W_{int}/kT) = \frac{c_0}{1 - c_0} \exp\left(-\frac{\beta \sin\theta}{r}\right), \tag{36}$$

where c_0 is the bulk equilibrium concentration and terms other than r and $\sin\theta$ in eq. (35) are included in β. Here it is important to use the Fermi–Dirac form, since for positive exponents $c \to 1$ as $r \to 0$, and the simpler Boltzmann form would artificially diverge. With δv positive, corresponding to a solute atom larger than the solvent atom, solute is depleted from the compressed region above the glide plane where $\sin\theta$ is positive, and attracted to the extended region (see fig. 2), as is physically reasonable. The solute of eq. (5) comprises the *Cottrell atmosphere*.

In addition, solute atoms are attracted to either screw or edge dislocation cores with a binding energy E_B for the appropriate case, forming a "core atmosphere". Also, interstitial atoms in bcc metals can form "Snoek atmospheres". These interstitials produce an asymmetric tetragonal strain in the octahedral or tetrahedral sites with the maximum principal strain aligned along $\langle 100 \rangle$ directions. A stress which produces extension along a [100] direction would then favor preferred occupation of that type of site with a corresponding negative interaction energy. All three atmospheres can produce dislocation drag and breakaway phenomena. As an example, the Cottrell atmosphere is considered here. The formation of the atmosphere is again a pure diffusion problem. For a discussion of the diffusion solutions, see BAIRD [1971].

The atmosphere-drag problem involves a diffusion solution, in a moving reference frame, that is quite lengthy and complex. However, quite good approximate results can be found by use of a simpler dissipation theorem (HIRTH and LOTHE [1968]; HILLERT and SUNDMAN [1976]). The diffusion flux J in a volume element δV is related to the chemical potential gradient $\nabla \bar{G}$ by the Einstein relation:

$$J = -\frac{D_c}{kT} \nabla \bar{G}. \tag{37}$$

The rate of dissipation of free energy in the element is:

$$-J \cdot \nabla \bar{G} \delta V = \frac{J \cdot J kT}{D_c} \delta V, \tag{38}$$

References: p. 1257.

with the use of eq. (37). However, for a problem where the driving force for diffusion is the Peach–Koehler force F, the total work done on the system per unit time is $F \cdot v$ where v is the dislocation velocity. Thus, the total force is given by:

$$F \cdot v = \frac{kT}{D} \int \frac{J \cdot J}{c} \, dV. \tag{39}$$

Diffusion can occur in response to a concentration gradient ∇c or to a potential gradient ∇W. At steady state in the moving reference frame, the total flux is zero,

$$-D \nabla c - (Dc/kT) \nabla W - v(c - c_0) = 0,$$

so the static frame flux J can be written

$$J = -D \nabla c - (Dc/kT) \nabla W = v(c - c_0). \tag{40}$$

The key to the use of the dissipation theorem is that, except for velocities so large that drag would be meaningless anyway, eq. (40) remains approximately valid with the *static* rest values substituted for J.

Let us consider a dislocation-type problem with a driving force σb in the x direction opposed by a drag force and with a dislocation velocity dx/dt. Equation (39) then reduces to

$$\frac{F}{L} = \sigma b = \frac{kT}{vD} \int \frac{J \cdot J}{c} \, dV. \tag{41}$$

As the dislocation moves through the solution, diffusion occurs to tend to establish the equilibrium concentration of eq. (36), producing the drag force of eq. (41). At low velocities, c is given by eq. (36) and the substitution of eq. (40) into (41) gives

$$\frac{F}{L} = \frac{kTv}{D} \int \frac{(c - c_0)^2}{c} \, dV. \tag{42}$$

In this regime F/L increases with increasing v. At high velocities $c \approx c_0$, $\nabla c \approx 0$ and one must use the form $J = (-Dc/kT) \nabla W$ giving:

$$\frac{F}{L} = \frac{D}{kTv} \int c_0 (\nabla W)^2 \, dV. \tag{43}$$

In this regime F/L decreases with increasing v.

The form of the stress–velocity curve is thus that shown in fig. 27. The maximum value $\sigma_c b \cong 17 \, c_0 \beta$ occurs at a critical velocity $v_c \cong 4 \, D \, kT/\beta$, according to several numerical solutions reviewed by HIRTH and LOTHE [1982]. In the region of positive slope in fig. 26, stable flow occurs. Large stresses can push the dislocation past the peak to velocities in the negative-slope region where breakaway occurs, with the dislocation accelerating until some other mechanism such as phonon damping controls its motion. At about v_c, breakaway can occur, but if the dislocation should be slowed down, repinning can occur, and the behavior can oscillate from pinned to unpinned motion.

These phenomena relate to macroscopic flow behavior in tension through the

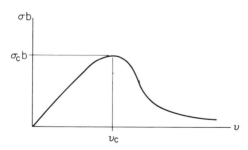

Fig. 27. Drag force–velocity relation for a dislocation with a Cottrell atmosphere of solute atoms.

Orowan relation:

$$\dot{\varepsilon} = m\rho bv, \qquad (44)$$

where $\dot{\varepsilon}$ is the strain rate, m is a strain resolution factor and ρ is the average mobile dislocation density. At temperatures $T \gg T_c$, D is large and the slope $d\sigma b/dv$ is small, eq. (42). Thus, for a given $\dot{\varepsilon}$, a large v can be attained with a small σ to satisfy eq. (44) and ρ will tend to be small. For $T \ll T_c$, D is small, the slope is large, and even the maximum stress σ_c is insufficient to produce a sufficiently large v to satisfy eq. (44). Then breakaway to another high velocity region will occur and again ρ will tend to be small. For $T \approx T_c$, $\sigma \approx \sigma_c$, $v \approx v_c$ will be small compared to the other cases, ρ will tend to be large, and oscillating breakaway–repinning behavior will occur. The breakaway and repinning correspond to macroscopic yield point and serrated flow phenomena. These effects are associated with low ductility which would correlate with the expected larger values of ρ needed to satisfy eq. (44).

6.   *Grain boundaries*

Grain boundaries represent a broad topic (see CHADWICK and SMITH [1976], also ch. 10B, §2.1), far beyond the scope of the present treatment. There are two aspects, however, that provide a connection between dislocations and other topics in this text. These are grain-boundary energy and grain-boundary dislocations.

Consider the simple edge-dislocation tilt boundary in fig. 28. The geometry of fig. 28 shows that the number of incomplete planes terminating to form the length h of tilt boundary is $n = (2h/b) \sin(\theta/2)$. Hence the mean spacing between dislocations in the boundary is:

$$D = \frac{h}{n} = \frac{b}{2 \sin(\theta/2)} \cong \frac{b}{\theta}, \qquad (45)$$

with the approximate form holding for small angles θ. The tilt array of fig. 28a is a minimum-energy array because the tension and compression fields of adjacent dislocations overlap and tend to cancel out over distances of the order of D

References: p. 1257.

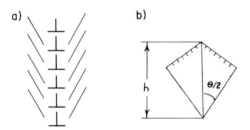

Fig. 28. A simple tilt boundary (a) and its representation in terms of planes ending at the boundary (b).

according to St. Venant's principle. The energy per dislocation is thus given by eq. (8) with $R \approx D \approx b/\theta$. Since there are $1/D = \theta/b$ dislocations per unit area of boundary, the strain-energy contribution to the boundary free energy per unit area (READ and SHOCKLEY [1950]) is:

$$\gamma = \frac{\mu b}{4\pi(1-\nu)} \theta (A - \ln \theta). \tag{46}$$

The term A contains the core energy, and while there are linear elastic estimates of A in terms of the core cutoff r_0, they are not expected to be very accurate for reasons discussed previously. Indeed, the measurement of A for small θ is a good way to estimate the core energy.

An equation with a similar θ-dependence holds for screw or mixed dislocation boundaries. Experiments by GJOSTEIN [1960] show that eq. (46) is valid up to $\theta \approx 5°$ for copper. Empirical fits of the form of eq. (46) are possible for much larger angles, but the work of Gjostein shows that these do not properly fit low-angle data.

Many high-angle grain boundaries can be regarded formally as superposed arrays of dislocations spaced atomic distances apart. This description has little advantage over a purely atomic disregistry view. However, for high-angle grain boundaries near twin boundaries, the coincidence-lattice, DSC lattice model of BOLLMANN [1970] is appropriate. These are illustrated for a first-order ⟨111⟩–⟨112⟩ twin in fcc in fig. 29a. The DSC is essentially the coarsest lattice containing all lattice sites from both constituent crystals of the twin. If a dislocation passes through the twin boundary, continuity of Burgers vector (§ 1) requires that a grain-boundary dislocation (GBD) is left in the boundary. As illustrated in fig. 29b, these GBDs have Burgers vectors corresponding to lattice vectors of the DSC lattice, small compared to those of perfect lattice dislocations.

Twin boundaries of relatively low order are expected also to correspond to low-energy configurations or cusps on a grain-boundary energy-orientation plot (CHADWICK and SMITH [1976]). Near these cusp orientations small deviations are accomplished by the superposition of small-angle boundaries composed of GBDs. Thus, the percentage of boundaries with dislocation structure is greatly increased if one includes the GBDs. Extensive observations of such GBDs in grain boundaries have been made with the use of electron transmission microscopy, see CHADWICK and SMITH [1976].

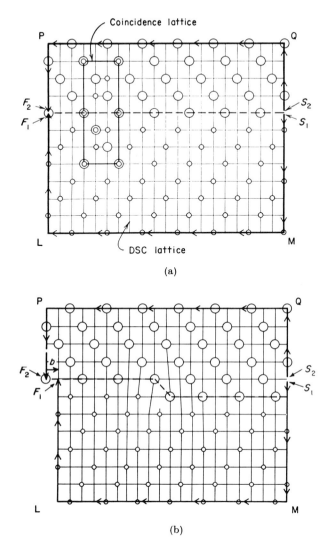

Fig. 29. View parallel to ⟨111⟩ of a first-order ⟨111⟩–⟨112⟩ twin in fcc showing the coincidence lattice and the DSC lattice (a). This view provides a reference lattice for the Burgers circuit in (b) which also shows the GBD in the DSC lattice. For faulted crystals, the Burgers circuit must start and end on the fault plane.

References

ALSHITS, V.I., and V.L. INDENBOM, 1975, Sov. Phys. Usp. **18**, 1.

AMELINCKX, S., 1979, in: Dislocations in Solids, ed. F.R.N. Nabarro (North-Holland, Amsterdam) vol. 2, p. 67.

BAIRD, J.D., 1971, Met. Rev. **16**, 1.

BALLUFFI, R.W., and A.V. GRANATO, 1979, in: Dislocations in Solids, ed. F.R.N. Nabarro (North-Holland, Amsterdam) vol. 4, p. 102.

BOLLMANN, W., 1970, Crystal Defects and Crystalline Interfaces (Springer, Berlin).

BULLOUGH, R., and V.K. TEWARY, 1979, in: Dislocations in Solids, ed. F.R.N. Nabarro (North-Holland, Amsterdam) vol. 2, p. 1.

CHADWICK, G.A., and D.A. SMITH, 1976, Grain Boundary Structure and Properties (Academic, New York).

CHERNS, D., P.B. HIRSCH and H. SAKA, 1980, Proc. Roy. Soc. (London) **371A**, 213.

CHOU, Y.T., and J.C.M. LI, 1969, in: Mathematical Theory of Dislocations, ed. T. Mura (Am. Soc. Mech. Eng., New York) p. 116.

DE WIT, G., and J.S. KOEHLER, 1959, Phys. Rev. **116**, 1113.

ESHELBY, J.D., 1953, J. Appl. Phys. **24**, 176.

FANTOZZI, G., C. ESNOUF, W. BENOIT and I.G. RITCHIE, 1982, Prog. Mater. Sci. **27**, 311.

FLEISCHER, R.L., 1964, in: Strengthening of Metals, ed. D. Peckner (Reinhold, New York) p. 93.

FRIEDEL, J., 1956, Les Dislocations (Gauthier–Villars, Paris) p. 123.

GALLAGHER, P.C.J., 1970, Metallurg. Trans. **1**, 2429.

GJOSTEIN, N.A., 1960, Acta Metall. **8**, 263.

GRANATO, A.V., and K. LÜCKE, 1956, J. Appl. Phys. **27**, 583.

HAASEN, P., 1983, ch. 21, this volume.

HAZZLEDINE, P.M., 1982, research in progress.

HILLERT, M., and B. SUNDMAN, 1976, Acta Metall. **24**, 731.

HIRSCH, P.B., 1962, Phil. Mag. **7**, 67.

HIRTH, J.P., 1972, Metallurg. Trans. **3**, 3047.

HIRTH, J.P., and J. LOTHE, 1982, Theory of Dislocations, 2nd Ed. (Wiley, New York); 1968, 1st Ed. (McGraw–Hill, New York).

HIRTH, J.P., T. JØSSANG and J. LOTHE, 1965, J. Appl. Phys. **37**, 110.

KOCKS, U.F., A.S. ARGON and M.F. ASHBY, 1975, Prog. Mater. Sci. **19**, 1.

KUHLMANN-WILSDORF, D., 1968, in: Work-Hardening, eds. J.P. Hirth and J. Weertman (Gordon and Breach, New York) p. 97.

LOHNE, O., 1974, Phys. Stat. Sol. (a) **25**, 209.

LOTHE, J., 1967, Phys. Norvegica **2**, 153.

MARCINKOWSKI, M.J., 1979, Unified Theory of the Behavior of Matter (Wiley, New York).

MOTT, N.F., and F.R.N. NABARRO, in: Report on Strength of Solids (Physical Society, London) p. 1.

NABARRO, F.R.N., 1947, Proc. Phys. Soc. (London) **59**, 256.

NABARRO, F.R.N., 1952, Adv. Phys. **1**, 269.

NABARRO, F.R.N., 1967, Theory of Crystal Dislocations (Oxford Univ. Press, Oxford).

OROWAN, E., 1934, Z. Phys. **89**, 605 and 634.

PEACH, M.O., and J.S. KOEHLER, Phys. Rev. **80**, 436.

PEIERLS, R.E., 1940, Proc. Phys. Soc. (London) **52**, 23.

POLANYI, M., Z. Phys. **89**, 660.

PULS, M.P., 1981, in: Dislocation Modeling of Physical Systems, eds. M.F. Ashby *et al.* (Pergamon, Oxford) p. 249.

READ, W.T., and W. SHOCKLEY, 1950, Phys. Rev. **78**, 275.

SEEGER, A., 1981, Z. Metall. **72**, 369.

SEEGER, A., and C. WÜTHRICH, 1976, Nuovo Chim. **33B**, 38.

SEIDMAN, D.N., and R.W. BALLUFFI, 1966, Phys. Stat. Sol. **17**, 531.

TAYLOR, G.I., 1934, Proc. Roy. Soc. (London) **A145**, 362.

TEUTONICO, L.J., 1964, Phil. Mag. **10**, 401.

THOMPSON, A.W., 1977, Work-Hardening in Tension and Fatigue (Met. Soc. AIME, Warrendale, PA).

THOMPSON, N., 1953, Proc. Phys. Soc. (London) **66B**, 481.

TIMPE, A., 1905, Z. Math. Phys. **52**, 348.

VESELY, D., 1972, Scripta Metall. **6**, 753.

VITEK, V., 1974, Cryst. Latt. Defects **5**, 1.

VOLTERRA, V., 1907, Ann. Ecole Norm. Super. **24**, 400.

WEERTMAN, J., and J.R. WEERTMAN, 1983, ch. 19, this volume.

CHAPTER 19

MECHANICAL PROPERTIES, MILDLY TEMPERATURE-DEPENDENT

J. WEERTMAN and J.R. WEERTMAN

Department of Materials Science and Engineering
Northwestern University
Evanston, IL 60201, USA

R.W. Cahn and P. Haasen, eds.
Physical Metallurgy; third, revised and enlarged edition
© *Elsevier Science Publishers BV, 1983*

1. Introduction

A piece of metal can be deformed permanently if it is pulled sufficiently hard in tension or is squeezed with a great enough force in compression or is twisted through a large enough angle in torsion. When the stress is removed, the dimensions of the piece of metal do not return to their original values as they would do if the deformation were elastic. The permanent distortion suffered by the metal specimen is called *plastic deformation*. The adjective *plastic* distinguishes this type of deformation from temporary elastic changes in dimensions.

The chief mechanism by which plastic deformation occurs is the motion of dislocations. Because there are an immense number of ways in which dislocations can bring about plastic deformation it is not surprising that this phenomenon is quite complex. Two metals can differ in their behaviour under plastic deformation if the crystal structures of the metals are not the same. The character of plastic deformation is a sensitive function of such variables as temperature, the strain rate of deformation, the past history of a sample, crystal size, and if the sample is a single crystal, the orientation of the crystal axes with respect to the stress system.

In this and the following chapter we shall present a rather simplified analysis of the subject of plastic deformation of metals. The reader who desires to have a more sophisticated knowledge should read the original articles on the subject (see *Further reading* at the end of this chapter).

At the present time we have a qualitative understanding of most of the phenomena of plastic deformation. Dislocations are so versatile in their behaviour that it is possible for a large number of independent theories to co-exist. As a result, although numerous quantitative or semi-quantitative theories have been proposed, there is no generally accepted theory for any plastic deformation phenomenon. The student is warned, therefore, that the theories discussed in this and the next chapter are by no means final or completely proven.

2. Stress–strain curves

Metal or, as a matter of fact, any crystalline material, work-hardens. The term *work-hardening* (or *strain-hardening*) simply describes the fact that if a crystalline specimen already has suffered plastic deformation, any further deformation will require a stress which exceeds that needed to produce the initial deformation. In order for work-hardening to occur the deformation temperature must be low relative to the melting point of the material. A piece of metal which has been deformed at a low temperature is said to have been cold-worked. The descriptive phrases *work-hardening, cold-working,* and *hot-working* (i.e. deforming at temperatures high enough to obviate work-hardening) are qualitative terms which arose years ago in the metal fabricating industry, and are used commonly by investigators in the field of plastic deformation.

The *stress–strain curve* of a metal is a source of quantitative information concerning the work-hardening properties of the metal. This curve is a plot of the stress applied to a specimen of the metal (at a constant strain rate) versus the strain produced by this stress. Unless stated to the contrary, it is assumed that the specimen has been annealed for some time at a temperature not very far below the melting point before the stress–strain curve was determined. High-temperature annealing removes any work-hardening that may have been introduced inadvertently during the preparation and shaping of the metal specimen.

The strains which are produced during plastic deformation can be large. The simple expressions which are used to describe the small strains treated by linear elasticity theory are not valid in the case of large plastic strains. For example, if a sample of initial length L_0 is pulled in tension to a length L, the strain ϵ is equal to the familiar term $(L - L_0)/L$ only if this ratio is small compared to one. An expression which is valid at large strains can be found as follows. The incremental strain $\delta\epsilon$ which is produced when the length of a specimen is changed from L to $L + \delta L$ is $\delta\epsilon = \delta L/L$. The total strain ϵ in a specimen which has been elongated from an initial length L_0 to a final length L is

$$\epsilon = \int_0^{\epsilon} d\epsilon = \int_{L_0}^{L} \frac{dL}{L} = \log\frac{L}{L_0}. \tag{1}$$

This reduces to the more familiar expression when L is almost equal to L_0. Equation (1) is valid for both tensile and compressive uniaxial stresses.

The stress–strain curve shown schematically in fig. 1 is typical for high-purity polycrystalline fcc or hcp metals which have been pulled to large strains. Note that the slope of the curve decreases as the strain increases. Thus in the case of polycrystalline metals the rate of work-hardening, which is defined as the slope of the stress–strain curve, decreases with increasing strain.

The stress–strain curve of a metal generally terminates at some finite strain because of the failure of the sample. The cause of failure varies from specimen to specimen and from metal to metal. It also depends on the type of loading. Work-hardening may raise the stress required for further deformation to such a high level that cracks open up and propagate across a specimen. In tensile tests of ductile

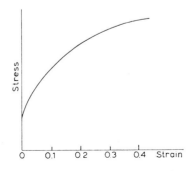

Fig. 1. Schematic diagram of the stress–strain curve of a polycrystalline fcc or hcp metal.

metals, failure occurs because in some local region the specimen *necks* down to a small cross-sectional area. Should the cross-sectional area in some region of a specimen be smaller than the average area, the stress in that region must be larger than the average stress. This increase in stress can lead to a greater amount of yielding in the region of smaller cross-section and thus produce a still smaller cross-sectional area. The process can become catastrophic and lead to failure.

The stress–strain curve of a polycrystalline material represents an average, of a sort, of the stress–strain relationships of the individual grains. However, the deformation of a polycrystalline mass produces non-uniform deformation within the individual grains. Thus the stress–strain curve for the sample as a whole is not a true average of the properties of the constituent crystals.

The behaviour of a metal under plastic deformation as revealed by its polycrystalline stress–strain curve obviously is important for practical applications. But for a fundamental understanding of the phenomenon of plastic deformation the experimental data contained in the stress–strain curves of single crystals are far more interesting. Such data are not subject to the loss of information which results from the averaging process associated with the deformation of polycrystalline samples. Furthermore, any complications which arise from the non-uniform deformation of the grains in a polycrystalline mass are eliminated.

2.1. Geometry of slip

Plastic deformation in crystals occurs by *slip* parallel to certain crystallographic planes, the *slip planes*. The *direction of slip* is along one of the crystallographic directions which lie in the slip plane. The slip mechanism is simply the motion in the slip plane of dislocations whose Burgers vectors lie parallel to the slip direction. The surface of a single-crystal specimen which originally is smooth becomes stepped as the result of deformation. The steps are produced where dislocations leave the crystal. Figures 2 and 3 show the appearance of the surface of deformed aluminium crystals. The lines on these pictures are steps on the surface where fifty or more dislocations which moved across one slip plane have left the crystal. These lines are called *slip lines*. If, as in fig. 3, they are clustered together they are called *slip bands*. Table 1 lists the more common slip planes and slip directions for a number of metals at relatively low temperatures (cf. ch. 18, § 3.4).

Because the deformation of crystals occurs by slip along crystallographic planes it is logical when plotting stress–strain curves of single crystals to resolve the applied stress into a shear stress acting on the slip plane and in the direction of slip. Similarly, it is desirable to resolve the strain into a shear strain across the slip plane and in the slip direction. The *resolved stress* and *strain* can be obtained from simple, easily derived formulae. Consider fig. 4. It shows a sample strained by a uniaxial tensile or compressive stress σ. The normal to the slip plane makes an angle ψ with the specimen axis. The slip direction makes an angle φ with the specimen axis. The resolved shear stress σ_r is given by the expression:

$$\sigma_r = \sigma(\cos \varphi) \cos \psi. \tag{2}$$

Fig. 2. Electron micrograph of an aluminium crystal deformed 20% at 90 K. Only fine slip (elementary slip) is observable. 12 500×. (From Van BUEREN [1960].)

The resolved shear strain ϵ_r is:

$$\epsilon_r = (\cos \varphi_0)^{-1} \left\{ \left[(L/L_0)^2 - \sin^2\psi_0 \right]^{1/2} - \cos \psi_0 \right\}. \tag{3}$$

Here φ_0 and ψ_0 are the values of φ and ψ at the start of slip, and L and L_0 are the final and initial lengths of the specimen. It is assumed in both eq. (2) and eq. (3) that slip takes place on only one set of slip planes and in one direction. Equation (3) is derived for a sample pulled in tension. If the sample is deformed in compression ϵ_r is the solution of the quadratic equation

$$(L_0/L)^2 = 1 + 2\epsilon_r(\cos \varphi_0) \cos \psi_0 + \epsilon_r^2 \cos^2\psi_0. \tag{4}$$

References: p. 1305.

Fig. 3. Aluminium crystal deformed 30% at room temperature. Most of the fine slip lines have clustered into slip bands. $12\,500\times$. (From Van BUEREN [1960].)

2.2. Stress–strain curves of hcp metals

The simplest stress–strain curves are obtained from high-purity single-crystal specimens of certain hexagonal close-packed (hcp) metals. The particular hcp metals are those which deform predominantly by slip on their basal planes. The basal planes, the $\{0001\}$ planes, are the close-packed planes of this structure. An hcp metal deforms primarily by basal slip at low temperatures if its c/a ratio is close to the ideal value of 1.6333 (e.g. Mg with $c/a = 1.624$) or exceeds it (Zn with $c/a = 1.856$ and Cd with $c/a = 1.885$). The hcp metals which have a c/a ratio appreciably smaller than the ideal value deform by slip on non-basal planes, usually the $\{10\bar{1}0\}$ prismatic planes or the $\{10\bar{1}1\}$ or $\{11\bar{2}2\}$ pyramidal planes. The slip direction on these

Table 1
The predominant slip planes and slip directions of various metals (after BARRETT and MASSALSKI [1980]).

Structure	Metal	Slip plane	Slip direction
fcc	Al	$\{111\}$	$\langle\bar{1}01\rangle$
	Cu	$\{111\}$	$\langle\bar{1}01\rangle$
	Ag	$\{111\}$	$\langle\bar{1}01\rangle$
	Au	$\{111\}$	$\langle\bar{1}01\rangle$
	Ni	$\{111\}$	$\langle\bar{1}01\rangle$
bcc	α-Fe	$\{110\}, \{211\}, \{321\}$	$\langle\bar{1}11\rangle$
	Mo	$\{211\}$	$\langle\bar{1}11\rangle$
	W	$\{211\}$	$\langle\bar{1}11\rangle$
	K	$\{321\}$	$\langle\bar{1}11\rangle$
	Na	$\{211\}\,\{321\}$	$\langle\bar{1}11\rangle$
	Nb	$\{110\}$	$\langle\bar{1}11\rangle$
hcp	Mg	$\{0001\}\,\{10\bar{1}1\}$	$\langle2\bar{1}\bar{1}0\rangle$
	Cd	$\{0001\}$	$\langle2\bar{1}\bar{1}0\rangle$
	Zn	$\{0001\}$	$\langle2\bar{1}\bar{1}0\rangle$
	Be	$\{0001\}$	$\langle2\bar{1}\bar{1}0\rangle$
	Re	$\{10\bar{1}0\}$	$\langle2\bar{1}\bar{1}0\rangle$
	Ti	$\{10\bar{1}0\}\,\{0001\}\,\{10\bar{1}1\}$	$\langle2\bar{1}\bar{1}0\rangle, \langle11\bar{2}0\rangle, \langle11\bar{2}2\rangle$
	Zr	$\{10\bar{1}0\}\,\{0001\}\,\{10\bar{1}1\}$	$\langle2\bar{1}\bar{1}0\rangle, \langle11\bar{2}0\rangle, \langle11\bar{2}2\rangle$
	Hf	$\{10\bar{1}0\}\,\{0001\}\,\{10\bar{1}1\}$	$\langle2\bar{1}\bar{1}0\rangle, \langle11\bar{2}0\rangle, \langle11\bar{2}2\rangle$

planes is either the $\langle11\bar{2}0\rangle$ direction, which also is the slip direction for the basal planes, or the $\langle11\bar{2}3\rangle$ direction. (Beryllium with $c/a = 1.568$ deforms primarily on basal planes and thus is an exception to these rules.)

When slip is confined to the basal planes there is only one set of parallel slip

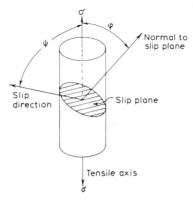

Fig. 4. Resolution of stress and strain onto the slip plane.

References: p. 1305.

planes. There are no intersections between sets of non-parallel slip planes. Obviously this is the simplest manner in which plastic deformation can take place. It is not surprising that hcp metals which deform through basal slip have the least complicated stress–strain curves. The schematic curve shown in fig. 5 is typical of the stress–strain curves of these hcp metals. The curve consists of an initial portion corresponding to the region of *elastic* deformation followed by a region of *plastic* deformation. As the shear stress (hereafter it is understood that the stresses and strains associated with single crystals are *resolved* shear stresses and shear strains) is increased from zero, the initial strains are given by the elastic law $\sigma = \mu\epsilon$, where μ is the shear modulus. These strains are recoverable. If a stress is brought back to zero, the strain likewise vanishes.

If the stress is raised to a value of the order of $10^{-5}\mu$–$10^{-4}\mu$, permanent plastic deformation occurs. The stress–strain curve bends at the onset of plastic deformation. At larger stresses the stress–strain curve again is linear. As indicated in fig. 5, its slope in the plastic region is approximately $10^{-4}\mu$. Only when the plastic strain reaches the magnitude of 1–2 (100% to 200%) does the work-hardening rate increase. The behaviour of the hardening rate at large strains is illustrated in fig. 6. This figure shows actual stress–strain curves of high-purity cadmium crystals for various orientations.

The hardening rate of $10^{-4}\mu$ per unit strain which is indicated in fig. 5 is a low value indeed. Even at a 100% strain the stress required to produce further plastic deformation is not much larger than that which caused the initial plastic deformation at the knee of the curve. This phenomenon of a very low rate of work-hardening has been given a special name: *easy glide*. The portion of the stress–strain curve in which it occurs is called the easy glide region.

The strain that is produced in the plastic region of fig. 5 is permanent. If the stress is removed after a sample has been deformed to the point O of the figure only an elastic strain is recovered. The dashed line from O to O′ is followed upon unloading. The slope of this line is about equal to the elastic modulus μ. The sample has suffered the permanent strain associated with the point O′. If the sample is reloaded the elastic curve is retraced to the point O. At this point plastic deformation will commence upon a further increase of stress. The same plastic stress–strain

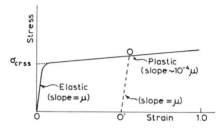

Fig. 5. Schematic plot of the resolved shear stress versus the resolved shear strain for an hcp crystal which deforms by basal slip.

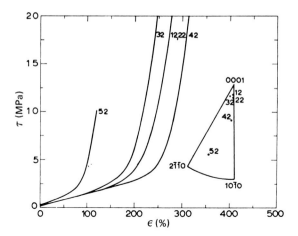

Fig. 6. Shear stress–shear strain curves of 99.9999% pure cadmium crystals deformed at 77 K (data of DAVIS [1963]).

curve that would have resulted had the specimen never been unloaded will be followed.

If very sensitive strain-measuring devices are used it generally is observed that the same paths are not followed between the points O and O′ during unloading and loading. The unloading curve is displaced slightly from the reloading curve. The strain is a double-valued function of stress. The loading and unloading curves form a *hysteresis loop* which is shown, greatly exaggerated, in fig. 7. This hysteretic behaviour is typical of any plastically deformed crystalline material which is unloaded and then stressed again. It represents a special case of the *Bauschinger effect*, which will be discussed in a later section.

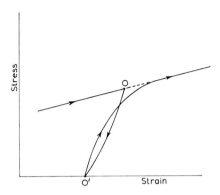

Fig. 7. Hysteresis loop upon unloading and reloading in the plastic portion of the stress–strain curve (schematic).

References: p. 1305.

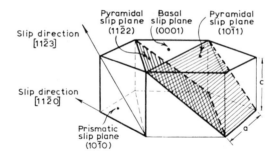

Fig. 8. A prismatic, pyramidal, and basal plane of a hexagonal crystal.

Those hcp metals which have c/a ratios appreciably lower than the ideal value generally deform on the prismatic planes and on the various pyramidal planes but sometimes on the basal planes too. These planes are illustrated in fig. 8. They are not all parallel to one other. Slip may occur on two or more sets of intersecting slip planes. Therefore the stress–strain curves of these hcp metals should differ from the stress–strain curves of hcp metals which deform by basal slip.

The form of the stress–strain curves of the hcp metals which have a low c/a ratio is not well established. Part of the problem in determining curves for these metals lies in the difficulty of obtaining high-purity material. Almost all the data on these metals have been obtained from samples which had an impurity content sufficiently high that the stress–strain curves showed typical impurity effects, e.g., the presence of yield points (see the section on bcc metals). Thus the stress–strain curves which have been obtained to date are not representative of the pure metals (HONEYCOMBE [1968]). Figure 9 shows stress–strain curves of zone-refined titanium single crystals

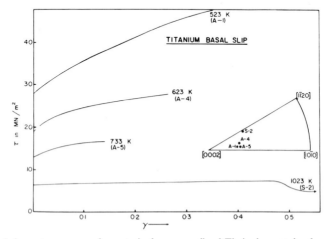

Fig. 9. Resolved shear stress versus shear strain for zone-refined Ti single crystals of an orientation that favours basal slip (AKHTAR [1975]).

of an orientation that favours basal glide (AKHTAR [1975]). At 250°C plastic deformation does not start until the stress is equal to $7 \times 10^{-4}\ \mu$ and the work-hardening coefficient is equal to $1.7 \times 10^{-3}\ \mu$. Both of these numbers are greater by over an order of magnitude than the corresponding values found for hcp metals with large c/a ratios. On the other hand, the work-hardening rate of Zn and Cd in the easy-glide range shows a strange temperature dependence with an intermediate maximum at $T = 0.3\ T_{\mathrm{m}}$ (WIELKE [1976]).

2.3. Critical resolved shear stress

The shear stress at which permanent plastic deformation first occurs in an annealed single crystal is known as the *critical* shear stress or the critical resolved shear stress. It is labelled σ_{crss} in fig. 5. Typical values of this stress range from 10^{-5} μ to $10^{-4}\ \mu$. Dislocation theory arose out of attempts to account for its magnitude.

It is difficult to define σ_{crss} precisely because the stress–strain curve bends in the transition region between elastic and plastic deformation. An unambiguous value for σ_{crss} may be obtained, in a somewhat arbitrary fashion, by extrapolating the linear portion of the plastic deformation curve to zero strain. Such an extrapolation is shown in fig. 5. Schmid's Law states that slip starts first on the slip plane with the greatest resolved shear stress.

2.4. Slip of face-centered cubic crystals

The fcc metal crystals deform by slip on any of the four $\{111\}$ close-packed planes, in any of the $\langle 1\bar{1}0 \rangle$ slip directions. A schematic plot of a typical stress–strain curve for single crystals of metals such as copper, silver, aluminium, or gold is given in fig. 10. In order to obtain such a curve it is necessary that the specimen be oriented initially so that the resolved shear stress in one direction on one slip plane is greater than the resolved shear stress in any other slip direction on that plane or in any slip direction on any other slip plane. That is, among all the resolved shear stresses which correspond to every possible slip direction on every possible slip

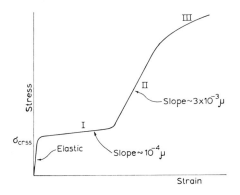

Fig. 10. Stress–strain curve of an fcc single crystal (schematic).

References: p. 1305.

plane, there must be a single resolved shear stress whose magnitude exceeds all the others. Single crystals of fcc metals which are so oriented deform initially by slip on only one set of parallel slip planes. This situation is analogous to the deformation by basal slip characteristic of certain hcp metals. It is to be expected that the initial portion of the stress–strain curve of a suitably oriented fcc crystal is similar to the stress–strain curve of an hcp crystal which deforms by basal slip. This similarity may be confirmed by comparing fig. 5 with the portion of the curve in fig. 10 labelled I. The slope of *stage* I in fig. 10 is about the same as the slope of the easy-glide region in fig. 5, namely, about $10^{-4} \mu$. Stage I of the stress–strain curves of fcc metals also is referred to as a region of easy glide. In the case of hcp crystals the easy-glide region extends out to strains of the order of 100% to 200%. With fcc crystals the easy-glide region terminates at strains of approximately 5% to 20%. A higher hardening rate sets in at larger strains. The stress–strain curve still is linear in this region, which is called *stage* II. The slope here is a factor of 30 or so larger than the slope in stage I. Stage II eventually passes into a third distinctive region of the stress–strain curve which is labelled *stage* III. In this last stage the stress–strain curve is often parabolic. The rate of hardening decreases as the strain increases. The onset of this third stage comes at strains of the order of 30% to 50%. The exact value depends on the test temperature. The higher the temperature, the smaller is the stress at which stage III starts. This stage can be eliminated by deforming the sample at a temperature close to absolute zero.

Experimentally, the most thoroughly studied of the stress–strain curves is that of the fcc metal single crystals. Almost all of the theoretical work which has been done on work-hardening has been directed to the problem of explaining this curve.

2.5. Slip of body-centred cubic crystals

The body-centred cubic metals deform by slip in any of the $\langle 111 \rangle$ directions. The slip planes are numerous. They include the {110}, {112}, and {123} planes. The slip lines that are seen at the crystal surface generally are very wavy. This indicates that the dislocations which produce slip are not confined to unique slip planes.

Fewer determinations have been made of the stress–strain curves of high-purity bcc single crystals than of fcc crystals. Figure 11 shows a stress–strain curve of zone-refined niobium (FOXALL et al. [1967]). Curves similar in shape to that shown in fig. 11 have been found for high-purity iron single crystals (JAOUL and GONZALES [1961]). The curves are very similar to those measured for fcc single crystals. Curves obtained for bcc single crystals of lesser purity tend to look "fuzzy" versions of fcc single-crystal curves. The different stages are smoothed out and are much less distinctive (ARSENAULT [1975]).

If a bcc metal contains traces of interstitial impurity atoms, such as carbon or nitrogen, its stress–strain curve for either single crystal or polycrystalline samples is similar to the schematic curve of fig. 12. This curve displays the phenomena of an upper yield point and a lower yield point. The deformation is elastic up to the stress which is labelled the *upper yield point*. Plastic deformation starts here. However once

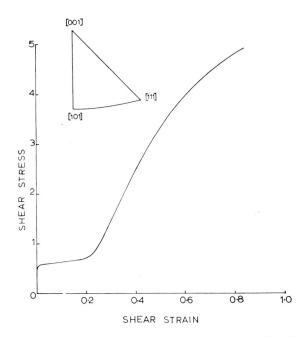

Fig. 11. Stress–strain curve of zone-refined niobium at room temperature (from FOXALL *et al.* [1967]).

plastic deformation does start it can continue at a stress level which is lower than the upper yield point. The stress level at which further plastic deformation proceeds is called the *lower yield point*. The stress–strain curve is horizontal during deformation at the lower yield point. This portion of the curve, which has a zero rate of work-hardening, may extend out to strains of the order of 0.1%. Beyond this strain

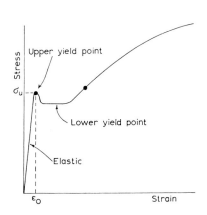

Fig. 12. Stress–strain curve of a slightly impure bcc crystal (schematic).

References: p. 1305.

the sample work-hardens and a normal stress–strain curve is followed.

The deformation is not uniform in the lower-yield-point section of the stress–strain curve. A kind of plastic wave of localized yielding passes across the specimen during this portion of the stress–strain curve. This region of localized yielding is known as a *Lüders band*. It will be discussed in a separate section.

The type of stress–strain curve shown in fig. 12 is not confined to bcc metals. It also is observed in the case of impure metal crystals of hcp and fcc structure (cf. ch. 21).

2.6. Asymmetric deformation of bcc metals

The bcc metal single crystals have the remarkable property that their deformation in tension is not symmetric with respect to deformation in compression (HIRSCH [1968], CHRISTIAN [1970,1983], ARSENAULT [1975]). Figure 13 presents a series of ψ–χ curves for tantalum and tantalum alloys that illustrate this asymmetry. Here χ, indicating the orientation of the crystal, is the angle between the (101) plane and the

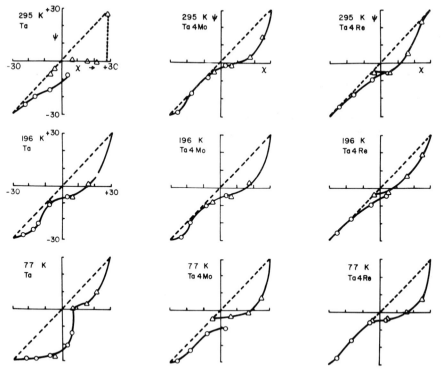

Fig. 13. ψ–χ curves for Ta, and alloys of Ta with 3.8 at% Mo and 3.8 at% Re, respectively. Triangles indicate tension data points, circles indicate compression data. (From ARSENAULT [1975] after a similar figure from ROGAUSCH and MORDIKE [1970].)

plane of maximum resolved shear stress, and ψ is the angle between the (101) plane and the observed slip plane.

The explanation of the asymmetry of flow in bcc metals and alloys cannot be very simple. The phenomenon is more complicated than a simple violation of Schmid's Law. For example, the difference in the tensile and compressive yield stresses of niobium single crystals near the [011] orientation varies with temperature and actually changes sign at 225 K (CHRISTIAN [1983], CHANG *et al.* [1983]). The ψ–χ curves differ for different bcc metals and alloys and change their shape as the temperature is varied. Explanations of the asymmetry effect generally are based on the behaviour and ease of movement of screw dislocations. Screw dislocations (or the core region of screw dislocations) can dissociate to a certain extent on two or three intersecting (112) planes, on three intersecting (110) planes, and combinations of these possibilities (cf. ch. 18, §3.2). The stress required to decrease the dissociation and thereby make the screw dislocation glissile can differ, depending on the character of the stress field. In terms of a Peierls mechanism (cf. ch. 18, §3.1), the Peierls stress can increase or decrease depending on whether there is a superimposed tensile or compressive stress across the slip plane.

Since the dissociation of the screw has to be removed before it is mobile on one or the other slip plane, the ease of cross-slip in the bcc structure right at the CRSS becomes understandable. Therefore stage II of the stress–strain curve is often missing (see ŠESTÁK and SEEGER [1978]).

3. *Effect of temperature on stress–strain curves*

The stress–strain curves of crystals exhibit a temperature effect. At present we shall assume that the upper limit of the temperature range under consideration is less than about one half of the melting temperature. We need not worry about the phenomenon of high-temperature creep, which is considered in the next chapter.

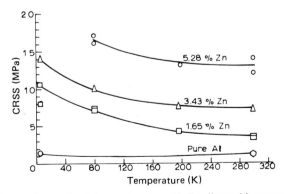

Fig. 14. Variation of σ_{crss} of pure aluminium and various Al–Zn alloys with temperature (data of DASH and FINE [1961]).

References: p. 1305.

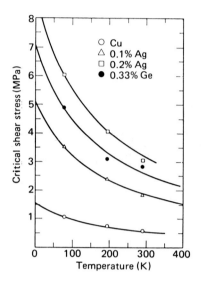

Fig. 15. Variation of σ_{crss} of pure copper and several copper alloys with temperature (data of GARSTONE and HONEYCOMBE [1957]).

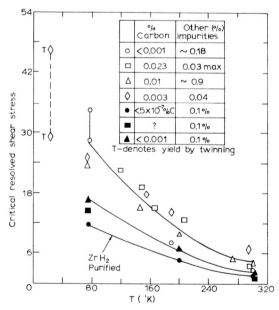

Fig. 16. Variation of σ_{crss} (in units of 7 KPa) of iron containing various amounts of carbon with temperature (data of STEIN et al. [1963]).

A change in the test temperature can alter the value of the critical resolved shear stress, the slope of the stress–strain curve, and the stress level at which stage III commences in fcc crystals. The variation of the critical shear stress with temperature for aluminium, copper and iron is shown in figs. 14, 15 and 16. Data on alloys of these metals also are plotted. These data show that the critical shear stress of pure metals is weakly temperature-dependent. Alloying is required to bring out an appreciable temperature dependence in σ_{crss} (cf. ch. 21).

The effect of changes in temperature on the appearance of the stress–strain curves of fcc crystals is shown in fig. 17. It can be seen that the higher the temperature the shorter is the extent of the easy-glide range and the smaller is the stress at the onset of stage III.

The rate of strain-hardening in the easy-glide range of fcc and hcp crystals is slightly temperature-dependent. It has been pointed out by Seeger *et al.* [1961] that the quantity θ_1/μ, where θ_1 is the slope of the stress–strain curve in the easy-glide range, is virtually temperature-independent. Therefore the temperature variation of the easy-glide strain-hardening arises chiefly through the temperature-variation of the shear modulus μ. The variation of the elastic constants with temperature usually is small.

3.1. Cottrell–Stokes Law

The stress–strain curves of a metal which are obtained at two different temperatures usually do not coincide. In general they can not be made to coincide by

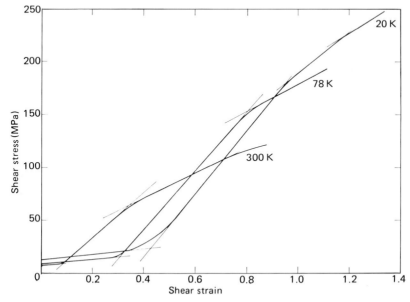

Fig. 17. Stress–strain curves of a high-purity nickel single crystal at 20 K, 78 K, and 300 K (data of Haasen [1958]).

dividing the stress by a temperature-dependent shear modulus. Figure 18 shows schematically two such divergent curves measured at a constant rate of deformation at the temperatures T_1 and T_2. It was pointed out by COTTRELL and STOKES [1955] in their now classic paper that this divergence can arise from two fundamentally different causes. The curves may differ if, at a given strain, the internal structure which has developed in each specimen as a result of plastic deformation depends on the test temperature. (The term *internal structure* covers such features as small-angle boundaries, subgrains and cells, the density and configuration of dislocations, dislocation tangles, etc.) We should expect to find a temperature variation in the stress–strain curves if different test temperatures lead to the development, at a given strain, of different internal structures.

It could happen that identical structures do develop in two metal crystals which are deformed at different temperatures. Yet their stress–strain curves need not coincide. The stress required to cause further plastic deformation (the flow stress) in crystals with identical internal structures could very well depend on temperature.

COTTRELL and STOKES [1955] attempted to measure the relative contribution of each of these fundamental causes to the observed variation with temperature of stress–strain curves. Their method was simple. Consider the schematic stress–strain curves of fig. 18. Let us suppose first that identical structures are produced at identical strains even though the test temperatures differ. In this case the divergence between the two curves of fig. 18 must be caused solely by the fact that the stress required to cause further plastic deformation in crystals with similar internal structures depends on temperature. If the temperature of a sample deformed at temperature T_2 to point B on the T_2 stress–strain curve is changed suddenly to temperature T_1, the flow stress jumps to the value associated with point D on the T_1 stress–strain curve. Further deformation will continue along the T_1 curve.

Now let us assume that the internal structures which develop during deformation do vary with deformation temperature, but that temperature has no effect on the flow stress of crystals with similar structures. Under these conditions a crystal which has been deformed to point B at temperature T_2 will continue along the dotted curve starting from B when the specimen temperature is changed to T_1. The slopes of the

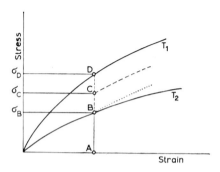

Fig. 18. Schematic stress–strain curves measured at a constant strain rate at temperatures T_1 and T_2 (after COTTRELL and STOKES [1955]).

dotted curve at B and the T_1 curve at D differ because, under the present assumptions, the internal structures of samples at points B and D are dissimilar.

If the divergence of the stress–strain curves arises from both of the causes just discussed, a sample whose temperature is changed from T_2 to T_1 at point B will continue along the dashed curve of fig. 18. The stress required to cause further plastic deformation will be raised to point C, which lies between B and D.

The ratio $(\sigma_C/\sigma_B)/(\sigma_D/\sigma_B)$ measures the relative contributions of the two fundamental sources of divergence to the observed variation in the stress–strain curves. If this ratio is unity, then all of the divergence arises from the temperature variation of the stress required to cause deformation in crystals with identical internal structures. If the ratio σ_C/σ_B is equal to unity, then all of the divergence is caused by a temperature variation in the internal structure which is developed with deformation. For aluminium it was found that when $T_1 = 78$ K and $T_2 = $ room temperature, the ratio σ_C/σ_B lies between 1.4 and 1.25. For copper measured at the same two tempertures σ_C/σ_B is between 1.25 and 1.1. The ratio $(\sigma_C/\sigma_B)/(\sigma_D/\sigma_B)$ can be as small as $1/2$ to $1/3$ in these metals. It would appear that at large strains, at least, a change in structure with test temperature is the predominant cause of the temperature dependence of the stress–strain curves of copper and aluminium.

COTTRELL and STOKES [1955] discovered an interesting relationship, the *Cottrell–Stokes Law*, which is concerned with the effect of temperature change on tests conducted along the lines of those illustrated in fig. 18. From experiments carried out on aluminium they deduced the following relationship:

$$\left(\frac{\sigma_2}{\sigma_3}\right)_u \left(\frac{\sigma_3}{\sigma_1}\right)_v = \left(\frac{\sigma_2}{\sigma_1}\right)_w = \text{const.} \tag{5}$$

The quantities appearing in the ratio $(\sigma_2/\sigma_3)_u$ are the flow stresses σ_2 and σ_3 which are measured on a specimen when its temperature is changed suddenly from T_2 to T_3. The test is carried out at a constant strain rate. The subscript u refers to the fact that the temperature change occurs when the strain of the specimen is u. Similarly the ratio $(\sigma_3/\sigma_1)v$ is obtained by changing the temperature of a specimen deformed to a strain v from T_3 to T_1, and $(\sigma_2/\sigma_1)_w$ by changing the temperature from T_2 to T_1 when the specimen strain is w. Equation (5) may be valid only beyond stage I in the stress–strain curves of fcc crystals. It is to be noted that eq. (5) is *independent* of the actual strain of a specimen.

The following is a more restricted form of the Cottrell–Stokes law:

$$\Delta\sigma/\sigma = \text{const.}, \tag{6}$$

where $\Delta\sigma$ is the change in flow stress caused by a sudden change in temperature when a sample is deformed at a constant strain rate. The stress σ can be chosen to be either the stress acting before or the stress acting after the change in temperature. The values of the initial and final temperatures remain fixed throughout the series of tests. Equation (6), like eq. (5), is independent of the strain. HAASEN [1958] has generalized eq. (6) to the following equation:

$$\sigma_1 - \sigma_2 = a + b\sigma_2, \tag{7}$$

References: p. 1305.

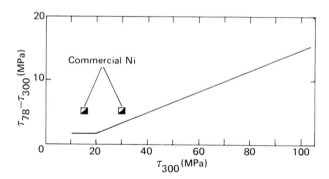

Fig. 19. Changes in the flow stress of pure nickel caused by a change in temperature between 78 K and room temperature, plotted against stress. Constant strain rate. (Data of HAASEN [1958].)

where a and b are constants and σ_1 and σ_2 are the flow stresses at temperatures T_1 and T_2. At large stresses this equation reduces to eq. (6). Figure 19 shows data obtained on nickel, which illustrate this last form of the Cottrell–Stokes law.

The Cottrell–Stokes law may not be valid for bcc metals. BASINSKI and CHRISTIAN [1960] have reported the following equation for iron,

$$\Delta\sigma \approx \text{const.,} \tag{8}$$

where $\Delta\sigma$ is the change in flow stress caused by a sudden temperature change. Again, this law is independent of the strain of the specimen.

In the case of hcp metals the Cottrell–Stokes law was found to be valid for cadmium (DAVIS [1963]) and magnesium (BASINSKI [1960]). The relationship holds in both the easy-glide region and the region of increased hardening at large strains. These results imply that the Cottrell–Stokes law should be valid in the easy-glide region of fcc crystals (see BASINSKI and BASINSKI [1979] and MECKING and KOCKS [1981]).

4.   Work-softening

In the course of their investigations STOKES and COTTRELL [1954] observed a curious effect in the experiments which involved changing the temperature from a lower to a higher value. (Because of this effect the data for the Cottrell–Stokes Law discussed in the last section always are obtained by lowering the temperature and never by raising the temperature.) They discovered the yield-point phenomenon illustrated in fig. 20. Evidently the dislocation structure which is introduced into a sample when it is deformed at a low temperature becomes unstable as the specimen temperature is increased. This instability results in the appearance of a yield point when the sample is further strained at the higher temperature. Stokes and Cottrell were able to prove that this phenomenon does not arise from the usual source of

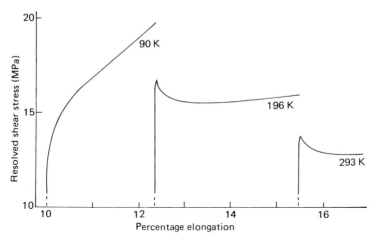

Fig. 20. Yield drops in aluminium observed at 196 K after strain at 90 K, and at 293 K after strain at 196 K (data of COTTRELL and STOKES [1955]).

yield point effects, namely, the diffusion of point defects or impurity atoms to dislocation lines and the subsequent pinning of these dislocation lines. They gave their yield-point effect the descriptive name of *work-softening*.

4.1. Haasen–Kelly effect

A phenomenon very similar to work-softening was discovered and thoroughly investigated by HAASEN and KELLY [1957]. They showed that a small yield-point effect can be introduced into a metal sample deformed at a relatively low temperature, merely by unloading, or partly unloading, the sample. When the stress is reapplied a yield point, illustrated schematically in fig. 21, is often observed.

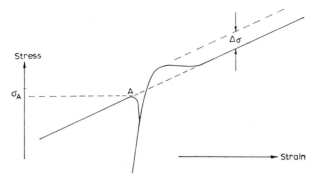

Fig. 21. The Haasen–Kelly yield point (from HAASEN and KELLY [1957]).

References: p. 1305.

4.2. Very large plastic deformation

The plastic deformation of a polycrystalline metal typically obeys the parabolic equation (see fig. 1)

$$\sigma = \sigma_0 \epsilon^m, \tag{9}$$

where ϵ is the plastic strain and σ_0 and m are constants. Typically m is about 0.3. This equation applies past the Lüders strain for material with an upper and lower yield point. It is quite simple to show that in a tensile test a specimen will neck down and fail once $\epsilon \geqslant m$. (The applied load reaches a maximum value at $\epsilon = m$. In an incipient neck, where the cross-sectional area is slightly smaller than elsewhere in the specimen, this strain value is reached sooner. Thus the neck region can continue to deform plastically as the stress in it increases while the load decreases. Elsewhere the stress is never quite large enough to continue the plastic deformation. For a discussion of the dependence of the necking instability on strain rate see GHOSH [1977] and LIN et al. [1981].) An ordinary tensile test terminates itself at a relatively modest strain of $\epsilon = m$. Much larger plastic strains can be attained by resorting to various tricks (machining periodically to remove incipient necks; machining away incipient barrels on compression tests). True strains as large as $\epsilon = 10$ have been reached. All the very large plastic strain data have been reviewed by HECKER et al. [1982] (see also GIL SEVILLANO et al. [1980]). A surprising feature of much of the data is that the strain region in which the parabolic eq. (9) holds is of limited extent. At very large strains the true stress–true strain curves become linear. Figure 22 illustrates this linear behavior for iron tested in effective tension. (Tensile specimens were made of wire cold-drawn to various reductions in area.)

Another feature of large strain deformation is that eventually the stress–strain curve levels out if the strain is made large enough. A saturation flow stress exists. In other words, despite the low temperature essentially a steady-state creep situation

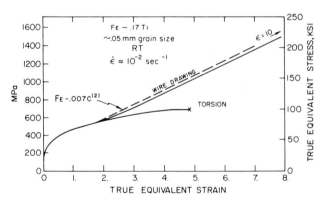

Fig. 22. Curves of true equivalent stress–true equivalent strain for polycrystalline iron tested in torsion and in tension (after wire drawing) at room temperature out to very large strains. Wire drawing data of LANGFORD and COHEN [1969]. (After YOUNG et al. [1973].)

exists. This steady-state creep normally is not seen because an unusually large plastic strain must first be produced before this flow regime is reached. The saturation flow stress is seen in fig. 22 for iron tested in torsion. Saturation is reached at a strain of about $\epsilon = 4$ in torsion. In tension no saturation occurs out to $\epsilon = 10$. This difference in the strain required to reach saturation most probably has its origin in the different grain-orientation textures that different deformation modes produce (GIL SEVIL-LANO et al. [1980], KOCKS and CANOVA [1983] and JONAS et al. [1983]).

5.   *Theories of work-hardening*

The reader learned in chapter 18 that the concept of dislocations in crystalline lattices was proposed independently by Taylor, Orowan and Polanyi. The paper of TAYLOR [1934] not only presented the concept of the dislocation but it also contained the first dislocation theory of work-hardening. A large number of work-hardening theories have been developed since that original theory appeared almost 50 years ago. The more recent workers have had the advantage of being able to profit from the results of actual electron microscope observations on dislocations. The theoreticians have addressed themselves to the task of explaining only a limited part of the work-hardening phenomenon. The main interest has been to explain the stress–strain curves of the fcc metals. The cause of this particular interest is the fact that more experimental data are available for these metals. In the sections that follow, an outline will be given of the types of work-hardening theories that have been considered.

5.1.   Long-range hardening

The theory of work-hardening developed by TAYLOR [1934] was the first in a series of such theories based on the long-range stress field of dislocations. The most detailed long-range work-hardening theories were developed in the late 1950s and early 1960s by Hirsch and by Seeger and their students and colleagues. This work is presented or summarized in the book "Work Hardening" and in the "Conference on the Deformation of Crystalline Solids" (see *Further reading*). Later theoretical developments in both long-range and short-range work-hardening theories have been summarized by HIRSCH [1975].

The stress field of a dislocation falls off as $\alpha\mu b/2\pi r$, where r is the radial distance from the dislocation line to any particular point in the crystal, μ is the shear modulus, b is the length of Burgers vector, and α is a constant. The value of α is 1 if the dislocation line has a screw orientation and is equal to approximately $3/2$ if the dislocation has an edge orientation. We shall assume that α is approximately 1.

If the distribution of the dislocations in a crystal is random with respect to both position and sign, the stress which all the dislocations in a crystal exert on any particular dislocation line must be of the order of $\mu b/2\pi r$, where r now is the mean spacing between dislocations. In other words, the stress field arising from all the dislocations except those closest to the particular dislocation must average out to zero. Since $r = 1/\sqrt{\rho}$, where ρ is the dislocation density in the crystal, the internal stress due to the dislocations can be rewritten as $\mu b(\rho)^{1/2}/2\pi$. Both the magnitude and the sign of the internal stress vary with position in the crystal. The wavelength of the internal stress is of the order of r.

Plastic deformation is produced by the motion of dislocations in crystals. In order to move the dislocations over appreciable distances, it is necessary to apply an external stress of the same magnitude as the maximum amplitude of the internal stress field of the crystal. If the maximum value of the internal stress field is approximately $\mu b(\rho)^{1/2}/2\pi$, the stress σ required to produce further plastic deformation is equal to

$$\sigma \approx \mu b(\rho)^{1/2}/2\pi. \tag{10}$$

This relationship can be checked experimentally. The dislocation density ρ can be determined from a count of etch pits. By carefully controlling the experimental conditions it is possible to produce an etch pit at each point on the surface of a crystal where a dislocation line terminates. The results of work by LEVINSTEIN and ROBINSON [1964] are shown in fig. 23. A plot is made in this figure of the square root of the dislocation density (i.e., the etch pit density) as a function of the flow stress of silver single crystals. The data points are taken from both stage I and stage II of the stress–strain curve. It can be seen that the functional form of eq. (10) is verified by this plot. Furthermore, if the measured values of μ and b are inserted into it, eq. (10) follows the experimental results very closely. For a critical evaluation of more recent data see BASINSKI and BASINSKI [1979].

As it stands, eq. (10) does not predict a stress–strain curve. Before a stress–strain curve can be obtained, the average distance a dislocation moves must be brought into the theory. Taylor assumed that each newly created dislocation moves a fixed distance L and then becomes stuck in the lattice. The dislocations in Taylor's theory are considered to be straight, parallel dislocations which extend from one side of a

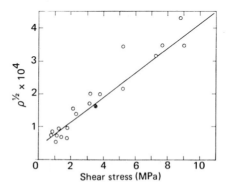

Fig. 23 Square root of dislocation density (in cm/cm³) versus shear stress in stages I and II for high-purity silver single crystals (data of LEVINSTEIN and ROBINSON [1964]).

crystal to the other. The distance L is assumed to be a constant throughout the course of deformation. The plastic strain ϵ is equal to $\rho b L$ and thus the dislocation density $\rho = \epsilon/bL$. Placing this expression in eq. (10) produces the following equation:

$$\sigma \approx (\mu b/2\pi)(\epsilon/bL)^{1/2}. \tag{11}$$

This equation predicts a parabolic relationship between stress and strain. Such a relationship is observed in the stress–strain curves of polycrystalline metals (see fig. 1).

Single crystals also can have a parabolic stress–strain curve when slip occurs on more than one set of slips planes. Figure 24 shows such a parabolic curve for aluminium. During deformation of polycrystalline samples slip generally occurs in each grain on more than one set of slip planes. Thus dislocations on one slip plane have to cut through dislocations on another active slip plane. This process has been described by saying that dislocations on one slip plane must cut through a *forest of dislocations* on other planes. The individual dislocations of the forest are referred to as *dislocation trees*. It requires energy for a dislocation to cut through a dislocation tree, just as cutting a real tree expends energy. A dislocation forest thus introduces a resistance to dislocation motion on other slip planes. The higher the density of the dislocation trees, the larger is this resistance. A crystal becomes harder as the dislocation content is increased on two or more sets of slip planes. This hardening does not depend on the long-range stress field of the dislocations.

It appears likely that the parabolic stress–strain curves of polycrystalline samples and of single crystals suitably oriented for slip on more than one plane may be explained by the dislocation forest mechanism. In such specimens the density of trees is high. The fact that eq. (11) also predicts a parabolic stress–strain curve points up the warning that a theory of work-hardening is not necessarily valid merely because it correctly predicts the shape of stress–strain curves.

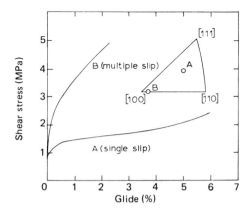

Fig. 24. Stress–strain curves of pure aluminium crystals (CUPP and CHALMERS's [1954] plot of data of LÜCKE and LANGE [1952]).

References: p. 1305.

5.2. Stage I hardening

Equation (11) does not explain the linear portions which comprise stage I and stage II of the stress–strain curves of fcc crystals. Neither does it explain the stress–strain curves of hcp metals. Seeger and his co-workers (the Stuttgart school of work-hardening) have attempted to explain these stress–strain curves by using the long-range hardening mechanism as the basis of their theories. The theory for stage I of fcc crystals or easy glide of hcp metals was developed as follows (SEEGER et al. [1961]). Let there be N dislocation sources per unit volume and let this number remain constant throughout easy glide. Let n be the average number of dislocation loops each source has produced when the stress level σ is reached. Because the stress level is low during easy glide the spacing between dislocation loops from the same source is large. Therefore these loops are not considered to be piled up.

An experimental fact now is introduced into the theory (which thus becomes partly phenomenological). It was observed with the electron microscope in studies made at Stuttgart that during easy glide the length of the slip lines and the spacing between them remains constant. On the basis of these observations Seeger et al. make the following assumption concerning the dimensions of the dislocation loops. They assume that the average linear dimension L of the ensemble of dislocation loops from a given source remains constant with time throughout easy glide.

Consider fig. 25. Shown here is a cross-section view of the slipped zones produced by the groups of n dislocation loops which have left each source marked by the letter S. Only a few of the dislocations are drawn in the figure. If the stress is raised by an increment $\delta\sigma$ each source will produce an additional number of loops δn. The resultant increase $\delta\epsilon$ in strains is given by

$$\delta\epsilon \sim bL^2 N\delta n \tag{12}$$

if the loops are taken to be roughly square in shape. All n loops will move a small amount when the stress is increased. The effect on the strain is the same as if the δn new loops moved a distance L while all the loops previously in existence remained stationary.

If d is the distance between slipped zones, by simple geometry the following approximate relationship must hold

$$N \sim 1/dL^2. \tag{13}$$

When δn new dislocation loops are created the back stress at the dislocation

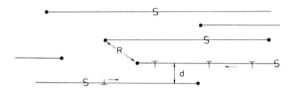

Fig. 25. Easy-glide dislocation model of SEEGER et al. [1961].

source is increased by an amount of the order of $(\mu b/2\pi L)\delta n$. When this increase in the back stress becomes equal to the stress increment $\delta\sigma$ no new dislocation loops can be produced at the source. Thus

$$\delta\sigma \sim \mu b\delta n/2\pi L. \tag{14}$$

Equations (12)–(14) may be combined to the following equation for the slope of the stress–strain curve in the easy-glide region:

$$\frac{d\sigma}{d\epsilon} = \frac{\mu}{2\pi}\left(\frac{d}{L}\right). \tag{15}$$

More refined calculations of Seeger *et al.* lead to a somewhat different expression for the slope

$$\frac{d\sigma}{d\epsilon} = \frac{8\mu}{9\pi}\left(\frac{d}{L}\right)^{3/4}. \tag{16}$$

For copper, d is of the order of 300 Å and L is approximately 600 μm. The substitution of these data into eq. (16) gives a value of 7.5 MPa for the slope of the stress–strain curve, a result which compares favourably with the measured value of 7.0 MPa.

In reviewing the theory of stage I hardening based on this model and similar models that were developed later, HIRSCH [1975] makes the point that no convincing hardening mechanism has been advanced which is based on interactions between the primary dislocations (in the case of low glide-plane friction stress). The dislocations of opposite sign in the model of fig. 25 form dipoles and multipoles. Hirsch points out that an excess of dislocations of one sign can easily move the dipoles and multipoles which presumably are locked in place. Other types of obstacles are needed to trap the multipoles. Hirsch concludes that stage I hardening still is not well understood.

5.3. Stage II hardening

The Stuttgart School (SEEGER [1957], SEEGER *et al.* [1957]) use the same dislocation model for stage II as for stage I. One change must be made however. It is assumed that the dislocations in stage II, unlike those of stage I, are piled up. The pile-ups are caused by the formation of Lomer–Cottrell locks (see ch. 18, §4.5). The stress field of n piled-up dislocations, as viewed from some distance away, is approximately the same as that of a single giant dislocation of Burgers vector nb located at the edge of the pile up. (Work of Hirsch and his students, published in "Work Hardening" and "Conference on Deformation of Crystalline Solids", has shown that the stress field of a piled-up group of dislocations is smaller than the value estimated by Seeger. Slip on secondary planes leads to a reduction of the long-range, stress field of a pile-up. The stress relief can be considered as arising from an "elastic polarization", a term used by NABARRO *et al.* [1964].) An experimental fact again is placed in the theory. It is the electron microscope observation

References: p. 1305.

that the average length L of a slip line in stage II changes according to the relationship

$$L = \Lambda/(\epsilon - \epsilon^*), \tag{17}$$

where Λ is a constant which generally has a value of about 4×10^{-4} cm, ϵ is the strain, and ϵ^* is a constant whose value is slightly smaller than the strain at the start of stage II.

Let R be the average distance between the ends of slipped zones (see fig. 25). At a distance R from the edge of a pile-up, the stress field of the pile-up is equal to

$$\mu nb/2\pi R \approx \mu nb (NL)^{1/2}/2\pi. \tag{18}$$

Here N is the density of active dislocation sources. It can be shown that the number n of dislocation loops in a slipped zone is proportional to the product of the applied stress and the length of the slipped zone (see ch. 18). Therefore n may be considered to remain constant as deformation proceeds since, according to eq. (17), the length of a slipped zone varies approximately as the inverse of the strain and thus approximately as the inverse of the stress. (We are anticipating the final result that the stress is proportional to strain.)

If the applied stress σ is set equal to the internal stress produced by a dislocation pile-up at a distance R, a change in strain $\delta\epsilon$ will produce a change in stress $\delta\sigma$ given by

$$\delta\sigma = (\mu nb/4\pi)\left[(dN/d\epsilon)(L/N)^{1/2} + (dL/d\epsilon)(N/L)^{1/2}\right]\delta\epsilon. \tag{19}$$

A change in strain $\delta\epsilon$ can be related to a change in the number of active dislocation sources through the equation

$$\delta\epsilon = bnL^2\delta N. \tag{20}$$

If eqs. (17), (18) and (20) are put into eq. (19), the following equation results

$$\frac{d\sigma}{d\epsilon} = \left(\frac{\mu^2 bn}{8\pi^2\Lambda}\right)\left(\frac{\epsilon - \epsilon^*}{\sigma}\right) - \frac{\sigma}{2(\epsilon - \epsilon^*)}. \tag{21}$$

An obvious solution of this equation is $\sigma = \theta(\epsilon - \epsilon^*)$, where θ is the (theoretical) slope of the stress–strain curve in stage II. If this solution is substituted into eq. (21) the value of θ can be found. It is

$$\theta = (\mu/2\pi)(bn/3\Lambda)^{1/2}. \tag{22}$$

The theoretical slope agrees with the experimental observations if the value of Λ mentioned previously is substituted into eq. (22) and if n is taken to be approximately 20–30. Since slip lines having step heights between $20b$ and $30b$ are found in stage II it is considered that there is satisfactory agreement between theory and experiment.

5.4. Stage III hardening

According to the Stuttgart school, stage III commences when the stress becomes sufficiently great for screw dislocations to transfer onto other slip planes. A perfect dislocation in a fcc crystal normally splits up into two partial dislocations with a stacking-fault ribbon between them (ch. 18, §3.3). Because of this extension a screw dislocation is confined to the slip plane which contains the stacking-fault ribbon. A screw dislocation can move onto another slip plane if a way can be found to transfer the stacking-fault ribbon to the new plane. The process of slip continued on a different slip plane is called *cross-slip*. Figure 26 shows one of several mechanisms which have been proposed to explain the start of cross-slip by a screw dislocation. A portion of an extended screw dislocation pinches together to form a segment of perfect dislocation line. This segment then separates into two partial dislocations on a different slip plane (see also ch. 18, §4 and fig. 21).

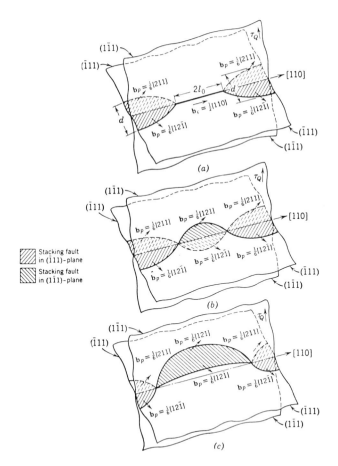

Fig. 26. Cross-slip by a screw dislocation (SEEGER [1957]).

References: p. 1305.

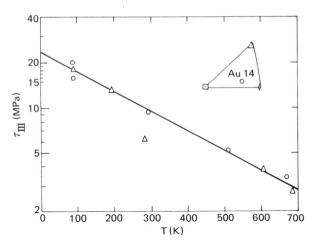

Fig. 27. Stress at start of stage III versus temperature for gold crystals (data of SEEGER *et al.* [1953]).

The work-hardening rate can be expected to decrease when cross-slip commences. The movement of screw dislocations onto cross-slip planes relaxes the high stress fields produced by dislocation pile-ups. The cross-slip process shown in fig. 26 and similar mechanisms involve an expenditure of energy. It is difficult for them to function at low stress levels. The cross-slip mechanisms require high stresses. Their operation is aided by thermal stress fluctuations. Consequently the stress required to initiate stage III should be temperature-dependent. Experimentally such is observed to be the case. The temperature dependence is shown in fig. 27. This figure is a semilog plot of the stress at the start of stage III versus temperature. It can be seen that the initial stress increases if the temperature is reduced.

5.5. Short-range hardening theories

Another viewpoint of the origin of work-hardening in metals will be presented in this section. The long-range stress field of dislocations no longer is considered to be of primary importance. Rather, hardening is attributed to much more local forces, that is, short-range interactions.

GILMAN's [1962] theory of hardening is an example of a short-range hardening theory. He believes that the dominant factor in the early part of a stress–strain curve is the "debris" left in the wake of moving screw dislocations. The debris consists of dislocation dipoles. A *dislocation dipole* is a pair of closely spaced parallel dislocations of opposite sign. Dislocation dipoles have been seen with the electron microscope in cold-worked crystals. A dislocation which is screw or partly screw in character produces dipoles at jogs. The jogs may have formed through the mechanism illustrated in fig. 28 (GILMAN [1962]). Dipoles also can be created at jogs which result from the intersection of two dislocations on different slip planes.

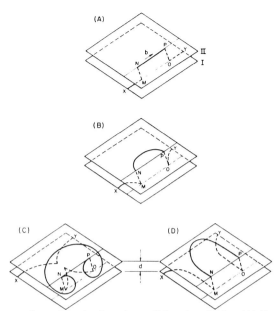

Fig. 28. Double cross-slip and production of two dislocation dipoles. (A) Screw dislocation cross-slips from I to II; (B) segment NP cross-slips onto II; (C) new loop forms if the stress is large but (D) dipoles form if the stress is not large enough to form the loop shown in (C). (After GILMAN [1962].)

The dipoles left behind by a screw- or partly screw dislocation interact with the dislocations which follow. Consequently these dislocations find it more difficult to move along the slip plane. The larger the amount of slip the greater is the amount of debris left on the slip planes and the more difficult does it become for further deformation to occur.

Another short-range hardening theory was developed by Kuhlmann-Wilsdorf (KUHLMANN-WILSDORF [1962, 1968] and KUHLMANN-WILSDORF *et al.* [1963]). She pictures the process of deformation in the easy-glide region as the gradual filling of a crystal with dislocations. The dislocation density is very non-uniform in this stage: some regions of the crystal have a high dislocation density while other regions are free of dislocations. She has electron microscope observations to support this picture. Stage I ends when all the previously empty regions of a crystal are filled with dislocations and the dislocation density is quasi-uniform throughout the sample.

She assumes that in stage II the dislocation density remains quasi-uniform but increases with increasing strain. If the dislocations are arranged in pile-ups at the end of stage I, during stage II they continue to be so arranged but the spacing between pile-ups decreases with increasing strain. On the other hand, if at the end of stage I the dislocations are disposed in irregular patterns called dislocation tangles (this is her preferred view of conditions at the end of stage I), the spacing between the tangles decreases during the course of stage II. Dislocation tangles commonly are

References: p. 1305.

observed in cold-worked metals. An electron micrograph of a tangle is shown in fig. 29. (The existence of dislocation tangles cannot be used as evidence that the long-range stress field is unimportant. Their origin also can be explained by a long-range stress theory of work-hardening, see DE WIT [1963], WEERTMAN [1963a].)

During stage II, work-hardening occurs because the segments of dislocation lines which act as Frank–Read sources become progressively smaller in length, i.e., the increase in dislocation density causes a reduction in the spacing between dislocations and thus in the free lengths of dislocation lines which can bow out to form new dislocation loops. The stress required to activate a dislocation source is inversely proportional to its length. Consequently, as stage II progresses, increasingly larger stresses are required to create new dislocations and cause further plastic deformation.

Kuhlmann-Wilsdorf's explanation of stage III is the same as Seeger's, namely, that it is dominated by cross-slip. She also would add the possibility that dislocation climb occurs in stage III. ALEXANDER and HAASEN [1961] have shown that stage III in germanium probably is caused by dislocation climb (ch. 21 – see also the large-strain behaviour of these crystals as recently observed by SIETHOFF and SCHRÖTER [1983] and BRION et al. [1981]).

Kuhlmann-Wilsdorf interweaves the development of her theory of work-hardening with a side theory on how dislocation tangles arise. She calls her mechanism for tangle formation a *mushrooming* process. According to her theory, a dislocation runs into the debris left by dislocations which move ahead of it. The debris consists of the dislocation dipoles considered by Gilman and any vacancies and small dislocation loops that may have been produced by the precipitation of vacancies. This debris attaches itself to straight dislocation lines and makes the lines crooked. Cross-slip also can aid the bending process. Eventually, initially straight dislocation lines become so deformed in shape that they are simply dislocation tangles. The hardening that occurs in stage I is thought to arise because tangle dislocations are difficult to move. This source of hardening is not considered to be important in stage II.

More recent work of Kuhlmann-Wilsdorf has focussed on dislocation-cell formation (KUHLMANN-WILSDORF [1977]). She has looked in detail into the geometric factors involved and how stress-strain curves can be explained through cell formation.

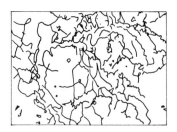

Fig. 29. Dislocation tangles in quenched, deformed aluminium (2% strain). Drawn from an electron micrograph of KUHLMANN-WILSDORF et al. [1963].

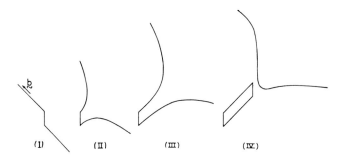

Fig. 30. Jog forming a dislocation dipole (after MOTT [1960] and SEGALL [1960]).

Another school of thought attributes a major portion of work-hardening in stage II to the presence of *jogs* on dislocation lines (ch. 18, §5.2). It is assumed that the jogs are formed when dislocation lines cut through a forest of immobile tangled dislocations. This theory has been developed by Hirsch and is described in detail by MOTT [1960]. HIRSCH [1962] has pointed out that in fcc metals jogs on moving dislocations may produce rows of point defects or dislocation dipoles in the manner shown in fig. 30. Only when it is moving in one of its two possible directions of motion does a jog create point defects or dipoles. The presence of jogs on a dislocation line decreases its mobility. As a result, jogs contribute to hardening. Since the jog density on a moving dislocation increases as the strain increases, the flow stress also increases with strain.

SAADA [1960] has proposed still another short-range hardening mechanism. Consider fig. 31. Saada points out that when a dislocation AA attempts to cross the tree dislocation BB, it is often energetically favourable for a dislocation reaction to occur. The two dislocations AA and BB may combine to form a segment of a third dislocation labelled CC in this figure. The tree dislocation must have some mobility for this process to take place. Reactions such as that shown in fig. 31 provide more effective obstacles to dislocation motion than do the intersections of two dislocations.

The most recent and ambitious short-range hardening theories have been developed by ARGON and by KOCKS, together with ASHBY. They are described at length

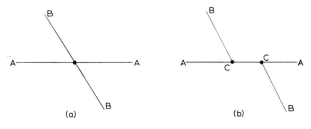

Fig. 31. (a) A dislocation AA on the slip plane crossing a tree dislocation BB. (b) The mobile and the tree dislocations have combined between CC to produce a third dislocation.

References: p. 1305.

in a book (KOCKS *et al.* [1975]) which also reviews the work of others on this and other topics. Statistical probability and thermodynamic arguments are featured strongly in the analysis.

5.6.   Dislocation cells

At large strains, dislocations in cold-worked metals take on a cell structure. Figure 32 is an electron micrograph of this structure obtained at a small strain where it is easily resolved. The dislocations are located primarily in the cell walls. The interior of the cells is relatively free of dislocations. KUHLMANN-WILSDORF [1977] has studied cell formation extensively. (See also references to her earlier work cited in this paper.)

Why dislocations seek to form cells is easy to understand from the original theory of their formation (HOLT [1970] and STAKER and HOLT [1972]). Suppose dislocations of density ρ were randomly distributed both spatially and by sign. Their average separation r is $1/\sqrt{\rho}$. The average self-energy per unit length of each dislocation thus is (cf. ch. 18, §2.1):

$$U \cong \left(\mu b^2/2\pi\right) \log(r/r_0),\tag{23}$$

where r_0 is the core radius. But suppose the dislocations were to move into cell walls

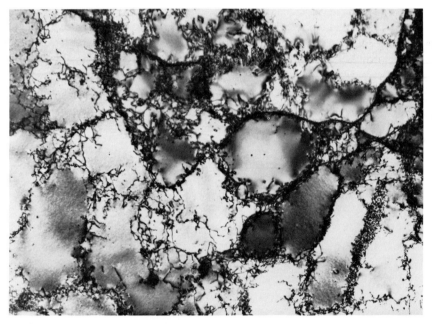

Fig. 32. Transmission electron micrograph of Al showing dislocation cells after a 5%×5% biaxial strain. Magnification 14000 (Courtesy D.L. ROHR, Los Alamos National Laboratory.)

without mutually annihilating each other. Let the width of a cell wall be w and the dimension of a cell be R. The average dislocation density of the dislocations in the cell walls, ρ_w, thus is approximately:

$$\rho_w \approx \rho(R/w) \qquad (24)$$

and the average dislocation spacing r_w of the dislocations in a wall is $r_w \approx r(w/R)^{1/2}$. Hence the average self-energy of a dislocation in a cell wall is:

$$U = (\mu b^2/2\pi) \log(rw^{1/2}/r_0 R^{1/2}) \qquad (25)$$

The average energy has indeed been reduced.

Obviously, there is an upper limit to the size of a dislocation cell at a given stress level. Were a cell to be too large, new dislocations could be nucleated in the cell interior, reducing the cell size. A reasonable value for the cell size is a dimension something like an order of magnitude larger than the average dislocation spacing $r = 1/\sqrt{\rho}$. The cell size becomes smaller as the stress is increased because the dislocation density ρ increases with increasing stress [see eq. (10)].

Dislocations in a cell wall can annihilate each other when dislocation-pipe diffusion (cf. ch. 20) permits climb of dislocations. WEERTMAN and HECKER [1983] have explained flow-stress saturation (fig. 22) at large strains as a pipe-diffusion controlled process that leads to a steady-state flow. They find that the saturation stress is given by

$$\sigma_s = \xi(Q_P - hkT)/\Omega \qquad (26)$$

where Q_P is the pipe diffusion activation energy, ξ is a dimensionless constant, and Ω is the atomic volume. The quantity h is stress-, temperature- and strain-rate dependent. The value of h also depends on the number of different types of Burgers vector the dislocations have in the cell walls. This number is the larger, therefore, the greater the number of active slip systems is. In tension the cell-wall dislocations come from more slip systems than in torsion. Hence the saturation stress is larger in tension that in torsion. (It is predicted that in tension the saturation stress in iron is reached near a strain of 10 at temperatures slightly greater than about 100°C.)

6.   *Theory of the temperature-dependence of the flow stress*

Figure 33 shows the results of a typical Cottrell–Stokes type of experiment. This experiment is designed to determine the temperature dependence of that part of the flow stress of a metal which is not caused by a change in internal structure. Each point was obtained by suddenly changing the temperature of the specimen from a standard initial value to a final value which varied from point to point. The temperature change was made at an arbitrary position on the stress–strain curve. The results were unaffected by the choice of this position. The data in fig. 33 are plotted as the ratio of the flow stress at the final temperature to the flow stress at the standard initial temperature versus the final temperature. Two curves are shown. In

References: p. 1305.

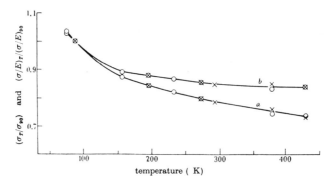

Fig. 33. Ratio of flow stress at temperature T to flow stress at 90 K for aluminium. The upper curve is compensated for the temperature variation of Young's modulus E (data of COTTRELL and STOKES [1955]).

the upper curve the data are corrected for the temperature-variation of the elastic constants. This curve is almost horizontal at the higher temperatures. In this temperature range, that part of the flow stress which does not depend on the change of structure with change in temperature is almost independent of temperature.

How are the data in fig. 33 to be explained? Hirsch's theory (MOTT [1960]) of work-hardening furnishes a simple explanation. There are two types of jog movement in Hirsch's theory. In one direction of motion a jog produces dipoles and point defects; in the other direction a jog can move without leaving defects in its wake (conservative motion). The force needed to move a dislocation in the former direction is larger than that required to produce conservative motion. At very low temperatures both types of jog motion harden the crystal. As the temperature is raised, thermal stress fluctuations aid the motion of the jogs. Since a lower force is required to move those jogs which do not produce point defects or dipoles, these are the first to be assisted in their motion by thermal stresses. The decrease in flow stress shown in fig. 33 reflects this assistance to the conservatively moving jogs. In the horizontal region of the curve the resistance offered by these jogs has become essentially zero, whereas the thermal stress fluctuations are still insufficient to move the defect-producing jogs. In this part of the curve, point defects are produced at jogs almost entirely by the applied stresses with very little aid from the thermal stresses. At still higher temperatures (not shown in the figure) the curve decreases again when thermal stress fluctuations are large enough to move both kinds of jogs. It can be seen that Hirsch's theory provides a natural explanation of the increase in hardening with decrease in temperature. (Hirsch concluded that in non-conservative jog motion only vacancies and no interstitials are formed. It can be shown that interstitials also may be produced, see WEERTMAN [1963b]. Hirsch's explanation can be generalized to take this later development into account.)

SEEGER's [1957] explanation of the temperature-dependence of the flow stress is based on an analysis of the contribution made by the forest dislocations to the total hardening. This contribution was ignored in the outline of Seeger's theory given above. Seeger first calculated the energy that is required to cut through the dislocation tree pictured in fig. 34. A mobile dislocation is shown hung up against the tree. The applied stress is not large enough to push the dislocation through the tree. Let Q be the additional energy which must be supplied by thermal stress fluctuations to the dislocation of fig. 34 in order to permit it to pass through the tree. Let the constant U_0 represent the value of this additional energy when no stress is applied to the crystal. If a stress σ assists in pushing the mobile dislocation, it does an amount of work equal to $l_0'd'b\sigma$ when the dislocation cuts the tree. In this expression, l_0' is the distance between trees as shown in the figure and d' is the distance the dislocation must move before the tree is cut. The energy Q can be set equal to

$$Q = U_0 - \nu\sigma, \tag{27}$$

where ν is $l_0'd'b$. The quantity ν may be regarded as an *activation volume*.

In a work-hardened crystal the stress which appears in eq. (27) must be replaced by an effective stress. The effective stress is equal to $\sigma - \sigma_b$, where σ is the applied stress and σ_b is the temperature-independent back stress which is produced by the long-range stress fields of the dislocations in the crystal.

The frequency with which trees are cut is proportional to $\exp(-Q/kT)$, which is the Boltzmann factor in any thermally activated process. In the case of a sample pulled at a constant strain rate $\dot\epsilon$, the strain rate is related to the effective flow stress through the following equation

$$\dot\epsilon = \dot\epsilon_0 \exp(-U_0/kT)\, \exp(\nu[\sigma - \sigma_b]/kT), \tag{28}$$

where $\dot\epsilon_0$ is a constant. This equation can be rewritten as

$$\sigma = \sigma_b - [U_0 - kT\log(\dot\epsilon_0/\dot\epsilon)]/\nu. \tag{29}$$

In the temperature range where $kT\log(\dot\epsilon_0/\dot\epsilon)$ is less than U_0, eq. (29) predicts that the flow stress will decrease as the temperature is increased. This temperature dependence matches the behaviour of the low-temperature end of the curve in fig. 33. At higher temperatures $kT\log(\dot\epsilon_0/\dot\epsilon)$ may be larger than U_0. Under these conditions the thermally activated stresses are so large that an applied stress no longer is needed to assist in cutting the forest dislocations. In this temperature range

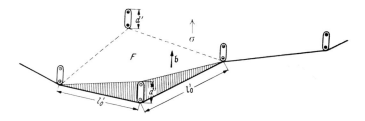

Fig. 34. A mobile dislocation hung up on dislocation trees (from SEEGER [1958]).

References: p. 1305.

the applied stress has only the back stress σ_b to oppose:

$$\sigma = \sigma_b. \tag{30}$$

This equation describes the right-hand side of the curve in fig. 33 where the flow stress is independent of temperature.

The temperature-dependent part of Seeger's theory also can be used to explain the temperature dependence of the critical resolved shear stress of well annealed crystals.

In many of the work-hardening theories, the Cottrell–Stokes Law is explained by establishing some relationship between the temperature-independent contribution σ_i to the flow stress and the temperature-dependent contribution σ_d. The total flow stress is the sum $(\sigma_i + \sigma_d)$. In Hirsch's theory, for example, the conservatively moving jogs contribute to σ_d and the non-conservatively moving jogs contribute to σ_i in the temperature range covered in fig. 33. If the ratio between the numbers of the two types of jogs is constant, the Cottrell–Stokes Law follows.

7. Lüders bands and yield points

The stress–strain curve obtained by pulling a single crystal or a polycrystalline sample of a slightly impure bcc metal, contains both an upper and a lower yield point (see fig. 12). This phenomenon of a *yield drop* is not unique to bcc metals. It has been observed in the stress–strain curves of impure metals with other crystal structures.

At the lower yield point, plastic deformation occurs through the propagation of a type of plastic wave called a *Lüders band* or Lüders strain. Figure 35 illustrates a Lüders band. It is a localized region of a sample which has undergone plastic deformation. As a consequence of this deformation, the cross-section of the sample

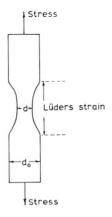

Fig. 35. Lüders strain in a specimen.

is smaller in the neighbourhood of the Lüders strain. The elongation which results from the spreading of a Lüders band throughout a specimen produces the lower yield-point portion of the stress–strain curve. Since a Lüders band grows at an almost constant stress level there is no indication of any strain-hardening in the stress–strain curve until the spread of the band is complete.

At the upper yield point the stress–strain relationship is essentially discontinuous. Measured on a specimen of macroscopic size, this discontinuity appears on the stress–strain curve as a region in which the strain is a multivalued function of the stress. It is this instability in the flow stress which makes possible the occurrence of a Lüders band. A sample that is pulled to the upper yield point supports a load σ_u per unit area, where σ_u is the upper yield stress. Upon any further increase in load instability sets in. Some small region of the sample undergoes a drop in flow stress to the lower yield stress and deforms. This deformed region, which is the beginning of a Lüders band, spreads into adjacent portions of the crystal. So long as the band continues to propagate, the specimen elongates without a change in load.

It is necessary that an upper yield point exist before a Lüders band can occur. The origin of an upper yield point is not difficult to explain. Upper yield points almost always are observed in deformation tests on impure metals or alloys (work-softening and the Haasen–Kelly effect are exceptions). This fact suggests that dislocations in impure metals are strongly pinned by impurity atoms which have diffused to the dislocations. Supporting evidence comes from the additional experimental observation that the upper yield point can be raised by ageing a sample at a temperature which is high enough to permit diffusion of impurity atoms to dislocation lines and yet is not so high as to cause dispersion of the impurity atoms which have segregated there.

The classic dislocation pinning mechanism was proposed by COTTRELL [1953] (see ch. 21). He showed that impurity atoms which differ in size from the host atoms of a lattice are attracted to a dislocation line. A force of attraction exists because the strain field of a dislocation can be partially cancelled by the strain field of the impurity atoms which arises because of their misfit in the lattice. The total elastic strain energy thus is reduced. The energy of interaction U of an impurity atom with a dislocation line is of the order of $4\mu|r_i - r|b^4/rR$, where r_i and r are the radii of the impurity and the host atoms, respectively, and R is the distance from the impurity atom to the dislocation line. The maximum value of U is found by setting $R = b$.

If c_0 is the average impurity concentration in a lattice, the equilibrium concentration c near a dislocation line reached by ageing at a temperature T is $c = c_0 \exp(U/kT)$. The excess of impurity atoms near a dislocation line is referred to as an *impurity cloud*. The impurity cloud pins the dislocation to it. At certain ageing temperatures or for certain values of U the impurity concentration close to a dislocation line actually will equal unity. The impurity atoms have precipitated out on the dislocation line. Such precipitates have been seen under the electron microscope (cf. also ch. 21).

Since the pinning effect of a Cottrell impurity cloud makes dislocation motion difficult, the presence of these clouds raises the flow stress of a crystal. The lower yield point of a crystal can be explained as the result of dislocations breaking away from their impurity clouds. Once freed from a cloud a dislocation can move at a lower stress level and thus produce plastic deformation at a stress which is less than the upper yield point. We see that the impurity-cloud concept leads to a straightforward explanation of the phenomenon of an upper and a lower yield point.

The development of another explanation for the upper and lower yield points was started by JOHNSTON and GILMAN [1959]. They noted that the stress–strain curves of lithium fluoride are characterized by both an upper and a lower yield point in the region where plastic deformation begins. However, the dislocations which already were in existence at the beginning of the tests were so strongly pinned that the applied stress never broke them away from their pinning impurity atoms. The pre-existing dislocations in LiF never moved during deformation. Newly created, unpinned dislocations produced the subsequent plastic deformation.

The Gilman–Johnston explanation for the upper and lower yield points was developed further by HAHN [1962]. The theory runs along these lines. Let a sample be pulled at a constant strain rate. The strain consists of an elastic component ϵ_e and a plastic component ϵ_p. The elastic part of the total strain rate is $\dot{\epsilon}_e = \dot{\sigma}/\mu$, where μ is an appropriate elastic constant and $\dot{\sigma}$ is the rate of change of the stress. The plastic contribution to the strain rate, $\dot{\epsilon}_p$, is equal to a term which is of the order of bLv, where L is the total length of unpinned dislocations within a unit volume of crystal and v is their average velocity of motion. From a study of moving dislocations in iron–silicon alloys, STEIN and LOW [1960] showed that in these alloys the stress dependence of the dislocation velocity v is $v = C\sigma^n$, where C and n are constants. The value of n is approximately 35. A similar dependence has been found in LiF.

The dependence of L on plastic strain can be established through the use of etch pit techniques (ch. 10). It is found that $L = C'\epsilon_p^\alpha + \lambda$, where C' is a constant, α is another constant which has a value between 0.7 and 1.5 for iron (HAHN [1962]), and λ is the length of any unpinned dislocations which are present initially.

The strain rate $\dot{\epsilon}$ of a sample thus can be written as

$$\dot{\epsilon} = \dot{\epsilon}_e + \dot{\epsilon}_p = (\dot{\sigma}/\mu) + bC(\lambda + C'\epsilon_p^\alpha)(\sigma - q\epsilon_p)^n, \tag{31}$$

where q is a constant. The term $q\epsilon_p$ is equal to the back stress acting on a dislocation. (It is assumed that the strain-hardening is linear.)

Because of the extremely high value of n ($n \approx 35$) the last term on the right-hand side of eq. (31) is either very much larger or very much smaller than the $\dot{\sigma}/\mu$ term. (N.B., a factor of 2 increase or decrease in the quantity $(\sigma - q\epsilon_p)$ changes the last term on the right-hand side by a factor of $10^{\pm 10.5}$.) Therefore the stress is either

$$\sigma \sim \mu\epsilon \tag{32a}$$

or

$$\sigma \sim q\epsilon + [\dot{\epsilon}/Cb(\lambda + C'\epsilon^\alpha)]^{1/n}. \tag{32b}$$

The variation of stress versus strain predicted by eq. (32) is shown in fig. 36. The total stress–strain curve is a combination of two curves. A lower yield point appears

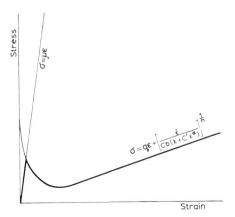

Fig. 36. Stress–strain curves given by eqs. (32) illustrating the explanation furnished by the theories of JOHNSTON and GILMAN [1959] and HAHN [1962] for the occurrence of an upper and lower yield point.

because the curve defined by eq. (32b) first decreases with increasing strain. The smaller the initial density λ of unpinned dislocations, the larger is the magnitude of the initial decrease (cf. also ch. 21, §4.2).

8. Dependence of the lower yield stress on grain size

The dependence of the lower yield stress on grain size is quite interesting. This dependence is illustrated in fig. 37. Here the lower yield stress of iron is plotted versus the inverse of the square root of the grain size.

The data follow the relationship

$$\sigma_{\text{lys}} = \sigma_0 + kd^{-1/2}, \tag{33}$$

where σ_{lys} is the lower yield stress, σ_0 and k are constants, and d is the average grain diameter. This law was discovered by HALL [1951], PETCH [1953] and LOW [1954]. It is known as the *Hall–Petch relationship* (see also ch. 10A, §4.2.2 and fig. 14). Equation (33) usually is valid for other bcc metals besides iron. It also is obeyed by fcc and hcp metals (GILBERT *et al.* [1963]). Sometimes the subgrain size rather than the grain size should be substituted into eq. (33) (WARRINGTON [1963]). For a recent survey see ARMSTRONG [1983].

PETCH [1953] has explained the grain-size dependence as follows. He assumes that all dislocations are pinned and immobile until an applied stress reaches the upper yield point. At the upper yield point plastic deformation commences first in a single grain. Yielding takes place in this grain when one dislocation source becomes unpinned and sends out dislocation loops. The dislocation loops move to the grain boundary, where they are stopped. These arrested dislocations are shown in fig. 38.

The dislocations which are piled up at a grain boundary produce a stress concentration in the adjacent grains. If this stress concentration is sufficiently large a

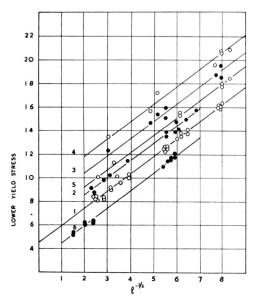

Fig. 37. Dependence of lower yield stress (in units of 15.2 MPa) on inverse of the square root of grain size (in mm$^{-1/2}$) for: (1) annealed mild steel, (2) nitrided, (3) quenched from 650°C, (4) quenched and aged 1 h at 150°C, (5) quenched and aged 100 h at 200°C, (6) annealed Swedish iron. (Data of CRACKNELL and PETCH [1955].)

source in a neighbouring grain will become activated and throw off dislocation loops. These new loops in turn will produce a stress concentration in another grain and cause the start of plastic deformation in it. Plastic yield thus spreads from one grain to the next and in this manner a Lüders band propagates across the sample.

Consider a dislocation pile-up such as pictured in fig. 38. It can be shown from dislocation theory (ch. 18) that the stress acting at a distance l from the head of a dislocation group piled up by an applied stress σ is roughly $\sigma(d/4l)^{1/2}$ when l is

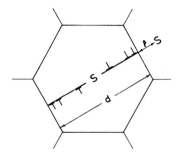

Fig. 38. Source in a grain of diameter d. The source has created dislocation loops which are piled-up at the grain boundaries.

smaller than $d/2$. The quantity $d/2$ is the width of the pile-up of dislocations of like sign. If a "frictional" force acts on the dislocation lines, the concentrated stress at the distance l is reduced to $\sigma_f + (\sigma - \sigma_f)(d/4l)^{1/2}$, where σ_f is the frictional stress. This frictional stress can arise from randomly spaced impurity atoms on the slip plane. It also can be produced by a Peierls stress. The frictional stress reduces the number of dislocations in a pile-up and thus also reduces the stress at the distance l from the pile-up.

Suppose σ' is the stress which must act on a pinned source in order to unlock it. Let l be the average distance from the head of a pile-up to the closest dislocation source in a neighbouring grain (see fig. 38). The stress acting on the pinned source is $\sigma_f + (\sigma - \sigma_f)(d/4l)^{1/2}$, which is the sum of the applied stress σ and the stress due to the pile-up. The source is unlocked when

$$\sigma' = \sigma_f + (\sigma - \sigma_f)(d/4l)^{1/2}. \tag{34}$$

Eq. (34) can be rearranged to give σ as a function of l:

$$\sigma = \sigma_f + (\sigma' - \sigma_f)(4l/d)^{1/2}. \tag{35}$$

This equation has the same form as the experimentally determined eq. (33).

9. *Bauschinger effect*

The stress–strain curves that have been considered up to this point were obtained by increasing the stress monotonically. Let us suppose now that at some point on a stress–strain curve the direction of the applied stress is reversed, that is, a tensile test is turned into a compressive test, or in a shear test the direction of shearing is reversed.

The plastic deformation which is produced by decreasing the stress on a sample to zero and reapplying it in the reverse direction differs from the deformation which results from simply unloading the sample and then reloading it to the original stress. The behaviour of the deformation under the latter conditions already has been described (see fig. 7). When the load on a sample is reversed, the metal deforms more easily in the reverse direction than it would have in the forward direction. A typical stress–strain curve during reversed stressing is shown in fig. 39. The dashed curve of this figure depicts the type of deformation previously shown in fig. 7 in which a specimen is unloaded and then reloaded in the same direction. The solid curve shows the course of deformation if the reloading is done in the opposite direction. Plotted as the ordinate of this curve is the absolute value of the stress. The strain coordinates indicate the total strain less the strain which exists when the specimen is unloaded. However, in order to facilitate comparison between the reloaded and the reversed-stress curves, the latter is plotted as a function of the negative of this strain difference. It can be seen that a metal sample deforms more easily if the stress is reversed than if it merely is reapplied in the same direction. This difference in plastic behaviour is known as the Bauschinger effect. The effect

References: p. 1305.

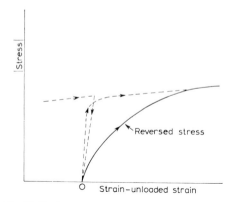

Fig. 39. Typical Bauschinger stress–strain curve.

generally is larger for single crystals than for polycrystalline material.

The existence of the Bauschinger effect obviously implies that in cold-worked crystals it is easier to move dislocations in the direction opposite to that in which they were pushed during the original plastic deformation. This behaviour is not surprising. If at least an appreciable fraction of cold-work arises from long-range interactions of dislocations, it is reasonable to expect that these interactions would help the dislocations move in a reversed direction of motion. The back stress of a dislocation pile-up is directed towards making the dislocation loops return to their source and thereby reversing the plastic strain. The Bauschinger effect is of particular importance in cyclic deformation (ch. 24) and in dispersion-hardened systems, where the dislocations stored in the particles carry the memory of the direction of shear (BROWN [1977]).

10. Mechanical twinning

Sometimes the stress–strain curves of metal single crystals are serrated. Figure 40 shows such a curve for a copper crystal deformed at 4.2 K. The stress required to cause further deformation suddenly falls, rises and then falls again. The stress–strain curve takes on the appearance of saw-tooth curve. These serrations in the deformation curve are caused by the phenomenon of *mechanical* or *deformation twinning*.

Normally deformation of a crystal occurs by the passage of perfect dislocations across slip planes. The movement of a single perfect dislocation across a slip plane shears the crystal by an atomic distance. A single crystal remains a single crystal despite the movement of one or of many perfect dislocations. Although atoms move with respect to each other across a slip plane, nevertheless they move into crystallographically equivalent positions.

When partial dislocations move across various slip planes during deformation, the atoms on either side of the slip planes are not moved into equivalent positions. If

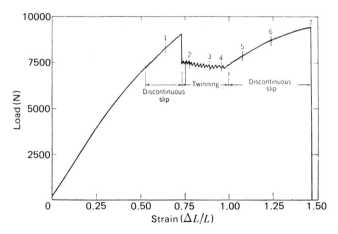

Fig. 40. Load versus strain for a copper crystal pulled in tension at 4.2 K. Curve shows deformation due to twinning. (Data of BLEWITT *et al.* [1957].)

one partial dislocation were to move across each individual slip plane of a set of parallel slip planes the orientation of the crystal actually would be changed. Figure 41 illustrates how the stacking sequence of close-packed planes in an fcc lattice is changed if the partial dislocations shown on the left-hand side of fig. 41a move across the planes and end up on the right-hand side of fig. 41b. The stacking sequence ABCABCABC is changed to ABCA/CBACB, where / indicates a fault in the stacking sequence. The orientation of the top half of the crystal in fig. 41b differs from that of the bottom half. Note that in the original state represented by fig. 41a the crystal was single.

The two crystals which form the bicrystal of fig. 41b are said to be in a *twin orientation* with respect to each other. A bicrystal with the same relative orientation between its component crystals can be formed from sections of the two identical fcc single crystals shown in fig. 42. To make a twinned crystal, the planes EFG and E′F′G′ are joined by superimposing E on E′, F on F′, and G on G′. The extra plane of atoms at the boundary then is removed. It can be seen that one crystal is the mirror image of the other.

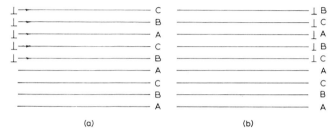

Fig. 41. Twinning in an fcc crystal by the passage of partial dislocations across the slip planes.

References: p. 1305.

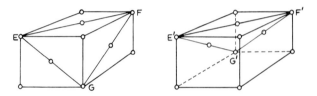

Fig. 42. A twin orientation is produced in a fcc crystal by joining planes EFG and E'F'G'.

The serrated part of the load–strain curve of fig. 40 is produced by the type of motion of partial dislocations illustrated in fig. 41. This motion changes the orientation of at least a part of the crystal into a twin orientation. The process of creating a crystal with a twin orientation by the application of stress is known as mechanical twinning or twin deformation.

Mechanical twins are made in fcc crystals by slip across {111} planes. The amount of slip across each plane is not equal to an atomic distance. In bcc crystals twins occur as the result of the slip produced by motion of partial dislocations on the {112} planes. In the case of hcp metals, the pyramidal {10$\bar{1}$2} planes are the most common, though not the only, twinning planes.

Mechanical twinning is an important mode of deformation both in bcc metals (particularly at low temperatures) and in hcp metals. It is not very important in the deformation of fcc metals. In fact it is very difficult to mechanically twin fcc crystals.

An interesting unresolved problem concerning mechanical twinning is the question of whether or not there exists a critical stress for the formation of deformation twins. We already have seen that a critical shear stress does exist for the start of ordinary deformation. It has not yet been conclusively shown whether there is an analogous critical shear stress for the start of mechanical twinning.

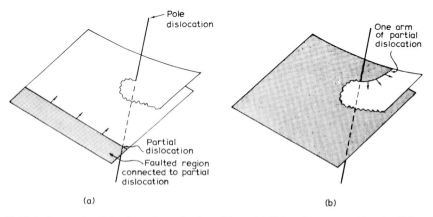

Fig. 43. Twinning produced by the pole mechanism: (a) partial dislocation approaching the dislocation pole; (b) two arms of the partial dislocation beginning to spiral around the pole.

In order to make a twin, a partial dislocation has to pass across each of a set of parallel crystallographic planes. This requirement presents the problem of how to accomplish the desired movement of partials. The pole mechanism illustrated in fig. 43 is a likely answer. As shown in this figure, a partial dislocation approaches a tree dislocation which is at least partially screw in character. The atomic planes form a ramp about the screw or partially screw tree dislocation. The tree dislocation itself acts as the pole for the spiral ramp. When the partial dislocation meets the pole dislocation one arm of the partial dislocation will move up one plane and the other arm will move down a plane. If the two arms are able to swing past each other they will spiral around the pole dislocation. Eventually one partial dislocation will have moved across each of the parallel planes in succession and a mechanical twin will have been created (cf. also ch. 21).

References

AKHTAR, A., 1975, Metallurg. Trans. **6A**, 1105.

ALEXANDER, H., and P. HAASEN, 1961, Acta Metall. **9**, 1001.

ARMSTRONG, R.W., 1983, in: Yield, Flow and Fracture of Polycrystals, ed. T.N. Baker (Appl. Science Publ., Barking, Essex) in press.

ARSENAULT, R.J., 1975, Treatise on Materials Science and Technology **6**, 1.

BARRETT, C.S., and T.B. MASSALSKI, 1980, Structure of Metals (Pergamon Press, Oxford).

BASINSKI, S.J., and Z.S. BASINSKI, 1979, in: Dislocations in Solids, vol. 4, ed. F.R.N. Nabarro (North-Holland, Amsterdam) p. 261.

BASINSKI, Z.S., 1960, Austr. J. Phys. 13, 284.

BASINSKI, Z.S., and J.W. CHRISTIAN, 1960, Austr. J. Phys. **13**, 299.

BLEWITT, T.H., R.R. COLTMAN and J.K. REDMAN, 1957, Dislocations and Mechanical Properties of Crystals (Wiley, New York) p. 179.

BRION, H.G., H. SIETHOFF and W. SCHRÖTER, 1981, Phil. Mag. **43**, 1505.

BROWN, L.M., 1977, Scripta Metall. **11**, 127.

CHANG, L.N., G. TAYLOR and J.W. CHRISTIAN, 1983, Acta Metall. **31**, 37.

CHRISTIAN J.W., 1970, in: 2nd Int. Conf. on the Strength of Metals and Alloys (ASM, Metals Park, OH) p. 29.

CHRISTIAN, J.W., 1983, Campbell Memorial Lecture, Sept. 1982, St. Louis, to be published in Metallurg. Trans.

COTTRELL, A.H., 1953, Dislocations and Plastic Flow in Crystals (Clarendon Press, Oxford).

COTTRELL, A.H., and R.J. STOKES, 1955, Proc. Roy. Soc. (London) **A233**, 17.

CRACKNELL, A., and N.J. PETCH, 1955, Acta Metall. **3**, 186.

CUPP, C.R., and B. CHALMERS, 1954, Acta Metall. **2**, 803.

DASH, J., and M.E. FINE, 1961, Acta Metall. **9**, 149.

DAVIS, K.G., 1963, Can. J. Phys. **41**, 1963.

DE WIT, R., 1963, Trans. Met. Soc. AIME **227**, 1443.

FOXALL, R.A., M.S. DUESBERRY and P.B. HIRSCH, 1967, Can. J. Phys. **45**, 607.

GARSTONE, J., and R.W.K. HONEYCOMBE, 1957, Dislocations and Mechanical Properties of Crystals (Wiley, New York) p. 391.

GHOSH, A.K., 1977, Acta Metall. **25**, 1413.

GILBERT, A., C.N. REID and G.T. HAHN, 1963, Refractory Metals and Alloys, vol. 2 (Interscience, New York) discussion, p. 16.

GILMAN, J.J., 1962, J. Appl. Phys. **33**, 2703.

GIL SEVILLANO, J., P. VAN HOUTTE and E. AERNOUDT, 1980, Prog. Mater. Sci. **25**, 69.

HAASEN, P., 1958, Phil. Mag. **3**, 384.

HAASEN, P., and A. KELLY, Acta Metall. **5**, 192.

HAHN, G.T., 1962, Acta Metall. **10**, 727.

HALL, E.O., 1951, Proc. Phys. Soc. **B64**, 747.

HECKER, S.S., M.G. STOUT and D.T. EASH, 1982, Proc. of Plasticity of Metals at Finite Strain: Theory, Computation and Experiment, eds. E.H. Lee and R.L. Mallett (Stanford Univ. & Rensselaer Polytechnic Inst.) p. 162.

HIRSCH, P.B., 1962, Phil. Mag. **7**, 67.

HIRSCH, P.B., 1968, Trans. Japan. Inst. Metals, **9**, Supplement, p. XXX.

HIRSCH, P.B., 1975, in: The Physics of Metals, 2. Defects, ed. P.B. Hirsch (Cambridge University Press) p. 189.

HOLT, D.L., 1970, J. Appl. Phys. **41**, 3197.

HONEYCOMBE, R.W.K., 1968, The Plastic Deformation of Metals (St. Martin's Press, New York).

JAOUL, B., and D. GONZALES, 1961, J. Mech. Phys. Solids **9**, 16.

JOHNSTON, W.G., and J.J. GILMAN, 1959, J. Appl. Phys. **30**, 129.

JONAS, J.J., G.R. CANOVA, S.C. SHRIVASTAVA and CHRISTODOULOU, 1982, Proc. of Plasticity of Metals at Finite Strain: Theory, Computation and Experiment, eds. E.H. Lee and R.L. Mallett (Stanford Univ. & Rensselaer Polytechnic Inst.) p. 206.

KOCKS, U.F., and G.R. CANOVA, 1981, Deformation of Polycrystals: Mechanisms and Microstructure, Second Risø International Symposium, Risø, Denmark, eds. N. Hansen, A. Horsewell, T. Leffers and H. Lilholt (Risø National Laboratory, Denmark) p. 35.

KOCKS, U.F., A.S. ARGON and M.F. ASHBY, 1975, Prog. Mater. Sci. **19**, 1.

KUHLMANN-WILSDORF, D., 1962, Trans. Met. Soc. AIME **224**, 1047.

KUHLMANN-WILSDORF, D., 1968, in: Work-Hardening, eds. J.P. Hirth and J. Weertman (Gordon and Breach, New York) p. 97.

KUHLMANN-WILSDORF, D., in: Work-Hardening in Tension and Fatigue, ed. A.W. Thompson (Metallurgical Society of AIME) p. 1.

KUHLMANN-WILSDORF, D., J. LEVINSTEIN, W.H. ROBINSON and H.G.F. WILSDORF, 1963, J. Austr. Inst. Metals **8**, 102.

LANGFORD, G., and M. COHEN, 1969, Trans. ASM **62**, 623.

LEVINSTEIN, H.J., and W.H. ROBINSON, 1964, The Relations Between Structure and the Mechanical Properties of Metals, Symposium at the National Physical Laboratory, Jan. 1963 (Her Majesty's Stationery Office) p. 180.

LIN, I.-H., J.P. HIRTH and E.W. HART, 1981, Acta Metall. **29**, 819.

LOW, J.R., Jr., 1954, Relation of Properties of Microstructure (ASM, Metals Park, OH) p. 163.

LÜCKE, K., and H. LANGE, 1952, Z. Metallk. **43**, 55.

MECKING, H., and U.F. KOCKS, 1981, Acta Met. 29, 1865.

MOTT, N.F., 1960, Trans. Met. Soc. AIME **218**, 962.

NABARRO, F.R.N., Z.S. BASINSKI and D.B. HOLT, 1964, The Plasticity of Pure Single Crystals, Adv. Phys. **13**, 193.

PETCH, N.J., 1953, J. Iron Steel Inst. **173**, 25.

ROGAUSCH, K.D., and B.L. MORDIKE, 1970, 2nd International Conference on the Strength of Metals and Alloys (ASM, Metals Park, OH) p. 16.

SAADA, G., 1960, Acta Metall. **8**, 841.

SEEGER, A., 1957, Dislocations and Mechanical Properties of Crystals (Wiley, New York) p. 243.

SEEGER, A., 1958, Encyclopedia of Physics, vol. 7/2 (Springer, Berlin).

SEEGER, A., R. BERNER and H. WOLF, 1953, Z. Phys. **155**, 247.

SEEGER, A., J. DIEHL, S. MADER and H. REBSTOCK, 1957, Phil. Mag. **2**, 323.

SEEGER, A., H. KRONMÜLLER, S. MADER and H. TRÄUBLE, 1961, Phil. Mag. **6**, 639.

SEGALL, R.L., 1960, Ph.D. thesis, Univ. of Cambridge.

ŠESTÁK, B., and A. SEEGER, 1978, Z. Metallk. **69**, 195, 355, and 425.

SIETHOFF, H., and W. SCHRÖTER, 1983, Scripta Metall. **17**, 393.
STAKER, M.R., and D.L. HOLT, 1972, Acta Metall. **20**, 569.
STEIN, D.F., and J.R. LOW, 1960, J. Appl. Phys. **31**, 362.
STEIN, D.F., J.R. LOW Jr. and A.U. SEYBOLT, 1963, Acta Metall. **11**, 1263.
STOKES, R.J., and A.H. COTTRELL, 1953, Acta Metall. **2**, 343.
TAYLOR, G.I., 1934, Proc. Roy. Soc. **A145**, 362.
VAN BUEREN, H.G., 1960, Imperfections in Crystals (North-Holland, Amsterdam).
WARRINGTON, D.H., 1963, J. Iron Steel Inst. **201**, 610.
WEERTMAN, J., 1963a, Trans. Met. Soc. AIME **227**, 1439.
WEERTMAN, J., 1963b, Phil. Mag. **8**, 967.
WEERTMAN, J., and S.S. HECKER, 1983, Mechanics of Materials, in press.
WIELKE, B., 1976, Phys. Stat. Sol. (a) **33**, 241.
YOUNG, C.M., L.J. ANDERSON and O.D. SHERBY, 1973, Metallurg. Trans. **5**, 519.

Further reading

Argon, A.S., ed., Physics of Strength and Plasticity (MIT Press, Cambridge, MA, 1969).
ARMSTRONG, R.W., 1983, see references.
Ashby, M.F., R. Bullough, C.S. Hartley and J.P. Hirth, eds., Dislocation Modelling of Physical Systems (Pergamon Press, Oxford, 1981).
Cottrell, A.H., Mechanical Properties of Matter (Wiley, New York, 1964).
Gifkins, R.C., ed., 6th Int. Conf. on Strength of Metals and Alloys, Melbourne, Australia (Pergamon Press, Oxford, 1982).
Haasen, P., V. Gerold and G. Kostorz, eds., 5th Int. Conf. on Strength of Metals and Alloys, Aachen, Germany (Pergamon Press, Oxford, 1979).
Hecker, S.S., A.K. Ghosh and H.L. Gegel, eds., Analysis, Modelling and Experimentation (Metallurgical Society of AIME, 1978).
Hirth, J.P., and J. Weertman, eds., Work-Hardening (Gordon and Breach, New York, 1968).
Honeycombe, R.W.K., The Plastic Deformation of Metals (St. Martin's Press, New York, 1968).
Kanninen, M.F., W.F. Adler, A.R. Rosenfield and R.I. Jaffee, Inelastic Behavior of Solids (McGraw–Hill, New York, 1970).
Meshii, M., ed., Mechanical Properties of bcc Metals (Metallurgical Society of AIME, 1982).
Nabarro, F.R.N., Z.S. Basinski and D.B. Holt, The Plasticity of Pure Single Crystals, Adv. Phys. 13 (1964) 193.
Pink, E., and R.J. Arsenault, Low-Temperature Softening in bcc Alloys, Prog. Mater. Sci. 24 (1979) 1.
Reed-Hill, R.E., J.P. Hirth and H.C. Rogers, eds., Deformation Twinning (Gordon and Breach, New York, 1964).
Proc. Int. Conf. on Crystal Lattice Defects, 1963, Tokyo Symposium, J. Phys. Soc. Japan **18** (1963) Suppl. 1.
Proc. Int. Conf. on the Deformation of Crystalline Solids, Can. J. Phys. 45 (1966) 453–1249.
Relations between Structure and Mechanical Properties of Metals, Symp. National Physical Laboratory, Jan. 1963 (Her Majesty's Stationery Office, London, 1963).

CHAPTER 20

MECHANICAL PROPERTIES, STRONGLY TEMPERATURE-DEPENDENT

J. WEERTMAN and J.R. WEERTMAN

Department of Materials Science and Engineering
Northwestern University
Evanston, IL 60201, USA

R.W. Cahn and P. Haasen, eds.
Physical Metallurgy; third, revised and enlarged edition
© *Elsevier Science Publishers BV, 1983*

1.   Creep of metals

It was assumed tacitly in the last chapter that the plastic strain which is produced by stressing a metal with a given load has a unique value, provided parameters such as temperature are held constant. This tacit assumption is not correct. There is no unique value of strain associated with a given stress. This is a consequence of the fact that plastic strain does not remain constant under a constant load. The strain actually increases with time. Figure 1 shows a schematic plot of strain versus time for a sample stressed under a constant tensile load at a temperature which is comparable to the melting point of the metal. At high temperatures the amount of time-dependent plastic strain which can occur is much larger than the plastic strain produced "instantaneously" during the time interval a load is placed on a specimen.

At low temperatures the amount of time-dependent plastic strain is relatively small, even after long time intervals. Because it is small it usually can be neglected when the stress–strain curves or the work-hardening properties of metals are being considered.

The time-dependent plastic strain which is observed in stressed materials is called *creep*.

1.1.   Stress relaxation

Creep tests on metals usually are carried out by keeping either the applied load or the stress constant and noting the specimen strain as a function of time. In another type of test, known as the stress-relaxation test, a sample first is deformed to a given strain and then the stress is measured as a function of time in such a manner that the total strain remains constant. Figure 2 shows a schematic plot of the results of a stress-relaxation test.

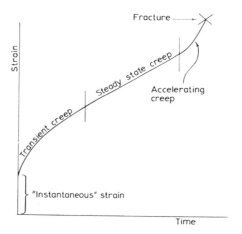

Fig. 1. Schematic creep curve for a metal stressed in tension under a constant stress at a relatively high temperature.

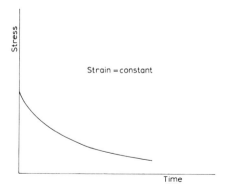

Fig. 2. Schematic plot of a stress-relaxation test. The total strain is held constant and the stress is allowed to vary as a function of time.

A stress-relaxation test is a creep test performed under a non-steady stress. When a specimen is deformed to a given strain, part of this strain is elastic in origin and part is plastic. Because of creep, the plastic strain increases with time. Since the total strain is held constant in a stress-relaxation test the elastic strain must decrease with time. The elastic strain can decrease only if the applied stress becomes smaller.

Stress-relaxation tests are more difficult to carry out than ordinary creep tests and also are more difficult to interpret. It is not surprising that most experimental work has been done by means of creep tests.

1.2. The three main types of creep and their locations on the creep diagram

The creep which is observed in a given test usually may be classified as belonging to one of three well-defined types. Which of the particular types it is depends on the test temperature and the applied stress. The temperature–stress diagram of fig. 3 indicates the region in which each of these types is to be found. A graph such as fig. 3 will be called a *creep diagram*. In order that a metal of low melting point and low elastic modulus, such as indium, may be treated on an equal footing with a metal of high melting point and high elastic modulus, such as tungsten, the temperature and stress are plotted as T/T_m and σ/μ rather than simply as T and σ. Here T is the test temperature, T_m is the melting point, σ is the applied stress, and μ is the shear modulus. Once temperature and stress are modified by dividing them by T_m and μ, the different types of creep fall roughly in the same regions on the creep diagram even though the metals tested differ widely in their melting points and elastic constants.

In the last chapter it was learned that large amounts of plastic deformation can occur only if the critical resolved shear stress σ_{crss} is exceeded. For stresses above σ_{crss} there is extensive dislocation multiplication and motion of dislocations. Below this stress the amounts of dislocation motion and multiplication are slight and thus

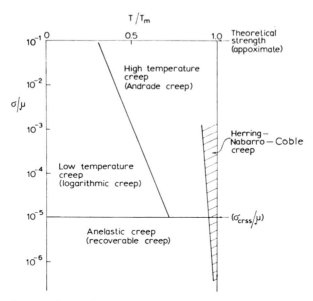

Fig. 3. The creep diagram: the conditions of temperature and stress which produce the three principal types of creep and, under special circumstances, Herring–Nabarro–Coble creep. The temperature and stress are plotted as T/T_m and σ/μ, where T = temperature (K), T_m = melting temperature of the metal, σ = applied resolved shear stress, μ = shear modulus, σ_{crss} = critical resolved shear stress for a well annealed crystal.

little plastic deformation takes place. Therefore it is not surprising that one of the boundaries on the creep diagram which divides a particular creep region from the others coincides with the line $\sigma/\mu = \sigma_{crss}/\mu$. The creep deformation which occurs at stresses smaller than σ_{crss} arises through mechanisms other than the large-scale motion of dislocations. Of course mechanisms that can give rise to creep strains in the region below σ_{crss} also contribute to the process of creep in the region above σ_{crss}. However, these contributions to the strain usually are negligible compared to the dislocation contribution.

At low temperatures, dislocations have no difficulty in moving in directions parallel to their slip planes. Motion in other directions is difficult for any dislocation which is edge or partly edge in character. Lattice vacancies or interstitial atoms must diffuse to or away from a dislocation with an edge or partly edge orientation when the motion of such a dislocation contains a component perpendicular to its slip plane. A motion perpendicular to the slip plane of a dislocation is known as a climb motion or *dislocation climb* (cf. § 1.8 and ch. 18, § 5). Diffusion is slow at low temperatures and the motion of dislocations in directions other than parallel to their slip planes likewise is slow. At high temperatures, diffusion permits relatively fast climb of dislocations. Consequently, dislocations can move with relative ease in all directions at high temperatures. We can consider the dislocations as having an extra "degree of freedom" at these temperatures.

Because of the extra degree of freedom available to dislocation motion at the higher temperatures, it is reasonable to expect that the portion of the creep diagram which corresponds to stresses larger than the critical shear stress is divided into two regions. One region embraces the higher temperatures in which dislocations have two degrees of freedom, whereas the other corresponds to temperatures which permit only one degree of freedom. The boundary between the two regions will shift to the right for very short-time tests and to the left for very long-time tests.

In the sections that follow, the characteristics of the creep found in each of the three regions of the creep diagram will be described. A unique type of creep labelled "Herring–Nabarro–Coble creep" in fig. 3 also will be considered.

1.3. Anelastic creep (recoverable creep)

It frequently is observed that below the critical shear stress the strain ϵ in a stressed specimen is well described by the following time-dependent equation:

$$\epsilon = \epsilon_e + \epsilon_0 \left[1 - \exp(-t/\tau) \right], \tag{1}$$

where ϵ_e is the purely elastic strain, t is time after loading, and ϵ_0 and τ are constants. The last term on the right-hand side of eq. (1) gives the non-elastic, time-dependent strain. This strain is described as *anelastic*.

Equation (1) predicts that the anelastic strain is zero immediately after loading. It then increases with time and exponentially approaches the limiting value ϵ_0. The constant τ is a characteristic time associated with the creep process. The anelastic creep strain saturates to the value ϵ_0 at times long compared to τ. A schematic plot of the relationship between strain and time predicted by eq. (1) is given in fig. 4.

The *anelastic creep strain* is recoverable. Should the applied load be removed after the strain ϵ' is reached, the strain at any time t after removal of the load is given by:

$$\epsilon = (\epsilon' - \epsilon_e) \exp(-t/\tau), \tag{2}$$

where ϵ_e is the elastic strain present just prior to the removal of the load. The strain given by this equation closely approaches zero at times which are long compared to the characteristic time (or decay time) τ. Therefore the anelastic creep is completely recoverable by removal of the load. This recoverable property of anelastic creep also is shown schematically in fig. 4.

Although anelastic creep can be studied by applying static loads to specimens, it is more convenient to investigate this phenomenon experimentally by applying cyclic stresses. Measurements are made of the damping loss per cycle of stress (the internal friction). Because of the appearance in eqs. (1) and (2) of the terms containing the decay time τ, the anelastic strain of a cyclically loaded sample in general is not in phase with the applied stress. If stress and strain are out of phase energy must be dissipated in damping losses. The subject of anelastic creep thus includes the whole field of the *internal friction* of metals.

One of the most celebrated mechanisms associated with internal friction and elastic creep is based on the motion of interstitial atoms in bcc iron. Interstitial

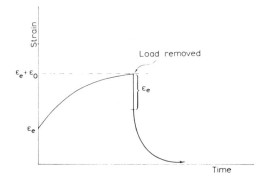

Fig. 4. Schematic creep curve typical of anelastic creep. Also shown is recovery creep after removal of load.

impurity atoms such as carbon and nitrogen reside on the three types of sites marked by an X in fig. 5a. These three sites are completely equivalent. Under equilibrium conditions, if no stress is applied to a crystal each type of site contains as many impurity atoms as does either of the others.

If, as indicated in fig. 5b, a tensile stress is applied to the crystal, one of the three interstitial sites assumes a preferred status. It is easier for an interstitial impurity atom to reside in a preferred site because the applied stress has slightly increased the separation between the two iron atoms on either side of the preferred site. If a very high stress were applied to the crystal at a low temperature the equilibrium positions of the interstitial atoms would be those indicated by X's in fig. 5b. At low applied stress and at higher temperatures under equilibrium conditions, the number of interstitial atoms in the preferred sites will be only slightly larger than in either of the two other types of sites. At low stresses the difference in energy between preferred and non-preferred sites is small. If the thermal energy kT, where k is Boltzmann's constant, is larger than this energy difference, thermal agitation will insure that the variation in the population of interstitial atoms on each of the three types of sites is slight.

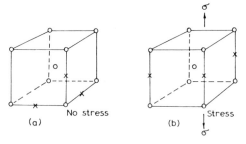

Fig. 5. (a) The three positions of interstitial atoms in the bcc lattice, and (b) favoured sites for interstitial atoms when the lattice is stressed as shown.

An interstitial atom on any site increases the separation between the two metal atoms on either side of it. If after application of a stress the number of interstitial atoms on the preferred sites increases at the expense of the number of interstitial atoms on the non-preferred sites, obviously the crystal must increase its length in the direction of the stress and decrease its dimensions in the perpendicular directions.

After application of a low tensile stress, there is a small bias favouring the sites shown in fig. 5b. The increase in the population of these sites is accomplished by diffusion of the impurity atoms. Thus immediately after application of a stress, the only strain produced in a sample is purely elastic. As diffusion of interstitial impurity atoms proceeds, the concentration of interstitial atoms on the preferred sites approaches its new equilibrium value.

The motion of interstitial atoms from one site to another takes place with a characteristic jump time τ. The concentration of impurities on the preferred sites will approach equilibrium through a term proportional to $[1 - \exp(-t/\tau)]$. This, or a similar expression, appears in eqs. (1) and (2). The decay time τ appearing in those creep equations can be identified with the jump time of an impurity atom.

The jump time of an impurity atom is proportional to the expression $\exp(Q/kT)$, where Q is the activation energy of diffusion. Thus the validity of the interstitial mechanism of anelastic creep may be checked by determining whether or not the activation energy of interstitial diffusion is the same as the activation energy derived from the observed decay time of anelastic creep. In the case of iron containing carbon these two activation energies have been found to be the same (see ch. 8). Another verification of the theory comes from the observation that the saturation strain ϵ_0 is proportional to stress. According to the theory, the proportionality arises because the larger the applied stress, the greater is the biasing favouring the preferred sites and hence the larger is the equilibrium population of interstitial atoms in these sites. The anelastic strain ϵ_0 also is proportional to the total concentration of interstitial atoms. Consequently the magnitude of the internal friction, determined by means of a simple torsion pendulum, frequently is used to measure very small concentrations of interstitial solutes in iron, tantalum, etc. Different solutes cause peaks in the internal friction at different temperatures and thus can be distinguished.

This example of the behaviour of interstitial impurity atoms in the bcc lattice is just one of a large number of mechanisms which can lead to an anelastic creep equation similar to eq. (1) or (2) or some modification of these equations. Dislocation motion over distances very much smaller than that required for dislocation multiplication also can lead to anelastic creep and internal friction. This source of internal friction is discussed in ch. 18. It is beyond the scope of this chapter to discuss all possible mechanisms of anelasticity.

1.4. Low-temperature creep (logarithmic creep)

Consider now a sample that has been deformed by a stress which exceeds the critical resolved shear stress. Assume that the temperature is relatively low, that is,

References p. 1339.

below about one half of the absolute melting point of the metal. Dislocations with an edge character will find it difficult to move perpendicular to their slip planes.

An applied stress can produce dislocation multiplication and move dislocations over long distances. The work-hardening mechanisms that were discussed in ch. 19 eventually cause the dislocation multiplication to stop. These mechanisms also make dislocation motion on slip planes more difficult. If all dislocation movement ceased immediately after the application of a load on a sample no time-dependent plastic strain would occur. However, we do not expect all motion of dislocations to stop. For example, we know that a slight increase in the applied stress will cause further plastic deformation because the stress–strain curve has a finite slope. If a small increase in the applied stress can cause further deformation, the momentary increases in stress produced by random thermal stress fluctuations present in any solid held at a finite temperature also can produce additional plastic deformation. (The thermal stress fluctuations arise from the thermal agitation which causes atoms in a lattice to vibrate about their mean equilibrium positions.) The probability that a stress fluctuation will have occurred within a small volume of a specimen is larger, the longer is the time interval during which the small volume is under observation. Plastic deformation does not cease immediately after a specimen is stressed under a static load. Thermal stress fluctuations which are added to the applied stress increase the plastic strain as time goes by. Thus without knowing anything about the work-hardening mechanisms one can predict that creep must occur in the low-temperature region of the creep diagram. It also can be predicted confidently that the creep found in this region of the creep diagram is of a transient nature. *Transient creep* is merely creep whose rate decreases with time. (In material with a low initial dislocation density, such as silicon, the creep rate *increases* with time during transient creep.) The creep curve of fig. 4 is an example of transient creep. In that figure the creep rate $\dot{\epsilon} = \mathrm{d}\epsilon/\mathrm{d}t$ approaches zero exponentially with time.

The transient nature of creep in the low temperature region may be explained as follows. Thermal stress fluctuations aid the process of deformation and thus produce a creep strain. However, by increasing the deformation, the stress fluctuations also increase the hardening of a sample. As a result the more a sample is deformed, the more difficult does further deformation become. Therefore the rate of creep deformation must decrease as time goes by.

Experimentally it is found that indeed low temperature creep is of a transient nature. The creep equations determined from measurements at low temperatures generally are of the type

$$\epsilon = \epsilon_e + \epsilon_p + \epsilon_0 \log(1 + \nu t), \tag{3}$$

which gives a creep rate

$$\dot{\epsilon} = \frac{\epsilon_0 \nu}{1 + \nu t}. \tag{4}$$

In these equations, ϵ_e is the elastic strain, ϵ_p is the instantaneous plastic strain, and ϵ_0 and ν are constants. The creep strain depends logarithmically on time and thus this type of creep is known as *logarithmic creep*. The creep is transient since the creep rate decreases to zero as the inverse of the time.

1.5. General theory of logarithmic creep

It is not difficult to set up a theory which accounts for eqs. (3) and (4) and yet does not require too specific a dislocation model. Imagine a crystal to be divided up into subvolumes of a length ℓ on a side. The length ℓ is the dimension of the smallest volume element that can be permanently deformed (by the passage of dislocations through it) independently of the deformation in any neighbouring volume element. That is, stresses within neighbouring volume elements cannot reverse the deformation of the particular volume element once the deformation has occurred. If σ is the applied stress and σ^* is a back stress due to the long-range contribution to the work-hardening, the additional energy W which must be supplied to increase the plastic deformation of the volume element by moving one dislocation through it is $W = b\ell^2(\sigma^* - \sigma) + U$. The quantity U is the energy required to overcome the resistance offered by short-range hardening mechanisms. The energy U could be, for example, the energy needed by a dislocation to cut through a dislocation forest if the mobile dislocations on the slip planes are hung up on the trees of the forest, or this energy could arise from the non-conservative motion of jogs. If an appreciable Peierls stress (cf. ch. 18, § 3.1) exists in a crystal, U could be the energy required to throw a segment of dislocation line over a "Peierls hill".

The long-range hardening stress σ^* can be rewritten as:

$$\sigma^* \equiv \sigma_0^* + h(\epsilon - \epsilon_e - \epsilon_p). \tag{5}$$

The constant h is approximately equal to the slope of the stress–strain curve if long-range hardening makes an appreciable contribution to the total work-hardening of the crystal. Equation (5) is merely a Taylor expansion of σ^* out to a first order term.

The creep rate $\dot{\epsilon}$ is proportional to the frequency with which thermal stress fluctuations cause deformation of each volume element having a volume $v = \ell^3$. According to chemical rate theory, the probability that a volume element will be deformed in a unit time is equal to $\nu' \exp(-W/kT)$, where ν' is some vibrational frequency. It is reasonable to identify ν' with the vibrational frequency of the dislocation lines. This frequency is somewhat lower than the vibrational frequency of atoms in a lattice. A reasonable estimate for ν' is a value in the range of $10^{10} - 10^{11}$ s^{-1}.

The strain produced per unit volume when one dislocation segment of length ℓ moves a distance ℓ is approximately $b\ell^2$. Therefore the creep rate is

$$\dot{\epsilon} = \nu' \frac{b\ell^2}{\ell^3} \exp\frac{-W}{kT} = \nu' \frac{b}{\ell} \exp\frac{-U_0}{kT} \exp\left(b\ell^2 \frac{\sigma - h\epsilon}{kT}\right), \tag{6}$$

where

$$U_0 = U + b\ell^2(\sigma_0^* - h\epsilon_e - h\epsilon_p).$$

If eq. (6) is integrated with respect to time the following equation is obtained

$$\epsilon = \epsilon_e + \epsilon_p + \epsilon_0 \log(1 + \nu t), \qquad (7)$$

where

$$\epsilon_0 = kT/hb\mathcal{L}^2$$

and

$$\nu \equiv \nu'(hb^2\mathcal{L}/kT)\exp(-U/kT).$$

It is assumed that $\sigma \cong \sigma_0^*$. Equation (7) is of the same form as eq. (3). The analysis just presented can be made much more specific by starting from a definite dislocation model. The creep equation derived from such a model usually is logarithmic.

Figure 6 presents results of measurements of the logarithmic creep of copper single crystals in stage I of the stress–strain curve. The log of the creep rate is plotted against the quantity $(\sigma - h\epsilon)$. The data were obtained at two temperatures over a range of values of the applied stress σ. All the experimental points taken at a given temperature fall more or less on one line. This result implies that U_0 of eq. (6) is independent of the applied stress. Similar results are found in the case of hcp metals in the easy-glide region and fcc metals in stage II of the stress–strain curve.

Let us consider how the activation energy of creep W may be determined experimentally. The activation energy can be obtained by measuring the creep rates immediately before and after an abrupt change of 5° or 10°C in the test tempera-

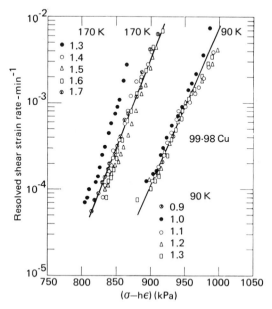

Fig. 6. Creep rate at two temperatures versus $(\sigma - h\epsilon)$ for copper single crystals in stage I. The values of the applied stress (in MPa) are shown behind the data symbols. (Data of CONRAD [1958].)

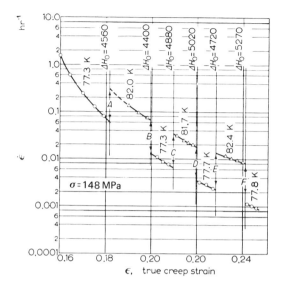

Fig. 7. Creep rate versus strain in temperature change tests used to obtain activation energies for high-purity aluminum. The ΔH_0 are given in units of 4.19 J/mole. (Data of SHERBY *et al.* [1957].)

ture. Typical results from such measurements are shown in fig. 7. This method has been used extensively by Dorn and Sherby and their school. From eq. (6), the activation energy of creep W is equal to:

$$W = \frac{kT_1T_2}{T_1 - T_2} \log(\dot{\epsilon}_1/\dot{\epsilon}_2), \tag{8}$$

where $\dot{\epsilon}_1$ is the creep rate at temperature T_1 and $\dot{\epsilon}_2$ is the creep rate at temperature T_2.

If eq. (6) is rearranged, the following expression for the activation energy of creep can be obtained:

$$W = kT \log(b\nu'/\dot{\epsilon}\ell). \tag{9}$$

It should be noted that the creep rate appears inside the log term in eq. (9). This fact has an interesting consequence. Because of practical restrictions, the creep rate can be measured conveniently only within certain limits (approximately between 10^{-8} and 10^{-2}/s). The logarithmic term of eq. (9) will remain reasonably constant throughout any experiment in which the creep rate can be measured. As a result, at low temperatures the activation energy of creep derived from eq. (9) appears to be approximately proportional to the absolute temperature. It vanishes as T approaches zero. This predicted temperature dependence at low temperatures can be observed in the creep data for aluminium presented in fig. 8. (The activation energy of creep at high temperatures is discussed in the following section.) Such a temperature dependence is usually found in polycrystalline specimens, but not always in single crystals.

References p. 1339.

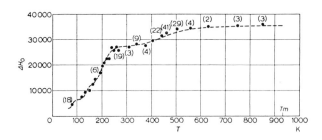

Fig. 8. Activation energy (in units of 4.19 J/mole) of creep for pure aluminum as a function of absolute temperature. Numbers in parentheses indicate number of tests made to obtain a point. (Data of SHERBY *et al.* [1957].)

In the liquid helium temperature range, creep activation energies as low as 105 J/mole have been reported (ARKO and WEERTMAN [1969]).

The fact that the measured creep activation energy increases with temperature does not imply that for a *given internal state* of a material the actual activation energy of creep changes with temperature. The observed temperature dependence means only that during creep the activation energy gradually increases from a value close to zero immediately after application of stress to a large value during that part of the creep curve in which $\dot{\epsilon}$ approaches zero. (As creep progresses, of course, the internal state of the material changes.) At the beginning of creep $\dot{\epsilon}$ is too fast to be measured, and at the end of a test $\dot{\epsilon}$ is too slow. Consequently the activation energy is measured during that part of a creep curve in which the value of W is such that measurable creep rates occur. Experimental observations are limited to a narrow range of values of W out of the wide field that exists. Since activated processes become slower with decreasing temperature it is not surprising that a smaller activation energy is measured at a lower temperature.

1.6. High-temperature creep (Andrade creep)

Around 1910 ANDRADE carried out a series of creep tests on metals in the region of the creep diagram which corresponds to high temperatures and moderate to high stresses (ANDRADE [1910]). He established that in this portion of the creep diagram, creep under constant stress and temperature follows the law

$$\epsilon = \epsilon_e + \epsilon_p + \beta t^n + Kt, \tag{10}$$

where ϵ_e is the elastic strain, ϵ_p is the instantaneous plastic strain, and n, β, and K are constants. Figure 1 shows a schematic plot of the strain predicted by *Andrade's equation*. The creep curve of fig. 1 shows an acceleration at the end of the test. This accelerated creep is not predicted by eq. (10). It usually is caused by a necking down of a sample made of a ductile metal or alloy, or the opening up of cracks or cavities in less ductile material.

It is strongly emphasized that the creep strain that occurs in high-temperature

(Andrade) creep is large. Strains of 100–300% can be produced in some tests without specimen failure, under stresses which produce only few per cent instantaneous plastic strain. These creep strains are very much larger than the relatively small strains found in the anelastic and low-temperature portions of the creep diagram. In those regions a creep strain of 1% is considered to be large.

The fact that the creep strain of high-temperature creep is much larger than the instantaneous plastic strain has an important implication. Unlike low-temperature creep, high-temperature creep cannot be considered to be the consequence of the effect of thermal stress fluctuations on the dislocation structure of a work-hardened metal. High-temperature creep deformation is unique and should be considered separately from the deformation produced at low temperatures.

The constant n appearing in eq. (10) was determined by Andrade to have the value $1/3$. Later workers have questioned whether its value is exactly equal to $1/3$. Results in the literature show a spread in the value of n from about $1/4$ to $2/3$ but many of the determinations lie in the neighbourhood of $1/3$.

If a time derivative is taken of eq. (10) the following creep rate is found

$$\dot{\epsilon} = n\beta/t^{1-n} + K. \tag{11}$$

The first term on the right-hand side of this equation represents transient creep, since it approaches zero as time goes to infinity (note that n is less than 1). The second term on the right gives rise to a constant or steady-state creep rate. According to eq. (11) the creep rate decelerates as time progresses and approaches a constant value. The initial part of the creep curve of fig. 1 is determined by the terms containing β in eqs. (10) and (11). The straight part of the creep curve is determined by the terms containing K.

The relative amount of transient creep that occurs in the high-temperature field of the creep diagram depends on the exact temperature and stress. Usually the lower the temperature and the higher the stress, the greater is the proportion of transient creep. It is possible to arrange conditions so that tests give predominately transient creep curves with little or no component of steady-state creep.

There is no sharp transition in creep behaviour between the high-temperature and the low-temperature regions of the creep diagram. The steady-state component of high-temperature creep becomes less important as the temperature is lowered. If the temperature is lowered still further the βt^n transient creep term of eq. (10) eventually must be replaced by the logarithmic time term of eq. (3). The transition from a βt^n to a $\log(1 + \nu t)$ time dependence occurs over a range of temperatures. WYATT [1953] found that in this transition region the creep curve is described by an equation which contains both of these time-dependent terms.

1.7. Temperature dependence of steady-state creep

Experimentally it is found that, at high temperatures and under constant stress, the steady-state creep rate K obeys the following law:

$$K = K_0 \exp(-Q/kT) \exp(-P\Delta V/kT), \tag{12}$$

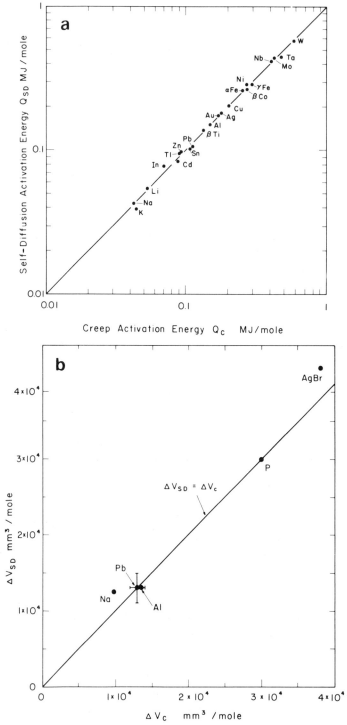

Fig. 9. (a) Comparison of the activation energies of creep and of self-diffusion for a number of metals. (b) The same plot as in (a), but for activation *volumes*. (From SHERBY and WEERTMAN [1979], data mainly compiled by POIRIER [1976,1978].)

where Q is the activation energy of creep, P is hydrostatic pressure and ΔV is the activation volume of creep. The quantity K_0 is almost independent of temperature but varies with stress. The activation energy of creep is independent of strain. The Dorn–Sherby school have shown that the activation energy of creep in the transient part of the creep curve (at high temperatures) usually is the same as that observed in the steady state part of the curve.

Dorn and Sherby have demonstrated that the activation energy of creep is identical, within experimental error, to the activation energy of self-diffusion (ch. 8). Later workers have confirmed this finding. (See reviews by SHERBY and BURKE [1967], WEERTMAN [1968] and TAKEUCHI and ARGON [1976], and the books by GITTUS [1975], LI and MUKHERJEE [1975] and POIRIER [1976].) The correlation between these two activation energies is shown in fig. 9a.

The activation volumes of creep and self-diffusion also are equal to each other as can be seen in fig. 9b. These striking correlations form the cornerstone of the dislocation-climb theory of high-temperature creep.

1.8. Stress dependence of steady-state creep

The stress dependence of the steady-state creep rate over the whole stress range above the critical stress σ_{crss} is described well by the empirical power-law hyperbolic-sine equation of GAROFALO [1965]. The constant K_0 of eq. (12) is

$$K_0 = C'(\sinh \beta'\sigma)^N. \tag{13}$$

The parameters C', β' and N are independent of stress. The value of β' varies with temperature. For pure metals and some alloys the power exponent N is approximately 4–5. Alloying often causes N to decrease to a value near 3. This change in N is shown in fig. 10 for a series of gold–nickel alloys. When the argument of the sinh function is small, eq. (13) becomes the simple power law

$$K_0 = C^*\sigma^N, \tag{14}$$

where C^* is a constant. Figure 11 is a log–log plot of the steady-state creep rate of aluminum versus stress. It is seen that the creep data follow a simple power law at moderate stresses and can be fitted to the hyperbolic-sine power law over the larger stress range.

Equations (12) and (14) can be combined to give the usual form of the power-law creep equation:

$$K = C(\sigma/\mu)^{N-1}(\sigma\Omega/kT)(D/b^2) = C(\sigma/\mu)^N(\mu\Omega/kT)(D/b^2), \tag{15}$$

where C is a dimensionless constant, Ω is the atomic volume and $D = D_0 \exp(-Q/kT) \exp(-P\Delta V/kT)$ is the diffusion constant. The forms of the terms in eq. (15) are those expected from a dimensional argument. The creep rate has the dimensions of inverse time. All other units must cancel out of the right-hand side of eq. (15). D has units of length squared and inverse time. A length unit can be eliminated by dividing by a characteristic distance of the problem. The interatomic

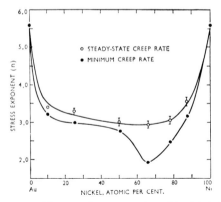

Fig. 10. Value of the stress exponent N versus alloy content for gold–nickel alloys at a test temperature of 860°C. (Note: a minimum creep rate is the smallest creep rate occurring during a creep experiment. The steady-state creep rate is the rate found after subtracting the transient component of the creep rate.) (Data of SELLARS and QUARRELL [1962].)

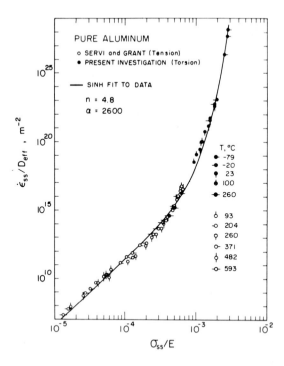

Fig. 11. Steady-state creep rate versus steady-state stress for high-purity aluminum. Stress is normalized with Young's modulus, and creep rate with effective diffusion constant. (From LUTHY et al. [1980].)

distance b is the obvious and only real choice for this distance. Similarly, to eliminate the units of stress the term σ can be divided by an elastic constant. The dimensionless term $\sigma\Omega/kT$ is introduced because whatever process controls the creep rate it is likely to go in a stress-assisted direction at a rate proportional to $\exp(\beta\sigma\Omega/kT)$ and against the the stress-assisted direction at a rate proportional to $\exp(-\beta\sigma\Omega/kT)$. Here β is a constant. The difference in rates is approximately proportional to $\sigma\Omega/kT$. We point out that when the activation energy of creep is determined, the temperature dependence of the terms μ and T is sufficiently large that it must be taken into account in the data analysis.

The activation energy of creep is equal to that of lattice self-diffusion only if the temperature is above a lower limit whose value ranges from about $0.5T_m$ for aluminium to about $0.8T_m$ for tin (POIRIER [1978]). Below this transition temperature the activation energy for steady-state flow is approximately equal to that measured or expected for pipe diffusion. Pipe diffusion is diffusion down a dislocation core. Figure 12 shows a plot of Q_{ss}/Q_L, where Q_{ss} is the steady-state creep activation energy and Q_L the lattice diffusion activation energy, versus T/T_m. The decrease in Q_{ss} at the lower temperatures can be seen in this figure. Very large strains ($\epsilon \approx 1-10$) are needed to reach the steady-state flow regime at low temperatures (cf. ch. 19). In the transient creep region at low temperatures, as pointed out in §1.5, the apparent activation energy is proportional to T, in agreement with prediction.

Over the entire temperature range eq. (15) is replaced with

$$K = C(\sigma/\mu)^N(\mu\Omega/kT)(D/b^2)\left[1 + \zeta(D_P/D)(\sigma/\mu)^2\right]. \tag{16}$$

where D_P is the pipe-diffusion constant and ζ is a dimensionless constant. At the

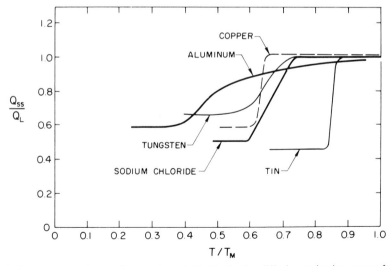

Fig. 12. Activation energy for steady-state flow divided by lattice diffusion activation energy for various materials as a function of T/T_m (from LUTHY *et al.* [1980]).

lower temperatures the stress exponent increases by a factor of two because of the increased importance of dislocation pipe diffusion (see ch. 19).

1.9. Theory of high-temperature creep

The theories of high-temperature creep are concerned primarily with explaining the steady-state part of the creep curve. Undoubtedly the transient part represents a transition period during which the dislocation structure changes from that produced during the instantaneous plastic deformation to a steady-state or quasi-steady-state configuration which is maintained during steady-state creep. The structure that occurs immediately after loading probably approximates the structure produced by deformation at low temperatures.

The creep of many solid solution alloys can be explained more easily than that of pure metals (ch. 21). When a metal is alloyed, a variety of mechanisms come into play which can hinder the motion of dislocations across their slip planes. At temperatures comparable to the melting point these mechanisms are similar to each other in that a dislocation pinned by any of them drifts at a constant velocity under a constant stress. Figure 13 depicts a number of these mechanisms. A dislocation may drift under action of an applied stress by dragging an impurity atmosphere with it (Cottrell–Jaswon). The impurity atmosphere moves by means of the diffusion of the constituent impurity atoms. The same drift motion occurs if the impurity or alloying atmosphere forms at the stacking fault separating two partial dislocations

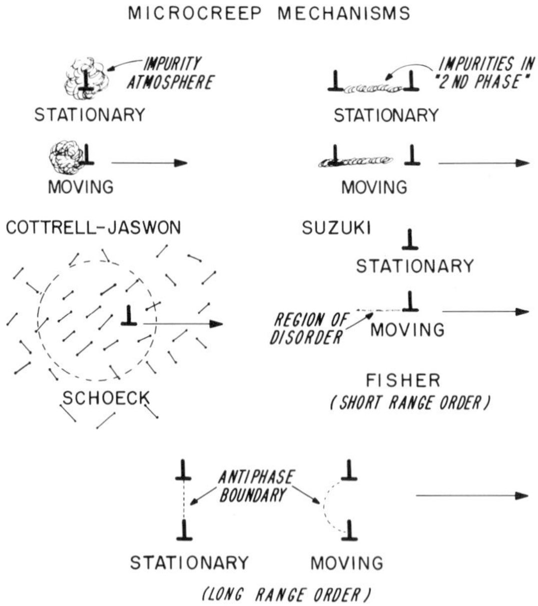

Fig. 13. Microcreep mechanisms (after WEERTMAN [1960]).

(Suzuki); or there may be a region of stress-induced order about a dislocation which can move at high temperatures (Schoeck); a dislocation located in an alloy with short-range order may create a strip of disorder behind it which diffusion heals up (Fisher). In an alloy with long-range order the dislocations may exist as pairs of perfect dislocations. Each pair must drag an antiphase boundary along with it (ch. 21). No matter which of these mechanisms is operative, the dislocation velocity v is given by the equation

$$v = \sigma b / A, \tag{17}$$

where A is a temperature-dependent constant whose exact value depends on the particular mechanism that controls the dislocation motion.

The steady-state creep rate always can be expressed by the equation

$$\dot{\epsilon} = K \equiv \varrho v b, \tag{18}$$

where ϱ is the dislocation density, v is the average dislocation velocity, and b is the length of the Burgers vector. In the previous chapter [ch. 19, eq. (9)] we estimated that the dislocation density in a stressed crystal is equal to $\varrho \cong (2\pi\sigma/\mu b)^2$, where μ is the shear modulus. If this value of the dislocation density is substituted into eq. (18), and eq. (17) is used to estimate the average dislocation velocity, the following equation is obtained for the creep rate:

$$K \cong \frac{4\pi^2\sigma^3}{A\mu^2}. \tag{19}$$

The creep rate is proportional to the stress raised to the third power. A stress dependence close to this is observed in a number of metal alloys. A satisfying feature of eq. (19) is the fact that the third power arises naturally, without resort to a specific dislocation creep model. It is not necessary to know how or where dislocations are created nor what kind of dislocation sources exist within a crystal.

A creep equation for pure metals cannot be derived so simply. The correlation which exists between the activation energies and activation volumes of creep and of self-diffusion in the case of pure metals (see fig. 9) strongly suggests that the mechanism of self-diffusion is involved in the creep process. Self-diffusion commonly takes place through the motion of lattice vacancies (ch. 8). Vacancies also play a role in the climb motion of edge dislocations and mixed dislocations. Figure 14 illustrates an edge dislocation in the process of climbing by the absorption of vacancies. Climb in the opposite direction occurs if the dislocation gives off rather than absorbs vacancies.

It is a logical step to postulate that the rate-controlling process in the high-temperature creep of pure metals is the climb of dislocations having an edge component. This suggestion was made by MOTT [1956]. (He concluded, however, that dislocation climb was a unimportant creep mechanism.) We also used dislocation climb as a rate-controlling process for creep (WEERTMAN [1955, 1957]). Our analyses led to the power law of eq. (14) as well as to a more general law that approximates eq. (13).

It is possible to develop a theory of creep in terms of climb from which a creep

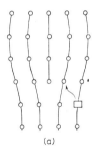
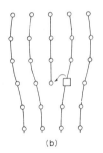
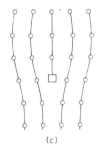

(a)   (b)   (c)

Fig. 14. The upward climb of an edge dislocation by the absorption of a vacancy (marked by the square). Downward climb occurs if the dislocation gives off a vacancy.

equation may be derived without referring to a specific dislocation model. We assume that a steady-state dislocation configuration develops within a sample after the transient creep has died out. In this steady-state configuration, dislocations are continuously being created and are continuously being destroyed. Destruction occurs when dislocations of opposite sign which originally were on different slip planes climb towards each other and suffer annihilation upon meeting. Some dislocations climb by creating vacancies, which then diffuse to other dislocations which climb by destroying vacancies. On the average, the number of dislocations which destroy vacancies equals the number of vacancy-creating dislocations. The creep rate is controlled by the diffusion of vacancies between dislocations.

Let L represent the average distance which a dislocation moves in a direction parallel to its slip plane between the time it is created and the time it is annihilated. Let d be the average distance a dislocation climbs in a direction normal to its slip plane before it is annihilated. The average amount of work done by the applied stress σ on a unit length of dislocation line from the time it is created until it is destroyed is equal to σbL. Suppose that during the whole motion history of an average dislocation line the dislocation moves only a negligible distance in an unstable, jerky manner between positions in which it is in unstable equilibrium to positions in which it is in stable equilibrium. This assumption enables us to neglect any work the applied stress does on a dislocation which becomes dissipated by the sound waves generated by the acceleration or deceleration of the dislocation.

Each edge dislocation creates or destroys approximately d/b^2 vacancies per unit length of dislocation line between the time the dislocation is created and the time it is destroyed. In an unstressed lattice a certain amount of energy, say Q_c, is required to make a vacancy. It takes this amount of energy to create a vacancy at a dislocation line which is not acted upon by any stress. However, the dislocations we are considering are under stress. The applied stress does an amount of work equal to σbL during the period a dislocation line climbs a distance d. Divided among the total number of vacancies created or destroyed during the lifetime of the dislocation, this energy amounts to approximately $\sigma b^3L/d$ per vacancy. Thus it can be expected that the average energy required to create a vacancy at a dislocation which gives off vacancies during creep is lowered from Q_c to $(Q_c - \sigma b^3 L/d)$. Similarly the energy of vacancy creation at a dislocation which climbs through the destruction of vacancies is raised to the value $(Q_c + \sigma b^3 L/d)$.

During steady-state creep two processes are occurring: dislocations are being created and destroyed and vacancies are being created and destroyed. At the moment of mutual annihilation, the self-energy of two long dislocation lines of opposite sign is zero because their combined stress field vanishes. The self-energy of a small, newly created dislocation loop is not large because the total length of dislocation line is small. Therefore approximately zero work is required to create a dislocation in a lattice and zero energy is recovered just at the moment when two dislocations annihilate each other. Therefore the applied stress can expend very little energy on the dislocation lines themselves during creep, provided the dislocations do not move in an unstable manner. Yet the work the applied stress does on a crystal must be dissipated somehow. This dissipation can be accomplished by assisting the climb motion of dislocations. Energy is expended to aid in creating vacancies on dislocation lines which climb by vacancy creation and in destroying vacancies on dislocations which must absorb vacancies in order to climb.

Towards the creation of a vacancy at one dislocation line and the destruction of another vacancy at a second dislocation, the applied stress expends an amount of energy equal to $2\sigma b^3 L/d$. It can be shown that the equilibrium number of vacancies c_v per unit volume of crystal in an unstrained metal is equal to $n_0 \exp(S_c/k)$ $\exp(-Q_c/kT) = n_0' \exp(-Q_c/kT)$, where n_0 is the number of lattice sites per unit volume, S_c is the entropy of formation of a vacancy, and $n_0' = n_0 \exp(S_c/k)$. The average difference Δc_v between the vacancy concentrations near climbing dislocations which are giving off vacancies and those which are absorbing vacancies is

$$\Delta c_v = n_0' \exp(-Q_c/kT)\left[\exp(\sigma b^3 L/dkT) - \exp(-\sigma b^3 L/dkT)\right]$$

$$= 2n_0' \exp(-Q_c/kT) \sinh(\sigma b^3 L/dkT). \tag{20}$$

At low stresses this equation reduces to

$$\Delta c_v = \left(2n_0'\sigma b^3 L/dkT\right) \exp(-Q_c/kT). \tag{21}$$

This difference in vacancy concentration gives rise to a gradient in the vacancy concentration between climbing dislocations.

The rate of vacancy flow between dislocations and thus the velocity of dislocation climb depends on the actual arrangement of the dislocations in a crystal. It is easy to estimate the average velocity of climb of dislocations which are disposed as shown in fig. 15. Here we see a group of $\mathfrak{N}$ straight dislocations all of the same sign and all climbing in the same direction. Out to a distance R_0 they maintain a vacancy concentration given by $c_v = c_0(1 \pm \sigma b^3 L/dkT)$, where $c_0 = n_0' \exp(-Q_c/kT)$. It is assumed that the value of the applied stress is low. At a distance R the vacancy concentration has changed to the average concentration c_0. The number of vacancies that flow to or away from the group of dislocations per unit time can be calculated from the diffusion equation. This number, per unit distance in the direction parallel

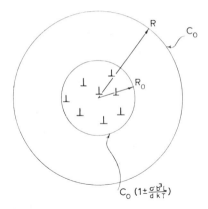

Fig. 15. Group of $\mathcal{N}$ climbing dislocations.

to the dislocation lines, is equal to $2\pi D_v c_0 (\sigma b^3 l/dkT)/\log(R/R_0)$, where D_v is the coefficient of diffusion for vacancies.

Thus the average velocity of climb v_c for each dislocation in the group is equal to

$$v_c \cong 2\pi D_v c_0 \frac{\sigma b^5 L/dkT}{\mathcal{N} \log(R/R_0)} \cong \frac{D_v c_0 \sigma b^5 L}{\mathcal{N} dkT}. \tag{22}$$

Let us suppose that under conditions which produce steady-state creep eq. (22) gives a reasonable estimate of the climb rate of a dislocation, provided the applied stress is sufficiently low that the vacancy concentration in the neighbourhood of a dislocation line is equal to $c_0(1 \pm \sigma b^3 L/dkT)$. The average velocity v of a dislocation in a direction parallel to its slip plane must equal

$$v = v_c L/d, \tag{23}$$

since a dislocation moves an average distance L parallel to its slip plane while climbing a distance d perpendicular to it. If this average velocity is substituted into eq. (18) and eq. (10) of the preceding chapter is used to estimate the dislocation density, the following equation is found for the steady-state creep rate of a pure metal:

$$\dot{\epsilon} = K \cong \frac{4\pi^2 \sigma^3 b^4 L^2 D_v c_0}{\mu^2 kTd^2 \mathcal{N}}. \tag{24}$$

This creep equation predicts that the activation energies of creep and of self-diffusion are identical. The product $D_v c_0$ is approximately equal to D, the self-diffusion constant. The self-diffusion activation energy thus is identical to the creep activation energy.

The stress dependence of creep predicted by eq. (24) is influenced by the stress dependence of each of the quantities d, L, and $\mathcal{N}$. The relationship between these quantities and the stress can be estimated as follows. The term $\mathcal{N}$ should approximate the number of dislocation loops which the applied stress can contain within a distance L. (These dislocation loops are not necessarily piled up.) Thus from dislocation theory, $\mathfrak{N} \sim \sigma L/\mu b$. The distance d might be expected to approximate the average distance between the $\mathfrak{N}$ dislocation loops from a given source which are contained within a radius L, or $d = \mu b/\sigma$. The distance L can be estimated if it is assumed that the number of dislocation sources is not a function of stress. If M is the density of sources per unit volume, and if each source sends out loops which grow to a radius L before they are annihilated by dislocations from other sources, a volume $\pi L^2 d$ will contain on the average only one source. Thus $L^2 \sim 1/\pi Md$.

If the dislocation structure that is developed during steady state creep is self-similar, a reasonable expectation, the ratio d/L and the term $\mathfrak{N}$ will be independent of stress. (Self-similar means that micrographs of the dislocation structure developed under different stress levels can all be made to look alike merely by changing the magnification.) Thus eq. (24) predicts a stress exponent $N = 3$ for creep controlled by dislocation climb. Numerous dislocation models have been analyzed in which the dislocation motion or the annealing of dislocation networks is controlled by lattice diffusion. Regardless of the details of the models they all predict similar creep rates and that $N = 3$ *if* no ad hoc assumptions are made that destroy the self-similar expectation (see papers in Li and Mukherjee [1975], Poirier [1976] and Gittus [1975]). But pure metals and many alloys have $N = 4$–5. Clearly the dislocation morphology must not be self-similar. No theory has yet predicted values of N larger than 3 without making ad hoc assumptions that destroy self-similarity. No theory explains why self-similarity should be destroyed. In fact, it is not established experimentally (because experimenters seem unaware that this is a significant problem) whether or not the dislocation morphology in steady-state creep is self-similar.

At lower temperatures account has to be taken of the enhancement of the climb velocity that is produced by diffusion paths through the cores of the dislocations themselves. When this is done it is found that at the lower temperatures the activation energy of creep is that of pipe diffusion and that the stress exponent is increased by 2 (Evans and Knowles [1977], Spingarn *et al.* [1979], and Sherby and Weertman [1979]). Equation (16) is of the form predicted by the theories. Experiments show that the power exponent increases from about 4–5 at the higher temperatures to values around 6–9 at the lower temperatures. Thus the discrepancy between theory and experiment involving the additional factor σ^2 persists as the temperature is lowered.

The reader should not be misled into believing that it has been proved conclusively that pipe-diffusion controls steady-state creep at lower temperatures. Other mechanisms not involving diffusion have been proposed for this region (Poirier [1978]). A recent series of high voltage electron microscope (HVEM) observations on aluminum made to creep at moderate temperatures indeed indicate that the process of a dislocation cutting through other dislocations may be rate controlling (Cailliard and Martin [1982,1983]). (In these experiments precrept specimens were made to creep further while in the microscope. Because of the resultant stress change the microscope observations must have been on a transient-type creep. In our earlier discussions in this chapter [see also ch. 19, eq. (26)] it was seen that at the lower temperatures a non-diffusion process can control the transient creep strain. But at the very large strains in the steady-state regime pipe-diffusion can be rate controlling. Thus Cailliard and Martin's observations do not disprove the pipe-diffusion mechanism.)

1.10. Nabarro–Herring–Coble creep

A special kind of steady-state creep can occur in specimens which are either very small or consist of very fine grain material. The dimensions of the grains or of the sample have to be approximately 0.01–0.001 cm or less. NABARRO [1948] pointed out that at high temperatures an appreciable number of vacancies can be transported from one side of a grain to another in a fine-grained specimen under stress. This mass transport is indicated in fig. 16. HERRING [1950] developed an analysis from which the creep rate produced by this mechanism can be calculated. The Nabarro–Herring creep is perhaps the best understood of all creep phenomena. The conditions of temperature and stress under which it can occur are indicated on the creep diagram of fig. 3.

The theory of Nabarro–Herring creep is simple. The energy which is required to form a vacancy at the top or bottom surface of the grain pictured in fig. 16 obviously is different from the energy required to form a vacancy at a side surface if a vertical tensile stress is applied to the grain. This energy difference is of the order of σb^3. If the stress σ is small compared to kT/b^3, the gradient in the vacancy concentration which exists between the top or bottom surface and a side surface is approximately equal to $(2c_0/L)(\sigma b^3/kT)$, where L is the dimension of the grain and c_0 is the equilibrium vacancy concentration when the applied stress is equal to zero. This gradient in the vacancy concentration gives rise to a vacancy flow per unit time equal to $D_v L^2(2c_0/L)(\sigma b^3/kT)$ in a grain of average dimension L.

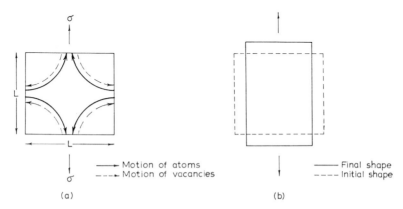

Fig. 16. (a) Mass motion of vacancies and atoms across a small grain of dimension L at a high temperature under an applied stress. (b) Change in shape of a grain under the mass motion shown in (a).

If there is a net flow of vacancies in one direction there must be an equal net flow of atoms in the opposite direction. Therefore atoms will be removed from the side faces of the grain and deposited on the top and bottom faces. In the course of time, the grain will change its dimensions in the manner indicated in fig. 16b. From a knowledge of the rate of flow of vacancies it is possible to calculate the rate of change of the grain dimension in the direction of the applied stress:

$$\dot{\epsilon} \approx \frac{5D}{L^2} \frac{\sigma \Omega}{kT}. \tag{25}$$

Note that the creep rate is proportional to the stress, and the activation energy of creep is identical to that of self-diffusion. Both of these predictions have been verified experimentally.

Figure 17 shows experimental data obtained on copper foils. In this figure the creep rate is plotted against the applied tensile stress. Because of surface tension, a thin foil actually suffers a compressive stress when the externally applied stress is zero. This compressive stress accounts for the negative creep rate of one point in the plot. A Nabarro–Herring-type of creep experiment thus can be used to make measurements of the surface tension of metals.

The diffusion path in Nabarro–Herring creep is not necessarily confined to the interior of a grain. COBLE [1963] pointed out this fact and calculated the creep rate enhancement produced by diffusion down the grain boundaries. If D_{gb} is the grain-boundary self-diffusion constant and δ the effective thickness of a grain boundary, eq. (25) becomes:

$$\dot{\epsilon} \approx \frac{5D}{L^2} \frac{\sigma \Omega}{kT} \left\{ 1 + \frac{\pi \delta D_{gb}}{LD} \right\} \tag{26}$$

when grain boundary diffusion is added to lattice diffusion. When the grain-boundary term in eq. (26) is large compared with unity, the activation energy of creep is that of grain-boundary diffusion (about one half that of lattice diffusion). The grain-size dependence then changes from L^{-2} to L^{-3}.

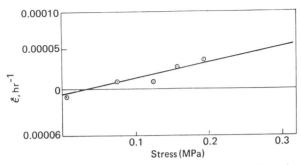

Fig. 17. Measured creep rates of copper foils 0.05 mm in thickness subjected to various stresses at 1002°C (data of PRANATIS and POUND [1955]).

2. Creep fracture and superplasticity

2.1. Superplasticity

In fig. 1 it is to be noted that eventually a tensile specimen fails in creep. Failure can occur simply because of a geometric effect. No specimen can be made with perfect dimensions. Some place along a length of specimen, the cross-sectional area $A_0 - \delta A_0$ is slightly smaller than the average cross-sectional area A_0. During creep a neck that pinches off may form at this point. For power-law creep it is simple to derive the value of the creep strain ϵ_P that exists in the region of initial area A_0 at pinch-off if it is assumed that the creep rate in the neck region is larger because the stress there is larger. (The effect of triaxial stresses is ignored.) The area $A - \delta A$ in the neck region after a strain ϵ is reached in the non-neck region, where the cross-sectional area is A, is given by:

$$\left[\frac{A - \delta A}{A}\right]^N = 1 + \left[\left(\frac{A_0 - \delta A_0}{A_0}\right)^N - 1\right]\exp(N\epsilon). \tag{27}$$

The neck pinches off when the right-hand side of this equation is equal to zero. If $\delta A_0/A_0 = 0.1$, the pinch off strain $\epsilon_P = 0.18$ when $N = 5$, $\epsilon_p = 0.27$ for $N = 4$, $\epsilon_p = 0.83$ for $N = 2$, and $\epsilon_p = 2.3$ for $N = 1$. Obviously, the smaller N, the larger will be the strain at failure. In practice, any material with a power exponent smaller than about $N = 2$ can be deformed to almost unlimited strain. Such material is called *superplastic*.

But how can N be made equal to or less than 2? Metals with moderately small to large grain size have values of N ranging from 3 to 5. However, if the grain size is very small (smaller than about 10 μm) and the grains do not grow appreciably during creep, N can be of the order of 2. A stable grain size can be produced by using two-phase material. Many superplastic metals have a eutectic or eutectoid composition or contain a dispersion of fine particles. A very fine microstructure can make an alloy superplastic. Hypereutectoid steels have been made superplastic (SHERBY *et al.* [1975]).

An important factor is that the grains in a superplastic material must be equiaxed. These equiaxed grains remain equiaxed throughout a deformation which may exceed 1000% (whereas in normal plastic deformation initially equiaxed grains become progressively elongated, just like the specimen itself which is stretched). The reason why N is small in superplastic alloys is not completely understood at present (MUKHERJEE [1979], TAPLIN *et al.* [1979], ARIELI and MUKHERJEE [1982] and LANGDON [1982]). It has to do with the strain-rate sensitivity of the flow stress (HOLT and BACKOFEN [1966]). One difficulty is that $N \approx 2$ only in an intermediate stress range. At higher stresses and usually at lower stresses as well N has a normal value. (A plot of log stress versus log strain rate has a sigmoidal shape because of this dependence of N on stress.) New creep mechanisms with smaller N values are possible in fine-grain material. The contribution of grain-boundary sliding to the

total creep strain is larger the smaller is the grain size. In superplastic alloys, 50–70%
of the total deformation is produced by grain-boundary sliding in the stress range in
which the alloys are superplastic. In both the lower and high stress ranges where they
are not superplastic, grain-boundary sliding accounts for only about 20–30% of the
total deformation (LANGDON [1982]). Various grain-boundary sliding theories of
creep do predict that $N = 2$ (LANGDON [1982]). (Nabarro–Herring–Coble creep
(with $N = 1$), that has been discussed in the last section, is important only when the
grain size and stress are small. If the stress is made sufficiently small this mechanism
must become dominant in fine-grain material. Thus at the very lowest stresses alloys
again become superplastic.) For a further discussion of mechanisms of superplasticity see ASHBY and VERALL [1973] and PADMANABHAN and DAVIES [1980].

2.2. Creep fracture

A creep specimen may fail in other modes before necking down. Deformation at a
temperature somewhere in the range of 0.4–0.7 times the (absolute) melting temperature can produce *cavitation* in a large number of metals and alloys. Small voids
nucleate on the grain boundaries, grow, and eventually coalesce to form long cracks
which may lead to failure (fig. 18). Even superplastic alloys can fail through cavity
growth (TAPLIN *et al.* [1979], MILLER and LANGDON [1979] and STOWELL [1980]).
The phenomenon of cavitation does not appear to be connected to a particular
crystal structure, as grain-boundary voids are found, for example, in copper, silver,
nickel, iron, zinc, and magnesium. Voids can form in high-purity, single-phase
material free of any grain-boundary particles (e.g., copper), in solid-solution alloys
such as α-brass and Cu–Al, and in more complex alloys such as ferritic steels,
austenitic stainless steels and various nickel-base superalloys. When present, hard
grain-boundary particles are favoured sites for the nucleation of voids. Whatever the
nucleation mechanism might be, it will be assisted by the large stresses which build
up around the particles during creep. These stresses at the particle–matrix interface
can arise from grain-boundary sliding (ARGON *et al.* [1980]) or from the presence of
differences in elastic and plastic behaviour between matrix and particles (SHIOZAWA
and WEERTMAN [1982]).

Once formed, voids grow by one of several mechanisms. (See SVENSSON and
DUNLOP [1981] for a review of the theories of void growth.) A tensile stress applied
normal to a grain boundary increases the equilibrium concentration of its vacancies.
Since the free surface of a void cannot support a normal tensile stress the concentration of vacancies in the grain boundary next to the void must be lower than that in
the grain boundary some distance away. Vacancies are driven into the void by the
concentration gradient and thus the void expands (HULL and RIMMER [1959] and
WEERTMAN [1974]). In this theory of void growth by vacancy diffusion it is assumed
that surface diffusion is so rapid that the void maintains its equilibrium shape of
spherical caps. However, as a void continues to enlarge eventually a point is reached
in which the surface diffusion no longer is fast enough to maintain the equilibrium
profile and the void becomes crack-like (CHUANG and RICE [1973] and GOODS and

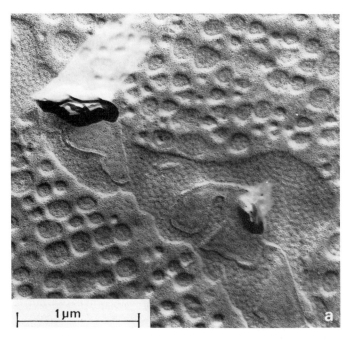

Fig. 18a. Voids on grain-boundary carbides in a specimen of the nickel-base superalloy Astroloy crept to failure at 750°C under a stress of 450 MPa. Shadowing is parallel to the stress axis. (From SAEGUSA *et al.* [1980].)

Fig. 18b. Voids coalesced into secondary cracks in a specimen of Astroloy crept to failure at 750°C under a stress of 450 MPa. Shadowing is parallel to the stress axis. (From SAEGUSA *et al.* [1980].)

NIX [1978]). Voids also can grow by plastic deformation (HANCOCK [1976], BEERE and SPEIGHT [1978]). The rate of void volume growth by stress-driven vacancy flow is proportional to the normal stress acting across the grain boundary, and is roughly independent of the size of the void, whereas in the case of growth by plastic deformation the growth rate is approximately proportional both to the void volume and to the strain rate. Therefore it is to be expected that low stresses and small size favour growth by vacancy diffusion, while large voids, or voids under high stress enlarge plastically. The rounded shape of very small voids often is taken as evidence of a diffusion mechanism at work, as is the fact that voids are most likely to be found on grain boundaries oriented transverse to the applied stress.

While grain-boundary cavitation has been studied extensively, both experimentally and especially theoretically, few accurate data are available on the kinetics of void nucleation and of void growth, and of the dependence of these processes on various deformation parameters such as temperature and creep stress. It has been noted by a number of workers that in the case of creep specimens which fail by cavitation, a Monkman–Grant relationship often holds between the time to failure, t_f, and the steady-state creep rate, $\dot{\epsilon}_{ss}$: $t_f \dot{\epsilon}_{ss} = C(T)$, where C is a constant which depends on the temperature T of deformation but not on the creep stress (PAVINICH and RAJ [1977]). However, this relationship may stem from the nucleation process, which can be rate-controlling in many cases. It is clear that grain-boundary cavitation is a complex phenomenon which is by no means completely understood.

2.3. Isoidionic maps

Figure 3, the creep diagram, presents only a very qualitative description of the conditions under which the different kinds of creep are dominant. Since the time this diagram was first presented, in the first edition of this book, the diagram has been improved vastly by being made quantitative. Ashby and his co-workers (ASHBY [1972], FROST and ASHBY [1982]) placed in the creep diagram curves of constant creep rate. Such plots, which Ashby called *deformation maps*, have become very popular. The creep rates are obtained either from experimental data or, where no data exist, from theoretical equations derived for different creep mechanisms. Figure 19a is one of the first plots. This way of presenting physical-mechanism information, experimental data and theoretical extrapolations is used now for many other physical properties of crystalline material besides steady-state creep rates. Such plots, as a general class, can be conveniently refered to as isoidionic maps. (We are indebted to Sandra V. v. Schlanger for this coinage.)

A further improvement (in ease of use as well as of construction) in isoidionic maps was made by LANGDON and MOHAMED [1978]. (See also RAJ and LANGDON [1982].) They chose scales for the axes such that the curves become straight lines. Figure 19b presents a simplified version of one such linearized plot of time-to-failure at high temperature.

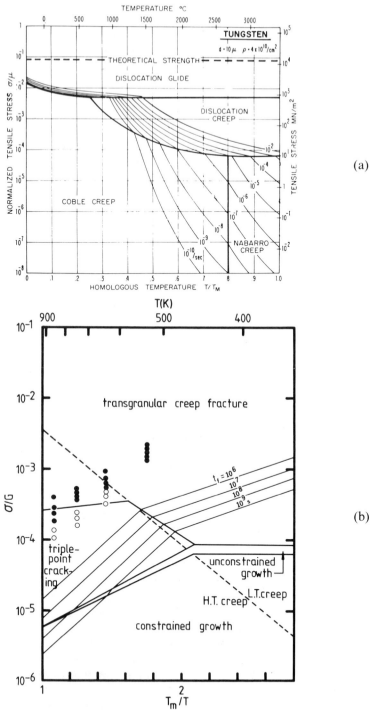

Fig. 19. (a) Isoidionic steady-state creep map for tungsten of grain size 10 μm (from Ashby [1972]). (b) Isoidionic fracture map for the theoretical fracture time of 99.3% pure aluminum which fails by constrained and unconstrained cavity growth as well as by other fracture mechanisms (from Raj and Langdon [1982]).

References

ANDRADE, E.N. DA C., 1910, Proc. Roy. Soc. **A84**, 1.

ARGON, A.S., I.-W. CHEN and C.-W. LAU in: Creep–Fatigue–Environment Interactions, eds. R.M. Pelloux and N.S. Stoloff (TMS–AIME, 1980) p. 46.

ARIELI, A., and A.K. MUKHERJEE, 1982 Metallurg. Trans. **13A**, 717.

ARKO, A.C., and J. WEERTMAN, 1969, Acta Metall. **17**, 687.

ASHBY, M.F., 1972, Acta Metall. **20**, 887.

ASHBY, M.F., and R.A. VERALL, 1973, Acta Metall. **21**, 149.

BEERE, W., and M.V. SPEIGHT, 1978, Met. Sci. **12**, 172.

CAILLIARD, D., and J.L. MARTIN, 1982, Acta Metall. **30**, 437 and 791.

CAILLARD, D., and J.L. MARTIN, 1983, Acta Metall. 31, in press.

CHUANG, T.J., and J.R. RICE, 1973, Acta Metall. **21**, 1625.

COBLE, R.L., 1963, J. Appl. Phys. **34**, 1679.

CONRAD, H., 1958, Acta Metall. **6**, 339.

EVANS, H.E., and G. KNOWLES, 1977, Acta Metall. **25**, 963.

FROST, H.J., and M.F. ASHBY, 1982, Deformation-Mechanism Maps (Pergamon Press, Oxford).

GAROFALO, F., 1965, Fundamentals of Creep Rupture in Metals (Macmillan, New York).

GITTUS, J., 1975, Creep, Viscoelasticity and Creep Fracture in Solids (Appl. Science, London).

GOODS, S.H., and W.D. NIX, 1978, Acta Metall. **26**, 739.

HANCOCK, J.W., 1976, Met. Sci. **10**, 319.

HERRING, C., 1950, J. Appl. Phys. **21**, 437.

HOLT, D.L., and W.A. BACKOFEN, 1966, Trans. Quart. ASM **59**, 755.

HULL, D., and D.E. RIMMER, 1959, Phil. Mag. **4**, 673.

LANGDON, T.G., 1982, Metallurg. Trans. **13A**, 689.

LANGDON, T.G., and F.A. MOHAMED, 1978, Mater. Sci. Eng. **32**, 49

LI, J.C.M., and A.K. MUKHERJEE, eds., 1975, Rate Processes in the Plastic Deformation of Materials (ASM, Metals Park, OH).

LUTHY, H., A.K. MILLER and O.D. SHERBY, 1980, Acta Metall. **28**, 169.

MILLER, D.A., and T.G. LANGDON, 1979, Metallurg. Trans. **10A**, 1869.

MOTT, N.F., 1956, Creep and Fracture of Metals at High Temperatures (Her Majesty's Stationery Office, London) p. 21.

MUKHERJEE, A.K., 1979, Ann. Rev. Mater. Sci. **9**, 151.

NABARRO, F.R.N., 1948, Report on a Conference on the Strength of Solids (Physical Society, London) p. 75.

PADMANABHAN, K.A., and G.J. DAVIES, 1980, Superplasticity (Springer, Berlin).

PAVINICH, W., and R. RAJ, 1977, Metallurg. Trans. **8A**, 1917.

POIRIER, J.P., 1976, Plasticité à Haute Température des Solides Cristallines (Eyrolles, Paris).

POIRIER J.P., 1978, Acta Metall. **26**, 629.

PRANATIS, A.L., and G.M. POUND, 1955, Trans. AIME **203**, 664.

RAJ, S.V., and T.G. LANGDON, 1982, Mech. and Corrosion Properties **B1**, 1.

SAEGUSA, T., M. UEMURA and J.R. WEERTMAN, 1980, Metallurg. Trans. **11A**, 1453.

SELLARS, C.M., and A.G. QUARRELL, 1962, J. Inst. Metals **90**, 329.

SHERBY, O.D., and P.M. BURKE, 1967, Prog. Mater. Sci. **13**, 325.

SHERBY, O.D., and J. WEERTMAN, 1979, Acta Metall. **27**, 387.

SHERBY, O.D., J.L. LYTTON and J.E. DORN, 1957, Acta Metall. **5**, 219.

SHERBY, O.D., B. WALSER, C.M. YOUNG and E.M. CADY, 1975, Scripta Metall. **9**, 569.

SHIOZAWA, K., and J.R. WEERTMAN, 1982, Scripta Metall. **16**, 735.

SPINGARN, J.R., D.M. BARNETT and W.D. NIX, 1979, Acta Metall. **27**, 1549.

STOWELL, M.J., 1980, Met. Sci. **14**, 267.

SVENSSON, L.E., and G.L. DUNLOP, 1981, Int. Met. Rev. **2**, 109.

TAKEUCHI, S., and A.S. ARGON, 1976, J. Mater. Sci. **11**, 1542.

TAPLIN, D.M.R., G.L. DUNLOP and T.G. LANGDON, 1979, Ann. Rev. Mater. Sci. **9**, 151.
WEERTMAN, J., 1955, J. Appl. Phys. **26**, 1213.
WEERTMAN, J., 1957, J. Appl. Phys. **28**, 1185.
WEERTMAN, J., 1960, Trans. Met. Soc. AIME **218**, 207.
WEERTMAN, J., 1968, Trans. ASM **61**, 681.
WEERTMAN, J., 1974, Metallurg. Trans. **5**, 1743.
WYATT, O.H., 1953, Proc. Phys. Soc. (London) **B66**, 495.

Further reading

Bhatia, M.L., Strengthening Against Creep, in: Alloy Design, eds. S. Ranganathan *et al.*, Proc. Indian Acad. Sci. (Eng. Sci.) 3 (1980) 255.
Frost, H.J., and M.F. Ashby, Deformation-Mechanism Maps (Pergamon Press, Oxford, 1982).
Garofalo, F., Fundamentals of Creep Rupture in Metals (Macmillan, New York, 1965).
Gifkins, R.C., ed., 6th Int. Conf. on Strength of Metals and Alloys, Melbourne, Australia (Pergamon Press, Oxford, 1982).
Gittus, J., Creep, Viscoelasticity and Creep Fracture in Solids (Appl. Science, London, 1975).
Haasen, P., V. Gerold and G. Kostorz, eds., 5th Int. Conf. on Strength of Metals and Alloys, Aachen, Germany (Pergamon Press, Oxford, 1979).
Jensen, R.R., and J.K. Tien, in: Metallurgical Treatises, eds., J.K. Tien and J.F. Elliott (Metallurgical Society of AIME, Warrendale, PA, 1981) p. 529.
Li, J.C.M., and A.K. Mukherjee, eds., Rate Processes in the Plastic Deformation of Materials (ASM, Metals Park, OH, 1975).
Mukherjee, A.K., Ann. Rev. Mater. Sci. 9 (1979) 191.
Myshlyaev, M.M., Ann. Rev. Mater. Sci. 11 (1981) 31.
Nowick, A.S., and B.S. Berry, Anelastic Relaxation in Crystalline Solids (Academic, New York, 1972).
Paton, N.E., and C.H. Hamilton, eds., Superplastic Forming of Structural Alloys (TMS–AIME, Warrendale, PA, 1980).
Poirier, J.P., Plasticité à Haute Température des Solides Cristallines (Eyrolles, Paris, 1976).
Sherby, O.D., and P.M. Burke, Progress in Materials Science, vol. 13 (Pergamon Press, Oxford, 1967) p. 325.
Takeuchi, S., and A.S. Argon, J. Mater. Sci. 11 (1976) 1542.
Taplin, D.M.R., G.L. Dunlop and T.G. Langdon, Ann. Rev. Mater. Sci. 9 (1979) 151.
Weertman, J. Trans. ASM, 61 (1968) 681.

CHAPTER 21

MECHANICAL PROPERTIES OF SOLID SOLUTIONS AND INTERMETALLIC COMPOUNDS

P. HAASEN

Institut für Metallphysik
Universität Göttingen
D-3400 Göttingen, FRG

R.W. Cahn and P. Haasen, eds.
Physical Metallurgy; third, revised and enlarged edition
© Elsevier Science Publishers BV, 1983

Introduction

This chapter deals with the influence of a second element on the mechanical properties of crystals, mainly metals. The foreign atom is assumed to *substitute* for an atom of the matrix. *Interstitial* solubility is in general limited and a second phase forms easily; the mechanical properties of this type of alloys are fully discussed in ch. 22. (See, however, also §2.4.2.) The observations to be described here were obtained mainly on single crystals. They are thus best suited for interpretation by means of dislocation theory (as presented in ch. 18). In the discussion of the mechanical properties of terminal solid solutions the reader is assumed to be familiar with the mechanical properties of pure metals (chs. 19/20) and with the structure of solid solutions (described in ch. 4). *Intermetallic compounds* are stoichiometric or nearly stoichiometric alloys whose atoms are arranged in a particular superlattice or structure (as is discussed in ch. 5). Another kind of non-random distribution of the component atoms may lead to the formation of precipitates (ch. 14) which greatly harden the alloy. A truly random distribution is certainly rare because of differences in atomic size, charge etc., which lead to attractive or repulsive interactions between solute atoms, precipitation or order. The dislocation is influenced by such inhomogeneities on a rather fine scale, determined by its interaction force with a "cluster" and the dislocation line tension. It is known that a dislocation is bent by a force f on its atomic length b to a minimum radius $R = \mu b^3/f$, where μ is the elastic shear modulus. In typical hardened alloys $R = 100b$ is reached.

A special kind of alloying and subsequent *hardening* is that by self point defects of the material, introduced in non-equilibrium concentrations by *irradiation* with fast particles (e.g. electrons, neutrons) or by *quenching* from high temperatures. In the former case self-interstitial atoms and vacancies are produced and especially associates between them: Frenkel pairs, vacancy clusters transforming into dislocation loops, stacking-fault tetrahedra and voids, interstitial–solute pairs. In the latter case vacancies and their associates are formed (see ch. 17). All of these defects interact with dislocations and therefore harden the base metal similarly to solute atoms and precipitates treated extensively in this chapter. There is of course a tendency for the point defects to annihilate on heating, either among themselves or with dislocations and at (internal) surfaces. In this respect their hardening effect differs with increasing temperature from true alloy hardening. There is an extensive literature on irradiation- and quench hardening (also due to their technological aspects) which has been summarized by DIEHL [1965], GITTUS [1978] and KIMURA and MADDIN [1971].

By mechanical properties we refer in the following mainly to the shear stress–shear strain (τ, ϵ) curve as measured at constant strain rate $\dot{\epsilon}$ and temperature T: $\tau = \tau(\epsilon, \dot{\epsilon}, T)$. Supplementary information is obtained from creep experiments under constant stress and temperature: $\epsilon = \epsilon(t, T, \tau)$, where t = time of creep. Much progress in understanding the mechanical properties has been made by combining these measurements with surface observations of slip lines, particularly with the electron microscope. Recently observations of dislocations in electron transmission, X-ray studies and atom probe measurements of the distribution of the component atoms and etch pit measurements of dislocation velocities have been used success-fully in this connection. Other types of measurements may also be of great help, but are less direct in the interpretation of alloy hardening.

1. Solid-solution hardening

Consider first the hardening effect of a second element in a nearly ideal solution. Most of the discussion is confined to binary substitutional alloy crystals with fcc structure, but hcp, bcc, diamond cubic or NaCl structures are briefly mentioned. Reports by COTTRELL [1954], SUZUKI [1957], HAASEN (1977, 1979a) and SUZUKI [1979a] describe results on polycrystals or on the *yield* stress of single crystals. Following the theoretical analysis of the stress–strain curve of pure fcc single crystals by SEEGER [1957,1958] it is interesting to discuss the effect of alloying on the different stages of this curve (see also HIRSCH [1975]). A first survey is presented in §1.1. After a description of relevant slip line and etch pit observations and electron transmission work in §1.2, the main theoretical ideas proposed for the explanation of solid solution hardening are presented in §1.3. This enables the effect of solid solution alloying on different parts of the τ–ϵ-curve to be discussed in §1.4. A discussion of similar effects in other structures follows in §1.5–7. Creep of solid solutions is investigated in §1.8, fatigue in §1.9.

1.1 Survey of stress–strain curves of fcc alloy single crystals

Stress–strain curves of typical solid solution crystals with a stacking-fault energy (SFE) lower than that of the base metal are shown in fig. 1 (Ni–Co) as a function of composition. Depending on strain, the flow stress of the alloy may be higher or lower than that of the base metal. This does not, however, imply softening by alloying. The three stages of the work-hardening curve of pure fcc crystals (cf. ch. 19) are still recognizable in the alloys, and thus the dislocation interactions leading to work-hardening must be much the same as in pure metals. But as the concentration of the solute is increased the critical shear stress increases, the easy glide range becomes longer, as does the stage II linear hardening range. Stage III, dynamic recovery, starts at higher stresses for the more concentrated alloys. The breaks in the Ni–Co curves indicate a delay in the usual change of slip system from primary to conjugate as a function of strain. As primary slip progresses, the orientation *overshoots* the side of the standard stereographic triangle as shown in fig. 31b (which implies that an abnormally high stress is needed to activate the conjugate slip system, cf. ch. 19). Horizontal serrated parts in the τ–ϵ-curves during which the crystals twin are observed in Cu–Ga and Cu–Ge (HAASEN and KING [1960]). Qualitatively, alloying with elements which lower the SFE changes the work-harden-ing behaviour in a similar way as does a decrease in temperature, as illustrated for Ni–Co alloys in fig. 2. If by alloying the SFE is not lowered, cross-slip starts already

References: p. 1403.

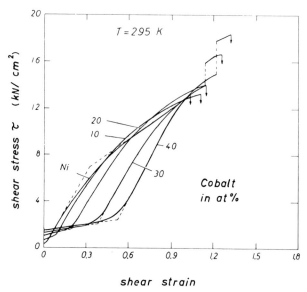

Fig. 1. Work hardening of Ni–Co single crystals of various compositions (% Co indicated) (after MEISSNER [1959]).

at the end of stage I, thereby reducing work-hardening in the next stage, see WILLE *et al.* [1980] for Cu–Mn.

1.2. Slip lines, etch pits, and electron transmission observations

Slip line measurements yield an authentic count of the number of (edge) disloca-tions that have left the crystal at a particular place. The intersection points of dislocations with the surface can frequently be detected by etch pits formed on them. By electron transmission micrography of thin foils cut from deformed crystals, dislocations left in the foil become visible. This powerful tool is particularly useful in alloy investigations.

1.2.1. Slip lines and etch pits

KUHLMANN-WILSDORF and WILSDORF [1953] have shown that slip lines on α-brass are more widely spaced and stronger than on pure metals (fig. 3). In between there is no fine slip. A strong line evidently does not correspond to a slip *band*, i.e. a packet of fine lines, but rather to a large single slip step on one atomic plane (WILSDORF and FOURIE [1956]). Deformed Ni–Co crystals, on the other hand, look more like pure metals in this respect although slip distribution is not quite as homogeneous in stage I (PFAFF [1962]). During easy glide the slip line density is found to remain constant, and steps grow. In stage II the density of lines increases and the new lines are shorter (inversely proportional to strain). Stage III is

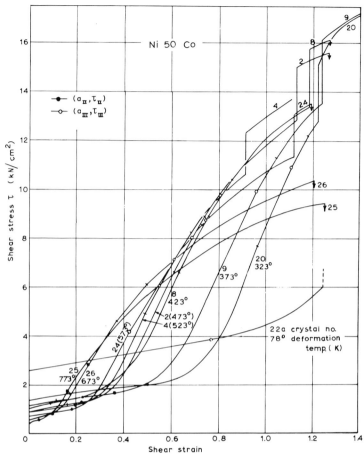

Fig. 2. Work-hardening curves of Ni–50 at% Co crystals as a function of temperature (after PFAFF [1962]).

characterized as in pure metals by intimate cross slip (ch. 19, fig. 3). α-brass shows prominent steps on cross-slip planes even during stage I. At 300°C and above, "fanning" slip lines are seen in Ni–50 at% Co, corresponding to climb of edge dislocations.

MEAKIN and WILSDORF [1960] studied etch pits on α-brass after "decorating" dislocations by adding 0.7 at% Cd. Annealed specimens had a dislocation density $N \approx 10^6/cm^2$ similar to that in good pure metal crystals. During easy glide N rose to $\sim 10^8/cm^2$. Dislocations on the primary slip planes were arranged in discrete groups of about 20 dislocations of like sign. The distances within a group agreed with those calculated for a dislocation pile-up (ch. 18). KLEINTGES [1980] has studied the operation of dislocation sources in Cu–0.6% Al activated by stress pulses. The technique was also used, most recently by KLEINTGES and HAASEN [1980], to

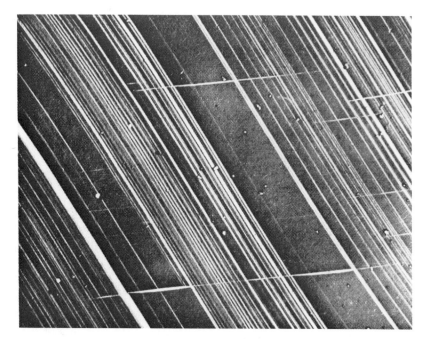

Fig. 3. Slip lines on Cu–30 at% Zn single crystal in stage I (from WILSDORF and FOURIE [1956]).

measure dislocation velocities in dilute alloys. Typical values are 1 μm/s at the critical resolved shear stress (CRSS), but increasing with a very high power of the stress. TRAUB et al. [1977] and NEUHÄUSER and FLOR [1979] have obtained dislocation velocities of the same order of magnitude by microcinematography from the length-wise and in-depth growth of surface slip lines in concentrated solid solutions.

MITCHELL et al. [1968] have studied the dislocation structure of narrow discrete slip bands on Cu–6% Al single crystals by optical and electron-microscopic techniques, and also via etch-pitting (HOCKEY and MITCHELL [1972]). These authors conclude that deformation proceeds by big avalanches of dislocations running right across the slip plane at the beginning of deformation. The residual dislocation distribution consists of trains of edge-dislocations of opposite sign on closely spaced slip planes which have activated some secondary slip systems. These slip bands resemble in their dislocation structure those of stage II or III of pure copper and do not further contribute to deformation.

1.2.2. Electron transmission pictures of fcc alloys

Dislocations in fcc alloys observed in transmission look somewhat different from those in pure metals. Instead of tangles and three-dimensional network structures, a planar arrangement of dislocations predominates. One reason for this change with alloying is probably the increasing difficulty of cross-slip. Also in non-homogeneous

Fig. 4. Cross-slip on Ni–50 at% Co single crystal deformed into stage III at 60°C (from MADER [1963]).

alloys preferential solute bonds are cut by the passage of the first dislocations and successive ones find an easier passage on the same slip plane. Dislocations in fcc crystals are split into partials. In between the partials there is a stacking fault. The width of the fault is proportional to $\mu b/\gamma$ where γ is the specific surface energy of the fault. WHELAN [1959] measured γ at extended dislocation nodes by electron transmission. γ decreases on alloying with an element of higher valency and cross-slip becomes more difficult (§1.4). Thus chances of retaining dislocation structures typical of bulk specimens during thinning to a foil are improved; see GALLAGHER [1970] for a recent review of measured SFE and CARTER and RAY [1977] for weak beam measurements of the SFE of copper alloys.

The changes with strain of dislocation arrangement in Cu–10 at% Al, Cu–15 at% Al single crystals were studied by HOWIE [1960], PANDE and HAZZLEDINE [1971] and KARNTHALER *et al.* [1972, 1975]. During stage I, "trains" of screw dislocations on primary slip planes are seen. Dislocations in neighbouring trains frequently have

References: p. 1403.

Fig. 5. Dislocations in Cu–10 at% Al single crystal foil deformed into stage II (from HOWIE [1960]).

opposite Burgers vectors. In stage II, interactions between trains and dislocations on other slip planes are observed (fig. 5), leading particularly to the formation of Lomer–Cottrell dislocations of ~ 1000 Å length. Large numbers of elongated loops are formed in the ⟨112⟩ direction. Howie has occasionally observed dislocation pile-ups on the primary slip plane. During stage III, faults widen and twins are formed. For comparison, in pure copper a ragged cell structure is observed during stages II and III. HOWIE (private communication) doubts whether the grown-in dislocation densities of pure metal and alloy differ by even as much as a factor of ten.

Dislocations in Ni–Co single crystals have been studied as a function of deformation and γ by MADER *et al.* [1963] and MADER [1963]. These authors etched thin foils from bulk crystals parallel to the primary slip plane, a tedious technique which, however, reduces interaction with the surface. The majority of the dislocations are found in edge orientation. Most of them are observed lying in dislocation "braids" parallel to ⟨112⟩ (fig. 6). Figure 7 shows two smaller overlapping dislocation pile-ups of opposite sign. These are rarely found, however, as are screw dislocations. Typical of stage II (except in pure Ni) is furthermore the formation of short Lomer–Cottrell dislocation segments (fig. 8). Little change of arrangement is observed with change of γ, although braids in pure nickel consist of short edge dislocation loops instead of groups of long dislocations of predominantly one sign.

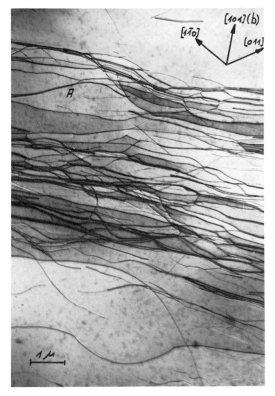

Fig. 6. Dislocation braid in Ni–40 at% Co single crystal deformed into stage I (from THIERINGER [1962]).

In stage I, most of the edge dislocations are paired with those of opposite sign and produce no long range stress field. This dislocation arrangement is certainly more complex than envisaged by any of the present theories, although Seeger's comes quite close to describing the uniaxial dislocation distribution with groups of predominantly one sign observed in the low stacking fault energy alloys.

1.3. Theories of solid-solution hardening

The last section shows that one reason for alloy-hardening is the change in the mechanism of work-hardening with alloying, i.e. the change in dislocation width, interaction, arrangement, density etc. The Hall–Petch relation for polycrystals describing the interaction of dislocations with grain boundaries is also changed in solid solutions (SUZUKI and NAKAMISHI [1975]). No great increase in grown-in dislocation density on addition of solute is observed although it could be expected for very dilute alloys (SEEGER [1957, 1958]). Most of the change of dislocation arrangement with alloying discussed so far can be explained in terms of a decrease in stacking-fault energy. Cross-slip then becomes more difficult. To form a jog in a

References: p. 1403.

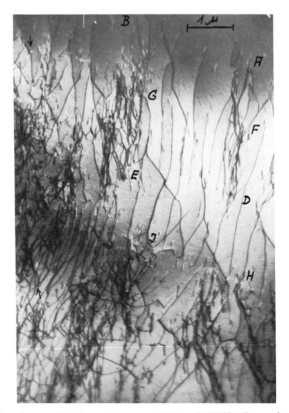

Fig. 7. Same as fig. 6, only stage II; pile-ups between ABDE and GFJH (of opposite sign).

widely split dislocation, a constriction must be produced. The formation and movement of jogs in extended dislocations in CuAl has been studied by CARTER [1980] by TEM. Twin nuclei are present in wide faults. In fact the stacking-fault energy would be expected to decrease in systems where alloying eventually produces a hexagonal phase (at low temperatures), as in the Hume–Rothery terminal solutions based on copper, silver, gold, and in Ni–Co. On the other hand adding nickel to copper etc., should raise γ (copper).

In addition to the *change of dislocation structure* on alloying, there are first in situ-observations of dislocation–solute interactions in specimens deformed in the TEM (KUBIN and MARTIN [1979]). Direct interactions between solute atoms and dislocations to be considered as mechanisms of solid solution hardening can be subdivided into two groups:

a) *Dislocation locking:* Solute atoms collect on dislocations at rest.
b) *Dislocation friction:* Solute atoms act on moving dislocations.

In the first case a pronounced yield point should be observed. Slip tends to

Fig. 8. Same as fig. 6. Lomer–Cottrell interactions in stage II.

concentrate by continuing where it has started. The effect of friction is simply to shift the whole stress–strain curve to higher stresses, although the friction may slowly decrease with increasing strain (if solute clusters are destroyed); a *gradual* yield point is then observed.

A few of the proposed elementary interactions between dislocations and solute will now be briefly described with respect to their overall dependence on solute concentration c and temperature of deformation T (not to be confused with the temperature T_e at which the alloy is in thermal equilibrium).

1.3.1. Dislocation locking mechanisms *

a) Chemical locking (SUZUKI [1957]). The stacking fault in the fcc lattice can be considered as a hexagonal layer a few atoms thick. Solubility in this hexagonal "phase" may differ from that in the fcc matrix, and solute can thus segregate in the fault. This normally increases the width of the fault. (The opposite may also occur.)

---

* See also ch. 18.

References: p. 1403.

Edge dislocations and (to a lesser extent) screw dislocations thus become pinned. For a dislocation to break loose an extra stress $\Delta\tau_c$ must be applied which depends on concentration c.

$\Delta\tau_c(c)$ describes a flat-topped parabola, but is approximately linear in c for small c. This stress is independent of temperature T below that of incipient diffusion, since the solute-enriched fault is too large for a correlated thermal fluctuation to help a dislocation to escape. The magnitude of $\Delta\tau_c$ depends on the thermochemical parameters characteristic of the alloy, see COTTRELL [1954] and FLINN [1958]. In typical cases $\Delta\tau_c \leqslant 10$ N/mm².

b) Elastic locking (COTTRELL [1954], SUZUKI [1979a]). Atoms of different size substituted in a crystal produce localized elastic strain fields which interact with those of dislocations. By compensation of their respective distortions, foreign atoms of different size find energetically favourable positions near dislocations. The elastic interaction energy between an edge dislocation and a substitutional atom may amount to several tenths of an eV. Since a screw dislocation has no hydrostatic stress field, its interaction with atoms of different size is much smaller. In the case of dissociated dislocations both partials will be locked separately, and since they in general will be of mixed character the strong locking difference between edges and screws disappears. Again a stress increment $\Delta\tau_c$ is needed to initiate dislocation movement which depends on c in a manner similar to that for mechanism **a)** (SUZUKI [1957]). In this case $\Delta\tau_c$ is proportional to the *misfit parameter*,

$$\delta = \frac{1}{a}\frac{\mathrm{d}a}{\mathrm{d}c},$$

where a is the lattice constant of the alloy. Again, $\Delta\tau_c$ is independent of temperature for an extended atmosphere, except for the case when an "atmosphere" of foreign atoms "condenses" along the dislocation core. Then $\Delta\tau_c$ should decrease with increasing T as T^{-3} (HAASEN [1959]). Besides the above size misfit or *parelastic interaction* there is a modulus misfit or *dielastic interaction* (FLEISCHER [1961,1963]). It is assumed that an individual solute atom interacts with a dislocation in the same way as a volume b^3 with a shear modulus differing from that of the matrix by $\Delta\mu$. Such elastically hard or soft spots in the matrix are felt by screw as well as edge dislocations. This interaction is called *dielastic*, as it is induced by the dislocation. It can be described by an "elastic polarizability" $\eta = 1/\mu(\mathrm{d}\mu/\mathrm{d}c)$ (KRÖNER [1964]).

c) Electrostatic locking (COTTRELL *et al.* [1953], FRIEDEL [1956]). The expanded dislocation core is negatively charged and interacts with the extra charge located near a solute atom of different valency. Although screening by free electrons is quite effective, there must be an appreciable change in the electron energy levels at such defects. This problem has not yet been adequately treated. Thus the estimated order of magnitude for the electrostatic interaction energy between dislocation and solute atom (about 1/5 of the elastic interaction) may not be correct. Apart from this the considerations of the previous paragraph also apply to this type of interaction.

d) Stress-induced order locking (SCHOECK and SEEGER [1959]). In the previously mentioned mechanisms a dislocation is locked when solute atoms collect on it. Even without this, the directionality of forces exerted by the dislocation on its environment may produce local order in the alloy around it. The dislocation "digs itself in" energetically in this way and a yield point may result. The effect should depend quadratically on concentration in the case of substitutional alloys. Schoeck and Seeger calculate it to be independent of temperature below the temperature of rapid diffusion. The effect acts differently on edge and screw dislocations.

1.3.2.   Summation of solute forces acting on moving dislocations

In the case of *locking* the solute atoms have the time to diffuse to positions of maximum interaction with a dislocation. In the other extreme called *friction* a stationary distribution of solute atoms, assumed to be a statistical distribution on both sides of the glide plane, interacts with a moving dislocation, and the question arises how many obstacles are in strong contact with the dislocation at any time. This is a summation problem. Its solution depends on geometrical parameters as solute spacing, $l_s = b/\sqrt{c}$, and range w of interaction as well as on the maximum elementary interaction force f_{max} in relation to the dislocation line tension $E_L = \frac{1}{2}\mu b^2$. LABUSCH [1970,1972] defines a dimensionless range parameter,

$$\eta_0 = \frac{w}{b}\left(\frac{cE_L^2}{f_{max}}\right)^{1/2}, \tag{1}$$

which differentiates between two cases of hardening, i.e., of summation:

$\eta_0 \ll 1$ (Fleischer–Friedel) means hardening by dilute, strong obstacles.

$\eta_0 > 1$ (Mott–Labusch) implies concentrated, weak obstacles. Figure 9 shows various states of dispersion with respect to the radius of curvature of the dislocations at the given stress.

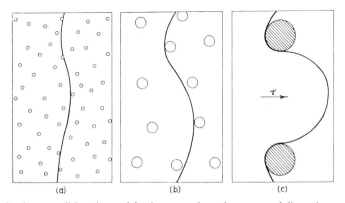

(a)   (b)   (c)

Fig. 9. Interaction between dislocation and foreign atoms in various states of dispersion.

References: p. 1403.

a) Dilute solutions. The mean spacing L between two solute atoms touched by a dislocation under a stress τ_c is (according to FRIEDEL [1964]):

$$L = (2E_L b/\tau_c c)^{1/3}, \tag{1a}$$

where c is the solute concentration per atom. The force balance at breakthrough reads:

$$\tau_c bL = f_{max}. \tag{1b}$$

Therefore:

$$\tau_c b^2 = f_{max}^{3/2} c^{1/2}/(2E_L)^{1/2}. \tag{1c}$$

This *Fleischer–Friedel yield stress* of dilute solid solutions has been confirmed (except for a numerical factor near one) by extensive computer simulations (KOCKS [1966], FOREMAN and MAKIN [1966], HANSON and MORRIS [1975], SCHWARZ and LABUSCH [1978]). Characteristic of dilute hardening is that obstacles are touched by the dislocations at full interaction force or not at all. This is no longer possible at higher solute concentrations (in the sense of $\eta_0 > 1$ above).

b) Concentrated solutions. LABUSCH [1970,1972] defines a distribution function $\tilde{\rho}(f)df = \rho(y)dy$ for the number of interactions with solute atoms which a unit length of dislocation (along the x-direction) has with strengths between f and $f + df$ or at spacings y, $y + dy$. The two quantities are uniquely related by the force–distance profile $f(y)$ of the obstacle (fig. 10). Then the force balance reads:

$$\tau_c b = \int \rho_c(y) f(y) \, dy, \tag{2}$$

where "c" refers to the state at breakthrough. The distribution function is calculated to

$$\rho_c = \frac{c}{b^2}[1 + G(0) \cdot f'(y)] \quad \text{or} \quad \rho_c = 0 \tag{2a}$$

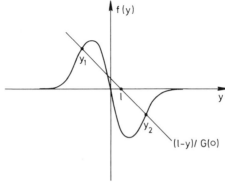

Fig. 10. Force–distance profile of elastic solute dislocation interaction, showing instable interval (y_1, y_2) given by mean dislocation position at l and compliance $G(0)$, see text.

for the intervals outside or inside (y_1, y_2) in fig. 10, respectively.
Here $G(0)^{-1} = 2\,(\varphi E_L)^{1/2}$, where

$$\varphi = \int \rho_c(y) f'(y)\, dy \tag{2b}$$

is the average curvature of the obstacle potential. It is plausible that most interactions ($\rho_c = \max$) take place where the curvature of the interaction potential has its maximum. If the "one" in eq. (2a) is neglected, eq. (2) can be integrated with eq. (2b) to yield

$$\tau_c b^2 = f_{max}^{4/3} c^{2/3} w^{1/3} / 2 (E_L b)^{1/3} (4I)^{1/3}, \tag{2c}$$

where

$$I = \int_0^1 \frac{\partial (f/f_{max})}{\partial (y/w)}\, d(f/f_{max})$$

is a pure number of order one. This is the *Labusch yield stress* for concentrated alloys ($\eta_0 > 1$). A rather similar result was obtained from completely different assumptions by RIDDHAGNI and ASIMOW [1968]. Computer simulations by SCHWARZ and LABUSCH [1978] have in the meantime confirmed the analytical solution for $\eta_0 > 1$, eq. (2c), and also allowed to interpolate between the Fleischer ($\eta_0 \ll 1$) and Labusch ($\eta_0 > 1$) situations as

$$\tau_c = \tau_c^{Fl}(1 + 2.5\eta_0)^{1/3}. \tag{2d}$$

Equation (2d) tends to τ_c^{Fl}, eq. (1c), for $\eta_0 \ll 1$ and to τ_c^{Lab}, eq. (2c), for $\eta_0 > 1$.

c) Effect of different obstacles on the same dislocation. One of the main advantages of Labusch's theory is that it allows to *superimpose*, by various distributions ρ_{ci}, the effects of different obstacles on the same dislocation. These might be obstacles at different spacings from the slip plane (where it follows that only the ones adjacent to the slip plane are important). Also solute atoms above and below the slip plane interact differently with the compression and dilation zones of an edge dislocation as far as the size misfit (parelastic interaction) is concerned: $f_{max}^P = \mu b^2 |\delta|$. The dielastic component $f_{max}^d = \alpha \mu b^2 |\eta|$ is of the same sign for solute atoms above and below the slip plane. Labusch shows that both effects and positions add up to an effective interaction force (see also HAASEN and KRATOCHVIL [1973]):

$$f_{max}^{eff} = \mu b^2 (\delta^2 + \alpha^2 \eta^2)^{1/2}. \tag{2e}$$

FLEISCHER [1963] in his theory had rather arbitrarily put:

$$f_{max}^{eff} = \mu b^2 |\delta - \alpha \eta'|, \quad \text{with} \quad \eta' = \eta/(1 + |\eta|/2),$$

see also KOSTORZ and HAASEN [1969] and SAXL [1964].

d) Short-range order destruction. Another possible flow stress component for a dislocation stems from the destruction of any existing *short-range order* in the alloy by slip. This creates a diffuse antiphase boundary of specific energy Γ per cm^2 and leads to (FISHER [1954]):

$$\tau_c b = \Gamma = f_{max} w c_p / b^2, \qquad (3)$$

where c_p is the fraction of solute pairs across the slip plane. c_p should depend on solute concentration as c^2 for small c (FLINN [1958]). This is an energy storing interaction, not an elastic one. The computer simulation (SCHWARZ and LABUSCH [1978]) interpolates in this case as $\tau_c \approx \tau_c^{Fl}(1 + \eta_0)$. The elastic interaction of solute clusters of i atoms can be included in Labusch's theory by putting $c_i = c/i$ and $f_{max,i} = f_{max}\sqrt{i}$ without changing the overall result of eq. (2c).

e) **Thermal activation.** The above calculations all refer to $T = 0$. Attempts have been made to include thermal activation into the theory of the friction stress due to elastic solute interactions (LABUSCH et al. [1975]). Labusch recently has described the motion of a dislocation in a field of obstacles under the action of random thermal forces, $b\tau_T$, by the equation

$$m\ddot{y} + B\dot{y} - E_L y'' = b\left[\tau_c + \tau_i(x - x_i, y - y_i) + \tau_T(x - x_n, t - t_K)\right]. \qquad (4a)$$

Here m is the dislocation mass per unit length, B the phonon and electron drag coefficient, i numbers the solute atoms acting on the dislocation while the thermal forces (phonons) hit at positions x_n, times t_K (there should be sums over all i, n, K). By clever scaling the equation can be written in new variables:

$$\xi = \frac{x}{b}\left(\frac{f_{max}c}{2E_L}\right)^{1/2}, \quad \eta_1 = \frac{y}{b}\left(\frac{c2E_L}{f_{max}}\right)^{1/2}, \quad \theta = \frac{t}{b}\left(\frac{f_{max}}{2m}c\right)^{1/2},$$

$$S = \tau_c/\tau_c^{Fl}; \quad S_i = \tau_i/\tau_c^{Fl}, \quad \gamma = Bb(2f_{max}mc)^{-1/2},$$

and then becomes

$$\frac{1}{2}\frac{\partial^2 \eta_1}{\partial \theta^2} + \gamma\frac{\partial \eta_1}{\partial \theta} - \frac{1}{2}\frac{\partial^2 \eta_1}{\partial \xi^2} = S + g((\eta_1 - \eta_i)/\eta_0)\delta(\xi - \xi_i)$$

$$+ \Delta\theta \cdot g_{n,K}\delta(\xi - \xi_n)\delta(\theta - \theta_K). \qquad (4b)$$

Solutions have been sought in three limiting cases:

1) Low temperatures ($T < 20$ K), no thermal activation, $\gamma \ll 1$, giving a lowering of the flow stress due to the dislocation inertial mass (especially in superconductors): GRANATO [1971], SCHWARZ and LABUSCH [1978], LABUSCH [1981].

2) Intermediate temperatures ($5T \leqslant T_{Debye}$), no inertial effects ($\gamma > 1$), giving athermal creep due to zero-point vibrations of the dislocation at the obstacles (NATSIK [1979], STARTSEV [1981]).

3) Normal temperatures, $\gamma > 1$, thermally activated flow. The dislocation is excited to oscillations with random amplitudes and phases, the energy contained in each mode being kT. Then Labusch calculates:

$$\langle g_{n,K}^2 \rangle = 2\gamma kT(2E_Lc)^{1/2}/bf_{max}^{3/2}, \quad \langle g_{n,K} \rangle = 0.$$

This leads to a scaling of the temperature as

$$\vartheta = kT(2E_L c)^{1/2}/bf_{max}^{3/2} = \frac{kTc}{\tau_c^{Fl}b^3},$$

while the yield stress at the temperature T scales with τ_c ($T = 0$). This is called the *stress equivalence* of solution hardening (BASINSKI *et al.* [1972]). NABARRO [1982] has recently shown that stress equivalence follows naturally from Labusch's theory but should not be observed in the low-concentration range of the Friedel theory.

Assuming a parabolic top of the function $f(y)$ in fig. 10 one obtains generally with an Arrhenius ansatz for the strain rate $\dot{\epsilon}$:

$$\tau_c(T, \dot{\epsilon}) = \tau_{c0}\left[1 - (T/T_0)^{2/3}\right]^{3/2}, \tag{4c}$$

with $kT_0 = f_{max}w/\ln(\dot{\epsilon}_0/\dot{\epsilon})$.

The problem now is to establish the contribution of the various mechanisms to the measured critical shear stress and yield stress increment according to their respective temperature- and concentration dependence. Various investigators have also tried to judge different models of solute solution hardening from the effect of varying the size misfit and valence. It is not always easy to define atomic "size" and "valency" in a solid solution (see ch. 4). A further difficulty arises in differentiating between locking and friction mechanisms: All models for dislocation locking might also lead to a friction drag if at high temperatures or low dislocation velocities solute enrichment moves with the moving dislocation (*microcreep* effect). KOCKS has recently worked out a theory (unpublished) for the temperature-independent plateau in the CRSS assuming fast jumps of solute atoms in the case of moving dislocations; see also SCHWARZ [1979]. Furthermore if the yield stress is attained successively in different parts of the specimen (by propagation of a *Piobert–Lüders band*) and work-hardening is rapid, a yield point maximum may not show and macroscopically the specimen will be characterized by a high friction stress. In both cases the stress–strain curve in the initial stages will appear serrated.

1.4   Solid-solution effect on fcc stress–strain curves

The different characteristic parameters describing a standard work-hardening curve of a fcc single crystal according to §1.1 will be discussed in this section in the light of direct observations of deformed alloy crystals (§1.2) and of theoretical considerations (§1.3).

1.4.1.   Critical shear stress τ_0 of fcc solid solutions

Without special treatment a yield point maximum is rarely observed at the beginning of plastic deformation in a tensile test at constant strain rate. Furthermore, the temperature dependence of the flow stress during easy glide as determined by temperature change on one crystal is found identical with that of the critical shear stress from a series of specimens (HAASEN and KING [1960], HENDRICKSON and FINE [1961b]). This implies that τ_0 is determined by a friction, rather than a

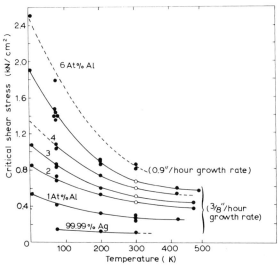

Fig. 11. Critical shear stress of Ag–Al single crystals (grown from the melt at two rates of solidification) versus temperature (after HENDRICKSON and FINE [1961b]).

locking mechanism. Typical examples for $\tau_0(T)$ are given in fig. 11. Below 300 K the critical shear stress increases more strongly with falling temperature in the alloy than in pure metals. Between 300 K and 500 K, $\tau_0(T)$ shows a nearly temperature-independent "plateau", the height of which increases rapidly for small concentrations. A second decline in $\tau_0(T)$ occurs at still higher temperatures, sometimes after a slight peak (ROGAUSCH, unpubl., see HAASEN [1964] and SEEGER [1958]). Examples of the concentration-dependence of τ_0 are given in fig. 12. The curvature of $\tau_0(c)$ for fcc · alloys is found to be negative at ~ 400 K with the exception of Al–Zn single crystals reverted after zone formation (DASH and FINE [1961]), which show a positive $d^2\tau_0/dc^2$ indicative of cluster hardening (see §1.3.2d, also §2 and fig. 23). At low temperatures τ_0 for Ag–Al increases more nearly linear with concentration. There is overwhelming evidence for many fcc alloys, particularly those based on Cu, Ag and Au, showing that the plateau stress τ_0 increases as the square root of the solute concentration, or as $c^{2/3}$, at least in the range $c \leqslant 0.1$ (HAASEN [1967a,1968], see also NIXON and MITCHELL [1981]). Figure 13 shows a plot of measured plateau stresses of copper alloy single crystals vs concentration to the 2/3 power.

To understand *plateau hardening*, i.e. the room temperature critical shear stress of the alloy, the forest intersection mechanism is of little help. Although the constriction energy will increase on alloying when γ decreases * and therefore dislocation intersection will become more difficult, the grown-in dislocation density does not change to the extent necessary to explain the strong decrease in activation volume observed for the alloys (see §1.2). HENDRICKSON and FINE [1961b] observe, however,

---

* There are, also, alloys like Cu–Ni in which γ should increase with concentration of solute.

a refinement of the substructure of Ag crystals with increasing Al concentration. HAMMAR *et al.* [1968] have investigated by etch pits the contribution to solution-hardening from changes in either the dislocation density or the dislocation arrangement due to solute additions. They found this contribution to be of minor importance in Ag–In and Ag–Sn crystals. In the former, a $c^{2/3}$ or $c^{3/4}$ law was found to fit the $\tau_0(c)$ results better than $c^{1/2}$. KLOSKE and FINE [1969] analyzed τ_0 in the Au–Ag system in terms of a c^2-component of the friction stress in addition to the $c^{1/2}$ term. The former may be interpreted in terms of local order hardening (mechanism **d** in §1.3.2). SVITAK and ASIMOW [1969] on the other hand found no effect of quenching on the CRSS of Ag–Au single crystals at 200 K except near the 75% Au composition. The plateau stress of all the investigated fcc alloys based on Cu, Ag, Au and Pb is, in its dependence on concentration and solute characteristics as well as in its absolute magnitude, fully explained by mechanism **b** (§1.3.2), eq. (2c), as fig. 13 proves for the case of copper alloys. In fig. 13b the slopes of the straight lines of fig. 13a are plotted to the power 3/4 versus the Labusch combined interaction parameter $\epsilon_L = (\delta^2 + \eta^2\alpha^2)^{1/2}$ as required by eq. (2e). This plot includes the effect of a second order parelastic interaction as proposed by JAX *et al.* [1970] for dissociated dislocations, see also TAKEUCHI [1968] and GYPEN and DERUYTTERE [1981]. In particular, size misfit (δ) alone is not able to explain plateau hardening as evidenced especially by the Cu–Si and Cu–Ni results. $\alpha^{-1} = 16$ (for edge dislocations) on the whole gives a better fit than $\alpha^{-1} = 3$ (for screw dislocations). There are

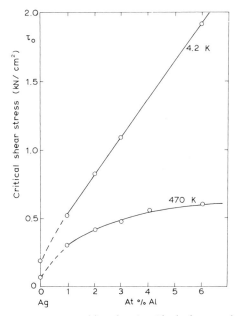

Fig. 12. Critical shear stress versus composition for Ag–Al single crystals at two temperatures (HENDRICKSON and FINE [1961b]).

References: p. 1403.

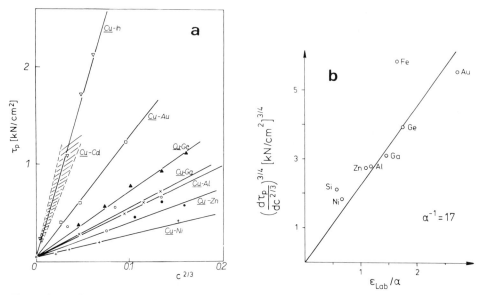

Fig. 13a,b. (a) Plateau stress of copper alloy single crystals versus solute concentration to power 2/3 (after KRATOCHVIL and NERADOVA [1971]). (b) Slopes of straight lines (to power 3/4) of fig. 13a versus the Labusch combined interaction parameter $\epsilon_L = (\delta^2 + \eta^2\alpha^2)^{1/2}$.

some alloy systems in which δ and η have the same sign, leading theoretically to partial compensation of dielastic and parelastic interactions if $\alpha > 0$. There is experimental evidence, however, that this does not happen in practice. Only the Labusch combination of parameters $(\delta^2 + \eta^2\alpha^2)^{1/2}$ is able to explain hardening in Ag–Al, Ag–Zn, Au–Zn within the frame-work of mechanism **b**. The dependence on the 4/3 power of ϵ_L in eq. (2c) fits the present experimental results better than a 3/2 power law. The localized obstacle mechanism, according to both theory and experiment, appears to be much more important in solution-hardening than the locking mechanisms (SUZUKI [1979]) at least for $c \leqslant 0.1$. It is interesting to note that solution-hardening is in fact controlled by atomic concentration of solute rather than by electron concentration as has once been proposed (HUTCHISON and HONEYCOMBE [1967]). PEISSKER [1965a] showed this by the dependence of τ_0 on c(Ga–Ge) for ternary Cu–Ga–Ge single crystals of constant electron concentration. Also the results in the Ag–Pd system (KRATOCHVIL [1970]) do not correlate with electron concentration, while they do correlate with atom concentration. Ternary hardening was investigated in Cu–Si–Ge by FRIEDRICHS and HAASEN [1975]. As expected from Labusch's theory the plateau stress τ_p follows from the binary hardening increments $\Delta\tau_{Ge}$ and $\Delta\tau_{Si}$ to the CRSS of pure Cu, τ_{Cu}, as

$$\left(\tau_p - \tau_{Cu}\right)^{3/2} = \Delta\tau_{Ge}^{3/2} + \Delta\tau_{Si}^{3/2}.$$

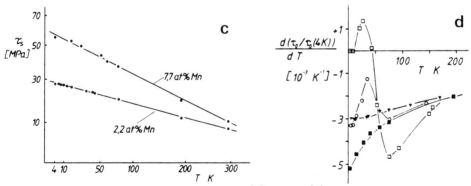

Fig. 13c. Thermal CRSS component $\tau_s = \tau_c - \tau_\mu$ as $\tau_s^{2/3}$ versus $T^{2/3}$ for CuMn solid solutions($\tau_\mu = \tau_{\mu 0}\mu(T)/\mu(0)$, μ = shear modulus, $\tau_{\mu 0}$ = 6 MPa (for 2,2% Mn), $\tau_{\mu 0}$ = 15MPa for 7,7% Mn (after WILLE and SCHWINK [1980]).

Fig. 13d. Slopes of normalized CRSS of CuMn alloys with respect to temperature. ■: 7,7%; ▼: 1,48%, ○:0.96%, □: 0,11% Mn. (After WILLE *et al.* [1982].)

A similar conclusion is reached by GYPEN and DERUYTTERE [1977]. The low temperature increase in the CRSS of fcc alloys is well described by the formal eq. (4c), see fig. 13c. There is however a low temperature anomaly at still smaller solute concentration, see fig. 13d (WILLE *et al.* [1982]). The CRSS goes through a maximum at 50 K, as expected from the theory of the inertial effect (GRANATO [1971] and SCHWARZ and LABUSCH [1978]). The activation energy obtained from the formal Arrhenius description, eq. (4c), often comes out rather high (0.5–1.5 eV), much higher than expected for the interaction of a dislocation with single solute atoms. It has been suggested (TRAUB *et al.* [1977], and KLEINTGES and HAASEN [1979]) that the dislocation on its way over the slip plane gets stuck in so-called hard configurations where it has to be thermally activated over *groups* of solute atoms. This would also explain the unexpectedly high stress exponents of the dislocation velocity (KLEINTGES and HAASEN [1979]). Furthermore the stress equivalence of solution hardening seems to get lost above about 2% solute concentration, perhaps as a consequence of clustering (SCHWINK and WILLE [1980] and WENDT and WAGNER [1982]). Another open question is that of strain aging in the plateau range under conditions of dynamic deformation (GLEITER [1968], NEUHÄUSER and FLOR [1978], SCHWARZ [1982]).

1.4.2. The yield phenomenon in fcc alloys

An initial maximum of the stress strain curve which can be recovered after deformation by heat treatment (strain aging) has been found in carefully treated fcc solid solutions. This yield point has been studied in particular on single crystals of Cu–Zn (ARDLEY and COTTRELL [1953]), Cu–Al (KOPPENAAL and FINE [1961]) and Ag–Al (HENDRICKSON and FINE [1961a]). A typical series of yield point observations on various Cu–Zn alloy crystals is shown in fig. 14. They fit the theoretical picture of pinning of dislocations by solute atoms described in §1.3.1. A time t_a is

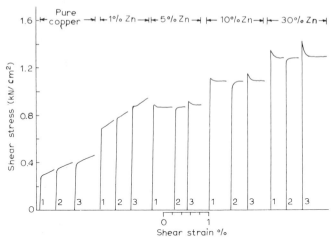

Fig. 14. Stress–strain curves of Cu and α-brass single crystals (1) first loading (2) immediately retested (3) after 2 h at 200°C. (after ARDLEY and COTTRELL [1953]).

necessary at a temperature T_e to permit solute atoms to diffuse to their energetically favourite positions at dislocations and thus establish a yield stress maximum (ch. 18). $\Delta\tau$ is the extra stress to free the dislocation from this interaction. t_a is surprisingly small in substitutional alloys after straining and depends exponentially on T_e with an activation energy E much smaller than expected for equilibrium solute diffusion. For Ag–6 at% Al, about 50% of $\Delta\tau_{max}$ is obtained after 2% straining and aging 1 min at 296 K, with $E \approx 0.55$ eV. Vacancies (and interstitials?) produced during deformation and/or moving in highly strained regions near dislocations could explain such rapid solute movement. If the aging is done under a stress approaching the flow stress it proceeds faster and reaches higher stresses $\Delta\tau_{max}$ (EVERS [1959]). This observation enables one to separate a strain-aging yield point as described above from an unloading yield point often observed in pure metals (HAASEN and KELLY [1957]) and alloys (PFEIFFER and SEEGER [1962]). For Cu–Zn and Ag–Al, $\Delta\tau_{max}$ varies linearly with concentration for small c and more gradually for larger c. In Cu–Al on the other hand, $\Delta\tau(c)$ is a quadratic function as expected for order-locking, although numerically too small (§1.3.1d). In this alloy the yield drop is rather gradual and shows a minimum as a function of the annealing temperature before quenching. There is a slight decrease of $\Delta\tau$ for increasing deformation temperatures for this alloy as well as Ag–Al, while $\Delta\tau$ is found to be independent of T for Cu–Zn. This is to be expected for chemical or order-locking, as well as for elastic locking in substitutional solid solutions. The latter is considered unlikely for Ag–Al because the size factor $\delta \approx 0$, and there are wide stacking faults. The hypothesis of chemical locking best fits the results for Ag–Al and Cu–Zn (see §1.3.1a). FLEISCHER [1966] concludes that the binding energy W_m between disloca-tions and solute atoms, see FIORE and BAUER [1968], correlates well with the combined elastic interaction $\epsilon_{FI} = |\eta' - 3\delta|$, for Cu-Si, Cu-Ge, Cu-Sn. W_m is smaller than 0.4 eV for copper alloys.

1.4.3. Easy glide and overshoot in fcc alloys

There are two possible mechanisms of initial slow work-hardening (stage I, ch. 19) in alloy crystals: (a) *single slip* occurs, the mean free path of dislocations on the primary plane approaches the specimen width; (b) *inhomogeneous slip* by Piobert–Lüders band propagation may simulate "easy glide". In the first case, the work-hardening rate θ_I is finite and decreases with decreasing temperature and increasing concentration. θ_I depends strongly on orientation and is small for single slip orientations only. The stress for onset of rapid hardening is $\tau_{II} \approx 2\tau_0$ (GARSTONE and HONEYCOMBE [1957]). The extent of easy glide a_{II} is therefore larger at low temperatures and for concentrated alloys. In the case of inhomogeneous slip $\theta_I \approx 0$ and $\tau_{II} \approx \tau_0$. Here a_{II} is fairly independent of orientation and temperature and increases with concentration. Examples for case (a) are found in Ni–Co (PFEIFFER and SEEGER [1962], fig. 1) and Ag–Al (HENDRICKSON and FINE [1961b]), while α-brass is typical for case (b) (PIERCY *et al.* [1955]). In brass, yield propagation has been observed directly. The Cu–Ga and Cu–Ge alloys appear to change from (a) to (b) with increasing concentration, τ_{II}/τ_0 decreases from 2 to 1 and a_{II} shows a flat maximum at small concentrations (HAASEN and KING [1960], PEISSKER [1965a]).

To understand the pronounced true easy glide behaviour of alloys [case (a)] it must be realized that dislocations of secondary slip systems find it difficult to move when dislocations in the primary system are widely extended (SEEGER [1958]). An exception is the activity of dislocations in the cross-slip plane (having the same Burgers vector as the primary dislocations) observed in α-brass. The same argument leads to an explanation of the *overshoot* phenomenon typical for alloys of low γ (see §1.1). Pinning of secondary slip dislocations on the other hand does not explain overshoot, as has been shown by a convincing experiment by PIERCY *et al.* [1955]. Overshooting is measured frequently by the ratio of the applied shear stresses $\varrho = \tau_{sec}/\tau_{prim}$ at the beginning of secondary (conjugate) slip ($\varrho = 1$ without overshoot, $\varrho > 1$ indicates overshoot). $\varrho = 1.28$ for Cu-30 at% Zn, independent of temperature and initial orientation of the crystal, while for Ni–Co ϱ is smaller at higher temperatures or unfavourable orientations (PFEIFFER and SEEGER [1962]).

1.4.4. Linear hardening

The work-hardening rate θ_{II} in stage II (ch. 19) of the stress–strain curve of fcc alloys is found to be remarkably independent of composition ($\theta_{II} \approx \mu/400$). This was predicted by SEEGER's [1957, 1958] theory of long-range elastic hardening which describes θ_{II} as proportional to the square root of the ratio of slip distance to slip-line length. A more recent treatment of stage II has been given by HIRSCH [1975]. Higher and longer surface steps have been observed in α-brass than in copper. The ratio of height to length is constant and thus the back stress remains unchanged. As we have seen in §1.2 the dislocation arrangement visualized by Seeger for stage II is realized best in alloy crystals of low γ.

References: p. 1403.

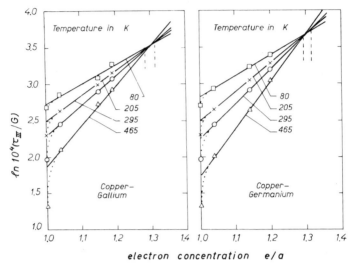

Fig. 15. Reduced stress at the beginning of stage III, logarithmically versus electron concentration $e/a = 1 + c(Z - 1)$ for copper with gallium ($Z = 3$) and germanium ($Z = 4$) and various temperatures (after HAASEN and KING [1960]).

1.4.5. Dynamic recovery

It was shown by slip-line observations by SEEGER [1958] and coworkers that the beginning of stage III (ch. 20) in pure metals coincides with the appearance of intimate cross-slip and fragmented slip bands. This stress-aided cross-slip is often described as dynamic *recovery* because it enables dislocations to circumvent obstacles. The same is true for Cu–Ga and Cu–Ge alloys (HAASEN and KING [1960]) and Ni–Co (PFEIFFER and SEEGER [1962]). A quantitative evaluation of the cross-slip process enabled HAASEN [1958] and SEEGER *et al.* [1959] to estimate stacking-fault energies γ from the stresses τ_{III} at the beginning of stage III as a function of temperature T and strain rate $\dot{\epsilon}$:

$$\ln\{\tau_{III}(T, \dot{\epsilon})/\tau_{III}^{\circ}\} = (\beta\gamma kT/\mu^2 b^4) \ln(\dot{\epsilon}/\dot{\epsilon}_0) \quad \text{for } 0{,}005 < \gamma/\mu b < 0.035. \quad (5)$$

Here τ_{III}°, $\dot{\epsilon}_0$ are constants to be evaluated from experiment and $\beta \approx 820$. (See SEEGER *et al.* [1959] for a more complete description and ESCAIG [1968] for an alternative theory.) As expected, the temperature- and strain-rate sensitivity of τ_{III} becomes zero and τ_{III} increases as the concentration of Ga and Ge approaches that of the fcc–hcp phase boundary (fig. 15) and the stacking faults widen. Figure 16 shows γ estimated for Ag alloys by TEM on extended nodes (GALLAGHER [1968,1970]) as a function of electron concentration. γ is a function of electron concentration only. While $\gamma(e/a)$ from τ_{III} appears to go through a maximum in Ag–Zn as well as in Cu alloys (ROGAUSCH [1967]) the node results indicate that γ does not change much with alloying, up to about $e/a = 1.05$ (cf. CARTER and RAY

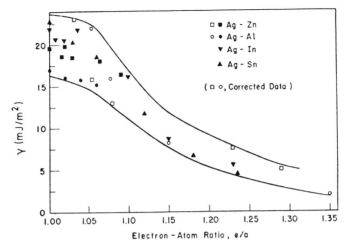

Fig. 16. Stacking-fault energy of copper alloys versus electron concentration, according to GALLAGHER [1970].

[1977]). In Ni–Co γ appears to decrease linearly towards the phase boundary (see PFAFF [1962]). Another theory to explain the increase of τ_{III} with concentration has been discussed by SEEGER [1958] for certain Al base alloys with strong size factor δ. Here dislocations must straighten themselves out before cross-slip can occur. A second mechanism of dynamic recovery is observed above 600 K in Ni–50 at% Co crystals (PFAFF [1962]) and has been identified with dislocation climb. A theory of τ_{III}^* under these conditions is presented in §1.8.

1.4.6. Deformation twinning in fcc alloys

Owing to the crystallographic similarity between stacking faults and twins in the fcc lattice, which are both determined by a $a/6 \langle 112 \rangle \{111\}$ shear, the low-γ alloys can be expected to twin easily during deformation. This has been verified by X-rays and metallography for Cu–Ga, Cu–Ge at temperatures up to 500 K. The relation between twinning and $\gamma(e/a)$ in Cu-Ga-Ge has been studied by PEISSKER [1965b]. In α-brass and generally in alloys deformed at low temperature the characteristic drop in load at the nucleation of the first twin lamellae is often missing. Cu–Zn nevertheless twins on a very fine scale (VENABLES [1961,1964], THORNTON and MITCHELL [1962]). In this case the propagation stress for the twin is thought to be higher than that for nucleation. Deformation twinning in Ag–Au crystals has been studied by SUZUKI and BARRETT [1958]. Silver twins at 0°C while gold has to be deformed at 77 K to twin. Twinning occurs when a critical shear stress is reached; this is found to depend on composition. A twinning stress maximum at intermediate concentrations indicates a solute hardening effect on twin propagation, in addition to the stress required to pull apart the partial dislocations bounding a stacking fault, which should vary with concentration linearly in γ(c). For a recent review see

MAHAJAN and WILLIAMS [1973], for new experiments on Ag alloys see NARITA and TAKAMURA [1974], on Cu–Al VERGNOL and VILLAIN [1979], STARTSEV et al. [1979] and MORI and FUJITA [1980], on nickel alloys NARITA et al. [1978].

1.5. Solid-solution hardening in hcp crystals

Little has been known until recently about how a second element in solution influences plastic deformation in hcp crystals. Here, in addition to the features of the cubic close packed structure (including stacking faults) the c/a axial ratio of hcp crystals is a variable with composition, and the contribution of non-basal slip may change with alloying. For $c/a > (c/a)_{ideal} = \sqrt{\frac{8}{3}}$ basal slip is favoured as in pure Zn and Cd, while for $c/a < \sqrt{\frac{8}{3}}$ slip often occurs also on prismatic planes or pyramidal planes. Mg is intermediate with a nearly ideal axial ratio. The slip direction in all cases is $\langle \bar{2}110 \rangle$.

The axial ratio of Mg increases slightly on addition of Al atoms which are ~ 11 at% larger. The critical shear stress τ_0 increases strongly with concentration up to about 7 at% Al as does the work-hardening rate, contrary to the behaviour of non-hcp alloys (SCHMID and SELIGER [1932]). Since pure Mg recovers rapidly even at room temperature, the work-hardening results of Mg alloys are difficult to interpret. The following discussion is therefore restricted to $\tau_0(c)$. The effect of alloying on the critical shear stress of Mg was investigated by SHEELY et al. [1959]. Of particular interest are the results for dilute Mg–In and Mg–Th alloys (fig. 17). Thorium atoms

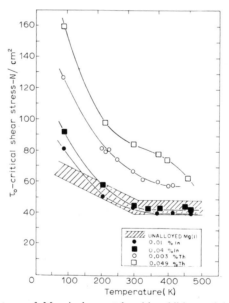

Fig. 17. Critical shear stress of Mg single crystals with additions of indium and thorium versus temperature (after SHEELY et al. [1959]).

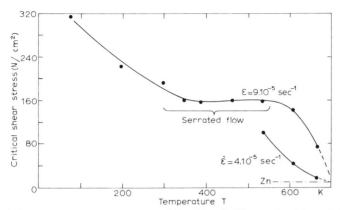

Fig. 18. Critical shear stress versus temperature of Zn–0.1 at% Cd crystals at two strain rates (after GILMAN [1956]).

are about 12% larger than Mg, indium only slightly so. The observed plateau-hardening thus seems to be due to size-misfit elastic interaction (§1.3.2b). On the other hand, In hardens Mg at low temperatures. A pronounced yield point has been observed in polycrystalline Mg–0.05 at% Th alloys at 200°C (COULING [1959]). Lithium markedly improves the ductility of Mg, as was observed in polycrystalline Mg–14.5 at% Li alloys by HAUSER *et al.* [1958]. Since the axial ratio decreases from 1.625 for pure Mg to 1.609 for this alloy it is likely that ductility is improved because additional prismatic slip can take place. Whereas pure Mg fails in a brittle manner below room temperature, it is ductile down to 78 K in the Mg–14.5 at% Li alloy.

It has been shown, however, that low temperature prismatic slip becomes easier too when Mg is alloyed with Zn or Al, although c/a does not decrease or even increases in these systems (AKHTAR and TEGHTSOONIAN [1968]). Alloying always increases the flow stress for basal slip. The plateau stress in Mg–Zn single crystals is found by these authors to follow eq. (1c) (beyond $c = 2.5 \times 10^{-4}$) with a reasonable order of magnitude for f_{max}. The same is observed for Mg–Cd single crystals by SCHARF *et al.* [1968]. Here again η and δ have the same sign and only $(\delta^2 + \alpha^2 \eta^2)^{1/2}$ gives a reasonable hardening from eq. (1c). A $c^{2/3}$ law fits these results better than $c^{1/2}$.

GILMAN [1956] investigated the influence of temperature and strain rate on τ_0 for Zn–0.1 at% Cd single crystals. Figure 18 shows the $\tau_0(T)$ curve typical for a dilute alloy as discussed in section 1.4.1. GILMAN finds prismatic slip to be less affected by the solute addition than basal slip. LUKÁČ and STULIKOVA [1974] have studied solution hardening in CdAg single crystals and found agreement with Labusch's theory. BOČEK *et al.* [1961] find for zinc single crystals with small additions of Cu, Pb, Cd etc. (of the order of 0.01 at%) that τ_0^2 is proportional to $A = \sum_i c_i \delta$, where c_i and δ_i are concentration and size factor, respectively, of the ith species of foreign atoms. According to a theory of TILLER [1958] the grown-in dislocation density N_0 should be proportional to A and thus $\tau_0 \infty \sqrt{N_0}$. Such an effect has been predicted by

SEEGER [1957,1958]. It plays no role, though, in Mg–Cd, Mg–Zn in the range investigated by AKHTAR and TEGHTSOONIAN [1968] and SCHARF et al. [1968]. It is noteworthy in this connection that HAESZNER and SCHREIBER [1957] observed no change in the stress–strain curve of Al with additions of Ag ($\delta \approx 0$) up to 0.12 at%.

1.6. Solid-solution hardening in the bcc structure

Metals of bcc structure are hardened much more by interstitial than by substitutional solute atoms. Interstitial atoms distort the lattice tetragonally and thus interact strongly with edge *and screw* dislocations. Also the crystal tolerates larger shear strains (from these distortions) than dilatations (from the symmetrical substitutional defects). If the interstitial foreign atoms are in solution like oxygen and nitrogen in the group V bcc metals, Nb and Ta, the hardening rate $d\tau_0/dc^{1/2}$ is well explained by a localized obstacle mechanism, see §1.3.2 (FLEISCHER [1964]). The same applies to the hardening of the hcp metals, Ti, Zr and Hf, by interstitials such as O, N, C (TYSON [1968]). The group VI bcc metals Cr, Mo, W as well as α-iron have very low interstitial solubilities at room temperature (for C in Fe, smaller than 10^{-7}). Small foreign atoms precipitation-harden these metals, as has been studied particularly in α-iron (see §2.4.2 and ch. 16, §5.5). Unfortunately interstitial and substitutional foreign atoms interact strongly, so their hardening effects are not additive (see ch. 16 and LÜTJERING and HORNBOGEN [1968]).

Fe–Cr single crystals have been studied by HORNE et al. [1963]. These authors observed a strong increase in the critical shear stress plateau (just above room temperature) with additions of chromium larger than 4 at% (fig. 19). The tempera-

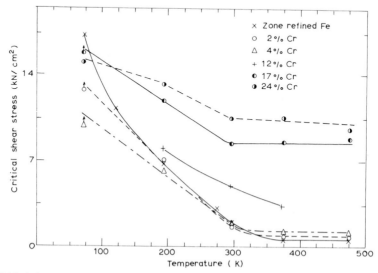

Fig. 19. Critical shear stress versus temperature for Fe–Cr single crystals (after HORNE et al. [1963]).

ture dependence of τ_0 in the $\langle 112 \rangle$, $\langle 110 \rangle \langle 111 \rangle$ systems below 300 K was actually reduced by the chromium addition. It is probable that chromium removed from solid solution some carbon which has been observed to influence the temperature-dependent part of the flow stress of α-iron + 0.003 at% C single crystals (MORDIKE and HAASEN [1962]). The size factor of Cr in Fe is very small. No stacking faults are expected in the alloy, although twins were observed to form. The almost quadratic increase of τ_0 (300 K) with concentration of chromium indicates friction due to cutting of favourable atom pairs by dislocations as discussed under §1.3.2d. A miscibility gap has been reported below 550°C, $c < 12$ at%. The work-hardening rate of the crystals is quite small so that in the 24 at% Cr alloy at room temperature necking started immediately after a pronounced yield point.

A solution-softening effect was also found at low temperatures in Ta-alloys, particularly in Ta–Re (fig. 20), by MITCHELL and RAFFO [1967]. It is doubtful whether in these alloys the scavenging effect of an interstitial–substitutional foreign atom interaction can be held responsible for this effect although scavenging can be shown to work in some alloys (GIBALLA and MITCHELL [1973]). Two other explanations have been forwarded: (a) anomalies in $\eta(c)$ (HARRIS [1967]); (2) easier kink

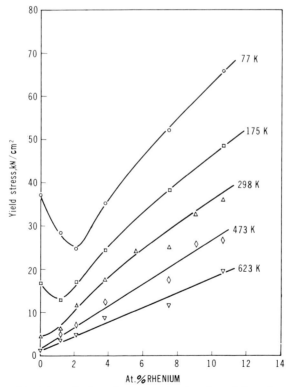

Fig. 20. Yield stress of tantalum–rhenium single crystals at various deformation temperatures (after MITCHELL and RAFFO [1967]).

References: p. 1403.

formation in the Peierls potential at substitutional atoms. It is clear now that screw dislocations in bcc metals are impeded by a high quasi-Peierls potential while edge dislocations are not. They interact more strongly with solute atoms (ŠESTAK [1979] and SUZUKI [1979b]).

It is surprising that assumption (2) has been disproved by SIETHOFF [1969] on Si–Ge alloys whose plasticity certainly is governed by the Peierls mechanism and which do not show solution softening. Neither do the Nb alloys investigated by KOSTORZ [1968]. The plateau stresses of Nb–W, Nb–Ta and Nb–V measured in this work can be fully explained by eq. (1c), the localized elastic interaction mechanism, with $\epsilon = |\eta' - 16\delta|$. Parelastic interaction alone would not explain the results on Nb–Ta ($\delta = 0$). The same conclusion is reached by MORDIKE et al. [1970] investigating Ta–Os, Ta–Re, Ta–Mo and Ta–Nb. KOSTORZ [1968] explains the deviation from the $\tau_0 \sim c^{1/2}$ curve above 3% W in Nb, simulating a linear concentration dependence (see also fig. 20), by an additional local-order term proportional to c^2 in τ_0 (see §1.3.2d).

LAWLEY and MADDIN [1962] studied single crystals of a solid solution of rhenium in molybdenium after polycrystalline studies had revealed the remarkable low-temperature ductility of these alloys. The solubility of Re in Mo is 42 at% at 2500°C. Deformation was by slip, except for the Mo–39 at% Re alloy which twinned extensively below 500 K. The critical shear stress on the $\{112\}\langle111\rangle$ or $\{110\}\langle111\rangle$

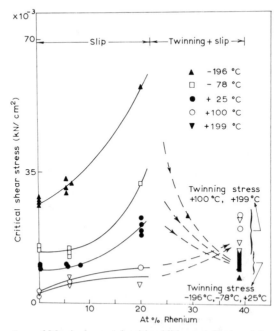

Fig. 21. Critical shear stress of Mo single crystals with additions of rhenium (after LAWLEY and MADDIN [1962]).

systems versus composition and temperature is shown in fig. 21. At room temperature and below, $\tau_0(c)$ rises with a positive curvature. The main temperature effect is present already in pure molybdenum. Above 30 at% Re, twinning becomes progressively easier (on $\langle 112\rangle\langle 111\rangle$) and is responsible for the fairly temperature-independent critical shear stress. A pronounced yield point (up to 8% drop in stress) was observed at the beginning of the tensile test and could be recovered by aging 2.5 h at 700°C, consistent with diffusion of C and N in Mo. The work-hardening rate is low (~ 70 kN/cm² average at 3% strain) and necking develops at 7% strain. The high ductility of the polycrystalline alloys with additions of Re is considered to be due both to the formation of a complex oxide (instead of MoO_2 in grain boundaries) and the promotion of twinning.

1.7. Hardening in the NaCl and diamond structures

From the classical work of Smekal and coworkers it is well known that divalent impurities M^{++} strongly harden NaCl crystals. In this case, a Na^+ vacancy has to be formed for every M^{++} introduced. Monovalent M^+ impurities have a much smaller effect on the yield stress. The dependence of the CRSS on temperature for doped *alkali halide* crystals and AgCl is shown schematically in fig. 22 (SKROTZKI and HAASEN [1981]). The upper curve is typical for, say, KCl with 70 ppm Sr^{++}, and the lower one for "pure" KCl, both in $\langle 110\rangle\langle 110\rangle$ slip. For $\langle 100\rangle\langle 011\rangle$ slip the curves look similar – only at 200 K higher temperatures. Four temperature regimes are visible. Regime I is controlled by the Peierls force and little influenced by doping. Regime II is that of Fleischer hardening by the divalent foreign ion associated with a vacancy (dipole). It follows eqs. (1c) and (4c) and the maximum interaction force is of parelastic as well as electrostatic origin. The latter is particularly strong for edge

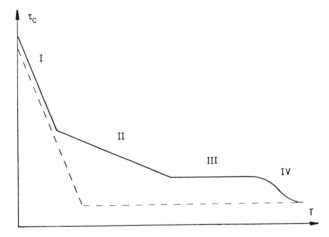

Fig. 22. Schematic regimes for the temperature dependence of CRSS of ionic crystals (dashed lines for pure, solid lines for doped crystals).

References: p. 1403.

dislocations on the {100} slip planes. In regime III the CRSS depends linearly on doping and little on temperature as predicted by the Snoek interaction with the rotating dipoles (§1.3.1d). Finally in regime IV dislocations are able to drag along the dipoles during their movement. Crystals with the *diamond cubic* structure like the semiconductors Si and Ge, but also those with partially ionic binding like the III–V and II–VI compounds become ductile only at elevated temperatures ($T > T_m/2$ for purely covalent bonding, while CdTe is ductile already at much lower temperatures (ALEXANDER and HAASEN [1968], GUTMANAS and HAASEN [1981]). The plasticity of these semiconductors is strongly enhanced by $\geqslant 10^{19}$ cm^{-3} n-dopants (V-valent additions to the IV-valent matrix). This *Patel-effect* is of electronic origin, not due to localized dislocation–solute interaction. It depends only on the position of the Fermi level for the particular temperature and doping relative to the occupation level of the neutral dislocation. If the two do not coincide the dislocation becomes charged and accordingly instable with respect to double-kink formation (HAASEN [1975]). This is an alloy-softening effect. On the other hand the charged dislocation interacts more strongly with charged point defects – this is the normal alloy-hardening (SIETHOFF [1969,1970]).

1.8. Creep of solid solutions

A creep test under constant stress yields essentially the same information as the dynamic (prescribed strain-rate) experiment. The creep test is of particular value at higher temperatures. Only steady state creep ($\dot{\epsilon} = $ const.) will be considered in the following and not transient creep (with decreasing rate). The reader is referred to reviews by SCHOECK [1957] and NIX and ILSCHNER [1979] and work by DORN [1957], WEERTMAN [1955,1957], COTTRELL [1953,1954] and MCLEAN [1962] for a more detailed account. The results of Dorn and coworkers for the stationary creep rate $\dot{\epsilon}(T, \tau)$ of metals and alloys are described by

$$\dot{\epsilon} = f(\tau) \exp(-H_{eff}/kT)$$

$$f(\tau) = \begin{cases} C_1\tau^n & \text{for small } \tau \\ C_2 \exp(B\tau) & \text{for large } \tau. \end{cases} \tag{6}$$

H_{eff} is determined by temperature change during creep and is found to approach at high temperatures H_d, the activation energy of diffusion of the material. For experimental reasons τ is chosen such that $\dot{\epsilon}$ keeps within certain limits. H_{eff} is then found to be appproximately proportional to temperature at low temperatures according to eq. (6). Relative to pure Al the measured activation energies around 600 K are somewhat larger in Al–Cu alloys. WALTON et al. [1961] attribute this to a modification of the cross-slip process caused by solute atoms. The region of strain-aging during creep will be discussed below (§1.8.2). In these class I alloys creep is controlled by solute drag. Above 800 K, the Al–0.05 at% Cu alloy crystal showed distinct sub-grain formation. This is expected if the mechanism of high-temperature creep of metals and dilute alloys (called class II) is dislocation climb.

Table 1
Stress-dependence of creep rate [a].

(Polycrystalline) Material	Temperature (°C)	Stress range (kN/cm^2)	Stress exponent n of creep rate
Al	371–593	0.06–2	4.4
Al–0.94 at% Mg	260–482	0.2–2	3.0
Al–1.9 at% Mg	260–482	0.2–2	3.0
Al–5.1 at% Mg	260–482	0.2–2	3.0
Al	480	0.03–0.3	3.9
Al–1.1 at% Mg	480	0.4–0.7	3.6
Al–2.1 at% Mg	480	0.2–0.7	3.4
Al–3.1 at% Mg	258–580	0.03–4	3.2
Al (coarse grain)	300	0.45–0.7	4.5
Al–2 at% Mg	250–380	0.5–3	4.0

[a] According to WEERTMAN [1955].

1.8.1. Steady-state creep through dislocation climb *

WEERTMAN [1955] considers dislocations which have been stopped in their glide planes by various obstacles. The mean shear stress on each dislocation is zero, but the normal stress has an average value of σ_i. Under the influence of this normal stress the dislocations will start to climb over the obstacles. Some dislocations give off vacancies, some absorb them. The local concentration of vacancies near the dislocations is thus changed and an extra chemical potential introduced which tries to pull the dislocations back against σ_i. In equilibrium the rate of climb is determined by the flux of vacancies from one dislocation to another. The solution of this diffusion problem results in a creep rate

$$\dot{\epsilon} = NAD/h\Lambda\left[\exp\left(\sigma_i b^3/kT\right) - 1\right], \tag{7}$$

where N is the density of dislocations per cm^2, Λ is the distance between obstacles touched by the dislocation, A the area of the slip plane the dislocation will cover after a successful climb, $D = D_0 \exp(-H_d/kT)$ is the coefficient of diffusion of the material, and h is the height of the obstacle the dislocation has to climb over. Depending on the ratio $\sigma_i b^3/kT$ and on the way N, h (and A, Λ) change with stress, the experimental relation (6) can be interpreted now with the help of eq. (7). In particular, for $\sigma_i \ll kT/b^3$ eq. (7) becomes

$$\dot{\epsilon} = \left(NAb^3/h\Lambda kT\right)\sigma_i D_0 \exp\left(-H_d/kT\right), \tag{8}$$

which is in agreement with most experimental data on creep rates of pure metals if the pre-exponential factor depends on stress as τ^n, $n \approx 4$–5 (see table 1). Concerning temperature dependence, SEEGER [1957,1958] has discussed a possible contribution of H_j, the energy of jog-formation (and constriction of the stacking fault), towards

---

* See also chs. 18 and 20.

References: p. 1403.

the measured activation energy H_{eff}, which would become noticeable when $H_j \approx H_d$ (particularly in alloys). For pure metals n can be explained by assuming $N \propto \tau^2/\mu^2$, i.e. the number of dislocations being determined by long range elastic interaction. Further $h \propto \mu b/\tau$ if the obstacle is a sessile dislocation and $\sigma_i \propto \tau$ as the mean internal stress in the crystal. Various other possibilities are discussed by WEERTMAN. For class II alloys again $n \approx 4$–5. For high stresses the exponent $\sigma_i b^3/kT$ in eq. (7) is found experimentally to be independent of temperature, compare eq. (6), which is difficult to reconcile with theory (see also TAKEUCHI and ARGON [1976a]). An application of (the inverted) eq. (8) in a dynamic experiment gives the stress (τ_{III}^*) above which climb contributes noticeably to the prescribed strain rate $\dot{\epsilon}$:

$$\tau_{III}^* \propto \dot{\epsilon}^{1/n} \exp(H_d/nkT). \tag{9}$$

This equation should be compared with high-temperature measurements on the dynamic recovery of low-γ alloys (see §1.4.5).

1.8.2. Creep limited by dislocation drag (class I) and the Portevin–Le Chatelier effect

Besides climb there is another mechanism to explain high-temperature creep of solid solutions at small stresses, by considering the glide movement of dislocations which drag solute "atmospheres" with them in a viscous manner (COTTRELL [1953,1954], BRION et al. [1971], TAKEUCHI and ARGON [1976b] and WEERTMAN [1977]). If by one of the pinning mechanisms considered in §1.3.1 a dislocation creates a dragging force F on a solute atom the stationary microcreep rate is

$$\dot{\epsilon} = Nbv, \quad \text{where} \quad v = (D_s/kT)F \tag{10}$$

and D_s is the diffusion coefficient of the solute atoms. The dislocation attains a velocity v in dynamic equilibrium between the applied stress τ and the dragging force F which is proportional to τ. $\dot{\epsilon}$ is thus proportional to stress for constant N, and proportional to τ^3 after dislocation multiplication and interaction has increased $N \propto \tau^2$ as above (WEERTMAN [1957]). $n = 3$ as found for class I alloys (see table 1) could also been interpreted by eq. (8) if h were independent of stress, i.e. if the obstacles are provided by heterogeneities of the alloy. On increasing the stress the dislocation reaches a critical velocity v_c just too high for solute diffusion and breaks loose from its atmosphere. Cottrell has pointed out that the velocity range near v_c is unstable since by sligthly accelerating the dislocation will free itself completely and by slowing down somewhat it will be caught by the atmosphere. (The material strain-ages during creep.) This phenomenon will give rise to a serrated stress–strain curve (jerky flow, *Portevin–Le Chatelier* (PL) *effect*) or a sharp bend in the creep curve at a certain strain rate and temperature. A detailed discussion of the mechanism of the PL effect has been given recently by SCHWARZ [1982] in connection with the strain-rate sensitivity of the flow stress. This effect coincides with the high-temperature end of plateau-hardening if the plateau is due to the friction with single solute atoms (cf. §1.4.1, fig. 18). In the calculation of the critical condition for the PL effect one has to consider enhancement of solute diffusion by vacancies produced during creep resulting in $\dot{\varepsilon} \sim \bar{c}_v \exp(-E_m/kT)$ at a critical strain. Here c_v is the excess vacancy concentration and E_m the activation energy of vacancy migration (COTTRELL [1954], MCCORMICK [1972] and VAN DEN BEUKEL [1980]).

1.9. Fatigue of solid solutions *

The low-amplitude fatigue-strength of fcc metals does not increase with solute addition in proportion to their unidirectional fracture strength. While in pure metals the long-life "fatigue limit" τ_{10^6} is about equal to the stress τ_{III} where thermally activated cross-slip starts (HAASEN [1965], see fig. 44 of ch. 19), it becomes much smaller than τ_{III} in solid solutions (RUDOLPH *et al.* [1965]). τ_{10^6} for ordered Cu_3Au is only 20% higher than for disordered single crystals, despite an enormous increase in τ_{III}. NEUMANN [1967] has recently explained this disappointing behaviour in a model which relates fatigue-crack nucleation and propagation to *the coarseness of slip*. Slip is indeed coarser in solid solutions (and alloys with short-range order or coherent precipitates) than in pure metals as is discussed in §1.2.1 for unidirectional deformation.

Cracks nucleate during fatigue often at *persistent slip bands* (PSBs) named so because their coarse surface features cannot be removed by intermediate polishing (MUGHRABI [1979]). Finally PSBs may turn into "extrusions" of material as an extreme case of surface topography developed during fatigue. PSBs are soft lamellae in the cyclically hardened matrix in which voluminous arrays of dislocation multipoles are replaced by rather sharp dislocation walls. In between those walls to- and fro-glide of dislocations produces an easy steady-state deformation indicated by a plateau in the cyclic stress–strain curve (i.e. the asymptotic stress versus plastic strain-amplitude of cyclic deformation). Cross-slip evidently is instrumental in the transformation of the dislocation arrangements into those of PSBs. This has been treated quantitatively for Cu–4%Ti solid solutions by SINNING and HAASEN [1981]. In alloys of low SFE therefore no PSBs form but slip is rather concentrated and planar; this by itself produces a rather sharp surface topography at slip steps on which cracks can nucleate (as on PSBs).

During the plateau of the cyclic stress–strain curve PSBs spread over the whole specimen. This can occur in single as well as in polycrystals where stress concentrations at grain boundaries sometimes help in the transformation of the dislocation microstructure (SINNING and HAASEN [1981]). The effect of precipitation on fatigue properties depends again on how this changes slip localization. This in turn seems to be favoured by a large size misfit (SINNING [1982]).

---

* For details see ch. 24.

References: p. 1403.

2. Precipitation-hardening

The formation of a precipitate greatly strengthens an alloy. The increase in yield stress depends principally on the strength, structure, spacing, size, shape and distribution of the precipitate particles as well as on the degree of misfit or coherency with the matrix and on their relative orientation. To understand precipitation-hardening one must study in detail the way in which dislocations interact with precipitate particles. In addition to mechanical testing (of mainly *single* crystals) the problem is investigated by the classical method of small-angle X-ray diffraction, by magnetic measurements, and more recently by transmission electron microscopy and by field ion microscopy. Very rarely, however, different methods have been compared in their application to one and the same specimen. This chapter is restricted to a review of some fundamental investigations of the dislocation–precipitate interaction in relative simple alloys as described in recent papers by ARDELL *et al.* [1976], MELANDER and PERSSON [1978], HAASEN and LABUSCH [1979], GRÖHLICH *et al.* [1982] and by GLEITER and HORNBOGEN [1968] rather than to discuss hardness versus aging time and X-ray diffraction results on alloys of technical importance. These studies are well documented in ch. 22 and in the reports of, e.g., BROWN and HAM [1971], BROWN [1979], KOCKS *et al.* [1975], GEROLD [1979], HARDY and HEAL [1954] and KELLY and NICHOLSON [1963].

2.1. Interactions between dislocations and precipitate particles

A dislocation moving on its slip plane containing a distribution of precipitate particles may cut through the particles or avoid them by moving out of its slip plane or by bending between the particles, leaving a dislocation ring around each precipitate. For all of these processes energy must be expended. Which takes place, however, depends on the applied stress and the nature of the precipitate.

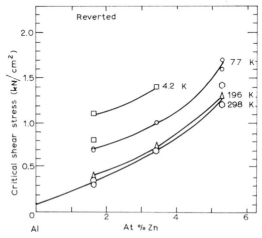

Fig. 23. Critical shear stresses of Al single crystals with various additions of zinc in solid solution at different temperatures (after DASH and FINE [1961]).

For a dislocation to *shear a particle*, sufficient energy must be supplied to break favourable bonds within the particle thus increasing its "surface" area. The energy Γ (J/m²) of the interface produced when a zone is sheared by one atomic distance is estimated from the heat of reversion for Al–Ag by KELLY and FINE [1957] to be about 0.1 J/m². In this case the precipitates are completely coherent with the matrix, there is little size misfit between Ag and Al. In the other extreme if the particles are completely incoherent and of a quite different crystal structure Γ will be of the order of the energy of a large-angle grain boundary, i.e. ~ 1 J/m². The stress to force an essentially straight dislocation through a distribution of precipitate particles ("zones") of distance Λ, and radius r is:

$$\tau_c = (\Gamma/b)(r/\Lambda). \tag{11}$$

This is rather similar to eq. (3) for the short-range order (cluster) friction mechanism (§1.3.2d) except for the factor $r/\Lambda = f^{1/2}$ where f = volume fraction of the precipitate (KELLY [1958]). In this case the unfavourable bonds formed by deformation only are concentrated in the interface of (spherical) zones instead of statistically distributed over the slip plane. If the composition of the zones (concentration c' of solute, assumed $c' \sim 1$) is independent of that of the alloy c, then Γ = constant and $f = c - c''$, where $c'' \ll 1$ is the concentration of solute in solution in equilibrium with the zones. Thus $\tau_c \sim (c - c'')^{1/2}$ for zone hardening whereas $\tau_c \sim \Gamma \sim c^2$ for cluster (favourable atom pair) hardening. This has been verified approximately for

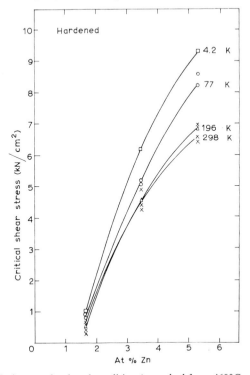

Fig. 24. As fig. 23, only for zone-hardened condition (quenched from 460°C and aged 24 h at 25°C).

References: p. 1403.

Al–Zn (DASH and FINE [1961]) (figs. 23 and 24). Equation (11) is a special case of energy-storing obstacles in the concentrated, Labusch limit (see §1.3.2). The corresponding equation in the Fleischer limit obtains by putting $f_{max} = \Gamma r$ into eq. (1c) for particles of volume fraction $f = cr^2/b^2$, where r is the particle radius, yielding:

$$\tau_c b = f_{max}^{3/2} f^{1/2}/(2E_L)^{1/2} r = \Gamma^{3/2}(rf)^{1/2}/(2E_L)^{1/2}. \tag{1d}$$

Another type of energy-storing obstacle is an ordered-phase particle which becomes disordered by the passage of a dislocation. With Γ_{APB} the antiphase boundary energy (per m²) then $f_{max} = \Gamma_{APB} \cdot r$ and eq. (1d) applies again.

The third case of an energy storing interaction is that of a difference in stacking-fault energy $\Delta\gamma$ between particle and matrix (HIRSCH and KELLY [1965]). Here one calculates $f_{max} \approx 2r \cdot \Delta\gamma$. In addition to this chemical hardening effect there may be a (short range) elastic one if the zones are surrounded by local coherency strains or have an elastic modulus different from that of the matrix as in Al–Cu ($\delta = -12\%$) which are to be overcome by a cutting dislocation. Temperature may assist in the shearing process if the dimensions of the zones are sufficiently small. In the parelastic case $f_{max} = \mu b\delta \cdot r$ and depending on the Labusch or Fleischer limits one obtains $\tau_c \sim f^{2/3} r^{1/3}$ or $\tau_c \sim f^{1/2} r^{1/2}$, respectively. In the dielastic case it is more complicated to calculate f_{max} (see WENDT and WAGNER [1982]), but it looks as if the modulus difference between particle and matrix does not determine the flow stress in any of the investigated alloy.

A dislocation can *avoid a particle* by climbing (§ 1.8) or cross-slipping (§ 1.4.5) around it. The minimum stress for this decreases strongly with increasing temperature; the slip lines are wavy as will be seen below. For a large enough particle spacing the dislocation may pass between two particles (fig. 9c) at an *Orowan* stress

$$\tau_m = \alpha\mu b/\Lambda, \tag{12}$$

where α is a constant of order of magnitude one (cf. ch. 18). This leaves dislocation rings around the particles which work-harden the crystal. τ_m depends on temperature through μ only. Comparing eqs. (11) and (12) one finds the largest size of particles to be sheared to be

$$r_{max} = \alpha b(\mu b/\Gamma), \tag{13}$$

where Γ is any of the above surface energies, or $\mu b|\delta|$ in the parelastic case. For a "critical dispersion", r_{max} is ~ 25 Å for $\Gamma = 1$ J/m², ~ 250 Å for $\Gamma = 0.1$ J/m².

A more detailed calculation of the Orowan stress, eq. (12) also has been made by ASHBY [1966], considering the orientation-dependence of the line tension, the dissociation into partials, and the interaction between the two dislocation segments on both sides of one particle (fig. 9c). If the particle is incoherent with the matrix, localized plastic flow may be nucleated at the particle–matrix interface by the approaching dislocation. The stress necessary for this is of the order of one hundredth of the shear modulus of the matrix and is decreasing with increasing particle size (ASHBY [1968]). Thus the Orowan process, eq. (12) and fig. 9c, truly applies only with large coherent particles.

A special problem arises in the interaction of dislocations with particles of a certain degree of internal order but with $\delta = 0$ (γ'-Ni$_3$Al precipitated from Ni–CrAl, GLEITER and HORNBOGEN [1965]). In this case dislocations move in pairs as superdislocations (§3.1) because single dislocations are partials of the ordered structure. The equilibrium shapes of these dislocations and the flow stress in these alloys have been calculated by GLEITER *et al.* in good agreement with electron transmission observations and polycrystalline deformation results. A dynamic theory of hardening by ordered particles is given by COPLEY and KEAR [1967].

SCHWARZ and LABUSCH [1978] have computer-simulated the movement of a dislocation in a dense array of energy-storing (or elastically interacting) obstacles. HAASEN and LABUSCH [1979] have compared the theory with experiments of ARDELL *et al.* [1976] on Ni–Ni$_3$Al. New experiments on this alloy have been compared with the Labusch–Schwarz theory by GRÖHLICH *et al.* [1982]. They make use of the superposition of order and parelastic hardening calculated by MELANDER and PERSSON [1978].

2.2. Direct observation of precipitate–dislocation interactions

The change in slip-line pattern of Al alloys due to the presence of various precipitates has often been studied (e.g. THOMAS *et al.* [1957], GREETHAM and HONEYCOMBE [1961]). In Al–2 at% Cu containing GP zones, slip lines are found to be wavy, indicating that dislocations cross-slip. In the presence of GP II and θ' precipitates well defined slip bands do not form and slip lines appear very "indistinct". Overaged alloys with a coarse θ precipitate show very fine multiple slip. At high resolution slip lines appear to bend around precipitate θ particles at room temperature whereas at heavy deformations partially coherent θ' plates are distorted by slip bands passing through them. Electron transmission observations on thin films of deformed precipitation-hardened Al alloys confirm that some dislocations do in fact pass through partially coherent precipitates, while others are held up, forming cusps (fig. 25). Large non-coherent θ precipitates are free of dislocations (NICHOLSON *et al.* [1960]). SWANN [1963] finds that dislocation walls connect large precipitates in deformed Al–2.3 at% Cu alloys. JAN [1955] and KELLY and SATO [1961] showed by small-angle scattering of X-rays that spherical precipitates in Al–Ag and Al–Zn alloys became elliptical by plastic deformation, as would be expected if slip were homogeneous within a volume of $(40 \text{ Å})^3$. The same conclusion was reached by LIVINGSTON and BECKER [1958] for precipitation-hardened Cu–2 at% Co alloys from the measurement of magnetic anisotropy. This is definitely not in agreement with the heterogeneous distribution of slip indicated by slip-line observations. It is possible that dislocations cross-slip in front of the particle to various parallel slip planes. This cross-slip may lead to the dislocation circumventing the particle. In agreement with this it was found that the precipitates were deformed somewhat less than the specimen as a whole.

GLEITER [1967b] has computed the formation of prismatic dislocation loops by cross-slip near large coherent particles. He also observed (fig. 24a) that large

References: p. 1403.

Fig. 25. Dislocations passing through θ' precipitate in Al–1.6 at% Cu aged 160 h at 190°C and strained 3% at room temperature (from BONAR and KELLY, unpublished).

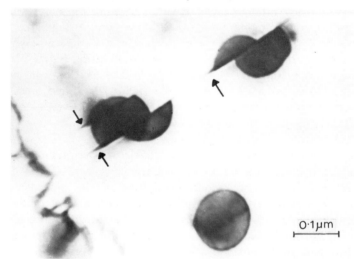

0·1μm

Fig. 24a. Sheared γ'-zones in Ni–19%Cr–6% Al aged 540 h at 750°C and deformed 2% (from GLEITER and HORNBOGEN [1965]).

γ'-zones in Ni–CrAl are cut by dislocations on a few slip planes only, in agreement with slip-line observations (GLEITER and HORNBOGEN [1965]).

2.3. Mechanical properties of precipitation-hardened Al alloys

As an alloy ages, a coarser and more stable precipitate is formed. The character of the stress–strain curve is then changed completely. After 2 h ageing at 190°C the precipitate in Al–1.6 at% Cu crystals is predominantly GP II; after 40 h some θ' has formed and after 300 h θ' only is present. For GP II alloys the critical shear stress is large and the rate of work-hardening low, but after 300 h ageing just the opposite is observed. The variations in τ_0 and $d\tau/d\epsilon$ (at $\epsilon = 2\%$) are shown in figs. 26 and 27, respectively, according to Kelly and Bonar, cf. KELLY [1963]. Between 30 and 100 h the mode of deformation changes from single slip, during which the specimen axis rotates towards the slip direction, to multislip showing extensive asterism without change in specimen axis-orientation and external specimen-shape.

2.3.1. Overaged alloys

After 300 h the minimum particle spacing in Al–1.6 at% Cu is about 2400 Å and appears to be controlling the initial flow stress. A critical shear stress of ~ 3 kN/cm² calculated from eq. (12) is of the same order of magnitude as the observed values (fig. 26). Similar conclusions were reached by DEW-HUGHES and ROBERTSON [1960] for Al–1.3 to 2.1 at% Cu overaged to precipitate the θ phase in various dispersions. Figure 28 shows a plot of the Orowan relation [eq. (12)] which is quite satisfactory for the 1.7 at% alloy except for a group of specimens which contained very coarse plate-shaped particles. For these, Orowan's particle distance may vary considerably with orientation. An uncertainty in the choice of Λ also arises for semi-coherent particles where coherency strain might produce a radius of influence larger than the particle radius. The high work-hardening rate of overaged alloys, about $\mu/10$, apparently is a consequence of multiple slip enforced by non-deforming particles and of the high dislocation density due to loops around particles. The work-hardening of overaged alloys depends strongly on deformation temperature (BYRNE *et al.* [1961]). This would not be expected from the theory by FISHER *et al.* [1953] who

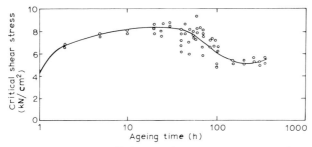

Fig. 26. Critical shear stress of Al–1.6 at% Cu crystals at room temperature after solution treatment, water quenching, and ageing at 190°C for various times. $\dot{\epsilon} = 3 \times 10^{-4}$ s⁻¹ (after KELLY and BONAR, cf. KELLY [1963]).

References: p. 1403.

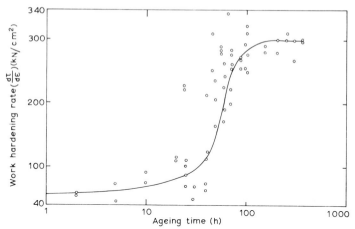

Fig. 27. Work-hardening rate (at 2% strain) of crystal as in fig. 26 versus aging time at 190°C.

explained work-hardening in terms of the back stress of planar bundles of loops around particles. Ashby [1968] analyzes the dislocation density as a function of strain in deformed dispersion-hardened alloys and concludes that the flow stress is best described by a "forest intersection" model (ch. 19, §5.1). According to this, diffuse dislocation walls connecting particles form by reaction between prismatic loops around particles in several slip systems. The stress necessary for glide dislocations to cut these walls with the help of thermal fluctuations is calculated to be

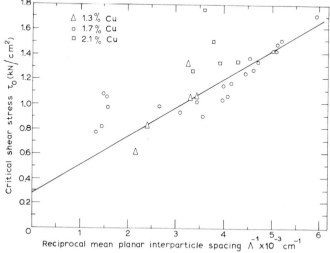

Fig. 28. Critical shear stress τ_0 of Al–Cu single crystals of various compositions aged to precipitate θ particles of different mean planar spacing Λ (data from Dew-Hughes and Robertson [1960]). For the solid solution $\tau_0 = 0.3$ kN/cm^2.

$$\tau(\epsilon) \approx \mu \left(\frac{fb}{r} \cdot \epsilon \right)^{1/2} \alpha(T, \dot{\epsilon}), \tag{15}$$

where the factor α decreases with temperature T and increases with strain rate $\dot{\epsilon}$ similar to the thermally activated flow-stress contribution τ_s for pure metals (ch. 19, SEEGER [1957]). The theory is in good agreement with experiments described by EBELING and ASHBY [1966], DEW-HUGHES and ROBERTSON [1960], JONES and KELLY [1968], and ASHBY [1964], see ch. 22.

2.3.2. Zone-hardened alloys

In zone-hardened alloys the precipitates are too closely spaced for dislocations to pass between them at the observed critical shear stresses [cf. eq. (13)]. These particles must be sheared during deformation in agreement with direct observations of the dislocation–precipitate interaction (§2.2). For aging times up to the peak value the yield stress then is determined by the work done in forcing the dislocations through the precipitates, and the rate of work hardening is similar to that of the matrix material ($\sim \mu/250$ for Al alloys). The parelastic interaction cannot in general explain the critical shear stress of zone hardened alloys. This is because no coherency strains are observed around silver-rich zones in Al–Ag and yet there is appreciable zone hardening. The stacking-fault hardening mechanism described above seems to work here. In the case of AlCu the interfacial strengthening mechanism is estimated to dominate hardening. For Al–1.7 at% Cu, the critical shear stress in the reverted and the GP I condition is strongly temperature-dependent below 200 K contrary to the behaviour of either pure Al or the alloys with larger precipitates (BYRNE *et al.* [1961]). In the reverted state small Cu clusters are expected to be present. The answer to the question which kind of precipitate provides strongest hardening thus depends on temperature. In Al–Ag τ_0 is very little temperature dependent as would be expected for spherical zones larger than a few Ångström units. In the case of Al–Cu on the other hand the monoatomic platelets of GP I can be cut with the help of thermal fluctuations. Much about the atomic mechanism of zone hardening can thus be learned from a "*thermal activation analysis*'" of the flow stress of alloy single crystals (BYRNE *et al.* [1961], KELLY [1963]).

2.4. Precipitation-hardening in other alloy systems

2.4.1. Copper–cobalt

In addition to the aluminium alloys aged Cu–1 to 3 at% Co alloys have been the subject of some fundamental mechanical studies, which are of particular interest to us here. In this system it is possible to determine independently the volume fraction f of precipitate, and the size and shape of the precipitated particles from the magnetic properties (LIVINGSTON and BECKER [1958], LIVINGSTON [1959]). It was found that during the first 1000 min at 600°C the yield stress at 300 K increased at apparently constant f, while the precipitate was coarsening, see eq. (1d). A broad maximum of the yield stress was reached at a particle radius of about 70 Å or a particle spacing of

500–1000 Å. This maximum stress is approximately proportional to $f^{1/2} = r/\Lambda$ as would be expected from eq. (11) describing the "chemical" shear strength of the particles. The constant of proportionality, Γ, is about 0.12 J/m^2. This can be interpreted as being caused by the parelastic and stacking fault mechanisms (HAASEN [1974]). Magnetic anisotropy measurements do in fact show the particles to be sheared during deformation (§2.2). The decrease of strength beyond the hardness peak can probably be explained by Orowan's mechanism although the observed particle spacing is somewhat too small for dislocations to pass between at the measured flow stresses. LIVINGSTON [1959] ascribes the decrease in strength above a critical particle diameter of 140 Å to a loss of complete coherency of the precipitate and the formation of interface dislocations. At a precipitate–matrix mismatch of $\delta = 1.6\%$ the interface dislocations should have a spacing about equal to the critical particle diameter. Electron transmission observations of *deformed* Cu–Co foils indicate, however, that particles maintain full coherency to several times the diameter corresponding to peak hardness [PHILLIPS 1964]. This is in agreement with KELLY [1963] who has pointed out that the hardening peak cannot in general be explained in terms of coherency strain (or particle size) in preference to particle spacing since coherency strains are only rarely observed in Al–1.6 at% Cu on both sides of the peak.

2.4.2. Iron–carbon

Interstitial carbon in α-iron will always be in excess of its solubility limit at room temperature. It will precipitate as carbide along dislocations and in the volume of the matrix especially if grain boundaries are absent. This gives rise to a pronounced yield point as well as an increase in the flow stress both of which depend on heat treatment (KÖSTER *et al.* [1932]). As an alternative process carbon atoms can form a more or less condensed solute atmosphere around dislocations, as considered by COTTRELL [1953] and coworkers. This would also explain a sharp yield point which in detail differs from the precipitate yield point. The Cottrell yield stress should be strongly temperature-dependent, as has been calculated by a number of workers and more recently by HAASEN [1959] using the assumption that the dislocation-line energy is not increased significantly in the yield process. On the other hand thermal fluctuations would not help much to free a dislocation from a row of precipitates. A precipitate yield point can be overaged but not the Cottrell yield maximum. Finally, the magnitude of the precipitate yield maximum depends on the curvature the dislocation has attained by ageing under load, i.e., on the line tension, contrary to the behaviour of a Cottrell-locked dislocation (HAASEN [1959]). Carefully aged iron–carbon specimens have been studied by thin film electron transmission (HALE and McLEAN [1961], LESLIE [1961]). Figure 29 shows a transmission photograph of an Fe–0.05 wt% C specimen thinned after water quenching from 690°C and ageing one month at room temperature. Lines of precipitates are seen along the dislocations as well as some carbide particles in the matrix. Leslie finds the precipitate formed in the matrix of Fe–0.014 wt% C on ageing below 60°C to be a voluminous, deformable carbide (probably $Fe_{16}C_2$), not the (brittle) ϵ-carbide. On overageing Hale and

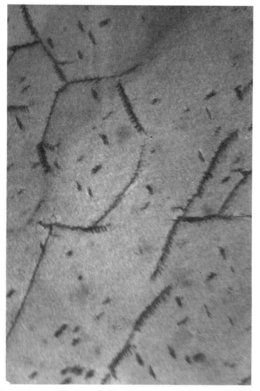

Fig. 29. Carbide precipitates in Fe–0.005 wt% C, water-quenched from 690°C and aged one month at room temperature (from HALE and McLEAN, unpublished).

McLean see loops around carbide particles after deformation.

EVERS [1961] studied the pronounced yield point of Fe–0.005 wt% C single crystals and found the difference between upper and lower yield stress to depend very little on temperature and strain rate, which the lower yield stress itself or the flow stress does very strongly (fig. 30). The constant in the Petch analysis of the dependence of yield stress on grain size should be proportional to the dislocation-carbon unlocking stress according to the theory of COTTRELL [1958] and yet in most materials is found to be independent of the temperature of deformation (HESLOP and PETCH [1956]). All this points towards a precipitate-locking yield point. The same conclusion was reached by MIGAUD [1960] in a study of mechanical properties and residual electric resistivity of zone-refined iron. Migaud finds that the yield stress decreases on overageing. The dependence of the yield stress on ageing stress (EVERS [1961]) also suggests precipitate and not atmosphere locked dislocations.

MORDIKE and HAASEN [1962] have investigated the flow stress of α-iron–0.003 wt% C single crystals by temperature- and strain rate changes between room temperature and -72°C. They measured a stress-free activation energy U_0 of 0.55

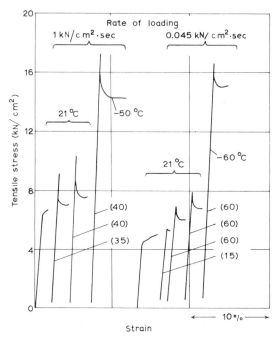

Fig. 30. Stress–strain curves for Fe–0.005 wt% C single crystals deformed at different rates and temperatures with intermediate ageing treatments at 21°C (times in ageing minutes given in brackets) (after EVERS [1961]).

eV and a very small activation volume v (corresponding to a distance between obstacles $v/b^2 \approx 7 \times 10^{-6}$ cm) which is sensitive to heat treatment. The authors identify the obstacles overcome by dislocations with thermal help as small carbide precipitates. CONRAD [1960] found that the temperature dependence of the lower yield stress τ_{ly} of polycrystalline iron depended on its purity, increasing particulary with the *total* carbon plus nitrogen content. HESLOP and PETCH [1956] have shown, on the other hand, that $\tau_{ly}(T)$ for iron with 0.16 wt% $(C + N)$ does not depend on the amount *in solution*. The work-hardening rate $d\tau/d\epsilon$ of Mordike and Haasen's single crystals showed a maximum of 30 kN/cm² at 350 K. A similar behaviour is observed for hardenable Al-alloys (§2.3). At low temperature $d\tau/d\epsilon$ was the same as in stage III of nickel single crystals (~ 10 kN/cm² at 200 K). On the whole, "pure" α-iron in its mechanical properties between 200 and 300 K resembles closely an age-hardened alloy (see however ŠESTÁK and SEEGER [1978]). For a more complete survey of the mechanical properties of α-iron see ch. 16.

2.4.3. Spinodally decomposed Cu–Ti alloys

WAGNER [1979] and KRATOCHVIL and HAASEN [1982] have recently studied the mechanical properties of alloys which have undergone spinodal decomposition and have developed modulated microstructures (Cu solid solution + Cu_4Ti) of various wavelengths and perfections. The CRSS shows a double-peak behaviour as a function of ageing (or a plateau-plus-peak curve). This is interpreted by the formation of a regular array of ordered particles at first which then by Ostwald ripening turns into a more statistical particle array.

3. *Order-hardening*

Some solid solution crystals develop long-range order in superlattice structures described in ch. 4. Of these, the fcc structures Cu_3Au, Ni_3Mn, fc tetr. CuAu (I) and bcc β'-CuZn, and Fe_3Al are considered here with respect to the influence of crystallographic order on the mechanical properties. Above the critical temperature for ordering T_c these properties can be influenced by short-range order (cf. §1.3.2). For a detailed review of this topic the reader is referred to surveys by WESTBROOK [1960,1967], STOLOFF and DAVIES [1966], STOLOFF [1971] as well as MARCINKOWSKI [1974].

The difference in stress–strain behaviour between the ordered and disordered state of a Cu_3Au crystal was demonstrated as early as 1931 by SACHS and WEERTS (fig. 31). The disordered and quenched crystal behaves like the random solid solutions discussed in §1.4. It has a high critical shear stress, long "easy glide" with little work-hardening, strong overshoot and shows coarse slip lines on the surface as does α-brass (TAOKA and SAKATA [1957]). The well annealed and long-range ordered

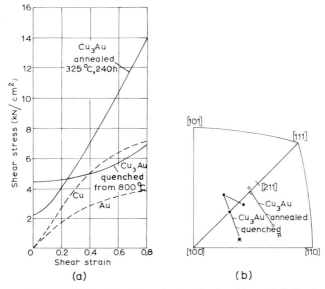

Fig. 31. (a) Stress–strain curves of disordered (quenched) and ordered (annealed) Cu_3Au single crystals at room temperature (after SACHS and WEERTS [1931]). Also curves for Cu and Au single crystals for comparison. (b) Change of orientation with strain of crystals (a).

References: p. 1403.

specimen on the other hand is similar in its stress–strain curve to a pure metal with a low stacking-fault energy. It shows strong linear hardening like pure copper, and a rather high τ_{III} while its critical shear stress is intermediate between pure copper and the quenched crystal. There is little overshoot and fine slip is observed for the ordered crystal.

Stress–strain behaviour of Ni_3Mn single crystals was found to differ considerably from that described above for Cu_3Au (VICTORIA and VIDOZ [1968]). The reason for that is the much smaller domain size of the ordered structure in the former (5×10^{-7} cm, compared with 10^{-5} cm in Cu_3Au). The critical shear stress of ordered Ni_3Mn is higher than that of disordered, the work-hardening rate lower for single slip orientations. Polyslip orientations work-harden very strongly though in ordered Ni_3Mn. This behaviour is explained by VIDOZ [1968] by the formation of antiphase-boundary tubes during intersecting slip. Jogs in super-dislocations will in general not be aligned, and disordered regions will so be formed leading to an extra flow stress component. HIRSCH and CRAWFORD [1979] have recently discussed work-hardening of ordered alloys.

3.1. The superdislocation

In order to understand the mechanical properties of ordered structures, one has to realize that a moving glide dislocation ($b = \frac{1}{2}a\langle 110\rangle$) in the fcc lattice creates disorder across its slip plane, leaving an *anti-phase boundary* (APB) behind. Order is restored, however, by a second identical dislocation following on the same plane. Thus in an ordered structure dislocation pairs of *superdislocations* are to be expected, bound together by an APB whose width is determined by the equilibrium between

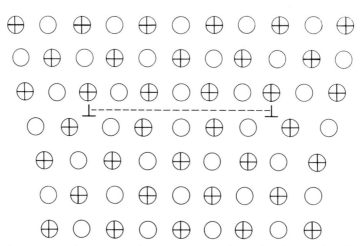

Fig. 32. Atomic arrangement in a cross-section through a superdislocation of edge orientation in a cubic primitive AB lattice, dashed line = APB (from BROWN [1960]).

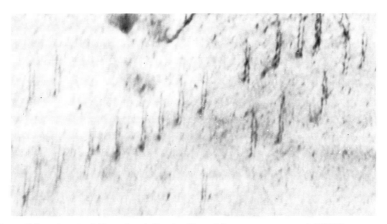

Fig. 33. Superdislocations in fully ordered Cu₃Au (from MARCINKOWSKI *et al.* [1961]).

the order fault energy Γ and the elastic repulsion between the bordering dislocations of like sign (cf. ch. 18, §6.3): fig. 32. The superdislocation is extended similarly to and independently of the stacking fault splitting. This explains the difficulty of cross slip and the high τ_{III}. In stage III slip then occurs by ordinary dislocations. Except for that, a superdislocation moving through a fully ordered crystal feels no solid-solution hardening (to a first approximation) in agreement with experiment. KEAR [1964] noted that since APBs in the Cu₃Au structure have lowest energies on {100} planes one partial of the superdislocation trailing the order fault would cross-slip

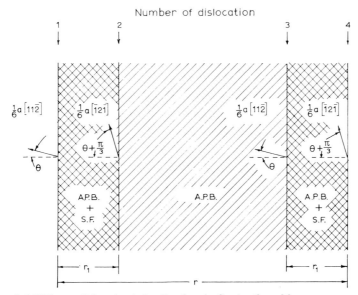

Fig. 34. Extended [011] superdislocation in its slip plane in Cu₃Au (from MARCINKOWSKI *et al.* [1961]).

References: p. 1403.

spontaneously onto this secondary type of slip plane. This explains the $\langle 110 \rangle$ alignment of dislocations seen in electron transmission in these ordered alloys as well as the increase in work-hardening rate, particularly in Cu_3Au, with temperature between 77 K and 373 K because spontaneous cross slip becomes easier (STOLOFF and DAVIES [1966]). The flow stress of a Ni_3Al-based alloy goes over a peak at 600 K, beyond which cubic slip is found to become prevalent (LALL et al. [1979]). VIDOZ [1968] is, however, able to explain $\theta_{11}(T)$ also by the formation of APB tubes as secondary slip becomes activated with increasing temperature. The extended super-dislocation has been observed directly in electron transmission by MARCINKOWSKI et al. [1961] in Cu_3Au (fig. 33) and Ni_3Mn (but not in CuZn, Fe_3Al, CuAu). These authors calculated the widths of the superdislocations in these structures by a quasichemical, nearest- (NN) and next-nearest (NNN) neighbour bond approach. In Cu_3Au and Ni_3Mn the combination of stacking- (SF) and order fault (APB) creates an extended superdislocation as shown in fig. 34. The widths r and r_1 are given by:

$$
\left.
\begin{aligned}
r_1 &= \frac{\mu b^2}{2\pi\left(\gamma + \frac{1}{2}S^2\Gamma\right)} f_1(\theta) \\[2mm]
r - r_1 &= \frac{\mu b^2}{2\pi\Gamma S^2} f(\theta)
\end{aligned}
\right\} \quad r_1 \ll (r - r_1),
\tag{16}
$$

where S is the long-range order parameter, $0 \leqslant S \leqslant 1$, f, f_1 are functions, of the order of magnitude one, of the angle θ as defined in fig. 34 (θ is $-30°$ and $60°$ for edge and screw dislocations, respectively). For the edge dislocation in Cu_3Au and for $S = 1$ these authors estimate $r = 102$ Å, $r_1 = 17$ Å from $\Gamma = 92$ mJ/m² and assuming $\gamma = 40$ mJ/m². Experimentally $r = 95$–130 Å (θ unknown). For Ni_3Mn they estimate $\gamma = 20$ mJ/m² and $\Gamma = 75$ mJ/m² and calculate, for $S = 1$, $\theta = -30°$, a width $r_1 = 35$ Å, $r \approx 176$ Å. The measurement gives $r = 110$–170 Å. In the case of β'-CuZn there should be no stacking, only an order fault which according to HERMAN and BROWN [1956] has a width too small to be observed, $r = 30b/S^2$, for the accessible values of S. For Fe_3Al the superdislocation in the DO_3 ordered structure must consist of four $1/2a\langle 111 \rangle$ (normal) dislocations in order to leave correct NN and NNN after slip. In this case Γ is so small and the width so large ($r \approx 1200$ Å for an edge dislocation) that only single (incomplete) dislocations are expected and in fact observed. The slip lines are wavy, the dislocations in electron transmission leave cross-slip trails. In CuAu(I) there are two nonequivalent types of dislocations in a (111) plane because of the tetragonal distortion of the fcc lattice. According to MARCINKOWSKI et al. [1961] the one type forms no superdislocation and the other an asymmetrical one.

As S decreases the widths r and r_1 in Ni_3Mn increase in agreement with observation (MARCINKOWSKI and MILLER [1961]). r_1 is determined by an effective stacking fault energy $\gamma_{eff} = \gamma + 1/2S^2\Gamma$. In the disordered specimens quenched from above T_c ($S = 0$) dislocations are observed to be single in Ni_3Mn (MARCINKOWSKI and MILLER [1961]). Increasing the equilibrium order temperature from $T = 0$ to T_c, i.e. decreasing S to zero generally results in the critical shear stress passing through a

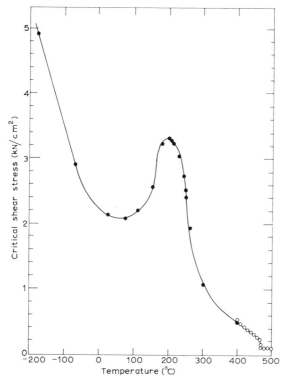

Fig. 35. Critical shear stress of β-brass, CuZn single crystals versus temperature (after ARDLEY and COTTRELL [1953] and BROWN [1959a]).

maximum: fig. 35 (ARDLEY and COTTRELL [1953], BROWN [1959a]). The low-temperature rise of τ_0 for CuZn is thought to be due to a phase transformation. A second, high-temperature peak of $\tau_0(T)$ for Cu_3Au has been ascribed by ARDLEY [1955] to strain-ageing during deformation.

3.2. Theoretical dependence of the yield stress on the degree of order

It is qualitatively evident that there will be a yield stress maximum at an intermediate degree of long-range order: Both the superdislocation in a completely ordered crystal and the single dislocation in a completely disordered structure will feel less friction than either of them at an intermediate S, see fig. (34a). Several mechanisms have been proposed to account for the order-hardening peak (e.g. BROWN [1960], LAWLEY et al. [1961] and STOLOFF [1971]).

3.2.1. Distortion of a partially ordered structure

When the ordered phase has a distorted structure relative to the disordered one like in AuCu (fc tetr. versus fcc) internal stresses are produced in fitting the two

References: p. 1403.

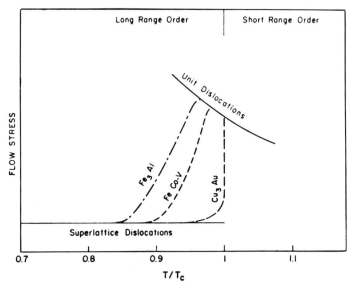

Fig. 34a. CRSS versus reduced temperature for several ordered alloys with quenched-in order (from STOLOFF [1971]).

structures together. This should impede dislocations particularly at an intermediate degree of order (or temperature if at low temperatures complete order is reached without misfit).

3.2.2. Change of lattice parameter with order

Even without changing the crystal structure long-range order will decrease the lattice parameter. This again can lead to misfit-hardening in a partially ordered structure. If only one phase exists at any time ordering may still appear near dislocations where atom distances are changed locally: At points where the disloca-tion strain is compressional, atom distances are decreased and the order increased and *vice versa* (SUMINO [1958]; see also §1.3.1d). The theory predicts a locking effect mainly for edge dislocations with a maximum near T_c.

3.2.3. Anti-phase domains of finite size

When a (super-)dislocation moves through an order domain structure it will increase the domain boundary area. COTTRELL [1954] has calculated the stress τ_c necessary for this as a function of domain diameter l.

$$\tau_c(\Gamma/l)[1 - \alpha(d/l)],   \qquad (17)$$

where d is the thickness of the domain boundary and α a geometrical factor of ~ 6. $\tau_c(l)$ passes through a maximum of $\tau_c(l_{max}) = \Gamma/4\alpha d$ for a diameter $l_{max} = 2\alpha d$. At smaller l the alloy is already disordered before slip and at larger l there are few boundaries for the dislocation to create disorder at: In both cases the alloy is softer than for l_{max}. A hardness maximum has in fact been observed by BIGGS and BROOM [1954] on Cu_3Au at $l_{max} = 50$ Å, giving $d = 2b$ and about the observed magnitude of τ_c from eq. (17). See also LOGIE [1957]. If order changes crystal symmetry (to non-cubic as in CoPt, IRANI and CAHN [1973]) then large internal stresses will develop in a microstructure of random small domains and contribute to order hardening.

3.2.4. Thickness of anti-phase boundary

BROWN [1959a] has drawn attention to the fact that the profile of order across an APB in a superdislocation must be different when this dislocation is in thermal equilibrium or in rapid movement. As the dislocation is pulled away from the equilibrium profile of its APB a *yield phenomenon* should be observed depending on the degree of order, i.e. on temperature. The calculated unlocking stress for CuZn has a maximum at $0.7T_c$ which has been related by Brown to the maximum observed for the *critical shear stress* of this material at about the same temperature (fig. 35). The agreement is good although this comparison needs justification.

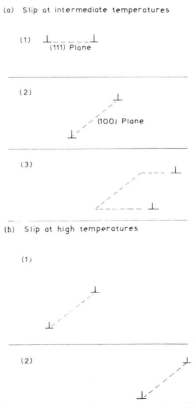

Fig. 36. Movement of a superdislocation at elevated temperatures (from FLINN [1960]).

References: p. 1403.

3.2.5. Climb of superdislocations

According to FLINN [1960] the minimum energy configuration of two dislocations connected by a strip of APB and forming a superdislocation may not be such that their common plane is the slip plane. If the temperature is high enough for some diffusion to occur, dislocations will climb to reach a lower-energy position (fig. 36a). They are then difficult to move unless the temperature is increased further such that diffusion becomes very rapid, possibly with the help of deformation-produced vacancies (fig. 36b). Thus again a maximum of the yield stress is expected at intermediate temperatures.

3.2.6. Short-range order hardening

This concept, due to FISHER [1954], has already been discussed in §1.3.2. ARDLEY [1955] has pointed out that a superdislocation moving through an *almost* perfect *super*lattice is restrained by the unfavourable bonds it creates during glide in just the same way as a *simple* dislocation moving through a *short-range* ordered lattice. The flow stress will increase in the first case as *disorder* increases, in the second as short range *order* increases. The maximum should appear for an intermediate degree of long-range and short-range order. RUDMAN [1962a] calculates $T = T_c$ for the position of the maximum.

3.2.7. Quench-hardening

If the order–disorder reaction is sluggish, the strength after quenching will depend on the degree of order at the quenching temperature. On the other hand vacancies may be trapped during the quench, leading to a faster rate of ordering than obtained by slow cooling. Dislocations may combine during the ordering process to form superdislocations of an equilibrium width corresponding to $S(T)$ [eq. (16)]. Such dislocation movement is expected to produce jogs and vacancies by intersection and climb. Both kinds of defects will harden the alloy (ch. 19). RUDMAN [1962] has pointed out that ordering due to excess vacancies will be inhomogeneous in the crystal since quenched-in vacancies decay more rapidly near dislocations. Thus dislocations will be locked by an order gradient in partially ordered crystals, and a yield point is to be expected.

In conclusion, several mechanisms have been proposed to account for the hardness peak at intermediate degrees of order and at temperatures of a certain fraction of T_c. Comparatively little experimental evidence is available so far to substantiate these theories.

3.3. Temperature dependence of the flow stress of ordered alloys

The initial flow stress of most of the ordered alloys considered here shows a maximum very near to the critical temperature, e.g. Cu_3Au, Ni_3Mn (MARCINKOWSKI and MILLER [1961]), and Fe_3Al (LAWLEY et al. [1961]). The degree of long-range order at the peak temperature is smaller than 0.7–0.4. An exception is CuZn (fig. 35) which shows a flow stress peak well below T_c, at $S > 0.9$. By initial flow stress is

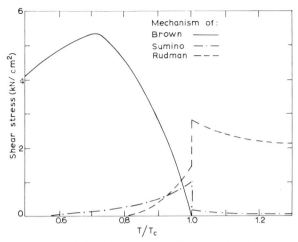

Fig. 37. Temperature dependence of the yield stress of CuZn as calculated by BROWN (§3.2.4), SUMINO (§3.2.2) and RUDMAN (§3.2.6) (after RUDMAN [1962a]).

meant the critical shear stress in the case of single crystals (Cu$_3$Au, CuZn) and the stress to produce a small defined strain for polycrystals. In addition, for a number of alloys under various conditions small pronounced yield points have been observed at the beginning of deformation. The dislocation-locking models of Sumino (§3.2.2), Brown (§3.2.4) and Rudman (§3.2.7) are well suited to explain them although some of the other mechanisms discussed above might also contribute. To explain the critical shear stress maximum of these three A$_3$B alloys the *local order–disorder friction* mechanism (§3.2.6) appears to be most suitable if a number of complicating factors can be eliminated. There should be no change in structure or lattice parameter with order and the domains should be larger than 50 Å. It is further assumed for the present discussion that the temperature is not high enough for extensive dislocation climb, and that the strength is measured at this temperature and not after quenching, which itself may produce hardening. Under these conditions the short-range order-disorder mechanism is used to explain the T_c flow stress peak in Cu$_3$Au by Ardley, Ni$_3$Mn by Marcinkowski *et al.* and in Fe$_3$Al by Lawley *et al.* The latter authors also see evidence for the Sumino locking mechanism for $T \leqslant T_c$, at the low-temperature flank of the peak in Fe$_3$Al. The same mechanism is invoked by BROWN [1959a] to explain the small step in τ_0 at T_c for CuZn (fig. 35). (The low temperature peak in CuZn has already been considered in §3.2.4. For a discussion of the possible contribution of an unlocking stress to the flow stress see §1.3.) In fig. 37, due to RUDMAN [1962a], the temperature dependence of the yield stress based on the short-range order, Sumino and Brown mechanisms is shown for the β' CuZn structure. On the basis of the available data no quantitative comparison between theory and experiment is attempted here.

Frequently the flow stress of ordered alloys is measured at room temperature after quenching from various temperatures T_Q near T_c. In as far as the degree of

References: p. 1403.

order at T_q is retained during quenching the yield stress should then correspond to that measured at temperature since the order-strengthening mechanisms of §3.2 depend little on temperature of deformation (with the exception of Flinn's climb mechanism which works only at elevated temperatures). In studies of order-hardening after quenching a possible contribution of quench-hardening must be considered. The latter can be investigated separately in CuZn which orders too rapidly for any disorder ever to be quenched in. BROWN [1959b] has shown that the yield stress of CuZn exhibits a maximum as a function of quenching temperature. The position of the maximum depends on the rate of quenching and coincides with a maximum in the density of vacancies retained after quenching. The yield stress after the quench anneals out rapidly with an activation energy of 64.5 kJ/mole.

3.4. Creep in ordered alloys

At high temperatures superdislocations in ordered alloys probably move by climb (§3.2.5). WEERTMAN [1956] has analyzed creep results of HERMAN and BROWN [1956] on CuZn as a function of temperature and stress within the framework of the theory presented in §1.8.1. The activation energy according to eqs. (6) and (8) in the ordered state is 160.5 kJ/mole, close to, although somewhat higher than, that for the diffusion of zinc in β' CuZn. The stress exponent is $n = 3.2$ in agreement with theory. The creep rate in the disordered state extrapolated to T_c is higher by a factor of 26 than in the ordered state. This could be due to additional glide movements by single dislocations in the disordered state.

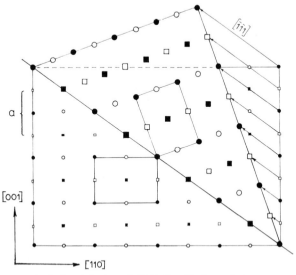

Fig. 38. Atomic movements during twinning of β' CuZn. Black symbols represent atoms in plane of projection, open symbols those on adjacent planes. Circles = Cu atoms, squares = Zn atoms. The cross-section of a unit cell is outlined. (From CAHN and COLL [1961].)

FLINN [1960] has investigated creep in an alloy based on the Ni_3Al structure (20 at% Al, 10 at% Fe, 70 at% Ni) which is ordered up to the melting point. He attributes a peak at 600°C in the initial flow stress of this alloy to the difficulty in moving superdislocations whose APB does not lie in their slip plane illustrated in fig. 36a. At high temperatures, diffusive movement of the APB may control the creep rate by a microcreep mechanism as described in §1.8.2. The APB is dragged by its bounding dislocations which glide on their regular slip planes (fig. 36b). The creep rate (according to §1.8.2) depends on stress and temperature in much the same way as in the climb theory, in agreement with the measurements. Flinn finds $n = 3.2$ and an activation energy of 327 kJ/mole close to the expected values. LAWLEY *et al.* [1960] observed anomalies in the creep behaviour of Fe–Al solid solutions near the composition Fe_3Al. The apparent activation energy for creep near T_c rises to about twice that of iron diffusion which controls creep in the disordered state [eq. (10)]. This is probably due to a rapid change of order parameter with temperature in this range. The authors also found that recovery periods during creep *hardened* the alloy while continuing creep *softened* it near T_c. Both of these creep anomalies could possibly be interpreted along the lines of §3.2.6 as due to the interaction of simple dislocations with short range order.

3.5. Twinning of ordered alloys

According to a classical prediction of LAVES [1952] an ordered cubic alloy should not twin mechanically. The twin shear would leave a certain fraction of the atoms in wrong places in the scheme of order (fig. 38). Diffusive movements are in general impossible at the low temperatures where twinning occurs. An alloy with partial long-range order will lose some of its favourable bonds during twinning. Below a critical degree of order S_c the applied stress will just be able to move the twinning dislocations against the friction resistance from the destruction of order. These predictions have been confirmed by BARRETT [1954] for β' CuZn and by CAHN and COLL [1961] in the case of Fe_3Al. The latter authors determined $S_c \approx 0.5$ corresponding to 5–10% of favourable nearest-neighbour bonds being destroyed during twinning.

4. Plasticity of intermetallic compounds

In this section the term intermetallic compound means intermediate phases in binary metal systems. Interstitial compounds will not be considered. Stoichiometry and the ordered structure play an important rôle in the classification of these alloys (ch. 5). They have a reputation of inherent brittleness which is related to the type of binding in these compounds. The present knowledge of the mechanical properties of these substances is summarized in symposia edited by WESTBROOK [1960,1967]. Recent technical interest in these compounds arises from their remarkable high-temperature strength and useful physical, e.g. semiconducting, properties. Furthermore their addition in fine dispersion to ductile metals causes strong hardening (ch. 22).

References: p. 1403.

4.1. Effect of temperature on strength

LOWRIE [1952] investigated the tensile elongation to fracture of various inter-metallic compounds as a function of a homologous temperature T/T_m, where T_m is the melting temperature. Most compounds showed some ductility for homologous temperatures larger than 0.65. Similar results were obtained by CHURCHMAN et al. [1956] for a number of diamond cubic and zincblende structures, including InSb. In this group of relatively similar substances slip started at $T/T_m = 0.44$ while cracking stopped at $T/T_m = 0.65$ as measured by indentation hardness. The sudden onset of ductility is not due to a loss of order but shows that thermal fluctuations are necessary for dislocations to move in their slip planes. The effect of temperature on hardness is particularly large for compounds with slight deviations from stoichiometry. In that case the crystal structure will contain "constitutional" vacancies (or interstitials). Figure 39 shows results of WESTBROOK [1960] on AgMg (CsCl structure) which exhibit a hardness maximum at the stoichiometric composition if the homologous temperature is large, a minimum when it is small. The constitutional point defects thus appear to assist slip at high temperatures where it is diffusion-con-

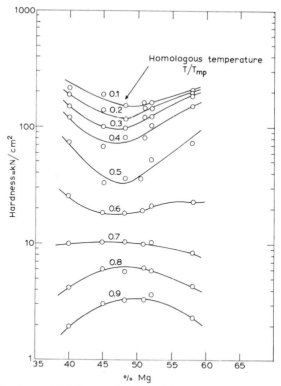

Fig. 39. Indentation hardness of AgMg at various homologous temperatures versus composition (after WESTBROOK [1960]). For flow-stress data on these alloys see WOOD and WESTBROOK [1962].

trolled. They can impede slip at low temperatures analogous to quench-hardening (§3.2.7). In addition, non-stoichiometric alloys could exhibit solid-solution hardening as discussed in §1. For the hexagonal close-packed compound Ag_2Al which has a nearly ideal axial ratio MOTE *et al.* [1961] found a strongly temperature-dependent critical shear stress in single crystals oriented for prismatic slip: fig. 40. Specimens oriented for basal slip exhibit a very pronounced upper and lower yield point which is independent of temperature. The prismatic yield stress presents a plateau at intermediate temperatures as described for solid solutions in §1. The dependence on composition is not known. The authors ascribe the plateau to short-range order in this (otherwise disordered) intermetallic phase (§1.3.2d). Stress-induced order locking could also explain the yield point in basal slip although chemical locking is expected for widely extended dislocations in the basal planes of this compound (§1.3.1). The steep low-temperature decrease of the prismatic yield stress cannot, however, be explained by any of the friction mechanisms discussed in §1.3.2. The corresponding activation volumes of $8b^3$ suggested to these authors a frictional movement of dislocations over the Peierls potentials by nucleation and growth of kinks.

4.2. Creep of indium antimonide

The mechanical properties of InSb have been studied extensively, especially by etch-pit techniques (STEINHARDT [1968]), dynamically (SCHÄFER *et al.* [1964]) and in bending creep tests (PEISSKER *et al.* [1962]). From the movement of etch-pitted

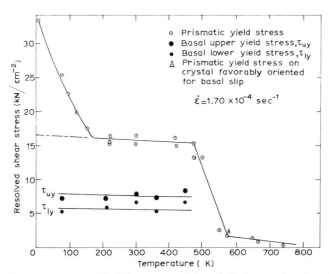

Fig. 40. Effect of temperature on critical shear stress of Ag_2Al single crystals oriented for basal and prismatic slip (after MOTE *et al.* [1961]).

References: p. 1403.

dislocations during a stress pulse the average dislocation velocity in InSb has been determined as a function of temperature and effective stress τ_{eff}:

$$v = K\tau_{eff}^m \exp(-Q/RT), \tag{18}$$

where K is a constant, $m = 1 \pm 0.2$ and $Q = 0.78$ eV (STEINHARDT [1968]). Equation (18) describes a quasi-viscous dislocation movement in the zincblende lattice. This behaviour originally proposed by HAASEN [1957] is expected if a microcrack exists in the core of a 60° dislocation (line direction $\langle 110 \rangle$), fig. 41a. For the core to move with the dislocation along its slip plane, atoms must make thermally activated jumps across the crack. This combined diffusive movement deviates slightly in InSb from that corresponding to a Newtonian viscosity ($m = 1$). (For comparison, in Fe–6.5 at% Si, $m = 40$ (STEIN and LOW [1960]).) A velocity–stress–temperature behaviour as expressed by eq. (18) is now interpreted in terms of the formation and movement of kinks in the Peierls potential for dislocations in the diamond and zincblende structure (HAASEN [1979b]). The creep rate can be calculated from $\dot{\varepsilon} = Nbv$, provided two assumptions are made. Firstly, a resonable dislocation multiplication rate is expressed by $(dN/dt)/Nv = c\tau_{eff}$ which is the number of new dislocation loops left behind a screw dislocation after moving 1 cm (GILMAN and JOHNSTON [1959]).

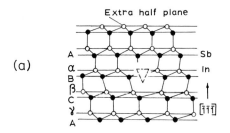

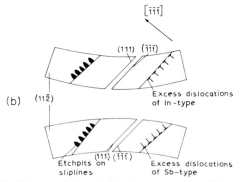

Fig. 41. (a) Model of the 60° dislocation in the zincblende structure (from HAASEN [1957]). (b) Schematic representation of etch pits and excess dislocations for InSb crystals bent in opposite directions (from PEISSKER et al. [1962]).

Secondly, work-hardening is introduced by

$$\tau_{\text{eff}} = \tau - \alpha \mu b \sqrt{N} \,, \tag{19}$$

where N is the dislocation density, and c and α are constants (HAASEN [1962]). This has been done and the result compared with measurements of plastic bending of InSb under various stresses and temperatures by PEISSKER *et al.* [1962]. A typical S-shaped creep curve of this material is shown schematically in fig. 42b. The creep rate $\dot\epsilon$ at any time corresponds to values of N, v calculated and plotted with $\dot\epsilon$ in fig. 42a. The creep rate increases as dislocations multiply; each dislocation then moves more slowly. After a steady-state creep period (rate $\dot\epsilon_w$) the rate decreases again because dislocation interaction now work-hardens the crystal, eq. (19). The agreement between the calculated and measured time laws of creep is quite good. The observed creep rate $\dot\epsilon_w$ is in reasonable agreement with the calculated one,

$$\dot\epsilon_w^{\text{obs}} = K_1 \tau^{3.3} \exp(-0.88 \text{ eV}/kT), \quad \dot\epsilon_w^{\text{calc}} = K_2 \tau^{(2+m)} \exp(-Q/kT), \tag{20}$$

if the m and Q from eq. (18) are used (K_1, $K_2 = \text{const}$). The dynamic yield behaviour of these compounds can be calculated in a similar way (ALEXANDER and HAASEN [1968]). The lower yield stress of InSb as a function of temperature and strain rate is obtained if one resolves eq. (20) for $\tau(T, \dot\epsilon)$, and is found in good agreement with dynamic measurements (SCHÄFER *et al.* [1964]).

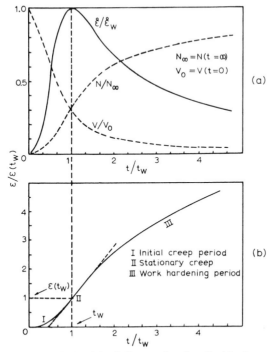

Fig. 42. Normalized creep curve calculated for InSb crystal as described in the text (from PEISSKER *et al.* [1962]).

References: p. 1403.

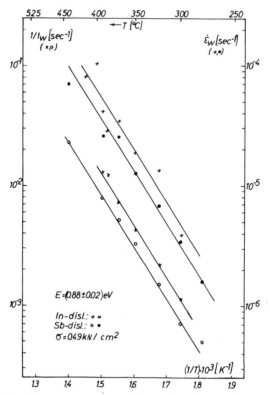

Fig. 43. Effect of bending direction on creep rate $\dot{\epsilon}_w$ and incubation time t_w of InSb for various temperatures (from PEISSKER *et al.* [1962]).

The polarity of stacking of the $\langle 111 \rangle$ slip planes in InSb causes an interesting anisotropy of the creep rate with respect to the bending direction as was predicted by HAASEN [1957] and is illustrated in fig. 41 (PEISSKER *et al.* [1962]). The extra half plane of a positive dislocation ends with a row of, say, Sb atoms (Sb dislocation) while that of a negative dislocation ends with a row of In atoms (In-dislocation). This assignment depends on the position of the slip plane in this structure: Between the widely spaced planes Aα as drawn in fig. 41a or – now more likely because of the existence of stacking faults – between the closely spaced planes αB (GOTT-SCHALK *et al.* [1978]). If the mobilities of these dislocations are determined by diffusive movements right in the core than these mobilities should differ. An effect of the dislocation charge on the rate of kink formation on dislocations in semiconductors was mentioned in §1.7. We expect the energy states and therefore the dislocation charges of In and Sb dislocations to differ. Similar effects occur with II–VI compounds like CdTe (GUTMANAS and HAASEN [1981]). A bending test introduces excess dislocations of one sign. By reversing the bending direction the sign of the excess dislocation is also changed. Thus the creep rate should depend on the direction of bending in InSb as has been in fact observed by PEISSKER *et al.* [1962] and is shown in fig. 43. The identification of the excess dislocations as In or Sb dislocations can be made by etching because of the polar nature of the zincblende lattice (see also SHIMIZU and SUMINO [1975]).

The plastic behaviour of intermetallic compounds may thus be better suited for a quantitative understanding than that of pure metals. This is a consequence of the quasi-viscous nature of dislocation motion in the former substances which strongly depends on temperature and less on the (difficult to control) stress parameter. Solid solutions appear to hold an intermediate position in this respect between pure metals and compounds. The micromechanical approach to plasticity *via* etch-pit studies described above for compounds has been shown to be a fruitful one also for solid solutions (KLEINTGES and HAASEN [1979]).

Acknowledgement

The author would like to thank Prof. Ch. Schwink and Dr. W. Skrotzki for critical comments on the third edition of this chapter.

References

AKHTAR, A., and E. TEGHTSOONIAN, 1968, Proc. Int. Conf. Strength of Metals and Alloys, Tokyo, Trans. Japan. Inst. Met. Suppl. **9**, 692.

ALEXANDER, H. and P. HAASEN, 1968, Solid State Physics, F. Seitz and D. Turnbull eds., vol. **22**, 28.

ARDELL, A.J., V. MUNJAL and D.J. CHELLMAN, 1976, Metallurg. Trans. **7A**, 1263.

ARDLEY, G.W., 1955, Acta Metall. **3**, 525.

ARDLEY, G.W., and A.H. COTTRELL, 1953, Proc. Roy. Soc. **A219**, 328.

ASHBY, M.F., 1964, Z. Metallk. **55**, 5.

ASHBY, M.F., 1966, Acta Metall. **14**, 679.

ASHBY, M.F., 1968, in: Strengthening Mechanisms in Solids, eds. A. Kelly and R.B. Nicholson (Elsevier, Amsterdam).

BARRETT, C.S., 1954, Trans. AIME **200**, 1003.

BASINSKI, Z.S., R.A. FOXALL and R. PASCUAL, 1972, Scripta Metall. **6**, 807.

BIGGS, W.D., and T. BROOM, 1954, Phil. Mag. **45**, 246.

BOČEK, M., P. KRATOCHVIL and P. LUKÁČ, 1961, Czech. J. Phys. **B11**, 674.

BRION, H.G., P. HAASEN and H. SIETHOFF, 1971, Acta Metall. **19**, 283,

BROWN, L.M., 1979, in: Strength of Metals and Alloys, eds P. Haasen *et al.* (Pergamon Press, Oxford) vol. 3, 1551.

BROWN, L.M., and R.K. HAM, 1971, in: Strengthening Methods in Crystals, eds. A. Kelly *et al.* (Elsevier, Amsterdam) p. 10.

BROWN, N., 1959a, Phil. Mag. **4**, 693.

BROWN, N., 1959b, Acta Metall. **7**, 210.

BROWN, N., 1960, in: Mechanical Properties of Intermetallic Compounds, ed. J. Westbrook (Wiley, New York) p. 177.

BYRNE, J.G., M.E. FINE and A. KELLY, 1961, Phil. Mag. **6**, 1119.

CAHN, R.W., and J.A. COLL, 1961, Acta Metall. **9**, 138.

CARTER, C.B., 1980, Phil. Mag. **A41**, 619.

CARTER, C.B., and J.L.F. RAY, 1977, Phil. Mag. **35**, 189.

CHURCHMAN, A.T., G.A. GEACH and J. WINTON, 1956, Proc. Roy. Soc. **A238**, 194.

CONRAD, H., 1960, Phil. Mag. **5**, 745.

COPLEY, S.M., and B.H. KEAR, 1967, Trans. AIME **239**, 984.

COTTRELL, A.H., 1958, Trans. AIME **212**, 192.

COTTRELL, A.H., C.S. HUNTER and F.R.N. NABARRO, 1953, Phil. Mag. **44**, 1064.

COULING, S.L., 1959, Acta Metall. **7**, 133.

DASH, J., and M.E. FINE, 1961, Acta Metall. **9**, 149.

DEW-HUGHES, D., and W.D. ROBERTSON, 1960, Acta Metall. **8**, 147 and 156.

DIEHL, J., 1965, in: Moderne Probleme der Metallphysik, ed. A. Seeger (Springer, Berlin) p. 227.

DORN, J.E., 1957, in: Creep and Recovery (ASM, Metals Park, OH) p. 255.

EBELING, R., and M.F. ASHBY, 1966, Phil. Mag. **13**, 805.

ESCAIG, B., 1968, in: Dislocation Dynamics, eds. A. Rosenfield *et al.* (McGraw–Hill, New York) p. 655.

EVERS, M., 1959, Z. Metallk. **50**, 638.

EVERS, M., 1961, Z. Metallk. **52**, 359.

FIORE, N.F., and C.L. BAUER, 1968, Prog. Mater. Sci. **13**, 85.

FISHER, J.C., 1954, Acta Metall. **2**, 9.

FISHER, J.C., E.W. HART and R.H. PRY, 1953, Acta Metall. **1**, 336.

FLINN, P.A., 1958, Acta Metall. **6**, 631.

FLINN, P.A., 1960, Trans. AIME **218**, 145.

FLEISCHER, R.L., 1961, Acta Metall. **9**, 996.

FLEISCHER, R.L., 1963, Acta Metall. **11**, 203.

FLEISCHER, R.L., 1966, Acta Metall. **14**, 1867.

FOREMAN, A.J.E., and M.J. MAKIN, 1966, Phil. Mag. **14**, 131.

FRIEDRICHS, J., and P. HAASEN, 1975, Phil. Mag. **31**, 863.

GALLAGHER, P.C.J., 1968, J. Appl. Phys. **39**, 160.

GALLAGHER, P.C.J., 1970, Trans. AIME **1**, 2429.

GARSTONE, J., and R.W.K. HONEYCOMBE, 1957, in: Dislocations and Mechanical Properties of Crystals, eds. J.C. Fisher *et al.* (Wiley, New York) p. 391.

GEROLD, V., 1979, in: Dislocations in Solids, ed. F.R.N. Nabarro (North-Holland, Amsterdam) vol. 4, p. 215.

GIBALLA, R., and T.E. MITCHELL, 1973, Scripta Metall. **7**, 1143.

GILMAN, J.J., 1956, Trans. AIME **206**, 1326.

GILMAN, J.J., and W.G. JOHNSTON, 1959, J. Appl. Phys. **31**, 331.

GITTUS, J., 1978, Irradiation Effects in Crystalline Solids (Appl. Sci. Publ., London).

GLEITER, H., 1968, Acta Metall. **16**, 455.

GLEITER, H., 1967b, Acta Metall. **15**, 1213.

GLEITER, H., and E. HORNBOGEN, 1965, Phys. Stat. Sol. **12**, 235.

GLEITER, H., and E. HORNBOGEN, 1968, Mater. Sci. Eng. **2**, 285.

GOTTSCHALK, G., H. PATZER and H. ALEXANDER, 1978, Phys. Stat. Sol. (a) **45**, 207.

GRANATO, A.V., 1971 Phys. Rev. Lett. **27**, 660; Phys. Rev. **B4**, 2196.

GREETHAM, G., and R.W.K. HONEYCOMBE, 1961, J. Inst. Met. **89**, 13.

GRÖHLICH, M., P. HAASEN and G. FROMMEYER, 1982, Scripta Metall. **16**, 367.

GYPEN, L.A., and A. DERRUYTTERE, 1981, Scripta Metall. **15**, 815.

GYPEN, L.A., and A. DERRUYTTERE, 1977, J. Mater. Sci. **12**, 1028, and 1034.

GUTMANAS, E.YU., and P. HAASEN, 1981, Phys. Stat. Sol. (a) **63**, 193.

HAASEN, P., 1957, Acta Metall. **5**, 598.

HAASEN, P., 1958, Phil. Mag. **3**, 384.

HAASEN, P., 1959, in: Internal Stresses and Fatigue in Metals, eds. G.M. Rassweiler and W.L. Grube (Elsevier, Amsterdam) p. 205.

HAASEN, P., 1962, Z. Phys. **167**, 461.

HAASEN, P., 1964, Z. Metallk. **55**, 55.

HAASEN, P., 1965, in: Alloying Effects in Concentrated Solid Solutions, ed. T.B. Massalski (Gordon and Breach, New York) p. 270.

HAASEN, P., 1967, Krist. u. Techn. **2**, 251.

HAASEN, P., 1968, Proc. Int. Conf. on Strength of Metals and Alloys, Tokyo, Trans. Japan. Inst. Met. Suppl. **9**, XL.

HAASEN, P., 1975, Phys. Stat. Sol. (a) **28**, 145.

HAASEN, P., 1977, in: Fundamental Aspects of Structural Alloy Design, eds. R.J. Jaffee *et al.* (Plenum, New York) p. 3.

HAASEN, P., 1979b, J. Physique **40**, C6-111.

HAASEN, P., and A. KELLY, 1957, Acta Metall. **5**, 192.

HAASEN, P., and A.H. KING, 1960, Z. Metallk. **51**, 722.

HAASEN, P., and P. KRATOCHVIL, 1973, Proc. Int. Reinststoff-Symp., Dresden (Akad. d. W., Berlin) p. 383.

HAASEN, P., and R. LABUSCH, 1979, in: Strength of Metals and Alloys, eds. P. HAASEN *et al.* (Pergamon Press, Oxford) vol. 1, p. 639.

HAESZNER, F., and D. SCHREIBER, 1957, Z. Metallk. **48**, 263.

HALE, K.F., and D. McLEAN, 1961, private communication; see also: Proc. Europ. Reg. Conf. Electron Microscopy, Delft, 1960, p. 425.

HAMMAR, R.H., W.C.T. YEH, T.G. OAKWOOD and A.A. HENDRICKSON, 1967, Trans. AIME **239**, 1692.

HANSON, K., and J.W. MORRIS, Jr., 1975, J. Appl. Phys. **46**, 2378.

HARDY, H.K., and T.J. HEAL, 1954, Prog. Met. Phys. **5**, 143.

HARRIS, B., 1967, J. Less-Common Met. **12**, 247.

HAUSER, F.E., P.R. LANDON and J.E. DORN, 1958, Trans. ASM **50**, 856.

HENDRICKSON, A.A., and M.E. FINE, 1961a, Trans. AIME **221**, 103.

HENDRICKSON, A.A., and M.E. FINE, 1961b, Trans. AIME **221**, 967.

HERMAN, M., and N. BROWN, 1956, Trans. AIME **206**, 604.

HESLOP, J., and N.J. PETCH, 1956, Phil. Mag. **1**, 866.

HIRSCH, P.B., 1975, in: The Physics of Metals, vol. 2, Defects (Cambridge Univ. Press) p. 189

HIRSCH, P.B., and R.C. CRAWFORD, 1979, in: Strength of Metals and Alloys, eds. P. Haasen *et al.* (Pergamon Press, Oxford) vol. 1, p. 89.

HIRSCH, P.B., and A. Kelly, 1965, Phil. Mag. **12**, 881.

HOCKEY, B.J., and J.W. MITCHELL, 1972, Phil. Mag. 26, 409.

HORNE, G.T., R.B. ROY and H.W. PAXTON, 1963, J. Iron Steel Inst. **201**, 161.

HOWIE, A., 1960, Thesis Cambridge; cf. Proc. Europ. Reg. Conf. on Electronic Microscopy Delft, p. 383.

HUTCHISON, M.M., and R.W.K. HONEYCOMBE, 1967, Met. Sci. J. **1**, 70

IRANI, R.S., and R.W. CAHN, 1973, Acta Metall. **21**, 575.

JAN, J.P., 1955, J. Appl. Phys. **26**, 1291.

JAX, P., P. KRATOCHVIL and P. HAASEN, 1970, Acta Metall. **18**, 237.

JONES, R.L., and A. KELLY, 1968, Proc. 2nd Bolton Landing Conf. on Oxide Dispersion Strengthening (Gordon and Breach, New York).

KARNTHALER, H.P., P.M. HAZZLEDINE and M.S. SPRING, 1972, Acta Metall. **20**, 459.

KARNTHALER, H.P., F. PRINZ and G. HASLINGER, 1975, Acta Metall. **23**, 155.

KEAR, B.H., 1964, Acta Metall. **12**, 555.

KELLY, A., 1958, Phil. Mag. **3**, 1472.

KELLY, A., and M.E. FINE, 1957, Acta Metall. **5**, 365.

KELLY, A., and R.B. NICHOLSON, 1963, Prog. Mater. Sci. **10**, 1.

KELLY, A., and S. SATO, 1961, Acta Metall. **9**, 59.

KIMURA, H., and R. MADDIN, 1971, Quench-Hardening in Metals (North-Holland, Amsterdam).

KLEINTGES, M., 1980, Scripta Metall. **14**, 993.

KLEINTGES, M., and P. HAASEN, 1980, Scripta Metall. **14**, 999.

KLOSKE, R.A., and M.E. FINE, 1969, Trans. AIME **245**, 217.

KOCKS, U.F., 1966, Phil. Mag. **13**, 541.

KOCKS, U.F., A.S. ARGON and M.F. ASHBY, 1975, Prog. Mater. Sci. **19**, 1.

KOPPENAAL, T.J., and M.E. FINE, 1961, Trans. AIME **221**, 1178.

KÖSTER, W., H. v. KÖCKRITZ and E.H. SCHULZ, 1932, Arch. Eisenhüttenw. **6**, 55.

KOSTORZ, G., 1968, thesis Göttingen; Z. Metallk. **59**, 941.

KOSTORZ, G., and P. HAASEN, 1969, Z. Metallk. **60**, 26.

KRATOCHVIL, P., 1970, Scripta Metall. **4**, 333.

KRATOCHVIL, P., and P. HAASEN, 1982, Scripta Metall. **16**, 197.

KRATOCHVIL, P., and E. NERADOVA, 1971, Czech. J. Phys. **B21**, 1273.

KRÖNER, E., 1964, Phys. Kondens. Materie **2**, 262.

KUBIN, L.P., and J.L. MARTIN, 1979, in: Strength of Metals and Alloys, eds. P. Haasen *et al.* (Pergamon Press, Oxford) vol. 3, 1639.

KUHLMANN-WILSDORF, D., and H.G. WILSDORF, 1953, Acta Metall. **1**, 394.

LABUSCH, R., 1970, Phys. Stat. Sol **41**, 659.

LABUSCH, R., 1972, Acta Metall. **20**, 917.

LABUSCH, R., 1981, Czech. J. Phys. **B31**, 165.

LABUSCH, R., G. GRANGE, J. AHEARN and P. HAASEN, 1975, in: Rate Processes in Plastic Deformation of Materials (ASM, Metals Park, OH) p. 26.

LALL, C., S. CHIN and D.P. POPE, 1979, Metallurg, Trans. **10A**, 1323.

LAVES, F., 1952, Naturwiss. **39**, 546.

LAWLEY, A., and R. MADDIN, 1962, Trans. AIME, **224**, 573.

LAWLEY, A., J.A. COLL and R.W. CAHN, 1960, Trans. AIME **218**, 166.

LAWLEY, A., A.E. VIDOZ and R.W. CAHN, 1961, Acta Metall. **9**, 287.

LESLIE, W.C., 1961, Acta Metall. **9**, 1004.

LIVINGSTON, J.D., 1959, Trans. AIME **215**, 566.

LIVINGSTON, J.D., and J.J. BECKER, 1958, Trans. AIME **212**, 316.

LOGIE, H.J., 1957, Acta Metall. **5**, 106.

LOWRIE, R., 1952, Trans. AIME **194**, 1093.

LUKAČ, P., and I. STULIKOVA, 1974, Czech. J. Phys. **B24**, 648.

LÜTJERING, G., and E. HORNBOGEN, 1968, Z. Metallk. **59**, 29.

MADER, S., 1963, in: Electron Microscopy and Strength of Crystals (Interscience, New York) p. 183.

MADER, S., A. SEEGER and H.M. THIERINGER, 1963, J. Appl. Phys. **34**, 3376.

MAHAJAN, S., and D.F. WILLIAMS, 1973, Int. Met. Rev. **18**, 43.

MARCINKOWSKI, M.J., and D.S. MILLER, 1961, Phil. Mag. **6**, 871.

MARCINKOWSKI, M.J., N. BROWN and R.M. FISHER, 1961, Acta Metall. **9**, 129.

McCORMICK, P.G., 1972, Acta Metall. **20**, 351.

McLEAN, D., 1962, Met. Rev. **7**, 481.

MEAKIN, J.D., and H.G. WILSDORF, 1960, Trans. AIME **218**, 737.

MEISSNER, J., 1959, Z. Metallk. **50**, 207.

MELANDER, A., and P.A. PERSSON, 1978, Met. Sci. J. **12**, 391.

MIGAUD, B., 1960, Thèse, Université de Paris.

MITCHELL, T.E., and P.L. RAFFO, 1967, Can. J. Phys. **45**, 1047.

MITCHELL, J.W., J. AHEARN, B.J. HOCKEY, J.P. MONAGHAN and R.K. WILD, 1968, Proc. Int. Conf. on Strength of Metals and Alloys, Tokyo, Trans. Japan. Inst. Met. Suppl. **9**, 769.

MORDIKE, B.L., and P. HAASEN, 1962, Phil. Mag. **7**, 459.

MORDIKE, B.L., A.A. BRAITHWAITE and K.D. ROGAUSCH, 1970, Met. Sci. J. **4**, 37.

MORI, T., and H. FUJITA, 1980, Acta Metall. **28**, 771.

MOTE, J.D., K.T. TANAKA and J.E. DORN, 1961, Trans. AIME **221**, 858.

MUGHRABI, H. 1979, in: Strength of Metals and Alloys, eds. P. Haasen *et al.* (Pergamon Press, Oxford) vol. 3, p. 1615.

NABARRO, F.R.N., 1982, Proc. Roy. Soc. **A381** 285.

NARITA, N., and J. TAKAMURA, 1974, Phil. Mag. **29**, 1001.

NARITA, N., A. HATANO, J. TAKAMURA, M. YOSHIDA and H. SAKAMOTO, 1978, J. Japan Inst. Metals **42**, 533.

NATSIK, V.D., 1979, Fiz. Nizkikh Temp. **5**, 400.

NEUHÄUSER, H., and H. FLOR, 1978, Scripta Metall. **12**, 443.

NEUHÄUSER, H., and H. FLOR, 1979, Scripta Metall. **13**, 147

NEUMANN, P., 1967, Z. Metallk. **58**, 780.

NICHOLSON, R.B., G. THOMAS and J. NUTTING, 1960, Acta Metall. **8**, 172.

NIX, W.D., and B. ILSCHNER, 1979, in: Strength of Metals and Alloys, eds. P. Haasen *et al.* (Pergamon Press, Oxford) vol. 3, p. 1503.

NIXON, W.E. and J.W. MITCHELL, 1981, Proc. Roy. Soc. **A376**, 343.

PANDE, C.S., and P.M. HAZZLEDINE, 1971, Phil. Mag. **24**, 1039 and 1393.

PEISSKER, E., 1965a, Acta Metall. **13**, 419.

PEISSKER, E., 1965b, Z. Metallk. **56**, 155.

PEISSKER, E., P. HAASEN and H. ALEXANDER, 1962, Phil. Mag. **7**, 1279.

PFAFF, F., 1962, Z. Metallk. **53**, 411 and 466.

PFEIFFER, W., and A. SEEGER, 1962, Phys. Stat. Sol. **2**, 668.

PHILLIPS, V.A., 1964, Trans. AIME **230**, 967.

PIERCY, G.R., R.W. CAHN and A.H. COTTRELL, 1955, Acta Metall. **3**, 331.

RIDDHAGNI, B.R., and R.M. ASIMOW, 1968, J. Appl. Phys. **39**, 4144 and 5169.

ROGAUSCH, K.D., 1967, Z. Metallk, **58**, 50.

RUDMAN, P.S., 1962a, Acta Metall, **10**, 253.

RUDMAN, P.S., 1962b, Acta Metall. **10**, 195.

RUDOLPH, G., P. HAASEN, B.L. MORDIKE and P. NEUMANN, 1965, Proc. 1st Intern. Conf. on Fracture, Sendai, ed. T. Yokobori (Japan. Inst. of Metals, Sendai) p. 501.

SACHS, G., and J. WEERTS, 1931, Z. Phys. **67**, 507.

SAXL, J., 1964, Czech, J. Phys. **B14**, 381.

SCHÄFER, S., H. ALEXANDER and P. HAASEN, 1964, Phys. Stat. Sol. **5**, 247.

SCHARF, H., P. LUKÁČ, M. BOČEK and P. HAASEN, 1968, Z. Metallk. **59**, 799.

SCHMID, E. and H. SELIGER, 1932, Metallwirtsch. **11**, 421.

SCHOECK, G., 1957, in: Creep and Recovery (ASM, Metals Park, OH) p. 199.

SCHOECK, G., and A. SEEGER, 1959, Acta Metall. **7**, 469.

SCHWARZ, R.B., 1979, in: Strength of Metals and Alloys, eds. P. Haasen *et al.* (Pergamon Press, Oxford) vol. 1, p. 953.

SCHWARZ, R.B., 1982, Scripta Metall. **16**, 385.

SCHWARZ, R.B., and R. LABUSCH, 1978, J. Appl. Phys. **49**, 5147.

SCHWINK, Ch., and Th. WILLE, 1980, Scripta Metall. **14**, 1093.

SEEGER, A., R. BERNER and H. WOLF, 1959, Z. Physik **155**, 247.

ŠESTAK, B., 1979, in: Strength of Metals and Alloys, eds. P. Haasen *et al.* (Pergamon Press, Oxford) vol. 3, 1461.

ŠESTAK, B., and A. SEEGER, 1978, Z. Metallk. **69**, 195, 355 and 425.

SHEELY, W.F., E.D. LEVINE and R.R. NASH, 1959, Trans. AIME **215**, 521, 693.

SHIMIZU, H., and K. SUMINO, 1975, Phil. Mag. **1**, 123 and 143.

SINNING, H.R., 1982, Acta Metall. **30**, 1019.

SINNING, H.R. and P. HAASEN, 1981, Z. Metallk. **72**, 807.

SIETHOFF, H., 1969, Thesis Göttingen.

SIETHOFF, H., 1969, Mater. Sci. Eng. **4**, 155.

SIETHOFF, H., 1970, Phys. Stat. Sol. **40**, 153.

SKROTZKI, W., and P. HAASEN, 1981, J. Physique **42**, C3-119.

STARTSEV, V.I., 1981, Czech. J. Phys. **B31**, 115.

STARTSEV, V.I., V.V. DEMIRSKI and S.N. KOMNIK, 1979, in: Strength of Metals and Alloys, eds. P. Haasen *et al.* (Pergamon Press, Oxford) vol. 2, p. 1043.

STEIN, D.F., and J.R. LOW, 1960, J. Appl. Phys. **31**, 362.

STEINHARDT, H., 1968, Dipl. thesis, Göttingen.

STOLOFF, N.S., 1971, in: Strengthening Methods in Crystals, eds. A. Kelly *et al.* (Elsevier, Amsterdam) p. 193

STOLOFF, N.S., and R.G. DAVIES, 1966, Prog. Mater. Sci. **13**, 1.

SUMINO, K., 1958, Sci. Rep. RITU, **A10**, 283.

SUZUKI, H., 1979a, in: Strength of Metals and Alloys, eds. P. Haasen *et al.* (Pergamon Press, Oxford) vol. 3, 1595.

SUZUKI, H., and C.S. BARRETT, 1958, Acta Metall. **6**, 156.

SUZUKI, H., and K. NAKAMISHI, 1975, Trans. Japan. Inst. Met. **16**, 17.

SVITAK, J.J., and R.M. ASIMOW, 1969, Trans. AIME **245**, 209.

SWANN, P.R., 1963, in: Electron Microscopy and Strength of Crystals (Interscience, New York) p. 131.

TAKEUCHI, S., 1968, Scripta Metall. **2**, 481.

TAKEUCHI, S., and A.S. ARGON, 1976a, J. Mat. Sci. **11**, 1542.

TAKEUCHI, S., and A.S. ARGON, 1976b, Acta Metall. **24**, 861 and 883.

TAOKA, T., and S. SAKATA, 1957, Acta Metall. **5**, 61.

THIERINGER, H.M., 1962, Thesis Stuttgart; cf. MADER [1963].

THOMAS, G., J. NUTTING and P.B. HIRSCH, 1957, J. Inst. Met. **86**, 7.

THORNTON, P.R., and T.E. MITCHELL, 1962, Phil. Mag. **7**, 361.

TILLER, W.A., 1958, J. Appl. Phys. **29**, 611.

TRAUB, H., H. NEUHÄUSER and Ch. SCHWINK, 1977, Acta Metall. **25**, 437 and 1289.

TYSON, W.R., 1968, Can. Met. Quart. **6**, 301.

VAN DEN BEUKEL, A., 1980, Acta Metall. **28**, 965.

VENABLES, J.A., 1961, Phil. Mag. **6**, 379.

VERGNOL, J., and J.P. VILLAIN, 1979, in: Strength of Metals and Alloys, eds. P. Haasen *et al.* (Pergamon Press, Oxford) vol. 3, p. 121.

VICTORIA, M.P., and A.E. VIDOZ, 1968, Phys. Stat. Sol. **28**, 131.

VIDOZ, A.E., 1968, Phys. Stat. Sol. **28**, 145.

WAGNER, R., 1979, in: Strength of Metals and Alloys, eds. P. Haasen *et al.* (Pergamon Press, Oxford) vol. 1, p. 645.

WALTON, D., L. SHEPARD and J.E. DORN, 1961, Trans. AIME **221**, 458.

WEERTMAN, J., 1955, J. Appl. Phys. **26**, 1213.

WEERTMAN, J., 1956, Trans. AIME **206**, 1409.

WEERTMAN, J., 1957, J. Appl. Phys. **28**, 1185.

WEERTMAN, J., 1977, Acta Metall. **25**, 1393.

WENDT, H., and R. WAGNER, 1982, Acta Metall. **30**, 1561.

WESTBROOK, J.H., ed., 1967, Intermetallic Compounds (Wiley, New York).

WHELAN, M.J., 1959, Proc. Roy. Soc. **A249**, 114.

WILLE, Th., and Ch. SCHWINK, 1980, Scripta Metall. **14**, 923.

WILLE, Th., B. WIELKE and Ch. SCHWINK, 1982, Scripta Metall. **16**, 561.

WILSDORF, H.G., and J.T. FOURIE, 1956, Acta Metall. **4**, 271.

WOOD, D.L., and J.H. WESTBROOK, 1962, Trans. AIME **224**, 1024.

Further reading

Brown, N., in: Mechanical Properties of Intermetallic Compounds, ed. J. Westbrook (Wiley, New York, 1960) p. 177.

Cottrell, A.H., Dislocations and Plastic Flow in Crystals (Clarendon, Oxford, 1953).

Cottrell, A.H., in: Relation of Properties to Microstructure (ASM, Metals Park, OH, 1954) p. 131.

Diehl, J., in: Moderne Probleme der Metallphysik, ed. A. Seeger (Springer, Berlin, 1965) p. 227.

Escaig, B., in: Dislocation Dynamics, eds. A. Rosenfield *et al.* (McGraw–Hill, New York, 1968) p. 655.

Fleischer, R.L., in: The Strengthening of Metals, ed. D. Peckner (Reinhold, New York, 1964) p. 93.

Friedel, J., Les Dislocations (Gauthier-Villars, Paris, 1956) Engl. Transl.: Dislocations (Pergamon, Oxford, 1964).

Gerold, V., in: Dislocations in Solids, ed. F.R.N. Nabarro (North-Holland, Amsterdam, 1979) vol. 4, 215.

Gittus, J., Irradiation Effects in Crystalline Solids (Applied Sci. Publ., London, 1978).

Haasen, P., Physical Metallurgy (Springer Verlag/CUP Cambridge, 1974, 1978).

Haasen, P., in: Dislocations in Solids, ed. F.R.N. Nabarro (North-Holland, Amsterdam, 1979a) vol. 4, p. 155.

Kelly, A., in: Electron Microscopy and Strength of Crystals (Interscience New York, 1963) p. 947.

Kelly, A., and R.B. Nicholson, Prog. Mater. Sci. 10 (1963) 151.

Kimura, H., and R. Maddin, Quench-Hardening in Metals (North-Holland, Amsterdam, 1971).

Kocks, U.F., A.S. Argon and M.F. Ashby, Prog. Mater. Sci. 19 (1975) 1.

Marcinkowski, M.J., in: Order–Disorder Transformations in Solids, ed. H. Warlimont (Springer, Berlin, 1974) p. 365.

McLean, D., Met. Rev. 7 (1962) 481.

Seeger, A., in: Dislocations and Mechanical Properties of Crystals, eds. J.C. Fisher *et al.* (Wiley, New York, 1957) p. 243.

Seeger, A., in: Encyclopedia of Physics, VII/2 (Springer, Berlin–Göttingen–Heidelberg, 1958) p. 1.

Suzuki, H., in: Dislocations and Mechanical Properties of Crystals, eds. J.C. Fisher *et al.* (Wiley, New York, 1957) p. 361; unpublished data.

Suzuki, H., in: Dislocations in Solids, ed. F.R.N. Nabarro (North Holland, Amsterdam, 1979b) vol. 4, p. 191.

Venables, J.A., in: Deformation Twinning, eds. R.E. Reed-Hill *et al.* (Gordon and Breach, New York, 1964) p. 77.

Westbrook, J.H., ed., Mechanical Properties of Intermetallic Compounds (Wiley, New York, 1960).

CHAPTER 22

MECHANICAL PROPERTIES OF MULTIPHASE ALLOYS

Jean-Loup STRUDEL

Centre des Materiaux
École Nationale Supérieure des Mines de Paris
91003 Evry, France

R.W. Cahn and P. Haasen, eds.
Physical Metallurgy; third, revised and enlarged edition
© *Elsevier Science Publishers BV, 1983*

1.   Introduction

A large majority of alloys used in everyday life are and have been for years or even centuries, knowingly or not, multicomponent materials. Except for pure noble metals such as platinum, gold, silver and copper which have been devoted to the manufacture of ornamental pieces since early periods of antiquity, high mechanical and thermal resistance have been expected from man-made metallic objects or engineering structures – blades, weapons, stoves, burners, piston- and jet-engine parts, nuclear reactors, etc. Strengthening a plastically deformable material can be achieved in a number of ways: reducing the grain size, solid-solution hardening of the matrix, strain-hardening, precipitate-, particle- or fiber-hardening. Each of these methods has its domain of stress, temperature, feasibility and cost where it is best suited but one must always keep in mind that grain-boundary resistance has to match the improved mechanical performance of the matrix.

This chapter will be devoted to polyphase alloys in which one phase, called the *matrix*, entirely surrounds on all sides the second phase, called *particle* or *precipitate*. This definition excludes dual-phase alloys such as certain types of steels which are examined in ch. 16, and titanium-base alloys. Fiber-composite materials are not considered either, although some of the theoretical tools developed in §3 can easily be extended to encompass this field of newly developed materials (ch. 29). Consequently the hardening obstacles considered in this chapter will be implicitly of polygonal convex shape, often described simply by spheres or roughly spheroidal objects with sizes ranging from a few nanometers to a few microns (10–20 μm at the most).

Macroscopic properties of materials are revealed to the metallurgist by experimental data such as stress–strain curves, fatigue, relaxation, or creep curves which depend on external parameters (temperature, strain rate or loading rate) and on the internal structure of the material. These basic manifestations of the alloy's response to various mechanical and thermal solicitations remain the ultimate framework of reference for the practical engineer on one hand, and for the theoretical metallurgist on the other hand. In the course of the last twenty years, innumerable transmission electron microscopy (TEM) observations have brought deep insights into microscopic mechanisms taking place during deformation. Complex dislocation interactions have enabled the metallurgist to develop microscopic models with a mass of geometric details which must be abandoned when an attempt is made to derive macroscopic flow rates or observed recovery kinetics. But the importance of these observations is of a qualitative nature – the halting of a dislocation in the glide plane in front of a particle, various alternatives for circumventing the obstacle, such as by-passing, cross-slipping, climbing, cutting, etc.

On the other hand, macroscopic models based on continuum mechanics effectively provide the modern engineer, surrounded by high-speed computers, with sophisticated equations capable of yielding precise quantitative answers in a good number of well-defined loading schemes: yield strength, fatigue curves, creep rates, hardening rates are well described, responses to abrupt thermal or stress transients and creep–fatigue interactions still need some improvements. Much too often these models are faced with the necessity of introducing numerical parameters whose physical significance is either obscure or often nonexistent. In the course of this chapter, only a few of these models are briefly described, and they have been chosen on the basis of the physical interpretation they can provide.

The first part of the chapter is concerned mostly with non-shearable particles such as oxides or carbides. The unusually high strength of multiphase materials is related to the origin and the structure of their internal stress pattern. Several approaches to the concept of *internal stress* are also presented in this section. (See also ch. 21.)

The second part is devoted to the behavior of alloys containing shearable precipitates. Nickel-base superalloys provide a large range of complex deformation processes also observed in other alloy systems. With origins ranging from powder-metallurgy manufacture to single-crystal growth, they tend to encompass all possible grain sizes and shapes and a large variety of precipitate diameters.

Recrystallization can either be stimulated or inhibited by the presence of a second phase in a plastically deformed matrix. This leads to patterns of recrystallization that are quite specific to multiphase materials and are described in a later section. At the end of the chapter a few examples are given of complex multiphase alloys or multiplex structures of importance in recent industrial developments of high-performance materials. (See also ch. 25.) A short review of grain-size effects in multiphase alloys is presented in the last section.

2. *Description and microstructure of dispersed-phase alloys*

The main goal achieved in two-phase alloys is to provide means of raising the yield stress in crystals above the value due to the lattice friction stress: either the Peierls–Nabarro stress alone, as in pure solids, or solid-solution hardening associated with solute atoms of odd sizes in the crystal lattice. In industrial alloys, both strengthening mechanisms are usually operating.

Hard particles such as carbides cannot be sheared by gliding dislocations and remain stable in size and distribution up to the highest service temperatures. In order to avoid the complications inherent in industrial alloys, model materials have been designed and abundantly studied. *Oxide-dispersion-strengthened* (ODS) alloys belong to this category and provide a good basis for the understanding of more complexe alloys.

The first-born among the synthetic alloys made of a metal oxide dispersed into a ductile metallic matrix seems to have been SAP (*sintered aluminium powder*) alloys obtained by LENEL *et al.* [1957] who fabricated bars by extrusion of a mixture of 1-10 vol% of Al_2O_3 powder in pure Al powders: ANSELL and WEERTMAN [1959], ANSELL and LENEL [1961] and later GUYOT [1962] and GUYOT and DEBEIR [1966] were intrigued by the internal structure of these alloys which exhibited interesting creep resistance. Observations by TEM (fig. 1) reveal a wide distribution of sizes (10 nm to 1 μm) and a lack of coherency with the matrix.

References: p. 1481.

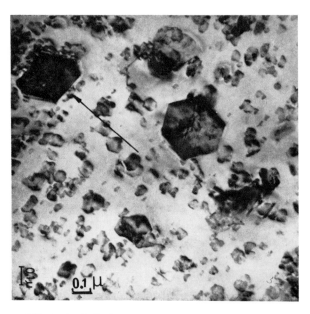

Fig. 1. Structure of Al–Al$_2$O$_3$ SAP alloy observed by TEM (after GUYOT and RUEDL [1967]).

Several types of crystallographic relationships between the lattice of the particle and that of the matrix can be met:

Coherency: either strict if both lattice parameters are equal and equivalent lattice directions are aligned, or quasi-coherency as described by fig. 2a if lattice parameters differ slightly. The lattice misfit is then described by the relative parameter difference

$$\delta = 2\frac{a_p - a_M}{a_p + a_M}, \tag{1}$$

where a_p and a_M are the precipitate and matrix parameters respectively. In fig. 2a the misfit parameter δ is positive. This situation is usually the result of precipitation from supersaturated solid solution, e.g., γ' precipitates in nickel-base alloys.

Semi-coherent precipitates (fig. 2b): the precipitate and the matrix lattices coincide along one set of crystallographic directions and the parameters are closely similar, but along another direction dislocations are needed in order to accommodate the large differences in lattice parameters. In aluminium-base alloys (Al–Cu), the θ' phase illustrates this situation.

Incoherent particles (fig. 2c): the lattice of the particle (Al$_2$O$_3$, SiO$_2$) has nothing in common with the lattice of the matrix (Al, Cu, Ni). Overaging of the θ' phase in Al–Cu stimulates the formation of θ particles which become incoherent with the matrix.

The average size of the population of dispersed particles plays a major role in the

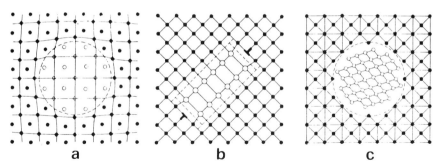

Fig. 2. Coherency of a precipitate: (a) quasi-coherent or coherent with positive misfit; (b) semi-coherent; (c) non-coherent. (From MARTIN [1980]).

mechanical and metallurgical properties of dispersed-phase alloys, which retain a dense network of subgrain boundaries even after long-term annealing treatments. There are several distinct situations:

Small particles (1–100 nm): they have the strongest effect on the yield strength and tend to assume approximately spherical shapes due to surface tensions. They are usually generated inside the matrix by internal oxidation techniques. A solid solution of Si, Al or Be in copper, for instance, is exposed to oxidizing conditions in the appropriate temperature range. Oxygen diffuses into the alloy and SiO_2, Al_2O_3 or BeO particles are formed inside the base metal. Single crystals containing fine dispersions of oxide particles can be produced and constitute an ideal model material for basic studies. Very low initial dislocation densities can easily be obtained in materials annealed before the oxidizing treatment.

Mechanical alloying consists in blending and grinding together oxide and metal powders in a vertical ball mill in which a rapidly rotating impeller agitates the steel balls. The initial oxide particles break into very fine debris and are cold-welded into the ductile matrix. The metal-oxide pellets thus formed (fig. 3) are then consolidated by hot extrusion, rolling and high-temperature annealing. Thoria-disperse nickel (TD-Nickel) and TD-nichrome have been successfully produced and the technique has been generalized by BENJAMIN and BROMFORD [1977] to manufacture complex ODS nickel-base superalloys and aluminium-base alloys. The end-result is a homogeneous distribution of finely dispersed small particles (fig. 4).

Secondary carbides (M_7C_3, $M_{23}C_6$) can precipitate in stainless steels after prolonged exposure at service temperature (fig. 5) and generate a finely dispersed colony of hardening particles which significantly alter the mechanical strength of the as-quenched material (DELEURY *et al.* [1981]). Since the nucleation of a misfitting precipitate is facilitated by the presence of a stress field they nucleate on dislocations.

Medium-size dispersoids (0,1–1 μm): this type of obstacle has a strong inhibiting effect on recrystallization and grain growth (§8). It appears as a losely dispersed population in aluminium-base alloys containing Cr, Zr or Mn additions.

More densely distributed, their strengthening influence is not so much felt on the

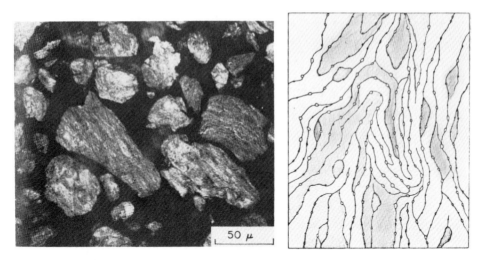

Fig. 3. Metal oxide pellets obtained by mechanical alloying (from BENJAMIN [1977]).

low-temperature yield strength as on the high-temperature creep resistance. They also have a tendency to induce cavitation on their tensile side during deformation.

Coarse inclusions (5–50 μm): because of their large size, they create problems of compatibility of deformation, and intense stress gradients are generated in their vicinity. They are widely recognized as the potential sources of weakness in commercial alloys. In steel and nickel-base alloys they are byproducts of deoxidation

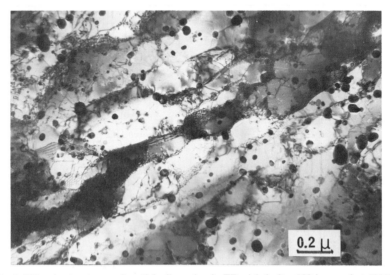

Fig. 4. Subgrain structure and particle dispersion in TD-nickel after 100 h anneal at 1200°C.

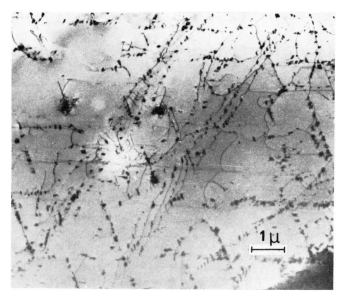

Fig. 5. Heterogeneous precipitation of $Cr_{23}C_6$ carbides in stainless steel after 5000 h at 600°C (from DELEURY *et al.* [1981]).

processes or may come from accidental slag bursts or debris of refractory materials used for casting the liquid melt.

Some of them are rather ductile, e.g., MnS inclusions in steels, or iron and silicon-rich inclusions in aluminium alloys. Their lack of cohesion with the surrounding matrix is always a preferential site for crack initiation and failure.

3.   Tensile properties of two-phase alloys

In engineering practice, the most basic criterion for a mechanical evaluation of a material still remains the yield strength and its decay with increasing temperature. The two striking features of multiphase materials are: (1) their rather high limit of elasticity at relatively low temperature (up to $\sim 0.2T_m$); (2) their tendency to preserve fairly high flow stresses up to temperatures close to their melting point $(0.90-0.95T_m)$. Let us successively examine each of these aspects in the following sections.

3.1.   Experimental results in macroscopic tests

A clean case of oxide dispersion strengthening is that of copper and copper–zinc single crystals containing alumina particles, as HIRSCH and HUMPHREYS extensively reported in 1969 and 1970.

Stress–strain curves obtained at various temperatures between that of liquid nitrogen and 200°C appear in fig. 6a for copper single crystals with 0.11% volume fraction (f) of alumina particles with average diameter $\bar{d}_s \approx 40$ nm and interparticle spacing $\bar{D}_s \approx 0.8$ μm and in fig. 6b for crystals with a higher volume fraction $f = 0.88\%$, $\bar{d}_s \approx 80$ nm and $\bar{D}_s \approx 0.5$ μm. Obviously the mechanical strength of these crystals (20–40 MPa) is one order of magnitude larger than that of pure copper (1–4 MPa).

The effect of the matrix composition was also examined by straining Cu–Zn single crystals with various solute concentrations (fig. 7). Solid-solution strengthening is superimposed on two-phase hardening.

3.1.1. Initial yield stress

Although not very sensitive to strain rate, the initial yield stress appears to be the sum of two contributions, the yield stress of the matrix, τ_m, which decreases with increasing temperature, and the particle by-passing stress, τ_p, which appears to be temperature-independent at least up to 500 K (HIRSCH and HUMPHREYS [1970]; JONES [1969] in Cu–BeO):

$$\tau_{ys} = \tau_m + \tau_p. \tag{2}$$

The friction stress due to solid-solution hardening is included in the first term, its

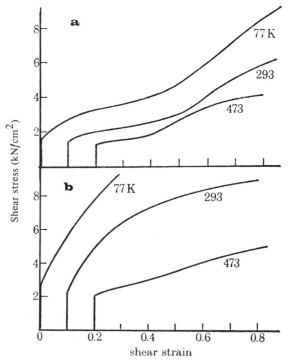

Fig. 6. Stress–strain curves at several temperatures of Cu–Al$_2$O$_3$ single crystals containing: (a) 0.11 vol% Al$_2$O$_3$; (b) 0.88 vol% Al$_2$O$_3$. (After HIRSCH and HUMPHREYS [1970].)

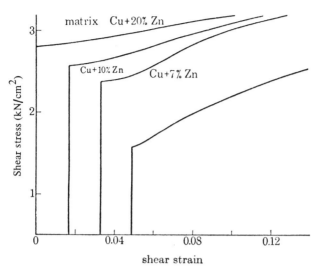

Fig. 7. Stress–strain curves at 293 K of Cu–Zn single crystals with 0.22 vol% Al$_2$O$_3$ (after HIRSCH and HUMPHREYS [1970]).

variation with solute concentration and temperature in single-phase alloys is described in ch. 21.

3.1.2. Stress–strain curves

All stress–strain curves are composed of three stages at room temperature, as in the case of single-phase crystals.

Stage I is characterized by an initial region ($\leq 1\%$) of low and quasi-linear hardening rate, followed by a short regime of rapidly increasing rate of hardening ($1\% \leq \epsilon \leq 3\%$) which merges into a long domain ($3\% \leq \epsilon \leq 15$–$20\%$) reputed for its parabolic hardening law. All the microscopic models presented in the literature are aimed at explaining each of these regions of the first stage. It is more developed at room temperature than at 77 K where it tends to disappear into stage II or even directly stage III to form an apparent single-stage hardening (fig. 6b). Orientations of slip lines, shape changes and crystal rotations examined by HIRSCH and HUMPHREYS [1970] and HUMPHREYS and MARTIN [1967] indicate that primary dislocations are responsible for the deformation. Solute additions tend to expand that stage, diminish its work-hardening rate (fig. 7) and drastically lower its temperature sensitivity (fig. 22, below), this latter effect being related to recovery of the microstructure.

Stage II is a domain of 10–20% extension in strain characterized by a linear rate of work-hardening. Lower temperatures and higher volume fractions of dispersed phase tend to make it vanish into stage I and the spreading stage III (fig. 6).

Stage III is parabolic with decreasing strain-hardening rates as in single-phase metals. It tends to overtake the whole curve when f is increased and T is low.

References: p. 1481.

3.1.3. Bauschinger effect

After tensile deformation, if the stress is reversed to compression, reverse strain-ing in two-phase alloys seems to start even before the externally applied stress reaches zero. This very strong *Bauschinger effect* typical of two-phase alloys (ch. 19, §10) can best be characterized by plotting the opposite of the compressive stress value versus cumulative strain (fig. 8) as suggested by ATKINSON *et al.* (1974). Notice in this figure that:

– the apparent reverse yield stress is negative (points A, A′);
– it is followed by a stage of 2% strain extension with a very high strain-hardening rate (zone B) as opposed to the forward loading curve;
– finally a "permanent softening" $\Delta\sigma_p$ appears in C where the reverse curve takes on nearly the shape of the forward curve.

When making several fatigue loops, the Bauschinger effect seems to decrease in amplitude. As reported by ATKINSON *et al.* [1974], Cu–SiO$_2$ crystals tested at 77 K show a lack of fatigue-hardening in the forward direction and progressive hardening in the reverse direction, so that the entire stress–strain loop tends to be more symmetrical with respect to the strain axis as the cumulative strain increases.

3.2. Microscopic mechanisms and models

An extensive description of dislocation mechanisms is outside the scope of this chapter and will be found mainly in the original papers published by the Cambridge and Oxford groups referred to in the text. A comprehensive review of micromecha-nisms in deformation and fracture processes of two-phase alloys has recently been published by MARTIN [1980]. Only the most striking features of this intricate problem of dislocation configurations in connection with the deformation of a mechanically heterogeneous medium will be presented.

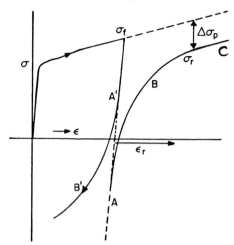

Fig. 8. Schematic stress–strain curve of an ODS material showing strong Bauschinger effect. The reverse loading curve ABC has been inverted to show the apparent softening $\Delta\sigma_p$. (After ATKINSON *et al.* [1974].)

3.2.1. Initial yield stress

An important contribution to the theory of yield stress of alloys containing non-shearable particles was made by OROWAN [1948] when he suggested that particles could be by-passed by the mechanism depicted in fig. 9, at a stress of the order of

$$\tau_p = \alpha \frac{Gb}{D_s}. \tag{3}$$

A more precise evaluation of the by-passing stress was obtained by taking into account the influence of the dislocation character (edge or screw) on the equilibrium shape of the loop and the interaction of the two arms of the dislocation on opposite sides of the particle (ASHBY [1968]), as well as the statistical nature of the effective spacing D_s between particles (KOCKS [1966] and FOREMAN and MAKIN [1966]). An excellent fit of experimental results is obtained with the expression:

$$\tau_p = K \frac{G_K b}{2\pi(D_s - d_s)} \ln \frac{d_s}{r_0}, \tag{4}$$

where K is a statistical factor and G_K is an average shear modulus combining both screw and edge contributions to the dislocation line tension.

When a dislocation is held up by point obstacles it bows out between them as described in ch. 18, §§4.2, 4.6. The force exerted by the dislocation line on the anchoring point is transmitted by the line tension. Strong obstacles, such as particles larger than 20 nm, do not release the dislocation before an Orowan loop is formed. Other, weaker, obstacles such as solute atoms, may release the dislocation at a larger angle ϕ (see fig. 23, ch. 18 for the definition of ϕ). The behavior of edge and screw dislocations encountering a statistical distribution of obstacles of various strengths

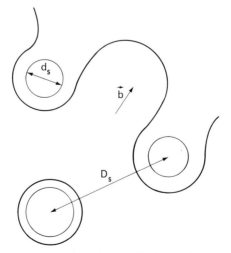

Fig. 9. Edge dislocation by-passing particles by the Orowan mechanism.

has been studied by several authors. BROWN and HAM [1971], using the assumption of constant line tension, show that the forward penetration of a dislocation following a path of easy movement is more pronounced in the case of strong obstacles ($0 < \phi < \pi/2$, fig. 10) than in the case of weak obstacles ($\pi/2 < \phi < \pi$, fig. 11) where it tends to sweep across the slip plane in a more rigid manner. The macroscopic effect of an array of randomly distributed obstacles is to lower the flow stress by a factor of 0.84 (KOCKS [1969]) with respect to a regular arrangement of similar obstacles. When particles tend to form clusters the mechanical properties are adversely affected further, as shown in fig. 12.

This topic is further discussed in ch. 29, §3.

3.2.2. Work-hardening at low temperature in alloys with small particles

One of the earliest model was imagined by FISHER, HART and PRY in 1953. They imagined that Orowan loops would pile up around particles, thereby creating such a back-stress in the matrix that the active dislocation sources would stop operating. The stress–strain curve then was found to be linear with strain:

$$\sigma = \sigma_0 + 6Gf^{3/2}\epsilon, \tag{5}$$

where f is the volume fraction of the particles. This model cannot explain the observed stage II of the work-hardening curves (fig. 6) where the dislocation density is extremely high, but may be relevant to the initial region of stage I at low temperature.

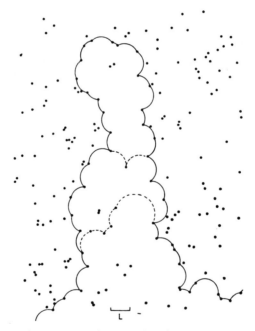

Fig. 10. Dislocation penetrating an array of strong obstacles ($0 < \varphi < \pi/2$) (from BROWN and HAM [1971]).

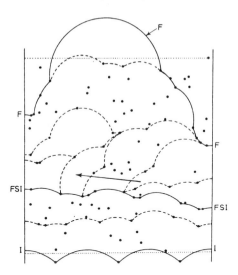

Fig. 11. Dislocation cutting through a population of weak obstacles ($\pi/2 < \varphi < \pi$) (from BROWN and HAM [1971]).

HART [1972] later refined the model and clearly stated that the friction stress of the pure matrix should be considered as an additional term to the general flow stress of the alloy rather than as an alternative strengthening mechanism: the very large Bauschinger effect of these alloys was even somewhat overestimated in this improved version, but the hardening rate was proportional to $\epsilon^{1/2}$, as found experimentally in the major part of stage I.

The retention of Orowan loops is simply a statement of the fact that the region within the loop has not undergone the plastic shear displacement of the rest of the surrounding material. One wonders then, how stable such a situation can be, especially when several loops are formed or when the initially applied stress is relaxed. HIRSCH and HUMPHREYS [1969,1970] reported from TEM observations of

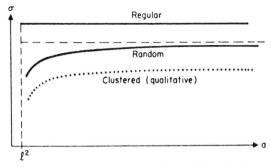

Fig. 12. Applied stress σ versus mean free slip area a of one dislocation for strong obstacles of area density $1/l^2$ as a function of obstacle distribution (from KOCKS [1969]).

References: p. 1481.

the early stages of plastic deformation (fig. 13) that Orowan loops may well be transformed into prismatic loops by cross-slip mechanisms such as those schematically drawn in fig. 14, and generalizing the mechanism originally proposed by HIRSCH [1957]. Screw segments can move in the cross-slip plane of the dislocation and recombine, thus leaving behind one or two loops with Burgers vector normal to their habit plane (prismatic loops). This event can be repeated a number of times and creates rows of prismatic loops (fig. 15) or be combined with further Orowan bowing of new dislocations as in figs. 14 or 15. The actual stability of coplanar stacks of Orowan loops in Cu–Al$_2$O$_3$ was studied in detail by HAZZLEDINE and HIRSCH [1974], who concluded that no more than four or five loops can be present around a particle at room temperature and that a hybrid model, in which both shear loops and prismatic loops are included, is necessary in order to account for the observed data: (1) the Bauschinger effect is entirely due to Orowan loops; (2) the

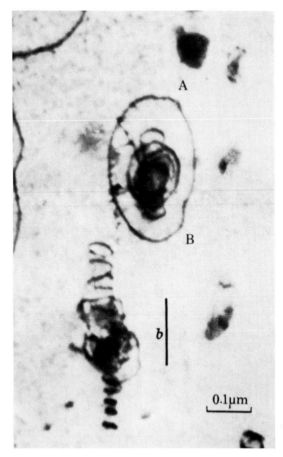

Fig. 13. Prismatic loop formation due to the cross-slip of inner Orowan loops. Segment AB is starting to cross-slip on the outer loop. Cu–30% Zn + 0.22% Al$_2$O$_3$. (From HUMPHREYS and HIRSCH [1970].)

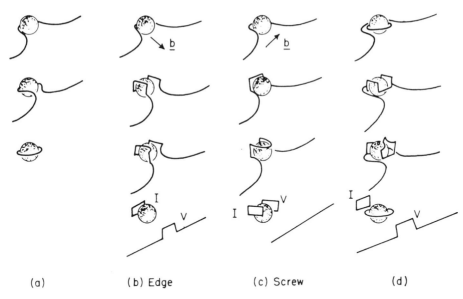

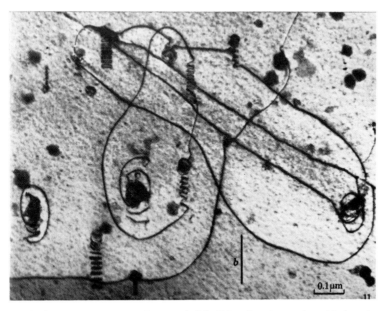

Fig. 14. By-passing of a hard particle: (a) a simple Orowan mechanism; (b) edge dislocation forming prismatic loops by double cross-slip; (c) screw dislocation; (d) multiple Orowan mechanism and cross-slip. (After ASHBY [1969].)

Fig. 15. Rows of prismatic loops, Orowan loops and glide dislocations interacting with the prismatic loops to form helices (same conditions as in fig. 13) (from HUMPHREYS and HIRSCH [1970]).

References: p. 1481.

low work-hardening rate of the initial region of stage I originates from the accumulation of Orowan loops, and the increasing proportion of prismatic loops progressively
dominates the rest of the domain.

BROWN and STOBB's model [1971] (also STOBBS [1973]) is based on ESHELBY's
[1961] technique of treating the mechanical problem of the stress and strain
distributions around an inclusion in a deformed matrix. They considered a local
plastic shear strain around each particle, of amplitude

$$\epsilon_p = \frac{nb}{4r_0},\tag{6}$$

in a matrix containing a volume fraction f of spherical inclusions of radius r_0
separated by an average distance L. The particles are assumed to be non-shearable
but to have the same elastic constants as the matrix. They are surrounded by n shear
loops of Burgers vector b which are smeared around the inclusion into infinitesimal
dislocations with infinitesimal Burgers vectors, in the usual fashion required by the
use of continuum mechanics.

It is interesting to note a general form of the solutions found for this model. The
various components of the stress tensor at a point located at a distance r from the
particle surface $(r > r_0)$ look like:

$$\sigma = G\epsilon_p r_0^3 \left(\frac{A}{r^3} + \frac{B}{r^5} \right).\tag{7}$$

The term in r^{-5} describes the *local stresses* prevailing in the immediate vicinity of
the particles, to which the first Orowan loops are directly exposed. The term falling
off as r^{-3} is related to the *source-shortening stress* which describes the increase in the
by-passing stress above the initial Orowan stress. It is a fluctuating component of the
internal stress with essentially zero mean value. The bowing dislocations will
experience it in their random movements across their glide plane as an effective
reduction of the interparticle spacing. ATKINSON et al. [1974] gave the following
simple estimate for it:

$$\sigma_{ss} = Gb\frac{f}{r_0} \left[\left(\frac{4r_0\epsilon_p}{b} \right)^{\frac{1}{2}} - 1 \right].\tag{8}$$

The concept of the source-shortening stress σ_{ss} also appears in the work-hardening
of fiber composite materials made of a copper matrix traversed by tungsten wires
(BROWN and CLARKE [1977] and LILHOLT [1977]).

On the other hand, the stresses within the particle $(r < r_0)$ are not fluctuating but
uniform and take on the value

$$\sigma(r < r_0) = 2G\nu'\epsilon_p,\tag{9}$$

where ν' is a numerical factor related to Poisson's ratio. When considering a finite
body with no externally applied stress, these stresses alone generate a uniform stress
throughout the matrix, with a magnitude:

$$\langle \sigma \rangle_M = -2G\nu'\epsilon_p,\tag{10}$$

where the subscript M implies averaging over the matrix containing a volume fraction f of particles. This is also called the *mean matrix stress* or *image stress*; it is the main contribution to the internal stress in two-phase alloys and has so far been treated generally as a scalar quantity rather than as a tensor.

Now the description of the stress–strain curves appears as:

$$\sigma = \sigma_0 + \alpha G f \epsilon_p + \frac{Gb}{r_0} f \left(\frac{4 r_0 \epsilon_p}{b} \right)^{1/2}, \qquad (11)$$

where the first term describes the initial yield under the Orowan stress $\sigma_0 = Gb/D$ [eq. (3)], the second term gives the short linear hardening domain of stage I and the third term yields the correct representation of the parabolic strain rate prevailing in the remainder of the stage. Notice also that the strengthening rate is proportional to f as expected from experimental results and not to $f^{3/2}$ as incorrectly found in earlier models [eq. (5)].

However, the relevance of a model cannot be established by comparison with the forward hardening curve alone, it has to be equally successful in its interpretation of related phenomena such as the reverse straining curve or the stress-relaxation curves. Considering the permanent softening effect denoted by $\Delta\sigma_p$ in fig. 8 as characteristic of the Bauschinger effect, one is interested in comparing it with twice the mean matrix stress:

$$\langle \sigma \rangle_M = \Delta\sigma_p / 2, \qquad (12)$$

since $\langle \sigma \rangle_M$ opposes flow in the forward direction and is additional to it in the reverse direction. ATKINSON *et al.* (1974) found excellent agreement between this expression and experimental results in a number of oxide-dispersion-hardened alloys.

Despite the obvious success of this model in describing stress–strain curves obtained at very low temperature (77 K) in Cu–SiO$_2$ for instance, it tends to grossly overestimate the hardening behavior around room temperature and above. Relaxation effects become essential at these temperatures; they lower the work-hardening rate and reduce the reversible part of the plastic flow taking place when unloading.

Another type of dislocation arrangement has been observed in Cu–Zn–SiO$_2$ alloys by HUMPHREYS and STEWART [1972]. They found very clear evidence of *secondary loops* of *prismatic* character (fig. 16) and interpreted it with the wire model shown in fig. 17. Shear loops in the primary glide plane are accompanied by primary prismatic loops with identical Burgers vectors. Outside the primary slip system, several sets of secondary dislocation loops are developed along $\langle 100 \rangle$ directions with non-primary Burgers vectors. A void tends to form on the dark side of the particle (fig. 17). The origin of this type of stress relief mechanism is best appreciated from the scheme originally drawn by ASHBY [1966] (fig. 18). The action of the tensile stress is to create a shear stress in the primary slip plane at the same time as it tends to separate the particle from the matrix along the tensile direction, thus generating a cavity. In the direction normal to the tensile axis, an excess of material does appear under the form of *prismatic punched loops* of interstitial character. Notice that the

References: p. 1481.

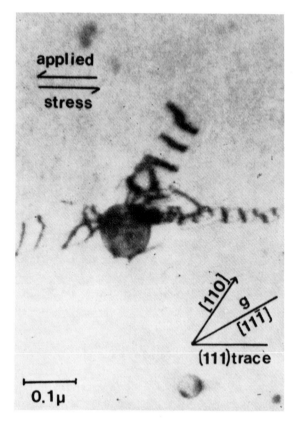

Fig. 16. Primary and secondary prismatic loops in α-brass + SiO₂ deformed at 25°C in simple glide (after HUMPHREYS and STEWART [1972]).

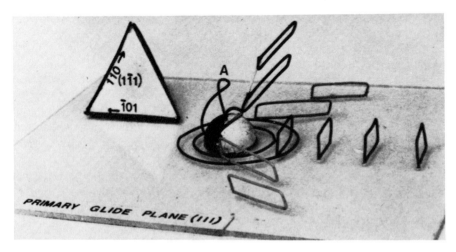

Fig. 17. A wire model describing prismatic loop formation along three systems and void formation on the dark side of the particle (after HUMPHREYS and STEWART [1972]).

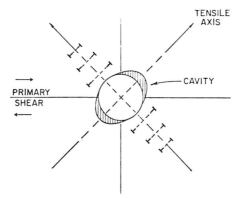

Fig. 18. Interstitial prismatic loops and void formation under tensile stress (after Ashby [1966]).

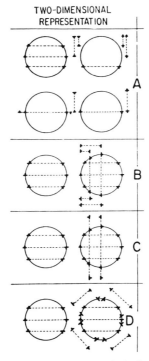

Fig. 19. Two-dimensional summary of the various modes of plastic relaxation: (A) formation of Orowan loops and cross-slip by Hirsch mechanism (small particles ≤ 300 nm); (B) activation of several slip systems, Orowan and prismatic loop formation, screw segments form twist boundaries, particle rotates (sizes ≥ 300 nm); (C) association of shear loops without rotation (not observed experimentally); (D) accumulation of prismatic loops and/or punched loops as in Asby's model (fig. 18). (From Brown and Stobbs [1971b].)

References: p. 1481.

exact geometry of a particular crystallographic system (fcc, bcc, etc.) is not considered in this scheme and will complicate the final picture.

Several types of dislocation arrangements have now been mentioned. They are best summarized by the two-dimensional representation proposed by BROWN and STOBBS [1971b] and shown in fig. 19.

Mechanism A describes the cross-slip of Orowan loops around small particles (< 300 nm) as suggested by HIRSCH's model [1957] (see also fig. 14), leading to the formation of prismatic loops having the Burgers vector of the primary slip system.

Mechanism B appears around somewhat larger particles (> 300 nm) and has been observed by CHAPMAN and STOBBS [1969] and reviewed by BROWN and STOBBS [1971b]. This involves Burgers vectors belonging to the cross-slip plane and shows very characteristic small rotations in the immediate surroundings of the particle with respect to the matrix. This can be easily deduced from the presence of two orthogonal sets of similar screw dislocations on the two facets of the particle contained in the projection plane in fig. 19.

Mechanism C is purely formal and has never been observed. Its particularity consists in containing only shear loops and no prismatic loops but it does not yield the correct compatibility of deformation.

Mechanism D is that of HUMPHREYS and STEWART [1972] and involves both primary and secondary prismatic loops of interstitial and vacancy characters, the latter type being able to condense on the particle–matrix interface and create a void along the tensile direction when the cohesive forces between the oxide and the metallic matrix are not very strong.

Before going further into the mechanisms of stress relaxation and recovery, let us look into more macroscopic models which are more relevant for the deformation of alloys with larger particles.

3.2.3. Heterogeneous deformation of alloys containing large particles

Instead of looking at detailed dislocation mechanisms in the immediate vicinity of the particle, one can envisage a more general approach to the deformation process of two-phase alloys and ask the question: how do crystals develop resistance to deformation? Of course, the density of forest dislocations in the path of gliding dislocations and the subsequent reduction of the mean link-length between dislocation nodes provide the correct answer. Following this approach, ASHBY [1971] draws a parallel between pure metals having a density of statistically stored dislocations, ρ^S, with a corresponding average slip distance λ^S, and two-phase materials where the multiplication of dislocations is greatly enhanced by the presence of the second phase. In the latter case, large strain gradients are induced in the matrix and *geometrically necessary dislocations* are created in order to accommodate these gradients and thus allow the two phases to deform in a compatible way. Their density ρ^G is definitely higher than in pure metals and its proposed value is given by the expression

$$\rho^G = \frac{f}{r}\frac{4\gamma}{b},\tag{13}$$

where γ is the shear strain. The average geometric slip distance, λ^G, for a volume fraction f of particles with radius r can be written as:

$$\lambda^G = \frac{r}{f} = \frac{1}{nr^2},\tag{14}$$

where n is the number of particles per unit volume. ρ^G takes on the form:

$$\rho^G = \frac{1}{\lambda^G}\frac{4\gamma}{b},\tag{15}$$

similar to the expression

$$\Delta\rho^S = \frac{1}{\lambda^S}\frac{4\Delta\gamma}{b}\tag{16}$$

in pure metals obtained by writing that the dislocation loops emitted by an active source stop expanding when they reach a diameter equal to λ^S.

When straining a two-phase crystal, the actual shear resistance is given by an expression similar to that of single-phase materials:

$$\tau = \tau_0 + \beta Gb\sqrt{\rho^G},\tag{17}$$

where τ_0 is the friction stress. This yields:

$$\tau = \tau_0 + \beta G\left(\frac{f}{2r}b\gamma\right)^{1/2}\tag{18}$$

which gives the correct parabolic dependence of the strain-hardening rate on both strain and volume fraction.

Another macroscopic continuum approach was followed by HUMPHREYS [1979] in trying to describe the deformation features produced at high strains (10–50%) in the neighborhood of micron-sized particles. A detailed account of dislocation arrangements can no longer be given, nevertheless a small volume of matrix containing a particle will tend to deform by a shearing mechanism involving several intersecting slip systems, the primary system being dominant (fig. 20a). If the particles were shearable as in fig. 20b no rotation would appear: this is the case in nickel-base alloys with excessively small γ' particles (see § 6.2). If, on the contrary, the particles

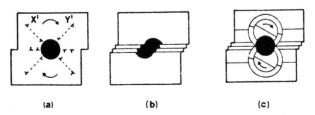

(a)   **(b)**   **(c)**

Fig. 20. The origin of a deformation zone at a particle: (a) Orowan loops on multiple slip systems; (b) when precipitates are sheared, as in γ'-hardened alloys, no rotation takes place; (c) continuum model of the rotation ariound a particle. (After HUMPHREYS [1977].)

are large and unshearable, the dense network of dislocations which develops in the volume adjacent to the particle will rearrange into subgrain boundaries. Each subgrain is slightly rotated with respect to its neighbor and the progressive misorientations will add up to the final rotation of the deformed zone of radius R (fig. 20c). The angle of rotation of a simple arrangement of similar dislocations separated by equal distances h is:

$$\tan \theta \approx b/h \approx \gamma. \tag{19}$$

Calling ρ^R the average dislocation density in the deformed zone, a more general expression applicable in the case where several types of dislocations are present in the deformation zone reads:

$$\tan \theta \approx \rho^R b R \approx \gamma. \tag{20}$$

This is a good approximation at small strains. Local lattice rotations can be estimated by X-ray techniques or by selected-area electron diffraction. This latter method was used by HUMPHREYS [1977] who examined an Al–Cu solid solution containing silicon particles of several μm size. Local misorientations of adjacent subgrains were measured and the cumulative misorientation was plotted as a function of the distance from the particle (fig. 21). We see that the cumulative rotation can attain 30° in a deformation zone of the order of the size of the initial particles. One can expect these regions of the material to be extremely active when recovery or recrystallization events start operating (see §8).

3.3.   Stress relaxation and recovery effects

All strengthening mechanisms described so far rest quite naturally on high densities of dislocations being developed in the matrix and particularly in the vicinity of the particles. If such structures are expected to be rather stable at low

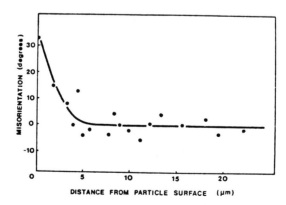

Fig. 21. Cumulative misorientation of the matrix in the vicinity of a 4 μm-size particle in cold-rolled aluminium (from HUMPHREYS [1977]).

temperature (77 K for instance), one wonders what will happen to them at room temperature and above. Experimentally, the strain-hardening coefficient of two-phase alloys is found to drop drastically in the temperature range 200–400 K and the rate of recovery exhibits a maximum in the same domain as shown in fig. 22, taken from the work of HIRSCH and HUMPHREYS [1970]. Note that higher strain rates tend to shift the drop to higher temperatures, as expected. Note also that the Cu–Zn solid solution is less affected by temperature (dashed curve, upper graph) than the pure matrix.

Turning to the point of micromechanisms responsible for recovery effects at about room temperature, the original value of about 1.1 eV (125 kJ/mole) measured by HIRSCH and HUMPHREYS [1970] most likely pertains to the local and limited rearrangement of dislocations by cross-slip and pipe-diffusion-aided climb. All mechanisms summarized in fig. 19 will benefit from limited climb by pipe diffusion and thus realize the equilibrium plastic relaxation state described by BROWN and STOBBS [1976]. Its main character however is of glissile nature. At still higher temperatures (550 K and above) diffusional relaxation starts playing a dominant role in the behavior of the material and most authors agree upon an apparent activation energy of 1.5–1.7 eV (140–160 kJ/mole). GOULD *et al.* [1974] consider

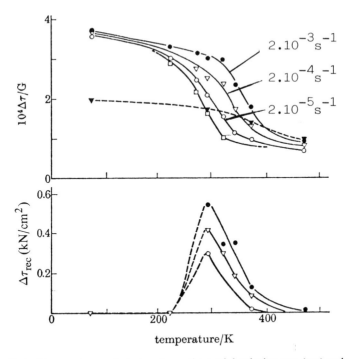

Fig. 22. The effect of temperature and strain rate on the work-hardening rate (top) and the extent of recovery (bottom) in Cu + 0.22% Al_2O_3. The Cu + 10% Zn + 0.22% Al_2O_3 crystals (dashed curve) are less sensitive to temperature.

References: p. 1481.

that climb is controlled by jog nucleation on the dislocations trying to by-pass the particles and interpret the rather high values of the activation energy as the sum of two terms: pipe-diffusion for copper plus twice the formation energy for a jog: $U_p + 2U_j \approx 0.9 + 2 \times 0.4 \approx 1.7$ eV. Experiments carried out on Cu–SiO$_2$ single crystals by LLOYD and MARTIN [1978] provide strong support for this model in the temperature range 470–1010 K. The interested reader will find an extensive discussion of creep recovery in original articles by LLOYD and McELROY [1975,1976] dealing with anelasticity. Anelastic strains are time-dependent recoverable components of deformation which accompany reduction or removal of stress during high temperature deformation.

Not only does the strain-hardening coefficient fall off rapidly around room temperature (HIRSCH and HUMPHREYS [1970]) and the initial yield stress in the temperature range of 400–500°C (LLOYD and MARTIN [1978]) but also the final dislocation structure is less disorganized as mentioned by HAUSSELT and NIX [1977] who observed by TEM the high temperature deformation of Ni–20Cr–2ThO$_2$. They also found that the final dislocation density is proportional to $\sigma^{1.6}$ rather than σ^2 as found in pure metals at low temperature. The threshold stress for by-passing of particles is significantly lowered ($\sim 0.43\sigma_0$, LLOYD and MARTIN [1978]) or even tends to vanish: the thermal fluctuations assist the dislocation in by-passing the particle by climb before the critical configuration for Orowan looping is reached (HOLBROOK and NIX [1974]). This effect is quantitatively considered in the model of SHEWFELT and BROWN [1973]. It becomes the key phenomenon in precipitation-hardened alloys at high temperature.

3.4. A continuum-mechanics approach to the internal stress

The problem of correctly describing the plastically deformed state of a single- or two-phase material is far from trivial. Metallurgists have been looking for a hardness state variable (HART [1970]) with the uneasy feeling that entrusting the effect of complex dislocation arrangements to a single scalar variable is far from satisfactory. Although on a microscopic scale ($< 1 \mu$m) the tensorial nature of dislocation stress fields (LI [1963]) and of strain gradients around particles (ESHELBY [1961] and BROWN and STOBBS [1971]) is duly acknowledged, this sophistication is lost in the equations describing stress–strain curves, creep curves or strain hardening rates.

On the other hand, the macroscopic approach of continuum mechanics has developed general constitutive equations for viscoelastic and viscoplastic behavior of isotropic and anisotropic materials, but seems to ignore the crystallographic nature of slip and dislocation mechanisms. Readers unfamiliar with basic descriptions of elasto–plastic flow equations will find them in McClintock and Argon's [1966] book (see *Further reading*).

Understanding and describing the behavior of crystalline solids under alternative loading (tension, compression, fatigue) and multiaxial stress conditions (tension–torsion) is forcing the mechanical and the metallurgical approaches towards a necessary point of convergence.

Since transient regimes are usually the most revealing test for a model, let us look at what happens to the creep curve when the applied stress σ_a is suddenly lowered slightly by $\Delta\sigma$ during the steady-state regime characterized by a strain rate $\dot{\epsilon}_1$. This so-called *strain transient dip test* was first proposed by SOLOMON and NIX [1970] in order to obtain an estimate of the internal stress σ_i. The various alternatives are schematically represented in fig. 23: for small stress reductions, forward creep continues with a reduced rate $\dot{\epsilon}_2 < \dot{\epsilon}_1$. For larger stress decrements the flow rate of the specimen is stopped for a period of time Δt which increases with $\Delta\sigma$. This has been called the *incubation period* by PARKER and WILSHIRE [1975], who observed it in Cu–Co and nickel-base superalloys (WILLIAMS and WILSHIRE [1973]). For still larger stress reductions, the instantaneous creep rate, just after the stress transient, is negative: macroscopic strain recovery is taking place. This effect in creep is the high-temperature equivalent of the Bauschinger effect in tensile tests. Reversing the stress is not necessary in order to reverse the flow: negative strain rates appear already after partial unloading.

In a more general fashion, if creep tests are carried out under multiaxial loading such as hollow cylinders tested in tension–torsion, experimental results such a those reported by OYTANA *et al.* [1979] and presented in fig. 24 confront us with the evidence that the material has developed an internal structure with an anisotropic character related to the temporal and spatial features of the applied stress. These experiments have been performed on α-brass and on XC48 mild steel. The tensile stress σ is plotted along the horizontal axis and the torsion stress $\tau_{z\theta}$ along the vertical axis so that each point fully describes a complex shear- and tensile stress state to which the specimen is submitted until a stationary creep rate is reached: the flow rate is represented in intensity and direction by the arrow attached to each point: the $\dot{\epsilon}_z$ component lies along the horizontal axis and $\dot{\epsilon}_{z\theta}$ along the vertical axis. Stress reductions affecting either the tensile load or the torque have been carried out as well as combined reductions, and the corresponding flow rates are indicated by arrows in fig. 24 for each new stress state. It is clear, here also, that a domain of zero creep rate does exist which is not centered around the origin of the diagram. The crystal is not hardened isotropically around the origin as a result of the strain it has undergone. There is, however, a region of *isotropic hardening* centered around the

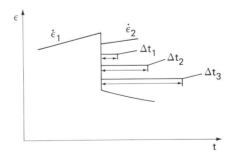

Fig. 23. Various possible alternative responses to a stress drop $\Delta\sigma$ during a creep test.

References: p. 1481.

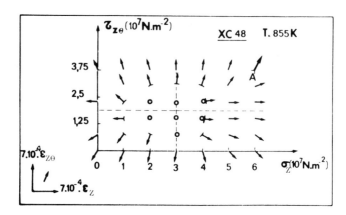

Fig. 24. Tension–torsion test. Steady-state stress with vectors indicating flow direction. Evolution of the plastic flow vector after various load decrements in tension and/or torque. (After OYTANA *et al.* [1979b].)

point defined by the vector **K** (fig. 25), and the stress shift **K** is called the *kinematical hardening* according to the suggestion first made by PRAGER [1955] who pioneered this new method of analyzing stress and strains in plastically work-hardening solids. Strain memory effects are adequately handled by this model. In the schematic Mohr representation (fig. 25), separating the shear and tensile components of the stress state, one would expect well-behaved materials to work-harden isotropically as shown by the dashed circle around the origin, but we find that most alloys and especially two-phase alloys develop strong kinematical hardening characteristics (large **K**) combined with rather weak isotropic hardening (small *R*).

    In order to develop a general formulation of the constitutive equations for plastic flow under complex stress states it is necessary to consider the referential coordinate system of the principal axis for the stress tensor and look at it in projection along the

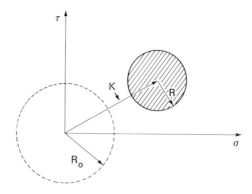

Fig. 25. Schematic representation in Mohr diagram of simple isotropic hardening R_0 and of kinematical + isotropic hardening $K + R$.

direction of the isostatic stress $\sigma = \frac{1}{3}(\sigma_1 + \sigma_2 + \sigma_3)$ where the usual yield criteria (Tresca, von Mises), are easily represented (fig. 26). Isotropic strain-hardening (dashed curves) leads to the absence of a Bauschinger effect, as shown schematically in fig. 26b. R_0 is the initial field strength of the material and $R(p)$ is the yield stress of the material after it has been hardened by the effect of the cumulative plastic strain identified by $p = |\epsilon_{ij}|$ and defined as:

$$\dot{p} = \frac{1}{3}\sqrt{2}\left[(\dot{\epsilon}_1 - \dot{\epsilon}_2)^2 + (\dot{\epsilon}_2 - \dot{\epsilon}_3)^3 + (\dot{\epsilon}_3 - \dot{\epsilon}_1)^2\right]^{1/2}, \tag{21}$$

which can be understood by analogy with eq. (22), below.

The yield criteria proposed by Von Mises (see McClintock and Argon's [1966] book, *Further reading*) simply expresses the fact that the second invariant of the strain deviator, the scalar quantity $J(\sigma)$, is the radius of a circle in the σ_1, σ_2, σ_3 plane (dotted circle in fig. 26):

$$J(\sigma) = \frac{1}{2}\sqrt{2}\left[(\sigma_1 - \sigma_2)^2 + (\sigma_2 - \sigma_3)^2 + (\sigma_3 - \sigma_1)^2\right]^{1/2}. \tag{22}$$

Now introducing a kinematical stress variable X of tensorial nature as suggested by Prager [1956] simply changes the last expression into a similar relation which can be written symbolically:

$$J(\sigma - X) = \frac{3}{2}\sqrt{2}\left[(\sigma'_{ij} - X'_{ij})(\sigma'_{ij} - X'_{ij})\right]^{1/2}, \tag{23}$$

where σ' and X' are the applied and kinematic stress deviators. The linear kinematic yield criterion reads:

$$\begin{cases} f(p) = J(\sigma - X) - R(p) \leqslant 0, & (24) \\ X = c\epsilon_p, & (25) \end{cases}$$

where the internal stress tensor X is proportional to the plastic strain tensor. Now the stress–strain curve (fig. 26b) exhibits a Bauschinger effect which the isotropic strain–hardening model could not account for. But the experimental curve (solid line

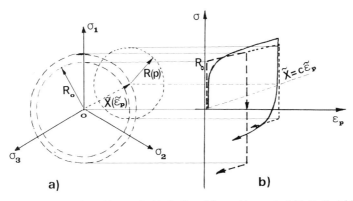

Fig. 26. Schematic representation of isotropic (dashed) and linear kinematical (dotted) yield criteria with associated stress–strain curve and reverse loading curve.

References: p. 1481.

in the figure) is not yet correctly represented by this elementary model because linear hardening is too crude an approximation. Confronted by the nonlinear strain-hardening behavior and the elasto-plastic memory effects of γ'-hardened nickel-base superalloys (IN 753 and Nimonic 80A) when tested in fatigue, ASARO [1975] already suggested the introduction of a kinematical hardening model with several successive linear regions. CHABOCHE [1977] has introduced the concept of nonlinear kinematical hardening by changing eq. (25) into

$$\dot{X} = ac\dot{\epsilon}_p - cX\dot{p} \tag{26}$$

and taking a general yield criterion with a set of convex hypersurfaces in σ_1, σ_2, σ_3 space (the dot symbolizes the time derivative). The general equation of an equidissipative surface can be written as:

$$\Omega[\sigma - X, R(p, T)] = K(p, T), \tag{27}$$

where R and K increase with the cumulative strain p and decrease with temperature T. The plastic flow rate $\dot{\epsilon}_p$ is found by writing that it is taking place in the direction normal to the equidissipative surfaces described above (fig. 27). So finally the constitutive equations for nonlinear kinematic hardening can be written formally as:

$$\begin{cases} \dot{\epsilon} = \dfrac{\partial \Omega}{\partial \sigma}, & (28) \\[2mm] d X = ac\,d\epsilon_p - cX\,dp - hX\,dt, & (29) \\[2mm] \sigma = X(\epsilon_p) \pm R(p, T) \pm K(p, T)|\dot{\epsilon}_p|. & (30) \end{cases}$$

Equation (29) is a more sophisticated version of eq. (26), now including two recovery terms. The first one, $-cX\,dp$, is related to cumulative strain and will tend to cancel kinematical hardening after repeated stress reversals so that a stabilized fatigue loop can be reached as experimentally observed in fatigue tests. The second

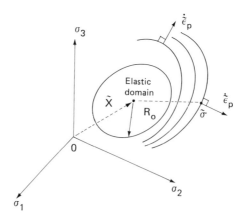

Fig. 27. Representation of successive yield surfaces and of the plastic flow rate which is perpendicular to them.

one, $-h\boldsymbol{X}\mathrm{d}t$, describes the time-dependent recovery effects taking place during relaxation tests, for instance, and previously referred to as *anelasticity*.

Finally, eq. (30) can closely describe experimental stress–strain curves with stress reversals such as in fig. 28: R_0 is the initial yield stress which can be a threshold stress as in the Orowan by-passing process; $R(p)$ is the isotropic strain hardening component of the material's resistance, the *latent hardening* produced by straining a crystal in one direction prior to deforming it by slip on a second slip system is an effect of this nature, it is also called *forest hardening* in dislocation theory; $\boldsymbol{X}(\boldsymbol{\epsilon}_p)$ on the other hand is of tensorial nature and describes the stress tensor associated with the *internal stress* field generated inside the material by the specific dislocation configuration associated with the strain $\boldsymbol{\epsilon}_p$; and the last term $K(p, T)|\dot{\boldsymbol{\epsilon}}_p|$ describes the strain-rate effects encountered in tensile tests. The elastic domain or zero-creep-rate domain after a stress reduction has a diameter equal to $2R(p)$ (see fig. 24, for instance).

Looking at equations (28)–(30), we can see that several numerical constants have been introduced. With the help of modern computers, they can be rapidly adjusted to obtain the closest possible fit with the various experimental curves, but such an approach does not provide any serious insight into the physical processes underlying the described mechanical behaviour. This model has been refined enough by CAILLETAUD and CHABOCHE [1979] to enable the authors to describe the macro-scopic effects of microstructural changes induced by varying the temperature in a series of cyclic tests performed on a γ'-hardened nickel base superalloy, namely

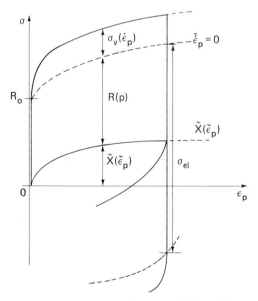

Fig. 28. Nonlinear kinematical hardening and strain-rate sensitivity of the stress–strain curve (after CHABOCHE [1977]).

References: p. 1481.

IN-100. More recently, CULIE *et al.* [1982] have made an interesting attempt to correlate microstructural changes observed by TEM in this alloy after various temperature and stress cycles with the parameters involved in the macroscopic equations describing the flow rate of the material. Dislocation rearrangements and γ'-coarsening during high-temperature anneals followed by the precipitation of fine γ' particles during rapid cooling, are correctly handled by this model. This successful attempt to bridge the gap between microscopic phenomena and the macroscopic equations describing the behavior of a two-phase industrial material under complex loading and thermal history needs to be developed further. Refinements in the description of elementary mechanisms at the scale of grains and subgrains in polycrystalline materials need to be introduced. Since the tensorial nature of the internal stress is given due attention, this quantitative approach can be used with confidence in polyphase materials which are known for their anisotropic response to stress, even at high temperature.

4. High-temperature behavior of dispersed-phase alloys

At temperatures low enough to keep climb processes down to a level where their contribution to the deformation rate is negligible, dislocations can only glide and possibly cross-slip when stress levels and orientations are favorable. At higher temperatures, however, ($T \gtrsim 0.5T_m$ or even $T > 0.4T_m$ under high local stresses) climb events start having a high enough frequency to contribute significantly to the material's efforts to relax local stress concentrations or even long-range stresses. High dislocation densities introduced during swaging, extrusion or cold work have raised the yield strength of the material to very high levels and the presence of the finely dispersed oxide particles has stabilized these disorganized tangles of dislocations. When the material is then heated up to temperatures where diffusion is operating, recovery can take place by climb-assisted glide. A new degree of freedom is thus introduced for dislocations which can now climb over small particles, recombine and annihilate or, at least, associate with each other to form better organized configurations or networks which are more energetically stable under local stress conditions.

4.1. High-temperature subgrains in polycrystalline oxide-dispersion-strengthened alloys

The various aspects of recovery in ODS alloys have been best explored by GREWAL *et al.* [1975] in a study of BeO-dispersed polycrystalline nickel. Samples were cold-worked by swaging at room temperature down to 70% reduction in area and then annealed for 1 h at various temperatures between 400 and 982°C. The resulting creep stress for 100 h rupture life at 982°C is plotted as a function of the intermediate anneal temperature in fig. 29. These authors report that unorganized and extremely dense tangles of dislocations formed during swaging at room temperature transform into small elongated subgrains 0.2–0.5 μm in width during inter-

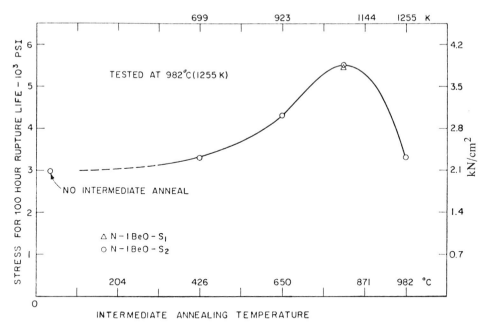

Fig. 29. Stress for 100 h rupture life at 982°C as a function of intermediate anneal temperature after swaging at 25°C of nickel-BeO (after GREWAL *et al.* [1975]).

mediate anneals and appear by TEM to be best organized for a 1 h treatment at 815°C, which corresponds to the highest creep resistance. At higher temperature the rearrangement of dislocations is too far advanced to preserve a good resistance of the material, and its stress rupture life drops again.

One can conclude that recovery in cold-deformed ODS alloys is a very active process as soon as temperatures are high enough to allow climb and rearrangement of dislocations into well organized subgrain boundaries (fig. 30). It is so active that recrystallization cannot take place even after large amounts of cold work. Rearrangements take place on a small scale (0.1–1 μm) owing to the stabilizing effect of the dispersed phase, and the formation of large-angle mobile boundaries is prevented. Existing grain boundaries on the other hand are also efficiently pinned by the particles and cannot absorb large amounts of dislocations by moving into deformed grains. Hence, recovery is extremely intense in these alloys until all necessary local rearrangements have taken place. Then again the structure appears frozen if no change in external conditions (σ, T, $\dot{\epsilon}$) takes place. If the material is exposed to stresses at high temperature, dynamic dislocation patterns will develop which look less stable than static arrangements (fig. 31). The average size of the dynamically stable subgrains is inversely proportional to the applied stress,

$$d = 4\frac{Gb}{\sigma_a}, \tag{31}$$

References: p. 1481.

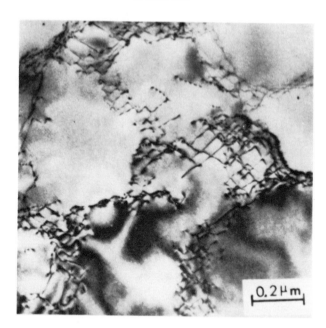

Fig. 30. Effect of 1 h recovery at 815°C in nickel–BeO after 60% reduction by swaging (from GREWAL *et al.* [1975]).

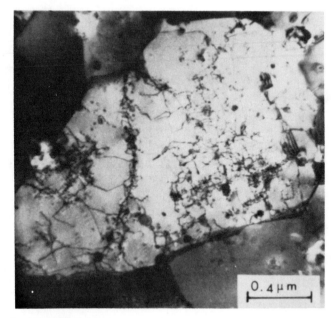

Fig. 31. Dislocation structure of the same material as in fig. 30 after 2 h creep at 982°C under 21 MPa (from GREWAL *et al.* [1975]).

as suggested by the observations of LASALMONIE and STRUDEL [1975a] in nickel and AHLQUIST and NIX [1971] in aluminium.

4.2. Apparent and effective creep parameters

Ever since the oxide-dispersion-strengthened alloys appeared on the market in the late fifties, scientists have been astonished and fascinated by the unusually high values of their apparent creep parameters (ANSELL and WEERTMAN [1959]). Experimentally measured activation energies for creep range from 400 to 800 kJ/mole and the apparent stress-sensitivity exponent can be as high as 10 and even 30 or 40 (WILCOX and CLAUER [1966]).

Another striking aspect of this type of material concerns its exceptionally high flow stress: at 600°C pure nickel will flow under a stress of about 20 MPa, whereas TD-nickel will strictly resist deformation up to an applied stress of 200 MPa. One is naturally led to inquire about the origin of such an unusual resistance: is there a stress inside the material itself which opposes the applied stress and thereby prevents individual dislocations from being exposed more or less directly to it? We are now aware of the fact that any kind of plastically deformable crystalline material develops particular dislocation arrangements whenever it undergoes deformation; it work-hardens in such a way as to oppose further straining in the same direction. In the case of monotonic uniaxial loading, a simple approach using scalar functions to deal with average values of stresses, the *internal stress* can be introduced as:

$$\sigma_a = \sigma_i + \sigma_e. \tag{32}$$

Here the applied stress, σ_a, is balanced by the internal stress, σ_i, plus the stress experienced by each individual dislocation, σ_e. This is of course a very direct and dramatic way of going from the macroscoppic scale (σ_a) to the level of the individual dislocation (σ_e) and it is probably much too coarse a model to give a precise description of the actual situation, except perhaps in single crystals.

Creep models using internal stresses are many in the literature and numerous controversies have given this potent concept a rather confusing image. We shall voluntarily restrict ourselves to one particular model which has developed enough self-consistency to be sketched rapidly.

If we accept eq. (32), despite its ambiguity, as a first approach to a definition of the internal stress, we can then ask: what variables does σ_i depend upon? Temperature, applied stress, internal structure of the material, strain, metallurgical history of the sample are probably some of them. Early models did not consider the internal stress to be temperature-dependent; most of them even used to call it the *athermal stress* as if it were an intrinsic property of a particular alloy.

The creep rate in the steady state is generally written as a function of applied stress σ_a and of temperature T as:

$$\dot{\epsilon}_s = A\sigma_a^{n_a} \exp(-Q_a/kT), \tag{33}$$

where the pre-exponential factor A includes entropy terms plus structural and metallurgical factors.

References: p. 1481.

The *apparent creep parameters* are defined by the following expressions:

$$n_a = \left(\frac{\partial \ln \dot{\epsilon}_s}{\partial \ln \sigma_a} \right)_T, \quad Q_a = -kT^2 \left(\frac{\partial \ln \dot{\epsilon}_s}{\partial (1/T)} \right)_{\sigma_a}. \tag{34}$$

Similarly, *effective creep parameters* can be introduced by reference to the effective stress:

$$n_e = \left(\frac{\partial \ln \dot{\epsilon}_s}{\partial \ln \sigma_e} \right)_T, \quad Q_e = -kT^2 \left(\frac{\partial \ln \dot{\epsilon}_s}{\partial (1/T)} \right)_{\sigma_e}. \tag{35}$$

These should relate directly to elementary mechanisms and thus be more easily interpreted physically. The creep equation can now be written as:

$$\dot{\epsilon}_s = B \left(\frac{\sigma_a - \sigma_i}{E} \right)^{n_e} \exp\left(-\frac{Q_e}{kT} \right), \tag{36}$$

but eqs. (33) and (36) are just two different ways of describing the same phenomenon, namely the steady-state creep rate. Apparent and effective creep parameters must therefore be related, as SAXL and KROUPA [1972] pointed out. Assuming that the internal stress is a function of applied stress σ_a and temperature only, $\sigma_i(\sigma_a, T)$, two partial derivatives must be introduced:

$$\sigma_i' = \left(\frac{\partial \sigma_i}{\partial \sigma_a} \right)_T, \quad \sigma_i^* = \left(\frac{\partial \sigma_i}{\partial T} \right)_{\sigma_a}. \tag{37}$$

They describe how the internal structure of the material is altered by changes in external conditions.

A complete derivation can be found in the original article (LASALMONIE and STRUDEL [1980]); it leads to the following relations:

$$n_a = n_e \frac{\sigma_a}{\sigma_a - \sigma_i} (1 - \sigma_i'), \tag{38}$$

$$Q_a = Q_e - n_e kT^2 \frac{dE}{EdT} - n_a kT^2 \frac{\sigma_i^*}{1 - \sigma_i'}, \tag{39}$$

which clearly indicates that the *apparent* creep parameters can be drastically different from their *effective* counterparts in particle-hardened alloys. Note that the temperature dependence of the elastic modulus is introduced by the second term of eq. (39).

4.3. Average internal stresses in dispersed-phase alloys

Some of the experimental data which have stimulated the development of the scalar but quantitative representation of the internal stress will now be examined.

In a **single-phase alloy** where no threshold stress is present, the strain-hardening can be described as a function of the applied stress by $\sigma_i = \alpha \sigma_a$, and eqs. (38) and

(39) reduce to:

$$n_a = n_e, \quad Q_a = Q_e - n_a kT^2 \frac{dE}{EdT}. \qquad (40)$$

where the corrective term for the activation energy comes from the decrease of Young's modulus with temperature alone. It is a significant correction, as pointed out by LUND and NIX [1975]. For instance, pure nickel at 900°C has a stress exponent of 4, $Q_e = Q_D \approx 350$ kJ/mole and the corrective term is about 42 kJ/mole, so that the measured activation energy for creep is about 390 kJ/mole.

In **dispersion-strengthened** alloys the situation is different since most authors seem to consider that one component of the back-stress is the threshold stress, either associated with the Orowan stress for by-passing the particles (LUND and NIX [1976]), or indirectly with the stress necessary for a dislocation to break away from well knitted networks of immobile dislocations (HAUSSELT and NIX [1977] and LASALMONIE and STRUDEL [1975a,1980]).

Considering the case of TD-nickel, for instance, the activation energy for plastic flow varies with temperature as shown in fig. 32. Similar results have been reported by GUYOT and RUEDL [1967] for Al–Al$_2$O$_3$ alloys. Several ranges of temperatures must be considered; in each of them a different dislocation mechanism is controlling the flow stress.

At *low temperatures* ($T < 100$°C) the activation energy is very low, ~ 0.4 eV, which is about twice the energy for the formation of a jog on a dislocation. At such low temperatures flow stresses are very high and decrease rapidly until a plateau is reached where the formation of jogs between repulsive dislocations can take place by thermal activation. The model for the formation of jogs on moving dislocations cutting repulsive trees of the forest dislocations is outlined in ch. 20, §5.5 (see also ch. 18, §3.5).

At *intermediate temperatures* (100°C $\leq T \leq 700$°C) activation energies are clearly in the range of the energy for self-diffusion in nickel (320–360 kJ/mole) but the stress exponents can become very large. The applied stresses are of the order of the threshold stress σ_0 and do not significantly modify the inner dislocation density of

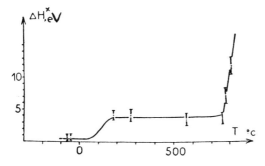

Fig. 32. Variation of the activation energy for plastic flow with temperature in TD-nickel (from LASALMONIE and SINDZINGRE [1971]).

References: p. 1481.

the material; they simply assist and bias the climb events into producing a strain in a given direction. Hence in eqs. (38) and (39), $\sigma_i \approx \sigma_0$ and σ_i' and σ_i^* are negligible:

$$n_a = n_e \frac{\sigma_a}{\sigma_a - \sigma_0}, \quad Q_a = Q_e. \tag{41}$$

This can be best observed in creep tests where σ_a is just above σ_0 and n_a takes on strikingly high values $(10-40)$.

At *high temperatures* $(T \geq 700°C)$ and for stresses above 180 MPa (LASALMONIE and STRUDEL [1980] the situation is more complex especially when the applied stress is larger than the threshold stress. This is clearly described by fig. 33, suggested by the work of LUND and NIX [1976] on single crystals of TD-nichrome. The total back-stress appears as the sum of the threshold stress (taken here as the Orowan stress) present in TD-nichrome and the strain-hardening stress as measured in the pure Ni–20% Cr matrix without ThO$_2$ particles. If we assume that a simple additive rule applies, then

$$\sigma_i = \sigma_0 + \alpha \sigma_a. \tag{42}$$

Now $\sigma_i' = \alpha$ as for pure metals and

$$n_a = n_e \frac{(1-\alpha)\sigma_a}{(1-\alpha)\sigma_a - \sigma_0}. \tag{43}$$

Since $\alpha \approx 0.1$–0.2, the denominator can be much smaller than the numerator. For

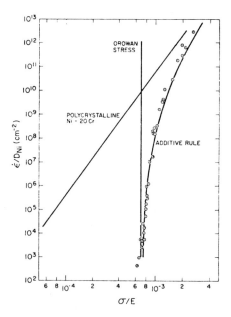

Fig. 33. Creep rates for polycrystalline single-phase Ni–20 Cr and TD-nichrome tested between 650°C and 1200°C (after LUND and NIX [1976]).

applied stresses just above the threshold stress, the apparent exponent reaches high values such as 10 or 40 and tends to infinity as the applied stress approaches the threshold stress. In fig. 33, n_a is simply the tangent to the line through the experimental points.

The other limiting case is achieved at high stress ($\sigma_a \gg \sigma_0$); equations (38) and (39) then become

$$n_a = n_e, \quad Q_a = Q_e - n_e kT^2 \left(\frac{\mathrm{d}E}{E\mathrm{d}t} + \frac{\sigma_i^*}{1 - \sigma_i'} \right), \tag{44}$$

which describes the situation in either tensile tests or short creep tests. The stress exponent is nearly that of the pure matrix (fig. 33) but the activation energy becomes very large (fig. 32): the dynamical dislocation arrangements which form inside the material are very sensitive to the high-stress regime ($\sigma_i' \approx 0.3$–0.6) and the dependence on temperature σ_i^* also becomes large since recovery is very effective and increases rapidly with a rise in temperature. In magnesium hardened by MgO particles, the apparent activation energy varies from 230 kJ/mol (CROSSLAND and JONES [1972]) to 420 kJ/mol (VICKERS and GREENFIELD [1968]) depending on experimental parameters, as compared to the 135 kJ/mol energy for volume self-diffusion. Similarly in TD-nickel, values of 800 kJ/mol (WILCOX and CLAUER [1966]) and ranging from 400 to 1100 kJ/mol (LASALMONIE and SINDZINGRE [1971]) are to be compared with the 295 kJ/mol of the energy for self-diffusion. Activation energies for creep in pure aluminium and in SAP exhibit the same kind of distortion (GUYOT and RUEDL [1967] and GITTUS [1975]).

5.   Composition and microstructure of precipitation-hardened alloys *

When the hardening phase is obtained by precipitation of an intermetallic compound dispersed inside the matrix and often at grain boundaries too, the particles are called precipitates. They are usually shearable although their crystallographic system may be entirely different from that of the matrix. As reviewed in detail by MARTIN [1980], the strengthening effects may come from one or several of the following sources:

Coherency hardening: coherency stresses developed in the matrix by misfitting particles interact with moving dislocations. This effect is very strong in Cu–Co alloys (THREADGILL and WILSHIRE [1974]) where it raises the yield stress at low temperature but does not bring any appreciable contribution to creep resistance.

Chemical hardening arises from the energy associated with the additional precipitate–matrix surface generated in the shear plane of the cutting dislocation. This term is only significant at low temperature or during rapid and repeated cyclic loading.

Order-hardening is related to the possible creation of a surface of antiphase

---

* See also ch. 21.

References: p. 1481.

boundary (APB) when a dislocation of the disordered matrix cuts across the precipitate.

Stacking-fault (SF) hardening is another example of a strengthening mechanism connected with the creation of a high-energy surface. These last two processes are very effective at high stresses at intermediate temperatures (0.5–0.7 T_m).

Modulus hardening originates from the difference of elastic moduli between matrix and precipitates.

Steel hardened by ferrite particles and various carbides and aluminium-base alloys hardened by GP zones or θ, θ' and θ'' precipitates have been examined in other chapters. Most of the examples presented in this chapter will be taken from modern nickel-base superalloys which are widely used in aircraft, marine and industrial gas turbines for their mechanical strength as well as in nuclear reactors, submarines or petrochemical equipment for their high resistance to corrosion and oxidation. The interested reader will find more specialized information in a book edited by Sims and Hagel (see *Further reading*).

The nickel atom has the striking property of being able to engage in bonds with a large number of elements and form solid solutions, intermetallic phases or chemical compounds of all sorts. This basic affinity for other elements stems from the partially filled 3d shell of nickel ($3d^8$, $4s^2$). This structure distinguishes it radically from its immediate neighbor, copper ($3d^{10}$ $4s^1$), which is more limited in its ways of forming bonds despite having one more electron. (See also ch. 5.)

The simplest way to start looking into the various ordered intermetallic compounds of type A_3B that nickel can form with Al, Ti, Nb and similar elements, is to consider the structure of the densely packed planes in the fcc crystal which is the basic crystallographic system for nickel. Two configurations are possible: in the first one the threefold symmetry is preserved (fig. 34a), in the second one a tetragonal symmetry appears (fig. 34b). Notice that each plane has the Ni_3M composition, with M atoms avoiding the nearest neighbor M–M configuration. In the dense packing of these planes in the third dimension, creating M–M bounds means generating an antiphase boundary. With the triangular scheme (fig. 34a), packing as indicated by ▲

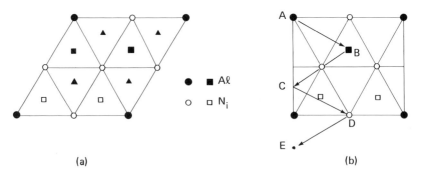

(a)   (b)

Fig. 34. Structure of the close-packed planes in A_3B ordered phases: (a) threefold symmetry of Ni_3Al, Ni_3Ti; (b) tetragonal or orthorhombic structures of Ni_3Nb, Ni_3Ta.

atoms will respect this rule and the sequence of planes can be called AB as in the fcc structure. If the third plane is located in such a way that M atoms are in ■ sites, an ABC stacking is obtained and can be repeated: this is the L1$_2$ structure (Strukturbericht notation) of Cu$_3$Au and Ni$_3$Al, which belong to the simple cubic system (Pm3m). When the stacking follows the simple ABABAB sequence, a hexagonal structure, D0$_{19}$, is created: it is that of MgCd$_3$ and only appears as locally faulted crystals in nickel-base intermetallic compounds. However, when the stacking sequence reads ABAC ABAC it belongs to the P6$_3$/mmc symmetry group as the ABAB stacking and it is the familiar arrangements of the η-phase, Ni$_3$Ti, with D0$_{24}$ structure. Finally with the rectangular pattern (fig. 34b) two basic types of stacking are possible: (1) ABAB when the ● and ■ positions of M atoms are alternately repeated. This is the body-centered tetragonal lattice, D0$_{22}$ of the γ″ phase represented by Ni$_3$Nb or Ni$_3$Ta. (2) A long sequence of six planes, ABCDEF, with orthorhombic structure, D0$_a$, in which Ni$_3$Nb and Ni$_3$Ta can also crystallize.

Optimal mechanical properties are obtained by combining precipitation-hardening with solid-solution hardening. In table 1 the major matrix elements appear in the first three columns (Fe, Ni, Co), next to the elements which tend to expand the matrix lattice and thereby increase the flow stress of the alloy. Large amounts of Cr are required to change the lattice parameter appreciably, whereas little Mo and/or W is needed and this latter element also hardens the γ′ phase. On the other hand, Ti is a very efficient strengthening factor when it remains in solid solution, as it does at very high temperature. At low temperature its solubility in the matrix is very low. The influence of several solutes on solid-solution hardening in nickel is shown in fig. 35.

For instance, Alloy 800, an iron-base alloy (table 1), is used, besides its excellent resistance to corrosive environments, in heat exchangers for nuclear power plants for its good creep resistance between 500 and 700°C. Chromium and iron provide the solid-solution strengthening, whereas the minute quantities of Al and Ti left in the alloy after the deoxidizing process are able to generate a very fine population (5–50

Table 1
Chemical composition of some alloys.

Trade name	Composition (wt%)									
	Fe	Ni	Co	Cr	Mo	W	Al	Ti	Nb	Ta
Alloy 800	44	34	–	21	–	–	0.3	0.4	–	–
INCO 718	7	53	–	18.5	3.1	–	0.4	2.5	–	–
Nimonic 80A	–	75	1	19	–	–	1.3	2.5	–	–
Waspaloy	–	58	13	19	4.5	–	1.3	3	–	–
Astroloy	–	55	17	15	5.3	–	4	3.5	–	–
IN100	–	60	15	10	3	–	5.5	4.5	–	–
MAR-M 200	–	60	10	9	–	12	5	2	1	–
CM SX2	–	65	5	8	0.5	8	5.5	1	–	6

References: p. 1481.

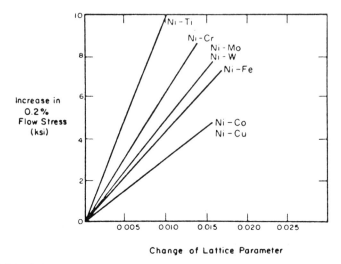

Fig. 35. Solid-solution hardening in nickel; 1 Ksi ≈ 15.2 kPa. (After PELLOUX and GRANT [1960]).

nm) of γ' precipitates. Their presence after 100 h at 500°C can be detected by hardness tests or relaxation tests. The γ' phase is fully precipitated after 1000 h at the same temperature and reaches about 2–3% volume fraction.

The volume fraction of the hardening precipitates is generally greater than 15%. Table 1 gives the composition of four typical wrought alloys, starting with INCONEL 718 which is hardened by γ' and γ'' simultaneously at a low volume fraction, and ending with Astroloy (or Udimet 700) containing 40–45% volume fraction. In the last three alloys of table 1, the volume fraction reaches 55–70%, the alloy can no longer be solutionized, even at temperatures close to the melting point of the alloy. These alloys cannot be forged and are used to cast highly critical engine parts such as turbine blades.

In most nickel-base superalloys the γ' phase is stabilized by the higher atomic proportion of Al atoms. Precipitation from the supersaturated solid solution is almost instantaneous. A recent investigation by electron diffraction and TEM of early stages of nucleation in a Ni–Ti alloy showed that formation of the ordered phase can result from a spinodal decomposition of the solid solution (LAUGHLIN [1976]). Even in the case where γ' precipitates may form by a nucleation and growth process, overcoming the low γ–γ' interfacial energy is easy and the growth rate is controlled by lattice diffusion during the ageing treatment. The expected growth rate of spherical precipitates of radius r, in this case, should obey an equation of the form:

$$r = A(Dt)^{1/2} \tag{45}$$

as a function of time t and diffusion coefficient D as described by MARTIN and DOHERTY [1976] in a recent review. During the course of prolonged ageing, the growth rate due to Ostwald ripening is slower:

$$r = B(Dt)^{1/3},\tag{46}$$

as given in the Wagner–Lifshitz model (see ch. 10B, §3.2.2).

Loss of coherency

After an extensive TEM study of several semi-coherent precipitates, WEATHERLY and NICHOLSON [1968] established that γ' precipitates in Al–Cu alloys lose coherency by the nucleation of dislocation loops within the precipitate, probably from collapsing point-defect clusters followed by climb of the loops into the particle–matrix interface.

On the other hand, γ' and β' (Al–Mg–Si) precipitates lose coherency by attraction of matrix dislocations initially stored in subgrain boundaries and pulled to the particle–matrix interface. They may move large distances by glide and reach their final equilibrium position by climb and interaction with other dislocations. LASALMONIE and STRUDEL [1975b] later described a three-dimensional model with all six $a/2 \langle 110 \rangle$ edge dislocations combined to form closed loops around each particle (fig. 36).

The dislocation spacing in the interface can be estimated by BROOKS's [1952] formula,

$$d = \frac{b}{|\delta|}\tag{47}$$

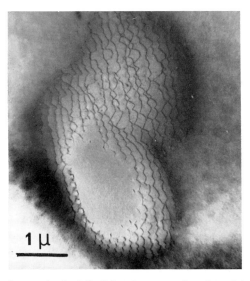

Fig. 36. Three-dimensional network of misfit dislocations around a γ' precipitate with $\delta = 0.3\%$ (from LASALMONIE and STRUDEL [1975b].

References: p. 1481.

where b is the Burgers vector and δ the misfit parameter defined by eq. (1).

A more accurate description of the problem of misfit dislocations in matrix–particle interfaces takes into account the differences in elastic constant between matrix and precipitate and the crystallographic anisotropy of each phase (see BONNET [1980] and DUPEUX and BONNET [1980]).

6. Tensile properties of precipitation-hardened alloys – behavior under high stress

Precipitation-hardened alloys are best appreciated at high temperature. While the 0.2% yield stress of pure nickel declines from about 20 down to 10 MN/m² between room temperature and 600°C and that of TD-nickel is fairly stable and of the order of 200 MN/m², it reaches 1000 MN/m² and more in most nickel-base superalloys in the same temperature range but decreases rapidly above 900°C.

6.1. Macroscopic properties

The stress–strain curves for polycrystalline materials depend strongly on temperature:

At low temperatures, below 650°C, the first-stage hardening is parabolic (up to 1–2% strain) and followed by a linear second stage which can extend from 5% for heavily hardened alloys to 20–30% for softer varieties. The hardening coefficient of stage II is of the order of 3000–4000 MN/m². Work-softening, constriction and rupture take place in stage III.

At intermediate temperatures (650–850°C) the linear stage is reduced at the expense of the parabolic stage at one end and the recovery stage at the other end.

At high temperatures (850°C and above) the initial yield stress is followed by a very short stage-I hardening before general decrease appears which extends until rupture (fig. 37). This mechanical instability at high temperature is also observed in single crystals tested in compression (LAW and GIAMEI [1976]) and has been attributed by CARRY and STRUDEL [1978] to the onset of massive dynamic recovery after the burst of gliding dislocations initially stored and locked in subgrain boundaries.

In single-crystal specimens, several different curves can be observed, depending on orientation:

For cube-oriented crystals the general features are similar to those of polycrystalline samples.

In crystals oriented for single slip, the stress–strain curve hardly rises above the initial yield strength; the crystal is progressively filled by slip bands which are channelled by the closely spaced precipitates until slip concentrates in one section of the specimen and induces terminal tearing of the material.

In $\langle 111 \rangle$ oriented crystals the yield strength is about 60% higher than in cube-oriented crystals at all temperatures. The elastic modulus is maximal in this

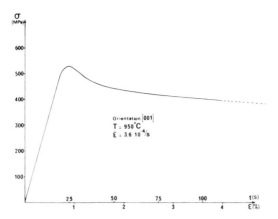

Fig. 37. Stress–strain curve of a cube-oriented nickel alloy single crystal showing mechanical instability at high temperature beyond the threshold stress (from CARRY and STRUDEL [1978]).

direction and the Schmid factor for all slip systems is minimal.

A detailed account of orientation and temperature effects on stress–strain curves of MAR M-200 single crystals will be found in the original publications of KEAR and PIEARCEY [1967] and COPLEY et al. [1972].

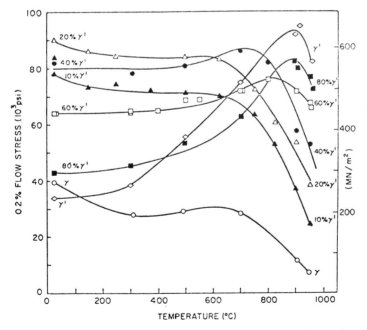

Fig. 38. Temperature dependence of flow stress in Ni–Cr–Al alloys with various volume fractions of γ′ precipitates (from BEARDMORE et al. [1969]).

References: p. 1481.

The influence of the volume fraction of γ' precipitates on the 0.2% flow stress of a series of Ni–Cr–Al alloys has been summarized by BEARDMORE et al. [1969] as shown in fig. 38. At first glance, it can already be seen that the strength of the γ' phase is not simply added to the strength of the γ matrix but it does become increasingly dominant at higher volume fractions, 60 and 80%. Since the γ' phase remains ordered up to temperatures close to its melting point (1100–1200°C) (COPLEY and KEAR [1967]) the prominent peak in the γ' flow stress around 800°C can only be attributed to slip mechanisms in the γ' phase itself. As suggested by MULFORD and POPE [1973] the anomalous temperature dependence of the yield stress is caused by the thermally activated formation of sessile dislocation segments which form by the cross-slip of screw dislocations from (111) slip planes on the (010) cube planes. Edge dislocations, on the other hand, retain high mobility (KURAMOTO and POPE [1978]). A ductility minimum of 3–5% elongation to rupture in tensile test and sometimes less than 1% rupture strain in creep adversely affects most nickel-base superalloys in the temperature range 750–800°C corresponding to the peak hardness of the γ' phase.

6.2. Deformation modes and hardening mechanisms

As seen in ch. 21, §3, ordered structures belonging to the Cu_3Au system (Ll_2) can be sheared by matrix dislocations travelling in pairs, thus forming a total Burgers vector $2 \times (a/2)$ [110] twice as long as in fcc crystals. The APB energy pulling the two similar dislocations together is of the order of 164 mJ/m² (COPLEY and KEAR [1967]) or may be less if due account is taken of the decrease in APB energy with ageing time. In complex alloys the final composition of the γ' phase varies as the matrix precipitates out other components. SHIMANUKI's recent investigations [1981] indicate 130 mJ/m².

The change in yield strength with particle size has been evaluated by GLEITER and HORNBOGEN [1965,1968] to be of the order of

$$\tau_c = \alpha\gamma_{APB}^{3/2} f^{1/3} r^{1/2}, \tag{48}$$

which has been successfully applied to a variety of austenitic alloys.

The parabolic $r^{1/2}$ dependence seems to indicate that larger precipitates will increase the critical stress. This is true until another mechanism requiring less energy becomes possible. Since the critical stress for Orowan by-passing by single dislocations falls off as $1/r$, the two curves meet (fig. 39) and the Orowan mechanism takes over for large γ' particles. The same kind of transition has been observed by REPPICH [1975] in iron-doped MgO where magnesia-ferrite particles precipitate in stress-free coherent octahedra inside the MgO matrix. In complex Al–Li alloys quasi-coherent ordered particles δ' (Al_3Li) with Ll_2 structure precipitate in the solid-solution hardened Al matrix. The microstructure of these alloys observed by TEM is strikingly similar to that of nickel-base superalloys. The transition from cutting by pairs of dislocations to by-passing is also observed (STARKE et al. [1981])

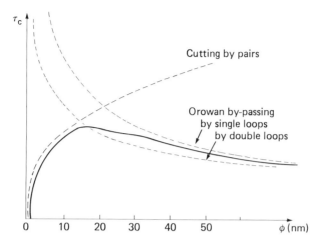

Fig. 39. Schematic representation of the transition from cutting to by-passing of ordered precipitates as a function of particle diameter ø. The solid line is a schematic experimental curve. (After STOLOFF [1972].)

and can be related to changes in the ductility of the material (SANDERS and STARKE [1982]). In actual crystals intermediate configurations can occur (fig. 40) where the first dislocation by-passes the γ' precipitate which is now surrounded by a shear loop of Orowan type. When a second dislocation sweeps across the same plane, shearing

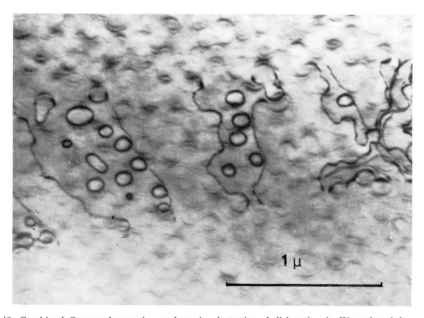

Fig. 40. Combined Orowan by-passing and cutting by pairs of dislocation in Waspaloy deformed at 750°C (from GUIMIER and STRUDEL [1970]).

References: p. 1481.

by pairs can take place under the increased stress of the combined dislocations: eventually the first Orowan loop is removed. This type of mechanism combined with the statistical size distribution of the γ' precipitates in the slip plane accounts for the flattened top of the schematic experimental curve in fig. 39 (BERGMAN [1975]; LAGNEBORG and BERGMAN [1976]).

The transition from cutting by pairs to Orowan by-passing actually depends on obstacle strength and volume fraction of the hardening phase. Although early models (GLEITER and HORNBOGEN [1965]; GUYOT [1971]) took these factors into account, they all assumed that obstacles remained rather far apart as compared to particle size, i.e. $f \leqslant 10\%$. For large volume fractions of the γ' phase, i.e., 20% to 60%, experimental data show major deviations from calculated curves (REPPICH et al. [1982]) unless the LABUSCH and SCHWARZ model [1978] is used. In this model the elastic interaction between the moving dislocation and the misfitting particle is represented by a particular profile, and the energy storing effect while the dislocation is approaching and cutting the particle is taken into account. HAASEN and LABUSCH [1979] have shown that the behavior of alloys with large γ' volume fractions can be rationalized well with this approach. REPPICH [1982], distinguishing between weakly coupled and strongly coupled dislocation pairs, has drawn theoretical strength vs particle diameter curves which give a close fit to experimental results. REPPICH et al. [1982] report an optimum particle size for strength above 650°C of about 30 to 40 nm for volume fractions up to 20% and of 80–160 nm for volume fractions of about 50%.

6.2.1. Superlattice stacking faults

In the temperature range 700–800°C, a two-phase alloy containing ordered coherent particles can deform by viscous shear under creep conditions, i.e., high stresses and/or slow dislocation motion enabling local atomic rearrangements.

The partial dislocation arrangement observed commonly in the Cu_3Au system (ch. 21, §3) creates high-energy surfaces connected either with complex faults (stacking fault + APB) or with APB. As first suggested by KEAR et al. [1968], simply adding an extra Shockley partial dislocation to the fcc scheme,

$$BA \rightarrow \delta A + B\delta, \tag{49}$$

yields, in the ordered phase, a *superlattice partial dislocation* (fig. 41):

$$2B\delta \rightarrow \delta A + B\delta + \delta C \tag{50}$$

and leads to the creation of a *superlattice intrinsic stacking fault* (S-ISF), a low-energy surface in the ordered phase of structure $L1_2$ since it is simply equivalent to four layers, ABAB, of the $D0_{19}$ hexagonal structure (see §5).

Taking the 2Bδ configuration as a superlattice Shockley partial dislocation, paired dislocation arrangements can be decomposed as:

$$2BA \rightarrow 2\delta A + 2B\delta = (C\delta + \delta A + B\delta) + (\delta A + B\delta + \delta C), \tag{51}$$

a decomposition which appears as dotted and dashed vectors in fig. 41. This scheme would describe a shearing mechanism taking place along a $\langle 110 \rangle$ direction and

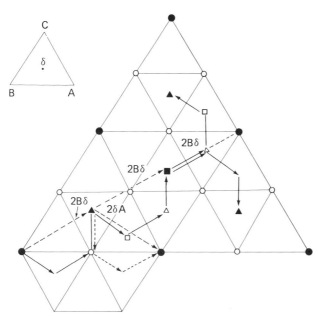

Fig. 41. Schematic directions of slip between dense planes in Ni_3Al ordered crystal, showing Shockley partial superdislocations $2B\delta$ and also δ shears associated with the formation of S-ISF/S-ESF pairs.

result in the creation of a S-ISF strip bounded by two superlattice partial dislocations. What is actually observed in nickel-base superalloys is much more complex and results in a general shear direction oriented along the $\langle 112 \rangle$ axis as described in detail by KEAR et al. [1970]. It is very similar to the intrinsic–extrinsic fault pairs observed by GALLAGHER [1970] in an fcc structure with low stacking-fault energy at the dissociated junction of two repulsive dislocations. If the dissociation described by eq. (49) takes place one more time on the plane just above (fig. 42a) two layers of S-ISF are superimposed, thus creating a *superlattice extrinsic stacking fault* (S-ESF) equivalent to seven layers, ABACABA, of the $D0_{24}$ structure, i.e., the crystal structure of Ni_3Ti. Finally a partial dislocation with a formal Burgers vector $2B\delta$ must bound the end of the S-ESF. The real atomic movements again take place in two successive planes and have actual Burgers vectors $2\delta A$ and $2\delta C$:

$$2\overline{B\delta} \begin{cases} 2\delta A = C\delta + \delta A + B\delta, \\ 2\delta C = B\delta + \delta C + A\delta. \end{cases} \tag{52}$$

Another possible decomposition of this superlattice stacking-fault pair leading to the same shear amplitude of $6B\delta$ is shown in fig. 42b: Shockley partials of the fcc lattice are coupled with dislocation dipoles, simply describing local lattice rearrangements (KEAR et al. [1969]). Such complex dislocation structures are mobile during creep at 750°C; their viscous glide motion requires short-range diffusion and is thermally activated.

References: p. 1481.

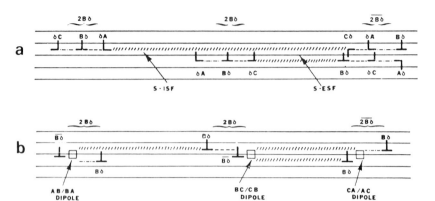

Fig. 42. Two possible configurations for superlattice intrinsic–extrinsic stacking-fault pairs with $6B\delta = a|211|$ net Burgers vector: (a) pure slip representation; (b) viscous slip including dipoles. (After KEAR *et al.* [1970].)

In two-phase alloys the complex γ' dislocation arrangements must be compatible with matrix shear vectors, i.e., the sum of all Burgers vectors in the matrix must be equal to the sum of the Burgers vectors in the precipitate. The combination of partial dislocations may be different in each phase. The correspondence

$$\begin{array}{ccc} 2\,BA + 2\,BC & \rightarrow & 6B\delta \\ \text{in the matrix} & & \text{in the precipitates} \end{array} \tag{53}$$

is explicitly drawn in fig. 43 and brings a correct interpretation of the widely observed stacking fault features in nickel-base alloys tested in creep under high stress (fig. 44). Simultaneous shearing on several successive planes or *synchro-shear* is known to take place in other ordered structures and also in corundum and spinels (ESCAIG [1974]).

6.2.2. Mechanical twinning of the ordered phase

As in fcc crystals, stacking-fault formation is associated with twinning for larger deformations. This is exemplified in Waspaloy (fig. 45) where the volume fraction of twinned increases with strain and also with temperature, isolated S-ISF being observed alone at 600°C and below. At higher temperatures, stacking faults first initiated in the γ' precipitates progressively invade the matrix lengthwise and then sideways by lateral thickening as deformation proceeds (fig. 46). Electron diffraction patterns reveal (GUIMIER and STRUDEL [1970]) that the twinned γ' phase remains ordered. The shear amplitude associated with twinning by single Shockley partial dislocations gliding in consecutive planes, as in fcc crystals, is very large: $(a/6|112|)/(a/3|111|) = \sqrt{2}/2 \approx 71\%$. In the $L1_2$ structure, repeated shearing by $2B\delta$ partials would double this value. It is therefore concluded that order is restored in the twinned γ' phase by local atomic rearrangements: hence the thermal dependence of its domain of occurrence (fig. 45). Mechanical twinning is also frequently observed in high-temperature fatigue tests (CLAVEL *et al.* [1980] and CLAVEL and

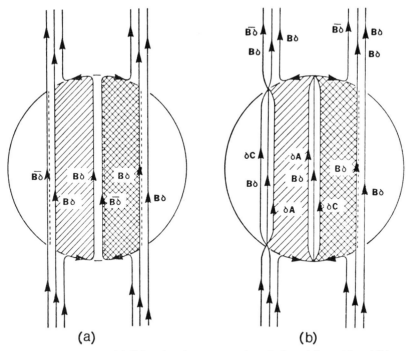

Fig. 43. Rearrangement of partial dislocations between matrix and γ' precipitate: (a) and (b) correspond to figs. 42b and 42a, respectively. (After KEAR *et al.* [1970].)

PINEAU [1982]) and in forged powder-metallurgy alloys (MENON and REIMAN [1975]).

Mechanical twining of fcc alloys at temperatures above 600°C and as high as 1080°C (DERMARKAR and STRUDEL [1979]) may seem incongruous since pure fcc metals resort to this deformation mechanism only at low temperature. Actually the general principle, that twinning only occurs when no other mechanism of lesser energy is available for plastic deformation, still holds in this case. Dislocations are too few at the start of the deformation, they are immobilized by the dense population of γ' precipitates, hence they cannot multiply. By-passing by climb requires low strain rates and time for recovery events to take place. The threshold stress for the Orowan by-passing mechanism is of the order of the critical stress for twin formation. The S-ISF energy of the γ' phase being lower than the SF energy of the matrix, twinning often originates in the ordered phase and spreads to the matrix.

7.   *High-temperature creep of precipitation-hardened alloys*

At high temperature and low stresses, the threshold stress associated with Orowan looping is never exceeded. Deformation processes are strongly climb-assisted although glide is not excluded but only restricted drastically.

References: p. 1481.

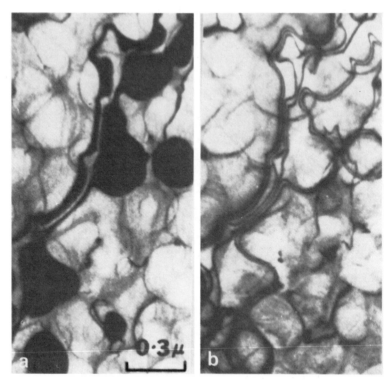

Fig. 44. Detail TEM observations of S-ISF/S-ESF pairs and associated partial dislocations in Mar-M 200 single crystals deformed at 760°C: (a) bright-field, (b) dark-field observation. (From KEAR and OBLAK [1974].)

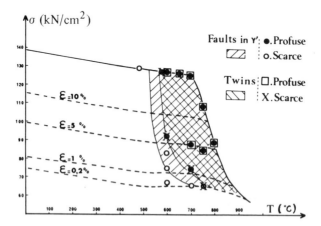

Fig. 45. Density of stacking faults and twins in deformed Waspaloy as a function of temperature and strain (after GUIMIER and STRUDEL [1970]).

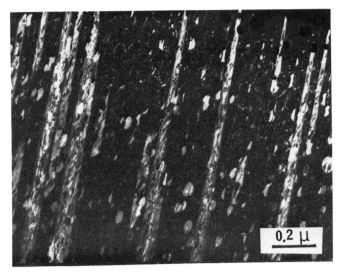

Fig. 46. Dark-field TEM observation of stacking faults and twins in Waspaloy deformed at 800°C.

7.1. Creep curves

The general aspect of high-temperature creep curves is radically different from that of the matrix alone (fig. 47). Primary creep is replaced either by a short sigmoidal stage at 850°C or by an incubation period (LEVERANT *et al.* [1973]). The

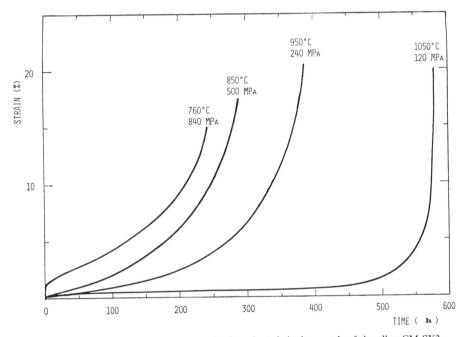

Fig. 47. High-temperature creep curves of cube-oriented single crystals of the alloy CM SX2.

References: p. 1481.

steady-state creep stage is short compared with an overwhelming third stage which either starts very early in the lower part of the temperature domain (850–1000°C) or contains most of the creep strain at higher temperature (1000–1100°C) and is soon followed by rupture. Comparing the forms of tensile (fig. 37) and creep (fig. 47) curves strongly points to the conclusion that two-phase alloys hardened by soluble quasi-coherent precipitates are mechanically unstable structures. The lack of initially mobile dislocations is the corner-stone of their mechanical resistance. The onset of dislocation movement, especially when it is dominated by climb, is the beginning of their degradation.

7.2. Deformation modes

In the lower part of the temperature domain (870°C), steady-state creep under high stresses can occur by viscous shearing of the γ' phase as evidenced by LEVERANT et al. [1973]. At higher temperature and lower applied stress, the hardening phase is no longer traversed by dislocations which remain confined to either:
 – thermally activated glide in the matrix
 – diffusion controlled climb in the γ–γ' interfaces.
Under conditions of low strain rate in single crystals oriented either for single slip or for multiple slip in the vicinity of the (001) pole, the dominant slip system in the fcc matrix is of type (110) $a/2$ [1$\bar{1}$0] as also observed by LE HAZIF et al. [1973] in work-hardened pure fcc metals at intermediate temperatures.

Dense arrays of edge dislocation dipoles are stored in the γ–γ' interfaces (fig. 48) and climb by stress-assisted pipe-diffusion (CARRY and STRUDEL [1978]). As they reach the edge of the γ' cuboids by climb, segments of opposite signs can glide toward one another and finally annihilate. A constant dislocation density is preserved by the glide of screw segments generating new edge dipoles in regions where the local internal stress has been lowered by intense recovery and the balance in the dislocation population as well as in the stress distribution is restored.

7.3 Internal stress

High-temperature creep in precipitation-hardened alloys takes place under stresses as low as 20–50% of the experimentally determined threshold stress or the calculated Orowan stress. It can only take place by a diffusion-controlled climb process around the unsheared particles as suggested by HOLBROOK and NIX [1974].

The tensorial structure of the internal stress field created inside the matrix by the closely spaced edge-dislocation dipoles (fig. 49) can be evaluated by summation of the stress fields of individual dislocations (CARRY et al. [1979]). In a cube-oriented single crystal tested in tension along the y axis, the σ_{yy} component of the stress tensor is the only component which exhibits a strong negative value in the matrix. All other components vary in sign and so does σ_{yy} in the γ' region. A substantial compression stress ($\sigma_{yy} < 0$) is created inside the matrix by the dislocations generated during deformation and it opposes a large fraction of the applied stress. This

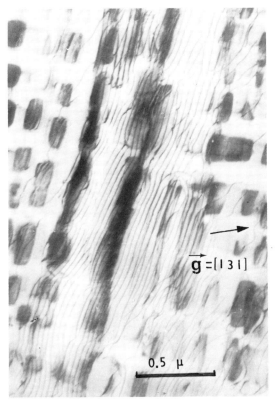

Fig. 48. Formation of edge-dislocation dipoles in the γ–γ′ interfaces by activated glide of screw segments bowing out between precipitates (after CARRY and STRUDEL [1975]).

kinematical hardening (§3.4) is the foundation on which the mechanical resistance rests at high temperature. It is anisotropic in nature and its degradation will accelerate the ruin of the crystal under unidirectional stress.

In order to write the simpler scalar equations for the flow rates, an average value of σ_{yy} is considered, usually in the form of an internal stress component $\sigma_i(\sigma_a, T)$ (§4.2). The apparent creep parameters are unusually high:

$$Q_a \simeq 400\text{–}700 \text{ kJ/mol}, \qquad n_a \simeq 6\text{–}12$$

according to most authors (KEAR and PIEARCEY [1967], PARKER and WILSHIRE [1975], CARRY and STRUDEL [1978] and STEVENS and FLEWITT [1981b]). But the temperature and stress dependence of σ_i can be experimentally measured. These introduce large corrective terms in eqs. (38) and (39). The effective creep parameters estimated from these equations are in the range of $Q_e \approx 250 \text{ kJ/mole}$ and $n_e \approx 1\text{–}2$. These values suggest that the elementary mechanism is effectively controlled by climb. In accord with the early predictions of Nabarro and Herring (ch. 20, §1.11), high-temperature creep of single crystals is taking place by an intense climb process.

References: p. 1481.

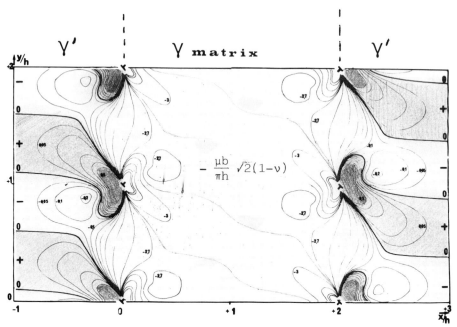

Fig. 49. Arrays of edge-dislocation dipoles generating a compression stress σ_{yy} along the tensile axis (after CARRY et al. [1979]).

The massive flux of atoms is not taking place in volume, however. It is channeled by the climbing arrays of dislocations held up in the γ–γ' interfaces. Similar conclusions drawn from a study of precipitation-hardened Al + 0.5% Fe alloys led BARRETT et al. [1972] to elaborate on the original creep model proposed by Nabarro and Herring.

A phenomenological approach to the internal stress problem has been taken by WHILSHIRE and co-workers (THREADGILL and WILSHIRE [1974] and PARKER and WILSHIRE [1975]), where the creep equations for precipitation-hardened alloys take on the form:

$$\dot{\epsilon}_s = C_0 (\sigma_a - \sigma_0)^4 \exp\left(- \frac{Q_e}{kT} \right) \tag{54}$$

similar to the flow equations for pure metals,

$$\dot{\epsilon}_s = C\sigma_a^4 \exp - \frac{Q_a}{kT}. \tag{55}$$

This type of analysis is found to be quite fruitful in interpreting the unusual creep behaviour of two-phase alloys. The friction stress σ_0 in this approach is connected with the internal stress distribution in the *relaxed state* after some amount of recovery has taken place, rather than with the instantaneous value of the internal stress as in the former approach.

7.4. Oriented coalescence of the hardening phase under strain

Several authors (TIEN and COPLEY [1971], CARRY and STRUDEL [1976], MIYAZAKI *et al.* [1979]) have observed that annealing of nickel-base alloys under stress can produce significant changes in the morphology of the γ' phase in relation to creep strain (CARRY and STRUDEL [1978]).

If the crystals are deformed in tensile creep along the (001) axis several slip systems are operating and two orthogonal sets of platelets appear along the cube planes parallel to the tensile axis when the γ–γ' misfit parameter δ is negative (fig. 50). When the stress is oriented along a direction of lower symmetry, (103) for instance, only one slip system is activated and γ' rafters form only along one plane. Similarly, if either the tensile stress is changed into a compression stress along the (001) direction, or the sign of δ is changed from negative to positive, γ' platelets extend in the direction normal to the stress (TIEN and COPLEY [1971]). The critical role of the lattice misfit is explained by fig. 51 (CARRY and STRUDEL [1976]): the matrix dislocations produced by the tensile stress aggravate the γ–γ' misfit along the vertical facets of the γ' cuboids when $\delta < 0$ and tend to compensate δ on the horizontal facets. Climb by pipe-diffusion along the dislocation line will follow the stress gradient and growth of horizontal facets will be favored (CARRY *et al.* [1981]). In order to restrain this deteriorating effect which destabilizes the strengthening configuration, δ must be minimized for the service temperature, and appropriate thermal treatment of the alloy can promote optimal resistance (STEVENS and FLEWITT [1981a]). The basic mechanism, however, can only be slowed down, not suppressed.

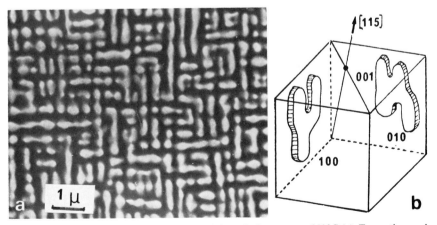

Fig. 50. Strain-induced coalescence of the hardening phase during creep at 950°C (a). Two orthogonal sets of rafters (b) are formed when several slip systems are operating.

References: p. 1481.

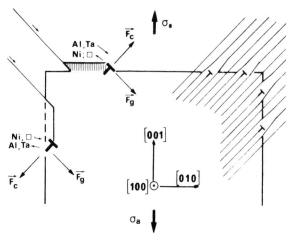

Fig. 51. Dislocation climb in the $\gamma - \gamma'$b interface with $\delta < 0$ (bold symbols: matrix dislocations; light symbols: misfit dislocations). (From CARRY et al. [1981].)

8. Recrystallization

Basic concepts concerning the recrystallization of alloys containing a second phase are set out in ch. 25. We shall simply emphasize and illustrate the main features of recrystallization phenomena specific to the materials with which we are here concerned. Three parameters play a major role in these alloys: *particle size, interparticle distances* and *degree of deformation.*

8.1 Particle size and amount of strain

Second-phase particles of radius r exert a retarding force F_Z, also called *Zener force*, on a migrating boundary as illustrated by HANSEN and JONES [1981] in the case of low-angle boundaries and subgrains (fig. 52). A rough estimate of this force was first proposed by ZENER [1948] as a function of volume fraction, f, and interfacial energy, γ_s, of the particles [see also eq. (8), ch. 25]:

$$F_z = \tfrac{3}{2} \frac{f}{r} \gamma_s. \tag{56}$$

Thus a high volume fraction of small particles can retard or even completely inhibit recrystallization. This is the case with TD-nickel which can retain its elongated grain structure (fig. 53) inherited from wire-drawing even after annealing at 1400°C.

JONES et al. [1979] and JONES and HANSEN [1980] have summarized the various physical causes for the inhibiting effect of small particles on the recrystallization process:

A fine dispersion of small particles homogenizes plastic deformation. Dislocations are more uniformly distributed in the matrix and subgrain boundary formation is

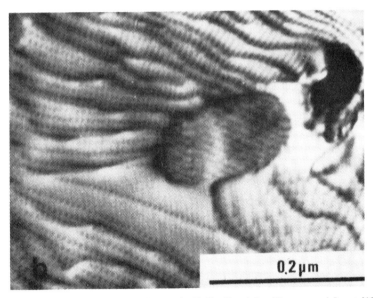

Fig. 52. Low-angle boundary pinning in an Al–Al$_2$O$_3$ alloy (after HANSEN and JONES [1981]).

more difficult. Local lattice rotations due to strain concentration are less frequent than in the absence of small particles, and the critical radius for recrystallization is thereby increased.

When subgrain boundaries do form, they are efficiently stabilized by the second phase. A good example is given by TD-nickel seen by TEM (fig. 4, above).

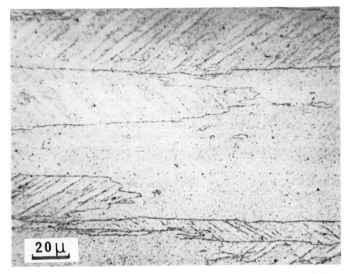

Fig. 53. TD-nickel wire-drawn and annealed 3 h at 1400°C (after LASALMONIE and SINDZINGRE [1971]).

References: p. 1481.

Rearrangement and coalescence of subgrains, leading to nucleation of recrystallized nuclei, is impeded.

As a consequence, nucleation of recrystallization embryos is more frequent at grain boundaries than inside the matrix.

These various aspects of the inhibiting effects of second-phase particles point to the excessive simplicity of Zener's approach whose expression tends to underestimate the retarding force experienced by each individual boundary or subgrain boundary developing inside the deformed matrix.

On the other hand, large second-phase particles tend to stimulate nucleation of recrystallization nuclei in their immmediate vicinity (see fig. 36, ch. 25). The critical size for nucleation enhancement depends directly on the amount of strain the alloy has been subjected to: let us call it $d(\epsilon)$. HUMPHREYS [1979b] gives a graphic representation of a $d(\epsilon)$ function established experimentally in the case of Al–Si alloys (see also ch. 25, §3). It is clear that in this material, particles less than 2 μm in diameter inhibit recrystallization whereas those of 4 μm and above act as strong nucleation sites. This effect of course is justified by the strain accumulation and large lattice rotations observed around larger particles as described in §§3.2 and 3.3, above.

8.2. Interparticle spacing

Kinetics of recrystallization are affected not only by particle size but also by interparticle spacing when their size is larger than the critical diameter $d(\epsilon)$. DOHERTY and MARTIN [1962] pointed out that both the growth rate (represented by the time for 50% recrystallization) (fig. 54a) and the nucleation rate (fig. 54b) are strongly dependent on interparticle spacing. They used an aluminium-base alloy hardened by various distributions of $CuAl_2$ precipitates obtained by a sequence of heat-treatments. A distance of at least 1 μm between precipitates is needed for the recrystallization process to overcome the inhibiting effects and to become stimulated by the presence of precipitates.

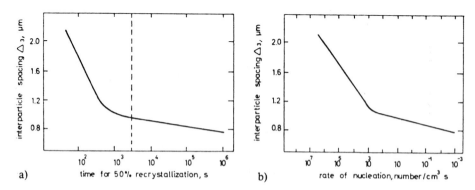

Fig. 54. Relation between interparticle spacing and: (a) time for 50% recrystallization (time for single-phase alloy shown by dashed line); (b) rate of nucleation. (After DOHERTY and MARTIN [1962].)

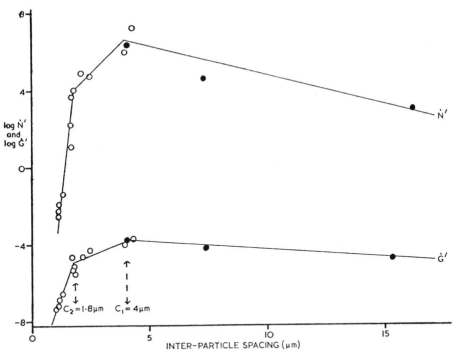

Fig. 55. Apparent nucleation rate ($\dot{N}$) and apparent growth rate ($\dot{G}$) as a function of interparticle spacing for Al–Cu alloys (after MOULD and COTTERILL [1967]).

MOULD and COTTERILL [1967] developed an elementary theory to describe similar results obtained in cold-drawn and annealed wires of an aluminium-base alloy hardened by Al_3Fe precipitates. The critical interparticle spacing they found (fig. 55) was of the order of 2–4 μm.

Finally, retardation and acceleration effects are best summarized by the two

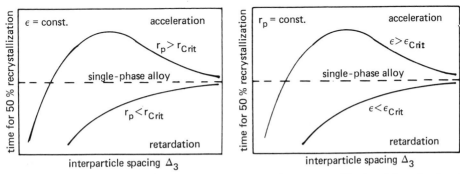

Fig. 56. Schematic diagram of the influence of dispersed particles on recrystallization: (a) acceleration of recrystallization occurs only if the particle size is large enough to give rise to local lattice curvature ($r_p > r_{crit}$); (b) for a given particle size, a critical strain is necessary for particle-simulated nucleation ($\epsilon > \epsilon_c$). (After HORNBOGEN and KÖSTER [1978].)

References: p. 1481.

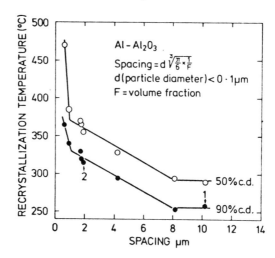

Fig. 57. Recrystallization temperatures of 50 and 90% cold-drawn (c.d.) Al–Al$_2$O$_3$ alloys as a function of particle spacing (after JONES and HANSEN [1980]).

schematics of fig. 56 first suggested by NES [1976] and later elaborated on by HORNBOGEN and KÖSTER [1978]. For a given deformation and various particle sizes the time for 50% recrystallization is plotted as a function of interparticle distance (fig. 56a) and for a given particle size one obtains a similar description when the amount of strain applied to the material is varied (fig. 56b).

8.3.   Effect of temperature

Since recrystallization is basically a thermally activated process, temperature has a strong effect on its kinetics even in two-phase alloys. JONES and HANSEN [1980] studied the variation of the recrystallization temperature in Al–Al$_2$O$_3$ alloys containing various volume fractions of small particles (0.1 μm). They observed that the recrystallization temperature was low and independent of interparticle spacing when this latter parameter was larger than 8 μm (fig. 57), but was slowly increasing when particles came as close as 8–2 μm apart and finally increased drastically for interparticle spacings in the submicron range.

8.4.   Micromechanisms

Micromechanisms have been observed in great detail by HUMPHREYS [1979a] who characterizes the immediate vicinity of a large particle as highly rotated with respect to the surrounding matrix and heavily deformed. Subgrain coalescence will therefore take place within a fraction of a micron from the particle. It will be followed by a rapid growth of the recrystallization nucleus into the deformed zone extending a few microns around the undeformed particle. Finally the recrystallization front will have grown rapidly through the more heavily strained region around the particle where the density of dislocations is much lower and the driving force much weaker: the growth rate therefore slows down considerably.

These three stages have been described more quantitatively in a recent model developed by SANDSTRÖM [1980b], which can be sketched as follows.

In stage I, associated with the coalescence of very small subgrains of average radii $\bar{R}$, the radius R of the growing subgrain increases by a coalescence mechanism:

$$\left(\frac{dR}{dt}\right)_I = \alpha m \tau \left(\frac{1}{\bar{R}} - \frac{1}{R}\right), \tag{57}$$

where m is the mobility of dislocations and τ their line tension.

In stage II, the growth rate in the deformed zone can be written as proportional to the difference between the energy associated with the high dislocation density in the deformed zone, ρ_{dz}, and that of the curved boundary:

$$\left(\frac{dR}{dt}\right)_{II} = \beta M \left(\tau \rho_{dz} - \frac{2\gamma_s}{R}\right), \tag{58}$$

where M is the mobility of the growing interface and γ_s the surface energy per unit area.

In stage III, growth outside the deformation zone can be described by the same equation as in stage II where ρ_{dz} is now replaced by ρ_M, the actual dislocation density in the matrix, which is much lower than in the deformed zone, and the growth rate drops drastically:

$$\left(\frac{dR}{dt}\right)_{III} = \beta' M \left(\tau \rho_M - \frac{2\gamma_s}{R}\right). \tag{59}$$

These three stages appear in fig. 58 where theoretical curves fit quite closely experimental points obtained in an Al–Al$_2$O$_3$ alloy. Extrapolating stage III growth rates backwards toward the origin may lead to apparently negative incubation periods. In practical terms, this means that the first two stages could take place so rapidly during a forging sequence at high temperature, that they may not be observed. This is the case in nickel-base superalloys, for instance, during forging sequences above 1000°C. Static recrystallization cannot take place when the material is water-quenched just after forging and yet micron-size recrystallized grains are observed in thin foils by TEM. If these grains would result from dynamic recrystallization, some of them would show traces of deformation such as dislocation bundles. Since these new grains are free of dislocations, DERMARKAR and STRUDEL [1980] suggest that they are the landmarks of a *metadynamic recrystallization* process as described by the above equations.

9. *Duplex structures and multiphase alloys*

At higher temperatures, little or no strain-hardening is available. On the other hand, plastic strain damages materials and tends to destabilize structures obtained

References: p. 1481.

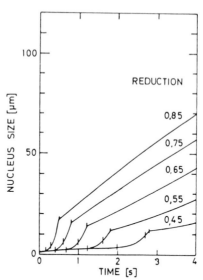

Fig. 58. Three-stage recrystallization process in forged nickel-base superalloys where metadynamic recrystallization is frequently observed (after SANDSTRÖM [1980a]).

by precipitation or thermomechanical treatments. As a consequence, engineers must imagine and implement ingenious stratagems to circumvent or retard the rapid mechanical and chemical degradation processes operating at high service temperatures.

9.1. Duplex structures

In nickel-base superalloys, the size of the γ' precipitates and to some extent the locations of the nucleation sites can be controlled by appropriate heat treatment: γ' precipitates will nucleate preferentially in grain boundaries and grow to large size if the first stage of an annealing treatment is carried out under conditions of low supersaturation, i.e., 20 K below the γ' solvus temperature, for instance. The alloy can now be forged at that temperature since the volume fraction of the precipitates remains small. Cooling it rapidly will preserve the deformed structure, and subsequent ageing at 100–200 K below the γ' solvus temperature will induce: (1) the precipitation of a dense population of 100 nm size γ' particles; (2) the recrystallization of the alloy. Owing to the stimulating effect of the large γ' particles, recrystallization nuclei will form near grain boundaries (fig. 59). Growth of these new grains will be inhibited by the dense population of submicronic γ' precipitates and recrystallization will only be of a very limited extent. A *duplex or necklace structure* is thus created (MENON and REIMAN [1975]). It is made of a necklace-like arrangement of 1–5 μm recrystallized grains located in the original grain boundaries of the material (fig. 60). The inside of each grain is partially recovered and contains a high density of small subgrains (0.1–0.5 μm) providing excellent tensile and creep

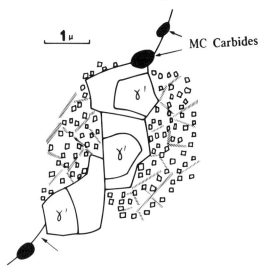

Fig. 59. Stimulated nucleation at large particles and inhibition of grain growth by small precipitates (after DERMARKAR and STRUDEL [1980]).

resistance to the complex structure. The detailed microstructure has been studied by BEE *et al.* [1980] and STRUDEL [1980]. The low-cycle fatigue properties of this duplex structure are higher than those of the heat-treated or conventionally forged material (MENON and REIMAN [1975] and SHAMBLEN *et al.* [1975]). The ductility is improved and the fatigue-crack growth rate is reduced owing to the multiple branching effect of the small grain size on the propagating crack tip as clearly shown by PEDRON and PINEAU (1982) for INCO-718 alloy at 650°C.

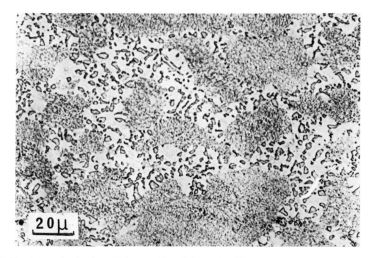

Fig. 60. Duplex-grain size in a PM superalloy deformed 22% at 1060°C and partially annealed.

References: p. 1481.

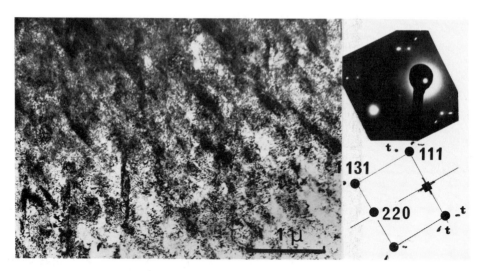

Fig. 61. Enhanced nucleation of twins on several systems. The slip dispersion effect is due to HfC particles finely distributed in the matrix of Waspaloy + 1% Hf. (After STRUDEL [1978].)

9.2. Multiphase precipitation-hardening

Nickel-rich austenites can not only precipitate γ' particles but also primary and secondary carbides (see the book by Sims and Hagel in *Further reading*). Deformed under high stresses, γ'-hardened alloys have a strong tendency to form twins (§6.2) which extend across the matrix grain and impinge upon the grain boundaries. Compatibility of deformation between grains cannot be maintained when only a small number of slip systems is activated, and cracks are therefore nucleated at the point where the mechanical twins meet grain boundaries. This mechanism is respon-sible for the very poor tensile ductility of most nickel-base alloys in the temperature range (750–800°C) where the strength of the γ' phase is maximum. The situation can be remedied however, not by inhibiting twinning which is not intrinsically detrimental to mechanical strength, but on the contrary by stimulating it. This can be implemented by adding 0.5–1% Hf to Waspaloy for instance (see table 1). Hafnium's avidity for carbon leads to the formation of micron-size primary HfC carbide particles with equiaxed morphology (STRUDEL [1978]). This undeformable phase acts as a strong dispersoid: it limits grain growth during ageing treatments and stimulates the nucleation of numerous small twins on the primary as well as on secondary slip systems (fig. 61). The homogeneity of plastic deformation is restored on the scale of the metallurgical grain because of the slip dispersion effect triggered by the HfC particles, and ductility is enhanced without loss of mechanical strength.

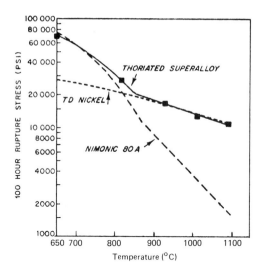

Fig. 62. Combined precipitation-hardening and OD strengthening in mechanically alloyed nickel-base superalloy IN-853 (1 psi ≈ 6.9 kPa) (after BENJAMIN [1970]).

9.3. Mechanical alloying of complex alloys

Encouraged by the unequalled mechanical properties of ODS nickel and nickel–chromium alloys, BENJAMIN [1970] extended mechanical alloying to more complex matrices which could, in turn, provide precipitation-hardening. This development was prompted by the rather poor mechanical performance of TD-nickel and TD-nichrome alloys in the temperature range 500–1000°C when compared with basic nickel-base superalloys such as Nimonic 80A (fig. 62). Alloy IN-853 (77 Ni–19 Cr–1.2 Al–2.4 Ti + 2.25 vol% Y_2O_3) combines the low-temperature yield strength and stress-rupture life of a precipitation-hardened alloy, with the high-temperature creep performance of an ODS alloy. The thermal stability of this ingenious mixture of materials can be characterized by its temperature of recrystallization: 1230°C (CAIRNS [1974]). The grain size of the material is in the micron range and the internal substructure starts recovering above 1100°C.

This sophisticated combination of metals and oxide cannot be achieved by melting and the appropriate blending of minute oxide particles cannot be manufactured by conventional powder metallurgy.

Recent progress has also been made in the development of aluminium-base alloys which combine solid-solution strengthening by Mg and OD and Al_4C_3 strengthening (BENJAMIN and SCHELLEN [1981]). Avoiding precipitation-strengthening improves the corrosion resistance of the material. The thermal stability of the fine-grained structure up to several hours at 350°C makes the forging and extrusion processes much easier.

References: p. 1481.

9.4. Grain-size effects in multiphase alloys

The effect of grain size on the mechanical properties of single-phase materials is treated in ch 19, §8, ch. 20, §§1.11 and 2.1, and also briefly referred to in ch. 28, §4.4. Recent aspects of the Hall–Petch relation between yield strength or flow stress and grain size or substructure size in single-phase metals and alloys have been reviewed by ARMSTRONG [1983].

Whatever the material, grain size effects are very sensitive to temperature and of course to strain rate. The case of manganese austenitic steels has been extensively documented experimentally by KUTUMBA RAO *et al.* [1975] and then discussed further by McLEAN (1976). Grain size has opposite effects on the flow stress at high and low temperatures; the finer the grain size, the higher the flow stress at low temperature and the lower the flow stress at high temperature (fig. 63). Therefore a *cross-over temperature* T_c can be defined where the effect is reversed. It lies in the temperature range 750–800°C in austenitic steels for a strain rate of 3.6/h and in the range 450–500°C for $\dot{\epsilon} = 3.6 \times 10^{-3}$/h when taking an activation energy for plastic flow and recovery of the order of 300 kJ/mole.

At *low temperature*, crystallographic slip in neighbouring grains leads to strain incompatibilities in the vicinity of grain boundaries and generates internal stresses that can be released only partially by the activation of secondary slip systems in that region (ASHBY [1970]). The theory conforms qualitatively to the experimental observation that the dislocation density increases with decrease in grain size. The yield strength of the material follows a Hall–Petch law,

$$\sigma_y = \sigma_0 + kl^{-1/2}, \tag{60}$$

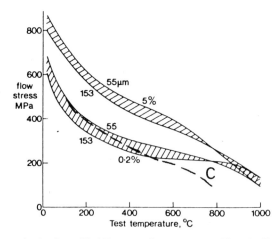

Fig. 63. Influence of a grain size from 55–153 µm on flow stress at 0.2% and 5% strain over a range of temperature straddling the crossover point C. Dashed curve corresponds to lower strain rate. (After McLEAN [1976].)

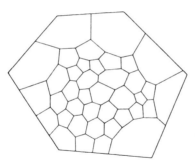

Fig. 64. Large dislocation meshes at grain boundaries caused by rapid grain-boundary diffusion – diagrammatic (from McLean [1976]).

where σ_0 is a friction or threshold stress and $k \simeq 0.16\sqrt{b}$ is the hardening coefficient. (l is used here for grain diameter, whereas d is used in ch. 19.)

At *high temperature*, or low strain rates, i.e., for $T > T_c$, stress concentrations near grain boundaries are removed by diffusion creep either in the grain boundary itself or by climb-controlled dislocation motion in its vicinity. The grain boundaries cease to act as barriers to deformation as they do at lower temperature; on the contrary, rapid grain-boundary transport accelerates the recovery of dislocation substructures and tends to create a softer zone in its immediate neighborhood as first suggested by McLean (fig. 64). A coarser subgrain structure near grain boundaries has been observed by Robert *et al.* [1981] in 316-type stainless steels after creep at 650°C. Hence the difficulty for single-phase materials of small grain size to retain a stable subgrain structure under creep conditions and also the absence of any dislocation substructure in materials having a grain size small enough to exhibit superplastic behavior (ch. 20, §2.1).

The role of grain size and grain elongation in strengthening of ODS alloys has been well documented by the work of Wilcox and Clauer [1972] who compared mechanical properties of various microstructures in pure nickel and nickel alloys hardened as solid solutions Ni–20 Cr and Ni–20 Cr–10 W with those of TD nickel, TD nichrome and Ni–20 Cr–10 W–2 ThO$_2$.

At *room temperature* all these materials follow a Hall–Petch relationship [eq. (60)] with somewhat lower k values for ODS alloys (fig. 65). The convergence of the lines in this figure indicates that the linear additivity rule for the various strengthening mechanisms (i.e., solid solution, particle, grain and subgrain refinement hardening) suggested by Hansen and Lilholt [1971] and implicitly assumed in the earlier parts of this chapter, is probably incorrect. Looking at linear distances between obstacles (i.e., oxide particles on one hand, grain or subgrain boundaries on the other hand), this plot indicates that boundaries are more than six times as potent in raising the room-temperature yield strength than are oxide particles, but of course this feature is not retained at high temperature.

At *elevated temperature* the yield strength of ODS alloys does not seem to correlate with $l^{-1/2}$ in the Hall–Petch manner but rather well with *grain aspect ratio*

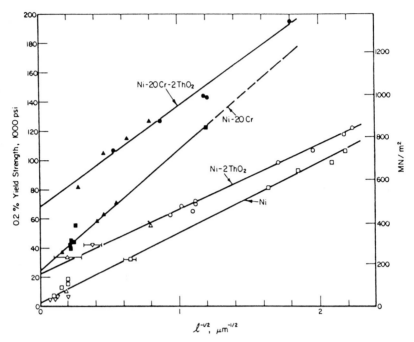

Fig. 65. Dependence of room-temperature yield strength (0.2% offset) of Ni, Ni–2ThO$_2$, Ni–20Cr and Ni–20Cr–2ThO$_2$ on grain size and cell size produced by drawing (from WILCOX and CLAUER [1972]).

(GAR), L/l, as suggested by WILCOX and CLAUER [1972]. This parameter is defined schematically in fig. 66b. A close fit of both tensile and creep properties at 1093°C

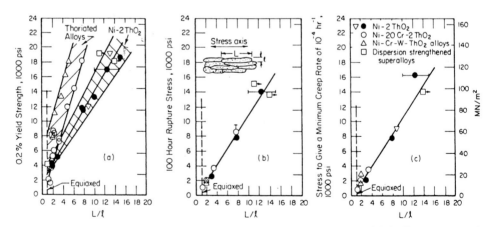

Fig. 66. The effect of grain aspect ratio, L/l, in dispersion-strengthened nickel alloys on strength properties at 1093°C: (a) 0.2% offset yield strength; (b) 100-hr rupture stress; (c) stress to give a minimum creep rate of 10^{-4}/h. Open points are recrystallized, closed points are non-recrystallized. (From WILCOX and CLAUER [1972].)

of these materials and microstructures is obtained when L/l is used (fig. 66). Considering these results, the authors suggest a dependence of the flow stress of the form:

$$\sigma = \sigma_e + K\left(\frac{L}{l} - 1\right), \tag{61}$$

where σ_e is the strength for equiaxed grains ($L/l = 1$) and K a hardening coefficient which increases with solid-solution hardening (fig. 65). The physical interpretation for the prominent role of this parameter in most materials designed for elevated temperature resistance rests on the important contribution of grain-boundary sliding in high-temperature yielding and strain accommodation (whenever a material is tested above the cross-over temperature T_c). A highly elongated microstructure tends to align grain boundaries along the main stress axis and thereby minimize the resolved shear and tensile stresses acting on them. These effects have been exploited in engineering applications where improved high-temperature ductility and creep rupture life are required: ODS tungsten filaments for lamps and electron guns, directionally solidified nickel-base superalloys for turbine blades, and thoriated platinum wire and strips for the chemical industry. Appropriate thermomechanical treatments can usually lead to strongly anisotropic mechanical properties based on large GAR values (5–20).

GRANT [1983] has recently reported comparative creep studies on single-phase

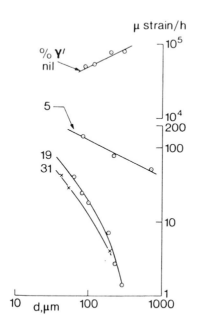

Fig. 67. Strain rate versus grain size for various γ' volume fractions in a nickel-base alloy tested at 750°C under 154 MN/m² (after GIBBONS and HOPKINS [1971]).

References: p. 1481.

type 316 stainless steel, tested at 650°C, and the same steel strengthened with a fine dispersion of TiC. The single-phase alloy resists creep much better with a 65 μm grain size (ingot processed) than with a 5 μm grain size (rapid solidification processed), whereas the dispersion-strenghtened alloy is actually somewhat more creep-resistant wit 5 μm grains. This behavior suggests that the cross-over temperature is probably lower for the single-phase material ($T_c < 650°C$) than it is for the TiC-strengthened steel. The basic reason for the shift of T_c to slightly higher temperatures in the two-phase material is probably to be related to the retardation of all recovery phenomena in the presence of a dispersed phase.

In precipitation-hardened alloys such as γ'-strengthened nickel, changes in mechanical properties are drastic from one side of the cross-over temperature T_c to the other. Increasing the γ' volume fraction enhances this tendency further as evidenced by the steeper slopes of the curves in fig. 67 as the γ' content of the alloy is increased.

Since the cross-over temperature is usually estimated to lie in the temperature range 650–700°C in nickel-base alloys, grain refinement by powder metallurgy or thermomechanical treatments is actively pursued only for jet engine parts such as turbine disks which require high strength and fatigue resistance up to 650°C (DAVIDSON and AUBIN [1982]). For parts exposed to temperatures above T_c, casting techniques are common practice and larger grain sizes yield longer creep rupture life in IN-100 (MOSKOWITZ et al. [1972]) as well as in IN 939 for instance (fig. 68).

From the many above examples, one would be tempted to conclude that the mechanical properties of ODS and precipitation-hardened alloys do not correlate

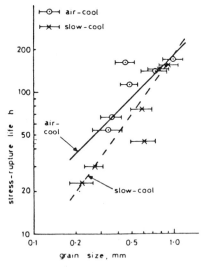

Fig. 68. Creep life versus grain size in IN939 for two different cooling rates from solution temperature, 1160°C. Creep test: 278 MN/m² at 870°C. (After CUTLER and SHAW [1979].)

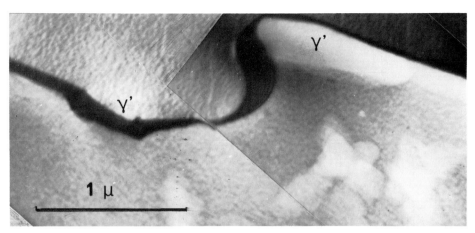

Fig. 69. An adequate cooling rate through the γ' solvus temperature yields wavy grain boundaries anchored by large γ' precipitates in a PM superalloy. Higher creep- and fatigue ductility is achieved.

with grain size in a simple manner. This difficulty often originates from the use of inappropriate parameters for characterizing the microstructure of the material. The significant structural distance is more likely to be the cell size or the subgrain size, which is directly related to the dislocation mean free path or the average dislocation link length. Subgrain boundaries in these materials are associated with strong local misorientations (as evidenced by electron diffraction in fig. 4, above); therefore the distinction between grain boundaries and large-angle subgrain boundaries tends to disappear, especially when cumulative misorientations are examined over several subgrains. This situation normally leads to continuous recrystallization during annealing (MONTHEILLET [1981]) but this process is strongly inhibited by precipitates and particles in multiphase alloys.

One further reason why grain boundaries often play a minor role in tensile tests as well as in creep in these composite materials is their wavy aspect, resulting from the pinning effect of the hardening particles (figs. 52, 53). In precipitation-hardened materials, an appropriate sequence of thermomechanical treatments can favor the development of interpenetrating grains (fig. 69) which provide the alloy with improved ductility and fatigue resistance (RUNKLE and PELLOUX [1979]). Hence both grains and subgrains are strongly pinned and distorted from the fairly straight polyhedral shape they would assume under the sole effect of interfacial tension. Their strength relies more on their morphology and on the anchoring effect of the hardening phases than on the absence of segregated elements or on their intrinsic structure.

References

ANSELL, G.S., and F.V. LENEL, 1961, Trans. AIME **221**, 452.
ANSELL, G.S., and J. WEERTMAN, 1959, Trans. Met. Soc. AIME **215**, 838.
AHLQUIST, C.N., and W.D. NIX, 1971, Acta Metall. **19**, 373.

ARMSTRONG, R.W., 1983, in: Yield, Flow and Fracture of Polycrystals, ed. T.N. Baker (Appl. Science Publ., Barking, Essex, UK) in press.

ASARO, R.J., 1975, Acta Metall. **23**, 1255.

ASHBY, M.F., 1966, Phil. Mag. **14**, 1157.

ASHBY, M.F., 1968, in: Proc. 2nd Bolton Landing Conf. on Oxide Dispersion Strengthening (Gordon and Breach, New York) p. 143.

ASHBY, M.F., 1969, in: Physics of Strength and Plasticity, ed. A.S. Argon (MIT Press, Cambridge, MA) p. 113.

ASHBY, M.F., 1970, Phil. Mag. **21**, 399.

ASHBY, M.F., 1971, in: Strengthening Methods in Crystals, eds. A. Kelly and R.B. Nicholson (Elsevier, New York) p. 137.

ATKINSON, J.D., L.M. BROWN and W.M. STOBBS, 1974, Phil. Mag. **30**, 1247.

BARRETT, C.R., E.C. MUEHLEISEN and W.D. NIX, 1972, Mater. Sci. Eng. **10**, 33.

BEARDMORE, P., R.G. DAVIES and T.L. JOHNSTON, 1969, Trans. Met. Soc. AIME **245**, 1537.

BEE, J.V., A.R. JONES and P.R. HOWELL, 1980, J. Mater. Sci. **15**, 337.

BENJAMIN, J.S., 1970, Metallurg. Trans. **1**, 2943.

BENJAMIN, J.S., 1977, Scientific American **234**, 40.

BENJAMIN, J.S., and M.J. BROMFORD, 1977, Metallurg. Trans. **8A**, 1301.

BENJAMIN, J.S., and G.R. SCHELLEN, 1981, Metallurg. Trans. **12A**, 1827.

BERGMAN, B., 1975, Scand. J. Metall. **4**, 97.

BONNET, R., 1980, Ann. Chim. Franç. **5**, 203.

BROOKS, H., 1952, in: Metal Interfaces (ASM, Metals Park, OH) p. 20.

BROWN, L.M., and D.R. CLARKE, 1977, Acta Metall. **25**, 563.

BROWN, L.M., and R.K. HAM, 1971, in: Strenghtening Methods in Crystals, eds. A. Kelly and R.B. Nicholson (Elsevier, New York), p. 9.

BROWN, L.M., and W.M. STOBBS, 1971a, Phil. Mag. **23**, 1185.

BROWN, L.M., and W.M. STOBBS, 1971b, Phil. Mag. **23**, 1201.

BROWN, L.M., and W.M. STOBBS, 1976, Phil. Mag. **34**, 351.

CAILLETAUD, G., and J.L. CHABOCHE, 1979, 3rd Int. Conf. on Mechanical Behavior of Materials (Cambridge Univ. Press).

CAIRNS, R.L., 1974, Metallurg. Trans. **5**, 1677.

CARRY, C., and J.L. STRUDEL, 1975, Scripta Metall. **9**, 731.

CARRY, C., and J.L. STRUDEL, 1976, in: Proc. 4th Int. Conf. on Strength of Metals and Alloys, Nancy (INPL, Nancy) p. 324.

CARRY, C., and J.L. STRUDEL, 1977, Acta Metall. **25**, 767.

CARRY, C., and J.L. STRUDEL, 1978, Acta Metall. **26**, 859.

CARRY, C., S. DERMARKAR, J.L. STRUDEL and B.C. WONSIEWICZ, 1979, Metallurg. Trans. **10A**, 855.

CARRY, C., C. HOUIS and J.L. STRUDEL, 1981, Mém. Sci. Rev. Métallurg. **78**, 139.

CHABOCHE, J.L., 1977, Bull. Acad. Polon. Sci. **25**, 33.

CHAPMAN,P.F., and W.M. STOBBS, 1969, Phil. Mag. **19**, 1015.

CLAVEL, M., and A. PINEAU, 1982, Mater. Sci. Eng. **55**, 157.

CLAVEL, M., C. LEVAILLANT and A. PINEAU, 1980, in: Proc. Conf. on Creep–Fatigue–Environment Interaction, eds. R.M. Pelloux and N.S. Stoloff (AIME, Warrendale, PA) p. 24.

COPLEY, S.M., and B.H. KEAR, 1967, Trans. Met. Soc. AIME **239**, 977.

COPLEY, S.M., B.H. KEAR and G.M. ROWE, 1972, Mater. Sci. Eng. **10**, 87.

CROSSLAND, I.G., and R.B. JONES, 1972, Met. Sci. J. **6**, 162.

CULIE, J.P., G. CAILLETAUD and A. LASALMONIE, 1982, La Récherche Aérospatiale **2**, 51.

CUTLER, C.P., and S.W.K. SHAW, 1979, in: Proc. 5th Int. Conf. on Strength of Metals and Alloys, Aachen, eds. P. Haasen, V. Gerold and G. Kostorz (Pergamon Press, Oxford) p. 1357.

DAVIDSON, J.H., and C. AUBIN, 1982, in: High Temperature Alloys for Gas Turbines, eds. R. Brunetaud *et al.* (D. Reidel, Dordrecht, Holland) p. 853.

DELEURY, M.J., J.R. DONATI and J.L. STRUDEL, 1981, Ann. Chim. Franç. **6**, 59.

DERMARKAR, S., and J.L. STRUDEL, 1979, Proc. Int. Conf. on Strength of Metals and Alloys, Aachen, eds. P. Haasen, V. Gerold and G. Kostorz (Pergamon Press, Oxford)) p. 705.

DERMARKAR, S., and J.L. STRUDEL, 1980, in: Recrystallization and Grain Growth in Multiphase and Particle-Containing Alloys, eds. N. Hansen *et al.* (Risø National Laboratory, Roskilde, Denmark) p. 139.

DOHERTY, R.D., and J.W. MARTIN, 1962, J. Inst. Metals, **91**, 332.

DUPEUX, M., and R. BONNET, 1980, Acta Metall. **28**, 721.

ESCAIG, B., 1974, J. Physique **35**, C7–151.

ESHELBY, J.D., 1954, J. Appl. Phys. **25**, 255.

ESHELBY, J.D., 1961, Prog. Solid Mech. **2**, 89.

FISHER, J.C., E.W. HART and R.H. PRY, 1953, Acta Metall. **1**, 336.

FOREMAN, A.S.E., and M.J. MAKIN, 1966, Phil. Mag. **14**, 911.

GALLAGHER, P.C.J., 1970, Metallurg. Trans. **1**, 2429.

GIBBONS, T.B., and B.E. HOPKINS, 1971, Met. Sci. J. **5**, 233.

GITTUS, J.H., 1975, Proc. Roy. Soc. (London) **A342**, 279.

GLEITER, H., and E. HORNBOGEN, 1965, Phys. Stat. Sol. **12**, 251.

GLEITER, H., and E. HORNBOGEN, 1968, Mater. Sci. Eng. **2**, 285.

GOULD, D., P.B. HIRSCH and F.J. HUMPHREYS, 1974, Phil. Mag. **30**, 1353.

GRANT, N.J., 1983, J. Metals **35**, 20.

GREWAL, M.S., S.A. SASTRI and N.J. GRANT, 1975, Metallurg. Trans. **6A**, 1393.

GUIMIER, A., and J.L. STRUDEL, 1970, Proc. 2nd Int. Conf. on Strength of Metals and Alloys, Asilomar (USA) (ASM, Metals Park, OH) p. 1145.

GUYOT, P., 1962, Acta Metall. **12**, 665.

GUYOT, P., 1971, Phil. Mag. **24**, 987.

GUYOT, P., and R. DEBEIR, 1966, Acta Metall. **14**, 43.

GUYOT, P., and E. RUEDL, 1967, J. Mater. Sci. **2**, 221.

HAASEN, P., and R. LABUSCH, 1979, Proc. 5th Int. Conf. on Strength of Metals and Alloys, Aachen, eds. P. Haasen, V. Gerold and G. Kostorz (Pergamon Press, Oxford) p. 639.

HANSEN, N., and A.R. JONES, 1981, in: Les Traitements Thermiques, 24e Colloque de Métallurgie de Saclay (CEN-Saclay, France), p. 95.

HANSEN, N., and H. LILHOLT, 1971, in: Modern Developments in Powder Metallurgy, ed. H.H. Hausner (Plenum, New York) vol. 5, p. 339.

HART, E.W., 1970, Acta Metall. **18**, 599.

HART, E.W., 1972, Acta Metall. **20**, 275.

HAUSSELT, J.H., and W.D. NIX, 1977, Acta Metall. **25**, 595.

HAZZLEDINE, P.M., 1974, Phil. Mag. **30**, 1327.

HAZZLEDINE, P.M., and P.B. HIRSCH, 1974, Phil. Mag. **30**, 1331.

HIRSCH, P.B., 1957, J. Inst. Metals, **20**, 275.

HIRSCH, P.B., and F.J. HUMPHREYS, 1969, in: Physics of Strength and Plasticity, ed. A.S. Argon (MIT Press, Cambridge, MA) p. 189.

HIRSCH, P.B., and F.J. HUMPHREYS, 1970, Proc. Roy. Soc. (London) **A318**, 45.

HOLBROOK, J.H., and W.D. NIX, 1974, Metallurg. Trans. **5**, 1033.

HORNBOGEN, E., and U. KÖSTER, 1978, in: Recrystallization of Metallic Materials, ed. F. Haessner (Dr. Riederer Verlag, Stuttgart) p. 159.

HUMPHREYS, F.J., 1977, Acta Metall. **25**, 1323.

HUMPHREYS, F.J., 1979a, Acta Metal. **27**, 180.

HUMPHREYS, F.J., 1979b, Met. Sci. **13**, 136.

HUMPHREYS, F.J., 1980, in: Recrystallization and Grain Growth of Multiphase and Particle-Containing Materials, eds. N. Hansen *et al.* (Risø National Laboratory, Roskilde, Denmark) p. 35.

HUMPHREYS, F.J., and P.B. HIRSCH, 1970, Proc. Roy. Soc. (London) **A318**, 73.

HUMPHREYS, F.J., and J.W. MARTIN, 1967, Phil. Mag. **16**, 927.

HUMPHREYS, F.J., and A.T. STEWART, 1972, Surf. Sci. **31**, 389.

JONES, A.R., and N. HANSEN, 1980, in: Recrystallization and Grain Growth of Multiphase and Particle-Containing Alloys, eds. N. Hansen *et al.* (Risø National Laboratory, Roskilde, Denmark) p. 13.

JONES A.R., B. RALPH and N. HANSEN, 1979, Met. Sci. **13**, 140.

JONES, R.L., 1969, Acta Metall. **17**, 229.

KEAR, B.H., and J.M. OBLAK, 1974, J. Physique 35, Colloque **C7**, 35.

KEAR, B.H., and B.J. PIEARCEY, 1967, Metallurg. Trans. **239**, 1209.

KEAR, B.H., A.F. GIAMEI, J.M. SILCOCK and R.K. HAM, 1968, Scripta Metall. **2**, 287.

KEAR, B.H., A.F. GIAMEI, G.R. LEVERANT and J.M. OBLAK, 1969a, Scripta Metall. **3**, 455.

KEAR, B.H., G.R. LEVERANT and J.M. OBLAK, 1969b, Trans. ASM **62**, 639.

KEAR, B.H., J.M. OBLAK and A.F. GIAMEI, 1970, Metallurg. Trans. **1**, 2477.

KOCKS, U.F., 1966, Phil. Mag. **14**, 1629.

KOCKS, U.F., 1969, in: Physics of Strength and Plasticity, ed. A.S. Argon (MIT Press, Cambridge, MA) p. 133.

KURAMOTO, E., and D.P. POPE, 1978, Acta Metall. **26**, 207.

KUTUMBA RAO, V., D.M.R. TAPLIN and P.R. RAO, 1975, Metallurg. Trans. **6A**, 77.

LAGNEBORG, R., 1969, Met. Sci. J. **3**, 161.

LAGNEBORG, R., 1972, Int. Metall. Rev. **17**, 130.

LAGNEBORG, R., and B. BERGMAN, 1976, Met. Sci. **10**, 20.

LASALMONIE, A., and M. SINDZINGRE, 1971, Acta Metall. **19**, 57.

LASALMONIE, A., and J.L. STRUDEL, 1975a, Phil. Mag. **31**, 115.

LASALMONIE, A., and J.L. STRUDEL, 1975b, Phil. Mag. **32**, 937.

LASALMONIE, A., and J.L. STRUDEL, 1980, Ann. Chim. Franç. **5**, 19.

LAUGHLIN, D.E., 1976, Acta Metall. **24**, 53.

LAW, C.C., and A.F. GIAMEI, 1976, Metallurg. Trans. **7A**, 5.

LE HAZIF, R., P. DORIZZI and J.P. POIRIER, 1973, Acta Metall. **21**, 903.

LENEL, F.V., G.S. ANSELL and E.C. NELSON, 1957, Trans. AIME **209**, 117.

LEVERANT, G.R., B.H. KEAR and J.M. OBLAK, 1973, Metallurg. Trans. **4**, 355.

LI, J.C.M., 1963, in: Electron Microscopy and Strength of Crystals, eds. G. Thomas and J. Washburn (Interscience, New York) p. 713.

LILHOLT, H., 1977, Acta Metall. **25**, 571 and 587.

LLOYD, G.J., and J.W. MARTIN, 1978, Scripta Metall. **12**, 217.

LLOYD, G.J., and R.J. McELROY, 1975, Phil. Mag. **32**, 231.

LLOYD, G.J., and R.J. McELROY, 1976, Acta Metall. **24**, 11.

LUND, R.W., and W.D. NIX, 1975, Metallurg. Trans. **6A**, 1329.

LUND, R.W., and W.D. NIX, 1976, Acta Metall. **24**, 469.

MARTIN, J.W., 1980, Micromechanisms in Particle-Hardened Alloys (Cambridge University Press).

MARTIN, J.W., and R.D. DOHERTY, 1976, Stability of Microstructure in Metallic Systems (Cambridge University Press).

McLEAN, D., 1962, in: Mechanical Properties of Metals (Wiley, New York; new edition 1977, Krieger, New York).

McLEAN, D., 1966, Rep. Prog. Phys. **29**, 1.

McLEAN, D., 1976, in: Proc. 4th Int. Conf. on Strength of Metals and Alloys (INPL, Nancy, France) p. 958.

MENON, M.N., and W.H. REIMAN, 1975, Metallurg. Trans. **6A**, 1075.

MIYAZAKI, T., R. NAKAMURA and H. MORI, 1979, J. Mater. Sci. **14**, 1927.

MONTHEILLET, F., 1981, in: Les Traitements Thermiques, 24e Colloque de Métallurgie de Saclay (CEN-Saclay, France) p. 57.

MOSKOWITZ, L.N., R.M. PELLOUX and N.J. GRANT, 1972, quoted by: DAVIDSON, J.H., and C. AUBIN, in: High Temperature Alloys for Gas Turbines, eds. R. Brunetaud *et al.* (D. Reidel, Dordrecht, Holland) p. 853.

MOULD, P.R., and P. COTTERILL, 1967, J. Mater. Sci. **2**, 241.
MULFORD, R.A., and D.P. POPE, 1973, Acta Metall. **21**, 1375.
NES, E., 1976, Acta Metall. **25**, 1323.
OROWAN, E., 1948, in: Symp. on Internal Stresses in Metals and Alloys, London (The Institute of Metals, London, p. 451).
OYTANA, C., P. DELOBELLE and A. MERMET, 1979a, J. Mater. Sci. **14**, 549.
OYTANA, C., MERMET, A. and P. DELOBELLE, 1979b, Proc. 3rd Int. Conf. on Strength of Metals and Alloys, Cambridge (The Institute of Metals, London) p. 203.
PARKER, J.D., and B. WILSHIRE, 1975, Met. Sci. J. **9**, 248.
PEDRON, J.P., and A. PINEAU, 1982, Mater. Sci. Eng. **56**, 143.
PELLOUX, R.M.N, and N.J. GRANT, 1960, Trans. Met. Soc. AIME **218**, 232.
PRAGER, W., 1955, Proc. Inst. Mech. Engrs **169**, 41.
REPPICH, B., 1975, Acta Metall. **23**, 1055.
REPPICH, B., 1982, Acta Metall. **30**, 87.
REPPICH, B., P. SCHEPP and G. WEHNER, 1982, Acta Metall. **30**, 95.
ROBERT, G., J. DESSUS and M.F. FELSEN, 1981, Ann. Chim. Franç, **6**, 209.
RUNKLE, J.C., and R.M. PELLOUX, 1979, Amer. Soc. Testing Mater. Spec. Tech. Publ. **675**, 501.
SANDSTRÖM, R., 1980a, Z. Metallk **71**, 681.
SANDSTRÖM, R., 1980b, in: Recrystallization and Grain Growth of Multiphase and Particle-Containing Materials, eds. N. Hansen *et al.* (Risø National Laboratory, Roskilde, Denmark) p. 45.
SAXL, I., and F. KROUPA, 1972, Phys. Stat. Sol. **A11**, 167.
SHAMBLEN, C.E., R.E. ALLEN and F.E. WALKER, 1975, Metallurg. Trans. **6A**, 2073.
SHEWFELT, R.S.W., and L.M. BROWN, 1973, Proc. 3rd Int. Conf. on Strength of Metals and Alloys, Cambridge (The Institute of Metals, London) p. 311.
SHIMANUKI, Y., 1981, Trans. Japan Inst. Met. **22**, 17.
SOLOMON, A.A., and W.D. NIX, 1970, Acta Metall. **18**, 863.
STARKE, Jr., E.A., T.H. SANDERS and I.G. PALMER, 1981, J. Metals **32**, 24.
STEVENS, R.A., and P.E. FLEWITT, 1981a, Acta Metall. **29**, 867.
STEVENS, R.A. and P.E. FLEWITT, 1981b, in: Creep and Fracture of Engineering Materials and Structures, eds. B. Wilshire and D.R.J. Owen (Pineridge Press, Swansea) p. 187.
STOBBS, W.M., 1973, Phil. Mag. **27**, 1073.
STOLOFF, N.S., 1972, in: The Superalloys, eds. C.T. Sims and W.C. Hagel (Wiley, New York) p. 79.
STRUDEL, J.L., 1978, J. Microsc. Spéctrosc. Electron. **3**, 337.
STRUDEL, J.L., 1980, in: Les Traitements Thermomécaniques, 24e Colloque de Métallurgie de Saclay (CEN-Saclay, France) p. 81.
THREADGILL, P.L., and B. WILSHIRE, 1974, Met. Sci. J. **8**, 117.
TIEN, J.K., and S.M. COPLEY, 1971, Metallurg. Trans. **2**, 543
VICKERS, W., and D.P. GREENFIELD, 1968, J. Nucl. Mater. **27**, 73.
WEATHERLY, G.C., and R.B. NICHOLSON, 1968, Phil. Mag. **17**, 801.
WILCOX, B.A., and A.H. CLAUER, 1966, Trans. Met. Soc. AIME **236**, 570.
WILCOX, B.A., and A.H. CLAUER, 1972, Acta Metall. **20**, 743.
WILLIAMS, K.R., and B. WILSHIRE, 1973, Met. Sci. J. **7**, 176.
ZENER, C., quoted by: C.S. SMITH, 1948, Trans. AIME **175**, 47.

Further reading

Argon, A.S., 1975, Constitutive Equations in Plasticity (MIT Press, Cambridge, MA).
Balakrishna Bhat, T., and V.S. Arunachalam, Strengthening Mechanisms in Alloys, in: Alloy Design, eds. S. Ranganathan *et al.,* Proc. Indian Acad. Sci. (Eng. Sci.) 3 (1980) 275.
Gittus, J., 1975, Creep, Viscoelasticity and Creep Fracture in Solids (Appl. Science Publ., London).

Hansen, N., A.R. Jones and T. Leffers, eds., 1980, Recrystallization and grain growth of Multiphase and Particle-Containing Materials (Risø National Laboratory, Roskilde, Denmark).

Hansen, N., A. Horwell, T. Leffers and H. Lilholt, 1981, Deformation of polycrystals: Mechanisms and Microstructures (Risø National Laboratory, Roskilde, Denmark).

Kelly, A., and R.B. Nicholson, 1971, Strengthening methods in crystals (Elsevier, Amsterdam).

Martin, J.W., 1980, Micromechanisms in Particle-Hardened Alloys (Cambridge University Press).

McClintock, F.A., and A.S. Argon, 1966, Mechanical behavior of materials (Addison–Wesley, Reading, MA).

Poirier, J.P., 1976, Plasticité à Haute Température des Solides Cristallines (Edit. Eyrolles, Paris).

Sims, C.T., and W.C. Hagel, 1972, The superalloys (Wiley, New York).

Wilshire, B., and D.R.J. Owen, 1981, Creep and Fracture of Engineering Materials and Structures (Pineridge Press, Swansea, UK).

CHAPTER 23

FRACTURE

Robb M. Thomson

Center for Materials Science
National Bureau of Standards
Washington, DC 20234, USA

R.W. Cahn and P. Haasen, eds.
Physical Metallurgy; third, revised and enlarged edition
© *Elsevier Science Publishers BV, 1983*

1.   *Introduction and fracture overview*

In this chapter, a fundamental approach to fracture will be followed in order to develop a few central ideas about why any given material is ductile or brittle, and how the basic lattice defects of the material cooperate to give rise to these two general classes of behavior. The analytical discussion will tend to focus on relatively brittle phenomena, partly because any idealization of fracture leads one first to the consideration of *brittle fracture*, and partly because the analysis for brittle fracture is better understood, and therefore more accessible. Nevertheless, the more ductile aspects of fracture, though more difficult to analyze quantitatively, will be kept in view throughout, because of the inherently ductile tendencies of most metals.

There are two extreme poles of material failure under stress. In the first, fig. 1, the material *necks down* under tension like taffy (single crystals of pure copper are an example) in a continuous plastic manner, until the last atoms in the narrowing neck come apart and the material splits into two pieces. This type of failure is usually termed *ductile rupture*. In the other case (fig. 2), the material suddenly and catastrophically fractures with no discernible plastic flow either preceding or during the failure event. Examples of the latter are glass and silicon at low temperature. These two poles are useful to keep in mind, because they illustrate how two crystal or lattice defects – dislocations and atomically sharp cleavage cracks – are responsible, ultimately, for the mechanical failure of crystalline materials. In the first instance, only dislocations are involved, while in the second only a sharp crack participates in the failure event. Materials scientists are not accustomed to thinking of brittle cracks (fig. 3) as one of the fundamental crystal defects which have proven so useful as constructs for understanding the structure-sensitive material properties. However, in this chapter, we will put such cracks on the same conceptual footing as dislocations because, like dislocations, they answer one of the two most fundamental questions in the mechanical behavior of materials. Dislocations were postulated to explain why the yield strength of a solid can be much lower than the theoretical shear strength of the atomic bonds of the solid. Likewise, cracks provide the answer

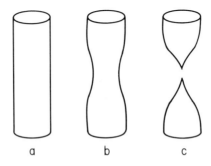

a   b   c

Fig. 1. A rod of ideally ductile material when pulled develops a region of plastic instability which finally thins uniformly down to a sharp point.

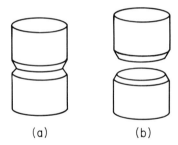

(a)   (b)

Fig. 2. An ideally brittle material when pulled separates suddenly by cleavage with no prior or simultaneous deformation. A notch is shown which localizes the plane of fracture.

to why fracture occurs in materials at average stress levels well below the theoretical tensile strength of the atomic bonds of the material. The reason is that a pre-existing crack in the material magnifies or concentrates the low applied stress to a value at the tip of the crack which may be many orders of magnitude higher. When this magnified stress at the tip reaches the maximum atomic bond strength, the crack will propagate through the solid and fracture it. From this chapter, we hope the reader will find that the otherwise very confusing subject of mechanical failure is best understood by viewing it as an interplay between, and sometimes a competition between dislocations and cracks.

The *stress concentration* referred to around a slit crack is shown in fig. 4 where the analogy of the magnetic lines of force around a similar slit in a magnet is striking. Physically, the crack must act as a stress concentrator, because every horizontal plane (line in two dimensions) in the figure must carry the total force imposed on the two external surfaces. Since the cut line representing the crack means there is less material on the cleavage plane defined by the cut to carry the applied force, the force

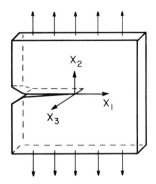

Fig. 3. A brittle crack in a material is viewed as a mathematical cut in the medium across which the atom bonds have been cut. For mathematical simplicity the crack line will be assumed to be straight and to lie along the x_3 axis. The cleavage plane is the negative $x_3 x_1$ plane. External stresses are also exerted on the external surfaces of the material in an arbitrary manner in order to hold the crack open.

References: p. 1548.

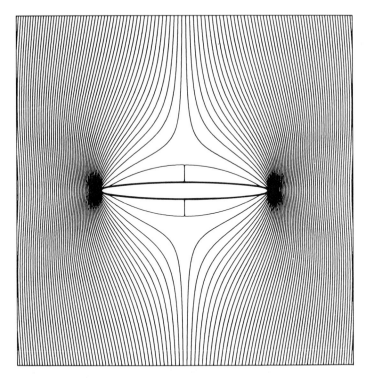

Fig. 4. Stresses are concentrated near the tip of a crack in a way suggestive of magnetic lines of force around a slit. In this diagram, the lines are computer-generated to simulate crack stress lines quantitatively. (Courtesy Prof. P. NEUMANN.)

per unit area (or stress) on this plane outside the crack must be higher. The stress is especially concentrated near the source of the "force transmission difficulty" at the crack tips, and indeed we will show that for a mathematically sharp cut, the stresses become singular at a tip. If the tip is rounded, the maximum stress is finite, and if the tip has a radius of atomic size, the maximum average stress which can be sustained in the material is that which begins to break the bonds at the tip in a progressive fashion as alluded to above (fig. 5).

An interesting contrast between dislocations and cracks follows from the fact that, unlike a dislocation, a crack cannot exist in a material without the presence of "external" stresses. This is easily seen from fig. 5, where the atomic bonds across the cleavage plane will close up the slit unless a counteracting set of external forces is applied. This "external" stress may of course be supplied by misfitting grains, precipitates, etc. in the solid, as well as by actual external forces applied to the surface of the specimen.

Although the atomically sharp brittle crack will play the dominant conceptual role in this chapter, other kinds of cracks also exist in materials. In particular, we

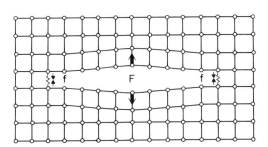

Fig. 5. Diagram of a crack in a two-dimensional square lattice with external forces exerted at the center of the crack. Bonds at the two crack tips are nonlinear. The nonlinear attractive forces at the tip are labeled f, and the external applied forces at the center of the crack are labeled F.

will find in the next section that in ductile materials, a sharp crack will open up under appropriate conditions into a wedge shape, with a finite opening angle at its tip. However, it is the sharp crack which possesses the well known dynamical properties of cleavage, while the wedge-shaped crack and its derivatives are associated with a complex form of ductile failure. Thus, the sharp crack is the only one which we shall elevate to the status of fundamental crystal defect.

Cracks are formed in materials in a variety of ways by any mechanism which causes a sufficiently high stress, locally. Examples are high-density dislocation structures formed near surfaces during fatigue, stress concentrations in brittle particles due to the strain incompatibility at the interface between the particle and the deforming metal matrix, grain-boundary triple points, etc. In their early stages these cracks in many cases are sharp.

The materials scientist will find in studying this chapter that the problem of fracture deeply involves several other disciplines, and that this fact is both the cause of stimulation because of the breadth of ideas involved, and also daunting because these ideas often involve unfamiliar subjects. In particular, fracture has from the first involved continuum fracture mechanics on a deep level. The fundamental reason why the mechanics of the problem cannot be ignored is that a crack cannot exist in a material without an external stress, and the way in which this stress is transmitted to the crack tip is the subject of *continuum mechanics*. The necessity of the external force is related to the long-ranged $1/\sqrt{r}$ stress field surrounding the crack, which means that the crack is not a localized defect in the same sense as a vacancy or even a dislocation is. Fortunately, for long cracks, the atomic configuration around the tip depends only on a quantity termed the *stress intensity*, K, and not on other mechanical and geometric details, but the way the plastic zone interfaces to the immediate region around the crack tip is both a materials problem and a mechanics problem. Likewise, fracture deeply involves surface physics and chemistry, because the tip region of the crack is an incipient surface, and reactions at the tip with external environments involve ideas and processes which are brought over from those fields. The concept of *fracture mechanics* is defined in §9.1.

References: p. 1548.

Finally, the question of the structure of the crack tip which determines such fundamental properties as whether dislocations can be emitted from it, or how it interacts with external chemical species, depends upon the kind of atomic bonding involved in the material. Unfortunately, the force laws of metals, and especially transition metals, are not understood for the type of extreme distortions typical of a crack tip. Force laws in metals are now under intensive study in solid state physics, and as answers becomes available, they will prove very useful to the materials scientist studying fracture. Thus, fracture is a subject which depends for its progress on the contributions from several adjacent fields, and the materials scientist cannot ignore these connections.

For other treatments of fracture on approximately the level observed here, the reader is referred to the books by LAWN and WILSHAW [1975], KNOTT [1973], and KELLY [1966]. The chapter on "Dislocations and Cracks" by SMITH [1979] in vol. 4 of Nabarro's series, "Dislocations in Solids", is also recommended. For an excellent modern, but qualitative, statement of the general status of *fracture mechanics* and its application in design and service, see the 1979 Discussion of the Royal Society, organized by H. FORD *et al.* [1981]. An up-to-date and very practical guide for the practicing metallurgist will be found in the book, "Applications of Fracture Mechanics for Selection of Metallic Structural Materials", by CAMPBELL *et al.* [1982]. More advanced treatments will be found in the various articles in vols. 1 and 2 of "Fracture" edited by LIEBOWITZ [1968] – see especially the article by BILBY and ESHELBY in vol. 1. An excellent review of fracture fundamentals applied to brittle materials has been written by LAWN [1983]. The various proceedings of the conference series, International Conference on Fracture and the International Conference on the Mechanical Behavior of Materials contain review articles in specific areas. The conference proceedings of the NATO Conference on the "Atomistics of Fracture", edited by LATANISION [1982], is also recommended. For a summary statement of elasticity theory, see Theory of Dislocations by HIRTH and LOTHE [1981].

2. *Qualitative and observational aspects of fracture*

Before digging into the quantitative discussions of later sections, a selection of fractographs and other material will be presented here to provide the reader with an observational basis for further discussion.

2.1. Fracture modes

We begin with a definition of terms which describe the macroscopic stress state of the fracture in terms of the three primary modes of fracture. A crack is defined by a two-step process. First, as in fig. 3, a cut is made in the medium, which becomes the *cleavage plane*, and then forces are exerted on the external surfaces of the specimen which define the particular fracture mode depicted in fig. 6. In *mode* I, the stress is a

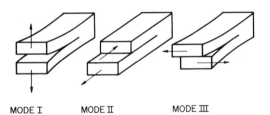

MODE I   MODE II   MODE III

Fig. 6. The three prototype modes of fracture. In each case, the crack is formed by making a planar slit in the material, with the crack line along the edge of the cut. In mode I, the opening mode, the force is exerted normal to the cleavage plane. In mode II, the force is in the plane of the crack normal to the crack line. In mode III, or antiplane strain, the force is in the plane along the crack line. Modes II and III are shear cracks.

tensile stress with the principal stress axis normal to the cleavage plane. Mode I is the only mode leading to physical fracture, because unless the external stress physically separates the two surfaces on the cleavage plane, rewelding would occur even after the stress is applied. In *mode* II, the stress is a *shear* parallel to the cut in the x_1-*direction*. (We retain in future the coordinate system of fig. 3.) In *mode* III, the stress is a *shear* parallel to the cut in the *antiplane* or x_3-direction. Mode III is important because antiplane strain is associated with a particularly simple analysis. We shall thus make extensive use of this analysis to describe results for cracks in a generic sense. Of course, most actual cracks are not pure cases; the fracture surfaces are not flat, and the crack fronts are not straight. Thus, a crack may be produced which is primarily mode II or III, but with enough mode I present to separate the cleavage plane. Needless to say, curved cracks possessing nonplanar fracture surfaces are much more difficult and often impossible to handle mathematically, at least analytically, so that the pure modes listed above will be featured almost exclusively in this discussion.

2.2. Fractographic observations

A desirable steel exhibits high *toughness*, which means that it requires a large energy to pull it apart in fracture. Such a steel has a fracture surface which exhibits a noncrystallographic "fibrous" quality on a microscopic scale as shown in fig. 7. The surface is composed of microdimples, which are the result of holes forming ahead of the main crack as shown in fig. 8. These holes are thought to initiate in practical alloy steels primarily at the site of precipitate particles in the matrix as illustrated in fig. 7. Often the large voids in the medium are connected by bands of intense shear, as shown in fig. 9. Within these shear bands, void formation also occurs, but typically at a much smaller average size than in the case of the larger precipitates. The initiating sites of the smaller voids are at the time of writing still a matter of controversy. In some cases, they are associated with precipitates – but in this case particles of very small size (100–1000 Å) (THOMPSON and WEIHRAUCH [1976]).

References: p. 1548.

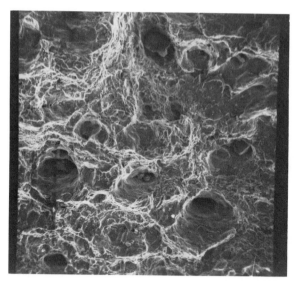

Fig. 7. Ductile fracture surface, showing the final dimpled structure of hole growth. Precipitate particles are visible in some of the dimples which served as nucleation sites for the holes. (Courtesy C. INTERRANTE.)

However, holes have been observed by GARDNER *et al.* [1977] to nucleate in very pure metals without the help of precipitates, and the small voids in impure steels may be produced homogeneously also. Figure 10 illustrates two forms of relatively brittle fracture in a high-strength steel under embrittling conditions. In one case, the fracture path follows the grain boundaries of the steel, while in the other, the cleavage path is transgranular.

In the next set of observations, phenomena characteristic of more ideal conditions are illustrated. Figure 11a shows results for completely brittle cracks by LAWN *et al.* [1980] in Si single crystals. The crack was injected at room temperature, where dislocations are immobile, by an indentation technique. In the figure, the crack is

Fig. 8. Side view of a growing fracture in a thin film of gold. Holes are shown nucleating ahead of the crack; see LYLES and WILSDORF [1975]. (Courtesy H.G.F. WILSDORF).

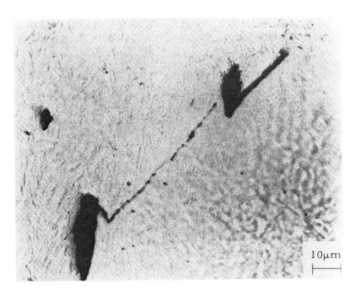

Fig. 9. Hole growth at precipitates in ductile fracture. Between the large holes, shear bands develop consisting of large numbers of much smaller holes. (After Cox and Low [1974]).

shown in cross-section. No dislocations whatsoever are present, proving that completely brittle atomically sharp cracks are possible. If cracks are injected into the same material at a temperature of 500°C, where the dislocations are mobile, then considerable dislocation activity is evident (fig. 11b). Whether these dislocations were emitted by the crack in a spontaneous fashion after it arrested, or whether the dislocations were dragged along in the stress field of the crack from elsewhere, is not yet established.

In fig. 12, photographs by VEHOFF *et al.* [1981] in Fe (+Si) show that the natural form of cracks in this material is wedge-shaped, and that propagation of the crack takes place by dislocation emission on the two complementary slip planes which interesect at the tip. When the crack is forced ahead either by raising the strain rate of the deforming machine or when hydrogen is introduced in the atmosphere, then the angle of the wedge narrows, which implies that cleavage at the crack tip can coexist with dislocation emission. Thus, in a material like iron, cleavage and emission at the crack tip are closely balanced processes, which can be tilted one way or the other by externally controlled circumstances.

Figure 13 shows results of KOBAYASHI and OHR [1981] in foils of Cu (similar results are obtained for Ni and steel). In this case, the foil "slides off" on its slip plane which is at an oblique angle to the plane of the foil. When the foil locally shears completely through its thickness, a cracklike artifact is generated (a mode III crack of zero length) which grows by dislocation emission and translation. This figure illustrates that the crack has a substantial region near its tip (termed an elastic enclave, or a dislocation-free zone) where no dislocations are present.

References: p. 1548.

Fig. 10. Appearance of relatively brittle fracture in steel. Top: the fracture is along the grain boundaries (intergranular). Bottom: the fracture is transgranular. (Courtesy C. INTERRANTE.)

2.3. The basis for fracture science

The illustrations selected above exemplify the major experimental findings to be used as a basis for a science of fracture. The reader is referred to vol. 9 of "The

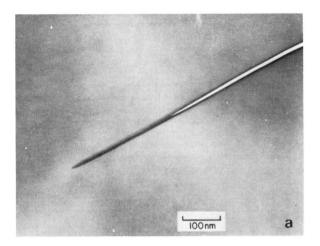

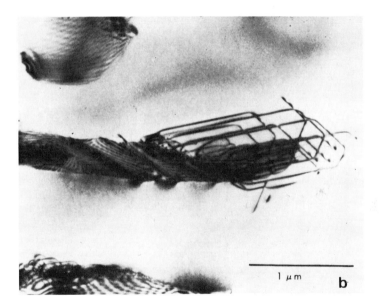

Fig. 11. (a) An electron microscope picture of a fully brittle crack in Si. In the photo, the crack is seen edge on. (b) Cracks formed in Si at 500°C are associated with dislocations as shown (see LAWN *et al.* [1980]). (Photos courtesy B. HOCKEY.)

Metals Handbook" (SHUBAT [1974]) and to vol. 1 of the "Fracture" series (LIEBO-WITZ [1958]) for a compendium and atlas of fractographic examples which go far beyond what we can display here. The illustrations demonstrate clearly enough, however, that fracture in materials is most often a very complex phenomenon which depends upon a large number of material and external variables, and that the task of

References: p. 1548.

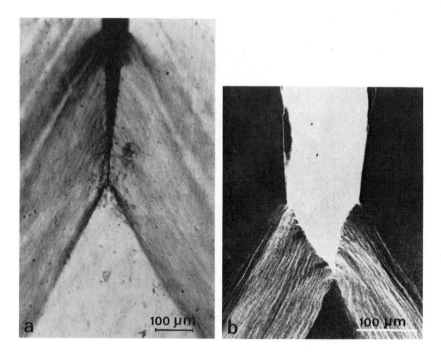

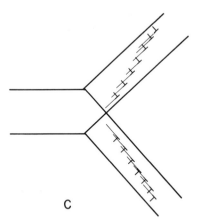

Fig. 12. In (a) sharp cracks are formed in a hydrogen atmosphere in Fe (+ Si). In (b) the crack opens entirely by dislocation slide-off at the crack tip. In both cases, dislocations are emitted from the crack tip, and form slip bands. In (a), cleavage must also be present. The angle of the wedge opening in (b) is defined by the intersection angle of the two slip planes with the crack plane. In (c) the dislocation configuration generating the crack is illustrated in terms of alternate emission of dislocations on the two complementary slip planes. (After VEHOFF and NEUMANN [1980], courtesy P. NEUMANN.)

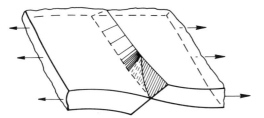

Fig. 13. (a) Dislocations emitted from the tip of a crack in a foil of copper. The crack has grown from the left. Immediately ahead of the crack is a region free of dislocations. The pile-up of dislocations is seen extending on the cleavage-slip plane to the right. In b) the crackdislocation geometrical configuration is illustrated. (After KOBAYASHI and OHR [1981], courtesy S.M. OHR.)

sorting out the fundamental phenomena which are critical to the fracture event is not an easy nor straightforward one. However, primarily on the basis of these observations, we shall state a number of qualitative conclusions as a point of departure for the sections which follow:

(a) In practice, cracks are most often flaws of very complex structure which only approximate the three idealized two-dimensional modes of fig. 6. However, because of the necessity of using models which can be analyzed and described in accessible mathematics, we are led to the study of these idealized modes.

(b) The two major poles, of brittle and ductile fracture, are discernible in the fractography of metals (figs. 7 and 10).

(c) The ubiquitous presence of dislocation sources in metals assures that copious dislocations are generated to accompany the fracture process in all cases, even when the fracture is brittle. Only in such exceptional (nonmetallic) cases as Si at low temperature are ideally brittle fractures found (fig. 11). In addition to considerations of the individual crack, since the stress fields of cracks and dislocations should interact strongly, the point of departure for a theoretical development in the next section is to investigate the extent to which these interactions may govern the response of the total configuration to an external stress. This development, for simplicity, will focus on the behavior of sharp cracks.

(d) Although the extreme pole of ductile failure is the macroscopic rupture by necking depicted in fig. 1, a more localized form of ductile failure is also possible which has the macroscopic appearance of a fracture. Pure forms of such ductile fracture occur when dislocation emission at the crack tip generates a mode I wedge-shaped crack (fig. 12) or a mode III shear crack (fig. 13). In both these cases

References: p. 1548.

again, the dislocation interactions with the underlying "crack" should be a key to the overall behavior. However, the criterion for dislocation emission, in the form of a local "fracture criterion" at the crack tip, will involve different physical ideas from the fracture criterion for a sharp cleavage crack as envisioned in the last paragraph. Such a treatment of ductile fracture has been developed for the mode III shear crack, and will be presented in the following sections, but the wedge case has not yet been analyzed in these terms.

(e) In practice, however, ductile fracture is characterized by void nucleation and growth as discrete events ahead of the main crack (figs. 7–9). The larger voids are initiated first at precipitates in practical alloys. Shear bands of intense localized shear associated and coincident with sheets of very small voids then connect the large voids. Final failure occurs by purely plastic rupture processes operating between the voids and the main crack. Clearly, this (mixed) form of failure is very complex, and we shall only be able to address it in a qualitative way, and only after development of ideas conforming to the simpler pure forms of brittle and ductile fracture.

(f) A given material can be influenced to respond over a range of the ductile–brittle axis by varying the external environment (fig. 12), the strain rate, or the temperature. Internal metallurgical variables also have the same effect. Thus, we shall be very sensitive in the following sections to the features in the models to be developed, which may lead to the breakdown of either the brittle or ductile poles in favor of the opposite pole of behavior.

3.   Elastic analysis of cracks

3.1.   Stress analysis

In this section we derive some of the principal results of elasticity as applied to slit cracks. We will derive some results for the antiplane strain (mode III), and quote results for mode I and II as appropriate. The analysis of mode III is much simpler than for the other two cases, and since many of the results which are possible to obtain quite easily in mode III can be taken over with obvious modifications to modes I and II, we shall restrict our formal analysis to this case. The difference between mode III and mode I cracks is analogous to the difference between screw dislocations (antiplane strain) and edge dislocations.

In antiplane strain, the displacement, u, is totally in the x_3 direction, and is a function of both x_1 and x_2, but not of x_3. In this case, the only nonzero stress components are σ_{32} and σ_{31} (and their symmetrical components σ_{23} and σ_{13}), and likewise for strain ϵ_{32} and ϵ_{31}. Hooke's Law is written as

$$\sigma_{3j} = \mu u_{3,j}, \quad j = 1,2, \tag{1}$$

where μ is the shear modulus (see also the table of symbols at the end of the chapter). In this and the following equations we shall define the comma in eq. (1) as

$u_{3,j} \equiv \partial u_3/\partial x_j$, and shall also use the "summation convention" wherein repeated indices in a product are summed over: $u_i u_i \equiv \Sigma_i u_i u_i$. Two-dimensional problems lend themselves to complex variable notation, and our expressions will be simplified by its use. Thus, writing $x_1 + i x_2 = z$, the stress tensor is written as a complex "vector":

$$\sigma(z) = \sigma_{32}(x_1, x_2) + i\sigma_{31}(x_1, x_2). \tag{2}$$

Since u has only a single component, it is convenient to introduce a somewhat redundant complex function, $w(z)$, related to u by the definition

$$\text{Im}[w(z)] = u_3(x_1, x_2). \tag{3}$$

Here "Im" means "the imaginary part of". Similarly, "Re" will mean "the real part of". In this notation, Hooke's Law is written:

$$\sigma(z) = \mu \frac{dw}{dz}. \tag{4}$$

The *strain energy density function*, W, takes the simple form

$$W = |\sigma^2|/2\mu. \tag{5}$$

The equations of elastic equilibrium become the simple Laplace equation

$$\nabla^2 u_3 = \frac{\partial^2 u_3}{\partial x_1^2} + \frac{\partial^2 u_3}{\partial x_2^2} \equiv u_{3,ii} = 0. \tag{6}$$

In the theory of complex functions, when eq. (6) is satisfied, it is known that a general solution is any regular complex function of z. It is then only necessary to choose from among these functions one which satisfies the appropriate boundary conditions of the particular problem. In the case at hand, a function which is finite at infinity is sought, which satisfies the boundary condition of elasticity at a free surface. This boundary condition states that no force is transmitted across the surface,

$$\sigma_{ij} n_j = 0, \tag{7}$$

for a surface, S, with normal components n_j. If the crack is an infinite cut along the negative x_1 axis, with the tip of the crack at the origin (see fig. 3), then in complex notation

$$\text{Re } \sigma = 0, \quad x_1 < 0, \quad x_2 = 0. \tag{8}$$

Square root functions are known to have special properties along the negative x_1 axis, and indeed a stress function proportional either to $\sqrt{z}$ or $1/\sqrt{z}$ would satisfy eq. (8). If we also require that σ be finite at infinity, then we choose

$$\sigma = \frac{K}{(2\pi z)^{1/2}} = \frac{K}{(2\pi r)^{1/2}} e^{-i\theta/2} \tag{9}$$

for the crack solution. That eq. (9) satisfies eq. (8) is easily verified. The alternative

References: p. 1548.

polar form of eq. (9) is often useful. K is a constant of proportionality, which is a measure of the magnitude of the stress-singularity at the crack tip, and is called the *stress intensity factor*. The displacement field, w, related to σ by eq. (4) is:

$$w = \frac{K}{\mu}\left(\frac{2z}{\pi}\right)^{1/2}.$$
(10)

The expression (9) demonstrates the basic result for cracks that they possess a stress concentration singularity of the form $1/\sqrt{z}$. The reader will note that the strain-energy density function, eq. (5), integrated over all space is linearly divergent at infinity, showing that the crack is not a localizable defect. The non-localizability is related to the fact that the crack must be associated with a long-range stress, such as a finite constant stress at infinity. Unfortunately, this long-range stress is not explicit in eq. (9), and the reason is that the cleavage plane of the crack is infinite in extent. To explore this further, it is necessary to discuss the case of a finite crack, explicitly.

Consider a finite cut on the x_1-axis from $x_1 = -a$ to $x_1 = +a$. Then we seek a complex function, w, which has the limiting form $(z - a)^{1/2}$ as $z \to a$ and $(z + a)^{1/2}$ as $z \to -a$. A simple complex function which has this property is the product of these two limiting forms,

$$w = A(z^2 - a^2)^{1/2}$$
(11)

Taking the derivative of eq. (11) to find the stress,

$$\sigma = \mu a \frac{z}{(z^2 - a^2)^{1/2}}.$$
(12)

The reader can verify directly that the boundary condition $\mathrm{Re}(\sigma) = 0$ is satisfied on the cut from $-a$ to $+a$. We make the connection to eq. (9) by requiring σ to take the form $\sigma(z) \to K/[2\pi(z - a)]^{1/2}$ near $z = a$ and write eq. (12) in terms of K:

$$\sigma = \frac{K}{(\pi a)^{1/2}} \frac{z}{(z^2 - a^2)^{1/2}}.$$
(13)

In the limit, the solution (13) has the desired finite uniform stress at $z \to \infty$, and the stress at infinity has the form

$$\sigma_\infty = \frac{K}{(\pi a)^{1/2}}.$$
(14)

Since in the limit for $z \to \infty$, σ is real, $\sigma_\infty = (\sigma_{32})\infty$. Equation (14) yields the desired relation between the externally applied stress and the stress-intensity factor. This relation involves the crack length in a characteristic way.

The solutions (11)–(14) represent a particular choice of complex functions which satisfy the boundary conditions on the crack, and which follow from the assumption of a particular form of externally applied stress. Other choices are possible; for example, the external stress may be applied at the midpoint of the crack itself, or at some other arbitrary place on the surface of the medium. In all cases, the stress at the crack tip is characterized by a $1/\sqrt{z}$ field with a stress intensity factor, K. By solving the elasticity problem, K can be related to the external force, resulting in some kind of relation such as eq. (14) which depends upon the geometry of the crack, the geometry of the external surfaces of the sample, and the distribution of the applied force. The case of eq. (14) represents a particularly simple choice for these geometry-dependent factors. Elastic calculations of K for a variety of special cases have been tabulated by TADA et al. [1973].

One other interesting fact is implicit in eq. (13). Although there is a $1/\sqrt{z}$ singularity at $\pm a$, the total stress field is not of the form $1/\sqrt{z}$, and from the standpoint of function theory should be represented as a power series expanded at, say $z = +a$, with $1/(z-a)^{1/2}$ as the first term in the expansion. This is a general result, and is forced by the fact that the stress must not only generate a singularity at the crack tip, but also describe the external loading arrangements at large distance from the crack tip, which vary from case to case.

Having derived elastic results for mode III, we summarize below the corresponding elastic equations for all modes. In these equations, the complex function format is dropped, and this set of equations, (15), reverts to the traditional real variables, x, y, and z or, correspondingly, r, θ and z.

Equations for Mode I:

$$\left.\begin{array}{c}\sigma_{xx}\\\sigma_{yy}\\\sigma_{xy}\end{array}\right\} = \frac{K_I}{(2\pi r)^{1/2}} \cdot \left\{\begin{array}{l}(\cos\theta/2)[1 - \sin(\theta/2)\sin(3\theta/2)]\\(\cos\theta/2)[1 + \sin(\theta/2)\sin(3\theta/2)]\ ;\\\sin(\theta/2)\cos(\theta/2)\cos(3\theta/2)\end{array}\right.$$

$$\left.\begin{array}{c}\sigma_{rr}\\\sigma_{\theta\theta}\\\sigma_{r\theta}\end{array}\right\} = \frac{K_I}{(2\pi r)^{1/2}} \cdot \left\{\begin{array}{l}(\cos\theta/2)[1 + \sin^2(\theta/2)]\\\cos^3(\theta/2)\\\sin(\theta/2)\cos^2(\theta/2)\end{array}\right. ;$$

$$\sigma_{zz} = \nu'(\sigma_{xx} + \sigma_{yy}) = \nu'(\sigma_{rr} + \sigma_{\theta\theta}),$$

$$\sigma_{xz} = \sigma_{yz} = \sigma_{rz} = \sigma_{\theta z} = 0;$$

$$u_z = -(\nu''z/E)(\sigma_{xx} + \sigma_{yy}) - (\nu''z/E)(\sigma_{rr} + \sigma_{\theta\theta});$$

$$\left.\begin{array}{c}u_x\\u_y\end{array}\right\} = \frac{K_I}{2E}\left(\frac{r}{2\pi}\right)^{1/2} \cdot \left\{\begin{array}{l}(1+\nu)[(2\kappa-1)\cos(\theta/2) - \cos(3\theta/2)]\\(1+\nu)[(2\kappa+1)\sin(\theta/2) - \sin(3\theta/2)]\end{array}\right. ;$$

$$\left.\begin{array}{c}u_r\\u_\theta\end{array}\right\} = \frac{K_I}{2E}\left(\frac{r}{2\pi}\right)^{1/2} \cdot \left\{\begin{array}{l}(1+\nu)[(2\kappa-1)\cos(\theta/2) - \cos(3\theta/2)]\\(1+\nu)[-(2\kappa+1)\sin(\theta/2) + \sin(3\theta/2)]\end{array}\right. \tag{15a}$$

References: p. 1548.

Equations for Mode II:

$$\left.\begin{array}{c}\sigma_{xx}\\\sigma_{yy}\\\sigma_{xx}\end{array}\right\}=\frac{K_{II}}{(2\pi r)^{1/2}}\cdot\left\{\begin{array}{l}-(\sin\theta/2)[2+\cos(\theta/2)\cos(3\theta/2)]\\\sin(\theta/2)\cos(\theta/2)\cos(3\theta/2)\\(\cos\theta/2)[1-\sin(\theta/2)\sin(3\theta/2)]\end{array}\right.\ ;$$

$$\left.\begin{array}{c}\sigma_{rr}\\\sigma_{\theta\theta}\\\sigma_{r\theta}\end{array}\right\}=\frac{K_{II}}{(2\pi r)^{1/2}}\cdot\left\{\begin{array}{l}(\sin\theta/2)[1-3\sin^2(\theta/2)]\\-3\sin(\theta/2)\cos^2(\theta/2)\\(\cos\theta/2)[1-3\sin^2(\theta/2)]\end{array}\right.\ ;$$

$$\sigma_{zz}=\nu'(\sigma_{xx}+\sigma_{yy})=\nu'(\sigma_{rr}+\sigma_{\theta\theta}),$$

$$\sigma_{xz}=\sigma_{yz}=\sigma_{rz}=\sigma_{\theta z}=0;$$

$$u_z=-(\nu''z/E)(\sigma_{xx}+\sigma_{yy})=-(\nu''z/E)(\sigma_{rr}+\sigma_{\theta\theta}).\tag{15b}$$

$$\left.\begin{array}{c}u_x\\u_y\end{array}\right\}=\frac{K_{II}}{2E}\left(\frac{r}{2\pi}\right)^{1/2}\cdot\left\{\begin{array}{l}(1+\nu)[(2\kappa+3)\sin(\theta/2)+\sin(3\theta/2)]\\-(1+\nu)[(2\kappa-3)\cos(\theta/2)+\cos(3\theta/2)]\end{array}\right.\ ;$$

$$\left.\begin{array}{c}u_r\\u_\theta\end{array}\right\}=\frac{K_{II}}{2E}\left(\frac{r}{2\pi}\right)^{1/2}\cdot\left\{\begin{array}{l}(1+\nu)[-(2\kappa-1)\sin(\theta/2)+3\sin(3\theta/2)]\\(1+\nu)[-(2\kappa+1)\cos(\theta/2)+3\cos(3\theta/2)]\end{array}\right.\ ;$$

Equations for Mode III:

$$\sigma_{xx}=\sigma_{yy}=\sigma_{rr}=\sigma_{\theta\theta}=\sigma_{zz}=0,\qquad\sigma_{xy}=\sigma_{r\theta}=0;$$

$$\left.\begin{array}{c}\sigma_{xz}\\\sigma_{yz}\end{array}\right\}=\frac{K_{III}}{(2\pi r)^{1/2}}\cdot\left\{\begin{array}{l}-\sin(\theta/2)\\\cos(\theta/2)\end{array}\right.\ ;$$

$$\left.\begin{array}{c}\sigma_{rz}\\\sigma_{\theta z}\end{array}\right\}=\frac{K_{III}}{(2\pi r)^{1/2}}\cdot\left\{\begin{array}{l}\sin(\theta/2)\\\cos(\theta/2)\end{array}\right.\ ;$$

$$u_z=(K_{III}/2E)(r/2\pi)^{1/2}[2(1+\nu)\sin(\theta/2)];$$
$$u_x=u_y=u_r=u_\theta=0.$$

For all modes:

$$\kappa=(3-\nu)/(1+\nu),\qquad\nu'=0,\qquad\nu''=\nu\quad\text{(for plane }\textit{stress}\text{)};$$
$$\kappa=(3-4\nu),\qquad\nu'=\nu,\qquad\nu''=0\quad\text{(for plane }\textit{strain}\text{)}.\tag{15c}$$

3.2. Eshelby's theorem and the J-integral

In addition to the stress-distribution problem, one other essential quantity in the elastic description of a crack is the force on the crack. By this is meant the analogous concept to the Peach–Koehler force on the dislocation (ch. 18, §2.2), or the driving force which moves the crack forward or backward on its cleavage plane. In the fracture mechanics literature, this quantity is referred to as the *crack extension force* or the *crack energy release rate*. It is also equal to the *J-integral* of RICE [1968]. For the material scientist, however, it is more appropriate to simply call it the "force" on the crack, and derive its value in a way which shows its close kinship to analogous dislocation concepts. We shall thus follow the development first given by ESHELBY [1956] in his discussion of the force on a general elastic singularity.

To derive the Eshelby theorem, imagine a finite body (fig. 14), on whose external surfaces an external stress is distributed. This discussion is fully general, so the full 3-d tensor notation will be used. The body will contain sources of self-stress, such as dislocations, cracks, etc., which will be stress-singularities in the medium. A strain-energy density function is assumed to exist, and since, except at the singularities, the medium is elastic, this function is given by:

$$W = \tfrac{1}{2}\sigma_{ij}u_{i,j}. \tag{16}$$

If all the singularities move a constant distance δx from their initial positions x_p to $x_p + \delta x$, the energy change is calculated in three stages.

In stage 1, the energy change, $\delta U^{(1)}$, arises from integrating the strain-energy density function over the solid. In this stage, when the sources translate, the stresses and displacements constituting the elastic solution are assumed to move rigidly with

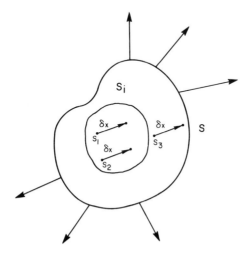

Fig. 14. Figure for deriving Eshelby's theorem. A body is contained within an external surface, S, and is subject to external stresses. Strain singularities exist at s_1, s_2, s_3..., which are translated by δx to new positions. S_i is an inner surface enclosing singularities s_1 and s_2.

References: p. 1548.

them in space. Thus, after the displacements occur, the new value of the strain energy function is

$$W^{(1)}(x) = W^{(0)}(x) - \frac{\partial W}{\partial x_i} \delta x_i. \tag{17}$$

The total energy change, integrated over the body is thus

$$\delta U^{(1)} = -\int \frac{\partial W^{(0)}}{\partial x_i} \delta x_i \, dV = -\int W \, dS_i \delta x_i. \tag{18}$$

In the second stage, the boundary condition on the surface of the body when the sources translate is considered. In stage 1, just above, when the elastic solutions translate rigidly with the sources, the stresses at the boundary translate in the same way. However, although the stress originally on the boundary satisfied the required boundary condition,

$$\sigma_{ij} n_j = 0 \tag{19}$$

on the boundary, where n is the normal to the boundary, after the rigid translation of the elastic solution the new stresses and displacements on the boundary no longer satisfy eq. (19). In stage 2, an additional surface stress is generated, such that when it is added to the translated stress, the total stress satisfies the required condition, eq. (19), on the surface. The added stress, $\Delta\sigma_{ij}$, generates a new set of displacements at the surface *after the translation has taken place*. Thus, if $u^{(0)}$ is the initial displacement before the sources are displaced, the changes generated in displacements during stage 2 are:

$$\Delta u_i = u_i^{\text{final}} - \left(u_i^{(0)} - \frac{\partial u_i^{(0)}}{\partial x_j} \delta x_j \right). \tag{20}$$

The work done on the body by the total stress acting over (Δu_i) is then, to first order $(\sigma + \Delta\sigma \sim \sigma)$,

$$\delta U^{(2)} \simeq \int \sigma_{ij}^{(0)} \Delta u_i \, dS_j$$

$$= \int \sigma_{ij}^{(0)} \left(u_i^{\text{final}} - u_i^{(0)} + \frac{\partial u_i}{\partial x_k} \delta x_k \right) dS_j. \tag{21}$$

The third contribution is the energy change in the external loading machinery caused by the total change in displacement at the surface, $u^{\text{final}} - u^{(0)}$. This is the total change in displacement from that before the source translates to that after the end of stage 2, when the boundary conditions are satisfied:

$$\delta U^{(3)} = -\int \sigma_{ij}^{(0)} \left(u_i^{\text{final}} - u_i^{(0)} \right) dS_j. \tag{22}$$

Addition of eqs. (18), (21) and (22) gives the final result:

$$f_m = -\frac{\delta U}{\delta x_m} = \int \left(W\delta_{im} - \sigma_{ij} u_{j,m} \right) dS_i. \tag{23}$$

f_m is the force on the singularity because it represents the negative of an energy derivative.

A subtlety arises when applying eq. (23) to a crack. The slit comprising the cleavage plane of the crack is, of course, a part of the external surface of the body. However, since on the free surface of the slit eq. (19) is satisfied, the second term in eq. (23) is zero. Also, by symmetry, W on the lower surface is equal to W on the upper. Because of the change in sign of dS_i when comparing the upper to the lower surface, the contribution of the first term is thus zero also. Hence, in making a contour around a crack, the contour may end on the crack cleavage surface.

The integration in eq. (23) is over the actual external surface of the body, and may enclose a number of singularities. The force in this case represents the total force on the entire set of singularities. If instead, the force on a single singularity, or subset of the original group, is desired, the following stratagem can be adopted. An inner boundary, S_i, is drawn surrounding the singularities in question (fig. 14), and the remainder of the body then serves as an "external driving system" on the inner surface. In carrying out stage 2 of the argument, however, a subtle point must be observed. The boundary conditions are established on the *real* surfaces of the body, and the Δu_i as calculated in eq. (20) are the Δu_i on the inner integration path S_i which are induced by satisfying the boundary conditions on the external surface, S. The rest of the argument then follows as before. In this case of course, the force on the singularities within S_i is not only a function of the external driving stresses, but also of the remaining singularities outside S_i. Equation (23) is quite general. In antiplane strain, a great simplification occurs. Using the complex variable notation, and Hooke's law, eq. (3.4), the simple result follows by direct substitution:

$$f^* = \frac{1}{i}\oint \frac{\sigma^2}{2\mu}\,dz. \tag{24}$$

In the complex plane $z = x_1 + ix_2$ the surface integral becomes the line integral of fig. 15. f^* is the complex conjugate of f. f is a complex number whose real part is the

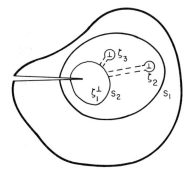

Fig. 15. Contours for Eshelby's theorem in two dimensions, in a body containing a crack and dislocations at $\zeta_1, \zeta_2, \zeta_3$. The inner surface, S_i, of fig. 24 becomes S_1. As S_1 shrinks to S_2, it passes ζ_2 and ζ_3. f^* is invariant if circles are left as shown, but not if S_2 encloses only ζ_1 and the crack.

x_1-component of the force, and whose negative imaginary part is the x_2-component.

Equation (24) has a familiar form in the theory of complex functions. Cauchy's residue theorem states that any such closed integral is zero unless there are singularities in the integrand of the form $1/z$. Then it has a finite value given by

$$f^* = \frac{\pi}{\mu} \sum_a \mathcal{R}_a(z^2), \qquad \mathcal{R} = \lim_{z \to a}(z - a) \sigma^2(z), \tag{25}$$

where $\mathcal{R}$ is the coefficient of the $1/(z - a)$ term in the Laurent expansion of σ^2 about the point $z = a$.

Equation (25) is a remarkable result. If we focus on a single singularity, it says that the force on that singularity is specified by the form of the stress as it goes to infinity at the singularity. For a screw dislocation, $\sigma \sim 1/z$, so there is *no* self-force on the dislocation, because $\sigma^2 \sim 1/z^2$, and σ^2 has no inverse linear term. If there is an external stress on the dislocation, then $\sigma^2 = (\sigma_0 + \mu b/2\pi z)^2 = \sigma_0^2 + \mu b \sigma_0/\pi z + \mu^2 b^2/4\pi^2 z^2$. Now there is a force $f^* = b\sigma_0$, as predicted by the Peach–Koehler result. If the singularity is a crack, then the force from eq. (14) is

$$f^* = \frac{K^2}{2\mu}. \tag{26}$$

This form of the force on a bare crack (mode III) was originally obtained by IRWIN [1957] by a different route, and in the fracture-mechanics community is called the *crack extension force*, or the *strain-energy release rate*.

The corresponding equation for the force in the x_1-direction on a mode I or mode II crack is

$$f_I = \frac{1 - \nu^2}{E} K_I^2, \qquad f_{II} = \frac{1 - \nu^2}{E} K_{II}^2, \tag{27}$$

where ν is Poisson's ratio and E is Young's modulus. K_1 refers to the mode I crack, etc. Equation (27) is calculated by a direct application of eq. (23) and the expressions for the appropriate stresses and displacements.

Equation (23) has been derived independently by RICE [1968], who called the integral quantity on the right-hand side (rhs) of eq. (23) the *J-integral*. *J* has been used extensively in the fracture-mechanics field for the analysis of fracture problems in plastic materials. The approach is to assume that the medium obeys a nonlinear stress–strain law which is a single-valued function of position. Then the strain-energy function can be defined everywhere, and *J* can be calculated from eq. (23). Such a stress–strain law adequately describes the loading history of a plastic material, and *J*, so calculated, can be used to characterize the stress condition of the material.

A further property of eq. (23) is that the integral on the rhs is to some degree path-independent. This result must be true if eq. (23) is indeed the force, because since any surface, S_i, can be used to calculate f, the force should not be a function of the surface of integration. Analytically, this result follows immediately from eq. (24), since from the Cauchy residue theorem, the integral is specified by the number and type of singularities enclosed in the path and not by its size or shape. If the path of the integral is thus deformed in such a way that it does not pass over one of the singularities, then, indeed, its value is unchanged. However, when the path is shrunk over one or more singularities so that it encloses fewer singularities than before, then the value of f^* is changed by the amount of the residue it bypasses during shrinking (fig. 15).

The quantity J, on the other hand, has a different meaning for a crack with a surrounding deformation field than does f. Since in J, the deformation field is not represented by dislocations, but by a continuous nonsingular nonlinear stress–strain function, no singularities are bypassed when the path is shrunk. Thus, J is completely independent of path throughout the deformation zone, and has the same value whether calculated within the deformation zone or outside it.

The complete path-independence of J gives it a great interpretive power for engineering purposes. The reason is that the critical value of K for which the material fails, known as K_{IC}, is an important engineering parameter for any material. However, sometimes K cannot be defined, but J can be. For example, in many cracked materials, the plastic region is confined to a small region near the crack tip (fig. 16). In these cases, the stresss in the elastic region is given by the simple result

$$\sigma = \frac{K}{(2\pi z)^{1/2}},$$
(28)

with the stress field centered at an effective crack tip position. In this case,

$$J = \frac{K^2}{2\mu}.$$
(29)

However, in a very tough material, and especially in "small" specimens, often the entire specimen undergoes plastic deformation, or a portion of the plastic zone overlaps an external surface. In this case, the stress is quite complicated, not given by eq. (28), and of course cannot be characterized by K. Nevertheless, even in these cases, J can be calculated by eq. (23). Because of its invariance properties, intuitively, the conditions near the crack tip which determine whether the crack grows or not should not depend strongly on the details of external specimen shape. The critical value of J when cracking occurs should still be a material property.

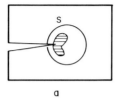

a   b

Fig. 16. The path for calculating J in (a) can be shrunk from a position where the crack and the shaded deformation zone are fully enclosed to any arbitrary size. In (b) there is no position of S enclosing the crack, which runs entirely through elastic material.

References: p. 1548.

Furthermore, J_{1C} should be equal to that calculated through eq. (29) for the larger specimens.

Thus, much materials testing effort has been expended on developing techniques for measurement of critical values of J, and establishing the limits of application for these measurements. Various publications of the ASTM are referred to for further details, e.q. BEGLEY and LANDES [1972] and RICE [1976].

A major conceptual problem with J, however, is that plastic materials when *unloaded* do not follow the same stress–strain law as when loaded. Thus, W is not a unique point function of position, and the J-integral theory breaks down. Nevertheless, if the crack is not allowed to move, and the stress is not reversed, the J can be used to characterize the not-yet-moving crack, and on this basis the engineering use of J as described above is well founded. The problems associated with *moving* cracks, where the J-integral does not exist, are still in an area of intense research. Another difficulty with using J to discuss the materials fundamentals of fracture is that it cannot easily describe the dislocation phenomena which ultimately control the actual response of a crack to external stress. For these two reasons, in this chapter, we shall be more concerned with f^* than J.

The elastic theory presented here treats the crack as an independent entity from the dislocation, and this approach is confirmed by the fact that the crack stress field goes as $1/\sqrt{z}$ and the dislocation as $1/z$. Nevertheless, there is a fundamental sense in which the distinction disappears. For example, by eq. (15a) in the mode I crack shown in fig. 17, the displacement on the negative x-axis cleavage plane is given by:

$$\Delta u_y \sim K\sqrt{z} \,. \tag{30}$$

If the space between the two opposite sides of the cut were filled with a continuous density of Burgers vector, which just fills the empty space with matter, then the external stress could be released, and the material around the crack would possess exactly the same elastic displacement, stress and strain as before. Thus a prismatic "pile-up" of dislocations is equivalent to the crack. This equivalence has been used to obtain analytic results for crack problems by using dislocation theory, and forms

Fig. 17. Near the tip of an elastic crack, the opening is parabolic (see eqs. (15). This opening can be made equivalent to the displacement of a distribution of prismatic dislocations as shown. The stress and displacement outside the crack is then given by a distribution of "crack dislocations".

the basis for the elastic theory of cracks presented by BILBY and ESHELBY [1968] and others. For the materials scientist who is more familiar with dislocations than with cracks, the equivalence is often an intuitive help.

4. Dislocation–crack interactions

The elastic interactions between cracks and dislocations can be derived from eqs. (24) and (25). General results are possible, but lead to very complicated expressions. Thus our treatment will address the simple dislocation–crack configuration illustrated in fig. 18 in mode III.

The first step is to find the stress for a dislocation in the presence of a crack. There are several techniques for doing this. One is to use a Green's function or source function for the elastic field (THOMSON and SINCLAIR [1982]) and another is to use the known solution for a dislocation in the presence of a plane surface and use complex function techniques to transform the plane surface into a slit surface (MAJUMDAR and BURNS [1981]).

Intuitively, however, we may obtain the solution by building on the known solution for a screw dislocation,

$$\sigma(z) = \frac{\mu b}{2\pi(z - \zeta)},$$

(31)

and modifying it to conform to the requirements of the crack. Here b is the Burgers vector of the dislocation, ζ the (complex) position of the dislocation, and z the point where the stress is measured. The origin is at the crack tip (fig. 18). The necessary modification is made by considering the form of the solution in the immediate neighborhood of the crack tip in a region where $z \ll \zeta$. If there is no external stress (and the cut is not allowed to heal) then the stress field of the dislocation will develop the characteristic crack stress-concentration around the crack tip just as an external stress would. That is, near the crack tip, the stress from the dislocation field must vary as $1/\sqrt{z}$. However, in a region near the dislocation, the stress field must look like eq. (31). Thus, if we multiply eq. (31) by the factor $(\zeta/z)^{1/2}$ the total stress

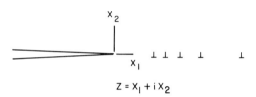

$$z = x_1 + i x_2$$

Fig. 18. For calculations of dislocation–crack configurations, the dislocations are all assumed to lie on the x_1-axis ahead of the crack tip. In this case, z and ζ are real.

References: p. 1548.

field has these required characteristics:

$$\sigma = \frac{\mu b}{2\pi(z-\zeta)}\left(\frac{\zeta}{z}\right)^{1/2} \tag{32}$$

The stress must also satisfy the boundary conditions, eq. (19), on the crack slit, and a direct check of eq. (32) shows that eq. (19) is indeed satisfied on the negative x_1 axis. Equation (32) corresponds to the dislocation–crack field without the presence of an external stress. If we add an external stress (i.e., a K field) to the dislocation field, and add other dislocations at positions ζ_j, then the final result for the stress *of symmetric distributions* is

$$\sigma(z) = \frac{K}{(2\pi z)^{1/2}} + \sum_j \frac{\mu b_j}{2\pi(z-\zeta_j)}\left(\frac{\zeta_j}{z}\right)^{1/2}. \tag{33}$$

The Burgers vector convention used in these expressions is that consistent with a positive sign in eq. (33). A geometric interpretation will be made later in terms of the absorption of dislocations by the crack surface.

In a calculation of f from eq. (25), it is only necessary to square eq. (33), and identify the $1/z$ singularities. Singularities in σ^2 appear at the crack tip and at each dislocation. Taking only those singularities of order (-1) at the cack tip, we find for the force on the crack, f_c:

$$f_c = \frac{k^2}{2\mu}, \qquad k = K - \sum_j \frac{\mu b_j}{(2\pi\zeta_i)^{1/2}}. \tag{34}$$

Thus, the force on the crack corresponds to a new K-field at the crack which we label k, and call the *local k*. Equation (34) means that the externally applied field is *shielded* from the crack tip by the presence of dislocations, and the contribution of each dislocation is in the amount $\mu b/(2\pi\zeta)^{1/2}$. It is important to note that the dislocation contribution is algebraic in b; that is, positive Burgers vectors shield the crack, while negative Burgers vectors enhance the external field. Thus, we speak of "shielding" dislocations and "antishielding" dislocations. Although eq. (34) has only been derived for the configuration of fig. 18, it is valid as it stands for an arbitrary symmetric distribution of screw dislocations around the crack; i.e., if there is a dislocation at ζ, there is also one at ζ^*.

The force on one of the dislocations, f_d, is equally straightforward. In the expression δ^2, there are terms which are of the form $1/(z-\zeta)$. These terms contribute to the residue, and after carrying out the necessary algebra, the result is

$$f_d^* = \frac{Kb}{(2\pi\zeta)^{1/2}} - \frac{\mu b^2}{4\pi\zeta} + \sum_j \frac{\mu b_j b}{2\pi}\frac{1}{\zeta-\zeta_j}\left(\frac{\zeta_j}{\zeta}\right)^{1/2}. \tag{35}$$

In this expression, the reference dislocation is unlabeled, but all other dislocations in the distribution are labeled by the index j. Thus the sum does *not* include the reference dislocation, b, at ζ. Again all b's are algebraic. The three terms in eq. (35) can be interpreted as follows:

The first term is the contribution from the direct crack stress field. Note that the appropriate K is the unshielded value.

The second term is an image term as if the crack were a vertical surface through its tip. Qualitatively, such an image term is expected because, after all, the cleavage plane of the crack is an open surface, and dislocations of both signs are attracted to surfaces as if by an image. The simplicity of the result and its equality to the image of a simple vertical surface is the only surprise.

The third term in eq. (35) is the dislocation–dislocation interaction, as modified by the crack. The $(\zeta_j/\zeta)^{1/2}$ factor must be present because, as explained earlier, a dislocation generates its own $(1/z)^{1/2}$ singularity at the crack tip, and a second dislocation used as a probe will experience the stress as a singular force near the crack tip. On the other hand, near the dislocation, the $(\zeta_j/\zeta)^{1/2}$ factor does not affect the normal dislocation–dislocation interactions. A final curiosity, however, results from an investigation of eq. (35) when $\zeta \gg \zeta_j$. In this case, the crack shields the dislocations! The reason is that the dislocation is near the open surface of the crack, and its stress field is partially cancelled by its image in the crack, though the cancellation is more complex than a simple image term.

Equation (35) is a simplification of the actual force on a dislocation at an arbitrary position off the crack plane. If the dislocation is not on the crack plane, additional image terms are present which lead to complex expressions for nonsymmetric distributions.

Equations (34) and (35) correspond to contours in the sense explained in fig. 15, drawn around each individual defect in question. If the contour is drawn around the entire distribution, then the force is that to translate the entire configuration rigidly. The residue is then composed of a sum of all the terms, eqs. (34) and (35), and the result is then simply

$$f^*_{\text{total}} = f_c + \sum_j f^*_d(j) = \frac{k^2}{2\mu} + \sum_j f^*_d(j) = \frac{K^2}{2\mu}. \tag{36}$$

f^*_{total} has the same value as the force on a bare crack sustaining a total external K-field value of K. f^*_{total} also has the same value as the J-integral defined in the last section for the crack with its deformation field eq. (29).

The discussions of crack–dislocation interactions do not yet reflect the important fact that the total dislocation Burgers vector is a conserved quantity. Suppose a configuration consisting of a crack and its associated screening dislocations is considered to be generated from an initially perfect crystal. The crack is constructed by making a cut in the presence of an external stress as explained in the introduction. The dislocations, however, will be assumed to be formed from dislocation sources in the neighborhood of the crack on the crystallographic slip planes specified by the crystal geometry. Although these multiplication processes in real materials are highly chaotic, for our two-dimensional purposes we may consider them to be formed in pairs of opposite Burgers vectors (fig. 19). The one with shielding Burgers vector will be repelled from the crack, and the other, with antishielding Burgers vector, will be attracted to the crack. The dislocations will move under the elastic

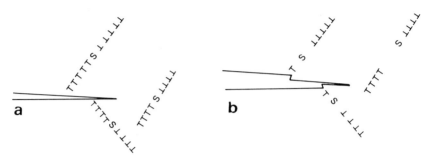

Fig. 19. Dislocation sources operate external to a crack and cause a crack-opening displacement as shown.

forces calculated in eq. (35) until they come into equilibrium with lattice friction forces (see the following section). Many of the antishielding dislocations will be attracted toward and be absorbed and annihilated by the open cleavage surface, producing steps on these surfaces, of height equal to the total absorbed Burgers vector. Of course, many of the antishielding dislocations are oriented to pile-up in the region ahead of the crack tip, and exert a force on the crack, forcing it to the right (fig. 18). The net result of these processes is to produce a crack with a new cloud of shielding dislocations (fig. 19), and a *crack opening displacement* (COD) equal to the antishielding dislocations which were absorbed by the cleavage plane. Figure 19 shows edge dislocations interacting with a mode I crack, whereas the analysis has been entirely in terms of screw dislocations in mode III. The qualitative similitude of the anaysis between the two cases allows us to do this. The crack opening displacement is quantitatively given by

$$\sum b_j = \text{COD} = Nb_\text{c},\tag{37}$$

where b_c is the component of the Burgers vector normal to the crack plane. N is the net remaining number of screening dislocations in the shielding configuration. Because many, and in fact most, of the antishielding dislocations created in the deformation zone around the crack do not find their way to the cleavage surface where they can be annihilated, the deformation zone is composed of dislocation networks whose Burgers vectors nearly cancel. Only the residual, or net Burgers vector of the distribution contributes to the screening effect.

As shown in fig. 19, those antiscreening dislocations whose slip planes pass in front of the crack are trapped in the vicinity ahead of the crack in a region of dislocation stagnation. Since the crack is attracted toward them, if sufficient pile-up takes place, the crack will be pulled ahead until absorption can occur on the cleavage surface. This effect has indeed been observed in NaCl by WIEDERHORN *et al.* [1970]. In metals, this stagnation region is also the region of hole formation.

5. Equilibrium configurations

5.1. The Griffith Crack

In terms of dislocation shielding, the fracture problem comes down to finding the equilibrium configuration of dislocations and cracks under a given external stress. The force represented by f^* is the elastic force exerted on a defect, and these elastic forces, which are partly interaction forces between the defects, and partly due to external stress, are balanced by other forces induced by the material. For the configuration to be in static equilibrium, each defect must experience a total net zero force.

The simplest crack configuration, of course, is an isolated crack, without dislocation screen. The force balance condition for the isolated brittle crack was first studied by GRIFFITH [1920], who suggested in 1920 that the elastic force is balanced by the surface tension of the two cleavage surfaces joined at the crack tip. That is, when the crack is opened, additional surface energy is created, and this energy residing in the surface exerts a force on the crack tip tending to close it. If γ is the surface tension, the equilibrium implies:

$$f_C^* = \frac{K_C^2}{2\mu} = 2\gamma. \tag{38}$$

γ is also equal to the surface free energy per unit area where K_C is the critical K. In modes I and II, the corresponding equations are

$$K_{IC}^2 = \frac{2\gamma E}{1 - \nu^2}, \qquad K_{IIC}^2 = 2\gamma E/(1 - \nu^2) \tag{39}$$

Equations (38) and (39) express the condition or criterion for brittle fracture. That is, for values of K greater than K_C, the crack grows, and catastrophic fracture ensues. K_C, K_{IC}, and K_{IIC} are also called *fracture toughness parameters*, because they relate to the onset of fracture in a material. In fact, the mechanical equilibrium expressed by eq. (38) is an unstable equilibrium, which can be seen from the expression for the energy of the system as a function of the crack length. The internal energy of the system, including the surfaces, is:

$$U = 2\int_0^a (-f^* + 2\gamma)\, dl. \tag{40}$$

In this equation, we use a finite crack configuration in which the crack extends from $x = -a$ to $x = +a$. The total crack length is $2a$. Using $f^* = K^2/2\mu$ and the relation (14) relating K to crack length for mode III, we get:

$$U = 4\gamma_a - \frac{\pi\sigma_\infty^2 a^2}{2\mu}. \tag{41}$$

This curve is an inverted parabola, with its maximum at the point of unstable static equilibrium. Thus if K is increased slightly above K_C, catastrophic crack growth occurs, as observed in such brittle materials as glass. The crack healing predicted for $K < K_C$ has been observed by LAWN et al. [1980] in very careful experiments, but requires the material to come back together in exact atomic registry, which is often difficult to ensure. Also, in practice, the cleavage surfaces are usually coated by oxides and other films, which makes crack growth an irreversible process.

Another useful relation for the "Griffith criterion" of eq. (38) is obtained by combining eq. (14) with eq. (38):

$$\sigma_C = \left(\frac{4\gamma\mu}{\pi a}\right)^{1/2}.\tag{42}$$

In this equation, σ_∞ given in eq. (14) (the externally applied stress) is renamed as the critical Griffith stress, σ_C.

Equation (41) and (42) have different forms when the loading geometry is changed and $K(a)$ is then different from the functional form, eq. (14), used in the derivation (see TADA et al. [1973]). The constant coefficients are also somewhat different in modes I or II, given the same crack geometry, and involve other elastic constants. In mode I plane strain, the Griffith relation corresponding to eq. (5.5) is:

$$\sigma_C = (2\gamma E/\pi a)^{1/2} \quad \text{for plane } strain,$$

$$\sigma_C = \left[2\gamma E/\pi a(1 - \nu^2)\right]^{1/2} \quad \text{for plane } stress.\tag{43}$$

The fracture criterion, or fracture toughness given in eqs. (38) and (39) for the pure brittle case describes only a few actual materials. Glass, silicon at room temperature, and mica are some examples. Historically, fracture mechanics as developed by Griffith in the 1920s thus had little applicability until Orowan and Irwin in the 1930s independently began the process of generalization to ductile materials when they realized that γ in eqs. (38) and (39) should be thought of as an effective surface energy which includes the plastic work done during fracture (see KNOTT [1973]).

5.2. Shielding by one dislocation

After the isolated crack, the next simplest configuration is a crack plus one dislocation on the x-axis, for which the equilibrium equations become:

$$f_C = 2\gamma = \frac{k^2}{2\mu} = \left(K - \frac{B\mu}{(2\pi x)^{1/2}}\right)^2 \frac{1}{2\mu},\tag{44}$$

$$f_d = \frac{KB}{(2\pi x)^{1/2}} - \frac{\mu B^2}{4\pi x} = B\sigma_f.\tag{45}$$

In eqs. (44) and (45), x is the distance along the crack plane, and all quantities are real. $B\sigma_f$ is the lattice resistance to dislocation motion expressed in terms of an effective friction stress, σ_f. B is the Burgers vector, written as a capital to note the possibility that the "dislocation" may be a superdislocation composed of many elementary crystal dislocations. If the external stress as specified through K and the distance, x, are independent variables, the solution is immediate. After eliminating x from eqs. (44) and (45),

$$K_C = \left(2B\mu\sigma_f + k^2\right)^{1/2} = \left(2B\mu\sigma_f + 4\mu\gamma\right)^{1/2}. \tag{46}$$

The local force balance on the core crack is satisfied by the Griffith relation as given in eq. (38), $k^2 = 4\mu\gamma$.

Toughness in a material is an engineering concept developed from the impact testing of specimens, and relates to the energy absorbed on fracture. It is thus also a measure of the resistance to fracture in the sense that a material requiring a high stress to fracture it is also a tough material. To be more precise, K_C is an alternative measure of toughness, and thus eq. (46) is a very simple or prototype toughness relation for a material. From the way it was derived, it is also true that high toughness is correlated with high dislocation shielding, and thus to the shielding "charge", B. Since B is also equal to the COD because of eq. (37), the COD is another alternative measure of toughness. Thus a second form of the toughness relation is

$$B = \text{COD} = \frac{K_C^2 - k^2}{2\mu\sigma_f} = \frac{K_C^2 - 4\mu\gamma}{2\mu\sigma_f}. \tag{47}$$

5.3. General shielding

Although eq. (46) illustrates the basic physics of the fracture toughness problem in a very transparent way, it does not address what the size of the shielding charge is – i.e., what is the size of the superdislocation Burgers vector, B. For this, a more complex theory is required which allows the dislocation distribution to arrange itself in a self-consistent fashion in the presence of sources in the material.

The general statement of the equilibrium problem for an arbitrary configuration of crack plus shielding dislocations is the generalization of eqs. (44) and (45) for an arbitrary number of dislocations:

$$f_c^* = 2\gamma, \qquad f_d^*(j) = b_j\sigma_f(\xi_j), \tag{48}$$

where f_c^* and f_d^* are given by the general relations (35) and (34), respectively. This prescription for solving the problem is intractable except by numerical methods for any but the simplest configurations. To make further progress, assume again that the dislocations are distributed along the x-axis as in fig. 18, so that the dislocation distribution is one-dimensional. The actual discrete distribution is also approximated by a continuous dislocation density.

Concentrating for the moment on the second equation of (48), from eq. (35) the force on the dislocation of Burgers vector b at x is

$$f_d = \frac{Kb}{(2\pi x)^{1/2}} - \frac{\mu b^2}{4\pi x} + \sum_{b_j} \frac{\mu b b_j}{2\pi} \left(\frac{x_j}{x}\right)^{1/2} \frac{1}{x - x_j}. \tag{49}$$

If the Burgers vector is a distributed quantity, then $db = \beta(x)dx$ and the sum becomes an integral:

$$\frac{1}{\beta}\frac{df_d}{dx} = \frac{K}{(2\pi x)^{1/2}} - \frac{\mu\beta dx}{4\pi x} + \frac{\mu}{2\pi}\int_c^d \sqrt{\frac{x'}{x}}\,\frac{\beta(x')}{x-x'}\,dx' \tag{50}$$

where c and d are the points at the beginning and end of the assumed dislocation distribution (fig. 20). The second term on the right is infinitesimal compared to the other terms, and should be dropped. The resulting equation is an integral equation for the unknown function β of a type which has been studied by HEAD and LOUAT [1955] in connection with dislocation pile-up theory.

A special case of eq. (50) has been studied for cracks in mode I by BILBY, COTTRELL, and SWINDEN [1963] (in what is called the BCS model) in which c is set equal to zero. The general equation was solved first by CHANG and OHR [1981] and in modified form by MAJUMDAR and BURNS [1983], and WEERTMAN et al. [1983]. In the approximation where $d > c$, very simple relations can be obtained:

$$\frac{k}{K} = \frac{3}{2\pi}\left(\frac{c}{d}\right)^{1/2}\left(\ln\frac{4d}{c} + \frac{4}{3}\right); \tag{51}$$

$$K = 2(2/\pi)^{1/2}\sigma_f\sqrt{d} ; \tag{52}$$

$$B = \int_c^d \beta(x)\,dx = \frac{2K}{\mu}\left(\frac{d}{2\pi}\right)^{1/2}. \tag{53}$$

These equations relate the external K to the local k, to the total dislocation screening charge, B, and to σ_f. The distribution β is a combination of elliptic integral functions, and is sketched in fig. 20. There is a strong maximum near the tip just ahead of the cut-off, c. As $c \to 0$, in the BCS limit, this maximum in fact diverges. It is important to note that as $c \to 0$ the crack becomes completely screened, and $k \to 0$. This result is an alternative expression of an important theorem by RICE [1965], that in a continuum treatment of the plastic field surrounding the crack, as the crack expands under an externally applied load, all the energy is absorbed by the plastic

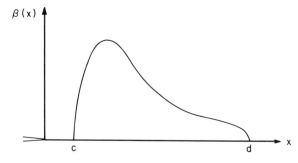

Fig. 20. Solution of eq. (50) for the continuous dislocation distribution, $\beta(x)$. The crack tip is at the origin, and the distribution begins at c and drops to zero at d. There is a strong maximum near c.

field, and none is absorbed by the crack itself. In such a theory, the crack is thus disconnected from the problem entirely. On the other hand, since dislocation densities in materials are almost never high enough to overlap the central nonlinear core region of the defect, there will be a region around the crack tip sufficient in size to establish its elastic identity, and under these conditions, a local k field at the crack tip can be defined. The distance c is the parameter which measures the average distance to the nearest dislocation in the configuration, and constitutes a lower cut-off for the application of continuum deformation theory. The relations (52) and (53) are relatively simple, and do not depend on c. If the field far from the crack and the dislocation screen is characterized by the K-field, then the stress is given there by $\sigma = K/(2\pi x)^{1/2}$. Where $x = d$, σ should be given by the yield stress, σ_f, on the slip plane at the far end of the dislocation screen. Hence, $\sigma_f \cong K/(2\pi d)^{1/2}$, which is a good approximation to eq. (52), and is a generally valid result for the size of the plastic zone for any shape of the plastic zone, and for all modes.

Equation (53) gives the total offset plastic displacement at the crack tip generated by the dislocations. In mode I, where a similar relation holds, this is the crack opening displacement known simple as the COD. Combining eqs. (52) and (53) then:

$$\text{COD} = \frac{K^2}{2\mu\sigma_f}, \tag{54}$$

which is the same as eq. (47) for $K_C^2 \gg k^2$. The corresponding result for mode I is:

$$\text{COD}_I = \frac{2K_I^2}{E\sigma_f}, \tag{55}$$

where E is Young's Modulus.

Expressions such as eqs. (47) and (54) have been derived in the literature in numerous ways, and have been widely used in applications of fracture mechanics. In particular, the COD which can be inferred experimentally is a popular measure of toughness, which is then related to K_C through eq. (54). In addition, attempts to correlate the critical COD with a critical strain in front of the crack have been made.

In the equations (51)–(53) however, the total number of dislocations is not determined. This question involves one of the major problems of dislocation theory, yet largely unsolved in a fundamental way, and we shall adopt one of the traditional heuristic approaches. Namely, a *constitutive relation* is introduced between stress and strain: in this case, between the number of dislocations produced, B, and the local stress. Remembering that σ_f is the negative of the elastic stress on the slip plane, because each dislocation experiences zero net force, such a constitutive relation can be written:

$$\sigma_f = \sigma_0 (B/b)^m. \tag{56}$$

This relation exhibits *work-hardening*, because a larger σ_f is required to achieve a larger B. m is the work-hardening power law parameter, and in such a law σ_0 is the *yield stress*, because it is the value of σ_f for the first dislocation, $B = b$.

References: p. 1548.

As a constitutive relation, however, eq. (56) is very limited, because in an adequate description σ_f should be a function of the *local* density of Burgers vector, not the total integrated value, B. Because the basic integral equations cannot be integrated for such a law however, eq. (56) is the best we can do for analytic results.

Using the constitutive relation, eqs. (51)–(53) can be combined to obtain a second approximation to a toughness relation:

$$K_C = \left(\frac{k^2}{c}\right)^{(1+m)/4m} \eta, \qquad \eta = \frac{\pi^3}{18} \frac{(2\mu)^{1/2}}{\sigma_0^{1/2m}(\ln 4d/c + 4/3)}. \qquad (57)$$

This relation is much improved over the primitive relation (46). Since the logarithm term in η is slowly varying, we may simply take η as a collection of material parameters. Then the only parameter in eq. (57) which cannot be determined from macroscopic measurements on materials is the distance c, the average distance from the crack tip to the closest shielding dislocation.

Another interesting relation can be obtained by eliminating K from eq. (51) and (52):

$$k = 3\left(\frac{2}{\pi^3}\right)^{1/2} \sigma_f\sqrt{c}\,(\ln 4d/c + 4/3). \qquad (58)$$

Except for numerical factors, this equation is approximately $k \sim \sigma_f\sqrt{c}$, which says that the stress at the radius of the dislocation-free zone is the local yield stress. This equation was in fact used by THOMSON [1978] and WEERTMAN [1978] as a cut-off in the standard plasticity solution for a crack embedded in a plastic zone, but with an elastic enclave around the core crack. In this work it was possible to incorporate work-hardening in the plastic zone in a more satisfactory manner than eq. (57), with the result given by THOMSON [1978]:

$$K^2 = 2\pi\left(\frac{2\gamma\mu}{\pi}\right)^{(1+m)/2m} \Big/ \left(c\sigma_y^2\right)^{(1-m)/2m}, \qquad \sigma = \sigma_y\left(\epsilon/\epsilon_y\right)^m, \qquad (59)$$

where σ_y is the initial yield stress, and m is the work hardening power coefficient. When the dislocation constitutive relations are included in the self-consistent solution of the total equilibrium configuration, as in eqs. (57) and (59), the toughness relations are seen to depend upon the intrinsic surface energy, γ, in an essential way. [Crudely, γ is simply the energy to break the bonds as the fracture is formed; see eq. (38).] The reason is that γ governs the local equilibrium of the core crack, and the required equilibrium between core crack and dislocation cloud thus involves γ in the expression of the overall toughness of the material. On the other hand, when $K^2 \gg k^2$ in (47) it would appear that γ could be ignored, as has often been claimed in the fracture literature. However, in the simple relation (46), and in the more complex equations (51)–(53), B (and σ_f) depend on γ in an implicit way because of the required overall equilibrium. Hence, it is in general an error to ignore the effect of γ on toughness even for tough materials.

5.4. Summary

In summary, the burden of this section has been to show that on a fundamental level, if a crack exists in a material, the material will fracture at a very low value of toughness unless the material can respond plastically to the high stress-concentration immediately surrounding the crack tip. If dislocations can be generated in this region, then the effect of these dislocations is to shield the crack from the external stress field after a portion of the antishielding dislocation population has been absorbed by the cleavage surface. The degree of shielding, or the level of toughness achievable, is then determined by the yield stress, work-hardening characteristics of the medium, and the size of the inner elastic enclave. It also, of course, depends upon the local critical k of the crack, which we have suggested is determined by the true surface energy of the solid, but may in fact be more complex for reasons discussed in the next section.

The descriptions and models used have, of course, been very simple, so the relations derived cannot be expected to describe in detail the complex fracture phenomena observed in a polycrystalline multiphase material such as a practical steel. Their major purpose, of course, is to demonstrate the materials science principles which underlie fracture toughness, and in particular the role of dislocations in multiplying the intrinsic lattice toughness of a crack by many orders of magnitude. Specifically, a typical value of k is of order 2×10^5 Pa$\sqrt{m}$, whereas, from eq. (59) one infers toughness values of order $K = 50$ MPa$\sqrt{m}$ for a high-strength steel.

There are several fundamental limitations of the model used which must be kept in mind, and which will suggest to the materials scientist further elaborations which may be used to achieve more satisfactory descriptions of cracks in materials:

a. It has always been assumed here that the fracture processes depend on the external shape of the sample and other macroscopic quantities, such as crack length, only through the stress intensity factor, K. The dependence of K on shape of sample, crack length, and the geometry of the stress application all lie in the domain of macroscopic stress analysis, and our approximation amounts to what is termed the two-dimensional small-scale yielding case, where the crack length is small compared to distances to all external surfaces in x_1 and x_2 directions, and the deformation zone is in turn small compared to the crack length. On the other hand, macrostructural inhomogeneities such as precipitates and grain boundaries are also assumed to be absent, and this is a much more serious assumption. To some degree, one can incorporate these effects in the dislocation constitutive law parameters, but we have already expressed reservations about the adequacy of the constitutive-law approach to deformation theory, especially in the region nearest the crack tip. A second difficulty concerns the interaction of the crack itself with microstructural inhomogeneities. By using interfacial surface energy ideas, one can make some progress in this direction by generalizing the Griffith argument for grain boundaries. However, the three-dimensional effects of cracks moving along faceted grain boundaries and

References: p. 1548.

precipitate interfaces, and the pinning effects of precipitates to cracks, introduce mathematical complexities which have not yet been worked out for cracks. In short, our understanding of three-dimensional crack interactions is very primitive when compared with the status of similar dislocation pinning theories. One promising approach would be to use the crack–dislocation analogue for these three dimensional crack configurations.

b. We have relied primarily on mode III in the analysis. This approach is analogous to studying dislocation theory by restricting oneself to screw dislocations. Nevertheless, the obvious geometrical generalizations are easy to make to modes I and II, and the analytical results are expected to be correct in a qualitative sense. In some cases we have given the mode I and II results in the text without proof.

c. The analysis leading up to the relations for f_c^* assumed the crack was sharp, and yet the requirement for COD $\neq 0$ is an essential aspect of a shielded crack as shown in eq. (37). Certainly, there is no difficulty so long as the COD is small, and the crack-opening angle is near zero. However, as the COD is increased, the changed geometry of the crack surface will lower the stress on the atom bonds at the crack tip. Indeed, if the material is sufficiently soft, plastically, then the stress at the crack tip may never be able to achieve a value sufficient to break the atomic bonds – in our terminology k_{critical} cannot be achieved because of progressively greater combined blunting and shielding – and the crack will never cleave. Only continued plastic blunting of the crack can occur. This process is, of course, precisely what happens during fully plastic failure. During fibrous fracture by hole growth, small cracks open ahead of the main crack and are progressively blunted out as soon as they are formed. In addition, the dramatic transition from relatively brittle fracture to ductile fracture observed when the temperature is varied in some steels and many other materials is also to be understood on these grounds (see KNOTT [1973]).

d. In fig. 19, antishielding dislocations are absorbed at the crack. Generally, of course, only a small fraction of the total dislocations formed in the dislocation screen will be absorbed. That is, the majority of dislocations in the screen cancel each other out, leaving only a small fraction of the total population (sometimes termed the geometric dislocation component) to perform the shielding function. Thus, in the analysis, sums over dislocations and Burgers vectors are in effect sums over the local *net* Burgers vector. The nonshielding dislocations are important, however, because they provide the essential work-hardening interactions which ultimately limit the total shielding dislocation density. They are also important, because as a crack moves through the material, the large tangled mass of non-shielding dislocations is left behind as a wake. Thus the energy required to form this wake, most of which is immediately converted into heat during the formation of the dislocation tangles, must be supplied by the crack and its effective screening charge of geometric dislocations as they move through the material. The rate of doing this work is the effective surface energy postulated by IRWIN [1957] and OROWAN [1949], and is given by the relation

$$\frac{K_C^2}{2\mu} = 2\gamma_{\text{eff}}.$$
(60)

6.   Atomic structure of cracks: thermal and chemical effects

6.1.   Theory

In the previous section, the importance of the local k of the crack tip was emphasized. Even when the energy required to produce plastic work in the wake of a moving crack is several orders of magnitude larger than the energy absorbed by the core crack, a balance of force is always required between the core crack and its shielding charge of dislocations. One can thus think of the small local k-field of the core crack as driving the entire fracture process because of the necessary force balance.

For this reason, and another given in the next section, the atomic structure of the crack tip is a crucial matter for understanding the behavior of materials in fracture, because it is the opening of atomic bonds at the crack tip which determines the critical k for cleavage. In the previous section, a surface tension argument was used to specify the critical value of k, and one of the tasks here will be to show under what circumstances the Griffith criterion for k_c is appropriate.

The easiest approach to atomic phenomena at crack tips is to consider an infinite lattice of atoms in which a crack is made by slitting the bonds between two rows in two dimensions (fig. 5). A force is then applied to the center atom pair which is required to hold the cracked atoms apart. For the first part of the argument, the force between the atoms corresponds to a simple linear spring which snaps at a critical stretch as in the left-hand part of fig. 21. Since the system is thus completely linear, its response to any external force is also linear. That is, if the displacement u_0 of the central atom is plotted as a function of the applied force, F, a straight line results:

$$u_0 = D_0(N)F.$$
(61)

Equation (61) is the compliance equation, and $D_0(N)$ is the compliance constant.

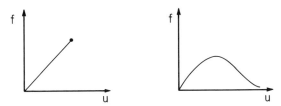

Fig. 21. Bond forces, f. A variety of bond forces are discussed. In the left-hand part, the force is a snapping spring. In the right-hand part, a more realistic nonlinear force is shown.

References: p. 1548.

The subscript corresponds to the displacement of the zeroth atom at the center of the crack. Since the compliance depends upon the crack length, N, D_0 is a function of N. The solution plotted for several values of the crack length is thus shown in fig. 22. If we now remember that the springs will snap when stretched to a critical displacement, and track one of the compliance lines (say for crack length $N = L$) as the force is increased, at some point the first atom at the tip with bonds connected ($N = L + 1$) will snap open because its critical force is exceeded. This point is indicated by the higher solid circles in fig. 22, and beyond this point the solution does not exist for crack length $N = L$. In fact, since the bond has snapped, the crack is now of length $N = L + 1$, and the solution has jumped to the next line corresponding to $N = L + 1$. Similarly, if the $N = L$ compliance line is tracked as the external force is decreased, it will eventually jump to a crack length $N = L - 1$ when the first open atom pair at the tip comes into the range where the atomic force snaps back "on". The total solution is then the set of solid lines plus dotted lines between the two limiting curves as drawn in fig. 22.

The striking result of this argument is that for a given crack length, the system is in static equilibrium over a range of external forces. Likewise, for a given external force, F, many crack lengths exist where static equilibrium is possible, as shown by the fact that the horizontal line in fig. 23 cuts many crack solutions in the allowed range. This result is contrary to the Griffith result, where a crack has a unique length where it is in equilibrium, and will catastrophically open or close at all other lengths. This phenomenon has been called lattice trapping because it is due to the discrete nature of the lattice. It is analogous to the Peierls barriers of a dislocation which are also caused by the lattice. We note that the trapping occurs only in a finite range of

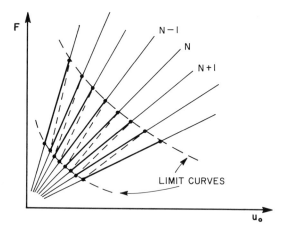

Fig. 22. Solution of eq. (61). A family of compliance curves is generated for various values of crack length, N. The curves are limited at a maximum displacement by the bond strength at the tip, and at a minimum displacement by reconnection of the first broken bond at the tip. At the maximum displacement for N, the solution snaps back to $N + 1$ at the minimum, as shown by the dotted lines, provided F can be varied at the same time.

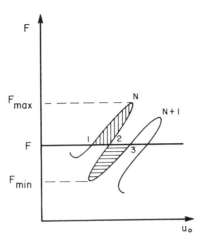

Fig. 23. Crack solution for nonlinear forces for N and $N + 1$. For a given F, several discrete solutions are possible. Positions 1 and 3 are stable. Position 2 is unstable. The cross-hatched areas correspond to trapping barriers. That from 1 to 2 is a forward barrier and that from 3 to 2 is a reverse barrier.

external forces from F_{max} to F_{min}, because if F_{max} in fig. 23 is exceeded, no equilibrium solution exists for any crack length and the system accelerates to complete failure. The sharp corners of the compliance lines in fig. 22 are smoothed off in fig. 23, because the forces between the atoms are assumed to be nonlinear as in the right-hand part of fig. 21, but the general results are preserved.

The energy of the trapping barriers can be calculated from the compliance curve of fig. 23. Since the internal energy of the lattice is given by the integral of the force–displacement curve for the atom pair where the external force is applied, the total energy change of the system is:

$$\Delta U = \int_{1}^{2} F(u_0) \, du_0 - F\Delta u_0. \tag{62}$$

The second term in eq. (62) is due to the change of the energy of the external force system as it induces Δu_0 displacement on the center atom pair, and is negative. The energy is thus given by the shaded region in fig. 23, and can be calculated if the compliance function (and the atomic force laws) is known. The trapping range of the external force is always finite. Thus, the energy barriers disappear at the ends of the trapping range, and are a maximum near its center. As a consequence, the energy barriers are a function of the external force, F. Recalling the smooth quadratic energy function derived for the continuum crack, eq. (41), the existence of the lattice barriers converts the smooth parabola into a "dinosaur back" function, as shown in fig. 24. It is an important result of the general theory that the energy barriers approach an asymptotic constant value as the crack length increases. Thus, even though the barriers are due to the atomic character of the crack tip, they do not disappear as the crack attains a macroscopic size.

References: p. 1548.

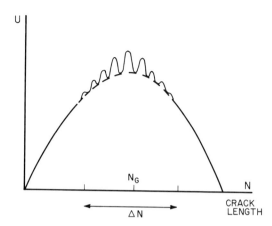

Fig. 24. Modified Griffith energy function. The smooth parabolic curve corresponding to eq. (41) (dashed), becomes a "dinosaur back" function. The trapping occurs over the range ΔN about the central maximum at N_G.

The important physical result following from the existence of lattice trapping is that thermal fluctuations which allow the barriers to be traversed will lead to a thermally generated slow crack growth. This crack growth has been observed in some glass systems (WIEDERHORN *et al.* [1980]) and probably exists in metal systems as well. Since the crack can move either forward or backwards by thermal fluctuations, the net velocity will be determined by the relative size of the barriers to forward and reverse movement. In particular, when the forward and reverse energy barriers are equal, the net motion will be zero. From fig. 24 the equality of energy barriers should occur at the top of the Griffith parabola, at N_G. In fact, it turns out that the energy barriers do not have their maximum value exactly at the top of the continuum parabola for subtle reasons, but nevertheless, a rigorous result (THOMSON [1980]) is that the crack is quiescent at the top of the Griffith parabola, and the condition for quiescence is the Griffith relation:

$$\frac{K^2}{2\mu} = 2\gamma(T).$$

(63)

In this expression γ is the thermodynamic surface work, also the surface free energy. In other words, γ contains temperature-dependent entropy terms in addition to the mechanical energy of breaking bonds. In eq. (63) the external force, F, is contained in K in the same way as for a standard macroscopic calculation.

On either side of the Griffith quiescent point, a stress- and temperature-dependent crack growth will occur. Model calculations by THOMSON and FULLER [1982] of crack growth in square lattices for a variety of force laws have shown that the crack velocity v is given by

$$v = v_0 \left\{ \exp\left[-U_0 (1 - K/K_+)^n / kT \right] - \exp\left[-U_0 (1 - K/K_-)^n / kT \right] \right\}.$$

(64)

Unless one is very close to the Griffith stress, only one of these exponentials is significant. K_+ is the value of K at the end of the lattice trapping range for forward growth. n was shown not to be very sensitive to the form of the force law, and has a value in the neighborhood of $3/2$; $n = 1$ is in fact a reasonable working assumption to use with experimental data.

One of the important applications of these ideas is to chemically enhanced fracture. In this case assume that an external chemical environment interacts with the stretched bonds at the crack tip, and becomes chemically adsorbed on the cleavage surface as the crack opens. At the crack tip, the force laws are modified, but the same arguments as before apply. In particular, the Griffith relation, eq. (63), is valid at the quiescent point, and the surface energy to be used is that corresponding to an adsorbed layer of the environmental species. Since chemical adsorption lowers the surface energy according to the Gibbs adsorption isotherm (ch. 6, §9):

$$\gamma(T, P) = \gamma_0 - \frac{kT}{b_0^2} \ln\left[1 + (P/P_0)^{1/2}\right]. \tag{65}$$

Thus, the local critical value for K will be shifted to lower values under chemical attack, and the material will be embrittled. In eq. (65) it is assumed that the external environment is a gas at pressure P. γ_0 and P_0 are relevant quantities at the "standard state", k is the gas constant, b_0 is the lattice parameter. If the external environment is a solution or liquid, an appropriate generalization is necessary.

The general result, eq. (65) and its application to eq. (63) and ultimately to toughness relations like eq. (59), are valid for the critical values of stress intensity k and K at the quiescent point, or the onset of fracture, because they relate to a thermodynamic condition at the crack tip. For this application, it is thus not necessary to know the details of the force-law changes induced by the chemical reaction at the crack tip. These changes are, of course, contained in γ, but γ is a macroscopic quantity *independent of any crack*, and can often be measured directly. If, however, it is desired to solve the problem of how fast thermal fluctuations cause the crack to grow on either side of the quiescent point, these questions lie in the realm of chemical kinetics, and force-law details are indeed then required. Provided the force laws are known, the problem is the same as that described in connection with fig. 23, and the activation energy is given by eq. (62).

An important insight into chemical effects can be gained by extending eq. (61) for nonlinear atom bonds at the crack tip. Suppose that the bonding force acting at the crack tip $n = L$ in fig. 5 is replaced by an external force, $-f$. Then, with two arbitrary external forces, f and F, the remaining *linear* system has the solution

$$u_0 = D_{00}F + D_{0L}f,$$
$$u_L = D_{L0}F + D_{LL}f. \tag{66}$$

The constants D are the compliance coefficients for the displacements of the central atom, u_0, and the tip, u_L in linear response to these two forces. If f is eliminated from eq. (66), then after taking differentials:

$$du_0 = \frac{D_{0L}}{D_{LL}} du_L + D_{0L}\left(\frac{D_{00}}{D_{0L}} - \frac{D_{L0}}{D_{LL}}\right) dF. \tag{67}$$

This equation can be substituted into the equation for the activation energy, eq. (62), with the result:

$$\Delta U = \int_1^2 \frac{D_{0L}}{D_{LL}} F \, du_L - F\Delta u_0. \tag{68}$$

The term $\int_1^2 F \, dF$ has been dropped from eq. (68) because states 1 and 2 in fig. 21 have the same value of F. Thus, $\int_1^2 F \, dF = 0$. Substituting for F from eq. (66) and for Δu_0 from eq. (67), we finally get:

$$\Delta U = -\frac{D_{0L}}{D_{L0}} \int_1^2 f(u_L) \, du_L + \left(\frac{D_{0L}}{2 D_{L0} D_{LL}}\right) \Delta(u_L^2) - \frac{D_{0L} F}{D_{LL}} \Delta u_L. \tag{69}$$

In eq. (69) we have finally inserted the fact that f is, after all, an interatomic bond force, and is given by the nonlinear force law shown in the right-hand part of fig. 21. Equation (69) is called the *decomposition theorem* and allows a very simple construction to be made which shows how external environments manifest themselves at the crack tip. The second and last terms of eq. (69) are completely independent of the external environment. Their contribution to the energy as a function of Δu_L is the parabola of fig. 25. In the figure the results of three examples of bond energy are shown as functions which are to be added to the parabola to give the total energy. If the bond function is weak and has a weak rise, as in (a), then the crack simply opens with little influence from the tip bond, and the crack is outside its trapping regime – in this case too long. If the bond is of medium strength as in (b), and operates in the region slightly to the left of the parabola minimum, a double valley is created with an activation energy corresponding to the crack being in its trapping regime. If the bond has, itself, an activation barrier, corresponding to an activation energy barrier for the chemical reaction of the molecular species at the crack tip as in (c), this additional chemical activation energy is added, roughly, to the lattice trapping barriers at the crack tip. Figure 25 and the decomposition theorem leading up to it depend only on the assumption that one atom pair at the tip contributes the major nonlinear effect to the system, which is expected to be a fairly good representation for actual brittle cracks.

When valid, the decomposition theorem is valuable because it then becomes possible to focus on the effects of the bond tip reaction independently of the complex interactions of this tip bond with the rest of the system. In this discussion, the term *tip bond* refers to the activated chemical complex composed of the two atoms originally at the tip combined with a molecule of the external chemical species. The bond energy function is then plotted as a function of the separation of the two original tip atoms in the complex.

Fracture in oxide glasses attacked by water and water-like molecules is the system in which the most detailed chemical picture has yet been worked out (see WIEDER-

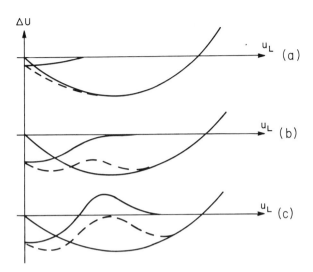

Fig. 25. Diagrams illustrating the decomposition theorem. The broad displaced parabola is composed of the second and third terms of eq. (69). The atomic bond function is displayed in each of three cases superimposed on the parabola. The resultant is the dotted curve. In (a) the bond function is weak and not near the parabola minimum; no activation energy. In (b), the bond is strong, and has its maximum slope close to the parabola minimum; activation energy. In (c), an additional chemical activation energy is present, which adds a large bump to the total activation energy.

HORN *et al.* [1980]). It is in general consistent with the decomposition theorem, and with the overall picture presented here.

6.2. Experimental observations

The question which immediately arises regarding lattice trapping is the size and magnitude of the effect. From calculations which vary the shape of the force laws, the trapping is found to be extremely sensitive to the "back side" of the potential function, where the interatomic bond weakens and the nonlinear aspect is predominant (see THOMSON and FULLER [1982]). In analogy to the Peierls calculation, for a soft bond release and long-range forces the trapping decreases rapidly to insensible values, while for a snapping type of release such as indicated in fig. 21 (l.h. part), the trapping is maximum, and the trapping stress range is a significant fraction of the Griffith stress. Because of the general lack of confidence in the numerical results to be obtained from atomic calculations, it is hard to say with any certainty, on the basis of theory alone, how significant the intrinsic barriers are in any particular material. When it is appreciated that the important parameter is the activation energy of a three-dimensional double kink on the crack, and not simply the two-dimensional energy barrier, the theoretical problem indeed becomes unmanageable. In a qualitative sense, just as in the analogous problems of the Peierls

References: p. 1548.

energy of a dislocation, metals probably have small barriers, while covalent materials probably have significant barriers. Calculations by SINCLAIR [1975] for the diamond structure do in fact suggest that kinks in that structure have energies which would be significant for thermal activation.

This question is more easily answered experimentally. The materials where completely brittle cracks are known to exhibit the most clear-cut example of thermally activated growth are the silica and silicate glasses. Table 1 shows the results for a number of glasses. The intrinsic stress dependence of crack velocity in vacuum is very steep, showing that the activation volume for crack growth is large, when compared with chemically enhanced slow crack growth. Slow crack growth has been looked for in Si, where one would expect on theoretical grounds to find it, but so far conclusive evidence for or against the barriers in this case is not available because the material has not been studied in the high vacuum necessary to control surface reactions. One of the possibilities for a material like silicon is that in fact the barriers are high, and that like the silica glasses, the activation volume is so high that the stress dependence is unobservably steep on a plot like fig. 26.

When it comes to chemically enhanced crack growth, the picture is more straightforward. Again, the clearest case is the completely brittle fracture in silica and silicate glasses. In these materials, water and water-like molecules give rise to a stress-dependent and temperature-dependent slow crack growth which is consistent with the theoretical ideas above. Figure 26 shows the effect of water on glass, with three different regions of crack growth. In region I, the growth is enhanced by chemical reaction with the water, depends on the water vapor pressure, and is stress dependent as predicted. In region II, the crack outruns the water reaction, and jumps to a faster growth curve, stage III. Stage III is similar to the steep curve of crack growth in vacuum, but is influenced by the dielectric properties of the environment.

MICHALSKE and FREIMAN [1982] have proposed that the water reaction features

Table 1
Summary of stress corrosion data in glass (WIEDERHORN *et al.* [1980]).

Environment	Glass	E kcal/mol (kJ/mol)	b $m^{5/2}/mol$	$\ln v_0$
Water	Silica	33.1 (139)	0.216	−1.32
	Aluminosilicate I	29.0 (121)	0.139	5.5
	Aluminosilicate II	30.1 (126)	0.164	7.9
	Borosilicate	30.8 (129)	0.200	3.5
	Lead-alkali	25.2 (106)	0.144	6.7
	Soda-lime–silica	26.0 (108)	0.110	10.3
Vacuum	61% lead	83.1 (348)	0.51	5.1
	Aluminosilicate III	176 (705)	0.77	14.1
	Borosilicate crown	65.5 (275)	0.26	6.6
	Soda-lime–silica	144 (605)	0.88	−5.7

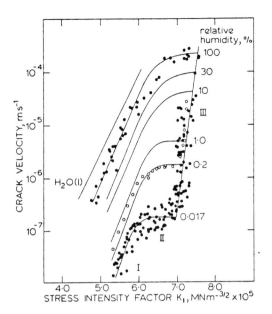

Fig. 26. Slow crack growth of glass in water. Three regions are observed. Region I corresponds to the chemically enhanced growth theory displayed here. In Region II, the external environment is outrun by the crack. Region III is the intrinsic growth region. (After WIEDERHORN *et al.* [1980], courtesy S. WIEDERHORN).

both hydrogen and electron donors in the reaction, leading to enhanced crack growth. Consistent with this proposal, they have also shown that molecules with a similar capability such as NH_3, N_2H_4, and CH_3NO have similar crack growth curves. The crucial role of hydrogen is confirmed by the fact that D_2O has a slower growth than H_2O. In other brittle materials, slow crack growth is also generated by external environments, but glass has been the most extensive case to be studied and provides the clearest picture.

In metals, the dislocation atmospheres becloud the situation. Chemical effects are certainly widespread, however. For example, hydrogen and other gases embrittle bcc metals. Also, fcc ductile metals can be made brittle by surface electrochemical reactions in solutions, a phenomenon which goes under the general title of stress corrosion. Finally, some low-melting-point metals in liquid form have dramatic embrittling effects on ductile metals. In all these cases, it is our opinion that the basic ideas of dislocation shielding and core crack chemical effects related to surface energy changes will ultimately be shown to play their obvious roles. However, in all these cases, partly because of the dislocation atmosphere the mechanisms have not yet been sorted out. We shall return to a limited discussion of some of the specific findings in a special section devoted to the metallurgy of fracture at the end of the chapter.

References: p. 1548.

7. Atomic structure of cracks: dislocation emission

In addition to the discrete lattice effects described in the last section, which lead to chemical and thermal fracture effects, the second major aspect of fracture which is a consequence of the atomic structure of the crack tip is the stability of a crack in a lattice.

KELLY et al. [1967] discussed the fact that an atomically sharp crack by definition has at its tip a stress equal to the theoretical tensile strength of the solid, and should therefore in the same region possess shear stresses of equal magnitude. Since, in general, the theoretical shear strength is lower than the theoretical tensile strength of a solid, one might suppose that materials with a sharp crack would easily break down in shear at the tip and become blunted and no longer atomically sharp.

In fact, such a shear breakdown is equivalent to the spontaneous emission of a dislocation from the crack tip (see fig. 27). A calculation on this basis has been made by RICE and THOMSON [1974] of the stability of a sharp crack in terms of the three forces acting on a dislocation as it is created out of the crack tip. The two principal forces are those represented in the first two terms in eq. (35), and a third weaker term is due to the fact that the emerging dislocation also interacts with the additional surface ledge created at the tip. This third term is very short-range and weak, so will be neglected in the following.

As seen in eq. (35), the K field for a positive blunting Burgers vector generates a repulsive force, and the image term is attractive. Because of the different ζ-dependence of these two terms, the image force is dominant at close distances, and the K field is dominant at larger distances. Thus, as shown in fig. 28, there is a balance point determined by setting $f = 0$ in eq. (35) with only one dislocation present, specified by the relation:

$$k = \frac{\mu b}{2(2\pi\zeta_e)^{1/2}}. \tag{70}$$

Carrying out an analogous argument for a mode I crack leads to a more complicated expression wherein eq. (70) is multiplied by a crystallography-and slip-plane dependent quantity. We shall use eq. (70) for the following qualitative discussion.

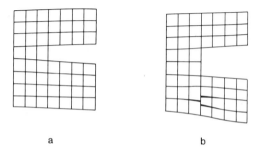

a   b

Fig. 27. Crack tip breakdown in shear by dislocation emission.

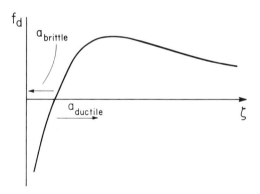

Fig. 28. Force between a crack and a dislocation as a function of the distance, ζ, between them. The core size of the dislocation, a, is shown in relation to the null point of the force for brittle and ductile cases.

When k is in the neighborhood of the Griffith value, the crossover, ζ_e, is a distance of atomic dimensions. On the other hand, unless the cross-over, ζ_e, is greater than the core size of the dislocation, the region of attractive force will not exist, and an emerging dislocation will always experience only a repulsive force. In this circumstance, the crack will spontaneously emit a dislocation. If the value of k at the crossover is labelled k_e, then

$$k_e = \frac{\mu b}{2(2\pi a)^{1/2}} .$$ (71)

In this expression, a is the core size of the dislocation.

Of course, cleavage will occur when the Griffith relation is satisfied, $k^2 \geqslant 4\mu\gamma$ in mode III, or in general:

$$k_c = (4\mu\gamma)^{1/2}, \qquad k_{Ic} = \left(\frac{2\gamma E}{1 - \nu^2} \right)^{1/2}$$ (72)

The subscript c refers to the critical k for cleavage using the Griffith criterion. When $k_c > k_e$, emission occurs before cleavage, and when $k_c < k_e$, vice versa.

In these considerations, the quantity $\mu b/\gamma$ obtained by setting $k_c = k_e$ is the major nondimensional materials factor which is important. When this factor is low, emission is favored, when high, cleavage is favored. The stability argument centered on dislocation emission is physically identical to the earlier proposal of Kelly *et al.* based on relative shear and tensile strengths. Using an estimate of the relation between γ and tensile strength originally given by OROWAN [1949], LIN [1982] has shown that the ratio $\mu b/\gamma$ also comes naturally out of the argument that the shear strength must not be exceeded at a crack tip in order for the crack to remain stable. Numerical estimates of the stability criterion have been given by Rice and Thomson which suggest that fcc metals tend to be unstable against dislocation emission, while such materials as diamond, hcp metals and ceramics are stable. These authors report

References: p. 1548.

iron as borderline, and tungsten as stable. Using different values of γ for W, OHR and CHANG [1982] claim Nb, Mo, and W among the bcc metals are also barely unstable.

In addition to the authors quoted above, the problem of spontaneous dislocation creation in the near vicinity of the crack has been studied by ARMSTRONG [1966], ATKINSON [1966] and WEERTMAN [1981] in the continuum approximation.

All these arguments relate to the inherent two-dimensional mechanical stability of a straight crack. The question also arises whether a dislocation loop can be emitted from a stable crack with a finite activation energy. The problem now becomes three-dimensional and a full three-dimensional calculation has yet to be performed, but estimates of the activation energy have been made by RICE and THOMSON [1974] for a number of materials. In these calculations, the activated state was assumed to be a semicircular dislocation ballooning out of the crack tip. Qualitatively, except in one case, the energies calculated were surprisingly large, which suggests that if a crack is stable at all in a material, it is quiescent relative to any thermally generated dislocations. The exception was bcc iron, where the activation energy was estimated to be about 2 eV. The accuracy of such estimates is of course low, and it is only suggested here that iron may be on the borderline between cleavage and emission.

In all this work, the properties of cracks and dislocations are inferred by elastic calculations down to within a few atomic spacings from the center of the defect, and recourse must be made there to cut-off techniques. Although the advantage of such elastic arguments is that broad classes of materials can be compared with one another for dicussion of trends, specific predictions must be viewed with caution when it comes to a particular material. Full scale atomistic calculations have the advantage that in principle, the stability of the structure could be addressed in a straightforward and complete way. Unfortunately, however, force laws for materials, and metals in particular, are not available for atomic configurations so distorted as a crack tip. This point is of far-reaching significance for finding the structure of any defect in a solid, but the crack is an especially severe test for any atomic configuration calculation attempt. Not only is the distortion very high in both tension and shear, but the crack also includes a surface which drastically alters the electron distribution (and thus the force laws) of the atoms. Most structural calculations of defects make use of two-body potentials, but again this approximation needs to be validated for the crack, and appropriate forms of the force law determined. At the time of writing, bonding in metals containing defects is an area of intense fundamental study (see NATO 1980 Conference discussions as reported by LATANISION [1982]), but the direction in which to look for answers to these questions is not yet clear.

Nevertheless, even without the needed fundamental guidance on what type of force laws to use, much can be learned from computer simulations. For example, using specific force laws for model solids, one can check the elastic criteria for stability based on eqs. (71) and (72) to see how bad the cut-off procedures are. This has been done, for example, by PASKIN et al. [1980] using a nearest-neighbor two-body potential function of type $V = A/r^{12} - B/r^{6}$. Their simulation indicates that this model solid only barely sustains a stable sharp crack, and that the elastic criterion from eqs. (71) and (72) as estimated from the potential function parameters agrees with the actual model. In other work on a simulated model of iron, again the computer results by KANNINEN and GEHLEN [1972] indicate that the model is very close to instability, in agreement with the elastic prediction for iron. However, these are isolated instances. One could gain considerable insight from computer model calculations in which crystal structure and force law forms were systematically varied, even though a direct connection to specific metals might be hard to make.

OHR and CHANG [1982] have developed a model of mode III cracking using the criteria (71) and (72) together with shielding calculations along the lines of the last section in comparison with experimental observations (see fig. 13) in a variety of materials (Fe, Cu, Ni, Al, Nb, Mo, W, MgO) to show that the crack will emit dislocations so long as the local k satisfies eq. (71). In all their cases, in mode III, cleavage does not intervene, and $k_e < k_c$ is implied. In this work, the role of the dislocation-free zone, as discussed for cleavage in eq. (59), and general validity of the dislocation screening and emission ideas is beautifully confirmed. The general conclusion from this work is that even such relatively "brittle" materials as niobium and tungsten can operate in a fully ductile fashion in mode III. Somewhat different observations by WILSDORF [1983], again in thin films, of fcc metals confirm the general picture of dislocation emission from cracks and sharp corners in fcc metals. Care must, of course, be taken in extending these results on thin films directly to bulk fracture.

In other observations by VEHOFF et al. [1981], confirmed by OHR et al. [1983], however, the bcc metals generally appear to be close enough to a stable crack configuration that simultaneous dislocation emission *and* cleavage can occur. This possibility is most strikingly demonstrated by VEHOFF et al. [1981] (fig. 12). They find that the normal fracture in silicon–iron is the ductile slide-off process first proposed by COTTRELL [1965], which produces a crack-opening angle equal to the angle between the active slip planes at the crack tip. When hydrogen gas is introduced, the fracture is altered to a mixed cleavage–emission mode in which the crack opening angle is any desired value from the fully ductile value down to zero. They also showed that by increasing the strain rate, the same effect on the crack opening angle can be induced.

The hydrogen experiments by Vehoff and Neumann show that the cleavage–emission competition occurs over a considerable range of hydrogen pressure, and that the Griffith point for cleavage quiescence is governed by the Gibbs adsorption relation, eq. (65).

By varying the pressure, P, one can vary the relative activation energies for cleavage over the lattice–chemical barriers, versus the thermal barriers for dislocation emission. Thus, these experiments suggest that in silicon–iron, both cleavage and dislocation emission processes are thermally activated.

The simultaneous emission and cleavage observed by both experimental groups in the body-centered metals suggests that very interesting rate effects occur at cracks in bcc metals. Apparently, the cleavage–dislocation emission competition at the crack

References: p. 1548.

tips is more complex than the either/or criterion enivisioned in the elastic calculations leading to the criterion (71)–(73). McMAHON *et al.* [1980] have been the first to
pursue this possibility, theoretically. In the view of these authors, interactions of the
dislocations with the crack going beyond the simple shielding and emission ideas
developed in the last two sections are at work; perhaps along the lines of an idea due
to KNOTT [1977] that dislocations may be reabsorbed in the cleavage plane after the
crack moves on; and perhaps due to an emission–cleavage reversal as suggested by
FINNIS and SINCLAIR [1982] after multiple dislocation emission has occurred.

8.   *Summary of basic ideas*

After the discussion of the fundamentals of fracture in the previous sections, and
before going on to some of the more applied aspects of fracture, it will be useful to
summarize here the general conceptual structure as it has been developed.

a. The material is intrinsically brittle if it can sustain an atomically sharp crack in
the material without shear breakdown and dislocation emission. The material is
intrinsically ductile if it cannot.

b. If the crack is sharp at the tip, then there is a local k-field, and the quiescent
point for the crack is governed by the macroscopic thermodynamic Griffith relation.
In general, and especially in the case of chemically enhanced fracture, lattice effects
and chemical reactions at the crack tip will lead to thermally activated stress-dependent slow crack growth. When significant dislocation screening fields are present, of
course, time-dependent dislocation processes will also contribute to the total rate
process, and these may be dominant in certain regimes of crack velocity. However,
the intimate balance always achieved between screening field and core crack must
always be kept in mind in considering the total time-dependent fracture growth.
Ultimately, local rates at the crack tip depend upon the activation barriers of the
chemical debonding and rebonding best thought of in terms of general chemical
reaction rate theory at the crack tip. Fracture in vacuum is a unimolecular decomposition of the lattice, while chemically enhanced fracture is a bimolecular (or higher
order) process. In the highly constrained configuration space of the crack tip, a full
scale fundamental theory will undoubtedly never be fully worked out. However,
semi-empirical chemical theories with guidance from fundamental quantum-mechanical bonding studies should be possible and ultimately very fruitful in piecing
together an adequate picture of the local crack growth law.

c. If the sharp crack is stable, then all metals, and nearly all other materials as
well, possess dislocation sources close enough to the crack to create a dislocation
screen which shields the crack from the external stress field. For a static crack in
equilibrium, detailed balance applies: the local core crack is in equilibrium with its
local fracture criterion (its simplest form is the Griffith relation). Each dislocation in

the cloud also takes an equilibrium position in which the overall configuration depends on the dislocation friction stress (or local yield stress), and the size of the elastic enclave at the crack tip. If the configuration is in motion, the same considerations apply, except that k and σ_f are velocity-dependent.

d. If a crack is not stable against spontaneous dislocation emission, the whole conceptual structure for quasi-brittle fracture in terms of a core cleavage crack shielded by dislocations breaks down, and a new set of ideas is necessary. The first interesting case is that of a mixed-mode response apparently realizable in the bcc metals in which both cleavage and emission may occur simultaneously. In this case, the Griffith relation governing brittle fracture must be modified to include the energy of the emitted dislocations. The rate of emission will be determined by the relative barriers for cleavage versus emission, but not all of the emitted dislocations will join the screening dislocation field if they fall back into the cleavage plane before the crack has moved completely away. The detailed analysis of this problem has only just begun.

e. If the crack is completely unstable, as in the fcc metals, then a sharp crack will break down under stress by continuous dislocation emission, and the crack will open with a notch of finite angle equal to the angle between the active slip planes (fig. 12). If the fracture is true mode III, then the notch angle is zero, and the emitted dislocations form a dislocation screen. The screening analysis developed in § 5 applies, and analytical predictions have been made for the fracture criterion in an analogous way as for the cleavage case (OHR and CHANG [1982]). If the notch angle is finite, however, although some numerical shielding analysis has been done, a complete picture has not yet been attempted. In all problems where major geometric deviations occur from the sharp crack, global compatibility constraints having nothing to do with crack tip phenomena will be important in determining the final geometrical shape of the crack, e.g., whether the far surfaces of the crack are parallel or wedge shaped. Although a wedge-shaped crack can grow continuously by successive dislocation emission as discussed, in general practice the fully ductile case is normally characterized by discontinuous hole nucleation and growth, so that the overall crack advance is too complex for the type of quantitative analysis envisioned here.

f. From a theoretical point of view, it is clear that the atomic structure of a crack tip is crucial in determining the fundamental criterion whether the material is intrinsically brittle (stable sharp crack) or intrinsically ductile (emitting crack). It also determines the cleavage criterion and rate laws. Major challenges exist in working out a semiquantitative understanding of this essential crack tip structure.

References: p. 1548.

9. Some practical implications

In this chapter, a very fundamental approach to fracture has been followed in order to develop a few central ideas regarding the interactions between cracks and dislocations. We have mostly focussed on sharp cracks and relatively brittle behavior, whereas engineering materials are deliberately designed to fall more in the middle of the brittle–ductile spectrum in order to optimize both high strength and high toughness. Consequently, engineering materials will show a much more complex mechanical response than can ever be quantitatively described by means of the fundamental properties of simple dislocations and cracks. Nevertheless, the fundamental ideas are useful as thinking tools when one is faced with the more complex systems, and in this short section, we shall tour briefly several areas of practical interest with these tools in hand. At the time of writing, this attempt will be more a promise and challenge for the future than a demonstration of current capability, however. This state of affairs exists partly because the fundamentals are presently in a state of intense investigation, with new insights still in the wings (especially in the area of dislocation-emission ideas as applied to ductile fracture), and partly because the application of these ideas to practical problems is still in a rudimentary and often controversial state.

9.1. Implications for final materials reliability

In the widest sense, fracture science and technology is a part of the overall concern for the reliability of a material in its ultimate use. This problem has been important to mankind since he first used tools of any kind, and is certain to remain with him as new materials are developed for new uses in new environments. Early engineering approaches to materials reliability made heavy use of safety factors and relatively ductile materials. More recently, the need for high materials performance and the heavy penalty paid for large safety factors in such areas as aerospace have driven the development of materials technology relentlessly, resulting in ever greater sophistication in materials engineering practice. One of the penalties paid for higher materials performance, unfortunately, is a greater exposure to the possibility of brittle failure, and the sudden catastrophe that often results from it, because of the rough correlation which exists between higher strength and lower toughness. Although there are ways to mitigate this tendency, it arises simply because when dislocation activity is constrained in order to obtain high strength, the ability of the material to shield its cracks diminishes as well.

The current approach to materials reliability is along two related paths. The straightforward first path is simply to optimize the strength–toughness durability through materials development. Two of the most fruitful approaches to this goal are: (i) Control of grain structure and second-phase morphology, usually by increasing uniformity and decreasing grain and precipitate size; the Hall-Petch relation [eq. (74)] is relevant here. (ii) Attention to embrittlement factors such as external chemistry, temperature, etc.

The second path is loosely called the *fracture mechanics* approach and works from the critical flaw concept. This concept connects the macroscopic stress in a part to the critical-sized crack which can be sustained, by means of eq. (14), and to the critical stress intensity, K_C, which a crack in the material can sustain before it advances. These relations can be expressed from eq. (14) as

$$\sigma_\infty = \frac{K_C}{(\pi a)^{1/2}},\tag{73}$$

where K_C is the same materials toughness parameter modeled in § 5. It is determined empirically by standardized measurements under appropriate conditions, and can be generalized to more ductile situations as explained in § 3.

In critical applications, the generalization of eq. (73) is derived from a computed stress-distribution in the part, so that a critical-sized flaw can be inferred for any portion of the structure or part for the assumed design maximum stress in the structure. The material toughness for the given service conditions must be known, of course. It then becomes a quality assurance problem during manufacture or construction to ensure that no flaws larger than the critical size are present, and a maintenance problem to ensure that the flaws do not grow to the critical size during service. Since cracks can grow owing to fatigue during service, the in-service inspection problem is often the most critical one. The technology for finding and evaluating flaws in materials is termed *nondestructive evaluation* (NDE) or *nondestructive testing* (NDT). Thus *materials reliability assurance* is a three-cornered technology involving materials technology through materials toughness considerations, mechanics through stress analysis, and NDE.

The materials technology "input" to materials reliability assurance in terms of the critical toughness parameter is decidedly complex. First, the flaws in materials are never the idealized cracks visualized in earlier sections, but voids and odd shapes associated typically with other inhomogeneities in the material. Hence the critical K for a given material must include a judgmental factor regarding how "flaws" relate to the idealized cracks studied in standardized measurements. Aside from this, however, all the materials science and technology already referred to are relevant to determining and enhancing toughness. Over and above all, however, ride concerns for a change of mode from ductile to brittle under conceivable service conditions.

In spite of all the complexity and implied uncertainty apparent to the reader, the use of the fracture mechanics approach to materials design and reliability assurance has been a great boon to designers in situations where reliability and performance must be optimized. It promises even more success for the future, as the materials aspects of fracture become better understood, and the NDE technology becomes better developed. For further perspective on the subject of materials reliability in general, the reader is referred to the excellent and mainly qualitative monograph covering the Royal Society Discussion of the subject, FORD *et al.* [1981].

9.2   Brittle crack initiation

Logically the next topic in this section should address the problem of how cracks are formed in a material. Generally speaking, of course, they exist in any material of technological interest as processing artefacts. But even when care is taken to reduce pre-existing cracks, they can be initiated by a variety of mechanisms associated with dislocations and twins. For example, when dislocations pile up against any obstacle such as a grain boundary, it has been proposed by STROH [1957] that the edge dislocations at the head of a sufficiently intense pile-up coalesce into a superdislocation with the lattice below the extra half-planes opening into a mode I crack (SMITH [1979]).

A review of the current status of the complicated subject of crack initiation in steels in general will be found in KNOTT [1977]. For the important special case of the initiation of transgranular brittle cleavage, it is found that the brittle carbide precipitate particles in the steel first crack because of the strain-incompatibility due to the brittle particle in the deforming matrix, or in other words, because of the pile-up of dislocations at the particle. This process is similar to the initial stage depicted for ductile hole growth in fig. 7. The initial size of the crack is then given by the size of the brittle particle, and the crucial question is whether the cleavage crack can then propagate through the ferrite matrix by means of the local effective stress on the crack. Provided the crack can propagate through the first ferrite grain, it can then generally propagate through the entire steel sample.

SMITH [1966] has published a successful, but qualitative model of the stress to nucleate and propagate a crack under these conditions which depends both on the size of the carbide particle and the effective shear stress caused by a dislocation pile-up. An important feature of this theory is the grain-size dependence of the fracture stress. Since the stress to form and propagate the crack depends on the effective shear stress, σ_{eff}, in the grain as given by the Hall–Petch relation (ch. 19, § 8),

$$\sigma_{eff}^2 = \lambda d, \tag{74}$$

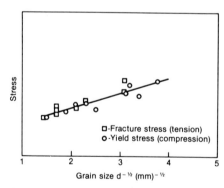

Fig. 29. Comparison of yield stress and fracture stress of steel at 77 K (after Low [1955]).

the fracture stress should also obey this relation. Here λ is a constant and d is the grain size. A striking verification of the same grain-size dependence of fracture and deformation is shown in fig. 29, after early work by Low [1955]. Later work and the comparison with the theory by Smith has been reviewed by KNOTT [1977].

9.3. Ductile fracture and hole growth

If a sharp crack is injected into a ductile specimen by some means such as that of §9.2, when a stress is applied, the first macroscopic observation is that the crack blunts plastically into a cigar-shaped cross-section with a significantly increased COD. This stage of the specimen compliance is termed the *stretch zone*. Continuum plastic studies have shown that such a cigar-shaped crack has maximum stress not at the surface of the tip, but at a point in the material in front of the crack, a distance about one COD from the tip (see KNOTT [1973]). There also exists a maximum negative pressure in the same region. As the stress is increased beyond the stretch zone, holes form in the region of maximum negative pressure, which under continued straining grow and eventually connect to one another and to the main crack (see fig. 8). Further straining simply continues this process. Most studies of this phenomenon (see KNOTT [1973,1977]; ARGON *et al.* [1975]) are made in multicomponent metals, in particular steel, and the holes are found to form at the sites of the minor-phase particles (fig. 7). Sometimes the particle interface is the origin of a hole. In other cases where the interface is strong, the particles themselves first crack in brittle fashion as discussed in §9.2, and this crack then blunts as it runs into the matrix. Even in pure metals containing no particles, such holes still form, apparently by nucleation of vacancy clusters in the dislocation cell walls (GARDNER *et al.* [1977]). One of the striking features of the hole growth fracture process is the development of bands of localized intense shear between the larger holes, within which large numbers of small holes are formed, as shown by Cox and Low [1974]. The origin of the shear instabilities is still being sorted out.

9.4. Ductile–brittle transitions – temperature and rate effects

As implied at various points in the chapter, aside from fatigue to which a whole chapter is devoted in this book, perhaps the most serious concern of the materials engineer when it comes to practical fracture control is the worry that a material may undergo a transition from its normal toughness to a much lower and dangerous value because of some unforeseen change of conditions, either in the material or in the way it is used. In general, the materials parameters which control toughness once the material has been fabricated are temperature, strain rate, and external chemical environment. The first two are usually linked effects in materials science for general reasons, so will be discussed together here, and discussion of chemical embrittlement will follow separately.

Much of the early study of fracture control in metals focussed on the transition which mild steels undergo between their high-temperature and low-temperature

References: p. 1548.

regimes, and the dramatic nature of this transition is perhaps best shown by a schematic example of these early tests (fig. 30). In this figure, the energy necessary to fracture a notched specimen is measured by a *Charpy apparatus*. In this measurement, a standard specimen is struck by a hammer pendulum, and the height of fall of the pendulum when the complete fracture occurs is converted to the energy plotted. The lower portion of the curve (the so-called *lower shelf*) corresponds to K_{1C}, or a dynamic version of it, rather straightforwardly. Above a rather indeterminant point, say near halfway in the rise to the *upper shelf*, the material exceeds its yield point through most of the specimen, and the K is no longer well defined. Nevertheless, the entire curve gives a good qualitative picture of material toughness, and the Charpy measurement is still used in many materials selection applications. As the diagram shows, the transition is also associated with a change in appearance of the fracture surface from a hole growth to a cleavage mode.

The temperature range of the ductile–brittle transition is a function of the strain rate as shown in fig. 31, and a function of microstructural variables, as shown in figs. 32 and 33.

Another extremely important factor in the brittle–ductile transition problem in plates is the effect of plate thickness. This effect is, strictly speaking, not a materials problem, but a mechanics effect associated with the size of the deformation zone around a crack in comparison with the plate thickness. Nevertheless, it is an important effect to the materials engineer. It is due to the change in stress state associated with the change from "plane strain" to "plane stress" (KNOTT [1973]) when the deformation zone decreases to the size of the plate thickness. Thus, the entire transition curve becomes a function of plate thickness. For thin plates, the transition temperature shifts to the left relative to thick plates, that is, thick plates are in a sense more brittle than thin ones.

This mixed bag of temperature, metallurgical variables, strain rate and specimen thickness leads to a very complex engineering problem in practice, which in critical

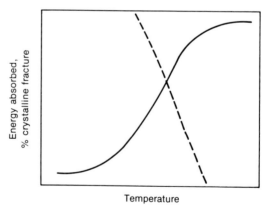

Fig. 30. Typical transition curve for mild steel. The solid curve gives energy absorption; the dashed curve shows "percent crystalline fracture" as estimated from surface appearance.

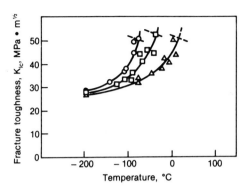

Fig. 31. Effect of strain rate on fracture toughness of ASTM A36 steel as a function of temperature. The curves are for strain rates of $10 \ s^{-1}$ (triangles), $10^{-3} \ s^{-1}$ (squares) and $10^{-5} \ s^{-1}$ (circles). (After Barsom [1979].)

applications is the subject of extensive data collection and code formulation efforts. A good introduction to the subject will be found in Knott [1973] and, beyond that, in the excellent and practical book by Campbell *et al.* [1982] and the various publications of the ASTM.

From the scientific point of view, the reason why a material is basically brittle or ductile is at the core of much of our discussion in this chapter, and most of the important fundamental questions about it still remain before us. One of the striking

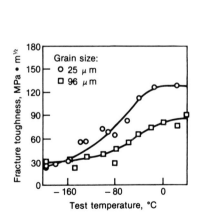

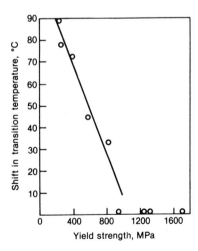

Fig. 32 (left). Effect of grain size on fracture toughness. Material is St37-3 steel in grain sizes shown. (After Dahl and Kretzchmann [1977].)

Fig. 33 (right). Effect of yield strength on shift in transition temperature between impact and slow-bend test loading rates. (After Barsom [1979].)

References: p. 1548.

features of both ductile and brittle fracture in steels is that in their earliest stages they are so similar. That is, in both cases a brittle crack exists either in the second-phase carbide particle, or at its interface, and the subsequent response of the ferrite matrix determines whether the material is ductile or brittle. As stated above, this response is both temperature- and rate-dependent; however, in view of the uncertainty concerning dislocation emission at sharp cracks in iron as explained in §7, and the problems associated with modeling cracks moving through a deforming matrix along the lines of §5, this area deserves much deeper study.

9.5. Chemical effects – hydrogen embrittlement

Chemical effects in metals are more complex than in brittle materials, because of the effects of chemical interactions on dislocations and dislocation emission as well as on the local k of the crack tip.

However, chemical effects are very widespread, and involve many different metals as well as different chemical interactions. We will mention here the effects of hydrogen, liquid metal embrittlement, and temper-embrittlement due to precipitate growth on grain boundaries. Stress corrosion is a further large field where chemical effects at cracks can embrittle a metal, but the electrochemical problems involved would carry us too far afield to discuss here.

Deleterious effects of hydrogen on steels have been observed since the 19th century, but the field has been characterized by controversy and confusion about the basic mechanisms involved. The reader is referred to the two excellent recent reviews by HIRTH [1980] and by ORIANI [1978] for a detailed discussion of hydrogen in iron and steel. In steels, the phenomenology is characterized by the facts that H diffuses very rapidly (10 μm in one second at room temperature), has a complex interaction with dislocations, has a very low (but stress-dependent) solubility of about 3×10^{-8} atom-fraction at room temperature and 1 atm pressure, but can be present in much larger amounts at traps in the medium. Further, the embrittlement in high-strength steels is much more drastic than in low-strength steels.

In high-strength 4340 steels, with a toughness value of 60 MPa$\sqrt{m}$, one atmosphere of hydrogen lowers the toughness to about 15 MPa$\sqrt{m}$, according to experiments by ORIANI and JOSEPHIC [1974]. The pressure-dependence of embrittlement over a pressure range of 0.1 to 1000 Torr can be fitted to a law $K \propto \ln P$, which is what one would expect if the primary effect is the change in the surface energy via the Gibbs adsorption equation, eq. (65). In these experiments, the hole-growth ductile fracture of the initial steel converts to a cleavage fracture on the prior austenite grain boundaries. Under these conditions, it is possible to model the toughness entirely in terms of the change in γ (THOMSON [1978]), or as ORIANI and JOSEPHIC [1974] have called it, by decohesion of the bonds at the crack tip. However, it is not possible to rule out the possibility of diffusion of hydrogen into the local region in front of the crack where discontinuous cracking might occur (see GERBERICH et al. [1975]). In either case, i.e., if the crack proceeds by hydrogen attack at the tip of the main crack or is nucleated at a small distance ahead of it, changes in surface energies or in interfacial energies are the crucial events.

In other cases of lower strength steels, LEE *et al.* [1979] have shown that macroscopically the cracking occurs at the surface when dynamic charging takes place, but it is usually believed that in lower-strength steels, diffusion of the hydrogen into the critical regions of high triaxial stress in front of a blunted crack is important. In low-strength steels, the action of the hydrogen has also been shown to enhance hole formation at precipitate particles (LEE *et al.* [1979]), presumably by lowering the interfacial energy. Moreover, an enhancement of plastic instability is also observed by the same authors, which is not yet understood. These effects can be engendered either by introducing hydrogen as an external gas, or by prior charging of the specimen to very high supersaturations of internally bound hydrogen.

The direct effect of hydrogen on a dislocation in pure iron is to lower the Peierls energy and thus soften the material (HIRTH [1980]). If the iron is impure however, addition of hydrogen hardens the material, presumably because of chemical interactions between hydrogen and oxygen, carbon, etc., impurities. Thus, the effect of hydrogen on the plastic properties of iron and steel is certainly complex, especially since the dislocations can transport hydrogen with them through the lattice. It is not surprising, then, that the full role of hydrogen in embrittling steels has not been sorted out and that opinion on which mechanisms are working is controversial.

Hydrogen also embrittles other metals. In bcc cases where, like iron, the solubility is low, similar effects as those in irons are observed, and similar mechanisms proposed. In numerous other systems, hydrogen is relatively soluble, and forms a stable hydride phase. In these metals, the hydride formation is generally aided by the stress in the vicinity of the crack. Since the hydrides are all found to be extremely brittle, the formation of the brittle phase at the crack tip drastically embrittles the material, presumably by inhibiting the formation of dislocations near the crack. These systems have been intensively studied and the embrittlement model verified by BIRNBAUM [1979], to whom we refer the reader for further details.

Hydrogen possesses a special place as an impurity in a metal, and as an embrittler, because of its small atom size and relative ease of diffusion. However, other impurities also embrittle metals because of their effect on metallic binding, and they can play this role either as agents in an external environment in contact with a crack tip, or as internal impurities. For example, segregation of impurities on grain boundaries often lowers the grain-boundary interfacial surface energy, which thus affects one of the crucial factors in fracture. (See ch. 13, §6.)

9.6. Grain-boundary impurities

Previous discussion in this chapter has not explicitly considered the possibility of cracking along a grain boundary, but a simple extension of the earlier argument in §6 would say that the quiescent point should be governed by a Griffith condition for the local k with an effective γ of the form

$$\gamma \to \gamma - \tfrac{1}{2}\gamma_{GB}, \tag{75}$$

References: p. 1548.

where γ_{GB} is the interfacial free energy of the grain boundary. An immediate issue arises in how to define the free-surface γ on the right-hand side of eq. (75). To be consistent with earlier use in §§5 and 6, γ is the equilibrium value for the material in contact with its external environment. When the grain boundary is cleaved, the impurity concentration present on the suddenly opened free surface will generally not correspond to that in equilibrium with the external environment, however. Thus a very slow crack growth during which chemical exchange can take place to bring about complete equilibrium at all times will have a different "γ" than one in which the crack grows fast compared to the time for chemical relaxation to equilibrium. The faster growth case will correspond to constrained thermodynamics in which vibrational contributions to surface entropy will be allowed, but not matter exchange. This effect has been discussed by HIRTH and RICE [1980].

On a deeper level, fracture along a boundary is not only one of the most important fracture paths in a material, but the special properties of boundaries add a new dimension to the topic of crack behavior. The reader is referred to a review of this subject by McMAHON et al. [1980]. One of the most important actions of the boundary is to act as a special site for the accumulation of impurities, and in this connection the chemical effects between impurities as they interact on a boundary is an important special case. It is especially important in engineering applications, for example, to recognize that Sn, Sb, P, and Si have attractive interactions with Ni in steels. But, the effect of such interactions may *in*crease or *de*crease embrittlement, depending on whether cosegregation occurs on the boundaries. Mo in steels, for example, interacts strongly with P, but remains in solution, and prevents segregation on the boundaries (see also ch. 13, §6.3). There is also evidence that hydrogen embrittlement may in some or even most cases be enhanced by hydrogen interactions with other impurities on the prior austenite boundaries in steels (TAKEDA and McMAHON [1982]).

Finally, we note that, on a fundamental level, the problem of dislocation emission by a crack on a boundary will require a different analysis than was used in the bulk. Cleavage on the boundary will be a function of the weakened cross-boundary bonds, whereas the dislocation after it is emitted will reflect the stronger bulk elastic properties. Thus, a boundary crack will be more "brittle" than one in the bulk. This argument is consistent with interpreting the brittle–ductile criterion in terms of the ratio $\mu b/\gamma$, in which μ is the bulk value, and γ is replaced by $\gamma - \gamma_{GB}/2$.

9.7. Liquid-metal embrittlement

Perhaps the most drastic chemical effect on fracture of a metal is caused by certain liquid metals. For example, simply painting a polycrystalline ductile fcc metal like Al with gallium changes the normally ductile metal into an easily fractured highly brittle material. This effect has a complicated temperature dependence, usually leads to intergranular fracture, and requires an external stress sufficient to cause some micro-yield. The effect is most dramatic with liquid metals, but vapors of the appropriate metallic agent also may be effective. The simplest and most straightforward explanation of this embrittlement is that γ is drastically lowered by the embrittling metal, so that dislocation emission cannot occur. The reader is referred to the review by NICHOLAS and OLD [1979] for details.

Although extensive use of surface- and interfacial energy ideas has been made in this chapter to explain the various chemical embrittlement effects, there is a very interesting and controversial suggestion that at least in some cases a very different process occurs. In liquid-metal embrittlement experiments by LYNCH [1981], and in hydrogen embrittlement studies in Ni by both LYNCH [1979] and BIRNBAUM [1979], it is claimed that the embrittlement process corresponds rather to an *enhancement* of dislocation emission at the crack tip, and that the overall effect is to localize the shear ahead of the crack on the crack plane. In the liquid-metal case, the adsorption must be highly localized in the external surface and crack tip. Thus, only cleavage processes can be affected by the adsorption. Since the dislocation is formed in the bulk, it should not be affected. Even though the dislocation is formed close to the surface, where the bonds are modified, in the ratio $\mu b/\gamma$ the average change in μ should not be as large as the change in γ. In the hydrogen case, it is conceivable that the internal dislocation sources are hardened, thus making localized shear in mode II easier than dislocations emitted from the crack. The liquid-metal case is more difficult to understand, but if the experimental interpretation holds up, the theory of dislocation emission will have to be reexamined in more atomistic detail.

9.8. Film formation

One additional result will be quoted regarding the effect of chemical environments on dislocation emission. SIERADZKI [1982] has shown that gaseous chlorine embrittles steel at pressures greater than 10^3 Pa. It is further shown that embrittlement does not occur until a film of the order of 10 μm forms on the surface. These results are only consistent with a model more complicated than one involving only changes in γ. In this case the dislocation emission takes place through a thin brittle film. In effect, the change in γ is insufficient to throw the material into a brittle mode, and this only occurs when a film of finite thickness builds up. It is not yet clear why the initial monolayer of adsorption shows no embrittlement, unless it is due to the inability of the Cl to affect the critical bonds in the iron because of steric effects.

9.9. Transformation-toughening

As the final application topic, a fundamentally different physical mechanism for crack shielding will be mentioned. When a local region of a material undergoes a phase transformation, the new phase, because it has a lattice with unit cell of different size and shape from the old, in general will generate compressive and shear stresses in both the old and the new lattice. By means of these stresses, transformed particles will interact with any cracks present for precisely the same reason that dislocations do. Also, just as a crack may generate a shielding plastic deformation in

References: p. 1548.

its vicinity, so a crack may induce phase transformations because of the shift in transformation free energy driving force caused by the k field of the crack. The transformation will take place in an orientation such that its reaction back on the crack is in a shielding configuration. There is not space here to develop fully the shielding theory for transformation analogous to that in §5 for dislocations. The straightforward way to do this, however, would be to assign a force multipole to each transformed particle, with a dilatation center to represent the net change in volume induced by the transformation, and a shear multipole to represent the shear component. The magnitude of the multipole is proportional to the local stress field with a constant of proportionality to be worked out on the basis of the transformation thermodynamics. The shielding theory then follows point-by-point that developed for dislocations in §5.

Toughening transformations have been studied in the ceramic, zirconia (see McMEEKING and EVANS [1981]) and in steels, where the so-called TRIP (TRansformation-Induced Plasticity) steels demonstrated a remarkably high toughness from room temperature to liquid nitrogen temperatures. In a TRIP steel, the relevant transformation is that from metastable quenched austenite to martensite, where the free energy of the transformation indeed depends on the local stress, and can thus be triggered by the presence of a crack. Thus the toughening potential due to this mechanism will be limited roughly to the temperature regime over which the martensite transformation can be stress-induced in the steel. This will, of course, depend on carbon content and other metallurgical variables. Estimates of the degree of toughening achievable by the transformation suggest that the energy absorbed by the transformation can be as much as five times as large as that contributed by deformation in a particular TRIP steel. (The increase in K_{IC} due to transformation-toughening is roughly a factor of two.) The reader is referred to GERBERICH et al. [1969,1971], and to ANTOLOVICH and SINGH [1970] for further details.

Acknowledgements

The author takes special pleasure in thanking his research colleagues at NBS for the many hours of discussion and collaboration on all aspects of the subject of fracture stretching over a period of ten years. Many of the points of view expressed in this chapter have indeed derived from this association. Thanks are also due to B. Lawn and N. Pugh for careful readings of the manuscript.

References

ANTOLOVICH, S., and B. SINGH, 1971, Metallurg. Trans. **2**, 2135.
ARGON, A., J. IM and R. SAFOGLU, 1975, Metallurg. Trans. **6A**, 825.
ARMSTRONG, R.W., 1966, Mater. Sci. Eng. **1**, 251.
ATKINSON, C., 1966, Int. J. Fract. Mech. **2**, 567.

BARSOM, J., 1979, Dynamic Fracture Toughness, ed. M. Davies (The Welding Institute, Abingdon Hall, Cambridge).

BEGLEY, J., and J. LANDES, 1972, Fracture Toughness (Am. Soc. Testing Mater., Spec. Tech. Pub. **560**) p. 187; see other articles in this publication.

BILBY, B.A., and J. ESHELBY, 1968, Fracture, vol. 1, ed. H. Liebowitz (Academic, New York) p. 99.

BILBY, B.A., A.H. COTTRELL and K. SWINDEN, 1963, Proc. Roy. Soc. (London) **272**, 304.

BIRNBAUM, H., 1979, in: Environmentally Sensitive Fracture of Engineering Materials, Proc. AIME Conference, ed. Z. Fourlis (Met. Soc. AIME, Warrendale, PA).

CAMPBELL, J., W. GERBERICH and J. UNDERWOOD, 1982, Application of Fracture Mechanics for Selection of Metallic Structural Materials (ASM, Metals Park, OH).

CHANG, S., and S. OHR, 1981, J. Appl. Phys. **52**, 7174.

COTTRELL, A.H., 1958, Trans. AIME **212**, 192.

COTTRELL, A.H., 1965, Proc. Roy. Soc. **A285**, 10.

COX, T. and T. LOW, 1974, Metallurg. Trans. **5**, 1457.

DAHL, W., and W. KRETZCHMANN, 1977, in: Fracture 1977, vol. 2a, ed. D. Taplin (Pergamon Press, New York) p. 17.

EASTMAN, J., T. MATSUMOTO, N. NARITA, F. HEUBAUM and H. BIRNBAUM, 1981, in: Hydrogen Effects in Metals, eds. I. Bernstein and A. Thompson (AIME, New York) p. 397.

ESHELBY, J.D., 1956, Solid State Physics vol. 3, eds. F. Seitz and D. Turnbull (Academic, New York) p. 79.

FINNIS, M., and J. SINCLAIR, 1982, in: Proc. NATO Advanced Study Institute, Atomistics of Fracture, ed. R. Latanision (Plenum, New York).

FORD, H., P.H. HIRSCH, J.F. KNOTT, A. WELLS and J. WILLIAMS, 1981, Fracture Mechanics in Design and Service, A Royal Society Discussion (Royal Soc. London); also in Phil. Trans. Roy. Soc., 1981, **A299**, 3–239.

GARDNER, R., T. POLLACK and H.G.F. WILSDORF, 1977 Mater. Sci. Eng. **29**, 169.

GERBERICH, W., P. HEMMING, V.F. ZACKAY and E.R. PARKER, 1969, Fracture 1969, ed. P.L. Pratt (Chapman & Hall, London) p. 288.

GERBERICH, W., P. HEMMING and V.F. ZACKAY, 1971, Metallurg. Trans. **2**, 2243.

GERBERICH, W., Y. CHEN and W. ST. JOHN, 1975, Metallurg. Trans. **6A**, 1485.

GRIFFITH, A.A., 1920, Phil. Trans. Roy. Soc. (London) **A221**, 163.

HEAD, A., and N. LOUAT, 1955, Austr. J. Phys. **8**, 1.

HIRTH, J.P., 1980, Metallurg. Trans. **11A**, 861.

HIRTH, J.P., and J. LOTHE, 1981, Theory of Dislocations (McGraw–Hill, New York).

HIRTH, J.P., and J. RICE, 1980, Metallurg. Trans. **11A**, 1501.

IRWIN, G., 1957, J. Mech. **24**, 361.

KANNINEN, J., and P. GEHLEN, 1972, in: Interatomic Potentials and Simulation of Lattice Defects, eds. P. Gehlen *et al.* (Plenum, New York) p. 713.

KELLY, A., 1966, Strong Solids (Oxford Univ. Press).

KELLY, A., A.H. COTTRELL and W. TYSON, 1967, Phil Mag. **29**, 73.

KNOTT, J.F., 1973, Fundamentals of Fracture Mechanics (Butterworths, London).

KNOTT, J.F., Fracture 1977, Proc. Fourth Int. Conf. on Fracture (Waterloo Univ. Press, 1977) vol. 1, p. 61.

KOBAYASHI, S., and S.M. OHR, 1981, Scripta Metall. **15**, 343.

LATANISION, R., ed., 1982, Atomistics of Fracture. Proc. NATO Advanced Study Institute (Plenum, New York).

LAWN, B.R., 1983, J. Am. Ceram. Soc. **66**, 83.

LAWN, B.R., and T.R. WILSHAW, 1975, Fracture of Brittle Solids (Cambridge Univ. Press).

LAWN, B.R., B. HOCKEY and S. WIEDERHORN, 1980, J. Mater. Sci. **15**, 1207.

LEE, T., T. GOLDENBURG and J. HIRTH, 1979, Metallurg. Trans. **10A**, 199.

LIEBOWITZ, H., 1968, Fracture, Vols. 1, 2 (Academic, New York).

LIN, I., 1982, to be published.

LOW, J.R., 1955, Proc. Conf. Intern. Union. Theor. & Appl. Mech., Madrid, **60**.

LYLES, R., and H. WILSDORF, 1975, Acta Metall. **23**, 269.

LYNCH, S., 1979, Scripta Metall. **13**, 1051.

LYNCH, S., 1981, Acta Metall. **29**, 325.

MAJUMDAR, B., and S. BURNS, 1981, Acta Metall. **29**, 579.

MAJUMDAR, B., and S. BURNS, 1983, Int. J. Fract. Mech. **21**, 229.

MCMAHON, C., V. VITEK and J. KAMEDA, 1980, in: Developments in Fracture Mechanics, ed. G. Chell (Appl. Sci. Publ., Barking, Essex, UK), 193.

MCMEEKING, J., and A. EVANS, 1981, J. Amer. Cer. Soc. **65**, 242.

MICHALSKE, T., and S. FREIMAN, 1982, Nature **295**, 511.

NICHOLAS, M., and C. OLD, 1979, J. Mater. Sci. **14**, 1.

OHR, S. and S. CHANG, 1982, J. Appl. Phys. **53**, 5645.

OHR, S., J. HORTON and S. CHANG, 1983, Proc. Int. Conf. on Defects, Fracture and Fatigue, ed. G.C. Sih, in press.

ORIANI, R.A., 1978, Ann. Rev. Mater. Sci. **8**, 327.

ORIANI, R.A., and P. JOSEPHIC, 1974, Acta Metall. **22**, 1065; 1977, Acta Metall. **25**, 979.

OROWAN, E., 1949, Prog. Phys. **12**, 185.

PASKIN, A., A. GOHAR and G. DIENES, 1980, Phys. Rev. Lett. **44**, 140; also DIENES, G., and A. PASKIN, 1982, Proc. NATO Advanced Study Institute on Atomistics of Fracture, ed. R. Latanision (Plenum, New York).

RICE, J., 1965, Proc. Int. Conf. on Fracture, eds. T. Yokohari *et al.* (Japan. Soc. Strength & Fracture, Sendai) vol. 1, 309.

RICE, J., 1968, J. Appl. Mech. **35**, 13.

RICE, J., 1976, The Mechanics of Fracture, ADM vol. 19 (ASME, New York) p. 23.

RICE, J., and R.M. THOMSON, 1974, Phil. Mag. **29**, 73.

SIERADZKY, K., 1982, Acta Metall., **30**, 973.

SINCLAIR, J., 1975, Phil. Mag. **31**, 647.

SHUBAT, G., *et al.* eds., 1974, Metals Handbook, vol. 9 (ASM, Metals Park, OH).

SMITH, E., 1966, Proc. Conf. Physical Basis of Yield and Fracture (Inst. of Phys. and Phys. Soc.) p. 36.

SMITH, E., 1979, in: Dislocations in Solids, vol. 4, ed. F.R.N. Nabarro (North-Holland, Amsterdam) p. 363.

STROH, A.N., 1957, Adv. Phys. **6**, p 418.

TADA, H., P. PARIS and G. IRWIN, 1973, The Stress Analysis of Cracks Handbook (Del. Rsch. Corp., Hillertown, PA).

TAKEDA, Y., and C. MCMAHON, 1981, Metallurg. Trns. **12A**, 1255.

THOMPSON, A., and P. WEIHRAUCH, 1976, Scripta Metall. **10**, 205.

THOMSON, R.M., 1978, J. Mater. Sci. **13**, 128.

THOMSON, R.M., 1980, J. Mater. Sci. **15**, 1027.

THOMSON, R.M., and E. FULLER, 1982, in: Fracture Mechanics of Ceramics, eds. R. Bradt *et al.* (Plenum, New York).

THOMSON, R.M., and J. SINCLAIR, 1982, Acta Metall. **30**, 1325.

VEHOFF, H., W. ROTHE and P. NEUMANN, 1981, Advances in Fracture Research **1**, 275; see also

VEHOFF, H., AND P. NEUMANN, 1980, Acta Metall. **28**, 265.

WEERTMAN, J., 1981, Phil. Mag. **A43**, 1103.

WEERTMAN J., I. LIN and R.M. THOMSON, 1983, Acta Metall. **31**, 473.

WIEDERHORN, S., R. MOSES and B. BEAN, 1970, J. Am. Ceram. Soc. **53**, 18.

WIEDERHORN, S., E. FULLER and R.M. THOMSON, 1980, Met. Sci. **14**, 450.

WILSDORF, H.G., 1982, Acta Metall. **30**, 247.

Further reading

1. Fracture Mechanics in Design and Service, A Royal Society Discussion, organized by H. Ford, P. Hirsch, J.F. Knott, A. Wells and J.Williams (Royal Society, London, 1981); also in Phil. Trans. Roy. Soc. **A299** (1981) 3–239.

2. Fundamentals of Fracture Mechanics, J. Knott (Butterworths, London, 1973).
3. Fracture of Brittle Solids, B.R. Lawn and T.R. Wilshaw (Cambridge Univ. Press, London, 1975).
4. Strong Solids, A. Kelly (Oxford Univ. Press, 1966).
5. Fracture, ed. H. Liebowitz (Academic, New York, 1968). See especially articles in vols. 1 and 2 by B.A. Bilby and J.D. Eshelby, J. Goodier; G. Sih and H. Liebowitz; J. Rice.
6. Dislocations in Solids, ed. F.R.N. Nabarro (North-Holland, New York, 1979) article by E. Smith, Dislocations and Cracks, vol. 4, p. 363.
7. Applications of Fracture Mechanics for Selection of Metallic Structural Materials, eds. J. Campbell *et al.* (ASM, Metals Park, OH, 1982)
 8. Atomistics of Fracture, eds. RM. Latanision and J.R. Pickens (Plenum, New York, 1982).
See comments on these and other sources in the Introduction.

Appendix A Table of symbols

σ	stress tensor
μ	shear modulus
u	displacement vector
$u_{ij} = \partial u_i / \partial x_j$	partial derivatives of displacement vector
x_1, x_2, x_3	rectangular coordinates
$z = x_1 + i x_2$	complex variable
W	strain energy function
w	u_3 displacement
a, N	half crack length
K	macroscopic stress intensity factor
k	local stress intensity
E	Young's modulus
K_C, K_{IC}, etc.	critical stress intensity for onset of fracture
b, B	Burgers vector, total Burgers vector
COD	crack opening displacement
c	core crack enclave radius
d	outer extent of plastic zone
β	dislocation density
σ_f	dislocation friction stress
m	work hardening parameter
D	compliance function of lattice
F	external force exerted on a lattice point
U	total energy of system plux external loading system
γ	surface free energy
f	interatomic force
ν	Poisson's ratio
f^*	complex vector force on an elastic singularity

CHAPTER 24

FATIGUE

P. NEUMANN

Max-Planck-Institut für Eisenforschung GmbH
Düsseldorf, FRG

R.W. Cahn and P. Haasen, eds.
Physical Metallurgy; third, revised and enlarged edition
© *Elsevier Science Publishers BV, 1983*

1. Introduction

The strength of a material in unidirectional testing is most simply characterized by the *ultimate tensile strength* (UTS), which is the maximum load in a tensile test divided by the initial cross-section. This definition does not comprise, however, that keeping the applied stress below the UTS would prevent failure. It is well recognized since more than a hundred years (WÖHLER [1870]), that much lower stresses lead to failure, if they are applied repetitively with alternating sign. The stress amplitude, τ_{10^6}, which leads to fracture after 10^6 loading cycles, is only 30–50% of the UTS for most commercial materials. This phenomenon, that most materials fail at lower stresses in cyclic loading if compared to unidirectional loading, is called *fatigue*.

Such behaviour is interesting enough from a fundamental point of view, but it has also enormous practical and economical consequences and makes the fatigue properties to be the most important mechanical design quantities of commercial materials. Most structures are designed –because of economical reasons – for a finite lifetime. Therefore there is a finite probability, that a given component may fail in service. This danger can be reduced effectively only by regular inspections. In the case of fatigue failure, cracks nucleate and grow up to a considerable size without any obvious shape change of the component. Sophisticated and costly methods of nondestructive testing have to be applied in order to detect fatigue cracks. The high inspection costs can be reduced only by improving the predictability of fatigue failure.

Reliable prediction of fatigue failure can be obtained only by a thorough understanding of the physical mechanisms involved. For commercial materials with their complicated microstructures we are, at the present, far away from a quantitative theory which relates elementary processes to the observable life under fatigue loading. Considerable progress has been made in the past twenty years, however, in the understanding of the physical mechanisms of fatigue in simple systems like pure metals or simple precipitation-hardened alloys, preferably in the form of monocrystals. This will be the subject of this chapter.

Fatigue loading of even the simplest materials produces a rich variety of phenomena. This justifies a treatment separate from that of mechanical properties which are observed in unidirectional loading. There are, of course, many similarities between unidirectional and fatigue loading. They become most pronounced if the strains within each cycle are large compared to the elastic strain amplitude. Then the fatigue test can be regarded as a succession of a number of tension and compression tests. In order to reveal typical fatigue properties, we shall not stress these similarities and refer only to fatigue experiments with more than 10 000 cycles and plastic strain amplitudes less than 1% (typically 10^{-4}). It is obvious that this type of loading has little resemblance to a tensile test and will show characteristics of its own.

2. *Fatigue tests and the representation of fatigue data*

In contrast to the tension test, fatigue tests can be conducted in many different ways, which may – if the details of the test are not specified – cause an unnecessary increase of the scatter of the results:

1. *Constant stress-amplitude tests:* In most commercial alloys, stresses of half their UTS (the size of typical fatigue stresses) result in small strains only. Therefore, genuine fatigue tests, with small strain amplitudes from the beginning, can be conducted under constant stress-amplitude control. The stress varies usually in a sinusoidal manner with time. If the material is in the annealed condition, the strain amplitude will be largest in the first cycle, will continuously decrease during an initial period and finally reach a saturation value after some 10 000 cycles.

2. *Increasing stress-amplitude tests:* In various soft commercial alloys and, more pronounced, in pure metal single crystals, the strains in the first cycle can become as large as 30%, at a stress amplitude equal to τ_{10^6}. In order to avoid this very large initial strain amplitude, the stress amplitude has to be raised continuously from zero to the desired final stress amplitude, which may then be kept constant for the rest of the experiment. Typically, the number of cycles spent to reach the final stress amplitude amounts to a few per cent of the total life.

3. *Constant plastic-strain-amplitude tests:* It is well known that plastic strains are necessary for the development of fatigue, whereas elastic strains, because of their reversibility, are less important. In typical fatigue situations, however, the plastic strain amplitude is a small fraction of the total strain amplitude and decreases due to cyclic hardening in the course of the experiment. Therefore, constant total strain-amplitude tests are rather ill-defined as far as the plastic strains are concerned and most experiments under strain control are performed nowadays in such a way, that the plastic strain amplitude is kept constant. Typical values of the strain amplitude are 5×10^{-5} to 10^{-2}.

4. For the study of the fatigue properties of *single crystals of pure metals*, however, a somewhat more complicated test seems to be best suited: these materials show a dramatic cyclic hardening of a factor of twenty or more (critical resolved shear stress to the saturation stress) and almost all fatigue phenomena depend strongly on how fast the cyclic hardening is done. This obviously favours a *two-step test*: First a high enough stress-amplitude level must be attained under identical conditions (preferably under increasing-stress-amplitude conditions), followed by the proper test with varying parameter values, e.g., constant plastic-strain-amplitude tests.

5. In order to simulate more realistically the loading history of members of real structures, tests are also performed with highly complicated loading sequences. Special techniques were developed to characterize such loading histories (WETZEL [1971]).

In all tests the *frequencies* range from 0.1–100 Hz. With some special resonance machines frequencies up to 500 Hz can be reached. Using the longitudinal resonance frequency of cylindrical specimens, fatigue experiments can also be performed with frequencies in the range of some 10 kHz (TSCHEGG and STANZL [1981]).

References: p. 1591.

A frequently used quantity which is specific for fatigue, is the plastic strain, which occurred in all previous cycles, summed up without regard of sign, Γ. Obviously we have:

$$\Gamma = 4 \sum_{n=1}^{N} \gamma_{p.n}, \tag{1}$$

where $\gamma_{p,n}$ is the plastic strain amplitude in the nth cycle. Sometimes this so-called *cumulative plastic strain* is taken as a damage parameter. It must always be kept in mind, however, that at small plastic strain amplitudes a considerable fraction of the plastic strain stems from reversible dislocation motion and thus does not contribute to any kind of fatigue damage.

The number of cycles at which fracture occurs, N_f, has been measured in constant stress-amplitude tests since more than a hundred years (WÖHLER [1870]) as a function of the applied stress amplitude. For historical reasons these data, which are the prime data for any design against fatigue, are usually plotted in the form shown in fig. 1, i.e. the independent variable, stress amplitude, is plotted on the ordinate versus the logarithm of the number of cycles to failure, N_f, on the abscissa. In many steels such a so-called *S–N curve* has a well defined knee at about 10^6 cycles, corresponding to a stress amplitude τ_{10^6}. This means, that below such a stress amplitude no failure will be observed. Such materials are said to have a genuine *fatigue limit*.

Most fatigue experiments are nowadays performed with the help of closed-loop hydraulic systems. These *testing machines* are capable of producing forces ranging typically from 5 kN to 1 MN by hydraulic pistons. The supply of high-pressure hydraulic fluid is controlled by a servo-valve, which controls the oil flux as a

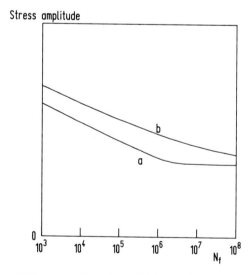

Fig. 1. Typical S–N curves (schematic) with (a) and without (b) endurance limit.

function of an electrical input. With modern valves a positional accuracy in the range of 0.1–0.5 µm at frequencies from 0–100 Hz can be realized. In order to control any combination of stress and strain as an arbitrary function of the number of cycles, usually a computer is employed. With additional measures (NEUMANN [1978]) the positional resolution can be improved to 20 nm.

3. Cyclic hardening

The natural experiment to measure cyclic hardening is the test with constant plastic strain amplitude. Figure 2 shows the development of the stress amplitude with the number of cycles, or with the cumulative plastic strain since both quantities are directly related in constant plastic strain tests via the equation

$$\Gamma = 4\gamma_p N. \tag{2}$$

If the cumulative plastic strain of the cyclic test is put in analogy to the strain in a tension test (as is done in fig. 2), it must be stated that *cyclic hardening is much slower than tensile work-hardening*. Obviously this is due to the Bauschinger effect or – in other words – due to the partial reversibility of plastic strain. For instance, in copper single crystals cumulative strains of the order of ten are necessary to reach a stress of 28 MPa, whereas in tension shear strains of 0.3 produce the same stress level. This effect is even more pronounced if the test is carried out at smaller plastic strain amplitudes, since the reversible fraction of the plastic strain amplitude increases with decreasing plastic strain amplitude.

The stress amplitude reaches a saturation value after a cumulative plastic strain of the order of 10 (see fig. 2). The value of the *saturation stress amplitude* does not depend on the plastic strain amplitude for $10^{-4} \leqslant \gamma_p \leqslant 10^{-2}$ (MUGHRABI [1978]). There is a close connection between the saturation stress amplitude and the

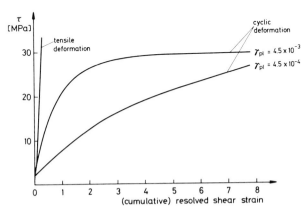

Fig. 2. The stress amplitude as a function of the cumulative plastic strain during constant plastic-strain-amplitude tests of copper crystals (from WILKENS *et al.* [1980]).

References: p. 1591.

formation of persistent slip bands. Thus the discussion of the physical sign ficance of the saturation stress amplitude is deferred to §4.

Physically the most striking differences between the cyclically and the unidirectionally hardened state is the *absence of long-range internal stresses* in the fatigued state. This was proved by electron microscopy (MUGHRABI [1981]) as well as X-ray studies (DEHLINGER [1931], HARTMANN and MACHERAUCH [1963] and WILKENS *et al.* [1980]). The dislocation structure after cyclic hardening is well defined and will be described in detail later.

If the hardening does not come about by long-range internal stresses, dislocation motion can be impeded in a pure metal only by mutual trapping of dislocations. This results in a metastable hardening state which tends to *instabilities and strain localization* in the form of coarse slip bands. These matters will be addressed in a later section.

In fatigue, huge amounts of cumulative plastic strain can be realized (up to 10^5, i.e. 10^7% (HUNSCHE and NEUMANN [1983]). Even if a large fraction of these strains are reversible, there will be numerous short-range interactions between dislocations, which may produce point defects. Thus a *large point-defect concentration* can be expected to be produced in the course of fatigue experiments. This high production of point defects has been verified by measurements of the electrical resistivity (POLAK [1970]) and is also reflected in a large temperature dependence of fatigue-hardening (BROOM and HAM [1959]). The problem of point-defect production will be addressed in more detail in connection with the development of persistent slip bands.

3.1. Matrix dislocation structure

In pure fcc metal single crystals, which are oriented for single slip, the dislocation structure after fatigue is well defined. Figure 3 shows a three-dimensional composite picture of the structure which is usually observed: There are areas of high dislocation density, the so-called *veins*, which have an elongated shape with the long axis parallel to the primary edge-dislocation line direction. The cross-section of these veins, perpendicular to the edge-dislocation line direction, is about equi-axed. The veins are separated by almost dislocation-free areas, the so-called *channels*. Their size is comparable to that of the veins. This *characteristic length* of the structure is about 1.5 μm in copper. Plastic strain amplitudes up to 10^{-4} can be maintained for more than 100 000 cycles before the occurrence of irreversible changes. For reviews see GROSSKREUTZ and MUGHRABI [1975] and MUGHRABI [1980,1981].

The *veins* consist predominantly of primary edge dislocation. The average Burgers vector within the veins is very close to zero, since no lattice rotation between the adjacent channels can be observed in the electron microscope. Thus the veins do not produce long-range internal stresses. Internal stresses were measured at zero applied stress by MUGHRABI [1981] and were found to be less than 20% of the saturation stress amplitude. The dislocation density can be as high as $\rho_{vein} = 10^{15}/m^2$, i.e., the mean dislocation separation is $d_{vein} = 30$ nm, which obviously is another characteris-

Fig. 3. Three-dimensional view of the dislocation structure in fatigued copper; the primary Burgers vector is denoted by b_p. (From MUGHRABI *et al.* [1979].)

tic length of the structure. At this high dislocation density individual dislocations can be resolved only with the help of the weak beam technique. Predominantly dislocation dipoles of the vacancy type were found in a limited amount of data (26 vacancy-type, 12 interstitial-type, 14 unidentified by ANTONOPOULOS *et al.* [1976] with a mean spacing of 5 nm between the dislocations of a dipole).

The dislocation density in the *channels* is three orders of magnitude smaller than the dislocation density in the veins, i.e., $\rho_{ch} = 10^{12}/m^2$. Thus the mean dislocation separation in the channels is about equal to the width of the channels, the first characteristic length.

The development of such a dislocation structure is a highly complicated process of some cooperative motion of many dislocations. However, some arguments can be given to explain most of the essential features of such a dislocation structure:

The *preponderance of primary dislocations* is understandable, since there is no orientation change during cyclic hardening as it is observed in a tension test.

References: p. 1591.

Therefore the primary slip system remains the most highly stressed slip system throughout the test. Multiplication of dislocations on the primary slip system (see below) produces strong latent hardening for the secondary slip systems. These two circumstances prevent secondary slip effectively.

It is obvious that during cyclic loading dislocations of both signs are effectively mixed because of their continuous to-and-fro motion. Therefore there will be frequent encounters of dislocations of different sign (because they attract each other). When the encounter is close enough, so that the attraction is stronger than the applied stress level, the dislocations will trap each other and stop moving over large distances. The resulting configurations are *dislocation dipoles*. Screw-dislocation dipoles are not stable because of the possibility of cross-slip and will annihilate. Thus only edge dislocation dipoles are left. Those dislocations (edge of screw), which did not have a close-enough encounter, continue to travel large distances through the crystal. Therefore they have a high probability of being trapped in the future. This process of mutual trapping continues until almost all dislocations are present in the form of edge-dislocation dipoles. In this way the average Burgers vector becomes zero.

The stresses due to the dipoles are much more short-ranged ($\propto 1/r^2$) than the stresses from dislocation pile-ups in a tension test ($\propto 1/\sqrt{r}$). Furthermore dislocation dipoles of the vacancy type as well as of the interstitial type are thoroughly mixed, so that not only the average Burgers vector becomes zero, but also the average dipole strength of the whole configuration, which may further reduce the internal stresses.

The above arguments hold only for vanishing applied load. At finite applied load the dipole patches become polarized and large internal stresses have to be expected. This agrees with the findings of MUGHRABI [1981], who measured internal stresses of the order of the saturation stress amplitude in the channels. HOLSTE et al. [1979] tried to calculate the internal stresses resulting from certain dislocation configurations, which were chosen according to the observed configurations. They obtained, however, too large stresses. This is most likely due to the fact, that the positions of their dislocations were prescribed at will and were not allowed to relax into equilibrium positions, which by definition would reduce the stresses.

The first characteristic length of the dislocation distribution, the *mean separation distance of dislocations within the veins*, is obviously related to the trapping distance of edge dislocations at the saturation stress, which is given by:

$$d_{\text{trapp}} \leqslant \frac{\mu b}{8\pi(1-\nu)\tau}. \tag{3}$$

With the data for copper ($\mu b/8\pi(1-\nu) = 0.67$ J/m^2 and $\tau = 28$ MPa) we obtain $d_{\text{trapp}} \leqslant 24$ nm. This is in reasonable agreement with the measured characteristic length of 30 nm.

The second characteristic length, the *size of the channels*, can be rationalized as follows: before the saturation stress is reached, the stress amplitude rises very slowly (0.03% of saturation stress per cycle or less). Therefore, at stress amplitudes smaller than the saturation stress, all the arguments for the dislocation configuration hold as well. Therefore the same dislocation configuration is expected – only with correspondingly increased dislocation distances. As a consequence the dislocation density (proportional to d^{-2}) must increase proportional to the square of the stress amplitude, which is also well documented experimentally. Therefore, during the process of cyclic hardening considerable dislocation-multiplication must happen. This is only possible by bowing out of dislocations. For this process, free areas with a diameter equal to the Orowan–Frank–Read length,

$$d_{\mathrm{Or}} = 1.5 \frac{\mu b}{\tau} \tag{4}$$

are necessary. With the data for copper ($\mu b = 11$ J/m^2 and $\tau = 28$ MPa) this yields $d_{\mathrm{Or}} = 0.6$ μm, which is in reasonable agreement with the observed channel width of 1.5 μm. An even better agreement can be achieved by taking into account the elastic anisotropy (BASINSKI *et al.* [1980]).

In order to understand the reversibility of plastic strains up to 10^{-4}, it is useful to estimate the average travelling distance, l, of dislocations at a plastic strain amplitude of 10^{-4}. This strain is produced by 10^{15}/m^2 dislocations in the veins, or 10^{12}/m^2 dislocations in the channels. According to the well known formula

$$\gamma_{\mathrm{p}} = \rho l b, \tag{5}$$

we obtain $l_{\mathrm{vein}} = 0.3$ nm $= 0.01$ d_{vein}, and $l_{\mathrm{ch}} = 0.3$ μm $= 0.2$ d_{ch}. Thus the travelling distances within the veins as well as in the channels are a small fraction of the corresponding average dislocation distances only. This calculation relies on the assumption that the plastic strain must be the same within the veins and in the channels. This is not necessary at all, since minor internal stresses of 4.2 MPa produce – via the shear modulus of 42 GPa – already elastic strains of 10^{-4}, which can compensate incompatibilities introduced by variations in the plastic strains. If, on the other hand, complete plastic incompatibility is assumed, the relevant parameters are changed by a factor of two only, since the volumes of veins and channels are about equal. The result $l \ll d$ for the veins as well as for the channels explains the reversibility of plastic strains up to 10^{-4}, because no further encounters between dislocations, and thus no irreversible changes, are possible with such small travelling distances.

Multiplication of dislocations and re-arranging them into a dipole structure with decreasing dislocation distances is a complicated process, however, and does not happen continuously but in local bursts of dislocation activity, which are discussed in the next section.

3.2. Instabilities during cyclic hardening

If an increasing-stress-amplitude fatigue test is performed with a single crystal oriented for single slip, the strain amplitude does not rise continuously with the applied stress amplitude, but passes through a succession of maxima and minima (fig. 4), which will be called strain bursts in the following (NEUMANN [1968]). They

References: p. 1591.

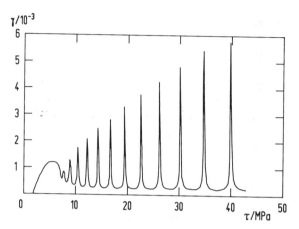

Fig. 4. Strain bursts in a copper crystal fatigued with a stress amplitude which increases linearly with the number of cycles at a rate of 7.1 kPa/cycle. Temperature: 90 K.

were observed in Cu, Ag, Mg, and Zn (NEUMANN and NEUMANN [1970]) as well as in CuAl alloys (DESVEAUX [1970]). The strain bursts are regularly spaced: They occur after a constant relative increment of the stress amplitude (11.5%) and are usually 50 cycles wide. Therefore they overlap, when the stress amplitude is increased by more than 11.5% within 50 cycles. Although the relative position of the bursts is well defined, the absolute values of the stress amplitude, at which strain bursts occur, is not defined at all.

These experimental results can be understood in the light of the above discussion of the dislocation arrangement in fatigued single crystals: Cyclic hardening in single-slip oriented single crystals is essentially a mutual trapping of dislocations of opposite sign. If enough loading cycles are performed at any given stress-amplitude level, the trapping will continue to reduce the plastic strain amplitude until the mean free path of the dislocation motion is smaller than their mean distance. If random encounters of dislocations are assumed, the probability of close encounters decreases proportional to the separation distances. Therefore the majority of dislocation dipoles will have a separation distance perpendicular to the slip planes, which is just below the passing distance at the current stress amplitude. Therefore, if – after trapping is complete – the stress amplitude is raised, a large number of dipoles will disintegrate. The resulting free dislocations will pile up against narrower dipoles, which are nominally stable against the current applied stress amplitude. With the help of these pile-ups, however, they will be decomposed too. This feedback leads to an avalanche of free dislocations and thus to a local strain burst. After the population of free dislocations has increased, the probability for trapping increases correspondingly and the strain amplitude will fall off again by mutual trapping. The result is a population of narrower dipoles, which are stable also at the current, increased, stress amplitude. This process was modelled quantitatively by NEUMANN [1974b].

These *local* strain bursts occur whenever the stress amplitude is raised slowly enough, i.e., also during the initial hardening in a constant plastic-strain-amplitude test. What is observed macroscopically, depends critically on the interaction of the local bursts: since dislocations are bound to move on the slip planes, the avalanches will develop and spread mainly along slip planes. Under favourable conditions (no inhomogeneities) the avalanche may also spread laterally due to dislocation interactions, which implies that all local microbursts in the gauge length happen *at the same time (coherently)*, yielding an observable macro-strain-burst. This is the case in a test with increasing stress amplitude. There is some experimental evidence for the global nature of the strain bursts: By measuring the plastic strain on both halves of the specimen independently, it could be shown [NEUMANN [1968]) that the strain bursts stem from the whole specimen gauge length.

During the initial hardening period in a test with constant plastic strain amplitude, the stress amplitude rises in the very same manner and local microbursts do also occur. But now coherent local bursts are obviously not compatible with the control, whereas incoherent local bursts are. Therefore the local bursts do also occur in a constant plastic-strain-amplitude test, but not synchronously (*incoherently*), such that they add up to a slowly varying macroscopic plastic strain amplitude. Besides differences in the loading history (the stress amplitude does not rise linearly in time in a γ_p = const. test) the amount of synchronism of the local strain bursts is the only difference between these two tests. In other words: strain bursts are not at all indicative of large local strains, but only of their perfect synchronism.

If the fatigue test is stopped above about half the saturation stress, and a tensile test is performed, a long zero-work-hardening region of up to 30% shear strain is found (PATERSON [1955] and BROOM and HAM [1959]). After fatigue tests which show bursts, the yield stress of the low-work-hardening region coincides with the value of the stress amplitude at which the next burst would have occurred if the fatigue test with increasing stress amplitude would have been continued (NEUMANN [1968]). The slip during this extended easy-glide region is extremely coarse with slip steps up to 0.3 μm high (BROOM and HAM [1959] and NEUMANN [1967]). In the light of these facts, the *coarse slip steps* may be taken as direct experimental evidence for the unidirectional version of the local strain bursts.

The bursts disappear if the crystal is *pre-strained* into stage II of the tensile stress–strain curve, indicating that secondary dislocations hinder the development of strain bursts. Accordingly, single crystals oriented for multiple slip also do not show strain bursts. The same is true for polycrystals (NEUMANN [1968]). These results fit well into the picture developed above, since all kinds of obstacles will oppose any synchronism of local avalanches.

4. Persistent slip bands

If fatigue tests are carried out under constant plastic strain amplitude, cyclic hardening is observed up to a certain stress level (see fig. 2). At large numbers of

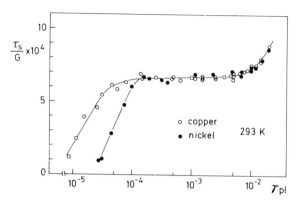

Fig. 5. The saturation stress amplitude of copper- and nickel crystals as a function of the plastic strain amplitude (from MUGHRABI *et al.* [1979]).

cycles the stress amplitude usually approaches a constant value, the stationary stress amplitude, or – at large strain-amplitude values – final fracture occurs. A plot of the stationary stress amplitude versus the constant plastic strain amplitude of the test (fig. 5) shows a marked *plateau* (MUGHRABI [1978]) which does not depend on the orientation (CHENG and LAIRD [1981b]). In the case of copper the plateau lies at approximately 30 MPa and extends over two orders of magnitude in the plastic strain amplitude ($10^{-4} \leqslant \gamma_p \leqslant 10^{-2}$), which makes the saturation stress amplitude the most important quantity of the fatigue properties. Its temperature dependence down to 4.2 K is slightly stronger than that of the yield stress of copper deformed in tension (BASINSKI *et al.* [1980]).

The surface of the specimen just before reaching saturation shows almost no evidence of slip in an optical microscope for $\gamma_p \leqslant 10^{-4}$. If, however, the specimen is inspected after reaching saturation, strong slip bands parallel to the primary slip plane can be detected even with the naked eye. In a well aligned tension–compression specimen these slip bands extend all the way through the cross-section. They are from two to one hundred micron wide and have a very rough surface. This surface topography will be discussed in §4.3. The surface roughness can be removed only with difficulty by electro-polishing. If the fatigue test is continued after the polishing, the slip activity re-occurs at the same locations. Furthermore, the slip band can be revealed deep inside the specimen by sectioning the specimen and subsequent etching. This behaviour led THOMPSON and WADSWORTH [1958] to name this strain-localization phenomenon *persistent slip bands (PSB)*.

4.1. Properties of persistent slip bands

It was soon proved by static straining after fatigue (BROOM and HAM [1959]) and by micro-hardness measurements (HELGELAND [1964]) that the persistent slip bands are *softer* than the surrounding matrix. The formation of PSB is usually connected with a yield drop (Broom and Ham [1959]). This can be observed most distinctly in the following way (Hunsche and Neumann [1983]): By an appropriate test (sufficiently small constant plastic strain amplitude or slowly increasing stress amplitude test) a crystal is hardened up to the saturation stress without formation of PSBs but with a well established matrix structure; if the plastic strain amplitude is then increased in order to form PSBs, the stress amplitude behaves a shown in fig. 6, which shows a well-defined *cyclic yield drop*. The first PSBs are observed at the maximum stress amplitude with an accuracy of ±15 cycles. After the following cyclic yield drop of 5% the stress amplitude stays constant for a large number of cycles. If crack initiation in the PSBs is delayed by an inert environment (vacuum) a very *slow secondary hardening* (Hunsche [1982] and Wang [1982]) can be observed after cumulative strains of the order of 1000 and more. This secondary hardening is most likely connected with dislocation cell formation in PSBs (Laufer and Roberts [1966] and Wang [1982]). When PSBs form, also a characteristic change of the shape of the hysteresis is observed (Mughrabi [1978]).

The volume fraction of the gauge length, which is occupied by PSBs, depends linearly on the plastic strain amplitude imposed (Roberts [1969], Winter [1974], Finney and Laird [1975] and Mughrabi [1978]). This implies, since the local plastic strain amplitude in the matrix is only 10^{-4}, that the local strain in the PSBs is independent of the applied plastic strain amplitude. The quantitative results yield for this *constant local plastic strain amplitude in the PSBs 10^{-2}*, i.e., 100 times the local plastic strain amplitude of the matrix!

The strong strain localization in the PSBs (Finney and Laird [1975], Cheng and

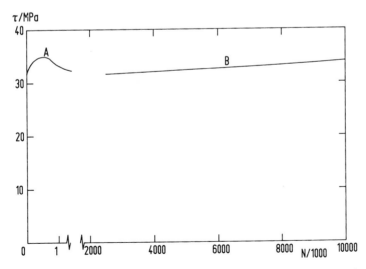

Fig. 6. Behaviour of the stress amplitude in a constant plastic-strain-amplitude test (after slow hardening into saturation). A: yield drop due to nucleation of PSBs. B: secondary hardening due to cell formation within the PSBs. (From Hunsche and Neumann [1983].)

References: p. 1591.

LAIRD [1981c] and LAIRD et al. [1981]) indicates that the dislocation structures of matrix and PSBs must be distinctly different. Figure 7 shows a TEM micrograph of a thin persistent slip band. The irregularly arranged veins of the matrix structure are replaced in the PSB structure by very regularly spaced edge-dislocation walls of high dislocation density (LAUFER and ROBERTS [1964,1966] and LUKAS et al. [1966,1968]). A {112} cross section through such a wall structure looks – as fig. 7 does – like a ladder, which is therefore commonly called *ladder structure*. Apart from the much more regular appearance, the features of the ladder structure are very similar to that of the matrix, and most of the arguments given in §3.1 hold for the ladder structure as well. Minor differences are as follows: The primary as well as secondary dislocation-densities in the walls are twice as high as in the veins. The dislocation densities in the PSB channels are ten times as high as in the matrix channels (WINTER [1974] and MUGHRABI [1978]).

The *ladder spacing*, which corresponds to the channel width of the matrix structure, has been determined also as a function of temperature. Detailed considerations by BASINSKI et al. (1980) showed that the absolute value as well as the temperature dependence agree well with the Orowan distance, if the temperature dependence of the saturation stress as well as the elastic anisotropy of copper is taken into account.

The regions of high dislocation density within the matrix as well as in the PSBs can be etched (WINTER [1973]). Such etching studies are much simpler to perform

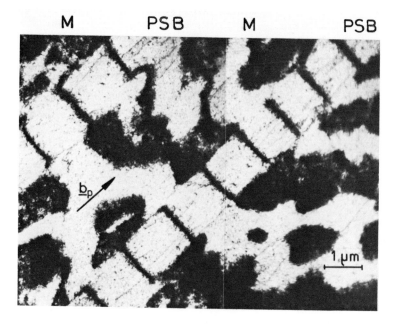

Fig. 7. Dislocation ladder structure of a PSB within the matrix structure (M) in copper; $\underline{b}_p$ denotes the primary Burgers vector. (From MUGHRABI et al. [1979].)

than TEM observations. In this way additional evidence was obtained, that the latter structure extends all the way through the specimen.

The ladder structure was also observed in hcp single crystals (KWADJO and BROWN [1978]), in fcc polycrystals near the surface (LUKAS *et al.* [1966]) and in the bulk (WINTER *et al.* [1981]), in impure bcc single crystals (MUGHRABI *et al.* [1979]), and in bcc polycrystals near the surface as well as in interior grains (POHL *et al.* [1980]). In the interior of polycrystals, RASMUSSEN and PEDERSEN [1980], CHARSLEY [1981] and WINTER *et al.* [1981] observed regularly spaced dislocation walls in two {100} planes (*labyrinth structure*). For a review see MUGHRABI and WANG [1981].

4.2. Dislocation models of persistent slip bands.

After discussing the most important experimental facts about persistent slip bands and their dislocation structure, it will be tried to explain the formation of PSBs and their most important properties on the basis of dislocation theory.

In a pure metal like copper, where no obstacles apart from dislocations and point defects can be expected, we must concentrate on the interactions between dislocations themselves. This task is facilitated to some extent by the fact that almost all dislocations are nearly parallel to each other in the high dislocation-density regions. Furthermore, long-ranging internal stresses are absent, because the average Burgers vector is close to zero everywhere. Therefore it is reasonable to model the dislocation structure in the high-density regions by straight edge dislocations, which are – for the sake of simplicity – arranged in a regular manner as shown in fig. 8.

Such an arrangement of edge dislocations – extended into infinity – was studied first by TAYLOR [1934] and NABARRO [1952]. Many properties of such *Taylor–Nabarro lattices* are known and were applied by KUHLMANN-WILSDORF [1979] to study plastic properties during cyclic loading. There is, however, one basic problem with infinite Taylor–Nabarro lattices: The interaction between those infinitely long edge dislocation walls, which are the constituents of Taylor–Nabarro lattices, is such (NABARRO [1952]) that they repel each other at any given distance. Therefore the infinite Taylor–Nabarro lattice is inherently instable and it is worthwhile to consider finite sections out of Taylor–Nabarro lattices and other regular arrays as models for the veins. The behaviour of finite sections of various shape cannot be described in general terms and numerical calculations must be carried out. This was done by NEUMANN [1983] with the following results:

1. Finite sections of Taylor–Nabarro lattices of very different shapes are *stable*, indeed. Figure 8 shows an example of a relaxed configuration, in which all dislocations are at equilibrium.

2. The *borders* of the sections are well defined in the equilibrium configuration, such that there is an abrupt change from high dislocation density to zero dislocation density.

3. The dislocations in the relaxed configurations form a *face-centered tetragonal lattice* to good accuracy. The tetragonality of the lattice ranges form 1.8 for small (36 dislocations), to 2.2 for large (324 dislocations) sections.

References: p. 1591.

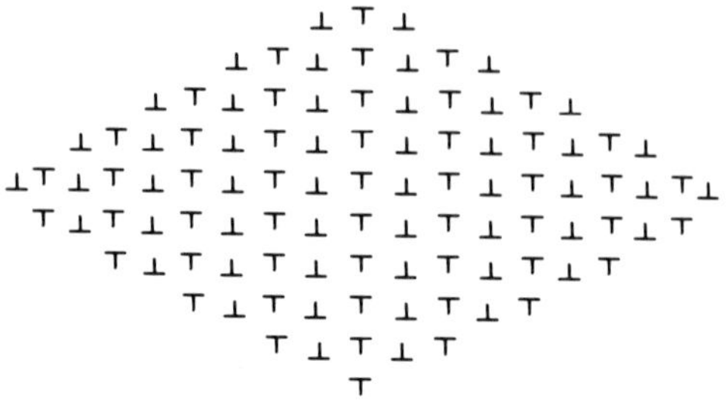

⊥ T ⊥
⊥ T ⊥ T ⊥ T ⊥
⊥ T ⊥ T ⊥ T ⊥ T ⊥ T ⊥
⊥ T ⊥ T ⊥ T ⊥ T ⊥ T ⊥ T ⊥
⊥T ⊥ T ⊥ T ⊥ T ⊥ T ⊥ T ⊥ T ⊥T⊥
T ⊥ T ⊥ T ⊥ T ⊥ T ⊥ T ⊥ T ⊥ T
T ⊥ T ⊥ T ⊥ T ⊥ T ⊥ T
T ⊥ T ⊥ T ⊥ T ⊥ T
T ⊥ T ⊥ T
T

Fig. 8. Finite section out of a Taylor–Nabarro lattice. All dislocations are shown at the calculated equilibrium positions ($\tau_{\text{applied}} = 0$). (From NEUMANN [1983].)

Under an applied load, which pushes the dislocations of opposite sign into opposite directions, the dislocation sections become *polarized* and dipole walls start to form, as is shown in fig. 9. At a certain applied stress level the whole configuration becomes unstable and decomposes into dipole walls, which start to move to infinity. This *decomposition stress*, τ_{dec}, characterizes the stability of the configuration and is plotted in fig. 10 as a function of the size of the section. The decomposition stress is given in units of the decomposition stress of the most narrow elementary dipoles of the configuration, $\tau_{\text{dec}}^{\text{dip}}$. In this way the results are valid for all values of d, the separation distance of the dislocations perpendicular to the slip planes. $\tau_{\text{dec}}/\tau_{\text{dec}}^{\text{dip}}$ decreases slowly with the number of dislocations in the section and is about 0.4.

Obviously, the value of the decomposition stress of such a configuration can be made as high as necessary by reducing d. But there is a lower limit, d_{min}, for the distance between edge dislocations under fatigue conditions. ESSMANN and MUGHRABI [1979] and ESSMANN [1982] have considered in detail the circumstances of annihilating dislocations and estimated the *maximum annihilation distance* to be about 1.6 nm in Cu at saturation. This is not of the order of the average distance of dislocations in the veins at saturation (see above). Based on TEM observations the authors solve this discrepancy essentially by postulating that the dislocations move

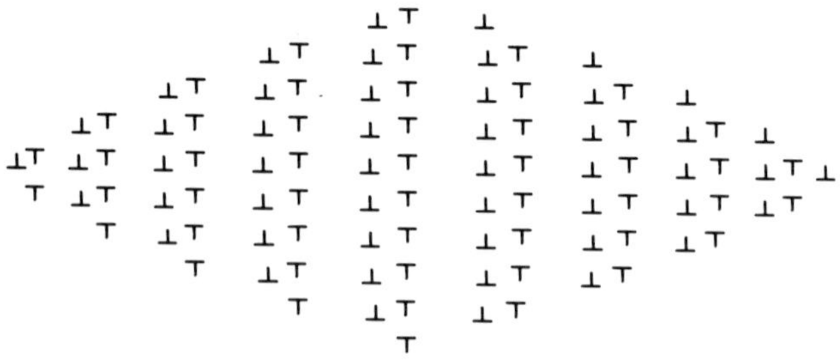

⊥ T   ⊥
⊥ T   ⊥ T   ⊥ T   ⊥
⊥T   ⊥ T.   ⊥ T   ⊥ T   ⊥ T   ⊥
⊥T   ⊥T   ⊥ T   ⊥ T   ⊥ T   ⊥ T   ⊥
⊥T ⊥T   ⊥ T   ⊥ T   ⊥ T   ⊥ T   ⊥ T   ⊥T ⊥
T ⊥T   ⊥T   ⊥ T   ⊥ T   ⊥ T   ⊥ T   ⊥ T ⊥T
T   ⊥T   ⊥ T   ⊥ T   ⊥ T   ⊥ T   ⊥T
T   ⊥ T   ⊥ T   ⊥ T   ⊥ T
T   ⊥ T   ⊥ T
T

Fig. 9. Calculated equilibrium positions of 100 dislocations in the Taylor–Nabarro lattice section shown in fig. 8, under an applied stress of $\tau_a = 0.52\,\tau_{\text{dec}}^{\text{dip}}$, where $\tau_{\text{dec}}^{\text{dip}}$ is the decomposition stress of the most narrow elementary dipole of the configuration (from NEUMANN [1983]).

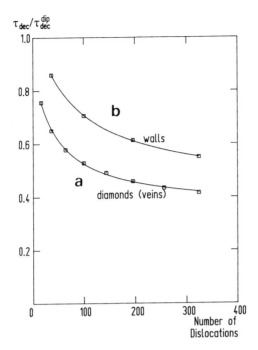

Fig. 10. τ_{dec}, the decomposition stress of various dislocation configurations in units of τ_{dec}^{dip} as a function of the total number of dislocations in the configuration: (a) Diamond-shaped Taylor–Nabarro lattice sections (see fig. 8 and 9) modelling the vein structure. (b) Wall structure (see fig. 11b). Wall structures are more stable than comparable (equal τ_{dec}^{dip}) veins. (From Neumann [1983].)

in groups of dislocations with the same Burgers vector. For details see the original publication. In any case there is overwhelming experimental evidence that at saturation decomposing veins cannot rearrange into veins with a smaller d. Instead, the dislocations rearrange into a ladder structure. This is possible only if there are wall structures, which are more stable than vein-type configurations with the same d.

Many stable ladder configurations were found indeed. Figure 11 shows two examples. The corresponding decomposition stresses are plotted in fig. 10, which shows that wall structures are indeed more stable than vein structures. All stable ladder structures which were calculated have a ladder spacing equal to the height of the walls, as is the case in the initial stages of PSB formation (see figs. 3 and 7). At large plastic strain amplitudes, however, many adjacent ladders join to form wide wall structures, in which the walls are much wider than their separation (Woods [1973] and Winter [1978]). Obviously these structures are not in such a complete equilibrium as the ladder structures. The calculations show that stable ladder structures are characterized by a small or even vanishing y-component of the dipole strength, D_y (see §4.4). If D_y is increased (e.g. by increasing d in fig. 11a to $d = p/2$) the stability is lost. This result still awaits experimental verification.

References: p. 1591.

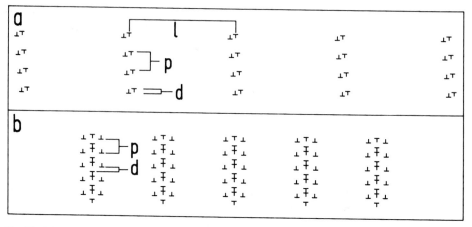

Fig. 11. Calculated equilibrium positions of two wall structures at $\tau_a = 0$. The D_y values of these two types of walls are: (a) $D_y = bd/lp$, (b) $D_y = 0$. (From NEUMANN [1983].)

Numerical calculations of this kind can obviously reproduce many features of dislocation structures observed in fatigue. They can also help to answer the important question, why the ladder structure can carry a hundred times the strain amplitude of the matrix. The above results imply that the borders of vein-type configurations are unexhaustable sinks for dislocations from the channels, since trapping of dislocations from within the channels would merely enlarge the vein, which – according to fig. 10 – reduces their decomposition stress only by a minimal amount. Such trapping of dislocations was indeed verified by numerical calculations. Therefore all free dislocations in the channels will be trapped efficiently, such that the channels finally become exhaustion-hardened.

The situation is different in the case of the ladder structure. Walls are the more stable the thinner they are. Thus additional dislocations are only loosely bound. Therefore the channels of the ladder structure cannot be hardened by dislocation-exhaustion. This agrees with the ten times larger dislocation densities found in the channels of the ladder structure than in those of the matrix structure (MUGHRABI [1981]).

4.3. Surface topography

Persistent slip bands are locations of high strain localization, which undergo cumulative strains of the order of 100–10 000. Therefore it is not astonishing that the surface of the PSBs shows an extreme roughness due to the large amounts of tension and compression slip. MAY [1960] was the first to calculate the expectation values of the difference between the highest and the lowest point in a slip band of given width under the assumption that a large number of elementary tension- and compression-slip steps are mixed randomly. Without further assumptions the maximum difference in elevation is proportional to the square root of the number of elementary steps, a result which is obvious in the context of the theory of random walk. This results, however, does not adequately describe the development of the surface topography in PSBs which show a profile far from random in many respects.

The profile of persistent slip bands was first measured by the so-called taper-sectioning technique (WOOD [1958]). In this technique, the specimen surface is sectioned along a plane which is almost parallel to the specimen surface. The angle between the section plane and the specimen surface is only about 5°. Therefore the profile of the surface is shown in the section plane exaggerated by a factor of $1/\cos 5° = 11.5$, the so-called taper magnification. Before the era of the SEM, this technique was applied because it extended the resolution of the optical microscope, although at the price of large distortions which can easily produce artefacts (see below). Similar profiles can be obtained nondestructively by using an interference microscope. The interference fringes appear along lines of constant elevation of the specimen surface. Thus, if the plane of focus is at a small angle with the specimen surface, the fringes form a taper-sectioning profile (NEUMANN [1967] and FINNEY and LAIRD [1975]).

With these techniques it was observed that narrow lips of material are extruded at the PSBs by cyclic deformation. In alloys these *extrusions* (FORSYTH [1953]) can have the quite astonishing length-to-thickness ratio of about 100, which has challenged many authors to try to explain their formation.

The material in the extrusions must come from somewhere, and therefore many investigators looked for holes or at least valleys, the *intrusions*, as counterparts for the extrusions. The search for holes was unsuccessful. Intrusions, however, are found quite frequently, although they are outnumbered by the extrusions and many "intrusions" are nothing but the gap between two neighbouring extrusions, and do not even reach below the level of the undeformed surface (see point 2 below).

More modern techniques like the scanning electron microscope resolve more details of the structure. Extrusions are shown in bright contrast due to the edge contrast. SEM stereo micrographs give an excellent impression of the rugged surface of PSBs.

WANG et al. [1982] have developed a simple technique to obtain in an nondestructive manner profiles from scanning electron micrographs: A *contamination line* is deposited across the persistent slip band by scanning with the electron beam along a single line for an extended period of time. The contamination line lies in the plane perpendicular to the surface but can be inspected from an inclined angle in the usual mode of operation of the SEM. The resolution is comparatively low, however, and the deep parts of the surface profiles, the intrusions and crack nuclei, are usually not accessible. In an alternative SEM method the specimen surface is viewed from a direction which is almost parallel to the trace of the PSB in the specimen surface (WANG et al. [1982]). Again, in this manner only an outer envelope of the profile can be obtained.

HUNSCHE and NEUMANN [1983] improved the sectioning technique in such a way that sections perpendicular to the surface can be obtained without any deformation

References: p. 1591.

of surface details down to the resolution of the SEM (20 nm). In this technique the original specimen surface is protected by an organic lacquer. The section is then produced by a micro-milling device, which, in the final stage, mills away layers as thin as 1 μm. After some ion-etching or electrolytic polishing a sharp edge between section plane and original surface is obtained, with an accuracy of the resolution of the SEM (20 nm). Figure 12 shows an example of such a section through a persistent slip band. Figure 13 shows the area near A in fig. 12 at a larger magnification, showing the resolution of the method.

In order to form the PSB out of a well developed matrix structure, the latter was formed by a slow enough cyclic hardening (30 000 cycles) to saturation. After reaching saturation the appropriate strain amplitude was enforced and kept constant afterwards in order to obtain the desired volume fraction of PSB. This *two-step test* was preferred over a constant plastic strain amplitude test, because in the latter test the initial hardening period and the resulting slip-distribution (CHENG and LAIRD [1981c]) depends strongly on the applied plastic strain amplitude before reaching saturation: Only a two-step test allows investigation of the plastic-strain-amplitude-dependence of PSB formation starting from the same structure. With this technique the following features of the surface topography in persistent slip bands have been verified:

1. The surface roughness within the PSB is composed of elementary extrusions which typically have triangular profile with a base width of 2 μm and a height of 2–3 μm.

2. In young PSBs there are no intrusions which reach below the original surface level.

3. If neighbouring extrusions touch each other, the resulting gap between them forms a V-shaped notch with a finite vertex angle of the order of 30° (see fig. 13).

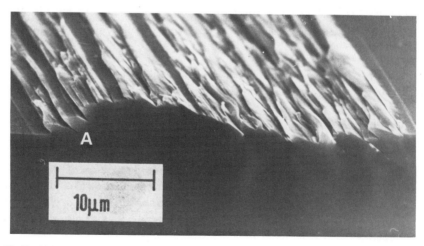

Fig. 12. Typical profile of a persistent slip band as revealed by the 90° sectioning technique (from HUNSCHE and NEUMANN [1983]).

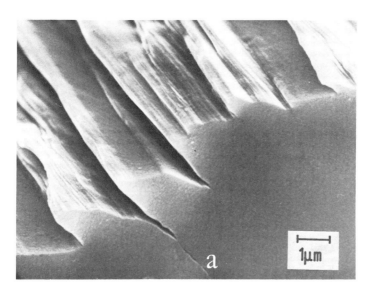

Fig. 13. Enlarged left hand side (area near A) of fig. 12, showing the shape of extrusions and intrusions and one crack nucleus at A (from HUNSCHE and NEUMANN [1983]).

4. In older PSBs the vertices of the so-called "intrusion" are even above the original surface level. Thus the PSB as a whole has a triangular profile with a base line of typically 20 μm and a height of 5 μm. It is proposed to call this bump which is much wider than a single extrusion a *protrusion* (THOMPSON and WADSWORTH [1958] and HUNSCHE and NEUMANN [1983]).

5. The growth rate of extrusions is 1 nm/cycle (HUNSCHE and NEUMANN [1983]) to 10 nm/cycle (MUGHRABI *et al.* [1982]), whereas the protrusion growth rate is at least ten times smaller.

6. Extrusions as well as protrusions do only grow in the direction of the primary Burgers vector. On those specimen side-faces which contain the primary Burgers vector, extrusion growth as well as protrusion growth is much slower (at least by a factor of twenty) than on those side-faces which form a large angle with the Burgers vector.

7. Protrusion growth leads to a *local lattice rotation* of the order of twenty degrees, which is evidenced by the direction of the individual extrusions.

8. It is important to mention that on opposite side faces of the specimen extrusions as well as protrusions are always growing away from the center of the specimen. If the other dimensions of the specimen were to stay constant, this would imply an apparent volume increase of up to 0.5%. Density measurements with a relative accuracy of 10^{-4} did not show, however, any evidence of such a large volume increase of 5×10^{-3} (HUNSCHE and NEUMANN [1983]).

9. Isolated, very high extrusions are sometimes observed (HUNSCHE and NEU-MANN [1983]).

References: p. 1591.

4.4. Models for extrusion formation

Extrusions form as soon as the matrix transforms into persistent slip bands. As discussed in §4.2, the triggering event for this transformation seems to be that the average dislocation distance approaches the annihilation distance. This leads to excessive annihilation of edge dislocations, and thus to heavy point-defect production. Electrical resisitivity measurements (POLAK [1970]) supply further evidence that there are high concentrations of point defects after fatigue. Therefore it seems natural to assume, that the formation of extrusions is somehow connected with dislocation annihilation and enhanced point-defect production in the PSBs. This point as well as most other physically relevant points of this section were discussed and developed into a detailed model by ESSMANN *et al.* [1981] and ESSMANN [1982]. In the following – formally different – presentation of this subject, simplification and unification are gained by using the concept of the dipole strength of dislocation configurations. Furthermore this presentation allows a discussion without reference to specific dislocation paths and clarifies the issue of dislocation layers.

Annihilation of edge dislocations on different slip planes can be accomplished only by *climb*. If we model the dislocation structure of the veins and of the walls in the ladders, as justified earlier, by arrays of straight edge dislocations with Burgers vectors $b_i = \pm b$ at the positions $r_i = (x_i, y_i)$, climb over y-distances Δy_i leads to a change of point defect densities

$$\Delta c_i - \Delta c_v = \sum b_i \Delta y_i L / V, \tag{6}$$

where c_i and c_v are the concentrations of interstitials and vacancies, L is the length of the dislocations, and V the volume of the configuration. The right-hand side is obviously the change of the y-component of the quantity

$$D = \sum b_i r_i / F, \tag{7}$$

where $F = V/L$ is the cross-section of the configuration. D, the so-called *dipole strength,* is characterized exclusively by the dislocation configuration (and not by the choice of the origin of the coordinate system), if and only if the net Burgers vector $B = \sum b_i$ vanishes. This is the case in fatigue dislocation structures, as discussed in §§ 3.1 and 4.1. Only after a large number of cycles, when the secondary hardening of PSBs sets in (fig. 6) the development of a cell structure with small orientation differences across the cell walls is observed (WANG [1982]). Thus the dipole strength is indeed an important characteristic of fatigue dislocation structures. A thorough discussion of the properties of this important quantity is necessary in this context in order to clarify the differences and implications of various models of processes within persistent slip bands.

For an *elementary dipole,* its strength is given according to (7) by $(b_2(x_2 - x_1), b_2(y_1 - y_1))$, in agreement with the popular notation. The dipole strengths of two more complex configurations are given in the caption of fig. 11. In the case of $B = 0$, D determines the *stress field* of the dislocation configuration at distances large compared to the size of the configuration.

In fatigued fcc metals, which have no assymmetry of slip, the x-component of the dipole strength is zero everywhere. If a completely random trapping of dislocations is assumed to be responsible for the matrix structure, it is very likely that also the y-component of the dipole strength D_y is zero in the veins of the matrix structure before the saturation stress is reached. At saturation, elastic nonlinearities may favour $D_y \geq 0$ (see below).

Only the x-components of the dislocation positions, x_i, not the y_i, can change by glide. Therefore the y-component of the dipole strength, D_y, cannot be changed by slip. This is an important statement and the reader is encouraged to verify this property with the help of some examples. The only way to change D_y is by climb, which requires changes of the point-defect densities as expressed by (6). Combining eqs. (6) and (7) yields:

$$\Delta c_i - \Delta c_v = \Delta D_y. \tag{8}$$

This equation can be rewritten as a conservation theorem as follows:

$$D_y + c_v - c_i = \text{const.} \tag{9}$$

This *conservation law* amplifies the importance of the dipole strength. It is valid as long as point-defect production and annihilation happen either via Frenkel pairs or via primary dislocations. These requirements are very likely fulfilled in the veins and ladders because of their high dislocation densities which make the formation of vacancy clusters and dislocation loops highly unlikely.

The *sign* of D_y defines whether there are excess vacancy-type dipoles or excess interstitial-type dipoles. If the Burgers vector is set positive by convention for a dislocation with an extra half plane extending to infinite positive y values, a vacancy dipole has a positive D_y.

The climb processes, characterized by the Δy_i, are non-conservative dislocation motions and give rise to *volume changes* or – more general – to internal strains. Their evaluation is very simple, if these processes happen with a constant density everywhere in a totally unconstrained piece of material. Then, by simply counting the missing extra half planes, one obtains immediately, if $\boldsymbol{b}_i = (\pm b, 0)$:

$$\gamma_{xx}^u = - \sum b_i \Delta y_i / F = -\Delta D_y, \quad \gamma_{ik}^u = \tau_{ik}^u = 0 \text{ for } ik \neq xx. \tag{10}$$

Equation (10) holds only under the prepositions that a vacancy occupies one atomic volume, which is a reasonable approximation, and that an interstitial has no volume, which is a bad approximation but does not matter since c_i is always extremely small.

If the climb processes take place in a *narrow strip of material* like the persistent slip bands (slip plane: xz-plane, Burgers vector in x-direction), the constraints by the neighbouring matrix enforce

$$\gamma_{xx} = \gamma_{xz} = \gamma_{zz} = 0. \tag{11}$$

From force equilibrium we have

$$\tau_{xy} = \tau_{yy} = \tau_{zy} = 0. \tag{12}$$

References: p. 1591.

Since (10), and (11) and (12) describe different elastic states of the same piece of material, the differences

$$\Delta\gamma_{ik} = \gamma_{ik} - \gamma_{ik}^{u} \quad \text{and} \quad \Delta\tau_{ik} = \tau_{ik} - \tau_{ik}^{u} \tag{13}$$

must fulfill Hooke's Law,

$$\Delta\tau_{ik} = \lambda\delta_{ik}\Delta\gamma_{ik} + 2\mu\Delta\gamma_{ik}, \quad \lambda = 2\mu\nu/(1 - 2\nu). \tag{14}$$

Inserting eqs. (11)–(13) into eq. (14) yields:

$$\tau_{xx} = 2\mu\frac{1 - 2\nu}{1 - \nu}\Delta D_y, \quad \tau_{yy} = 2\tau_{zz} = -2\mu\frac{2\nu}{1 - \nu}\Delta D_y, \quad \gamma_{yy} = -\frac{\nu}{1 - \nu}\Delta D_y. \tag{15}$$

Thus under this type of constraint (PSB in a matrix structure) no net change of the external shape (extrusion formation) results. Instead, a *uniform stress* develops (ANTONOPOULOS *et al.* [1976]).

This very simple stress state is modified near the stress-free surface down to a depth of approximately the thickness of the PSB (*St. Venant's principle*). Only if the rather unrealistic assumption is made that the free surface disturbs the stresses, but not the dislocation distribution, logarithmic singularities result where the interfaces between matrix and PSB intersect the surface (BROWN [1981]). In a more realistic treatment the maximum stresses must be about equal to the yield stress, i.e., the saturation stress. This subject is important for crack nucleation and will be discussed further in § 5.1.

It is worthwhile to estimate the size of the quantities involved in the above equations: An upper limit for the vacancy concentration far away from equilibrium is of the order of $c_v = 10^{-3}$. The dislocation densities are $10^{15}/m^2$ in the high density areas, and the *maximum dipole strength* – at a given dislocation density – is obtained in configurations like fig. 11 with $p = 2d = 1$. In such a configuration the dipole strength is $D_y = b/8d$. With $b = 0.3$ nm and $d = 30$ nm we obtain $D_y = 1.25 \times 10^{-3}$. This agrees with the experimental result of ANTONOPOULOS *et al.* [1976] for the local value of D_y within the walls, and is – via eq. (8) – comparable indeed to the upper limit of the vacancy concentration. The stresses which can develop are of the order of $\mu c_v = \mu \times 10^{-3}$, which is of the order of the saturation stress.

It is important to know whether more vacancies are produced than interstitials, because this would result in volume changes, which might be related to the formation of extrusions. Effects of *nonlinear elasticity* become important at high local stresses and can produce such an asymmetry in the production of point defects. It is generally accepted that the lattice is harder in compression than in tension. Therefore the self-energy of a vacancy dipole with its high dilatational stress is smaller than the self-energy of an interstitial dipole. ANTONOPOULOS *et al.* [1976] have argued on these grounds, that the tendency for annihilation of vacancy dipoles should be smaller. This would result in more vacancy dipoles than interstitial dipoles at saturation, as was observed experimentally by ANTONOPOULOS *et al.* [1976].

On the other hand, the *self-energy* of the resulting vacancies is smaller than that of the resulting interstitials. This stresses the opposite argument: vacancy dipoles can

more readily disintegrate, because the resulting vacancies have lower energy. Ess-
MANN *et al.* [1981] follow the latter argument and assume, that the majority of point
defects are vacancies. From the conservation law (8) it follows that the majority of
the dipoles must be interstitial, which contradicts the only existing direct measurement of dipole densities (ANTONOPOULOS *et al.* [1976]). As a consequence, both
authors predict different signs for the stresses within the PSBs, which will be
discussed further in § 5.1.

As stated earlier, climb processes on their own can produce stresses in the PSBs,
but no external shape-change (extrusions), because of the constraints due to the
neighbouring matrix. Thus *slip is a necessary requirement for extrusion formation.*
ESSMANN *et al.* [1981] considered a specific combination of slip and climb and
deduced the resulting changes in point defect densities and external shape. The
reader is referred to the original publications for details. Here we shall discuss only
net effects in general terms as in eqs. (8), (10) and (15).

Only the net flux of dislocations out of the PSB is relevant for its net shape-change.
Since dislocations of either sign are equally abundant ($B = 0$) in all known fatigue
dislocation structures, the net flux of escaping dislocations must have a vanishing net
Burgers vector, too, because of elementary statistics: $B^{esc} = \sum b_i^{esc} = 0$. Therefore it
is reasonable to consider the dipole strength of the escaping dislocations, D_y^{esc}. It is
easy to verify that escaping dislocations form an extrusion if and only if $D_y^{esc} < 0$
(interstitial-type dipoles). $D_y^{esc} > 0$ (vacancy-type dipoles) is synonymous to intru-
sion-formation. As in the case of $B^{esc} = 0$, random dislocation egress will enforce
that D_y^{esc} and D_y are of the same sign. Thus, the *sign of D_y is decisive whether
extrusions or intrusions are formed.* This was realized in less general terms by
PARTRIDGE [1965].

Thus, the above discussed assumptions by ANTONOPOULOS *et al.* [1976] lead to
formation of intrusions instead of extrusions or protrusions (contrary to experiment),
but agree with the experimentally found abundance of vacancy dipoles. The assumption of ESSMANN *et al.* [1981] lead to the opposite results. There may be a way
(ANTONOPOULOS *et al.* [1976]) that both assumptions hold in successive stages during
the development of the PSB, assuming that $D_y^{PSB} - D_y^{Matrix}$ is first negative and
becomes positive after further cycling. Then the stress concentrations at the PSB
interfaces would also change sign, thus switching preferred crack-nucleation sites as
observed experimentally (see § 5.1).

As long as climb and slip of primary edge dislocations are considered only, the
excess volume of the extrusions and protrusions comes from the point-defect
densities inside the PSB, which are left over after all reactions with primary
dislocations are considered [see eq. (10)]. Thus from definition they are those point
defects, which did not precipitate inside the walls of the PSB at the most abundant
sinks – primary edge dislocations. Also in the channels between the walls the
probability for a precipitation without a reaction with primary edge dislocations is
very small because of the sweep-up by primary screw dislocations (ESSMANN *et al.*
[1981]). Neither can the point defects stay unprecipitated inside the PSBs, since in
this case their extra volume should show up in density measurements, which is not

References: p. 1591.

the case, as discussed in §4.3, point 8. Thus these point defects must flow out of the PSBs and precipitate there. This introduces other difficulties, since extrusions are also observed at 4 K (Mc CAMMON and ROSENBERG [1957]). It is felt that these questions point to an important problem, which is not resolved by current theories of extrusion formation. Reconsideration of these questions may even rejuvenate the simplest explanation, that extrusions are soft PSB material which was literally squeezed out (LYNCH [1979]). Careful measurements of D_y as well as internal stresses are required to shed more light on these problems.

4.5. PSBs in various materials

The features of PSBs discussed in the above section were mainly observed in copper, but most of them apply equally well in other pure fcc metals. Depending on the impurity content, temperature, and strain rate, *bcc metals* can also behave like fcc metals – as far the dislocation structure is concerned (MUGHRABI et al. [1979]). In this case also persistent bands with a regular ladder structure have been observed (POHL et al. [1980]). If, however, the temperature is low enough, the preponderance of screw dislocations changes the picture considerably. Especially all those features which depend on the existence of edge dislocations are absent. The shape change due to the asymmetry of slip is another complication which drastically changes the events in single crystals (NINE [1973], R. NEUMANN [1975] and MUGHRABI and WÜTHRICH [1976]).

In *precipitation-hardened alloys* the localization of strain into very thin persistent slip bands is much more extreme. Accordingly, very long extrusions of extreme shape are found in these alloys (FORSYTH [1953] and VERPOORT [1980]). Electron microscopy shows, however, no indication of any ladder structure within the PSBs. On the other hand there is evidence (VOGEL et al. [1982a]) that the precipitates are dissolved within the PSBs and/or the precipitates are made less effective by disordering (CALABRESE and LAIRD [1974] and LAIRD [1975]). Since these changes in the precipitate structure may soften the material considerably, very large amounts of cyclic softening (VOGEL et al. [1982a,b]) can be observed in these alloys. It is generally accepted, that this is the reason for the high strain localization (CHENG et al. [1981]) within the PSBs of precipitation-hardened alloys.

In a *super alloy*, VERPOORT [1980] was able to study the interior of large and uniform extrusions with transmission electron microscopy. The dislocation structure within the extrusions, however, does not reveal details about their formation.

SINNING and HAASEN [1981] found PSBs in polycrystals but not in monocrystals of Cu–4 at% Ti and related this to a change of the annihilation distance in pile-ups. This idea agrees with observations of PARTRIDGE [1965] who preferentially found extrusions near twin boundaries in Mg.

5. Crack initiation and stage-I crack growth

Following the ideas of MAY [1960], there is a widespread belief that crack nuclei are a continuously evolving result of random slip within PSBs. For a review see LAIRD and DUQUETTE [1972]. Detailed observations of the surface topography (see §4.3) open the possibility for a more distinctive classification. Firstly, the surface topography of a PSB with its triangular extrusions and its protrusion is far from random. Secondly, an intrusion in a PSB is in most cases the valley between two extrusions and has a finite angle at the vertex, which is of the order of 20–30° (fig. 13). This distinguishes the intrusions uniquely from real crack nuclei, which are almost perfectly closed and just visible in the SEM (fig. 13 at A). Thus, at the resolution of the SEM, they are characterized by a vanishing crack tip angle. This is a qualitative difference which shows – together with the different dependence on environment (§5.2) – that *intrusion formation and crack nucleation are two different processes.*

5.1. Crack initiation in persistent slip bands

After one quarter of the total life, which corresponds to a cumulative plastic strain in the persistent slip band of 1200, the number of intrusions as well as crack nuclei within a PSB are proportional to its width (fig. 14). The data of fig. 14 show that the intrusions inside a PSB are, on the average, 2 μm apart, independent of the width of the PSB. This distance agrees with the average width of the dislocation ladder structure in the PSBs as well as the mean width of extrusions.

All crack nuclei less than 5 μm deep are parallel to the primary slip plane and are therefore called *stage-I cracks.* TEM observations show that they can propagate parallel to coexisting ladders (KATAGIRI *et al.* [1977]).

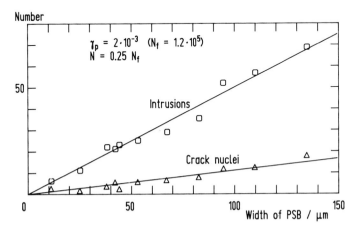

Fig. 14. Number of intrusions and crack nuclei in PSBs of varying width, yielding a constant intrusion density (one per 2 μm) and a constant crack-nuclei density (one per 9 μm at $N = 0.25 N_f$) (from HUNSCHE and NEUMANN [1983]).

References: p. 1591.

About 20% of the intrusions contain a crack nucleus after $N = 0.25 N_f$ (fig. 14). Later in the life, almost every intrusion contains a crack nucleus. Thus the intrusions provide crack nucleation sites at a constant lateral spacing of 2 μm. It is reasonable, therefore, to calculate the average length of cracks as an *average crack length per nucleation site*. In other words, an intrusion without a crack nucleus is counted as a crack of zero length. In this way the calculated average crack length is less sensitive to the number of crack nuclei which are below the detection limit. The results of such measurements are shown in fig. 15 as a function of the cumulative plastic strain in the PSB, Γ_{PSB}. From the double logarithmic plot a power law dependence like

$$a_{av} = 4 \times 10^{-12} \Gamma_{PSB}^{1.7} \text{ m} \tag{16}$$

can be deduced.

Sequential sectioning shows that the length of the crack nuclei along the trace of the persistent slip band on the specimen surface is usually less than 100 μm. These crack nuclei are less than 10 μm deep. The larger cracks, which form preferentially at position 2 in fig. 16, extend over several millimeters along the PSB on the specimen surface. They finally become macroscopic cracks and lead to failure.

Within a PSB the frequency of crack nuclei varies from position to position. In fig. 16 the typical PSB profile is divided into six sections and the number of crack nuclei found in each section is plotted as a histogram. It is evident from the histogram that early in the life position 6 is uniquely preferred, but later on, when larger cracks develop, the preference is inverted: on the other interfaces between the PSBs and the matrix, at position 2, most of the long cracks are found. These results

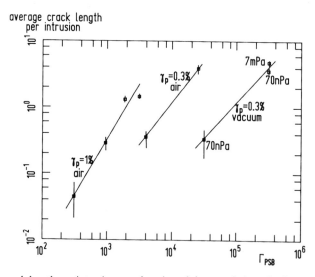

Fig. 15. Average crack length per intrusion as a function of the cumulative plastic strain within the PSB, Γ_{PSB}, for various environments and local plastic-strain-amplitude values in the PSB, γ_p (from HUNSCHE and NEUMANN [1983]).

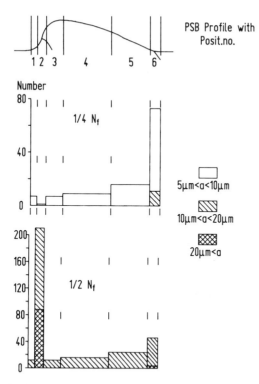

Fig. 16. Number of cracks of length a found at various locations within PSBs in copper during a constant plastic strain test ($\gamma_p = 2 \times 10^{-3}$, $N_f = 1.2 \times 10^5$) after slow hardening into saturation (from HUNSCHE and NEUMANN [1983]).

clarify the contradiction between older observations which claim crack nuclei preferably at position 6 (NEUMANN [1967] and HAHN and DUQUETTE [1978]) as well as at position 2 (BOETTNER and McEVILY [1965] and DÖNCH and HAASEN [1971]). ANTONOPOULOS *et al.* [1976] have proposed that the stress concentrations at the interface between the PSB and matrix due to the discontinuity of D_y (§4.4) may be the reason for the preferred crack nucleation there. For $D_y^{PSB} - D_y^{matrix} > 0$ tensile stresses concentrate only at position 6 and for $D_y^{PSB} - D_y^{matrix} < 0$ at position 2. This, together with the data of fig. 16, may perhaps be taken as a hint that the sign of $D_y^{PSB} - D_y^{matrix}$ changes during the fatigue life (ANTONOPOULOS *et al.* [1976]).

    Another feature makes the PSB–matrix interfaces unique within the specimen: As a plane of discontinuity in D_y, it necessarily must contain a layer of primary edge dislocations of the same sign with an area density

$$\rho^{area} = \left(D_y^{PSB} - D_y^{matrix} \right) / 2b. \tag{17}$$

This edge-*dislocation layer* does not produce lattice rotations. Therefore it does not produce striking contrast in the TEM, but it may weaken the interface, perhaps only

References: p. 1591.

in conjunction with the environment, which is very important for crack nucleation (see below). ESSMANN *et al.* [1981] did consider such dislocation layers as a consequence of their specific dislocation paths and proposed gas-stabilized voids in these dislocation layers as possible weakening mechanism, which may favour crack nucleation.

HUNSCHE and NEUMANN [1983] pointed out that position 2 may be a preferred crack nucleation site because there are *large local lattice rotations* (§4.3, point 7) which may introduce compatibility problems for primary slip. For a review of crack-nucleation sites in polycrystals see MUGHRABI and WANG [1981].

5.2. Effect of environment on crack initiation and early growth

It is well known that the fatigue life can be extended by performing the test in an inert environment. Until recently, there was little evidence whether this effect of environment comes about via the crack nucleation or crack propagation. With the help of the 90° sectioning technique it is possible to answer this question unequivocally. HUNSCHE and NEUMANN [1983] have performed tests in various environments including ultra high vacuum and have measured with the 90° sectioning technique the progress of crack-nuclei formation at intrusions. It was found that the formation of persistent slip bands as well as their surface topography is not altered by an inert environment if the comparison is based on identical values of the cumulative plastic strain in the PSBs, Γ_{PSB}. Crack nucleation is considerably delayed however, such that much larger values of $\Gamma_{PSB} = 10\,000$ can be reached, which is impossible in air. Under these conditions the surface topography of the PSBs is more rugged than in experiments in air with their smaller values of Γ_{PSB}.

Normal fatigue failure by crack initiation and growth cannot be obtained in vacuum. Instead, plastic buckling occurs even in perfectly aligned specimens with a length-to-thickness ratio of 2.5. Crack nuclei can be observed, however. Figure 15 shows the experimental data for the average crack length as a function of Γ_{PSB}. These data show clearly that crack nucleation, detected with a resolution of 20 nm, is deferred in vacuum by 1–2 orders of magnitude in the value of Γ_{PSB}.

These experimental data show no difference for a vacuum of 1 mPa (10^{-5} mbar) and a ultra-high vacuum of 100 nPa (10^{-9} mbar). This indicates that at vacuum conditions (10^{-5} mbar) the effects of the environment have been eliminated already. The finite crack nucleation measurements in ultra-high vacuum indicate that crack nucleation without the help of environment is possible, but only after a large amount of cumulative plastic strain of 10 000. The latter fact makes it very likely that there is the possibility to form crack nuclei just by slip irreversibilities. Because of the large amounts of cumulative strain a very small amount of slip irreversibility would be sufficient to produce such an effect. Therefore it is very difficult to specify which kind of irreversibility may be responsible for crack nucleation.

Because of the strong environmental effect on crack nucleation, the most important mechanism of crack nucleation seems to be that proposed by THOMPSON *et al.* [1956], which is sketched in fig. 17. According to this mechanism, every slip step will be covered by adsorbed atoms or molecules from the environment. After slip-reversion this adsorption layer will partially prevent annihilation of the just formed surface of the slip step. This leads to a closed crack or at least an area with imperfect cohesion. According to such a model, crack nucleation should happen where the slip concentrations are strongest, i.e., at the roots of the instrusions.

The most useful *phenomenological law* which describes the crack initiation life in terms of a loading parameter, was proposed by COFFIN [1954] and MANSON [1954] for constant plastic strain amplitude tests:

$$N_f = C\Delta\gamma_p^{-m}. \tag{18}$$

This relationship was tested for a large number of materials and was found to be highly successful, especially under low-cycle fatigue conditions, where the plastic strain amplitudes are large. Figure 18 shows some data of constant plastic strain experiments and constant total strain tests. Even copper crystals show this behaviour in constant plastic-strain-amplitude tests (CHENG and LAIRD [1981a]).

In spite of this success the stress amplitude has maintained its importance, since in high-cycle fatigue, crack nucleation often occurs at brittle inclusions. For this type of crack nucleation, also the applied stress range besides the applied plastic strain range is a controlling parameter. Best results are obtained therefore with *composite parameters,* which contain $\Delta\gamma_p$ as well as $\Delta\tau$ (HEITMANN et al. [1983]).

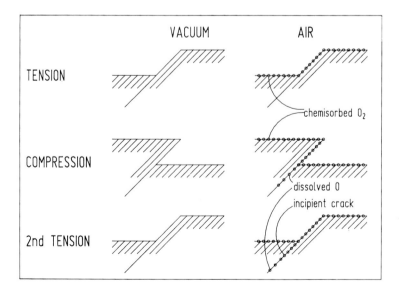

Fig. 17. Mechanism of crack nucleation and stage-I crack growth by the interaction of slip and reactive environment (after THOMPSON et al. [1956]).

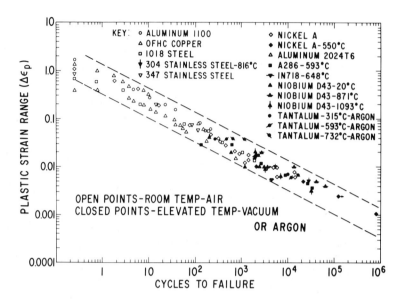

Fig. 18. The relation between fatigue life and plastic strain range according to the Coffin–Manson law (from COFFIN [1972]).

6.   Stage-II crack propagation

Crack nuclei below a depth of 5 μm are usually parallel to the primary slip plane. They are called stage-I cracks. When such crack nuclei link along the persistent slip band and grow deeper, they soon deviate from the orientation of the primary slip plane and propagate approximately perpendicular to the stress axis. This type of propagation is called *stage II*. TEM observations (AWATANI *et al.* [1978]) indicate, that stage-II cracks propagate through dislocation cells. The cell size decreases only slightly in the vicinity of the crack. The crack advance per cycle in stage-II can usually be resolved with TEM replica or even with SEM observations. In many cases fracture surface markings can be observed with a periodicity of the crack advance per cycle. Figure 19 shows an example with extremely long and wide-spaced *striations*. These striations can be used effectively to measure the local crack-advance rate as well as the exact shape of the crack front. Under certain conditions, especially in inert environments, the visibility of these striations can be poor or the striations are completely absent. Most likely, striations are also formed under these conditions, but when the crack grows larger, the crack faces, which are repetitively pressed onto each other during the compression phases of the loading cycles, tend to reweld because of the inert invironment, and are torn appart in the tension phases of the loading cycles. Such processes will, of course, destroy striations which may have existed initially.

Fig. 19. Fracture surface in a single crystal with the tensile axis parallel to [100], and crack propagation parallel to [011], with plane fracture surface and straight striations (parallel to [0,1, − 1]). Crack growth from top to bottom. (From NEUMANN [1974a].)

6.1. Mechanism of stage-II crack propagation

The mechanism of stage-II crack propagation is well understood. LAIRD [1967] has proposed that the increment of crack extension within each loading cycle is due to plastic blunting of the crack tip (fig. 20): Such a *crack-tip blunting* enlarges the surface area of the crack tip. If in the succeeding compression-phase the crack tip is closed again, it will *re-sharpen*, but it is highly unlikely, that the new surface which has formed during the blunting, will be annihilated during the compression. Thus after completion of every cycle the crack becomes a little longer. Furthermore it is very unlikely that the compressive strains are the exact reverse of the tensile strains which have blunted the crack tip. Any such imperfection in the back slip will produce a modulation of the fracture surface with a periodicity of the crack advance per cycle and give rise to striations. Therefore all kinds of striations may be expected and the exact profile of the striations in a specific case may depend on many imponderabilities.

Laird's model applies because of its general nature to a wide range of materials including commercial polycrystalline alloys. In the case of single crystals, however,

References: p. 1591.

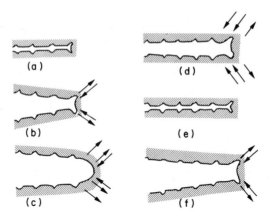

Fig. 20. Plastic blunting model of stage-II crack growth (from LAIRD [1967]).

the slip processes at stage-II crack tips are known in greater detail. NEUMANN [1967] and PELLOUX [1969] have proposed that alternating slip on two slip planes which intersect along the crack front is responsible for the crack extension. The details of this mechanism are shown schematically in fig. 21. The activation of two slip systems alternates, because the currently active slip plane work-hardens and the crack tip moves away from the formerly activated and work-hardened slip plane. In tension this mechanism can produce only cracks with large crack tip angles, since crack advance and crack opening are about the same. In fatigue, however, this drawback does not apply, because the large crack-tip opening as well as the large strains which are produced during the tension phase are reversed during the compression phase,

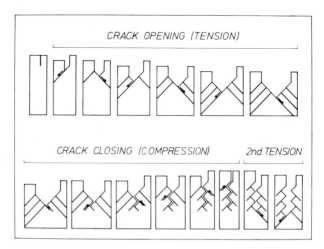

Fig. 21. Mechanism of stage-II crack growth by alternating slip on two slip systems (after NEUMANN [1967]).

thus producing a sharp crack on the average. Furthermore the irreversibility of surface production can be understood: If the back-slip process in compression (see fig. 21) is supposed to be such that a perfect surface annihilation is achieved, an ideal correlation of the slip events on two intersecting slip planes down to the motion of single dislocations is necessary. Since no such correlation of slip exists on independent slip systems surface production by alternating slip is highly irreversible.

Many details of this model can be verified experimentally (NEUMANN [1974a]). According to the model (see fig. 21) two slip planes – i.e. {111} planes in fcc metals – intersect along the crack front. Therefore the crack front should always be parallel to the intersecting line of two {111} planes, which are $\langle 011 \rangle$ directions. If the tensile axis is parallel to a $\langle 100 \rangle$ direction, the two slip planes are symmetrically loaded and very regular fatigue crack propagation can be obtained. Figure 19 shows an example of such a fracture plane. The dark area without striations is the surface of the spark cut, which was used to initiate crack growth in the desired orientation. Large crack-growth rates per cycle (150 μm) were used in order to produce easily visible striations. The striations, which mark the position of the crack front, are straight and parallel to $\langle 110 \rangle$ all the way through the specimen due to the special triangular shape of the ligament. In a rectangular ligament the crack front would lag behind near the specimen surface (by forming a zigzag line, see below). This triangular ligament shape fulfils the same purpose as the side grooves in compact tension specimens.

It is interesting to observe how a fatigue-crack front behaves if it is forced into a forbidden direction, e.g. into the [010] direction on the (100) plane. This can be achieved by bending a suitably notched specimen. The resulting fracture surface (fig. 22) shows striations which are strongly zigzagged in such a way that most of the crack front is parallel to either [011] or [01$\bar{1}$] as required by the alternating-slip model. These experiments strongly support the view that *the activation of two slip planes is necessary and sufficient for fatigue-crack growth.* In bcc metals, which have many crystallographically different slip planes, the requirements for fatigue-crack growth are less stringent, and microscopically straight crack fronts can be realized in many directions on many fracture planes (VEHOFF and NEUMANN [1979] and RIEUX *et al.* [1979]).

The *plastic zone* of such a geometrically simple fatigue crack can be observed on the side faces of the specimen. FUHLROTT [1978] has examined the slip distribution in detail. First, there are the main slip bands, which are responsible for the crack-tip opening. They contain strains of the order of a hundred per cent as required by the geometry of the process (NEUMANN [1974a]). In addition to these main slip bands, however, small amounts of slip are found on slip planes which extend from the main slip bands into the region ahead of the crack tip. This is in agreement with calculations by RIEDEL [1976] of the slip on inclined slip planes through a crack tip.

The events directly at the crack tip can be observed under high magnification, if the experiment is performed in situ within a SEM. Best results are obtained if the resultant Burgers vector of the slip in the main slip bands is contained in the specimen side face. Such a configuration is possible only in bcc metals, and fig. 23

References: p. 1591.

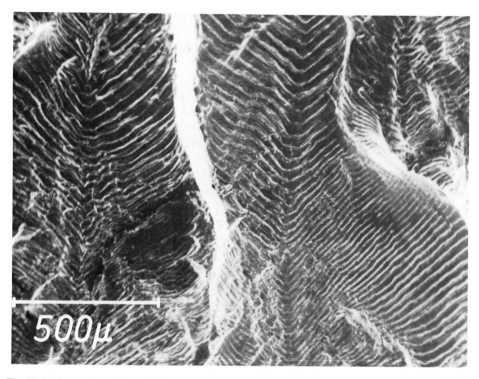

Fig. 22. A zigzagged crack front is obtained in copper, if the average crack front is not parallel to ⟨110⟩. Crack growth from top to bottom. (From NEUMANN [1974a].)

shows a sequence of SEM micrographs which directly show the alternating activation of slip planes, exactly as proposed in the model.

Such in situ experiments also show that *plastic crack closure* starts at the current crack tip and proceeds backwards. Schematically, the slip processes of fig. 21 are reversed in space in the reverse time sequence. These experiments also show how striations are formed by inexact backslip, illustrating the experimental result that many different and hardly reproducible striation profiles can be produced, depending on the more or less stochastic differences between forward- and back-slip.

The appearance of fatigue-fracture surfaces depends strongly on the orientation. Therefore the crack-growth resistance depends on the average fracture plane, as found by RICHARDS [1971], RIEUX et al. [1979] and HIGO and NUMOMURA [1981].

The behaviour of cracks with slip on inclined slip planes has also been considered *theoretically* on the basis of the Cottrell–Bilby–Swinden concept of treating cracks as arrays of dislocations. RIEDEL [1976] and VITEK [1976] calculated the continuous dislocation distribution on inclined slip planes and the stress distribution near the crack tip. Two singular slip planes (and not sets of parallel slip planes) were considered only, and the resulting crack-tip opening was neglected. A more involved

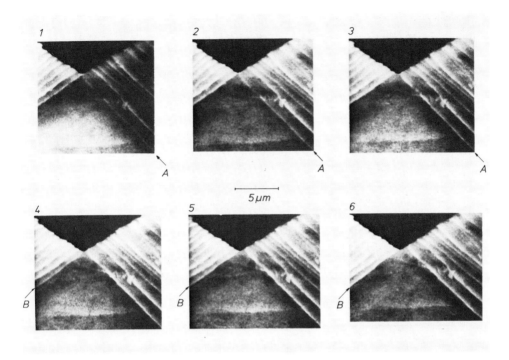

Fig. 23. Crack tip in Fe–3% Si at successive increments of the crack-tip opening displacement ($\Delta\delta = 0.4$ μm), showing the successive activation of slip plane A (23.1–23.3) and slip plane B (23.4–23.6), according to the alternating-slip model (see fig. 21). Note the changing relative position of crack tip and slip lines due to the changing slip displacements. (From VEHOFF and NEUMANN [1979].)

model for fatigue-crack growth, which on the other hand represents slip on two slip planes by just two singular dislocations, was presented by ATKINSON and KANNINEN [1977]. WEERTMAN [1981] considered fatigue-crack tips in the context of fracture-mechanical concepts with special consideration of an elastic enclave around the crack tip. The result of such calculations is usually a crack-growth law of the kinds discussed in the next section.

6.2. Fracture-mechanical concepts

Fracture mechanics is a well developed discipline of continuum mechanics concerned with the stress- and strain fields around cracks in isotropic continua (RICE [1968a]; ch. 23, §9.1). The main purpose is to characterize the stress situation at a crack tip in an arbitrarily shaped specimen or structural member by a single parameter. If the plastic zone size is small compared to the crack length as well as to all characteristic dimensions of the specimens, the major part of the surrounding material of the crack tip behaves elastically. In this case the linear elastic fracture mechanics applies and the *stress-intensity factor* is the only necessary parameter to

References: p. 1591.

characterize the stress field around the crack tip. For an infinitely extended specimen in plane strain, which is loaded at infinity by a stress τ^∞, K_I is given by

$$K_I = \tau^\infty (\pi a)^{1/2},\tag{19}$$

where a is the crack length. For more complicated geometries the values for the stress-intensity factor are tabulated by Rooke and Cartwright [1976]. For slow fatigue-crack growth the requirements of linear elastic fracture mechanics are usually fulfilled. Therefore the stress-intensity factors, which are produced by the loading cycle, determine the crack advance per cycle. According to Laird's model the crack growth depends on the blunting action in tension, but also on the re-sharpening action in the compression phase. Therefore the range of the stress-intensity factor, ΔK_I, is the most reasonable one-parameter characterization of fatigue loading.

There is, however, a problem involved in the choice of the minimum value of K_I. If this minimum value corresponds to compression, the crack faces will definitely touch somewhere far away from the crack tip, because fatigue cracks are quite serrated and very narrow compared to their length. This *crack closure* may reduce the effective crack length, which is necessary to calculate $K_{I,min}$ from the external load. The effect of crack closure has been discussed thoroughly by many authors (Taplin [1977]), because crack closure seems to be important even in tension–tension fatigue tests.

From the mechanism of crack advance, which is essentially a plastic blunting in the tensile phase and a re-sharpening in the compression phase, the crack advance per cycle must be roughly equal to the crack-tip opening during one cycle. In linear fracture mechanics the plastic crack tip opening, δ, is given by

$$\delta = K_I^2 / E.\tag{20}$$

If we replace K_I by $\Delta K_{I,eff}$ (the suffix "effective" indicating that crack closure should be taken into account), we expect proportionality between crack-growth rate per cycle and the square of the stress-intensity factor. This is also the result of more detailed considerations by Atkinson and Kanninen [1977] and Weertman [1981]:

$$\frac{da}{dN} = C\Delta K_{I,eff}^m,\tag{21}$$

This *Paris law* (Paris et al. [1961]) is very well observed in many materials for crack-growth rates ranging between 1 nm/cycle to 10 μm/cycle with an exponent, m, of 2–5.

Deviations from the Paris law are only observed at small stress-intensity ranges, since below a certain finite stress-intensity range, $\Delta K_{I,thr}$, no crack propagation is possible in most materials. This *threshold* behaviour (Tschegg and Stanzl [1981]) of cracks (among other effects) complicates the quantitative description of crack growth, and many modifications of eq. (21) have been proposed. For a review see McEvily [1982].

If the limits of linear elastic fracture mechanics are exceeded, there is another very promising one-parameter characterization of crack tips. Eshelby [1956] considered the driving force on a crack due to the stress field in a nonlinear elastic medium. In two dimensions the driving force for crack extension to the positive x-direction can be expressed by a path-independent line integral (RICE [1968b]), the so-called *J-integral*,

$$J = \int_L (W \, \mathrm{d}y - \partial \boldsymbol{u}/\partial x \cdot \boldsymbol{\tau} \cdot \mathrm{d}\boldsymbol{l}); \quad W = \int \boldsymbol{\tau} \cdot \cdot \, \mathrm{d}\boldsymbol{\gamma}, \tag{22}$$

where $\boldsymbol{u}$ is the displacement vector. The path L (line element $\mathrm{d}\boldsymbol{l} = (\mathrm{d}x, \mathrm{d}y)$ must run from one side of the crack to the other, avoiding the process zone ahead of the crack tip, where other than nonlinear elastic behaviour must be assumed. The limitation of such a characterization of the crack tip is given by the requirement that the material must behave in an nonlinear elastic manner. It is well known from deformation theory, that plasticity can be considered as nonlinear elastic behaviour as long as the stresses at every volume element increase monotonically under the additional constraint of "proportional loading". Both requirements can be cast into the equation

$$\boldsymbol{\tau}(\boldsymbol{r}, t) = h(t) \cdot \boldsymbol{\tau}_0(\boldsymbol{r}), \tag{23}$$

where $h(t)$ is a monotonic function of time.

It is obvious that plasticity behaviour deviates considerably from nonlinear elastic behaviour if unloading is encountered. Therefore an additional assumption about the material behaviour is necessary to apply the *J*-integral to cyclic loading: If the cyclic material behaviour can be approximated by the so-called Masing behaviour (MASING [1925]), stresses and strains in the *J*-intregral may be replaced by the corresponding ranges (WÜTHERICH [1982]). This modified *J*-integral was called *Z-integral* by Wütherich, and he showed that all appealing properties of the *J*-integral, especially its path-independence, are also attributes of the *Z*-integral. Also the experimental procedures for measuring *J* (DOWLING and BEGLEY [1976]) may be modified to measure *Z* in cyclic deformation and may be used to characterize the cyclic loading of crack tips (HEITMANN *et al.* [1983]).

References

ANTONOPOULOS, J.G., L.M. BROWN and A.T. WINTER 1976, Phil. Mag. **34**, 549.
ATKINSON, C., and M.F. KANNINEN, 1977, Int. J. Fracture **13**, 151.
AWATANI, J., K. KATAGIRI and H. NAKAI, 1978, Metallurg. Trans. **A9**, 111.
BASINSKI, Z.S., A.S. KORBEL and S.J. BASINSKI, 1980, Acta Metall. **28**, 191.
BOETTNER, R.C., and A.J. MCEVILY, 1965, Acta Metall. **13**, 937.
BROOM, T., and R.K. HAM, 1959, Proc. Roy. Soc. **A251**, 186.
BROWN, L.M., 1981, in: Proc. Int. Conf. on Dislocation Modelling of Physical Systems, Gainesville, FL (Pergamon Press, Oxford) p. 51.
CALABRESE, C., and C. LAIRD, 1974, Mater. Sci. Eng. **13**, 141.
CHARSLEY, P., 1981, Mater. Sci. Eng. **47**, 181.
CHENG, A.S., and LAIRD, C., 1981a, Mater. Sci. Eng. **51**, 55.
CHENG, A.S., and LAIRD, C., 1981b, Mater. Sci. Eng. **51**, 111.

CHENG, A.S., and LAIRD, C., 1981c, Fatigue Eng. Mater. Str. **4**, 331.

CHENG, A.S., J.C. FIGUERO, C. LAIRD and J.K. LEE, 1981, in: Deformation of Polycrystals: Mechanisms and Microstructures, Proc. 2nd Risø Int. Symp. on Metallurgy and Materials Science, eds. N. Hansen, A. Horsewell, T. Leffers and H. Lilholt (Risø National Laboratory, Roskilde, Denmark) p. 405.

COFFIN, L.F., 1954, Trans. ASME **76**, 931.

COFFIN, L.F., 1972, Metallurg. Trans. **3**, 1777.

DEHLINGER, U., 1931, Metallwirtschaft **10**, 26.

DESVAUX, M.P.E., 1970, Z. Metallk. **61**, 206.

DÖNCH, J., and P. HAASEN, 1971, Z. Metallk. **62**, 780.

DOWLING, N.E. and J.A. BEGLEY, 1976, in: Mech. of Crack Growth, Amer. Soc. Testing Mater. Spec. Tech. Publ. **590**, 82.

ESHELBY, J.D., 1956, in: Prog. Solid State Phys. **3**, 79.

ESSMANN, U., 1982, Phil. Mag. **45**, 171.

ESSMANN, U., and H. MUGHRABI, 1979, Phil. Mag. **A40**, 731.

ESSMANN, U., U. GÖSELE and H. MUGHRABI, 1981, Phil. Mag. **A44**, 405.

FINNEY, J.M., and C. LAIRD, 1975, Phil. Mag. **31**, 339.

FORSYTH, P.J.E., 1953, Nature **171**, 172.

FUHLROTT, H., 1978, Doctorate thesis, Technical Univ. Aachen.

GROSSKREUTZ, J.C., and H. MUGHRABI, 1975, in: Constitutive Equations in Plasticity, ed. A.S. Argon (MIT Press, Cambridge, MA) p. 251.

HAHN, H.N., and D.J. DUQUETTE, 1978, Acta Metall. **26**, 279.

HARTMANN, H.J., and E. MACHERAUCH, 1963, Z. Metallk. **54**, 161. 197 and 282.

HEITMANN, H.-H., H. VEHOFF and P. NEUMANN, 1983, Fatigue Eng. Mater. Str., to be published.

HELGELAND, O., 1964, J. Inst. Metals **93**, 570.

HIGO, Y., and S. NUMOMURA, 1980, in Advances in Fracture Research, eds. D. Francois *et al.* (Pergamon Press Oxford) vol. 1, 291.

HOLSTE, C., G.K. SCHMIDT and R. TORBER, 1979, Phys. Stat. Sol. (a) **54**, 305.

HUNSCHE, A., 1982, Doctorate thesis, Technical Univ. Aachen.

HUNSCHE, A., and P. NEUMANN, 1983, Acta Metall., to be published.

KATAGIRI, K., A. OMURA, K. KOYANAGI, J. AWATANI, T. SHIRAISHI and H. KANESHIRO, 1977, Metallurg. Trans. **8A**, 1769.

KUHLMANN-WILSDORF, D., 1979, Mater. Sci. Eng. **39**, 231.

KWADJO, R., and L.M. BROWN, 1978, Acta Metall. **26**, 1117.

LAIRD, C., 1967, in: Fatigue Crack Propagation, Amer. Soc. Testing Mater. Spec. Tech. Publ. **415**, 131.

LAIRD, C., 1975, in: Treatise on Materials Science and Technology, vol. 6: Plastic Deformation of Materials, ed. R.J. Arsenault (Academic, New York) p. 101.

LAIRD, C., and D.J. DUQUETTE, 1972, in: Corrosion Fatigue (National Association of Corrosion Engineers, Houston, TX) p. 88.

LAIRD, C., J.M. FINNEY and D. KUHLMANN-WILSDORF, 1981, Mater. Sci. Eng. **50**, 127.

LAUFER, E.E., and ROBERTS, W.N., 1964, Phil. Mag. **10**, 883.

LAUFER, E.E., and ROBERTS, W.N., 1966, Phil. Mag. **14**, 65.

LUKAS, P., M. KLESNIL, J. KREJCI and P. RYS, 1966, Phys. Stat. Sol. **15**, 71.

LUKAS, P., M. KLESNIL and J. KREJCI, 1968, Phys. Stat. Sol. **27**, 545.

LYNCH, S.P., 1979, Amer. Soc. Testing Mater. Spec. Tech. Publ. **675**, 174.

MANSON, S.S., 1954, NACA Technical Note **2933**.

MASING, G., 1925, Z. Tech. Phys. **6**, 569.

MAY, A.N., 1960, Nature (London) **185**, 303.

MCCAMMON, R.D., and H.M. ROSENBERG, 1957, Proc. Roy. Soc. **A242**, 203.

MCEVILY, A.J., 1982, ASTM Int. Conf. on Quantitative Measurement of Fatigue Damage, May 10–11, Dearborn, MI.

MUGHRABI, H., 1978, Mater., Sci. Eng. **33**, 207.

MUGHRABI, H., 1980, in: Proc. 5th Int. Conf. on Strength of Metals and Alloys, Aachen, 1979, eds. P. Haasen, V. Gerold and G. Kostorz (Pergamon Press, Oxford) vol. 3, 1615.

MUGHRABI, H., 1981, in: Proc. 4th Int. Conf. on Continuum Models of Discrete Systems, Stockholm, eds. O. Brulin and R.K.T. Hshieh, (North-Holland Amsterdam) p. 241.

MUGHRABI, H., and R. WANG, 1981, in: Deformation of Polycrystalls: Mechanisms and Microstructures, Proc. 2nd Risø Int. Symposium on Metallurgy and Materials Science, eds. N. Hansen, A. Horsewell, T. Leffers, and H. Lilholt (Risø National Laboratory, Roskilde, Denmark) p. 87.

MUGHRABI, H., and C. WÜTHRICH, 1976, Phil. Mag. **33**, 963.

MUGHRABI, H., F. ACKERMANN and K. HERZ, 1979, in: Fatigue Mechanisms, Proc. ASTM-NBS-NSF Symp., May 22–24, 1978, Kansas City, MO, Amer. Soc. Testing Mater. Spec. Tech. Publ. 675, 69.

MUGHRABI, H., R. WANG, K. DIFFERT and U. ESSMANN, 1982, ASTM Int. Conf. on Quantitative Measurement of Fatigue Damage, May 10–11, Dearborn, MI.

NABARRO, F.R.N., 1952, Adv. Phys. **1**, 269 (see p. 327–328).

NEUMANN, P., 1967, Z. Metallk. **58**, 780.

NEUMANN, P., 1968, Z. Metallk. **59**, 927.

NEUMANN, P., 1974a, Acta Metall. **22**, 1155 and 1167.

NEUMANN, P., 1974b, in: Constitutive Equations in Plasticity, ed. A.S. Argon (MIT Press, Cambridge, MAss) p. 449.

NEUMANN, P., 1978, Exp. Mech. **18**, 152.

NEUMANN, P., 1983, Acta Metall., to be published.

NEUMANN, R., 1975, Z. Metallk. **66**, 26.

NEUMANN, R., and P. NEUMANN, 1970, Scripta Metall. **4**, 645.

NINE, H.D., 1973, J. Appl. Phys. **44**, 4875.

PARIS, P.C., M.P. GOMEZ and W.E. ANDERSON, 1961, The Trend in Engineering (Univ. of Washington) **13**, no. 1, 9.

PARTRIDGE, P.G., 1965, Acta Metall. **13**, 517.

PATERSON, M.S., 1955, Acta Metall. **3**, 491.

PELLOUX, R.M.N., 1969, Trans. ASM, **62**, 281.

POHL, K., P. MAYR and E. MACHERAUCH, 1980, Scripta Metall. **14**, 1167.

POLAK, J., 1970, Scripta Metall **4**, 761.

RASMUSSEN, K.V., and O.B. PEDERSEN, 1980, Acta Metall. **28**, 1467.

RICE, J.R., 1968a, in: Fracture, ed. H. Liebowitz (Academic, New York) vol. 2, 191.

RICE, J.R., 1968b, J. Appl. Mech., Trans. ASME **35**, 379.

RICHARDS, C.E., 1971, Acta Metall. **19**, 583.

RIEDEL, H., 1976, J. Mech. Phys. Solids **24**, 277.

RIEUX, P., J. DRIVER and J. RIEU, 1979, Acta Metall. **27**, 145.

ROBERTS, W.N., 1969, Phil. Mag. **20**, 675.

ROOKE, D.P., and D.J. CARTWRIGHT, 1976, Compendium of Stress-Intensity Factors (Her Majesty's Stationery Office, London).

SINNING, H.-R., and P. HAASEN, 1981, Z. Metallk. **72**, 807.

TAPLIN, D.M.R., ed., 1977, Proc. Int. Conf on on Fracture 1977 (Univ. of Waterloo, Canada) vol. 2, 1009 ff.

TAYLOR, G.I., 1934, Proc. Roy. Soc. **A145**, 362.

THOMPSON, N., and N.J. WADSWORTH, 1958, Adv. Phys. **7**, 72.

THOMPSON, N., N.J. WADSWORTH and N. LOUAT, 1956, Phil. Mag. **1**, 113.

TSCHEGG, E., and S. STANZL, 1981, Acta Metall. **29**, 33.

VEHOFF, H., and P. NEUMANN, 1979, Acta Metall. **27**, 915.

VERPOORT, C., 1980, Dissertation, Univ. of Bochum.

VITEK, V., 1976, J. Mech. Phys. Solids **24**, 263.

VOGEL, W., M. WILHELM and V. GEROLD, 1982a, Acta Metall. **30**, 21.

VOGEL, W., M. WILHELM and V. GEROLD, 1982b, Acta Metall. **30**, 31.

WANG, R., 1982, Doctorate thesis, Stuttgart Univ.

WANG, R., B. BAUER and H. MUGHRABI, 1982, Z. Metallk. **73**, 30.

WEERTMAN, J., 1981, IUTAM Symp. on Three-Dimensional Constitutive Relations and Ductile Fracture, ed. S. Nemat-Nasser (North-Holland, Amsterdam) p. 111.

WETZEL, R.M., 1971, Ph. D. thesis, Univ of Waterloo, Canada.

WILKENS, M., K. HERZ and H. MUGHRABI, 1980, Z. Metallk. **71**, 376.

WINTER, A.T., 1973, Phil. Mag. **28**, 57.

WINTER, A.T., 1974, Phil. Mag. **30**, 719.

WINTER, A.T., 1978, Phil. Mag. **37**, 457.

WINTER, A.T., O.B. PEDERSEN and K.V. RASMUSSEN, 1981, Acta Metall. **29**, 735.

WÖHLER, A., 1870, Z. Bauwesen **20**, 73.

WOOD, W.A., 1958, Phil. Mag. **3**, 692.

WOODS, P.J., 1973, Phil. Mag. **28**, 155.

WÜTHRICH, C., 1982, Int. J. Fracture **20**, R35.

CHAPTER 25

RECOVERY AND RECRYSTALLIZATION

Laboratoire de Métallurgie Physique
Université de Paris-Sud
91405 Orsay, France

R.W. Cahn and P. Haasen, eds.
Physical Metallurgy; third, revised and enlarged edition
© *Elsevier Science Publishers BV, 1983*

1. Classification of phenomena

We shall be concerned in this chapter primarily with the mechanisms by which metals and alloys repair the structural damage caused by mechanical deformation, and incidentally with the resulting changes in physical and mechanical properties. These repair mechanisms are thermally activated and thus the deformed material has to be heated: any heat-treatment intended to reduce or eliminate deformation-induced damage is termed *annealing*.

The first distinction to be made is between *recovery* and *recrystallization*. The term *recovery* embraces all changes which do not involve the sweeping of the deformed structure by migrating high-angle grain boundaries. The deformed crystal (or polycrystalline structure) thus retains its identity, while the density of crystal defects and their distribution changes. A special form of recovery occurs when *residual stresses* resulting from metal-working processes are removed by heat-treatment. Where such stresses are *long-range* (i.e. are approximately uniform over distances large compared with the grain size) then their removal is termed *stress-relief*. (We do not discuss this topic further.)

In *recrystallization*, the crystal orientation of any region in the deformed material is altered, perhaps more than once. This results from the passage through the material of high-angle grain boundaries.

In *primary recrystallization*, a population of new grains is *nucleated*, often at the grain boundaries of the deformed material, and these then *grow* at the expense of the deformed structure until this is all consumed. Thereafter, grain boundaries continue to migrate, but more slowly: this stage of cannibalism among the new population of grains is termed *grain growth*. Usually all boundaries migrate at roughly equal rates, with the result that at any stage the grains are roughly uniform in size: sometimes, however, migration is restricted to a minority of boundaries only, so that a few grains grow very large at the expense of all the rest. This is termed *secondary recrystallization*; alternative terms in use are *coarsening* and *exaggerated grain growth*.

Boundaries of individual grains in the original deformed structure are sometimes observed to migrate over short distances, leaving in their wake a "healed" crystal of the same orientation as the growing grain. The regions swept by the moving grain-boundary are substantially free of dislocations. This process is called *strain-induced boundary migration*. Another process which can take place in the deformed structure is *subgrain coalescence*: here small-angle (sub) boundaries between sub-grains disappear progressively by the climb and migration of the individual dislocations constituting the sub-boundaries.

The term *recovery* is also often applied in a different sense to the gradual return of physical and mechanical properties to the values characteristic of the undeformed condition, irrespective of the mechanism by which this return is effected. In this sense, the yield stress, say, may recover principally as a result of primary recrystallization. The reference state is always the completely recrystallized (*fully annealed*) condition; the material then retains none of the original deformed grains.

2. Recovery

2.1. Recovery of electrical properties

Plastic deformation slightly increases the electrical resistivity. A great many investigations have been devoted to the stages by which the electrical resistivity returns to its fully annealed value. This is of interest both because it helps to disentangle the separate contributions made to the resistivity-increase by dislocations and by deformation-induced vacancies, and because it helps to cast light on the complex mechanism of the damage caused by neutron irradiation in nuclear reactors; this damage also causes resistivity changes which anneal out in a different manner from those caused by plastic deformation. Further details will be found in ch. 17, §§ 2.2.7.4 and 2.3.

2.2. Recovery of stored internal energy

When a piece of metal is plastically deformed, a certain amount of external work has to be expended in the process. A small fraction of this work is retained as stored energy, and on annealing the metal much or all of this is progressively released in the form of heat. The measurement of this released heat requires highly sensitive differential calorimeters, in which the specific heat of a deformed sample is compared over a range of temperature with that of an undeformed sample. These calorimeters either operate at steadily rising temperatures (CLAREBROUGH *et al.* [1955]) or isothermally (e.g. GORDON [1955], BELL and KRISEMENT [1962]). Similar instruments are now commercially available as *scanning differential calorimeters*; they are beginning to be used to study retained energy as a function of strain and strain rate (SCHÖNBORN and HAESSNER [1982]).

CLAREBROUGH *et al.* [1955, 1956] in a series of classic experiments determined the stages of the release of stored energy for copper and nickel. Figure 1 shows some of their results and correlates the energy release with the change of other physical properties. Pure copper gives off little energy during its recovery stage; only 3–10% of the total stored energy is released, depending on the amount of prestrain; during this stage, most of the vacancies generated by deformation have already diffused out. Most of the energy release and resistivity drop, and all of the hardness drop, are associated with recrystallization. When the copper contains impurities, the recrystallization temperature is raised and more energy is released during recovery. The temperature spectrum of energy release is also altered.

An alternative, very sensitive technique for measuring stored energy (not, however, capable of determining a temperature spectrum) is liquid-solution calorimetry. As an example of the application of this technique, SMITH and BEVER [1968] showed that the stored energy in gold after deformation at 78 K is considerably higher than the energy stored after room-temperature deformation to the same strain. The difference is predominantly due to the fact that at 78 K point defects are retained, while at room temperature they anneal out in this metal. (The 78 K-sample is held at

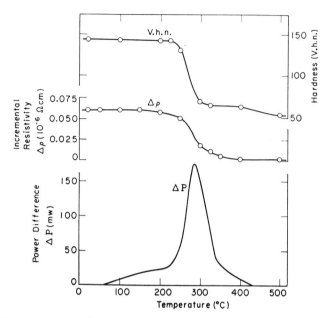

Fig. 1. Power differential, representing released energy, during the uniform heating of plastically twisted pure copper rod. Recovery of resistivity and hardness are also shown. (After CLAREBROUGH *et al.* [1955].)

that temperature until the moment of its dissolution in the liquid metal bath of the calorimeter.)

The most recent review of calorimetric studies of stored energy of cold work and its release on annealing is by BEVER *et al.* [1973]. Since then, RÖNNPAGEL and SCHWINK [1978] have developed a sensitive new calorimetric method which operates *during* plastic deformation.

In nickel, vacancies do not diffuse as readily as in copper, and here the disappearance of vacancies is represented by the region below about 200°C. The second small peak represents a form of recovery of uncertain origin; BELL and KRISEMENT [1962] regard it as the beginnings of recrystallization. Evidence summarized by CLAREBROUGH *et al.* [1963], however, suggests that this stage is connected in some way with the presence of impurities which slow down the release of vacancies.

The energy released during the recovery of aluminium must be due entirely to dislocation rearrangement and the formation of a cell structure (§ 2.4.2), since vacancies will have migrated out of the lattice at room temperature. Two recovery stages have been found in this metal (VANDERMEER and GORDON [1963]) and have been identified with the processes termed meta- and ortho-recovery (§ 2.3).

Experimental results such as those in fig. 1 can be combined with theoretical values for density, resistivity and stored energy associated with individual vacancies and dislocations and estimates made of the concentration of these defects in the original deformed metal; the overlapping data permit useful cross-checks. Estimates vary somewhat according to the assumptions made. Thus Clarebrough *et al.* found that nickel compressed by 70% contained about 0.01% vacancies and over 10^{11} dislocations/cm^2. Bell and Krisement deduced a higher vacancy concentration and lower dislocation concentration. An uncertainty by a factor of 5–10 must be accepted in this kind of calculation; BROOM and HAM [1957] discuss the matter in detail.

It is probable (CLAREBROUGH *et al.* [1963]) that at least part of the energy release during the recovery stage is due to changes in the arrangement of dislocations, with or without the formation of a cell structure (§2.4.2). This consideration complicates the calculation of vacancy and dislocation concentrations from calorimetric measurements. In nickel, the energy, E_{ry}, released in recovery is about equal to (Clarebrough *et al.*) or three times greater than (Bell and Krisement) the energy, E_{rn}, released during recrystallization. In aluminium, the proportion of energy released during recovery is again about equal to that released in recrystallization. In fact, the fraction E_{ry}/E_{rn} is an inverse function of the specific energy of stacking faults in the (111) plane of face-centred cubic metals (chapter 19, §3.3); the lower the stacking-fault energy, the more difficulty dislocations have in *climbing* out of their glide planes (chapter 19, §5.3) – an important process in recovery – and the more the recovery process is inhibited. In metals with very low stacking fault energy, recovery is limited to the diffusion of vacancies out of the deformed metal; dislocations are not substantially rearranged or eliminated until recrystallization begins. (Nevertheless, it has recently been established by CHERNS *et al.* [1979] that after intense irradiation, when a large supersaturation of point defects is present, dissociated dislocations *can* climb by a special mechanism without the prior formation of a constriction – i.e., without the local disappearance of the stacking fault. This matter is also discussed in ch. 18, §§3.5 and 5.4.)

2.3. Recovery of mechanical properties

Figure 1, and tests on nickel, shows that no drop in hardness accompanies stress relief or recovery in brass, copper or nickel. These are all metals with low stacking-fault energy, and therefore little climb and rearrangement of dislocations can take place, especially in brass or copper. Since the flow stress and hardness are a function of the concentration and disposition of dislocations, this immutability of dislocation structure prior to recrystallization implies that the mechanical properties also remain fixed. Slight dislocation movements have been postulated to account for considerable stored-energy release during the recovery stage in some metals such as silver (§2.2), but such rearrangement must be sufficiently slight to leave the yield stress unaffected. In other metals, of which aluminium and α-iron are the most important, dislocations can climb easily, and correspondingly these soften during a recovery anneal. Under favourable circumstances, the whole of the work-hardening may be recovered without the intervention of recrystallization. Thus a weakly deformed silicon-iron crystal * (DUNN [1946]), and likewise a very slightly deformed iron-

---

* Iron–silicon alloy is often used instead of pure iron in experiments on recovery and recrystallization since this alloy does not undergo a phase transformation on heating.

References: p. 1665.

aluminium polycrystal (LAWLEY *et al.* [1961]) recovers *completely* at 700–800°C; after heavier deformation, however, iron will recover only partially (fig. 2). Slightly deformed aluminium crystals can recover almost completely at 400–600°C (MASING and RAFFELSIEPER [1950]) while various investigators have found that aluminium single or polycrystals will recover about half the work-hardening at 300–400°C. The general rule is: the larger the deformation, the smaller is the fraction R of the work-hardening that can be recovered by a standard recovery anneal. Crystals of hexagonal metals such as zinc or cadmium are exceptions: these can recover completely even after very large tensile strains by *easy glide* (ch. 19, §2.2).

Figure 2 depicts softening isothermals for polycrystalline iron prestrained in tension. These show the form characteristic of recovery: the *rate* of softening is highest at the beginning and then steadily diminishes. Recrystallization-softening isotherms, on the other hand, have a sigmoidal shape: softening (when proportional to fraction recrystallized) is slow at first, accelerates and then slows once more (fig. 13, below). The kinetics of recovery in their dependence on prestrain, prestraining temperature and annealing temperature are reviewed by PERRYMAN [1957].

Because of the considerable softening of aluminium by recovery, *macro*hardness measurements cannot be used to estimate recrystallization in this metal; the technique *can* however be used for copper, in which there is virtually no recovery-softening. For both metals, it is possible to make arrays of *micro*hardness impressions; the histogram of microhardness-values then consists of two groupings, one due to unrecrystallized (but possibly recovered) grains and the other to recrystallized ones (GORDON [1955] and LÜCKE and RIXEN [1968]). When this technique was applied to explosively deformed copper, which is intensely hardened but sluggish to recrystallize, then after partial recrystallization it was found that the unrecrystallized grains had somewhat reduced (i.e., recovered) microhardness (fig. 3) (CHOJNOWSKI and

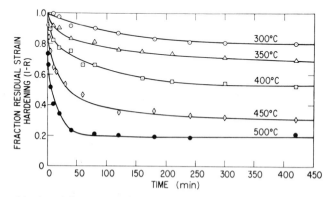

Fig. 2. Recovery kinetics of flow stress of identically strained iron samples at various temperatures. $(1 - R)$ represents the fraction of flow stress *increment* which remains after recovery, i.e., $R = (\sigma_m - \sigma_r)/(\sigma_m - \sigma_o)$, where σ_m is flow stress after deformation, σ_r is recovered flow stress, σ_o is fully annealed flow stress. (After LESLIE *et al.* [1962].)

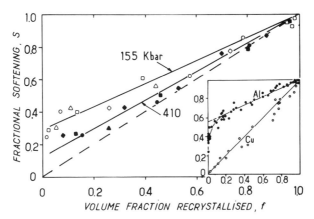

Fig. 3. Softening of copper previously subjected to shock deformation at two different levels of explosive pressure. Fractional softening, $(H_s - H_t)/(H_s - H_o)$, as a function of volume fraction recrystallized. H_s, H_t and H_o are hardness values as shock-deformed, annealed for t and before shock-deformation, respectively. Increasing temperatures in sequence: ◇ ○ □△, solid symbols at 410 kbar, open symbols at 155 kbar. (After CHOJNOWSKI and CAHN [1973].)
Inset: Curve similar to the above for cold-worked copper and aluminium (after LÜCKE and RIXEN [1968]).

CAHN [1973]); the delay in recrystallization gives the copper time to experience some recovery.

Recovery of metals such as aluminium and iron can be considerably accelerated, and the total degree of recovery enhanced, if stress is applied during the anneal (e.g. THORNTON and CAHN [1961]). The stress must be large enough to generate some plastic creep strain, but not so large that the resultant workhardening outweighs the enhanced recovery. Figure 4 shows stress-enhanced recovery of aluminium; the maximum creep strain generated during the stress-anneal was about 4% in these experiments. Even copper can be made to soften very slightly under these conditions, but the recovery is negligible in comparison with aluminium. This effect of stress is due to its accelerating influence on the climb of dislocations, which is an essential part of the recovery mechanism.

Recovery can also proceed *during* hot-working, whether it be rapid or slow. In this case it is denoted *dynamic recovery* *. Under conditions of rapid deformation, this process is again most pronounced in aluminium and α-iron, metals of high stacking-fault energy, in which dislocation climb is rapid. The details of subgrain morphology as affected by dynamic recovery have been extensively studied. MC-QUEEN and JONAS [1975] have shown that because of continuous growth of subgrains, or cells (§2.4.2) during hot-working (the subgrain equivalent of normal grain growth), some sub-boundaries disappear while others are born, and the outcome is

---

* The term "dynamic recovery" is also applied to the reduction of the rate of work-hardening in a stress–strain curve, at high strain (even at room temperature) (ch. 21, §1.4.5).

References: p. 1665.

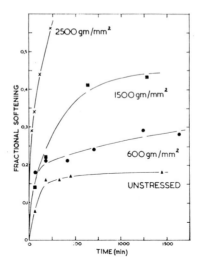

Fig. 4. Fractional recovery of polycrystalline aluminium prestrained 30% and annealed at 225°C (after THORNTON and CAHN [1961]).

that the mean subgrain misorientation remains constant at a few degrees, right up to high strains. TWISS [1977] has made a general study of mean subgrain diameter, d, during hot working, and found that d is related to the applied stress, σ, by:

$$\sigma = kd^{-n},$$

where $n \approx 1$ for many materials. (Twiss has proposed the subgrain size as a means of estimating the pressures that have acted upon rocks in the geological past – a form of 'palaeopiezometer'; hence the publication in a geological journal.) Dynamic recovery, and its uses, have been reviewed by McQUEEN [1977].

If a specimen undergoing creep is allowed to recover by temporary removal of the stress, not only is the subsequent creep-rate enhanced but there is evidence that the rupture life may be restored to that characteristic of virgin metals (EVANS and WADDINGTON [1969]; HART and GAYTER [1968]). It appears that this "healing" is associated with the sintering-up and consequent disappearance of grain-boundary cavities formed by creep.

Not only creep strain but also fatigue strains (chapter 24) can assist recovery. Figure 5 shows the change of flow stress when a succession of predetermined strain cycles is imposed on an aluminium sample which had previously been unidirection-ally cold-worked. The cold-worked samples recovered extensively under the influence of the fatigue cycles. Presumably, fatigue strain also assists dislocation climb.

Normally, recovery is measured in terms of the flow stress or hardness. A more fundamental approach is to measure the entire stress–strain curve both before and (on a duplicate sample) after recovery. Thus, fig. 6 shows measurements on aluminium polycrystals by CHERIAN et al. [1949]. The stress–strain curves, after recovery at different temperatures and for different periods, of samples prestrained to a stan-

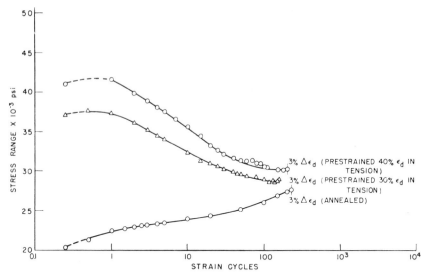

Fig. 5. Aluminium prestrained 0, 30 or 40% in tension and recovered at room temperature under simultaneous imposed fatigue cycles over a 3% range of strain, $\Delta\epsilon_a$. The stress range required to maintain this strain range adjusts itself automatically as the material softens or hardens. (After COFFIN and TAVERNELLI [1959].)

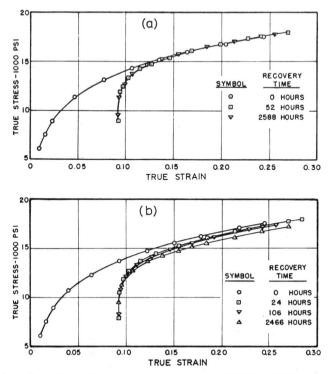

Fig. 6. Aluminium alloy deformed in tension and recovery-annealed (a) at 100°C (metarecovery) and (b) at 150°C (orthorecovery) (after CHERIAN *et al.* [1949]).

References: p. 1665.

dard strain, are compared with the stress–strain curves of samples whose straining has not been interrupted for a recovery anneal. Low-temperature recovery, termed *metarecovery*, is transient in the sense that eventually the sample resumes work-hardening characteristics identical to those of the sample which has not been annealed; the two stress–strain curves join up. When the temperature of the recovery anneal is sufficiently raised, its effect becomes permanent and the stress–strain curves remain distinct. This is *orthorecovery*. Generally, the work-hardening rate of a recovered metal is greater than it was before recovery at the same total strain. (See also § 2.4.2, end.)

Mechanical properties of deformed metals or alloys such as flow stress, hardness and ductility normally recover monotonically towards the values characteristic of the fully annealed state. However, there are exceptions: thus brass, copper–aluminium solid solutions (KOPPENAAL and FINE [1961]) and some other similar alloys are subject to slight *anneal-hardening* when annealed at temperatures too low to cause recrystallization; this phenomenon has been thought to be connected with the creation of short-range crystallographic order in the solid solution (ch. 4, § 10.3; ch. 14, § 3.2; chapter 21, § 1.3.2). Recently, WARLIMONT [1979] has found large levels of anneal-hardening in a range of nickel-base solid solutions and found that the magnitude of the effect is well correlated with atomic-size misfit between solvent and solute. He deduces that anneal-hardening (negative recovery as it were) is due to segregation of solute to dislocations, and he holds this to be true for copper alloys also.

2.4. Recovery of microstructure

2.4.1. Polygonization and subgrains
An especially simple form of structural recovery is observed when a crystal is bent in such a way that only one glide system operates, and is subsequently annealed. The crystal breaks up into a number of strain-free *subgrains*, each preserving the local orientation of the original bent crystal, and separated by plane *sub-boundaries* which are normal to the glide vector of the active glide plane. This process is termed *polygonization*, because a smoothly curved lattice vector in the crystal turns into part of a polygon.

The polygonized structure can be seen particularly clearly in fig. 7a, which shows subgrains in a sharply bent and polygonized sapphire (aluminium oxide) crystal, as seen by transmitted polarized light. The different shades of the subgrains indicate their differing orientations: only those subgrains are seen which have sufficient "extinction". It should be noted that in spite of the sharp curvature of the original crystal, the sub-boundaries are straight. This is in accord with the elastic theory of *bend–gliding* (the process by which a crystal simultaneously glides and becomes bent), which specifies that the bent glide planes have the form of the involute of a cylinder, as indicated in fig. 7b.

In polygonized metals, the sub-boundaries can be revealed by etching; they are normally etched in the form of dense rows of pits (fig. 8) (CAHN [1949]). To

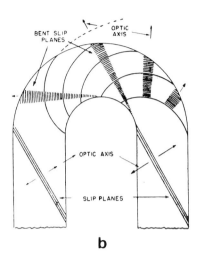

a   b

Fig. 7. (a) Single crystal of sapphire bent at 1800°C and then annealed 10 min at 2000°C. Photographed by transmitted polarized light. (b) Diagrammatic representation of plane bending, by glide on a single set of planes. Normals to the curved planes define the positions of the optic axis. (After KRONBERG [1955].)

understand this, it is necessary to consider the polygonization process in terms of the dislocation distribution. When a crystal undergoes plane gliding, then it is possible for all dislocations, both positive and negative, to pass right through the crystal and

Fig. 8. Bent, annealed and etched single crystal of aluminium, showing sub-boundaries, 70×. (After CAHN [1949].)

References: p. 1665.

out at the surface. To a first approximation this is what happens in easy glide (ch. 19, §2.2). However, in bend–gliding as exemplified in fig. 7b, a number of excess dislocations of one sign must remain in the crystal to accommodate the plastic curvature. This is most readily understood by considering the large difference in dimension between the outer and inner surfaces of a bent crystal. This strain implies the presence of many extra lattice planes near the outer surface, which terminate at edge dislocations within the crystal. The density of these excess dislocations is $1/Rb$, where R is the mean radius of curvature and b is the Burgers vector (ch. 18, §5.1). When now the bent crystal is annealed, these dislocations rearrange themselves in walls or *tilt boundaries* normal to the Burgers vector, because in this position they largely relieve each other's elastic strain field. This can best be appreciated in terms of the Read–Shockley equation (SHOCKLEY and READ [1950]) for the energy, E_b, of a tilt boundary:

$$E_b = E_0\theta(A - \ln \theta).$$

This equation, based on the theory of the elastic interaction of edge dislocations, and valid for values of θ up to several degrees, shows that dislocations constituting a tilt boundary progressively relieve each other's energy as θ increases; if there were no such relief then, since the dislocation density in a small-angle boundary $\propto \theta$, E_b would be proportional to θ and there would be no driving force for polygonization. (See also ch. 18, §5.4 and ch. 10B, §3.3).

Figure 9a shows the structure of a tilt boundary in more detail.

An incipient tilt boundary exerts a small force on isolated dislocations some distance away, sufficient to cause them to glide into the vicinity of the boundary plane. In fact, incipient *glide-polygonization* has been observed even before annealing begins. Each dislocation must then *climb* (ch. 18, §5.3) sufficiently to find a niche in the boundary. This process requires thermal activation and determines the rate of polygonization. The metastable end state is a tilt boundary with a uniformly spaced array of edge dislocations whose spacing h is given by the relation: $b/2h = \tan(\theta/2)$ or, approximately, $b/h = \theta$, where θ is the relative tilt angle of the two subgrains (fig. 9a). This relation has been verified by counts of etch pits on tilt boundaries, particularly in germanium and silicon, where each etch pit locates an individual dislocation.

Later stages of polygonization take place by the progressive merging of pairs of adjacent sub-boundaries; the driving force for this comes from the progressive reduction of the boundary energy *per dislocation* in the boundary, as the misorientation angle θ increases. The speed of this process (sketched in fig. 9b) is also limited by dislocation climb, since the dislocations in the two merged boundaries will not be uniformly spaced unless some dislocations climb.

A detailed survey of the mechanism and kinetics of polygonization is found in a review by HIBBARD and DUNN [1957]. The elastic theory of tilt boundaries and the consequential properties of such boundaries are treated in detail by LI [1961].

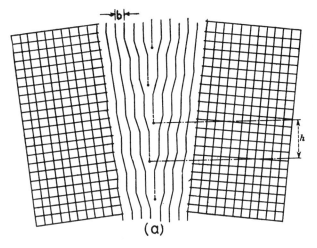

Fig. 9a. Structure of a tilt boundary. The two subgrains are mutually tilted through an angle θ.

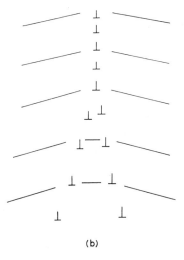

(b)

Fig. 9b. Titl boundaries meeting at a "triple point". The tilt axis relating the orientations of the two subgrains is normal to the plane of the figure.

2.4.2. Cell formation

The distribution of dislocations in a piece of plastically deformed metal is not uniform. This was first recognized in a classic research by GAY *et al.* [1954]: they deduced from X-ray microbeam diffraction experiments that grains in various deformed polycrystalline metals consist of subgrains or *cells*, the interiors of which have a comparatively low dislocation concentration while the small-angle cell boundaries are packed with dislocations. During subsequent recovery treatments, the cell boundaries (often called *cell walls*) sharpen and then the cells progressively grow larger while their interiors become further drained of dislocations. The cells, both before and after recovery, are normally roughly equiaxed – i.e., their dimensions are nearly the same in all directions – although under some circumstances, especially in the deformation of single crystals, cells may be elongated and arranged in bands. Cell formation is characteristic of metals that glide on more than one glide system, so that dislocations with several different Burgers vectors are available to form cell walls.

The sharpness of the cell walls formed during plastic deformation varies from metal to metal. There is a clear correlation between stacking-fault energy – and hence the degree of dissociation of dislocations and their ability to climb – and the sharpness of cell walls. Alloys such as α-brass or austenitic stainless steel after deformation have only the most rudimentary cell structure (dislocation tangles with regions of low dislocation density), copper and silver have recognizable cells with thick tangled boundaries, nickel and especially aluminium have better defined cells which rapidly sharpen and grow during annealing after deformation.

The mechanism by which the diffuse cells present after deformation sharpen and grow is a complex form of polygonization. Since dislocations are present on several glide planes (or, if on a single glide plane, with different glide vectors), the cell walls cannot be simple tilt boundaries. However, where the misorientation across cell walls has been examined, as in zinc (CAHN et al. [1954]), it turns out that they are effectively very similar to tilt boundaries. In zinc, each wall contains at least two families of mixed edge-cum-screw dislocations, but so arranged that the misorienting effects of the screw components cancel and the tilt axis is still parallel to the dislocation lines lying in the wall. There is some evidence that cell walls in aluminium have a similar structure. Since a variety of dislocations are available, cell walls have a number of different orientations relative to the crystal axes and can thus completely enclose individual cells.

Further evidence that cell boundaries are similar to sub-boundaries produced by polygonization is that the former, like the latter, often appear in the form of rows of etch pits on an etched section. Thus, fig. 10 shows a family of etched cells in annealed aluminium; these have been formed by slight accidental distortion during heating. When aluminium is substantially cold-worked and then given a recovery-anneal, again sharply delineated cells are formed. Since the cold-worked lattice is more sharply distorted, misorientations across cell walls are larger and it becomes convenient to reveal them by an etching process which produces colour contrast by white polarized light (ch. 10A, § 1.3.4). Figure 11 shows such a cell structure in cold-rolled and annealed aluminium.

A number of clear electron micrographs of sub-boundaries in α-iron have been published; sub-boundaries in this metal frequently consist of two-dimensional networks of dislocations.

Cell structures are formed most effectively during creep (chapter 20, § 1.6) and fatigue. Here the applied stress both produces the deformation and helps the dislocations to climb into the cell walls; since this is the rate-limiting factor,

Fig. 10. Subgrains in annealed aluminium sheet. 100×. (After Lacombe and Beaujard [1947].)

cell-formation is accelerated *. The growth of some cells at the expense of others, which follows on the first formation of sharply defined cell walls, is akin to the growth of polygonized subgrains by the merging of pairs of tilt boundaries (fig. 9b) and is similarly rate-limited by dislocation climb; here again, the effect of stress is to accelerate cell growth.

Cell walls are far more mobile than ordinary large-angle grain boundaries. Lacombe and Beaujard [1947] verified that in aluminium the cell morphology was completely rearranged during an anneal too short to alter the *grain* morphology detectably. This well established fact is paradoxical, because measurements of the mobility of tilt boundaries of different misorientations, θ, in the range 2–32° (Viswanathan and Bauer [1973a]) have established that the smaller θ, the lower the mobility. It may be that mobility reaches a minimum for some value of θ smaller than 2° and again becomes large for misorientations of a few minutes, although no reason for this is apparent. An alternative possibility is that in copper, with its low SFE, the migration of small-angle boundaries is severely inhibited by sluggish climb; whereas in aluminium, with high SFE, small-angle boundaries might move much more easily. This has not been systematically explored.

---

* The fact that climb is accelerated by stress is generally accepted (see also fig. 4, above) but the mechanism of this acceleration is still debated. A stress, to be effective, must produce some permanent strain, so the action of the stress in generating glide motion must be relevant. The stress also aids the weak elastic forces attracting dislocations to cell walls. Finally, the plastic strain generates vacancies, which again accelerate the self-diffusion required for climb. The theory of climb is discussed in ch. 18, §§ 5.1, 5.3.

References: p. 1665.

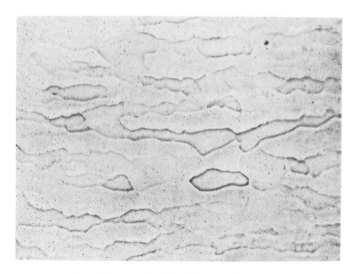

Fig. 11. Cell structure in an aluminium crystal cold-rolled to 20% reduction and annealed at 350°C for 20 s (from PERRYMAN [1954]).

It may be that cell walls, in any metal, move easily under the influence of internal stresses in each grain; it is well established that small-angle tilt boundaries move readily under the influence of an appropriate shear stress (creating thereby a small plastic strain). This notion has been discussed by EXCELL and WARRINGTON [1972] and is consistent with the observation by CAHN et al. [1954] on subgrain behaviour in polycrystalline zinc. On this view, the mobility of very small-angle sub-boundaries (cell walls) under purely thermal activation (i.e. in the absence of stress) may indeed be vanishingly small, but the presence of a suitable small stress may increase apparent mobility by orders of magnitude. This would only be true for pure tilt boundaries, and not for other kinds of small-angle boundaries.

The recovery of the yield stress or hardness of a metal during progressive annealing after plastic deformation is almost certainly correlated with the formation and growth of a sharp cell structure. The interiors of the cells become partially drained of dislocations at an early stage of the process, so it is most unlikely that it is the dislocation concentration in the cells, rather than the cell size, which determines the strength.

We have seen above (§ 2.3) that in hot-worked materials, where the mean mutual misorientation of neighbouring subgrains remains approximately invariant with strain, the subgrain diameter, d, is related to the applied stress, σ, by $\sigma = kd^{-n}$, where $n \approx 1$ for many materials. This relationship was first observed by BALL [1957], who studied aluminium deformed in creep. It seems that in creep or hot-working, the subgrain size does not define the strength but rather the applied stress defines the subgrain size.

For cells formed by recovery from cold work, the subgrain size *does* apparently determine the recovered flow stress. A general relationship has been established:

$$\sigma = \sigma_0 + k_y d^{-m}.$$

When $m = 0.5$, this reduces to the Hall–Petch relationship (chapter 19, §8). In the cold-worked, unrecovered state, when cells are ill defined, $m \approx 1$, but as recovery proceeds, subgrains grow and misorientations increase, m drops towards 0.5. By contrast, in hot-worked metals, m remains close to 1 throughout, as its also does for very small subgrains in recovered metals (McQueen [1977] and Thompson [1977]).

Other aspects of the relationship between yield stress and cell structure include:

(a) Metals of low stacking-fault energy, in which dislocations cannot readily climb, do not form sharp cell structures (though diffuse structures are observed). Correspondingly, no recovery of yield stress is observed.

(b) A stress applied during recovery anneal accelerates recovery of mechanical properties and, as we have seen, it also accelerates the growth of the cell structure.

(c) For aluminium, Vandermeer and Gordon [1973] have proposed that meta-recovery corresponds to the stage at which many dislocations have left the cell interiors but the cell walls have not yet begun to sharpen, while orthorecovery corresponds to the stage when cell walls become more sharply defined and the cells grow; more recent work by Hasegawa and Kocks [1979] on aluminium has attributed metarecovery to the sharpening of the cells already created during cold work, whereas orthorecovery is wholly associated with cell growth. A similar correlation has been observed in the case of iron (Leslie *et al.* [1963]).

(d) A further complication in relating cell size to properties derives from the observation (Dillamore *et al.* [1972]) that the mean subgrain size and misorientation within various grains in rolled iron depends on the orientation of individual grains relative to the rolling plane and direction. This was found to have a pronounced effect on the subsequent formation of annealing textures (§3.6).

A general survey of subgrain structures in metals subjected to diverse treatments has been published by Doherty [1978]; a comprehensive review of recovery, in regard to both properties and microstructure, has been published by Li [1966].

3. Primary recrystallization

3.1. Laws of recrystallization

A very large body of experimental information concerning primary recrystallization, spanning more than seventy years, can be summarized by the following six laws (Burke and Turnbull [1952]):

1) Some minimum (critical) deformation is necessary to initiate recrystallization.

2) The smaller the degree of deformation, the higher is the temperature required to initiate recrystallization.

References: p. 1665.

3) Increasing the annealing time decreases the temperature required for recrystallization.

4) The final grain size depends chiefly on the degree of deformation and to a lesser degree on the annealing temperature, normally being smaller the greater the degree of deformation and the lower the annealing temperature.

5) The larger the original grain size, the greater the amount of deformation required to give equivalent recrystallization temperature and time.

6) The amount of deformation required to give equivalent deformational hardening increases with increasing temperature of working and, by implication, for a given degree of deformation a higher working temperature entails a coarser recrystallized grain size and a higher recrystallization temperature.

To this a seventh law can be added:

7) New grains do not grow into deformed grains of identical or slightly deviating orientation, or into grains close to a twin orientation (TIEDEMA et al. [1949]).

An eighth law, not strictly concerned with primary recrystallization, is:

8) Continued heating, after primary recrystallization is complete, causes the grain size to increase.

These qualitative laws can be understood in a general manner if they are examined in terms of the component processes of primary recrystallization, i.e. nucleation of the new grains and their growth. Metallographic evidence has established that nuclei are formed in regions where the dislocation-concentration and strain-hardening are greatest. A certain local concentration of elastic energy is required to bring about a nucleus, hence the requirement of a threshold strain (law 1). Since the process of nucleation is thermally activated, a longer annealing time and higher temperature increase the probability of producing a nucleus, and correspondingly reduce the threshold strain (laws 2 and 3). The need for thermal activation also accounts for the *induction period* often observed before recrystallization becomes detectable.

The grain size depends on the balance between the nucleation- and growth rates; normally the nucleation rate increases faster with increased work-hardening, but the growth rate increases faster with rising temperature (hence law 4). Law 5 is attributed to the fact that most nuclei form at or near grain boundaries, so that a smaller original grain size enhances nucleation, while law 6 results from the reduced dislocation concentration resulting from deformation at elevated temperature. In fact, the acceleration of recrystallization at a given annealing temperature, as the temperature of previous deformation is lowered, can be dramatic. (A general survey of the influence on recrystallization of the conditions of deformation can be found in the book by Cotterill and Mould, see *Further reading* list.)

If the threshold strain for recrystallization is only just exceeded then very few nuclei will form, and this can be exploited for the preparation of quite large metal single crystals. Normally, a lightly strained fine-grained sample is heated at a very slowly rising temperature, or alternatively passed slowly through a sharp temperature gradient. In either case, one nucleus will form and grow rapidly enough to consume other potential nucleation sites before they become activated. This method has been applied principally to aluminium, lead, silicon–iron, and some other less common metals. It cannot normally be applied to the hexagonal close-packed metals since the twins which are produced on deformation are too effective as nucleation sites. Copper, silver and gold cannot be grown, because the formation of annealing twins leads to multiple orientations. High purity may be an obstacle, because rapid recovery may impede recrystallization; this is especially a problem with zone-refined iron.

The preparation of single crystals by recrystallization is reviewed by AUST [1963] (metals) and THORNTON [1968] (alloys).

The variation in grain size of a recrystallized metal as a function of prestrain and temperature is apt to be quite complex. This information is sometimes gathered in a single perspective figure, a *recrystallization* (or *Czochralski*) *diagram*. Figure 12 shows such a diagram for electrolytically refined iron (BURGERS [1941]); some other diagrams are a good deal more complex. A recrystallization diagram is at best approximate: for instance, where new grains nucleate at the boundaries of the deformed grains (VANDERMEER and GORDON [1959] and ENGLISH and BACKOFEN [1964]), the final grain size must evidently depend on the grain size before deformation. Perhaps because of such inadequacies, recrystallization diagrams have fallen out of use in recent years.

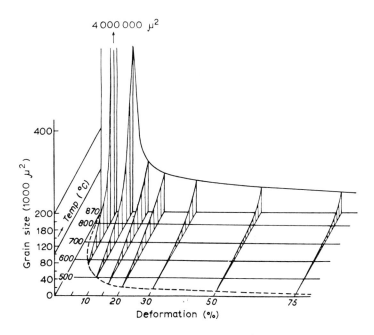

Fig. 12. Recrystallization diagram for electrolytically refined iron annealed 1 h (after BURGERS [1941]).

References: p. 1665.

3.2. Kinetics of primary recrystallization

A great deal of effort has been expended on the analysis of the isothermal kinetics of recrystallization, in the hope that this would cast indirect light on the mechanisms involved. In particular, it was hoped in this way to derive separate values for the nucleation rate, $\dot{N}$, and the growth rate, G. Since these quantities themselves are often a function of time and G may be anisotropic, and moreover the analysis must allow for the mutual interference of growing grains in the later stages of recrystallization, this type of analysis has rarely been fruitful. Most investigators agree on a resultant equation of the form

$$X_v = 1 - \exp(-Bt^k) \tag{1}$$

where X_v is the volume fraction recrystallized, t is time and B and k are constants. Values of k are most commonly in the range 1 to 2; the exact value is sure to vary with the experimental material and circumstances. A low value of k implies one- or two-dimensional recrystallization, i.e. new grains growing in the form of rods or platelets. Figure 13 shows typical recrystallization kinetics for isothermal recrystallization, obtained here by the use of quantitative metallography (chapter 10A, §4).

Some kinetic experiments have indicated a nucleation rate which is low at first and then increases, passing through a maximum, while others show a constant rate or else indicate that all nuclei form at zero annealing time; the last of course implies the presence of preformed nuclei. There is more agreement about growth rates; these either remain constant or else slowly diminish as growth continues. Here again, the variation with time and temperature of the constituent processes depends on the purity of the material, degree of deformation and grain size, and generalizations are dangerous.

One study has been particularly informative, in that the purity of the sample was systematically varied and careful metallographic and calorimetric studies were combined. VANDERMEER and GORDON [1959, 1963] studied the isothermal recrystallization of zone-refined polycrystalline aluminium, with or without small additions of copper, after deformation by 40%. The kinetics for the pure metal followed

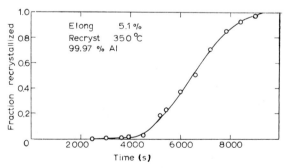

Fig. 13. Recrystallization kinetics for aluminium deformed 50% and annealed at 350°C (after ANDERSON and MEHL [1945]).

eq. (1), with $k = 2$. All new grains were nucleated in clusters at sites along grain boundaries of the deformed structure. Only a small proportion of the total grain-boundary area produced nuclei, but all those nuclei began to grow at the very beginning of the anneal; there was no induction period. New grains, long and narrow, all grew at a steady rate into the adjoining grains.

If impurity was added or the annealing temperature lowered, or both, deviations were found from the kinetics of eq. (1); these were attributed to a slowing down of the growth of new grains, and this was confirmed by direct measurement. Figure 14 shows typical results. The retardation was attributed to recovery: the authors' own isothermal calorimetric experiments showed that recovery and recrystallization normally overlap in aluminium. Figure 15 shows schematically the kind of calorimetric results obtained, and how they can be decomposed into constituent processes. When the component due to recrystallization has been separated out, it can be used to calculate a recrystallization isotherm like that of fig. 13; the metallographically and calorimetrically derived isotherms then agree well. (If a prolonged low-temperature recovery anneal precedes the recrystallization anneal, the two stages of fig. 15 no longer overlap; there is less energy left to recover when the test temperature is

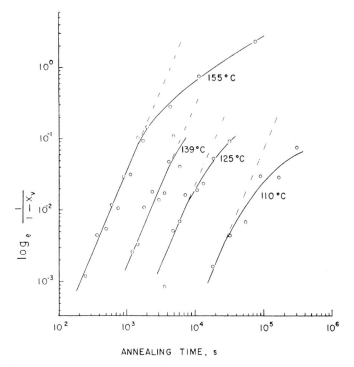

Fig. 14. Ln[$1/(1 - x_v)$] plotted as a function of isothermal annealing time at various temperatures for zone-refined aluminium containing 0.004% copper, rolled 40% at 0°C (after VANDERMEER and GORDON [1963]).

References: p. 1665.

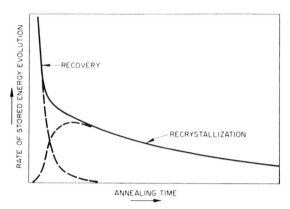

Fig. 15. Schematic illustration of the overlapping of recovery and recrystallization in the isothermal evolution of stored energy (after VANDERMEER and GORDON [1963]).

reached, and moreover recrystallization is retarded since the driving force for it has been reduced by the preceding recovery.)

For a given annealing temperature, the deviations due to added impurity are the more pronounced, the greater the impurity content, while for a given composition they are greater for lower annealing temperatures. This correlation is interpreted in terms of the different activation energies for recovery and recrystallization, and the differing effect of copper additions on these activation energies; the activation energy for recrystallization is increased more than that for recovery. As copper is added, therefore, both recovery and recrystallization are slowed down, but recrystallization is slowed down more; as the temperature is lowered, recrystallization is slowed more than recovery since it always has the higher activation energy.

The retarding influence of recovery on recrystallization may go so far as to inhibit growth altogether, especially at the low temperatures which must be used for starting the growth of single crystals. Difficulty has been reported for very pure aluminium, and single crystals of iron cannot be made at all by straining and annealing if the iron is too pure; an intentional carbon contamination must then be added before growth, and removed by gas-phase reaction after the crystal has been grown (TALBOT et al. [1957]). Similarly, MONTUELLE [1955] found that superpure (zone-refined) aluminium can only be recrystallized, to make single crystals, if it is doped with 0.04 wt% iron or lithium: but the high purity cannot then be restored. If these observations are taken in conjunction with Vandermeer and Gordon's results, which show that under appropriate experimental conditions impurities depress recovery *less* than recrystallization, it can be seen how complex the interaction of recovery and recrystallization is.

It is not possible to generalize as to recrystallization mechanisms from a single experimental analysis such as Vandermeer and Gordon's. For instance, ANDERSON and MEHL [1945], studying polycrystalline aluminium rather less pure and much less heavily deformed than in the work just described, reached the conclusion that the nucleation rate varied with time, that there was an induction period for nucleation, that the growth rate did not diminish with time and that growth of new grains was isotropic. If the amount of strain is even smaller, no proper nuclei arise at all and some existing grains grow into their neighbours; this is the phenomenon of strain-induced boundary migration, described in § 3.3.

3.3. Nucleation of primary grains

Few topics in physical metallurgy have generated so much and such prolonged argument as has the nature of the nucleation process in recrystallization; the debate is about sixty years old. Only studies undertaken in the past twelve years have at length been able to resolve the problem. The difficulty in resolving it has stemmed from the great difficulty of actually observing a nucleus (or even a grain which has barely progressed beyond the nuclear stage) under a microscope. Optical microscopy lacks resolution and the means of determining orientation, and electron microscopy normally suffers from a small field of view, so that the statistical chance of hitting on

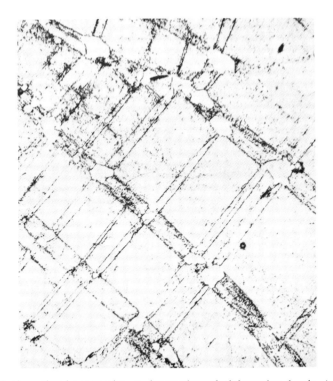

Fig. 16. Nucleation of primary grains at intersection of deformation bands, in a crystal of titanium–molybdenum alloy lightly deformed and annealed. The deformation bands are regions of high dislocation density. 75×. (After SCHOFIELD and BACON [1961].)

References: p. 1665.

a nucleus is small. This fundamental difficulty was finally overcome by the use of three techniques:

1) Optical microscopy coupled with the transmission Kossel method of X-ray diffraction (FERRAN et al. [1971]).

2) Transmission electron microscopy at 100 kV, using a sample thinned in such a way as to provide a large thinned area, combined with the use of Kikuchi lines as a means of determining local orientation (FAIVRE and DOHERTY [1979]).

3) High-voltage transmission electron microscopy, which provides a large field of view in three dimensions (RAY et al. [1975]).

From these studies it became clear that the mechanism of nucleation is intimately linked with the structure of the deformed metal, and specifically with the degree of heterogeneity of orientation *within* each deformed grain. This crucial factor was left out of account during the long debate, which had tended to concentrate on the relationship between the (supposedly uniform) orientation of a deformed grain and that of the recrystallized grain born in it.

Any theory of nucleation must account for the following experimental generalizations:

a) Nuclei form preferentially in regions where the local degree of deformation is highest, such as grain boundaries, deformation bands, inclusions, twin intersections and free surfaces. The rather unexpected conclusion, that strain density and work-hardening just below the free surface of a deformed sample is particularly high and leads to a high nucleation rate, has been proposed (KIMURA et al. [1964]) but more recently has been questioned (MUGHRABI [1977]). Figure 16 illustrates nucleation in a region of high dislocation concentration, seen at low magnification.

b) The nucleation rate increases sharply with increasing strain, above a minimum critical amount *.

c) There is extensive and conclusive evidence that the orientations of nuclei have a statistical connection with the orientations of the deformed regions in which they form, at any rate for moderate degrees of deformation. Formerly, this relationship was deduced indirectly from a comparison of deformation and annealing textures (a process fraught with pitfalls); recently, it has been established by direct metallo-graphic/diffraction methods.

d) Since, as we have seen, small-angle boundaries are apt to have a very small mobility, a nucleus cannot grow into material of nearly identical orientation.

The older models of nucleation are as follows:

1) *The classical nucleation (or fluctuation) model*, either homogeneous or heteroge-neous. This is now recognized to be impracticable, because of the small driving forces available and the high energy of the large-angle boundary which has to be created. Local elastic strains of the order of 20% over a region some nanometres in

---

* However, the proclivity to recrystallization depends not only on the amount of strain but also on its nature. RICHARDS and WATSON [1969] have found that cold-swaged iron recrystallizes much more sluggishly than iron cold-rolled to the same strain, and they attribute the difference to the more uniform distribution of stored energy after swaging.

diameter would be needed, and this level of strain is not credible.

2) *The growth of a locally polygonized region of the deformed structure* (a model worked out by CAHN [1950], COTTRELL [1953] and BECK [1954]). The idea was that polygonization would remove the strain energy from a small region which is more severely bent than its environment. The viable nucleus so created would be misoriented relative to its surroundings and can thus grow.

3) A related model based on the hypothesis of *coalescence* of neighbouring subgrains, by the "evaporation" of the dislocations constituting the sub-boundaries between them. (This process must involve both glide and climb.) The enlarged subgrain thus formed, much larger than its neighbours, could grow at their expense – i.e., act as a nucleus. This model was originally advanced by HU [1962, 1963].

4) A model based on the process of *strain-induced boundary migration* (SIBM), in which a subgrain within a deformed grain grows into its neighbour, forming a bulge which has the orientation of the source grain, but is largely free of dislocations. This ballooning subgrain can act as a nucleus. SIBM was first reported by BECK and SPERRY [1949, 1950] and subsequently studied by BAILEY and HIRSCH [1962].

5) A form of *martensitic mechanism* which produces a nucleus by shear. It was proposed by VERBRAAK and BURGERS [1957] and has been pursued by SLAKHORST [1975].

This last model was put forward to explain the puzzle of the creation of cube textures on annealing (§ 3.4). However, no direct experimental evidence has ever been advanced to substantiate it.

Model 1 having been eliminated, this leaves four among the older contenders. It is still too early to reach a definite conclusion about the field of applicability of model 5. Recent work has proved the validity of models 3 and 4. Model 2 is only applicable in an indirect sense: subgrains (cells), first produced by a dislocation reaction which is equivalent to polygonization, are the necessary precursor to nucleation, and furthermore, regions in which the orientation varies rapidly with position (i.e., sharp lattice curvature) are essential for the creation of a viable nucleus. The sharply bent lattice region invoked in model 2 is, in fact, most commonly the edge of a deformation band.

Studies of compressed aluminium polycrystals by FERRAN *et al.* [1971] and BELLIER and DOHERTY [1977] by the Kossel X-ray method and of rolled or compressed iron by INOKUTI and DOHERTY [1977, 1978], have shown the extreme fluctuation of orientation in the deformed grains. Variations of up to about 45° within a single grain were found after only 40% strain, over distances of often only a few μm. *Deformation bands* are formed, regions which begin by rotating differently from neighbouring regions when deformation starts; such rotational instability, once under way, continues to progress. Figure 17a, due to DILLAMORE and KATOH [1974], shows theoretically predicted orientation changes during uniaxial compression of iron for grains with initial orientations close to the (dashed) instability locus; figure 17b shows corresponding orientations of regions inside and outside deformation bands, as observed by Inokuti and Doherty. While the rotations found by the latter

References: p. 1665.

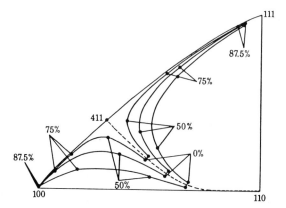

Fig. 17a. Slip rotation paths for axisymmetric compressive deformation of a body-centred cubic metal, deforming by pencil glide. Three different initial orientations are indicated, with rotations corresponding to various compressive strains for each initial orientation. (After DILLAMORE and KATOH [1974].)

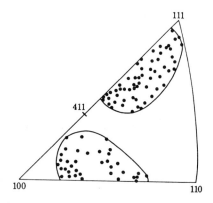

Fig. 17b. Orientations of various positions in a single grain of polycrystalline iron deformed 40% in compression. Orientations were determined by individual X-ray Kossel diffraction photographs. The two groups of orientations correspond to the matrix and to a series of deformation bands. (After INOKUTI and DOHERTY [1977].)

are greater than the computed rotations (presumably individual grains differ in their degree of mean strain), yet the pattern of orientation change is as predicted. Even in the absence of such deformation bands, with their sharp boundaries and extreme orientation gradients, there are still very large but less heterogeneous orientation gradients, more especially in the vicinity of grain boundaries.

The studies of annealed specimens, reported in the papers cited at the beginning of the preceding paragraph, showed that new grains are nucleated either at grain boundaries or at deformation-band limits (which may be characterized as "artificial grain boundaries"). Figure 18 shows a large grain in a compressed iron sample, and

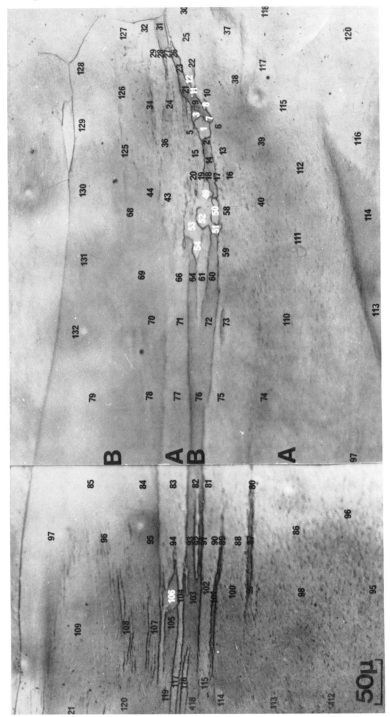

Fig. 18. Part of a large iron grain compressed 40%. The compression axis was vertical in the plane of the micrograph. The sample, which was annealed for 2 h at 500°C and nital-etched, has broken up into two interpenetrating matrix bands, A and B. New grains, numbered in white, have formed at the deformation bands. (After INOKUTI and DOHERTY [1978].)

References: p. 1665.

fig. 19 is the corresponding stereogram. This shows the distinct orientation clusters in the matrix and deformation bands (black numbers in fig. 18), and of the recrystallized grains after partial recrystallization (white numbers). Figure 20 is a micrograph showing new grains growing from a grain boundary in compressed

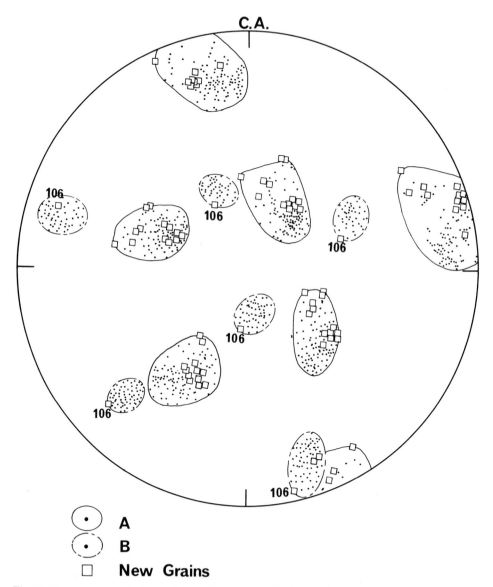

Fig. 19. Stereographic projections of the positions marked in fig. 18. New grain 106 has an orientation within the spread of matrix band B while all the other new grains have an "A" orientation. (After INOKUTI and DOHERTY [1978].)

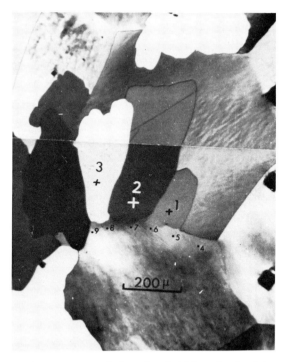

Fig. 20. Cluster of new grains seen at a grain boundary in high-purity aluminium compressed 20% and annealed for 11 min at 395°C. The three new grains were shown by transmission Kossel X-ray diffraction to come from within the orientation spread of the lower grain (4–9), into which they have *not* grown. (After BELLIER and DOHERTY [1977].)

aluminium. Here again, the original stereogram shows that the new grains have orientations within the spread of the neighbouring deformed region.

Extensive studies of this kind (reviewed by DOHERTY and CAHN [1972] for the earlier studies and, very fully, by DOHERTY [1978] for the later ones) all showed the same thing. A subgrain grows slowly, either at a grain boundary or at the limit of a deformation band, and if it becomes large enough it will grow fast: it will be able to do so if (and only if) its immediate environment differs in orientation by about 12° or more (the different studies in fact established a critical misorientation which ranged from 9–18°, according to the metal and degree of deformation).

The tendency of a subgrain to grow if it is larger than its neighbours arises from the associated changes in total sub-boundary energy of the configuration. This can be understood by reference to fig. 21, due to DILLAMORE *et al.* [1972]. This represents the limit region of a deformation band with its long narrow subgrains, and at the centre is a larger-than-average subgrain. The orientation gradient vertically is small, while that horizontally, across the band limit, is large. Therefore the specific interfacial energy of the vertical subboundaries (σ_t) is greater than that of the short horizontal boundaries (σ_r).

References: p. 1665.

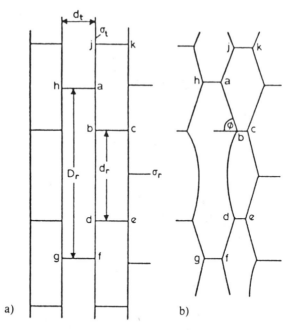

Fig. 21. Schematic representation of a deformation band, (a) as formed by deformation, (b) relaxed to the equilibrium angles on annealing (after DILLAMORE et al. [1972]).

On annealing, the structure changes to that shown in fig. 21b. The equilibrium configuration at 'b' is given by

$$\cos \phi = \frac{\sigma_r}{2\sigma_t}$$

The large subgrain, abdfgh, can continue to grow if 'b' and 'c' come into contact before ϕ has decreased sufficiently for the above equilibrium condition to be satisfied. For this to happen, the inequality

$$D_r > \tfrac{4}{3}\left[d_r + d_t\left(\frac{4\sigma_t^2}{\sigma_r^2} - 1 \right)^{1/2} \right]$$

must be satisfied; that is to say, the central subgrain must be sufficiently long.

Just this type of growth was observed in early recrystallization studies from deformation bands (also termed 'transition bands') in silicon–iron (WALTER and KOCH [1963], HU [1963]). Figure 22 shows an instance of an early stage of nucleus formation at the limit of such a band. Another kind of deformation band which, unlike those in iron, runs across a number of grains, are those found in brass and other copper alloys (GREWEN et al. [1977] and DUGGAN et al. [1978]). These are likewise the site of nucleation during annealing (HUBER and HATHERLY [1979]).

The question now remains by what mechanism the crucial subgrains in fig. 21 and

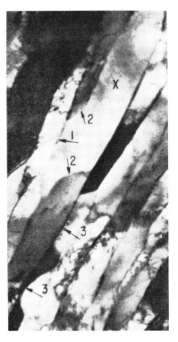

Fig. 22. Electron transmission micrograph showing origin of a recrystallized grain within a transition band, in a silicon-iron crystal rolled 70% and annealed 5 min; at 700°C subgrains are growing to form nuclei at 1, 2 and 3. 50000×. (After WALTER and KOCH [1963].)

22 achieve their initial growth, so that they become larger than their neighbours. Two rival models have been in competition: *strain-induced boundary migration* (SIBM) and *coalescence*. A micrograph showing coarse SIBM in aluminium is shown in fig. 23, and its interpretation by BECK [1954] in fig. 24. An example from more recent work is shown in fig. 25 (some of the 'tongues' originate from above or below the plane of section and thus appear as islands). The tongues were demonstrated to share the orientation of a nearby region of the deformed structure, below the central grain boundary; other similar tongues were also seen to proceed from deformation band limits.

Coalescence, as originally proposed by Hu, has in the past few years been repeatedly been observed by TEM, mostly in aluminium. Sub-boundaries within a subgrain have been photographed in a state of partial disappearance. Clear examples are to be found in papers by FAIVRE and DOHERTY [1980] and by JONES *et al.* [1979]. HU [1981] has recently reviewed the state of knowledge concerning coalescence. The second of these papers established very clearly that sub-boundary dislocations link continuously with dislocations in an adjacent high-angle grain boundary; the disappearance of the sub-boundary entails a detectable rearrangement of the dislocations in the grain boundary, and indeed, it was found that grain boundaries act more

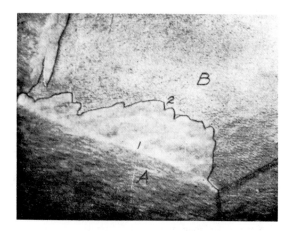

Fig. 23. Strain-induced boundary migration. 75×. (After BECK and SPERRY [1950].)

efficiently as sinks than as sources of dislocations. This is the precondition of coalescence – i.e., the 'evaporation' of sub-boundaries.

Figure 26 shows a model version, due to JONES *et al.* [1979] of nucleus formation based on coalescence. – It is to be noted that, as between SIBM and coalescence, it is only the earliest stage that is different; the later stages of nucleus growth are indistinguishable in the two models.

The HVEM studies of recrystallization in heavily rolled copper (RAY *et al.* [1973]), however, showed no signs of subgrain coalescence; as so often, copper and aluminium behave differently. Long, thin subgrains grow steadily, as indicated in fig. 21, into a strongly misoriented neighbourhood – a pure instance of SIBM. The process was observed in situ, by means of a hot stage.

In summary, from studies of aluminium, iron and copper, it would appear that moderate degrees of deformation favour coalescence as the initial process for turning

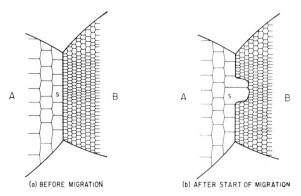

(a) BEFORE MIGRATION   (b) AFTER START OF MIGRATION

Fig. 24. Model for strain-induced boundary migration (after BECK [1954]).

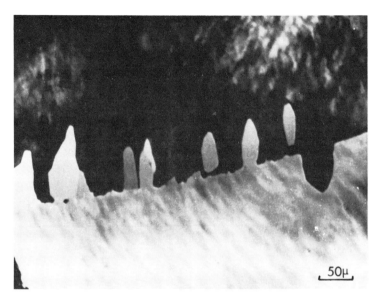

Fig. 25. Aluminium compressed 40% and annealed for 1 h at 328°C, showing strain-induced boundary migration into both grains. The direction of growth is not always parallel to the plane sectioned, as can be seen from the two white 'bulges' that do not contact the lower parent grain in the section seen. (After BELLIER and DOHERTY [1977].)

a subgrain into a nucleus, whereas high strain and low SFE favour SIBM. The reason for this distinction is not clear. What *is* now established beyond reasonable doubt is that nuclei are always *preformed* – i.e., an existing subgrain turns into a nucleus. Therefore, the orientation of a new grain is always already present in the previous deformed microstructure.

The necessary and sufficient condition for nucleus formation is that one subgrain should first grow substantially larger than its neighbours, and that, immediately adjacent, there should be a steep orientation gradient; for this reason, nucleation takes place predominantly at grain boundaries and at deformation bands, i.e., artificial grain boundaries. Where steep orientation gradients are absent, recrystallization is sluggish or impossible. This applies, at one extreme, to crystals of metals such as aluminium and zinc which after very large strains in single glide (as much as 200%) will not recrystallize, for lack of misorientation; and at the other extreme, to explosively shocked metals which, although much work-hardened, apparently lack orientation gradients and so resist recrystallization (HIGGINS [1971], CHOJNOWSKI and CAHN [1973]). – Different forms of plastic deformation apparently produce quite different misorientation gradients. On present evidence, these gradients diminish in the sequence: compression, rolling, wire drawing or swaging, shock deformation. Correspondingly, recrystallization becomes more sluggish in the same sequence.

References: p. 1665.

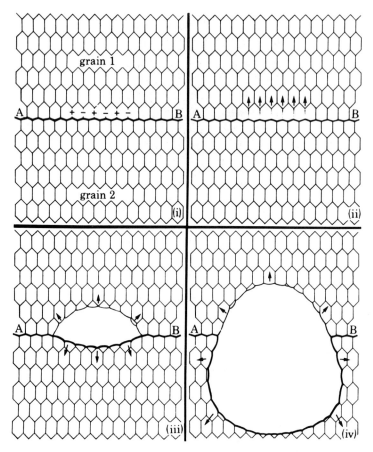

Fig. 26. Nucleation involving subgrain coalescence (after JONES *et al.* [1979a]).

One suggested nucleation mechanism has very recently been discussed which is quite distinct from any of those discussed above. Evidence has gradually accumulated that in certain alloys of low stacking-fault energy, nucleation takes place from very small *twins* which form during the recovery stage and then grow by a SIBM mechanism. HUBER and HATHERLY [1980] were the first to recognize this clearly on the basis of TEM of rolled brass, and they review earlier observations. They propose the term *recovery twin* for this kind of nucleus, to distinguish them from annealing twins formed during grain growth. JONES [1981] provided further micrographic evidence on copper alloys and stainless steels. The formation of such recovery twins at a very early stage of subgrain growth can be rationalized in terms of the misorientation-dependence of the energies of both sub- and grain boundaries: it is possible for a growth accident, creating a twin, to be energetically favoured, if the resulting total boundary energy is thereby reduced; in fact, this is the way

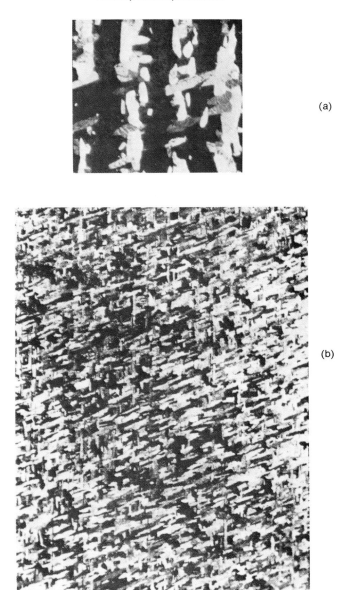

(a)

(b)

Fig. 27. (a) Incipient new grains on abraded side of rolled aluminium crystals after 5 s at 750°C. Etched 50×. (b) New grains on far side of rolled aluminium crystal, after 600 s at 350°C. Etched. 6×. (After KOHARA *et al.* [1958].)

annealing twins also form at a later stage (§4.2).

A related study is that of WILBRANDT and HAASEN [1980], based on HVEM of tensile-deformed and annealed copper single crystals. They claim that the orienta-

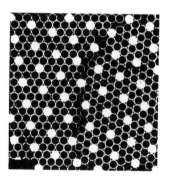

Fig. 28. Coincidence lattice for two grains both of which have [111] parallel to the plane of the figure but which are mutually rotated by 38° about an axis normal to this (111) plane. Only one lattice plane is shown for each grain. (After KRONBERG and WILSON [1949].)

tions of recrystallized grains found in this study can be interpreted on the basis of multiple twinning (i.e., successive generations of twins), up to at least five generations. The twinning generations were not observed in the HVEM but are deduced from consideration of initial and final orientations. The authors assert that the recrystallized orientation is *not* present in the deformed microstructure, having been modified by twinning at a very early stage of nucleus development. A later paper by WILBRANDT [1980] comments further on the reliability of the deductive arguments used to work out this multiple twinning mechanism, and points out that the successive twinning generations cannot be on random {111} planes, but must in some way be biased so that only particular twinning variants actually operate. This notion is further discussed in a recent paper by RAE *et al.* [1981], in which it is suggested that twinning accidents tend to survive if thereby the velocity of a sluggishly moving sub-boundary is increased.

Related evidence comes from the study of recrystallization of non-cubic metals containing *mechanical* twins. A scratch on a zinc single crystal generates many twins, and new grains grow from these in orientations closely related to the twin orientations (CAHN [1964]). In deformed uranium, the orientation of new grains is also very closely related to the twin orientations; one group of recrystallized grains had an orientation accurately identical with the twin, another had a different but closely reproducible orientation (CALAIS *et al.* [1959]). In all these instances, the new grains are evidently produced by oriented nucleation.

3.4. Growth of primary grains and the role of impurities

3.4.1. Impurity drag

We have already discussed in §3.2 how recovery of the deformed matrix can slow the growth of new grains into it. Here we shall discuss in particular the role of dissolved impurities in controlling the migration rate of boundaries and summarize present theoretical ideas about the process.

It is clear that the rate of growth of a new grain into a deformed structure is sensitive to the mutual orientation of the two grains, and moreover, that for a given mutual orientation, the growth rate may be very anisotropic. This first became clear from a classic study by KOHARA *et al.* [1958]. A crystal of aluminium was rolled down by 80% (producing a very sharp deformation texture, almost a single orientation); one side was then abraded lightly with a fine abrasive, and the strip then lightly annealed at 350°C. Copious recrystallization took place in the thin abraded layer (fig. 27a). After a much longer anneal at 350°C, the new grains penetrated to the far side of the strip (fig. 27b). When the grains of fig. 27b were examined by X-ray diffraction, they proved to be related to the texture of the rolled crystal by a rotation of about 38° about several axes common to the deformed orientation and the new grains. This special relationship is best understood by reference to fig. 28; it has so often been found to relate deformation and annealing textures in face-centred cubic metals that it has been given a name of its own, the *Kronberg–Wilson rotation* (KRONBERG and WILSON [1949]). This represents a special instance of a *coincidence lattice*, and the white atoms of fig. 28 are *coincidence sites* (see also § 3.4.2). A more general version of these concepts was developed by CHALMERS and GLEITER [1971], and a fuller account will be found elsewhere in this book (ch. 10B, § 2.2.1.3). Other rotation angles about a common [111] axis also turn up but ~ 38° is the most common. There is always some scatter about this ideal orientation relationship, but in the sample of fig. 27 the scatter was found to be less on the far side of the strip (fig. 27b) than for the freshly nucleated grains (fig. 27a). A copious supply of randomly oriented nuclei must have been formed in the abraded layer, since abrasion is an intense but isotropic mode of deformation. Among these nuclei, only the favourably oriented grew to the stage seen in fig. 27a, the others were consumed by the faster-growing grains. Continuing competition resulted in a sharpening of the texture by the time the grains had grown through to the other side of the strip. The fact that the grains of figs. 27a and b are plate-shaped (seen in cross-section) shows that for the most favourably oriented grains the growth rate was very anisotropic; analysis showed that growth was slow normal to the common (111) plane and fast in directions parallel to this plane. From this it follows further, if fig. 27 is closely examined, that out of the four ⟨111⟩ axes of the deformed crystal, only two acted as rotation axes to produce new grains; only two families of platelets appear in the micrograph. This can most simply be explained by postulating that the nuclei were not in fact randomly distributed in orientation; an alternative explanation is that the growth rate of a new grain depends not only on the mutual orientation of new and deformed grains but also on the inclination of the growing grain's boundaries to the sample surface.

Kohara's experiment has recently been repeated and extended by HELLER *et al.* [1981]: although grains nucleated after surface abrasion did again show a strong texture, the authors' other evidence indicates that this texture could not be due primarily to selective growth, but rather to oriented nucleation (see below).

References: p. 1665.

A number of investigators have performed experiments similar in conception to the experiment by KOHARA et al. [1958], but so arranged that the migration of individual primary grain boundaries could be measured over a period of time. This work was first done with strained aluminium monocrystals, squeezed or cut at one end to generate a moderately plentiful supply of nuclei. These were then allowed to compete and the migration rate of the victor determined (fig. 29). These spontaneously selected grains usually had a [111] axis in common with the parent crystal and were rotated with respect to the latter by roughly 38° about the common axis (the Kronberg–Wilson rotation). The 60° value represents the established fact that an interface between twin-related crystals has zero or very low mobility. These experiments therefore confirm the qualitative conclusions drawn from the work of Kohara et al.

Other investigators found no such distinction in migration rates between differently oriented boundaries. This contradiction was eventually traced to the state of the impurities, especially iron, in the aluminium (GREEN et al. [1959]). Growth selectivity according to the Kronberg–Wilson rotation was not found when iron and silicon (0.1% each) were in solution; when some had been precipitated then growth selectivity returned but did not survive for complete precipitation. It appears that when substantial amounts of solute are present they exert such powerful control over boundary motion that orientation differences lose their influence; in a two-phase alloy, again, growth selectivity is prevented.

A classic series of experiments on boundary migration kinetics in lead was performed by AUST and RUTTER [1959, 1960]. The metal was zone-refined (ch. 9, §5.3) and thus exceedingly pure. The driving force for boundary migration was provided by *striation substructure* in the lead single crystals into which artificially

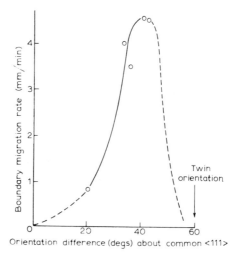

Fig. 29. Growth rates of new grains into a plastically stretched aluminium crystal at 615°C. The common [111] axis is always the same one in the deformed crystal. (After LIEBMANN and LÜCKE [1956].)

nucleated new grains were allowed to grow. These striations are a form of sub-boundary introduced during growth of the crystals from the melt; they represent a reproducible dislocation concentration and therefore a small but reproducible driving force. Figure 30 shows some of the results. Two striking conclusions can be drawn: (i) Very small impurity concentrations dramatically retard boundary migration; one part per million of tin will reduce the rate by a factor of four. (ii) Grains with a Kronberg–Wilson or other "special" orientation relationship to the crystal which is being consumed grow much faster than do grains of arbitrary orientation, *but only if* impurity is present. Later experiments with silver or gold as solute led to similar conclusions, except that these solutes were even more effective than tin in slowing down boundary migration. A full account of these important experiments and also of their possible interpretation will be found in a review by AUST and RUTTER [1963] and in the discussion following it *. More recent reviews on the same subject have been published by AUST [1969, 1974]. In the second of these, grain-boundary mobility as a function of small deviations from coincidence-orientation relationships is discussed in detail, for various degrees of purity. It must be recognized that these studies refer to experiments with very small driving forces, much smaller than in normal primary recrystallization experiments, and there is no assurance that the orientation-dependence (and purity-dependence) of migration rates is the same for the small and large driving forces. RATH and HU [1972] have assembled comparative data and themselves performed experiments with very small driving forces, for zone-refined aluminium, which show a very steep dependence of boundary velocity on driving force.

A few measurements have been made on iron alloys. LESLIE *et al.* [1963] describe experiments in iron–molybdenum and iron–manganese alloys. 0.01% solute had a substantial effect on migration rates. Unfortunately the base metal was not zone-refined.

The implication of experiments such as those set out in fig. 30 is that a metal entirely devoid of dissolved impurity should have a grain-boundary mobility wholly independent of misorientation (except, presumably, for quite small misorientations as found in the previously discussed experiments of VISWANATHAN and BAUER [1973a]). Fig. 30a, based on studies by FRIDMAN *et al.* [1975], is consistent with this expectation: a reduction of impurity level in aluminium from 8 ppm to 0.5 ppm greatly reduces the orientation-dependence of mobility. DEMIANCZUK and AUST [1975] find that the activation energy for migration in aluminium containing 40 ppm of copper changes sharply above a certain critical temperature; they suggested that beyond this temperature, boundary motion is no longer controlled by segregated impurity and the boundaries become 'intrinsic'.

The matter is further complicated by the fact that migration rates during primary recrystallization progressively decrease, during an anneal, because of recovery of the deformed matrix (VANDERMEER and GORDON [1959], RATH *et al.* [1979]). The

---

* It will be interesting to discover whether large impurity additions in lead will once again destroy growth selectivity, as indicated by the experiments on aluminium of GREEN *et al.* [1959].

References: p. 1665.

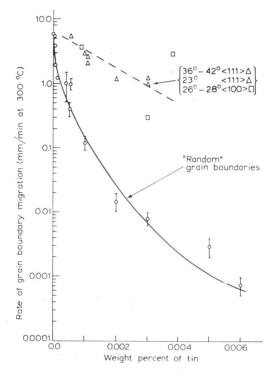

Fig. 30. Plot of log grain-boundary migration rate at 300°C into zone-refined lead crystals doped with various amounts of tin (after AUST and RUTTER [1959]).

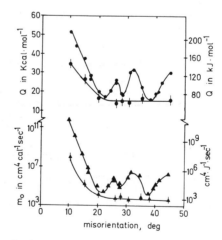

Fig. 30a. Variation of the activation energy, Q, and the pre-exponential factor of the mobility, m_0, with misorientation for the motion of $\langle 100 \rangle$ tilt boundaries in aluminium samples of different purities. ●, ▲: Al 99.99995 at%; ●, ▲: Al 99.9992 at%. (After FRIDMAN et al. [1975].)

theoretical modelling of this effect is discussed by HU [1981]).

The whole topic of the dependence of migration rates on purity, misorientation and driving force is very fully treated by HAESSNER and HOFMANN [1978].

The first attempt to interpret measurements of grain-boundary mobility was due to LÜCKE and DETERT [1957]; improved versions of this theory were published independently by CAHN [1962] and by LÜCKE and STÜWE [1963]. These are all based on the postulate that *dissolved* impurities retard a moving boundary through an elastic attraction of impurity atoms towards the open structure of the boundary (ch. 13); the boundary must then either drag the impurity atoms along – so that its speed is limited by the diffusion rate of these atoms – or else *break away* if the impurity concentration is small enough, or the driving force or temperature high enough.

In Cahn's revised version of the theory, we have two extreme cases:

(a) For low velocities of the boundary (low driving force P or high solute concentration C_0, or both) the velocity V is given by

$$V = \frac{P}{\lambda + \alpha C_0} , \tag{2}$$

where λ is the reciprocal of the intrinsic mobility (which is determined by the grain-boundary self-diffusion of the solvent, since this is the rate-limiting process in boundary migration in an entirely pure material). α is a parameter depending on the variation of the interaction energy $E(x)$ between solute atoms and the boundary, and of the solute diffusivity $D(x)$, as a function distance x from the boundary. The theory predicts that $V \propto P$ at constant composition, that $(1/V)$ is a linear function of C_0, and that a lower diffusivity of the solute will lead to a stronger drag (as would be intuitively expected).

(b) For high boundary velocities, the velocity is given by

$$V = \frac{P}{\lambda} - \left(\frac{\alpha}{\beta^2} \right) \frac{C_0}{P} , \tag{3}$$

where α/β^2 is a different function of $E(x)$ and $D(x)$. Here the velocity is no longer proportional to P unless P is very large or C_0 very small, and in particular a *higher* solute diffusivity generates a stronger drag on the boundary. This is because at high velocities the solute atoms can move very quickly. This last conclusion in particular, that a rapidly diffusing solute will exert more drag on a fast boundary, is in complete agreement with Aust and Rutter's results; the relative efficacy of the solutes, tin, silver and gold, can be explained in terms of this conclusion.

The transition behaviour is complicated, and for certain values of the parameter a breakaway is to be expected: that is, increase of the driving force will cause only gradual change in the boundary velocity until a critical value is reached, when the velocity will suddenly jump to a much larger value; in effect, the solute atmosphere around the boundary evaporates. This kind of abrupt change has been observed in aluminium by DEMIANCZUK and AUST [1975]. In fig. 31 a comparison is made between some of Aust and Rutter's results for tin in lead and the formulae derived from Lücke and Stüwe's version of the migration theory, with two alternative

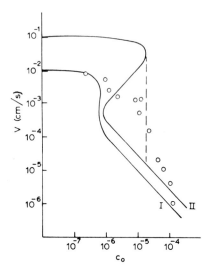

Fig. 31. Comparison of observed boundary migration rates in tin-doped lead as a function of fractional tin concentration c_0 with theory (after LÜCKE and STÜWE [1963]).

reasonable values for the pre-exponential term D in the diffusivity equation $D = D_0 \exp(-Q/RT)$. An increase in the slope of the experimental relation occurs at about 10^{-3} cm/s.

Recent ideas concerning the role of vacancies in grain-boundary migration are discussed in § 3.4.3, below.

3.4.2. Special orientations

It remains to discuss the reason for the remarkable difference in migration rates between the arbitrary or "random" boundaries and the "special" boundaries of fig. 30, and at the same time to interpret the strongly anisotropic migration velocity of such boundaries (fig. 27). Figure 28 shows the structure of an idealized Kronberg–Wilson boundary. The white atoms form a *coincidence lattice*, which continues without disturbance across the boundary. It is intuitively obvious that any boundary which involves the existence of a coincidence lattice will have a less disturbed grain-boundary structure than a random boundary: the average fit between the contiguous grains will be better. The higher the ratio of coincidence lattice sites to total lattice sites (this ratio is $\frac{1}{7}$ in fig. 28), the better the boundary-fit; for a given coincidence site ratio but for different orientations of the same grain boundary, the fit will be the better, the higher the density of coincidence sites in the plane of the boundary. Thus in fig. 28, the density of coincidence sites is higher along the boundary orientation shown (normal to (111), the plane of the figure) than it would be if the boundary lay *in* the plane of the figure; that is, the (111) plane normal to the common [111] axis has a comparatively disturbed structure.

The question of the structural disturbance at grain boundaries in relationship to

the coincidence lattice structure has been discussed in mathematical terms by
BRANDON *et al.* [1964]. Figure 32, from their paper, shows for a *body-centred cubic*
bicrystal with a coincidence lattice ratio of $\frac{1}{11}$ how the boundary fit varies with
boundary orientation. In this connection, selective growth of special orientations,
akin to the Kronberg–Wilson grains in aluminium, was observed in niobium (a bcc
metal) by STIEGLER *et al.* [1963].

AUST [1969] has reported a joint experiment with RUTTER on zone-refined lead, in
which they directly confirmed the lower energy of "coincidence boundaries" as
compared with random boundaries, in agreement with the lesser structural disturbance of the former. The variation of grain-boundary energy in aluminium with
misorientation has been measured over a wide range of angles by HASSON *et al.*
[1972] and found to match closely with values computed from the coincidence lattice
model (see also ch. 10B, §2.2.1.3).

The observed facts as to both growth selectivity of special grains (those with a
high coincidence-site ratio) and of the growth-rate anisotropy of each grain, can be
explained if it is postulated that a high coincidence-site ratio, by improving lattice fit
at the boundary, reduces the interaction energy, E, of solute atoms with the
boundary and also the diffusivity of the atoms across the boundary. The reduction
of E is to be expected; the reduction of D is not so obvious, since we are concerned
with diffusion *across* the boundary, not in its plane. However, the actual steps across
the boundary will be impeded if the diffusing impurity atom encounters a host atom

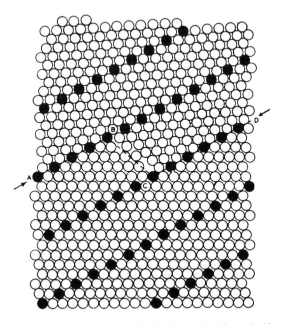

Fig. 32. Two-dimensional model for a body-centred cubic bicrystal with a coincidence site ratio of 1 in 11.
Arrows indicate ends of the boundary. (After BRANDON *et al.* [1964].)

References: p. 1665.

in a coincidence site; it will have to wait for a wandering vacancy to remove this atom, whereas elsewhere it can jump into a hole in the boundary itself. If a boundary deviates slightly from an ideal coincidence relationship, the boundary can be regarded as containing some dislocations superimposed on the coincidence sites (BRANDON et al. [1964]) and such a boundary should still be unusually mobile so long as its mobility is impurity-controlled.

Field-ion microscopic observations by FORTES AND RALPH [1967] have established directly that oxygen atoms segregate to random grain boundaries in iridium, but that "special" boundaries with a high density of coincidence sites are free of oxygen segregation. This was the first direct demonstration of a difference in purity between the two types of boundary. (See also ch. 13, §4.2.4.)

GLEITER [1969] has published a novel theory of grain-boundary migration in terms of the movement of kinked lattice steps at the boundary (akin to the process of crystal growth from the melt). He is able neatly to relate the migration rate to the degree of misorientation; he obtains good agreement with experiment for small misorientations, but the local variations near critical misorientations do not emerge satisfactorily from the theory as worked out for pure metal.

3.4.3. Vacancies in grain boundaries

An ordinary high-angle grain boundary is sufficiently imperfect to be regarded as a source of lattice vacancies; indeed, it must be regarded as always containing a certain concentration of vacancies per unit area. However, the effective vacancy concentration in boundaries is strictly limited, and this in turn limits the rate of any process which depends on grain-boundary diffusion (ch. 8, §7). The pre-exponential term in the diffusivity, which depends directly on the vacancy concentration, is increased if the vacancy concentration is increased. A boundary, as well as being a vacancy source, is also a vacancy sink, and it has in the past been tacitly assumed that the boundary structure is not altered when vacancies are absorbed. There is indirect evidence, however, from grain-boundary migration measurements, that vacancy absorption can so alter a boundary as to increase its migration rate drastically.

MULLINS [1956] first observed a gradual slowing down of grain boundaries in bismuth moving under a very small driving force; the circumstances of the experiment were such that there was no possibility of a reduction of driving force through recovery, and he concluded that vacancies were diffusing out along the boundary to the free surface and thus reducing the boundary mobility. The bismuth had not been deformed and thus the moving boundary could not sweep up any substantial fresh supply of vacancies.

When the driving force is due to plastic deformation, a moving boundary must gather in a steady supply of excess vacancies from the lattice; whether or not it slows down will depend upon the balance between the input and outflow of vacancies. IN DER SCHMITTEN et al. [1960] have carried out experiments on deformed aluminium bicrystals to test this idea. Pieces of a deformed crystal were etched down to various diameters and growth rates of primary grains measured for a standard annealing treatment. Figure 33 shows the results. The growth rate varies steeply with diameter; to explain this, the authors developed a theory based on the following structural hypotheses:

(i) Grain-boundary diffusion operates through the migration of vacancies.

(ii) Vacancies above the thermal-equilibrium concentration, and also dislocations, are absorbed by a sweeping boundary so as to produce a steady supply of excess vacancies in the boundary.

(iii) These excess vacancies diffuse along the boundary for a critical distance R_{cr} before they are annihilated by an unspecified mechanism, unless they disappear at a free surface first.

(iv) The degree of disorder of the boundary is not increased during a diffusional vacancy jump, i.e. the entropy of activation is zero.

The quantitative theory shows that below a certain critical sample radius, vacancies will be annihilated at the free surface and migration velocity will depend on the sample radius. Beyond this radius, the velocity should be independent of radius. R_{cr} was estimated by fitting the curve of fig. 33 to a theoretical expression developed by the authors, and a critical radius R_{cr} of 2.4×10^{-2} cm adopted. With this value, the theoretical dashed curve was drawn in fig. 33. Rough agreement is found for the form of the curve; in view of the simplifications inherent in the derivation of the equations, this offers some support to the underlying idea.

Further evidence of the vital role of vacancies in promoting boundary migration comes from attempts to observe primary recrystallization of thin metal films in situ in an electron microscope. DIMITROV [1960] and his colleagues sought to observe recrystallization in films of heavily rolled zone-refined aluminium, even though this metal in massive form recrystallizes completely below room temperature. Recrystallization detectable by electron microscopy proved to be restricted to film regions which were more than 1500 Å in thickness. Below this thickness, even nuclei which

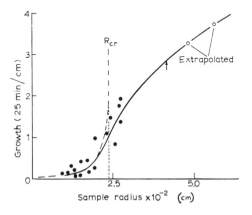

Fig. 33. Aluminium single crystal (normal purity) deformed 10% in tension. Pieces were etched down to different diameters and each annealed 25 min at 616°C. Growth distances of nuclei produced artificially at sample tips were measured. (After IN DER SCHMITTEN *et al.* [1960].)

References: p. 1665.

had been previously formed by brief annealing before the metal had been thinned still would not grow. The most straightforward explanation is that both any boundaries present, and the unconsumed cold-worked matrix, are entirely drained of excess vacancies by diffusion to the adjacent surface. Admittedly, the critical half-thickness, 750 Å, is orders of magnitude smaller than the value of R_{cr} derived from the experiment on stretched crystals of aluminium, which was however much less pure than Dimitrov's metal. It is possible that surface grooving where boundaries intersect the surface (§4.1) complicates the interpretation of such experiments.

CHRIST et al. [1965] have observed a substantial acceleration of recrystallization in heavily rolled copper if the metal had been irradiated with neutrons before rolling. The acceleration was attributed to the cutting of vacancy clusters by the moving boundary and the redistribution of the vacancies. Presumably the transfer of vacancies from the clusters to the boundaries enhances the mobility of the boundaries (HAESSNER and HOFMANN [1971]).

Later experiments (HAESSNER and HOLZER [1974]) on rolled copper crystals which were neutron-irradiated at intervals interspersing the anneal showed a jump in boundary-migration velocity after each irradiation. This finding is again consistent with the idea that vacancies (or, more generally, point defects) aid migration.

More recently, evidence has been obtained which proves that the vacancy concentration just behind a *moving* grain boundary is strongly enhanced. An important study by GOTTSCHALK et al. [1980] showed that Ni_3Al precipitates from a solution supersaturated in vacancies much faster than within a vacancy-equilibrated grain. Figure 34 shows the ratio of precipitation rates, and these indicate that at the lower temperatures, the local vacancy concentration behind the moving boundary is enhanced thousandfold. However, BALLUFFI [1982] has questioned the validity of the conclusions deduced from fig. 34.

GLEITER [1969, 1981] proposed that these findings, and others pointing in the same direction, imply that a moving boundary *generates* vacancies by means of successive growth accidents, and that these vacancies in turn create a vacancy drag,

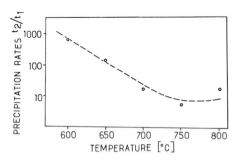

Fig. 34. Measured relative precipitation rates, t_2/t_1 of a Ni_3Al precipitate in a recrystallized (t_2) and in a vacancy-equilibrated (t_1) Ni–12.5 at% Al alloy as a function of the recrystallization temperature. It is seen that the Ni_3Al precipitates form in a recrystallized material up to 1000 times faster than in a vacancy-equilibrated material of the same composition and at the same temperature. (After GOTTSCHALK et al. [1980].)

analogous to an impurity drag. Although there is no experimental evidence to test the orientation dependence of the influence of vacancies on migration, in his 1981 paper Gleiter advances arguments for the hypothesis that coincidence boundaries contain localized vacancies whereas random boundaries contain 'distributed' vacancies, rather as in metallic glasses (chapter 28, § 3.2.1). The implication seems to be that the migration of random boundaries should be most strongly affected by the presence of vacancies.

Since the work of In der Schmitten *et al.*, discussed above, suggests that vacancies need to be *absorbed* from outside to 'lubricate' grain-boundary migration, and Gleiter's arguments indicate that a moving boundary also *rejects* vacancies, it appears that such a boundary must be regarded as a sort of vacancy pump!

The motion of a grain boundary in relation to its intake and rejection of vacancies and of the concentration of vacancies *in* the boundary, has been analyzed quantitatively in a series of theoretical papers by LÜCKE and coworkers (the fifth in the series is by ESTRIN and LÜCKE [1982], and here the new theory is also related to the other ideas mentioned above). In this treatment, the moving boundary is shown to be enriched in vacancies, and the supply of vacancies from the grain being consumed is shown to be an essential feature of rapid migration. This last conclusion is consistent with a number of experimental results cited above.

VAIDYA and EHRLICH [1983] have observed enhancement of recrystallization in irradiated austenite stainless steel (deformed 20%); they observed that irradiation-induced voids were dissolved by migrating grain boundaries, thereby releasing vacancies which were presumed to be responsible for accelerating recrystallization.

3.5. Recrystallization during hot-working

In §2.5, we saw that recovery is accelerated if metal is subjected to strain and high temperature simultaneously. This *dynamic recovery* can be followed by *dynamic recrystallization* if the temperature and strain rate are high enough. Dynamic recrystallization takes place most readily in metals of low stacking-fault energy, in which (because of slow climb of dislocations) dynamic recovery is sluggish.

During dynamic recrystallization, nuclei at first grow rapidly but the concurrent deformation steadily increases the dislocation density within the growing grains, so that the driving force for boundary migration is progressively reduced. The new grains reach a limiting size and then stop growing, and a new cycle of nucleation begins. Figure 35 indicates the correlation between this repeated recrystallization and the instantaneous creep rate. For the circumstances illustrated here, the strain ϵ_x, during the time required for almost complete recrystallization, is less than the critical strain for renewed nucleation, so that there is a necessary interval of further strain before the next cycle of recrystallization can start.

Precise control of dynamic recrystallization is crucial in thermomechanical processes such as controlled rolling. Thus CUDDY [1981] explains how to achieve the finest possible ferrite grain size in Nb-bearing microalloyed steels by controlled rolling: high-temperature deformation with repeated recrystallization, as in fig. 35, is

References: p. 1665.

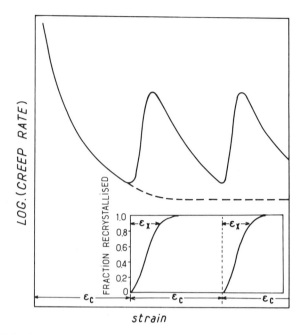

Fig. 35. Schematic behaviour of a metal recrystallizing repeatedly during hot working (after LUTON and SELLARS [1969]).

used to achieve a uniform fine austenite grain size, followed by further rolling at a temperature below the austenite recrystallization range; the flattened, unrecrystallized austenite grains generate very fine ferrite grains.

Repeated dynamic recrystallization neutralizes incipient grain-boundary cracks in materials of limited ductility, by separating them repeatedly from their associated boundaries, and thus enhances hot-workability (which is likewise enhanced by the repeated softening of the grain interiors).

The subject, including microstructural aspects, is reviewed by MCQUEEN and JONAS [1975] and by MECKING and GOTTSTEIN [1978]. Both reviews also treat dynamic recovery. Secondary recrystallization associated with hot-working has been investigated by SINGER and GEISSINGER [1982].

3.6. Annealing textures

When a piece of metal is deformed by some directional process such as wire-drawing or rolling, the constituent grains acquire a *preferred orientation* or *texture*; the grains approximate, generally with a good deal of scatter, to an *ideal orientation*. In the extreme case the whole sample may turn into a pseudo-single crystal. Sometimes several ideal orientations coexist, so that the total scatter is greater. The actual orientation distribution of grains is termed a *deformation* texture. When such material is recrystallized it again acquires a texture, which may be identical to the previous deformation texture but more often is quite different; this is an *annealing texture*. We shall be concerned here mainly with annealing textures, though it is impossible to understand these properly without reference to their precursor textures. In this outline, we shall be concerned only with the *mechanism* of texture formation.

A texture can be described in terms of the ideal orientation or group of orientations to which it approximates. Thus for a rolled sheet the ideal orientation is quoted as $(hkl)[UVW]$; here (hkl) is the lattice plane which tends to lie in the rolling plane and $[UVW]$ the lattice vector which tends to lie along the rolling direction. (For a drawn wire only a direction $[UVW]$ need be specified; the distribution around the wire axis is random). For instance, brass heavily rolled and annealed gives a texture approximating to $(113)[21\bar{1}]$. Other metals, such as aluminium and iron, have a mixed texture (several ideal orientations coexisting); the texture of rolled and annealed aluminium has been described as a mixture of $(100)[001]$ (the *cube texture*), $(110)[011]$ and $(236)[53\bar{3}]$; the proportion of the various components varies with purity and annealing temperature.

A clear quantitative expression of the ideal orientation and statistical spread around it, and of the relative proportions of several constituent textures, is provided by a graphical device, the *pole figure*, which is plotted directly from X-ray diffraction data obtained from the sample. To produce a pole figure, imagine all grains concentrated at the centre of a sphere and then construct normals to all the lattice planes of a particular family [say $\{110\}$] in all grains, so that each forms a spot where it cuts the sphere. If there is a texture, the spot density varies in different parts of the spherical surface, and contours of equal spot density are drawn to define areas of high and low concentration. The contours on one hemisphere are projected stereographically into a plane parallel to the rolling plane. Figure 36 is a typical pole figure of an annealing texture.

The orientation of a crystallite in a polycrystalline specimen is determined by three orientation angles (α, β, γ). Their choice is somewhat arbitrary; one can, for instance, choose the Eulerian angles. The orientation distribution of all crystals in the specimen, i.e. the texture, may then be defined by the *orientation distribution function* (ODF),

$$\frac{dV}{V} = f(\alpha\beta\gamma)\,d\alpha\,d\beta\,d\gamma, \tag{4}$$

giving the relative frequency or volume fraction of crystals having an orientation with respect to the specimen geometry specified by the ranges $\alpha/\alpha + d\alpha$; $\beta/\beta + d\beta$, $\gamma/\gamma + d\gamma$. Straightforwardly one may determine f by X-ray Laue or electron diffraction photographs (LÜCKE *et al.* [1964]). The statistical task is however formidable. It is more promising to start from a pole figure p for a low-index plane hkl, which is a two-dimensional projection of the ODF:

$$p_{hkl}(\alpha\beta) = \frac{1}{2\pi} \int f(\alpha\beta\gamma)\,d\gamma. \tag{5}$$

References: p. 1665.

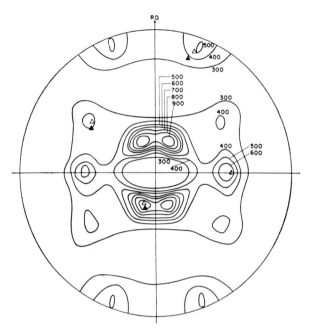

Fig. 36. (111) pole figure of the primary recrystallization texture of rolled brass, showing contours of equal pole density. The normal to the rolling plane is at the centre, RD represents the rolling direction. ▲ represents pole positions for the ideal (113)[$\bar{2}$11] texture; △, (225)[73$\bar{4}$] is a slightly better approximation to the observed texture. There are four symmetrical ideal orientations which are mutually equivalent. (After BECK and HU [1952].)

From many such projections for different *hkl* one then can construct the three-dimensional ODF. Often this is done by series expansions of both functions *p* and *f* (BUNGE [1981]), but care has to be taken to avoid "ghosts" in this transformation. This error in earlier ODFs has been pointed out by MATTHIES [1979].

An enormous volume of work has been done on annealing textures, on the details of which we need not dwell. A very full though somewhat dated account of the subject can be found in the book by WASSERMANN and GREWEN [1962]; this also discusses the consequences of annealing texture in generating anisotropy of physical and mechanical properties, which have important technological implications, especially for deep drawing. Later additions can be found in a conference report by the same authors (GREWEN and WASSERMANN [1969]). We shall return to one particular instance of this anisotropy in a later section (§4.3.2). The most recent, concise but very informative review is by GREWEN and HUBER [1978], and a number of up-to-date papers will also be found in the proceedings of the latest texture conference (NAGASHIMA [1981]).

There has been much argument concerning the physical origin of recrystallization textures. Opinion has divided into two camps: One group hold that nuclei form only in particular orientations in relation to the deformation texture, and the non-random nuclei then grow to form a preferred orientation of primary grains. This is the *oriented nucleation* hypothesis. The other view is that such nuclei are so plentiful in a heavily deformed metal that all orientations are present, but only those favourably oriented with respect to the deformation texture grow to a large size. This is the *oriented growth selection* hypothesis. The specific models for oriented nucleation and oriented growth as separate processes are discussed in § 3.3 and § 3.4, respectively.

LÜCKE [1963] described experiments on copper and brass. Pure copper rolled at − 196°C (A) produces almost exactly the same texture as a copper-6% zinc alloy rolled at room temperature (B), whereas pure copper rolled at room temperature produces quite a different texture (C) *. On annealing, A and B produced quite different recrystallization textures. If growth selection alone were responsible for determining the texture, then the similar deformation textures of samples A and B should impose a similar growth selection on the new grains growing into these two samples. The difference actually observed can only be attributed to a difference in the supply of nucleus orientations available in the two cases; this may well be due to a difference in fine structure of the twins in samples A and B, or to a variation in the spatial distribution of the different orientations in the two samples.

Oriented nucleation, as we have seen in § 3.3, is now well established. Leaving aside the complications which arise from multiple twinning during the early stages of growth (end of § 3.3), it is clear that the orientations of nuclei are always present in the deformed structure (which generally, however, has extensive orientation gradients in each grain).

Likewise, we have seen that there is clear evidence of preferential growth of certain orientations only, when a plentiful supply of nuclei is provided by artificial means (fig. 27).

The present consensus, as expressed by GREWEN and HUBER [1978], is that for moderately deformed single crystals (as in the case of Kohara's experiment shown in fig. 27) preferred growth undoubtedly is a determining factor of the resultant texture. But as one goes to a polycrystal, the deformation texture necessarily becomes more scattered and the microstructure more complex, and all the indications are that then the textures are determined largely by preferred nucleation; but there may be a residual element of growth selection, acting on available orientations already determined by preferred nucleation.

In the past, attempts were made to draw conclusions about the mechanism involved in texture formation from a critical comparison of deformation and recrystallization textures. This is however a blunt approach, especially if pole figures rather than ODFs are used. It is much more informative to map the behaviour of

---

* This difference is probably due to the incidence of mechanical twinning in the A and B cases and its absence in C (WASSERMANN [1963]).

References: p. 1665.

individual grains both during deformation and subsequent annealing; when this is done for instance by Kossel point X-ray diffraction, stereograms like figs. 17b and 19 are obtained, and if this is done for many grains, as for instance was done by INOKUTI *et al.* [1980] for silicon-iron, then the genesis of an annealing texture can be identified in detail. This approach confirms the primacy of oriented nucleation.

A topic which has caused much difficulty is the interpretation of the very pronounced *cube texture*, (100)[001], obtainable by annealing heavily rolled copper, aluminium and some alloys; it was long thought that only selective growth of cube-oriented grains could explain this texture, but it was not at all clear why just these grains should grow preferentially. Now, evidence marshalled by GREWEN and HUBER [1978] shows that small amounts of material in 'cube orientation' have been detected by TEM in rolled sheet, and these can develop by oriented nucleation in the special circumstances that lead to this texture.

One of the most impressive investigations into the origin of an annealing texture is the study by DILLAMORE *et al.* [1967], set out more fully by HUTCHINSON [1974], to account for the annealing texture of rolled mild steel (of great industrial importance). They showed that differently oriented *deformed* grains incorporated different amounts of strain energy (greatest in (110)[001] and (111)[UVW] grains – but the former are only present in trace amounts and do not contribute much to subsequent annealing texture). If a precisely critical dispersion of a second phase such as AlN is present, this can just inhibit the components with smaller stored elastic energy from growing into large recrystallized grains, while allowing the technologically desirable grains with (111) parallel to the sheet plane to grow. The matter is critically discussed by GREWEN and HUBER [1978].

3.7 Primary recrystallization of two-phase alloys

A dispersed second phase exerts two opposed influences on the progress of primary recrystallization of a deformed alloy:

(i) Nucleation of new grains may be accelerated, especially if the second-phase particles are comparatively large. If they are very small, nucleation is retarded or prevented altogether.

(ii) Growth of new grains is always impeded, because of the drag exerted on a migrating grain boundary by dispersed particles, especially if these are small and numerous.

When dispersed particles are quite coarse (several microns across) it appears that they generate a local concentration of lattice distortion caused by the applied deformation, which in turn enhances the nucleation rate in the matrix close to the inclusions. This has been clearly demonstrated by LESLIE *et al.* [1963], who examined a series of dilute iron–oxygen alloys containing a second phase in various degrees of dispersion. Fig. 37 shows clusters of newly nucleated grains near oxide inclusions in cold-rolled iron, briefly annealed, and fig. 38 shows recrystallization isotherms for various alloys. Increase of oxygen content enhances nucleation and thus accelerates the entire recrystallization process, and correspondingly the recrystallized grain size

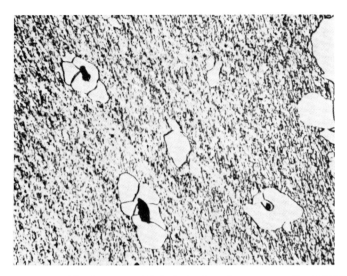

Fig. 37. Initiation of recrystallization at an oxide inclusion in an iron–oxygen alloy (0.033% oxygen). 60% rolling reduction, 2 min at 540°C. 100×. (After LESLIE *et al.* [1963].)

also becomes smaller. BLADE (quoted by DOHERTY and MARTIN [1963]) similarly observed that precipitation of FeAl$_3$ particles of about 1 μm diameter in aluminium–iron alloys accelerated nucleation more than it retarded growth of new grains, so that recrystallization as a whole was accelerated. DOHERTY and MARTIN [1963] observed the same in their careful study of the recrystallization of two-phase aluminium–copper alloys. Here the nucleation rate was critically sensitive to particle spacing, while growth rate was much less sensitive to this parameter. Figure 39 illustrates how strongly this extreme sensitivity of nucleation rate on the state of dispersion affects the recrystallization kinetics. The steep softening here is due to

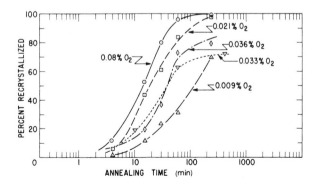

Fig. 38. Effect of oxygen in iron on recrystallization kinetics at 540°C. 60% rolling reduction. (After LESLIE *et al.* [1963].)

References: p. 1665.

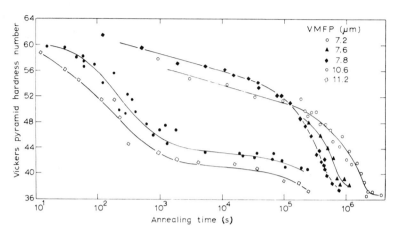

Fig. 39. Hardness–log-time isotherms for aluminium–5% copper alloy heat-treated to different states of dispersion. VMFP is the volumetric mean free path between $CuAl_2$ particles. (After DOHERTY and MARTIN [1963].)

recrystallization. There is evidence in Doherty and Martin's work that nucleation becomes very difficult when the particle spacing becomes so small that each developing subgrain collides with a particle before it becomes a viable nucleus.

A particularly detailed and thorough study of recrystallization in a two-phase alloy was undertaken by MOULD and COTTERILL [1967], on aluminium–iron alloys heat-treated to give different precipitate morphologies. Nucleation rate was a maximum for a mean particle separation of about 4 μm (the smallest separation studied); taken in conjunction with Doherty and Martin's results, the results suggest that this is the optimum spacing in an aluminium matrix. The authors believe that nucleation becomes difficult when inter-particle spacing becomes of the order of twice the mean subgrain diameter in the cold-worked matrix. The same authors discuss the matter in greater detail in their book (see Bibliography).

GAWNE and HIGGINS [1969] studied recrystallization kinetics in iron containing dispersed Fe_3C particles with a mean separation of about 2 μm, and found a pronounced acceleration compared with unalloyed iron rolled to the same reduction. It thus appears that in iron, the critical interparticle spacing is less than in aluminium (the rolling reductions were similar in the two experiments). Further information on the recrystallization of two-phase steels will be found in ch. 16, §4.2.

A very detailed TEM study of the role of large particles in stimulating nucleation in deformed two-phase alloys of aluminium and copper has been undertaken by HUMPHREYS [1977, 1979a] and he compares his findings with many other observations in another paper [1979b]. He finds that large particles stimulate nucleation: they do this not only by concentrating deformation locally, but especially by creating surprisingly large local misorientations which can be as large as 45° for 2 μm particles. As strain increases, substantial misorientations, and the associated stimulation of nuclei, were found for particle sizes down to 0.1 μm, for isolated particles; the smaller the particle, the less the limiting misorientation which was observed. This study has established in detail the mechanism of particle-stimulated nucleation in deformed metals. (See also ch. 22, §8.)

RYAN [1967] has studied the recrystallization of chromium containing a fine tantalum carbide dispersion, with a mean interparticle spacing of less than 1 μm, and found that recrystallization was severely inhibited. The presence of the precipitates produced an intractable dislocation substructure and the particles interfered with the migration both of sub-boundaries (fig. 40a) and of high-angle boundaries (fig. 40b).

Extensive work has been done on the annealing characteristics of deformed metals containing dispersed oxide. The extreme case of this is SAP (sintered aluminium powder), a very stable material containing closely-spaced oxide inclusions. These apparently are far too close together to permit any nuclei to develop.

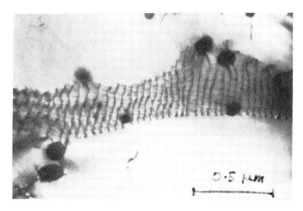

Fig. 40a. Sub-boundaries held up by precipitate in deformed chromium annealed at 1300°C (courtesy N.E. RYAN).

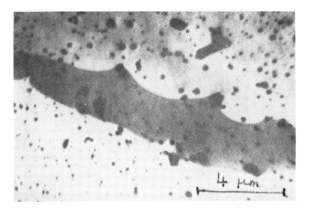

Fig. 40b. High-angle boundaries held up by precipitate in deformed chromium annealed at 1300°C (courtesy N.E. RYAN).

References: p. 1665.

Fine oxide dispersions in many other alloy systems behave similarly; thus, internally oxidized alloys (dilute metallic solid solutions in which oxide dispersions are produced in situ by diffusing in oxygen) generally resist not only nucleation, but also the growth into the two-phase region of new grains nucleated in an adjacent region which is still single-phase. These findings are reviewed by BONIS and GRANT [1960] (see also DESALVO and NOBILI [1968]), and JONES *et al.* [1979b].

The problem of the migration of a grain boundary in the presence of finely dispersed particles was first systematically examined in connection with grain growth (§4.1), and the fundamental theory was first presented by SMITH [1948], who followed an unpublished treatment due to ZENER. This must be the most quoted unpublished theory in metallurgical history!

A great deal of work has been done on the optimum dispersion of a second phase to achieve specific objectives, especially (for aluminium alloys) the production of the finest possible grain size (e.g., FURRER [1980] and many other papers in the same conference proceedings). Special interest attaches to the behaviour of alloys with a bimodal distribution of particle sizes (a few large, many very fine particles). The complications of recrystallization in such alloys are reviewed by NES [1980].

Figure 41 represents a grain boundary migrating upwards under the influence of an unspecified driving force, and intersecting a spherical inclusion or pore. The drag exerted by the inclusion on the boundary, resolved in the y direction, is given by:

$$F = \pi r \lambda \sin 2\theta, \tag{6}$$

where λ is the specific interfacial energy of the boundary; this is effectively equivalent to a surface *tension*. The maximum value $F_{max} = \pi r \lambda$, for $\theta = 45°$.

If now there are N similar particles per unit volume, randomly distributed, their volume fraction f is $\frac{4}{3}\pi r^3 N$. A boundary of unit area will intersect all particles within a volume of $2r$, that is, $2Nr$ particles. Hence the number, n, of particles intersecting unit area of a boundary is given by:

$$n = \frac{3f}{2\pi r^2}. \tag{7}$$

Now suppose that the boundary is migrating purely under the influence of its own interfacial tension; if the grain periphery has a minimum radius of curvature R, the driving force is simply $2\lambda/R$ (§4.1). For a homogeneous grain structure (i.e., one where no grain greatly exceeds its neighbours in size) R approximates to the mean grain diameter D. As the grains grow, R increases and the driving force diminishes,

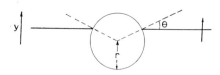

Fig. 41. Spherical inclusion resisting migration of a grain boundary.

until it is balanced by the drag due to inclusions, and growth suddenly ceases. Assuming that $R \sim D$, the critical diameter, D_{crit}, is given by the condition

$$nF_{max} \sim 2\lambda/D_{crit},$$

or

$$D_{crit} = 2\lambda \frac{2\pi r^2}{3f} \frac{1}{\pi r\lambda} = \frac{4r}{3f}. \tag{8}$$

It is obvious that a given volume fraction of inclusions is much more effective in retarding grain growth when the particles are very small. Thus for $f = 0.01$ and $r = 10^{-4}$ cm (just visible in the *optical* microscope), $D_{crit} \sim 1.3$ cm; in effect, the particles exert no constraint, since homogeneous grain growth does not normally progress to such a large grain size. With $f = 0.01$ but $r = 10^{-6}$ cm (just visible in the *electron* microscope), $D_{crit} \sim 0.013$ cm, which represents effective grain size control. In primary recrystallization, where the driving force is much greater than $2\lambda/D$, the particle drag will be less effective.

HAZZLEDINE et al. [1980] have published a more sophisticated version of the Zener treatment, allowing for boundary flexibility and for the fact that not all particles hinder boundary motion. The result is of the same order as Zener's estimate. LOUAT [1981] developed this further and found drag values for $f > 0.001$ which substantially exceed Zener's estimate.

HILLERT [1965] and GLADMAN [1966] have worked out more sophisticated analyses of grain growth in metals; the former considers pure metals, while the latter computes the critical particle size for unpinning for a given initial geometry (see next section).

HAROUN and BUDWORTH [1968] have made an experimental check of Zener's formula by studies of grain growth in magnesia, and conclude that it predicts a terminal grain size about an order of magnitude too large. They recognize that not every boundary is necessarily impeded by an inclusion, and if this is allowed for, better agreement is obtained.

OLSON et al. [1982] have recently established that MnS formed in an ultrafine dispersion by rapid solidification processing (ch. 28) leads to exceptional resistance to grain coarsening in a steel even at a temperature as high as 1200°C. They show thermodynamically that other dispersoids such as TiN should be even more effective than MnS in this capacity.

Small bubbles or cavities can act like inclusions in pinning boundaries. Thus, sintered and hot-worked pure platinum (MIDDLETON et al. [1949]) and copper and silver (BHATIA and CAHN [1978]) recrystallize more slowly than the cast metal, similarly deformed. The general problem of interactions between pores and grain boundaries has been reviewed by CAHN [1980]. Small voids form around ThO_2 particles in thoria-disperse nickel which has been heavily cold-worked, and these voids have been found to hold up grain boundaries and thus impede recrystallization (WEBSTER [1968]).

A further complication in connection with particle drag is that small (< 0.5 μm)

References: p. 1665.

particles, especially if their structure is amorphous and has low viscosity, can be dragged bodily through the metal by a moving boundary (ASHBY and CENTAMORE [1968]). The general problem of particle drag has been reviewed by ASHBY [1980].

The general problem of grain-boundary migration in a two-phase alloy in which the precipitate phase is itself liable to form, coarsen or dissolve during the anneal, has received a good deal of attention lately and has received a masterly survey by HORNBOGEN and KREYE [1969a, b] (see also HORNBOGEN and KÖSTER [1978]). The most important variables are solute concentration and temperature of anneal. At a high concentration or low temperature, precipitate will form *during* recrystallization and strongly impede boundary migration.

In many heat-treatment programmes, precipitation and recrystallization compete with each other; the extent (and mechanism) of recrystallization depends on the temporal balance between the two, and is therefore very sensitive to temperature. Figure 42 (LOMBRY et al. [1980]) shows an instance of this, for a stainless steel which precipitates NbCN during heat-treatment. The TTT curve shows that, for 44% strain, recrystallization in almost complete at 1100°C but does not proceed far at 1000°C. This kind of balance is classified and reviewed by HORNBOGEN and KÖSTER [1978], who also deal with other aspects of two-phase recrystallization, and by DOHERTY [1982].

Special problems arise if the precipitate is coherent, as for instance in Cu–Co (PHILLIPS [1966]; TANNER and SERVI [1966]), Ni–Al (PHILLIPS [1967]; HAESSNER et al. [1966]; OBLAK and OWCARSKI [1968], KREYE and HORNBOGEN [1970]) or Al–Cu (HORNBOGEN and KREYE [1969a, b]). Under these conditions a special form of particle coarsening may be observed: the coherent particles dissolve as the boundary passes and re-precipitate behind the boundary in coarser form. This process can act as equivalent to a driving force, since the interface area is reduced. According to the particle-size distribution, temperature, solute concentration and equilibrium structure, such particles may either retard or accelerate boundary migration, though the former is more frequently observed. Hornbogen and Kreye for the first time have succeeded in interpreting in an orderly way the complex situations which arise, and RALPH et al. [1980] survey the different ways particles react to passing grain boundaries in superalloys. HOTZLER and GLASGOW [1982] have shown how the

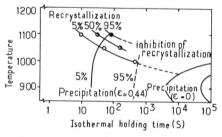

Fig. 42. Interaction of recrystallization and precipitation of an austenitic type-347 stainless steel. Plastic deformation to $\epsilon = 44\%$ or $\epsilon = 0$. (After LOMBRY et al. [1980].)

dissolution of coherent γ'-phase in a superalloy can be rate-determining for secondary recrystallization (see below). (See also ch. 10B, § 3.4.1.)

4. Grain growth and secondary recrystallization

4.1. Mechanism and kinetics of grain growth

When primary recrystallization is complete, the grain structure is not yet stable. The main driving force, associated with the retained energy of deformation, is spent, but the material still contains grain boundaries which have finite interfacial energy. This situation is at best metastable, and ideal thermodynamic stability is only attained when the sample has been converted into a monocrystal. The situation is closely analogous to a froth of soap bubbles, which will gradually coarsen and may finish as a single bubble; indeed, such froths, under reduced pressure to accelerate growth, have been used as quantitative analogue of an unstable grain structure (SMITH [1964]).

Figure 43 serves to explain the instability of grain structures on geometrical grounds. Suppose first that, in two dimensions, the grains were to exist as an array of perfectly regular hexagons. The sides of the grains would then be flat and all triple points would be in equilibrium, because grain boundaries (in two dimensions, strictly grain edges) then all meet at 120°. Since all boundaries are assumed to be high-angle boundaries and thus of equal energy, the triangle of forces at the triple point is stable. If however a "rogue grain" has only, say, five sides, the average internal angle exceeds 120° unless the sides are curved convex outwards; then the sides must be unstable, and in seeking to shorten themselves by straightening, they will disturb the triple-point equilibria at the apices; thus the grain is gradually consumed. The smaller the number of sides, the sharper is the curvature of the sides

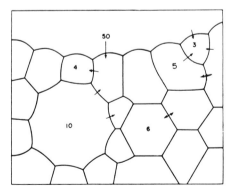

Fig. 43. Schematic diagram showing effect of the number of sides of grains on the curvature of grain boundaries. Six-sided grains are in two-dimensional equilibrium; smaller grains disappear, larger grains grow. (After COBLE and BURKE [1963].)

References: p. 1665.

for a given grain size, and the faster is the process of absorption. Once a rogue grain has vanished, the new neighbours find themselves out of equilibrium, and the process continues. Grains with more than six sides grow, those with fewer vanish. In three dimensions, the topological details are more complex (SMITH [1964]) but the principles are similar. The only entirely stable configuration (never encountered in practice) consists of an array of non-regular 14-sided polyhedra with doubly curved faces. In practice, arrays of grains are always unstable and coarsen continuously, until limited by some counterforce. The detailed grain morphologies that arrive in the course of grain growth are discussed by MCLEAN [1957], in more detail by HILLERT [1965] and most completely by RHINES and PATTERSON [1982].

In a soap-bubble array, it has been experimentally confirmed that the bubble diameter D follows the time law

$$\frac{dD}{dt} = \frac{1}{D}. \tag{9}$$

This law follows directly from the hypothesis that the rate of boundary migration is inversely proportional to the mean radius of curvature and that this in turn is proportional to D. For this hypothesis to be acceptable, the boundary *mobility* (as distinct from the driving force acting upon it) must be invariant with time; this is reasonable, for the mobility is determined by the diffusivity of gas through the bubble walls, while the driving force is determined by the gas pressure in each bubble, which in turn is fixed by the radius of curvature of its walls. In a metal, correspondingly, either self-diffusion or else impurity diffusion will determine the mobility, while the surface tension of the curved boundary provides the driving force.

From eq. (9) it follows that

$$D - D_0 = Kt^{1/2}. \tag{10}$$

Experimentally, this has been confirmed for a number of pure metals or alloys, though n is usually slightly less than $\frac{1}{2}$. Frequently a value near 0.3 is found.

The simple law can be disturbed by a number of factors. In particular, impurities slow down and may eventually stop grain growth: for instance, a fraction of one per cent of molybdenum or manganese in *solution* in iron reduces the grain growth rate at a given mean grain size and temperature by several orders of magnitude (LESLIE *et al.* [1962]). Here impurities exert their effect through segregation to grain boundaries but without forming precipitates *; indeed, grain growth offers a good way of studying the impurity drag effect, because the driving force is so well-defined and reproducible.

Again, a disperse phase, even in very small quantities, can retard boundary motion as explained in §3.6, and eventually bring grain growth to a halt. Here the

---

* This is why it was long believed that cast structures were intrinsically incapable of grain growth. Geometrically there is no reason why this should be so, but impurities, which segregate at grain boundaries during solidification, slow down the process to vanishing point. Zone-refined cast polycrystalline metals do indeed undergo rapid grain growth (HOLMES and WINEGARD [1960]).

growth law should be controlled not by the instantaneous grain size, but by the difference between this and the final limiting grain size; this, and the experimental evidence generally, are discussed by BURKE and TURNBULL [1952]. This influence is termed *disperse phase inhibition*.

The geometrical theory of grain growth in the presence of a dispersion of second phase particles has been taken to a very detailed level by the work of GLADMAN [1966, 1980] and HELLMAN and HILLERT [1975]. The treatment is based on an idealized grain-shape distribution, and it is assumed that the disperse particles are all the same size.

Only a grain larger than its neighbour, by about 30% at least, can consume the latter with a decrease of total boundary energy. The shrinkage of small grains is energetically favourable, the more so the smaller they become. There are distinct critical particle sizes (for a given volume fraction) to just inhibit growth of large grains and shrinkage of small ones. These critical sizes depend also on the degree of instantaneous heterogeneity of grain sizes. The critical particle size for inhibiting growth of large grains is proportional to the mean grain size, so that it can be seen that grain growth must eventually stop so long as the inhibiting particles do not simultaneously coarsen. An alternative outcome (see next section) is *secondary recrystallization*.

The most remarkable recent study of grain morphology during grain growth is that by RHINES and PATTERSON [1982]: it is also an excellent demonstration of the power of quantitative metallography (chapter 10A, §4). They demonstrated that the amount of deformation determines the width of the statistical distribution of grain sizes after primary recrystallization. This persists during subsequent grain growth. (The reduced heterogeneity of grain sizes after heavier strain is shown to be consistent with the hypothesis that nucleation centres have a minimum distance of approach.) A highly heterogeneous grain-size distribution leads to relatively more small (triangular) grain faces; grains containing these shrink very rapidly, and grain-growth kinetics are thus intimately linked with the initial heterogeneity of grain sizes – and thus with the degree of deformation imparted in the first place.

VARIN and TANGRI [1980] have established that the characteristics of extrinsic boundary dislocations in a stainless steel were different after primary recrystallization and after grain growth. This implies that grain-boundary structure may not reach a stable state when a boundary migrates rapidly. This might affect relative grain-boundary energies, but has not been investigated.

Grain growth can be limited, not only by a particle distribution but also by thickness of the sample when this is in sheet form. The rate of grain growth diminishes sharply once a substantial number of grains span the thickness of the sheet, so that the process becomes two-dimensional, and normally growth stops altogether when the mean grain diameter (as measured in the plane of the sheet) reaches between two and three times the sheet thickness (BURKE [1949]). This is due to the reduced driving force associated with boundaries which are cylindrically rather than spherically curved, and also with the drag caused by the grooves formed on a free surface where grain boundaries emerge; these grooves are formed by a

References: p. 1665.

process of surface diffusion and diffuse along with the migrating grain boundaries. This influence is termed *thickness inhibition*.

Finally, grain growth is seriously impeded if the grains resulting from primary recrystallization have a sharp texture (BECK and SPERRY [1949]). This is due to the much reduced grain boundary energy λ associated with grains having mutual misorientations of less than $15°$ (e.g., MCLEAN [1957]); this in turn reduces the driving force for grain growth, which for a given grain size is proportional to λ. This influence is termed *texture inhibition*.

4.2. Formation of annealing twins

A commonly observed feature of recrystallized structures in certain face-centred cubic metals (particularly the copper-group metals and their solid solutions, lead and austenitic steels) is the presence of copious parallel-sided *annealing twin* lamellae, as sketched in fig. 44a. These lamellae are always bounded by {111} planes or *coherent boundaries*, apart from short incoherent steps and terminations, which are crystallo-graphically more complicated. The orientation of the lattice within a twin lamella is related to that of the parent grain by reflection in the coherent plane. Immediately after primary recrystallization there are few lamellae per grain, but their number increases with the progress of grain growth. However, this received truth has been cast into doubt by GINDVAUX and FORM [1973] who established by direct observa-tion that most annealing twins form during primary recrystallization and few additional twins appear during subsequent grain growth.

The genesis of these twin lamellae has been clarified by Fullman, Burke and others (BURKE and TURNBULL [1952], SMITH [1964]). It rests upon the fact that the interfacial energy of coherent interfaces in the above-mentioned metals and alloys is only about 5% of the energy of an arbitrary high-angle grain boundary. If now a region of a grain structure goes through the configurational sequence shown in fig. 44b, then at stage (2) a new grain contact is established between grains C and D.

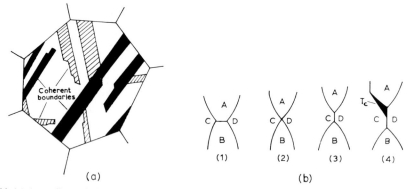

Fig. 44. (a) Annealing twin lamellae in a face-centred cubic metal grain. (b) Stages in the development of a twin during grain growth.

Now it may happen, according to the accident of the particular orientation of these grains, that the twin T_c has an orientation close to that of grain D. Then the T_c–D interface is of lower energy than the C–D interface. C–A and T_c–A on the other hand will in general be of roughly similar energies. In that case the configuration (4) will be of lower energy than configuration (3), in spite of the extra area of interface represented by the coherent twin interface C–T_c; this is only possible because the latter has a very low specific interfacial energy. As growth proceeds, a *growth fault* will sooner or later be generated and a twin will appear at the advancing A–C–D triple point, and such a fault will be stable so that the twin grows as shown in stage (4). When the apex of the growing twin makes contact with a fresh grain, it may become unstable and the original orientation C will again take over; the result will be a parallel-sided twin lamella. On this model, the number of twin lamellae per unit grain boundary area should depend only on the number of new grain contacts (as at stage (2) of the figure) which have been made during growth, irrespective of absolute grain size and temperature; experiments by Hu and Smith [1956] have shown that this is indeed so. This study has established the model beyond doubt. The earlier observations on the balance of grain-boundary energies were extended by Viswanathan and Bauer [1973b]; they showed in particular that, as predicted by the theory, the migration of relatively small-angle boundaries does *not* give rise to any twins. A recent review, taking into account the newer observations, is by Gleiter [1981].

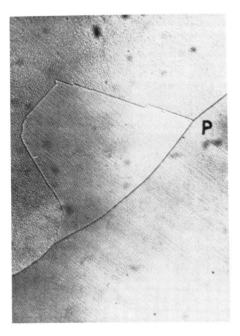

Fig. 45. Annealing twin in an iron–aluminium alloy after prolonged grain growth. 170×. (After Cahn and Coll [1961].)

References: p. 1665.

When the coherent twin interfacial energy, γ_T, is higher in relation to the energy of ordinary grain boundaries, γ_B, as in aluminium where $\gamma_T/\gamma_B \sim 0.2$, then annealing twins are very rare.

Annealing twins have also been found in body-centred cubic metals and alloys; here the crystallography is more complicated, and one twin can have three distinct coherent interfaces. Figure 45 shows an example. The low energy of the coherent boundary can be deduced from the fact that the normal grain boundary is only slightly deflected where a coherent boundary abuts on it (at P).

4.3. Secondary recrystallization

4.3.1. General features

When the annealing of an initially deformed sample is continued long beyond the stage when primary recrystallization is complete, the even tenor of grain growth may be interrupted by the sudden very rapid growth of a few grains only, to dimensions which may be of the order of centimetres, while the rest of the grains remain small and are eventually all swallowed by the large grains. This is *secondary recrystallization*, or *coarsening*. Figure 46 shows an instance of incipient secondary recrystallization. The process has the following general characteristics:

(a) The large grains are not freshly nucleated; they are merely particular grains of the primary structure grown large.

(b) The first stages of growth of the large grains are sluggish; there is an induction period before secondary recrystallization gets under way.

(c) The factors governing the choice of the victorious primary grains which are to grow large, and the mechanism of the early stages, are the least understood parts of the process. It is generally agreed that secondary-grains-to-be must be appreciably

Fig. 46. Incipient secondary recrystallization in zinc sheet. 100×.

larger than the mean primary grain size, and they must have orientations which diverge from the main primary texture.

(d) Something must inhibit normal uniform grain growth; it is only when grain growth is very slow that large secondary grains can effectively grow. Inhibition by a dispersed phase, by primary texture or by sheet thickness, may all play a part.

(e) The secondary structure, when complete, sometimes has a pronounced texture. Such a texture always differs from the previous primary texture.

(f) A well-defined minimum temperature must be exceeded for secondary recrystallization. The largest grains are normally produced just above this temperature; at higher annealing temperatures smaller secondary grains result.

(g) The driving force of secondary recrystallization, once the large grains are well launched, normally arises from grain-boundary energy (just as does normal grain growth); under special circumstances, the surface energy of the grains can contribute.

Figure 47 exemplifies several of these regularities.

Suppose a primary structure contains a grain of roughly twice the average diameter such as the lower left-hand grain of fig. 43. If we assume that the interfacial energy of the boundaries separating this grain from its smaller neighbours (γ_s) is the same as the energy of the boundaries between these neighbours themselves (γ_p), then the sides of the large grain must be convex inwards as shown in the figure, if the

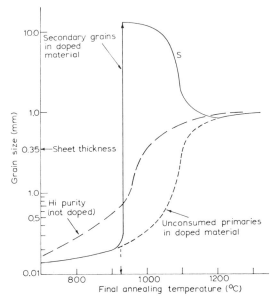

Fig. 47. Grain size (log scale) as a function of temperature for pure and MnS-doped 3% silicon-iron, cold-rolled 50% to 0.35 mm thickness, annealed 1 h. The drop in curve S is due to an increase in the number of secondary grains at higher temperatures, until above 1100°C all grains grow equally to a size limited by the sheet thickness. (After MAY and TURNBULL [1958].)

References: p. 1665.

triple grain junctions are to be in equilibrium. This configuration, which is clearly recognizable in fig. 46, is evidently unstable, since the curved boundaries will seek to become straight and in the process will upset the equilibrium at the triple points, which thus migrate outwards. The larger the grain grows, the more sharply will its bounding faces become curved and the process is thus self-sustaining. At a certain size a grain has reached "breakaway point", and thereafter grows rapidly, quickly reaching the stage of the grain at the top of fig. 43. There is no mystery, then, about the continuation of secondary growth once it has started; only the first stages are hard to interpret.

Consider a grain such as the ten-sided one in fig. 43. Let its diameter be D while its smaller neighbours have diameter d. Whether or not this grain has reached breakaway point depends on the boundary-energy ratio γ_s/γ_p and on the volume fraction f and radius r of the dispersed-phase particles. When γ_s/γ_p is large (i.e. a grain of divergent orientation surrounded by primary grains with a strong texture) then the critical ratio D_{crit}/d is large (because the equilibrium angle between adjacent sides of the large grains is then greater than 120°). The larger is f and the smaller r, the stronger is the boundary drag. It is only the curvature of the sides of the large grains which drives them to grow and the drag counteracts this driving force. Thus a large f and small r (i.e. an effective dispersion) increase D_{crit}/d. May and Turnbull [1958] have analyzed this problem in detail and conclude that:

$$\frac{D_{crit}}{d} = \frac{\frac{3}{4}(4\gamma_s/\gamma_p - 1)^{1/2}}{1 - (\alpha\gamma_s/3\gamma_p)f/r}. \qquad (11)$$

Once the large grain has somewhat exceeded the critical diameter D_{crit}, then the driving force ΔF per unit volume for boundary migration settles down to a value

$$\Delta F \sim 2\left(\frac{\gamma_p}{d} - \frac{3}{8}\gamma_s\frac{f}{r}\right) \qquad (12)$$

and the rate of growth G of the secondary grains is derived by multiplying this term by the mean boundary mobility. This expression for ΔF is most simply regarded as containing an energy term depending on the total area of primary grain boundaries destroyed per unit distance moved by the secondary boundary, and a drag term which counteracts this.

This analysis is qualitatively in agreement with the known facts about secondary recrystallization, but it has not been subjected to quantitative test. However, the notion of a growth rate determined essentially by a grain-boundary energy and a drag term has received quantitative confirmation in a special situation, as shown below.

Three processes suggest themselves that might enable a primary grain to grow to the point of breakaway:

(a) There is a sharp primary texture with only a few sharply deviating primary grains. All or many of these grains grow, not because they have a strong tendency to do so (the large value of γ_s/γ_p indeed discourages them) but only because the small value of γ_p is even more effective in discouraging the normal growth of grains belonging to the sharp texture. Such a situation will not produce a sharply defined secondary texture. This is believed to be the most common source of secondary grains.

(b) There are many deviant primary grains and among them a few are so oriented that their grain-boundary mobility M is particularly high. Then during normal grain growth such grains will forge ahead, not because of a higher driving force but only because of their higher mobility, and eventually they reach the critical diameter for breakaway. KRONBERG and WILSON's [1949] original discovery of "special orientations" stemmed from a study of secondary grains in copper; here, presumably, growth selectivity operated merely during the very early stages of growth. A sharp texture results from this process.

(c) The successful grains may have a more perfect lattice than their neighbours and thus possess a larger driving force for growth. This has often been suggested but never substantiated. It is hard to see how it can produce a secondary texture.

From the foregoing, it is evident that the nature of the disperse phase is vital in controlling secondary recrystallization; unless there is a very sharp primary texture with a few rogue grains, a properly dispersed second phase is vital for secondary recrystallization. If the dispersion drag is insufficient, too much normal grain growth occurs; if it is too great, the secondary grains cannot grow at all (the right-hand side of eqs. (11) and (12) is negative). Thus in fig. 47 the primary grain size of the pure alloy increases more rapidly than that of the alloy doped with a manganese sulphide dispersion; the latter undergoes secondary recrystallization, the former does not. GLADMAN [1980] has analyzed the role of the fine-particle dispersion in the light of his (and Hillert's) theory of grain growth in the presence of a dispersion, outlined in §4.1. When a distribution of grains is *just* inhibited from growing, then a slight degree of coarsening of the disperse particles by Ostwald ripening (chapter 10B, §3.2.2) will unpin the few largest grains only. These can then act as 'nuclei' for secondary recrystallization.

Secondary recrystallization can take place during or after *hot*-working, if normal grain growth is inhibited. A recent, very detailed study on the hot-working of a superalloy (SINGER and GESSINGER [1982]) showed that for large strains and low strain rates, normal grain growth is enhanced by the applied strain and this prevents the 'take-off' of secondary grains.

The control of dispersions, especially of manganese sulphide, is evidently of vital importance in producing strip for transformer cores made of iron–3% silicon alloy. This material, when properly treated, consists of secondary grains with the *Goss texture*, (110)[001]. This texture is beneficial because the [001] direction is one of easy magnetization and thus a smaller mass of transformer laminations is required than if the texture were random. While much of this work is subject to commercial secrecy, WASSERMANN and GREWEN [1962] give an extensive survey of the procedures published up to that date. FIEDLER [1964] has illustrated the role of manganese sulphide inclusions in controlling secondary recrystallization in this material. More recent developments concerning texture in iron–silicon alloys are discussed in ch. 26, §4.2.3.3.

References: p. 1665.

The analysis by DILLAMORE *et al.* [1967] to interpret the primary annealing texture of steel containing and AlN dispersion (§ 3.6) can also serve to explain the development of the Goss texture from a primary annealing texture containing a weak (110)[001] component. That component is weakly represented, because in the deformed structure grains of (110)[001] orientation were few in number, but they did have the largest stored elastic energy; therefore, though few, they began to recrystallize first when the rolled sheet was annealed, and hence grew largest. When the AlN dispersion eventually coarsens, all the largest grains (which then become unpinned) have the (110)[001] orientation. The fact that the secondary texture is so sharp derives from the very sharp variation, in the cold-worked polycrystal, of stored elastic energy with small deviations from the (110)[001] orientation, which in turn implies that the size of the largest primary grains is very sensitive to their exact orientations.

A recent study by INOKUTI *et al.* [1980,1983], in which orientations of many individual grains were mapped, is consistent with Dillamore's model. (110)[001] grains grow from the surface regions, where they are particularly large after primary recrystallization (because of a variation, through the thickness, of the degree of dispersion of second-phase particles). This is a particularly comprehensive investigation.

In recent years, major improvements have been achieved in Japan in the secondary recrystallization of iron–silicon transformer alloys: very fine control of dispersed phases has permitted a sharper (110)[001] texture to be obtained (e.g., INOKUTI *et al.* [1980], TAGUCHI and SAKAKURA [1969]; ch. 26, §§ 4.2.3.3, 4.2.4).

A striking illustration of the importance of even very small amounts of disperse phase in promoting secondary recrystallization is provided by the work of CALVET and RENON [1960]. They heat-treated a number of aluminium–copper solid solutions at temperatures close to the solid solubility limit. Effective inhibition of normal grain growth and very rapid secondary recrystallization was observed whenever the annealing temperature was as little as 1–2°C below the temperature at which all the copper present just entered into solid solution; at or above this temperature, only rapid normal grain growth took place. The dividing line between the two modes of recrystallization in alloys of different compositions accurately traced out the known solubility curve. The volume fraction of disperse phase just below the solubility line must be very small, less than 0.1% but this is apparently enough to retard normal grain growth quite brusquely and thus permit secondary recrystallization. This behaviour is the analogue, for secondary recrystallization, of the behaviour in primary recrystallization of an Al–Fe alloy that precipitates and recrystallizes, sluggishly, at the same time (MIKI and WARLIMONT [1968]).

Especially complex behaviour is observed when secondary recrystallization is controlled both by a stable oxide dispersion and by a coherent γ'-dispersion which must dissolve and reprecipitate to allow a grain boundary to pass (HOTZLER and GLASGOW [1982]).

4.3.2. Surface-controlled secondary recrystallization

In 1958 it was discovered independently in several laboratories that after prolonged annealing in the range 1000–1200°C, thin sheets of 3% silicon-iron (the alloy used for transformer cores) sometimes developed a secondary (001)[100] or *cube texture*, as distinct from the normal (110)[001] Goss texture referred to above. DETERT [1959] and WALTER and DUNN [1959], again independently, clarified the necessary conditions: The sheet has to be thin (normally below 0.1 mm in thickness) and the annealing has to be done in an atmosphere containing traces of oxygen. If, after cube texture has been established, the atmosphere is replaced by very dry hydrogen or by high vacuum, the texture disappears and is replaced by the Goss texture. This process has been termed *tertiary recrystallization*; indeed, by repeated changes of atmosphere the texture can be changed repeatedly.

It has been convincingly established that the development of these textures is governed by an anisotropy in surface energy γ_s. The surface energy (surface tension) of a grain at a free surface depends on its orientation; various indirect experiments on copper, silver, iron and other metals show that there is a measurable difference between, for instance, γ_s (100) and γ_s (111); according to the type of experiment, differences of 3–30% have been inferred (SHEWMON and ROBERTSON [1963]). There is also some evidence, summarized in the same review, that adsorbed impurities (which are known to change the crystal structure of the surface) considerably reduce the anisotropy. It appears that when the strip surface is comparatively clean, (110) faces have the lower surface energy, while in the presence of adsorbed oxygen, (100) has the lower energy; other orientations are probably intermediate. If the strip is thin enough, this anisotropy of surface energy is sufficient to cause the most favourably oriented grains to grow at the expense of all the others.

If this model is correct, and if moreover the surface energy varies rather sharply with small changes in orientation of a grain, then one would expect the (001) planes to be accurately parallel to the surface, while the parallelism of [100] to the rolling direction might be subject to considerable scatter. This is what is observed: (001) is always parallel to the strip within 5–7° at most, while scatter of up to 40° is observed for the [100] vector in the plane of the strip. This scatter is determined purely by the range of orientations in the primary texture.

FOSTER *et al.* [1963] studied the kinetics of the growth of (001)-oriented secondary grains in silicon-iron as a function of primary grain size and sheet thickness. They assumed that the growth rate G is given by the expression

$$G = M\Delta F = M\left(\frac{2\gamma_p}{r} + \frac{2\Delta\gamma_s}{t} - C\right), \tag{13}$$

where M is the mobility of the secondary boundaries, ΔF is the total driving force per unit volume, γ_p is the energy of the boundaries between primary grains, $\Delta\gamma_s$ is the difference between the surface energy of (001)-oriented grains and an "average" orientation, $2r$ is the mean primary grain diameter, t is the sheet thickness and C is the drag force associated with disperse particles. When a secondary grain is big, the driving force for secondary growth is essentially the energy ΔF_p per unit volume associated with primary grain boundaries. For roughly cylindrical primary grains (diameter d) spanning the sheet thickness, $\Delta F_p = 2\pi r\gamma_p/\pi r^2 t = 2\gamma_p/r$. The driving

References: p. 1665.

force ΔF_s from surface energy anisotropy, per unit volume (i.e. the difference in surface energy per unit volume of metal in a sheet of thickness t, which is composed entirely of cylindrical (001)-oriented grains, as compared with grains of random orientation) is $\Delta F_s = 2\pi r^2 \Delta\gamma_s / \pi r^2 t = 2\Delta\gamma_s / t$. Since $\Delta F = \Delta F_p + \Delta F_s - C$, eq. (13) follows.

From measurements of G in sheets with various r and t, the correctness of the relationship was established. For instance, with a given t, G should vary as $1/r$. Figure 48 shows that this relationship is obeyed, and shows how sensitive secondary grain growth is to primary grain size. Extrapolation shows that above a limiting grain size of about 1 mm, secondary recrystallization could not continue at all; since, however, primary grain size is restricted by thickness inhibition to a maximum of about $3t$, this limiting primary grain size is never reached. Small thicknesses favour surface-induced secondary recrystallization both because it restricts primary grain size and because the surface/volume ratio of secondary grains is greater.

From their measurements, Foster, Kramer and Wiener derived values of $\Delta\gamma_s$ and C in terms of the primary grain boundary energy γ_p. Typically, for $t = r = 0.3$ mm, $\Delta F_p \sim 33\gamma_p$, $\Delta F_s \sim 2\gamma_p$ and $C \sim 14\gamma_p$. Thus $\Delta F = (33 + 2 - 14)\gamma_p = 21\gamma_p$ mJ/m². Thus the surface energy term is only 10% of the total resultant driving force. This term decides the selection of the orientation which is to grow, but ΔF_p and C together determine the rate of growth of the secondary grains.

Cube texture in silicon-iron would be very desirable for transformer laminations, in spite of the scatter in the texture (ch. 26, §§4.2.3.3 and 4.2.4), and this is the reason for the considerable effort which has been devoted to attempts to manufacture this material commercially. Up to the present, however, it has not been possible to achieve the necessary rigorous control of processing variables at an economic cost.

It is probable that surface-energy-controlled secondary recrystallization is a good deal more common than has been recognized. Thus DUNN and WEBSTER [1965] have been able to grow (110)-oriented grains in thin tantalum strip by this mechanism;

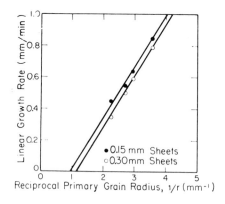

Fig. 48. Effect of primary grain size on growth rate of (001)-oriented secondary grains. Strips of two different thicknesses were used. (After FOSTER et al. [1963].)

this orientation is particularly favourable if the metal is to be used as an electron emitter in thermionic energy converters. Again, MᴄLᴇᴀɴ and Mʏᴋᴜʀᴀ [1965] have found that (111)-oriented grains grow preferentially in thin platinum sheets, and (0001)-oriented grains were found to grow in sheets of zinc (Mɪʟʟᴇʀ and Wɪʟʟɪᴀᴍs [1964]).

The environmental influence in surface-controlled secondary recrystallization appears to be very subtle in detail. Thus Mᴇᴇ [1968] has found that oxygen and sulphur both have a major effect on the phenomenon, and that moreover the metallic impurities in the alloy and their state of heat-treatment also influence the texture.

The situation concerning secondary recrystallization has been reviewed in detail by Dᴜɴɴ and Wᴀʟᴛᴇʀ [1966], Wᴀʟᴛᴇʀ [1969] and Dᴇᴛᴇʀᴛ [1978].

4.3.3. Secondary recrystallization and sintering

When a compressed compact of metal or ceramic particles is heated, the pores between the particles gradually contract and eventually most of them disappear; this is *sintering*. It has been established (Cᴏʙʟᴇ and Bᴜʀᴋᴇ [1963]) that the essential process is the diffusion of excess vacancies from the immediate neighbourhood of a pore into a nearby grain boundary and rapidly away along the boundary by grain-boundary diffusion (ch. 8, §7). Complete sintering (i.e., total removal of the pores), which is highly desirable in many applications, thus depends on a plentiful supply of boundaries near all pores. If slow, steady grain growth is maintained, each pore is repeatedly crossed by a migrating grain boundary; a simple statistical calculation (Kᴏᴏʏ [1962]) shows that a tenfold increase in grain size by normal grain growth implies that each pore is on the average crossed by seven migrating grain boundaries. This gives plenty of opportunity for substantial sintering.

If, however, secondary recrystallization takes place, then all pores within a secondary grain have been crossed only once by a moving boundary, and moreover one that is moving rapidly. Such pores thus remain stranded within their large grains, far away from the nearest boundary; vacancies then cannot diffuse rapidly away from the pores and sintering virtually comes to a halt. For effective sintering, then, secondary recrystallization must be prevented; this insight has been put to good use by doping aluminium oxide with small quantities of magnesium oxide. This produces a powerful boundary drag, slowing down grain growth without stopping it entirely, and preventing secondary recrystallization. The compressed powder can be completely sintered and the dense product so formed has valuable properties for practical applications.

The relation between grain structure and sintering characteristics is further discussed in ch. 30, §1.4.

References

Aɴᴅᴇʀsᴏɴ, W.A. and R.F. Mᴇʜʟ, 1945, Trans. AIME **161**, 140.

Asʜʙʏ, M.F., 1980, in: Recrystallization and Grain Growth of Multi-phase and Particle-Containing Alloys, eds. N. Hansen *et al.* (Risø National Laboratory, Denmark) p. 325.

ASHBY, M.F., and R.M.A. CENTAMORE, 1968, Acta Metall. **16**, 1081.

ASSMUS, F., K. DETERT and G. IBE, 1958, Z. Metallk. **48**, 344.

AUST, K.T., 1963, in: Art and Science of Growing Crystals, ed. J.J. Gilman (Wiley, New York) p. 452.

AUST, K.T., 1969, in: Textures in Research and Practice, eds. J. Grewen and G. Wassermann (Springer, Berlin) p. 160.

AUST, K.T., 1974, Can. Met. Quart. **13**, 133.

AUST, K.T., and C.G. DUNN, 1957, Trans AIME **209**, 472.

AUST, K.T., and J.W. RUTTER, 1959, Trans. AIME **215**, 119 and 608.

AUST, K.T., and J.W. RUTTER, 1960, Trans. AIME **218**, 682.

AUST, K.T., and J.W. RUTTER, 1963, Recovery and Recrystallization of Metals (Wiley, New York) p. 131.

BAILEY, J.E., and P.B. HIRSCH, 1962, Proc. Roy. Soc. **A267**, 11.

BALL, C.J., 1957, Phil. Mag. **7**, 1011.

BALLUFFI, R.W., 1982, Metallurg. Trans. **13A**, 2069.

BECK, P.A., 1954, Adv. Phys. **3**, 245.

BECK, P.A., and H. HU, 1952, Trans. AIME **189**, 83.

BECK, P.A., and P.R. SPERRY, 1949, Trans. AIME **180**, 240.

BECK, P.A., and P.R. SPERRY, 1950, J. Appl. Phys. **21**, 150.

BELL, F., and O. KRISEMENT, 1962, Z. Metallk. **53**, 115; also F. BELL, Acta Metall. **13** (1965) 363.

BELLIER, S.P., and R.D. DOHERTY, 1977, Acta Metall. **25**, 521.

BEVER, M.B., D.L. HOLT and A.C. TITCHENER, 1973, Prog. Mater. Sci., vol. 17 (Pergamon Press, Oxford).

BHATIA, M.L., and R.W. CAHN, 1978, Proc. Roy. Soc. **362A**, 341.

BONIS, L.J., and N.J. GRANT, 1960, Trans. AIME **175**, 151.

BRANDON, D.G., B. RALPH, S. RANGANTHANAN and M.S. WALD, 1964, Acta Metall. **12**, 813.

BROOM, T., and R.K. HAM, 1957, Vacancies and other Point Defects in Crystals (Inst. Metals, London) p. 41.

BUNGE, H.J., ed., 1981, Quantitative Texture Analysis (Deutsche Gesellschaft für Metallkunde, Oberursel).

BURGERS, W.G., 1941, Rekristallisation, Verformter Zustand und Erholung (Akademische Verlagsges., Leipzig) p. 356.

BURKE, J.E., 1949, Trans. AIME **180**, 73.

BURKE, J.E., and D. TURNBULL, 1952, Prog. Met. Phys. **3**, 220.

CAHN, J.W., 1962, Acta Metall. **10**, 789.

CAHN, R.W., 1949, J. Inst. Metals **76**, 121.

CAHN, R.W., 1950, Proc. Phys. Soc. **A63**, 323.

CAHN, R.W., 1964, Deformation Twinning (Wiley, New York) p. 1.

CAHN, R.W., 1980, in: Recrystallization and Grain Growth of Multi-phase and Particle-Containing Alloys, eds. N. Hansen *et al.* (Risø National Laboratory, Denmark) p. 77.

CAHN, R.W., and J.A. COLL, 1961, Acta Metall. **4**, 683.

CAHN, R.W., I.J. BEAR and R.L. BELL, 1954, J. Inst. Metals **82**, 481.

CALAIS, D., P. LACOMBE and N. SIMENEL, 1959, J. Nucl. Mater. **1**, 325.

CALVET, J., and C. RENON, 1960, Mém. Sci. Rev. Métallurg. **57**, 347.

CHALMERS, B., and H. GLEITER, 1971, Phil. Mag. **23**, 1541.

CHERIAN, T.V., P. PETROKOWSKY and J.E. DORN, 1949, Trans. AIME **185**, 948.

CHERNS, D., P. HIRSCH and H. SAKA, 1979, Proc. ICSMA **5**, 295.

CHOJNOWKSI, E.H. and R.W. CAHN, 1973, in: Metallurgical Effects at High Strain Rates, eds. Rohde *et al.* (Plenum, New York) p. 631.

CHRIST, W., K. GSCHWENDTER and F. HAESSNER, 1965, Phys. Stat. Sol. **10**, 337.

CLAREBROUGH, L.M., M.E. HARGREAVES and G.W. WEST, 1955, Proc. Roy. Soc. **A232**, 252.

CLAREBROUGH, L.M., M.E. HARGREAVES and G.W. WEST, 1956, Phil. Mag. **1**, 528.

COBLE, R.L., and J.E. BURKE, 1963, Progr. Ceramic Sci. **3**.

COFFIN, JR., L.F., and J.F. TAVERNELLI, 1959, Trans. AIME **215**, 794.

COTTRELL, A.H., 1953, Prog. Met. Phys. **4**, 255.

CUDDY, L.J., 1981, Metallurg. Trans. **12A**, 1313.

DE LIBANATI, N.A., and S. VOLMAN DE TANIS, 1962, Comp. Rend. **252**, 2710.

DEMIANCZUK, D.W., and K.T. AUST, 1975, Acta Metall. **23**, 1149.

DESALVO, A., and D. NOBILI, 1968, J. Mater. Sci. **3**, 1.

DETERT, K., 1959, Acta Metall. **7**, 589.

DETERT, K., 1978, in: Recrystallization of Metallic Materials, 2nd Ed., ed. F. Haessner (Dr. Riederer-Verlag, Stuttgart) p. 97.

DILLAMORE, I.L., and H. KATOH, 1974, Met. Sci. **8**, 21 and 73.

DILLAMORE, I.L., C.J.E. SMITH and T.W. WATSON, 1967, Met. Sci. **1**, 49.

DILLAMORE, I.L., P.L. MORRIS, C.J.E. SMITH and B.W. HUTCHINSON, 1972, Proc. Roy. Soc. **329A**, 405.

DIMITROV, O., 1960, Nouvelles Propriétés Physiques et Chimiques des Métaux de très Haute Pureté (CNRS, Paris) 89 and 103.

DOHERTY, R.D., 1978, in: Recrystallization of Metallic Materials, 2nd Ed., ed. F. Haessner (Dr. Riederer-Verlag, Stuttgart) p. 23.

DOHERTY, R.D., 1982, Met. Sci. **16**, 1.

DOHERTY, R.D., and R.W. CAHN, 1972, J. Less-Common Met. **28**, 279.

DOHERTY, R.D., and J.W. MARTIN, 1963, J. Inst. Metals, **91**, 332.

DUGGAN, B.J., M. HATHERLY, B.W. HUTCHINSON and P.T. WAKEFIELD, 1978, Met. Sci. **12**, 342.

DUNN, C.G., 1946, Trans. AIME **167**, 373.

DUNN, C.G., and J.L. WALTER, 1966, in: Recrystallization, Grain Growth and Textures, ed. H. Margolin (ASM, Metals Park, OH) p. 461.

DUNN, C.G., and H.F. WEBSTER, 1964, Trans. AIME **230**, 1567.

ENGLISH, A.T., and U.A. BACKOFEN, 1964, Trans. AIME **185**, 501.

ESTRIN, Y., and K. LÜCKE, 1982, Acta Metall. **30**, 983.

EVANS, H.E., and J.S. WADDINGTON, 1969, J. Nucl. Mater. **30**, 337.

EXCELL, S.F., and D.H. WARRINGTON, 1972, Phil. Mag. **26**, 1121.

FAIVRE, P., and R.D. DOHERTY, 1979, J. Mater. Sci., **14**, 897.

FERRAN, G.L, R.D. DOHERTY and R.W. CAHN, 1971, Acta Metall. **19**, 1019.

FIEDLER, H.C., 1964, Trans. AIME **230**, 95.

FORTES, M.A., and B. RALPH, 1967, Acta Metall. **15**, 707.

FOSTER, K., J.J. KRAMER and G.W. WIENER, 1963, Trans. AIME **227**, 185.

FRIDMAN, E.M., T.W. KOPEZKII and L.S. SHVINDLERMAN, 1975, Z. Metallk. **66**, 533.

FURRER, P., 1980, in: Recrystallization and Grain Growth of Multi-phase and Particle-Containing Alloys, eds. N. Hansen *et al.* (Risø National Laboratory, Denmark) p. 109.

GAWNE, D.T., and G.T. HIGGINS, 1969, in: Textures in Research and Practice, eds. J. Grewen and G. Wassermann (Springer, Berlin) p. 319.

GAY, P., P.B. HIRSCH and A. KELLY, 1954, Acta Cryst. **7**, 41.

GINDVAUX, G, and W. FORM, 1973, J. Inst. Met. **101**, 85.

GLADMAN, T., 1966, Proc. Roy. Soc. **294A**, 298.

GLADMAN, T., 1980, in: Recrystallization and Grain Growth of Multi-phase and Particle-Containing Alloys, eds. N. Hansen *et al.* (Risø National Laboratory, Denmark), p. 183.

GLEITER, H., 1969, Acta Metall. **17**, 853.

GLEITER, H., 1979, Acta Metall. **27**, 1754.

GLEITER, H., 1981, Prog. Mater. Sci. (Chalmers Anniversary Volume) p. 125.

GORDON, P., 1955, Trans. AIME **203**, 1043.

GOTTSCHALK, CHR., K. SMIDODA and H. GLEITER, 1980, Acta Metall. **28**, 1653.

GREEN, R.E., B. LIEBMAN and H. YOSHIDA, 1959, Trans. AIME **215**, 610.

GREWEN, J., and J. HUBER, 1978, in: Recrystallization of Metallic Materials, 2nd Ed., ed. F. Haessner (Dr. Riederer-Verlag, Stuttgart) p. 111.

GREWEN, J., and G. WASSERMANN, eds., 1969, Textures in Research and Practice (Springer, Berlin).

GREWEN, J., T. NODA and D. SAUER, 1977, Z. Metallk. **67**, 260.

HAESSNER, F., and S. HOFMANN, 1971, Z. Metallk. **62**, 807.

HAESSNER, F., and S. HOFMANN, 1978, in: Recrystallization of Metallic Materials, 2nd Ed., ed. F. Haessner (Dr. Riederer-Verlag, Stuttgart) p. 63.

HAESSNER, F. and H.P. HOLZER, 1974, Acta Metall. **22**, 695.

HAESSNER, F., E. HORNBOGEN and M. MUKHERJEE, 1966, Z. Metallk. **57**, 270.

HAROUN, N.A., and D.W. BUDWORTH, 1968, J. Mater. Sci. **3**, 326.

HART, R.V., and H. GAYTER, 1968, J. Inst. Metals **96**, 338.

HASEGAWA, T., and U.F. KOCKS, 1979, Acta Metall. **27**, 1705.

HASSON, C., J.T. BOOS, I. HERBEVAL, M. BISCONDI and G. GOUX, 1972, Surf. Sci. **31**, 115.

HELLER, H.W.F., J.H. VAN DORP, G. WOLFF and C.A. VERBRAAK, 1981, Met. Sci. **15**, 333.

HELLMAN, P., and M. HILLERT, 1975, Scand. Metall. J. **4**, 211.

HAZZLEDINE, P.M., A.B. HIRSCH and N. LOUAT, 1980, in: Recrystallization and Grain Growth of Multiphase and Particle–Containing Alloys, eds. N. Hansen et al. (Risø National Laboratory, Denmark) p. 159.

HIBBARD JR., W.A., and C.G. DUNN, 1957, Creep and Recovery (ASM, Metals Park, OH) p. 52.

HIGGINS, G.T., 1971, Metallurg. Trans. **2**, 1271.

HILLERT, M., 1965, Acta Metall. **13**, 227.

HOLMES, E.L., and W.C. WINEGARD, 1960, J. Inst. Metals **88**, 468.

HORNBOGEN, E., and H. KREYE, 1969a, in: Textures in Research and Practice, eds. J. Grewen and G. Wassermann (Springer, Berlin) p. 274.

HORNBOGEN, E., and H. KREYE, 1969b, J. Mater. Sci. **4**, 944.

HORNBOGEN E., and U. KÖSTER, 1978, in: Recrystallization of Metallic Materials, 2nd Ed., ed. F. Haessner (Dr. Riederer-Verlag, Stuttgart) p. 159.

HOTZLER, R.K. and T.K. GLASGOW, 1982, Metallurg. Trans. **13A**, 1665.

HU, HSUN, 1963, Recovery and Recrystallization of Metals (Interscience, New York) p. 311.

HU, HSUN, 1962, Acta Metall. **10**, 112; Trans. Met. Soc. AIME **224**, 75.

HU, HSUN, 1981, in: Metallurgical Treatises, eds. J. Tien and J.F. Elliott (Met. Soc. AIME) p. 385.

HU, HSUN, and C.S. SMITH, 1956, Acta Metall. **4**, 638.

HUBER, J., and M. HATHERLY, 1979, Met. Sci. **13**, 665.

HUBER, J., and M. HATHERLY, 1980, Z. Metallk. **71**, 15.

HUMPHREYS, F.J., 1977, Acta Metall. **25**, 1323.

HUMPHREYS, F.J., 1979a, Acta Metall. **27**, 1801.

HUMPHREYS, F.J., 1979b, Met. Sci. **13**, 136.

HUTCHINSON, W.B., 1974, Met. Sci. **8**, 185.

IN DER SCHMITTEN, W., P. HAASEN and F. HAESSNER, 1960, Z. Metallk. **51**, 101.

INOKUTI, Y., and R.D. DOHERTY, 1977, Texture **2**, 143.

INOKUTI, Y., and R.D. DOHERTY, 1978, Acta Metall. **26**, 61.

INOKUTI, Y., Y. SHIMIZU, C. MAEDA, and H. SHIMANAKA, 1980, in: Recrystallization and Grain Growth of Multi-phase and Particle-Containing Alloys, eds. N. Hansen et al. (Risø National Laboratory, Denmark) p. 71. Fuller account (same authors): Trans. Iron & Steel Inst. Japan 23 (1983) 440.

JONAS, J.J., C.M. SELLARS and W.J. McG. TEGART, 1969, Metallurg. Rev. **14**, 1.

JONES, A.R., 1981, J. Mater. Sci. **16**, 1374.

JONES, A.R., B. RALPH and N. HANSEN, 1979a, Proc. Roy. Soc. **368A**, 345.

JONES, A.R., B. RALPH and N. HANSEN, 1979b, Met. Sci. **13**, 149.

KIMURA, A., R. MADDIN and H. KIMURA, 1964, Acta Metall. **12**, 1167.

KOEHLER, J.S., J.W. HENDERSON and J.H. BREDT, 1957, Creep and Recovery (ASM, Metals Park, OH) 1.

KOHARA, S., M.N. PARTHASARATHI and P.A. BECK, 1958, Trans. AIME **212**, 875.

KOOY, C., 1962, Science of Ceramics (Academic, London) p. 21.

KOPPENAAL, T.J., and M.E. FINE, 1961, Trans. Met. Soc. AIME **221**, 178; J. Appl. Phys. **32**, 1781.

KREYE, H., and E. HORNBOGEN, 1970, J. Mater. Sci. **5**, 1.

KRONBERG, M.L., 1955, Science **122**, 599.

KRONBERG, M.L., and H.F. WILSON, 1949, Trans. AIME **185**, 501.

LACOMBE, P., and L. BEAUJARD, 1947, J. Inst. Metals **74**, 1.

LAWLEY, A., A.E. VIDOZ and R.W. CAHN, 1961, Acta Metall. **9**, 287.

LESLIE, W.C., J.T. MICHALAK and F.W. AUL, 1963, Iron and its Dilute Solid Solutions (Interscience, New York) p. 119.

LI, J.C.M., 1961, J. Appl. Phys. **32**, 525.

LI, J.C.M., 1966, in: Recrystallization, Grain Growth and Textures, ed. H. Margolin (ASM, Metals Park, OH) p. 45.

LIEBMANN, B., and K. LÜCKE, 1956, Trans. AIME **209**, 427.

LOMBRY, R., C. ROSSARD and B. THOMAS, 1980, in: Recrystallization and Grain Growth of Multi-phase and Particle-Containing Alloys, eds. N. Hansen *et al.* (Risø National Laboratory, Denmark), p. 257.

LOUAT, N., 1981, Acta Metall. **30**, 1291.

LÜCKE, K., 1963, Ecrouissage, Restauration, Recristallisation (Presses Universitaires de France, Paris) p. 1.

LÜCKE, K., and K. DETERT, 1957, Acta Metall. **5**, 628.

LÜCKE, K., and R. RIXEN, 1968, Z. Metallk. **59**, 321.

LÜCKE, K., and H.P. STÜWE, 1961, Recovery and Recrystallization of Metals (Interscience, New York) p. 131.

LÜCKE, K., H. PERLWITZ, and W. PITSCH, 1964, Phys. Stat. Sol. **7**, 733.

LUTON, M.J., and C.M. SELLARS, 1969, Acta Metall. **17**, 1033.

MASING, G., and J. RAFFELSIEPER, 1950, Z. Metallk. **41**, 65.

MATTHIES, M., 1979, Phys. Stat. Sol. (b) **92**, K135.

MAY, J.E., and D. TURNBULL, 1958, Trans. AIME **212**, 769.

McLEAN, D., 1957, Grain Boundaries in Metals (Oxford Univ. Press, Oxford) p. 44 and 87.

McLEAN, M., and H. MYKURA, 1965, Acta Metall. **13**, 1291.

McQUEEN, H.J., 1977, Metallurg. Trans. **8A**, 807.

McQUEEN, H.J., and J.J. JONAS, 1975, in: Treatise on Materials Science and Technology, vol. 6, ed. R. Arsenault (Academic, New York) p. 393.

MECKING, H., and G. GOTTSTEIN, in: Recrystallization of Metallic Materials, 2nd Ed., ed. F. Haessner (Dr. Riederer-Verlag, Stuttgart) p. 195.

MEE, P.B., 1968, Trans. Met. Soc. AIME **242**, 2155.

MIDDLETON, A.B., L.B. PFEIL and E.C. RHODES, 1949, J. Inst. Metals **75**, 595.

MIKI, I., and H. WARLIMONT, 1968, Z. Metallk. **59**, 408.

MILLER, W.A., and W.M. WILLIAMS, 1964, J. Inst. Metals **93**, 125.

MONTUELLE, J. 1955, Compt. Rend. Acad. Sci. (Paris) **241**, 1304.

MOULD, P.R., and P. COTTERILL, 1967, J. Mater. Sci. **2**, 241.

MUGHRABI, H., 1977, in: Surface Effects in Crystal Plasticity, eds. R.M. Latanision and J.F. Fourie (Noordhoff, Leyden).

MULLINS, W.W., 1956, Acta Metall. **4**, 431.

NAGASHIMA, S., ed., 1981, Proc. 6th Int. Conf. on Textures of Materials (Iron and Steel Institute of Japan).

NES, E., 1980, in: Recrystallization and Grain Growth of Multi-phase and Particle-Containing Alloys, eds. N. Hansen *et al.* (Risø National Laboratory, Denmark) p. 85.

OBLAK, J.M., and W.A. OWCARSKI, 1968, Trans. Met. Soc. AIME **242**, 1563.

OLSON, G.B., H.C. LING, J.S. MONTGOMERY, J.B. VANDER SANDE and M. COHEN, 1982, in: Rapidly Solidified Amorphous and Crystalline Alloys, ed. B.H. Kear (North-Holland, Amsterdam) p. 355.

PERRYMAN, E.C.W., 1954, Acta Metall. **2**, 26.

PERRYMAN, E.C.W., 1957, Creep and Recovery (ASM, Metals Park, OH) p. 111

PHILLIPS, V.A., 1966, Trans. Met. Soc. AIME **236**, 1302.

PHILLIPS, V.A., 1967, Trans. Met. Soc. AIME **239**, 1955.

RAE, C., C.R.M. GROVENOR and K.M. KNOWLES, 1981, Z. Metallk. **72**, 798.

RALPH, B., C. BARLOW, B. COOKE and A. PORTER, 1980, in: Recrystallization and Grain Growth of Multi-phase and Particle-containing Alloys, eds. N. Hansen *et al.* (Risø National Laboratory, Denmark) p. 229.

RATH, B.B. and HSUN HU, 1972, in: The Nature and Behaviour of Grain Boundaries, ed. Hsun Hu (Plenum, New York) p. 405.

RATH, B.B., R.T. LEDERICH, C.F. YOLTON and F.H. FROES, 1979, Metallurg. Trans. **10A**, 1013.

RAY, R.K., B.W. HUTCHINSON and B.J. DUGGAN, 1975, Acta Metall. **23**, 831.

RHINES, F.N., and B.P. PATTERSON, 1982, Metallurg. Trans. **13A**, 985.

RICHARDS, C.E., and T.W. WATSON, 1969, J. Iron and Steel Inst. **207**, 582.

ROBINSON, P.M., and P.N. RICHARDS, 1965, Phil. Mag. **11**, 407.

RÖNNPAGEL, D., and C.H. SCHWINK, 1978, Acta Metall **26**, 319.

RYAN, N.E., 1967, Metallurgy Report ARL/Met 64 (Aeronautical Research Laboratories, Australia).

SCHMIDT, W., K. LÜCKE and J. POSPIECH, 1974, in: Texture and Properties of Materials, eds. G.J. Davies *et al.* (The Metals Society, London).

SCHÖNBORN, K.H., and F. HAESSNER, 1982, Z. Metallk. **73**, 739.

SCHOFIELD, T.H., and A.E. BACON, 1961, Acta Metall. **9**, 653.

SHEWMON, P.G., and W.M. ROBERTSON, 1963, Metal Surfaces (ASM, Metals Park, OH) p. 67.

SHOCKLEY, W., and T.W. READ, 1950, Phys. Rev. **78**, 275.

SINGER, R.F. and GESSINGER, G.H., 1982, Metallurg. Trans. **13A**, 1463.

SLAKHORST, J.W.H.G., 1975, Acta Metall. **23**, 301.

SMITH, C.S., 1948, Trans. AIME **175**, 151.

SMITH, C.S., 1964, Met. Rev. **9**, 1.

SMITH, J.H., and M.B. BEVER, 1968, Trans. Met. Soc. AIME **242**, 880.

STIEGLER, J.O., C.K.H. DUBOSE, R.E. REED and C.J. MCHARGUE, 1963, Acta Metall. **11**, 851.

TAGUCHI, S., and A. SAKAKURA, 1969, J. Appl. Phys. **40**, 1539.

TALBOT, J., P. ALBERT and G. CHAUDRON, 1957, Compt. Rend. Acad. Sci. (Paris) **244**, 1577.

TANNER, L.E., and I.S. SERVI, 1966, Mater. Sci. Eng. **1**, 153.

THOMPSON, A.W., 1977, Metallurg. Trans. **8A**, 833.

THORNTON, P.H., 1968, in: Techniques of Metal Research, ed. R.F. Bunshah (Interscience, New York) vol. 1, part 2, p. 1069.

THORNTON, P.H., and R.W. CAHN, 1961, J. Inst. Metals **89**, 455.

TIEDEMA, T.J., W. MAY and W.G. BURGERS, 1949, Acta Cryst. **2**, 151.

TWISS, R.J., 1977, Pure and Appl. Geophys. **115**, 227.

VAIDYA, W.V., and K. EHRLICH, 1983, J. Nucl. Mater. **113**, 149.

VANDERMEER, R.A., and P. GORDON, 1959, Trans. AIME **215**, 577.

VANDERMEER, R.A., and P. GORDON, 1963, Recovery and Recrystallization of Metals (Interscience, New York) p. 211.

VARIN, R.A., and K. TANGRI, 1980, Scripta Metall. **14**, 337.

VERBRAAK, C.A., and W.G. BURGERS, 1957, Acta Metall. **5**, 765.

VISWANATHAN, R., and C.L. BAUER, 1973a, Acta Metall. **21**, 1099.

VISWANATHAN, R., and C.L. BAUER, 1973b, Metallurg. Trans. **4**, 2645.

WALTER, J.L., 1969, in: Textures in Research and Practice, eds. J. Grewen and G. Wassermann (Springer, Berlin) p. 227.

WALTER, J.L., and C.G. DUNN, 1959, Acta Metall. **7**, 424; J. Metals **11**, 599.

WALTER, J.L., and G.F. KOCH, 1963, Acta Metall. **11**, 923.

WARLIMONT, H., 1979, Proc. ICSMA **5**, 1055.

WASSERMANN, G., 1963, Z. Metallk. **54**, 61.

WASSERMANN, G., and J. GREWEN, 1962, Texturen Metallischer Werkstoffe (Springer, Berlin). (See also Dillamore and Roberts in *Further reading* list.)

WEBSTER, D., 1968, Trans. Met. Soc. AIME **242**, 640.

WILBRANDT, P.-J., 1980, Phys. Stat. Sol. (a) **61**, 411.

WILBRANDT, P.-J. and P. HAASEN, 1980, Z. Metallk. **71**, 273 and 385.

Further reading

Burgers, W.G., in: The Art and Science of Growing Crystals, ed. J.J. Gilman (Wiley, New York, 1957). (The older work is summarized here.)

Haessner, F., ed., Recrystallization of Metallic Materials, 2nd Ed. (Dr. Riederer-Verlag, Stuttgart, 1978). (The most up-to-date survey of the whole field.)

Cotterill, P., and P.R. Mould, Recrystallization and Grain Growth in Metals (Wiley, New York, 1976).

Maddin, R., ed., Creep and Recovery of Metals (ASM, Metals Park, OH, 1957).

Martin, J.W., and R.D. Doherty, Stability of Microstructure in Metallic Systems (Cambridge University Press, 1976) ch. 3: Instability due to strain energy.

Dillamore, I.L., and W.T. Roberts, Preferred Orientations in Wrought and Annealed Metals, Met. Rev. **10** (1965) 271.

Hatherly, M., and W.B. Hutchinson, An Introduction to Textures in Metals (Monograph No. 5; Institution of Metallurgists, London, 1979).

Margolin, H., ed., Recrystallization, Grain Growth and Textures (seminar volume, ASM, Metals Park, OH, 1966).

Sellars, C.M., and W.J.McG. Tegart, Hot-Workability of Metals, Internat. Met. Rev. **17** (1972) 1.

McQueen, H.J., and J.J. Jonas, chapter on dynamic recovery and recrystallization, in: Treatise on Materials Science and Technology,, ed. R. Arsenault (Academic, New York, 1975) vol. 6.

CHAPTER 26

MAGNETIC PROPERTIES OF METALS AND ALLOYS

F.E. LUBORSKY and J.D. LIVINGSTON

Corporate Research and Development,
General Electric Company
Schenectady, NY 12301, USA

G.Y. CHIN

Bell Telephone Laboratories
Murray Hill, NJ 07974, USA

R.W. Cahn and P. Haasen, eds.
Physical Metallurgy; third, revised and enlarged edition
© *Elsevier Science Publishers BV, 1983*

1. Origins of fundamental magnetic properties

In general, there are three kinds of magnetic effects: diamagnetism, paramagnetism, and cooperative magnetism. Cooperative magnetism includes the most important phenomena we have to describe in this chapter, e.g., ferromagnetism, antiferromagnetism and ferrimagnetism. The principal cause of magnetism is to be found in the interactions of electrons with magnetic fields and interaction of electrons with each other. Apart from its orbital motion, the electron possesses a spin which is equivalent to a magnetic moment of the electron itself. Electrons are arranged in energy states of successive order, and for each energy state there can only be two electrons, which must be of opposite spin. In atoms with closed shells of electrons, therefore, the spin magnetism is entirely self-cancelling and the application of a magnetic field has the effect of distorting the electron orbits so that the internal magnetic field remains unchanged. This effect of realignment of electron orbits is very small and is called *diamagnetism*. The substance behaves as though an internal rearrangement opposing the external magnetic field took place. Although this diamagnetism can be used for the examination of materials, it is not of great importance in this connection. It must, however, be noted that the effect is fundamental and occurs in all substances, i.e., diamagnetism is superimposed on any paramagnetic or ferromagnetic effect, but can normally be neglected in comparison.

If within the atomic shells there are some shells which contain electrons whose spin is not compensated by others of corresponding opposite spin, as for instance in the d-shell of the transition metals, we can have atoms which carry a resultant magnetic moment, μ, due to unpaired or uncompensated electron spins (μ is measured by Bohr magnetons). Such atoms are called magnetic atoms, and a dilute assembly of such atoms when exposed to a magnetic field shows a certain degree of magnetic orientation of the atomic magnetic moments. This orientation is opposed by thermal agitation and the law governing the magnetization, M, of the assembly as a function of the strength of the applied magnetic field H is known as the *Langevin law:*

$$M = N\mu \coth\frac{\mu H}{kT} - \frac{kT}{\mu H},\tag{1}$$

where N is the number of magnetic carriers per cm^3 and k and T have their conventional significance. For low values of $\mu H/kT$ the change of magnetic moment with field is linear ($M = N\mu^2 H/3kT$) and the *magnetic susceptibility*, which is the proportionality constant between field and magnetization, is a constant (*Curie law*):

$$\frac{M}{H} = \chi = \frac{N\mu^2}{3kT} = \frac{\text{const.}}{T}.\tag{2}$$

If we apply very high magnetic fields, this *paramagnetic* magnetization reaches a saturation value which is equal to $N\mu$, the total sum of all the magnetic moments of the carriers. In dilute assemblies, the paramagnetic susceptibility can serve as a measure of the magnetic moment of the individual carriers and can thus give information about the magnetic properties of the atoms of which the assembly is composed. However, in solids containing many paramagnetic atoms or ions there are usually interactions of the magnetic electrons with either the magnetism arising from neighboring atoms or with the electrostatic fields arising from neighboring atoms. These influences disturb the ideal behaviour of the magnetic carriers, and although magnetic effects of a similar nature to those described for the dilute assembly occur, these effects can no longer be used to give simple information about the number of magnetic electrons per atom. Thus, in general, the paramagnetic susceptibility, whilst giving useful information about the substance, is not a direct measure of the metallurgically important factors contributing to the material structure. However, discontinuities which occur in the paramagnetic susceptibility as the composition of a system is systematically changed are of value in the examination of metals and alloys.

If the magnetic atoms are in sufficiently close contact with each other so that the magnetic electrons can exchange between neighboring magnetic atoms, a cooperative phenomenon may occur which spontaneously aligns the spins of all the magnetic carriers in the lattice and binds their moments very strongly. This spontaneous magnetization is characteristic for *ferromagnetic* materials. The spins of neighboring magnetic atoms are aligned by *exchange forces* which are equivalent to magnetic fields of the order of $8-80 \times 10^7$ A/m (1–10 million Oe). However, the coupling is not itself magnetic but is due to a quantum-mechanical interaction between the electrons of neighbouring atoms. The exchange interaction between magnetic atoms may often be indirect and take place via an intermediate nonmagnetic atom such as oxygen or sulphur. Under some circumstances, the coupling between magnetic atoms in a substance can cause antiparallel alignment of the spins of neighboring atoms. This is called *antiferromagnetism*. Thus we have substances which consist of two interpenetrating lattices of similar magnetic ions or atoms with magnetizations in exactly opposite directions; these cancel out so that there is no resultant magnetic moment. Finally, the two interpenetrating lattices can have antiparallel magnetizations of unequal magnitude so that there remains a resultant magnetic moment in the direction of the stronger magnetic sub-lattice. This uncompensated antiferromagnetism has been called *ferrimagnetism*.

The saturation magnetization of ferromagnetic materials depends on the number of uncompensated spins in the magnetic atoms. In the ferromagnetic metals – iron, cobalt and nickel – the magnetic moment per atom that is measured is 2.2, 1.7 and 0.6 spin moments (electron units or Bohr magnetons), respectively. The very simple rigid band model (ch. 4, §5.1) calculation gives values of 2.6, 1.6 and 0.6. The electronic band theory of magnetism is discussed in some detail in ch. 3, §8.

Ferromagnetic and ferrimagnetic materials are characterized by their high saturation magnetization, M_s, which decreases slowly as the temperature is increased and fairly abruptly disappears at a characteristic temperature known as the *Curie temperature* T_c (fig. 1). The ferrimagnetic substances behave in most respects very similarly to the ferromagnetic substances in showing field-independent magnetization and a definite Curie point. (In what follows the term ferromagnetism will be

References: p. 1730.

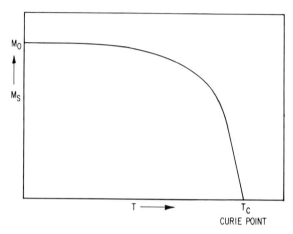

Fig. 1. Variation of saturation magnetic moment M_s with temperature for a ferromagnetic substance.

used for both ferro- and ferrimagnetism.) Antiferromagnetics behave in many respects like paramagnetic substances because the two oppositely directed magnetic lattices compensate each other, showing only a small external magnetic effect which depends on the applied magnetic field. At a given temperature, however, the antiferromagnetic coupling is lost and this is accompanied by an anomaly in the specific heat and by an anomaly in the magnetic susceptibility. The temperature at which antiferromagnetic substances lose their antiparallel alignment and become paramagnetic is called the *Néel point*.

In the paramagnetic region, for a ferromagnet, the variation of susceptibility χ with temperature follows a *Curie–Weiss law* which contains the Curie temperature T_c as the additional constant, $\chi = C/(T - T_c)$. For an antiferromagnetic material above the Néel temperature, T_N, χ is instead given by $\chi = C/(T + T_N)$.

The exchange interaction between the electrons of neighboring magnetic atoms in ferromagnetic and ferrimagnetic materials causes the individual magnetic moments of all atoms in such materials to be aligned, and the material possesses a *spontaneous magnetization* at zero field, M_s. This is in contradiction with the observation that even ferromagnetic materials do not show external polarization under normal circumstances. This apparent paradox is resolved by the fact that a ferromagnet is always subdivided into a number of microscopic *domains*. Inside the domains the magnetization that prevails is equal to M_s but the domains are oriented in different directions so that externally their magnetic moments cancel and no magnetization is observed outside the body. *Domain walls*, i.e., regions which separate domains of different directions of magnetization from each other, can be observed by a variety of methods, and the study of the spontaneous magnetization M_s can give interesting structural information about the materials. The position and density of domain walls as well as their special features can again be of great importance to the study of the structure of materials. The temperature at which the energy of thermal agitation is sufficiently large to overcome the effect of the exchange interaction is the Curie temperature, T_c. The saturation magnetization M_s and the Curie temperature are not dependent on structural imperfections of the material but are characteristic of the atomic constitution and therefore can be used to measure or determine that constitution. In the examination of metals and alloys, observation of M_s and its variation with temperature can be used for phase diagram investigations.

The exchange interaction between electrons of neighboring atoms which causes the magnetization of all magnetic atoms to be aligned is the principal contribution to the magnetic energy of a specimen. In addition, for crystalline materials the ease with which the magnetization to saturation is achieved differs as we magnetize along different crystal directions (fig. 2). Thus the measurement of this *magnetocrystalline anisotropy energy* can be a sensitive test of the crystal structure. The magnetocrystalline anisotropy energy changes with temperature and, of course, at the Curie point the anisotropy disappears together with the magnetization.

An elastic deformation of the lattice will additionally exert an influence on the magnetization and its symmetry. This effect is called *magnetostriction* and can be expressed as the relation between elastic constants of a material and the strain dependence of the anisotropy energy. The magnetostriction is direction-dependent and can be expressed as a harmonic function of the directions of both the magnetization and the crystal symmetry. The constants of this harmonic series are called *magnetostriction constants* and an accurate analysis of magnetostriction as a function of both field direction and temperature can help in the investigation of structural properties of magnetic materials.

An example of a ferromagnetic material divided into domains to produce no net magnetization is shown at the bottom of fig. 3. Subdivision into domains removes most of the *magnetostatic energy* otherwise associated with surface magnetic poles. The boundaries between domains with different directions of magnetization, called *domain walls*, have a finite thickness and surface energy dependent primarily on the strengths of the exchange interaction and the magnetocrystalline anisotropy energy. For given specimen shape and dimensions, equilibrium domain sizes are determined by a balance between magnetostatic and domain-wall energies. Typical domain sizes

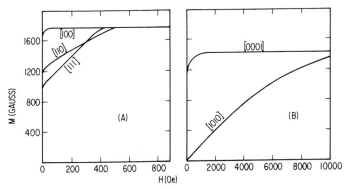

Fig. 2. Magnetization along different crystallographic axes in single crystals of (a) iron and (b) cobalt.

References: p. 1730.

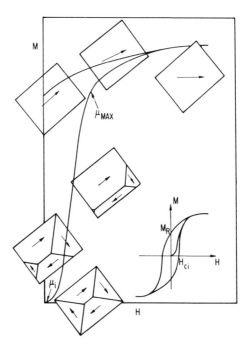

Fig. 3. Schematic diagram of the magnetization process and corresponding changes in domain configuration of a single crystal (inset: hysteresis loop).

range from several μm to several mm. Domain-wall surface energy increases with increasing exchange and anisotropy constants. Domain-wall width increases with increasing exchange constant, but decreases with increasing anisotropy constant, ranging from several hundred nm in low-anisotropy materials down to interatomic distances in very high-anisotropy materials.

Let us now describe the conditions in a magnetic material in low fields where the domain walls play an important role. At first, consider a single crystal. In the demagnetized state, the specimen is sub-divided into domains, each domain magnetized along one of the easy directions of magnetization and the various domains separated by domain walls. The external demagnetizing energy and the domain wall energy as well as the contributions from magnetocrystalline and magnetostrictive energy are minimized by the existence of domain walls. An example of such a demagnetized specimen is shown in fig. 3 where the magnetization process is shown schematically. Upon the application of a magnetic field, initially domain walls will tend to move reversibly, so that the domains which are magnetized with a component in the direction of the field grow at the expense of domains magnetized in opposite directions. The increase in magnetization with field in this region is reversible and the *initial permeability* $\mu_i = (dB/dH)_{H=0}$ and the *initial susceptibility* $\chi_0 = (dM/dH)_{H=0}$. Note that this curve is often plotted as B versus H rather than

M versus H, where $B = 4\pi M + H$. Continuing to apply an increasing field the domain walls begin to move from their original positions irreversibly into new positions. Usually in this state the increase in magnetization with field is steeper and reaches a maximum $B/H = \mu_{max}$. Finally, the domain walls have largely moved out of the specimen and the specimen is magnetized in that easy direction which is nearest to that of the applied field; further increase of the field has only the result of rotating the direction of magnetization until it finally coincides with the field direction. This final rotational change is reversible and the final state is the saturated magnetic state.

Reduction and subsequent reversal of the field produces the *hysteresis curve*, as shown by the inset in fig. 3. The magnetization left in a specimen when the applied field has been removed after full saturation is called *residual magnetization* (M_r) or sometimes *remanence*, and the field required after full saturation to reduce the magnetization M to zero is called the *intrinsic coercivity*, H_{ci}, while the field required to reduce B to zero is called simply the *coercivity*, H_c. Another quantity, commonly used in permanent-magnet technology, is the *maximum energy product*, $(BH)_m$, the maximum of $B \cdot H$ in the second quadrant of the $B-H$ hysteresis curve. The various parameters of the $B-H$ hysteresis curve – χ, M_r, H_c and $(BH)_m$ – are very much dependent on the structural properties of the material and can be used in some instances to obtain information about crystallinity, internal stresses, inclusions, etc., in magnetic materials.

In soft materials, the *core loss* (§ 2.7), or energy loss associated with each cycle of magnetization (as in a transformer core) is of interest. The loss is traditionally separated into two components: a *hysteresis loss* proportional to frequency, and an *eddy-current loss* proportional to the square of the frequency. The latter is, of course, most important in the higher frequency range. It can be reduced by reducing the physical dimensions, for example by using thin films or small particles. Another approach is to increase the resistivity, for example by using ferrites instead of metals. Magnetic aftereffect also gives rise to losses especially at higher frequency. *Magnetic aftereffect* is the delayed change in magnetization after a change in magnetic field. This does not include delays due to eddy currents, or to changes in structure. This aftereffect may arise from many different physical processes; for example, it is due to ion diffusion in ferrites occurring at frequencies of a few Hz or to electron diffusion between Fe^{+2} and Fe^{+3} at a frequency of several hundred MHz.

2. *Magnetic measurements*

This discussion is limited to strongly magnetic materials, especially materials with engineering applications with emphasis on the physical principles involved and on the relative ease and accuracy of the various methods. For more details see CULLITY [1972] or ZIJLSTRA [1967].

References: p. 1730.

2.1. Magnetization

The maximum value of the magnetic moment per unit volume, or *magnetization*, is called the *saturation magnetization, M_s*; it is temperature-dependent, and its value at absolute zero is M_0. In SI units, magnetization may be measured in A/m, M, or in tesla, J. In this convention, $J = \mu_0 M$, where μ_0 is the magnetic constant or the permeability of free space, and J is known as polarization. In cgs units, magnetization is measured in erg oersted/cm^3, or emu/cm^3. (Both kinds of units are used in this chapter, reflecting current practice among specialists.) Absolute measurements of magnetization are notoriously difficult, so that almost always the measuring apparatus is calibrated with a standard sample of nickel or iron.

There are two basic methods for determining magnetization. The first depends on the fact that a magnetic moment experiences a force in a field gradient: $F = M$ grad(H). The resulting force is measured with a chemical balance or equivalent force detector. This method is especially useful for measurements on very small or very weakly magnetic samples; it is also good for measurements at high and low temperatures, since the only connection to the sample is a single supporting wire.

The second basic method relies on the voltage generated in a coil by a changing magnetic flux: $E = -N \, d\phi/dt$, giving the magnetic moment per unit volume or equivalently lines of magnetic flux per unit area. This voltage, proportional to the time derivative of the flux in the sample, can be integrated to give the change in flux. Its disadvantages are that only changes in flux can be measured, not the magnetization in constant field; and the measured quantity is actually the sum of the sample flux plus the flux from the field acting on the sample. At high fields, the flux from the applied field is large, so the method is usually used for materials that do not require high magnetizing fields.

Measurement of the field at a fixed distance from the sample can also be used to measure the moment of the sample. The sample is normally oscillated at a fixed frequency (usually below 100 Hz) through a small amplitude. The field at a coil fixed in space near the sample then varies at the frequency of the oscillation, producing an ac voltage in the coil. The ac signal is amplified, normally with a lock-in amplifier, and converted to a dc signal which is proportional to the magnetic moment of the sample. This *vibrating-sample magnetometer* can be made very sensitive, and can operate over a wide temperature range. It is limited to small samples.

A problem that may be severe in high-field measurements is the *image effect*. The local magnetic field around a magnetized sample is distorted if there are nearby masses of high-permeability material, such as the pole pieces of an electromagnet. At high fields, the iron of the pole pieces becomes saturated so that its permeability decreases and the image weakens. The calibration constant relating the sample magnetization to the output signal will thus vary with field. A similar problem exists in a superconducting solenoid. The only solution is to calibrate with a sample whose moment is known as a function of field.

2.2. Magnetic field

The most convenient way to measure magnetic fields is with a *Hall-effect probe*. This instrument gives a dc signal proportional to the field component perpendicular to a semiconductor plate. Fields less than the earth's field (25 A/m ≈ 0.3 oersted) and up to the highest steady fields attainable in the laboratory (8 MA/m ≈ 100 kOe) can be measured, with an accuracy generally of the order of 1%.

Time-varying fields can be measured easily by observing the output voltage from a coil in the field. In practice, this is most useful for sinusoidally varying fields. Steady fields can be measured with a coil by rotating the coil to generate an ac voltage. This device, known as a *rotating-coil gaussmeter,* provides a signal that is accurately linear with field, but the equipment is bulky and inconvenient for routine use.

Very accurate field measurements can be made using nuclear magnetic resonance. The resonance frequency of a nucleus, usually hydrogen, is directly proportional to the applied field, and the proportionality constant is known with great accuracy. The field, however, must be highly uniform and stable, so this method is mainly used for the calibration of other field-measuring instruments.

2.3. Demagnetizing field

Any magnetized sample not infinitely long sets up a field inside the sample as well as outside it. In the sample it is always directed opposite to the magnetization. It is therefore called the *demagnetizing field*.

There are four basic ways to deal with the demagnetizing field. First, one can try to make it negligibly small, by using a very long sample, or, more commonly, a ring-shaped sample. Second, the demagnetizing field can in principle be calculated if the magnetization is known and the sample is ellipsoidal in shape, on the assumption that the magnetization is uniform throughout the sample. Ellipsoidal samples are difficult to make, so often a simpler shape, such as rod or disk, is assumed to be ellipsoidal. Third, one can attempt to measure the field as near as possible to the surface of the sample. Since the tangential component of the field does not change across the surface, this measurement gives the true field acting on the sample, including the demagnetizing field. Finally, one can determine the demagnetizing field experimentally. A sample of the same shape as the real sample, but made from a material with negligibly small coercivity, is measured. The demagnetizing field is equal and opposite to the applied field and the sample magnetization is linear with applied field. The slope of this line is inversely proportional to the demagnetizing factor, which should be the same for all samples of the same shape.

2.4. Curie temperature

The Curie temperature is the temperature at which ferromagnetic behavior disappears. Its approximate value can be obtained from the temperature dependence

References: p. 1730.

of the magnetization measured at any field. At high fields, however, the transition is often not sharply defined, so that low-field determination is more accurate, or at least more reproducible.

2.5. Magnetic anisotropy

Magnetic anisotropy is expressed in terms of a series of *anisotropy constants*. Magnetization curves measured with the field applied in different crystallographic directions can be interpreted to give anisotropy data, but the best method is to use a spherical or disk-shaped crystal and measure the torque exerted on the sample by a magnetic field as a function of the angle between the field and a major crystallographic direction. The major difficulty is that the measured anisotropy constants usually depend slightly on the field, and the proper extrapolation is not clear on theoretical grounds. The dependence of the ferromagnetic resonance frequency on crystallographic direction can also be interpreted to give anisotropy data.

2.6. Magnetostriction

Magnetostriction refers to changes in dimensions with state of magnetization. The term is applied to two rather different phenomena. In engineering usage, magnetostriction means the change in length when a demagnetized polycrystalline sample is magnetized. Scientifically, magnetostriction refers to the change in dimensions resulting from the rotation of the saturation magnetization from one crystallographic direction to another in a single crystal. The strains are usually small, typically 10^{-5}, and the standard measuring technique makes use of resistance strain gages. For routine ac measurements on commercial materials, where long samples are available, phonograph pickups or other displacement transducers can be used.

2.7. Core loss

The principal quantity of interest for soft magnetic materials is the *power loss* under ac excitation. This is specified at a particular operating frequency, at a particular maximum flux density. The sample is usually made into the core of a small transformer; if the material is available in sheet form, a standard test geometry called the *Epstein test* is used. The procedure is equivalent to determining the area of the ac hysteresis loop, which can now also be done by digital data recording and computation. (See also § 4.2.3.)

3. Permanent-magnet materials

Permanent-magnet or "hard" magnetic materials, as their name implies, strongly resist demagnetization once magnetized. As used in motors, loudspeakers, meters, holding magnets, etc., permanent-magnet materials usually have coercivities ranging

Table 1
Representative permanent-magnet properties.

Material	H_c		B_r		$(BH)_m$	
	(kA/m)	(Oe)	(T)	(G)	(kJ/m³)	(MGOe)
ESD * Fe–Co	70	870	0.8	8 000	25	3.2
Alnico 5	58	620	1.25	12 500	42	5.3
8	130	1 600	0.83	8 300	40	5.0
9	120	1 450	1.05	10 500	68	8.5
$Fe_{65}Cr_{32}Co_3$	40	500	1.25	12 500	34	4.3
$Fe_{63}Cr_{25}Co_{12}$	50	630	1.45	14 500	61	7.7
Cunife	44	550	0.54	5 400	12	1.5
Co_5Sm	760	9 500	0.98	9 800	190	24
$(Co, Fe, Cu, Zr)_8Sm$	800	10 000	1.20	12 000	260	33
Ba ferrite	170	2 100	0.43	4 300	36	4.5
Sr ferrite	250	3 100	0.42	4 200	36	4.5
Mn–Al–C	220	2 700	0.61	6 100	56	7.0
Co ~ Pt	360	4 500	0.65	6 500	73	9.2

* Elongated Single Domain (§ 3.3.1).

from about 10 kA/m to over 100 kA/m (several hundred to many thousand Oersteds). Properties of some representative materials are shown in table 1. More detailed data and information have been presented by McCaig [1977]. Some applications require coercivities in the 800–8000 A/m (10–100 Oe) range, and such materials are sometimes called "semi-hard" materials. Magnetic-recording materials are permanent-magnet materials whose direction of magnetization can be varied over short distances, thereby recording information. These have typically consisted of dispersions of magnetic oxide particles, but there has been growing interest in metallic recording media.

3.1. Reversal mechanisms and coercivity

To understand the connection between coercivity and metallurgical microstructure, it is necessary to consider the microscopic mechanisms of magnetization reversal. Consider a region of magnetic material initially magnetized along an "easy" or low-energy direction and then subjected to a reverse magnetic field. Figure 4 shows several alternative mechanisms whereby the material can reverse its magnetization. These include *coherent rotation* of all the atomic moments (fig. 4a), *non-coherent rotation* modes such as curling (fig. 4b), and *nucleation and growth of reverse domains* (fig. 4c). Coercivity will be determined by the easiest of these reversal mechanisms.

To impede coherent rotation, the material must possess magnetic anisotropy that will provide an energy barrier to rotation of the magnetization. One possibility is the shape anisotropy of non-spherical particles. A rod-shaped particle has a much lower demagnetizing energy (magnetostatic energy associated with surface magnetic poles)

References: p. 1730.

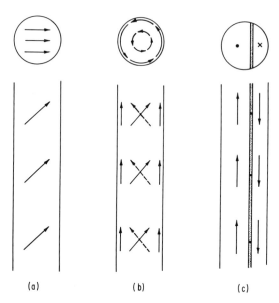

Fig. 4. Schematic representation of magnetization reversal mechanisms. Top and side views of cylindrical region changing from upward to downward magnetization: (a) by coherent rotation; (b) by incoherent rotation (curling); (c) by domain-wall motion.

when magnetized along its length than when magnetized across its width (because the demagnetizing field is lower). The resulting energy barrier can be overcome by coherent rotation only for a reverse field given by $(N_t - N_l)M_s$, where N_t and N_l are the transverse and longitudinal demagnetizing factors, respectively. In the limit of infinite length/width ratio, $N_t = 2\pi$ and $N_l = 0$, leading to a predicted coercivity of $2\pi M_s$. A second possibility to provide the energy barrier is *crystal anisotropy*. The simplest case is that of a *uniaxial* crystal anisotropy given by an energy per unit volume of $K \sin^2\theta$, where θ is the angle between the magnetization and the low-energy crystal direction. For this case, a coercivity of $2K/M_s$ is predicted for coherent rotation. Real magnetic materials generally have coercivities far below those predicted for coherent rotation, indicating that other reversal mechanisms are dominant.

Non-coherent rotation modes such as curling (fig. 4b) can significantly lower the energy barrier associated with *shape* anisotropy by lowering the transverse demagnetizing fields. The non-parallel magnetization configurations are opposed by the exchange forces between neighboring moments, but this constraint becomes less important as the size of the independent magnetic regions increases. Thus these theories predict coercivities that decrease with increasing particle size, and curling or other non-coherent rotation modes are believed to control coercivity in some shape-anisotropy materials. However, where *crystal* anisotropy is dominant, non-coherent rotation cannot contribute to lowering the energy barrier or coercivity.

It is now believed that most crystal-anisotropy materials reverse primarily through nucleation and growth of reverse domains. In this process, magnetization rotation is discontinuous, confined to the narrow region within the domain wall. Coercivities for this process can be far lower than for continuous rotation processes, and will be determined by the more difficult of the two necessary steps, reverse-domain nucleation or domain-wall motion. The former becomes more difficult in finer particles, leading to a size-dependence of coercivity when that step is dominant.

In discussing size dependence of coercivity, there are three different size parameters of significance. These are $\delta = \pi(A/K)^{1/2}$, $b_c = 2A^{1/2}/M_s$, and $D_c = 1.4\gamma/M_s^2$, where A is the exchange constant and $\gamma = 4(AK)^{1/2}$ is the domain-wall energy per unit area.

The parameter δ is the width of a domain wall. The parameter b_c is the scaling factor in curling theory, and represents the cylinder diameter below which coherent rotation is favored over curling. It is related to the other parameters through $b_c \approx \frac{1}{2}(\delta D_c)^{1/2}$. The ratio $D_c/\delta \approx 2K/M_s^2$ is a measure of the relative importance of crystal and shape anisotropies. Where $D_c \gg \delta$, the parameter D_c represents the diameter of a sphere below which a single-domain structure is of lower energy at zero field than a two-domain structure. Although particles both above and below D_c may reverse by domain nucleation and growth, domain nucleation becomes increasingly difficult in particles approaching this size range.

In high-anisotropy compounds like Co_5Sm, typical values may be $\delta \approx 5$ nm and $D_c \approx 1$ μm. In pure cobalt, all three size parameters are of the same order of magnitude, about 15–30 nm. In iron and nickel, $\delta > D_c$.

The increase in coercivity with decreasing particle size expected from curling or domain-nucleation theories does not continue indefinitely. For extremely small particles, magnetization rotation over the energy barrier can be thermally activated, leading to *superparamagnetism* (BEAN and LIVINGSTON [1959]) and a decrease of coercivity at small sizes. For a given anisotropy and temperature, there may thus be an optimum particle size for peak coercivity. A more detailed review of coercivity models has recently been prepared by ZIJLSTRA [1983].

3.2. Microstructure and properties

Since dimensions and morphology play such a direct role in reversal mechanisms, coercivity can be extremely sensitive to metallurgical microstructure (LIVINGSTON [1981a, b]). However, satisfactory quantitative correlation of observed microstructure–coercivity relationships and theoretical models of reversal has been achieved only in a few model materials.

Model shape-anisotropy materials have been produced by multiple drawing of compacts of ferromagnetic wires in a nonmagnetic ductile matrix. By drawing down nickel wires in a silver matrix, NEMBACH et al. [1977] were able to demonstrate an increase in coercivity with decreasing wire diameter, consistent with the theory of magnetization reversal by *curling* (fig. 5).

In single-phase materials dominated by crystal anisotropy, coercivity is usually

References: p. 1730.

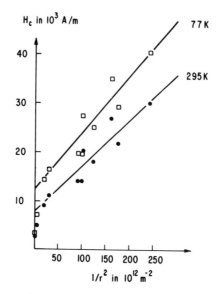

Fig. 5. Variation of coercivity (10^3 A/m = 4π Oe) of drawn composite of Ni filaments in Ag with inverse square of filament radius. Linear dependence predicted by theory of magnetization-reversal by curling. (From NEMBACH *et al.* [1977].)

determined by *reverse-domain nucleation* at defects such as low-anisotropy regions produced by chemical inhomogeneity. Coercivity can be increased by minimizing the size and density of nucleating defects and by subdivision into fine regions in which the probability of such defects is small. Thus, coercivity in such materials generally increases with decreasing particle size or, in fully dense materials, with decreasing grain size. Apparently grain boundaries frequently provide sufficient magnetic isolation from neighboring grains that separate nucleation events are required for magnetic reversal in each grain. Although theories of the size dependence of coercivity based on domain nucleation have been developed (e.g., BROWN [1962], MCINTYRE [1970], ROWLANDS [1976]), quantitative comparison of theory and experiment has been hampered by inadequate characterization of the nucleating defects.

In many hard and semi-hard materials, coercivity is not determined by curling or by domain nucleation, but by microstructural resistance to domain-wall motion, i.e., by *pinning*. Any local variation in magnetic properties can produce local variations in domain-wall energy and thereby produce forces that resist wall motion. The case most studied by theory and experiment is that of domain-wall pinning by second-phase particles, usually nonmagnetic. Optimum pinning is believed to occur for particle sizes comparable with the domain-wall width δ. For particles smaller than δ, a linear dependence of coercivity on particle diameter has been observed for Co_3Ti precipitates in a Co–Fe–Ti alloy (fig. 6), with the experimentally measured slope in good agreement with that predicted by theory (SHILLING and SOFFA [1978]).

For high-anisotropy materials, domain walls become very narrow, and finer

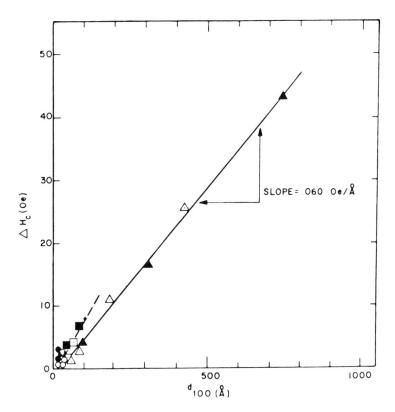

Fig. 6. Variation of coercivity-increment with average diameter of δ' precipitates in Co–Fe–Ti alloy. Linear dependence predicted by theory of domain-wall pinning by nonmagnetic precipitates. (From SHILLING and SOFFA [1978].)

defects become important for pinning. In such cases, planar defects such as grain boundaries and anti-phase boundaries can produce pinning (HILZINGER [1977]), as can point defects and even, in some cases, the discrete nature of the crystal lattice itself (HILZINGER and KRONMÜLLER [1973]). Domain-wall pinning by the lattice structure itself can be viewed as analogous to the Peierls resistance to dislocation motion.

More generally, magnetic "hardening" through domain-wall pinning by defects is analogous in several ways to mechanical hardening by dislocation pinning, (and "hardening" of type II superconductors through vortex pinning) (HAASEN [1972, 1977] and ch. 27, §4.3). In many alloys, thermal or mechanical treatments that lead to increases in coercivity, e.g. by precipitation, also lead to increases in yield stress. However, because of the different dependence of domain-wall pinning and dislocation pinning on defect size, magnetic and mechanical hardness are not always concurrent. For example, solid-solution alloying elements in iron or nickel can produce mechanical hardening without producing magnetic hardening. An extreme

References: p. 1730.

example is that of amorphous alloys, which are mechanically very hard but magnetically very soft.

The general effect of defects depends on which mechanism is controlling coercivity. Defects may *lower* coercivity by serving as nucleation sites for reverse domains, or *raise* coercivity by pinning domain walls. Whereas a defect size of the order of δ is effective for pinning, the defect size effective for reverse-domain nucleation is closer to D_c, i.e., much larger for the case of high-anisotropy materials.

The other magnetic properties are much less sensitive to microstructure than is coercivity. Saturation magnetization depends of course on the volume fraction of the ferromagnetic phase. The remanence-to-saturation ratio depends on the degree of alignment of the easy magnetic axes. Where crystal anisotropy is dominant, it will depend therefore on crystallographic texture. Where shape anisotropy is dominant, it will depend on the alignment of the rod axes. Materials with a high degree of alignment, sometimes called "oriented" magnets, are generally highly anisotropic in properties, with a greatly reduced remanence and energy product in the transverse direction. This is undesirable for some applications, for which "isotropic" magnets, with a random distribution of magnetic easy axes, are more suitable. In such materials, magnetic reversal may occur by domain nucleation and growth or by non-coherent rotation, but full saturation will require higher fields and occurs by coherent rotation from the various easy-axis directions to the field direction.

3.3. Shape-anisotropy materials

Iron and its solid-solution alloys generally have moderate to low crystal anisotropy and usually serve as soft, rather than hard, magnetic materials. However, when processed to produce fine microstructures with elongated morphologies, useful coercivities can be achieved with these materials through shape anisotropy. Commercial permanent-magnet materials based on shape anisotropy include *elongated-single-domain* (ESD) magnets and various alloys, like the alnico series, produced by spinodal decomposition.

3.3.1. ESD magnets

Development of these materials (LUBORSKY [1961]) was inspired by the theoretical predictions of high coercivity ($2\pi M_s$) for coherent rotation in rod-shaped particles. Elongated particles of iron–cobalt alloy about 10–20 nm in diameter are formed by electrodeposition from an aqueous electrolyte into a mercury cathode. Further processing steps include aging at ~ 200°C to remove dendritic branches from the particles, addition of lead–antimony to provide the matrix metal in which particles are dispersed, removal of the mercury, pressing, grinding, etc. For oriented magnets, a magnetic field is applied during the final processing steps to align the particles.

The coercivities and energy products of ESD magnets, although commercially useful, fall far short of the predictions of coherent-rotation theory. The maximum coercivities attained (about 160 kA/m = 2 kOe) are believed to be limited by a non-coherent mode of rotation, influenced by shape irregularities of the particles. Further, coercivity decreases with increasing volume fraction of the iron–cobalt because of magnetic interactions between particles, thereby limiting energy products.

In later work (LUBORSKY and MORELOCK [1964a, b]), vapor-grown whiskers of iron, iron–cobalt, and cobalt, were produced with more perfect shapes than the electrodeposited particles. Higher coercivities were achieved, but these whiskers could not be developed into a commercial product.

3.3.2. Spinodal alloys

Although the important *alnico* series of permanent-magnet alloys was originally developed empirically, it is now understood in terms of fine (~ 20 nm) elongated particles produced by spinodal decomposition (ch. 14, §3.1) of a high-temperature body-centered-cubic (bcc) phase into two coherent bcc phases. The earliest alloys in this series contained only Fe, Ni, and Al, and the decomposition resulted from a miscibility gap that developed at low temperatures between the Fe-rich bcc phase (α_1) and the NiAl-rich ordered bcc phase (α_2). Alloys were later improved by major additions of Co and minor additions of Cu, Ti, and other elements.

The Co raised the saturation magnetization and Curie temperature of the α_1 phase, and made the alloy more susceptible to property improvement through *magnetic annealing*, i.e. annealing in a magnetic field. The phases usually are elongated in $\langle 100 \rangle$ directions, and magnetic annealing causes preferential growth along the $\langle 100 \rangle$ direction nearest the field direction. Most of this preferential alignment appears to occur during the coarsening of the precipitate. The development of $\langle 100 \rangle$ rods from coherent precipitation is believed to be related to lattice-misfit strains and anisotropy in the elastic moduli. This $\langle 100 \rangle$ alignment appears to be accentuated by increases in the Co and Ti content. Alloying additions also alter the volume fractions of the α_1 and α_2 phases and their magnetic properties. For maximum alignment of the microstructure, directional solidification has been used to produce a strong $\langle 100 \rangle$ fibre texture. The relations between microstructure and properties in alnico alloys have been reviewed by DE VOS [1969], PFEIFFER [1969], and McCURRIE [1982].

After initial decomposition of alnico alloys, both α_1 and α_2 phases are ferromagnetic. However, after subsequent lower-temperature aging to increase coercivity, the composition difference between the two phases increases, and the α_2 phase usually becomes nonferromagnetic at room temperature. The microstructure is often modelled as consisting of isolated ferromagnetic α_1 rods in a nonferromagnetic α_2 matrix, and coercivity is interpreted by non-coherent rotation theory, usually curling, modified by interparticle interaction. However, micrographs in some alloys indicate a more complex microstructure, with both α_1 and α_2 phases continuous and interconnected, suggesting the possibility of extended domain walls in the α_1 phase and a coercivity controlled by domain-wall pinning. A similar uncertainty of the operative reversal mechanisms exists in the analogous Fe–Cr–Co alloys.

Guided by the growing understanding of alnico alloys in terms of spinodal decomposition and the Fe–Ni–Al miscibility gap, KANEKO et al. [1971] initiated the

References: p. 1730.

development of a series of permanent-magnet alloys based on the Fe–Cr miscibility gap. Additions of Co were found to raise the decomposition temperatures and increase the composition spread between the Fe-rich α_1 phase and the Cr-rich α_2 phase (fig. 7). Other alloying additions have been used to avoid the formation of the unwanted γ (fcc) and σ phases, and to modify precipitate morphology. Although precipitates are often nearly spherical, $\langle 100 \rangle$ elongated morphology can be enhanced, for example, by Mo additions (HOMMA *et al.* [1980]).

As with the alnicos, magnetic annealing to improve alignment, and subsequent aging at lower temperatures to increase the composition difference between α_1 and α_2 phases, improve the magnetic properties. A major advantage of the Fe–Cr–Co alloys over the alnico series is that most compositions retain some ductility even after spinodal decomposition has occurred. This allows enhancement and alignment of the shape anisotropy by uniaxial plastic deformation, which can produce significant increases in coercivity and energy product (KANEKO *et al.* [1976]). JIN [1979] developed a "deformation-ageing" technique in which this particle elongation is accomplished before the final ageing. The dependence of magnetic properties on this intermediate deformation is shown for one alloy composition in fig. 8.

Another advantage of the Fe–Cr–Co alloys over the alnicos is that they can achieve comparable magnetic properties at lower percentages of cobalt (HOMMA *et al.* [1981], CHIN *et al.* [1981]). Recent price and supply problems with cobalt have made this an important consideration. Work has also progressed on other elements that raise the Fe–Cr miscibility gap, and promising Fe–Cr–V alloys have recently been reported (INOUE and KANEKO [1981]).

In some Fe–Cr–Co alloys at peak coercivity, both α_1 and α_2 phases appear to be

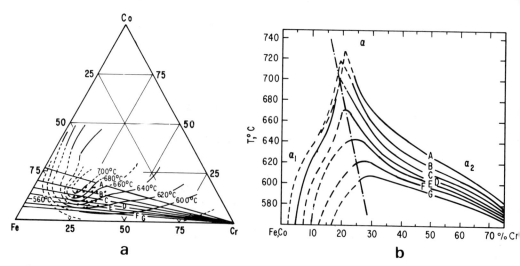

Fig. 7. Miscibility gap in Fe–Cr–Co system: (a) at various temperatures; (b) at various Co/Fe ratios. (From MINOWA *et al.* [1980].)

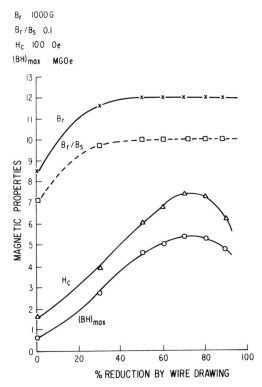

Fig. 8. Magnetic properties of Fe–Cr–Co alloy versus degree of plastic deformation. Alloy was aged before deformation to produce precipitation, and after deformation (at a lower temperature) to increase the composition difference between the phases. (From JIN [1979].)

ferromagnetic, and domain walls have been seen by Lorentz electron microscopy (BELLI *et al.* [1978], MAHAJAN *et al.* [1978]). Such results suggest that, at least for some compositions and heat treatments, domain-wall pinning is a more appropriate model for magnetization reversal than rotation processes in single-domain particles.

Face-centered-cubic spinodal alloys have also found use as permanent-magnet materials. In the Cu–Ni–Fe and Cu–Ni–Co systems, a high-temperature fcc phase decomposes on ageing into two coherent fcc phases, one Cu-poor and ferromagnetic and the other Cu-rich and nonferromagnetic. Phases are elongated along ⟨100⟩ directions, and the extensive ductility of these alloys allows enhancement and alignment of the shape anisotropy. Magnetization reversal is believed to occur by curling (KIKUCHI and ITO [1972]).

3.4. Crystal-anisotropy materials

Crystal anisotropy provides an energy barrier that impedes both coherent and non-coherent rotation processes, independent of particle size and volume fraction.

References: p. 1730.

Permanent-magnet materials based on crystal anisotropy generally possess hexagonal or tetragonal crystal symmetry with the hexagonal or tetragonal axis the magnetic easy axis. Since experimental coercivities are always well below the $2K/M_s$ predicted by coherent or non-coherent rotation, magnetization in these materials clearly occurs by nucleation and growth of reverse domains. The materials differ, however, in the relative importance of domain nucleation and domain-wall pinning in determining coercivity.

3.4.1. Cobalt–rare earths

The permanent-magnet materials with the highest coercivities and energy products are the cobalt–rare earth compounds, which have crystal anisotropy constants as high as 10^7 J/m^3 (10^8 erg/cm^3). These magnets can be divided into four types, depending on whether the primary phase is of the Co_5R (R = rare earth, usually Sm), or $Co_{17}R_2$ type, and whether the magnet has a predominantly single-phase or two-phase microstructure. All four types generally consist of a sintered compact of magnetically aligned powders of the order of 10 μm in diameter. Coercivities are usually higher in Co_5R-based magnets, because of the higher anisotropy constants associated with this phase. Magnetizations are generally higher in $Co_{17}R_2$-based magnets, leading to higher remanence and the potential of higher energy products. Coercivity mechanisms in these materials have been reviewed by several authors (LIVINGSTON [1973], MENTH et al. [1978], KRONMÜLLER [1978]).

The two-phase magnets generally contain Cu, which enhances the formation of fine coherent precipitates, either of Co_5R phase in a $Co_{17}R_2$ matrix or vice versa. Another common alloying element is Fe. The precipitates sometimes form a cellular microstructure, as in fig. 9. As with the alnicos and Fe–Cr–Co alloys, optimum properties are often obtained by achieving phase separation at one temperature and subsequently ageing at a lower temperature (or temperatures) to increase the composition difference between the two phases. Both phases remain ferromagnetic, and coercivity is controlled by the pinning of domain walls. Lorentz electron microscopy of magnets has directly demonstrated the pinning of domain walls by the cellular microstructure (MISHRA and THOMAS [1979]).

In single-phase Co_5R or $Co_{17}R_2$ magnets, coercivity is controlled largely by the nucleation of reverse domains, although domain wall pinning by grain boundaries is also important. The difference between pinning-controlled and nucleation-controlled behavior can be seen clearly by observing the initial magnetization curve of thermally demagnetized samples. Thermal demagnetization generally leaves the samples with several domains per grain. In the two-phase, pinning-controlled magnets, the existing domain walls cannot move easily, and permeability remains low until the applied field approaches the coercivity (solid curve of fig. 10). In the single-phase, nucleation-controlled magnets, the existing walls move easily within the grains and initial permeability is high, the magnet approaching saturation at low fields (dashed curve of fig. 10). However, once saturated, i.e., once reverse domains are removed, this magnet acquires considerable coercivity. The variation of coercive force with magnetizing field for the two magnets is shown as an inset in the figure.

Fig. 9. Transmission electron micrograph of cellular precipitation structure within single-grain-sintered Co–Cu–Fe–Sm aged to peak coercivity (560 kA/m = 7 kOe). Section normal to easy magnetic axis (hexagonal axis). Cell interiors have 17–2 structure, cell boundaries have 5–1 structure, fully coherent. Coercivity controlled by domain-wall pinning by cell boundaries. (From LIVINGSTON and MARTIN [1977].)

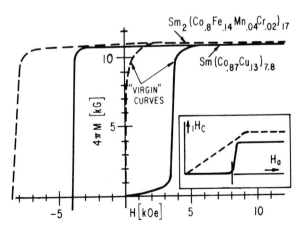

Fig. 10. "Virgin" magnetization curves for precipitation-hardened (solid line) and single-phase (dashed line) $Co_{17}R_2$ permanent magnets. Characteristic, respectively, of pinning-controlled and nucleation-controlled coercivity. Insert shows schematic dependence of coercive field on magnetizing field for the two coercivity mechanisms. (From MENTH et al. [1978].)

References: p. 1730.

Reverse-domain nucleation is generally believed to occur at regions of lowered crystal anisotropy, such as $Co_{17}R_2$ precipitates. Co_5Sm, for example, is unstable to eutectoid decomposition below about 800°C (DEN BROEDER and BUSCHOW [1980]), and ageing at too low a temperature can produce large $Co_{17}Sm_2$ particles, particularly along grain boundaries. (It is interesting that multi-micron $Co_{17}Sm_2$ particles decrease coercivity in Co_5Sm through domain nucleation, whereas coherent ~ 10 nm $Co_{17}Sm_2$ particles formed in Cu-containing Co_5Sm alloys increase coercivity through domain-wall pinning.) In some cases, experimentally-defined "nucleation" may not correspond to the initial formation of a reverse domain, but to its irreversible breakaway from its nucleation site, generally to reverse the entire grain. Thus residual reverse domains resulting from incomplete saturation can serve as "nuclei" (BECKER [1973]). Preferential oxidation in grain-boundary regions has been suggested as a source of low-anisotropy nucleation sites (BARTLETT and JORGENSON [1974]).

Domain studies of magnetic reversal in sintered Co_5Sm magnets (LIVINGSTON [1973b], DEN BROEDER and ZIJLSTRA [1976]) show that grain boundaries serve to block the propagation of magnetic reversal from grain to grain. This pinning of domain walls by grain boundaries is important, because otherwise one nucleating defect would lead to the reversal of an entire magnet. The effectiveness of grain boundaries as domain-wall pins is probably sensitive to the physical and chemical state of the boundary. Annealing at too high a temperature has been shown to decrease the pinning effectiveness of grain boundaries in Co_5Sm (DEN BROEDER and ZIJLSTRA [1976]).

3.4.2. Hard ferrites

The hexagonal barium and strontium ferrites are today the most used permanent-magnet materials. These oxides are ferrimagnetic rather than ferromagnetic, i.e., some of the atomic moments are oriented opposite to the remaining moments. This significantly limits the magnetizations and attainable energy products. However, they have a significant crystal anisotropy (about 3×10^5 J/cm^3 = 3 $\times 10^6$ erg/cm^3), and high coercivities (160–400 kA/m = 2–5 kOe) can be achieved by sintering aligned particles about 1 μm in diameter. Their development and properties have been reviewed by VAN DEN BROEK and STUIJTS [1977].

These magnets appear to be predominantly single phase, and magnetization curves of thermally-demagnetized samples resemble the behavior of single-phase cobalt–rare earth magnets (dashed curves, fig. 10). These results plus direct domain observations of magnetic reversal indicate that coercivities are controlled by domain nucleation, with grain-boundary pinning isolating grains from their neighbors (CRAIK and HILL [1977]).

3.4.3. Mn–Al–C

None of the equilibrium phases in the Mn–Al alloy system are ferromagnetic. The ferromagnetic near-equiatomic τ-phase is metastable, formed from the high-temperature hexagonal phase by an ordering followed by a martensitic shear transformation

(KOJIMA *et al.* [1974]). It can be formed by quenching and ageing, or by cooling at moderate rates. At long ageing times or slow cooling rates, it decomposes eutectoid-ally into the low-temperature equilibrium phases.

The ferromagnetic phase has the face-centered-tetragonal CuAu(I) structure, which imparts a uniaxial crystal anisotropy of about 10^6 J/m^3 (10^7 erg/cm^3). If grain size is 1 μm or finer, coercivities of 240 kA/m (3 kOe) or more can be attained. However, production of an aligned magnet is difficult, because the instabil-ity of the τ-phase at elevated temperatures precludes the sintering of aligned powder. Additions of carbon have been formed to stabilize the τ-phase up to 700°C for extended times. This permits high-temperature extrusion, which produces a texture with significant alignment of the [001] magnetic easy axes. This alignment enhances remanence and energy product (OHTANI *et al.* [1977]).

Transmission electron microscopy reveals an essentially single-phase microstruc-ture, suggesting that, as in the hard ferrites and single-phase cobalt-rare earths, coercivity is controlled by reverse-domain nucleation, with grain-boundary pinning serving to magnetically isolate grains from their neighbors. Micrographs also reveal a significant volume fraction of twins, which is probably a major factor limiting alignment.

The high coercivities of "silmanal" (Ag–Mn–Al) magnets apparently derive from fine precipitates of MnAl τ-phase dispersed in a nonmagnetic Ag-rich matrix (McCURRIE and HAWKRIDGE [1975]).

3.4.4. Co–Pt and related alloys

Below 825°C, Co–Pt alloys near the equiatomic composition undergo an ordering reaction from a disordered fcc solid solution to an ordered fc tetragonal CuAu(I) phase. This structure imparts a crystal anisotropy of 5×10^6 J/m^3 (5×10^7 erg/cm^3), and Co–Pt alloys with coercivities over 500 kA/m (6 kOe) and energy products over 95 kJ/m^3 (12 MGOe) have been produced (KANEKO *et al.* [1968]). Peak coercivity appears to occur before ordering is complete, and the microstructure consists of a fine coherent mixture of ordered and disordered phases. Coercivity appears to be governed by domain-wall pinning (GAUNT [1966]).

Fe–Pt and Fe–Pd alloys, and related ternaries, also go through an ordering reaction to a CuAu(I) tetragonal phase with high crystal anisotropy. Since each disordered fcc grain usually yields three tetragonal variants, Co–Pt and related magnets are usually isotropic. However, application of a stress or magnetic field during the order reaction can favor one of the three variants, leading to alignment of the [001] magnetic easy axes. Stress-assisted ordering has been used to produce an 64 kJ/m^3 (8 MGOe) FePd magnet (YERMAKOV *et al.* [1978]).

4. *Soft magnetic materials*

Soft magnetic materials are characterized by high permeability, low coercive force, and low core loss. They are widely used industrially as pole pieces in electric

References: p. 1730.

motors and generators, laminations in transformer and inductor cores, and in a variety of electromagnetic devices. There are six major groups of commercially important soft magnetic materials: iron and low-carbon steels, iron–silicon alloys, iron–aluminium and iron–aluminium–silicon alloys, nickel–iron alloys, iron–cobalt alloys, and ferrites. Ferrites are oxide ceramics and hence will not be discussed in the present context of metals and alloys. Because of their high electrical resistivity, ferrites are used at high frequencies ($\geqslant 100$ kHz) where eddy-current losses limit the usefulness of metallic magnets. An emerging group of new soft magnetic alloys are the amorphous magnetic alloys which will be discussed in § 5.

In general, the magnetic behavior of soft magnetic materials is governed by domain-wall pinning at heterogeneities such as grain boundaries, surfaces, precipitates and inclusions. Hence, a major common goal in the metallurgy of soft magnetic materials is to minimize such heterogeneities through the use of high-purity starting materials, improved melting and casting practice as well as subsequent fabrication. In addition, eddy-current loss is minimized through alloying additions which increase the electrical resistivity. Initial permeability, a useful parameter in the design of electronic transformers and inductors, is improved by minimizing all sources of magnetic anisotropy energy. Indeed, as will be discussed in § 4.4, the successful development of the high-permeability nickel–iron alloys is primarily the result of the scientific understanding and technical exploitation of the various types of magnetic anisotropy energy and their minimization based on composition and degree of atomic order. On the other hand, a high maximum permeability, useful for power transformer applications, is increased through the alignment of a strong anisotropy. In iron–silicon alloys, where magnetocrystalline anisotropy is dominant, crystallographic texture control is most important. In nickel–iron alloys, where several types of anisotropy can be dominant, the situation is much more complex and the interplay between structure and processing becomes more intricate. Finally, since most soft magnetic materials are processed to thin strips or fine wires, ductility control often becomes as important as magnetic control.

Thus a discussion of soft magnetic materials, as constituted in this section, must necessarily be linked to each material in accordance with its intended applications, with production and processing playing a major role. For a more complete treatment, the reader is referred to the article by CHIN and WERNICK [1980] and the book by CHEN [1977].

4.1. Iron and low-carbon steels

Commercially pure iron, generally of 99.9 + % purity, is used in dc applications such as pole pieces of electromagnets. The value of saturation magnetization is high, and coercivity low, but the electrical resistivity is also low. Low-carbon steel has higher resistivity and is thus more suitable for ac applications such as small motors.

Chief impurities in iron are C, Mn, Si, P, S, O and N, with C, O, S, and N having the greatest detrimental effect on the magnetic properties as they enter the iron lattice interstitially. These impurities strain the lattice and interfere with domain wall motion. By annealing in H_2 and in vacuum much of these impurities can be removed and the permeability consequently much improved. Although best results are obtained at very high temperatures (1200–1300°C), the usual commercial practice is to anneal at ~ 800°C.

The carbon and nitrogen contents can also be lowered by treating molten iron with titanium and aluminium. The remaining carbon and nitrogen are then tied up as second phase compounds of titanium and aluminium. Such treatment has the further beneficial effect of "stabilizing" the iron, since any appreciable dissolved amounts of carbon and nitrogen can lower the permeability with time by precipitation of carbides and nitrides, resulting in magnetic ageing.

Systematic investigations by SWISHER et al. [1969] and SWISHER and FUCHS [1970] have clarified the influence of impurities on the magnetic properties and susceptibility to ageing of low-carbon steels. Carbon and nitrogen have similar effects on initial magnetic properties, but only nitrogen promotes ageing at 100°C. Second phases also increase the coercivity significantly. However, the morphology of the second phase is important. In SAE1010 steel, for example, cementite (Fe_3C) is generally present as fine lamellae or as a grain-boundary network, resulting in high coercivity. On the other hand, a spheroidizing heat treatment can markedly decrease the coercive force (SWISHER et al. [1969] and SWISHER and FUCHS [1970]). Such treatment could be used to advantage to upgrade the magnetic properties of a normally low-cost, low-grade steel.

In addition to dissolved interstitials, inclusions and second phase, grain size and crystallographic texture have a substantial influence on the magnetic behavior of low-carbon steels. Coercivity generally decreases with increasing grain size, which is obtained by annealing at high temperatures, in higher-purity starting materials, and under conditions where impurities are removed (e.g., decarburization). For iron, $\langle 100 \rangle$ is the easy magnetic direction. Here sheet material with a $\{110\}$ sheet texture has superior soft magnetic properties as compared with that having a $\{111\}$ texture. In the investigation by RASTOGI [1977], an increased Mn/S ratio was found to increase the $\{110\}$ component at the expense of $\{111\}$ in cold-rolled and decarburized low-C steels, resulting in an improvement in permeability at the 1.5–1.8 T level. A previous study (RASTOGI [1976]) indicated that, under otherwise similar conditions, a rephosphorized 0.1% carbon steel exhibits superior permeability and core loss as compared to a 0.06% C steel as a result of a more favorable $\{110\}$ texture.

In general, permeability at the 1.5–1.8 T level is quite well correlated with crystallographic texture, while the core loss is a more complex parameter which depends on sheet thickness, electrical resistivity, grain size, inclusion and second-phase content and morphology, and texture. For commercial low-C steels subjected to standard decarburization treatments, RASTOGI and SHAPIRO [1973] have provided an empirical relationship for the 1.5 T/60 Hz core loss as:

$$P_T = 11.33 \frac{t^{1.35} G^{0.065}}{(B_{30}/B_s)^{3.09} \rho^{0.40}} \text{ (W/kg)},$$

References: p. 1730.

where t is the thickness (m), G is the average number of grain-boundary intercepts/mm, ρ is the resistivity ($\mu\Omega$ cm), and B_{30}/B_s is a texture-related ratio of induction at 30 Oe (24 A/cm) to that at saturation.

4.2. Iron–silicon alloys

Iron–silicon alloys containing up to about 4% Si used for magnetic applications are known as *silicon steels*. Higher-loss non-oriented silicon steels are generally used in small motors and generators, relays, and small power transformers where efficiency is of less concern, while the low-loss oriented grades tend to be used in large generators and power- and distribution-transformers where weight and efficiency considerations are of paramount importance.

As a result of intense research and development efforts over the years, the quality of silicon steels has been improved continually. Intense studies on the mechanisms of recrystallization textures and of magnetic core loss have led to the development of new grades of high-induction low-loss material of exceptionally sharp $\{110\}\ \langle 001\rangle$ texture.

4.2.1. Phase diagram and intrinsic magnetic properties

The technical magnetic properties of silicon steels are intimately related to the phase diagram and intrinsic magnetic properties of the Fe–Si system. According to the most recently drawn iron-rich portion of the Fe–Si phase diagram (HULTGREN et al. [1973]), the high-temperature γ-loop extends from 912°C to 1394°C, with the tip at 2.5% Si. Thus one of the benefits of Si addition to Fe is to enable high-temperature heat treatment for grain-orientation control without the deleterious effect of the α–γ (bcc–fcc) phase transformation.

The size of the γ-loop, however, is highly sensitive to small additions of carbon. For this reason, carbon is generally suppressed to below 0.01% through a combination of melting control and decarburization heat treatment.

Silicon additions in the 5–6% range lead to the formation of the DO_3 ordered α_1 phase, resulting in a loss of ductility. Thus practically all commercial grades of silicon steels contain less than 3.5% Si.

The four important intrinsic magnetic parameters of Si–Fe, saturation induction, Curie temperature, magnetocrystalline anisotropy constant K_1, and saturation magnetostriction λ_{100}, are shown in fig. 11 (LITTMANN [1971]). The addition of silicon lowers the saturation induction and Curie temperature and is thus undesirable magnetically. On the other hand, the decrease in K_1 is beneficial. The near-zero value of magnetostriction at 6% Si is highly attractive, but as already noted, material, of this composition is also brittle.

Another benefit of silicon addition, from the magnetic viewpoint, is the increase in electrical resistivity, which helps lower the eddy-current loss.

4.2.2. Magnetic permeability

In general, the value of permeability near the knee of the magnetization curve,

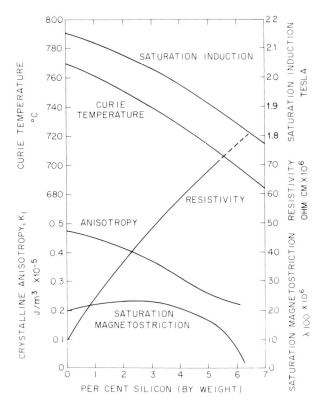

Fig. 11. Effect of silicon on properties of iron.

typically measured at 10 Oe (800 A/m), is a reasonably sensitive measure of grain orientation and insensitive to other factors. However, experiments by SHILLING *et al.* [1978] on a 3% Si–Fe (110) [001] single crystal revealed that at small misorientations such as a tilt of [001] at 2° from the sheet plane, the magnetization is aligned in the field direction rather than along [001]. Thus in this case $\mu = 2030$ (at $H = 800$ A/m), the same value as for a zero tilt angle.

4.2.3.   Core loss

The single most important technical magnetic property of silicon steels is the core loss. The major variables affecting core loss are: composition, impurities, grain orientation, stress, grain size, thickness, and surface condition.

   4.2.3.1.   Composition. It is well known that the addition of silicon decreases the core loss. The primary reasons for the decrease are the decrease in magnetocrystalline anisotropy, the ability to eliminate harmful impurities through annealing at higher temperature due to the vanishing γ-loop, and the increase in electrical resistivity.

   4.2.3.2.   Impurities. As stated in the earlier section on iron and low carbon steels, interstitial impurities elastically distort the lattice and impede domain wall motion, and hence cause an increase in core loss. For these reasons, C, S, N and O are generally kept below 0.01% in today's silicon steels. The widespread use of various modern liquid-steel refining techniques has permitted the reduction of carbon levels to as low as 0.005%.

Similarly, insoluble inclusions such as SiO_2 and Al_2O_3 also increase the loss by pinning the moving domain walls and by creating spike domains.

Of the substitutional solid-solution elements, manganese and aluminium are commonly encountered in silicon steels. Manganese increases the loss only slightly and about 0.25% is added to improve workability in non-oriented grades, although the content is kept below 0.1% in oriented grades. Aluminium is also present to about 0.3% in non-oriented grades for improved ductility, but again kept low in oriented grades.

4.2.3.3. Grain orientation. Grain orientation, or texture, is another important variable that affects core loss. The breakthrough by Goss [1934] in the development of the $\langle 110 \rangle \langle 001 \rangle$, or *cube-on-edge* (COE) texture, is now classic history. It has been estimated (LITTMANN [1971]) that for COE-textured 3.15% Si–Fe, a 1° smaller average misorientation would improve the total core loss at 1.5 T/60 Hz by 5%.

The relationship of core loss to orientation can be traced to the magnetic domain structure. SHILLING and HOUZE [1974] observed that there are two components of the domain structure in (110) [001] Si–Fe: a main structure consisting of large flux-carrying slab domains with magnetizations along [001], and a supplementary structure which flux-closes along the grain surface with magnetizations along [010] and [001]. Core loss is mainly associated with rearrangement of the supplementary structure during an ac cycle. With increases in misorientation, there is an increase in the supplementary structure and hence an increase in loss.

The above is true for tilts of [001] out of the surface greater than 2°. For lesser tilts, however, the trend is reversed, core loss increasing with decreasing tilt (SHILLING et al. [1978] and NOZAWA et al. [1978]). This is shown in fig. 12. Here it was found that the 180° domain wall spacing of the main structure increases rapidly with decreasing tilt, a factor which increases the core loss. Detailed measurements are given in fig. 13, and domain observations are shown in fig. 14.

4.2.3.4. Stress. Since λ_{100} is positive for 3% Si–Fe, the application of a tensile stress near [001] is expected to decrease the core loss by suppressing the supplementary domain structure. The data of fig. 12 show that core loss is indeed lowered by a tensile stress, but the effect is greater for smaller tilts. The domain observations of SHILLING et al. [1978] indicate that in addition to suppressing the supplementary structure, the applied stress was also effective in refining the spacing of the main structure. Both effects were greater for smaller tilts – hence the larger lowering of core loss there. A stress-inducing surface coat may be applied (§ 4.2.4).

4.2.3.5. Grain size. Everything else being equal, the core loss goes through a minimum with grain size at a grain diameter of about 0.5 mm (LITTMANN [1967]). Domain observations indicate that in the larger grain size range ($\sim$ 5 mm), a decrease in grain size results in a decrease in the 180° main domain spacing – hence a decrease in loss. In the finer grain size range ($\sim$ 0.05 mm), large magnetostatic

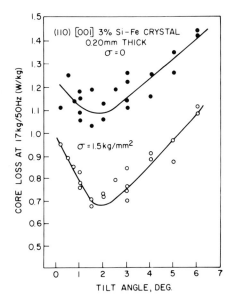

Fig. 12. Core loss versus tilt angle for (100) [001] 3% Si–Fe crystals of 0.20 mm thickness (from NOZAWA *et al.* [1978]).

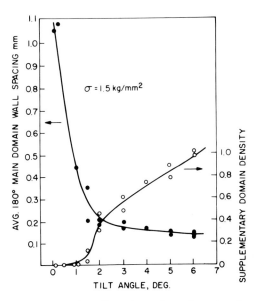

Fig. 13. Dependence of 180° main domain wall spacing and supplementary domain density (normalized at 1 for 6°) on tilt angle for crystals of fig. 12 (from NOZAWA *et al.* [1978]).

References: p. 1730.

Fig. 14. Domain structure of (100)[001] 3% Si–Fe crystals in the 60 Hz demagnetized and unstressed state: (a) tilt angle $\theta = 0$; (b) $\theta = 2°$; (c) $\theta = 5°$. (From SHILLING *et al.* [1978].)

energy is expected at the grain boundaries, causing an increase in loss owing to domain-wall pinning there.

4.2.3.6. Thickness. A decrease in thickness generally leads to a decrease in loss by reducing the eddy-current effects. At very small thicknesses, surface pinning of domain walls becomes dominant and core loss rises again. The core loss minimum, however, is shifted from a thickness of about 0.25 mm for unoriented material to 0.12 mm for oriented material (LITTMANN [1967]). Thus the reduced surface pinning associated with grain orientation drives the thickness minimum to smaller values.

4.2.3.7. Surface morphology. It is well known that surface pinning is more severe at rough spots. Hence it is expected, and generally observed, that a smooth surface, such as that obtained by chemical polishing, leads to lower losses. In commercial production of grain-oriented material a separator such as MgO is applied to the steel surface prior to the final high-temperature ($\sim 1200°C$) hydrogen purification cycle. Depending on the separator material and the details of the heat treatment, a smooth or rough surface can result, leading to changes in core loss (SWIFT *et al.* [1975]).

One conclusion to be drawn from the discussion of the effect of various parameters on core loss is the clear importance of grain orientation. If the grain orientation is sharp, the effects of grain size, thickness, and stress are much more dramatic. Exploitation of such synergistic effects is largely responsible for the design and production of new high-induction steels.

4.2.4. Metallurgy of silicon steels

A major development in silicon steels has been the high-induction material with an exceptionally sharp COE texture. The average misorientation is $3°$ instead of $7°$ for conventional oriented grades. The new material exhibits an induction of 1.92 T at $H = 800$ A/m versus 1.83 T for conventional grades, and a 20% lower core loss at high induction (1.7 T).

The new material was first introduced by Nippon Steel in 1968 under the trade name HI-B. Kawasaki Steel introduced its RG-H grade in 1974 and Allegheny Ludlum its Silectperm grade in 1977. Armco Steel produces the TRAN-COR H version of Hi-B in the U.S.

The new high-induction grades differ from conventional grades in the use of grain-growth inhibitors (ch. 25, § 3.7 and 4.1), the processing procedure, and the type of glass coating used.

The conventional process for manufacturing regular grain-oriented material involves hot-rolling a cast ingot near $1370°C$ to a thickness of about 2 mm, annealing at $800–1000°C$, and then cold-rolling to a final thickness of 0.27–0.35 mm in two steps, with a recrystallization anneal ($800–1000°C$) in-between. The final cold reduction is about 50%. The cold-rolled material is first decarburized to $\sim 0.003\%$ C at about $800°C$ in wet $H_2 + N_2$, a step which also results in primary recrystallization. It is then coated with MgO as separator and annealed in dry hydrogen at $1100–1200°C$ for about 24 h to form the COE texture via secondary recrystallization (ch. 25, §4.3). During this anneal the impurities are further removed and inclusions of the grain-growth inhibitors are dissolved and absorbed in the glass film formed on the surface. Manganese sulfide has been the most common grain-growth inhibitor used.

The HI-B process (TAGUCHI [1977]) differs from the conventional process in two major areas: (i) the use of AlN in addition to MnS as grain-growth inhibitor and (ii) a one-stage cold rolling with large thickness reduction (80–85%). Usually a large final cold reduction leads to a sharper COE texture in the finished strip, but it also enhances the undesirable grain growth of the primary grain structure. A strong

References: p. 1730.

inhibitor such as AlN thus restrains primary grain growth and permits the large reduction. For the AlN to be effective, however, the anneal after hot-rolling is done at 950–1200°C, higher than conventional, followed by rapid cooling. In addition, the glass film which contains the absorbed aluminium has a smaller coefficient of thermal expansion than the usual glass film, resulting in a larger tensile stress on the strip and a lowering of the core loss.

In addition to these major processing differences, the HI-B process takes into account the effects of silicon content, grain size, inclusion content, etc.

In the Kawasaki RG-H process (GOTO *et al.* [1975]), antimony was added along with MnSe or MnS as a grain-growth inhibitor. It is thought that Sb segregates to the primary recrystallized grain boundaries. The cold rolling is two-stage as in the conventional process, but the final stage is closer to a 60–70% reduction rather than 50%. Finally, the last anneal step is carried out first at 820–900°C for 5–50 h to promote a sharper orientation of the secondary grains, and then at about 1100°C for purification. A low-thermal-expansivity inorganic phosphate coating was also developed to impart a large tensile stress to the finished steel (SHIMANAKA *et al.* [1975]).

In the Allegheny Ludlum Silectperm process (MALAGARI [1977]), boron and nitrogen together with sulfur or selenium are used as grain-growth inhibitors. These are also thought to segregate to the primary recrystallized grains. In addition, since the Mn and S contents are quite low, the MnS solubility temperature is lowered, permitting the hot-rolling operation at 1250°C instead of above 1300°C. This is advantageous in terms of lower fuel costs, added mill life, etc. As in the HI-B process, cold rolling is done in one stage, the reduction exceeding 80%. The Silectperm process is a commercial realization of the laboratory technique developed at General Electric (FIEDLER [1977], GRENOBLE [1977]).

It may be pointed out that the core-loss advantage of the new high-induction material over the conventional oriented material is in the high-induction range only. Below an induction level of 1.5 T, the core-loss difference becomes minimal. In recent years, the sharp increase in energy costs has prompted electric utilities to incur a first-cost penalty for decreased core loss over the working life of transformers. As a result, transformer designers have reduced design induction to the 1.3 T range to minimize the loss penalty, even at the expense of a larger transformer. Consequently, the anticipated large-scale introduction of the high-induction material has been slowed, despite a price premium of only ~ 10%. There is a general tendency in industry to use a thinner-gauge conventional oriented material, at 0.22 mm rather than at 0.27 mm, for lower core loss at the lower design inductions (LITTMANN [1982]). (See also ch. 25, §4.3.1, and ch. 28, §4.1 and fig. 28.)

4.3. Iron–aluminium and iron–aluminium–silicon alloys

The iron-rich portion of the Fe–Al phase diagram is similar to that of the Fe–Si diagram. Both Al and Si suppress the formation of the fcc γ-phase and both elements increase the electrical resistivity significantly. The Fe–Al alloys are more

References: p. 1730.

ductile than the Fe–Si alloys. Hence Al is added to silicon steel in the unoriented grades to improve ductility.

There has been no extensive commercial development of binary Fe–Al alloys. However, since the magnetocrystalline anisotropy constant K_1 goes to zero at 12–14% Al and the magnetostriction constant λ_{100} is near maximum in this range, alloys near these compositions have some interest as magnetostrictive transducers. In the 14–16% Al range, both K_1 and λ_{100} can be small and thus exhibit high permeability. These alloys have found limited use as recording-head material on account of the additional benefits of high hardness and electrical resistivity.

For recording-head applications, Fe–Al–Si alloys centered around the *Sendust* composition (5.4% Al, 9.6% Si) are more widely used than the binary Fe–Al alloys. Here the $K_1 = 0$ and $\lambda_s = 0$ curves intersect, leading to high values of permeability. The addition of 2–4% Ni, referred to as *Super Sendust* (YAMAMOTO and UTSUSHIKAWA [1976]), reportedly increases the permeability and lowers the coercivity even further.

Because Sendust is brittle, much effort has been expended to shape the alloy for recording-head use. In addition to normal powder-metallurgy techniques, developments have included "squeeze casting" whereby molten Sendust is forced into a die of predetermined shape (SENNO et al. [1977]), sputtered films (SHIBAYA and FUKUDA [1977]), and rapid quenching from the melt (TSUYA and ARAI [1978]). The last technique, patterned after that for producing amorphous metal ribbons from the melt, is interesting in that it provides a ribbon with sufficient ductility to punch out cores for recording-head applications. The enhanced ductility presumably comes from the extremely small grain size ($\sim 10~\mu$m) of the as-quenched material.

4.4. Nickel–iron alloys

The nickel–iron alloys in the *Permalloy* range, from about 35 to 90% nickel, are probably the most versatile soft magnetic alloys in use today. With suitable alloying additions and proper processing, the magnetic properties can be controlled within wide limits. Some exhibit high initial permeability up to $100\,000\mu_0$ useful for high-quality electronic transformers. Others display a square hysteresis loop ideal for inverters, converters and other saturable reactors. Still others combine low remanence with high constant permeability, ideal for unbiased unipolar pulse transformers. And since these alloys are extremely ductile, material can be precision cold-rolled to thin tapes useful for some magnetic-core components.

4.4.1. Phase diagram and intrinsic magnetic properties

In the nickel–iron system, the fcc γ-phase exists in the 30–100% Ni range at room temperature. At the low-nickel end, the alloy is bcc α, with the α–γ transformation exhibiting considerable hysteresis. Near the Ni_3Fe composition the alloys undergo a long-range-ordered $L1_2$ transformation below about 500°C. This transformation has a significant influence on the intrinsic magnetic parameters and is exploited in the processing of commercial alloys.

Figure 15 shows the major magnetic parameters in the Ni–Fe system: saturation induction B_s, Curie temperature T_c, magnetocrystalline anisotropy constant K_1, and saturation magnetostriction λ_s. Values of B_s, K_1 and λ_s are for room temperature. Highest B_s occurs near 48% Ni and highest T_c is at 68% Ni. The solid K_1 curve is for material quenched from above 600°C to retain a disordered structure. The large dip near 75% Ni in the dashed K_1 curve, for slowly cooled material, is a result of the L1$_2$ ordering.

Magnetic theory indicates that high initial permeability is achieved by minimizing both K_1 and λ_s. Thus a major aim in the development of high-permeability Ni–Fe alloys is to search for compositions and processing conditions whereby K_1 and λ_s are minimized. In the Ni–Fe binary, this occurs at 78.5% Ni in the rapidly-cooled condition, as indicated in fig. 15. With the addition of molybdenum, the kinetics of ordering is slowed and simultaneous attainment of zero K_1 and λ_s is possible with moderate cooling rates for alloys near 4% Mo–79% Ni–17% Fe. The addition of

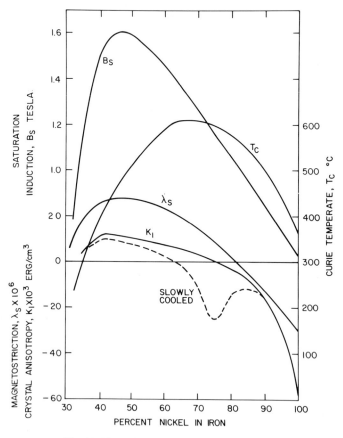

Fig. 15. Magnetic parameters of Ni–Fe alloys.

molybdenum confers the additional benefit of increased resistivity, desirable for high frequency usage, but it lowers B_s and T_c. Copper has also been found useful as an addition to minimize K_1 and λ_s. Commercial alloys near the 4% Mo–5% Cu–77% Ni–bal Fe range have been so developed.

In addition to magnetocrystalline anisotropy and magnetostriction, two other sources of magnetic anisotropy are important for Ni–Fe alloys. One is *thermomagnetic anisotropy*, developed by annealing below T_c (GRAHAM [1959]; CHIKAZUMI and GRAHAM [1969]). Thermomagnetic anisotropy is most pronounced near 68% Ni where T_c is highest, and is utilized commercially to develop square or skewed hysteresis-loop shaped materials by applying a magnetic field during annealing. If the field is absent, the hysteresis loop becomes constricted. The other is *slip-induced anisotropy*, obtained by cold-working, particularly when the material is atomically ordered prior to deformation (CHIKAZUMI and GRAHAM [1969]; CHIN [1971]).

Both thermomagnetic and slip-induced anisotropies have their origin in the short-range directional ordering of nearest-neighbor atom pairs (NÉEL [1954]). Since the pseudo-dipolar magnetic coupling energy of a pair of atoms depends on the identity of the atoms, e.g., Fe–Fe, Fe–Ni, annealing below T_c in the presence of a magnetic field tends to align in the field direction those pairs with minimum coupling energy. Fast cooling to a low temperature then freezes-in the directional order structure and gives rise to a uniaxial magnetic anisotropy. Similarly, when material with an atomically ordered structure is deformed, the slip process creates planar anti-phase domain boundaries and establishes a directional-order structure with unequal distribution of atom pairs. For Ni–Fe alloys, the thermomagnetic anisotropy energy is ~ 0.1 kJ/m^3 and the slip-induced anisotropy energy is ~ 10 kJ/m^3, both overlapping the values of K_1 and λ_s.

Thus all four types of anisotropy can be, and have been, manipulated in custom-engineering the magnetic properties of Ni–Fe alloys.

4.4.2. Metallurgy of nickel–iron alloys
Commercial Ni–Fe magnetic alloys are predominantly used for high-permeability and square-loop applications, but some have also been developed for use with skewed-loop characteristics and some for wear-resistant recording-head applications.

4.4.2.1. High-permeability alloys. There are two broad classes of high-permeability Ni–Fe alloys. The low-Ni alloys, near 50% Ni, are characterized by moderate permeability ($\mu_{40} \approx 10\,000\mu_0$, measured at 40 G/60 Hz) and high saturation ($B_s \approx 1.5$ T). This region corresponds to high B_s, low K_1 and λ_{100}. The high-Ni alloys, near 79% Ni and usually combined with Mo and/or Cu, are characterized by high permeability ($\mu_{40} \approx 50\,000$) but lower saturation ($B_s \approx 0.8$ T). This region corresponds to near-zero values of K_1 and λ_s.

Commercial 50% Ni alloys generally range from 45–50% Ni and are of two grades: rotor grade, of $\mu_{40} \approx 6000\mu_0$, useful for rotors and stators where non-direction properties are desired, and transformer grade, of $\mu_{40} \approx 10\,000\mu_0$, where high permeability is achieved both parallel and perpendicular to the rolling direction. To achieve the essentially random texture of the rotor-grade material, the final cold

References: p. 1730.

reduction is held low (50–60%). Annealing is done in dry H_2 at 1200°C, followed by furnace cooling (~ 150°C/h). The transformer grade, on the other hand, is produced by severe cold reduction (~ 95%) followed by high-temperature annealing to produce a secondary recrystallization texture of $\langle 210 \rangle \langle 001 \rangle$ type.

A large increase in permeability can be achieved by annealing the alloys, particularly in the 56–58% range, in the presence of a magnetic field after the usual high-temperature anneal. This is the result of the establishment of a thermomagnetic anisotropy by the magnetic annealing.

In the 79% Ni regime, modern alloys contain Mo and Cu as stated earlier. There is a standard grade with $\mu_{40} \approx 30\,000$–$40\,000$, and a very-high-permeability grade with $\mu_{40} \approx 60\,000$. The latter is based on the development of high-purity *Supermalloy* (BOOTHBY and BOZORTH [1947]) with $\mu_{40} > 100\,000$. Today's Supermalloy composition is close to 5% Mo–80% Ni–15% Fe. The attainment of superior permeability requires a simultaneous achievement of zero K_1 and λ_s, which parameters are highly sensitive to composition and heat-treatment conditions. One example is shown in fig. 16 illustrating the effects of annealing temperature and nickel content for alloys containing 4% Mo and 5% Cu (SCHOLEFIELD *et al.* [1967]). It can be understood that as the nickel content is increased, for a given Mo and Cu composition, less atomic ordering is required to achieve $K_1 = 0$ (ENGLISH and CHIN [1967]); hence a higher annealing temperature (i.e., less order) is needed for optimum initial permeability. Figure 16 also shows that the absolute value of μ_i goes down at both the high- and the low-Ni end as a result of deviation from the $\lambda_s = 0$ composition. For a more extended discussion of these complex effects, the reader is referred to the article by CHIN and WERNICK [1980].

In addition to composition selection and heat-treatment control, melting practice and microstructure control are of paramount importance. Today all high-quality Ni–Fe alloys are prepared by vacuum-melting, which minimizes harmful ingredients such as oxygen and sulfur. Figure 17 shows the effect of these two elements on the initial permeability of the two classes of Ni–Fe alloys (COLLING and ASPEN [1969, 1970] and AMES [1970]). In this connection, deoxidation by the use of strong deoxidizers such as Ca, Si and Al has not met with success, primarily because the fine oxide particles left in the matrix restrict grain growth and block domain-wall motion. Figure 18 shows, for example, the initial permeability μ_5, i.e., at $H = 5$ mOe (4 mA/cm), as a function of grain diameter for a 50% Ni alloy with and without Al additions (HOFFMANN [1971]). The lower value of μ_5 for Al-added samples is attributed to the presence of oxide precipitates. A quantitative study by ADLER and PFEIFFER [1974] led to the following expression for the coercivity of 47.5% Ni alloys:

$$H_c = H_{c0} + H_{ci} + H_{cK} = 0.8 + 2.8 \times 10^{-4} N_F + 0.29/d_K \; (\text{A/m}),$$

where H_{c0} is the residual coercivity of the pure alloy, H_{ci} is due to inclusions with N_F as the number of particles per mm² in the 0.02–0.5 μm range, and H_{cK} represents the grain size effect with d_K as the grain diameter in mm.

A similar study was done on a Mo–Cu–Ni–Fe alloy (KUNZ and PFEIFER [1976]). For a more detailed discussion, see PFEIFER and RADELOFF [1980] and CHIN and WERNICK [1980].

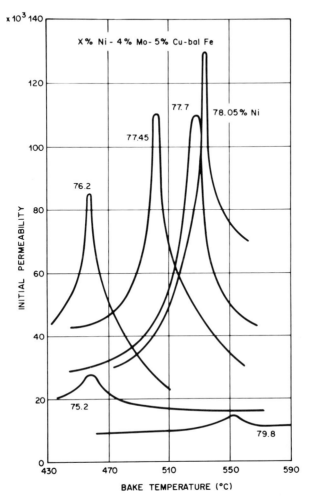

Fig. 16. Influence of annealing ("bake") temperature on initial permeability for alloys of Mumetal type (from SCHOLEFIELD *et al.* [1967]).

4.4.2.2.   Square-loop alloys. A square hysteresis loop is obtained in a material by aligning a nonzero magnetic anisotropy with the easy axis in the direction of magnetization. In 50% Ni alloys, this is primarily achieved through the development of a cube-on-face, $\{100\}$ $\langle 001 \rangle$, texture, since $K_1 > 0$ and $\langle 001 \rangle$ is the easy direction. The texture is obtained by extensive prior rolling ($> 95\%$) followed by primary recrystallization in the vicinity of 1100–1200°C. Among factors favoring texture formation are small penultimate grain size, large cold reduction, and a lowering of impurity elements such as Si and C. For 79% Ni alloys, the situation is more complicated. For weakly textured material normally encountered in commercial processing, a heat treatment to achieve $K_1 < 0$ has been found to provide "quasi-

References: p. 1730.

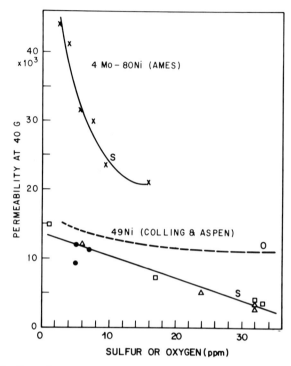

Fig. 17. Influence of sulfur and oxygen on initial permeability (after COLLING and ASPEN [1969, 1970] and AMES [1970]).

squareness" of the hysteresis loop (PFEIFER [1966a]). This is done by lowering the isothermal annealing temperature or decreasing the cooling rate, as compared to the heat treatment for high μ_i where $K_1 = 0$ is desired. The results are illustrated in fig. 19 for a series of Mo-bearing alloys in which Ni is held at 80%.

The other major technique for achieving a square hysteresis loop is to establish a thermomagnetic anisotropy by magnetic annealing. Here alloys in the 55–65% Ni

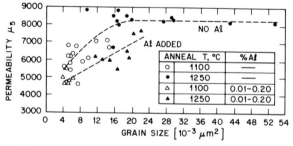

Fig. 18. Dependence of initial permeability on grain size for a commercial 50% Ni alloy with and without Al added as deoxidizer (from HOFFMANN [1971]).

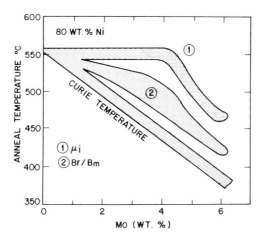

Fig. 19. Optimum annealing temperatures for high μ_i and B_r/B_m in Mo-Permalloys (PFEIFER [1966a]).

range are particularly susceptible to magnetic annealing effects as a result of high T_c in this region. The nature of the response, however, is highly dependent on the texture and on annealing conditions under which both magnetocrystalline and thermomagnetic anisotropies compete with each other. For example, to produce maximum squareness in unoriented material, it is important that the magnetic annealing be done such that $K_u \gg K_1$, i.e., aligning the thermomagnetic anisotropy at the expense of crystal anisotropy. Such combination is obtained at a relatively low annealing temperature ($\sim 430°C$) (PFEIFER [1966b]).

4.4.2.3. Skewed-loop alloys. While the alignment of a uniaxial anisotropy, such as by magnetic annealing, results in a square hysteresis loop in the direction of alignment, it also results in a skewed loop in the perpendicular direction. The result is a low-remanence, constant-permeability characteristic highly desirable in loading coil and unipolar pulse transformer applications. Commercial alloys of 2.5% Mo–65% Ni–bal Fe and 4% Mo–79% Ni–bal Fe compositions have been developed with such characteristics. Again there is a complex relationship among texture, K_1 and K_u, which is reasonably well understood.

4.4.2.4. Wear-resistant alloys. For applications such as recording heads, there is a need for high mechanical hardness to reduce wear without sacrifice of magnetic softness. Standard Permalloys have poor wear resistance. It has been found, however, that alloys near 79% Ni with Nb and Ti additions (MIYAZAKI *et al.* [1972]) can greatly enhance the hardness while retaining the high permeability. Figure 20 shows how the mechanical and magnetic parameters change as a function of ageing time at 630°C, for a 3% Ti–2.8% Nb–3.8% Mo–10.5% Fe–bal Ni alloy. It may be noted that after ageing for 4 h, the permeability goes through a maximum and the coercivity a minimum, but the Vickers hardness is still increasing. This observation is explainable in terms of precipitation where the optimum particle size for blocking dislocation motion differs from that for domain-wall motion.

References: p. 1730.

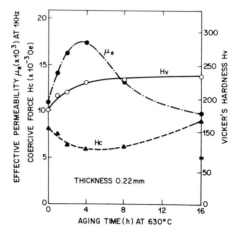

Fig. 20. Magnetic and mechanical properties of a 3% Ti–2.8% Nb–3.8% Mo–10.5% Fe–bal Ni alloy as function of ageing time at 630°C (MIYAZAKI et al. [1972]).

4.5. Iron–cobalt alloys

Iron–cobalt alloys have the highest value of saturation induction at room temperature, at 2.45 T for the 35% Co alloy. Somewhat lower values of saturation induction but substantially higher permeability are obtained in equiatomic alloys because of their low magnetocrystalline anisotropy. Commercial magnetic alloys are known as *Hiperco, Permendur* and *Supermendur*. Because they are relatively costly, these alloys are limited to specialized applications such as aircraft generators where weight and size are important considerations.

The Fe–Co alloys have an fcc structure above 912–986°C to ∼ 70% Co. The low-temperature form is bcc. At ∼ 50% Co, ordering to the B2 structure sets in below about 730°C and the alloy becomes brittle. Vanadium additions retard the ordering reaction and improve the ductility. An additional benefit is an increase in resistivity.

The magnetocrystalline anisotropy constant K_1 decreases with Co addition to a value near zero at 50% Co, but is sensitive to the degree of order. The magnetostriction constants λ_{100} and λ_{111} increase with Co content, λ_{111} reaching a high value of ∼ 150×10^{-6} at 50% Co.

4.5.1. Metallurgy of equiatomic Fe–Co alloys

Since 2V-Permendur and Supermendur, both containing 2% V–49% Fe–49% Co, are the most important Co–Fe soft magnetic alloys, their metallurgy may be discussed in more detail. Preparation of thin sheets of 2V-Permendur involves hot-working in the fcc state, followed by quenching to room temperature for further cold reduction. Optimum magnetic properties are developed by annealing at ∼ 850°C, below the fcc–bcc transformation temperature but above the B2 ordering temperature. However, for aerospace high-speed rotor applications, there is a need to develop a high-strength ductile alloy at the sacrifice of maximum magnetic properties. THORNBURG [1969] showed that annealing cold-rolled sheet at 695°C yielded material of high strength and reasonable ductility and magnetics. The microstructure of this material indicated only partial recrystallization. A similar study by THOMAS [1975] resulted in annealing at 650°C for rotor material and 875°C for stator material, where the high-strength requirement is of less importance.

By adding 4.5% Ni to the 49/49/2 Fe–Co–V alloy, MAJOR *et al.* [1975] were able to enhance the ductility and strength over a wider heat-treating temperature range, thus eliminating the need for relatively close temperature control of the regular alloy. X-ray examination showed that after heat-treatment in the 650–740°C range, some γ-phase and ordered α-phase are present. It is suggested that the enhanced properties stem from the restriction of grain growth owing to the presence of these phases.

Supermendur (GOULD and WENNY [1957]) is essentially of the 2V-Permendur composition, except that the starting ingredients are of higher purity. In addition, the heat-treatment to develop the soft magnetic properties is carried out in a magnetic field of 80 A/m or greater. The result is an extremely square hysteresis loop with a small coercivity, ideal for saturable reactor type applications.

5. Amorphous magnetic materials

5.1. Introduction

The amorphous magnetic alloys most investigated, and of most interest from the applications viewpoint, are of the type $(Fe, Ni, Co)_{80}$ (metalloid)$_{20}$. The metalloids may be B, Si, C, P or Ge. In the past 20 years research and development on amorphous metals and alloys has increased drastically (ch. 28, §3). The reasons for this increase are twofold. Firstly, they are a new state of matter, with no long-range order, with many opportunities for scientific investigations. Secondly, many of the properties are of interest in potential applications. For example, the magnetic losses and exciting power are extremely low in alloys with high magnetization, making them of interest in both large and small transformer application. Their hardness, high permeability and zero magnetostriction make them useful for recording heads. The ductility of certain intermetallic alloys makes them extremely useful for brazing alloys. They have excellent corrosion resistance, making them potentially useful for coatings. Their high permeability and toughness make them useful for magnetic-shielding applications. Their exceptional magnetomechanical coupling factor and their change in Young's modulus with applied magnetic field make them exceptional electronically controlled delay lines. Finally, the high stress-sensitivity of some of the alloys makes them useful as stress transducers. In addition, there are many other potential applications reported in the literature.

In this section we will describe the preparation and properties of amorphous magnetic materials.

References: p. 1730.

5.2. Preparation

All methods for producing a non-crystalline, i.e. an amorphous or glassy solid, depend on achieving high enough quenching rates so as to prevent crystallization. The quench rate required depends critically on the alloy and alloy composition and its properties, e.g. its melting point and glass temperature. The maximum thickness of ribbon, or splat, which can be cast then is determined by the heat-withdrawal rates achievable through the alloy and across the interface between the alloy surface and quenching surface.

There are a variety of methods available for producing the very rapid quenching necessary to prevent crystallization of a suitable alloy. For example, methods which have been used include sputtering and evaporation methods, chemical and elec- trodeposition (for a bibliography of glasses prepared by these four methods see DIETZ [1977]), ion implantation (BORDERS [1979]), and various splat-quenching techniques which involve the rapid spreading of a liquid droplet on a surface, or pair of surfaces, by propelling the droplet onto the surface or by propelling the surfaces together with the droplet in between. There are also a variety of methods for the preparation of amorphous metal powders, e.g., gas atomization, centrifugal atomiza- tion, or spark erosion. These powders may then be compacted by pressing or extrusion to near-final shapes to obtain bulk amorphous parts (ch. 28, § 4).

Most interest in commercial preparations of amorphous alloys has been con- cerned with the preparation of continuous tapes or ribbons. Narrow ribbons are prepared by rapidly expelling the molten alloy from a circular orifice onto the surface of a spinning wheel. This is normally called *chill-block melt-spinning* (LIEBER- MANN [1980] (ch. 28, § 1). A puddle builds up on the spinning wheel and serves as a reservoir from which ribbon is formed by the solidification of the melt next to the surface of the wheel. The ribbon typically adheres to the wheel for some time before being released, or thrown off, as a continuous filament. A puddle which remains stable in size results in a ribbon with smooth edges and uniform width. Normally casting is done onto the outside rim of a wheel, but casting onto the inside of a wheel, between two rollers or onto belts have also been done in an attempt to improve the ribbon in various respects. For example, casting between two rollers results in ribbon with both surfaces much smoother than casting on a single wheel. This is of particular importance in recording-head applications where it is necessary to achieve the highest permeability and highest packing fraction. However it is a much more difficult process to control and thus is not used by many. Casting onto the inside of a rim results in longer contact with the wheel and therefore a quench to a lower temperature. This results in a ribbon structure with a larger atomic free volume which in term results in more rapid stress-relief, for example. Other methods have also been used to increase the quench rate, for example, by using a gas stream impinging on the ribbon to hold the ribbon down, or by using a secondary roller to hold the ribbon down on the wheel.

The above methods are only suitable for preparing narrow ribbons, i.e., ribbon less than about 2 mm in width. To make wider ribbon an elongated orifice must be used. In order to reduce or eliminate the perturbations caused by the impingement of the air on the molten metal stream, the casting from wide nozzles must be done at a reduced atmospheric pressure or with the nozzle very close to the wheel. This latter configuration is called *planar flow-casting* and is characterized by the molten metal filling the space between the crucible and wheel. This is the method now used to prepare all wide ribbons (LIEBERMANN [1979]).

Some novel variations of planar flow-casting have been described for particular applications. For example, preparations of wide tapes with teeth or other geometries have been made without the need to punch the teeth out with a die. The teeth were formed by casting onto a wheel with a low-thermal-conductivity pattern. These low-conductivity regions result in slower cooling of the ribbon, causing the ribbon to be brittle and easily separated from the very tough non-brittle remainder of the ribbon.

Another novel variation is the preparation of helical rather than straight ribbons (LIEBERMANN [1981]). The helical ribbon can be wound without producing the degradation in properties associated with the bending stresses produced on winding up straight ribbon into a core. By combining this helical-ribbon technique with the previously described technique for producing a cut-out pattern, electric motor stators or rotors can be prepared directly from the cast ribbon without need for punching.

Another variation of planar flow-casting is the formation of composite layers of ribbons by casting successive layers on top of amorphous ribbons. The problem is to balance the heat transfer against the crystallization of the amorphous alloy being cast on. In this way thick sections of ribbons and combinations of amorphous/crystalline and crystalline/crystalline ribbons have been prepared.

Modeling of the chill-block melt-spinning process has been done both empirically and theoretically (LIEBERMANN 1980, KAVESH [1978], PAVUNA [1982], and VINCENT *et al.* [1982]). Empirical relations have been established, with little scatter, for the ribbon geometry as a function of melt ejection pressure, temperature, flow rate, wheel velocity, atmosphere etc. The theoretical models are all based on the propagation of a solidification boundary which determines the ribbon thickness. Several approaches have evolved, depending whether thermal or momentum boundary layers are the controlling factors. In the thermal boundary-layer propagation, while temperature varies continuously through the melt, there is an abrupt velocity change in the solidification front. This abrupt velocity change is taken as the glass-transition isotherm. A momentum-transfer mode has more recently been proposed to be the dominant process and a combined thermal–momentum transfer process has been worked out. All of these approaches are successful to some extent in describing the experimental results. However, the complex nature of the ribbon-formation process and the factors which control it are still to be described by a physically consistent model.

The modeling of the planar flow-casting process proceeds along the same lines as for chill-block melt-spinning. Empirical relations have also been established using the same parameters as for chill-block melt-spinning. It should be noted that

References: p. 1730.

perturbation of the melt on the wheel by the gas boundary-layer and by surface roughness of the wheel, etc., are substantially reduced because of the constraint of the melt imposed by the crucible. The specific effect of the gas on melt-puddle stability and interfacial contact has been studied in detail. The mathematical modeling has been described by characterizing the fluid dynamics as a laminar-flow process. This can be approximated by lubrication theory. It has been calculated that a momentum, rather than a thermal, boundary layer controls ribbon dimensions.

5.3. Properties

The magnetic properties reflect the short-range order existing in the amorphous alloy. The short-range order, in general, reflects the structure existing at the same composition in the crystalline alloys. The same rules that govern the behavior of the crystalline alloys, appropriately averaged for randomness, also govern the behavior of amorphous alloys. The various magnetic properties will next be discussed in the above terms.

5.3.1. Curie temperature

The simplest system of amorphous alloys is probably the binary iron, cobalt or nickel, FeX, CoX, and NiX alloys, where X is a single metalloid or a combination of metalloid atoms such as B, P, Si, Ge and C. The Curie temperature of these alloys varies with X in a systematic manner as shown in fig. 21a (LUBORSKY *et al.* [1980b], GRAHAM and EGAMI [1978]). It should be noted here, however, that T_c decreases with increase in Fe content, as shown in fig. 21b, suggesting that the structure is behaving like close-packed, rather than bcc iron.

When two transition metal, TM, atoms are present in the amorphous alloy, the change in T_c with variation in the TM ratio can be modeled phenomenologically. On the basis of the coherent potential approximation it can be shown (LUBORSKY [1980a]) that:

$$\alpha^2 T_c^2 + \left\{\alpha(T_{AA} + T_{AB}) - \alpha(1 + \alpha)\left[x^2 T_{AA} + 2x(1 - x)T_{AB}\right]\right\}T_c$$
$$+ \left[(1 - \alpha)(1 - x)^2 - \alpha\right]T_{AA}T_{AB} = 0, \tag{3}$$

where x is the atom fraction defined by the alloy $A_{x-1}B_x$, T_{AA}, T_{AB}, and T_{BB} are so-called interaction temperatures, $\alpha = (z/2) - 1$, z is the number of nearest neighbors and T_c is the Curie temperature. An *interaction temperature* is an effective temperature for the strength of the interaction of an atom pair. These temperatures involve the exchange coefficients and the atomic moments. Using this relation, the curves in fig. 22 were calculated, giving excellent agreement with the data using only one adjustable parameter, T_{AB}, to fit the data.

5.3.2. Saturation magnetization
5.3.2.1. Dependence of magnetization on alloy composition. The saturation magnetization of $Fe_x B_{100-x}$ varies almost linearly with x at low temperatures, as shown

Magnetic properties

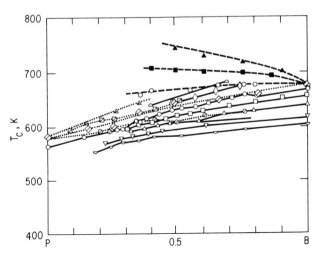

Fig. 21a. Open symbols are for Curie temperatures as a function of P/B fraction for $Fe_x(PB)_{1-x}$ amorphous alloys: ⌒, $Fe_{75}(PB)_{25}$; ◇, $Fe_{77}(PB)_{23}$; ○, $Fe_{79}(PB)_{21}$; □, $Fe_{80}(PB)_{20}$; △, $Fe_{81}(PB)_{19}$; ▽, $Fe_{82}(PB)_{18}$; ▿, $Fe_{83}(PB)_{17}$ (after GRAHAM and EGAMI [1978]). Solid symbols are for addition of M to $Fe_{80}B_{20-x}M_x$: ▲, Ge; ■, Si; ●, C. Slashed symbols are for additions of N to $Fe_{80}P_{20-x}N_x$: ◭, Ge; ⧄, Si; ⦸, C; ◈, B.

in fig. 23 (LUBORSKY *et al.* [1978]), but at room temperature and above σ_s reaches a peak and then drops with increasing x. This decrease is due simply to the approach of T_c as x increases since T_c is decreasing sharply.

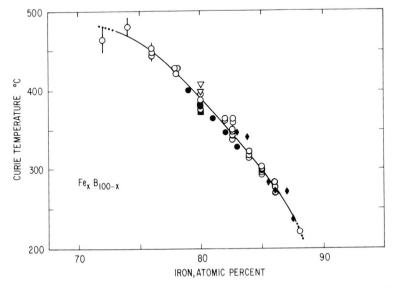

Fig. 21b. Curie temperatures for Fe_xB_{100-x} amorphous alloys (from LUBORSKY *et al.* [1978]).

References: p. 1730.

On the replacement of B by another metalloid the magnetization typically decreases as shown by the results in fig. 24 (LUBORSKY [unpublished]) for ternary Fe–B–Si alloys. KAZAMA et al. [1978] examined the effect of various metalloids, M, in $Fe_{80}B_{20-x}M_x$ on σ_s and found that the magnetization in Bohr magnetons per metallic atom, μ_B, varied with P < C < Si < Ge. However, LUBORSKY et al. [1978]

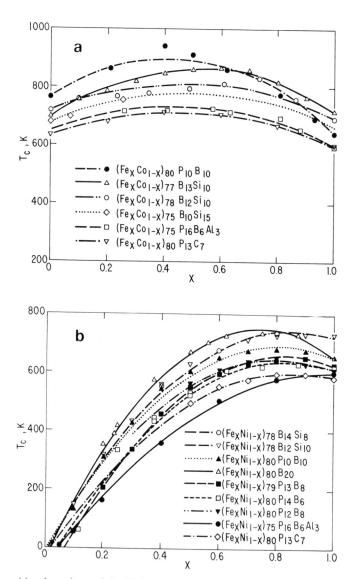

Fig. 22. Composition dependence of the Curie temperature of amorphous alloys. Curves calculated using an equation based on a phenomenological model using the coherent potential approximation: (a) Fe–Co based alloys; (b) Fe–Ni based alloys; (c, on facing page) other alloys.

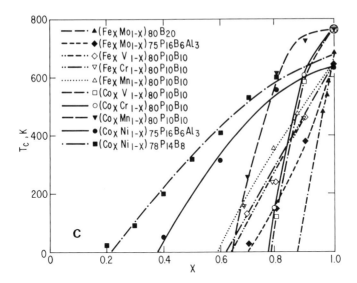

found P < Si < Ge < C. The reasons for the differences are not known; however, the differences between Si, Ge, and C are quite small.

It would be expected on the basis of charge transfer that P < Si = Ge = C since the effective valency of the metalloid, defined as the number of outer electrons, is 1 for B, 2 for C, Si, and Ge and 3 for P. This is approximately true since the differences between Si, Ge, and C are small.

The effect of changing transition-metal content is shown in fig. 25 (LUBORSKY [1980b]) for a variety of metalloid combinations. To a first approximation the μ_B are independent of the metalloids used varying from μ_B for Fe to μ_B for Co or Ni. However, on close examination it is found that alloys containing only boron have slightly higher moments than those containing phosphorus, as would be expected on the basis of the charge transfer model.

5.3.2.2.   Temperature-dependence of magnetization and spin waves. The magnetization–temperature curve has been measured for a variety of amorphous alloys. In all cases the decrease in σ is always more rapid than in crystalline metals. This might be predicted from the distribution of exchange interactions resulting from the distribution in atomic spacings. The mean-field model is generally used with a parameter that is a measure of the rms deviation of the nearest-neighbor exchange from its mean value. Values from 0.1 to 0.5 are generally obtained.

The decrease in σ with increasing temperatures can be fitted to the *Bloch law*, $\sigma/\sigma_0 = 1 - \beta(T/T_c)^{3/2}$. For amorphous alloys this law holds up to temperatures as high as $0.5T_c$, much higher than for crystalline alloys. Furthermore, the higher-order terms, e.g. $T^{5/2}$, are much smaller than in crystalline alloys. The reason for these differences is not known. The value of β can be directly related to the quadratic coefficient of the spin-wave dispersion relation, called the *spin-wave stiffness, D*, by

References: p. 1730.

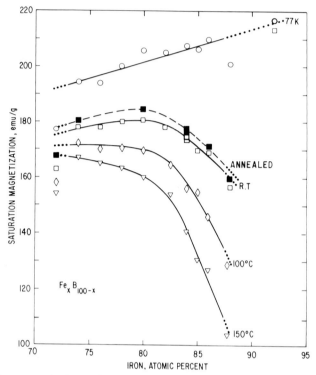

Fig. 23. Saturation magnetization measured at various temperatures for $Fe_x B_{100-x}$ amorphous alloys. Solid symbols measured after annealing at ~30° below crystallization temperature for 2 h. (From LUBORSKY and BECKER et al. [1979].)

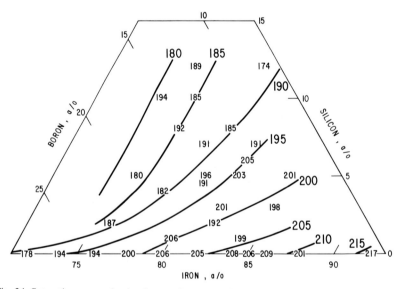

Fig. 24. Saturation magnetization in emu/g measured at 77K for Fe–B–Si amorphous alloys.

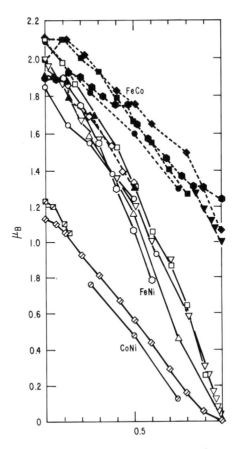

Fig. 25. Magnetization in Bohr magnetons per metallic atom versus atomic percent transition metal. Open symbols for FeNi amorphous alloys: □, $(FeNi)_{80}P_{10}B_{10}$; ○, ◇, $(FeNi)_{80}B_{20}$; ▽, $(FeNi)_{79}P_{13}B_8$; △, $(FeNi)_{80}P_{20}$; ◯, $(FeNi)_{80}P_{14}B_6$. Solid symbols for FeCo amorphous alloys: ◆, $(FeCo)_{80}P_{10}B_{10}$; ■, $(FeCo)_{80}B_{20}$; ●, $(FeCo)_{78}Si_{10}B_{12}$; ●, $(FeCo)_{80}P_{20}$; ▲, $(FeCo)_{80}P_{13}C_7$; ▼, $(FeCo)_{75}Si_{15}B_{10}$. Slashed symbols for CoNi amorphous alloys: ▨, $(CoNi)_{78}Si_{10}B_{12}$; ◈, $(CoNi)_{78}P_{14}B_8$; ⊘, $(CoNi)_{80}P_{20}$. (From Graham and Egami [1978].)

an equation $\beta \propto (1/D)^{3/2}$, in which the proportionality constant can be calculated from the material constants. The spin-wave stiffness constant is a measure of the stiffness of the spin-wave system. Experimentally it has been found, in the Fe_xB_{100-x} series, that D is proportional to x, as seen in fig. 26 (Luborsky et al. [1980]). Furthermore, the replacement of B by C, Si, or Ge increases the value of D in the order C < Si < Ge. It has been shown, on the basis of theoretical arguments, that D should be linearly related to T_c and extrapolate to $D = 0$ at $T_c = 0$. Many of the results reported to date do not extrapolate to zero, as shown in fig. 27 (Luborsky et al. [1980a]). It has therefore been suggested that the amorphous Fe–B–X alloys do not behave as described by the model but may be regarded as strong ferromagnets,

References: p. 1730.

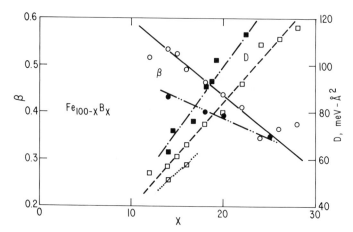

Fig. 26. Slope of the spin wave equation β and the exchange stiffness constant D for $Fe_x B_{100-x}$ amorphous alloys (from LUBORSKY et al. [1980a]).

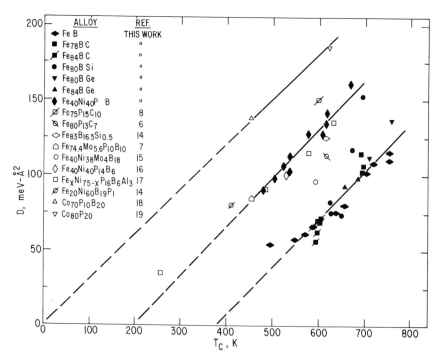

Fig. 27. The exchange stiffness constant D as a function of Curie temperature for various alloys. Solid symbols for D derived from magnetization data; open symbols for D derived from neutron measurements. (From LUBORSKY et al. [1980a].)

thus predicting that the values of D decrease with increasing T_c and extrapolate to finite T_c.

5.3.3. Anisotropy

Amorphous solids usually are assumed to have no long-range order and thus should be isotropic on a macroscale. However, magnetic anisotropic behavior is observed. Its origin is varied and sometimes not completely understood. In many cases, it reflects the existence of short-range order in the amorphous alloy. In addition, it has become something like folklore that in an ideal amorphous alloy the atomic-scale anisotropy would be averaged away so that there would be no anisotropy effects visible. CHI and ALBEN [1976] have shown from a simple model calculation that when the local random anisotropy is small the coercivity is indeed small but high coercivities are abruptly obtained as the local anisotropy increases and dominates the behavior. In the following subsections the various anisotropies existing in amorphous alloys will be described.

5.3.3.1. Structural and compositional anisotropy.
Anisotropic microstructures can arise in amorphous alloys owing to their preparation. These microstructures may involve density or compositional fluctuations which produce internal shape effects, that is, variations in magnetization. The resultant anisotropy field can then be no greater than that for long rods, namely $2\pi M_s$. The amorphous electro-deposited Co–P and CoNiP alloys have been studied in some detail. These alloys show a weak perpendicular anisotropy whose origin is believed to be due to such compositional fluctuations. Such fluctuations are often on the scale that can be seen by microprobe analysis or small-angle X-ray scattering, i.e., in the range of 1–1000 nm (10–10000 Å). Small-angle X-ray-scattering observations on these electrodeposited Co–P films have been interpreted by CHI and CARGILL [1976]) as showing the presence of oriented ellipsoidal scattering regions, assumed to be of high Co concentration. Annealing at temperatures well below crystallization reduced the perpendicular anisotropy and the small-angle scattering, suggesting that the annealing resulted in some homogenization of the alloy. In some amorphous alloys of Gd–Co, anisotropy fields greater than $2\pi M_s$ are observed even with no external stresses. The magnitude of K_u depends on the composition and conditions of preparation which influence the quantity of argon incorporated into the film.

5.3.3.2. Strain–magnetostriction anisotropy.
Most amorphous ferromagnetic materials have nonzero magnetostriction, λ. Internal stresses, σ, which may be uniform or non-uniform, arise from the original solidification or from subsequent fabrication. These strains couple with λ to produce an anisotropy, K_λ. Uniform strains are often induced in evaporated, sputtered or electrodeposited films due to the differential thermal expansion between the film and the substrate. The magnitude of λ and the direction and magnitude of σ will then determine the direction and magnitude of K_λ. An important example of non-uniform strains is observed in drum-quenched alloys of the $(TM)_{80}(P, B, Al...)_{20}$ type. The non-uniform strains develop during the preparation of the ribbon and result in a periodic fluctuation in the perpendicular component of anisotropy along the length of the tape. Thermal

References: p. 1730.

annealing removes the internal strains, causing the anisotropy to disappear. The domain structure and its disappearance on annealing reflect this perpendicular K_λ and its removal.

5.3.3.3. Directional-order anisotropy. Like crystalline alloys, amorphous alloys also order under the influence of a magnetic field or stress, applied at temperatures below the Curie temperature. This results in an *induced uniaxial anisotropy* arising from the ordering of both the magnetic and nonmagnetic atoms in relation to the applied field direction. *Directional-order* theory predicts the dependence of the magnitude of the directional order anisotropy, K_u, on various parameters. It will be seen that the behavior of K_u in amorphous alloys is very similar to the behavior in thin films or bulk alloys of Ni–Fe. Typical results are shown in fig. 28 (LUBORSKY [1978]).

In the case of magnetic ordering, the magnitude of K_u is expected to be a function of the anneal temperature, θ, the magnetization at the anneal temperature, M_θ, at the measurement temperature, M_T, and at 0 K, M_0, and the concentrations of the ordering atoms, c_A and c_B, respectively, as given by:

$$K_u + Af(c)(M_T/M_0)^2(M_\theta/M_0)^2/\theta, \tag{4}$$

where A is a constant which depends on the atomic arrangement and the range of interactions considered. For dilute solutions and *monatomic directional order* (i.e., where only one species takes up preferential positions) $f(c) = c$ where c is the concentration of the ordering species. For *di-atomic directional order* (where pairs of distinct atoms arrange themselves in preferential directions), $f(c) = c_a^2$, where c_a is the concentration of the dilute species. If neither constituent is dilute, but assuming

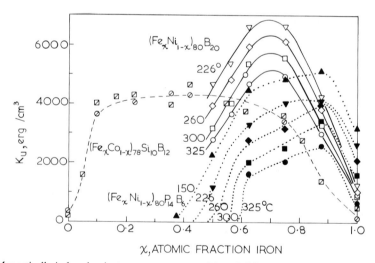

Fig. 28. Magnetically induced anisotropy versus composition. Solid and open symbols both annealed to equilibrium; slashed symbols obtained by heating at successively higher temperatures and on cooling from 380°C at 1.6 K/min.

ideal solutions, i.e., for negligible nonmagnetic interaction energy, then

$$f(c) = Nnc_a^2 c_b^2 / 2,\tag{5}$$

where N is the number of atoms per unit volume and n is the number of possible orientations of each pair referred to a crystal lattice. For non-ideal solutions the effective nonmagnetic interaction energy must be included in the expression for $f(c)$. In amorphous alloys we expect the interaction energy to be negative, i.e., leading to precipitation. These relations for amorphous Fe–Ni–B alloys were studied in detail by LUBORSKY and WALTER [1977a,b]. The reported values of K_u can be changed reversibly by changing the field direction and anneal temperature. The nonzero value of K_u at $x = 1$ is interpreted as due to ordering of the boron. The temperature-dependence observed fits the theoretically expected dependence on anneal temperature. The dependence of K_u on Fe–Ni composition follows the theoretical dependence for non-ideal solid solutions with a negative interaction energy, i.e., of the type leading to precipitation, as expected.

There have also been studies of the kinetics of the reorientation of K_u (LUBORSKY and WALTER [1977c]). The time constants for this reorientation for various amorphous alloys show that the amorphous alloys are more closely related in their behavior to the quenched crystalline alloys than to the annealed crystalline alloys. In the quenched crystalline alloys, the rate-determining step in the reorientation has been associated with the excess vacancies present. It has thus been suggested that the disordered structure of the amorphous alloys has produced a similar atomic environment and thus similar kinetics for the reorientation. The $Fe_{40}Ni_{40}P_{14}B_6$ alloy surprisingly exhibited simple first-order kinetics in the reorientation of its anisotropy, as measured by the changes in its remanent-magnetization ratio M_r/M_s. This suggests that a uniform atomic environment exists around the reordering species. On the other hand, the reorientation kinetics of the $Fe_{40}Ni_{40}B_{20}$ alloy did not exhibit simple first-order kinetics; the kinetics could be fitted to a distribution of time constants or by second-order kinetics with equal concentrations of the two species.

Amorphous alloys are also susceptible to stress-induced ordering. As in crystalline alloys, this ordering presumably occurs via the interactions of the strain produced with the magnetostriction. The activation energy for stress-induced ordering is about twice as large as the values for magnetic ordering. It is thus concluded that stress-induced directional ordering involves different atomic species or motions than those involved in magnetically induced ordering. In addition, the final state produced by the two ordering processes must be different.

5.3.4. Magnetostriction

Since amorphous alloys are isotropic on a macroscale, the magnetostriction is expressed as a single saturation constant, λ_s. Measured values of λ_s are comparable to those in the corresponding crystalline phase. Some results are shown in fig. 29, for various series of amorphous alloys. Most measurements were obtained using strain gages but the small size of the ribbons may mean that the gage-backing and cement may not be negligible, meaning that the measurements represent only lower limits.

References: p. 1730.

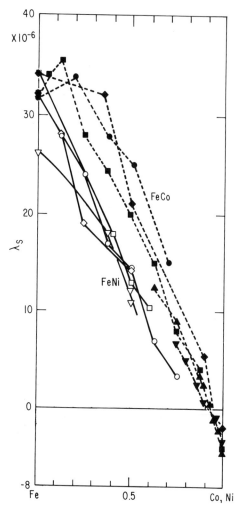

Fig. 29. Saturation magnetostriction measured at room temperature versus atomic percent transition metal. Open symbols for FeNi amorphous alloys; ◇, $(FeNi)_{80}P_{13}C_7$; ○, $(FeNi)_{80}B_{20}$; □, $(FeNi)_{78}B_{14}Si_8$; ▽, $(FeNi)_{80}P_{14}B_6$. Solid symbols for FeCo amorphous alloys: ■, $(FeCo)_{80}B_{20}$; ▲, $(FeCo)_{80}B_{20}$; ◆, $(FeCo)_{78}B_{14}Si_8$; ▼, $(FeCo)_{75}S_{15}B_{10}$; ●, $(FeCo)_{80}P_{13}C_7$. (From GRAHAM and EGAMI [1978].)

TSUYA et al. [1975] have used a capacitance technique to avoid this problem. Note that both positive and negative values of λ_s are achievable.

O'HANDLEY [1977] has explored a wide range of temperatures. Band models which adequately describe the λ_s of crystalline Fe–Ni alloys do not apply to amorphous alloys. A combination of the single-ion and two-ion models does account for the temperature dependence λ_s of amorphous alloys.

5.3.5. Low-field properties

The low-field properties include the coercivity, remanence, losses, and permeability. The very low coercivities and losses make amorphous alloys particularly attractive in many applications, e.g., in transformers of all sizes. The low-field properties are usually the controlling parameters in the applications.

At least five pinning effects have been identified and discussed by KRONMÜLLER [1981] as contributing to the total coercivity. In order of increasing importance these are:

a) *Intrinsic fluctuations* of exchange energies and local anisotropies (10^{-3}–1 mOe), $H_c(i)$.

b) Clusters of *chemical short-range-ordered* regions (< 1 mOe), $H_c(SO)$.

c) *Surface irregularities* (< 5 mOe), $H_c(surf)$.

d) *Relaxation effects* due to local structural rearrangements (0.1–10 mOe), $H_c(rel)$.

e) *Volume pinning* of domain walls by defect structures in magnetostrictive alloys (10–100 mOe), $H_c(\sigma)$.

Within the framework of the statistical potential theory (KRONMÜLLER *et al.* [1979]), on the basis of which the coercivity has to be determined, the contributions of the different statistically distributed pinning centers must be added quadratically. However, the effect of $H_c(rel)$ has to be added linearly because each domain wall sees a stabilizing energy due to the relaxation with a wavelength of $2\delta_0$, where δ_0 is related to the domain wall width. This wavelength is of the same order of magnitude as the wavelength of the potential due to statistical fluctuations. Thus, the resultant total coercivity is:

$$H_c(total) = \left[H_c^2(\sigma) + H_c^2(surf) + H_c^2(SO) + H_c^2(i) \right]^{1/2} + H_c(rel). \qquad (6)$$

In the special case where the contribution of the pinning due to surface irregularities is larger than all other contributions the various terms add linearly, i.e.:

$$H_c(total) = H_c(\sigma) + H_c(rel) + H_c(surf) + H_c(SO) + H_c(i). \qquad (5.5)$$

This is so because the wavelength of the surface-pinning is much larger ($\sim 10~\mu m$) than that of the intrinsic fluctuations ($< 0.5~\mu m$). Explicit mathematical expressions have been derived for each of these contributions, giving the limits on H_c shown above.

The question now is how we can minimize the coercivity. Defect structures which give rise to the maximum contribution to the coercivity cannot be completely suppressed. These defect structures are generated during the rapid quenching process by an agglomeration of the free volume. This free volume is stable with respect to annealing and thus cannot be removed completely. Thus, the interaction of domain walls with defects can only be reduced significantly by using nonmagnetostrictive alloys. Then short-range order effects and relaxation effects become important. These latter effects may be minimized by a suitable anneal treatment. The intrinsic coercivity may only be detectable in ribbons with extremely smooth surfaces, with $\lambda = 0$, and after an anneal arranged to avoid relaxation and short range order effects.

It is well established that amorphous alloys with high saturation magnetization

References: p. 1730.

Table 2

Some typical magnetic properties of amorphous alloys compared to commercial alloys after optimum magnetic anneals.

Alloy	60 Hz, 1.4 T		50 kHz, 0.1T	
	Core loss W/kg	Exciting power VA/kg	Core loss W/kg	Exciting power VA/kg
$Fe_{81}B_{13.5}Si_{3.5}C_2$ (METGLAS 2605 SC)	0.20	0.31	6	26
$Fe_{79}B_{16}Si_5$ (METGLAS 2605 S3)	–	–	9	21
Fe–3.2% Si (0.28 mm thick)	0.90	1.1	–	–
Supermalloy	–	–	9.4	–
Ferrite, H7A	–	–	11	–

have losses at 60 Hz that are smaller than the best commercial grain-oriented Fe–$3\frac{1}{4}$% Si by a factor of three to twenty (LUBORSKY and BECKER [1979]) as indicated in table 2. Since amorphous alloys have four times the resistivity and are about 1/5 the thickness of Fe–$3\frac{1}{4}$% Si we would predict, using a simple classical model, that the eddy-current losses would be reduced by a factor of ten; i.e., eddy-current losses are proportional to the inverse of resistivity and the square root of thickness. Thus the remaining losses at 60 Hz in amorphous alloys, for example in $Fe_{81}B_{13.5}Si_{3.5}C_2$, are due to hysteresis losses associated with domain wall motion. Refining the domain structure, for example by inducing an easy axis at some angle to the length of the ribbon, leads to reduction in losses (LUBORSKY and LIVINGSTON [1982]). The exciting power necessary to achieve a given level of induction remains relatively constant for field angles out to about 30° and then rises rapidly.

The same annealing which minimizes losses for the iron-rich metallic glasses for 60 Hz applications does not minimize losses in the higher frequency region, e.g., from 1 to 100 kHz. Minimum losses in this frequency region are achieved by a higher-temperature anneal which induces a precipitation of fine crystallites (DATTA et al. [1982]). These precipitates provide nucleation sites for domain walls, thereby refining the domain structure and reducing losses. Typical results are shown in table 2.

6. Magnetic measurements in metallurgy

When alloys contain one or more ferromagnetic phases, magnetic measurements can provide significant metallurgical information (HOSELITZ [1952, 1960] and BERKOWITZ [1969]). Curie temperatures are characteristic of particular phases and compositions. Therefore, measurement of magnetization versus temperature at fixed field, or thermomagnetic analysis, is often used to identify the Curie temperatures of

the various phases present, and thereby to identify those phases and their compositions. In special cases, changes in the magnetic symmetry of phases with temperature
can lead to discontinuities in magnetization–temperature curves that can also be
used to identify the presence of particular phases (BERKOWITZ *et al.* [1979]).
Thermomagnetic analysis, of course, is only useful if no metallurgical changes are
produced by the heating.

If only one ferromagnetic phase is present, *saturation magnetization* can be used
as a measure of the volume fraction of that phase. This is frequently used to study
phase transformations, particularly in steels, in which Fe-rich austenite is not
ferromagnetic, but ferrite and martensite are. For example, saturation-magnetization
measurements have been used to determine the amounts of retained austenite
(KRAWIARZ *et al.* [1979]) or strain-induced martensite (CHANANI *et al.* [1971],
LARBALESTIER and KING [1973] and COLLINGS [1979b]) in steels. Measurement of
saturation magnetization may require larger fields than are available in some
situations, and magnetic quantities that are simpler to measure are often used
instead to determine the volume fraction of ferromagnetic phase. *Permeability* is
commonly used, as is the *force* exerted by the sample on a standard permanent
magnet. Force or permeability measurements are often used, for example, to
determine the volume fraction of ferrite in duplex steel castings or welds (RATZ and
GUNIA [1969]). Such measurements can be useful, but depend not only on saturation
magnetization but also on sample microstructure and on experimental details,
making determination of volume fractions less reliable.

Since measurements of Curie temperature and saturation magnetization can be
used to identify compositions and volume fractions of ferromagnetic phases, they
have frequently been used in phase-diagram determination. It is less well appreciated
that the free energies involved in the ferromagnetic (or antiferromagnetic) state can
influence phase equilibria and thereby affect the boundaries in the phase diagram
(MIODOWNIK [1982]).

The structure-sensitivity of various features of the hysteresis loop, such as
permeability, coercivity, approach to saturation, etc., has led to use of these magnetic
parameters as measures of microstructural parameters such as grain size, internal
stresses, dislocation densities, etc. Coercivity, for example, has been used as a
measure of microstructural refinement in Co–WC composites (FISCHMEISTER and
EXNER [1966] and FREYTAG *et al.* [1978]). The noise signals produced by irregular
domain-wall motion ("Barkhausen noise") have been used to determine internal
stresses (GARDNER *et al.* [1977] and JAMES and BUCK [1980]). Many earlier examples
have been discussed by HOSELITZ [1952, 1960]. However, such measurements must
be used with great care, since these quantities usually depend on more than one
microstructural variable.

One special case in which details of the magnetization curve can provide considerable metallurgical information is that of an alloy containing "superparamagnetic"
particles (§ 3.1). These are ferromagnetic particles that are sufficiently fine that the
assembly is in thermal equilibrium with an applied field. This yields a non-hyster-
etic, temperature-dependent magnetization curve that can be used to determine not

References: p. 1730.

only volume fraction but also average particle size, distribution of sizes, and particle shape (BEAN and LIVINGSTON [1959]).

Another measurement sometimes used to provide metallurgical information is that of *magnetic anisotropy*, frequently determined by *torque* measurements in an applied field. Where the magnetization curve is determined largely by crystal anisotropy, this can be a useful means of *texture* determination. In other cases, shape, or stress, or directional-ordering anisotropies may be dominant, and torque measurements can provide quantitative information about the dominant anisotropy.

Direct metallographic observation of the distribution of ferromagnetic phases can be enhanced with the use of ferromagnetic colloids (GRAY [1974] and GRAY et al. [1977]). Magnetic-powder techniques are also used for flaw detection in steels. In special cases, magnetic domain patterns can be revealed using polarized light, aiding phase identification and orientation determination (LIVINGSTON [1981c]).

The hyperfine magnetic fields experienced by the nuclei in alloys can be analyzed through Mössbauer spectra, nuclear magnetic resonance, and related measurements. Since they are characteristic of particular phases and compositions, they have been used in phase-diagram determination in several alloy systems (RENO et al. [1978]) and in determining ferrite content in duplex steels (SCHWARTZENDRUBER et al. [1974]).

Finally, magnetic measurements can be used in metallurgical studies even where the phases are paramagnetic rather than ferromagnetic, provided the paramagnetic susceptibilities of the constituent phases are known (COLLINGS [1975, 1979a]). The disturbing effect of ferromagnetic impurities on such measurements is usually surmounted by assuming that the ferromagnetic material saturates in moderate fields, and that the high-field susceptibility is caused by the paramagnetic phase alone. The presence of superparamagnetism or of high anisotropies in the ferromagnetic phase may make this assumption questionable in some cases.

References

ADLER, E., and H. PFEIFFER, 1974, IEEE Trans. Magn. **MAG-10**, 172.
AMES, S.L., 1970, J. Appl. Phys. **41**, 1032.
BARTLETT, R.W., and P.J. JORGENSON, 1974, J. Less-Common Met. **37**, 21.
BEAN, C.P., and J.D. LIVINGSTON, 1959, J. Appl. Phys. **30**, 120S.
BECKER, J.J., 1973, IEEE Trans. Magn. **MAG-9**, 161.
BELLI, Y., M. OKADA, G. THOMAS, M. HOMMA and H. KANEKO, 1978, J. Appl. Phys. **49**, 2049.
BERKOWITZ, A.E., 1969, in: Magnetism and Metallurgy, vol. 1, eds. A.E. Berkowitz and E. Kneller (Academic, New York) p. 331.
BERKOWITZ, A.E., J.D. LIVINGSTON, B.D. NATHAN and J.L. WALTER, 1979, J. Appl. Phys. **50**, 1754.
BOOTHBY, O.L., and R.M. BOZORTH, 1947, J. Appl. Phys. **18**, 173.
BORDERS, J.A., 1979, Ann. Rev. Mater. Sci. **9**, 313.
BROWN, W.F., Jr., 1962, J. Appl. Phys. **33**, 3022.
CHANANI, G.R., V.F. ZACKAY and E.R. PARKER, 1971, Metallurg. Trans. **2**, 133.
CHEN, C.W., 1977, Magnetism and Metallurgy of Soft Magnetic Materials (North-Holland, Amsterdam).
CHI, G.C., and R. ALBEN, 1976, AIP Conf. Proc. **34**, 316.
CHI, G.C., and G.S. CARGILL, 1976, Mater. Sci. Eng. **23**, 155.

CHIKAZUMI, S., and C.D. GRAHAM Jr., 1969, in: Magnetism and Metallurgy, vol. 2, eds. A.E. Berkowitz and E. Kneller (Academic, New York) ch. 12.

CHIN, G.Y. 1971, Adv. Mater. Res. **5**, 217.

CHIN, G.Y., and J.H. WERNICK, 1980, in: Ferromagnetic Materials, vol. 2, ed. E.P. Wohlfarth (North-Holland, Amsterdam) ch. 2.

CHIN, G.Y., S. JIN, M.L. GREEN, R.C. SHERWOOD and J.H. WERNICK, 1981, J. Appl. Phys. **52**, 2536.

COLLING, D.A., and R.G. ASPEN, 1969, J. Appl. Phys. **40**, 1571.

COLLING, D.A., and R.G. ASPEN, 1970, J. Appl. Phys. **41**, 1040.

COLLINGS, E.W., 1975, J. Less-Common Met. **39**, 63.

COLLINGS, E.W., 1979a, Metallurg. Trans. **10A**, 463.

COLLINGS, E.W., 1979b, Cryogenics **19**, 215.

CRAIK, D.J., and E.W. HILL, 1977, J. Physique **38**, C1-39.

CULLITY, B.D., 1972, Introduction to Magnetic Materials (Addison–Wesley, Reading, MA).

DATTA, A., N. DeCRISTOFARO and L.A. DAVIS, 1982, in: Proc. Fourth Int. Conf. on Rapidly Quenched Metals, Sendai, Japan (1981) eds. T. Masumoto and K. Suzuki (Japan. Inst. Metals, Sendai) pp. 1007 and 1031.

DEN BROEDER, F.J.A., and K.H.J. BUSCHOW, 1980, J. Less-Common Met. **70**, 289.

DEN BROEDER, F.J.A., and H. ZIJLSTRA, 1976, J. Appl. Phys. **47**, 2688.

DE VOS, K.J., 1969, in: Magnetism and Metallurgy, vol. 1, eds. A.E. Berkowitz and E. Kneller (Academic, New York) p. 473.

DIETZ, G., 1977, J. Magn. Magn. Mater. **6**, 47.

ENGLISH, A.T., and G.Y. CHIN, 1967, J. Appl. Phys. **38**, 1183.

FIEDLER, H.C., 1977, IEEE Trans. Magn. **MAG-13**, 1433.

FISCHMEISTER, H., and H.E. EXNER, 1966, Arch. Eisenhüttenw. **37**, 499.

FREYTAG, J., P. WALTER and H.E. EXNER, 1978, Z. Metallk. **69**, 546.

GARDNER, C.G., G.A. MATZKANIN and J. LANKFORD, 1977, Int. Adv. Nondestructive Testing **5**, 201.

GAUNT, P., 1966, Phil. Mag. **13**, 579.

GOSS, N.P., 1934, U.S. Patent **1965**, 559.

GOTO, I., I. MATOBA, T. IMANAKA, T. GOTOH and T. KAN, 1975, Proc. 2nd EPS Conf. on Soft Magnet Materials (Wolfson Centre for Magnetic Technology, Cardiff, Wales) p. 262.

GOULD, H.L.B., and D.H. WENNY, 1957, AIEE Spec. Publ. T-97, 675.

GRAHAM, C.D., Jr., 1959, Magnetic Properties of Metals and Alloys (ASM, Metals Park, OH) ch. 13.

GRAHAM, C.D., Jr., and T. EGAMI, 1978, Ann. Rev. Mater. Sci. **8**, 423.

GRAY, R.J., 1974, in: Microstructural Science, vol. 1, eds. R.J. Gray and J.L. McCall (American Elsevier, New York) p. 159.

GRAY, R.J., R.S. CROUSE, V.K. SIKKA and R.T. KING, 1977, in: Microstructural Science, vol. 5, eds. J.D. Braun, H.W. Arrowsmith and J.L. McCall (Elsevier, New York) p. 65.

GRENOBLE, H.E., 1977, IEEE Trans. Magn. **MAG-13**, 1427.

GROGER, B., 1979, J. Magn. Magn. Mater. **13**, 53.

HAASEN, P., 1972, Mater. Sci. Eng. **9**, 191.

HAASEN, P., 1977, Contemp. Phys. **18**, 373.

HILZINGER, H.R., 1977, Appl. Phys. **12**, 253.

HILZINGER, H.R., and H. KRONMÜLLER, 1973, Phys. Stat. Sol. (b) **59**, 71.

HOFFMANN, A., 1971, Z. Angew. Phys. **32**, 236.

HOMMA, M., E. HORIKOSHI, T. MINOWA and M. OKADA, 1980, Appl. Phys. Lett. **37**, 92.

HOMMA, M., M. OKADA, T. MINOWA and E. HORIKOSHI, 1981, IEEE Trans. Magn. **MAG-17**, 3473.

HOSELITZ K., 1952, in: Ferromagnetic Properties of Metals and Alloys (Oxford Univ. Press, London) ch. 4.

HOSELITZ, K., 1960, in: The Physical Examination of Metals, 2nd Ed., eds. B. Chalmers and A.G. Quarrell (Arnold, London) p. 225.

HULTGREN, R., P.D. DESAI, D.T. HAWKINS, M. GLEISER and K.K. KELLEY, 1973, Selected Values of the Thermodynamic Properties of Binary Alloys (ASM, Metals Park, OH) p. 871.

INOUE, K., and H. KANEKO, 1981, U.S. Patent 4 273 595.

JAMES, M.R., and O. BUCK, 1980, CRC Reviews **9**(1), 61.

JIN, S., 1979, IEEE Trans. Magn. **MAG-15**, 1748.

KANEKO, H., M. HOMMA and K. SUZUKI, 1968, Trans. Japan. Inst. Metals **9**, 124.

KANEKO, H., M. HOMMA and K. NAKAMURA, 1971, AIP Conf. Proc. **5**, 1088.

KANEKO, H., M. HOMMA, M. OKADA, S. NAKAMURA and N. IKUTA, 1976, AIP Conf. Proc. **29**, 620.

KAVESH, S., 1978, in: Metallic Glasses, eds. J.J. Gilman and H.I. Leamy (ASM, Metals Park, OH) p. 36.

KAZAMA, N.S., M. MITERA and T. MASUMOTO, 1978, in: Proc. Third Int. Conf. on Rapidly Quenched Metals, Univ. of Sussex, England, 1978, ed. B. Cantor (The Metals Society, London) vol. 2, p. 164.

KIKUCHI, S., and S. ITO, 1972, IEEE Trans. Magn. **MAG-8**, 344.

KOJIMA, S., T. OHTANI, N. KATO, K. KOJIMA, Y. SAKAMOTO, I. KONNO, M. TSUKAHARA and T. KUBO, 1974, AIP Conf. Proc. **24**, 768.

KRAWIARZ, J., T. MALKIEWICZ and A. MAZUR, 1979, Mém. Sci. Rev. Métallurg. **76**, 5.

KRONMÜLLER, H., 1978, in: Proc. 2nd Int. Symp. on Magnetic Anisotropy and Coercivity in Rare-Earth–Transition-Metal Alloys, ed. K.J. Strnat (Univ. of Dayton, OH) p. 1.

KRONMÜLLER, H., 1981, J. Appl. Phys. **52**, 1859; 1981, J. Magn. Magn. Mater. **24**, 159.

KRONMÜLLER, H., M. FÄHNLE, N. DOMANN, H. GRIMM, R. GRIMMAND and B. GROGER, 1979, J. Magn. Magn. Mater. **13**, 53.

KUNZ, W., and F. PFEIFER, 1976, AIP Conf. Proc. **34**, 63.

LARBALESTIER, D., and H.W. KING, 1973, Cryogenics **13**, 160.

LIEBERMANN, H.H., 1979, IEEE Trans. Magn. **MAG-15**, 1393.

LIEBERMANN, H.H., 1980, Mater. Sci. Eng. **43**, 203.

LIEBERMANN, H.H., 1981, Mater. Sci. Eng. **49**, 185.

LITTMANN, M.B., 1967, J. Appl. Phys. **38**, 1104.

LITTMANN, M.B., 1971, IEEE Trans. Magn. **MAG-7**, 48.

LITTMANN, M.B., 1982, J. Appl. Phys. **53**, 2416.

LIVINGSTON, J.D., 1973a, AIP Conf. Proc. **10**, 643.

LIVINGSTON, J.D., 1973b, Phys. Stat. Sol. (a) **18**, 579.

LIVINGSTON, J.D., 1981a, Prog. Mater. Sci., Chalmers Anniversary Volume, 243.

LIVINGSTON, J.D., 1981b, J. Appl. Phys. **52**, 2544.

LIVINGSTON, J.D., 1981c, J. Appl. Phys. **52**, 2506.

LIVINGSTON, J.D., and D.L. MARTIN, 1977, J. Appl. Phys. **48**, 1350.

LUBORSKY, F.E., 1961, J. Appl. Phys. **32**, 171S.

LUBORSKY, F.E., 1976, AIP Conf. Proc. **29**, 209.

LUBORSKY, F.E., 1978, J. Magn. Magn. Mater. **7**, 143.

LUBORSKY, F.E., 1980a, J. Appl. Phys. **51**, 2808.

LUBORSKY, F.E., 1980b, Amorphous Ferromagnets, in: Ferromagnetic Materials, vol. 1, ed. E.P. Wohlfarth (North-Holland, Amsterdam) ch. 6.

LUBORSKY, F.E., and J.J. BECKER, 1979, IEEE Trans. Magn. **MAG-15**, 1939.

LUBORSKY, F.E., and J. LIVINGSTON, 1982, IEEE Trans. Magn. **MAG-18**, 908.

LUBORSKY, F.E., and C.R. MORELOCK, 1964a, J. Appl. Phys. **35**, 2055.

LUBORSKY, F.E., and C.R. MORELOCK, 1964b, in: Proc. Int. Conf. on Magnetism, Nottingham (The Institute of Physics and The Physical Society, London) p. 763.

LUBORSKY, F.E., and J.L. WALTER, 1977a, IEEE Trans. Magn. **MAG-13**, 1635.

LUBORSKY, F.E., and J.L. WALTER, 1977b, IEEE Trans. Magn. **MAG-13**, 953.

LUBORSKY, F.E., and J.L. WALTER, 1977c, Mater. Sci. Eng. **28**, 77.

LUBORSKY, F.E., H.H. LIEBERMANN, J.J. BECKER and J.L. WALTER, 1978, in: Proc. Third Int. Conf. on Rapidly Quenched Metals, Univ. of Sussex, England, 1978, ed. B. Cantor (The Metals Society, London) vol. 2, p. 188.

LUBORSKY, F.E., J.L. WALTER, H.H. LIEBERMANN and E.P. WOHLFARTH, 1980a, J. Magn. Magn. Mater. **15-18**, 1351.

LUBORSKY, F.E., J.L. WALTER and E.P. WOHLFARTH, 1980b, J. Phys. **F10**, 959.

MAHAJAN, S., E.M. GYORGY, R.C. SHERWOOD, S. JIN, S. NAKAHARA, D. BRASEN and M. EIBSCHUTZ, 1978, Appl. Phys. Lett. **32**, 688.

MAJOR, B.V., M.C. MARTIN and M.W. BRANSON, 1975, in: Proc. 2nd EPS Conf. on Soft Magnetic Materials (Wolfson Centre for Magnetics Technology, Cardiff, Wales) p. 103.

MALAGARI, F.A., 1977, IEEE Trans. Magn. **MAG-13**, 1437.

MCCAIG, M., 1977, Permanent Magnets in Theory and Practice (Wiley, New York).

MCCURRIE, R.A., 1982, in: Ferromagnetic Materials, vol. 3, ed. E.P. Wohlfarth (North-Holland, Amsterdam) p. 107.

MCCURRIE, R.A., and D.G. HAWKRIDGE, 1975, Phil. Mag. **32**, 923.

MCINTYRE, D.A., 1970, J. Phys. **D3**, 1430.

MENTH, A., H. NAGEL and R.S. PERKINS, 1978, Ann. Rev. Mater. Sci. **8**, 21.

MINOWA, T., M. OKADA and M. HOMMA, 1980, IEEE Trans. Magn. **MAG-16**, 529,.

MIODOWNIK, A.P., 1982, Bull. Alloy Phase Diagrams **2**, 406.

MISHRA, R.K., and G. THOMAS, 1979, in: Proc. 4th Int. Workshop in Rare-Earth–Cobalt Magnets (Society for Non-Traditional Technology, Tokyo) p. 301.

MIYAZAKI, T., E. SAWADA and Y. ISHIJIMA, 1972, IEEE Trans. Magn. **MAG-8**, 501.

NÉEL, L., 1954, J. Phys. Radium **15**, 225.

NEMBACH, E., C.K. CHOW and D. STOCKEL, 1977, Physica B **86–88**, 1415.

NOZAWA, T., T. YAMAMOTO, Y. MATSUO and Y. OHYA, 1978, IEEE Trans. Magn. **MAG-14**, 252.

O'HANDLEY, R.C., 1977, in: Amorphous Magnetism II, eds. R.A. Levy and R. Hasegawa (Plenum, New York) p. 379.

OHTANI, T., N. KATO, S. KOJIMA, K. KOJIMA, Y. SAKAMOTO, I. KONNO, M. TSUKAHARA and T. KUBO, 1977, IEEE Trans. Magn. **MAG-13**, 1328.

PAVUNA, D., 1982, J. Mater. Sci. **16**, 2419.

PFEIFER, F., 1966a, Z. Metallk. **57**, 295.

PFEIFER, F., 1966b, Z. Metallk. **57**, 240.

PFEIFER, F. and C. RADELOFF, 1980, J. Magn. Magn. Mater. **19**, 190.

PFEIFFER, I., 1969, Cobalt **44**, 115.

RASTOGI, P.K., 1976, AIP Conf. Proc. **34**, 61.

RASTOGI, P.K., 1977, IEEE Trans. Magn. **MAG-13**, 1448.

RASTOGI, P.K., and J.M. SHAPIRO, 1973, IEEE Trans. Magn. **MAG-9**, 122.

RATZ, G.A., and R.B. GUNIA, 1969, Met. Prog. **95**(1), 76.

RENO, R.C., L.J. SCHWARTZENDRUBER, G.C. CARTER and L.H. BENNETT, 1978, in: Applications of Phase Diagrams in Metallurgy and Ceramics, ed. G.C. Carter (NBS Special Publication 496) p. 450.

ROWLANDS G., 1976, J. Phys. **D9**, 1267.

SCHOLEFIELD, H.H., R.V. MAJOR, B. GIBSON and A.P. MARTIN, 1967, Brit. J. Appl. Phys. **18**, 41.

SCHWARTZENDRUBER, L.J., L.H. BENNETT, E.A. SCHOEFER, W.T. DE LONG and H.C. CAMPBELL, 1974, Welding J. **53**, 1-S.

SENNO, H., Y. YANAGIUCHI, M. SATOMI, E. HIROTA and S. HAYAKAWA, 1977, IEEE Trans. Magn. **MAG-13**, 1475.

SHIBAYA, H., and I. FUKUDA, 1977, IEEE Trans. Magn. **MAG-13**, 1029.

SHILLING, J.W., and G.L. HOUZE Jr., 1974, IEEE Trans. Magn. **MAG – 10**, 195.

SHILLING, J.W., and W.A. SOFFA, 1978, Acta Metall. **26**, 413.

SHILLING, J.W., W.G. MORRIS, M.L. OSBORN and P. RAO, 1978, IEEE Trans. Magn. **MAG-14**, 104.

SHIMANAKA, H., I. MATOBA, T. ICHIDA, S. KOBAYASHI and T. FUNAHASHI, 1975, in: Proc. 2nd EPS Conf. on Soft Magnetic Materials (Wolfson Centre for Magnetic Technology, Cardiff, Wales) p. 269.

SWIFT, W.M., W.H. DANIELS and J.W. SHILLING, 1975, IEEE Trans. Magn. **MAG-11**, 1655.

SWISHER, J.H., and E.O. FUCHS, 1970, J. Iron Steel Inst. **208**, 777.

SWISHER, J.H., A.T. ENGLISH and R.C. STOFFERS, 1969, Trans. ASM **62**, 257.

TAGUCHI, S., 1977, Trans. Iron Steel Inst. Japan **17**, 604.

THOMAS, B., 1975, in: Proc. 2nd Conf. EPS on Soft Magnetic Materials (Wolfson Centre for Magnetics Technology, Cardiff, Wales) p. 109.

THORNBURG, D.R., 1969, J. Appl. Phys. **40**, 1579.

TSUYA, N., and K.I. ARAI, 1978, Solid State Phys. **13**, 237.

TSUYA, N., K.I. ARAI, Y.Y. SHIRAGA, M. YAMADA and T. MASUMOTO, 1975, Phys. Stat. Sol. (a) **31**, 557.

VAN DEN BROEK, C.A.M., and A.L. STUIJTS, 1977, Philips Tech. Rev. **37**, 157.

VINCENT, J.H., J.G. HERBERTSON and H.A. DAVIES, 1982, in: Proc. 4th Int. Conf. on Rapidly Quenched metals, Sendai, Japan, 1981, eds. T. Masumoto and K. Suzuki (Japan. Inst. Metals, Sendai) p. 77.

YAMAMOTO, T., and Y. UTSUSHIKAWA, 1976, J. Japan. Inst. Metals **40**, 975.

YERMAKOV, A.Ye., N.I. SOLOKOVSKAYA, U.A. TSURIN, G.V. IVANOVA and L.M. MAGAT, 1978, Phys. Met. Metallogr. **46**(4), 46.

ZIJLSTRA, H., 1967, Experimental Methods in Magnetism, vol. 2 (North-Holland, Amsterdam; Wiley–Interscience, London).

ZIJLSTRA, H., 1983, in: Ferromagnetic Materials, vol. 3, ed. E.P. Wohlfarth (North-Holland, Amsterdam) p. 37.

Further reading

R.M. Bozorth, Ferromagnetism (Van Nostrand, New York, 1951).

K. Hoselitz, Ferromagnetic Properties of Metals and Alloys (Oxford Univ. Press, London, 1952).

E. Kneller, Ferromagnetismus (Springer, Berlin, 1962).

A.E. Berkowitz and E. Kneller, eds., Magnetism and Metallurgy, vols. 1 and 2 (Academic, New York, 1969).

R.S. Tebble and D.J. Craik, Magnetic Materials (Wiley–Interscience, London, 1969).

B.D. Cullity, Introduction to Magnetic Materials (Addison–Wesley, Reading, MA, 1972).

C.W. Chen, Magnetism and Metallurgy of Soft Magnetic Materials (North-Holland, Amsterdam, 1977).

E.P. Wohlfarth, ed., Ferromagnetic Materials, vols. 1 and 2 (North-Holland, Amsterdam, 1980; vol. 3, 1982).

G. Conderchon and J.F. Tiers, Some Aspects of Magnetic Properties of Ni–Fe and Co–Fe alloys, J. Magn. Magn. Mater. 26 (1982) 196.

R. Hasegawa, ed., Glassy Metals: Magnetic, Chemical and Structural Properties (CRC Press, Boca Raton, FL, 1983) chs. 4–6.

F.E. Luborsky, ed., Amorphous Metallic Alloys (Butterworths, London, 1983) chs. 14–20.

CHAPTER 27

SUPERCONDUCTING MATERIALS

W.L. JOHNSON

M.W. Keck Laboratory of Engineering Materials
California Institute of Technology
Pasadena, CA 91125, USA

R.W. Cahn and P. Haasen, eds.
Physical Metallurgy; third, revised and enlarged edition
© *Elsevier Science Publishers BV, 1983*

1. Introduction

This chapter concerns the nature and occurrence of the superconducting state in metals, alloys, and intermetallic compounds. It deals with the microscopic material parameters which determine the superconducting transition temperature, magnetic behavior, and other characteristic features of superconductors. Particular emphasis will be placed on the properties of actual materials and the study of superconductivity in the context of a metallurgical tool.

2. Background and theory

This section outlines the theory and phenomenological background essential to understanding the sections that follow. The later sections deal with the properties of actual superconducting materials.

2.1. Phenomenology

Superconductivity is a rather ubiquitous phenomenon occurring in 27 pure metallic elements with observed superconducting transition temperatures, T_c, ranging from 0.026 K for Be to 9.25 K for Nb. The number of known alloys and intermetallic compounds exhibiting superconductivity numbers in the thousands. An intermetallic compound, Nb_3Ge having the A15 structure, exhibits the highest known transition temperature of 23.2 K.

Superconductors are characterized by the absence of dc electrical resistance below the transition temperature, as first observed by KAMERLINGH ONNES [1911]. In addition, most pure superconducting metals exhibit perfect diamagnetism in an applied external magnetic field ranging up to a critical field $H_c(T)$ which depends on temperature. This is expressed by

$$B = H + 4\pi M = 0 \quad \text{for } H < H_c(T),$$
$$M = 0 \qquad\qquad \text{for } H > H_c(T), \tag{1}$$

so that $\chi = M/H = -1/4\pi$ in the superconducting state. Above $H_c(T)$, the sample becomes normal. The above behavior is called the *Meissner effect* (MEISSNER and OCHSENFELD [1933]). In contrast, for nearly all alloys and intermetallic compounds, perfect diamagnetism is observed up to a lower critical field H_{c1}. Above this field, partial diamagnetism ($-1/4\pi < \chi < 0$) persists. The superconducting state persists in the form of a mixed state (GINZBURG and LANDAU [1950]) up to an upper critical field H_{c2} where $\chi \to 0$ and the sample becomes normal. The behavior of the magnetization together with an (H, T) phase diagram is shown in fig. 1 illustrating the two types of behavior outlined above. Pure metals exhibiting behavior described by eq. (1) are referred to as *type-I superconductors* while impure metals and alloys exhibiting only partial diamagnetism between H_{c1} and H_{c2} are referred to as *type-II*

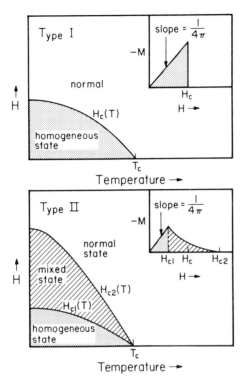

Fig. 1. Phase diagrams for type-I and type-II superconductors showing homogeneous superconducting state, inhomogeneous superconducting state, and normal state in the (H, T) plane.

superconductors. The microscopic parameters which determine type-I and type-II behavior will be discussed in the next section.

In addition to perfect conductivity and the Meissner effect, superconductors exhibit other interesting properties. The transition to the superconducting state is a second-order phase transition in the absence of an external field H. The transition becomes first order in an applied field. For $H = 0$, the specific heat exhibits a discontinuity (or λ-anomaly) at T_c. An example is shown in fig. 2. This type of anomaly is common to second-order phase transitions. Notice that the electronic heat capacity vanishes exponentially at low temperatures, that is, $c_v \propto \exp(-B/T)$ as $T \to 0$. This behavior is similar to that observed in semiconductors where it is attributed to thermal activation of carriers over an energy gap in the density of electronic states. This suggests that superconductors possess an energy gap in their electronic excitation spectrum. Direct evidence of a gap was first provided by electron tunneling experiments (GIAEVER [1960]). The existence of an energy gap has other important consequences. It leads, for example, to a reduction in the attenuation of sound waves below T_c. These and other phenomena, though interesting, cannot be adequately discussed within the scope of this chapter. The interested

References: p. 1776.

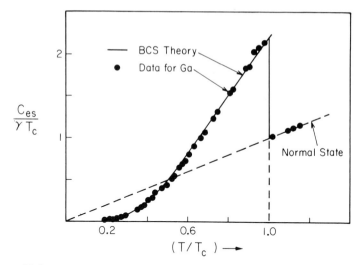

Fig. 2. The specific heat anomaly as predicted by the BCS theory and measured for the metal Ga (after PHILLIPS [1964]).

reader is referred to the references for further information. Several textbooks dealing with the physics of superconductivity are also particularly readable and complete. (TINKHAM [1975], DE GENNES [1966] and SCHRIEFFER [1962]). In the following section, a brief outline of the theory of superconductivity is given.

2.2. Microscopic theory

A successful microscopic theory of superconductivity was first proposed by BARDEEN, COOPER and SCHRIEFFER [1957]. The *BCS theory* of superconductivity is based on the existence of an attractive interaction between pairs of electrons. This concept may seem at first a bit puzzling since the normal bare Coulomb interaction between electrons is repulsive. In a metal, the electrostatic field of a fixed point charge Z induces a rearrangement of the electronic charge in the metal. In particular, a simple semi-classical argument shows that the Coulomb potential of a fixed point charge, Z, in an electron gas takes the form $V(r) = Ze^2e^{-\lambda r}/r$, where λ is the reciprocal Thomas–Fermi screening length and r the distance from the point charge. This screened Coulomb potential is much weaker and of shorter range than the bare Coulomb potential. A proper quantum-mechanical treatment of the screening problem yields the somewhat more complicated form of $V(r)$ as discussed in several textbooks (ZIMAN [1972], see also ch. 3), but the same conclusion regarding the strength and range of the potential applies. Electronic screening reduces the strength and range of the potential of charged particles, in particular that of electrons. The Coulomb repulsion between two electrons is reduced in metals by the collective screening effect of other electrons. However, this is not a complete picture. Metals also contain a lattice of ions, each carrying a positive charge. The ions can also respond to internal fields. Each ion is situated in a three-dimensional potential well determined by the requirement that the equilibrium metal forms a stable lattice. A given ion cannot move translationally but rather can only exercise vibrations about its equilibrium position. The ions behave like a coupled set of simple harmonic oscillators. The coupling is due to the fact that a displacement of any ion produces a change in the crystal potential as a whole and thus produces a force on neighboring ions. The normal modes of these coupled oscillators are called lattice waves or phonons as discussed in ch. 3. It is the interaction of electrons with phonons that leads to an effective attraction between electrons.

A simple schematic illustration, fig. 3, shows how electron attraction comes about. A given electron in a Bloch state $\psi_K(r)$ has an associated electronic charge density given by $e\psi_K^*(r)\,\psi_K(r)$. Ions are attracted to regions where this electronic charge is concentrated. The ions attempt to screen the charge density of the electron, and the ionic lattice becomes polarized. A second electron initially in a state $\psi_{K'}(r)$ is now perturbed by this polarization. It can lower its energy by changing its wavefunction so as to concentrate charge in those regions where ionic charge density is already higher. Two electrons can lower their mutual energy by following ionic motion in a phase-coherent manner. A more detailed description of the mutual interaction

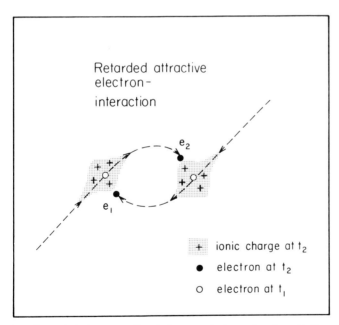

Fig. 3. Illustration of retarded phonon-mediated electron–electron attraction. At time t_1, the two fast moving electrons are located at the open circles and attract ionic charge. At the later time t_2, the slow moving ionic charge is maximized near the open circles and the two electrons are attracted toward these regions of excess ionic charge.

References: p. 1776.

between electrons due to lattice polarization and the dynamic aspects of this interaction can be found in several textbooks (ZIMAN [1972]). For the present purposes, the following points are important:

1. The collective screening effect of electrons reduces the magnitude and range of the normal Coulomb repulsion between pairs of electrons in metals.

2. Through their mutual interaction with ions, electrons can experience an interaction which lowers their mutual energy and is thus attractive.

3. The net combined effect of the above interactions can, under certain conditions, lead to a weak net attraction between pairs of electrons. As a result, the electrons may tend to form a collective bound state at low temperatures.

The BCS theory gives a detailed *microscopic description* of how the above mentioned bound state is formed and shows that it indeed has the properties of the superconducting state. It treats the problem of determining the superconducting ground state of the electron–ion systems as well as the question of thermally excited states. It predicts the thermodynamic properties of a superconducting electron–ion system, in particular, the highest temperature for which the superconducting state exists before being destroyed by thermal disorder. The essential results of the BCS theory are as follows. The superconducting transition temperature is given in terms of microscopic material parameters as (BARDEEN *et al.* [1957]):

$$T_c = 1.14 \langle \omega \rangle \exp\left[-1/D(\epsilon_F)V\right] = 1.14 \langle \omega \rangle \exp(-1/\lambda), \qquad (2)$$

where

$D(\epsilon_F) =$ the density of electronic states at the Fermi level of the metal;

$\langle \omega \rangle$ = a typical lattice vibration frequency expressed as a temperature (i.e., $\langle \omega \rangle = \hbar\omega/k_B$, k_B = Boltzmann's constant);

V = effective strength of the attractive electron–electron interactions expressed in units of interaction energy.

The parameter $\lambda = D(\epsilon_F)V$ is dimensionless and is called the superconducting coupling constant. The BCS theory assumes that $0 < \lambda \ll 1$ and is not strictly valid when λ becomes of order unity. The theory predicts an energy gap in the density of electronic states of the superconductor. This gap vanishes at T_c, $\Delta(T_c) \rightarrow 0$, and approaches a constant given by $\Delta(T = 0) = \Delta(0) = 1.76\, k_B T_c$ at low temperatures. The thermal properties of the superconductor are determined by this gap and by the nature of electron-like excitations with energies above the gap. The low-temperature specific heat and other properties predicted are found to be in good agreement with experiment as seen, for example, in fig. 2.

The magnetic properties of superconductors also follow from the BCS theory which predicts the Meissner effect, and the temperature-dependence of the critical field $H_c(T)$ for type-I superconductors. The interested reader is referred to the original BCS paper and to TINKHAM's [1975] particularly clear discussion of the theory for further details.

The BCS theory of superconductivity employs a simplified model of the phonon spectrum, electron–phonon interaction, and electronic states of a metal. In real metals, the detailed features of these characteristic properties are more complex.

Furthermore, the BCS theory is valid only in the limit where $\lambda \ll 1$, the *weak coupling limit*. The theory has been extended (ELIASHBERG [1960]) to include the details of the electron–phonon interaction as well as the *strong-coupling case* where $\lambda \approx 1$. The more general theory is capable of describing real metals with arbitrary λ. The full strong-coupling Eliashberg theory cannot be dealt with here. Rather, we outline some of the significant results which follow from the theory. McMILLAN [1968] has solved the Eliashberg energy-gap equation using a simple but realistic approximation. In his solution, McMillan used the phonon density of states for bcc-Nb. From his solution, he derived an approximate equation for T_c which is valid both in the weak and strong-coupling regimes. His result can be expressed as:

$$T_c = \frac{\theta_D}{1.45} \exp\left[\frac{-1.04(1+\lambda)}{\lambda - \mu^*(1 + 0.62\lambda)}\right] \qquad (3)$$

where

θ_D = Debye temperature (or other suitable measure of the characteristic phonon frequency);

λ = electron–phonon coupling constant;

μ^* = effective repulsive coupling interaction constant for electrons.

The constant μ^* takes into account the repulsive bare Coulomb interaction between electrons. McMillan finds that a value of $\mu^* \approx 0.13$ is suitable for many metals. Equation (3) resembles the BCS equation (2). For many purposes one can approximate eq. (3) as:

$$T_c \approx \frac{\theta_D}{1.45} \exp\left[-(1+\lambda)/\lambda\right], \qquad (3b)$$

which is identical to eq. (2) in form if one replaces λ by $\lambda/(1+\lambda)$. This replacement is required by the strong-coupling theory and is often referred to as renormalization of the coupling constant. The same renormalization is also required for several other electronic characteristics of strong-coupling superconductors. For example, the electronic energy itself is renormalized. This has important consequences. For example, the electronic density of states of the Fermi level $D(\epsilon_F)$ is enhanced by the factor $(1+\lambda)$. One can determine $D(\epsilon_F)$ from a low-temperature specific heat experiment using the well known relation

$$C_{el} = \gamma T = \tfrac{2}{3}\pi^2 D(\epsilon_F) k_B^2 T, \qquad (4)$$

where C_{el} is the contribution of electrons to the low-temperature specific heat, which is characterized by a linear temperature dependence with coefficient γ, and where $D(\epsilon_F)$ is the density of states for electrons in one spin state (up or down). When electron–phonon coupling is taken into account, this must be replaced by

$$C_{el} = \tfrac{2}{3}\pi^2 D(\epsilon_F)(1+\lambda) k_B^2 T. \qquad (5)$$

Finally, McMillan expressed λ in terms of directly calculable properties of a metal. He obtained the expression

$$\lambda = \frac{D(\epsilon_F)\langle I^2 \rangle}{M\langle \omega^2 \rangle_M}, \qquad (6)$$

References: p. 1776.

where $\langle I^2 \rangle$ is a suitably averaged matrix element for change in potential seen by an electron as an ion is moved, $\langle \omega^2 \rangle_M$ is a suitably defined mean square phonon frequency of the metal, and M is the ionic mass. In later sections, we will have occasion to refer to this formula for λ in attempting to understand the variation of λ in real materials. For a derivation of the formula, the reader should consult McMillan's original paper.

2.3. Inhomogeneous superconducting states: Ginzburg–Landau theory

The BCS theory briefly described above gives an excellent description of homogeneous type-I superconductors. On the other hand, there are many practical situations in which the first-principles microscopic theory is difficult to apply. The case of the mixed state of type-II superconductors referred to in §1.1 is such a situation. Between H_{c1} and H_{c2}, only partial diamagnetism exists, and the superconducting state turns out to be characterized by an inhomogeneous spatial density of paired electrons as well as a spatially varying internal magnetic field B. The microscopic theory can be extended to include such cases (GORKOV [1959]), however this treatment is mathematically complex and does not emphasize the physical phenomena involved. A semi-classical approach to superconductivity based on the general theory of second-order phase transitions was developed by GINZBURG and LANDAU [1950], which provides an eloquent and rather simple framework in which to treat inhomogeneous superconducting states. The *GL theory* is based on the concept, first proposed by LANDAU [1938], that the difference in free energy between the low-temperature "ordered" phase and the high-temperature "disordered" phase of a second-order phase transition can be expanded in a power series based on an *order parameter*. In the case of superconductivity, the order parameter can be taken as the wavefunction, ψ, of paired or superfluid electrons. The free energy of the homogeneous superconducting state can be written as (TINKHAM [1975]):

$$f_s = f_{no} + \alpha|\psi|^2 + \tfrac{1}{2}\beta|\psi|^4, \tag{7}$$

where

f_s = free energy in the superconducting state;
f_{no} = free energy of the normal metal at low temperatures;
ψ = wavefunction for superfluid electrons;

and α and β are coefficients, possibly temperature dependent, introduced phenomenologically. Symmetry requirements guarantee the absence of odd terms in this expansion. Treating ψ like a quantum-mechanical wavefunction leads one to introduce a kinetic-energy term which describes the spatial variation of ψ. In the absence of a magnetic field, one follows the usual prescription where kinetic energy is written as:

$$\frac{p^2}{2m} \rightarrow -h^2/2m^*|\nabla\psi|^2, \tag{8a}$$

where m* is the mass of the superfluid particles. In the presence of a magnetic field, the momentum operator must be replaced by

$$P \rightarrow \left[\frac{\hbar}{i} \nabla - \frac{e^*}{c} A \right],$$ (8b)

where e^* is the charge of the superfluid particles and A is the vector potential for the magnetic field. The kinetic energy density becomes:

$$\frac{1}{2m^*} \left| \left(\frac{\hbar}{i} \nabla - \frac{e^*}{c} A \right) \psi \right|^2.$$ (9)

The total free-energy density, including the energy stored in the magnetic field becomes:

$$f_s = f_{no} + \alpha |\psi|^2 + \tfrac{1}{2} \beta |\psi|^4 + \frac{1}{2m^*} \left| \left(\frac{\hbar}{i} \nabla - \frac{e^*}{c} A \right) \psi \right|^2 + \frac{|h|^2}{8\pi}.$$ (10)

Here, we can use h to represent the local microscopic magnetic flux in the sample. We allow h to vary with position x. The GL theory is based on this free-energy expansion and a variational principle which minimizes the integral of the above free-energy density over the volume of the superconductor. This leads to the famous *Ginzburg–Landau (GL) equations* which can be written:

$$\alpha \psi + \beta |\psi|^2 \psi + \frac{1}{2m^*} \left(\frac{\hbar}{i} \nabla - \frac{e^*}{c} A \right)^2 \psi = 0$$ (11)

and $J = (c/4\pi) \operatorname{curl} h = \dfrac{e^*\hbar}{2m^*i} (\psi^* \nabla \psi - \psi \nabla \psi^*) - \dfrac{e^{*2}}{m^*c} \psi^* \psi A,$ (12)

where J is the superfluid current or supercurrent. These equations are a pair of coupled nonlinear second-order differential equations for the electromagnetic vector potential A (which determines B and h) and the superconducting order parameter ψ. A general solution to these equations is not possible. On the other hand, solutions can be obtained in certain physically important limiting cases. Below, a few of these cases are discussed which are relevant to our later discussion.

2.4. Special solutions to the GL equations

Case I: $A = 0$, $\nabla \psi = 0$. This case describes a homogeneous superconductor. Equation (12) has the trivial solution $J = 0$ and the order parameter can be taken as real. The solution of eq. (11) becomes $|\psi|^2 = -\alpha/\beta$. If ψ is to have a real solution, then we must have $\alpha/\beta < 0$. In order that the superconducting state ($\psi \neq 0$) be stable just below T_c and unstable just above T_c, α must change sign at T_c. Expanding α about T_c requires the expansion to be of the form (TINKHAM [1975]):

$$\alpha \approx \alpha_0 (T_c - T) + [\text{higher order terms}], \quad \text{with } \alpha_0 < 0.$$ (13)

A similar expansion of β admits a constant to lowest order so that

$$\beta \approx \beta_0 + [\text{higher order terms in } (T_c - T)], \quad \beta_0 > 0.$$ (14)

Thus, we find $\psi \equiv [|\alpha_0|(T_c - T)/\beta_0]^{1/2}$ for a homogeneous superconductor. ψ vanishes like $(T_c - T)^{1/2}$ near T_c. This homogeneous solution is often referred to as ψ_∞.

Case II: $A = 0$, $\nabla\psi \neq 0$. In this case, we define a new function $f = \psi/\psi_\infty$. Then eq. (11) can be written as:

$$(\hbar^2/2m^*|\alpha|)\frac{d^2f}{dx^2} + f - f^3 = 0, \tag{15}$$

where *only* the x-component of $\nabla\psi$ is assumed nonvanishing. We see that a natural characteristic length emerges which can be written as:

$$\xi^2(T) = \frac{\hbar^2}{2m^*|\alpha_0|(T_c - T)} \propto (T_c - T)^{-1}, \tag{16}$$

where $\xi(T)$ is called the *coherence length*. In terms of $\xi(T)$, eq. (15) becomes:

$$\xi^2(T)\frac{d^2f}{dx^2} + f - f^3 = 0. \tag{17}$$

Assuming that deviations from homogeneity are small allows one to linearize eq. (17). Letting $f(x) = 1 + g(x)$, where $g(x) \ll 1$, gives $d^2g/dx^2 = (2/\xi^2)g$ to lowest order in g. This linearized equation admits solutions of the form $g(x) \sim \exp[\pm(\sqrt{2})x/\xi(T)]$. We see that the coherence length ξ is a natural scale of length over which g, and therefore ψ, varies in space. From the microscopic theory, the coefficients α_0 and β_0 can be identified with microscopic parameters (GORKOV [1959]). One finds:

$$\xi(T) = 0.74\xi_0\left(\frac{T_c - T}{T_c}\right)^{-1/2} = 0.74\frac{\hbar v_F}{\pi\Delta(0)}\left(\frac{T_c - T}{T_c}\right)^{-1/2}, \tag{18}$$

where $\Delta(0)$ is the superconducting energy gap (at $T = 0$) and v_F the Fermi velocity of electrons. This expression holds only when scattering of electrons in the normal metal is weak, as in the case of pure metals. When normal electron scattering is strong, as in the case of disordered alloys, one must introduce an *electron mean free path* l. In the limit where $l \ll \xi_0$, the coherence length becomes:

$$\xi(T) \to \xi_d(T) = 0.855(\xi_0 l)^{1/2}\left(\frac{T_c - T}{T_c}\right)^{-1/2}. \tag{19}$$

Roughly speaking, $\xi_d(T)$, the coherence length in the *dirty limit*, is the geometric mean of the coherence length in the *clean limit* [eq. (18)] and the electron mean free path l. The two expressions for ξ are important in that they determine the natural length scales for variations of ψ in space in real materials. It is worthwhile to estimate this scale of length for real materials. Using the simple BCS result for $\Delta(0)$ together with $T_c \approx 5$ K for a typical superconductor, and taking v_F of a typical free-electron metal (see ch. 3) as $v_F \sim 10^8$ cm/s, we find for a "clean" metal:

$$\xi(T) \approx 0.74 \left(\frac{\hbar v_F}{\pi \Delta(0)} \right) \left(\frac{T_c - T}{T_c} \right)^{-1/2} \approx (2 \times 10^{-5} \text{ cm}) \left(\frac{T_c - T}{T_c} \right)^{-1/2}. \qquad (20)$$

We see that ξ is typically of the order of 10^{-5} cm in pure metals. In very "dirty" alloys, the electron scattering mean free path can be as small as a few angstroms. In this case, we see that

$$\xi_d(T) \approx 0.855(\xi_0 l)^{1/2} \left(\frac{T_c - T}{T_c} \right)^{-1/2} \sim (10^{-6} \text{ cm}) \left(\frac{T_c - T}{T_c} \right)^{-1/2}. \qquad (21)$$

The characteristic length for variation of ψ can thus range from $\sim 10^{-6}$ cm in alloys to $\sim 3 \times 10^{-5}$ cm in pure metals. Note that ξ always diverges as the temperature T approaches T_c.

Case III: $A \neq 0$, $\nabla \psi$ small. This case corresponds to "clean" superconductors (ξ large, will make $\nabla \psi$ small) in a weak magnetic field. In this case, eq. (12) reduces to:

$$J = (c/4\pi) \text{ curl } h = (e^{*2}/m^*c) \psi^* \psi A. \qquad (22)$$

Taking the curl of both sides and using standard vector identities gives:

$$\text{curl}[(c/4\pi) \text{ curl } h] = (c/4\pi)[\nabla(\nabla \cdot h) - \nabla^2 h] = (c/4\pi)\nabla^2 h = (e^{*2}/m^{*c})\psi^* \psi h, \qquad (23)$$

since $\nabla \cdot h = 0$ by Maxwell's equations. This yields an equation for h which can be written as:

$$(4\pi e^{*2}/m^*c^2)\psi^* \psi h = -\nabla^2 h. \qquad (24)$$

This equation predicts the Meissner Effect, as can be seen as follows. Taking $\psi^* \psi$ to represent the density of superfluid electrons, $n_s = \psi^* \psi$, allows us to define another characteristic length:

$$\lambda^2(T) = [4\pi e^{*2} n_s/m^*c^2]^{-1},$$

and eq. (24) becomes:

$$-\nabla^2 h = -\frac{1}{\lambda^2(T)} h. \qquad (25)$$

This describes the behavior of the microscopic magnetic field inside a superconductor, which must behave like a solution to eq. (25), $h(x) \sim \exp[\pm x/\lambda(T)]$. As an example, consider the case of an external field, H, applied to a superconducting slab with boundaries in the yz-plane as shown in fig. 4. The microscopic field h well inside the interior must approach zero as predicted by eq. (1). Outside, h must be equal to the external field H. The only physically admissable solution is one which decays exponentially in the interior, with characteristic length λ. We obtain the solution for $h(x)$ shown in fig. 4. Thus we see that λ is the natural scale of length for

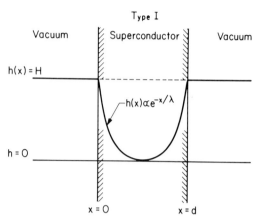

Fig. 4. A slab of type-I superconducting material located in an external field **H** parallel to the slab. The field penetrates the slab as shown by $h(x)$ a distance of order λ.

variations in the microscopic magnetic field **h**. We see, for example, that the Meissner effect is confined to the bulk of the sample and is incomplete near a surface layer of thickness $\sim \lambda$. Again using the microscopic theory, one can express λ in terms of microscopic parameters. In the clean limit described above, we find (TINKHAM [1975]):

$$\lambda_L(T) = \frac{\lambda_L(0)}{[2(1-t)]^{1/2}}, \quad \text{with} \quad t = \left(\frac{T_c - T}{T_c}\right), \tag{26}$$

$$\text{and} \quad \lambda_L(0) = \frac{mc^2}{4\pi ne^2},$$

the subscript L refers to LONDON and LONDON [1935] and n is the total density of conduction electrons. $\lambda_L(T)$ is called the *London penetration depth* since it measures the depth of penetration of a magnetic field in a type-I superconductor. In the dirty limit, the penetration depth becomes

$$\lambda_d(T) = \lambda_L(T)\left(\frac{\xi_0}{1.33l}\right)^{1/2}, \tag{27}$$

where l is the electron mean free path. Again, we can estimate the size of λ in real materials. In pure metals, the typical density of conduction electrons is 10^{23} electrons/cm^3 so that $\lambda_L(0) \sim 10^{-6}$/cm. For very impure (dirty) metals and alloys, from eq. (18) one has $(\xi_0/l) \approx 100$ and $\lambda_L(0) \sim 10^{-5}$ cm. We see that the introduction of electron scattering by impurities and alloying increases λ while it decreases ξ. As we shall now see, this leads to a change-over from type-I to type-II behavior as the ratio (λ/ξ) becomes larger on going from pure metals to dirty materials.

2.5. Type-I and type-II superconductors

Having looked at several special limiting solutions to the GL equations, we now outline a few general results. Through the *Meissner effect*, one can relate the thermodynamic critical field $H_c(T)$ to total magnetic work done when a type-I superconductor is driven normal by an applied field. We find

$$f_{no}(T) - f_s(T) = \int_0^{H_c(T)} H \, dM = \frac{H_c^2(T)}{8\pi}. \tag{28}$$

This must be equal to $f_{no} - f_s$ as determined by our solution to the GL equations in Case I. Thus, we find for the coefficients α and β the relation

$$(-\alpha/\beta) = \frac{H_c^2(T)}{8\pi}. \tag{29}$$

Using our other previous results, we can express α and β as (TINKHAM [1975]):

$$\alpha(T) = \frac{-e^{*2}}{m^*c^2} H_c^2(T)\lambda^2(T) = -\frac{2e^2}{mc^2} H_c^2(T)\lambda^2(T), \tag{30}$$

$$\beta(T) = \frac{4\pi e^{*4}}{m^{*2}c^4} H_c^2(T)\lambda^4(T) = \frac{16\pi e^4}{m^2c^4} H_c^2(T)\lambda^4(T), \tag{31}$$

where we have now made the identification of the superfluid particles as paired electrons with $e^* = 2e$ and $m^* = 2m$.

We now consider the question of distinguishing type-I and type-II superconductors. It turns out that the answer is simply determined by the ratio of the two characteristic lengths $\xi(T)$ and $\lambda(T)$. This ratio, $\kappa = \lambda(T)/\xi(T)$, is called the *Ginzburg–Landau parameter*. For sufficiently large fields H, one can assume that ψ is very small compared with ψ_∞. In this case, one can to a good approximation drop the cubic term in ψ in eq. (11). This equation becomes a linear differential equation:

$$\alpha\psi + \frac{1}{2m^*}\left(\frac{\hbar}{i}\nabla - \frac{e^*}{c}A\right)^2\psi = 0, \tag{32}$$

$$\text{or} \quad \left(\frac{\nabla}{i} - \frac{2\pi A}{\phi_0}\right)^2\psi = \frac{\psi}{\xi^2(T)},$$

where $\phi_0 = hc/2e$ is called the fundamental *flux quantum*. The solution to this equation is outlined by TINKHAM *. For example, when H is along the z-direction a solution of the form $\psi = e^{iK_y y}e^{iK_z z}f(x)$ exists, where $f(x) = \exp[-(x-x_0)^2/2\xi^2]$ is "localized" near x_0. Notice that ξ is the characteristic "size" of this localized solution. The maximum field for which this solution exists is given by

$$H_{\max} = H_{c2} = \frac{\phi_0}{2\pi\xi^2(T)}, \tag{33a}$$

---

* Tinkham [1975], p. 128.

References: p. 1776.

or equivalently

$$H_{c2}(T) = H_c(T)\kappa\sqrt{2} = \left[\frac{\lambda(0)}{\xi(0)}\right] H_c(T)\sqrt{2} . \tag{33b}$$

Using eq. (2.21), it can also be shown (ORLANDO et al. [1979]) that as $T \to T_c$,

$$\left.\frac{dH_{c2}}{dT}\right|_{T_c} = 4.48\rho\gamma \times 10^5, \tag{33c}$$

with γ in J/m³ K², ρ in Ωcm and H_{c2} in gauss. We see that when $\kappa > 1/\sqrt{2}$, we have $H_{c2} > H_c$. Stated differently, an inhomogeneous solution to the GL equations exists for fields $H_c < H < H_{c2}$ whenever $\kappa > 1/\sqrt{2}$. No homogeneous (Case I) solution exists for such fields. This defines the upper critical field H_{c2}. The condition for type-II (inhomogeneous) superconductivity is simply $\kappa > 1/\sqrt{2}$. When $\kappa < 1/\sqrt{2}$, the superconductor is homogeneous for $H < H_c(T)$ and becomes normal for $H > H_c(T)$. This describes type-I superconductors.

In actuality, the lowest-energy inhomogeneous state of a three-dimensional super-conductor with H along the z-axis turns out to be a linear combination of solutions which form a regular two-dimensional array. Each solution is localized in both the x and y directions. Such a solution is illustrated in fig. 5. It turns out, in fact, that this type of solution has a lower free energy than the homogeneous (Case I) solution not just for $H_c < H < H_{c2}$ but for a range of fields below H_c as well. The lowest field for which a *mixed state* is more stable than a homogeneous state is called H_{c1}, the lower critical field. It can be shown that the magnetic flux through each unit cell of the periodic xy array shown in fig. 5 is quantized in units of ϕ_0. Such solutions are often referred to as *flux lattices* or mixed states. It will be seen in the following sections that the mixed state is of extreme practical importance in that it can often persist to very high fields $H_{c2}(T)$. The existence of the mixed state is the basis for applications of superconductivity to high-field magnets.

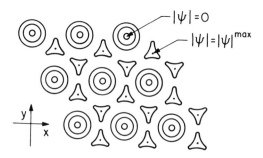

Fig. 5. Illustration of the variation of the order parameter ψ in the mixed state of a type-II superconductor.

3.   Pure metals

The systematics of superconductivity among the metallic elements in the periodic table are surveyed in this section. In simple metals, the Fermi surface is formed by electronic states having a predominantly s- or p-electron character. The nearly-free-electron model is used as a basis for discussing the microscopic origins of superconductivity in this case. In contrast, transition metals contain narrow partially filled d-bands which can make a large contribution to the area of the Fermi surface and to the electronic density of states at the Fermi level $D(\epsilon_F)$. Superconductivity in transition metals is dominated by d-electrons. The free-electron model does not give an adequate description of superconductivity in this case. The tight binding method (TBM) gives a more reasonable description of d-band superconductors. Simple metals and transition metals are thus discussed separately, below.

3.1.   Simple metals

A list of transition temperatures, critical fields $H_c(0)$, and other characteristic parameters of superconducting simple metals is given in table 1. Included are $D(\epsilon_F)$

Table 1

Properties of simple metals. Data from ROBERTS [1976], GESCHNEIDER [1965] and MCMILLAN [1968]. Original references in these articles.

Metal	Z	T_c (K)	θ_D (K)	γ mJ/mole K^2	λ	$D(\epsilon_F)$ (states/ eV atom spin)	$H_c(0)$ (kG)
Na	1	< 0.08	89.4	2.11	–	–	–
K	1	< 0.09	157	1.38	–	–	–
Rb	1	< 0.011	54	2.52	–	–	–
Cu	1	< 0.02	342	0.693	–	–	–
Ag	1	< 0.002	228	0.659	–	–	–
Au	1	< 0.002	165	0.748	–	–	–
Be	2	0.026	1160	0.21	0.23	0.032	–
Mg	2	< 0.002	396	1.30	–	–	–
Ca	2	< 0.017	234	2.90	–	–	–
Sr	2	< 0.017	147	3.64	–	–	–
Zn	2	0.850	310	0.66	0.38	0.098	310
Cd	2	0.517	209	0.69	0.38	0.106	209
Hg	2	4.154	87	1.81	1.00	0.146	411
Al	3	1.175	420	1.35	0.38	0.208	104.9
α-Ga	3	1.083	325	0.60	0.40	0.091	59.2
β-Ga	3	6.2	–	–	–	–	560
γ-Ga	3	7.0	–	–	–	–	950
In	3	3.408	109	1.672	0.69	0.212	281
Tl	3	2.38	78.5	1.47	0.71	0.182	178
Sn	4	3.722	195	1.78	0.60	0.238	305
Pb	4	7.196	96	3.10	1.12	0.276	803

References: p. 1776.

and θ_D (the Debye temperature) as determined from low-temperature specific heat measurements. Data for several non-superconducting simple metals are also given for comparison and further analysis. The determination of $D(\epsilon_F)$ from specific heat measurements is accomplished using eq. (5). The values given have been corrected by the factor $(1+\lambda)$ to properly account for the effect of the electron–phonon interaction on electron excitations (McMillan [1968]). The values of λ used are also given in the table. These values were deduced from McMillan's T_c equation using $\langle\omega\rangle = \theta_D/\sqrt{2}$, and $\mu^* = 0.13$ as suggested by McMillan. With this information, one can attempt to analyze the microscopic parameters which determine T_c and its variation in the periodic table. It is natural to examine the ratio T_c/θ_D. In particular, $-\ln(T_c/\theta_D)$ is inversely proportional to λ [see eqs. (2) and 2(3b)]. In McMillan's theory $-\ln(T_c/\theta_D) = f(\lambda)$ where $f(\lambda) \approx (1+\lambda)/\lambda$. Among the simple metals, certain trends are apparent. For example, there is a clear connection between superconductivity and valency. No monovalent metals are known to be superconducting. The alkali-earth divalent metals are also nonsuperconducting while the group-IIB divalent metals (i.e. Zn, Cd, and Hg) are superconducting albeit with somewhat low T_c's. All trivalent and quadrivalent simple metals are superconductors. A plot of $\ln(T_c/\theta_D)$ versus valency is shown in fig. 6 which illustrates this trend. One might be tempted to attribute this correlation to the effect of Z on $D(\epsilon_F)$. However, a plot of $\ln(T_c/\theta_D)$ versus $D(\epsilon_F)$ shows no well defined correlation. In the BCS theory, $\lambda = D(\epsilon_F)V$ where V is an effective electron–electron interaction parameter. We see that the correlation of λ with valence must result from an interplay between the dependence of both $D(\epsilon_F)$ and V on valence. It is not surprising that V increases with increasing valence since the electron–phonon interaction is directly related to

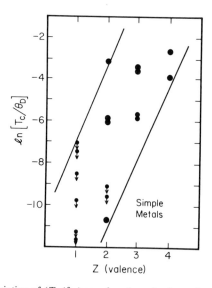

Fig. 6. The variation of (T_c/θ_D) as a function of valency for simple metals.

the strength of the ionic potential (ZIMAN [1960]) which increases with ionic charge. It is also clear from the scatter of the points in fig. 6 that Z alone is not sufficient to provide a reliable estimate of λ. A successful theory of λ must account for the details of the electronic band structure, phonon spectrum, and electron–phonon interaction. In fact, these parameters are all interdependent and must be treated together in a self-consistent manner. Such calculations have recently been carried out (VARMA and WEBER [1979]) with considerable success. The details of these calculations are too difficult to adequately outline here. It can at least be said that one must go beyond the free-electron model in order to predict T_c from first-principles calculations. On the other hand, the trend of T_c with valence discussed above seems to dominate the other physical effects.

3.2. Transition metals

A summary of the superconducting and electronic parameters of transition metals is given in table 2. The parameters given are the same as those given for simple metals in the previous section. One notices that the values of $D(\epsilon_F)$ determined from specific heat data are often rather larger for transition metals than for simple metals.

Table 2
Properties of transition metals. Data from ROBERTS [1976], GESCHNEIDER [1965] and McMILLAN [1968]. Original references are in these articles.

Metal	Group No.	T_c (K)	θ_D (K)	γ (mJ/mole K^2)	λ	$D(\epsilon_F)$ (states/ eV atom spin)	$H_c(0)$ (kg)
Sc	3	< 0.032	470	10.8	–	–	–
Ti	4	0.40	415	3.3	0.38	0.51	56
V	5	5.40	383	9.82	0.60	1.31	1408 [a]
Y	3	< 0.005	268	10.1	–	–	–
Zr	4	0.61	290	2.77	0.41	0.42	47
Nb	5	9.25	276	7.80	0.82	0.91	2060 [a]
Mo	6	0.915	460	1.83	0.41	0.28	96
Tc	7	7.8	411	6.28	0.66	–	1410 [a]
Ru	8	0.49	580	2.8	0.38	0.46	69
Rh	9	< 0.086	480	4.6	–	–	–
Pd	10	< 0.10	283	10.0	–	–	–
α-La	3	4.88	151	9.8	0.82	–	800
Hf	4	0.128	256	2.40	0.34	0.34	–
Ta	5	4.47	258	6.15	0.65	0.77	829 [a]
W	6	0.0154	383	0.90	0.28	0.15	1.15
Re	7	1.697	415	2.35	0.46	0.33	200
Os	8	0.66	500	2.35	0.39	0.35	70
Ir	9	0.1125	425	3.19	0.34	0.51	16
Pt	10	< 0.10	234	6.68	–	–	–

[a] These pure transition metals are type-II superconductors ($\kappa > 1/\sqrt{2}$). The values given are $H_{c2}(0)$ values. See ROBERTS [1976].

References: p. 1776.

The large values of $D(\epsilon_F)$ are due to the contribution of d-electron states to $D(\epsilon_F)$. The d-electron states form a set of narrow bands (typically a few eV wide). In contrast, s-electrons (and p-electrons) tend to form broad (typically ~ 10 eV wide) bands which are nearly parabolic in shape as predicted by the free-electron model. Furthermore, the d-bands can be occupied by up to 10 electrons per atom while s-bands (p-bands) may contain a maximum of 2 (6) electrons per atom.

The basic difference between d-electrons and nearly-free (s-, p-) electrons can be appreciated by examining a plot of the free-atom radial wavefunctions of a transition-metal atom. As an example, the s and d radial atomic wavefunctions from atomic-orbitals calculations for the valence electrons of molybdenum are shown in fig. 7. (The 5p states of molybdenum lie in energy well above the 5s and 4d states and are not considered here.) Also shown in the figure is the atomic (Wigner–Seitz) radius of molybdenum. This atomic radius is defined as half the distance between nearest-neighbor molybdenum atoms in metallic molybdenum while the Wigner–Seitz radius gives a sphere of volume equal to the average volume per metal atom. In the atomic ground state, the valence electron configuration of a molybdenum atom is $4d^5 5s^1$ so that both types (d- and s-) of orbitals are partially filled with electrons. From the figure, it is obvious that the 5s states have large amplitude at distances extending well beyond the atomic radius whereas the more localized 4d states have an amplitude which rapidly decays with distance from the nucleus, becoming quite small at the atomic radius. When a collection of atoms are brought together to form a metallic crystal, the atomic s-states of neighboring atoms will be strongly overlapping, creating a more or less high density of s-electron charge everywhere in the crystal as suggested in the figure. This leads to free-electron-like behavior with a

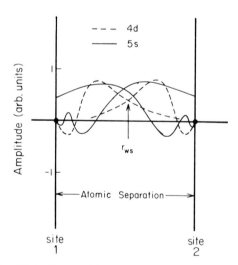

Fig. 7. Atomic 4d and 5s orbitals of two neighboring Mo atoms, showing the degree of overlap of the orbitals on two sites separated by twice the Wigner–Seitz cell radius observed in pure Mo. Notice the large overlap of the 5s orbitals compared with the smaller overlap of 4d orbitals.

broad parabolic s-band. On the other hand, the atomic d-states on neighboring atoms are only weakly overlapping. This is also illustrated in the figure. In the limit of weakly overlapping orbitals, it is reasonable to use a linear combination of the atomic wavefunctions as a first approximation to electronic states in the metal (FRIEDEL [1969], see ch. 3). This is called the LCAO, *linear combination of atomic orbitals, approximation* and is found to be a reasonable approach to understanding d-electrons in transition metals. Using Bloch's theorem, an appropriate linear combination is found to be:

$$\psi_{d,K}(r) = \sum_{a,i} \exp(iK \cdot r_i) c_\alpha \phi_\alpha(r - r_i),$$ (34)

where α runs over the five degenerate atomic d-orbitals, the r_i label the atomic positions in the metallic crystal and the c_α are coefficients to be determined. Using this as a trial wavefunction in a variational calculation, one determines the c_α's and a set of energy dispersion relations $\epsilon_{K,\beta}$ for d-bands. The subscript β labels the five d-bands which arise from linear combinations of the five independent atomic d-orbitals. This technique for calculating d-bands was first employed by Friedel. By methods outlined in ch. 3, one can use the calculated energy bands to compute the s- and d-electron contribution to the density of states function for electrons. Figure 8 shows the function $D(\epsilon)$ calculated for tungsten by a more sophisticated method employing accurate APW, *augmented plane wave*, calculations of the energy bands (MATTHEISS [1972]). The main features of $D(\epsilon)$ can be reproduced by the simple

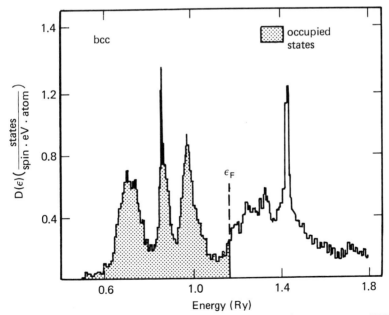

Fig. 8. The electronic density of states for bcc tungsten as calculated by MATTHEISS [1972].

References: p. 1776.

LCAO calculation, showing that the main features of the d-band contribution to $D(\epsilon)$ are in fact well described by this simple approach.

The Fermi level of tungsten is shown in fig. 8 and is observed to lie near a deep minimum in the $D(\epsilon)$ function. Thus the d-band contribution to $D(\epsilon_F)$ is small for tungsten. Figure 9 shows $D(\epsilon)$ calculated by the APW method for the neighboring element niobium (MATTHEISS [1972]) which also forms the bcc structure. One notices a remarkable similarity in the $D(\epsilon)$ functions of W and Nb. In particular, the $D(\epsilon)$ functions have the same shape but differ in total width. The width change is characteristic of going from 4d (Nb) to 5d (W) metals. On the other hand, the Fermi level of niobium is located at a somewhat lower energy with respect to the d-band structure. In fact, ϵ_F is located near a sharp maximum in $D(\epsilon)$ so that the d-band contribution to $D(\epsilon_F)$ is very large in niobium. A simple model which describes the above facts is obtained by assuming that the function $D(\epsilon)$ is roughly the same for bcc transition metals of a given row. As one moves along a given transition metal series from left to right in the Periodic Table, one simply increases the total number of d-electrons, and thus increases the position of the Fermi level with respect to $D(\epsilon)$. This is called the rigid-band model. The function $D(\epsilon)$ is assumed to be "rigid" and determined only by crystal structure (e.g. bcc). The Fermi level moves progressively higher in the band structure as one moves along a transition series and adds additional d-electrons. This model has a natural extension to alloys as well. For a solution of two metals having the same crystal structure, $D(\epsilon)$ is taken to be a rigid band with ϵ_F determined by the average number of electrons per atom. For example, a bcc alloy of composition $Nb_{50}Mo_{50}$ is assumed to have the same $D(\epsilon)$ function as

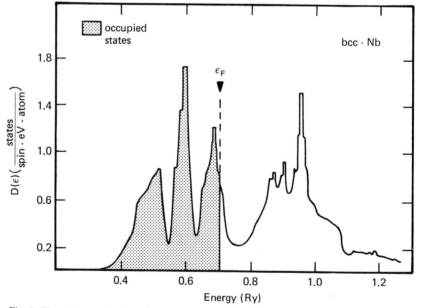

Fig. 9. The electronic density of states for bcc niobium as calculated by MATTHEISS [1972].

Nb and Mo with ϵ_F located so as to give 5.5 valence electrons per atom in the filled part of the band. Molybdenum is taken to have 6 valence electrons, and niobium 5 valence electrons. We will use this concept in the next section.

For the present purposes, we see that the formation of narrow d-bands leads to a density of states which varies rapidly over a wide range with energy. The exact position of ϵ_F determines $D(\epsilon_F)$ which can take on a large range of values. By examining table 2, we see that the large value of $D(\epsilon_F)$ experimentally found for Nb (1.82 states/eV atom) and the small value for W (0.30 states/eV atom) are both consistent with the calculated band structure. In fact, the results of calculations for a large number of transition metals give reasonable agreement with experiment.

In looking at the variation of T_c among transition metals (table 2) one notices a definite correlation between T_c and $D(\epsilon_F)$ for metals of a given structure. From McMillan's equations, we see that the electron–phonon coupling constant is given by eq. (6). Considerable controversy has arisen over the relative importance of $D(\epsilon_F)$, $\langle I^2 \rangle$, and $\langle \omega^2 \rangle$ in determining λ. HOPFIELD [1969] first pointed out that the product $D(\epsilon_F)\langle I^2 \rangle = \eta$ is an atomic-like parameter determined by the nature of the atomic potential. As such, it would be expected to vary only slowly across a given transition series. MCMILLAN [1968] tested this hypothesis for bcc transition metals and found that η is roughly constant for Mo, Nb and bcc solid solutions of Zr–Nb, Nb–Mo, Mo–Ru, and Mo–Re. Thus McMillan concluded that the average squared phonon frequency $\langle \omega^2 \rangle$ was primarily responsible for variations in T_c. Later on, GOMERSALL and GYORFFY [1974] noted that $\langle \omega^2 \rangle$ itself is strongly influenced by d-electron screening of the electrostatic fields caused by ionic displacements involved in phonon vibrations. They concluded that a reciprocal relation also exists between $D(\epsilon_F)$ and $\langle \omega^2 \rangle$. They conclude that $D(\epsilon_F)$ in fact is the most important parameter in determining λ. VARMA and DYNES [1976] have more recently carried out a general analysis using the LCAO description of d-electrons together with the ideas of Gomersall and Gyorffy. This confirms the prediction that $D(\epsilon_F)$ is an important factor governing the T_c of transition metals and alloys. Although these issues are far from entirely resolved, several statements can nevertheless be made with reasonable confidence. Firstly, it is clear that the d-electrons play a key role in determining the superconductivity of transition metals. Secondly, it is clear both empirically and theoretically that a large value of $D(\epsilon_F)$ arising from d-band contributions to $D(\epsilon)$ leads to favorable conditions for a large λ and consequently to high T_c's in d-band alloys. In §4.2, we will return to the interrelationship of $D(\epsilon_F)$, $\langle I^2 \rangle$, and $\langle \omega^2 \rangle$ and will examine a possible connection between large λ and "soft" phonons (small $\langle \omega^2 \rangle$) in transition metals and alloys.

Finally we note that more detailed experimental studies of the electron–phonon interaction and its role in the superconductivity of transition metals have been carried out using the method of superconductive tunneling (MCMILLAN and ROWELL [1969]). These data and their comparison with the theory of superconductivity have established beyond any doubt that the present theory of superconductivity based on electron–phonon coupling correctly describes the origin of superconductivity in transition metals.

References: p. 1776.

4. Superconducting alloys and compounds

In this section, we examine several selected examples of superconducting alloy materials of interest in current research and technology. In addition to the variation of T_c, we also consider the critical fields, critical currents, and magnetic properties of these materials. In particular we examine materials which are of practical interest in the generation of high magnetic fields. We also examine some features which seem to be common to high-T_c superconductors. One of these features, the presence of low-frequency or "soft" phonon vibrations, is examined in some detail. The relationship of soft phonons to instabilities of the crystal lattice and resulting allotropic phase transitions is discussed with reference made to examples. In the last part of this section, we briefly examine non-crystalline superconducting materials and show that these materials can be described within the same theoretical framework as crystalline materials.

4.1. Selected examples of materials

We divide the discussion of high-T_c superconductors into three parts. These are respectively a discussion of solid solutions, intermetallic compounds, and non-crystalline materials. We choose representative examples for each class of materials.

4.1.1. Solid solutions

When two metals are alloyed to form a solid solution having the same crystal structure as one of the parent metals, the resulting superconducting properties are often very different from those of the parent metals. Alloying has several effects. First it destroys the perfect periodicity of the crystal since equivalent lattice positions are now occupied by chemically different atomic species. This is called chemical disorder. One immediate result is that Bloch's theorem is no longer strictly valid. The electron states no longer have a well defined crystal momentum. This leads to scattering of electrons and an increase in the electrical resistivity compared with that of pure metals. A simple argument (NORDHEIM [1931]) shows that the excess resistivity varies roughly like $\Delta\rho x(1-x)$, where x is the composition of the alloy $A_{1-x}B_x$ and $\Delta\rho$ is characteristic of the two elements in the alloy. As mentioned in §2.3, one can characterize electron scattering by a mean free path l. When l becomes small compared with the clean-limit coherence length ξ_0 [eq. (19)] one obtains type-II superconducting behavior with $\xi \propto (\xi_0 l)^{1/2}$. This is generally the case with concentrated alloys where electron scattering by the chemical disorder occurs each time an electron propagates over a few atomic distances. Thus, solid solutions tend to be type-II superconductors. The Ginsburg–Landau parameter $\kappa = \lambda/\xi$ for solid solutions tends to be rather large. The values of κ are most frequently much larger than the critical value $(1/\sqrt{2})$ which separates type-I from type-II behavior. As will be seen, this leads to very large values of the upper critical field H_{c2} in certain alloys.

A second effect of alloying is to vary the average microscopic parameters $D(\epsilon_F)$,

$\langle I^2 \rangle$, and $\langle \omega^2 \rangle$ which enter in the determination of T_c and λ. For d-band alloys, the variation of $D(\epsilon_F)$ can be roughly understood using the rigid-band model provided that the pure metals have d-band structures with absolute energies not too different. This amounts to requiring that the alloy constituents be neighbors or second neighbors in the Periodic Table. The alloys we consider below generally satisfy this requirement reasonably well. Experimentally one can determine the variations of $D(\epsilon_F)$ and θ_D using low-temperature specific heat data. Assuming that $\langle \omega^2 \rangle \propto \theta_D^2$, one tries to analyze which of the microscopic parameters are important in determining the λ in alloys. A more accurate analysis of the microscopic parameters determining λ can be obtained from tunneling data although such analysis is beyond the scope of the present discussion.

As a first example of solid solutions, we consider alloys of Nb with elements in the neighboring column to the left in the Periodic Table. We look particularly at the Nb–Ti and Nb–Zr systems. The phase diagram of Nb–Ti is shown in fig. 10. One sees that a broad range of bcc solid solutions exists. Table 3 lists several measured properties of bcc Nb–Zr and Nb–Ti alloys. Figure 11 shows the temperature dependence of the upper critical field H_{c2} of $Nb_{44}Ti_{56}$ with $T_c = 9.4$ K. Below $T = 4$ K, values of H_{c2} are of the order of 100 kOe, far larger than the critical field $H_c(0)$ of either pure metal (table 2). Using eq. (33) one can estimate the coherence length near $T = 0$ of this alloy to be $\xi(0) = 50$ Å. The clean-limit coherence length could be estimated by taking the BCS expression for $\Delta(0)$ and eq. (18) with $v_F \approx 10^7$ cm/s appropriate to a d-band alloy. We find $\xi_0(0) \approx 200$–300 Å and notice that this material is a type-II superconductor in the dirty limit since $\xi \ll \xi_0$.

The variation of T_c in Nb–Ti and Nb–Zr alloys can be attributed to a combination of two effects. First, the values of $D(\epsilon_F)$ increase on alloying Zr (or Ti) with Nb

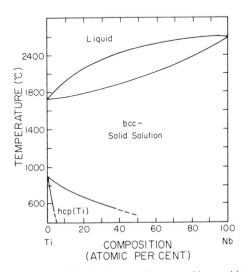

Fig. 10. The phase diagram of Nb–Ti showing the range of bcc and hcp solid solutions.

References: p. 1776.

Table 3

Properties of solid solutions of transition metals in neighboring columns of the periodic table. Data from ROBERTS [1976], MCMILLAN [1968] and HEINIGER *et al.* [1966].

Alloy	x	T_c (K)	$H_{c2}(0)$ (kG)	θ_D (K)	γ mJ/mole K^2	λ
$Nb_{1-x}Ti_x$	0.1	9.61	35	–	–	–
	0.25	10.0	90.5	–	–	–
	0.45	9.4	108 [a]	–	–	–
	0.78	7.5	77 [a]	–	–	–
$Nb_{1-x}Zr_x$	0.50	9.3	92	238	8.3	0.88
	0.25	10.6	82	246	8.9	0.93
$Nb_{1-x}Mo_x$	0.15	5.85	–	265	6.3	0.70
	0.40	0.60	–	371	2.87	0.41
	0.60	0.05	–	429	1.62	0.31
	0.70	0.016	–	442	1.46	0.29
	0.80	0.095	–	461	1.49	0.33
	0.90	0.36	–	487	1.67	0.36
$Mo_{1-x}Re_x$	0.05	1.5	–	450	2.2	0.45
	0.10	2.9	–	440	2.6	0.51
	0.20	8.5	–	420	3.8	0.68
	0.30	10.8	–	395	4.1	0.76
	0.40	12.6	19 [a]	340	4.4	0.86
		11.5	23–33 [a]	320	4.4	0.85

[a] at $T = 4.2$ K.

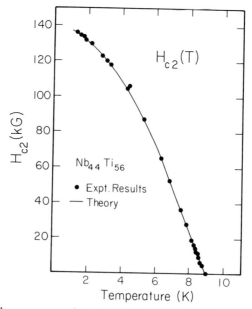

Fig. 11. The temperature dependence of the upper critical field of $Nb_{44}Ti_{56}$.

before falling rapidly as one approaches the hcp or bcc phase of the pure metals Ti or Zr. The experimental values of $D(\epsilon_F)$ for the alloys can be compared with those of a calculation of $D(\epsilon)$ for bcc Nb by assuming the rigid band model (see fig. 9). Alloying with Zr lowers ϵ_F of Nb so that ϵ_F lies closer to a peak in the $D(\epsilon)$ function for bcc transition metals. Thus $D(\epsilon_F)$ is increased by alloying; in addition, also note that θ_D is decreased on alloying by comparison with θ_D for either pure Zr or pure Nb. This suggests that $\langle\omega^2\rangle$ is decreased on alloying. Both the change in $D(\epsilon_F)$ and $\langle\omega^2\rangle$ should have the result of increasing λ [see eq. (6)]. Thus we can understand the enhancement of T_c by alloying. In §4.2 it will be noted that Nb–Zr alloys are prone to an allotropic phase transformation at elevated temperatures. The allotropic phase is referred to as the ω-phase. The general softening of phonons evidenced by the minimum in θ_D for these alloys may be in part due to the softening of a particular set of phonon modes involved in the ω-phase transformation (see ch. 15).

A second example of an alloying effect is also given in table 3. When one alloys Mo with Re to form a bcc solid solution, one observes a dramatic increase in T_c. Again, this can be correlated with a rise in $D(\epsilon_F)$ and a lowering of the Debye temperature as seen in the table. Again, these effects are characterized by a destabilization of the bcc lattice and accompanying softening of phonon frequencies. Figure 12 illustrates the composition dependence of T_c, $D(\epsilon_F)$, θ_D^2, and λ. This figure shows very clearly the reciprocal relationship between λ and $\langle\omega^2\rangle \sim (\theta_D^2)$ as well as the direct correlation of both T_c and λ with $D(\epsilon_F)$.

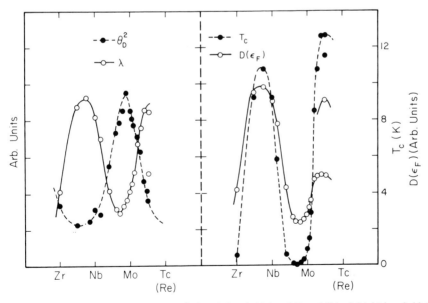

Fig. 12. The variation with group number of θ_D^2, λ (left-hand side) and T_c and $D(\epsilon_F)$ (right-hand side) for bcc transition metals and solid solutions of neighboring metals of the 4d-series. Data for Mo–Re solutions (group VI–group VII) are used in place of Mo–Tc alloys since data of Tc alloys is lacking.

References: p. 1776.

Below, the properties of Nb–Ti solutions will be further discussed in conjunction with high-field applications of superconductivity. The nature of the type-II state will be described in more detail there.

4.1.2. A15 compounds and other special structures *

No discussion of superconducting materials would be complete without a survey of several special crystal structures which have been found to be favorable for high-temperature superconductivity. Foremost among these special structures is the β-W or *A15 structure*. Figure 13 shows the unit cell of this crystal. The prototype compound having this structure is Nb_3Sn which has $T_c = 18.3$ K. Other examples of A15 superconductors are listed in table 4. Among these is Nb_3Ge with $T_c = 23.2$ K. This is the highest transition temperature observed for any superconductor. Table 4 also gives data from specific heat measurements and values of H_{c2} for these materials. The H_{c2} values range as high as 430 kG and are typically in the range 100–300 kG.

The origin of the high T_c in these materials has been the subject of considerable research and an equal amount of controversy. Early work on the electronic structure of A15 compounds focused on the chain-like structure of the Nb atoms (WEGER [1964]). These chains are illustrated in fig. 13. It was noted that the Nb atoms form a three-dimensional array of such chains. Along each chain the interatomic Nb distance is substantially less ($\sim 15\%$) than that observed in bcc Nb. This leads to a band structure with quasi-one-dimensional features and to an unusually large value of $D(\epsilon_F)$. Friedel and his coworkers (LABBÉ et al. [1966]) used the LCAO approach and the quasi-one-dimensional nature of the chain structure to further understand the large $D(\epsilon_F)$ and other anomalous properties of A15 compounds such as V_3Si ($T_c = 17$ K). Tunneling studies (GEERK et al. [1982] and KIHLSTROM and GEBALLE [1981]) have revealed that in addition to a large $D(\epsilon_F)$, the A15 materials Nb_3Al and Nb_3Ge also exhibit an anomalous phonon softening with unusually low values of $\langle \omega^2 \rangle$.

Several of the A15 compounds are prone to a type of lattice instability which leads to a martensitic phase transformation from a cubic to a tetragonally distorted lattice at low temperatures. This transformation is believed to be related to the existence of *soft shear modes* in the phonon spectrum and is discussed further in a later section (see also ch. 15).

The upper critical field of Nb_3Sn as a function of temperature is shown in fig. 14. At low temperatures, it approaches a value of ~ 250 kG. Estimates of ξ and λ reveal (ORLANDO et al. [1979]) that Nb_3Sn is a large-κ type-II superconductor. The large critical field results from the combined effect of the high T_c [large $\Delta(T)$] and small value of v_F which derives from the narrow d-bands ($v_F \propto dE_K/dK$) which contribute to $D(\epsilon_F)$. This leads to a small coherence length even in the clean limit. The degree of atomic ordering in Nb_3Sn has been studied in detail and found to be extremely high. The electron mean free path l could thus be rather large since little

---

* See also ch. 5.

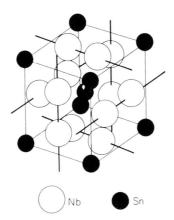

Nb   Sn

Fig. 13. Illustration of the unit cell of the A15 crystal Nb_3Sn.

or no electron scattering is expected from chemical disorder. These facts have raised the question of whether the "clean" or "dirty" limits appropriately describe ξ and H_{c2} in these materials.

Another interesting class of high T_c materials are the refractory metal borides, carbides, and nitrides having the *B1 (NaCl-type) structure*. The properties of some selected examples of these materials are shown in table 5. The compound NbN ($T_c = 17$ K) was one of the first known of these high-T_c materials. The members of this family of superconductors can be viewed as interstitial compounds in which the small atoms B, C, and N occupy the octahedral interstitial sites in a close-packed transition metal lattice. The electronic structure of the materials has been analyzed in detail. The $D(\epsilon)$ function contains many contributions in the vicinity of the Fermi level (MATTHEISS [1972]). These include a large d-band contribution from the transition-metal atoms as well as a significant contribution from the s and p-orbitals of the B, C, and N. The superconducting properties of these materials are sensitively

Table 4

Properties of some selected A15 superconducting materials. Data from SINHA [1972], ROBERTS [1976] and STEWART [1978].

Alloy	T_c (K)	θ_D (K)	γ mJ/atom K^2	$H_{c2}(0)$ (kG)	$D(\epsilon_F)$ (states/eV atom)
Nb_3Al	18.8	290	4.49	295 [a]	1.4
Nb_3Au	10.6	280	5.71	–	1.1
Nb_3Sn	18.0	290	8.96	180–240	1.4
V_3Si	17.0	330	11.1	250	2.4
V_3Ga	16.8	310	14.6	200–270	3.0
Nb_3Ge	23.0	302	7.6	370 [a]	1.2
$Nb_3(Al_{0.8}Ge_{0.2})$	20.7	278	8.75	~ 430	1.4

[a] at $T = 4.2$ K.

References: p. 1776.

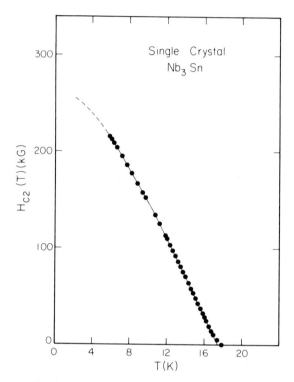

Fig. 14. The temperature dependence of the upper critical field of Nb$_3$Sn. Data from ORLANDO et al. [1979].

Table 5

Properties of selected superconducting materials having the B1 structure. T^{Obs} is the temperature at which H_{c2} was observed. References for data in ROBERTS [1976].

Material	T_c	H_{c2}	T^{Obs}
MoC$_{0.60}$	12.1	98.5	1.2
NbC$_{0.98}$	11.56	–	–
NbC$_{0.8-0.99}$	< 1.5–11.56	–	–
HfB	3.1	–	–
ZrB	3.1	–	–
NbN$_{0.84}$	12.9–13.5	–	–
NbN$_{0.92}$	16.3	130	0.0
NbN$_{0.96}$	15.0	250	0.0
TiN$_{0.6-0.99}$	< 1.2–4.35	–	–
TiN$_{0.84}$	1.2	–	–

dependent on the concentration of the smaller interstitial atoms (B, C and N). A perfectly ordered B1 crystal requires the equiatomic composition. In reality, these compounds tend to form over a homogeneity range with compositions deficient in the interstitial atom. For example, a partial binary phase diagram of the Nb–C system is shown in fig. 15. Only at very high temperatures does the B1 structure form near the equiatomic composition. At lower temperature, one has a C-deficient structure Nb_1C_x $(x < 1)$. Figure 16 shows that the superconducting transition temperature is strongly depressed by vacancies on the carbon sites. The tendency to form vacancies on the interstitial atom sites in B1 superconductors can be viewed as another type of instability of the lattice.

The high T_c values observed for B1 superconductors have been attributed to a modest value of $D(\epsilon_F)$ coupled with large values of $\langle I^2 \rangle$, the Fermi surface averaged electron–phonon matrix element. Soft phonons (low $\langle \omega^2 \rangle$) are also thought to play a possible role in giving a large electron-phonon coupling constant λ. The upper critical fields H_{c2} for a number of B1 materials are quite high as seen in table 5. This is presumed to be due (as in the A15 compounds) to small values of v_F (associated with d-band electrons) and the large values of T_c $[\Delta(T)]$. As seen in eq. (18) this leads to a small clean-limit coherence length ξ_0. On the other hand, there is also evidence that atomic disorder and electron scattering can decrease the electron mean free path l and lead to a large enhancement of H_{c2} as one moves toward the dirty limit [eq. (19)]. Disorder decreases T_c but increases ρ and thus increases $(dH_{c2}/dT)_{T_c} \propto \rho\gamma$ [see eq. (33)]. The combined effects of lowering T_c but concur-

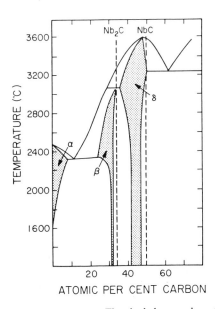

Fig. 15. The phase diagram of the Nb–C system. The shaded areas show the homogeneity range of the various carbide compounds.

References: p. 1776.

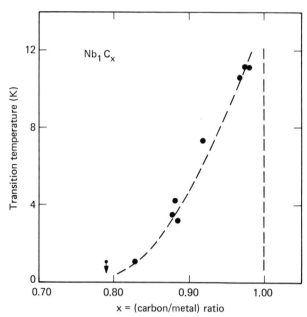

Fig. 16. The superconducting transition temperature of Nb_1C_x with the B1 (NaCl-type) structure. For $x < 1$, the structure contains carbon vacancies which depress T_c. (After GIORGI *et al.* [1962].)

rently raising (dH_{c2}/dT) gives for the case of NbN an increase in H_{c2} at low temperature (ASHKIN *et al.* [1972]). This behavior is illustrated in fig. 17 for a "clean" highly ordered bulk sample of NbN compared with a "dirty" thin film sample. The figure shows the $H_{c2}(T)$ curves for both samples. Although T_c is lower in the "dirty" sample, one observes a much steeper $H_{c2}(T)$ curve which rapidly overtakes the $H_{c2}(T)$ curve of the "clean" sample at sufficiently low temperatures. This illustrates the dual role that disorder plays in decreasing T_c while increasing dH_{c2}/dT.

In recent years, research on superconducting compounds has turned to a variety of structurally unique materials. Among these are materials of lower dimensionality. Strictly speaking, all superconducting materials are three-dimensional. On the other hand, the atomic and electronic structures of many materials are highly anisotropic to the extent that they can be thought of as quasi-two dimensional (2D) or quasi-one-dimensional (1D). Among the quasi-2D compounds are several layered compounds. One of the most extensively studied classes of such materials are the *Chevrel phases*. These are layered compounds with a highly two-dimensional electronic structure. The best studied subgroup of the Chevrel phases are the intercalation compounds of the form $Mo_{3-y}S_6I_{1+x}$ where I = Pb, Sn, Cu etc. In these materials, the Mo atoms are arranged in puckered 2D nets. Each Mo atom is surrounded by an octahedron of S atoms. The narrow d-bands formed by the 4d-electrons of Mo have a highly two-dimensional character, and these d-bands are

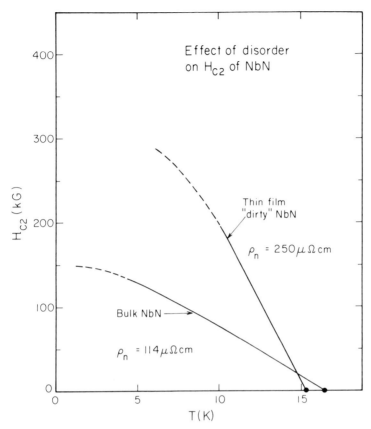

Fig. 17. The effect of disorder (as evidenced by electrical resistivity) on the upper critical field and T_c of NbN.

presumed responsible for the superconducting properties of these materials. As in the "NaCl-type" superconductors, the narrow d-bands result in a small v_F and thus a small clean-limit coherence length. Table 6 lists a few properties of some selected examples of these materials. One unusual characteristic of these materials is directional dependence of H_{c2} *in single crystals*. H_{c2} measured along the direction parallel to the layers considerably exceeds that measured along the direction normal to the layers. This reflects the two-dimensional character of the electronic structure. Table 6 gives an example of one such measurement. From the table, it is also seen that the parallel upper critical field *for polycrystalline samples* is extremely large, ranging up to 600 kG. These are the highest values of H_{c2} observed for any superconducting material. The corresponding coherence length can be estimated from eq. (33) and is found to be $\xi(0) \approx 23$ Å for $Pb_{1.0}Mo_{5.1}S_6$. This suggest a combination of very narrow d-bands (small v_F) and considerable electron scattering (small l) in these high-field materials.

References: p. 1776.

Table 6
The properties of some selected Chevrel phases.

Compound	T_c (K)	$H_{c2}(T)$ (kG)	T^{obs} (K)	Ref.
$Cu_{1.5}Mo_{4.5}S_6$	10.3	130	0	ODERMATT et al. [1975]
$Cs_{0.3}MoS_2$	6.8	12	4.2	WOOLHAM et al. [1974]
		30	5.8	
$PbMo_6S_8$	15.2	600	0	FISCHER et al. [1980]
$Pb_{0.9}Mo_{5.1}S_6$	11.7	486	0	FONER et al. [1974]
		390	4.2	
$Pb_{1.0}Mo_{5.1}S_6$	14.4	598	0	FONER et al. [1974]
		510	4.2	
$Sr_{0.2}MoS_2$	5.6	9	2.6	WOOLHAM et al. [1974]
		25	3.2	

There are a number of other interesting crystal types exhibiting high-temperature superconductivity. Time and space do not permit a detailed discussion of all of these and so we only mention a few. The hydrides and deuterides of several metals and alloys exhibit superconductivity at rather high temperatures. For example, the stoichiometric *hydride (deuteride)* of Pd having the NaCl structure has been prepared by ion implantation of hydrogen at low temperature (STRITZKER and BUCKEL [1972]) and found to become superconducting at $T_c = 9$ K. The origin of the superconductivity is thought to involve the reduction in spin fluctuations of Pd and the coupling of electrons to high-frequency optical phonon modes involving hydrogen motion. The deuteride, PdD, has $T_c = 11$ K showing that the superconductivity is altered by changing the nuclear mass of hydrogen. The hydrides formed from Pd–Ag and Pd–Cu fcc solid solutions have an even higher transition temperature with a maximum T_c of 16 K for $Pd_{1-x}Ag_xH_y$ hydrides and 16.6 K for $Pd_{1-x}Cu_xH_y$ hydrides (STRITZKER [1974]).

Several *ternary compounds* have been found to have unusual superconducting properties. These include such materials as the ternary oxide $LiTi_2O_4$ with $T_c = 13.7$ K (JOHNSTON et al. [1973]) and the ternary borides of the form $(RE)Rh_4B_4$ where RE stands for rare-earth elements. The compound $ErRh_4B_4$ becomes superconducting at $T_{c1} = 9$ K but reverts to normal metal behavior at a second, lower temperature $T'_{c2} = 1$ K below which the Er atoms order ferromagnetically. These latter compounds form a fertile area for the study of the coexistence of superconductivity with magnetic ordering. Figure 18 shows a phase diagram indicating the regions where superconductivity, and various types of magnetic ordering exist in the alloys $(Er_{1-x}Ho_xRh_4B_4)$. The richness of electronic and magnetic behavior observed in these materials provides the opportunity to test theories of phase transitions in magnetically ordered superconductors. Unfortunately, these subjects cannot be dealt

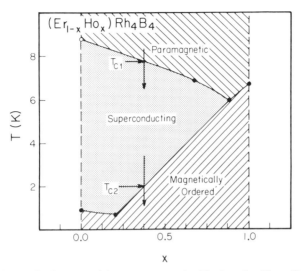

Fig. 18. A phase diagram in the composition–temperature (x, T) plane for $(Er_{1-x}Ho_x)Rh_4B_4$ showing paramagnetic, magnetically ordered, and superconducting regions.

with here without developing considerable theoretical background. The reader is therefore referred to the many review articles in the literature (e.g. MAPLE [1978]).

4.1.3. Amorphous alloys

In the BCS theory of superconductivity, the electronic states which are involved in Cooper pairing are assumed to have well defined crystal momentum. In imperfect crystals and chemically disordered alloys, electron scattering occurs and Bloch's theorem is no longer valid. Nevertheless, superconductivity still occurs in such materials. This was first explained by ANDERSON [1959] in a paper dealing with "dirty" superconductors. Anderson showed that in spite of the failure of Bloch's theorem, one can still find suitable electronic states (time-reversed pairs of states) from which to form a superconducting ground state. The ultimate test of this concept is the case of an amorphous material. Here, electrons are so strongly scattered that the electron mean free path is reduced to interatomic distances and the notion of eigenstates of momentum breaks down entirely. Nevertheless, superconductivity was found to occur in amorphous metals (BUCKEL and HILSCH [1954]). The earliest experiments on these materials were carried out on thin films produced by quenching metallic vapors onto a cryogenically cooled substrate. Particularly fascinating was the observation of superconductivity in amorphous Bi (a semimetal in its crystalline form) with $T_c = 6$ K. Later studies revealed that most simple metals have enhanced T_c when prepared in amorphous form. A few examples are given in table 7. The data are taken from a review by BERGMANN [1976]. A similar survey of transition metals reveals that T_c is raised or lowered depending on the metal

References: p. 1776.

Table 7

Properties of some amorphous superconducting materials. Data from BERGMANN [1976] and JOHNSON [1981].

Material	T_c (K)	Preparation method [a]	$(dH_{c2}/dT)_{T_c}$ (kG/K)
Bi	5	VQ	–
$Pb_{90}Cu_{10}$	6.5	VQ	–
Ga	8.4	VQ	3.04
$Sn_{90}Cu_{10}$	6.76	VQ	4.4
$La_{80}Au_{20}$	3.5	LQ	23.0
$Mo_{48}Ru_{32}B_{20}$	6.05	LQ	23.6
$Zr_{75}Rh_{25}$	4.55	LQ	26.3
$Nb_{60}Rh_{40}$	4.8	LQ	31
Nb_3Ge	3.6	S	25

[a] VP = vapor quenched; LQ = liquid quenched; S = sputtered.

(COLLVER and HAMMOND [1973]). Figure 9 shows the variation of T_c found as one traverses a given row of transition metals in the Periodic Table. The T_c's shown are for pure amorphous metals and alloys of neighboring metals of the 4d transition series. Data for the corresponding crystalline alloys are also shown. The crystalline alloys show high T_c values for specific group numbers or valence-electron-per-atom ratios $e/a \approx 4.5$ and $e/a \approx 6.5$ as first pointed out by MATTHIAS [1957]. This presumably reflects the variation of $D(\epsilon_F)$ with e/a as discussed in §§ 3.2 and 4.1.1.

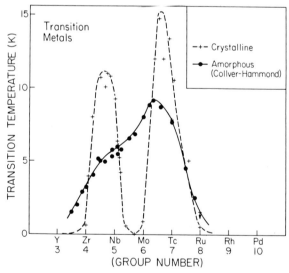

Fig. 19. Illustration of the variation of T_c with group number for crystalline and amorphous 4d transition metals and alloys of neighboring metals. Data for amorphous alloys from COLLVER and HAMMOND [1973].

In contrast, amorphous alloys show a slowly varying T_c which peaks near $e/a \approx 6.8$ and falls slowly as one moves away from this e/a value. This suggests that $D(\epsilon_F)$ varies slowly with e/a in amorphous alloys, having a broad maximum when the d-electron states are roughly half filled. This is discussed in more detail by JOHNSON [1981].

Amorphous alloys have also been produced by rapid quenching of liquid alloys (JOHNSON [1981]) to form metallic glasses. Some examples are given in table 7. In contrast to thin cryoquenched films, metallic glasses are generally stable against crystallization at room temperature and considerably above. The trends in T_c for metallic glasses based on transition metals are roughly the same as those observed by Collver and Hammond. The variation of T_c in these alloys can be qualitatively understood in terms of variations in $D(\epsilon_F)$ as determined by electron photoemission spectra and from specific heat data (JOHNSON and TENHOVER [1983]). The reader is referred to this reference for a detailed discussion. Table 7 gives selected data for several metallic glasses.

The high degree of electron scattering in amorphous alloys leads, as mentioned above, to small values of l. This in turn suggests that these should be "dirty" superconductors. Experiments confirm this expectation and yield large values of H_{c2} and small values of $\xi = \xi_d$ as predicted by eq. (19). The upper critical field for several metallic glasses is shown as a function of temperature (both in reduced units) in fig. 20. In addition to the large H_{c2} values, one notes that the $H_{c2}(T)$ curves for these cases are rather linear over a broad range of temperatures. This contrasts with the prediction of the microscopic theory of H_{c2} (WERTHAMER *et al.* [1966], and MAKI [1966]) in the dirty limit. The most nearly linear curve predicted by this theory is shown in the figure for comparison. A similar discrepancy has also been observed

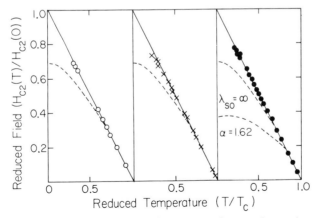

Fig. 20. The upper critical field H_{c2} as a function of temperature for several amorphous alloys, plotted in reduced units $[T/T_c]$ and $[H_{c2}/H_{c2}(0)]$. Dashed curves are nearest-to-linear theoretical predictions. From left to right: $Mo_{40}Ru_{40}P_{20}$, $T_c = 5.76$ K, $H_{c2}(0) = 133.6$ kG; $Mo_{52}Ru_{34}B_{16}$, $T_c = 6.88$ K, $H_{c2}(0) = 159.6$ kG; $Mo_{30}Re_{70}$, $T_c = 8.39$ K, $H_{c2}(0) = 240.8$ kG.

References: p. 1776.

for some very "dirty" crystalline alloys. In the case of amorphous alloys, it is not clear whether the theory fails or whether inhomogeneities in the amorphous state are responsible. This is but one of several problems which are raised by non-crystalline superconductors. A recent review is by POON [1983].

4.2. High T_c's, lattice instabilities, and soft phonons

In this section, the problem of the ultimate limit for T_c in real materials is considered. It is argued that materials with strong electron–phonon coupling (large λ) have an inherent tendency to become unstable. This instability can take on a variety of forms. For example, certain high-T_c superconductors are frequently observed to undergo distortive phase transformations of the crystal lattice as a function of temperature. In other systems, the crystal lattice is observed to possess *soft* (low-frequency) phonon modes. Phonon softening can ultimately lead to static distortions. The underlying origin of these instabilities is closely linked to the strength of the electron–phonon interaction itself. As such, the tendency toward instability tends to limit the ultimate T_c of real materials. The discussion that follows begins with some simple theoretical arguments illustrating the above ideas. It is followed by examples of instabilities observed in real materials.

The total effective interaction between two electrons in a solid can be written as $V^{\mathrm{eff}}(r_1 - r_2)$, where r_1 and r_2 are the electron position vectors and $r = r_1 - r_2$. In free space, this interaction is simply the bare Coulomb repulsion:

$$V^C(r_1 - r_2) = \frac{e^2}{r}, \tag{35}$$

with $r = |r|$. The Fourier components of this potential are given by

$$V_q^C = \int e^{iq \cdot r} \frac{e^2}{r} \, dr = \frac{4\pi e^2}{q^2}, \tag{36}$$

with $q = |q|$.

Using linear response theory of screening (ZIMAN [1972]) one can define a total dielectric response function $\epsilon(q, \omega)$ for a solid. This function describes the combined screening response of ions and electrons in the solid to a perturbation of wavevector q and frequency ω. For a static perturbation, one uses $\epsilon(q, \omega = 0) = \epsilon(q, 0)$. In particular the bare static Coulomb interaction of two electrons in a metal will be modified by the combined screening effects of the ions and other electrons. Each Fourier component will become:

$$V_q^{\mathrm{eff}} = \frac{V_q^C}{\epsilon(q, 0)} = \frac{4\pi e^2}{\epsilon(q, 0) q^2}. \tag{37}$$

In order for an attractive static interaction of two electrons to occur in a metal one would require $\epsilon(q, 0) < 0$. But then it would follow that a periodic static charge wave in the metal with wavevector q would have an associated energy density

$$\mathscr{E} = \tfrac{1}{2} \epsilon(q, 0) |E_q|^2 < 0, \tag{38}$$

where E_q is the electric field associated with the charge wave. This means that one could lower the energy of the metal by simply creating any static charge wave of wavevector q. The crystal would *spontaneously* develop such waves and would thus be unstable. Thus, we must have $\epsilon(q, 0) > 0$ in stable crystals.

On the other hand, as mentioned earlier for the case of superconductivity the attractive interaction between electrons is in fact a phonon-mediated dynamic interaction which does not violate the above condition. We see that one can only have $\epsilon(q, \omega) < 0$ for $\omega > 0$. When $\epsilon(q, \omega) < 0$ for small ω, one approaches an unstable condition.

Using McMillan's expression for the electron–phonon coupling constant [eq. (6)] with

$$\lambda = \frac{D(\epsilon_F)\langle I^2 \rangle}{M\langle \omega^2 \rangle} = \frac{\eta}{M\langle \omega^2 \rangle}$$

we follow the arguments mentioned earlier that η is roughly constant for metals of a given type. The parameter $\eta = D(\epsilon_F)\langle I^2 \rangle$ was analyzed by HOPFIELD [1969] and shown to be an "atomic-like" parameter determined mainly by valence and atomic volume. BARASIC *et al.* [1970] analyzed the variation of η for d-band metals and found that $\eta \propto q_0^2 E_c$ where q_0 is a characteristic length for spatial decay of atomic d-orbitals (a Slater coefficient) and E_c is the cohesive energy of the transition metal contributed by the d-electron bonding. Again it was found that η is expected to vary slowly over a given transition metals series in the Periodic Table. All of the above work leads to the conclusion that the mean square phonon frequency $\langle \omega^2 \rangle$ is a primary factor controlling λ. Strong-coupling (high-T_c) superconductors tend to have a soft or low-frequency phonon spectrum. In fact, combining the above expression for λ ($\eta \approx$ constant) with eq. (3b), one can obtain an upper bound on T_c (McMILLAN [1968]):

$$T_c \sim \langle \omega \rangle \exp\left[-\left(\frac{M\langle \omega^2 \rangle}{\eta} - 1\right)\right], \tag{39}$$

which has a maximum value with respect to $\langle \omega \rangle$ (ignoring the difference between $\langle \omega^2 \rangle$ and $\langle \omega \rangle^2$ for purposes of argument) of

$$T_c^{max} = (\eta/2M)^{1/2} e^{-3/2} \tag{40}$$

when $\langle \omega \rangle = (\eta/2M)^{1/2}$. We see that a maximum T_c exists as a function of $\langle \omega \rangle$ when $\lambda = 2$ in the present argument. On the other hand, in order to obtain λ values so large, one must have rather low $\langle \omega^2 \rangle$. This in turn would suggest $\epsilon(q, \omega) < 0$ for small ω which would tend to lead to an unstable crystal.

There are many interesting examples of soft phonons and lattice instabilities observed in real high-T_c superconducting materials. A few of these are discussed below. We first consider the bcc transition metals and solid solutions of neighboring metals discussed earlier. One would expect the phonon frequencies to be related to the bulk elastic moduli of the metals and thus with the cohesive energy of the metals. These parameters in turn vary through a broad maximum as one traverses a series of

References: p. 1776.

d-band alloys. (GESCHNEIDER [1964], FRIEDEL [1969]). In contrast, fig. 12 shows that the variation of θ_D^2 is rather extreme and inversely correlated to that of λ for 4d bcc transition metals Zr, Nb, Mo, and bcc solid solutions Zr–Nb, Nb–Mo, Mo–Tc and Mo–Re. Neglecting the variation of atomic mass M within this series (which is small), we expect that $1/\theta_D^2$ roughly gives the variation of $1/M\langle\omega^2\rangle$. There is thus indeed an obvious correlation between λ, T_c, and $1/M\langle\omega^2\rangle$. The variation of $1/\theta_D^2$ does not follow a simple pattern as might be expected from the variation of cohesive energy and bulk modulus. This is precisely the soft phonon behavior alluded to above. This analysis can be taken one step further. For example, the actual experimental phonon dispersion curves of Nb and Mo and a $Nb_{0.25}Mo_{0.75}$ alloy along high-symmetry directions in the Brillouin zone have been measured. These data are obtained from inelastic neutron scattering experiments (POWELL *et al.* [1968]). One observes a general lowering in the phonon frequencies of Nb compared with either Mo or with an alloy of Nb and Mo. In addition, one sees several anomalous low-frequency dips in the Nb dispersion curves. In particular one notices a pronounced dip in the longitudinal phonon curves of Nb along the (1 1 1) direction in the Brillouin zone. These anomalous features can be reproduced by calculations (WEBER [1980]) and are directly attributable to the electron–phonon interaction. The soft phonons in Nb play an important role in determining the high T_c (large λ) of Nb. We see a clear difference between the phonons of the low-λ (low-T_c) metal Mo, and the high-λ (high-T_c) metal Nb.

As a second example of soft phonons and lattice instability, we turn to the A15 compounds. Among the high-T_c member of this family, several are observed to undergo a distortive martensitic transformation at low temperatures. This transformation lowers the symmetry of the unit cell from cubic to tetragonal. This transition has been observed in V_3Si (GORINGE and VALDRE [1965]), Nb_3Sn (BATTERMANN and BARRETT [1964]), and V_3Ga (NEMBACH *et al.* [1970]). The temperature at which the transition occurs is generally low (< 100 K) but above the superconducting T_c. The cubic modification is metastable at low temperature but can often be stabilized by strains or defects. This permits measurement of the superconducting transition temperature of both phases, and the tetragonal modification is invariably observed to have a depressed T_c. TESTARDI and BATEMAN [1967] have shown that this transition is associated with an anomalous softening of the alloy shear modulus which is arrested in the tetragonal phase.

Direct evidence of soft phonons in high-T_c A15 compounds has been obtained from superconductive tunneling experiments (GEERK *et al.* [1982] and KIHLSTROM and GEBALLE [1981]). In these experiments, a direct correlation between high T_c and large λ, and the existence of enhanced contributions to λ arising from low-frequency phonon modes was found for a series of $Nb_{3+x}Al_{1-x}$ and $Nb_{3+x}Ge_{1-x}$ A15 phases. For off-stoichiometric alloys with $x > 0$, one finds small λ, low T_c, and an absence of soft phonons. As stoichiometry is approached ($x \to 0$), both λ and T_c increase while the average squared phonon frequency $\langle\omega^2\rangle$ decreases.

Earlier we pointed out the tendency of "NaCl-type" superconductors (e.g. NbN) to form off stoichiometry. The deficiency of small atoms (e.g. N and C) is accommodated by the formation of small-atom vacancies. As seen in an earlier section, these deviations from ideal stoichiometry result in a dramatic lowering of T_c (and λ). Furthermore, the equilibrium concentration of vacancies increases as the temperature is lowered (see fig. 15), indicating that the thermodynamic ground state of this system tends to contain vacancies. This is another example of a type of lattice instability. In this case, the lattice of the high-T_c material is unstable against defect formation (e.g., vacancies). The defective lattice is characterized by lower values of λ and T_c.

There are many other examples of soft phonons and lattice instability among high-T_c superconducting materials. They are too numerous to discuss here. Instead, we summarize by saying that a general correlation between high T_c and small $\langle \omega^2 \rangle$ seems to exist for a variety of superconductors. There is substantial evidence that phonon softening and lattice instability (which frequently accompanies it) tend to be a dominant factor in determining the highest achievable T_c of real materials. High-T_c superconductors tend to have a lattice which is prone to transformation to structures of lower symmetry and lower T_c. As such, high-T_c materials tend to be difficult to synthesize.

4.3. Flux pinning, critical current, and applications

The persistence of the superconducting state to high magnetic fields in type-II superconductors was recognized about twenty years ago to have practical application in the technology of magnets. The first high-field superconducting magnets were constructed at Bell Laboratories over twenty years ago and utilized the superconducting solid solutions of Mo–Re and the A15 compound Nb_3Sn. It was soon recognized that the flux line lattice (§ 2.5) of mixed-state superconductors interacts with material inhomogeneities in a manner which determines the *current-carrying capacity* of type-II superconductors in high magnetic fields. An understanding of the mechanism of this flux-pinning interaction was essential to the applications of type-II superconductors in technology. In the following discussion, the mechanism of flux pinning and its effect on the critical current density of type-II superconductors is described. The last part of the section briefly outlines some of the technological developments in this field.

In § 2.5, a description of the mixed state of type-II superconductors was given. This state can be described as a two-dimensional periodic array of cells normal to the applied field (see fig. 5), each containing a quantum ϕ_0 of magnetic flux. The order parameter ψ vanishes in the center or core of each cell while the microscopic magnet field $h(x, y)$ (which is directed along the z-axis normal to the page in fig. 5) is maximum at the core and decreases to a minimum value in the interstices between cores. The magnetic flux in each cell is often referred to as a flux line, or *vortex* (SAINT-JAMES *et al.* [1969]). When a current flows in a mixed-state superconducting sample along a direction normal to an applied magnetic field, the individual vortices are acted upon by a Lorentz force. This is illustrated in fig. 1. The magnitude of this

References: p. 1776.

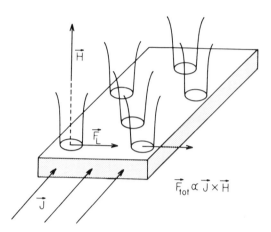

Fig. 21. Illustration of the Lorentz force F_L exerted on a single vortex in the presence of a flowing current. The total force F_{tot} per unit volume on the entire flux lattice is proportional to $J \times H$.

force per unit length of a single isolated vortex is given by (TINKHAM [1975]):

$$F_{\phi_0} = J \times z_1 (\phi_0/c),\tag{42}$$

where z_1 is the unit vector in the z-direction; the force per unit volume of superconductor, exerted on the flux lattice as a whole is given by

$$F_L = J \times B/c,\tag{43}$$

where B is the macroscopic average field in the sample and J is the current density flowing through the sample normal to the field. This behavior is illustrated in fig. 21. The Lorentz force tends to produce acceleration of the flux line lattice in a direction normal to both H and J. However, it can also be shown that the motion of the flux lattice is opposed by a viscous drag. The viscous drag produces an opposing force $-\eta v_L$ were η is an effective viscosity and v_L the velocity of the flux lines. For a single vortex, the equation

$$|F_L| = J \frac{\phi_0}{c} = \eta |v_L|\tag{44}$$

expresses the fact that steady-state equilibrium requires that the magnitude of the Lorentz force and viscous drag force be equal. This equation gives the steady-state flow velocity. The viscous flow results in power dissipation. The power dissipated per unit length of each vortex is

$$W = -F \cdot v_L = \eta v_L^2.\tag{45}$$

Strictly speaking, the sample is no longer superconducting under these conditions. In practical superconductors in high fields and carrying a high current density (typically $H \approx 100$ kG and $J \approx 10^4 - 10^5$ A/cm^2) the energy is dissipated as heat which warms the sample and leads to a thermal run-away effect whereby superconductivity is destroyed altogether. In order to prevent this catastrophe, a means of preventing flux flow is required. Fortunately, nature provides a variety of means to accomplish this. Real superconductors contain crystal defects such as dislocations as well as other types of inhomogeneities such as surfaces, second-phase precipitates, etc. The local superconducting properties of the material, such as ξ, λ, T_c etc. are all locally altered by the presence of the defects and inhomogeneities in the material. Under these conditions, the free energy of a vortex becomes a function of the vortex location with respect to these material inhomogeneities. As a result, vortices tend to have preferred arrangements in inhomogeneous superconductors. A finite force is required to move vortices from their preferred locations. The force which opposes motion is referred to as a *pinning force*. The presence of pinning forces implies that the Lorentz force must exceed a critical value before flux motion will occur. The sample can then carry a finite current density before dissipative loss sets in. The maximum loss-free current density is called the *critical current density* J_c. The problem of calculating J_c in real inhomogeneous superconducting materials is immensely complex. Only a few general considerations are outlined here. The reader is referred to reviews by CAMPBELL and EVETTS [1972], FREYHARDT [1974], ULLMAIER [1975] and KRAMER [1978].

The problem of calculating the total pinning-force density on the flux lattice can be divided into two parts. First one considers the interaction of an individual vortex with the inhomogeneities of the material. This interaction is referred to as the *elementary pinning interaction*. Once this is known, one proceeds to attempts to sum the pinning forces. The *summation problem* must be dealt with statistically (LABUSCH [1969], CAMPBELL and WEBB [1972] and MATSUSHITA and YAMAFUJI [1981]). In fact, it is easy to see that for a random distribution of inhomogeneities, a strictly rigid flux line lattice could not be pinned at all. This follows from the fact that for random inhomogeneities and a nondeformable flux lattice, each vortex sees an elementary pinning force directed with equal probability in any direction. The sum of these elementary forces on all vortices must necessarily be near zero, giving a vanishing average pinning-force density (ALDEN and LIVINGSTON [1966]). Thus, the existence of a net average pinning force requires that the flux lattice be deformable. One is thus led to consider the elastic deformation behavior of the flux lattice. For a more detailed discussion, the reader is referred to above references.

The problem of flux pinning by inhomogeneities in type-II superconductors ("hardening" of the superconductor) resembles those of alloy hardening (ch. 21) and magnetic hardening (ch. 26). A flexible "carrier" (flux line, dislocation, Block wall) which changes the state of the material interacts with "obstacles" of suitable dimensions distributed in the material and becomes pinned up to a maximum level of the applied "force" (critical current density, critical shear stress, coercive magnetic field). For a unified discussion of the three subjects see HAASEN [1970, 1977].

In real superconductors, the elementary pinning interaction can be comparatively strong. This allows substantial deformation of the flux lattice and results in a total pinning-force density which is a substantial fraction of what one would expect if all the elementary pinning forces acted in concert.

References: p. 1776.

As an example, the critical current density of a sample of the A15 superconductor Nb_3Sn or the solid solutions of Nb–Ti can be as large as 10^5–10^6 A/cm^2 for magnetic fields $H \approx 100$–150 kG. These are examples of materials with strong flux-pinning interactions. In a case such as this, it is possible to use the superconductor for the generation of high magnetic fields.

Superconducting magnets based on strongly pinning type-II superconductors such as NbTi alloys and Nb_3Sn have now become commonplace in many research laboratories and have made possible the widescale availability of high magnetic fields for research applications. Magnets based on NbTi can be routinely used to generate fields ranging from 50–100 kG while magnets based on a Nb_3Sn superconductor stabilized with copper to form a composite can produce stable dc fields ranging up to 150 kG or higher. The application of type-II superconductors in the technology of high-field magnets is a striking example of how the development of a basic understanding of the phenomena of superconductivity has led to important technological advances.

References

ALDEN, T.H., and J.D. LIVINGSTON, 1966, J. Appl. Phys. **37**, 3551.
ALEKSEEVSKI, N.E., N.M. DOBROVOL'SKII and V.I. TSEBRO, 1974, JETP Lett. **20**, 25.
ANDERSON, P.W., 1959, J. Phys. Chem. Solids **11**, 26.
ASHKIN, M., D.W. DEIS, J.R. GAVALER and C.K. JONES, 1972, in: Superconductivity in d- and f-band Metals, ed. D.H. Douglas (AIP, New York) p. 204.
BARASIC, S., J. LABBÉ and J. FRIEDEL, 1970, Phys. Rev. Lett. **25**, 919.
BARDEEN, J., L. COOPER and J.R. SCHRIEFFER, 1957, Phys. Rev. **108**, 1175.
BATTERMANN, B.W., and C.S. BARRETT, 1964, Phys. Rev. Lett. **13**, 390.
BERGMANN, G., 1976, Phys. Rep. **27C**, 161.
BUCKEL, W., and R. HILSCH, 1954, Z. Phys. **138**, 109.
CAMPBELL, A.M., and J.E. EVETTS, 1972, Adv. Phys. **21**, 199.
CAMPBELL, A.M., and W.W. WEBB, 1972, Critical Currents in Superconductors (Taylor and Francis, London).
COLLVER, R., and R. HAMMOND, 1973, Phys. Rev. Lett. **30**, 92.
* DE GENNES, P.G., 1966, Superconductivity in Metals and Alloys (Benjamin, New York).
ELIASHBERG, G.M., 1960, Zh. Eksp. Teor. Fiz. **38**, 966 [Sov. Phys. JETP 11 (1960) 696].
FISCHER, Ø., B. SEEBER and M. DECROUX, 1980, in: Superconductivity in d- and f-band Metals, eds. H.Suhl and M.B. Maple (Academic, New York) p. 485.
FONER, S., E.J. McNIFF and E.J. ALEXANDER, 1974, Phys. Lett. **49A**, 269.
FREYHARDT, H.C., 1974, Proc. Int. Disc. Meeting Sonnenberg (Gött. Akad. Wiss.) p. 98.
FRIEDEL, J., 1969, in: Physics of Metals, ed. J.M. Ziman (Cambridge University Press) p. 340.
GEERK, J., J.M. ROWELL, P.H. SCHMIDT, F. WÜCHNER and W. SCHAVER, 1982, in: Proc. IV Int. Conf. on d- and f-band Metals, eds. W. Buckel and W. Weber (KKG, Karlsruhe) p. 23.
GESCHNEIDER, K., 1964, in: Solid State Physics, vol. 16, eds. H. Ehrenreich et al. (Academic, New York) pp. 333–371.
GIAEVER, I., 1960, Phys. Rev. Lett. **5**, 147 and 464.
GINZBURG, V.L., and L.D. LANDAU, 1950, Zh. Eksp. Teor. Fiz. **20**, 1064.
GIORGI, A.L., E.G. SZKLARZ, E.K. STORMS, A.L. BOWMAN and B.T. MATTHIAS, 1962, Phys. Rev. **125**, 837.
GOMERSALL, I.R. and B.L. GYORFFY, 1974, Phys. Rev. Lett. **33**, 1286.

GORINGE, M.J., and U. VALDRE, 1965, Phys. Rev. Lett. **14**, 823.

GORKOV, L.P., 1959, Zh. Eksp. Teor. Fiz. **36**, 1918 [Sov. Phys. JETP 9 (1959) 1364].

HAASEN, P., 1970, Nachr. Akad. Wiss. Göttingen no. 6.

HAASEN, P., 1977, Contemp. Phys. **18**, 373.

HEINIGER, F., E. BUCHER and J. MULLER, 1966, Phys. Kond. Mater. **5**, 243.

HOPFIELD, J., 1969, Phys. Rev. **186**, 443.

JOHNSON, W.L., 1981, in: Glassy Metals, I, ed. H.J. Guntherodt (Springer, Heidelberg) ch. 9.

JOHNSON, W.L., and M.A. TENHOVER, 1983, in: Magnetic, Chemical, and Structural Properties of Glassy Metallic Alloys, ed. R. Hasegawa (CRC Press, Boca Raton, FL) ch. 3.

JOHNSTON, D.C., H. PRAKASH, W.H. ZACHARIASEN and P. VISWANATHAN, 1973, Mater. Res. Bull. **8**, 777.

KAMERLINGH ONNES, H., 1911, Leiden Comm. 120b, 112b and 124c.

KIHLSTROM, K.E., and T.H. GEBALLE, 1981, Phys. Rev. **B24**, 4101.

KRAMER, E.J., 1978, J. Nucl. Mater. **72**, 5; J. Appl. Phys. **49**, 742.

LABBÉ, J., S. BARASIC and J. FRIEDEL, 1966, Phys. Rev. Lett. **19**, 1039.

LABUSCH, R., 1969, Cryst. Lattice Defects **1**, 1.

LANDAU, L.D., 1938, Nature **141**, 688.

LONDON, F., and H. LONDON, 1935, Proc. Roy. Soc. **A149**, 71.

MAKI, K., 1966, Phys. Rev. **148**, 362.

MAPLE, M.B., 1978, J. Physique, Suppl. **3**, C6-1374.

MATSUSHITA, T., and K.YAMAFUJI, 1981, J. Phys. Soc. Japan **50**, 38.

MATTHEISS, L.F., 1972, in: Superconductivity in d- and f-band Metals, ed. D.H. Douglass (AIP, New York) p. 57.

MATTHIAS, B.T., 1957, Progress in Low Temperature Physics (Interscience, New York) vol. 2, p. 138.

MCMILLAN, W.L., 1968, Phys. Rev. **B167**, 331.

* MCMILLAN, W.L., and J.M. ROWELL, 1979, in: Superconductivity, ed. R.A. Parks (Marcel Dekker, New York) vol. 1, ch. 11.

MEISSNER, W., and R. OCHSENFELD, 1933, Naturwiss. **21**, 787.

MOOK, H.A., *et al.*, 1982, in: Superconductivity in d- and f-band Metals, eds. W. Buckel and W. Weber (KKG, Karlsruhe) p. 201.

NEMBACH, E., K. TACHIKAWA and S. TOKANO, 1970, Phil. Mag. **21**, 869.

NORDHEIM, L, 1931, Ann. Phys. **9**, 607.

ODERMATT, O., O. FISHER, H. JONES and G. BONGI, 1975, J. Phys. **C7**, L13.

ORLANDO, T.P., E.J. MCNIFF, JR., S. FONER and M.R. BEASLEY, 1979, Phys. Rev. **B19**, 4545.

PHILLIPS, N.E., 1964, Phys. Rev. **134**, 385.

POON, S.J., 1983, in: Amorphous Metallic Alloys, ed. F.E. Luborsky (Butterworths, London) p. 432.

POWELL, B.M., P. MARTEL and A.D.B. WOODS, 1962, Phys. Rev. **171**, 727.

ROBERTS, B.W., 1976, J. Phys. Chem. Ref. Data **5**, 581.

SAINT-JAMES, D., G. SARMA and E.J. THOMAS, 1969, Type-II Superconductivity (Pergamon Press, Oxford) ch. 5.

SCHRIEFFER, J.R., 1962, The Theory of Superconductivity (Benjamin, New York).

SINHA, A., 1972, Prog. Mater. Sci. **15**, 79.

STEWART, G., 1982, Superconductivity in d- and f-band Metals (KKG, Karlsruhe) p. 81.

STRITZKER, B., 1974, Z. Phys. **268**, 261.

STRITZKER, B., and W. BUCKEL, 1972, Z. Phys. **257**, 1.

TESTARDI, L.R., and T.B. BATEMAN, 1967, Phys. Rev. **154**, 402.

* TINKHAM, M., 1975, Introduction to Superconductivity (McGraw–Hill, New York).

* ULLMAIER H., 1975, Springer Tracts in Modern Physics, vol. 76 (Springer, Berlin).

VARMA, C.M., and R.C. DYNES, 1976, in: Superconductivity in d- and f-band Metals, ed. D.H. Douglass (Plenum, New York).

VARMA, C.M., and W. WEBER, 1979, Phys. Rev. **B19**, 6142.

WEBER, W., 1980, in: Superconductivity in d- and f-band Metals, eds. H. Suhl and M.B. Maple (Academic, New York) p. 131.

WEGER, M., 1964, Rev. Mod. Phys. **36**, 175.

WERTHAMER, N.R., E. HELFAND and P.C. HOHENBERG, 1966, Phys. Rev. **147**, 295.

WOOLHAM, J.A., R.B. SOMOARO and P.O'CONNOR, 1974, Phys. Rev. Lett. **32** 712.

ZIMAN J.M, 1960, Electrons and Phonons (Clarendon Press, Oxford) ch. 5.

ZIMAN, J.M., 1972, Principles of the Theory of Solids (Cambridge University Press) ch. 5.

Further reading

For further reading we refer to the publications marked with an asterisk in the list of references.

CHAPTER 28

ALLOYS RAPIDLY QUENCHED FROM THE MELT

R.W. CAHN

Laboratoire de Métallurgie Physique
Université de Paris-Sud
91405 Orsay, France

R.W. Cahn and P. Haasen, eds.
Physical Metallurgy; third, revised and enlarged edition
© *Elsevier Science Publishers BV, 1983*

1.   General features of rapid quenching from the melt

1.1   Experimental methods

All casting methods involve a chill zone close to the mould surface; the metal in the chill zone cools more rapidly than the interior of the casting, and its grain morphology and composition are accordingly different. Such a zone will normally cool at a rate of up to some hundreds of K/s. The modern techniques of *Rapid Solidification Processing* (RSP), with which this chapter is principally concerned, however, involve cooling rates in the range 10^4–10^7 K/s, and require an approach quite distinct from the traditional casting methods.

Atomization of liquid metal to form small granules – one of the methods widely used today – goes back to the last century; and as recounted by JONES [1981] in a historical survey, several investigators in the period 1925–1955 developed variants of chill-casting which gave estimated cooling rates up to 10^5 K/s, while several studied extension of solid solubilities resulting from such cooling rates. However, the remarkable present-day developments in RSP were initiated by Duwez in 1959–60. Duwez, in California, set out to establish whether a *continuous* metastable series of solid solutions could be created in the Cu–Ag system, to bring it into line with the Cu–Au and Au–Ag systems which both show continuous solid solubility, as required according to the familiar Hume-Rothery rules. For this purpose, he argued, the melt would have to be frozen fast enough to inhibit nucleation of two distinct fcc phases. Having unsuccessfully tried a number of other approaches, he designed the *Duwez gun*, a device in which a gaseous shock wave atomizes a drop, ~ 10 mg, of molten alloy and projects the microdroplets into contact with a copper substrate, or *chill block*, to produce small foils, or *splats*. The technique was by some given the onomatopoeic nickname *splat-quenching*, now falling out of use together with the original gun as improved methods have been developed.

In 1960, Duwez and his collaborators were successful in making a continuous, metastable series of Cu–Ag alloys without any two-phase region, and in the same year also discovered the first *metallic glasses* made by rapid quenching, in the Au–Si and Au–Ge systems (DUWEZ et al. [1960]). The essential condition for rapid quenching from the melt, as realized in the Duwez gun, was that small particles of liquid metal should rapidly be flattened into a thin sheet in intimate contact with a good heat sink. More generally – and this was the real importance of the innovation wrought by Duwez – only by starting from the molten metal was it possible to achieve really high rates of cooling in the *solid*. The importance of RSP arises *both* from the changes brought about in the freezing process *and* from the more rapid cooling of the solid so formed.

The Duwez gun was used for a decade to investigate the crystallographic and calorimetric characteristics of numerous metastable splat-quenched crystalline phases and of a few glasses, notably $Pd_{80}Si_{20}$ which was for a number of years a standard composition for those interested in the properties of metallic glasses. DUWEZ [1967] has described these pioneering days, and also the background to his original experiments.

The next stage came when the first methods of making *continuous* rapidly quenched ribbons were invented. All these depend on contact between a thin liquid ribbon and a *moving* chill block. The old method of *single-roller melt-spinning* was re-invented by Pond in 1958 and later published (POND and MADDIN [1969]); *twin-roller melt-spinning* is due to CHEN and MILLER [1970] and BABIĆ *et al.* [1970] and was improved by MURTY and ADLER [1982]; *melt-extraction* was developed by MARINGER and MOBLEY [1974]. The fourth major technique, much used in basic research, is the *drop-smasher*, also known as *piston-and-anvil quencher*, in which an alloy drop is levitation-melted inside a conical induction coil, released and quenched between two moving copper surfaces to form a disc (HARBUR *et al.* [1969], BEGHI *et al.* [1969] and CAHN *et al.* [1976]). The principles of these four methods are indicated in fig. 1. These and various subsidiary methods, as well as the powder-generating methods and the surface treatments to be discussed below, are surveyed in more

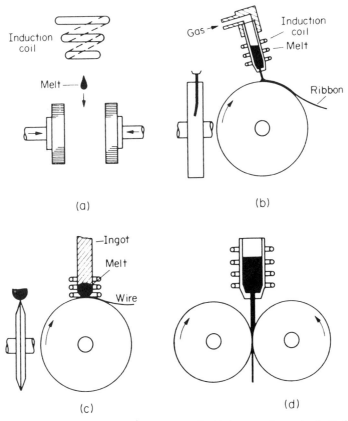

Fig. 1. Principal methods of rapid quenching from the melt: (a) drop-smasher, using levitation-melting by induction; the pistons are pneumatically or electromagnetically accelerated and come into action when the falling drop breaks a light beam; (b) melt-spinning; (c) pendant-drop melt-extraction (there is also a version with the wheel dipping into a melt surface); (d) twin-roller quenching device.

References: p. 1845.

detail by JONES [1981] and by LIEBERMANN and WALTER [1982]. Melt-spinning and its derivatives are also discussed in ch. 26, §5.2; in particular, that section includes an account of the *planar flow-casting* variant of single-roller melt-spinning, used to make wide sheets of glass.

Plasma-spraying is a method which in a sense represents a return to Duwez's microdroplet approach: an electric discharge between two electrodes in argon provides energy to melt alloy globules which are then projected by an argon jet on to a substrate to build up a deposit. The starting material is pre-alloyed powder, 50–100 μm in diameter. Special procedures (CAHN [1978]) are needed to ensure rapid cooling of the impinging droplets. GAGNÉ and ROY [1982] have succeeded in plasma-depositing layers of a Ni–Cr–Si–B glass on to a metallic substrate. A full account of the use of this technique has recently been published (SAFAI and HERMAN [1981]). The method can also be used to solidify fine droplets in flight for subsequent compaction, though it has been little used in this mode. A simple flame-heated metal-spraying gun has also been used to deposit glassy layers of $Fe_{40}Ni_{40}P_{14}B_6$ (SHINGU [1979]).

A number of different methods have been used for the production of RSP *wires*. One is *free-jet melt-spinning* of a molten alloy, usually a steel, through a nozzle into a gaseous quenching medium (MOTTERN and PRIVOTT [1978]). The problem here is to avoid Rayleigh instabilities which tend to break up the free jet: this is best achieved by alloying the steel with aluminium, quenching into carbon monoxide and relying on the surface oxide layer formed on the freshly formed jet to resist its break-up. Other alloys can be correspondingly treated. An alternative approach (OHNAKA and FUKUSAKO [1978], OHNAKA et al. [1981], INOUE et al. [1982]) relies on a variant of melt-spinning: the alloy jet impinges on the inner surface of a rotating vessel containing water which is held in an annular shape by centrifugal force. This, like the free-jet method, produces wires 100–200 μm in diameter. Rayleigh instabilities are avoided by carefully matching the speed of the liquid metal jet and the rotational speed of the water annulus. There is evidence from comparative relaxation studies (INOUE et al. [1983a]) that a glass in the form of wire is more drastically quenched than the same glass in the form of a melt-spun ribbon several times thinner!

Following an early preview (GRANT [1970]), LEBO and GRANT [1974] published a key paper on production of RSP aluminium alloy powder by atomization and also of splats against a cold metal surface, followed by consolidation of the powder at a temperature and for a time designed to prevent deterioration of the desirable properties induced by RSP. Another key paper, describing the compaction and properties of splat-quenched Al–Fe alloys, was by THURSFIELD and STOWELL [1974]. All subsequent research and development of industrial uses of RSP as applied to crystalline alloys has been based on this approach. Atomization methods in current use (reviewed by GRANT [1978a]) include ultrasonic atomization (see also ANAND et al. [1980]), centrifugal atomization in helium (GLICKSTEIN et al. [1978]) and quenching of droplets between rollers to make flakes (SANKARAN and GRANT [1980]). Consolidation is either by warm extrusion or by hot isostatic pressing (ch. 30, §2). Even cold explosive compaction can be used (MORRIS [1981]). An alternative approach which is increasingly finding favour is to pulverize a brittle crystalline melt-spun ribbon. A small-scale variant, capable of producing extremely fine, submicron powders, is electrostatic extraction of particles from a liquid-alloy tip drawn out into a point by a strong electric field (PEREL *et al.* [1978]). This apparatus has been used for a detailed metallographic study of the microstructures of RSP aluminium alloy globules (LEVI and MEHRABIAN [1982]; the globules were fine enough for TEM examination without further thinning).

All of the rapid-quenching methods described above have been used to make not only crystalline alloys but also metallic glasses. Several glass-making methods which do not depend on melt-quenching will be outlined here, though on a strict interpretation they do not fall within the province of this chapter. These methods may collectively be termed *pseudo-RSP*.

One group of such methods depends on the deposition of vapour on cold substrates (often held at liquid-nitrogen or even liquid-helium temperature). These methods are in fact of older ancestry than true RSP methods and were used to make various amorphous and supersaturated crystalline materials in the 1930s (KRAMER [1934,1937]) and 1950s (BUCKEL and HILSCH [1956]), generally in the form of thin films. Sometimes thermal evaporation is used to produce the vapour (MADER [1965] and SINHA *et al.* [1976]), but cathodic sputtering is more usual; if a powerful high-rate sputtering apparatus is used, glass samples several mm thick can be made, though at very high cost (see review by DAHLGREN [1978]). These techniques are now waning, in spite of the fact that more alloy compositions and wider composition ranges can be made glassy by vapour methods than by liquid-quenching (TURNBULL [1981]). Perhaps loss of interest in these methods is a recognition that in metallurgy, as opposed to microcircuitry, thin films have only a modest role.

Another approach is heavy irradiation (to a dose of at least one displacement per atom) of a crystalline alloy, using either neutrons or ions (RECHTIN *et al.* [1978], ELLIOTT *et al.* [1980] and ELLIOTT and KOSS [1981]). As judged by calorimetric tests, such "artificial" glasses are structurally very similar to the same alloys vitrified by RSP. A related "solid" approach to making a glass is by *implantation* into a pure metal of ions of a different element, most usually metalloid ions. Phosphorus has been most effectively used. The bombarding ion both destroys crystalline order and, by occupying strategic locations in the glass structure, stabilizes it. Self-ion bombardment (e.g., Ni^+ into Ni) never generates a stable glassy structure. The technique is reviewed by BORDERS [1979] and more recent research by RAUSCHEN-BACH and HOHMUTH [1982]. The most recent observations are on *electron*-induced amorphization. MORI and FUJITA [1982] have established the temperature dependence of the critical dose for amorphizing NiTi with 2 MeV electrons, while MORI *et al.* [1983] have observed the surprising phenomenon of initial amorphization preferentially at dislocations, in the same alloy. The theoretical background, including a discussion of critical irradiation doses, is presented by BOURGON [1979] and by CARTER and GRANT [1982]. It is important to recognize that glassy layers made in this way are less than one micron thick.

MENDOZA-ZELIS *et al.* [1982] describe the amorphization of nickel films by

References: p. 1845.

phosphorus ion implantation. Electrical resistivity indicates that the glass thus prepared is structurally identical to glasses of the same compositions made by evaporation or electrodeposition. (The last two approaches to glass-making, together with the related technique of electroless plating, with a partial but extensive list of compositions prepared in these ways, are outlined by DIETZ [1977].) Closely related to ion implantation is the technique of *ion-mixing*, in which two discrete layers at a surface are mixed on an atomic scale by bombardment with 'neutral' ions such as A^+ or Xe^+. Thus, Au–Si glasses have been made by ion-mixing of a gold-coated silicon substrate (LIU *et al.* [1982]); this same glass species has been made by RSP of liquid alloy, evaporation on to a cold substrate, laser-treatment and ion-mixing.

A very recently discovered way of making a glass is by means of a chemical reaction involving a crystalline intermetallic compound. YEH *et al.* [1983] found that a compound, Zr_3Rh, prepared in a metastable crystalline form by freezing at a moderate rate, can be turned into a glass by reacting with hydrogen at a fairly low temperature ($\leqslant 200°C$) to produce an amorphous hydride, $\sim Zr_3RhH_{5.5}$. (Zr_3Rh can also be made into a glass by rapid quenching from the melt and *then* hydriding: the structure, as assessed by X-ray diffraction, is the same in both versions.) The discoverers of amorphization by hydrogenation attribute the tendency to the presumed fact that the glass is stabler than the metastable crystalline form, whereas the fact that a glass forms in preference to the *stable* crystalline species is due to *chemical frustration*, in that there is no time for the necessary segregation of Zr and Rh which would be needed to form stable crystalline ZrH or ZrH_2, plus some Rh-rich crystalline phase. This experiment may well prove to be the first of a substantial family of glasses made by *reaction-amorphization*. The rapid interdiffusion of two crystalline metals, Au and La, in thin-film form, was found (SCHWARZ and JOHNSON [1983]) to generate an amorphous solid solution. The process is driven by the large negative heat of mixing in the amorphous alloy; it takes place below T_{cryst}.

Finally, a whole family of techniques has been developed over the past six years on the basis of surface-treatment of alloys and of silicon-based semiconductors by lasers or electron beams. This family of techniques is discussed in §1.5.

1.2. Forms of metastability resulting from rapid solidification processing

Alloys rapidly solidified or prepared by pseudo-RSP methods may show one or other of four forms of metastability:

1) An extension of solid solubility beyond the equilibrium value, which may be partial or (as in the Ag–Cu system) complete.

2) The formation of one or more metastable crystalline phases.

3) Lowering of the M_s temperature when a martensitic transformation takes place, sometimes to such an extent that the martensite is replaced by a different phase.

4) The formation of a metallic glass.

Consequences 1–3 are treated in §4, while the formation and properties of glasses are discussed in §3. The broad topic of metastability has recently been examined in masterly fashion by TURNBULL [1981].

During the first decades of RSP research, emphasis was heavily on crystalline phases, especially nonferrous metastable phases. From the beginning of the 1970s, interest began to spread to steels, but the principal characteristics of the 1970s was a growth in emphasis on metallic glasses. This probably reached its zenith at the time of the 1981 Rapid Quenching Conference, when over 380 papers out of 400 presented were concerned with metallic glasses. Shortly before that time, a strong revival of interest in RSP crystalline alloys became evident and has continued since (CAHN [1982]).

The intense interest in metallic glasses has a double basis: firstly, the remarkable scientific interest of these (until recently) unknown materials, and in particular their structure, unusual mechanical properties (very high strength combined with toughness), liquid-like electrical properties, and their ferromagnetic properties; secondly, these magnetic properties (fully treated in ch. 26, §5) are in process of finding practical applications.

The recent recrudescence of attention to RSP crystalline alloys, including the surface treatments designed to generate metastable structures, is almost entirely due to the perceived practical promise of the RSP approach: the benefits include lesser heterogeneity of composition, finer grain, cell and dendrite sizes, the ability to incorporate more solute than by conventional methods (and thus permit more age-hardening on subsequent heat-treatment) and the consequent improvement of mechanical properties, at both ambient and elevated temperatures.

1.3. Cooling rates in rapid solidification processing

The rate of cooling of the melt before, and of the solid after solidification can in principle be assessed by direct measurement. Alternatively, the cooling rate at the freezing stage can only be deduced by measurement of a microstructural feature of the resulting solid, most commonly the secondary dendrite arm spacing or (where apposite) eutectic lamellar spacing; but such a microstructural approach is not available when the solid product is a glass.

The relationship between cooling rate, $\dot{T}(= G_L V$, where G_L is the temperature gradient in the liquid and V is the velocity of the solid–liquid interface) and primary- (λ_1) and secondary- (λ_2) dendrite arm spacings is discussed in ch. 9, §6.1.3.4. Primary dendrite spacings are not well suited to an estimate of $\dot{T}$, since $\lambda_1 = A_1 G_L^{-m} V^{-n}$, where in general $m \neq n$. However, for *secondary* dendrite arms, $\lambda_2 = B_1 (G_L V)^{-n}$, where B_1 is a constant and $n \approx \frac{1}{3}$. So long as dendrite-coarsening during the cooling of the solid can be neglected (by no means a secure assumption for the finest spacings), λ_2 can be used to calibrate very fast cooling rates by extrapolation from slower, directly measurable cooling rates. Figure 2 shows a composite plot for Al–Cu and Al–Si alloys, including data from several investigators. The fastest cooling rates were measured directly for piston-and-anvil quenching of an Al–Si alloy by means of a microthermocouple projecting into the melt (ARMSTRONG and JONES [1979]). It was found that the $\dot{T}^{1/3}$ relationship is obeyed very well for these alloys, though for Ni and Fe-base alloys n is found to be

References: p. 1845.

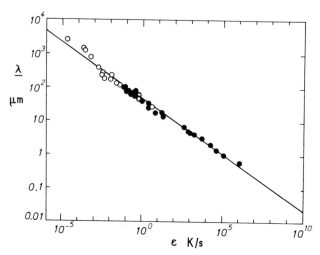

Fig. 2. Dendrite spacing, λ, as a function of cooling rate near the melting-point, ϵ, for Al–4 to 5 wt% Cu (open circles) and Al–7 to 11 wt% Si (solid circles). The plot refers to secondary dendrite arms. The line shown represents $\lambda \epsilon^{1/3} = \text{const.} = 50 \ \mu\text{m (K/s)}^{1/3}$. (After JONES [1982b].)

somewhat variable (JONES [1982b]). The eutectic-lamellae approach, used for a melt-spun aluminium alloy ribbon and extrapolating from slower, measured cooling rates, gave an estimate of 3×10^6 K/s in the neighbourhood of the freezing temperature (CHATTOPADYAY and RAMACHANDRARAO [1980].

The limitations of such indirect methods are: (1) they are not applicable to glass formation; (2) they can give an estimate only of the effective cooling rate (related to the solidification time and specimen thickness) in the temperature range in which solidification takes place; (3) each family of alloys requires separate calibration; (4) the dendrites may coarsen after solidification and so yield misleading results. For all these reasons, direct measurement is preferable, difficult though it is.

Most of the direct measurements of liquid-quenching rates which have been published have referred to the piston-and-anvil technique. Thus, HARBUR *et al.* [1969] made use of the metals (suitably insulated) of two dissimilar pistons as the legs of an intrinsic thermocouple, whereas DUFLOS and CANTOR [1982] used a minute intrinsic thermocouple, electrically shorted by the melt itself. Pyrometry of the droplet has also been employed (e.g., KATTAMIS *et al.* [1973]). Melt-spinning poses a more difficult measurement problem altogether: HAYZELDEN [1983] developed a technique of calibrated colour photography to establish temperature profiles along the ribbon while WARRINGTON *et al.* [1982] used calibrated black-and-white photography.

Figure 3 shows the results of Duflos and Cantor. The various measurements have been critically compared by CANTOR [1982]. It is quite clear that the cooling rate varies substantially between the freezing point and ambient temperature. Thus at 1500°C, near the freezing-point of a steel, the cooling rate for piston-quenched steel

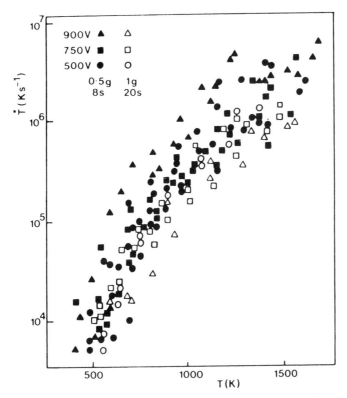

Fig. 3. Data from many cooling curves obtained by microthermocouple and oscilloscope traces during piston-and-anvil drop-smashing of iron, replotted as cooling-rate, $\dot{T}$, versus temperature, T. The voltages cited are electromagnetic accelerating voltages; higher voltages imply faster impact. The weights refer to specimen sizes; larger specimens, heated for longer periods, had greater superheat. (After DUFLOS and CANTOR [1982].)

or iron samples 25–50 μm in *half*-thickness (the proper measure) is 10^5–10^7 K/s; the higher values apply to the very rapid-acting piston-and-anvil apparatus designed by CAHN *et al.* [1976]. The cooling rate at 1000°C is ~ 10 times slower than at 1500°C, and at 500°C, 2–3 times slower again. It is evident that the custom of characterizing the cooling rate by a single figure is unacceptable, unless the temperature range to which such a figure applies is specified. Single-roller melt-spinning turns out to be slower than piston-and-anvil quenching (HAYZELDEN [1983]): thus, for a 40 μm-thick steel ribbon, the measured cooling rate is 5×10^5 K/s at 1500°C and only 10^4 K/s at 1000°C (because the ribbon has by then left the wheel). It is generally accepted that dual-roller quenching is even slower than single-roller melt-spinning overall, in spite of the presence of double chill block, because the duration of contact between the ribbon and the chill block is so brief. However, the initial cooling rate appears to be very high (DAVIES *et al.* [1978]).

In the case of rapid quenching by atomization, cooling rates have been estimated

References: p. 1845.

from dendrite spacings: thus, for a gas-atomized superalloy, cooling rates ranging from 10^3–10^6 K/s have been estimated as particle diameters range from 400 to 10 μm (DUFLOS and STOHR [1982]). The preferred approach at present, however, is by means of heat-transfer calculations, as exemplified for conventional casting in ch. 9, §2.2. The crucial parameter here is the *Nusselt* (or *Biot*) *number*, N, of both specimen and "chill block" (which last, in the case of a droplet solidifying in flight, is gaseous). $N = hd/k$, where h is the *heat-transfer coefficient*, d is the effective specimen or chill-block thickness and k is the thermal conductivity. As h increases, cooling changes from Newtonian, via intermediate to "ideal" cooling, with progressively larger temperature gradients in the specimen. h can be estimated from measured cooling rates in the case of piston-and-anvil quenching (CANTOR [1982]), but for powder atomization it is necessary to compute h from first principles for specified boundary conditions (h very low for pure radiative transfer and several orders of magnitude larger for favourable convective cooling). The theory has become extremely complex: full treatments have been published by LEVI and MEHRABIAN [1982], MEHRABIAN [1982] and CLYNE [1984]; reference should be made to these papers.

The cooling rate varies of course as a function of droplet diameter and depth within the droplet, and is perturbed by the release of latent heat in a way which depends on the undercooling that preceded the start of freezing. A particularly important regime is with heat-transfers rapid enough to permit *hypercooling*, that is, suppression of nucleation till the sample has undercooled so far that the release of the entire latent heat is insufficient to reheat the sample to the equilibrium freezing point: freezing can then go to completion without any further heat extraction.

In the theory of atomization, it turns out to be more useful to compute solid–liquid interface velocities, V, rather than cooling rates, $\dot{T}$, since these relate more closely to interface stability theory. Mehrabian cites values for 50 μm aluminium droplets of 4 m/s at the start of freezing, dropping to 0.04–0.13 m/s in the intermediate stages of freezing. Corresponding estimates of V for melt-spinning or piston-and-anvil quenching do not appear to have been made yet.

1.4. Process physics of rapid quenching

In the production of RSP foils by melt-spinning or planar flow-casting, as outlined in ch. 26, §5.2, it is important to be able to predict the width, thickness and smoothness of foil in terms of wheel velocity, jet speed and diameter, jet attack angle and atmosphere in which solidification is carried out. Similar problems exist with respect to free-jet spinning. This range of problems has been approached both empirically and theoretically (KAVESH [1978], HILLMAN and HILZINGER [1978], VINCENT et al. [1982], LIEBERMANN and WALTER [1982], PAVUNA [1982] and LIEBERMANN [1978,1983]). Pavuna's paper gives the most detailed empirical results on ribbon geometry, while Vincent et al. and Liebermann and Walter outline present views on modelling; the latter include an especially varied list of citations. LIEBERMANN [1978] was the first to establish the major importance of the aerodynamic properties of the atmosphere in determining ribbon quality: helium is much preferable to argon or air because of its low molecular weight and high viscosity. In such modelling, the unresolved disagreement is between *heat-transfer control*, where foil thickness is in effect controlled by the velocity of the liquid–solid interface, and *momentum-transfer control*, according to which friction between the chill-block and the melt drags out a liquid film whose thickness is hydrodynamically controlled. A recent analysis (HUANG and FIEDLER [1981]) takes into account both the high viscosity of severely undercooled alloy melt in contact with the wheel, and the observed dependence of heat-transfer on wheel surface speed (because of better melt–wheel contact at higher speeds): the process model here is a hybrid of the momentum-transfer and heat-transfer models. It is plain that modelling of melt-spinning is not yet in final form, and several variables (such as gas environment) have not yet been incorporated in any model.

The process physics of twin-roller quenching is discussed by MURTY and ADLER [1982] and by MIYAZAWA and SZEKELY [1981].

1.5. Rapid solidification processing of surfaces

The RSP techniques presented in the preceding sections, other than ion-beam implantation and mixing, and radiation damage, involve complete melting of the alloy before rapid quenching begins. An alternative which began to receive serious attention only quite recently is the very rapid melting and refreezing of a thin surface layer on a massive alloy object: this can be done either without modification of composition or else in the presence of a thin surface layer of different composition, which becomes alloyed with the substrate during the rapid melting and freezing cycle. Chemical modification of the surface by means of a reactive atmosphere is also possible. The most usual instrument used is the *laser*.

The motive for such a procedure is twofold: (1) it requires possibly expensive alloying elements to be present only in a very thin layer; (2) it offers an escape from the obligation to work with thin foils or wires which is the normal price to be paid for very rapid cooling. To these strictly metallurgical considerations one must add the many ways in which surface-RSP has proved able to improve the structure of doped silicon for microelectronic devices; this has been the principal motive for such research in industry until very recently.

1.5.1. Laser surface treatment

Focused laser beams offer a means of concentrating extremely high power densities in a small spot (typically 0.1–1.0 mm diameter) at an alloy surface, because of the very high parallelism of the beam emerging from a laser. Two kinds of lasers are available: *pulsed* ruby, neodymium-glass or other solid-state lasers, and *continuous-wave* gas (CO_2) lasers. The former are used in stationary mode, whereas the latter are used in conjunction with a rapid scanning of the alloy surface perpendicularly to the beam (e.g. by rotation of a circular sheet, or purely optical scanning). If such scanning is combined with slow transverse translation of the laser beam, then successive strips of melted material can cover the whole surface.

References: p. 1845.

The earlier studies used laser-treatment as an alternative to other methods of RSP, and in this way, metastable solubility extension (ELLIOTT et al. [1973]) and metastable phase formation (LARIDJANI et al. [1972]) was achieved with pulsed lasers, the substrate acting as built-in chill-block. Later, it also proved possible to make glassy layers at alloy surfaces, although at first only with "easy" glass-formers, a Pd–Cu–Si alloy in particular. Surface-hardening of laser-treated tool steels was another early application (TULI et al. [1978]). These studies all had in common that the alloys were used without addition of 'dopants' at the surface.

Figure 4 shows the variables involved in the treatment of a surface by a continuous-wave laser. Heat-flow calculations (which, for sufficiently large heated spots, can be simplified by neglecting sideways heat flux and treating the problem as a one-dimensional heat flux) have been published by MEHRABIAN [1982]. He assumes that the heat which can usefully be absorbed is limited by the onset of vaporization; on this basis, increasing laser power (which entails faster scanning of the specimen) yields a shallower melt, faster refreezing and a steeper thermal gradient. Typical figures, for aluminium, are 650 μm or 6.5 μm/melt depth, for 5×10^8 or 10^{10} W/m^2 heat flux, respectively. The estimated solid–liquid interface velocities are 0.035 or 3.5 m/s, respectively. KEAR et al. [1981] estimate that, again taking vaporization as limiting factor, quenching rates up to 10^6 K/s can be achieved, for a melted layer ~ 10 μm thick. The fact that, even with cooling rates of this order, it is so difficult to turn an alloy surface into a glass, can be attributed to the presence of a crystalline substrate capable of acting as a crystalline catalyst.

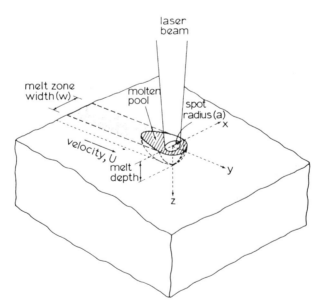

Fig. 4. Schematic illustration of laser beam–substrate geometry during rapid surface melting and refreezing by laser treatment (after MEHRABIAN [1982]).

The nomenclature in the field of laser-treatment is distinctly confusing: the semiconductor community has adopted the term *laser-annealing*, because the original aim was to remove radiation damage from silicon into which dopants had been ion-implanted; unlike ordinary metallurgical annealing, however, this process is now widely believed to involve the melting and epitaxial re-crystallization of a thin surface layer. (Nevertheless, this new orthodoxy is already being challenged by VAN VECHTEN [1982], who has adduced evidence that the silicon does *not* melt; this challenge, however, is not accepted by most researchers.) This whole field, which carries many lessons for metallurgists, has been admirably reviewed by POATE and BROWN [1982] and more recently by BOYD and WILSON [1983]. Metallurgists refer to *laser surface alloying*, even though, as we have seen, laser treatment does not necessarily entail a change in surface composition. BREINAN and KEAR [1978] have proposed to denote surface melting and refreezing by the term *laser-glazing*; this is unsatisfactory because it implies that the surface is turned into a glass, which is rarely the case. They confusingly introduced the further term *layer-glazing*, which denotes a process of laser-aided repeated deposition of alloy on the circumference of a rotating disc, so that a thick spiral layer of RSP material can be deposited. It would avoid confusion if metallurgists, for their purposes, were to retain the neutral term *laser surface treatment*, which implies no particular mechanism or outcome.

No attempts will be made here to review the very large literature which has accumulated over the past six years in regard to "laser-annealing" of silicon, and the reader is referred to Poate and Brown's review and bibliography, and to several conference proceedings cited by them. A few important findings of this body of research will however be outlined, since they have important metallurgical implications. A Q-switched (ultrarapid) ruby-laser pulse can lead to an interface velocity during recrystallization of the molten silicon spot exceeding 3 m/s, and this could be directly measured by time-dependent measurements of local electrical conductivity. Above 3 m/s, the silicon melt re-crystallizes with a high degree of epitaxial perfection – hence the semiconductor industry's interest. It is possible to trap ion-implanted dopants, dissolving much more than the equilibrium solubility, by again depending on rapid freezing. The degree of supersaturation can be assessed by using ion-back-scattering methods which are sensitive to both amounts and lattice-site locations of solute atoms (ch. 10A, §3.7.2; ch. 13, §3.2; POATE *et al.* [1978]). According to measurements by WHITE *et al.* [1980], the solubility of arsenic in silicon can be enhanced fourfold, that of gallium ten-fold and that of bismuth by a factor of 500. If the quench rate is fast enough (pulses in the nanosecond range or shorter), the molten silicon cannot crystallize at all and an amorphous silicon layer forms at the surface. On the other hand, if an amorphous or microcrystalline silicon layer is laser-treated by a scanning technique, local crystallization takes place and elongated grains are produced. Variants on this technique are described by Poate and Brown. A detailed critical survey of *short-time annealing* of silicon, covering both transient surface melting and solid-state changes, has been published by SEDGWICK [1983].

By combining local melting by laser with chemical reactions induced by gaseous reagents, both preferential local metal removal and material deposition on silicon

References: p. 1845.

can be achieved (CHRISTENSEN [1982]). A related metallurgical use of lasers offers a way of selective "laser-machining" of aluminium (EHRLICH et al. [1981]). Atomic zinc is generated near an aluminium surface by means of an UV laser acting on a zinc compound, while an optical laser is used to melt a strip of aluminium in which the zinc vapour dissolves. This Al–Zn alloy is then differentially etched by acetic acid.

The very varied applications of lasers in metallurgy are comprehensively reviewed, with a full bibliography, by DRAPER [1982], by KEAR et al. [1981] and also in a recent book edited by MUKHERJEE and MAZUMDER [1981]. (The review by Kear et al. also covers the various ion-beam methods of surface treatment, already outlined in § 1.1.)

Apart from functions such as welding, cutting, pulse-annealing in the solid state and shock-hardening, which are not directly relevant to RSP, the most fruitful development in recent years has been *laser-aided surface alloying*. Draper discusses in detail the *uniformly* alloyed surface layers, several microns thick, which can be achieved by laser-treatment, for instance, of nickel with a deposited gold layer. Considerably thicker uniform layers can be obtained by laser-treatment than by ion-implantation or by ion-mixing. Such surfaces can be highly corrosion-resistant without the use of a large amount of precious metal.

BERGMANN and MORDIKE [1980,1982] were able to make several metastable glasses by laser-treating alloys. The problem of easy epitaxially induced crystallization was circumvented by changing the composition locally: thus, $Ni_{60}Nb_{40}$ glass was made on a laser-alloyed Nb single crystal, and a glassy ferrous layer by laser-alloying an austenitic Fe–Cr–C steel with boron which had previously been plasma-sprayed on to the surface. More generally, several glasses in familiar systems (Pd–Cu–Si, Au–Si) have been made by nanosecond laser-pulsing, using alloys containing less solute (5 at% Si in Pd–Cu–Si, 10 at% Si in Au–Si) than is found necessary in normal RSP (TURNBULL [1981]), and LIN and SPAEPEN [1982], by picosecond laser treatment of a finely layered Fe/Fe_3B sandwich, were able to make a Fe–B glass with as little as 5 at% B.

BERGMANN and MORDIKE [1980,1982] also describe a number of instances of surface-hardening of steels by laser-treatment without amorphization of the surfaces, both with and without addition of dopants, depending on steel composition. Because of considerable surface-supersaturation of solute, subsequent heat-treatment of laser-treated steel can produce extremely hard surfaces.

1.5.2. Electron-beam surface treatment

An alternative to the use of lasers is an electron beam, focused magnetically. Very high energy densities can be obtained, but the inconvenience of a vacuum chamber must be accepted. The technique has been extensively applied to steels, especially tool steels (LEWIS and STRUTT [1982]).

2. Solidification theory of rapid quenching

In spite of a great deal of theoretical activity, beginning with a now classic paper by BAKER and J.W. CAHN [1971], it is still by no means clear just how solute distribution in a rapidly melt-quenched crystalline alloy is determined. Three distinct criteria are involved, and each predicts a different requirement to achieve partitionless freezing:

(a) A *thermodynamic criterion*, defined in terms of a composition-dependent temperature T_0, which, for an alloy of composition c, is the temperature at which solid and liquid each of composition c have the same free energy.

(b) A *morphological criterion*, which determines for what interface velocities and thermal gradients the interface is morphologically stable – i.e., no cells or dendrites form and partitionless solidification ensues.

(c) A *heat-flow criterion*, defined in terms of the requirement for *hypercooling* which, as already defined, is that degree of liquid undercooling preceding solid nucleation which ensures that all the melt can solidify, even in the absence of any further heat extraction, without release of latent heat causing the sample to reheat to the equilibrium melting temperature.

There is a further question as to the thermal conditions needed to suppress crystallization altogether and permit a glass to be formed. This question involves:

(d) *a kinetic criterion for glass formation*, i.e., the question of what cooling rate will prevent any crystalline nuclei from forming for purely kinetic reasons. When this criterion has been determined for a number of alloy systems, the further question arises of the structural and chemical factors that determine the critical cooling rate, that is,

(e) *a structural criterion for glass formation.*

2.1. The thermodynamic criterion

BAKER and CAHN [1971] were among the first * to show theoretically that T_0 is the maximum solid–liquid interface temperature at which the partition coefficient k_0 (ch. 9, §5.1) can be unity – i.e., solidification can proceed without any solute segregation. Such solidification is also referred to as *partitionless*; another term widely used is *solute trapping*. Another way of putting this is to point out that only below T_0 is there a thermodynamic driving force for a liquid to freeze into a solid of the same composition. BOETTINGER [1982a], in a comprehensive survey of kinetic and thermodynamic limitations on solidification mechanisms, explains that in practice it is necessary to undercool somewhat below the thermodynamically imposed limiting temperature T_0 before solute trapping in fact becomes possible. The solute trapping conditions which, as mentioned in §1.5.1, have been directly measured in laser-treated silicon containing Bi imply a further *kinetic undercooling*, $\Delta T'$, below T_0, sufficient to generate an interface velocity exceeding 3 m/s, but this velocity

---

* Earlier treatments of the role of T_0 are discussed by BILONI and CHALMERS [1965].

cannot be experimentally determined as a function of undercooling. Independent theories cited by Boettinger suggest that $\Delta T'$ is less than 1 K (below T_0) for an interface velocity of 1 cm/s – i.e., not a serious limitation on solute trapping.

The only conditions under which undercooling can be directly measured at all is for slow cooling of droplets in a microcalorimeter following the Turnbull–Cech approach, as described in ch. 9, § 3.3.1. This approach has been fruitfully applied to several compositions which had also been studied by RSP (PEREPEZKO and PAIK [1982] and PEREPEZKO and SMITH [1981]), but such experiments yield only the supercooling required for homogeneous or hetereogeneous nucleation (which of these, in any particular system, is an open question), not the kinetic undercooling required for solute trapping. In principle, however, this could be achieved by subsequent local chemical analysis of droplets if, in some way, each droplet could be tagged with its individual measured undercooling!

Nevertheless, the T_0 criterion does have qualitative explanatory power. Thus, figs. 5a and b show (neglecting the small kinetic undercooling), according to Baker and Cahn's theory, what solid compositions can form from a liquid of composition C_L^*

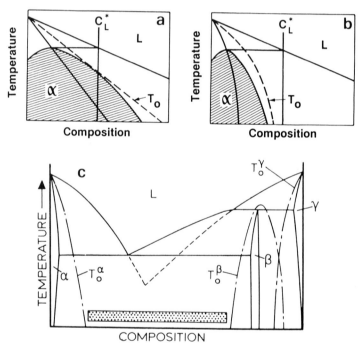

Fig. 5. Hypothetical phase diagrams showing different variations of T_0 with composition. In (a) and (b), shaded regions show possible solid compositions which can form from liquid of composition C_L^* at various interface temperatures, for a plane interface. Partitionless solidification is possible for this liquid composition in (a) but not in (b). In (c), T_0 loci relating the three solid phases to liquid are shown. In the shaded composition range, partitionless crystallization to a single-phase solid is impossible. (After BOETTINGER [1982a] and MEHRABIAN [1982].)

for a plane interface, as a function of solid–liquid interface temperature. (The diagrams show, for reasonable assumed variation of free energies with composition, the solid temperature–composition range for which there exists a positive thermodynamic driving force for solidification.) For the system of fig. 5a, solute trapping is possible at C_L^*, for that of fig. 5b it is not. In fact, whenever T_0 varies steeply with temperature, solute trapping is limited to quite dilute alloys. Figure 5c shows an imaginary phase diagram with a range of compositions over which crystalline freezing with solute trapping is thermodynamically feasible. PEREPEZKO and SMITH [1981] performed microcalorimetric measurements on slowly cooled droplets of various Cu–Te alloys; here the thermodynamic criterion is favourable and, between 60 and 80 at% Te, glasses form when nucleation is suppressed by subdividing the melt finely. The T_0 criterion is also discussed in some detail in ch. 14, §2.2.1.

Figure 6 shows a computer-calculated phase diagram for Cu–Ag alloys: the T_0 curve never dips to a temperature at which solid diffusion is very slow: in such a system, while the formation of a series of solute-trapped metastable crystalline solid solutions may be achieved (as is indeed the case in Cu–Ag), glasses cannot be formed. A different way of expressing this is to say that suppression of crystal nucleation is kinetically more difficult than is solute trapping, in a crystalline context.

MASSALSKI [1982] discusses in general terms how glass-forming tendency relates to the form of binary or ternary phase diagrams.

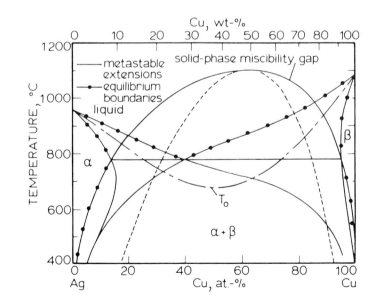

Fig. 6. Shallow eutectic with a continuous T_0 curve, for Cu–Ag alloys. (Computation by MURRAY [1983]. Very similar plots were obtained by ISHIHARA and SHINGU [1982] and much earlier by KRISHTAL [1960].)

References: p. 1845.

2.2. The morphological criterion

For *dilute* alloys, both very slow and very fast interface velocities can lead to planar solidification; for low velocities there is only equilibrium segregation, for high velocities (e.g., 0.1–1 m/s for dilute Al–Cu) there is solute trapping. The theory is explained in ch. 9, §6.3 and figs. 20, 21, while experimental confirmation is presented in ch. 9, §6.3.1.7 and fig. 25. The reader is referred to these sections and figures.

Morphological stability theory can also be applied to the freezing of *concentrated* (eutectic) compositions, and here also there has been progress. The standard theory is presented in ch. 9, §7.2: when this theory is complemented by an analysis of liquid diffusion kinetics, it is possible to extrapolate theoretical eutectic growth rates to temperatures well below the eutectic temperature. Figure 7 shows an example of such a calculation. It follows that if the specimen is forced by imposed heat flow to exceed the maximum feasible growth rate, then either a metastable solid solution or a glass must form. Such an eutectic–glass transition at a critical forced growth rate (here only 2.5 mm/s) was found experimentally for a Pd–Cu–Si alloy, an easy glass-former (BOETTINGER [1982b]).

It should be noted that even if solute trapping is not feasible in a particular instance of freezing, yet RSP leads to fine dendrites and therefore to a fine scale of microsegregation. This implies that homogenization then is much faster than with a slowly cooled alloy: fig. 8 presents some estimated figures.

This figure is based on the simplified model of initially sinusoidal composition

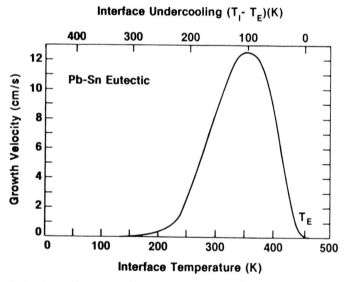

Fig. 7. The calculated growth rate of the Pb–Sn eutectic as a function of interface temperature, extrapolated from low-velocity experimental data using a measured activation energy (after BOETTINGER [1982b]).

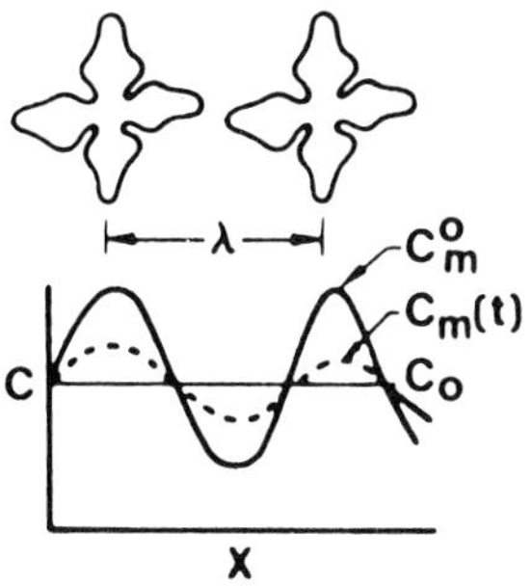

$$\delta = \frac{C_m(t) - C_0}{C^0_m - C_0} = \text{microsegregation ratio}$$

- From Fick's Law

$$t = \frac{-\lambda^2 \ln \delta}{4\pi^2 D}$$

$$= \frac{0.12\lambda^2}{D} \quad \text{for 99\% decay}$$

- For W in Ni

$$D_W^{Ni} = 2.08 \times 10^{-10} \text{ cm}^2/\text{s} \quad \text{at 1600 K}$$

$$t = 5.8 \times 10^8 \, \lambda^2 \text{ s}$$

TYPICAL VALUES OF HOMOGENIZATION TIMES

	Conventional solidification		Rapid solidification	
$\lambda \, (\mu m)$	1000	100	10	1
t	10 weeks	16 h	10 min	6 s

Fig. 8. Effect of diffusion distance, λ, on the expected homogenization kinetics of a Ni–W alloy solidified at different cooling rates in the dendritic mode (after COHEN *et al.* [1980]).

variations between dendrite arms; the times needed for the sinusoidal amplitude to decay to 1% of its original value are estimated for different dendrite arm spacings, for the case of W diffusing in nickel at 1600 K. Homogenization may even go a long way during solid-state cooling, after freezing, and likewise it has been shown that dendrites can coarsen appreciably during cooling in RSP. In spite of this last complication, it remains true that, with or without solute trapping, RSP aids the formation of homogeneous solid solutions, stable or metastable, and this is one of its most crucial advantages as a technique.

2.3. The heat-flow criterion

Even if solute-trapping is achieved according to one or other of the preceding criteria, the solute will be released to segregate after all if the supercooling is not large enough to prevent recalescence to the liquidus temperature. JONES [1982a] points out that this heat-flow criterion is the most demanding: according to his simple model, the necessary supercooling to prevent last-ditch segregation is at least L/c, where L and c are the latent heat of solidification and the specific heat of the liquid, respectively; for Al, this implies $\Delta T_{crit} \geqslant 330$ K. This would be a *heat-flow criterion*. Nevertheless, Jones points out that in alloys such as Al–Cr or Al–Zr, where the solidus rises slowly with solute concentration (and T_0 presumably rises slowly also), considerable supersaturation is achieved even at modest cooling rates such as chill-casting provides, where certainly the heat-flow criterion is not obeyed. The role of the heat-flow criterion awaits further analysis.

2.4. Glass-forming criteria

2.4.1. The kinetic criterion

The prime condition for glass formation is that no nuclei of crystals should be formed during quenching. Theoretical treatments recently have all been based on UHLMANN's [1972] approach to combined nucleation and growth under isothermal conditions, and the standard criterion for glass formation is that a volume fraction of no more than 10^{-6} should be crystalline when cooling is complete. DAVIES [1976,1978] developed a detailed theory of the critical cooling rate, R_c, and found that R_c is approximately determined by the ratio T_g/T_f, where T_f is the equilibrium freezing temperature (sometimes referred to, below, as T_m, the equilibrium melting temperature) and T_g is the glass-transition temperature (§3.2); the higher T_g/T_f, the smaller R_c – i.e., the easier it is to form a glass.

Theories such as Davies's need to be corrected for the fact that TTT (isothermal time–temperature–transformation curves, such as are presented for steels in ch. 16), obtained by a combination of experiment and theory for glass-forming alloys, require modification for the circumstance of steady continuous cooling, to give a *continuous transformation* (CT) *curve*. Figure 9 gives examples of such CT curves, computed by two alternative correction methods from the TTT curve (MACFARLANE [1982]). Effectively, in terms of quantities marked in fig. 9, the critical cooling rate is then given by $(T_f - T_p)/t_p$, which is less demanding than an R_c computed directly from a TTT curve as used to be done. For the Pd–Si alloy of fig. 9, $R_c = 6-8 \times 10^3$ K/s, which compares with a measured value of 8×10^2 K/s (DREHMAN and TURNBULL [1981]).

The value of T_g usually varies less with composition than does T_f (especially in the neighbourhood of deep eutectics) – fig. 11, below, offers an illustration – so that a *deep eutectic forms a favourable composition for glass formation*, with a high value of T_g/T_f and a low value of R_c. The most important physical variable is the viscosity of the melt, and a low freezing temperature implies a higher viscosity at the freezing point, which in turn retards nucleation. However, it is not possible to measure viscosity at all the temperatures at which it needs to be known so that extensive

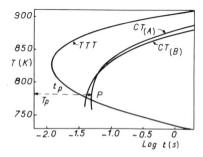

Fig. 9. Continuous-cooling transformation curves for a $Pd_{82}Si_{18}$ melt–glass, calculated from isothermal TTT characteristics by two distinct methods (after MACFARLANE [1982]).

interpolation of viscosity (fig. 18b, range B) is unavoidable, which means that a TTT or CT curve cannot be calculated very accurately. An order-of-magnitude disparity between calculated and observed R_c, as cited above, is not at all surprising. The viscosity problem is fully discussed by DAVIES [1976] and by RAMACHANDRARAO *et al.* [1977a].

Figure 10 shows estimated critical quenching rates for a number of glass-forming alloys as a function of T_g/T_f, and corresponding limiting thicknesses (since the attainment of R_c throughout the thickness of a foil depends on a sufficiently small foil thickness during RSP).

It is now clear that the kinetic criterion, as expressed by the prediction that glasses form most readily at and near the compositions of deep eutectics, is inadequate to rationalize most glass-forming systems. It works well for some of the best-known metal–metalloid glass-forming systems, such as Pd–Si (see fig. 11a), but not for many metal–metal glass-forming systems, such as Cu–Zr (to judge from fig. 11b) or Ca–Cu–Mg. Some alloys expected on the kinetic criterion to produce glasses do not do so, or only at extreme quenching rates. Thus near-eutectic binary Fe–C alloys can only partly be turned to glass (SHINGU *et al.* [1976]); this is probably due to very rapid diffusion of the small C atoms in the liquid, encouraging rapid crystallization. SUZUKI *et al.* [1983] have been able to turn an aluminium-based alloy, $Al_{69.6}Fe_{13}Si_{17.4}$, entirely into glass by the gun method (the first aluminium-rich alloy to be totally vitrified). T_g/T_f for this alloy is only 0.47, and $R_c > 10^6$ K/s.

It appears that the thickness limitations of fig. 10 can be circumvented if heterogeneous nucleation of crystals can be prevented. DREHMAN *et al.* [1982] point out that easy glass-formers with high T_g/T_f tend to have a low homogeneous nucleation rate. The glass $Pd_{40}Ni_{40}P_{20}$ has a particularly high T_g/T_f and is an

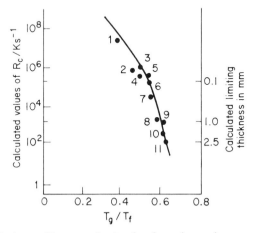

Fig. 10. Calculated critical quenching rates, R_c, for glass formation, and corresponding estimated limiting ribbon thickness for wholly glassy ribbons. Key to alloys: (1) $Fe_{89}B_{11}$; (2) $Au_{78}Ge_{14}Si_6$; (3) $Fe_{83}B_{17}$; (4) $Fe_{41.5}Ni_{41.5}B_{17}$; (5) $Co_{75}Si_{15}B_{10}$; (6) $Fe_{79}Si_{10}B_{11}$; (7) $Fe_{80}P_{13}C_7$; (8) $Pd_{82}Si_{18}$; (9) $Ni_{63}Nb_{37}$; (10) $Pd_{77.5}Cu_6Si_{16.5}$; (11) $Pd_{40}Ni_{40}Pd_{20}$. (After DAVIES [1978].)

References: p. 1845.

especially easy glass-former, with a critical thickness (fig. 10) of about 2.5 mm. Drehman *et al.* found that molten ingots up to 1 cm in diameter could be turned into glass if heterogeneous nucleation was discouraged by careful etching of the ingots before remelting and cooling at a rate of only 2 K/s, some fifty times less than the normal R_c. TURNBULL [1981] points out that careful fluxing with molten silicate is another possible way of removing surface nucleating sites comprehensively enough to allow large blocks of metallic glass to be made.

2.4.2. Structural glass-forming criteria

Table 1 classifies the categories of alloys which have been successfully turned into glasses. The most important categories are the metal–metalloid glasses (category 1 in the table) and the binary transition-metal glasses (category 2). Hundreds of different compositions, binaries, ternaries and more complex, have been successfully melt-spun to form glasses, and for many the limits of glass-forming compositions have been established. Figure 11 shows examples of such limits. The precise limits depend on the efficacy of quenching – ribbon thickness in particular – and the variations between the five limit curves in fig. 11c are thought to be attributable to differences in the quenching conditions used in the various studies. TAKAYAMA [1976] reviews numerous binary and ternary glass-forming system, citing glass-forming limits, atomic-radius ratios and other relevant information.

Many have attempted to extract some order from the mass of empirical information now available on glass-forming systems. The luxuriance of rival hypotheses is such that only a bare outline can be offered here. The author has recently published a critical review of the field (CAHN [1983]). The criteria for good *glass-forming ability* (GFA) include:

(1) *Steric hindrance.* It is simply proposed that a minimum disparity of atomic radius between two constituents is necessary to achieve good GFA. Experiments with two-dimensional model systems – hard spheres or soap bubbles (MADER [1965], SIMPSON and HODKINSON [1972], ARGON and KUO [1978] and ARGON

Table 1
Classification of glass-forming alloy systems [a].

Category	Representative systems	Typical composition range, at%
1. T^2 or noble metal + m	Au–Si, Pd–Si, Co–P, Fe–B, Fe–P–C, Fe–Ni–P–B, Mo–Ru–Si, Ni–B–Si	15–25 m
2. $T^1 + T^2$ (or Cu)	Zr–Cu, Zr–Ni, Y–Cu, Ti–Ni, Nb–Ni, Ta–Ni, Ta–Ir	30–65 Cu or T^2, or smaller range
3. A-metal + B-metal	Mg–Zn, Ca–Mg, Mg–Ca	Variable
4. T^1-metal + A-metal	(Ti, Zr)–Be	20–60 Be
5. Actinide + T^1	U–V, U–Gr	20–40T^1

[a] Key to symbols: A-metal: Li, Mg groups. B-metal: Cu, Zn, Al groups. T^1: early transition metal (Sc, Ti, V groups); T^2: late transition metal (Mn, Fe, Co, Ni groups). m: metalloid.

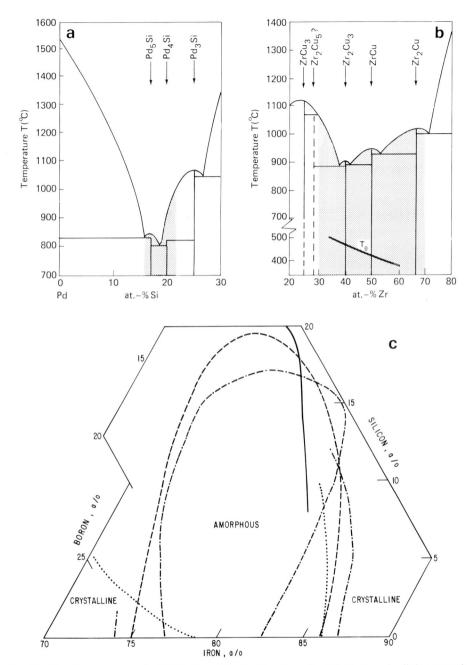

Fig. 11. Glass-forming ranges in (a) Pd–Si and (b) Cu–Zr alloys. For Pd–Si, T_g varies little across the glass-forming range; for Cu–Zr, the variation of T_g is shown. Also (c) five independent experimental estimates of the glass-forming range in Fe–Si–B alloys. For Fe–B alloys, see fig. 24c. (After LUBORSKY *et al.* [1979].)

[1982]) – agreed with empirical indications that a radius mismatch approaching 15% is necessary. SOMMER [1982] and others have combined this criterion with some measure, X, of bond strength, such as the mean vaporization enthalpy of the constituent metals, and an empirical diagram with radius mismatch as one axis and X along the other axis is drawn up, on which putative binary glass-forming systems are plotted; it turns out in all such studies that radius mismatch plays a considerably greater role than bond strength. Plainly, category-1 glasses, combining large metal atoms with small metalloid ones, are all favoured by this criterion. (On the other hand, *very* small solute atoms diffuse so easily that a favourable radius ratio may be thereby annulled; this is probably why carbon so rarely features among glass-forming additives, and hydrogen can evidently only feature as a glass stabilizer in extremely large concentrations in view of its ultrafast diffusion.) POLK [1972] originally put forward the idea that small metalloid atoms fill and thus stabilize the larger holes in a glass structure.

The studies of RAUSCHENBACH and HOHMUTH [1982] on amorphization of pure metals by ion-implantation suggest that for successful amorphization, the implanted atom should have a radius in the range $0.59R_A–0.88R_A$, where R_A is the matrix atom radius. It is obvious that metalloid projectiles are ideal for this purpose. The steric criterion here is related to Hägg's long-established criterion for the formation of interstitial crystalline structures (ch. 5, §5).

(2) *Hole formation enthalpy.* BUSCHOW [1982] has shown for a variety of category-2 glasses that stability, as measured by crystallization temperature, T_x, of a glass, scales linearly with the computed enthalpy of formation of a hole (equivalent to a vacancy) in a glass (fig. 12). His paper must be consulted concerning the method of computation, which has been well tested for crystalline materials. The underlying idea is that a high enthalpy implies a small hole concentration, which implies a high

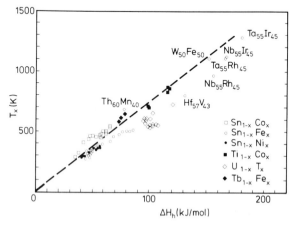

Fig. 12. Dependence of the measured crystallization temperature, T_x, of various binary metallic glasses on the formation enthalpy of a hole the size of the smaller of the constituent atoms; the hole formation enthalpy is calculated from thermochemical data. (After BUSCHOW [1982]).

viscosity and therefore a hindrance to crystallization. The hole formation enthalpy was also estimated for several glasses by RAMACHANDRARAO *et al.* [1977b] from diverse experimental data, by a statistical–mechanical computation distinct from that used by Buschow, and also found to vary linearly with T_x (or T_g, which is always close to T_x). However, this type of theory has been criticized on thermodynamic grounds by HILLERT [1982] and its theoretical respectability is at present uncertain.

In this approach, *glass stability*, or resistance to crystallization, T_x, is considered instead of GFA (which is tested by the critical cooling rate, R_c). It is widely assumed that glass stability and GFA are themselves closely correlated, but this assumption has never been properly tested.

(3) *An electronic criterion.* NAGEL and TAUC [1977] postulated a relationship, $2k_F = K_p$, between the wave vector at the electronic Fermi level, k_F, and the wave vector at the first peak in the diffuse X-ray scattering curve, K_p; this is, loosely, analogous to the well-known Hume-Rothery criterion for electron phase formation in crystalline solid solutions (ch. 4, §7.1). The equality $2k_F = K_p$ can be shown to be equivalent to the location of the crystalline Fermi energy at the peak of an electronic density-of-states curve (ch. 4, §5.1), and in both cases the Fermi level (for a given electron concentration), and thus the mean electron energy, is thereby minimized.

This controversial criterion fits many glasses but a number of exceptions have now been found; BUSCHOW [1983] in particular has advanced detailed evidence of the inadequacy of the criterion, for instance for the Ni–Ti glasses. The electronic criterion is critically reviewed in its theoretical context by HAFNER [1981].

(4) *Short-range order criterion.* The notion here is that many metallic melts have substantial "chemical" short-range order (CSRO), i.e., a tendency to unlike or like nearest neighbours. SOMMER [1982] has proposed that this CSRO is heterogeneous and can be assessed indirectly by thermodynamic measurements. If the clusters of strongly ordered atoms resemble, in local composition, a crystalline compound which tends to crystallize in equilibrium, then crystallization is facilitated and GFA is poor; if the local composition differs from any stable crystalline species, then GFA is high. This model fits particularly well with the observed variation of GFA in the Cu–Ti alloys, which have been very fully studied. The composition of maximum SRO, as measured by neutron diffraction, coincides with maximum T_x (see fig. 17). There is some independent physical evidence in other systems for what is, in effect, a sharp variation of CSRO character as a function of composition (HOPKINS and JOHNSON [1982]). This criterion offers some prospect of interpreting the *composition range* of good GFA, as opposed merely to the combination of elements which form glasses at all. Apart from the semi-empirical T_g/T_f criterion discussed above, this problem has received scarcely any theoretical attention to date.

(5) *Thermodynamic criterion.* We have already seen that the variation of T_0 with composition (fig. 5, above), based on Baker and Cahn's ideas, is theoretically related to GFA. MASSALSKI [1982] analyzes the validity of this criterion in relation to a number of alloy systems, and concludes that it has good interpretative power.

(6) *Free volume criterion.* This is closely related in conception to criterion (2), above. RAMACHANDRARAO [1980] and YAVARI *et al.* [1983] have independently

References: p. 1845.

published different versions of this criterion: a zero (or negative) change of volume on melting of a crystalline species favours glass formation on subsequent RSP. This criterion, according to both papers cited, is mostly consistent with the relative GFA of a number of alloys tested, with apparently no exceptions, although glass-forming ability is not found to vary necessarily in the same sequence as volume change on melting.

The concept of *free volume* is explained in §3.2.1; the larger the free (unoccupied) volume in a liquid, the lower is the viscosity and the faster is diffusion. The approach of Yavari and his coworkers is that if a crystal is denser than the melt from which it grows, then in growing it effectively rejects free volume into the adjacent melt, reduces its viscosity and thus encourages crystal growth. They establish that good glass-forming systems are those in which the crystalline solid is *not* denser than the liquid. (It is however puzzling why it is the liquid and not the solid–liquid interface which should act as a sink for free volume.)

It is not yet possible to see clearly which of these various criteria for either GFA or glass stability has the greatest predictive power, and indeed up to now the criteria have generally only been used for ex-post-facto interpretation of findings. As with crystalline phases (ch. 5), different criteria may prove to be apposite according to the types of atoms and bonds concerned. In particular, so many quite different metal–metal glasses are now known that it is improbable that their stability ranges can be rationalized on any single criterion.

3. Metallic glasses

3.1. Structure

The problem of determining the structure of a crystal consists merely of identifying the coordinates of all the atoms constituting a unit cell; though the task is very difficult for large and complex unit cells, yet in principle (if the "phase problem" can be solved) the structure can be determined precisely. Conversely, for a glass, the structure can only be described on a statistical basis: there are no unit cells and the environments of different, chemically identical atoms will necessarily vary. Herein lies the difficulty of determining and describing the structure of any glass (or liquid), and the intellectual challenge no doubt accounts for the quite remarkable amount of attention which has been devoted to the problem in the past twenty years.

The task of determining glassy structures resolves itself into two constituent tasks – determining a *radial* (or *pair*) *distribution function* (RDF), and *modelling*. An example of a radial distribution function, $G(r)$, is shown in fig. 13a; it is a measure of the probability of finding an atom centre at a distance r from an average central atom. Specifically, $G(r)$ as shown is a *reduced* RDF, given by $G(r) = 4\pi r[\rho(r) - \rho_0]$, where ρ_0 is the glass density. $\int_{r_1}^{r_2} 4\pi r^2 \cdot \rho(r)\, dr$ is the number of atom centres located in a spherical shell between radii r_1 and r_2. $\rho(r)$ or its variant $G(r)$ are determined by Fourier inversion from normalized X-ray scattering curves of the kind shown in

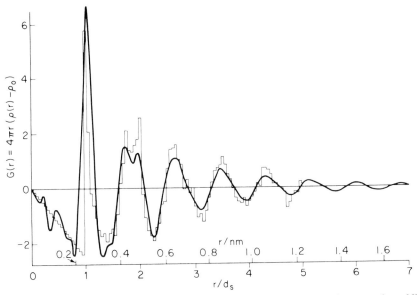

Fig. 13. Experimentally estimated 'reduced' radial distribution function, $G(r)$, for amorphous $Ni_{76}P_{24}$ (continuous curve) compared with Finney's version of the dense random packing (hard sphere) model, taking into account nickel atoms only (histogram) (after CARGILL [1979]).

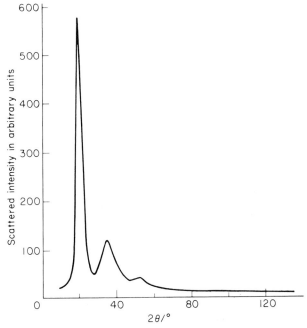

Fig. 13b. Scattered X-ray intensity versus scattering angle, θ, for amorphous $Ni_{81}P_{19}$ (after CARGILL [1979]).

References: p. 1845.

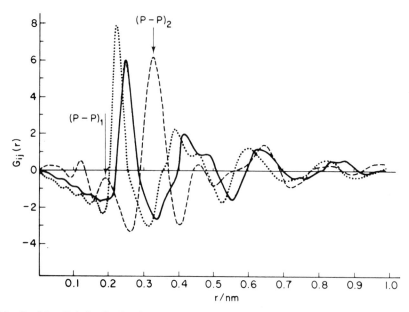

Fig. 13c. Partial radial distribution function, $G_{ij}(r)$, for glassy $Co_{80}P_{20}$. Solid curve, Co–Co; dotted curve, Co–P; dashed curve, P–P. (After DIXMIER et al. [1965].)

ch. 12; however, all the curves in that chapter refer to crystalline materials, whereas the scattering curves from glasses have no Bragg peaks. Figure 13b shows an example of such a scattering curve before normalization, for a glass similar to that denoted by fig. 13a.

Scattering curves such as fig. 13b can be obtained with X-rays, neutrons or electrons (though the precision obtainable with electrons is too poor so this approach is not useful). If both X-rays and neutrons are used, or X-rays showing anomalous atomic scattering factors near an absorption edge (ch. 12, §2.2 and fig. 3), then *partial* RDF's can be obtained for binary glasses (fig. 13c); these indicate separately the probable separations of specified pairs of atomic species. For details of the theory underlying the derivation of RDF's from scattering curves and the various experimental and computational errors to which both are subject, the reader should consult the very full review by CARGILL [1975]. It is worth pointing out here that until very recently no attempt has been made to obtain a figure of merit for the expected accuracy of RDF's (analogous to the R values currently used to assess the quality of a crystal structure determination). LIVESEY and GASKELL [1982] have taken the first steps towards the definition of such a figure of merit. Until this has been done systematically, it is impossible to decide with conviction between some rival models since the differences between the idealized RDF's computed from them are often less than the uncertainty in the 'observed' RDF.

To determine a glass structure (just as with a crystal structure) it is necessary to guess a structural model, compute its RDF (total or partial) and compare this with the RDF deduced from the observed, normalized scattering curve. Such modelling leans on as much incidental information as can be gleaned about the sizes, bonding and coordination of the constituent atoms: observations of density, atomic radii in crystalline forms, interaction potentials (Lennard–Jones, Morse, etc), interbond angles where appropriate are all taken into account in constructing a model. The actual construction was initially done physically, with balls and springs, or just steel balls, but now is always done by computer simulation. Models for covalently bonded glasses such as silicon involve distinct problems from metallic glass models (only the former incorporate directed atomic bonds) and we discuss only the latter here.

In the very extensive literature on glass modelling, there are so many technical difficulties to discuss (e.g., the surface errors resulting from the finite size of a model) that it is very hard to see the wood for the trees: a survey of the underlying principles that is both clear and comprehensive has however just been published by FINNEY [1983]. WAGNER [1983] expounds the study of glasses by diffraction.

The matching of RDF's and models is fully discussed in Cargill's review; however, much has changed since 1975 and a somewhat more up-to-date detailed review is by GASKELL [1979a], which covers all varieties of glass. For metallic glasses there are in effect two rival models, the *Dense Random Packing* (DRP) model originally due to BERNAL [1964] and initially intended for liquid structure, and the alternative model recently proposed by GASKELL [1979].

Bernal's empirical model originally referred to the DRP of hard spheres all of the same diameter and was based on physical experiments with steel balls kneaded in a rubber bag. The packing is statistically reproducible and consists of a number of different kinds of polyhedra, of which the principal varieties are shown in fig. 14. The small polyhedra, tetrahedra and octahedra, make up the great majority, and there is still some dispute about the precise frequency ratios between the different polyhedra. For some metallic glasses, there is fairly good accord between the computed RDF and that deduced from diffraction data (fig. 13a). Note, however, that the computed RDF of this Ni–P glass takes no account at all of the P atoms, and the spheres used in modelling are infinitely hard. BOUDREAUX and GREGOR [1977] took the important step of correcting these deficiencies: P atoms are now placed in the structure in such a way that no P atoms are nearest neighbours of each other (as required by the negligible $(P–P)_1$ peak in fig. 13c), and sequential computer relaxation of *soft* spheres was used to densify the modelled structure. The original hard-sphere DRP model yields a packing fraction of ~ 0.63, while soft-sphere relaxed DRP gives packing fractions of 0.69–0.70, in agreement with deductions from measured densities. The agreement between modelled and "observed" RDF's is distinctly improved.

This bare outline hides numerous computational approximations and difficulties which are well dealt with in Finney's review. FROST [1982] also treats in detail the distribution of different kinds of cavity in DRP structures, which have now altered somewhat from the classic forms shown in fig. 14.

GASKELL [1979b] emphasized one of the principal deficiencies of the DRP model

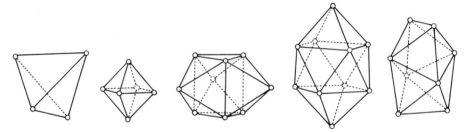

Fig. 14. Polyhedra formed by dense random packing of hard spheres, according to Bernal.

(even more pronounced for the soft-sphere than for the hard-sphere version): it is difficult to make sense of the common metal–metalloid glasses (category 1 in table 1) with typically ~ 20% of metalloid atoms. POLK [1972] proposed that such glasses are stabilized by the insertion of small atoms into the larger structural voids in the assembly of the large atoms, but to accommodate *without distortion* 20% of metalloid atoms even in the largest voids then (using one of the more recent refined versions of DRP) Gaskell shows that the metalloid diameter must be ⩽ 0.48 of the metal atom diameter. This is satisfied only for carbon and boron, not for phosphorus, silicon or arsenic (all common glass stabilizers). The force of Gaskell's observation about the insufficiency of the space available in DRP voids to accommodate the larger metalloid atoms is however blunted by Turnbull's intriguing empirical observation of the mean atomic volumes of both metallic glasses and interstitial–crystalline-metallic solid solutions: the mean atomic volume is almost entirely determined by the metallic skeleton and the larger metalloids are constrained to reduce their size to fit the available interstitial sites. This is explicable, according to Turnbull, in terms of a particularly "soft" (i.e., easily variable) interatomic separation in the easy glass-forming combinations (TURNBULL [1981]).

To overcome the difficulty he had identified in forcing metalloids into available voids, GASKELL [1979b] proposed a model as shown in fig. 15, based on a computer-

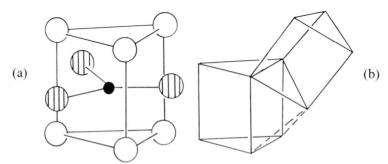

(a)   (b)

Fig. 15. (a) Trigonal prismatic packing of transition metal atoms (large circles) and a metalloid atom (small circle), as found in crystalline Fe_3C or Fe_3P. In a glass, such prisms can be connected in various ways after relaxation, of which (b) is the most important example. (After GASKELL [1979b].)

modelled packing of prisms which themselves mimic the smallest identifiable units of a stable crystalline phase. This particular version of what might be termed, by analogy with oxide glass structures, the *network model*, refers to Pd_3Si; the linkage of the prisms among themselves is refined by successive computer approximations to minimize atomic overlap. RDF's and densities agree well, and Gaskell proposes that such a model, mutatis mutandis, can apply to many metal–metalloid glasses. Recent developments in the network model are described by GASKELL [1982]; here he adduces reasons for regarding the formation of this structure as a compromise between the tendency to the closest possible packing and the need for space to insert metalloid atoms.

Striking confirmation of this network model has come from a recent study by MEHRA *et al.* [1983] of $(Ru_{84}Zr_{16})_{1-x}B_x$ glasses with $0.40 < x < 0.53$ – i.e., an unusually high metalloid content. They show that no DRP model can give even moderate agreement with the diffraction data from such glasses, while a form of irregular prismatic packing based on the Ru_2B crystal structure gives good, though not perfect agreement of RDF's. This kind of structure can accommodate much higher metalloid concentrations than can RDP, hard or soft.

As exemplified in fig. 11, most glass-forming alloy systems are able to form glasses over a range of compositions. In many such systems, physical properties vary substantially with composition. In some systems, there are clear discontinuities in such properties as electrical resistivity, Debye temperature, mean atomic volume and crystallization temperature as the composition varies. JOHNSON [1982], who has studied several such glass-forming systems, analyzes this behaviour in terms of sharp changes in the electronic density-of-states (which can be deduced from transport or XPS measurement): such changes in turn are associated with structural changes which themselves are very difficult to detect. Figure 16 shows three examples of such property discontinuities.

The $Fe_{1-x}B_x$ system in particular (fig. 24c) has been intensively studied, in the range $x \approx 0.12$–0.28. There are discontinuities in density (or atomic volume), radial distribution function, magnetic properties and crystallization characteristics, all near $x = 0.18$ (figs. 16c, 24a). Other physical properties have also uncovered pronounced discontinuities at exactly $x = 0.25$ (WARLIMONT and GORDELIK [1982]). Mössbauer spectroscopy, in particular, is well suited to disentangle the origin of such discontinuities, since the method is sensitive to the nature and number of nearest neighbours of 'Mössbauer atoms', of which an iron isotope is one. DUBOIS and LE CAER [1982,1983] have used this technique to good effect in the study of $Fe_{1-x}B_x$ glasses. They conclude that for $x > 0.20$, the glasses are well described in terms of the network model, in the form of trigonal iron prisms containing interstitial boron. When $x < 0.20$, the glass undergoes *phase separation* into boron-rich and boron-poor regions. This conclusion is well supported by the results of PILLER and HAASEN [1982] who used field-ion microscopy (ch. 10a, §3.2.2) to examine the fine compositional variations in space of a $Fe_{40}Ni_{40}B_{20}$ glass. The glass was found to be inhomogeneous on a 2–4 nm scale (a typical scale of glass inhomogeneity, according to Johnson): one phase is boron-poor, the other contains ~ 25% B. Piller and

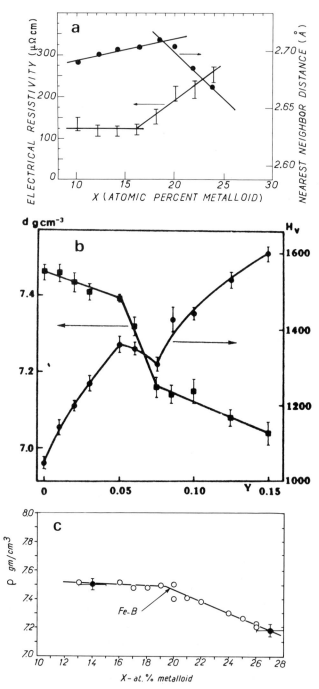

Fig. 16. Instances of discontinuities in the variation of properties with composition in metallic glasses: (a) electrical resistivity and nearest-neighbour distance in $(Mo_{0.6}Ru_{0.4})_{1-x}B_x$ glasses (after JOHNSON [1982]); (b) density, d, and Vickers microhardness, H_v, in $Fe_{0.85-y}Si_yB_{0.15}$ glasses (after AL BIJAT et al. [1982]); (c) densities of Fe–B glasses (after RAY et al. [1977]).

Haasen interpret this second phase as representing the transition from DRP to network glass structure.

The observed variation of density with composition for $F_{1-x}B_x$ glasses, and particularly the discontinuity in slope at $x = 0.18$ (as shown in fig. 16c) matches remarkably well with that calculated from a close analysis of a computer model constructed on soft DRP principles (BOUDREAUX and FROST [1981]). It thus appears that a DRP $\leftrightarrow$ network transition can emerge spontaneously from a model which does not explicitly try to mimic crystal structures.

Curiously, the X-ray or neutron small-angle scattering technique which has proved so effective in studying phase separation in oxide glasses (ZARZYCKI [1974]) has been applied very little to metallic glasses, and then mostly to study plastically deformed materials. An exception is the study by NOLD et al. [1982], by small-angle neutron scattering, of $Fe_{80}B_{20}$. Their results might have been interpreted in terms of structural inhomogeneity, but were not; this reinforces the general principle that, with metallic glasses, it is best to combine several techniques of investigation before conclusions are drawn.

The model derived from the various studies on Fe–B and Fe–Ni–B glasses has been discussed by DUBOIS and LE CAER [1982] with respect to metal–metal glasses. Field-ion microscopy has failed to uncover any heterogeneity in Ni–Nb glass but did confirm the somewhat disputed TEM observations of TANNER and RAY [1980] of heterogeneities in the glass $Ti_{50}B_{40}Zr_{10}$ (ÖHRING [1983]).

A question which has come very much to the fore recently as more attention has focused on metal–metal glasses is the matter of *chemical short-range order* (CSRO), the tendency of nearest-neighbour identities not to be random. It is in fact a crucial issue for all kinds of glasses. CSRO can be assessed by obtaining partial RDFs on the basis of multiple diffraction measurements (e.g., for $Ni_{40}Ti_{60}$, RUPPERSBERG et al. [1980]), by EXAFS, a technique for analyzing the *fine structure* of an *X*-ray *absorption* edge which is highly sensitive to the electronic character of nearest neighbours ($Pd_{80}Ge_{20}$ etc., WONG [1981] and GURMAN [1982]) or by nuclear magnetic resonance ($Co_{75}B_{25}$ etc., DURAND [1981]). These experiments are difficult and subject to considerable error, especially those by EXAFS where the conclusions are highly sensitive to the details of the structural model employed (so that the influence of atom positions and atom identities is hard to separate). It is nevertheless now quite clear that considerable deviations from chemical randomness in fact exist in a number of glasses. The system which has been most thoroughly studied is Cu–Ti: SAKATA et al. [1982] have used neutron diffraction to establish the variation of CSRO with composition over a wide composition range; they also found that the crystallization temperature – one measure of glass stability – is linked with the degree of CSRO (fig. 17), and this behaviour of the Cu–Ti series in turn was interpreted by SOMMER [1982] in terms of his short-range-order criterion for glass stability, outlined above. The diffraction identification of strong CSRO in Cu_2Ti has been confirmed by several recent EXAFS studies (e.g., FLANK et al. [1982]).

One feature concerning which measurements are entirely lacking up to now is the variation of CSRO with temperature *in the glassy state* (though there is some

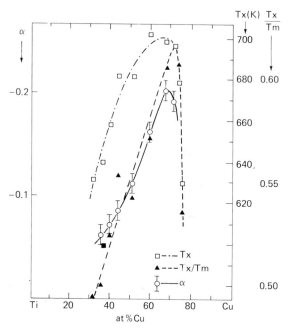

Fig. 17. The variation with composition of the chemical short-range order parameter, α, and the crystallization temperature, T_x (measured at 80 K/s), together with the reduced crystallization temperature, T_x/T_m, for Cu–Ti glasses (after SAKATA et al. [1982]).

evidence that CSRO increases on cooling from the equilibrium melt to the glass). CAHN [1983] explains in some detail why it is important to have this kind of information: a number of properties of metal–metal glasses in particular, especially in connection with reversible thermal relaxation, are likely to be linked with it. One recent EXAFS study (SADOC et al. [1983]) has indeed established that heat-treatment below the glass transition temperature modifies the CSRO as well as the atomic packing in a $Cu_{46}Zr_{54}$ glass (whereas other glasses, including Cu_2Ti, do not show such changes). The investigators did not examine the changes as a function of annealing time or temperature and they did not look for changes reversible with change of annealing temperature. Such reversible changes have however been elegantly confirmed, albeit indirectly, by detailed resistometric measurements on $Cu_{50}Ti_{50}$ and $Ni_{35}Ti_{65}$ (BALANZAT and HILLAIRET [1982]). CSRO of metallic glasses is reviewed by EGAMI [1983].

3.2. The glass transition and free volume

When an alloy is rapidly quenched from the melt, various physical and thermodynamic properties change with temperature as shown in fig. 18. At the higher temperatures, range A, the melt is in equilibrium. It is extremely difficult (though possible under certain circumstances, PEREPEZKO and GLEITER [1983]) to superheat a

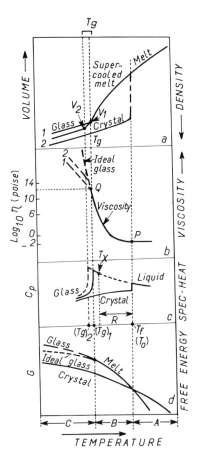

Fig. 18. The glass transition: variation of several properties with temperature. See text.

crystal above the thermodynamic melting point, so that metastable states are very rare in range A. In defining the boundary between ranges A and B, we take T_f (the equilibrium freezing temperature) for the more interesting case of falling temperature – T_f is not necessarily the same as T_m, the melting temperature, for alloys with a solidus–liquidus gap, but we ignore here the complexities arising from an interposed solid + liquid range. In effect, the diagram is based on the simple hypothesis that $T_f = T_0$.

In range B, the melt is thermodynamically supercooled; if, however, a critical nucleus is provided, crystal growth is easy because the viscosity is sufficiently low for easy atom transport through the melt. At all temperatures as the melt is cooled through range B, it remains in *configurational equilibrium* even for the high cooling rates in the RSP category – that is, the atomic packing and translational motion of the atoms are always in equilibrium. Crystallization is avoided only because fast cooling gives no time for nucleus formation, not because atomic transport in the melt is congealed. In range B, the viscosity varies over some fifteen orders of magnitude, and though most of the range is not accessible to direct measurement (because crystallization would intervene during the measurement), the curve can be estimated by interpolation between experimentally accessible points in the neighbourhood of P and Q (e.g., RAMACHANDRARAO et al. [1977a]).

Range C is defined by a temperature, T_g – termed the *glass transition* – below which the configuration is congealed in a pattern which corresponds to equilibrium at T_g. T_g is close to the temperature at which the viscosity reaches 10^{13} poise. Below T_g, translational (i.e., exchange) motions of atoms in the melt become so slow that, during the rapid quench presupposed in drawing fig. 18, the configuration no longer changes below T_g. The mobile melt has become a rigid glass.

It is important to recognize that T_g is *not* a thermodynamically defined temperature; its location is determined wholly by kinetic considerations. If the quenching rate is somewhat reduced (but not so much as to permit crystallization in range B), then $(T_g)_1$ is lowered to, say, $(T_g)_2$. This slower quench entails a smaller volume (higher density) for the glass and also a transiently higher specific heat; this enhanced specific heat is associated with the rapid change of configuration just above T_g. Figure 18b moreover shows that the viscosity–temperature variation shows a sharp discontinuity at T_g: for an *ideal glass*, in which configurational equilibrium is imagined maintained even at low temperatures at which atomic diffusion rates are negligible, the viscosity curve would continue to steepen. Instead, it veers off and varies much more slowly with temperature. The viscosity branches labelled "1" and "2" in fig. 18b correspond to two quenching rates ($\dot{T}_1 > \dot{T}_2$), and are each *isoconfig-urational*, i.e., below T_g the configuration or atomic packing no longer varies appreciably with temperature. (The factors that govern the temperature variation of viscosity in the isoconfigurational state, such as variations in vibration amplitude of atoms, are still subject to discussion.)

A *glass* can be adequately defined as a supercooled liquid congealed into a rigid, isoconfigurational condition.

The difference in specific heat, c_p (fig. 18c), between supercooled melt and crystal falls as the temperature rises from T_g. The entropy change for the liquid–solid transformation decreases as the temperature falls towards T_g: the detailed reasons for this are much disputed, but crudely we can say that the configurational entropy of the melt diminishes with falling temperature as the density increases and the packing approaches more nearly to close packing.

Theorists of the glassy state postulate that there exists a Platonic "ideal glass", denser and more perfect and with a lower free energy than any actually attainable glass (fig. 18d). A number of thermodynamic arguments induced Kauzmann, in 1947, to announce the paradox now named after him: below some temperature well above 0 K, i.e., at the temperature T_K of the *Kauzmann catastrophe*, the entropy of an *ideal* glass should become lower than that of the perfect crystal. It is a paradox because one cannot see how a perfectly periodic structure can have a higher entropy than a relatively disordered, non-periodic one! The paradox is still the subject of lively discussion by theorists and especially those interested in polymeric glasses (e.g., WOODCOCK [1981]). It seems that the paradox must be resolved by asserting that if it *were* possible to keep a glass in equilibrium down to T_K, it would inevitably crystallize however severe the kinetic hindrance.

3.2.1. Free volume

Returning again to fig. 18a, we see that a faster-quenched glass (1) has a lower density than a slower-quenched one (2) or, in other terms, "1" has a larger *free volume* than "2". Free volume is a somewhat shadowy concept, originally introduced and analyzed by polymer physicists (FOX and FLORY [1951]). RAMACHANDRARAO *et al.* [1977a,b] discuss the background to the concept. In its simplest terms, free volume is the unfilled space between atoms or molecules (which at once poses the problem of defining the contours of these entities); it can also be defined in terms of thermal expansion starting from an ideal glassy state, or else in terms of the hole theory of liquids, in which free volume is postulated to consist of identifiable holes with a rather narrow size distribution.

The crucial point is that, the smaller the free volume, the more sluggish is any structural rearrangement. It is in any case quite clear that the viscosity, η, varies with free volume approximately according to the *Doolittle equation*, originally derived for polymers: $\eta = A \exp(B/f_T)$, where f_T is the *relative free volume* defined by $(V_T - V_0)/V_0$, where V_T, V_0 are the volume of the material at temperatures T and 0 K, respectively. The equation is valid only for relatively high fluidities (not below T_g). The contentious question whether the free volume is more or less uniformly distributed or else concentrated in holes which are probably at least the size of the smallest atoms in the glass, is still open and can be pursued in the papers by Ramachandrarao *et al.* and in a review by CAHN [1979]. Certainly, one theory of glass stability (BUSCHOW [1982], fig. 12 and RAMACHANDRARAO *et al.* [1977b]) is based self-consistently on the notion that a stabler glass has a higher formation energy for holes, therefore fewer holes form, the total free volume is smaller and thus crystallization more sluggish. It has been proposed (FOX and FLORY [1951]) that the relative free volume for different glass-forming materials is always the same at T_g; this does not stand up to critical examination as an overall generalization, but it *is* in the range 0.025–0.03 for a number of glass-formers. Yet another model, relating T_g via hole fraction to the molecular weight of the glass unit (a concept which can be adapted to metallic mixtures) has been proposed by HIRATA [1979]).

The reason for the sharp change of slope of the viscosity curve at T_g (fig. 18b) is simply that above T_g, the relative free volume changes with temperature, remaining always more or less in thermal equilibrium, whereas below T_g, the relative free volume is congealed and no longer changes with temperature; viscosity then varies with temperature primarily because of the temperature dependence of vibrational amplitudes, which affect the probability of atomic interchanges.

The pronounced peak in specific heat at T_g has been a puzzle, and in general terms has been attributed to a variation of defect concentration with temperature. This generalized notion (which has recently been used by HILLERT [1982] as the basis of a novel thermodynamic theory of the glass transition) has independently been particularized on the basis of the hard-sphere DRP model (CANTOR and RAMACHANDRARAO [1982]). This important quantitative treatment starts from the idea that the relative numbers of Bernal tetrahedra, octahedra, tetragonal dodecahedra, etc. (as shown in fig. 14) must alter with rising temperature, in the sense that tetrahedra become less numerous and the other polyhedra more so. This implies that the number of larger interstitial sites increases – a hypothesis shared by Hillert. From the quantitative statistical mechanics in Cantor and Ramachandrarao's analysis, the variation of specific heat and thermal expansion coefficient below and above T_g are very well accounted for, especially bearing in mind the unavoidable approximations employed – Properly, this analysis should be repeated for soft-sphere DRP, especially since FINNEY and WALLACE [1981] have shown by computer-modelling that, in such a structure, the larger polyhedra tend to be preferentially "squeezed out". The statistical mechanics of non-crystalline assemblies was first approached in the Ramachandrarao papers with respect to postulated holes, then treated more rigorously but only for two dimensions by KAWAMURA [1979] and now has been taken a step further by being specifically tied to DRP; further developments will no doubt need to progress from there, since it is difficult to see how Gaskell's network model can be used as the basis of a statistical–mechanical analysis.

COHEN and GREST [1979,1980,1981] have recently proposed what is in quantitative terms the most subtle model available for the glass transition. The underlying notion is that each "molecule" or *cell* (the polymeric parentage of the free-volume concept leaves traces) has a temperature-dependent volume, v, part of which is free when $v \geqslant v_c$, a critical value. A cell is *liquid-like* if its free volume exceeds another critical value, and matter transport is only possible if there is a continuous network of liquid-like cells in contact in such a way that each liquid-like cell has more than one liquid-like neighbour. The main feat of the theory is to have solved the percolation problem implied by this condition, and a number of observations, such as the $\dot{T}$ dependence of T_g and the effect of anneals near T_g on the height of the C_p peak at T_g, are neatly accounted for. CHEN [1981] has proposed a somewhat related model based on his relaxation studies (§ 3.3).

Nothing has been said yet about the determination of T_g. This is nowadays always done by microcalorimetry (§ 3.5); a slow calorimetric run, done at 20–100 K/min, determines both T_g and the crystallization temperature T_x (fig. 18c), though it is not always possible to separate them.

3.3. Relaxation

It follows from what has been said in the preceding section that the properties of a glass are a function of quenching rate, and that even the fastest-quenched glass is far removed from the structure of the "ideal glass". The structure of any glass, therefore, if it is annealed just below T_g where atomic mobility is not negligible, should relax towards the unattainable state of ideality, all its properties changing as it does so. This process is termed *relaxation* and it has been very extensively investigated. There is room here only for an outline and the citation of some key publications.

Table 2 lists the physical and mechanical properties modified by relaxation. No explicit attempt has been made to study the change of T_g itself on relaxation, though the calorimetric studies by CHEN [1981] on $Pd_{48}Ni_{32}P_{20}$ glass indicated that T_g does *not* sensibly change on relaxation.

Table 2 refers to both reversible and irreversible changes. (The terms "recoverable" and "irrecoverable" have been proposed, but have not found favour). If relaxation were exclusively a matter of striving towards the structure of the ideal glass, then all relaxational changes would be irreversible, and the reversible variety when first observed was a major surprise. Figure 19a shows an example of an almost perfectly *irreversible* change: this is an isothermal dilatometric run on a $Fe_{40}Ni_{40}B_{20}$ glass, and shows a reduction in length equivalent to an increase in density of $\sim 0.12\%$ in 1500 s at 350°C, assuming isotropic contraction. If the sample is cycled between, say, 350°C and 250°C, only a tiny, probably not significant reversible change of length is found (when the instantaneous thermal expansion or contraction is allowed for). Figure 19b, on the other hand, shows what is, on a short-term basis at least, perfectly reversible behaviour: it refers to the elastic (Young's) modulus of a Co-rich glass cycled between 350°C and 450°C. Figure 19c shows isothermal elastic

Table 2

Changes in physical and mechanical properties during structural relaxation. (After EGAMI *et al.* [1982], slightly modified.)

Properties	Direction of change during relaxation [a]	Reversible (R) or Irreversible (I) below T_g
Volume	D	I
Specific heat	I	I
	I/D	R
Young's modulus	I	I
	I/D	R
Internal friction [b,c]	D	I
Electrical resistivity	I/D	I
	I/D	R
Diffusivity [b]	D	I
Viscosity [b]	I	I
Embrittlement [b]	I	I
Thermal resistivity	D	I
Curie temperature	I/D	I/R
Coercive field	I/D	I/R
Magnetic anisotropy, field-induced	I/D	R
Superconductive transition temperature	D	I

[a] I denotes increase, D decrease.

[b] These properties are conventionally (for crystalline solids) considered to be structure-sensitive, and indeed do show large changes in amorphous solids upon annealing.

[c] This property is itself a reversible relaxation phenomenon.

References: p. 1845.

modulus changes for the same glass at various temperatures, and fig. 19d, for the glass $Fe_{80}B_{20}$, shows how the Curie temperature T_c, is modified by annealing isothermally (T_c can be measured very rapidly so no difficulty arises even if the annealing temperature is close to T_c). EGAMI [1978a] was the first to establish that T_c also behaves almost reversibly, though it was later found that an irreversible rising trend is superimposed on oscillations akin to those of fig. 19b when a glass is cycled, for instance between 250° and 350°C. Figure 19e shows roughly reversible behaviour for the magnetic coercive field.

The relationship between irreversible and reversible relaxation has been most fully discussed by SCOTT and KURŠUMOVIĆ [1982]. The *irreversible* behaviour has been attributed to what EGAMI [1978b] first termed *topological short-range order* (TSRO), i.e., the more efficient packing of the constituent aroms irrespective of their chemical identity, whereas *reversible* behaviour was attributed to changes in CSRO. SCOTT [1981] adduced strong though indirect evidence from isothermal micro-calorimetry on thermally cycled $Fe_{40}Ni_{40}B_{20}$ glass that the degree of CSRO varies reversibly with change of temperature; similar calorimetric evidence was obtained for other glasses by DRIJVER *et al.* [1982], while BALANZAT [1980] has obtained similar evidence for thermally cycled $Fe_{40}Ni_{40}B_{20}$ glass by monitoring electrical resistivity. (As already indicated, however, the change of CSRO reversibly with temperature has not yet been verified by direct X-ray scattering or EXAFS experiments.) Differential scanning calorimetry (§3.5) can also be used to monitor reversible structure changes induced by annealing just below T_g though it is not yet clear whether these should be attributed to CSRO (SUZUKI *et al.* [1982]). This, and many other aspects of relaxation are discussed in detail by CHEN [1983].

What the above-mentioned hypothesis effectively amounts to is the postulate that those properties which are largely determined by nearest-neighbour distances and identities should behave largely reversibly, while those which depend on the collective packing of the entire structure should behave largely irreversibly. CAHN [1983] has proposed a tentative ordering of properties on a scale along which I/R, the ratio of irreversible to reversible components of relaxation, increases; it is: Curie temperature, elastic modulus, field-induced magnetic anisotropy (CHAMBRON and CHAMBEROD [1980]), coercive field, electrical resistivity (BALANZAT *et al.* [1980]), density.

The form of relaxation kinetics, especially for the irreversible variety, has been the subject of much theoretical debate. As exemplified in fig. 19a, the central part of an sothermal irreversible relaxation curve is often linear in log(time). This log(time) functional dependence has been asserted to be incompatible with the operation of a singly activated process, while the commonly observed *crossover effect* (a special variant of thermal cycling test, see GREER and SPAEPEN [1981], SCOTT and KURŠUMOVIĆ [1982]) can *only* be interpreted on the hypothesis of at least two different activation energies. GIBBS *et al.* [1983] have proposed a detailed theory postulating a continuous spectrum of activation energies for relaxation (valid for both irreversible and reversible varieties), the different constituent processes being, as it were, successively exhausted during a long anneal; whereas VAN DEN BEUKEL

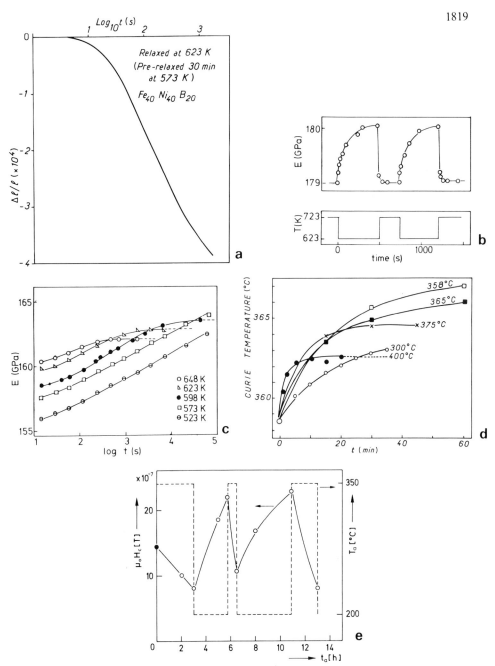

Fig. 19. Manifestation of relaxation of metallic glasses: (a) irreversible length change at 623 K of $Fe_{40}Bi_{40}B_{20}$ glass, measured by isothermal dilatometry (after SINNING and CAHN [1983]); (b) short-term reversible changes of Young's modulus of $Co_{58}Fe_5Ni_{10}B_{16}Si_{11}$ between 623 K and 723 K (after KURŠUMOVIĆ et al. [1980]); (c) long-term changes of Young's modulus of $Co_{58}Fe_5Ni_{10}B_{16}Si_{11}$, isotherm-ally held at various temperatures, from as-fabricated condition (after KURŠUMOVIĆ et al. [1980]); (d) changes of Curie temperature of $Fe_{80}B_{20}$, as fabricated, with time of isothermal holding at various temperatures, substantially reversible (after GREER and SPAEPEN [1981]); (e) coercive field, u_0H_c, during cyclic annealing of low-magnetostriction glass $Co_{58}Ni_{10}Fe_5Si_{11}B_{16}$ substantially reversible (after DONG et al. [1982b]).

References: p. 1845.

and RADELAAR [1983] have countered with a theory incorporating but a single activation energy. This disagreement has yet to be resolved.

Another theoretical approach to relaxation which also offers an interpretation of log(time) kinetics is due to EGAMI et al. [1982]. So far as irreversible relaxation goes, they associate it with a progressive mutual annihilation of postulated localized regions of higher-than-average and lower-than-average density in the glass which they had previously analyzed by computer simulation (EGAMI et al. [1980]); this interpretation is shown to be compatible with the measured changes, during irreversible relaxation, of the pair distribution function (EGAMI [1981]), which incidentally is a remarkable experimental achievement.

It is gradually coming to be realized that one and the same glass can vary substantially in its structure and properties according to the quenching rate used during fabrication. Thus, DINI et al. [1982] have shown that several separately quenched samples of $Fe_{83}B_{17}$ glass had significantly different X-ray structure factors, while HARMELIN et al. [1982] made a detailed study of the calorimetric behaviour of $Cu_{60}Zr_{40}$ glass samples and found T_x varying much more than T_g; (this confirms Chen's finding that T_g, while sensitive to cooling or heating rate during testing, is not sensitive to cooling rate during fabrication or to relaxation anneals after fabrication). They emphasize the important conclusion that it is not possible to standardize the structural state of samples of a metallic glass, following different rates of quench, by a relaxation anneal below T_g. Finally, GREER [1982b] found that the kinetics of reversible relaxation are sensitive to the rate of quenching during fabrication (as influenced by varying ribbon thickness), just as CHAMBRON and CHAMBEROD [1980] have shown that preceding irreversible relaxation affects the kinetics of subsequent reversible relaxation.

3.4. Diffusion in metallic glasses

The processes of relaxation and crystallization are intimately linked with self-diffusion and chemical diffusion in metallic glasses. The underlying concepts of diffusion are fully set out in ch. 8 and will not be repeated here.

Diffusion in metallic glasses, an experimentally very difficult topic, has only very recently received serious attention, and most investigations have been done during the past five years. The difficulties stem entirely from the fact that diffusion times and temperatures are limited by the need not to initiate crystallization; this mostly entails diffusional penetration depths of the order of 0.1 μm and never exceeding 1 μm. Very special methods are needed for such shallow profiles.

There are in fact two quite distinct approaches to the measurement of diffusion coefficients in metallic glasses: (1) direct measurements of self or solute diffusion, and (2) indirect methods.

Indirect methods include deduction from the following features: (a) the isothermal growth rate of crystals in the glass (see §3.5); (b) the changes in a concentration-modulated layer sandwich, as assessed by the gradual decay of X-ray diffraction peaks due to the sandwich spacings (GREER et al. [1982]); (c) equilibrium segregation, of the kind discussed in ch. 13, to the surface layer of glass (BAER *et al.* [1981]); (d) measurement of the Gorsky internal friction peak in an oscillating hydrogen-charged glass strip, due to rapid thermoelastic diffusion of hydrogen through the strip thickness (BERRY and PRITCHET [1981]); (e) electrolytic pulsing, i.e., discharge of a pulse of hydrogen through a foil and measurement of the time it needs to traverse the strip (KIRCHHEIM *et al.* [1982]); (f) motional narrowing in NMR, used for hydrogen diffusing in a Cu–Ti glass (BOWMAN and MAELAND [1981]); (g) reorientation of field-induced magnetic anisotropy, which is known to involve exchange of nearest-neighbour atoms (LUBORSKY and WALTER [1977] and CHAMBRON and CHAMBEROD [1980,1982]); this was used by the authors of the last-cited paper to demonstrate the particularly rapid diffusion of carbon in an iron-base glass, though no exact D value was obtained; finally, (h) measurement of changes of diffusion coefficient, D, by examining the change of viscosity, η, of a glass by using creep measurements and relying on the *Stokes–Einstein* relationship, $D = (kT/3\pi a)/\eta$, a being the ionic diameter; this relationship is exact for liquids but its applicability for very high viscosities is still in doubt (LIMOGE *et al.* [1983]).

In *direct methods*, the diffusing species is evaporated on or, better, ion-implanted in the surface, the sample is annealed and then one of the following profiling methods can be used: (a) measurement of either solute or stable isotope profile using secondary ion mass spectroscopy (ch. 10A, §3.7.2; ch. 13, §3.2); this may involve preparing an 'artificial' glass layer at the surface before diffusion-annealing to incorporate the tracer isotope (CAHN *et al.* [1980]); (b) use of a nuclear reaction (such as the $^{11}B + p \rightarrow {}^8Be + \alpha$ reaction) which allows a shallow profile to be estimated from analysis of the ejected α particles as a function of incident proton energy (KIJEK *et al.* [1981]); (c) use of ion (Rutherford) back-scattering (ch. 10a, §3.7.1.; ch. 13, §3.2) to measure profiles of heavy atoms in a light matrix; this is the same technique as has been used in connection with alloying studies in ion-implanted and laser-treated glasses (§1.4); (d) use of radioactive tracers, combined with a sputtering-away of the surface and counting of the sputtered atoms (e.g., VALENTA *et al.* [1981]).

The results from all the direct methods, including hydrogen diffusion, have been collected and analyzed by CANTOR and CAHN [1983], while the results of indirect methods were also included in another review by LIMOGE *et al.* [1983]. A frequent drawback to the various indirect methods is that either an activation energy but no pre-exponential (D_0) value is obtained so that no absolute D values can be obtained, or else, as in the crystallization-based methods, one is not always clear as to which of two or more species is the one to which the derived diffusion coefficient refers. Nevertheless, sometimes indirect values obtained from crystallization kinetics can be useful confirmation when taken in context with other information. Thus fig. 20a shows results for P diffusion in $Fe_{40}Ni_{40}P_{14}B_6$ glass, from FERNENGEL *et al.* [1982]. The points at the lower end of the graph are from direct measurements (with radioactive tracers) whereas the three sets of data at the high-temperature end are all deduced from crystal-growth measurements. A similar close relationship was established for diffusion of B in the binary $Fe_{86}B_{14}$ alloy, examined by crystal growth measurements (NUNOGAKI et al. [1982]) and compared with the direct measurements of CAHN et al. [1980]. Agreement was excellent, though over a narrow temperature range; yet the crystal-growth results are ambiguous (CANTOR and CAHN [1983]).

It is best to concentrate attention on direct methods. Figure 20 reproduces assembled direct measurements of diffusion in metal–metalloid glasses (b), including several instances of metalloid or metal self-diffusion, and in metal–metal glasses (c). The temperature scale is normalized with respect to the T_g of each glass, to make comparisons more significant.

For many metal–metalloid glasses, D values depend only on the diffusing species and increase as the diffusing atoms become lighter and smaller; for metal–metal glasses, however, there is less of a trend and in particular, there is no obvious correlation with the size of the diffusing atoms. For many diffusion couples straight Arrhenius plots were obtained within the (rather large) experimental errors, but for B in $Ni_{59.5}Nb_{40.5}$ which, thanks to the ultrasensitive nuclear-reaction method used covers a very wide temperature range, a very distinct curvature can be noted in the Arrhenius plot.

In some experiments, a pronounced reduction of diffusion rate was observed following a relaxation anneal near T_g; in others, there was no significant effect. This disagreement was resolved in terms of the effective quenching rate, and therefore the

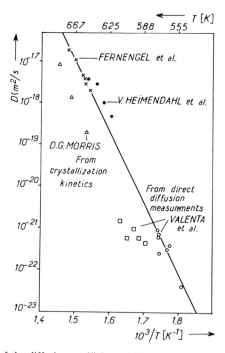

Fig. 20a. Arrhenius plot of the diffusion coefficient of P in $Fe_{40}Ni_{40}P_{14}B_6$ glass, from direct diffusion measurements and from measurements of growth rates of (eutectic) crystals at different temperatures (after FERNENGEL et al. [1982]).

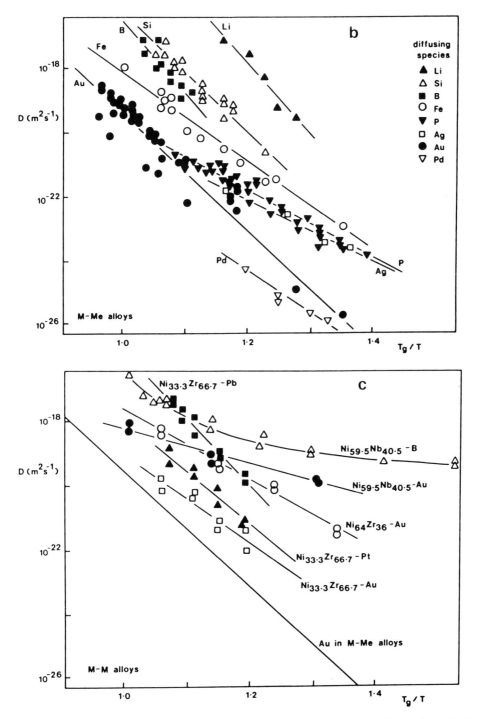

Fig. 20(b,c). Normalized Arrhenius plots of directly measured diffusion coefficients for metal–metalloid glasses (b) and for metal–metal glasses (c) (after CANTOR and CAHN [1983]).

References: p. 1845.

initial free volume, of the as-quenched glass used in the various experiments; this variation led to a difference in the scope for relaxation in the different experiments. For details, Cantor and Cahn's review should be consulted.

In this connection, it is also desirable to cite the very careful work of TAUB and SPAEPEN [1980]; using creep measurements under isoconfigurational conditions, they were able to accurately measure the increase of viscosity, η, as a function of time with isothermal anneal of a $Pd_{82}Si_{18}$ glass: a linear increase of η with time was always found, and if the Stokes–Einstein law remains valid at these temperatures, this implies a decrease of D of more than one order of magnitude for the lowest relaxation temperature. A direct comparison of $\eta(T)$ with directly determined $D(T)$ casts doubt on the validity of the Stokes–Einstein law for a metallic glass below T_g (LIMOGE et al. [1983]).

Hydrogen, the smallest atom, diffuses very much faster than all the others; this, together with the results of fig. 20b, does suggest that there is a form of interstitial diffusion, faster for smaller atoms. However, as explained by Cantor and Cahn, the theory of diffusion in metallic glasses is still in an inchoate state, especially since the migration energy of a diffusing atom varies from one site to another and simple random-walk theory cannot apply.

Figure 21 shows an especially interesting observation. The line shows hydrogen diffusion (at unknown concentration) in a Pd–Si glass (BERRY and PRITCHET [1981]) and also in Pd–Cu–Si glass at various known concentrations (KIRCHHEIM et al.

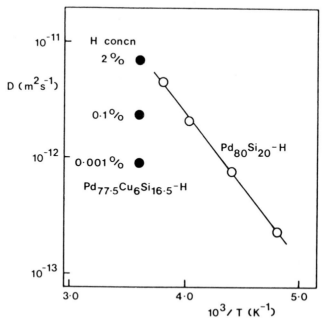

Fig. 21. Arrhenius plot of hydrogen diffusion coefficients in amorphous Pd–Si (BERRY and PRITCHET [1981]) and Pd–Cu–Si (KIRCHHEIM et al. [1982]), related to hydrogen concentration.

[1982]). It is clear that diffusivity at a fixed temperature increases with hydrogen concentration, and this has been quantitatively interpreted by a statistical–mechanical analysis: the deepest trapping sites are filled first, thereupon hydrogen goes to shallower traps from which it can jump more easily. This (KIRCHHEIM [1982]) is the first successful application of atomistic theory to the difficult problem of diffusion in metallic glasses.

3.5. Crystallization of metallic glasses

The study of crystallization of metallic glasses consists of four components: (a) determination of crystallization temperatures, (b) identification of nucleation and growth rates of crystals, (c) identification of crystalline species and morphologies, and (d) examination of the role of surface nucleation, including the possible role of surface-active fluids. This branch of the study of metallic glasses has expanded very rapidly and the literature is already large, so that only an outline is possible here. It is however a topic of potentially great practical importance, because there are recent indications (RAY [1981,1982]) that crystallization of glasses specially designed for the purpose (e.g., compositions with only 5–13% metalloids, boron preponderating) can give rise to microcrystalline alloys, grain size < 1 μm, stabilized by extremely finely dispersed borides, with impressive high-temperature properties.

For objective (a), above, it is desirable to use a *differential scanning calorimeter* (DSC), which is a highly sensitive and precise commercial instrument in which the power required to heat the sample through any small temperature interval is automatically compared with the power necessary to heat a reference sample (undergoing no transformations) through the same temperature interval. The instrument can be used in constant $\dot{T}$ or isothermal modes. Samples are very small (typically 3 mg) and the technique is thus ideally suited to metallic glass foils; however, the instrument was developed by polymer physicists and so the upper temperature limit is still restricted to ~ 700°C, which can be an irksome limitation for metallurgists. Specific and latent heats can be accurately determined, which is not the case with conventional differential thermal analysis. The mode of operation, calibration and error sources of DSC's are very clearly explained by GREER [1980], and some of the subtle ways in which the technique can be used to analyze details of the crystallization process are exemplified, with respect to $Fe_{80}B_{20}$, by GREER [1982].

Figure 22 shows an example of part of a DSC curve for a $Pd_{77.5}Cu_6Si_{16.5}$ glass. As can be seen from fig. 18c, at T_g the specific heat peaks; the sample here absorbs more power and so T_g is associated with a steep negative slope, corresponding to rapidly increasing heat absorption as the temperature rises. T_g is usually identified with the point of inflexion. Contrariwise, crystallization leads to the evolution of a large latent heat; the peak temperature, T_x, is strongly sensitive to heating rate (this can be seen most strikingly by comparing the T_x values for the three heating rates cited in figs. 24a, b). One way to determine the activation energy for crystallization is to determine how T_x varies with $\dot{T}$; another is to study crystallization kinetics isothermally at various temperatures. It is noteworthy that T_g and T_x are clearly

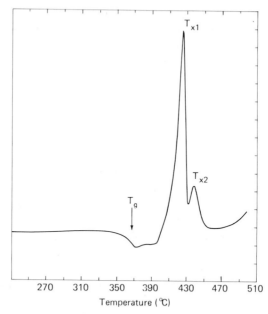

Fig. 22. Differential scanning calorimeter record during the heating, at 20 K/min, of $Pd_{77.5}Cu_6Si_{16.5}$ glass. 6 mg sample. The ordinate represents power released (exotherm, positive displacements) or absorbed (endotherm, negative displacements). T_g is clearly differentiated from the first crystallization peak, T_{x1}.

separated on DSC records only if the glass crystallizes with difficulty (easy glass-formers) or the heating rate is very rapid, or both. Otherwise T_x follows so closely on T_g that the T_g endotherm overlaps the crystallization exotherm.

To return to the objectives listed at the start of this section, objective (b) is achieved by a combination of DSC and TEM, (c) involves X-ray diffraction and selected-area electron diffraction, while (d) again involves microscopy.

For most glasses, T_x lies in the range 0.4–0.65 T_m, for normal DSC heating rates of 10–100 K/s, where T_m is the equilibrium melting temperature of the crystalline alloy. The absolute range hitherto is 400–1200 K. Most iron-base glasses have T_x near 700 K.

The glass used to prepare fig. 22 has *two* crystallization peaks. This is not uncommon: the separate peaks, where they occur (e.g., for $Fe_{40}Ni_{40}P_{14}B_6$, $Fe_{40}Ni_{40}B_{20}$, $Ni_{58}Zr_{42}$) are always rather close together. Double crystallization peaks, it appears, can be due to initial primary crystallization (see below) during which the residual glass changes in composition and then crystallizes at a higher temperature to form a distinct phase; or, in principle, it can be due to a crystallization without change of composition followed by a subsequent crystalline polymorphous transition (but no instance of this has been unambiguously established as yet). It can also, it appears, be due to crystallization into micrograins which then coarsen at a slightly higher temperature (THOMAS [1983]).

The temperature of the (first) crystallization peak is sensitive not only to heating rate but also to the free volume in the glass. GREER [1982] has made a careful study of the variation of T_x with conditions of preparation for $Fe_{80}B_{20}$ glass, as part of a study of crystallization kinetics. He found that if a ribbon is spun ~ 50% thicker, then T_x at constant heating rate actually increases ~ 15 K. This must be attributed to the presence of some quenched-in nuclei in the thicker, more slowly quenched ribbon, and this factor evidently outweighs the slower diffusion to be expected in a more slowly quenched ribbon (which ipso facto must contain less free volume).

Four distinct crystallization mechanisms have been identified:

(1) *Polymorphous crystallization*, in which there is no change of composition. Crystal shape depends only on crystallographic anisotropy of growth rates.

(2) *Eutectic crystallization*, in which the glass transforms to two phases growing in closely coupled form. There is again no overall composition change between the glass and the eutectic colony. Figure 23a, by way of example, shows such a colony in $Fe_{70}Ni_{10}B_{20}$ at high magnification while fig. 23b shows a random surface distribution of such colonies at low power in $Fe_{40}Ni_{40}B_{20}$. The rate of growth is presumably controlled by solute diffusion just ahead of the interface, but no experimental evidence is available. The lamellar spacing, curiously enough, varies little with crystallization conditions. MORRIS [1982] has shown that in the glass $Fe_{40}Ni_{40}P_{14}B_6$, growth of the dominant eutectic colonies is initiated by homogeneous nucleation of the boride phase, after which the iron-rich phase (which forms the interleaving lamellae) nucleates without difficulty.

(3) In *primary crystallization*, the first crystals to form have a composition different from the glass (e.g., nearly pure Fe from Fe–B glass) and the rate of growth

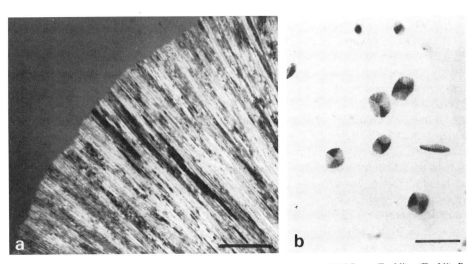

Fig. 23. (a) Eutectic colony in $Fe_{40}Ni_{40}B_{20}$ glass, annealed 1 h at 390°C: γ-(Fe, Ni)+(Fe, Ni)$_3$B. Marker = 0.1 μm. (b) Distribution of eutectic cells at the free surface of $Fe_{30}Ni_{50}B_{20}$ glass, annealed 120 s at 400°C. Marker = 10 μm. (Both after HEROLD [1982].)

References: p. 1845.

of primary crystals is controlled by the kinetics of solute rejection in the glass. The enriched glass subsequently undergoes a polymorphous transition to another crystalline species. Only one study has been published on the use of analytical TEM to examine the solute profile in the glass ahead of a growing primary crystal, in $Cu_{58}Zr_{42}$ glass into which $Cu_{10}Zr_7$ crystals were growing (FREED and VANDER SANDE [1980]).

Measurement of isothermal growth rates of primary crystals has been used to estimate diffusion rates of metalloid atoms in glasses (KÖSTER and HEROLD [1981]).

(4) Some glasses such as $Zr_{36}Ti_{24}B_{40}$ (TANNER and RAY [1980]) appear to *separate into* very fine but distinct *glassy phases* below T_x, each glassy phase then crystallizing separately. This has been confirmed by FIM observations (BÖHRING [1983]).

Crystallization kinetics generally follow the familiar Avrami equation, $x(t) = 1 - \exp(-kt^n)$; the exponent n varies from 1.5 to 4, and activation energies vary from 40 kJ/mol to 400 kJ/mol. SCOTT [1978b] explains experimental procedures which can be used to analyze the kinetics and obtain activation energies.

There has been much discussion of nucleation kinetics. In many instances, nucleation appears to be homogeneous and $\dot{N}$ is approximately invariant with time at temperature. A good recent example of a detailed study over a range of compositions in a single glass system, incorporating the study of nucleation rates, growth rates of crystals and crystallographic identification, is the study of the Ni–Zr system by DONG *et al.* [1981, 1982a]. Few attempts have been made to establish whether nucleation in particular cases is heterogeneous and to relate this to crystal morphology, or to look for preferred orientation. Both kinds of nucleation, and the existence of preferred orientation also, have been noted in a series of $Fe_{80}C_{20-x}B_x$ glasses (SHIMOMURA *et al.* [1980]). THOMPSON *et al.* [1984] have applied kinetic analysis to obtain indications of homogeneous nucleation in a Au–Cu–Si–Ge glass.

The most detailed information, however, has been assembled for the important $Fe_{1-x}B_x$ series of glasses. Figure 24 is a composite figure showing the principal facts. Figure 24a shows crystallization temperatures for two heating rates (there have been many such measurements), and it can be seen that there are two T_x temperatures below 17.5 at% B (which is near the critical composition at which there is a structural discontinuity, fig. 16c). Figure 24b shows results from one recent study (KHAN *et al.* [1982]) in which a Debye–Scherrer diffraction pattern was continuously recorded on a moving film during heating of the glass at 17 K/h. (The characteristics of one of the phases found, the unfamiliar Fe_3B, first discovered by WALTER *et al.* [1978], were further clarified (KHAN *et al.* [1982]) and these findings are incorporated in the phase diagram shown in fig. 24c.) Figure 24d shows the lattice parameter of the primary α-iron as a function of temperature for one composition of glass: the iron continues to crystallize out with changing composition over a range of temperature.

It can be seen that, at low B contents, the first phase to crystallize (at $T_{x,1}$) is α-Fe; fig. 24d indicates that at first this iron is B-rich but gradually approaches pure iron. The residual glass is enriched in boron, till at $T_{x,2}$ the disordered high-temper-

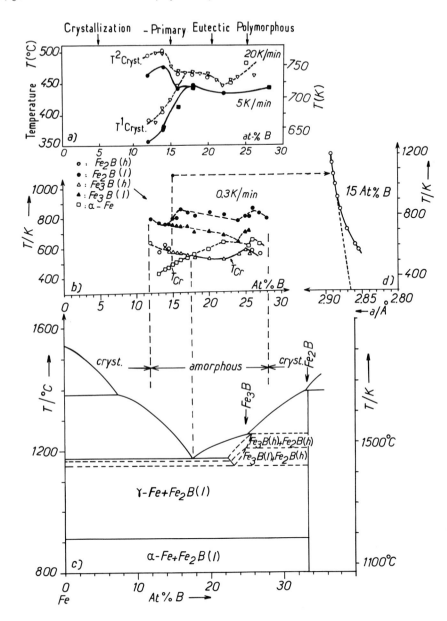

Fig. 24. Crystallization of $Fe_{1-x}B_x$ glasses: (a) crystallization temperature for two heating rates; (b) phases identified by X-ray diffraction; (c) equilibrium diagram of Fe–B system, also showing glass-forming range; (d) α-iron lattice parameter as function of temperature during heating of $Fe_{85}B_{15}$. (After KHAN *et al.* [1981].)

ature (ht) form of Fe_3B crystallizes, then orders (1t form) at a higher temperature, ~ 750 K, which is still lower than the equilibrium 1t–ht transition temperature, which is ~ 1500 K. At higher B contents, eutectic (~ 20 at% B) or polymorphous (25 at% B) crystallization is found micrographically and this is consistent with the X-ray findings.

KHAN et al. [1981] stress that the first phase to crystallize during heating is invariably that which would first form in equilibrium during slow *cooling* of a liquid alloy. This, together with much other evidence, indicates that crystallization generally proceeds from nuclei quenched in during glass fabrication. GREER [1982], by an ingenious approach, found that a thin surface layer of $Fe_{80}B_{20}$ glass is free of quenched-in nuclei, while the main thickness of the ribbon contains them. However, there is also evidence in some cases of homogeneous nucleation during crystallization, as in the glass $Fe_{40}Ni_{40}P_{14}B_6$ (MORRIS [1982]).

Recent studies on Cu–Zr glasses (CHEVALIER et al. [1983]) indicate that small amounts of dissolved oxygen cause a transition from single-stage to two-stage crystallization, so it seems that the role of impurities must be kept in mind in evaluating information from DSC runs.

There is a range of observations concerning the behaviour of free surfaces. In most cases the surface in contact with the chill-block, which has a larger free volume (but, as indicated, may be free of quenched-in nuclei) crystallizes more sluggishly than the interior, while the converse is true of the other surface; but exceptions have repeatedly been reported. Thus, TANIWAKI et al. [1982] have found that oxidation at the chill-surface of a $Fe_{75}Si_{15}B_{10}$ glass foil led to preferred crystallization there. Removal of the surface by electrolytic polishing may lead to enhanced crystallization, because of change of composition induced by the removal process, whereas ion-bombardment etching does not lead to anomalies.

Overall, surface behaviour is not yet clearly understood: existing knowledge is reviewed by KÖSTER and HEROLD [1981]. Further information on crystallization kinetics and mechanisms generally will be found in the comprehensive reviews by Köster and Herold and by SCOTT [1983].

3.6. Plastic deformation and fracture

The principal plastic characteristics of metallic glasses are, first, their great strength (the flow stress is typically as much as $1/50$ of the shear modulus); second, their fracture toughness, which in the normal state is very much greater than that of comparably strong fibrous materials such as graphite; and third, the mechanisms of flow and fracture which are entirely distinct from those found in crystalline materials.

Table 3 compares the flow stresses of a number of ductile metallic glasses, both on an absolute basis and normalized for density, with those of some other ultra-high-strength wires or fibres. Metallic glasses can be seen to be among the strongest practical materials known (the adjective is to distinguish them from single-crystal whiskers).

Table 3
Tensile properties of various continuous filaments and ribbons.

Material	Yield stress σ_y/GPa	Relative density d	Young's modulus		
			$\sigma_y d^{-1}$/GPa	E/GPa	Ed^{-1}/GPa
S-glass (SiO$_2$–Al$_2$O$_3$–MgO)	5.0	2.5	2.0	85	34
C fibre (high-yield PAN type) [a]	3.2	2.6	1.7	490	153
B filament (on W core) [a]	2.5–4.5	2.6	1.0–1.7	380	146
SiC microcrystalline filament [a]	3.5	2.6	1.4	200	77
Kevlar fibre (organic polymer) [b]	2.8	1.5	1.9	135	90
High-C steel wire	4.2	7.9	0.55	210	27
Fe$_{80}$B$_{20}$ met. glass [b]	3.6	7.4	0.5	170	23
Ti$_{50}$Be$_{40}$Zr$_{10}$ met. glass	2.3	4.1	0.55	105	26
Ti$_{60}$Be$_{35}$Si$_5$ met. glass [b]	2.5	3.9	0.65	~110	~28
Cu$_{50}$Zr$_{50}$ met. glass [b]	1.8	7.3	0.25	85	12

[a] Indicates heat-resistant materials.
[b] Indicates materials with some ductility.

Metallic glasses deform plastically by two entirely different regimes. Under tension, at high stresses near their fracture limit, they deform extremely inhomogeneously: shear bands form on planes close to the orientation of maximum resolved shear stress. These are very thin (typically 10–20 nm thick according to DONOVAN and STOBBS [1981]) and high-speed cinematography and acoustic emission indicate that the bands form extremely rapidly. (NEUHÄUSER [1978] found a formation time for individual bands in Pd$_{80}$Si$_{20}$ < 2 ms.) At low stresses, there are no shear bands and deformation is by homogeneous creep. A schematic deformation map separating the two deformation regimes is shown in fig. 25. The ductility in inhomogeneous deformation is usually only about 1% in tension, whereas many glasses can be rolled to reductions exceeding 50% if sufficiently hard and stiff miniature rolls are used. Creep rates in most creep experiments have been so low that total creep strains in reasonable times are less than 1%, but it is possible at temperatures close to T_g to hot-press a metallic glass foil such as Fe$_{40}$Ni$_{40}$P$_{14}$B$_6$ to strains of the order of unity (PATTERSON *et al.* [1978]).

The mode of ductile fracture of a metallic glass is unique. Fracture following inhomogeneous flow always takes place by separation along a previously active shear band – i.e., it is strictly a shear failure – and the fracture surface shows a "river pattern" of fine ridges which can be exactly mimicked by simultaneously shearing and pulling apart a pair of glass slides coated with vaseline. Accordingly, it has been proposed that the material in a shear band which is about to fail behaves in a liquid-like manner. There is no record of the appearance of a fracture in creep;

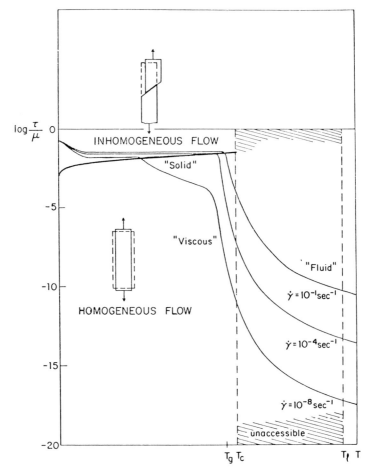

Fig. 25. Deformation map for an idealized metallic glass. T_g is the glass transition, T_c the crystallization temperature, T_ℓ the liquidus temperature, τ is shear stress, μ is shear modulus, $\dot{\gamma}$ is shear rate. (After SPAEPEN [1977].)

apparently glasses subject to slow homogeneous flow never reach their ductility limit. The complex fracture mechanics of metallic glasses are reviewed by KIMURA and MASUMOTO [1983].

The ductility of some glasses, especially iron-based ones, can be destroyed by a low-temperature anneal, typically in the range 100–200°C. This was first observed by LUBORSKY and WALTER [1976]. Figure 26 illustrates this behaviour for isochronal anneals at different temperatures for three different glasses: an intermetallic glass which is immune to embrittlement, a $Fe_{80}B_{20}$ glass which is at the border of embrittling behaviour and $Fe_{40}Ni_{40}P_{14}B_6$ which embrittles readily. The test involves simple bending to failure, recording a flow stress en route if flow precedes failure.

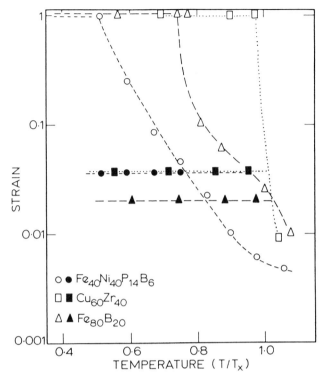

Fig. 26. Ductility of several metallic glasses (max. strain in bending) versus annealing temperature for 2 h anneals. Solid symbols represent elastic limit (i.e., flow stress); open symbols are fracture stresses. Temperatures normalized to crystallization temperatures. (After SCOTT [1978a].)

One investigator (ARGON [1982]) asserts that all glasses are subject to some degree of thermal embrittlement. Glasses which have been thermally embrittled or are brittle as fabricated fracture in a normal way on a plane normal to applied tension.

One recent study has established that a bimetallic glass not subject to thermal embrittlement undergoes "quench-ageing", if the as-spun glass is annealed near T_g (BRESSON *et al.* [1982]). In fact, flow stress and Young's modulus increased in the same proportion, indicating that it is the *strain* at yield which remains constant. This interesting finding is as yet without an interpretation. In similar tests on an iron-base glass, ageing increased the elastic limit at 77 K but not at room temperature (KRENITSKY and AST [1979]).

The aforementioned generalizations apart, a large amount of experimental information about elastic properties, fatigue, fracture toughness and the fracture mechanism is to be found in two major reviews by PAMPILLO [1975] and DAVIS [1978]. Internal friction is reviewed by BERRY [1978] and a recent detailed experimental study is by HETTWER and HAESSNER [1982]. The theory of plastic deformation in metallic glasses is treated by SPAEPEN [1977, 1978] and by ARGON [1979]. The

References: p. 1845.

last-named has also used to good effect two-dimensional simulation of plastic flow in glasses by means of soap-bubble rafts incorporating two bubble sizes, and has recently (ARGON [1982]) reviewed in great detail the insights gained in this way into the plastic flow mechanisms.

Spaepen's and Argon's versions of plasticity theory both hinge on the behaviour of small variable regions of free volume, which is not assumed to be quantized into holes. A very short-range migration of an element of free volume (so long as it exceeds a critical size), accompanied by rearrangement of the atoms surrounding the element, constitutes a "unit" of flow. Such rearrangement leads to a reduction of viscosity. There are then two possibilities: in the first case deformation is slow enough and the temperature high enough so that recovery of the "damage" (i.e., a steady stream of rearrangements) can keep pace with damage creation. This leads to *homogeneous* deformation, and the deformation map of fig. 25 was computed on this basis. In the alternative case, deformation is too fast, viscosity continues to fall and strain becomes localized in *shear bands*; this is *heterogeneous* deformation. Similar behaviour has been observed in certain glassy polymers (BOWDEN and RAHA [1970]).

The implication of such localization, as POLK and TURNBULL [1972] and SPAEPEN and TURNBULL [1974] first pointed out, is a progressive loss of order (both chemical and topological, in today's terminology): this notion is confirmed by the fact that, on a repolished surface, the sites of former shear bands are etchable by chemical attack, whereas a subsequent heat-treatment destroys the etchability. Further, the concentration of free volume in a shear band leads to a large local density decrease – as much as 20% has been suggested (DONOVAN and STOBBS [1982] and MEGUSAR et al. [1982]) – and a consequent overall density decrease of ~ 0.3% for a heavily deformed sample (PRATTEN [1980]). This density loss is recoverable on heat-treatment.

No metallic glass has been recorded as showing any work-*hardening*, but further, the concentration of shear in bands implies work-*softening* within the bands. This is very difficult to quantify, but a recent study in bending of the plastic flow of a Ni-based glass by ZIELINSKI and AST [1983] has produced results quantitatively consistent with a substantial degree of *progressive* work-softening, which in turn implies that the interior of a shear band becomes steadily more disordered as flow proceeds.

LI [1982] has recently published a comprehensive review of the characteristics of shear bands in metallic glasses.

Homogeneous creep has been studied in precise detail in a series of important papers beginning with one by TAUB [1980]. The principal innovation in this research was to study the kinetics of isothermal relaxation (which was found to lead to a linear viscosity-increase with time) so that thermal programmes could be devised to ensure *isoconfigurational creep* – deformation in which the configuration, and there-fore the free volume, stays constant. Very substantial changes of isoconfigurational creep rate result from modest changes of free volume. One unexpected result of the detailed measurements was that change of isoconfigurational creep rate with temper-ature for fixed stress is found to entail reversible changes of the concentration of "defects" effectively available to 'carry' the strain, whereas the jump frequency of such defects does *not* vary steeply with temperature.

The concept of isoconfigurational creep and its implications are fully explained by TSAO and SPAEPEN [1982] and its relation to relaxation and anelastic behaviour by GREER and SPAEPEN [1981] and TAUB and SPAEPEN [1981]. TAUB and LUBORSKY [1981, 1982] have taken this investigation a step further to relate homogeneous creep to stress relaxation (under constant strain), a phenomenon of great practical importance in that magnetic characteristics, coercive field in particular, are closely linked to residual stress in wound toroids. The original papers must be consulted for details of the involved relations between the various linked phenomena. (See also a review by SPAEPEN and TAUB [1983].)

The origins of the thermal embrittlement have given rise to much controversy, outlined by SCOTT [1978a]. It is clear that phosphorus, the prime embrittling element in glassy as in crystalline steels, segregates to form small local clusters. This form of heterogeneity and alternatively decomposition into two glassy phases have been held responsible for embrittlement. PILLER and HAASEN's [1982] FIM experiments on the effect of heat-treatment on heterogeneity in $Fe_{80}B_{20}$ glass support this hypothesis. There is also now extensive evidence that quite small fractions (0.01–0.02) of crystalline phase in a heat-treated glass leads to rapid weakening and embrittlement (e.g., FREED and VANDER SANDE [1980]). A very recent hypothesis, backed by extensive observations by several techniques of investigation, is that in susceptible glasses a low-temperature anneal causes a very thin surface layer to become imperfectly crystallized, so that the glass in effect becomes surface-embrittled (JANOT *et al.* [1983]). ARGON [1982], on the basis of his comprehensive soap-bubble simulations, attributes thermal embrittlement to an alteration in the *distribution* of free volume, while GERLING *et al.* [1983], on the basis of detailed irradiation studies on $Fe_{40}Ni_{40}B_{20}$ glass, conclude that embrittlement is due to a critical degree of cluster formation or a critical reduction in free volume; either of these alone can induce embrittlement. The question of embrittlement mechanisms in general is still very much open.

An excellent and very complete review of the elastic properties of metallic glasses, with an extensive summary and comprehensive tables of mechanical and physical properties of many glasses has been published by LI [1981]. Temperature-invariant elastic moduli (the *Elinvar* effect), linked with temperature-dependent magnetostriction, have been observed in some ferromagnetic glasses (FUKAMICHI *et al.* [1978]).

Technological applications of the special mechanical properties of metallic glasses have not hitherto been conspicuous. However, possibilities of use of glass ribbons as reinforcement in composite materials and, by themselves, for the construction of large flywheels, have been reviewed by DAVIS *et al.* [1981].

3.7 Other properties of metallic glasses

Electrical transport in metallic glasses is beyond the scope of this chapter: the most recent surveys are by COTE and MEISEL [1981] and RAO [1983]. Superconduction in metallic glasses is treated in ch. 27, §4.1.3, while the crucially important ferromagnetic properties are treated in extenso in ch. 26, §5. Low-temperature physical properties are discussed in Güntherodt and Beck's book [see *Further reading* (G)]. Electronic properties (Fermi level, electronic specific heat, etc.) are comprehensively reviewed by MIZUTANI [1984]. Some metallic glasses, especially those rich in Cr and P, are very highly corrosion-resistant (MASUMOTO and HASHIMOTO [1978]), and certain glasses have also been shown to be highly selective catalysts in organic hydrogenation reactions (BROWER *et al.* [1983]), especially in the partly crystallized condition.

4. Rapid-solidification-processed crystalline alloys

4.1. General characteristics

The principal changes which can be brought about in crystalline alloys consequent on RSP include:

(1) Extension of solid-state solubility;

(2) refinement of grain size, with possible modification of grain shapes and textures;

(3) reduction or elimination of microsegregation;

(4) formation of metastable phases;

(5) achievement of high point-defect concentrations.

All of these have been surveyed in a general review by HONEYCOMBE [1978].

We have already seen, in §2, how thermodynamic and kinetic considerations, the variation of T_0 with composition in particular, affect solute trapping – that is, metastable extension of solid solubility. Aluminium alloys have been far more studied than any others, especially those with transition-metal solutes, and fig. 27. taken from MIDSON and JONES [1982], summarizes available results. It is plain that the solubility enhancement factor is in general much larger for solutes with a small equilibrium solubility. Midson and Jones analyze these results in terms of the morphological criterion and Jones has also sought to interpret the lattice parameters of some greatly supersaturated solid solutions in terms of rival physical models (JONES [1983]). An early but substantial collection of experimental data on extended solubility in many different solvents was assembled by ANANTHARAMAN and SURYANARAYANA [1971].

WOOD and HONEYCOMBE [1981] review the available information on extension of solubility in iron-base alloys. Their figures, together with some additional ones assembled by CANTOR [1982] are collected in table 4. In addition, C solubility in martensites is enhanced up to 2.0 wt% and in austenites up to 3.5 wt% before cementite is detected; but results are very variable and dependent on the measurement technique used. Boron, which is almost insoluble in equilibrium in pure α-iron (~ 50 ppm) was forcibly dissolved by RSP up to 1 wt%; as with aluminium alloys, it appears that the least soluble elements have the highest enhancement factor.

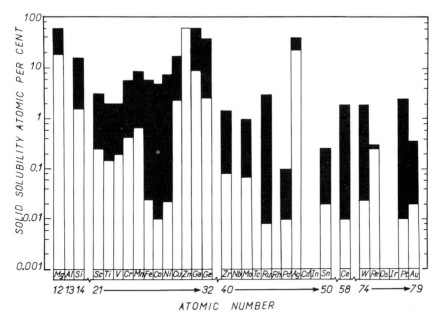

Fig. 27. Solubility extensions of various solutes in aluminium. The black bars indicate the increase beyond equilibrium room-temperature solubilities obtained by RSP. (After MIDSON and JONES [1982].)

Extended metastable solid solubility of course leads to greater volume fractions of pre-precipitation structures (GP zones) and consequently enhanced hardening. Much of the work on this has been purely empirical and there is a dearth of basic studies of pre-precipitation and precipitation mechanisms in RSP alloys. The rejection of solute from the matrix and the identity and density of zones and phases need to be studied and compared with those for solid-quenched alloys of the same type. The rejection of solute from supersaturated RSP Al–12 at% Cu alloy has been studied by

Table 4
Extensions of solubility for substitutional solutes in rapidly solidified binary iron-base alloys.

Solute	Max.equilibrium solubility (at%)	Extended solubility (at%)
Cu	7.2	15.0
Ga	47.0	50.0
Ge	20.0	25.0
Mo	26.0	40.6
Rh	50.0	100.0
Sn	9.8	20.0
Ti	9.8	16.0
W	13.0	20.8, 18.0

References: p. 1845.

KRISHNANAND and CAHN [1976] and the decomposition products of RSP Al–17 at% Cu by SCOTT and LEAKE [1975]. The behaviour of supersaturated RSP Al–8 wt% Fe alloys was examined by JACOBS et al. [1974], Al–Fe–Cr alloys by YEARIM and SHECHTMAN [1982], while a range of supersaturated Al–Zr alloys have recently been examined in detail by SAHIN and JONES [1978] and HORI et al. [1982].

It is important to emphasize that to obtain the full benefits of improved mechanical properties from RSP of alloys, it is necessary to develop special solute-rich compositions and not to restrict oneself to standard compositions used for normal casting.

Grain size is always refined by RSP, to the point where many commentators somewhat confusingly use the terms *microcrystalline* and *metastable* as alternatives in denoting the results of rapid quenching on crystalline materials. Grain sizes in the range 1–5 μm are quite common and the presence of solute may facilitate creation of grain sizes less than one micron. BOSWELL and CHADWICK [1977] have attempted to rationalize micrograin sizes in aluminium in terms of homogeneous nucleation theory. Consequences include "Petch-strengthening" at ambient temperature (DUFLOS and CANTOR [1978]) as well as an enhanced liability to superplasticity in RSP alloys (e.g., CHEESE and CANTOR [1980]), in the sense that superplasticity persists to higher strain rates. Quite generally, fine grains enhance hot-formeability and thus RSP alloys are easier to shape as well as often having better mechanical properties after forming and heat-treatment, than slowly solidified analogues.

Fine grains also have consequences for martensitic transformations in steels and for creep in superalloys; these are discussed below.

Grains are not necessarily equiaxed. Thus studies of piston-and-anvil quenched discs often show slender columnar grains growing normal to the chill-blocks, in the surface region of the solidified metal, or even right through the thickness (fig. 28a). LIEBERMANN et al. [1983] confirmed the prediction by COHEN et al. [1980] that, in RSP, hyperquenching is needed to produce columnar grains.

Only two studies have been reported of preferred orientations in as-cast structures produced by RSP. BLAKE et al. [1982] found remarkably perfect alignment of (0001) planes of Zn and Cd parallel to the surface of melt-spun ribbons; several other metals formed weaker textures of various types. The texture of high-purity (99.999%) zinc was found to deteriorate with time, and this was tentatively attributed to selective grain growth at ambient temperature. KAN et al. [1982] have melt-spun ribbons of highly alloyed Fe–Si (6.5 wt% Si compared with the usual 3%) and examined the preferred orientation. Grains with (100) [0kl] orientations grew preferentially to give a sharp planar texture (fig. 28b), leading to much reduced ac losses. The texture-sharpening is no doubt an instance of surface-controlled grain growth (ch. 25, §4.3.2).

The general technological implications of the combination of extended solubility, microcrystallinity and reduced segregation have been lucidly reviewed by GRANT [1978a]. Recent work has been summarized by CAHN [1982].

A particularly important practical objective is the long-standing requirement of substantially increasing the elastic stiffness of aluminium alloys for airframe applica-

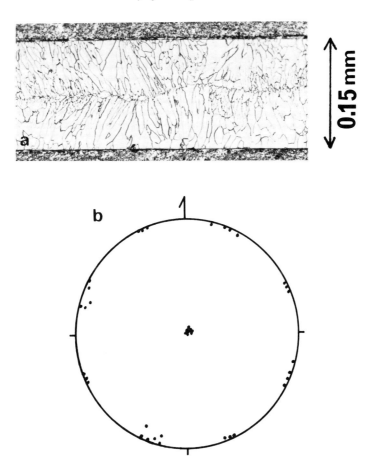

Fig. 28. Iron–6.5 wt% Si sheet made by twin-roller quenching: (a) cross-section, etched; (b) typical (200) pole figure of as-cast ribbon after annealing at 1100–1200°C. (The texture becomes close to an ideal (001) [100] texture if the ribbon is repeatedly rolled and annealed.) (After KAN *et al.* [1982].)

tions. Elastic moduli are not structure-sensitive and so this objective is best approached by introducing large amounts of dispersed intermetallic compounds, e.g. one based on Mn. When this was first attempted (DUDZINSKI *et al.* [1947–8]), the slow freezing and associated segregation led to coarse, uneven particle distributions and the alloys were hopelessly brittle. The quest has been renewed with alloys based on Al–Li (with further consituents); if cast in the ordinary way they have unsatisfactory toughness, but in RSP alloys the fine particle distribution and unsegregated, fine grain structure can lead to still modest, but acceptable toughness as well as specific stiffness enhanced by over 25% (e.g., SANKARAN and GRANT [1980]).

All such RSP aluminium alloys have to be prepared by atomization or splat-quenching followed by consolidation. The many techniques for consolidation and

References: p. 1845.

the literature of aluminium powder metallurgy generally are reviewed by PICKENS [1981]. He points out that RSP could with advantage be combined with other strengthening techniques such as mechanical alloying to introduce a dispersed oxide phase.

The formation of metastable phases was very extensively studied in the 1960s and early 1970s. The very varied results do not lend themselves to any comprehensive generalizations or rationalization; the most recent detailed review of the observed facts and related theories (SINHA et al. [1976]) was preceded by a full account of such phases by ANANTHARAMAN and SURYANARAYANA [1971], and the reader is referred to these.

Point-defect concentrations are enhanced in RSP metals and alloys. HONEYCOMBE [1978] explains the consequences of this for precipitation reactions in steels, while JONES [1977] sets out the evidence for vacancy enhancement in aluminium.

4.2. Microcrystalline alloys from glasses

RAY [1981, 1982] has pioneered the development of practically useful, high-strength crystalline alloys from metallic glasses, by a process akin to the formation of fine-grained pyrocerams (which are now a successful industrial category). His large range of alloys – based on Fe, Ni, Cr, Mo, Co and W in various combinations with metalloid contents not exceeding 12 at%, boron preponderantly – are brittle when glassy; the melt-spun glass tape (melt-spinning is preferred to atomization as giving uniformly cooled, even-sized particles) is brittle and easily powdered, and the powder is hot-extruded or hot isostatically pressed and simultaneously crystallized into a microcrystalline structure stabilized by minute boride particles. If carbon is present, the alloys are heat-treatable. Such alloys are very hard and wear-resistant and can be used like high-speed steels. There is a great deal of research to be done on the family of alloys made in this way, which might well, by analogy with "pyrocerams", be called *pyromets*. OHNAKA et al. [1982] have shown that spherical metallic glass powders of controllable, fairly homogeneous size distribution can be made by quenching a liquid alloy jet into a rotating water annulus, with sufficiently high jet velocity. The production and consolidation of metallic glass powders is reviewed by MILLER [1983].

4.3. Steels

The quite extensive work on microstructure of RSP steels has been fully reviewed by WOOD and HONEYCOMBE [1981]. Extension of solid solubilities was outlined in §4.1. The findings concerning the precipitation morphology of various carbides, both in simplified "model" steels and in commercial compositions, do not lend themselves to summary and the review should be consulted. CANTOR [1982], citing a good deal of unpublished work, analyzes in quantitative detail the conditions which lead to partitionless solidification of concentrated Fe–Ni alloys and compares the cooling rates and estimated undercoolings found necessary to achieve this with the predictions of the thermodynamic and kinetic theories outlined in §2.

The principal driving force for research in this field has been the desire to improve the characteristics of tool ("high speed") steels. Much of the research has centred on simplified model alloys such as Fe–W–C. The most striking finding (see for instance SARE and HONEYCOMBE [1978] and RAYMENT and CANTOR [1981]) is the fact that for the fastest quenches and for solute-rich steels, δ-iron is retained; it is only at intermediate quench rates that austenite is retained in such steels, and at conventional cooling rates, martensite is the major phase. The kinetics of hardening during heat-treatment are not much affected though the pattern of carbide precipitation is altered; the peak hardness is however raised for optimum treatments (RAYMENT and CANTOR [1978]). The recent practical developments based on RSP of tool steels are surveyed by JONES [1982b].

In scientific terms, one of the most striking sets of findings concerns the modification of the martensitic transformation by RSP. INOKUTI and CANTOR [1976] found a variation of microhardness values for a series of RSP Fe–Ni alloys, as shown in fig. 29a; values for the solid-quenched alloys are shown for comparison. This was traced to a very pronounced change in martensite start (M_s) temperatures on cooling (fig. 29b.) This curve, and also the $\alpha \rightarrow \gamma$ start temperatures on reheating, were obtained calorimetrically: the martensitic alloy was heated slowly in a DSC until it had all transformed back to γ, and then slowly cooled again; M_s ($\gamma \rightarrow \alpha$) was marked by a pronounced exothermic peak, $\alpha \rightarrow \gamma$ by an endotherm (INOKUTI and CANTOR [1982]). The striking feature was that the depressed M_s caused by the original rapid quench from the melt does *not* disappear on repeated cycling between α and γ; only a prolonged γ anneal restores the alloy to normal behaviour (labelled "solid quench" in fig. 29b). It is therefore not the rapid cooling per se which reduces the M_s, so it can only be the austenite structure resulting from that quench. In fact, various earlier studies not involving RSP have shown a relationship between austenite grain size, d, and M_s (e.g., for Fe–32% Ni, $\Delta M_s = 40$–100 K for d reduced from 200 to 10 μm). In the RSP alloys, d = 4–5 μm and this leads to ΔM_s up to 200 K for Fe–17% Ni. Thermal cycling, via the reverse shear transformation, restores the fine austenite grains, and only a long γ anneal gives the γ grains a chance to grow substantially. Such austenite grain growth seems often to be particularly sluggish in RSP steels (SUGA *et al.* [1980]); a recent study suggests that stable, extremely fine sulphide dispersions at austenite grain boundaries in RSP steels are responsible for such retardation of grain growth. (KELLY and VANDER SANDE [1980]).

The work cited in the preceding paragraph leaves it uncertain whether the M_s depression is wholly due to austenite grain size refinement or whether dislocations generated in the austenite by piston impact during RSP also play a part. Recently, HAYZELDEN [1983] has succeeded in separating the effects of austenite grain size and dislocation density and was able to prove convincingly that M_s depression in a Fe–Ni–C steel is due to austenite grain refinement alone; in fact an increase in dislocation density alone was found to *raise* M_s.

Figure 29a shows that pure iron can also form martensite. The structure, formation conditions and crystallography of this martensite were studied by DUFLOS and CANTOR [1982]; hardness goes up with martensite/ferrite ratio, and for 100%

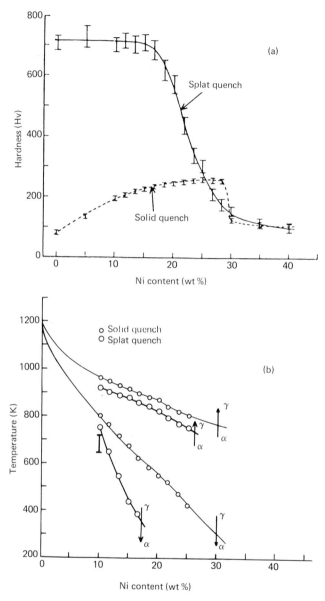

Fig. 29. (a) Microhardness values for liquid-quenched and solid-quenched Fe–Ni alloys. (b) γ → α (M$_s$) and α → γ transformation temperatures for a range of Fe–Ni alloys. (Both after INOKUTI and CANTOR [1982].)

martensite, the microhardness of pure iron is 600–700! A Hall–Petch plot (ch. 19, §8) for pure iron is shown in fig. 29c; the "grain size" for martensitic iron was taken as the width of the martensite laths, ~ 0.35 μm. It can be seen that the remarkably

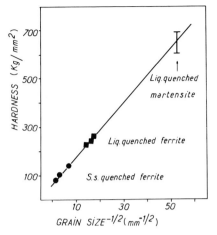

Fig. 29c. Microhardness versus (grain size)$^{-1/2}$ for pure iron (after DUFLOS and CANTOR [1978]).

high hardness of martensitic iron can be largely, or even wholly, attributed to "grain-size" strengthening.

Depressions of the M_s temperature connected with a reduced β grain size, and other modifications of transformation behaviour, were found in a melt-spun Cu–Zn–Al memory alloy (PERKINS *et al.* [1981]).

A recent study by MAWELLA *et al.* [1982] of a melt-spun ultrahigh-strength alloy steel indicated that the considerable strength increase generated by melt-spinning was primarily due to refinement of the martensite structure. It has also been established that high-carbon cast iron (2.4–3 wt% C) on RSP by centrifugal atomization consists largely of retained austenite. Subsequent annealing produces a microcrystalline (α + carbide) structure. Even at 650°C this microcrystalline structure is so stable that the material behaves superplastically (EISELSTEIN *et al.* [1983]). The conversion of brittle cast iron into a superplastic material is a particularly striking indicator of the changes in metallurgical characteristics obtainable by RSP.

4.4. Superalloys

Apart from tool steels and aluminium alloys, the other principal category of alloys to which RSP has been extensively applied is that of the superalloys, both those used for turbine discs and those destined for blades. The two Reston Conferences cited under *Further reading* at the end of this chapter are the principal source of information up to now on these studies; some other publications are by TRACEY and CUTLER [1981] and by SMUGERESKY [1982]. JONES [1982b] also provides a lucid summary.

Most investigators up to now have used atomized powders, especially those made by centrifugal atomization in helium. However, DUFLOS and STOHR [1982] have

References: p. 1845.

demonstrated that melt-spinning can provide higher quench rates and thus comminuted melt-spun ribbon provides an alternative material for subsequent hot consolidation, most commonly done by hot isostatic pressing. The benefits of RSP include the possibility of forcing more solute into the γ and γ′ phases, producing alloys fine-grained and homogeneous enough to permit easy hot-working, even superplastic forming in extreme cases, and uniformity of structure throughout a large piece. The microstructural implications of a change from gas atomization to melt-spinning of superalloys have been examined in some detail by LIEBERMANN *et al.* [1983]. SHOHOJI *et al.* [1982] have also examined the structure of melt-spun superalloys and estimated cooling rates.

INOUE *et al.* [1983b] have melt-quenched a number of superalloys consisting of γ′ only. They found that these RSP alloys, instead of being almost perfectly ordered with large antiphase domains, were imperfectly ordered with APDs well below 0.1 μm in diameter, and were free of grain-boundary segregation. These structural modifications lead, in several alloys, to greatly enhanced ductility and some increase in strength.

There is, however, one major problem. The fine grain sizes produced by RSP, while they facilitate forming, leave the alloy very liable to creep at high temperatures but low stresses. This can be overcome by a process of *directional recrystallization* of the RSP alloy after it has been consolidated and formed: the process consists of a slow passage through a steep thermal gradient, to form large elongated grains rather like those generated by directional solidification. However, as SMUGERESKY [1982] indicates, the prolonged period spent at high temperatures during consolidation, forming and directional recrystallization may damage the desirable phase structure generated by RSP, and this problem will require further attention.

Concluding note

The breakneck speed at which the field of rapid solidification processing has been advancing during the past four years has made it essential to provide a substantial number of citations of recent work, since in the space here available no extensive experimental details or theoretical proofs can be set out. The examples given to illustrate a variety of general principles need to be complemented by reading recent publications. The nature of this field of research is best appreciated when it is noted that more than half the papers cited were published in the period 1981–1983.

I should like to express my thanks to the many colleagues who sent copies of papers in advance of publication; without their help this chapter could not have been written.

References

To save space, references to articles in the surveys and conference proceedings cited under *Further reading* at the end of the chapter are denoted by appropriate capital letters.

AL BIJAT, S., R. IRALDI, J.M. DUBOIS, G. LE CAER and C. TETE, 1982, in: **T**, p. 375.

ANAND, V., A.J. KAUFMAN and N.J. GRANT, 1980, in: **P**, p. 273.

ANANTHARAMAN, T.R., and C. SURYANARAYANA, 1971, J. Mater. Sci. **6**, 1111.

ARGON, A.S., 1979, Acta Metall. **27**, 47.

ARGON, A.S., 1982, J. Phys. Chem. Solids **43**, 945.

ARGON, A.S., and H.Y. KUO, 1978, in: **M**, vol. 2, p. 269.

ARMSTRONG, G.R., and H. JONES, 1979, in: Solidification and Casting of Metals (The Metals Soceity, London) p. 456.

BABIĆ, E., E. GIRT, R. KSNIK, B. LEONTIĆ and I. ZORIĆ, 1970, in: **K**, paper 2.

BAER, D., L.R. PEDERSEN and M.T. THOMAS, 1981, Mater. Sci. Eng. **48**, 283.

BAKER, J.C., and J.W. CAHN, 1971, in: Solidification (ASM, Metals Park, OH) p. 23.

BALANZAT, M., 1980, Scripta Metall. **14**, 1245.

BALANZAT, M., and S. HILLAIRET, 1982, J. Phys. **F12**, 2907.

BALANZAT, M., C. MAIRY and S. HILLAIRET, 1980, in: **Q**, p. 871.

BEGHI, G., R. MATERA and G. PIATTI, 1969, J. Nucl. Mater. **31**, 259.

BERGMANN, H.W., and B.L. MORDIKE, 1980, Z. Metallk. **71**, 658.

BERGMANN, H.W., and B.L. MORDIKE, 1982, in: **V**, p. 467 and 497.

BERNAL, J.D., 1964, Proc. Roy. Soc. **A284**, 299.

BERRY, B.S., 1978, in: **A**, p. 161.

BERRY, B.S., and W.C. PRITCHET, 1981, Phys. Rev. **B24**, 2299; Scripta Metall. **15**, 632.

BILONI, H., and B. CHALMERS, 1965, Trans. Met. Soc. AIME **233**, 373.

BLAKE, N.W., F.A. HAMES and R.W. SMITH, 1982, in: **V**, p. 363.

BOETTINGER, W.J., 1982a, in: **V**, p. 15.

BOETTINGER, W.J., 1982b, in: **T**, p. 99.

BORDERS, J.A., 1979, Ann. Rev. Mater. Sci. **9**, 313.

BOSWELL, P.G., and G.A. CHADWICK, 1977, Scripta Metall. **11**, 459.

BOUDREAUX, D.S., and H.J. FROST, 1981, Phys. Rev. **B23**, 1506.

BOUDREAUX, D.S., and J.M. GREGOR, 1977, J. Appl. Phys. **48**, 152 and 5057.

BOURGON, J.C., 1979, in: **O**, p. 351.

BOWDEN, P.B., and S. RAHA, 1970, Phil. Mag. **22**, 463.

BOWMAN, JR., R.C., and A.J. MAELAND, 1981, Phys. Rev. **B24**, 2328.

BORD, I.W., and J.I.B. WILSON, 1983, Nature (London) **303**, 481.

BREINAN, E.M., and B.H. KEAR, 1978, in: **N**, p. 87.

BRESSON, L., J.P. CHEVALIER and M. FAYARD, 1982, Scripta Metall. **16**, 499.

BROWER, JR., W.E., M.S. MAZYJASZCZYK, T.L. PETTIT and G.V. SMITH, 1983, Nature (London) **301**, 497.

BUCKEL, W., and R. HILSCH, 1956, Z. Phys. **146**, 27.

BUSCHOW, K.H.J., 1982, Solid State Commun. **43**, 171.

BUSCHOW, K.H.J., 1983, J. Phys. **F13**, 563.

CAHN, R.W., 1978, in: **N**, p. 129.

CAHN, R.W., 1979, in: **O**, p. 137.

CAHN, R.W., 1982, Ann. Rev. Mater. Sci. **12**, 51.

CAHN, R.W., 1983, in: **W**, p. 55.

CAHN, R.W., K.D. KRISHNANAND, M. LARIDJANI, M. GREENHOLZ and R. HILL, 1976, Mater. Sci. Eng. **23**, 83.

CAHN, R.W., J.E. EVETTS, J. PATTERSON, R.E. SOMEKH and C. KENWAY-JACKSON, 1980, J. Mater. Sci. **15**, 702.

CANTOR, B., 1982, in: **V**, p. 317.

CANTOR, B., and R.W. CAHN, 1983, in: **J**, p. 487.

CANTOR, B., and P. RAMACHANDRARAO, 1982, in: **T**, p. 291.

CARGILL, G.S., 1975, in: Solid State Physics, eds. H. Ehrenreich *et al.* (Academic, New York) p. 227.

CARGILL, G.S., 1979, private communication.

CARTER, G., and W.A. GRANT, 1982, Nucl. Instr. and Meth. **199**, 17.

CHAMBRON, W., and A. CHAMBEROD, 1980, Solid State Commun. **35**, 61.

CHAMBRON, W., A.F. LANÇON and A. CHAMBEROD, 1982, J. Physique Lettres **43**, L–55.

CHATTOPADYAY, K., and P. RAMACHANDRARAO, 1980, J. Mater. Sci. **15**, 685.

CHEESE, R., and B. CANTOR, 1980, Mater. Sci. Eng. **45**, 83.

CHEN, H.S., 1981, J. Non-Cryst. Solids **46**, 289.

CHEN, H.S., 1983, in: **J**, p. 169.

CHEN, H.S., and C.E. MILLER, 1970, Rev. Sci. Instr. **41**, 1237.

CHEVALIER, J.P., V. CALVAYRAC, A. QUIVY, M. HARMELIN and J. BIGOT, 1983, Acta Metall. **31**, 465.

CHRISTENSEN, C.P., 1982, Science **218**, 113.

CLYNE, T.W., 1984, Metallurg. Trans. **A**, in press.

COHEN, M., B.H. KEAR and R. MEHRABIAN, 1980, in: **P**, p. 1.

COHEN, M.H. and G.S. GREST, 1979, Phys. Rev. **B20**, 1077.

COHEN, M.H. and G.S. GREST, 1980, Phys. Rev. **B21**, 4113.

COHEN, M.H., and G.S. GREST, 1981, in: Structure and Mobility in Molecular and Atomic Glasses, Ann. N.Y. Acad. Sci. **371**, 199.

COTE, P.J., and L.V. MEISEL, 1981, in: **G**, p. 144.

DAHLGREN, S.D., 1978, in: **M**, vol. 2, p. 36.

DAVIES, H.A., 1976, Phys. Chem. Glasses **17**, 159.

DAVIES, H.A., 1978, in: **M**, vol. 1, p. 1.

DAVIES, H.A., B.G. LEWIS and I.W. DONALD, 1978, in: **N**, p. 78.

DAVIS, L.A., 1978, in: **A**, p. 196.

DAVIS, L.A., N.J. DECRISTOFARO and C.H. SMITH, 1981, in: **S**, vol. 1, p.1.

DIETZ, G., 1977, J. Magn. Magn. Mater. **6**, 47.

DINI, K., N. COWLAM and H.A. DAVIES, 1982, J. Phys. **F12**, 1557.

DIXMIER, J., K. DOI and A. GUINIER, 1965, in: Physics of Non-Crystalline Solids (North-Holland, Amsterdam) p. 67.

DONG, Y.D., G. GREGAN and M.G. SCOTT, 1981, J. Non-Cryst. Solids **43**, 463.

DONG, Y.D., G. GREGAN and M.G. SCOTT, 1982a, in: **T**, p. 671.

DONG, X.Z., W. FERNENGEL and H. KRONMÜLLER, 1982b, Appl. Phys. **A28**, 103.

DONOVAN, P.E., and W.M. STOBBS, 1981, Acta Metall. **29**, 1419.

DRAPER, C.W., 1982, J. Metals (AIME) **34** (June) 24.

DREHMAN, A.J., and D. TURNBULL, 1981, Scripta Metall. **15**, 543.

DREHMAN, A.J., A.L. GREER and D. TURNBULL, 1982, Appl. Phys. Lett. **41**, 718.

DRIJVER, E., L. MULDER and S. RADELAAR, 1982, in: **T**, p. 535.

DUBOIS, J.M., and G. LE CAER, 1982, Nucl. Instr. and Meth. **199**, 307.

DUBOIS, J.M., and G. LE CAER, 1983, in: **W**, p. 67.

DUBOIS, J.M., M. BASTICK, G. LE CAER and C. TETE, 1980, Rev. Phys. Appl. **15**, 1103.

DUDZINSKI, N., J.R. MURRAY, B.W. MOTT and B. CHALMERS, 1947–8, J. Inst. Metals **74**, 291.

DUFLOS, B., and B. CANTOR, 1978, in: **M**, vol. 1, p. 110.

DUFLOS, B., and B. CANTOR, 1982, Acta Metall. **30**, 323.

DUFLOS, B., and J.F. STOHR, 1982, J. Mater. Sci. **17**, 3641.

DURAND, J., 1981, in: The Magnetic, Chemical and Structural Properties of Glassy Metallic Alloys, ed. R. Hasegawa (CRC Press, Boca Raton, FL).

DUWEZ, P., 1967, Trans. ASM **60**, 607; Prog. Solid-State Chem. **3**, 377.

DUWEZ, P., R.H. WILLENS and W. KLEMENT, 1960, Nature (London) **187**, 869; J. Appl. Phys. **31**, 1136 and 1500.

EGAMI, T., 1978a, Mater. Res. Bull. **13**, 575.

EGAMI, T., 1978b, J. Mater. Sci. **13**, 2587.

EGAMI, T., 1981, in: **G**, p. 25.

EGAMI, T., 1983, in: **J**, p. 100.

EGAMI, T., K. MAEDA and V. VITEK, 1980, Phil. Mag. **A41**, 883.

EGAMI, T., V. VITEK and D. SROLOVITZ, 1982, in: **T**, p. 517.

EHRLICH, D.J., R.M. OSGOOD, JR. and D.F. DEUTSCH, 1981, Appl. Phys. Lett. **38**, 399.

EISELSTEIN, C.E., O.A. RUANO and O.D. SHERBY, 1983, J. Mater. Sci. **18**, 483.

ELLIOTT, R.O., and D.A. KOSS, 1981, J. Nucl. Mater. **97**, 339.

ELLIOTT, R.O., D.A. KOSS and B.C. GIESSEN, 1980, Scripta Metall. **14**, 1061.

ELLIOTT, W.A., F.P. GAGLIANO and G. KRAUSS, 1973, Metallurg. Trans. **4A**, 2031.

FERNENGEL, W., H. KRONMÜLLER, M. RAPP and Y. HE, 1982, Appl. Phys. **A28**, 137.

FINNEY, J.L., 1983, in: **J**. p. 42.

FINNEY, J.L., and J. WALLACE, 1981, J. Non-Cryst. Solids **43**, 165.

FLANK, A.M., P. LAGARDE, D. RAOUX, J. RIVORY and A. SADOC, 1982, in: **T**, p. 393.

FOX, T.G., and P.J. FLORY, 1951, J. Appl. Phys. **21**, 581.

FREED, R.L., and J.B. VANDER SANDE, 1980, Acta Metall. **28**, 103.

FROST, H.J., 1982, Acta Metall. **30**, 889.

FUKAMICHI, K., M. KIKUCHI, S. ARAKAWA and T. MASUMOTO, 1978, Solid State Commun. **23**, 955.

GAGNÉ, H., and C. ROY, 1982, High Temp. Technol. **1**, 93.

GASKELL, P.H., 1979a, J. Phys. **C12**, 4337.

GASKELL, P.H., 1979b, J. Non-Cryst. Solids **32**, 207.

GASKELL, P.H., 1982, Nucl. Instr. and Meth. **199**, 45.

GERLING, R., F.P. SCHIMANSKY and R. WAGNER, 1983, Scripta Metall. **17**, 203.

GIBBS, M.R.J., J.E. EVETTS and J.A. LEAKE, 1983, J. Mater. Sci. **18**, 278.

GLICKSTEIN, M.R., R.J. PATTERSON II and N.E. SHOCKEY, 1978, in: **N**, p. 46.

GRANT, N.J., 1970, in: **K**, paper 16.

GRANT, N.J., 1978a, in: **N**, p. 230.

GRANT, N.J., 1978b, in: **M**, vol. 2, p. 172.

GREER, A.L., 1980, Thermochim. Acta **42**, 193.

GREER, A.L., 1982a, J. Mater. Sci. **17**, 1117.

GREER, A.L., 1982b, Acta Metall. **30**, 171.

GREER, A.L., and F. SPAEPEN, 1981, in: Structure and Mobility in Molecular and Atomic Glasses, Ann. N.Y. Acad. Sci. **371**, 218.

GREER, A.L., C.J. LIN and F. SPAEPEN, 1982, in: **T**, p. 567.

GURMAN, S.J., 1982, J. Mater. Sci. **17**, 1541.

HAFNER, J., 1981, in: **G**, p. 93.

HARBUR, D.R., J.W. ANDERSON and W.J. MARAMAN, 1969, Trans. Met. Soc. AIME **245**, 1055.

HARMELIN, M., Y. CALVAYRAC, A. QUIVY, J.P. CHEVALIER and J. BIGOT, 1982, Scripta Metall. **16**, 707.

HAYZELDEN, C., 1983, D. Phil. thesis, University of Sussex.

HEROLD, U., 1982, Doctoral thesis, Ruhruniversität, Bochum.

HETTWER, K.J., and F. HAESSNER 1982, Mater. Sci. Eng. **52**, 147.

HILLERT, M., 1982, in: **V**. p. 3.

HILLMAN, H., and H.R. HILZINGER, 1978, in: **M**, vol. 1, p. 22.

HIRATA, T., 1979, Scripta Metall. **13**, 555.

HONEYCOMBE, R.W.K., 1978, in: **M**, vol. 1, p. 73.

HOPKINS, S.T., and W.L. JOHNSON, 1982, Solid State Commun. **43**, 537.

HORI, S., S. SAJI and A.TAKEHARA, 1982, in: **T**, p. 1545.

HUANG, S.C., and H.C. FIEDLER, 1981, Mater. Sci. Eng. **51**, 39.

INOKUTI, Y., and B. CANTOR, 1976, Scripta Metall. **10**, 655.

INOKUTI, Y., and B. CANTOR, 1982, Acta Metall. **30**, 343.

INOUE, A., M. HAGIWARA and T. MASUMOTO, 1982, J. Mater. Sci. **17**, 580.

INOUE, A., H.S. CHEN, T. MASUMOTO and M. HAGIWARA, 1983a, Scripta Metall. **17**, in press.

INOUE, A., H. TOMIOKA and T. MASUMOTO, 1983b, Metallurg. Trans. **14A**, in press.

ISHIHARA, K.N., and P.H. SHINGU, 1982, Scripta Metall. **16**, 837.

JACOBS, M.H., A.G. DOGGETT and M.J. STOWELL, 1974, J. Mater. Sci. **9**, 1631.

JANOT, C., B. GEORGE, D. TEIRLINCK, G. MARCHAL, C. TETE and P. DELCROIX, 1983, Phil. Mag. **47**, 301.

JOHNSON, W.L., 1982, in: U, p. 183.

JONES, H., 1977, in: Vacancies 1976, eds. R.E. Smallman and J.E. Harris (The Metals Society, London) p. 175.

JONES, H., 1981, in: E, p. 1.

JONES, H., 1982a, in: T, p. 71.

JONES, H., 1982b, Rapid Solidification of Metals and Alloys (Institution of Metallurgists, London).

JONES, H., 1983, Scripta Metall. **17**, 97.

KAN, T., Y. ITO and H. SHIMANAKA, 1982, J. Magn. Magn. Mater. **26**, 127.

KATTAMIS, T.Z., W.E. BROWER and R. MEHRABIAN, 1973, J. Cryst. Growth **19**, 229.

KAVESH, S., 1978, in: A, p. 36.

KAWAMURA, H., 1979, Prog. Theor. Phys. **61**, 1584.

KEAR, B.H., J.N. MAYER, J.M. POATE and P.R. STRUTT, 1981, in: Metallurgical Treatises, eds. J.K. Tien and J.F. Elliott (TMS–AIME, Warrendale, PA) p. 321.

KELLY, T.F., and J.B. VANDER SANDE, 1980, in: P, p. 100.

KHAN, Y., E. KNELLER and M. SOSTARICH, 1981, Z. Metallk. **72**, 553.

KHAN, Y., E. KNELLER and M. SOSTARICH, 1982, Z. Metallk. **73**, 624.

KIJEK, M., M. AHMADZADEH and B. CANTOR, 1981, in: S, vol. 2, p. 397.

KIMURA, H., and T. MASUMOTO, 1983, in: J, p. 187.

KIRCHHEIM, R., 1982, Acta Metall. **30**, 1069.

KIRCHHEIM, R., F. SOMMER and G. SCHLUCKBIER, 1982, Acta Metall. **30**, 1059.

KÖSTER, U., and U. HEROLD, 1981, in: G, p. 225.

KRAMER, J., 1934, Ann. Phys. **19**, 37.

KRAMER, J., 1937, Z. Phys. **106**, 675.

KRENITSKY, D.J., and D.J. AST, 1979, J. Mater. Sci. **14**, 275.

KRISHNANAND, K.D., and R.W. CAHN, 1976, in: L, part 1, p. 67.

KRISHTAL, YU. A., 1960, Izv. Vysshikh Uchernykh Zaved. (Chern. Met.), no. 3, 110.

KURŠUMOVIĆ, A., M.G. SCOTT, E. GIRT and R.W. CAHN, 1980, Scripta Metall. **14**, 1303.

LARIDJANI, M., P. RAMACHANDRARAO and R.W. CAHN, 1972, J. Mater. Sci. **7**, 627.

LEBO, M., and N.J. GRANT, 1974, Metallurg. Trans. **5**, 1547.

LEVI, C.G., and R. MEHRABIAN, 1982, Metallurg. Trans. **13A**, 13.

LEWIS, B.G., and P.R. STRUTT, 1982, J. Metals (AIME) **34** (Nov.) 37.

LI, J.C.M., 1981, in: E, p. 325.

LI, J.C.M., 1982, in: V, p. 267.

LIEBERMANN, H.H., 1978, in: M, vol. 1, p. 34.

LIEBERMANN, H.H., 1983, in: J. p. 26.

LIEBERMANN, H.H., and J.L. WALTER, 1982, in: V, p. 111.

LIEBERMANN, H.H., R.E. MAXWELL, R.W. SMASHEY and J.L. WALTER, 1983, Metallurg. Trans. **14A**.

LIMOGE, Y., G. BRÉBEC and Y. ADDA, 1983, in: DIMETA 82, Proc. Conf. on Diffusion in Metals and Alloys, Tihany, Hungary, Sept. 1982 (Hungarian Academy of Sciences) in press.

LIN, C.-J., and F. SPAEPEN, 1982, Appl. Phys. Lett. **41**, 721.

LIU, B.X., L.S. WIALUNSKI, M. MAENPAA, M.A. NICOLET and S.S. LAM, 1982, in: U, 133.

LIVESEY, A.K., and P.H. GASKELL, 1982, in: T, p. 335.

LUBORSKY, F.E., and J.L. WALTER, 1976, J. Appl. Phys. **47**, 3648.

LUBORSKY, F.E., and J.L. WALTER, 1977, IEEE Trans. Magn. **MAG-13**, 953 and 1635.

LUBORSKY, F.E., J.J. BECKER, J.L. WALTER and H.H. LIEBERMANN, 1979, IEEE Trans. Magn. **MAG-15**, 1146.

MACFARLANE, D.R., 1982, J. Non-Cryst. Solids **53**, 61.

MADER, S., 1965, J. Vac. Sci. Technol. **2**, 35.

MARINGER, R.E., and C.E. MOBLEY, 1974, J. Vac. Sci. Technol. **11**, 1067.

MASSALSKI, T.B., 1982, in: **T**, p. 203.

MASUMOTO, T., and K. HASHIMOTO, 1978, Ann. Rev. Mater. Sci. **8**, 215; in: **M**, vol. 2, p. 435.

MAWELLA, K.J.A., R.W.K. HONEYCOMBE and P.R. HOWELL, 1982, J. Mater. Sci. **17**, 2850.

MEGUSAR, J., A.S. ARGON and N.J. GRANT, 1982, in: **V**, p. 283.

MEHRA, M., A.R. WILLIAMS and W.L. JOHNSON, 1983, Phys. Rev. **B28**, in press.

MEHRABIAN, R., 1982, Int. Metallurg. Rev. **27**, 185.

MENDOZA-ZELIS, L., L. THOME, L. BROSSARD, J. CHAUMONT, Q. KROLAS and H. BERNAS, 1982, Phys. Rev. **B26**, 1306.

MIDSON, S.P., and H. JONES, 1982, in: **T**, p. 1539.

MILLER, S.A., 1983, in: **J**, p. 506.

MIYAZAWA, K., and J. SZEKELY, 1981, Metallurg. Trans. **12A**, 1047.

MIZUTANI, U., 1984, Prog. Mater. Sci. **28**, in press.

MORI, H., and H. FUJITA, 1982, Jap. J. Appl. Phys. **21**, L484.

MORI, H., H. FUJITA and M. FUJITA, 1983, Jap. J. Appl. Phys. **22**, L94.

MORRIS, D.G., 1981, in: **P**, p. 372.

MORRIS, D.G., 1982, Scripta Metall. **16**, 585.

MOTTERN, J.W., and W.J. PRIVOTT, eds., 1978, Spinning Wire from Molten Metal (Amer. Inst. Chem. Engrs., New York) A.I.Ch.E. Symp. Series **74**, no. 180.

MUKHERJEE, K., and J. MAZUMDER, eds., 1981, Lasers in Metallurgy (TMS–AIME, Warrendale, PA).

MURRAY, J.L., 1983, Metallurg. Trans. **14A**, in press.

MURTY, Y.V., and R.P.I. ADLER, 1982, J. Mater. Sci. **17**, 1945.

NAGEL, S.R., and J. TAUC, 1977, Solid State Commun. **22**, 129.

NEUHÄUSER, H., 1978, in: **M**, vol. 2, 431.

NOLD, E., S. STEEB, P. LAMPARTER and G. RAINER-HARBACH, 1980, in: **Q**, p. C8-186.

NUNOGAKI, K., Y. KABAO and M. KIRITANI, 1982, in: **T**, p. 509.

ÖHRING, M., 1983, Thesis, Universität Göttingen.

OHNAKA, I., and T. FUKUSAKO, 1978, J. Japan Inst. Metals **42**, 415.

OHNAKA, I., T. FUKUSAKO and T. OHMICHI, 1981, J. Japan Inst. Metals **45**, 751.

OHNAKA, I., T. FUKUSAKO and H. TSUTSUMI, 1982, J. Japan. Inst. Metals **46**, 1095.

PAMPILLO, C.A., 1975, J. Mater. Sci. **10**, 1194.

PATTERSON, J., A.L. GREER, J.A. LEAKE and D.R.W. JONES, 1978, in: **M**, vol. 2, p. 293.

PAVUNA, D., 1982, J. Mater. Sci. **16**, 2419.

PEREL, J., J.F. MAHONEY, B.E. KALENSHER, K.E. VICKERS and R. MEHRABIAN, 1978, in: **N**, p. 258.

PEREPEZKO, J.H., and H. GLEITER, 1983, to be published.

PEREPEZKO, J.H., and J.S. PAIK, 1982, in: **V**, p. 49.

PEREPEZKO, J.H., and J.S. SMITH, 1981, J. Non-Cryst. Solids **44**, 65.

PERKINS, J., J.J. RAYMENT, B. CANTOR and R.W. CAHN, 1981, Scripta Met. **15**, 771.

PICKENS, J.R., 1981, J. Mater. Sci. **16**, 1437.

PILLER, J., and P. HAASEN, 1982, Acta Metall. **30**, 1.

POATE, J.M., and W.L. BROWN, 1982, Physics Today, June, p. 24.

POATE, J.M., K.N. TU and J.N. MAYER, eds., 1978, Thin Films – Interdiffusion and Reactions (Wiley, New York).

POLK, D., 1972, Acta Metall. **20**, 485.

POLK, D., and D. TURNBULL, 1972, Acta Metall. **20**, 493.

POND, R., and R. MADDIN, 1969, Mater. Sci. Eng. **23**, 87.

PRATTEN, N.A., 1980, D. Phil. thesis, University of Sussex.

RAMACHANDRARAO, P., 1980, Z. Metallk. **71**, 177.

RAMACHANDRARAO, P., B. CANTOR and R.W. CAHN, 1977a, J. Non-Cryst. Solids **24**, 109.

RAMACHANDRARAO, P., B. CANTOR and R.W. CAHN, 1977b, J. Mater. Sci. **12**, 2488.

RAO, K.V., 1983, in: **J**, p. 401.

RAUSCHENBACH, B., and K. HOHMUTH, 1982, Phys. Stat. Sol. (a) **72**, 667.

RAY, R., 1981, J. Mater. Sci. **16**, 2924 and 2927.

RAY, R., 1982, in: **V**, p. 435.

RAY, R., R. HASEGAWA, C.P. CHOU and L.A. DAVIS, 1977, Scripta Metall. **11**, 973.

RAYMENT, R.R., and B. CANTOR, 1978, Met. Sci. **12**, 156.

RAYMENT, R.R., and B. CANTOR, 1981, Metallurg. Trans. **12A**, 1557.

RECHTIN, M.D., J.B. VANDER SANDE and P.M. BALDO, 1978, Scripta Metall. **12**, 639.

RUPPERSBERG, H., D. LEE and C.N.J. WAGNER, 1980, J. Phys. **F10**, 1645.

SADOC, A., A.M. FLANK, D. RAOUX and P. LAGARDE, 1983, in: **W**, p. 43.

SAFAI, S., and H. HERMAN, 1981, in: **E**, p. 183.

SAHIN, E., and H. JONES, 1978, in: **M**, vol. 1, p. 138.

SAKATA, M., N. COWLAM and H.A. DAVIES, 1982, in: **T**, p. 327.

SANKARAN, K.K., and N.J. GRANT, 1980, Mater. Sci. Eng. **44**, 213.

SARE, I.R., and R.W.K. HONEYCOMBE, 1978, J. Mater. Sci. **13**, 1991.

SCHWARZ, R.B., and W.L. JOHNSON, 1983, Phys. Rev. Lett. **51**, in press.

SCOTT, M.G., 1978a, in: **M**, vol. 1, p. 198.

SCOTT, M.G., 1978b, J. Mater. Sci. **13**, 291.

SCOTT, M.G., 1983, in: **J**, p. 144.

SCOTT, M.G., and A. KURŠUMOVIĆ, 1982, Acta Metall. **30**, 853.

SCOTT, M.G., and J.A. LEAKE, 1975, Acta Metall. **23**, 503.

SEDGWICK, T.D., 1983, J. Electrochem. Soc. **130**, 484.

SHIMOMURA, K., P.H. SHINGU and R. OZAKI, 1980, J. Mater. Sci. **15**, 1175.

SHINGU, P.H., 1979, Trans. Japan. Inst. Metals **20**, 33.

SHINGU, P.H., K. SHIMOMURA, K. KOBAYASHI and R. OZAKI, 1976, Mater. Sci. Eng. **23**, 183.

SHOHOJI, N., H.A. DAVIES, H. JONES and D.H. WARRINGTON, 1982, in: **T**, p. 1529.

SIMPSON, A.W., and P.A. HODKINSON, 1972, Nature (London) **237**, 320.

SINHA, A.K., B.C. GIESSEN and D.E. POLK, 1976, in: Treatise on Solid-State Chemistry, ed. N.B. Hannay (Plenum, New York) vol. 3, p. 1.

SINNING, H.R., and R.W. CAHN, 1983, submitted to Z. Metallk.

SMUGERESKY, J.E., 1982, Metallurg. Trans. **13A**, 1535.

SOMMER, F., 1982, in: **T**, p. 209.

SPAEPEN, F., 1977, Acta Metall. **25**, 407.

SPAEPEN, F., 1978, in: **M**, vol. 2, p. 253.

SPAEPEN, F., and A.I. TAUB, 1983, in: **J**, p. 231.

SPAEPEN, F., and D. TURNBULL, 1974, Scripta Metall. **8**, 563.

SUGA, M., J.L. GOSS, G.R. OLSEN and J.B. VANDER SANDE, 1980, in: **P**, p. 364.

SUZUKI, R., K. SHIBUE, K. OSAMURA, P.H. SHINGU and Y. MURAKAMI, 1982, J. Mater. Sci. Lett. **1**, 127.

SUZUKI, R.O., Y. KOMATSU, K. KOBAYASHI and P.H. SHINGU, 1983, J. Mater. Sci, in press.

TAKAYAMA, S., 1976, J. Mater. Sci. **11**, 164.

TANIWAKI, M., M. MAEDA, S. UMEYAMA and Y. ISHIDA, 1982, in: **T**, p. 699.

TANNER, L., and R. RAY, 1980, Scripta Metall. **14**, 1657.

TAUB, A.I., 1980, Acta Metall. **28**, 633.

TAUB, A.I., and F.E. LUBORSKY, 1981, Acta Metall. **30**, 1939.

TAUB, A.I., and F.E. LUBORSKY, 1982, Mater. Sci. Eur. **56**, 157.

TAUB, A.I., and F. SPAEPEN, 1980, Acta Metall. **28**, 1871.

TAUB, A.I., and F. SPAEPEN, 1981, Acta Metall. **29**, 1939.

THOMAS, M., 1983, D. Phil. Thesis, Univ. of Sussex.

THOMPSON, C.V., A.L. GREER and F. SPAEPEN, 1984, Acta Metall., in press.

THURSFIELD, G., and M.J. STOWELL, 1974, J. Mater. Sci. **9**, 1644.

TRACEY, V.A., and C.P. CUTLER, 1981, Powder Metallurg. **24**, 37.

TSAO, S.S., and F. SPAEPEN, 1982, in: **T**, p. 463.

TULI, M., P.R. STRUTT, H. NOWOTNY and B.H. KEAR, 1978, in: **N**, p. 112.

TURNBULL, D., 1981, Metallurg. Trans. **12A**, 695.

VALENTA, P., K. MEIER, H. KRONMÜLLER and K. FREITAG, 1981, Phys. Stat. Sol. **105**, 537; **106**, 129.

UHLMANN, D.R., 1972, J. Non-Cryst. Solids **7**, 337.

VAN DEN BEUKEL, A., and S. RADELAAR, 1983, Acta Metall. **31**, 419.

VAN VECHTEN, J., 1982, in: Laser and Electron Beam Annealing, eds. B.A. Appleton and C.K. Cellars (Elsevier, New York) p. 49.

VINCENT, J.H., J.G. HERBERTSON and H.A. DAVIES, 1982, in: T, p. 77.

WAGNER, C.N.J., 1983, in: J, p. 58.

WALTER, J.L., S.F. BARTRAM and R.R. RUSSELL, 1978, Metallurg. Trans. **9A**, 803.

WARLIMONT, H., and P. GORDELIK, 1982, private communication.

WARRINGTON, D.H., H.A. DAVIES and N. SHOHOJI, 1982, in: T, p. 69.

WHITE, C.W., S.R. WILSON, B.R. APPLETON and F.W. YOUNG, Jr., 1980, J. Appl. Phys. **51**, 738.

WILLIAMS, M.F., R.F. LANDEL and J.D. FERRY, 1955, J. Amer. Chem. Soc. **77**, 3701.

WOOD, J.V., and R.W.K. HONEYCOMBE, 1981, in: E, p. 117.

WOODCOCK, L.V., 1981, in: Structure and Mobility in Molecular and Atomic Glasses, Ann. N.Y. Acad. Sci. **371**, 274.

WONG, J., 1981, in: G, p. 43.

YAVARI, A.R., P. HICTER and P. DESRÉ, 1983, J. Chim. Phys. **79**, 572.

YEARIM, R., and D. SHECHTMAN, 1982, Metallurg. Trans. **13A**, 1891.

YEH, X.L., K. SAMWER and W.L. JOHNSON, 1983, Appl. Phys. Lett. **42**, 242.

ZARZYCKI, J., 1974, J. Appl. Crystallogr. **7**, 200.

ZIELINSKI, P.G., and D.G. AST, 1983, Phil. Mag., in press.

Further reading

a. Surveys

(A) Metallic Glasses, Proc. Sem. on Metallic Glasses, 1976, eds. H.J. Leamy and J.J. Gilman (ASM, Metals Park, OH, 1978).

(B) R.W. Cahn, Metallic Glasses, Contemp. Phys. **21** (1980) 43.

(C) H.S. Chen, Glassy Alloys, Rev. Mod. Phys. **43** (1980) 353.

(D) F.E. Luborsky, in: Ferromagnetic Materials, vol. 1, ed. P.E. Wohlfahrt (North-Holland, Amsterdam, 1980) ch. 6, p. 451. (Detailed quantitative data on magnetic properties, densities, etc. of many metallic glasses.)

(E) H. Herman, ed., Ultrarapid Quenching of Liquid Alloys, Treatise on Materials Science and Technology, vol. 20, (Academic, New York, 1981). (The chapter by Li, on mechanical properties, also tabulates T_g, T_x, densities, elastic moduli, etc., of many metallic glasses.)

R. Wang, Thermal Stability Data of Amorphous Binary Alloys, Bull. Alloy Phase Diagrams **2** (1981) 269.

(F) Japanese work (articles in English): Amorphous Materials I–VI, Sci. Rep. Res. Inst. Tohoku Univ., series A, 26 (June 1976), 27 (March 1977), 27 (March 1979), 28 (March 1980), 29 (March 1981) and 31 (March 1983). Also: Structures and Properties of Amorphous Metals, Proc. Fukushima Symposium, 1977: Suppt. to Sci. Rep. Res. Inst. Tohoku Univ., Series A, June 1978 (no volume number).

(G) H.-J. Güntherodt and H. Beck, eds., Glassy Metals I: Ionic Structure, Electronic Transport and Crystallization (Springer, Berlin, 1981).

(H) H. Jones, Rapid Solidification of Metals and Alloys (Monograph No. 8, The Institution of Metallurgists, London, 1982) (also review paper in J. Mater. Sci. (1983) in press).

I.S. Miroshnichenko, Zakalka iz Zhidkovo Sostoyaniya (Quenching from the Liquid State) (Metallurgiya Publ. House, Moscow, 1982).

T.R. Anantharaman, ed., Metallic Glasses: Production, Properties and Applications (Trans Tech SA, Aedermannsdorf, Switzerland, 1983).

(I) S.R. Nagel, Metallic Glasses, Adv. Chem. Phys. **51** (1982) 227.

(J) F.E. Luborsky, ed., Amorphous Metallic Alloys (Butterworths, London, 1983). The most complete survey.

R. Hasegawa, ed., Glassy Metals: Magnetic, Chemical and Structural Properties (CRC Press, Boca Raton, FL, 1983).

b. Conference proceedings

(**K**) Proc. 1st Int. Conf. on Metastable Metallic Alloys, Brela, 1970, Fizika (Yugoslavia) **2** (Suppl. 2).

(**L**) Proc. 2nd Int. Conf. on Rapidly Quenched Metals, Cambridge, MA, 1975, eds. N.J. Grant and B.C. Giessen, part I (MIT Press, Cambridge, MA, 1976); part II, Mat. Sci. Eng. 23 (1976).

(**M**) Proc. 3rd Int. Conf. on Rapidly Quenched Metals, Brighton, 1978, ed. B. Cantor (Metals Society, London, 1978). 2 vols.

(**N**) Rapid Solidification Processing: Principles and Technologies, Proc. Conf. on Rapid Solidification Processing, Reston, 1977, eds. R. Mehrabian *et al.* (Claitor's, Baton Rouge, LA, 1978).

(**O**) Alliages et Matériaux Amorphes, 21ᵉ Colloque de Métallurgie, Saclay, 1978 (Publ. Centre d'Etudes Nucleaires de Saclay, Gif-sur-Yvette, 1979).

(**P**) Rapid Solidification Processing: Principles and Technologies II., Proc. (2nd) Conf. on Rapid Solidification Processing, Reston, 1979, eds. R. Mehrabian *et al.* (Claitor's, Baton Rouge, LA, 1980).

(**Q**) Proc. Int. Conf. on Liquid and Amorphous Metals, Grenoble, J. Physique, Colloq. C-8 (1980).

(**R**) Proc. Int. Conf. on Amorphous Metallic Phases, Smolenice, 1978, Physics and Applications, vol. 5 (VEDA, Publishing House of the Slovak Academy of Sciences, Bratislava, 1980).

(**S**) Proc. Int. Conf. on Metallic Glasses: Science and Technology, Budapest, 1980, 2 vols. (Central Research Institute for Physics, Budapest, 1981, distrib. by Kultura, PO Box 149, Budapest 1389, Hungary).

(**T**) Proc. 4th Int. Conf. on Rapidly Quenched Metals, Sendai, 1981, eds. T. Masumoto and K. Suzuki (Japan Inst. of Metals, Sendai, 1982) 2 vols.

(**U**) Metastable Materials Formation by Ion Implantation, Proc. Materials Research Soc. Ann. Meeting, Boston, 1981, eds. S.T. Picraux and W.J. Choyke (North-Holland, Amsterdam, 1982).

(**V**) Rapidly Solidified Amorphous and Crystalline Alloys, Proc. Materials Research Soc. Ann. Meeting, Boston, 1981, eds. B.H. Kear, B.C. Giessen and M. Cohen (North-Holland, Amsterdam, 1982).

(**W**) Proc. 5th Int. Conf. on Physics of Non-Crystalline Solids, Montpellier, 1982, ed. J. Zarzycki (Les Editions de Physique, Orsay, 1983) (J. Physique, Colloque C-9, Suppt. au no. 12.)

(**X**) Lasers in Metallurgy, Proc. Symp., eds. H. Mukherjee and J. Mazumder (The Metallurgical Society of AIME, Warrendale, PA, 1981).

c. Classified bibliographies of rapid solidification papers

(**Y**) H. Jones and C. Suryanarayana, Annotated Bibliography 1958–72, J. Mater. Sci. 8 (1973) 705.

(**Z**) C. Suryanarayana, Rapidly Quenched Metals: A Bibliography, 1973–79 (Plenum, New York, 1981). (Includes 187 papers published in 1980.)

CHAPTER 29

METALLIC COMPOSITE MATERIALS

G. FROMMEYER

Department of Materials Science and Engineering
Stanford University,
Stanford, CA 94305, USA

R.W. Cahn and P. Haasen, eds.
Physical Metallurgy; third, revised and enlarged edition
© *Elsevier Science Publishers BV, 1983*

1. Classification

Metallic composite materials are combinations of two or more different metals, intermetallic compounds, or second phases having particular microstructural arrangements within a metallic matrix. They are produced by mixing the components in such a way that the morphologies of the constituents are controlled so as to achieve optimum properties. The manifold possibilities of combining metallic and nonmetallic materials allow the creation of a new class of metallic alloys with superior or unusual properties.

The essence of composite materials and their technology is the ability to modify materials with hard particles, strong stiff fibres or layers in the proper distribution, orientation, and volume fraction. Ceramics and polymers alone or in combination with metals cover the complete spectrum of properties for all composite materials. However, in the interest of confining the topic to the realm of physical metallurgy, only metal and metal-matrix composites shall be discussed.

These composites may be classified in many ways, depending on the concept of interest. One useful classification system is based on microstructural aspects, differentiating between composite materials according to the *morphology* of the constituent phases. Using this system, composite materials fall into the following categories:
 – dispersion- and particle-strenghtened composites;
 – fibre-reinforced composites;
 – layer-composite materials.

Dispersion and *particle composites* * contain the second phase as fine particles randomly distributed in the matrix (fig. 1a). In *fibre-composites*, the second phase is mainly one-dimensional, and may be continuously or discontinuously aligned in the matrix (fig. 1b). In *layered* or *laminated composites*, the second components and phases are two-dimensional. They can also be continuously or discontinuously arranged (fig. 1c). These different types of microstructures can be obtained on submicroscopic, microscopic, and macroscopic levels.

Another classification system may be based on the *size and distribution* of the second phases in the composite. Under this system the following categories are obtained:
 – submicrocomposite materials (particle size, fibre diameter: $d \ll 1\ \mu m$), e.g. dispersion-strengthened alloys and fibre composites with extremely thin fibres.
 – microcomposite materials (particle size, fibre or layer thickness: $d \approx 1\ \mu m$), e.g. particle-, fibre- and laminate-reinforced materials, particle-reinforced composites and directionally solidified eutectics.
 – macrocomposite materials (macroscopic dimensions of the components: $d \gg 100$ μm), e.g. tungsten- and steel wire-reinforced copper and aluminium alloys.

Macrocomposite materials belong more to the category of technological structures

---

* The term *composite* often refers to fibre composites only in the literature. Here we use it in the more general sense.

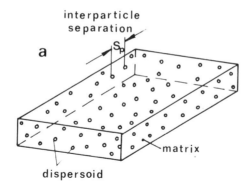

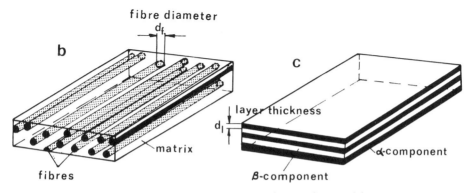

Fig. 1. Schematic representation of different microstructures of composite materials.

and structural engineering products, and are discussed in the literature of mechanical engineering. This chapter will discuss the relation between structure and properties of metallic micro- and submicrocomposite materials with particular reference to the properties of the fibres, particles, matrix phases, and interfaces. In addition, the predominant production methods will be described shortly. For a more detailed discussion see BROUTMAN and KROCK [1967].

Two widely used approaches for preparing metallic composites are:
– powder-metallurgical techniques (see ch. 30);
– directional solidification (see ch. 9).

Powder-metallurgical techniques and their variations use metals and/or hard second phases in powder form as starting materials. The diameter of the powder particles ranges from 0.05 μm to 100 μm. For alloy preparation, the components are mixed or mechanically alloyed and compacted by pressing. This can be carried out at room temperature or higher temperature by sintering. Mass transport by diffusion decreases the volume fraction of cavities, thus simultaneously increasing the density of the compacted composites. Hot isostatic pressing (HIP) and extrusion allow the density of the material to increase to its theoretical limit. WAHL and WASSERMANN [1970], WASSERMANN [1973] and FROMMEYER and WASSERMANN [1975] showed that

References: p. 1882.

extensive plastic deformation by rolling or wire-drawing enables very thin layers or fibres to be produced in the matrix. A promising technique especially for producing laminate composites with sandwich structure by superplastic bonding was described by SHERBY et al. [1979]. All dispersion- and particle-strenghened, as well as fibre- and layer-reinforced composites can be prepared by powder-metallurgical techniques. In addition to this, such modified techniques as internal oxidation (in situ oxide-particle formation in a metallic matrix by oxidation), and mechanical alloying enable the preparation of high-temperature creep and oxidation resistant superalloys. Readers who are interested in this topic are referred to the treatment by HAUSNER [1966] and LENEL [1980]. Other methods, such as chemical and physical vapor deposition, and explosive plating may be used, especially for preparing high-melting-point bcc- and hcp-metal composites with layer structure.

An elegant method for producing metallic composites with fibre and layer structures from the melt is directional solidification of binary or ternary eutectic alloys. This process can be used to take advantage of the simultaneous formation of the separate phases which form from a eutectic upon cooling. The same principles are valid for directionally transformed eutectoids. For *directional solidification*, it is necessary to solidify the eutectic melt at a controlled rate, with the temperature gradient perpendicular to the liquid–solid interface. This can be accomplished by removing a crucible containing the molten eutectic alloy from a furnace with a specific temperature gradient. The solidification rate of the phases is then controlled by the velocity of the crucible.

Figure 2 represents the fibrous microstructure of the unidirectionally solidified eutectic $Al–Al_3Ni$ composite material in transverse section (a) and longitudinal section (b) shown by HERTZBERG et al. [1965]. The fibres are of the order of 1 μm in diameter for most binary and ternary eutectic alloys. The morphology of the fibres is characteristic of many alloy systems, e.g. Cr–NiAl, Cu–Cr, $Al–Al_3Ni$, Sb–MnSb,

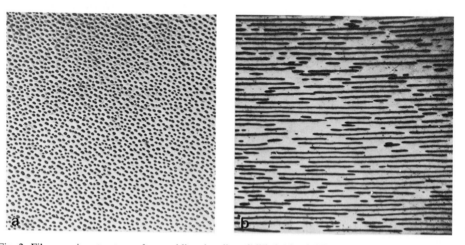

Fig. 2. Fibrous microstructure of an unidirectionally solidified $Al–Al_3Ni$ eutectic alloy; (a) transverse section and (b) longitudinal section. 500×. (From HERTZBERG et al. [1965].)

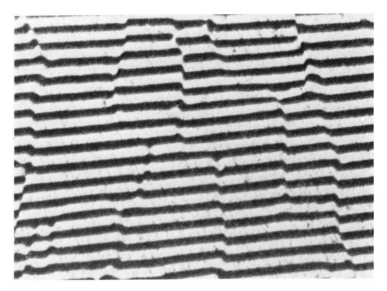

Fig. 3. Lamellar microstructure of an unidirectionally solidified Al–CuAl$_2$ eutectic (transverse section). 1000×. (From HERTZBERG *et al.* [1965].)

etc. Many factors influence the fibrous shape and morphology. The most important parameters are the solidification rate, V_c, and the volume ratio of the second phase in the eutectic. If this ratio is very small, e.g. $c_0 < 0.1$, then the eutectic alloy will show anomalous directional-solidification characteristics with irregular shapes of the fibres (e.g. Ag–Si eutectic).

Another common morphology of directionally solidified composites is the lamellar structure (fig. 3). Here the two phases form as alternating platelets. This microstructure is characteristic of eutectics with approximately equal volume fractions, e.g. Al–CuAl$_2$, Cd–Sn, Al–Zn, Co–CoBe, see e.g. KURZ and SAHM [1975]. It can also occur in systems with unequal volume fractions if the energy associated with the interface is particularly low for a particular crystallographic orientation. The controlled solidification of eutectic alloys is discussed more thoroughly in ch. 9.

2. Dispersion-strengthened composite materials

In dispersion-strengthened composite materials (particle size $d \ll 1\ \mu$m) and also in particle-reinforced composites (particle diameter $\simeq 1\ \mu$m) the matrix is the major load-bearing constituent. The fine dispersion of a second phase is present primarily to impede the motion of dislocations. The metallic matrix will be strengthened or reinforced in proportion to the effectiveness of the dispersion as a barrier to the motion of dislocations.

References: p. 1882.

The prime structure parameters determining the effectiveness of a dispersion are the mean free path between particles in the dispersion and the interparticle separation. The *mean free path between particles*, $\bar{L}_p$, is defined as the mean separation of particles along a randomly drawn straight line across a random section of a specimen. The *interparticle separation* of randomly distributed particles, S_p, may be evaluated as the radius of the smallest circle about one particle in a random plane of finding a second particle, for which the probability is unity, see e.g. ASHBY [1964]. These parameters are related to the particle diameter, d, and volume fraction of particles, V_p, by the following equations:

$$\bar{L}_p = \frac{2}{3} d \frac{1 - V_p}{V_p}, \tag{1}$$

$$S_p = \frac{d}{\sqrt{V_p}} \sqrt{\frac{2}{3}}. \tag{2}$$

For a matrix dislocation to pass through a dispersion of fine particles, the applied stress must be sufficient to bend the dislocation into a semicircular loop. The smallest radius of curvature to which a dislocation can be bent under the influence of an internal stress field, τ_i, is:

$$R = G_m |\boldsymbol{b}| / 2\tau_i, \tag{3}$$

where G_m is the shear modulus and $\boldsymbol{b}$ is the Burgers vector of the matrix. For a given interparticle spacing S_p, the stress required to bow dislocations around this dispersion is:

$$\tau_i = G_m |\boldsymbol{b}| / S_p. \tag{4}$$

If the applied stress is large enough to create a dislocation loop of radius $R = S_p/2$, then the loop can be expanded without a further stress increment. The limits for S_p can be predicted by using the matrix yield strength, $\tau_m \approx G_m/1000$, as the minimum stress to expand a dislocation loop between the particles. For the maximum stress, the theoretical matrix shear stress, which for fcc-metals is $\tau_{th} \approx G_m/30$ and for bcc metals $\tau_{th} \approx G_m/10$ is used. Substituting these values of the stress into eq. (3), and using $|\boldsymbol{b}| \approx 3$ Å for the Burgers vector, we see that the interparticle separation for effective dispersion strengthening should be $0.01 \leq S_p \leq 0.3$ μm. Many properties of the metallic matrix, such as ductility, electrical and thermal conductivity and impact strength, should be preserved. Exceptions are strength improvement and creep resistance. The former restrictions require that the volume fraction of the dispersoid should be kept small.

Figure 4 shows clearly that the particle diameter must be less than 0.1 μm to keep the mean free matrix path in the desired range, 0.01 μm $< \bar{L}_p < 0.3$ μm, and to maintain an optimal volume fraction of $V_p \approx 15\%$ of the second phase. The described stress-induced dislocation bowing creates a loop of slipped area around the dispersed particles. This circular slipped area effectively reduces the interparticle spacing. The additional increase in stress required to move dislocations through the disper-

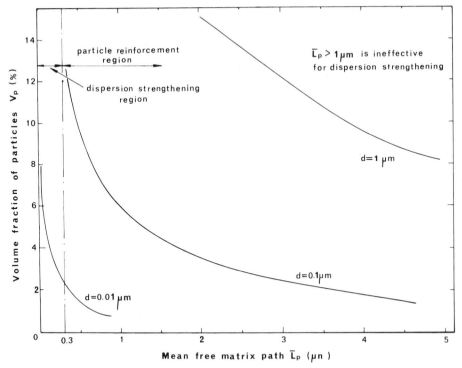

Fig. 4. Relationship between volume fraction and mean free matrix path for dispersion-strengthened composites of different particle size (after BROUTMAN and KROCK [1967]).

sion produces work-hardening of dispersion-strengthened composite materials.

The main advantage of dispersion-strengthened composites is not their ability to improve the room-temperature yield strength or work-hardening rate, but more the ability to maintain this yield-strength increase and attendant *creep resistance* increase over a wide temperature range, up to $T \approx 0.8\ T_m$ of the melting point of the metallic matrix, see e.g. ANSELL et al. [1966]. The effectiveness of the dispersion and dispersoids is determined by their insensitivity to high temperatures. This distinguishes dispersion-hardened composites from precipitation-hardened alloys, which soften with increasing temperature. The dispersoids in composites may be oxides, carbides, silicides, nitrides, borides and refractory metal particles, etc., which are insoluble and incoherent with the matrix. PRESTON and GRANT [1961] showed that these microstructural effects cause excellent hardness retention of, e.g., dispersion strengthened $Cu-SiO_2$ and $Cu-Al_2O_3$ composites over a wide temperature range, represented in fig. 5.

Table 1 shows some dispersion-strengthened alloy systems: The dispersion-strengthened composites with higher melting point and good oxidation behavior are used as high-temperature and creep resistant materials. The silver- and copper-based

References: p. 1882.

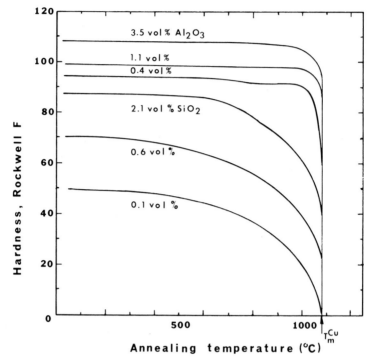

Fig. 5. Hardness retention as a function of temperature for various Cu–SiO$_2$ and Cu–Al$_2$O$_3$ dispersion-strengthened composite materials (from PRESTON and GRANT [1961]).

alloys, including tungsten doped with ThO$_2$, show good electrical and thermal conductivities, and have applications as electrical wires and contact materials (SCHREINER [1964]). The improved creep behavior of conventional superalloys by the

Table 1
Some dispersion-strengthened alloy systems.

Composite system		Volume fraction, V_p	Particle diameter, d (μm)
type	constituents		
metal matrix–oxide	W–ThO$_2$	0.03	0.01
	Ni–Al$_2$O$_3$	0.10	0.012
	NiCrAlTi–Y$_2$O$_3$	0.02	0.015
	Cu–SiO$_2$	0.10	0.08
	Cu–Al$_2$O$_3$	0.08	0.05
	Ag–Al$_2$O$_3$	0.10	0.05
metal matrix–carbide	Ni–TiC	0.05	0.03
metal–metal	Cu–W	0.05	0.015
	Ag–W	0.08	0.015

addition of fine dispersoids is the most important achievement of research on dispersion-strengthened composites, see e.g. SIMS and HAGEL [1972]. Dislocation theory has been used as the primary method for predicting the creep rate. This topic is discussed in detail in chs. 20 and 22.

3. *Particle-reinforced composites*

The important role of the matrix phase in particle-reinforced composite materials is to transmit load to the reinforcing particles. These types of composites continue where dispersion-strenghened composites leave off. In this case the metallic matrix and the particles share the applied load. The volume fraction of the second phase exceeds $V_p = 0.25$, and the particle diameter as well as the mean free path in the matrix is of the order of 1 μm. Therefore, particle-reinforced composites are intermediate between dispersion-strengthened and fibre-reinforced materials.

Hardening and strengthening of particle-reinforced composites occur initially when the dispersed particles restrict matrix deformation by mechanical restraint. The amount of the restraint effect is complex, and depends on structural factors. In general, it is a function of the ratio of the particle separation distance to the particle diameter, S_p/d, and the ratio of the elastic properties of the matrix and the second phase. The elastic modulus of particle-reinforced composites can be described using the *rule of mixtures*

$$E_c = V_m E_m + V_p E_p,$$ (5)

where V_m and V_p are the volume fractions of the matrix and the particles, and $E_{c/m/p}$ are the elastic moduli of the composite and the components, respectively. KROCK [1966] showed that if the constraint is severe enough, this bound may be exceeded, the isostrain conditions will not be met and deviations from eq. (5) occur. Some typical data are represented in fig. 6. The *elastic properties* of the iron–copper composite follow the rule of mixtures under the isostrain condition. In contrast to this, the tungsten-particle reinforced Cu–W alloy shows remarkable deviations from the predicted upper bound modulus given by eq. (5).

In this case, the elastic moduli of particle composites should follow the isostress criterion:

$$\sigma_c = \sigma_m = \sigma_p.$$ (6)

The elastic properties can then be described as

$$\frac{1}{E_c} = \frac{V_m}{E_m} + \frac{V_p}{E_p}.$$ (7)

All particle-reinforced metallic composites investigated exhibit positive deviation from eq. (7) and signify matrix constraint. This has been observed in systems such as: W–Cu, Mo–Cu, Mo–Ag, Ni–Ag, Fe–Cu, Cu–Cr, Ni–TiC, WC–Co, etc. Figure 7 represents positive deviations of the elastic modulus of Cu–W particle composites

References: p. 1882.

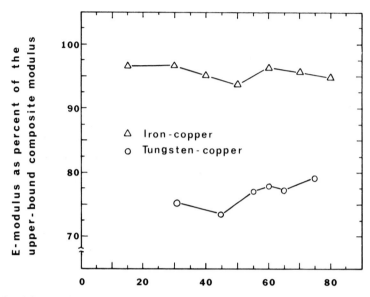

Fig. 6. Ratio of the experimental composite modulus to the theoretical upper-bound modulus in percent as a function of the volume fraction of dispersed particles for Cu–Fe, and Cu–W composites (from Krock [1966]).

from the lower bound isostress curve described by Krock [1965]. Therefore a certain amount of hydrostatic stress is generated in all particle-reinforced metallic composites and the obtained elastic values are located between the upper-bound iso-strain and lower-bound isostress curve. The amount of deflection by the degree of hydrostatic constraint depends on the mechanical and microstructural parameters.

The deformation behavior beyond the elastic range of particle reinforced composites may be classified into categories according to whether the particles themselves undergo *plastic deformation* before fracture occurs. Under an applied stress, the softer matrix phase is restrained from deformation by the hard particles. A hydrostatic stress field, identical to that in the elastic range, is built up. With continued load increase, the magnitude of the hydrostatic stress increases. When the hydrostatic stress component reaches about three times the unconstrained matrix yield strength, fracture is likely. If a stress in the particles is about 3–3.5 times the unconstrained matrix flow stress, fracture through the matrix occurs, initiated by *particle cracking*. Such behavior is typical for carbide and cermet materials, where the hard dispersion of carbide and oxide particles cannot be deformed, see e.g. Bergmann and Wassermann [1978]. Under load, the internal shear stress, τ_i, is equal to the product of the applied stress and the number of dislocation loops piled up against the particles:

$$\tau_i = n\sigma. \tag{8}$$

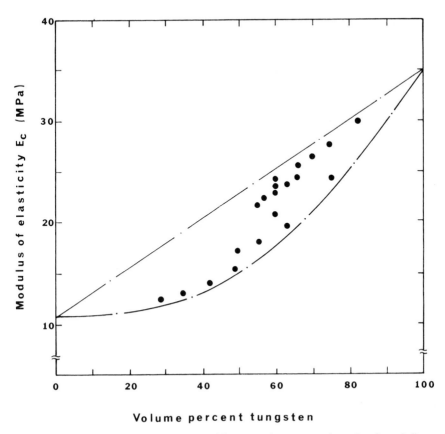

Fig. 7. Elastic modulus of tungsten–copper particle composite versus volume fraction of dispersoid. Experimental values fall between the rule-of-mixture upper (isostrain) and lower (isostress) bounds. (from KROCK [1963].)

The number of piled-up dislocations, n, is proportional the interparticle spacing, S_p, by

$$n = \sigma S_p / G_m |b|. \tag{9}$$

Combining eqs. (8) and (9) we obtain the internal shear stress:

$$\tau_i = \sigma^2 S_p / G_m |b|. \tag{10}$$

When the shear stress acting on the particles is equal to their fracture stress σ_{fp}, the particles will fracture, nucleating a crack which will cause yielding of the composite:

$$\tau_i = \sigma_{fp} = G_p / C \quad \text{at yield,} \tag{11}$$

where C defines the strength of the composite as:

$$\sigma_{cy} = \left(G_m G_p |b| / C S_p \right)^{1/2}. \tag{12}$$

References: p. 1882.

This expression predicts that the composite yield strength is proportional to the inverse square root of the interparticle spacing, S_p, when the particles do not yield under load. This was shown by ROBERTS *et al.* [1952] for a number of carbide-reinforced iron–carbon alloys (fig. 8) and for numerous cemented carbide composites, such as Co–WC and Ni–TiC cermets. When the interparticale spacing is $S_p > 0.5$ µm, the strength drops due to crack-nucleation at weak particle–particle boundaries.

In cases where the constrained matrix creates stresses high enough to *deform the dispersed particles*, the yield criterion is:

$$\tau_i = \tau_{py} = \sigma^2 S_p / G_m |b|. \tag{13}$$

Under the hydrostatic stress, the particle yield strength is reduced from its uncon-

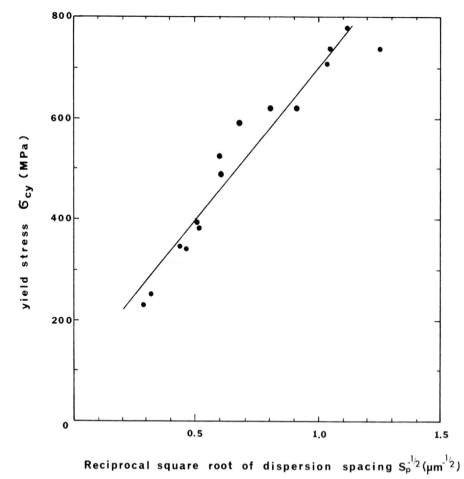

Fig. 8. Yield point of several hypoeutectoid, eutectoid, and hypereutectoid steels as a function of the reciprocal square root of dispersion spacing, S_p, of cementite (after ROBERTS *et al.* [1952]).

strained yield strength by a an amount proportional to the constraint:

$$\tau_{py} = \frac{G_p S_p}{C^*}.\tag{14}$$

Combining eqs. (13) and (14) we obtain the yield strength relation for deformable particle composites as

$$\sigma_{cy} = \left(G_p G_m |b|/C^*\right)^{1/2}.\tag{15}$$

This expression describes the independence of the yield stress of the composite of the interparticle spacing. Such behavior has been observed by KROCK and SHEPARD [1963] in W–Ni–Fe cemented alloys, where the tungsten particles undergo large plastic deformation before fracture of the particle-reinforced composite occurs (fig. 9).

Under these circumstances the Fe–Ni matrix is constrained to a sufficiently high stress that the dispersed tungsten particles deform, and extensive plastic deformation in the composite occurs. The stress necessary to impart extensive ductility to the composite is identical to the flow stress of the W-dispersoid under the present stress (reduced applied stress and hydrostatic compression stress). The flow stress of this W–Ni–Fe particulate material is independent of the volume fraction and the mean free matrix separation of the dispersed phase as predicted by eq. (15). The unusual behavior has contributed much information to an understanding of the deformation and strengthening behavior of particulate composites in general.

For further detailed treatment of these topics, see ch. 22, §§ 3 and 4.

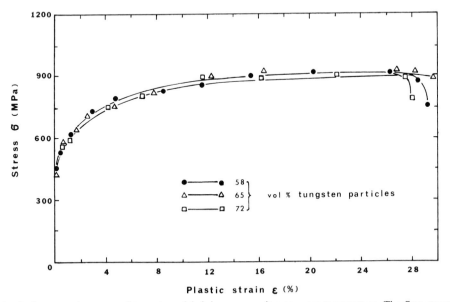

Fig. 9. Stress–strain curves of tungsten–nickel–iron composites at room temperature. The flow stresses are independent of the volume fraction of the dispersoids. (After KROCK and SHEPARD [1963].)

References: p. 1882.

4. Fibre- and laminate-reinforced composites

4.1. Continuous fibres and layers

In microcomposite materials with fibre- and lamellar structure, the second components are dispersed throughout the metallic matrix. In laminated composites, the phases are sequentially arranged, as shown in fig. 1c. For this discussion, we assume that the second phase is uniform, continuous, and unidirectional, and is bound by the matrix so that no slippage can occur at the matrix–fibre or matrix–lamellae interface. Under the isostrain condition, $\epsilon_c = \epsilon_m = \epsilon_f$, the total load, F_c, that is applied to the composite, is shared between the fibre load, F_f, and the load on the matrix, F_m. Therefore:

$$F_c = F_m + F_f \quad \text{or, in terms of stresses,} \qquad \sigma_c = \sigma_m V_m + \sigma_f V_f, \tag{16}$$

where V_m and V_f represent the volume fractions of matrix and fibres, respectively. When slippage does not occur at the interface, the isostrain criterion is fulfilled, and eq. (16) may be rewritten as:

$$\sigma_c = E_m \epsilon_c V_m + E_f \epsilon_c (1 - V_m), \tag{17}$$

since $V_m + V_f = 1$. The ratio of the load carried by the fibres or layers to the load carried by the matrix is:

$$\frac{F_f}{F_m} = \frac{E_f}{E_m} \left(\frac{1 - V_m}{V_m} \right). \tag{18}$$

This ratio is plotted in fig. 10 as a function of the ratio of the elastic moduli E_f/E_m and the volume fraction of the second phase. To obtain optimum reinforcement by the fibres and lamellae, the modulus of the second phase should be higher than the modulus of the metallic matrix. Under this condition, the reinforcement carries the main contribution of the total applied load and attains the maximum stress compared to the matrix. The theoretical maximum of the volume fraction of a fibrous second component is almost $V_f = 0.91$. However, above a volume fraction of the fibres of about $V_f \gtrsim 0.8$, most artifically prepared composites exhibit binding problems at the interfaces and many superior properties of the composites begin to deteriorate. When a composite with uniaxially aligned and continuous fibres or laminates is stressed in an isostrain mode, the deformation behaviour can be classified into *four different stages:*
– the fibres or lamellae *and* the matrix deform elastically;
– The matrix deforms plastically and the fibres or lamellae continue to deform elastically;
– the matrix *and* the fibres or lamellae deform plastically;
– fibre fracture is followed by complete fracture.
These four stages are shown schematically in fig. 11. In stage I the elastic modulus of the composite can be predicted accurately using the rule of mixtures,

$$E_c = V_p E_f + V_m E_m. \tag{19}$$

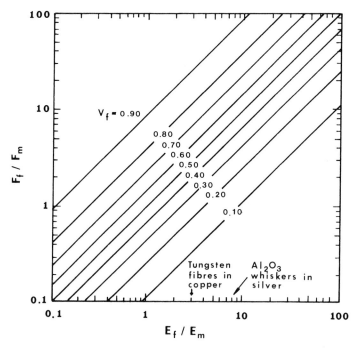

Fig. 10. Ratio of the fibre-to-matrix load versus ratio of the elastic moduli for various volume fraction of the fibres (from BROUTMAN and KROCK [1967]).

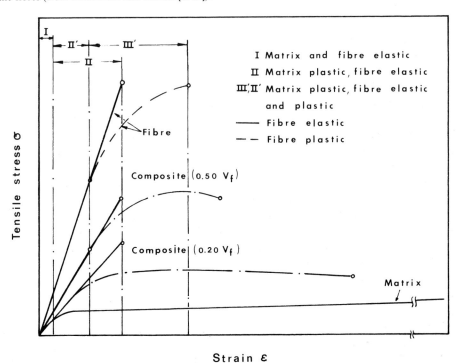

Fig. 11. Stress–strain curves of a ductile metal reinforced with ductile and/or brittle fibres of equal fracture stress (after KELLY and DAVIS [1965]).

References: p. 1882.

The largest portion of the composite stress–strain curve belongs to stage II, where the matrix stress–strain curve is not linear. The composite modulus may be determined at different stress–strain levels by

$$E_c = E_f V_f + \left(\frac{d\sigma_m}{d\epsilon_m} \right)_{\epsilon_f} \cdot V_m, \tag{20}$$

where $(d\sigma_m/d\epsilon_m)_{\epsilon_f}$ is the slope of the matrix stress–strain curve at different strains ϵ_f of the fibres. In stage III the modulus of the composites with ductile fibres must be modified in a similar way to the reduction of the matrix modulus as

$$E_c = \left(\frac{d\sigma_f}{d\epsilon_f} \right)_{\epsilon_f} \cdot V_f + \left(\frac{d\sigma_m}{d\epsilon_m} \right)_{\epsilon_f} \cdot V_m. \tag{21}$$

In composites with ductile fibres which fracture by necking during a larger amount of plastic deformation, additional factors such as a hydrostatic stress state exerted by the ductile matrix to prevent necking of the fibres cause deviations from the rule of mixtures.

For metallic composites containing more than a minimum volume fraction, V_{min}, of continuous fibres or lamellae, the ultimate tensile strength will be ideally reached at a total strain of the fibres at their ultimate tensile strength. The following expression describes the ultimate tensile strength of a composite:

$$\sigma_c^u = \sigma_f^u V_f + (\sigma_m)_{\epsilon_f} (1 - V_f), \tag{22}$$

where $V_f > V_{min}$ and σ_f^u is the ultimate tensile strength of the reinforcing fibres. $(\sigma_m)_{\epsilon_f}$ is the stress of the metallic matrix when the fibres are strained to their ultimate tensile strain. The value of $(\sigma_m)_{\epsilon_f}$ for metallic matrices is equal to the flow stresses at the ultimate tensile strain of the fibres. This value can be determined from the stress–strain curve of the metallic matrix. The extensive *work-hardening rate* of the matrix can be explained by dislocation pile-up at the matrix–fibre interfaces. This was shown by NEUMANN and HAASEN [1971] on tungsten-fibre reinforced copper composites. Equation (22) predicts a linear increase of the ultimate tensile strength and also of the yield strength of the composites with increasing volume fraction of the reinforcing components. This relation has been verified for many systems, such as aluminium–steel and silver–steel fibres, copper–tungsten fibres, silver– and NiCr–Al$_2$O$_3$ whiskers etc. Some of the data are represented in fig. 12, and show good agreement with the rule of mixtures curve. However, the exception is the silver–steel-fibre composite, where the experimental values fall above the predicted linear relationship. In this case the silver matrix provides a hydrostatic stress component on the ductile steel fibres and thus restrains necking. This effect leads to enhanced strength, and to a positive deviation form the predicted composite strength.

It should be noted that for very small volume fractions of the second phase, V_f, the mechanical properties of the composite may not follow the prediction by eq. (22). This is because there is an insufficient number of fibres and lamellae to restrain

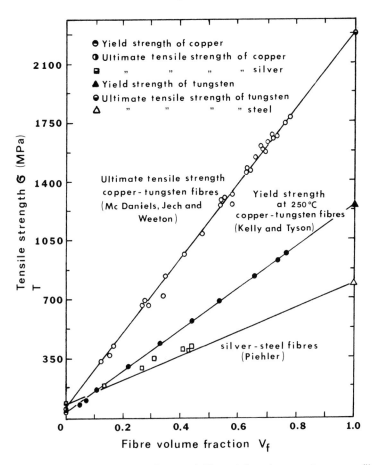

Fig. 12. Strength of copper–tungsten and silver–steel fibre-reinforced composites versus fibre volume fraction (KELLY and TYSON [1965], MC DANIELS *et al.* [1965], and PIEHLER [1965]).

the elongation of the matrix phase, so that the fibres are rapidly stressed to their fracture point.

For effective reinforcement of the matrix phase (e.g. in work-hardened or precipitation-hardened metallic matrices), the strength of the composite must exceed the *ultimate tensile strength of the matrix*, σ_m^u, so that $\sigma_c^u \geq \sigma_m^u$. This condition defines a critical volume fraction of the reinforcing second component which must be exceeded for effective strengthening, see KELLY and DAVIS [1965]. This volume fraction can be determined by the expression

$$V_f > V_{crit} = \frac{\sigma_m^u - (\sigma_m)_{\epsilon_f}}{\sigma_f^u - (\sigma_m)_{\epsilon_f}}, \tag{23}$$

which is graphically represented in fig. 13. The minimum volume fraction, indicated

References: p. 1882.

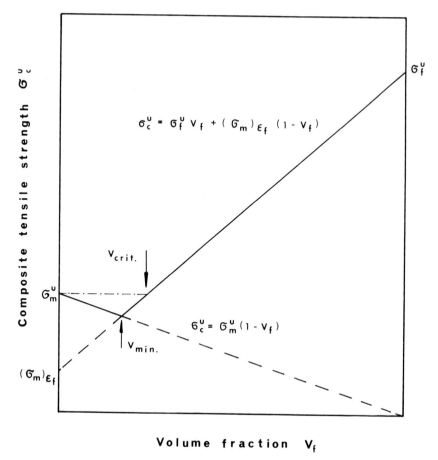

Fig. 13. Variation of composite strength, σ_c^u, with volume fraction, V_f, for reinforcement with continuous brittle fibres (after KELLY and DAVIS [1965]).

also in fig. 13, is given by the following expression:

$$V_{min} = \frac{\sigma_m^u - (\sigma_m)_{\epsilon_f}}{\sigma_f^u + \sigma_m^u - (\sigma_m)_{\epsilon_f}}.\qquad(24)$$

In all cases V_{min} is less than V_{crit}. Table 2 presents some selected values of V_{crit} for composites of strong fibres in ductile metal matrices at room temperature. It is evident that the critical volume fraction increases with increasing work-hardening in the matrix $(\sigma_m^u - (\sigma_m)_{\epsilon_f}) \Rightarrow$ max., and also as the strength of the metallic matrix phase approaches that of the fibres $(\sigma_f^u - (\sigma_m)_{\epsilon_f}| \Rightarrow$ min.. Thus, reinforcement of a strong matrix by fibres of comparable strength requires a large critical volume fraction of the second component.

An interesting example of a composite with a continuous layer structure is the

Table 2
Values of V_{cit} for fibres of various strengths, σ_f^u, in ductile metallic matrices.

Matrix metal	$(\sigma_m)_{\epsilon_f}$ (MPa)	σ_m^u (MPa)	Critical fibre volume fraction V_{crit}			
			$\sigma_f^u = 700$ MPa	$\sigma_f^u = 1750$ MPa	$\sigma_f^u = 3500$ MPa	$\sigma_f^u = 7000$ MPa
Aluminium	28	84	0.083	0.033	0.016	0.008
Copper	42	210	0.255	0.098	0.047	0.024
Nickel	63	315	0.396	0.150	0.073	0.036
Stainless steel (18/8)	175	455	0.534	0.178	0.084	0.041

directionally solidified Al–CuAl$_2$ eutectic, which consists of alternating ductile Al and reinforcing intermetallic CuAl$_2$ single crystal platelets. The volume fractions of the two phases are approximately equal, and the thickness of the platelets is about 1 to 2 μm (see fig. 3). The strong anisotropy of the mechanical properties of this system was investigated by HERTZBERG *et al.* [1965]. Other composite systems like this also show strongly anisotropic properties. The elastic stiffness and the strength of the composite decrease as the Al and Al$_2$Cu lamellae are rotated away from the maximum applied normal stress direction ($\theta = 0$). The influence of the microstruct-

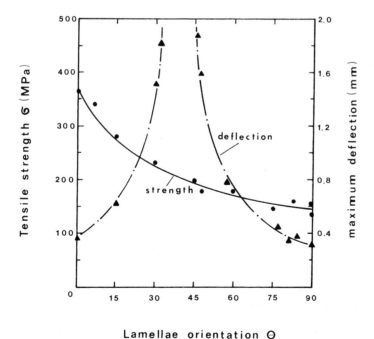

Fig. 14. Tensile strength and lamellae deflection versus lamellae orientation relative to tensile axis, θ, of a directionally solidified Al–CuAl$_2$ eutectic alloy (after HERTZBERG *et al.* [1965]).

References: p. 1882.

ural anisotropy on the tensile strength of this laminated composite materials is illustrated in fig. 14. As mentioned above, the tensile strength decreases continuously with increasing degree of misorientation of the lamellae over the whole range from the isostrain to the isostress limits. A particularly surprising feature of the plastic derformation of the platelets is the unusual deflection, which develops a maximum for lamellar orientations between $30° \leq \theta \leq 45°$ to the tensile axis, see fig. 14. This extraordinary increase in ductility of the composite may be caused by slip systems activated in the Al_2Cu lamellae. This behavior of the laminated composites indicates both microstructural and crystallographic anisotropy.

4.2. Discontinuous fibre composites

The properties of composite materials reinforced with continuous fibres are relatively simple to describe using the rule of mixtures, since the stress profile along any single fibre is nearly uniform. The rule of mixtures must be modified for describing the properties of discontinuous fibre-reinforced composites, see e.g. CHRISTENSEN [1979]. The fibre ends will not support the same load as the middle part, hence there is a critical length, l_c, called the fibre transfer length, which is the smallest fibre length for which the maximum fibre stress, σ_f^{max}, may be achieved (fig. 15).

For fibres longer than the critical length l_c, a certain fraction will support the maximum fibre stress, while a fixed fibre length $l_{c/2}$ at each end has a stress profile which decreases from σ_f^{max} to zero. When these *fibre-end effects* are taken into account, an equation analogous to eq. (22) may be developed. The quantity $\tilde{\sigma}_f$ is defined as the ratio of the area under the stress curve to that part which supports σ_f. Under this assumption, the ineffective fibre ends contribute a reduced average stress $\alpha\sigma_f^{max}$ to the transmitted maximum stress σ_f^{max}. The average stress, $\bar{\sigma}_f$, in the

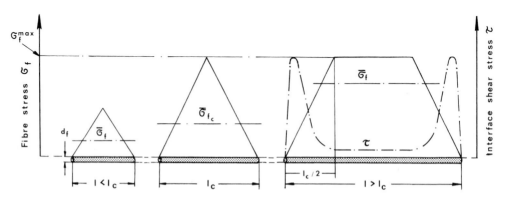

Fig. 15. Schematic representation of fibre tensile stress (σ_f) and interfacial shear stress (τ) when the matrix exhibits elastic deformation (from KELLY and DAVIS [1965]).

discontinuous fibres is given by expression:

$$\bar{\sigma}_f = \frac{1}{l}\int_0^l \sigma_f \, dl = \sigma_f\left[1 - (1 - \alpha)\frac{l_c}{l}\right]. \tag{25}$$

In the case of an ideal plastic matrix with yield stress τ_m^0, α is equal to 0.5 and the average fibre stress can be described as

$$\bar{\sigma}_f = \sigma_f\left(1 - \frac{l_c}{2l}\right). \tag{26}$$

The critical ratio of the fibre length, l_c, to fibre diameter, d_f, is defined as

$$\frac{l_c}{d_f} = \frac{\sigma_f^{max}}{2\tau_m^0}. \tag{27}$$

For a work-hardened metallic matrix, τ_m^0 is identical with the ultimate shear strength of the matrix, and in the case of interfacial failure τ_m^0 must be considered as the interfacial shear strength between the fibre and the matrix. However, in an aligned short fibre composite two important failure mechanisms are possible, depending on the relative values of τ^0, σ_f^u, and σ_m^u. In one case fibre fracture occurs before fracture in the matrix takes place. The build-up in stress in the fibres, as the load on the composite is increased, can be described by eq. (26). If the average stress in the fibre reaches the stress for fibre fracture, then the ultimate tensile strength of the composite can be described by

$$\sigma_c^u = \left(1 - \frac{l_c}{2l}\right)\sigma_f^u V_f + (\sigma_m)_{\epsilon_f}(1 - V_f). \tag{28}$$

The second failure condition, illustrated in fig. 15, occurs when the build-up of stress in the fibre is insufficient to cause fibre fracture and the average fibre stress,

$$\bar{\sigma}_f = \tau\frac{l}{r}, \tag{29}$$

reaches the fracture stress of the matrix. This relationship is valid for $l < l_c$. The tensile strength of an aligned short fibre composite is given by

$$\sigma_f^u = \left(\frac{l}{2r}\tau\right)V_f + \sigma_m^u(1 - V_f). \tag{30}$$

Equations (28) and (30) allow the prediction of the variation in strength as a function of the geometry of the discontinuous fibres if σ_f^u and τ are known. Eq. (28) indicates that if the discontinuous fibre length, l, is equal to the critical fibre length, l_c, then 50% of the maximum fibre strength will be realized. When $l \geq l_c$, then 90% or more of the tensile strength of the fibres will be reached. Therefore if the fibre strength or the aspect ratio, l/d_f, is sufficiently large, the composite with reinforced discontinuous fibres approaches the strength of the composites with continuous fibres. This case is illustrated in fig. 16 for a tungsten-fibre reinforced copper matrix.

　　KELLY and TYSON [1965] have shown that for fibres having an aspect ratio $l/d_f < \infty$ a deviation from the linear upper bound line, predicted by the rule of

References: p. 1882.

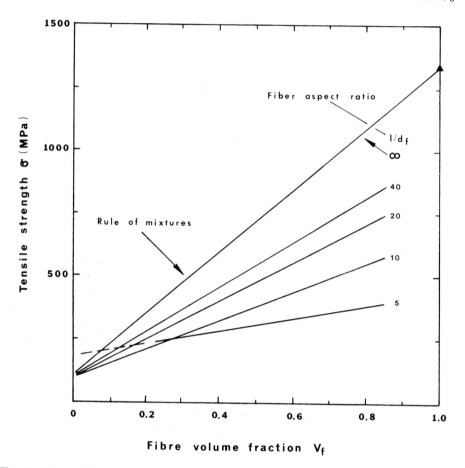

Fig. 16. Effect of fibre aspect ratio (l/d_f) and fibre volume fraction (V_f) on tensile strength of a copper–tungsten composite; tested at 250°C. (From KELLY and TYSON [1965].)

mixtures, occurs. This deviation increases with decreasing values of the fibre aspect ratio. For $l/d_f = 40$ and fibre volume fraction $V_f = 0.8$, the tensile strength of the composite materials reaches 80% of that with continuous fibres. CRATCHLEY [1963] reported similar data which are obtained from other systems, such as aluminium–steel fibres, silver–Al_2O_3 whiskers etc. For fibres with a small fibre aspect ratio, misalignment is the strongest factor which contributes to a decrease in the elastic modulus and also the tensile strength of composite materials. KELLY and DAVIS [1965] described the effect of the fibre orientation on the elastic properties and the tensile strength, that is qualitatively represented in fig. 17.

A misalignment of about $\Phi_{crit} \lesssim 8°$ has no influence on the discussed mechanical properties of the composites. Beyond this critical angle, tensile strength and elastic modulus diminish rapidly with increasing orientation angle, and at $\Phi = 45°$ the

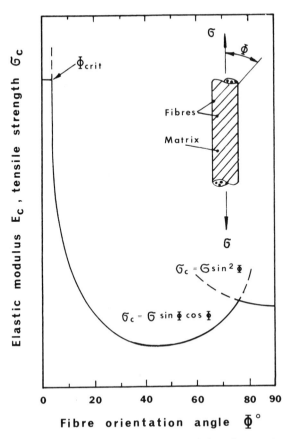

Fig. 17. Variation in elasticity and tensile strength of a fibre-reinforced composite as a function of fibre misorientation (after KELLY and DAVIS [1965]).

curve passes its minimum. With further increase of the Φ values tensile strength and elastic modulus increase slightly until the curve reaches the intersection with the isostress-condition curve at $\Phi = 80°$. When the fibres are oriented perpendicularly to the load direction, then the isostress condition is exactly fulfilled, and the properties of the composite material can be described by the lower-bound isostress rule of mixtures.

KOPPENAAL and PARIKH [1962] determined the tensile strength as a function of the volume fraction of the second phase for different orientations of steel fibres in a silver matrix. The obtained results are represented in fig. 18. It is evident that composite materials with discontinuously aligned fibres of varying orientation relative to the stress axis have lower strengths than composites with continuously aligned fibres. Composites with randomly distributed fibres exhibit a minimum reinforcement effect. Similar behaviour is observed in directionally solidified

References: p. 1882.

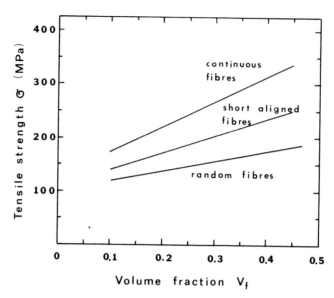

Fig. 18. Effect of fibre volume fraction and fibre orientation on the strength of silver–steel fibre composite (from BOUTMAN and KROCK [1967]).

eutectics, where stiff whiskers of pure metallic and intermetallic compounds are embedded in a ductile metallic matrix.

The first example of a fibre-reinforced eutectic alloy prepared by controlled

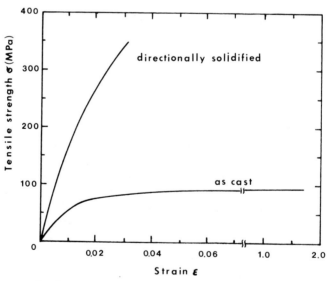

Fig. 19. Tensile properties of as-cast and unidirectionally solidified Al–Al$_3$Ni eutectic alloy (after LEMKEY et al. [1965]).

solidification was the Al–Al$_3$Ni system. This eutectic alloy contains 10% volume fraction of Al$_3$Ni whiskers, surrounded by the ductile Al matrix. The reinforcement effect of the unidirectionally aligned intermetallic Al$_3$Ni whiskers is compared with that of randomly distributed Al$_3$Ni phase in the as-cast condition in fig. 19.

All composite materials containing misaligned fibres are weaker than those with an aligned fibre structure. This graph shows that the directionally solidified fibre structure causes a marked improvement in strength to about $\sigma_c^u = 350$ MPa. In contrast to this, the as-cast eutectic alloy exhibits a tensile strength of $\sigma^u = 90$ MPa and an elongation to failure of $\epsilon_{tot} = 15\%$. Calculations of the ultimate tensile strength of the controlled-solidified Al–Al$_3$Ni composite, using eq. (28), yield $\sigma_c^u = 335$ MPa, in good agreement with the experimentally determined values. Improvements in stiffness and strength by controlled eutectic structures have been obtained in many other systems, such as Cu–Cr, Cu–Zr–Si, Al–CuAl$_2$, Sn–Cu$_6$Sn$_5$, Ag–ZrAg, Ta–C, etc., see e.g. KURZ and SAHM [1975].

4.3. Composites with extremely thin fibres and layers

Metallic composites with a fibre or a layer structure exhibit anomalous mechanical and physical properties due to size and interface effects when the fibres or the layers become extremely thin ($d_f \lesssim 100$ nm). The properties of these types of composites show strong deviations from the rule of mixtures. In addition to that, structural parameters as fibre diameter, interface concentration, and lattice defect density must be taken into account to describe this anomalous behavior properly. Very thin fibres or layers can be obtained by extensive cold deformation of the composites by wire drawing or rolling. Composites with such submicrostructural arrangements of the components show marked changes in their defect structure, which influence the physical properties strongly. For example, the extraordinary changes in the electrical and thermal properties are caused by electron and phonon interactions with the closely spaced interfaces. Ferromagnetic hardening, and also the stability of the magnetic flux lattice of type II-superconductors are caused by pinning processes at the interfaces. Some characteristic properties of submicrostructural composites will be discussed with special consideration of their structure.

Tensile strength and hardness of fibre-reinforced composites with extremely thin fibres show a remarkable positive deviation from the rule of mixtures of the isostrain condition. This has been observed by WAHL and WASSERMANN [1970], FROMMEYER and WASSERMANN [1975], SEIFERT and WASSERMANN [1981] and FROMMEYER [1977] on several composite systems: iron–silver, copper–iron, silver–copper, iron–glass, nickel–silver, aluminium–glass, etc. The effect of different fibre diameters, obtained by certain degrees of deformation of the composite, on the tensile strength of an iron–silver system is illustrated in fig. 20. The ultimate tensile strength of this composite increases with increasing degree of cold deformation and decreasing fibre diameter of the components. In the Fe–Ag system, maximum values are obtained for the composition Fe–40 vol% Ag-fibres and the fibre diameter $d_f^{Ag} = 165$ nm. The high strength is caused by marked changes in the defect densities of the thin fibres.

References: p. 1882.

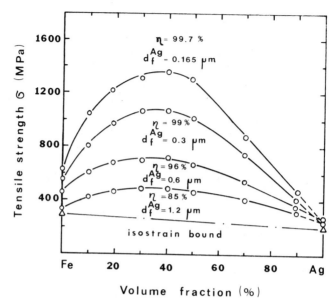

Fig. 20. Tensile strength versus volume fraction of an Fe–Ag composite system for different degrees of deformation η and fibre diameters d_f (from WAHL and WASSERMANN [1970]).

This has been shown on a directionally solidified and heavily deformed eutectic Ag–30 vol% Cu fibre-composite by electron microscopic investigation.

The increase in the dislocation density in the copper and silver fibres by extensive cold deformation is about two orders of magnitude greater than in heavily deformed bulk metals and alloys. At the largest reduction of cross-section ($\eta \gtrsim 99.9\%$) and for very thin Cu-fibres ($d_f < 60$ nm) a rapid decrease in dislocation density seems to occur, and extended fibre areas appear almost free of dislocations, showing a whisker-like structure, fig. 21. In this stage, the fibre diameter of the components is comparable to the mean free path of dislocation. In addition, the section of Frank–Read dislocation sources is restricted by small fibre diameters of $d_f \lesssim 10$ nm. This leads to a tensile strength which is comparable to the theoretical strength, τ_{th}, of metals, which can be described by the relation:

$$\left(\frac{G_{\langle hkl \rangle}}{10} \right)_{bcc} \gtrsim \tau_{th} \gtrsim \left(\frac{G_{\langle hkl \rangle}}{30} \right)_{fcc}, \tag{31}$$

where $G_{\langle hkl \rangle}$ is the shear modulus of the fibres for the crystallographic fibre orientation parallel to the stress axis.

Anomalous *changes in physical properties* are to be expected if the physical parameters, e.g. the mean free path of the conduction electrons or of the phonons and the size of ferromagnetic domains, strongly interact with the fibre distances. It is well known that lattice defects such as vacancies, impurities, dislocations and particularly phase boundaries influence the electrical and thermal conductivity as

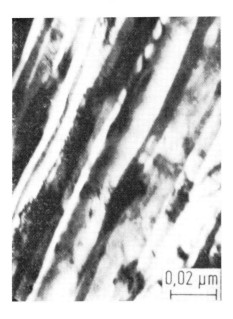

Fig. 21. Transmission electron micrograph of highly deformed Ag–Cu composite wires, longitudinal section; reduction in area $\eta = 99,99\%$.

well as the thermoelectric properties, including the Hall effect of bulk metals and alloys. For example, the electrical resistance of thin metallic fibres and films increases markedly with decreasing specimen thickness. This is caused by the restriction of the mean free path of electrons and their enhanced scattering at phase boundaries.

Figure 22 represents the ratio of the d.c. electrical conductivities of bulk metal and thin layer (or fiber), $\mathcal{K}_\infty/\mathcal{K}_l$, plotted versus the ratio of fibre (layer) thickness and mean free path l_0 of the conduction electrons for various test temperatures and composite systems. All investigated materials show identical curves similar to a hyperbola. When the fibre or layer thickness is much smaller than the mean free path of the conduction electrons, the ratio of the conductivities tends to infinity. If the fibre or layer thickness approaches the dimension of a bulk component, then the conductivity ratio becomes unity, see FROMMEYER and BRION [1981]. The electrical conductivity of these composites can be described satisfactorily by the rule of mixtures for the limiting cases of series and parallel connection of the components by the expressions:

$$\rho_c = \rho_f^A V_f^A + \rho^B\left(1 - V_f^A\right) \quad \text{(series)} \tag{32}$$

and

$$\mathcal{K}_c = \mathcal{K}_f^A V_f^A + \mathcal{K}^B\left(1 - V_f^A\right) \quad \text{(parallel)}, \tag{33}$$

when the mean free path of the conduction electrons is much less than the fibre or

References: p. 1882.

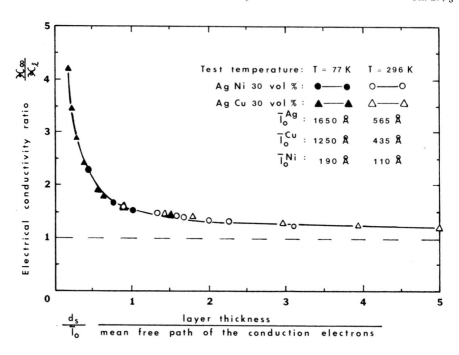

Fig. 22. Ratio of the electrical d.c. conductivity (bulk metal to thin metal layers) for Ag–Ni and Ag–Cu composites at different test temperatures as a function of relative layer thickness.

layer thickness ($\bar{l}_0^l \ll d_{f/l}$). $\mathcal{K}_f^A$ and ρ_f^A are the specific electrical conductivity and resistance of the fibre or layer component A and V_f^A is the volume fraction. For $\bar{l}_0 \gtrsim d_{f/l}$ a quantitative theory has been worked out by FUCHS [1938], DINGLE [1950] and SONDHEIMER [1952], which yields values of $\mathcal{K}_f$ in terms of the mean free path, $\bar{l}_0$ and the scattering factor, p, of the conduction electrons as:

$$\mathcal{K}_f = \mathcal{K}_\infty \left[1 - \frac{\pi}{8} \frac{\bar{l}_0}{d_f} (1 - p) \right]. \tag{34}$$

This equation describes the hyperbolic function of the experimentally obtained values in fig. 22, and the calculated values $\bar{l}_0$ and p of the components are listed for the test temperatures 77 K and 297 K. The mean free path of the electrons increases with decreasing test temperature. The reflection coefficient $p \approx 0.15$ signifies that about 15% of the conduction electrons will be reflected at the phase boundaries and the rest will be diffusely scattered.

A promising technological application of fibre composites for electrical purposes are the type-II-filament superconductors, see TSUEI et al. [1974] as well as BORMANN et al. [1979]. The high-magnetic-field superconductors with relatively high transition temperatures may play an important role in the construction of strong electromagnets in the fusion reactor technology and in powerful electric generators. Multifilament superconductors contain thin Nb_3Sn or V_3Ga fibres of A15 structure with good superconducting properties, embedded in a ductile copper or copper–tin matrix. The intermetallic superconducting fibres are brittle, and therefore a mechanically and electrically stable matrix phase is necessary to achieve optimum properties for technologically useful superconductors. These type II-superconductors have the ability to carry a high electric current in a high magnetic field with normally no energy losses. This is caused by pinning processes of magnetic flux lines by lattice defects and particularly by the fibre–matrix interfaces (HAASEN and FREYHARDT [1974], HARBISON and BEVK [1977]). For a detailed discussion see ch. 27.

Ferromagnetic properties, such as the coercive force, the remanence and the hysteresis loss of submicrostructural composites exhibit anomalous behavior when the very thin ferromagnetic fibres or layers have simple magnetic domain structures. Strong magnetic hardening will be expected by interface pinning processes. This was first noted in iron–silver composites with fibre structure (WAHL and WASSERMANN [1970]). Figure 23 illustrates the magnetic hardening effect of a copper–iron, and a silver–nickel layer-composite in terms of the coercive force (FROMMEYER [1978]). With decreasing layer thickness of the components, a marked increase in the relative coercive force, $H_c/H_{c\infty}$, is observed. The maximum values of the coercive force of the submicrostructural composite are about 30–40 times greater than the coercive force of the composite with bulk components. The ferromagnetic remanence as a function of the layer thickness shows similar behavior. The magnetization of these types of composites is strongly influence by interactions of microstructural defects and residual stresses with Bloch-wall movement and with curling and buckling of the magnetization vectors. Below the critical layer thickness of about $d_1 \approx 100$ nm, these

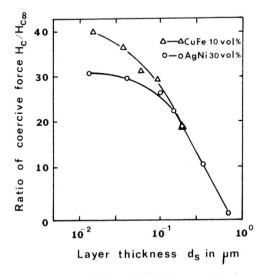

Fig. 23. Influence of the layer thickness of the embedded ferromagnetic phases on the relative coercive force.

References: p. 1882.

mechanisms become less significant and a single-domain structure in layers dominates with a minimum of free energy. The saturation of the relative coercive force at $d \approx 10$ nm is also observed in thin metal films, see e.q. CHOPRA [1969], and indicates a strong interaction of Fe and Ni single domains with the interfaces of the nonferromagnetic components.

References

ANSELL, G.S., T.D. COOPER and F.V. LENEL, 1966, Oxide Dispersion Strengthening, AIME Conference, vol. 47 (Gordon and Breach, New York, London, Paris).

ASHBY, M., 1964, Z. Metallk. **55**, 5.

BERGMANN, H.W., and G. WASSERMANN 1978, Z. Metallk. **69**, 126.

BERGMANN, H.W., G. PHILLIPI and G. WASSERMANN, 1979, Z. Metallk. **70**, 802.

BORMANN, R., L. SCHULTZ, H. FREYHARDT and B.L. MORDIKE, 1979, Z. Metallk. **70**, 467.

BROUTMAN, J. and R.H. KROCK, 1967, Modern Composite materials, vol. 1–5 (Addison–Wesley, Reading, MA).

CHOPRA, K.C., 1969, Thin Film Phenomena (McGraw–Hill, New York).

CHRISTENSEN, R.M., 1979, Mechanics of Composite Materials (Wiley, New York).

CRATCHLEY, D., 1963, Powder Metallurg. **11**, 59.

DINGLE, R.G., 1950, Proc. Roy. Soc. **A202**, 545.

FROMMEYER, G., 1977, Metall **31**, 732.

FROMMEYER, G., 1978, Ber. Bunsen Gesell. Phys. Chem. **82**, 323.

FROMMEYER, G., and H.G. BRION, 1981, Phys. Stat. Sol. (a) **66**, 559.

FROMMEYER, G., and G. WASSERMANN, 1975, Acta Metall. **23**, 619.

FUCHS, K., 1938, Proc. Cambridge Phil. Soc. **34**, 100.

HAASEN, P., and H.C. FREYHARDT, eds., 1974, Proc. Int. Disc. Meeting on Flux Pinning in Superconductors, Sonnenberg, West Germany (Akad. Wiss., Göttingen).

HARBISON, J.P., and J. BEVK, 1977, J. Appl. Phys. **48**, 5180.

HAUSNER, H.H., ed., 1966, Modern Developments in Powder Metallurgy, vol. 1–3 (Plenum, New York).

HERTZBERG, R.W., and R.W. KRAFT, 1963, Trans AIME **227**, 530.

HERTZBERG, R.W., F.D. LEMKEY and J.A. FORD, 1965, Trans. AIME **233**, 342.

KELLY, A., and G.J. DAVIS, 1965, Metallurg. Rev. **10**, 1.

KELLY, A., and W.R. TYSON, 1965, in: High Strength Materials, ed. V.F. Zackay (Wiley, New York) p. 578.

KOPPENAAL, T.J., and N.M. PARIKH, 1962, Trans. AIME, **224**, 1173.

KROCK, R.H., 1965, in: Plansee Proceedings 1964 – Metals for the Space age, ed. F. Benesovsky (Metallwerk Plansee AG, Reuttle/Tyrol) p. 256.

KROCK, R.H., 1966, in: Modern Developments in Powder Metallurgy, vol. 3, ed. H. Hausner (Plenum, New York) p. 105.

KROCK, R.H., and L.A. SHEPARD, 1963, Trans. AIME **227**, 1127.

KURZ, W., and SAHM, P.R., 1975, Gerichtet erstarrte eutektische Werkstoffe (Springer, Berlin, Heidelberg, New York).

LEMKEY, F.D., R.W. HERTZBERG and J.A. FORD, 1965, Trans. AIME **233**, 334.

LENEL, F.V., 1980, Powder Metallurgy (Principles and Applications) (Metal Powder Industries Federation, Princeton, NJ).

MC DANIELS, D.L., R.W. JECH and J.W. WEETON, 1965, Trans. AIME **233**, 636.

NEUMANN, P., and P. HAASEN, 1971, Phil. Mag. **23**, 285.

PIEHLER, H.R., 1965, Trans. AIME **233**, 12.

PRESTON, O., and J.J. Grant, 1961, Trans. AIME, **221**, 164.

ROBERTS, G.S, R.C. CARRUTHERS and B.L. AVERBACH, 1952, Trans. Amer. Soc. Met. **77**, 1150.

SCHREINER, H., 1964, Pulvermetallurgie elektrischer Kontakte, in: Reine und angewandte Metallkunde in Einzeldarstellung, vol. 20, ed. W. Köster (Springer, Berlin).

SEIFERT, R., and G. WASSERMANN, 1981, Z. Metallk. **72**, 381 and 854.

SHERBY, O.D., J. WADSWORTH, R.D. CALIGIURI, L.E. EISELSTEIN, B.C. SNYDER, and R.T. WAHLEN, 1979, Scripta Metall. **13**, 941.

SIMS, G.T., and W.C. HAGEL, 1972, The Superalloys (Wiley, New York).

SONDHEIMER, E.H., 1952, Adv. Phys. **1**, 1.

TSUEI, C.C., M. SUENAGA and W.B. SAMPSON, 1974, Appl. Phys. Lett. **25**, 318.

WAHL, W.P., and G. WASSERMANN, 1970, Z. Metallk. **61**, 326.

WASSERMANN, G., 1973, Z. Metallk. **64**, 844.

CHAPTER 30

SINTERING PROCESSES

H.E. EXNER and E. ARZT

Max-Planck-Institut
Institut für Werkstoffwissenschaften
7000 Stuttgart 1, FRG

R.W. Cahn and P. Haasen, eds.
Physical Metallurgy; third, revised and enlarged edition
© *Elsevier Science Publishers BV, 1983*

Introduction

The technique of forming metal parts from powders by pressing and sintering dates back to the beginning of human civilization. Almost every metal or ceramic material was initially made via the powder route. Modern applications of sintering in materials technology are widespread: powder-metallurgical production of structural steel parts, self-lubricating bearings, porous metals for filtering, tungsten wires for lamp filaments, soft and hard magnetic materials, electrical contacts, cemented carbides for cutting tools and a large variety of ceramic components are only a few of the many technical production processes involving sintering as an important step. A recent book by LENEL [1980] gives a comprehensive review. Sintering processes are also important in a number of other fields of materials technology. One example is sintering of finely ground ores into pellets, thereby preparing them for smelting. On the other hand, sintering of finely dispersed catalyst materials at operating temperatures is undesirable since it lowers their activity.

The consolidation of powders and densification of porous solids is possible by pressing and subsequent pressureless heat-treatment (*solid-state sintering*), by simultaneous application of pressure and heat (*hot-pressing* or *pressure-sintering*) or with the aid of a limited amount of melt (*liquid-phase sintering*). The technology of sintering and hot-pressing has been thoroughly developed down to minute details. Physicists, chemists, metallurgists and ceramists have worked together to establish the theoretical basis for understanding the complex interaction of geometrical and thermodynamic factors and the effects of a number of mechanisms which occur simultaneously or in sequence. Some of these results are discussed in the following.

1. Solid-state sintering

In the majority of technical applications, powders are compacted into shaped parts which then are heat-treated in order to give them the required mechanical and physical properties. The essential difference between a powder (or a porous body) and a dense solid body of exactly the same material and identical mass is, from the energetic point of view, the excess free energy due to the broken atomic bonds at the surface. Stressing this fundamental aspect, we can define *pressureless sintering* as material transport driven by surface energy or, in other words, by capillary forces. This material transport involves several basic mechanisms, essentially of diffusional-flow type. By filling in the necks at the points of contact between the powder particles and, at a later stage, the pore space, sintering increases density and strength of the compacted powders.

Figure 1 shows neck formation between spherical copper particles, and fig. 2 shows the geometrical changes taking place during sintering of tungsten powder. There are numerous experimental methods for studying the progress of sintering in laboratory experiments, reviewed recently by EXNER [1980]. Table 1 gives a summary of the frequently used techniques. In the following we will focus on experimental

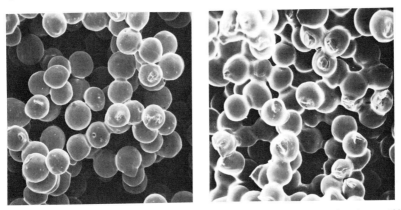

Fig. 1. Neck formation between loosely packed spherical copper particles during sintering at 1300 K for 1 h (left) and 8 h (right). 150×.

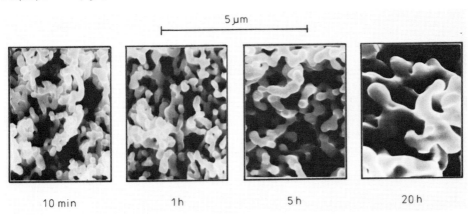

5 µm

10 min   1 h   5 h   20 h

Fig. 2. Fine tungsten powder sintered in hydrogen at 1470 K for 10 min, 1 h, 5 h, and 20 h (from left to right). (After Exner [1979a].)

facts elucidating the basic phenomena and on the theoretical concepts for quantifying sintering mechanisms.

1.1. Driving energy

Surface energy provides the motivation of material transport during pressureless sintering. Its magnitude can be estimated theoretically or determined experimentally in various ways and ranges for metals, alloys, intermetallic compounds and non-metallic crystalline solids from a tenth to a few J/m^2 (for references see Roth [1975] and Exner [1982]). The most reliable and instructive technique of visualizing the effect and measuring the value of surface energy is the so-called *zero-creep* technique

References: p. 1908.

Table 1
Experimental method for deriving quantitative information on changes of pore geometry during sintering.

Method	Quantity measured	Remarks
Dilatometry	Length change	Shrinkage may vary in different directions. Relative precision approximately 10^{-6} of sample length.
Buoyancy	Density	Impregnation or pore sealing necessary. Relative precision approximately 10^{-3}.
Gas	Solid–pore interface	Only for high specific interfaces (> 0.1 m^2/cm^3). Closed pores not included. Relative precision approximately 10^{-2} for total areas > 0.5 m^2.
Porosimetry	Accessible pore volume	For open and fine pore systems. Interpretation of pressure–volume diagrams difficult.
Indirect methods	Physical properties	Exact relationships between pore geometry and properties usually not known.
Quantitative microscopy	Direct geometric parameters	Tedious but most effective method for complete characterization of pore geometry.

first described by UDIN *et al.* [1949]. This technique is described in ch. 13, §3.1.

Depending on the size of the powder particles or the amount and dispersion of porosity in a compact, the total excess energy of the surface amounts to 0.1–100 J per mole of solid where the smaller number applies to coarse powders (~ 100 μm diameter) and to low-porosity material and the larger number to submicron powders or highly dispersed porosity. Grain-boundary energy usually provides a back-driving force because, at least in the early stages, new grain boundaries are formed while the particle contacts are being filled in (fig. 3). For most materials (especially metals)

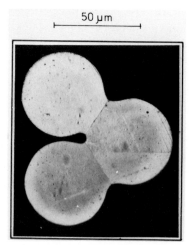

50 µm

Fig. 3. Metallographic cross-section through an arrangement of three copper particles sintered at 1300 K for 8 h. Note the grain boundaries in the necks. (From EXNER [1979a].)

grain-boundary energies are lower (approximately by a factor of three) than surface energies, and the dihedral angle is of the order of 160°. Therefore the back-driving effect is not very pronounced. However, as pointed out by HOGE and PASK [1977], material transport during sintering may cease because of the establishment of local equilibrium between grain-boundary- and surface energy in cases where the grain-boundary energy is high (dihedral angle < 120°).

If a volume element dV is removed from (dV < 0) or added to (dV > 0) a surface with the principal radii of curvature r_1 and r_2, the energy change, dE, involved is

$$\frac{dE}{dV} = \gamma \frac{dA}{dV} = \gamma \left(\frac{1}{r_1} + \frac{1}{r_2} \right), \tag{1}$$

where γ is the (isotropic) surface energy and dA is the change in surface area. dE/dV formally corresponds to a stress, σ, which leads to the well-known Laplace equation

$$\sigma = \gamma \left(\frac{1}{r_1} + \frac{1}{r_2} \right). \tag{2}$$

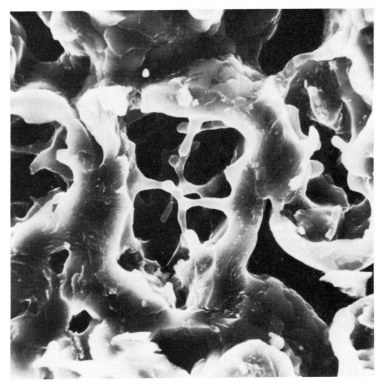

Fig. 4. High-porosity beryllium alloy produced from cobalt-coated beryllium particles by sintering. 1000×. (From ALDINGER [1974].)

References: p. 1908.

Under a concave surface, this stress is tensile and under a convex surface, compressive. Therefore, a stress gradient exists between the particle interior and the neck. Similarly, a gradient in chemical potential can be defined which, in turn, corresponds to a gradient of vapour pressure (Kelvin equation) or vacancy concentration (Thomson–Freundlich equation, see ch. 6, §9.3). Owing to the stress gradient, the difference in vapour pressure, or the gradient in vacancy concentration, material is transported by viscous flow (in amorphous materials), by plastic flow, evaporation/vapour transport/condensation, diffusion along the surface and grain boundaries, or volume diffusion.

Chemical driving forces due to nonequilibrium composition of powders (e.g. mixture of elemental powders which react at sintering temperature) are usually much higher than capillary forces. This fact, which becomes apparent by merely comparing the molar surface energies with energies of mixing or compound formation, has been demonstrated in models, e.g. by sintering a cobalt sphere to a nickel plate (THÜMMLER and THOMMA [1966]). The same effect causes the well-known Kirkendall porosity (ch. 5, §5.3) and has been used to prepare highly porous bodies from nickel or cobalt-covered beryllium spheres, as shown in fig. 4. However, in special cases chemical driving forces can be overcome by capillary forces: owing to the higher diffusivity of indium, an intermetallic compound is formed at the neck region between two wires of an originally homogeneous solid solution of copper–indium (KUCZYNSKI et al. [1960]), and silver is enriched in the neck region between initially homogeneous Ag–Au spheres (MISHRA et al. [1975]). However, rehomogenization occurs as soon as the sharp curvatures at the neck surface are filled in and capillary forces are reduced. Chemical effects, which play a major role in most real systems, must clearly be differentiated from ideal surface-energy controlled sintering.

1.2. Material sinks and sources

Possible sinks and sources for material (or vice versa for vacancies) are the surface (pore–solid interface), grain boundaries, and dislocations. The role of the surface is clear: Since surface energy is reduced when concave regions are filled in, these regions (in the early stages predominantly the highly concave neck surface, in the later stages the concave parts of the pore–solid interface) are the sinks for material (or the sources of vacancies). The convex part of the particle surface or solid–pore interface is one of the material sources. It is obvious that redistribution of material over the surface by surface and/or volume diffusion will not result in shrinkage of the compact or reduction of pore space (i.e. densification) but can only increase strength by enlarging the contact areas and reducing the notch effect of sharp pore contours.

Densification occurs when material is removed from the volume between the particle contacts. In numerous experiments, the decisive role of grain boundaries as sinks for vacancies arriving from the neck surface or pores has been demonstrated (ALEXANDER and BALLUFFI [1957], ICHINOSE and KUCZYNSKI [1962]; for further references see GEGUZIN [1973], USKOKOVIĆ and EXNER [1977], and EXNER [1979a]).

Pores shrink only when attached to or located very close to grain boundaries, and particle centres approach each other only when the particles are separated by at least one grain-boundary. A most convincing experiment is the comparison of shrinkage rates of sintered copper with a high density of grain boundaries, resulting in pronounced shrinkage, with that of dezincified brass with a low density of grain boundaries, showing virtually no shrinkage (BRETT and SEIGLE [1963]).

Grain boundaries are usually assumed to be perfect vacancy sinks at the supersaturation levels caused by capillarity, but recent models for the structure of grain boundaries (ASHBY *et al.* [1978], BALLUFFI [1980], ch. 10B, § 2.2) assume a fairly well-defined structure that would be disrupted if atoms were removed. This leads to the suggestion (ASHBY [1972], ARZT *et al.* [1982]) that a divergence of the diffusive flux of matter can occur only at dislocation-like defects (grain-boundary dislocations as observed, for example, by GLEITER [1969], SCHOBER and BALLUFFI [1970], NILSSON *et al.* [1979] and KING and SMITH [1980]). The details of these dislocation structures (ch. 10B, § 2.2) can depend on the misorientation of the adjacent grains. By emitting atoms, these defects move in a non-conservative way in the boundary plane. The effectiveness of a grain boundary as a source for atoms or sink for vacancies is thus determined by the mobility of the grain-boundary dislocations, which may be reduced by solute atoms exerting a viscous drag, or by grain-boundary particles pinning the dislocations. The pronounced dependence of neck growth upon orientation between a zinc single crystal sphere and a zinc plate (NUNES *et al.* [1971]), the variation of neck sizes between simultaneously sintered copper spheres (EXNER [1979a]), the low apparent diffusivities observed in some sintering experiments (for references see EXNER [1979a]) or the complete suppression of shrinkage in metals containing dispersed particles (ASHBY *et al.* [1980]) can be understood in these terms, as well as the effect of grain-boundary structure in zero-creep experiments (JAEGER and GLEITER [1978]). (In this connection see also ch. 25, § 3.4.1.)

There is more or less general agreement that lattice dislocations cannot be generated under the action of capillary stresses prevailing during sintering. Nevertheless, high densities of lattice dislocations have been observed in regions close to the grain boundary of a particle sintered to a plate and it has been suggested that these dislocations may contribute to the shrinkage in various ways (SCHATT *et al.* [1982]). For example, diffusion can take place from one dislocation to another, which can result in removal of atomic planes in the direction of compressive stress and insertion of those under tension. Lattice dislocations then act both as vacancy sources and sinks, and move non-conservatively by climb (NABARRO [1967], SCHATT [1981]). Obviously, increased dislocation density in the neck region would also provide short-circuits for diffusion and thus reduce the effective diffusion distance between material sinks and sources.

1.3 Neck growth and shrinkage equations

There have been a large number of attempts to quantify the kinetics of neck growth and shrinkage. Early work dating back to the time between the two World

References: p. 1908.

Table 2
Constants appearing in eqs. (3) and (5) (after EXNER [1979a]).

Transport Mechanism	Range		Plausible Values [b]		
	n	m	n	m	C
Viscous flow	2	1	2	1	$\dfrac{3\gamma}{2\eta}$
Evaporation and recondensation [a]	3 – 7	2 – 4	3	2	$3 \cdot \sqrt{\dfrac{2}{\pi}} \dfrac{\nu\gamma p M^{3/2}\rho^{1/2}}{(RT)^{3/2}}$
Grain-boundary diffusion	6	4	6	4	$\dfrac{48\gamma \cdot D_b \cdot b \cdot M}{RT}$
Volume-diffusion from the grain boundary	4 – 5	3	5	3	$\dfrac{16 D_V \cdot \gamma \cdot M}{RT}$
Volume diffusion from the surface [a]	4	3	4	3	$\dfrac{20 D_V \cdot \gamma \cdot M}{RT}$
Surface diffusion [a]	3 – 7	2 – 4	7	4	$\dfrac{23\gamma \cdot D_s \cdot w \cdot M}{RT}$

[a] No centre approach, $h/a \equiv 0$.
[b] With the symbols in the expressions for C denoting:

D_b, D_S, D_V = diffusion coefficients for grain boundary, surface and volume diffusion; M = molar volume; R = gas constant; T = absolute temperature; b = grain-boundary width; p = gas pressure; w = width of surface atom layer; γ = specific surface energy; η = viscosity constant; ν = accomodation coefficient for gas transport; ρ = specific gravity.

Wars (TAMMANN [1926], HEDVALL and HELIN [1927] and BALSHIN [1936]) is characterized by intuition rather than by physical reasoning (GEGUZIN [1973]). Some of the important fundamental ideas can be found in papers published around 1950 (e.g. FRENKEL [1945], JONES [1946], PINES [1946], HÜTTIG [1948], SHALER and WULFF [1948], MACKENZIE and SHUTTLEWORTH [1949] and CLARK and WHITE [1950]). In particular, the work by KUCZYNSKI [1949] and HERRING [1950] marks the beginning

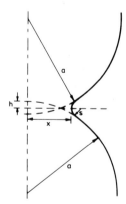

Fig. 5. Geometry of an idealized neck contour between two spherical particles (a = particle radius, x = neck radius, s = radius of neck curvature for a tangent circle, h = half of the centre approach and particle interpenetration).

of a quantitative modelling of sintering processes. Several research schools have developed since then in the various parts of the world (for reviews see FISCHMEISTER and EXNER [1964], THÜMMLER and THOMMA [1967], GEGUZIN [1973] and EXNER [1979a]). The classical approach has been the study of two coalescing particles (Frenkel–Kuczynski model), usually spheres or rods, in order to reduce problems inherent in the highly complicated geometry of powder compacts. Using idealized neck shapes (cylindrical or toroidal necks which result in contours formed by tangent circles, see fig. 5), analytical relationships were derived for the time- and particle-size dependence of neck size, x, during isothermal sintering. These equations are generally of power-law type

$$(x/a)^n = Ca^{-m}t \tag{3}$$

where a is the radius of the spherical or cylindrical particles in contact which each other, t is the time of isothermal sintering, and n, m and C are constants which are typical for the individual transport mechanism. Table 2 (see also COBLENZ *et al.* [1980], and GERMAN [1982]) lists the range of values of n and m given in the literature and plausible values for n, m and C following derivations by FRENKEL [1945] for viscous flow, by KINGERY and BERG [1955] for evaporation and condensation, and by ROCKLAND [1967] for the diffusional transport mechanisms. From the neck growth the amount of interpenetration of the two particles, and thus the centre approach between the two particles, can be estimated. Evaporation and condensation, volume diffusion from the surface, and surface diffusion cannot produce centre approach. For the other mechanisms it is assumed that all the material filling the neck comes from the grain boundaries in contact regions, and using again the idealized geometry and neglecting second-order terms, the approach of the two particle centres, $2h$, (see fig. 5) can be estimated from the neck size, x:

$$h = x^2/4a. \tag{4}$$

Using eq. (3), the time- and temperature dependence of the relative centre approach, h/a, is:

$$(h/a)^{n/2} = 2^{-n}Ca^{-m}t, \tag{5}$$

with C, n and m shown in table 2 for the relevant mechanisms.

Viscous flow and volume diffusion can act on their own to remove material from regions between the particle centres while grain-boundary diffusion obviously needs the cooperation of another process to distribute the material reaching the surface at its intersection with the grain boundary, e.g. surface- or volume diffusion, which may then be rate-controlling (COBLE [1958], GESSINGER [1970], JOHNSON [1970] and SWINKELS and ASHBY [1980]). The major role of surface diffusion is to reshape the surface in such a way that the curvature gradient (and thus the chemical potential gradient) changes continuously over the surface and sharp curvatures are reduced quickly. This effect, termed *undercutting* (NICHOLS and MULLINS [1965]), is clearly revealed in fig. 6; it is particularly pronounced at the early stages of contact formation, producing a bulb-shaped neck contour.

References: p. 1908.

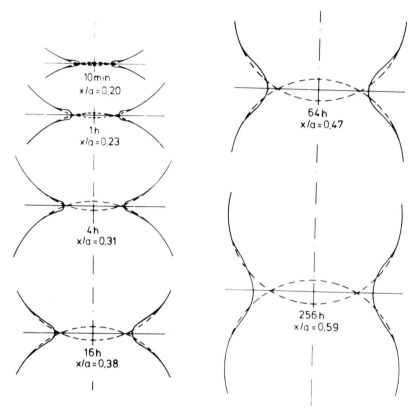

Fig. 6. Neck shapes developed between two copper spheres during sintering at 1200 K. The redistribution of material along the surface close to the neck causes undercutting. The dashed lines show the circular contours of the original particles, the fully drawn lines are taken from shadow-graphs. (From EXNER [1979a].)

There have been numerous attempts to derive equations for neck growth and centre approach for more realistic neck geometries and for simultaneously acting sintering mechanisms. Reviews and detailed discussions have been published recently by COBLENZ et al. [1979] and EXNER [1979a]. Analytical solutions are still possible for neck geometries with elliptical or cartenoid-shaped contours (SWINKELS and ASHBY [1980] and GERMAN and MUNIR [1975]), but numerical techniques must be applied for assumption-free modelling of neck shapes (for recent solutions and references see BROSS and EXNER [1979], EXNER [1979a], NICHOLS [1980] and ROSS et al. [1982]). Figure 7 shows, as an example, the geometry for simultaneous grain-boundary and surface transport obtained by computer simulation, which, compared to the simple tangent circle approximation, conforms much more closely to the experimental geometry (compare fig. 6).

The results of these complex calculations are not as easily visualized or applied as the simple power laws [eqs. (3) and (5)] and, therefore, have not gained much popularity. Usually, however, application of the simple equations is taken too far in interpreting experimental work quantitatively, considering the severe geometric simplifications and the problems connected with superimposed mechanisms (in addition to chemical effects due to impurities, among other divergences between theoretical assumptions and real systems). The same restrictions must apply to *sintering maps* delineating the temperature and neck-size regions in which a specific material transport mechanism is predominant (ASHBY [1974] and SWINKELS and ASHBY [1981]). Unlike the *hot-pressing maps* discussed below (see figs. 12 and 13), these sintering-mechanism diagrams are rather limited for quantifying the effects of various sintering parameters for practical purposes owing to the severe simplifications they are based on. Nevertheless, they provide a convenient means of visualizing the results of theoretical calculations.

In spite of these principal deficiencies, the theoretical treatments and experiments based upon the two-particle approach have been highly successful in categorizing the active material-transport processes. There is general agreement that sintering of most metals and ceramics can be understood on the basis of *diffusional or Herring–Nabarro–Coble creep* (ch. 20, § 1.11) with surface diffusion playing a major role, while vapour transport and plastic flow play a minor part which, however, cannot be neglected in all cases. Most of the basic phenomena occurring in the early

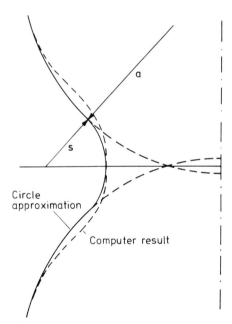

Fig. 7. Computer simulation of neck formation between two spherical particles for superimposed surface- and grain-boundary diffusion (*a* and *s*: compare fig. 5. $x/a = 0.58$). Note the agreement with experimental results shown in fig. 6. (From BROSS and EXNER [1979].)

stages of solid-state sintering are now well understood, and the general findings give
a sound basis for a qualitative understanding of the later stages and more complex
geometries as, for example, for contacts between pressed spheres (WELLNER *et al.*
[1974]), spheres of different diameter (COBLE [1973]), for nonspherical geometries
(USKOKOVIĆ and EXNER [1977]) and for models consisting of three or more particles
(EXNER [1979a]).

Nearly unsurmountable problems are present when a concise quantitative description of the sintering behaviour of real powder compacts is attempted by extrapolating the results derived for two-particle models to multiparticle systems. Shrinkage
equations have been derived in a straightforward way by simply assuming that the
relative linear shrinkage, $\Delta L/L$, is equal to the ratio of centre-approach and particle
diameter, h/a. Thus, the well-known and frequently applied power-law shrinkage
relationship dating back to KINGERY and BERG [1955] is obtained:

$$\Delta L/L = h/a = ct^n, \tag{6}$$

where: $\Delta L/L$ = relative linear shrinkage, c, n = constants, t = sintering time. However, this extrapolation has no relevance and determination of the exponent n from
shrinkage experiments with powder compacts (which was highly popular during the
last three decades) is of no use for identifying the dominant sintering mechanism. To
prove this fact, fig. 8 shows the pronounced difference of individual relative centre
approach and shrinkage values measured for irregular planar arrays of copper

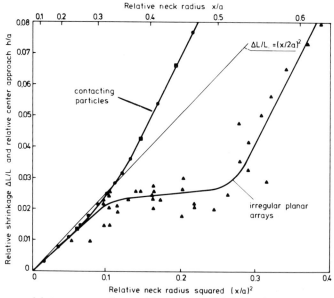

Fig. 8. Centre approach between contacting particles and overall shrinkage in an irregularly packed planar
array of equally sized copper spheres sintered at 1300 K. The straight line corresponds to eq. (2). Note the
pronouncedly lower shrinkage for the irregular array than predicted from centre approach. (From EXNER
[1979a].)

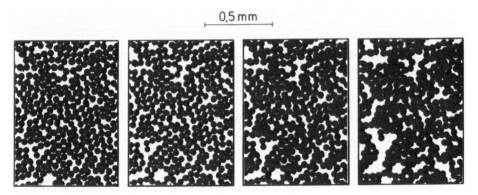

Fig. 9. Planar array of spherical copper particles sintered at 1300 K. Identical field at different stages (from left to right: presintered for 1 min, sintered for 2, 5 and 25 h). Note the local shrinkage and opening of large void regions. (From EXNER [1979a].)

spheres. The reason is that particles shift positions, contacts form or break up, and pores open up owing to rearrangement processes (fig. 9). Though uniform packing and pressing reduce these differences (EXNER *et al.* [1973] and EXNER [197a]), particle shapes typical for technical powders, local variations in packing, asymmetric contact geometry, and stresses induced by nonuniform particle size and packing may cause pronounced rearrangement and deviations from uniform shrinkage and thus invalidate eq. (6) for most practical applications. EVANS [1982] has shown that the stresses induced by inhomogeneous shrinkage may be large enough to open cracks as frequently observed in sintered ceramic compacts.

The decisive part of shrinkage takes place during a sintering stage where the individual necks and powder particles grow together and cannot be clearly identified any longer. The pore space then is still a continuous network embedded in an (obviously) continuous solid. During this stage, porosity is attached to the grain boundaries (or vice versa). Simple geometric, statistical and stereological models have proven useful for describing this situation (see EXNER [1980]). The most successful model, designed by COBLE [1961], has been a geometrically simple arrangement of cylindrical pores situated along the edges of regularly polyhedral grains. The flow of vacancies to each of these cylindrical pores is independent of its radius, r, because the higher curvature of a thin pore channel (proportional to $1/r$) is compensated by its smaller area for the arriving atoms (proportional to r). A cubic time-dependence for grain growth (increase of edge length of the polyhedral grains) and diffusional creep as rate-controlling mechanisms are assumed. This model yields a logarithmic dependence of porosity, P, upon sintering time t:

$$P_0 - P = k \ln t/t_0,\tag{7}$$

where P_0 and t_0 are porosity and time at the beginning of isothermal sintering, respectively. The cubic relationship for grain growth has repeatedly been confirmed for porous materials while, as pointed out in ch. 25, §4.1, a quadratic relationship is typical for pure dense metals. The linear dependence of porosity on the logarithm of time has been found to fit many of the experimental results obtained with powders of alumina, copper, silver, iron, cobalt and nickel (for references see EXNER [1979a]). More recently, refined versions of this model have been published (e.g. BEERE [1975], KUCZYNSKI [1975, 1978], EADIE *et al.* [1978] and WONG and PASK [1979]). Its interesting aspects are the very clear connection between grain growth and densification and the extreme simplicity of the geometric and thermodynamic reasoning.

Table 3 shows a systematic survey of these and other popular concepts for deriving equations for isothermal shrinkage (linear or volume shrinkage, increase of

Table 3

Basic concepts for deriving quantitative relationships for shrinkage kinetics of powder compacts (after EXNER [1980]).

Basic concept	Authors	Critical remarks
Extrapolation of idealized models:		
Two-particle model	KINGERY AND BERG [1955] COBLE [1958, 1970] ROCKLAND [1967]	Nonrealistic assumptions on neck shape. Effects of asymmetry and rearrangement not considered. Explicit equations valid only for single mechanism. Not applicable to later sintering stages.
Simple-pore model	COBLE [1961]	Geometry greatly oversimplified. Empirical grain-growth relationship.
Statistical approach	KUCZYNSKI [1978]	Simplifying geometric assumption on pore shape (no convex interfaces).
Rheological approach	SKOROKHOD [1972]	Defect geometry and interaction porosity–defects not clearly defined. Phenomenological definition of viscosity.
Sterological approaches	JOHNSON [1972] AIGELTINGER and DROLET [1974]	Integration of rate equation not possible. Phenomenological introduction of geometric parameters. Valid for early or late sintering stage, respectively.
Empirical approaches	TIKKANEN and MÄKIPIRTTI [1965]	Posterior motivation of empirical equation on basis of vacancy– dislocation interaction.
	IVENSEN [1973]	Posterior motivation of ad hoc equation based on non-defined defects. 3 adjustable parameters.
	Other authors (for references see EXNER [1979a])	Mathematical derivations or phenomenological and purely pragmatic equations without reasonable physical support.

density, or decrease of porosity) together with some major objections (EXNER [1980]). Each of these concepts has merits in focusing attention on particular aspects of the sintering process. The equations derived analytically or empirically were often applied to a variety of materials with little awareness of the critical assumptions implied. Nevertheless, excellent agreement with experimental data has been obtained. This is due to the fact that most equations contain more than one adjustable parameter, which makes it difficult to reach conclusive differentiation of the quality of different sintering equations (and the underlying ideas) on the basis of quality of fit (PEJOVNIK *et al.* [1979]).

In the very late stages of sintering (below 10% porosity), when isolated and geometrically well defined pores are present, a more realistic modelling of the behaviour of real materials is possible. Especially in ceramics, where high density values are striven for, breakaway of grain boundaries from the pores is the pre-eminent barrier to complete densification (see next section). In the light of the extensive efforts to give a proper theoretical basis to sintering processes, it is interesting to note that the techniques for improving sintering behaviour of materials which are difficult to densify have been developed essentially empirically. To date, the effects of dopants, such as transition metals in tungsten or MgO in Al_2O_3, which activate sintering dramatically, or the influence of the composition of sintering atmospheres, are still a matter of controversy (see, for example, GERMAN and MUNIR [1982]).

In addition, the effects of compacting, density distribution, contact geometry, structural defects, impurities and other factors of influence are understood for specific cases but not in terms of a general theory, and further work will be needed to close some of these gaps.

1.4. Development of microstructure and grain growth

The initial state of microstructure in a pressed compact is characterized by the microstructure of the individual particles and the shape of the pore space which, in turn, is determined by the shape and the arrangement of the particles. During the initial stage of sintering, grain boundaries develop at the contact regions, and the grain structure in the particles changes quickly by recrystallization if deformation during pressing has exceeded a critical value. The microstructure is then characterized by a continuous pore space with grain boundaries located at the small cross-sections of the solid. Further development of microstructural geometry (i.e. shape, dimensions and topological arrangement of porosity and grain-boundary network) can be assessed by quantitative microscopy (ch. 10A, §2.4). Descriptive work using stereological and topological parameters (DE HOFF and AIGELTINGER [1970], AIGELTINGER and EXNER [1972], JERNOT *et al.* [1980] and CHERMANT *et al.* [1981]) shows that the geometric changes related to the reduction of pore volume are fairly similar for most materials. During the intermediate stage, the pore space forms a continuous network attached to grain boundaries, and the dimensions of the pore cross-section and the grains are related in much the same way as predicted by the

References: p. 1908.

Zener relationship (KUCZYNSKI [1975]):

$$R = k\frac{r}{P}, \tag{8}$$

where R and r are typical average dimensions of the grains and the porosity, respectively (e.g., mean linear intercepts or area-equivalent radii of planar cross-sections). The constant k is of the order of unity and related to the shape and to the size-distribution of the grains and the cross-sections of the pore network (see also ch. 25, § 3.7).

Only later in the sintering process, when the continuous pore system breaks down, are more or less equilibrium-shaped isolated pores formed at the grain boundaries (fig. 10). Theoretical studies show how the grain-growth kinetics are influenced by pores (MOCELLIN and KINGERY [1973], CARPAY [1977], CAHN [1980], HSUEH et al. [1982], SPEARS and EVANS [1982] and WEI and GERMAN [1982]). Pores can be dragged by grain boundaries and coalesce when meeting at grain edges or corners until they break away from the grain boundaries. The separation limits final densification since volume diffusion is usually much slower than grain-boundary diffusion. Additionally, local breakaway results in exaggerated grain growth or secondary recrystallization (ch. 25, § 4.3). Thus, the separation process is a decisive step in final-stage sintering, determining the microstructural details as well as the remaining porosity.

Grain-boundary motion, pore drag, pore shrinkage, pore coalesence, and pore detachment from grain corners, edges and facets have been modelled in order to explain the effects or process variables as heating rate, sintering time, sintering temperature, dopants etc. (COBLE and CANNON [1978], HSUEH et al. [1982] and SPEARS and EVANS [1982]). Figure 11 shows schematically the dependence of pore size on grain size and the conditions under which the pores become separated from the grain boundaries. Two pore-grain trajectories are indicated, one of which indicates a technically advantageous sintering route leading to full densification and

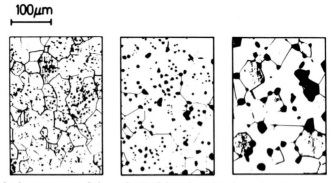

Fig. 10. Typical microstructures of sintered materials at late sintering stages. In carbonyl iron (left) the pores are mostly separated from grain boundaries, in magnesia (middle) small pores lie inside grains and the larger ones on grain boundaries, and in zinc oxide (right) most pores are on grain boundaries and corners. (From EXNER [1979b].)

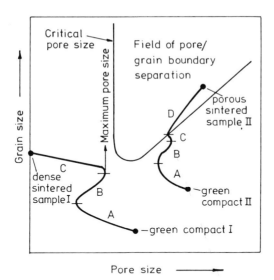

Fig. 11. Schematic plot of theoretical results indicating the relationship between pore and grain size in the final sintering stages. Sample I: favourable path for high final density. Sample II: path for porous product. A: rearrangement and densification by contact formation. B: intermediate stage shrinkage and coarsening. C: normal grain growth and final densification. D: pore–grain boundary separation, exaggerated grain growth, Ostwald ripening of isolated pores and cease of densification. (After SPEARS and EVANS [1982].)

a fine grain microstructure, while the second one enters the pore-detachment region resulting in residual porosity and coarse microstructure. In spite of the detailed models and derivations, equations describing these processes and their interactions, care must be taken not to carry the conclusions too far for practical cases. There are still many assumptions unproven by experimental facts (see, for example, CAHN [1980]). On the other hand, however, one should be able to optimize sintering conditions on this basis in order to obtain taylor-made microstructures in sintered materials, and to better understand the pronounced effects of special heating schedules such as rate-controlled heating (HUCKABEE *et al.* [1978]), quasi-isothermal heating (SOERENSEN [1980]), rapid sintering (BROOK [1969]), etc., by comparing the experimental findings with theoretical approaches and ideas.

2. *Hot pressing (pressure-sintering)*

Powders which are difficult to sinter can be consolidated to high density and strength by simultaneous application of heat and pressure. Such a process may be thought of as sintering enhanced by a pressure (*pressure-sintering*) or as a pressing operation activated by high temperatures (*hot-pressing*). The pressure allows lower temperatures than in pressureless sintering to be employed; thus normal grain growth can be reduced and abnormal or exaggerated grain growth, caused by breakaway of grain boundaries from pores, may be avoided. If the pressure is applied uniaxially, the process is commonly referred to as hot-pressing, while *hot Isostatic pressing* (HIP, for short) utilizes hydrostatic pressure. The HIP process in particular has attained the position of the leading hot consolidation process; presses are commercially available which allow hydrostatic pressure, transmitted by gas, to be applied to a large heated volume (HANES *et al.* [1977] and FISCHMEISTER [1978]). Examples of its successful application are the production of tool steels, of superalloys, of alumina (e.g. for nuclear waste encapsulation), and of ceramic cutting tools.

2.1. Stresses and mechanisms

When an external pressure is applied to a hot powder compact, it is transmitted through the powder bed as a set of forces acting across the particle contacts. The exact force distribution depends on the stress state: in uniaxial hot-pressing some of the pressure is dissipated by die-wall friction, while HIP produces a uniform distribution resulting in more homogeneous densification. The local contact force per unit contact area (the "effective pressure") may exceed the capillary stress by several orders of magnitude, especially in the early stages of densification. This high stress not only enhances the diffusional processes contributing to pressureless sintering, but also introduces new densifying mechanisms. When the pressure is first applied, the contacts between the particles are small and the effective pressure will be sufficient to cause instantaneous plastic yielding in the contact zone. The resulting contact flattening leads to a rapid attenuation of the effective pressure until yielding stops. Then time-dependent deformation mechanisms determine the rate of further densification: power-law creep and stress-enhanced diffusion from a grain-boundary source to the neck surface, as in pressureless sintering. The contribution of vapour transport is even more insignificant than in pressureless sintering because it is not enhanced by the applied pressure.

It is convenient to divide the densification process into two stages (sometimes a third, intermediate stage is assumed which bridges the two mentioned here). During the initial stage the individual particles, which are commonly assumed to be spherical, can still be distinguished. The densification is determined by the deformation of the particle contacts caused by the local effective pressure acting on the contact area. In isostatic compaction, this effective pressure, σ, is proportional to the applied pressure, p (KAKAR and CHAKLADER [1968], COBLE [1970] and MOLERUS [1975]), e.g.:

$$\sigma = \frac{4\pi a^2}{AZ\rho} p, \tag{9}$$

where a is the particle radius, ρ the relative density of the compact (volume fraction of the solid), and A and Z are the average contact area and number of contacts per particle, respectively. Under certain assumptions concerning the structure of the particle-packing and the contact geometry, A and Z can be expressed as functions of ρ (ARZT [1982] and FISCHMEISTER and ARZT [1983]).

During the final stage, at relative densities greater than roughly 90%, the compact is usually modelled as a homogeneous solid containing isolated spherical pores. The effective pressure causing densification is then identical with the applied pressure, unless gas trapped in the pores causes a back-pressure which may prevent the compact from reaching full density.

2.2. Densification models

Numerous theoretical models have been developed for the mechanisms leading to densification in pressure-sintering. The initial densification upon application of the pressure is due to *plastic yielding*, provided the effective stress [eq. (9)] exceeds the stress necessary for plastic flow of two contacting spheres. This stress is higher than the yield stress in simple compression, because the material around the contact zone acts as a constraint for plastic deformation. The appropriate slip-line field is similar to that of a punch indenting a flat surface (or a hardness indentation) for which the yield criterion requires (HILL [1960]):

$$\sigma \approx 3\sigma_y, \tag{10}$$

where σ is the indentation stress and σ_y the yield stress, at temperature, of the powder material in compression. Yielding enlarges the contact areas and, as a consequence of the resulting densification, increases the number of contacts per particle. Once the effective pressure [eq. (9)] drops below the indentation stress [eq. (10)], yielding stops.

If the applied pressure is high, the compact may instantaneously reach densities above 90%. It has then entered the "final stage" by yielding alone and its behaviour is better modelled as the plastic collapse of a thick spherical shell (TORRE [1948] and HEWITT *et al.* [1973]), which requires:

$$p = \tfrac{2}{3}\sigma_y \ln(1 - \rho). \tag{11}$$

Densification by *power-law creep* has been considered only recently, although it is often the dominant densifying mechanism (WILKINSON and ASHBY [1975], MATTHEWS [1980] and ARZT *et al.* [1983]). Exact solutions for this case are difficult to obtain; an approximation for the densification rate whose validity has been confirmed (SWINKELS *et al.* [1983]) is given by:

$$\frac{d\rho}{dt} = f(\rho, G)\frac{x}{a}\dot{\epsilon}_0\left(\frac{\sigma}{3\sigma_0}\right)^n, \tag{12}$$

where σ_0 and n are material properties, $\dot{\epsilon}_0$ is proportional to the volume diffusivity, x is the radius of the contact area and $f(\rho, G)$ is a function of the density and the geometry only.

As a model for final-stage densification, the creep of a thick spherical shell can be analyzed. This leads to the following equation for the densification rate:

$$\frac{d\rho}{dt} = f(\rho)\dot{\epsilon}_o\left(\frac{3}{2n}\frac{\sigma}{\sigma_0}\right)^n, \tag{13}$$

References: p. 1908.

where $f(\rho)$ is a complicated function of the relative density ρ.

A common feature of densification by plastic flow and power-law creep is their independence of the particle size. Further, both mechanisms are highly stress-sensitive, and for this reason the calculation of the effective stress is critical. Second-order effects, such as the increasing number of particle contacts, which at constant external pressure implies diminishing contact forces, can become important (ARZT [1982]). Also the question of whether the pores remain cusp-like or sintering necks are formed can be of relevance (SWINKELS *et al.* [1983]).

Diffusion, which is the most important densifying mechanism in pressureless sintering, contributes to pressure-sintering, too. It is enhanced by an applied pressure, because the additional energy $\sigma\Omega$ is gained on removing an atom with the volume Ω from a grain boundary acted on by a compressive traction σ (HERRING [1950], COBLE [1970]). The total chemical-potential difference between the grain boundary and the neck surface is

$$\Delta\mu = \sigma\Omega + \gamma\Omega\left(\frac{1}{r_1} + \frac{1}{r_2}\right). \tag{14}$$

Under normal pressure-sintering conditions, the second term (the capillary stress) is negligible and the driving force for densification is provided only by the external pressure. Hence pressure-sintering by diffusion, unlike pressureless sintering, is not sensitive to the exact shape and curvature of the sintering necks, and simplifying assumptions about the contact geometry will be less critical.

The rate of densification by diffusion is proportional to the number of atoms deposited, per second, on the surface of the sintering necks:

$$\frac{d\rho}{dt} = g(\rho, G)\frac{bD_b + sD_v}{kT a^3}\Delta\mu. \tag{15}$$

Here D_b and D_v are the boundary and volume diffusivities, b is the boundary thickness, k is Boltzmann's constant, T the absolute temperature, and s is the radius of curvature of the neck surface. $g(\rho, G)$ is a function of the relative density and the geometry. Here the particle radius a is a critical variable: fine powders densify much faster by diffusion than do coarse powders. Since a stress biases the chemical potential in a linear manner [eq. (14)], the rate of diffusional densification is directly proportional to the effective stress. A similar equation describes diffusional densification during the final stage.

There are several complicating features of diffusional densification. One is the limited efficiency of grain boundaries as vacancy sinks, which may result in a quadratic stress-dependence of the densification rate or a complete suppression of densification below a threshold stress (ARZT *et al.* [1982], see also §1.2). Another is the influence of impurity segregation at grain boundaries on the boundary diffusivity (ch. 13, §6.2). Further, the redistribution, by surface diffusion, of atoms arriving at the neck by grain-boundary diffusion can become the rate-limiting process. This has been recognized in pressureless sintering (§1.3). Because a pressure will not enhance this redistribution process, the effect should be even more pronounced in pressure-sintering.

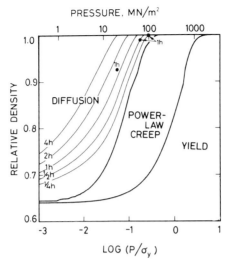

Fig. 12. A theoretical density–pressure diagram for hot-isostatic pressing (HIP) of a tool-steel powder with a particle radius of 25 μm, at a temperature of 1200°C. σ_y is the yield stress of the powder material. Data points correspond to typical industrial HIP cycles (with time marked in hours). Follow a vertical line (at a given external pressure p) in the direction of increasing density: the initial densification is due to plastic yielding of the particle contacts; at intermediate densities power-law creep in the contact zone dominates; diffusion may finally achieve full density. The thin lines are contours of constant time. (After ARZT et al. [1983].)

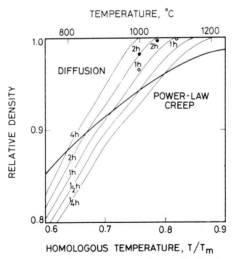

Fig. 13. The high-density portion of a density–temperature diagram for hot-isostatic pressing of the same powder as in fig. 12, at an applied pressure of 100 MN/m² (after ARZT *et al.* [1983]).

References: p. 1908.

2.3. Hot-pressing maps

This short treatment of the densifying mechanisms has established different dependences of the rate of densification on pressure, temperature, and particle size. For a given particle size, there are therefore distinct pressure–temperature regimes in which one mechanism is dominant, i.e. produces more densification than the others. This competition can be illustrated in theoretical diagrams (ARZT *et al.* [1983]), two of which are shown in figs. 12 and 13 for a tool-steel powder. The heavy lines bound the fields of dominance of one mechanism, while the superimposed thin lines are contours of constant time, which predict the extent of densification. It is seen, for example, that at a temperature of 1200°C and an applied pressure equal to a tenth of the yield stress of the powder material, power-law creep dominates up to a density of about 85%. Unless very high pressures are employed, the final densification is always due to diffusion. Therefore particle size is a critical variable. If the effective particle size is increased by grain growth, for instance, the dominance of power-law creep, which is insensitive to such a microstructural change, will be extended to higher densities.

Experimental data points typical of industrial HIP cycles are included in both figures; they are in promising agreement with the theoretical predictions. But it should be borne in mind that these and similar "maps" are based on a set of rate equations derived from simplifying assumptions as outlined above and depend on the knowledge of the appropriate material data which may vary considerably even within a specific grade of a technical alloy. At present, these maps are intended rather as a rational basis for designing HIP cycles, which in view of the high cost of HIP equipment is a worthwhile purpose. The maps also summarize conveniently our understanding of the pressure-sintering process: it is not one but several mechanisms which contribute to the densification. These mechanisms reflect the ways in which a stressed crystalline solid can deform. Pressure-sintering provides an example of a technical process in which the dominant deformation mechanism may change with time.

3. Sintering with a liquid phase

High-quality technical products can be made from mixed powders by cold compacting and then heating above the melting temperature of the lower-melting, low-volume-fraction (1–40 vol%) component. The shape of the compact is maintained as in solid state sintering, but higher densification levels are obtained as a rule.

Heavy metals (W with Cu, Fe, Ni), cemented carbides (WC, TiC with Co), ceramics containing glassy phases, aluminium, superalloys and cobalt–rare-earth magnets, and a variety of sintered steels are prominent examples (LENEL [1980]). The technical or economic reasons for applying this production process vary largely for these materials; reasons include the incapacity of other processes to produce optimized microstructures with specified compositions and density, a need for reduced shaping cost, or a need for improved homogeneity. Usually, the original powder mixtures are not in chemical equilibrium at sintering temperature, at least in early stages, and chemical driving forces are present. Owing to the very low stresses needed for activating the flow of liquids, capillary forces play a major role even in non-equilibrium systems.

The shrinkage processes occurring during isothermal liquid state sintering are usually divided into three stages (HUPPMANN [1975] and LENEL [1980]):

(i) Rearrangement by liquid flow, retarded by friction between the solid particles. This stage requires good wetting since capillary forces acting at the liquid bridge are determined by the wetting angle (CAHN and HEADY [1970]).

(ii) Dense packing by shape-accommodation of particles separated by liquid films through solution and reprecipitation. This stage requires that the solid phase has a finite solubility in the liquid. In most systems this stage is connected with pronounced structural coarsening.

(iii) Skeleton formation and solid-state sintering. After the liquid is squeezed out between the particles, pores are removed by shrinkage of the solid skeleton. After full density is reached locally, interfacial and surface energy drive further shrinkage of the solid, and liquid is squeezed out into large void spaces (KWON and YOON [1980], KAYSSER *et al.* [1982]) or to the surface (RIEGGER *et al.* [1980]).

Neglecting all complicating factors (e.g. chemical driving forces, size and shape distribution of particles, bridging and coarsening effects), KINGERY [1959] has derived a quantitative description of shrinkage during isothermal liquid-phase sintering which results in equations of the type

$$\Delta V / V \approx 3 \Delta L / L = k t^n, \tag{18}$$

where ΔV and V are the volume change and the initial volume, ΔL and L length change and initial length of the sample and k and n constants typical for each stage and mechanism involved. Application of this equation to compacts of metal and ceramic powder mixtures showed reasonable agreement, and conclusions on the rate-controlling steps were made on the basis of experimental values of n (e.g. 0.5 for reaction-controlled and 0.33 for diffusion-controlled solution/reprecipitation). In the light of later work (for references see HUPPMANN and PETZOW [1980] and LENEL [1980]) this agreement must be considered fortuitous or, by analogy with observations made with powders and solid-state sintering equations, due to a convenient fitting of two-parametric equations rather than to physical relevance.

In systems with appreciable solubility of the solid in the liquid phase (as in virtually all materials produced by liquid-phase sintering), the liquid penetrates particle bridges and grain boundaries in the early stages, which causes swelling if shrinkage by rearrangement is not rapid enough as, for example, in iron–copper alloys (BERNER *et al.* [1974], KAYSSER *et al.* [1980], TABESHFAR and CHADWICK [1982]). According to the classical second-stage mechanism (KINGERY [1959]), the capillary pressure at the contact points of the solid particles increases solubility and causes solution, transport through the liquid and reprecipitation at pressure-free solid–liquid interfaces. Additional and in most cases dominating driving forces for material transport through the liquid are caused by differences in surface curvature when irregular powders are concerned, and variations of defect densities or composition not only due to variations in the original powder particles but also due to the low defect density and the equilibrium composition of the reprecipitated parts (HUPPMANN and PETZOW [1980]). Though these effects do not directly affect the driving force for shrinkage exerted by porosity, they may contribute to densification and particle accommodation indirectly (HUPPMANN and PETZOW [1980] and LENEL [1980]). Finally, when a solid skeleton forms, shrinkage depends on solid state sintering mechanisms and follows the lines described in § 1.

Owing to the rapid material transport in liquids, coarsening of the microstructure is usually pronounced in liquid-phase sintering. Theories developed for coarsening (ripening) in dispersed systems (Wagner–Lifshits–Ardell theory for Ostwald ripening, see ch. 10B, § 3.2.2) apply to systems in which the amount of liquid is high (e.g., EXNER [1973]). In the typical range for liquid-phase sintering (volume fractions < 30%) or for systems with incomplete wetting of the solid by the liquid experienced with most combinations used in technical materials, the particles are in contact and coarsening will be effected by coalescence, i.e. grain-boundary movement assisted by reprecipitation of material into the energetically unfavourable grooves left behind. A concise theoretical treatment of liquid-phase densification and microstructural development is difficult owing to the complex interaction of mechanisms and processes involved. The basic phenomena are well understood, however, and a quantitative description of liquid-phase sintering for adequately realistic models seems feasible in the near future.

References

AIGELTINGER, E.H., and J.P. DROLET, 1974, in: Modern Developments in Powder Metallurgy, vol. 6, eds. H.H. Hausner and W.E. Smith (Metal Powder Industries Federation, Princeton, NJ) p. 323.

AIGELTINGER, E.H., and H.E. EXNER, 1972, J. Mater. Techn. **3**, 425.

ALDINGER, F., 1974, Acta Metall. **22**, 923.

ALEXANDER, B.H., and R.W. BALLUFFI, 1957, Acta Metall. **5**, 666.

ARZT, E., 1982, Acta Metall. **30**, 1883.

ARZT, E., M.F. ASHBY and R.A. VERRALL, 1982, Technical Report CUED/C/MATS/TR60 (Cambridge University, Cambridge).

ARZT, E., M.F. ASHBY and K.E. EASTERLING, 1983, Metallurg. Trans. **14A**, 211.

ASHBY, M.F., 1969, Scripta Metall. **3**, 837.

ASHBY, M.F., 1972, Surf. Sci. **31**, 498.

ASHBY, M.F., 1974, Acta Metall. **22**, 275.

ASHBY, M.F., F. SPAEPEN and S. WILLIAMS, 1978, Acta Metall. **26**, 1647.

ASHBY, M.F., S. BAHK, J. BEVK and D. TURNBULL, 1980, Prog. Mater. Sci. **25**, 1.

BALSHIN, J.M., 1936, Vestn. Metalloprom. **16**, No. 17, 87.

BALLUFFI, R.W., 1980, Grain-Boundary Structure and Kinetics (ASM, Metals Park, OH).

BEERE, W., 1975, Acta Metall. **23**, 139.

BERNER, D., H.E. EXNER and G. PETZOW, 1974, in: Modern Developments in Powder Metallurgy, vol. 6, eds. H.H. Hausner and W.E. Smith (Metal Powder Industries Federation, Princeton, NJ) p. 237.

BRETT, J., and L.L. SEIGLE, 1963, Acta Metall. **11**, 467.

BROOK, R.J., 1969, J. Amer. Ceram. Soc. **52**, 56 and 339.

BROSS, P., and H.E. EXNER, 1979, Acta Metall. **27**, 1013.

CAHN, J.W., and R.B. HEADY, 1970, J. Amer. Ceram. Soc. **53**, 406.

CAHN, R.W., 1980, in: Recristallisation and Grain Growth of Multi-Phase and Particle Containing Materials, eds. N. Hansen, A.R. Jones and T. Leffers (Risø National Laboratory, Roskilde, Denmark) p. 77.

CARPAY, F.M.A., 1977, J. Amer. Ceram. Soc. **60**, 82.

CHERMANT, J.L., M. COSTER, J.P. JERNOT and J.L. DUPAIN, 1981, J. Microsc. **121**, 89.

CLARK, P.W., and J. WHITE, 1950, Trans. Brit. Ceram. Soc. **49**, 305.

COBLE, R.L., 1958, J. Amer. Ceram. Soc. **41**, 55.

COBLE, R.L., 1961, J. Appl. Phys. **32**, 787.

COBLE, R.L., 1970, J. Appl. Phys. **41**, 4798.

COBLE, R.L., 1973, J. Amer. Ceram. Soc. **56**, 461.

COBLE, R.L., and R.M. CANNON, 1978, in: Processing of Crystalline Solids, Materials Science Research, vol. 11, eds. H. Palmour, R.F. Davis and T.M. Hare (Plenum, New York) p. 151.

COBLENZ, W.S., J.M. DYNYS, R.M. CANNON and R.L. COBLE, 1980, in: Sintering Processes, Materials Science Research vol. 13, ed. G.C. Kuczynski (Plenum, New York) p. 141.

DEHOFF, R.T., and H.E. AIGELTINGER, 1970, in: Advanced Techniques in Powder Metallurgy, Perspectives in Powder Metallurgy, vol. 5, eds. J.S. Hirschhorn and K.H. Roll (Plenum, New York) p. 81.

EADIE, R.L., G.C. WEATHERLY and K.T. AUST, 1978, Acta Metall. **26**, 759.

EVANS, A.G., 1982, J. Amer. Ceram. Soc. **65**, 497.

EXNER, H.E., 1973, Z. Metallk. **64**, 273.

EXNER, H.E., 1979a, Rev. Powder Metallurg. Phys. Ceram. **1**, 7.

EXNER, H.E., 1979b, J. Microsc. **116**, 25.

EXNER, H.E., 1980, Powder Metallurg. **23**, 203.

EXNER, H.E., 1982, Met. Sci. **16**, 451.

EXNER, H.E., G. PETZOW and P. WELLNER, 1973, in: Sintering and Related Phenomena, Materials Science Research, vol. 6, ed. G.C. Kuczynski (Plenum, New York) p. 351.

FISCHMEISTER, H.F., 1978, Powder Metallurg. Int. **10**, 119.

FISCHMEISTER, H.F., and E. ARZT, 1983, Powder Metallurg., in press.

FISCHMEISTER, H.F., and H.E. EXNER, 1964, Metall **18**, 932; 1965, Metall **19**, 113.

FRENKEL, J., 1945, J. Phys. USSR **9**, 385.

GEGUZIN, J.A., 1973, Physik des Sinterns (VEB Deutscher Verlag für Grundstoffindustrie, Leipzig).

GERMAN, R.M., 1982, Sci. Sint. **14**, 13.

GERMAN, R.M., and Z.A. MUNIR, 1975, Metallurg. Trans. **6A**, 2223.

GERMAN, R.M., and Z.A. MUNIR, 1982, Rev. Powder Metallurg. Phys. Ceram. **2**, 9.

GESSINGER, G.H., 1970, Scripta Metall. **4**, 673.

GLEITER, H., 1969, Acta Metall. **17**, 565.

GLEITER, H., 1979, Acta Metall. **27**, 187.

HANES, H.D., D.A. SEIFFERT and C.R. WATTS, 1977, Hot Isostatic Processing (Metals and Ceramics Information Center, Columbus, OH).

HEDVALL, J.A., and E. HELIN, 1927, Jernkont. Ann. **82**, 265.

HERRING, C., 1950, J. Appl. Phys. **21**, 437.

HEWITT, R.L., W. WALLACE and M.C. DE MALHERBE, 1973, Powder Metallurg. **16**, 88.

HILL, R., 1960, The Mathematical Theory of Plasticity (Clarendon Press, Oxford) p. 254.

HOGE, C.E., and J.A. PASK, 1977, Ceramurgia Int. **3**, 95.

HSUEH, C.H., A.G. EVANS and R.L. COBLE, 1982, Acta Metall. **30**, 1281.

HUCKABEE, M.L., T.M. HARE and H. PALMOUR, 1978, in: Processing of Crystalline Solids, Materials Science Research, vol. 11, eds. H. Palmour, R.F. Davis and T.M. Hare (Plenum, New York) p. 205.

HÜTTIG, G.F., 1948, Arch. Metallk. **2**, 93.

HUPPMANN, W., 1975, in: Sintering and Catalysis, Materials Science Research, vol. 10, ed. G.C. Kuczynski (Plenum, New York) p. 359.

HUPPMANN, W., and G. PETZOW, 1980, in: Sintering Processes, Materials Science Research, vol. 13, ed. G.C. Kuczynski (Plenum, New York) p. 189.

ICHINOSE, H., and G.C. KUCZYNSKI, 1962, Acta Metall. 10, 205.

IVENSEN, V.A., 1973, Densification of Metal Powders during Sintering (Consultants Bureau, New York, London).

JAEGER, W., and H. GLEITER, 1978, Scripta Metall. 12, 675.

JERNOT, J.P., J.L. CHERMANT, P. MAIRE and P. GILES, 1980, Mikroskopie 37 (Suppl.), 331.

JOHNSON, D.L., 1970, Scripta Metall. 4, 677.

JOHNSON, D.L., 1972, in: Powder Metallurgy for High Performance Applications, eds. J.J. Burke and V. Weiss (Syracuse Univ. Press) p. 139.

JONES, W.D., 1946, Metal Treatment 13, 265.

KAKAR, A.K. and A.C.D. CHAKLADER, 1968, Trans. AIME 242, 1117.

KAYSSER, W.A., W.J. HUPPMANN and G. PETZOW, 1980, Powder Metallurg. 23, 86.

KAYSSER, W.A., O.J. KWON and G. PETZOW, 1982, in: P/M-82 in Europe, Preprints Intern. Powder Metallurgy Conference (Assoc. Italiana di Metallurgia, Milano) p. 23.

KING, A.H., and D.A. SMITH, 1980, Phil. Mag. 42, 495.

KINGERY, W.D., and M. BERG 1955, J. Appl. Phys. 26, 1205.

KINGERY, W.J., 1959, J. Appl. Phys. 30, 301.

KUCZYNSKI, G.C., 1949, Metallurg. Trans. AIME 185, 169.

KUCZYNSKI, G.C., 1975, in: Sintering and Catalysis, Materials Science Research, vol. 10, ed. G.C. Kuczynski (Plenum, New York) p. 325.

KUCZYNSKI, G.C., 1978, Science of Sintering, Monograph 17 (The Serbian Academy of Science and Arts, Beograd).

KUCZYNSKI, G.C., G. MATSUMURA and B.B. CULLITY, 1960, Acta Metall. 8, 205.

KWON, O.J., and D.N. YOON, 1980, in: Sintering Processes, Materials Science Research, vol. 13, ed. G.C. Kuczynski (Plenum, New York) p. 203.

LENEL, F.V., 1980, Powder Metallurgy – Principles and Applications (Metal Powder Industries Federation, Princeton, NJ).

MACKENZIE, J.K., and R. SHUTTLEWORTH, 1949, Proc. Phys. Soc. 62, 833.

MATTHEWS, J.R., 1980, Acta Metall. 28, 311.

MISHRA, A., F.V. LENEL and G.S. ANSELL, 1975, in: Sintering and Catalysis, ed. G.C. Kuczynski (Plenum, New York) p. 339.

MOCELLIN, A., and W.D. KINGERY, 1973, J. Amer. Ceram. Soc. 56, 309.

MOLERUS, O., 1975, Powder Techn. 12, 259.

NABBARRO, F.R.N., 1967, Phil. Mag. 16, 231.

NICHOLS, F.A., 1980, Scripta Metall. 14, 951.

NICHOLS, F.A., and W.W. MULLINS, 1965, J. Appl. Phys. 36, 1826.

NILSSON, J.-O., P.R. HOWELL and G.L. DUNLOP, 1979, Acta Metall. 27, 179.

NUNES, J.J., F.V. LENEL and G.S. ANSELL, 1971, Acta Metall. 19, 107.

PEJOVNIK, S., V. SMOLEJ, D. SUSNIK, and D. KOLAR, 1979, Powder Metallurg. Int. 11, 22.

PINES, B.J., 1946, Zh. Tekh. Fiz. 16, 737.

RIEGGER, H., J.A. PASK and H.E. EXNER, 1980, in: Sintering Processes, Materials Science Research, vol. 13, ed. G.C. Kuczynski (Plenum, New York) p. 219.

ROCKLAND, J.G.R., 1967, Z. Metallk. 58, 467 and 564.

ROSS, J.W., W.A. MILLNER and G.S. WEATHERLY, 1982, Z. Metallk. 73, 391.

ROTH, T.A., 1975, Mater. Sci. Eng. 18, 183.

SCHATT, W., ed., 1981, Einführung in die Werkstoffwissenschaft (VEB Deutscher Verlag für Grundstoffindustrie, Leipzig).

SCHATT, W., H.E. EXNER, E. FRIEDRICH and G. PETZOW, 1982, Acta Metall. 30, 1367.

SCHOBER, T., and R.W. BALLUFFI, 1970, Phil. Mag. 21, 109.

SHALER, A.J., and J. WULFF, 1948, Ind. Eng. Chem. **40**, 838.
SKOROKHOD, V.V., 1972, Rheologicheskie Osnovy Teorii Spekania (Rheological Theory of Sintering) (Izd. Naukovaya Dumka, Kiew).
SOERENSEN, O.T., 1980, in: Thermal Analysis (Birkhäuser, Basel) p. 231.
SPEARS, M.A. and A.G. EVANS, 1982, Acta Metall. **30**, 1281.
SWINKELS, F.B., and M.F. ASHBY, 1980, Powder Metallurg. **23**, 1.
SWINKELS, F.B., and M.F. ASHBY, 1981, Acta Metall. **29**, 259.
SWINKELS, F.B., D.S. WILKINSON, E. ARZT and M.F. ASHBY, 1983, to be published.
TABESHFAR, K., and G.A. CHADWICK, 1982, in: P/M-82 in Europe, Preprints Intern. Powder Metallurgy Conference (Assoc. Italiana di Metallurgia, Milano) p. 693.
TAMMANN, G., 1926, Z. Anorg. Allg. Chem. **157**, 321.
THÜMMLER, F., and W. THOMMA, 1966, in: Modern Developments in Powder Metallurgy, vol. 12, ed. H.H. Hausner (Plenum, New York) p. 3.
THÜMMLER, F., and W. THOMMA, 1967, Metallurg. Rev. **12**, 69.
TIKKANEN, M.H., and S.A. MÄKIPIRTTI, 1965, Int. J. Powder Metallurg. **1**, No. 1, 15.
TORRE, C., 1948, Berg- u. Hüttenm. Mh. **93**, 62.
UDIN, H., A.J. SHALER and J. WULFF, 1949, Metallurg. Trans. AIME **185**, 186.
USKOKOVIC, D., and H.E. EXNER, 1977, Phys. Sint. **9**, 265.
WEI, T.S., and R.M. GERMAN, 1982, in: P/M-82 in Europe, Preprints Intern. Powder Metallurgy Conference (Assoc. Italiana di Metallurgia, Milano) p. 239.
WELLNER, P., G.H. GESSINGER and H.E. EXNER, 1974, Z. Metallk. **65**, 602.
WILKINSON, D.S., and M.F. ASHBY, 1975, Acta Metall. **23**, 1277.
WONG, B., and J.A. PASK, 1979, J. Amer. Ceram. Soc. **62**, 138.

Further Reading

Burke, J.J., and V. Weiss, eds., 1972, Powder Metallurgy for High-Performance Application (Syracuse University Press, Syracuse).
Exner, H.E., 1978, Grundlagen von Sintervorgängen (Gebr. Borntraeger Verlag, Stuttgart), in German, English translation by V.F. Lenel published 1980, Rev. Powder Metallurg. Phys. Ceram **1**, 7.
Geguzin, J.E., 1973, Physik des Sinterns (VEB Deutscher Verlag für Grundstoffindustrie, Leipzig).
Hanes, H.D., D.A. Seiffert and C.R. Watts, 1977, Hot Isostatic Processing (Metals and Ceramics Information Center, Columbus, Ohio).
Lenel, V.F. 1980, Powder Metallurgy – Principles and Applications (Metal Powder Industries Federation, Princeton, NJ).
Schatt, W., ed., 1977, Pulvermetallurgie, Sinter- und Verbundwerkstoffe (VEB Deutscher Verlag für Grundstoffindustrie, Leipzig).
Thümmler, F., and W. Thomma, 1967, Metallurg. Rev. **12**, 69.

The following journals and book series are specialized in powder metallurgy and topics treated in this chapter are frequently discussed:

Journals

Science of Sintering (edited by the International Institute for the Science of Sintering, Beograd, Yugoslavia).
Powder Metallurgy International (edited by Schmid-Verlag, Freiburg, FR-Germany).
Powder Metallurgy (edited by The Metals Society, London, UK).
International Journal of Powder Metallurgy (edited by Metal Powder Industries Federation, Princeton, NJ, USA).
Soviet Powder Metallurgy and Metal Ceramics (translated from Russian by Consultants Bureau, New York, USA).

Reviews in Powder Metallurgy and Physical Ceramics (edited by G. Freund Publishing House, Tel Aviv, Israel).

Planseeberichte für Pulvermetallurgie (Metallwerk Plansee AG, Reutte) discontinued 1981.

Book series

Materials Science Research (Plenum Press, New York, London):

Vol. 6, 1973, G.C. Kuczynski, ed., Sintering and Related Phenomena.

Vol. 9, 1975, A.R. Cooper and A.H. Heuer, eds., Mass Transport Phenomena in Ceramics.

Vol. 10, 1978, G.C. Kuczynski, ed., Sintering and Catalysis.

Vol. 13, 1980, G.C. Kuczynski, ed., Sintering Processes.

Vol. 14, 1981, J. Pask and A. Evans, eds., Surfaces and Interfaces in Ceramic and Ceramic Metal Systems.

Modern Developments in Powder Metallurgy, Proceedings of the International Powder Metallurgy Conferences (Metal Power Industries Federation, Princeton, NJ, USA).

Perspectives in Powder Metallurgy – Fundamentals, Methods and Applications (Plenum, New York, London).

New Perspectives in Powder Metallurgy (Metal Powder Industries Federation, Princeton, NJ, USA).

Progress in Powder Metallurgy, Proceedings of the National US Powder Metallurgy Conferences (Metal Powder Industries Federation, Princeton, NJ, USA).

AUTHOR INDEX

SUBJECT INDEX